New Advances in High-Entropy Alloys

New Advances in High-Entropy Alloys

Editor

Yong Zhang

MDPI • Basel • Beijing • Wuhan • Barcelona • Belgrade • Manchester • Tokyo • Cluj • Tianjin

Editor
Yong Zhang
University of Science and
Technology Beijing
China

Editorial Office
MDPI
St. Alban-Anlage 66
4052 Basel, Switzerland

This is a reprint of articles from the Special Issue published online in the open access journal *Entropy* (ISSN 1099-4300) (available at: https://www.mdpi.com/journal/entropy/special_issues/High-Entropy_Alloys).

For citation purposes, cite each article independently as indicated on the article page online and as indicated below:

LastName, A.A.; LastName, B.B.; LastName, C.C. Article Title. *Journal Name* **Year**, *Article Number*, Page Range.

ISBN 978-3-03943-619-4 (Pbk)
ISBN 978-3-03943-620-0 (PDF)

Contents

About the Editor

Yong Zhang has been a university professor of Materials Science at USTB since 2004. He has published over 200 papers, including two comprehensive review papers published in the journal of *Progress in Materials Science* (*PMS*). He authored a book entitled "High-Entropy Materials: A Brief Introduction", published by Springer Nature. His contributions include proposing a balance parameter to evaluate the configurational entropy effect over the enthalpy effect in liquid state, which has been verified effective to predict the phase formation of multicomponent materials.

Recent publications:

[1] Zhang, Y.; Yan, X.-H.; Liao, W.-B.; Zhao, K. Effects of Nitrogen Content on the Structure and Mechanical Properties of (Al0.5CrFeNiTi0.25)Nx High-Entropy Films by Reactive Sputtering. Entropy 2018, 20, 624.

[2] Wen-Jie Sheng, Xiao Yang, Jie Zhu, Cong Wang, Yong Zhang, Amorphous phase stability of NbTiAlSiNX high-entropy films. Rare Met. (2018) 37(8):682–689.

[3] Zhang, Y.; Yan, X.H.; Ma, J.; Lu, Z.P.; Zhao, Y.H.; Compositional gradient films constructed by sputtering in a multicomponent Ti–Al–(Cr, Fe, Ni) system. JMR, 2018.

[4] Sheng, W.; Yang, X.; Wang, C.; Zhang, Y. Nano-Crystallization of High-Entropy Amorphous NbTiAlSiWxNyFilms Prepared by Magnetron Sputtering. Entropy 2016, 18, 226.

[5] X.H. Yan, J.S. Li, W.R. Zhang, Y. Zhang, A brief review of high-entropy films, Materials Chemistry and Physics 210 (2018) 12–19.

[6] Zhang, W., Liaw, P.K. & Zhang, Y., Science and technology in high-entropy alloys, Sci. China Mater. (2018) 61: 2.

[7] Rui Xuan Li, P. K. Liaw, Yong Zhang, Synthesis of Al x CoCrFeNi High-entropy Alloys by High-gravity Combustion from Oxides, Materials Science and Engineering A 707, 668-673.

[8] Michael C. Gao, Daniel B. Miracle, David Maurice, Xuehui Yan and Yong Zhang, Jeffrey A. Hawk, High-entropy functional materials, J. Mater. Res., Vol. 0, No. 0, 2018.

[9] Xia S, Lousada CM, Mao H, Maier AC, Korzhavyi PA, Sandström R, Wang Y and Zhang Y (2018) Nonlinear Oxidation Behavior in Pure Ni and Ni-Containing Entropic Alloys. Front. Mater. 5:53. doi: 10.3389/fmats.2018.00053.

[10] Christopher M. Barr, James E. Nathaniel II, Kinga A. Unocic, Junpeng Liu, Yong Zhang,YongqiangWang, Mitra L. Taheri, Exploring radiation induced segregationmechanisms at grain boundaries in equiatomic CoCrFeNiMn high entropy alloy under heavy ion irradiation, Scripta Materialia 156 (2018) 80–84.

[11] Shao Lei, et al., A Low-Cost Lightweight Entropic Alloy with High Strength, Journal of Materials Engineering and Performance, ASM International https://doi.org/10.1007/s11665-018-3720-0.

[12] Xing Q, et al., High-Throughput Screening Solar-Thermal Conversion Films in a Pseudobinary (Cr, Fe, V)–(Ta, W) System, ACS Combinatorial Science, 2018.

[13] Li R., et al., Graded microstructures of Al-Li-Mg-Zn-Cu entropic alloys under supergravity, SCIENCE CHINA Materials, 2019.

[14] Xing Q, et al., Mechanical properties and thermal stability of (NbTiAlSiZi)Nx high-entropy ceramic films at high temperature, JMR, 2018.

[15] Li D., et al., AnnealingeffectfortheAl0.3CoCrFeNihigh-entropyalloyfibers, JACs, 2018.

[16] Y. Zhang*, J. Liu, S. Chen, X. Xie, P. Liaw, K. Dahmen, J. Qiao, Y. Wang, Serration and noise behaviors in materials. Progress in Materials Science. 2017, 90:358–460.

[17] Y. Zhang, Authored Book, ≪High-entropy Materials: A Brief Introduction≫. Springer-Nature Publishing. 2019.

[18] X. Yan, J. Ma, Y. Zhang*, High-throughput screening for biomedical applications in a Ti-Zr-Nb alloy system through masking co-sputtering. Science China: Physics, Mechanics and Astronomy. 2019, 62(9):996111.

entropy

MDPI

Editorial

New Advances in High-Entropy Alloys

Yong Zhang [1,2,3,*] and Ruixuan Li [1]

1 Beijing Advanced Innovation Center of Materials Genome Engineering, State Key Laboratory for Advanced Metals and Materials, University of Science and Technology Beijing, Beijing 100083, China; liruixuan1208@163.com
2 Qinghai Provincial Engineering Research Center of High Performance Light Metal Alloys and Forming, Qinghai University, Xining 810016, China
3 Shunde Graduate School, University of Science and Technology Beijing, Foshan 528399, China
* Correspondence: drzhangy@ustb.edu.cn

Received: 6 October 2020; Accepted: 13 October 2020; Published: 15 October 2020

Keywords: high-entropy alloys; mechanical behaviors; high-entropy film; low-activation alloys

---

Exploring new materials is an eternal pursuit in the development of human civilization. In recent years, people have tended to adjust the order/disorder degree to explore new materials. The order/disorder can be measured by entropy, and it can be divided into two parts: the topological ordering and the chemical ordering. The former mainly refers to the ordering in the spatial configuration, e.g. amorphous alloys [1] with short-range ordering but without long-range ordering; the latter mainly refers to the ordering in the chemical occupancy, e.g. high-entropy alloys (HEAs) where components can replace each other. HEAs, in sharp contrast to traditional alloys based on one or two principal elements, have one striking characteristic: the unusually high entropy of mixing. They have not been too much noticed until a review paper entitled: "Microstructure and properties of high entropy alloys" published in 2014 on the journal of Progress in Materials Science [2]. Lots of reports showing they exhibit five recognized performance characteristics, namely strength-plasticity trade-off breaking, irradiation tolerance, corrosion resistance, high impact toughness within a wider temperature range, and high thermal stability [3]. So far, the development of HEAs has gone through three main stages: 1. quinary equal-atomic single-phase solid solution alloys; 2. quaternary or quinary non-equal-atomic multiphase alloys; 3. medium-entropy alloys, high entropy fibers, high entropy films, lightweight HEAs, etc. Nowadays, deeper research on HEAs is urgently needed.

This Special Issue includes forty-two research articles and six review papers dedicated to the frontiers of high entropy materials, from exploring the microstructures by experiment to revealing the structure-performance relationship by simulation. These innovative studies will provide new methods to solve challenging problems and open up new possibilities for emerging research fields.

Configuration entropy plays an important role in the microstructures and properties of HEAs, which is believed to stabilize disordered solid solution phases over intermetallic compounds by lowering the Gibbs free energy. Traditionally, configuration entropy was computed by the empirical formula. In Fernández-Caballero's work [4], a new formalism based on a hybrid combination of the Cluster Expansion Hamiltonian and Monte Carlo simulations is developed to predict the configuration entropy as a function of temperature from multi-body cluster probability in a multi-component system with arbitrary average composition. Experimentally, Haas et al. [5] used differential scanning calorimetry to determine the thermal entropy and compared it to the configurational entropy. The contributions of entropy and enthalpy to the Gibbs free energy was calculated and examined by them.

In addition to the above basic research on the entropy of the alloy system, research on HEAs mainly focuses on their composition design, preparation and processing technology, mechanical properties, and physical and chemical properties. Other main contents of this issue are listed as follows.

1. Compositional Design

Because HEAs are characterized by multiple components and high contents of alloying elements, the complexity of their compositional design is greatly increased compared with traditional alloys. Some empirical rules are used early on to realize simple prediction of phase structure. With the development of computer technology and simulation technology, phase diagram and first-principles methods are now widely used to help alloy design. Now, machine learning is also popular by using various algorithms. Here, our Special Issue contains three articles using phase diagram and first principles for compositional design.

A phase diagram is the geometric description of the system under a thermal equilibrium and it is the basis for the study of solidification, phase transformation, crystal growth, and solid-phase transformation. Since the 1970s, the calculation of phase diagrams (CALPHAD) based on the thermodynamic theory has become a new trend. Gorsse et al. [6] and Klaver et al. [7] conducted an in-depth study on the use of CALPHAD methods for composition design. On the other hand, Zheng et al. [8] used first principles to calculate the elastic properties of seventy different compositions in the refractory V–Mo–Nb–Ta–W system. Effects of different elements to the modulus are precisely calculated so as to optimize the composition in this refractory system.

2. Preparation and Processing

Traditional preparation methods of bulk HEAs mainly include vacuum arc melting, vacuum induction melting, mechanical alloying, and so on. In order to obtain a certain crystal orientation, directional solidification can also be employed. There are also some new preparation methods for bulk HEAs, such as high-gravity combustion synthesis and additive manufacturing. Chen et al. [9] reviewed the application of additive manufacturing methods in the fabrication of HEAs. Compared with the casting counterparts, HEAs prepared by additive manufacturing are found to have a superior yield strength and ductility as a consequence of the fine microstructure formed during the rapid solidification in the fabrication process. As a result, this is an effective method to improve their comprehensive properties.

Different processing methods are closely related to performance. Our Special Issue contains two research papers on the use of annealing to optimize alloys and two review papers on alloys using high pressure and welding.

Annealing is considered to be an effective method to improve the microstructure and properties of alloys, and different annealing temperature and time are closely related to comprehensive properties. Zhuang et al. [10] investigated the effect of annealing temperature on the microstructure, phase constituents, and mechanical properties of $Al_{0.5}CoCrFeMo_xNi$ (x = 0, 0.1, 0.2, 0.3, 0.4 and 0.5) HEAs at a fixed annealing time (10 h). They found that the alloys annealed at 80 °C showed relatively fine precipitations and microstructures and, therefore, higher hardness and yield stress. Sathiyamoorthi et al. [11] put the high-pressure torsion-processed CoCrNi alloy with a grain size of ~50 nm in different annealing conditions and they explored the optimal processing technology. The sample annealed at 700 °C for 15 min exhibited a remarkable combination of tensile strength (~1090 MPa) and strain to failure (~41%).

Pressure, as another fundamental and powerful parameter, has been introduced to the experimental study of HEAs. Many interesting reversible/irreversible phase transitions that were not expected or otherwise invisible before have been observed by applying high pressure. Zhang et al. [12] reviewed recent results in various HEAs obtained using in situ static high-pressure synchrotron radiation X-ray techniques and provided some perspectives for future research.

Welding is an important emerging area with significant potential impact to future application-oriented research and technological developments in HEAs. The selection of feasible welding processes with optimized parameters is essential to enhance the applications of HEAs. Guo et al. [13] reviewed the key recent works on welding of HEAs in detail, focusing on the research of main HEA systems when applying different welding techniques. They also highlighted the future challenges and main areas to research.

3. Mechanical Behaviors

Research on the mechanical properties of HEAs is the most extensive. The main research hotspots are three types of alloys: 1, 3D transitional-group-element HEAs; 2, 3D transitional-group-element HEAs with Al or Ti added; 3, refractory high-entropy alloys with excellent high-temperature properties. In addition, other trace alloying elements, such as Mo, Nb, Zr, etc., have also been added to the alloy in order to study the effect of their content on the microstructures and properties of the alloy.

HEAs based on 3D transition metals, such as CrCoNi and CrMnFeCoNi alloys, are revealed to have remarkable mechanical properties. Especially the CoCrFeMnNi alloy is famous as the Cantor alloy for its high plasticity and comparable strength. The tensile creep behavior of the CoCrFeNiMn HEA was systematically investigated by Cao et al. [14] over an intermediate temperature range (500–600 °C) and applied stress (140–400 MPa). The alloy exhibited a stress-dependent transition from a low-stress region to a high-stress region, which was characterized by the dynamic recrystallization and the grain-boundary precipitation. This alloy, deformed at high strain rates (900 to 4600 s^{-1}), was investigated by Wang et al. [15]. The yield strength was sensitive to the change of high strain rates, and serration behaviors were also observed on the flow stress curves, which could be predicted by the Zerilli–Armstrong constitutive equation. Bu et al. [16] performed an in situ atomic-scale observation of deformation behaviors in a nano-scaled CoCrCuFeNi alloy. Exceptional strength was realized, and the deformation behaviors, including nano-disturbances and phase transformations, were distinct from those of corresponding bulk HEAs. The same result was confirmed in Mridha's [17] experiments. From the perspective of theory and simulation, Ikeda et al. [18] investigated the impact of compositional fluctuations in the vicinity of stacking faults on CrCoNi and CrMnFeCoNi by employing first-principles calculations. Their research and experimental results complement and perfect each other.

Adding other alloying elements is one of the research hotspots, in order to further improve the strength. The addition of Nb promoted the precipitation of nano-phases and thus increased the strength of the alloy, and Mo addition in the CoCrFeNi alloy effectively helped to increase the corrosion resistance. Detailed work was carried out by Han et al., Wang et al., and Tsau et al. [19–21]. The effect of Zr addition on the CoCrFeNiMn alloy was studied by Zhang et al. [22]. They prepared the alloy by using the ZrH_2 powders and a mechanical alloying technique. Multi-phase microstructures formed in the alloys, which can be attributed to the large lattice strain and negative enthalpy of mixing, caused by the addition of Zr. Sun et al. [23] also employed the mechanical alloying technique and prepared the CoCrNiCuZn alloy. Pd addition promoted the local- and long-range lattice distortions in CoCrFeNi alloy and also had effect on the phase stability and phase transformation. This phenomenon was revealed by Zhang et al. [24]. Tan et al. [25] then studied the effect of Mn addition on the microstructures and mechanical properties of this CoCrFeNiPd alloy.

Alloying Al and Ti in 3D transitional Co-Cr-Fe-Ni HEAs shows a strong influence on microstructure and phase composition, as well as the ability to decrease the density. They tend to strengthen the alloy by increasing the lattice distortion, as confirmed by Wu et al. [26]. Löbel et al. [27] studied the influence of Ti in the $AlCoCrFeNiTi_x$ alloy, and Zhang et al. [28] characterized the partially recrystallized $CoCrFeNiTi_{0.2}$ alloy. It seems that alloying with Ti leads to an increase in microstructural heterogeneity. Similarly aimed at the Al-Co-Cr-Fe-Ni-Ti alloy after the composition adjustment, Manzoni et al. [29] studied the microstructure and properties of the $Al_{10}Co_{25}Cr_8Fe_{15}Ni_{36}Ti_6$ alloy and Haas et al. [30] studied those of the $Al_{10}Co_{25}Cr_8Fe_{15}Ni_{36}Ti_6$ alloy. Both kinds of alloy showed beneficial particle-strengthened microstructures. Riva et al. [31] investigated the effect of Sc alloying on the $Al_2CoCrFeNi$, $Al_{0.5}CoCrCuFeNi$, and $AlCoCrCu_{0.5}FeNi$ HEAs. It caused grain refinement as well as hardness and electrical conductivity increases (up to 20% and 14%, respectively). Furthermore, the solid-state phase transformation kinetics of as-cast and cold rolling deformed $Al_{0.5}CoCrFeNi$ HEAs have been investigated by Wang et al. [32] using the thermal expansion method. Lightweight AlCrMoTi and AlCrMoTiV HEAs were designed by Kang et al. [33] and they were confirmed as solid solutions with minor ordered B2 phases. They also have superb specific hardness compared to that of commercial alloys.

With the help of their excellent mechanical properties, high-entropy thin film developed by a magnetron sputtering technique has attracted attention, which has exciting potential to make small-structure devices and precision instruments with sizes ranging from nanometers to micrometers. Liao et al. [34] fabricated $Al_{0.3}CoCrFeNi$ film, and Zhang et al. [35] prepared $(Al_{0.5}CrFeNiTi_{0.25})N_x$ high-entropy films. It showed that the phase structure changes from the amorphous to the face-centered cubie (FCC) structure with increasing N content, which was proved to be related with the atomic size difference in the alloy system. Similarly, $(AlCrTiZrV)Si_xN$ high entropy film was prepared and characterized by Niu et al. [36]. Then, an $Al_{0.5}CoCrCuFeNi$ coating was successfully prepared on an AZ91D magnesium alloy surface [37], and an AlCoCrFeNi coating was prepared on the stainless steel substrate [38]. The coating obviously had significant high hardness, high wear resistance, and corrosion resistance, and can integrate the properties of the substrate.

The development and preparation of composites have realized the complementary advantages of a variety of different component materials, which can further improve the overall performance of alloys. Zhang et al. [39] prepared a high-entropy ceramic composite by using HfC, Mo_2C, TaC, and TiC in pulsed-current processing. Furthermore, Li et al. [40] used a mixture of carbides and oxides in the preparation of an NbMoCrTiAl HEA. With the presence of approximately 7.0 vol. % Al_2O_3 and 32.2 vol. % TiC-reinforced particles, the compressive fracture strength of the composite reached 1542 MPa, and this was increased by approximately 50% compared with that of the as-cast NbMoCrTiAl HEA. Furthermore, two types of novel HEA/diamond composites were reported, providing a new model for material development [41,42].

It can be found that nano-precipitation phases appear widely in HEAs, which play a vital role in improving their strength and plasticity. In this case, Wang et al. [43] summarized precipitation behavior and precipitation strengthening in HEAs comprehensively, including the morphology evolution of second-phase particles and precipitation strengthening mechanisms. They claimed that the challenge in the future is to design a stable, coherent microstructure in different solid-solution matrices.

Besides room-temperature mechanical properties, HEAs tend to exhibit great high-temperature properties for their high-entropy stabilization effect. The Hf-Nb-Ta-Ti-Zr refractory HEA shows great high-temperature properties and room-temperature properties. Zýka et al. [44] presented investigations of the room-temperature tensile mechanical properties of selected three- and four-element medium entropy alloys derived from the Hf-Nb-Ta-Ti-Zr system, and they found that it is the five-element HEA alloy that exhibits the best combination of strength and elongation. Tseng et al. [45] aimed to reveal the effects of Mo, Nb, Ta, Ti, and Zr on mechanical properties of equiatomic Hf-Mo-Nb-Ta-Ti-Zr alloys. Another $VCrFeTa_xW_x$ HEA, which has great high-temperature properties and excellent heat-softening resistance, was studied and recommended by Zhang et al. [46].

4. Physical and Chemical Properties

In addition to mechanical properties, research on the physical and chemical properties of HEAs mainly focuses on soft magnetic properties, radiation resistance, electrical properties, and corrosion resistance.

Because HEAs contain more principal elements, including magnetic elements, such as Fe, Co, Ni, and Mn, and these magnetic elements are mixed with other elements, HEAs exhibit different magnetic properties compared to traditional alloys. Li et al. [47] exhibited the effects of Sn addition on the soft magnetic properties of dual-phase $FeCoNi(CuAl)_{0.8}Sn_x$ ($0 \le x \le 0.10$) HEAs. They showed that saturation magnetization of the alloy increased greatly, while the remanence (Br) decreased after the addition of Sn. Furthermore, thermomagnetic curves indicated that the phases of the alloy will transform from FCC with low Curie temperatures (Tc) to the body-centered cubic (BCC) phase with high Tc at temperatures of 600–700 K. For the FeSiBAlNi alloy, the effects of Co and Gd addition combined with subsequent annealing on microstructures were investigated by Zhai et al. [48]. FeSiBAlNi-based HEAs also possessed soft magnetism, especially the as-annealed FeSiBAlNiGd alloy, which can be ascribed to the formation of Gd-oxides.

Some HEA systems also have good radiation resistance. Disordered solid-solution phase structure is mainly formed in HEAs. The biggest feature of their structure is the large lattice distortion caused by the difference in atomic size and the high configuration entropy. Therefore, atomic-level stress may be formed. In addition, the sluggish diffusion effect is caused by the interaction of multiple components. As a result, HEAs have shown excellent performance in radiation resistance, which provides new ideas for the development of nuclear materials. CoCrFeCuNi alloys were irradiated with a 100 keV He^+ ion beam by Wang et al. [49]. It was indicated that Cu-rich phases were favorable sites for the nucleation and gathering of He bubbles. At ion doses of 2.5×10^{17} ions/cm^2 and 5.0×10^{17} ions/cm^2, the HEAs showed obvious hardening, which could be attributed to the formation of large amounts of irradiation defects.

5. Outlook

The development of HEAs has only been occurring for no more than ten years until now, and people's understanding of this material is still in its infancy. This Special Issue summarizes some of the latest developments in HEAs. Even though not all the questions related to HEAs have been given a reasonable explanation in this issue, it does pay attention to some typical research hotspots, and significant progress can, indeed, be seen. We believe this special issue will guide further research and promote the property breakthrough of HEAs.

Funding: Y.Z. would like to thank the financial support from the National Science Foundation of China (Grant No. 51671020), the Fundamental Research Funds for the Central Universities (Grant No. FRF-MP-19-013), and the project of Al-Mg based medium-entropy alloys with Dongguan Eontec company.

Acknowledgments: We express our thanks to the authors of the above contributions, and to the journal *Entropy* and MDPI for their support during this work.

Conflicts of Interest: The author declare no conflict of interests.

References

1. Chen, S.; Wang, J.; Xia, L.; Wu, Y. Deformation Behavior of Bulk Metallic Glasses and High Entropy Alloys under Complex Stress Fields: A Review. *Entropy* **2019**, *21*, 54. [CrossRef]
2. Zhang, Y.; Zuo, T.T.; Tang, Z.; Gao, M.C.; Dahmen, K.A.; Liaw, P.K.; Lu, Z.P. Microstructures and properties of high-entropy alloys. *Prog. Mater. Sci.* **2014**, *61*, 1–93. [CrossRef]
3. Derimow, N.; Abbaschian, R. Liquid Phase Separation in High-Entropy Alloys—A Review. *Entropy* **2018**, *20*, 890. [CrossRef]
4. Fernández-Caballero, A.; Fedorov, M.; Wróbel, J.; Mummery, P.; Nguyen-Manh, D. Configurational Entropy in Multicomponent Alloys: Matrix Formulation from Ab Initio Based Hamiltonian and Application to the FCC Cr-Fe-Mn-Ni System. *Entropy* **2019**, *21*, 68. [CrossRef]
5. Haas, S.; Mosbacher, M.; Senkov, O.; Feuerbacher, M.; Freudenberger, J.; Gezgin, S.; Völkl, R.; Glatzel, U. Entropy Determination of Single-Phase High Entropy Alloys with Different Crystal Structures over a Wide Temperature Range. *Entropy* **2018**, *20*, 654. [CrossRef]
6. Gorsse, S.; Senkov, O. About the Reliability of CALPHAD Predictions in Multicomponent Systems. *Entropy* **2018**, *20*, 899. [CrossRef]
7. Klaver, T.; Simonovic, D.; Sluiter, M. Brute Force Composition Scanning with a CALPHAD Database to Find Low Temperature Body Centered Cubic High Entropy Alloys. *Entropy* **2018**, *20*, 911. [CrossRef]
8. Zheng, S.; Wang, S. First-Principles Design of Refractory High Entropy Alloy VMoNbTaW. *Entropy* **2018**, *20*, 965. [CrossRef]
9. Chen, S.; Tong, Y.; Liaw, P. Additive Manufacturing of High-Entropy Alloys: A Review. *Entropy* **2018**, *20*, 937. [CrossRef]

10. Zhuang, Y.-X.; Zhang, X.-L.; Gu, X.-Y. Effect of Annealing on Microstructure and Mechanical Properties of Al0.5CoCrFeMoxNi High-Entropy Alloys. *Entropy* **2018**, *20*, 812. [CrossRef]
11. Sathiyamoorthi, P.; Bae, J.; Asghari-Rad, P.; Park, J.; Kim, J.; Kim, H. Effect of Annealing on Microstructure and Tensile Behavior of CoCrNi Medium Entropy Alloy Processed by High-Pressure Torsion. *Entropy* **2018**, *20*, 849. [CrossRef]
12. Zhang, F.; Lou, H.; Cheng, B.; Zeng, Z.; Zeng, Q. High-Pressure Induced Phase Transitions in High-Entropy Alloys: A Review. *Entropy* **2019**, *21*, 239. [CrossRef]
13. Guo, J.; Tang, C.; Rothwell, G.; Li, L.; Wang, Y.-C.; Yang, Q.; Ren, X. Welding of High Entropy Alloys—A Review. *Entropy* **2019**, *21*, 431. [CrossRef]
14. Cao, C.; Fu, J.; Tong, T.; Hao, Y.; Gu, P.; Hao, H.; Peng, L. Intermediate-Temperature Creep Deformation and Microstructural Evolution of an Equiatomic FCC-Structured CoCrFeNiMn High-Entropy Alloy. *Entropy* **2018**, *20*, 960. [CrossRef]
15. Wang, B.; Yao, X.; Wang, C.; Zhang, X.; Huang, X. Mechanical Properties and Microstructure of a NiCrFeCoMn High-Entropy Alloy Deformed at High Strain Rates. *Entropy* **2018**, *20*, 892. [CrossRef]
16. Bu, Y.; Peng, S.; Wu, S.; Wei, Y.; Wang, G.; Liu, J.; Wang, H. Unconventional Deformation Behaviours of Nanoscaled High-Entropy Alloys. *Entropy* **2018**, *20*, 778. [CrossRef]
17. Mridha, S.; Komarasamy, M.; Bhowmick, S.; Mishra, R.; Mukherjee, S. Small-Scale Plastic Deformation of Nanocrystalline High Entropy Alloy. *Entropy* **2018**, *20*, 889. [CrossRef]
18. Ikeda, Y.; Körmann, F.; Tanaka, I.; Neugebauer, J. Impact of Chemical Fluctuations on Stacking Fault Energies of CrCoNi and CrMnFeCoNi High Entropy Alloys from First Principles. *Entropy* **2018**, *20*, 655. [CrossRef]
19. Han, B.; Wei, J.; He, F.; Chen, G.; Wang, Z.; Hu, A.; Zhou, W.; Kai, J.-J. Elemental Phase Partitioning in the γ-γ'' Ni2CoFeCrNb0.15 High Entropy Alloy. *Entropy* **2018**, *20*, 910. [CrossRef]
20. Tsau, C.-H.; Tsai, M.-C. The Effects of Mo and Nb on the Microstructures and Properties of CrFeCoNi(Nb,Mo) Alloys. *Entropy* **2018**, *20*, 648. [CrossRef]
21. Wang, W.; Wang, J.; Yi, H.; Qi, W.; Peng, Q. Effect of Molybdenum Additives on Corrosion Behavior of (CoCrFeNi)100−xMox High-Entropy Alloys. *Entropy* **2018**, *20*, 908. [CrossRef]
22. Zhang, H.; Zhang, L.; Liu, X.; Chen, Q.; Xu, Y. Effect of Zr Addition on the Microstructure and Mechanical Properties of CoCrFeNiMn High-Entropy Alloy Synthesized by Spark Plasma Sintering. *Entropy* **2018**, *20*, 810. [CrossRef]
23. Sun, Y.; Ke, B.; Li, Y.; Yang, K.; Yang, M.; Ji, W.; Fu, Z. Phases, Microstructures and Mechanical Properties of CoCrNiCuZn High-Entropy Alloy Prepared by Mechanical Alloying and Spark Plasma Sintering. *Entropy* **2019**, *21*, 122. [CrossRef]
24. Zhang, F.; Tong, Y.; Jin, K.; Bei, H.; Weber, W.; Zhang, Y. Lattice Distortion and Phase Stability of Pd-Doped NiCoFeCr Solid-Solution Alloys. *Entropy* **2018**, *20*, 900. [CrossRef]
25. Tan, Y.; Li, J.; Wang, J.; Kou, H. Effect of Mn Addition on the Microstructures and Mechanical Properties of CoCrFeNiPd High Entropy Alloy. *Entropy* **2019**, *21*, 288. [CrossRef]
26. Wu, C.-S.; Tsai, P.-H.; Kuo, C.-M.; Tsai, C.-W. Effect of Atomic Size Difference on the Microstructure and Mechanical Properties of High-Entropy Alloys. *Entropy* **2018**, *20*, 967. [CrossRef]
27. Löbel, M.; Lindner, T.; Mehner, T.; Lampke, T. Influence of Titanium on Microstructure, Phase Formation and Wear Behaviour of AlCoCrFeNiTix High-Entropy Alloy. *Entropy* **2018**, *20*, 505. [CrossRef]
28. Zhang, H.; Liu, P.; Hou, J.; Qiao, J.; Wu, Y. Prediction of Strength and Ductility in Partially Recrystallized CoCrFeNiTi0.2 High-Entropy Alloy. *Entropy* **2019**, *21*, 389. [CrossRef]
29. Manzoni, A.; Haas, S.; Daoud, H.; Glatzel, U.; Förster, C.; Wanderka, N. Tensile Behavior and Evolution of the Phases in the Al10Co25Cr8Fe15Ni36Ti6 Compositionally Complex/High Entropy Alloy. *Entropy* **2018**, *20*, 646. [CrossRef]
30. Haas, S.; Manzoni, A.M.; Krieg, F.; Glatzel, U. Microstructure and Mechanical Properties of Precipitate Strengthened High Entropy Alloy Al10Co25Cr8Fe15Ni36Ti6 with Additions of Hafnium and Molybdenum. *Entropy* **2019**, *21*, 169. [CrossRef]
31. Riva, S.; Mehraban, S.; Lavery, N.; Schwarzmüller, S.; Oeckler, O.; Brown, S.; Yusenko, K. The Effect of Scandium Ternary Intergrain Precipitates in Al-Containing High-Entropy Alloys. *Entropy* **2018**, *20*, 488. [CrossRef]
32. Wang, J.; Yang, H.; Guo, T.; Wang, J.; Wang, W.Y.; Li, J. Effect of Cold Rolling on the Phase Transformation Kinetics of an Al0.5CoCrFeNi High-Entropy Alloy. *Entropy* **2018**, *20*, 917. [CrossRef]

33. Kang, M.; Lim, K.; Won, J.; Lee, K.; Na, Y. Al-Ti-Containing Lightweight High-Entropy Alloys for Intermediate Temperature Applications. *Entropy* **2018**, *20*, 355. [CrossRef]
34. Liao, W.-B.; Zhang, H.; Liu, Z.-Y.; Li, P.-F.; Huang, J.-J.; Yu, C.-Y.; Lu, Y. High Strength and Deformation Mechanisms of Al0.3CoCrFeNi High-Entropy Alloy Thin Films Fabricated by Magnetron Sputtering. *Entropy* **2019**, *21*, 146. [CrossRef]
35. Zhang, Y.; Yan, X.-H.; Liao, W.-B.; Zhao, K. Effects of Nitrogen Content on the Structure and Mechanical Properties of (Al0.5CrFeNiTi0.25)Nx High-Entropy Films by Reactive Sputtering. *Entropy* **2018**, *20*, 624. [CrossRef]
36. Niu, J.; Li, W.; Liu, P.; Zhang, K.; Ma, F.; Chen, X.; Feng, R.; Liaw, P. Effects of Silicon Content on the Microstructures and Mechanical Properties of (AlCrTiZrV)-Six-N High-Entropy Alloy Films. *Entropy* **2019**, *21*, 75. [CrossRef]
37. Huang, K.; Chen, L.; Lin, X.; Huang, H.; Tang, S.; Du, F. Wear and Corrosion Resistance of Al0.5CoCrCuFeNi High-Entropy Alloy Coating Deposited on AZ91D Magnesium Alloy by Laser Cladding. *Entropy* **2018**, *20*, 915. [CrossRef]
38. Cui, W.; Karnati, S.; Zhang, X.; Burns, E.; Liou, F. Fabrication of AlCoCrFeNi High-Entropy Alloy Coating on an AISI 304 Substrate via a CoFe2Ni Intermediate Layer. *Entropy* **2018**, *21*, 2. [CrossRef]
39. Zhang, H.; Akhtar, F. Processing and Characterization of Refractory Quaternary and Quinary High-Entropy Carbide Composite. *Entropy* **2019**, *21*, 474. [CrossRef]
40. Li, T.; Liu, B.; Liu, Y.; Guo, W.; Fu, A.; Li, L.; Yan, N.; Fang, Q. Microstructure and Mechanical Properties of Particulate Reinforced NbMoCrTiAl High Entropy Based Composite. *Entropy* **2018**, *20*, 517. [CrossRef]
41. Zhang, M.; Zhang, W.; Liu, F.; Peng, Y.; Hu, S.; Liu, Y. Effect of Binding and Dispersion Behavior of High-Entropy Alloy (HEA) Powders on the Microstructure and Mechanical Properties in a Novel HEA/Diamond Composite. *Entropy* **2018**, *20*, 924. [CrossRef]
42. Zhang, W.; Zhang, M.; Peng, Y.; Liu, F.; Liu, Y.; Hu, S.; Hu, Y. Effect of Ti/Ni Coating of Diamond Particles on Microstructure and Properties of High-Entropy Alloy/Diamond Composites. *Entropy* **2019**, *21*, 164. [CrossRef]
43. Wang, Q.; Li, Z.; Pang, S.; Li, X.; Dong, C.; Liaw, P. Coherent Precipitation and Strengthening in Compositionally Complex Alloys: A Review. *Entropy* **2018**, *20*, 878. [CrossRef]
44. Zýka, J.; Málek, J.; Veselý, J.; Lukáč, F.; Čížek, J.; Kuriplach, J.; Melikhova, O. Microstructure and Room Temperature Mechanical Properties of Different 3 and 4 Element Medium Entropy Alloys from HfNbTaTiZr System. *Entropy* **2019**, *21*, 114. [CrossRef]
45. Tseng, K.-K.; Juan, C.-C.; Tso, S.; Chen, H.-C.; Tsai, C.-W.; Yeh, J.-W. Effects of Mo, Nb, Ta, Ti, and Zr on Mechanical Properties of Equiatomic Hf-Mo-Nb-Ta-Ti-Zr Alloys. *Entropy* **2018**, *21*, 15. [CrossRef]
46. Zhang, W.; Liaw, P.; Zhang, Y. A Novel Low-Activation VCrFeTaxWx ($x = 0.1, 0.2, 0.3, 0.4$, and 1) High-Entropy Alloys with Excellent Heat-Softening Resistance. *Entropy* **2018**, *20*, 951. [CrossRef]
47. Li, Z.; Wang, C.; Yu, L.; Gu, Y.; Pan, M.; Tan, X.; Xu, H. Magnetic Properties and Microstructure of FeCoNi(CuAl)0.8Snx ($0 \leq x \leq 0.10$) High-Entropy Alloys. *Entropy* **2018**, *20*, 872. [CrossRef]
48. Zhai, S.; Wang, W.; Xu, J.; Xu, S.; Zhang, Z.; Wang, Y. Effect of Co and Gd Additions on Microstructures and Properties of FeSiBAlNi High Entropy Alloys. *Entropy* **2018**, *20*, 487. [CrossRef]
49. Wang, Y.; Zhang, K.; Feng, Y.; Li, Y.; Tang, W.; Wei, B. Evaluation of Radiation Response in CoCrFeCuNi High-Entropy Alloys. *Entropy* **2018**, *20*, 835. [CrossRef]

entropy

MDPI

Review

Deformation Behavior of Bulk Metallic Glasses and High Entropy Alloys under Complex Stress Fields: A Review

Shunhua Chen [1,2,*], Jingyuan Wang [1], Lei Xia [3] and Yucheng Wu [2,4,*]

1 School of Mechanical Engineering, Hefei University of Technology, Hefei 230009, China; wjyyyyuan@163.com

2 National-Local Joint Engineering Research Centre of Nonferrous Metals and Processing Technology, Hefei 230009, China

3 Laboratory for Microstructures, Shanghai University, Shanghai 200444, China; xialei@shu.edu.cn

4 School of Materials Science and Engineering, Hefei University of Technology, Hefei 230009, China

* Correspondence: shchen@hfut.edu.cn (S.C.); ycwu@hfut.edu.cn (Y.W.)

Received: 1 December 2018; Accepted: 9 January 2019; Published: 11 January 2019

Abstract: The plastic deformation of bulk metallic glasses (BMGs) depends significantly on applied stress states, and more importantly, in practical applications of BMGs as structural materials, they always deform under complex stress fields. The understanding of deformation behavior of BMGs under complex stress fields is important not only for uncovering the plastic deformation mechanisms of BMGs, but also for developing BMG components with excellent mechanical performance. In this article, we briefly summarize the recent research progress on the deformation behavior of BMGs under complex stress fields, including the formation and propagation of shear bands, tunable macroscopic plasticity, and serrated plastic flows. The effect of complex stress fields on the plastic deformation mechanisms of BMGs is discussed from simple stress gradient to tailored complex stress fields. The deformation behavior of high entropy alloys (HEAs) under complex stress states has also been discussed. Challenges, potential implications and some unresolved issues are proposed.

Keywords: bulk metallic glass; complex stress field; shear band; flow serration; deformation mechanism; high entropy alloy

1. Introduction

As a new class of structural materials, bulk metallic glasses (BMGs) are considered poised for widespread engineering applications [1]. With non-ordered atomic structures, plastic deformation in BMGs is accommodated by initiation and propagation of shear bands, rather than dislocation movements as in crystalline alloys [2–4]. Under applied stress, plastic strain is localized within a thin layer of atoms with a width of about 2–210 nm, forming a shear band [5–10]. Extensive studies have shown that the initiation and propagation of shear bands are significantly dependent on applied stress fields. For example, the burst of a shear band relies on applied shear stress, and a shear band can be initiated only when the applied shear stress exceeds a critical value [5,11–14]. In addition, in practical applications of BMGs as structural materials, they always deform under complex stress states [15]. Studies have shown that BMGs can demonstrate more plastic deformation under complex stress states [16–18]. This implies that although some BMGs are brittle under conventional compression/tension tests with relatively-uniform stress fields, they can still have large capability to deform plastically in practical structural applications. For example, BMG foams can exhibit large nominal plasticity, which is much larger than solid BMG specimens [19–21]. BMG honeycombs and cellular BMG structures also demonstrate similar behavior [22–24]. The understanding of the deformation behavior of BMGs under complex stress fields is therefore important not only for uncovering the plastic deformation mechanisms of BMGs, but also for developing BMG components with excellent mechanical performance. Although many papers have reviewed the atomic structures, mechanical and physical properties, and functional and structural applications of BMGs [2–4,25–40], a comprehensive review on the plastic deformation behavior of BMGs under complex stress fields has not been reported. In this work, the research progress on the plastic deformation behavior of BMGs under complex stress fields is reviewed, including the initiation and propagation of shear bands, macroscopic plastic deformation behavior, criticality of plastic flows, and transition of deformation modes. The recent contributions on the deformation behavior of high entropy alloys (HEAs) under complex stress fields have also been discussed. The challenges, unresolved issues as well as future research directions are proposed.

2. Initiation of Shear Bands under Complex Stress Fields

Driven by applied shear stress, the dilatancy of atoms, associated with excess free volume, results in the decrease of viscosity localized in thin layers of shear bands [41,42]. The localization process causes local heating/softening, which was evidenced by experimental observations through fusible coatings [43–45]. The initiation of shear bands in BMGs was studied extensively during the past decades. Greer et al. [4] have summarized the formation mechanisms into three scenarios: the percolation of homogeneous nucleated shear transformation zones (STZs), stemming from the intrinsic structural fluctuations; the nucleation from the sites with extrinsically introduced stress concentrators, such as casting defects; and a two-consecutive-stage formation, including the burst of a viable band from activated STZs at stress-concentrated sites and a following rapid sliding process. Although the shear-banding mechanisms in BMGs are still under debate, all three kinds of scenarios share a common point that the initiation of shear bands is significantly dependent on applied stress fields. Johnson and Samwer [46] have shown that the potential barrier for activating STZs at a given shear stress, τ, can be expressed as $W_\tau = 4RG_0\gamma_C{}^2[(\tau_C - \tau)/\tau_C]^{3/2}\zeta\Omega$, where τ_C is the critical shear stress of the BMG, γ_C is the critical shear strain, R is a constant parameter, G_0 is the shear modulus of unstressed BMGs at 0 K, Ω is the volume of STZs, and ζ is a correction factor. With the change of applied shear stress (τ) under complex stress fields, the activation process of STZs will be varied accordingly, resulting in differences in the initiation of shear bands. On the other hand, the annealing of stressed MGs can be used to shape MG films without embrittlement, where the change of free volume in stressed BMGs may also cause different deformation behavior [47]. Hufnagel et al. [48] have indicated that although many previous studies have focused on the deformation behavior of BMGs under uniaxial stress states, the understanding of the deformation behavior as well as structural evolution under more complex stress states is urgently needed to uncover the shear-banding mechanisms, such as the shear localization process. The research progress on the formation mechanisms of shear bands in BMGs has been summarized in many review papers [4,27,30,34,35,48–54], however, the formation of shear bands under complex stress fields has rarely been emphasized. Here, the initiation of shear bands under more complex stress fields is summarized.

With well-known size effect, MGs demonstrate different deformation behavior at submicron scales, for example, larger elastic limit, higher ductility, and more homogeneously plastic deformation with necking effect [55–59]. The introducing of stress concentrators in submicron-sized specimens will facilitate the initiation of shear bands, giving better insight into the shear band formation mechanisms. A simple method to introduce stress concentrators is to create notches in testing specimens. Based on MD simulations of some notched CuZr MGs, Sha et al. [60] have reported that shear bands emanate from the notch root when the plastic zone size ahead of the notches increases beyond a critical value. As shown in Figure 1, the notches can serve as stress concentrators to facilitate the localization of plastic zones, leading to the initiation of shear bands. Following studies have examined the effect of notch geometries and sizes on the formation of shear bands [61–63]. They have shown that the initiation of shear bands can be tailored by introducing notches with appropriate geometries and sizes, which can serve as stress concentrators for highly localized shear strains. By creating surface roughness in MG nano-pillars can also achieve similar localized shear strains [64]. It can then be concluded that under complex stress fields, stress concentrators serve as origin sites for the initiation of shear bands due to higher localized shear strains. The authors would like to point out that at sub-micron scales, the formation of localized shear bands can also be suppressed till fracture due to the change of deformation modes [65]. This phenomenon has also been observed in notched MG specimens with sub-micron sizes, where the change of notch sizes can tune the deformation modes from localized shear-banding to homogeneous softening and necking [61–63]. A summary on the homogeneous deformation behavior of BMGs under complex stress fields will be given in Section 5 in this paper, and thus not described in detail here.

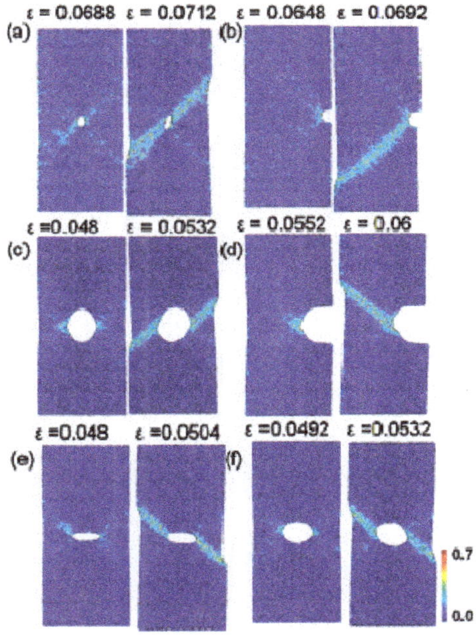

Figure 1. MD simulation results showing the initiation and propagation of shear bands in some notched $Cu_{50}Zr_{50}$ MG specimens, where the color denotes the localized shear strains. Reprinted from [60] with permission of 2013 AIP Publishing LLC.

Despite the efforts to understand the initiation of shear bands in BMGs under complex stress fields, some problems are still unresolved. For example, how to control the propagation of shear bands through the design of complex stress distributions? With well-known size effect, what's the difference between the mechanisms on the formation of shear bands at submicron scales and macroscopic scales? Further understandings of the initiation and propagation of shear bands under complex stress fields could add more knowledge to the plastic deformation mechanisms of BMGs.

3. Macroscopic Deformation Behavior under Gradient Stress Distribution

The macroscopic plastic deformation in BMGs is accommodated by the initiation and propagation of shear bands. The rapid propagation of shear bands causes catastrophic failures, while the confinement of the propagation of shear bands can enhance the macroscopic plasticity. Moreover, the impeding of the propagation of shear bands is also beneficial for the initiation and bifurcation of more shear bands, leading to more plastic deformation. The applied complex stress fields in BMGs not only bring to stress concentrators, serving as sites for the initiation of shear bands, but also influence the propagation of shear bands, even under simple gradient stress distributions. Studies on the deformation behavior of BMGs under gradient stress distributions, for example, the specimens with tailored sample geometries and surface treatments, are discussed here.

3.1. Stress Gradient Resulting from Tailored Sample Geometry (Loading Angle)

Under compression tests of BMGs, by tailoring a tilted angle between the end surface and loading platen, as shown in Figure 2a inset, significant improvement in the macroscopic plasticity, associated with work-hardening-like behavior, was observed [16]. Based on the scanning electron microscopy (SEM) observations and Finite Element Modeling (FEM) results, Chen et al. [16] have shown that the propagation of shear bands can be stopped in the regions with relatively lower stress concentration orders, leading to the multiplication of more shear bands associated with greatly enhanced macroscopic plasticity. Similar large nominal plasticity was also observed in some other BMG samples with tilted surface angles [17,66,67]. The tilted geometry constrained the initiation of shear bands in the regions with localized strains, and at the same time hindered the propagation of shear bands, leading to the multiplication of more shear bands. Despite the brittleness, some BMG communities still have certain plasticity under compression tests. However, almost all communities of BMGs are brittle under tensile loadings. The investigations on the plastic deformation behavior of the tensile side of bending BMG specimens have indicated that more plastic deformation was found due to the presence of tensile stress gradient [68,69]. The gradient stress distributions may also be possible to be used to improve the plastic deformation behavior of BMGs under tensile loadings. Chen et al. [70] have then introduced gradient stress distribution into some tensile specimens by tailoring tilted angles in Z-shaped BMG specimens, as shown in Figure 2b. It is interesting to find that the tilted BMG specimens have demonstrated more plastic deformation in the stress-concentrated regions, similar to compressive testing results. This implies that although BMGs are brittle in conventional compression/tension tests with relatively uniform stress distributions, they may still demonstrate more plastic deformation in practical engineering applications, where they always deform under complex stress fields. Additionally, with the presence of stress gradients, BMGs can also exhibit attractive performance for engineering applications, such as higher reliability in nominal strains [71] and less dependence on the change of loading rates [72]. The understanding of the deformation behavior of BMGs under stress gradients opens a new window for the practical applications of BMGs.

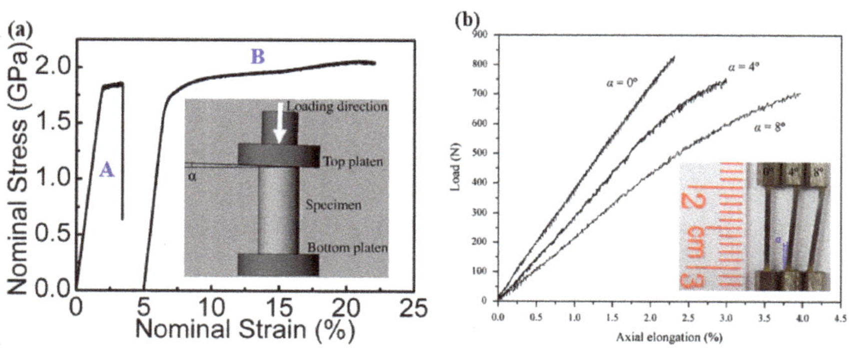

Figure 2. (**a**) Effect of stress gradient on the compressive plastic deformation behavior of BMGs, where the gradient stress distribution was introduced by tailoring a loading angle between the plateau and the specimen surface. A and B are macroscopic nominal stress-strain curves of the specimens without and with stress gradient, respectively (adapted from [16] with permission of 2008 AIP Publishing LLC.); (**b**) Effect of stress gradient on the deformation behavior of BMGs under tensile loadings, where the gradient stress distribution was introduced by designing Z-shaped specimens (adapted from [70] with permission of Elsevier).

3.2. *Effect of Surface Residual Stress*

In engineering applications of structural materials, residual stress is usually created to improve the mechanical performance. In BMG communities, the introducing of surface residual stress using techniques, such as shot peening, laser melting and mechanical attrition, has also been investigated to improve the macroscopic plastic deformation behavior. For example, Zhang et al. [73] have treated some Zr-based BMGs using shot peening, and enhanced the bending plasticity significantly (Figure 3). A maximum surface plastic strain of about 0.35% was obtained while the as-cast BMG specimen only has a limited value of about 0.15%. They have shown that the compressive residual stress on the surface of shot-peened specimens can suppress the propagation of shear bands and cracks. Combined with the pre-existed shear bands nucleated underneath the surface, more shear bands will be initiated to achieve larger macroscopic plasticity. By laser surface melting, gradient distribution of both tensile and compressive residual stresses as well as atomic structural changes can also be employed to improve the macroscopic plasticity of BMGs [74,75]. In addition, Wang et al. [76] have obtained an obvious enhancement in the tensile plasticity of BMGs by surface mechanical attrition treatment. Gradient atomic structures were formed during the mechanical attrition process, resulting in a gradient distribution of residual stress with both compressive and tensile stresses. They have shown that the gradient distribution of residual stress can cause the formation of more shear bands, and at the same time, delay the cavitation effect, resulting in a work-hardening mechanism associated with enhanced plasticity [76].

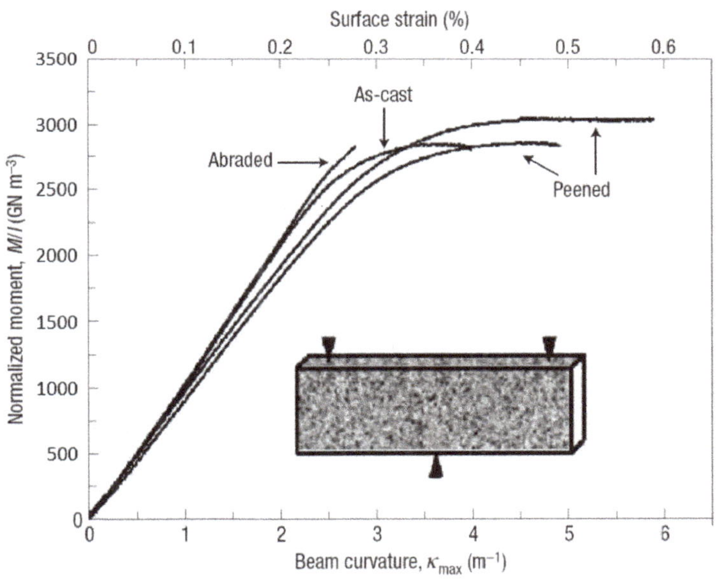

Figure 3. Three-point bending behavior of BMGs with surface treatment by shot peening. Reprinted from [73] with permission of Springer Nature.

Despite the use of different methods/techniques to introduce gradient stress distributions to BMGs, it can be concluded that the localized regions with high orders of stress concentration can promote the initiation of more shear bands, and the regions with less stress concentrations impede their propagations. This will further induce the bifurcation and multiplication of shear bands, leading to better macroscopic plasticity. By design of appropriate distributions of more complex stress fields, it is able to tailor the formation and propagation of shear bands, achieving controllable plastic deformation behavior. In fact, many studies have been devoted to achieving tunable plastic deformation behavior of BMGs by designing complex stress fields, which are given in the following section.

4. Tunable Plastic Deformation Behavior under Tailored Complex Stress Fields

During the past decade, tunable plastic deformation behavior has been widely achieved by extensive studies based on the design of complex stress fields, for example, to guide the propagation of shear bands and cracks, to tune the criticality of the plastic-flow dynamics, and to obtain large apparent plasticity/elongations. The studies are of significance for giving more insight into the shear-banding mechanisms of BMGs, and obtaining better macroscopic plastic deformation performance for engineering applications.

4.1. Guiding the Propagation of Shear Bands and Cracks

The initiation of shear bands emanating from notch roots was widely studied to evaluate the fracture behavior of BMGs [77,78]. Tandaiya et al. have shown that by changing the mixity of loading modes (I/II) of bending specimens can tune the plastic zone sizes as well as their distributions [79,80].

Under mixed mode loading conditions, the notch root deforms as that one part sharpens and the other part blunts. The increase of mode II component can enlarge the plastic zone sizes and enhance the localized strain levels ahead of the notch tips, resulting in controllable directions for shear-banding and micro cracks. For some notched BMGs, Yi et al. [81] have demonstrated that by creating pre-existed shear bands can change the plastic deformation behavior as well as the crack propagations around the notch root. To guide and deflect the cracks in those notched BMGs can improve the fracture resistance behavior, which is useful for toughening BMG components with stress concentrations in engineering applications. The controlled shear-banding behavior of notched BMGs was understood by Yang et al. from a perspective of "multiple shear band deformation mechanisms", differing from the conventional materials fracture mechanics [82]. The proposed fracture mechanisms agree well the fracture morphologies of notched BMGs, giving more theoretical understandings on the controlling of shear-banding in notched BMGs. Li et al. [83] have also conducted theoretical analysis on the controlling of shear-banding in BMGs using an instability theory. As shown in Figure 4, the formation of shear bands in notched specimens with mode mixities of 0.5 and 0.75 was successfully predicted by instability analysis, which is highly in line with the FEM results and experimental observations [83]. Such a theory may be further used to predict the shear-banding behavior of BMGs under complex stress fields in practical engineering applications.

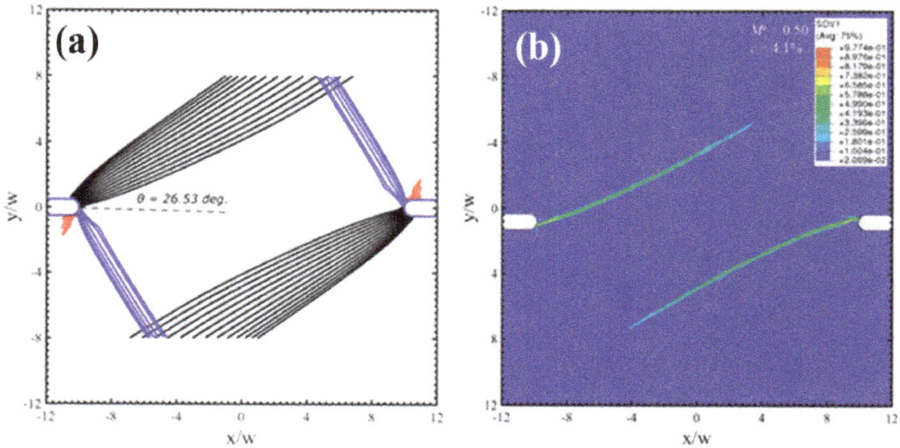

Figure 4. *Cont.*

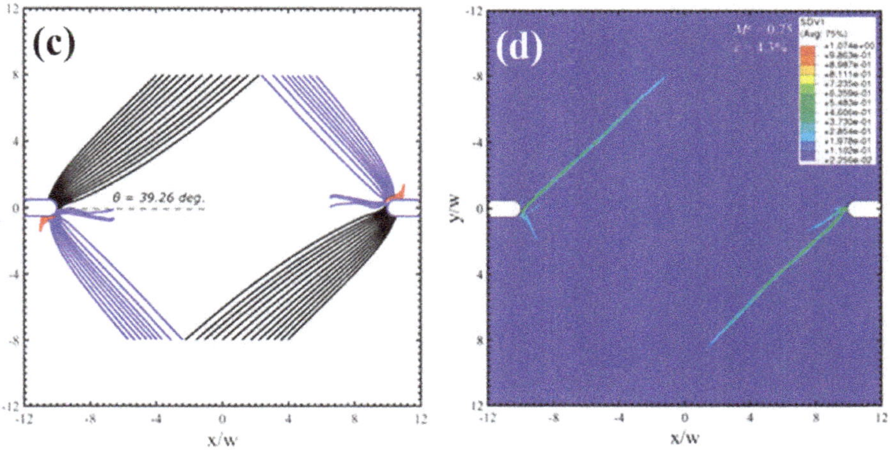

Figure 4. Shear bands formation in notched specimens with mode mixity (M^e) of 0.5 (**a**,**b**) and 0.75 (**c**,**d**), respectively (M^e = 1 and 0 stand for mode-I and mode-II loading conditions, respectively). (**a**,**c**) show the predictions by an instability theory, and (**b**,**d**) are the FEM results [83].

4.2. Tunable Criticality in Flow Serrations

The discontinuous plastic flows in crystalline alloys display serrations or jerky flows in stress-strain curves, which are known as the Portevin-Le Chatelier (PLC) effect [84,85]. Although BMGs have vanished crystalline lattices, the plastic deformation in BMGs is limited to inhomogeneously-localized shear bands, which also results in discontinuous stress-strain behavior similar to the PLC effect in crystalline alloys, i.e., serrated plastic flows. A great consideration of research has shown that the serrated plastic flows of BMGs can evolve to a power-law scaling criticality [40,86–94]. The criticality of the plastic-flow dynamics is usually correlated to larger macroscopic plasticity [86]. The criticality of the flow serrations can be attributed to the multiplication and intersections of shear bands, which suggests that the burst of shear bands may be intrinsically correlated [86,95]. However, for BMGs under tensile loadings, the catastrophic failures lead to limited serrations where the understanding of the plastic-flow dynamics is very challenging. By designing complex stress fields through double-side notches, Chen et al. [96] have reported tunable criticality in the plastic flows of BMGs under mixed mode (I/II) loading conditions. As shown in Figure 5, by the multiplication of shear bands within the regions with complex stress fields, a stable plastic flow stage was observed, resulting in the delay of catastrophic failures [96]. The two-stage plastic flows were also observed under compression tests, where the formation of new shear bands tends to delay catastrophic failures [97]. For the notched specimens with complex stress fields (Figure 5), the flow serrations during the stable plastic flow stages have smaller magnitudes, and evolve to a power-law scaling. The findings suggest that it may be possible to obtain more flow serrations in tensile BMGs by introducing of complex stress fields, where enough serration data could be collected to give more insight into the plastic-flow dynamics of BMGs under tensile loadings.

Thereafter, by tailoring single-side notches with varying radii, complex stress fields with varying stress concentration factors have been introduced to some tensile BMG specimens [76]. Despite different trends in the evolution of amplitudes from compression tests, the flow serrations of the single-side notched BMG specimens also demonstrate power-law criticality within the stable plastic flow stages, similar to the results under compression and nano-indentation tests [76]. The power-law criticality in BMGs might be a universal rule for all kinds of loading conditions. Based on such assumption, the catastrophic failures in BMGs can then be delayed or avoided in practical applications by designing complex stress fields, regardless of loading conditions. Additionally, under complex stress fields, BMGs can also have high uniformity for accumulating elastic energy during the stress-arising process of the flow serrations [98]. By tailoring of complex stress fields is helpful for uncovering the mechanisms of the plastic flow serrations in BMGs and to achieve better plastic performance.

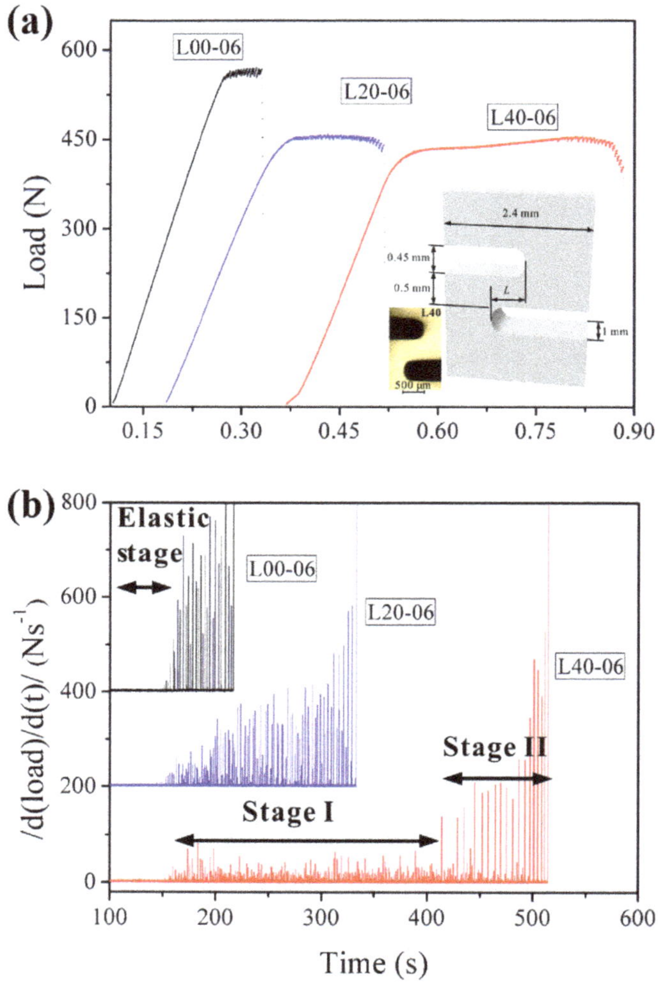

Figure 5. (**a**) Serrated plastic flows of three notched BMG specimens at a loading rate of 0.06 mm/min., where L = 0, 0.2, and 0.4 mm for L00, L20 and L40 specimens, respectively. (**b**) The |d(load)/d(t)| versus time relationships showing the observation of two-stage plastic flows [96].

4.3. Achieving Large Macroscopic Plasticity/Axial Elongation

The improved macroscopic plastic deformation behavior of BMGs under complex stress fields can be observed in some specimens with casting defects, such as voids [99]. In order to create complex stress fields to block the propagation of shear bands, Zhao et al. [100–103] have introduced notches into BMGs and obtained greatly enhanced nominal strains. An example showing the improvement of nominal strains of notched BMGs was shown in Figure 6a. The introducing of symmetrical notches can result in stable plastic flows confined within the regions between notches. With high orders of stress concentrations, shear bands are easier initiated from the notch bottoms. However, only the stress fields between two symmetrical notches can effectively confine the propagation of shear bands. In the specimens with single notches or two asymmetric notches, significant improvement in nominal plasticity was not observed due to the rapid propagation of shear bands [100–103]. By tailoring the distribution of more notches is useful for achieving larger macroscopic plasticity of BMGs, especially in practical applications of BMGs [104]. Moreover, such a strategy may be particular useful for BMGs due to the unique atomic orders. The notched high-strength steel and ceramic specimens exhibited no obvious improvement or even decreased nominal strains (Figure 6b) [102].

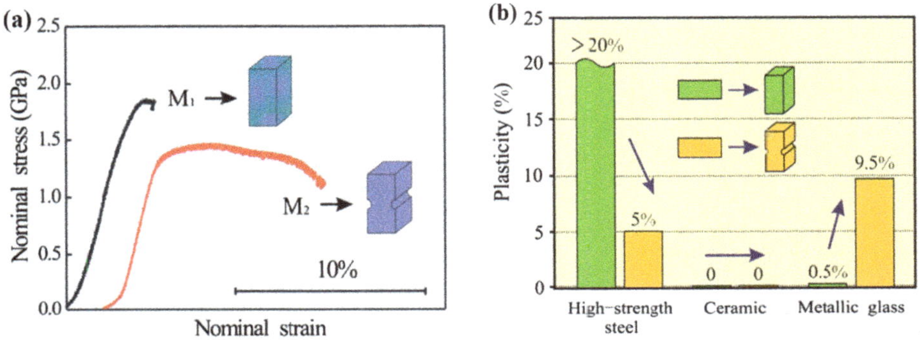

Figure 6. Enhanced nominal strain in a notched BMG specimen under compression (**a**), and the comparison of the effect in different materials (**b**). Adapted from [102] with permission of Elsevier.

As compared with the plastic deformation behavior of BMGs under compressive loadings, it is challenging to obtain plasticity under tensile loadings due to catastrophic failures. With vanishing of crystalline lattices and boundaries, the rapid propagation of shear bands cannot be impeded, and leads to brittle failures. Nevertheless, it is fortunately to find that the introducing of complex stress fields can be employed to improve the plastic deformation behavior of the stress-concentrated regions. Qu et al. [105,106] have reported that the complex stress field created by double-side notches can prevent the unstable propagation of shear bands, and result in a stable plastic zone. Li et al. [107] have further examined the effect of notch sizes and shapes on the formation and propagation of shear bands in the stress-concentrated regions, based on FEM simulations, and identified the notch conditions which are beneficial for plastic deformation. By tailoring the plastic deformation in stress-concentrated regions, tunable axial elongations have also been obtained in some curved specimens, as shown in Figure 7 [108]. With both compressive and tensile stresses, the curved segment forms a shear band multiplication stage during tensile loading process [18,108]. Although the large plastic deformation was localized in the stress-concentrated regions, the straightening of the curved segments results in large axial elongations along the loading directions. Despite the brittleness under uniform stress distributions, large axial elongations can still be achieved in BMGs and BMG structures through the geometry design, laying a sound foundation for the practical applications of BMGs as structural materials.

Additionally, tensile ductility in BMGs has also been achieved by introducing complex stress states onto the surface of BMG specimens, such as imprinting method [109,110] and laser surface treatment [111,112]. The enhancement of macroscopic tensile ductility was attributed to the heterogeneities on the specimen surface, such as the mechanical properties and geometries, resulting from tailored complex stress fields [109,110]. The heterogeneities cause the formation of plastic zones with multiple shear bands. The multiplication and intersection of shear bands limit the rapid propagation of shear bands, and therefore enhance the macroscopic ductility. Gao et al. [111] have found that with complex stress fields introduced by laser surface treatment, the shear band propagation was impeded, leading to a relatively homogeneous deformation mode. Dong et al. [112] have further shown that the tensile ductility may be ascribed to the large scale flow, driven by the complex stress fields introduced by laser engraving. Although the mechanisms on the enhancement of macroscopic ductility of BMGs under complex stress fields are still being debated, in fact, complex stress fields have been widely employed to investigate the deformation mechanisms of BMGs, i.e., under indentation tests. For example, indentation tests have been used to investigate the shear band formation and propagation mechanisms [113,114], and the transition of deformation modes from serrated flows to homogeneous deformation [115,116]. However, since the specimens do not fracture during indentation tests and the studies mainly focused on the corresponding deformation mechanisms, these studies on the indentation tests of BMGs are therefore not discussed in detail here.

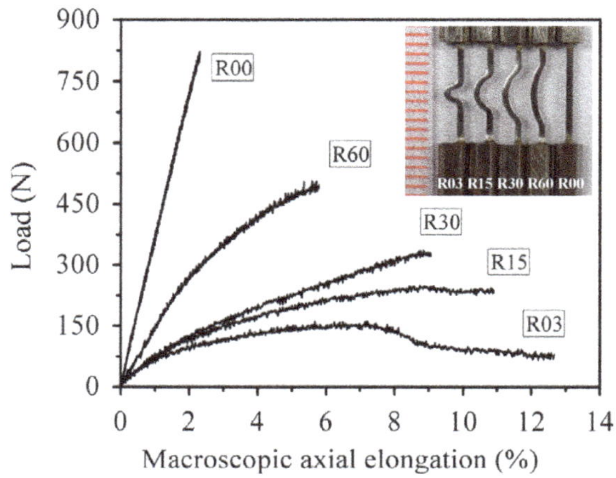

Figure 7. Tunable large axial elongations in curved BMG specimens under tensile loadings, where the inset image showing the reduced sections of the curved specimens. Adapted from [108] with permission of Elsevier.

Up to date, complex stress fields have been used to characterize the shear band formation and propagation mechanisms, as well as the global deformation behavior, such as the plasticity and fracture. However, theories on how to control the propagation of shear bands and cracks, and subsequently control the macroscopic plastic deformation behavior of BMGs are still urgently needed. In the practical applications of BMGs as structural materials, most of the parts of BMG structures may deform under complex stress fields. The understanding of the macroscopic deformation behavior of BMGs under complex stress fields is also useful for designing BMG structures with better mechanical performance, for example, the development of BMG foams [20], BMG honeycombs [22] and cellular BMGs [117].

5. Transition of Deformation Modes under Complex Stress Fields

Attributed to the amorphous atomic structures, homogeneous plastic flow in BMGs is usually observed at high temperature [42]. However, at room temperature, the introducing of complex stress fields can also result in the transition of deformation modes from highly localized shear-banding to relatively homogeneous deformation. A pioneered work by Flores and Dauskardte [118] on the strain localization behavior of some notched BMG bars has suggested that the stress states may affect the deformation and failure behavior of BMGs. Want et al. [119] have then examined the plastic deformation behavior of a $Zr_{64.13}Cu_{15.75}Ni_{10.12}Al_{10}$ (at. %) BMG under multi-axial tensile stress states, introduced by circumferential deep notches. It was surprising to find that the notched BMG specimens demonstrate strain hardening behavior at room temperature, as can be seen in Figure 8. This unusual phenomenon was attributed to the diffusional relaxation driven by multiaxial stress states, where obviously shear-banding behavior was not observed. Further studies have shown that the change of the notch dimensions can cause BMGs to deform plastically through the nucleation and coalescence of voids/cavies, where shear-banding behavior is suppressed [120]. Since the formation of shear bands occurs from very localized regions into sub-micron scales, the examination of homogeneous plastic deformation at sub-micron scales brings more understandings to the transition of deformation modes from localized shear-banding to homogeneous deformation.

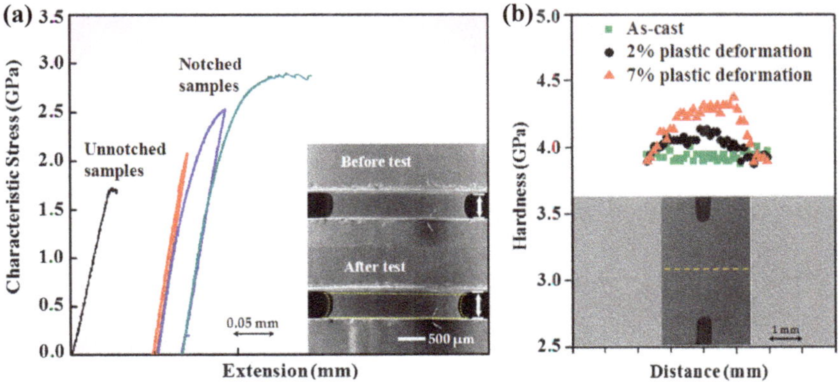

Figure 8. (**a**) Strain hardening effect in a notched BMG at room temperature, and (**b**) shows the trace of mircohardness at the notched regions. Adapted from [119] with permission of American Physical Society.

Gu et al. have examined the shear-banding and fracture behavior of some notched Ni-P/Fe-P MGs of about 70 nm in diameter, as shown in Figure 9a [121]. Combined with experimental observations and molecular dynamics (MD) simulations, they showed that plastic deformation in these notched MG specimens initiated from the notch root by forming microscopic voids, and the coalescence of voids resulted in cavitation and final brittle failures. Narayan et al. [122] have studied the deformation mechanisms of some double-side-notched CuZr MGs with varying sharpness (Figure 9b). They have found that the specimens with sharper notches can delay the formation of shear bands, resulting from a higher degree of triaxiality in stress distributions. When the cavitation stress reaches a threshold value, plastic deformation in MGs can transit from shear-banding to microscopic voids coalescence, similar to the MD simulation results reported by Gu et al. [121]. On the other hand, Narayan et al. [122] have also examined the effect of notch depth on the plastic deformation mechanisms of MGs. It was shown that a deeper notch tends to facilitate homogeneously activated STZs and then suppress shear-banding. However, to date the experimental studies on the formation of STZs and shear bands in notched MG specimens are not sufficient enough to make a consensus conclusion on the formation and evolution mechanisms of STZs/shear bands. Since STZs only involve a small cluster of atoms, it is still challenging but necessary to investigate the initiation of STZs under complex stress fields directly through in-situ TEM observations, and how these STZs evolve to shear bands/cracks. MD simulations are therefore more feasible to be employed to investigate the formation of shear bands under complex stress fields.

Based on MD simulations, many previous studies focused on the effect of notch sizes and geometries on the plastic deformation behavior of nanoscaled MG specimens, aiming to shed more light into the transition of deformation modes. Sha et al. have shown that the design of notches with increased depth and sharpness can suppress shear-banding and result in more homogeneous defamation with presence of necking [61]. Similar transition of deformation modes can also be observed in nanopillars with tailored surface roughness [64]. Pan et al. [123] have investigated the physical origin of the homogenous deformation, and found that voids and cavitations were initiated at the notch root when the stress triaxiality exceeds a critical value. On the other hand, Dutta et al. [62] have also observed the transition of deformation modes from shear-banding (Figure 10a,b) to homogeneous deformation with necking (Figure 10c,d). With a relatively sharper notch (Figure 10c), plastic zones first initiated from the notch root due to stress concentrations, and then evolved to incipient shear bands (Figure 10d). However, due to the rapid expansion and coalesce of plastic zones, the formation of shear bands was finally suppressed, showing a necking effect (Figure 10e,f). It is reasonable to see the initiation of some incipient shear bands, where stress concentrators can also serve as sites for the nucleation of STZs, as discussed in Section 2. There may exist a competing process for the formation of localized shear bands and the coalescence of plastic zones, where the rapid coalesce of plastic zones can suppress the initiation of shear bands and result in different plastic deformation mechanisms. This phenomenon is also highly in line with the in-situ TEM observations on the plastic deformation behavior of some nanosized MG specimens [122]. The change of atomic packing in a shear band was observed in a Ni-based MG [124]. More recently, Cui et al. [63] have examined the structural evolution of CuZr MGs during the transition of deformation modes from shear-banding to homogeneously necking in notched MGs. They have shown that the Voronoi volume recovery can be dominant in the localized regions with triaxial stress state, differing from the unnotched specimens. Nevertheless, more effort is still needed for revealing the underlying physical origins of the homogeneous plastic deformation within the localized regions with complex stress fields, such as the movements of atoms, atomic structural evolutions as well as the formation and coalesce of microscopic voids.

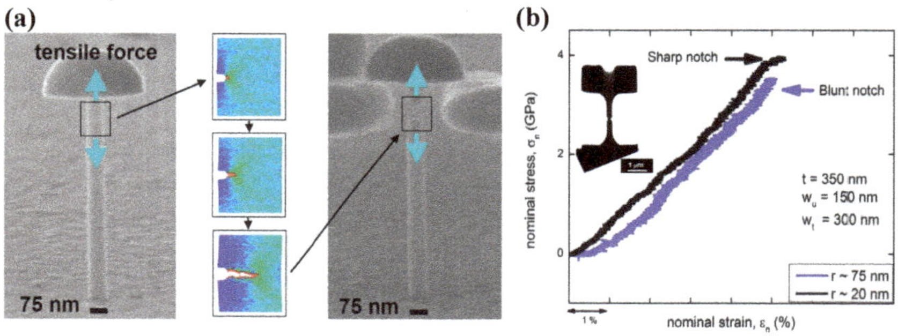

Figure 9. (**a**) Deformation behavior of single-side-notched Ni-P/Fe-P MGs (Reprinted from [121] with permission of ACS Publications); (**b**) in-situ TEM observation of the deformation behavior of double-side-notched CuZr MG specimens with varying sharpness (adapted from [122] with permission of Elsevier).

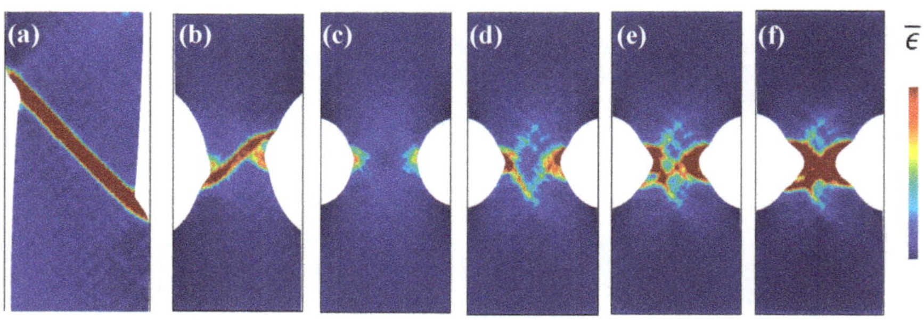

Figure 10. MD simulation results of the Mises strain plots of unnotched (**a**) and notched (**b–f**) specimens, where (**c–f**) shows the evolution of plastic zones in a specimen with a sharper notch than (**b**). Adapted from [62] with permission of Elsevier.

6. Deformation Behavior of HEAs under Complex Stress Fields

HEAs are a new class of alloys, having at least five elements with equal or near equal atomic percentages, where the solvent and solute elements are not easily distinguished [125]. The mixing of multi component in the solution states results in very high entropy associated with unique properties. For example, they can have high strength comparable to BMGs [125]. More importantly, in conventional metallurgical methods, the increase of strength leads to the decrease of ductility, and it seems that the strength and ductility are usually mutually exclusive in conventional alloys [126]. However, some HEAs can exhibit both high strength and ductility [127]. The plastic deformation behavior of HEAs shares similar characteristics as compared with BMGs. For example, differing from conventional alloys, HEAs also have serrated plastic flows, resulting in challenges to accurately predict and control the plastic deformation behavior [40,128]. The serrated plastic flows of HEAs are also significantly dependent on the change of strain rates, temperature and sample dimensions [125,128,129]. For instance, Zou et al. [129] have examined the plastic deformation behavior of some HEAs with Ar+ ion beam-assisted deposition, where flow serrations were obviously observed in the pillar of 70 nm diameter, while the pillars with larger diameters tend to have smooth plastic flows (Figure 11).

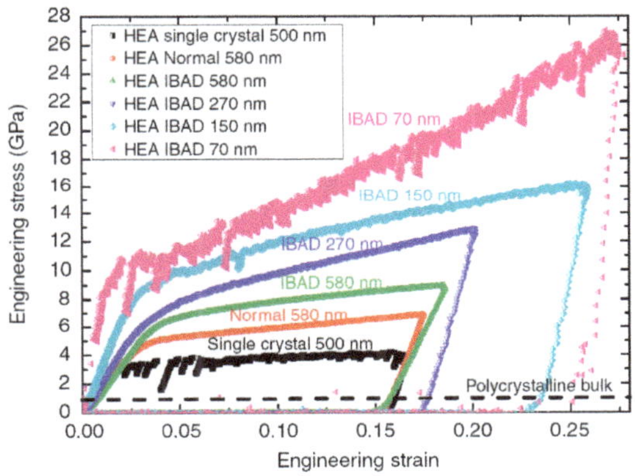

Figure 11. The compressive stress-strain curves of NbMoTaW HEAs, where IBAD stands for the specimens with Ar^+ ion beam-assisted deposition [129].

Recent research has shown that the residual strains may cause the instability of phases and result in the transition of phases [130]. This may further change the mechanical properties of HEAs, for example, transition induced plasticity [127]. Joseph et al. [131] have reported that some laser-fabricated $Al_{0.3}$CoCrFeNi HEAs exhibited tension/compression asymmetric deformation behavior, without/with mechanical twining. Therefore, the change of applied stress fields could also affect the evolution of microstructures in HEAs, such as the phase transition and mechanical twining, and the resultant flow serrations and mechanical properties. The tailoring of complex stress fields can be used to improve the plasticity and tune the criticality of plastic flows in BMGs, while some research has also shown that the notches can reduce the nominal strains of crystalline high-strength steels [102]. Whether the introducing of complex stress fields is helpful for improving the plasticity and tuning the criticality of flow serrations in HEAs is still a mystery. Up to date, a study focused on the deformation behavior of HEAs under complex stress fields has yet been reported. Regarding that the evolution of microstructures of HEAs is significantly affected by applied stress states, it should be worthy of further investigations on the deformation behavior/mechanisms of HEAs under complex stress fields.

7. Conclusions and Future Directions

The change of applied stress fields can significantly affect many aspects of the plastic deformation in BMGs, such as the shear band initiation and propagation, evolution of flow serration, and macroscopic plastic performance. Varying complex stress fields have been employed to investigate the mechanisms on the burst of shear bands, guiding the propagation of shear bands/cracks, tunable plastic-flow dynamics, transition of deformation modes and better plastic performance. On one hand, complex stress fields play a significant role for elucidating the mechanisms on the plastic deformation in BMGs. For example, how STZs are initiated and when the formation of STZs is suppressed? On the other hand, complex stress fields can be tailored to achieve controllable plastic deformation behavior for practical engineering applications. A case in point is to design BMG structures/devices with enhanced mechanical properties. However, the understanding on the deformation behavior of BMGs under complex stress fields still faces challenges, and some unresolved issues are summarized and given below:

- Although studies have been devoted to investigating the formation of shear bands under complex stress fields, how to control the formation of a shear band under a given complex stress fields is still challenging, especially under experimental observations. The localization of plastic deformation in BMGs at submicron scales involves size effect and transition of deformation modes. Due to different mechanical/physical properties, differences may exist during the formation of shear bands when characterized in specimens with varying sample dimensions. With complex stress fields, the formation and propagation of shear bands can be tailored, and even be eliminated by homogeneous deformation. The mechanisms on the plastic deformation of BMGs under complex stress fields are worthy of further attention to uncover the fundamental deformation/fracture mechanisms of BMGs.
- Extensive studies have shown that the burst of shear bands is not an independent event and affected by previously existing shear bands [73,96]. The flow serrations in BMGs, which are related to the formation of shear bands, may also have intrinsic links. Despite the well-known tunable power-law criticality, how to predict and control the serrated plastic flows in BMGs is still very difficult and challenging. The control of the initiation and propagation of shear bands under complex stress fields could be helpful for shedding more light into the underlying relationships among the bursts of flow serrations.
- The engineering applications of BMGs still have many challenges due to catastrophic failures and metastable microstructures, associated with uncertainty in mechanical properties. The achievement of controllable plastic deformation behavior under complex stress fields may not only improve the macroscopic mechanical performance, but also lead to more reliable behavior. Combining with tailored complex stress fields, BMG devices/structures with enhanced performance as well as predictable properties can be further developed, exploring the engineering applications of BMGs.
- HEAs have some characteristics similar to BMGs, for example, the serrated plastic flows which are difficult to predict/control. The introducing of complex stress fields can result in the evolution of microstructures, such as phase transition and mechanical twining, which may also be beneficial for uncovering the deformation mechanisms of HEAs. Furthermore, with controllable evolution of microstructures, the outstanding mechanical properties of HEAs could be further improved and optimized.

Author Contributions: S.C. prepared the manuscript. S.C. and J.W. collected the data. S.C., L.X. and Y.W. designed the scope of the paper. All authors discussed the conclusions and reviewed the manuscript.

Funding: This research was funded by National Natural Science Foundation of China grant number 51801049.

Conflicts of Interest: The authors declare no conflict of interest.

References

1. Plummer, J.; Johnson, W.L. Is metallic glass poised to come of age? *Nat. Mater.* **2015**, *14*, 553–555. [CrossRef] [PubMed]
2. Wang, W.H.; Dong, C.; Shek, C.H. Bulk metallic glasses. *Mater. Sci. Eng. R* **2004**, *44*, 45–89. [CrossRef]
3. Yavari, A.R.; Lewandowski, J.J.; Eckert, J. Mechanical properties of bulk metallic glasses. *MRS Bull.* **2007**, *32*, 635–638. [CrossRef]

4. Greer, A.L.; Cheng, Y.Q.; Ma, E. Shear bands in metallic glasses. *Mater. Sci. Eng. R* **2013**, *74*, 71–132. [CrossRef]
5. Chen, M.W.; Inoue, A.; Zhang, W.; Sakurai, T. Extraordinary plasticity of ductile bulk metallic glasses. *Phys. Rev. Lett.* **2006**, *96*, 245502. [CrossRef] [PubMed]
6. Hieronymus-Schmidt, V.; Rosner, H.; Wilde, G.; Zaccone, A. Shear banding in metallic glasses described by alignments of eshelby quadrupoles. *Phys. Rev. B* **2017**, *95*, 134111. [CrossRef]
7. Wang, J.G.; Pan, Y.; Song, S.X.; Sun, B.A.; Wang, G.; Zhai, Q.J.; Chan, K.C.; Wang, W.H. How hot is a shear band in a metallic glass? *Mater. Sci. Eng. A* **2016**, *651*, 321–331. [CrossRef]
8. Dai, L.H.; Jiang, M.Q.; Wang, W.H. Prediction of shear-band thickness in metallic glasses. *Scripta Mater.* **2009**, *60*, 1004–1007. [CrossRef]
9. Joshi, S.P.; Ramesh, K.T. Stability map for nanocrystalline and amorphous materials. *Phys. Rev. Lett.* **2008**, *101*, 025501. [CrossRef]
10. Liu, C.; Roddatis, V.; Kenesei, P.; Maass, R. Shear-band thickness and shear-band cavities in a Zr-based metallic glass. *Acta Mater.* **2017**, *140*, 206–216. [CrossRef]
11. Han, Z.H.; Yang, W.; Wu, F.; Li, Y. Invariant critical stress for shear banding in a bulk metallic glass. *Appl. Phys. Lett.* **2008**, *93*, 231912. [CrossRef]
12. Cao, A.J.; Cheng, Y.Q.; Ma, E. Structural processes that initiate shear localization in metallic glass. *Acta Mater.* **2009**, *57*, 5146–5155. [CrossRef]
13. Ketov, S.V.; Louzguine-Luzgin, D.V. Localized shear deformation and softening of bulk metallic glass: Stress or temperature driven? *Sci. Rep.* **2013**, *3*, 02798. [CrossRef] [PubMed]
14. Joo, S.H.; Kato, H.; Gangwar, K.; Lee, S.; Kim, H.S. Shear banding behavior and fracture mechanisms of $Zr_{55}Al_{10}Ni_5Cu_{30}$ bulk metallic glass in uniaxial compression analyzed using a digital image correlation method. *Intermetallics* **2013**, *32*, 21–29. [CrossRef]
15. Kumar, G.; Desai, A.; Schroers, J. Bulk metallic glass: The smaller the better. *Adv. Mater.* **2011**, *23*, 461–476. [CrossRef] [PubMed]
16. Chen, L.Y.; Ge, Q.; Qu, S.; Jiang, Q.K.; Nie, X.P.; Jiang, J.Z. Achieving large macroscopic compressive plastic deformation and work-hardening-like behavior in a monolithic bulk metallic glass by tailoring stress distribution. *Appl. Phys. Lett.* **2008**, *92*, 211905. [CrossRef]
17. Wu, W.F.; Zhang, C.Y.; Zhang, Y.W.; Zeng, K.Y.; Li, Y. Stress gradient enhanced plasticity in a monolithic bulk metallic glass. *Intermetallics* **2008**, *16*, 1190–1198. [CrossRef]
18. Chen, S.H.; Chan, K.C.; Xia, L. Deformation behavior of a Zr-based bulk metallic glass under a complex stress state. *Intermetallics* **2013**, *43*, 38–44. [CrossRef]
19. Brothers, A.H.; Dunand, D.C. Ductile bulk metallic glass foams. *Acta Mater.* **2005**, *17*, 484–486. [CrossRef]
20. Brothers, A.H.; Dunand, D.C. Plasticity and damage in cellular amorphous metals. *Adv. Mater.* **2005**, *53*, 4427–4440. [CrossRef]
21. Louzguine-Luzgin, D.V.; Inoue, A.; Wada, T. Improved mechanical properties of bulk glassy alloys containing spherical pores. *Mater. Sci. Eng. A* **2007**, *471*, 144–150. [CrossRef]
22. Sarac, B.; Ketkaew, J.; Popnoe, D.O.; Schroers, J. Honeycomb structures of bulk metallic glasses. *Adv. Funct. Mater.* **2012**, *22*, 3161–3169. [CrossRef]
23. Chen, W.; Liu, Z.; Robinson, H.M.; Schroers, J. Flaw tolerance vs. performance: A tradeoff in metallic glass cellular structures. *Acta Mater.* **2014**, *73*, 259–274. [CrossRef]
24. Chen, S.H.; Chan, K.C.; Wu, F.F.; Xia, L. Pronounced energy absorption capacity of cellular bulk metallic glasses. *Appl. Phys. Lett.* **2014**, *104*, 111907. [CrossRef]
25. Wang, W.H. Bulk metallic glasses with functional physical properties. *Adv. Mater.* **2009**, *21*, 4524–4544. [CrossRef]
26. Ma, E.; Cheng, Y.Q. Atomic-level structure and structure-property relationship in metallic glasses. *Prog. Mater. Sci.* **2011**, *56*, 379–473. [CrossRef]
27. Takeuchi, S.; Edagawa, K. Atomistic simulation and modeling of localized shear deformation in metallic glasses. *Prog. Mater. Sci.* **2011**, *56*, 785–816. [CrossRef]
28. Nieh, T.G.; Yang, Y.; Lu, J.; Liu, C.T. Effect of surface modifications on shear banding and plasticity in metallic glasses: An overview. *Prog. Nat. Sci.* **2012**, *22*, 355–363. [CrossRef]
29. Yu, H.B.; Wang, W.H.; Samwer, K. The beta relaxation in metallic glasses: An overview. *Mater. Today* **2013**, *16*, 183–191. [CrossRef]

30. Louzguine-Luzgin, D.V.; Louzguina-Luzgina, L.V.; Churyumov, A.Y. Mechanical properties and deformation behavior of bulk metallic glasses. *Metals* **2013**, *3*, 1–22. [CrossRef]
31. Liu, L.; Zhang, C. Fe-based amorphous coatings: Structures and properties. *Thin Solid Films* **2014**, *561*, 70–86. [CrossRef]
32. Lin, Y.C.; Tsai, Y.C.; Ono, T.; Liu, P.; Esashi, M.; Gessner, T.; Chen, M.W. Metallic glass as a mechanical material for microscanners. *Adv. Funct. Mater.* **2015**, *25*, 5677–5682. [CrossRef]
33. Sarac, B.; Sopu, D.; Park, E.; Hufenbach, J.K.; Oswald, S.; Stoica, M.; Eckert, J. Mechanical and structural investigation of porous bulk metallic glasses. *Metals* **2015**, *5*, 920–933. [CrossRef]
34. Sun, B.A.; Wang, W.H. The fracture of bulk metallic glasses. *Prog. Mater. Sci.* **2015**, *74*, 211–307. [CrossRef]
35. Trexler, M.M.; Thadhani, N.N. Mechanical properties of bulk metallic glasses. *Prog. Mater. Sci.* **2010**, *55*, 759–839. [CrossRef]
36. Chen, S.H.; Cheng, H.Y.; Chan, K.C.; Wang, G. Metallic glass structures for mechanical-energy-dissipation purpose: A review. *Metals* **2018**, *8*, 689. [CrossRef]
37. Schroers, J. Processing of bulk metallic glass. *Adv. Mater.* **2010**, *22*, 1566–1597. [CrossRef] [PubMed]
38. Jafary-Zadeh, M.; Kumar, G.P.; Branicio, P.S.; Seifi, M.; Lewandowski, J.J.; Cui, F. A critical review on metallic glasses as structural materials for cardiovascular stent applications. *J. Funct. Biomater.* **2018**, *9*, 19. [CrossRef]
39. Khan, M.M.; Nemati, A.; Rahman, Z.U.; Shah, U.H.; Asgar, H.; Haider, W. Recent advancements in bulk metallic glasses and their applications: A review. *Crit. Rev. Solid State* **2018**, *43*, 233–268. [CrossRef]
40. Zhang, Y.; Liu, J.P.; Chen, S.Y.; Xie, X.; Liaw, P.K.; Dahmen, K.A.; Qiao, J.W.; Wang, Y.L. Serration and noise behaviors in materials. *Prog. Mater. Sci.* **2017**, *90*, 358–460. [CrossRef]
41. Argon, A.S.; Shi, L.T. Development of visco-plastic deformation in metallic glasses. *Acta Metall.* **1983**, *31*, 499–507. [CrossRef]
42. Spaepen, F. A microscopic mechanism for steady-state inhomogeneous flow in metallic gasses. *Acta Metall.* **1977**, *25*, 407–415. [CrossRef]
43. Lewandowski, J.J.; Greer, A.L. Temperature rise at shear bands in metallic glasses. *Nat. Mater.* **2006**, *5*, 15–18. [CrossRef]
44. Leamy, H.J.; Chen, H.S.; Wang, T.T. Plastic-flow and fracture of metallic glass. *Metall. Trans.* **1972**, *3*, 699–708. [CrossRef]
45. Steif, P.S.; Spaepen, F.; Hutchinson, J.W. Strain Localization in Amorphous Metals. *Acta Metall.* **1982**, *30*, 447–455. [CrossRef]
46. Johnson, W.L.; Samwer, K. A universal criterion for plastic yielding of metallic glasses with a $(T/T\text{-}g)^{(2/3)}$ temperature dependence. *Phys. Rev. Lett.* **2005**, *95*, 195501. [CrossRef] [PubMed]
47. Yavari, A.R.; Aljerf, M.; Georgarakis, K. Shaping of metallic glasses by stress-annealing without thermal embrittlement. *Acta Mater.* **2011**, *59*, 3817–3824. [CrossRef]
48. Hufnagel, T.C.; Schuh, C.A.; Falk, M.L. Deformation of metallic glasses: Recent developments in theory, simulations, and experiments. *Acta Mater.* **2016**, *109*, 375–393. [CrossRef]
49. Chen, M.W. Mechanical behavior of metallic glasses: Microscopic understanding of strength and ductility. *Annu. Rev. Mater. Res.* **2008**, *38*, 445–469. [CrossRef]
50. Schuh, C.A.; Hufnagel, T.C.; Ramamurty, U. Mechanical behavior of amorphous alloys. *Acta Mater.* **2007**, *55*, 4067–4109. [CrossRef]
51. Wang, W.H.; Yang, Y.; Nieh, T.G.; Liu, C.T. On the source of plastic flow in metallic glasses: Concepts and models. *Intermetallics* **2015**, *67*, 81–86. [CrossRef]
52. Maaß, R.; Loffler, J.F. Shear-band dynamics in metallic glasses. *Adv. Funct. Mater.* **2015**, *25*, 2353–2368. [CrossRef]
53. Qiao, J.W.; Jia, H.L.; Liaw, P.K. Metallic glass matrix composites. *Mater. Sci. Eng. R* **2016**, *100*, 1–69. [CrossRef]
54. Tian, L.; Wang, X.L.; Shan, Z.W. Mechanical behavior of micronanoscaled metallic glasses. *Mater. Res. Lett.* **2016**, *4*, 63–74. [CrossRef]
55. Guo, H.; Yan, P.F.; Wang, Y.B.; Tan, J.; Zhang, Z.F.; Sui, M.L.; Ma, E. Tensile ductility and necking of metallic glass. *Nat. Mater.* **2007**, *6*, 735–739. [CrossRef] [PubMed]
56. Luo, J.H.; Wu, F.F.; Huang, J.Y.; Wang, J.Q.; Mao, S.X. Superelongation and Atomic Chain Formation in Nanosized Metallic Glass. *Phys. Rev. Lett.* **2010**, *104*, 215503. [CrossRef] [PubMed]
57. Jang, D.C.; Greer, J.R. Transition from a strong-yet-brittle to a stronger-and-ductile state by size reduction of metallic glasses. *Nat. Mater.* **2010**, *9*, 215–219. [CrossRef] [PubMed]

58. Tian, L.; Cheng, Y.Q.; Shan, Z.W.; Li, J.; Wang, C.C.; Han, X.D.; Sun, J.; Ma, E. Approaching the ideal elastic limit of metallic glasses. *Sci. Rep.* **2012**, *3*, 609. [CrossRef]
59. Liontas, R.; Jafary-Zadeh, M.; Zeng, Q.S.; Zhang, Y.W.; Mao, W.L.; Greer, J.R. Substantial tensile ductility in sputtered Zr-Ni-Al nano-sized metallic glass. *Acta Mater.* **2016**, *118*, 270–285. [CrossRef]
60. Sha, Z.D.; Pei, Q.X.; Sorkin, V.; Branicio, P.S.; Zhang, Y.W.; Gao, H.J. On the notch sensitivity of CuZr metallic glasses. *Appl. Phys. Lett.* **2013**, *103*, 253104. [CrossRef]
61. Sha, Z.D.; Pei, Q.X.; Liu, Z.S.; Zhang, Y.W.; Wang, T.J. Necking and notch strengthening in metallic glass with symmetric sharp-and-deep notches. *Sci. Rep.* **2015**, *5*, 10797. [CrossRef] [PubMed]
62. Dutta, T.; Chauniyal, A.; Singh, I.; Narasimhan, R.; Thamburaja, P.; Ramamurty, U. Plastic deformation and failure mechanisms in nano-scale notched metallic glass specimens under tensile loading. *J. Mech. Phys. Solids* **2018**, *111*, 393–413. [CrossRef]
63. Cui, W.; Pan, J.; Blackwood, D.J.; Li, Y. Voronoi volume recovery during plastic deformation in deep-notched metallic glasses. *J. Alloys Compd.* **2019**, *776*, 460–468. [CrossRef]
64. Adibi, S.; Branicio, S.P.; Liontas, R.; Cheng, D.Z.; Greer, J.R.; Scrolovitz, D.J.; Joshi, S.P. Surface roughness imparts tensile ductility to nanoscale metallic glasses. *Extreme Mech. Lett.* **2015**, *5*, 88–95. [CrossRef]
65. Deng, Q.S.; Cheng, Y.Q.; Yue, Y.H.; Zhang, L.; Zhang, Z.; Han, X.D.; Ma, E. Uniform tensile elongation in framed submicron metallic glass specimen in the limit of suppressed shear banding. *Acta Mater.* **2011**, *59*, 6511–6518. [CrossRef]
66. Tariq, N.H.; Akhter, J.I.; Hasan, B.A.; Hyder, M.J. Design induced plastic deformation in Zr-based bulk metallic glass. *J. Alloys Compd.* **2010**, *507*, 414–418. [CrossRef]
67. Wu, W.F.; Li, Y.; Schuh, C.A. Strength, plasticity and brittleness of bulk metallic glasses under compression: Statistical and geometric effects. *Philos. Mag.* **2008**, *88*, 71–89. [CrossRef]
68. Chen, S.H.; Chan, K.C.; Xia, L. Deformation evolution of a Zr-based bulk metallic glass under three-point bending tests. *Adv. Mater. Res.* **2014**, *939*, 31–38. [CrossRef]
69. Chen, S.H.; Chan, K.C.; Xia, L. Fracture morphologies of Zr-based bulk metallic glasses under different stress states. *Adv. Eng. Mater.* **2015**, *17*, 366–373. [CrossRef]
70. Chen, S.H.; Chan, K.C.; Xia, L. Effect of stress gradient on the deformation behavior of a bulk metallic glass under uniaxial tension. *Mater. Sci. Eng. A* **2013**, *574*, 262–265. [CrossRef]
71. Fan, J.J.; Yan, Y.F.; Chen, S.H.; Ng, C.H.; Wu, F.F.; Chan, K.C. Reliability of the plastic deformation behavior of a Zr-based bulk metallic glass. *Intermetallics* **2016**, *74*, 25–30. [CrossRef]
72. Chen, S.H.; Yue, T.M.; Tsui, C.P.; Chan, K.C. Effect of external disturbances on the strain-rate dependent plastic deformation behavior of a bulk metallic glass. *Mater. Sci. Eng. A* **2016**, *669*, 103–109. [CrossRef]
73. Zhang, Y.; Wang, W.H.; Greer, A.L. Making metallic glasses plastic by control of residual stress. *Nat. Mater.* **2006**, *5*, 857–860. [CrossRef] [PubMed]
74. Chen, B.Q.; Li, Y.; Yi, M.; Li, R.; Pang, S.J.; Wang, H.; Zhang, T. Optimization of mechanical properties of bulk metallic glasses by residual stress adjustment using laser surface melting. *Scripta Mater.* **2012**, *66*, 1057–1060. [CrossRef]
75. Wu, G.J.; Li, R.; Liu, Z.Q.; Chen, B.Q.; Li, Y.; Cai, Y.; Zhang, T. Induced multiple heterogeneities and related plastic improvement by laser surface treatment in CuZr-based bulk metallic glass. *Intermetallics* **2012**, *24*, 50–55. [CrossRef]
76. Chen, S.H.; Yue, T.M.; Tsui, C.P.; Chan, K.C. Flaw-induced plastic-flow dynamics in bulk metallic glasses under tension. *Sci. Rep.* **2016**, *6*, 36130. [CrossRef]
77. Kimura, H.; Masumoto, T. Plastic constraint and ductility in tensile notched specimens of amorphous $Pd_{78}Cu_6Si_{16}$. *Metall. Trans. A* **1983**, *14*, 709–716. [CrossRef]
78. Demetriou, M.D.; Launey, M.E.; Garrett, G.; Schramm, J.P.; Hofmann, D.C.; Johnson, W.L.; Ritchie, R.O. A damage-tolerant glass. *Nat. Mater.* **2011**, *10*, 123–128. [CrossRef]
79. Tandaiya, P.; Ramamurty, U.; Narasimhan, R. Mixed mode (I and II) crack tip fields in bulk metallic glasses. *J. Mech. Phys. Solids* **2009**, *57*, 1880–1897. [CrossRef]
80. Tandaiya, P.; Narasimhan, R.; Ramamurty, U. On the mechanism and the length scales involved in the ductile fracture of a bulk metallic glass. *Acta Mater.* **2013**, *61*, 1558–1570. [CrossRef]
81. Yi, J.; Wang, W.H.; Lewandowski, J.J. Guiding and deflecting cracks in bulk metallic glasses to increase damage tolerance. *Adv. Eng. Mater.* **2015**. [CrossRef]

82. Yang, G.N.; Shao, Y.; Yao, K.F. The shear band controlled deformation in metallic glass: A perspective from fracture. *Sci. Rep.* **2016**, *6*, 21852. [CrossRef] [PubMed]
83. Li, W.D.; Gao, Y.F.; Bei, H.B. Instability analysis and free volume simulations of shear band directions and arrangements in notched metallic glasses. *Sci. Rep.* **2016**, *6*, 34878. [CrossRef] [PubMed]
84. Kubin, L.P.; Estrin, Y. The Portevin-Le Chatelier effect in deformation with constant stress rate. *Acta Metall.* **1985**, *33*, 397–407. [CrossRef]
85. Lebyodkin, M.A.; Brechet, Y.; Estrin, Y.; Kubin, L.P. Statistics of the catastrophic slip events in the Portevin—Le Châtelier effect. *Phys. Rev. Lett.* **1995**, *74*, 4758–4761. [CrossRef] [PubMed]
86. Wang, G.; Chan, K.C.; Xia, L.; Yu, P.; Shen, J.; Wang, W.H. Self-organized intermittent plastic flow in bulk metallic glasses. *Acta Mater.* **2009**, *57*, 6146–6155. [CrossRef]
87. Sarmah, R.; Ananthakrishna, G.; Sun, B.A.; Wang, W.H. Hidden order in serrated flow of metallic glasses. *Acta Mater.* **2011**, *59*, 4482–4493. [CrossRef]
88. Ren, J.L.; Chen, C.; Liu, Z.Y.; Li, R.; Wang, G. Plastic dynamics transition between chaotic and self-organized critical states in a glassy metal via a multifractal intermediate. *Phys. Rev. B* **2012**, *86*, 134303. [CrossRef]
89. Bian, X.L.; Wang, G.; Chan, K.C.; Ren, J.L.; Gao, Y.L.; Zhai, Q.J. Shear avalanches in metallic glasses under nanoindentation: Deformation units and rate dependent strain burst cut-off. *Appl. Phys. Lett.* **2013**, *103*, 101907. [CrossRef]
90. Wang, Z.; Qiao, J.W.; Wang, G.; Dahmen, K.A.; Liaw, P.K.; Wang, Z.H.; Wang, B.C.; Xu, B.S. The mechanism of power-law scaling behavior by controlling shear bands in bulk metallic glass. *Mater. Sci. Eng. A* **2015**, *639*, 663–670. [CrossRef]
91. Li, J.J.; Wang, Z.; Qiao, J.W. Power-law scaling between mean stress drops and strain rates in bulk metallic glasses. *Mater. Des.* **2016**, *99*, 427–432. [CrossRef]
92. Antonaglia, J.; Antonaglia, J.; Wright, W.J.; Gu, X.J.; Byer, R.R.; Hufnagel, T.C.; LeBlanc, M.; Uhl, J.T.; Dahmen, K.A. Bulk metallic glasses deform via slip avalanches. *Phys. Rev. Lett.* **2014**, *112*, 155501. [CrossRef] [PubMed]
93. Krisponeit, J.O.; Pitikaris, S.; Avila, K.E.; Kuchemann, S.; Kruger, A.; Samwer, K. Crossover from random three-dimensional avalanches to correlated nano shear bands in metallic glasses. *Nat. Commun.* **2014**, *5*, 3616. [CrossRef] [PubMed]
94. Antonaglia, J.; Xie, X.; Schwarz, G.; Wraith, M.; Qiao, J.; Zhang, Y.; Liaw, P.K.; Uhl, J.T.; Dahmen, K.A. Tuned critical avalanche scaling in bulk metallic glasses. *Sci. Rep.* **2014**, *4*, 4382. [CrossRef] [PubMed]
95. Sun, B.A.; Pauly, S.; Tan, J.; Stoica, M.; Wang, W.H.; Kuehn, U.; Eckert, J. Serrated flow and stick-slip deformation dynamics in the presence of shear-band interactions for a Zr-based metallic glass. *Acta Mater.* **2012**, *60*, 4160–4171. [CrossRef]
96. Chen, S.H.; Chan, K.C.; Wang, G.; Wu, F.F.; Xia, L.; Ren, J.L.; Li, J.; Dahmen, K.A.; Liaw, P.K. Loading-rate-independent delay of catastrophic avalanches in a bulk metallic glass. *Sci. Rep.* **2016**, *6*, 21967. [CrossRef]
97. Louzguine-Luzgin, D.V.; Zadorozhnyy, V.Y.; Chen, N.; Ketov, S.V. Evidence of the existence of two deformation stages in bulk metallic glasses. *J. Non-Cryst. Solids* **2014**, *396*, 20–24. [CrossRef]
98. Tang, H.H.; Cai, Y.C.; Zuo, Q.; Chen, S.H.; Liu, R.P. Achieving high uniformity of the elastic strain energy accumulation rate during the serrated plastic flows of bulk metallic glasses. *Mater. Sci. Eng. A* **2018**, *736*, 269–275. [CrossRef]
99. Bei, H.; Lu, Z.P.; Shim, S.; Chen, G.; George, E.P. Specimen size effects on Zr-based bulk metallic glasses investigated by uniaxial compression and spherical nanoindentation. *Metall. Mater. Trans. A* **2010**, *41A*, 1735–1742. [CrossRef]
100. Zhao, J.X.; Qu, R.T.; Wu, F.F.; Li, S.X.; Zhang, Z.F. Deformation behavior and enhanced plasticity of Ti-based metallic glasses with notches. *Philos. Mag.* **2010**, *90*, 3867–3877. [CrossRef]
101. Zhao, J.X.; Qu, R.T.; Wu, F.F.; Li, S.X.; Zhang, Z.F. Enhanced plastic deformation in a metallic glass induced by notches. *Philos. Mag. Lett.* **2010**, *90*, 875–882. [CrossRef]
102. Zhao, J.X.; Wu, F.F.; Qu, R.T.; Li, S.X.; Zhang, Z.F. Plastic deformability of metallic glass by artificial macroscopic notches. *Acta Mater.* **2010**, *58*, 5420–5432. [CrossRef]
103. Zhao, J.X.; Zhang, Z.F. Comparison of compressive deformation and fracture behaviors of Zr- and Ti-based metallic glasses with notches. *Mater. Sci. Eng. A* **2011**, *528*, 2967–2973. [CrossRef]

104. Zhao, J.X. Achieving the desirable compressive plasticity by installing notch cluster in metallic glass. *Mater. Sci. Eng. A* **2015**, *634*, 134–140. [CrossRef]
105. Qu, R.T.; Calin, M.; Eckert, J.; Zhang, Z.F. Metallic glasses: Notch-insensitive materials. *Scr. Mater.* **2012**, *66*, 733–736. [CrossRef]
106. Qu, R.T.; Zhao, J.X.; Stoica, M.; Eckert, J.; Zhang, Z.F. Macroscopic tensile plasticity of bulk metallic glass through designed artificial defects. *Mater. Sci. Eng. A* **2012**, *534*, 365–373. [CrossRef]
107. Li, W.D.; Bei, H.B.; Gao, Y.F. Effects of geometric factors and shear band patterns on notch sensitivity in bulk metallic glasses. *Intermetallics* **2016**, *79*, 12–19. [CrossRef]
108. Chen, S.H.; Chan, K.C.; Xia, L. Deformation behavior of bulk metallic glass structural elements. *Mater. Sci. Eng. A* **2014**, *606*, 196–204. [CrossRef]
109. Qu, R.T.; Zhang, Q.S.; Zhang, Z.F. Achieving macroscopic tensile plasticity of monolithic bulk metallic glass by surface treatment. *Scripta Mater.* **2013**, *68*, 845–848. [CrossRef]
110. Scudino, S.; Bian, J.J.; Shahabi, H.S.; Sopu, D.; Sort, J.; Eckert, J.; Liu, G. Ductile bulk metallic glass by controlling structural heterogeneities. *Sci. Rep.* **2018**, *8*. [CrossRef] [PubMed]
111. Gao, M.; Dong, J.; Huan, Y.; Wang, Y.T.; Wang, W.H. Macroscopic tensile plasticity by scalarizating stress distribution in bulk metallic glass. *Sci. Rep.* **2016**, *6*, 21929. [CrossRef]
112. Dong, J.; Gao, M.; Huan, Y.; Feng, Y.H.; Liu, W.; Wang, W.H. Enhanced tensile plasticity of Zr based bulk metallic glasses by a stress induced large scale flow. *J. Alloys Compd.* **2017**, *727*, 297–303. [CrossRef]
113. Wu, F.F.; Zhang, Z.F.; Shen, J.; Mao, S.X. Shear deformation and plasticity of metallic glass under multiaxial loading. *Acta Mater.* **2008**, *56*, 894–904. [CrossRef]
114. Jana, S.; Ramamurty, U.; Chattopadhyay, K.; Kawamura, Y. Subsurface deformation during vickers indentation of bulk metallic glasses. *Mater. Sci. Eng. A* **2004**, *375*, 1191–1195. [CrossRef]
115. Schuh, C.A.; Nieh, T.G. A nanoindentation study of serrated flow in bulk metallic glasses. *Acta Mater.* **2003**, *51*, 87–99. [CrossRef]
116. Schuh, C.A.; Argon, A.S.; Nieh, T.G.; Wadsworth, J. The transition from localized to homogeneous plasticity during nanoindentation of an amorphous metal. *Philos. Mag.* **2003**, *83*, 2585–2597. [CrossRef]
117. Chen, S.H.; Chan, K.C.; Wu, F.F.; Xia, L. Achieving high energy absorption capacity in cellular bulk metallic glasses. *Sci. Rep.* **2015**, *5*, 10302. [CrossRef] [PubMed]
118. Flores, K.M.; Dauskardt, R.H. Mean stress effects on flow localization and failure in a bulk metallic glass. *Acta Mater.* **2001**, *49*, 2527–2537. [CrossRef]
119. Wang, Z.T.; Pan, J.; Li, Y.; Schuh, C.A. Densification and strain hardening of a metallic glass under tension at room temperature. *Phys. Rev. Lett.* **2013**, *111*, 135504. [CrossRef] [PubMed]
120. Pan, J.; Wang, Y.X.; Li, Y. Ductile fracture in notched bulk metallic glasses. *Acta Mater.* **2017**, *136*, 126–133. [CrossRef]
121. Gu, X.W.; Jafary-Zadeh, M.; Chen, D.Z.; Wu, Z.X.; Zhang, Y.W.; Srolovitz, D.J.; Greer, J.R. Mechanisms of failure in nanoscale metallic glass. *Nano Lett.* **2014**, *14*, 5858–5864. [CrossRef] [PubMed]
122. Narayan, R.L.; Tian, L.; Zhang, D.L.; Dao, M.; Shan, Z.W.; Hsia, K.J. Effects of notches on the deformation behavior of submicron sized metallic glasses: Insights from in situ experiments. *Acta Mater.* **2018**, *154*, 172–181. [CrossRef]
123. Pan, J.; Zhou, H.F.; Wang, Z.T.; Li, Y.; Gao, H.J. Origin of anomalous inverse notch effect in bulk metallic glasses. *J. Mech. Phys. Solids* **2015**, *84*, 85–94. [CrossRef]
124. Lesz, S.; Griner, S.; Nowosielski, R. Deformation mechanisms and fracture of Ni-based metallic glasses. *Arch. Metall. Mater.* **2016**, *61*, 791–795. [CrossRef]
125. Zhang, Y.; Zuo, T.T.; Tang, Z.; Gao, M.C.; Dahmen, K.A.; Liaw, P.K.; Lu, Z.P. Microstructures and properties of high-entropy alloys. *Prog. Mater. Sci.* **2014**, *61*, 1–93. [CrossRef]
126. Ritchie, R.O. The conflicts between strength and toughness. *Nat. Mater.* **2011**, *10*, 817–822. [CrossRef] [PubMed]
127. Li, Z.M.; Pradeep, K.G.; Deng, Y.; Raabe, D.; Tasan, C.C. Metastable high-entropy dual-phase alloys overcome the strength-ductility trade-off. *Nature* **2016**, *534*, 227–230. [CrossRef]
128. Carroll, R.; Lee, C.; Tsai, C.W.; Yeh, J.W.; Antonaglia, J.; Brinkman, B.A.W.; LeBlanc, M.; Xie, X.; Chen, S.Y.; Liaw, P.K.; et al. Experiments and model for serration statistics in low-entropy, medium-entropy, and high-entropy alloys. *Sci. Rep.* **2015**, *5*, 16997. [CrossRef]

129. Zou, Y.; Ma, H.; Spolenak, R. Ultrastrong ductile and stable high-entropy alloys at small scales. *Nat. Commun.* **2015**, *6*, 7748. [CrossRef]
130. Ye, Y.F.; Liu, C.T.; Yang, Y. A geometric model for intrinsic residual strain and phase stability in high entropy alloys. *Acta Mater.* **2015**, *94*, 152–161. [CrossRef]
131. Joseph, J.; Stanford, N.; Hodgson, P.; Fabijanic, D.M. Tension/compression asymmetry in additive manufactured face centered cubic high entropy alloy. *Scr. Mater.* **2017**, *129*, 30–34. [CrossRef]

entropy

MDPI

Review

Liquid Phase Separation in High-Entropy Alloys—A Review

Nicholas Derimow and Reza Abbaschian *

Department of Materials Science and Engineering, University of California, Riverside, CA 92521, USA; nderimow@engr.ucr.edu
* Correspondence: rabba@engr.ucr.edu; Tel.: +1-951-827-6374

Received: 24 October 2018; Accepted: 16 November 2018; Published: 20 November 2018

Abstract: It has been 14 years since the discovery of the high-entropy alloys (HEAs), an idea of alloying which has reinvigorated materials scientists to explore unconventional alloy compositions and multicomponent alloy systems. Many authors have referred to these alloys as multi-principal element alloys (MPEAs) or complex concentrated alloys (CCAs) in order to place less restrictions on what constitutes an HEA. Regardless of classification, the research is rooted in the exploration of structure-properties and processing relations in these multicomponent alloys with the aim to surpass the physical properties of conventional materials. More recent studies show that some of these alloys undergo liquid phase separation, a phenomenon largely dictated by low entropy of mixing and positive mixing enthalpy. Studies posit that positive mixing enthalpy of the binary and ternary components contribute substantially to the formation of liquid miscibility gaps. The objective of this review is to bring forth and summarize the findings of the experiments which detail liquid phase separation (LPS) in HEAs, MPEAs, and CCAs and to draw parallels between HEAs and the conventional alloy systems which undergo liquid-liquid separation. Positive mixing enthalpy if not compensated by the entropy of mixing will lead to liquid phase separation. It appears that Co, Ni, and Ti promote miscibility in HEAs/CCAs/MPEAs while Cr, V, and Nb will raise the miscibility gap temperature and increase LPS. Moreover, addition of appropriate amounts of Ni to CoCrCu eliminates immiscibility, such as in cases of dendritically solidifying CoCrCuNi, CoCrCuFeNi, and CoCrCuMnNi.

Keywords: high-entropy alloys; liquid phase separation; immiscible alloys; HEAs; multicomponent alloys; miscibility gaps; multi-principal element alloys; MPEAs; complex concentrated alloys; CCAs

---

1. Introduction

1.1. Liquid Phase Separation

Liquid phase separation (LPS), a widely-observed phenomenon in metals, is related directly to the Gibbs free energy of the system, and the most prevailing cases are often two distinct immiscible liquids of varying compositions. Although there is often some degree of solubility between the alloying elements in a metallic system exhibiting LPS, each liquid will have its own equilibrium vapor pressure, such that the vapor pressures of both phases are the same, with a positive deviation from Raoult's law. When positive deviations from Raoult's law are large, phase segregation tends to occur.

The occurrence of liquid phase separation in an alloy can lead to heterogeneous microstructures, which may or may not be desirable depending on the intended application. For example, an alloy exhibiting liquid phase separation would not be suited for use as a structural material due to the heterogeneity of the microstructure; however, it may have potential use as a self-lubricating bearing material, such as the case with Cu-Pb. There have been several comprehensive reviews of immiscible metal systems of common alloys about the phenomenon [1–5]. Therefore, the scope of this review will focus particularly on the liquid phase separation in the high-entropy alloy (HEA), complex concentrated alloy (CCA), and multi-principal element alloy (MPEA) systems.

1.2. Thermodynamics of Liquid Phase Separation

Factors such as positive deviations from Raoult's law, positive heat of mixing, and atomic size mismatch in some cases do not overcome the entropy term in the overall Gibbs free energy and cause overall immiscibility in the liquid, as is the case of miscibility between Au-Bi [5]. B. Mott in the late 1950s put together a review of the immiscible liquid metal systems, as well as the corresponding thermodynamic data for each material at the time. The immiscible alloys Mott compiled in the study contained many of the known immiscible binary monotectic alloys of the time [1]. Nearly ten years later, Mott compiled another review detailing the thermodynamics of these metal systems, as well as provided models for predictions of immiscibility in metals [2].

In Table 1, we provide a non-exhaustive table of binary alloys with miscibility gaps in the liquid state. The table expands the tables from Mott's reviews [1,2] by adding immiscible alloys from binary phase diagrams provided by the Centre for Research in Computational Thermochemistry using the FactSage thermochemical software databases [6].

Table 1. Binary systems that contain a stable miscibility gap in the liquid state.

Ag-B	Au-Ru	Bi-V	Ce-Cr	Cs-Fe	Fe-Na	K-Mo	Li-Tb	Na-Y	Sr-Tm
Ag-Co	B-Ge	Bi-Zn	Ce-Eu	Cu-K	Fe-Pb	K-Nd	Li-Ti	Na-Yb	Sr-V
Ag-Cr	B-Sn	C-Cu	Ce-K	Cu-Mo	Fe-Rb	K-Ni	Li-V	Na-Zn	Sr-Y
Ag-Fe	Ba-Ce	C-Sn	Ce-Li	Cu-Na	Fe-Sn	K-Pb	Li-Yb	Na-Zr	Sr-Zr
Ag-Ir	Ba-Cr	Ca-Cd	Ce-Mo	Cu-Pb	Fe-Sr	K-Pm	Li-Zr	Nd-Sr	Tb-Ti
Ag-K	Ba-Fe	Ca-Ce	Ce-Na	Cu-Ru	Fe-Tl	K-Pr	Lu-Na	Nd-Ti	Tb-V
Ag-Mn	Ba-Gd	Ca-Cr	Ce-Sr	Cu-Se	Ga-Hg	K-Sc	Lu-Sr	Nd-V	Te-Tl
Ag-Nb	Ba-K	Ca-Dy	Ce-Ti	Cu-Tl	Ga-Pb	K-Sm	Lu-V	Nd-Yb	Th-U
Ag-Nb	Ba-La	Ca-Er	Ce-U	Cu-Tu	Ga-Te	K-Sr	Lu-Yb	Ni-Pb	Ti-Yb
Ag-Ni	Ba-Mn	Ca-Fe	Ce-V	Cu-U	Ga-Tl	K-Tb	Mg-Mn	Ni-Sr	Tl-Zn
Ag-Os	Ba-Nd	Ca-Gd	Ce-Zr	Cu-V	Ge-Tl	K-Ti	Mg-Mo	Ni-Tl	Tm-V
Ag-Os	Ba-Pm	Ca-Ho	Co-In	Cu-W	Gd-K	K-Tm	Mg-Na	Pb-Se	V-Y
Ag-Rh	Ba-Pr	Ca-K	Co-K	Cr-Pb	Gd-Li	K-V	Mg-Nb	Pb-Si	V-Yb
Ag-Rh	Ba-Ru	Ca-La	Co-Li	Cr-Sn	Gd-Mo	K-Y	Mg-Ru	Pb-Zn	W-Zn
Ag-Se	Ba-Sc	Ca-Lu	Co-Pb	Dy-K	Gd-Na	K-Yb	Mg-Ru	Pb-Zr	
Ag-Ta	Ba-Sm	Ca-Mn	Co-Tl	Dy-Li	Gd-Sr	K-Zn	Mg-Ta	Pm-Sr	
Ag-U	Ba-Ti	Ca-Na	Cr-Dy	Dy-Na	Gd-Ti	K-Zr	Mg-Ti	Pm-Ti	
Ag-V	Ba-Y	Ca-Nd	Cr-Er	Dy-Sr	Gd-V	La-Li	Mg-W	Pm-V	
Ag-W	Ba-Zr	Ca-Pm	Cr-Eu	Dy-Ti	Gd-Yb	La-Mn	Mg-Zr	Pr-Sr	
Al-Bi	Be-K	Ca-Pr	Cr-Gd	Dy-V	Hg-Nb	La-Na	Mn-Na	Pr-Ti	
Al-Cd	Be-Li	Ca-Ru	Cr-K	Er-K	Hg-Si	La-Sr	Mn-Pb	Pr-V	
Al-In	Be-Mg	Ca-Sc	Cr-La	Er-K	Hf-Mg	La-Ti	Mn-Sr	Pr-Zr	
Al-K	Be-Na	Ca-Sm	Cr-Li	Er-Na	Ho-K	La-V	Mn-Tl	Sc-V	
Al-Na	Be-Se	Ca-Tb	Cr-Mg	Er-Sr	Ho-Mo	La-Zr	Mn-Yb	Sc-Sr	
Al-Pb	Be-Sn	Ca-Ti	Cr-Na	Er-V	Ho-Na	Li-Cs	Mo-Na	Sc-V	
Al-Tl	Be-Sr	Ca-Tm	Cr-Nd	Eu-Li	Ho-Sr	Li-Fe	Na-Nd	Se-Tl	
As-Tl	Be-Zn	Ca-V	Cr-Pb	Eu-Mn	Ho-Ti	Li-K	Na-Ni	Si-Tl	
Au-B	Bi-Co	Ca-Y	Cr-Pm	Eu-Na	Ho-V	Li-Na	Na-Pm	Sm-Sr	
Au-Ir	Bi-Cr	Ca-W	Cr-Pr	Eu-T I	In-Te	Li-Nd	Na-Pr	Sm-Ti	
Au-Mo	Bi-Fe	Ca-Zr	Cr-Sm	Eu-V	In-V	Li-Ni	Na-Sc	Sm-V	
Au-Rh	Bi-Ga	Cd-Cr	Cr-Sn	Eu-Zr	K-La	Li-Pm	Na-Sm	Sn-V	
Au-Rh	Bi-Mn	Cd-Fe	Cr-Sr	Fe-In	K-Li	Li-Pr	Na-Tb	Sn-W	
Au-Se	Bi-Rb	Cd-Ga	Cr-Tm	Fe-K	K-Lu	Li-Rb	Na-Ti	Sn-Zr	
Au-W	Bi-Se	Cd-K	Cr-Y	Fe-Li	K-Mg	Li-Sc	Na-Tm	Sr-Tb	
Au-Ru	Bi-Si	Cd-Na	Cr-Yb	Fe-Mg	K-Mn	Li-Sm	Na-V	Sr-Ti	

The molar Gibbs free energy of a system of stable unmixed liquids is represented additively via the atomic fraction of the free energies of the constituent liquids,

$$G^{L}_{A+B+C...} = \sum_{i=A,B,C...} x_i G^{L}_{i} \tag{1}$$

where $x_{i\,=\,A,B,C...}$ are the molar fractions of elements A, B, C, etc. The molar Gibbs free energy of mixing is classically defined as,

$$\Delta G_{\text{mix}} = \Delta H_{\text{mix}} - T\Delta S_{\text{mix}} \tag{2}$$

where the entropy of mixing is given as,

$$\Delta S_{\text{mix}} = -R\sum_{i} x_i \ln x_i \tag{3}$$

and the enthalpy of mixing is:

$$\Delta H_{\text{mix}} = \sum_{i=1,i\neq j}^{n} \Delta H^{\text{mix}}_{x_i,x_j} \tag{4}$$

where $\Delta H^{\text{mix}}_{x_i,x_j}$ is the interatomic interaction between concentrations of "*i*" and "*j*" elements in the system. Immiscible alloys typically have a positive value of ΔH_{mix}, which implies a preference of nearest neighbors of similar atoms as opposed to compound formation with different atoms. Many of the immiscible binary systems can be categorized by their liquid state miscibility gaps and positive enthalpy of mixing, ΔH_{mix}, of which extensive thermodynamic treatments are presented in [3,5].

The region of a phase diagram where there is non-mixing of the constituents is defined as a miscibility gap. The liquid miscibility gap in many of the monotectic binary systems assumes a dome-like shape; the shape and location of which may shift with the addition of more alloying elements. For example, one of the most well-studied ternary systems with a stable liquid miscibility gap is the Co-Cu-Fe system [7–17], while with equiatomic additions of Cr and Ni, the CoCrCuFeNi high-entropy alloy solidifies dendritically from a single-phase liquid, as observed by Yeh et al. in 2004 [18].

A generalized equilibrium monotectic phase diagram is presented in Figure 1, where the miscibility gap in the liquid state is present as a dome with label $L_1 + L_2$. The size and width of the immiscibility gap varies from system to system; however, the concept is the same. That is, cooling the alloy system from a liquid state in the concentrations that fall within the miscibility gap will lead to the liquid decomposing into two compositionally different liquids, the temperature of which is known as the critical temperature (labeled T_c in Figure 1).

As the temperature decreases to T_1 in Figure 1, the entropy term $T\Delta S$ is smaller than the enthalpy of mixing ΔH_{mix} in the free energy of the system (Figure 2); therefore, the free energy of the liquid G_L with respect to concentration of the B element in A will also assume a dome shape, presented in Figure 1. If the temperature T_1 is held, the equilibrium phases will be L_1, $L_1 + L_2$, or L_2 dependent on composition X_B. Cooling the system through T_2 until the monotectic temperature, T_3, the monotectic reaction will take place, and we will start to see α precipitate out of the liquid as L_1 is no longer stable until we reach T_4, where the remaining equilibrium phases are the solidified α and liquid L_2. Cooling through the eutectic temperature at T_5 to reach T_6, we are ultimately left with $(\alpha + \beta)$ solid phases.

Due to the lack of experimental data for mixing enthalpies of many binary alloys, a model for generating approximate mixing enthalpies was first developed by Miedema et al. in 1973 [19], which uses the electron density at the Wigner–Seitz cell boundary and the chemical potential of electronic charge of pure metals as input and can be written as $\Delta H_{\text{mix}} = \sum_{i=1,i\neq j}^{n} \Delta H^{\text{mix}}_{c_i,c_j}$. This model was used by Takeuchi et al. in 2005 for the classification of bulk metallic glasses by atomic size difference and heat of mixing [20] and later revisited by Takeuchi in 2010 [21] for mixing enthalpies of binary alloys, which includes an additional model for sub-regular solutions [22]. The ΔH_{mix} of the binary alloys from [21] serve as a starting point for many of the recent calculations of ΔH_{mix} for HEAs, MPEAs, and CCAs. Using the calculated binary mixing enthalpies (ΔH_{mix}) from Takeuchi et al. [21], ΔH_{mix}, much of the values used for determining the mixing enthalpies for HEAs/CCAs/MPEAs were calculated using Equation (4) where $\Delta H^{mix}_{x_i,x_j} = 4\Omega_{0_{ij}} x_i x_j$ for the ith and j^{th} elements at $A_{0.50}B_{0.50}$ concentrations from the tables in [21]. The values for c_i and c_j are the normalized atomic concentrations in the multicomponent alloy.

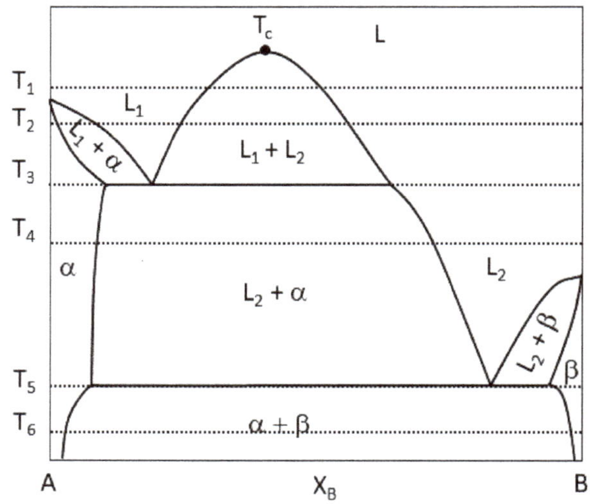

Figure 1. Generalized equilibrium monotectic binary phase diagram.

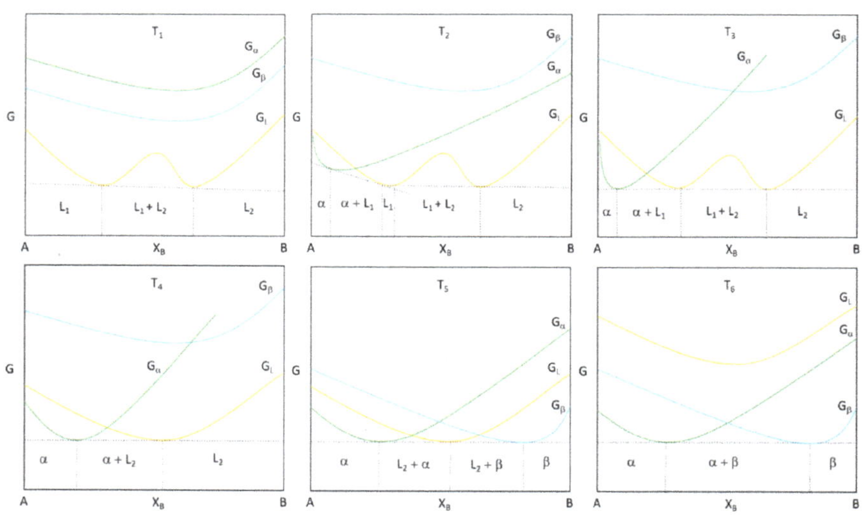

Figure 2. Gibbs free energy corresponding to the monotectic phase diagram.

1.3. Metastable Liquid Phase Separation

Unlike the stable liquid state immiscibility observed in the monotectic binary alloys, there are certain cases where a completely miscible liquid alloy can de-mix in the presence of impurities or when supercooled below the freezing temperature of the alloy, as demonstrated for Co-Cu and Cu-Fe by Nakagawa in 1958 [23]. Since then, there has been an enormous amount of LPS studies on metastable Co-Cu [8,15,16,24–34] and Cu-Fe [15,35–43], as well as the stable LPS that occurs in the combination of all three elements in Co-Cu-Fe [7–17]. Metastable liquid phase separation is defined as the liquid phase separation that occurs when undercooling an alloy such that it enters a miscibility gap that would not have been observed if solidified via conventional methods, presented in the phase diagram from [25] in Figure 3. These studies have shown that when undercooling past freezing, the single-phase alloy liquid will then split into two liquids (L1 + L2), specifically in these cases, into Cu-rich and Cu-lean liquids, which solidify often as spherical globules trapped in the frozen regions of the other liquid (the microstructure of such will be discussed later). This metastable LPS implies that there exists a dome shape similar to the monotectic alloys beneath the liquidus curves in their respective equilibrium phase diagrams.

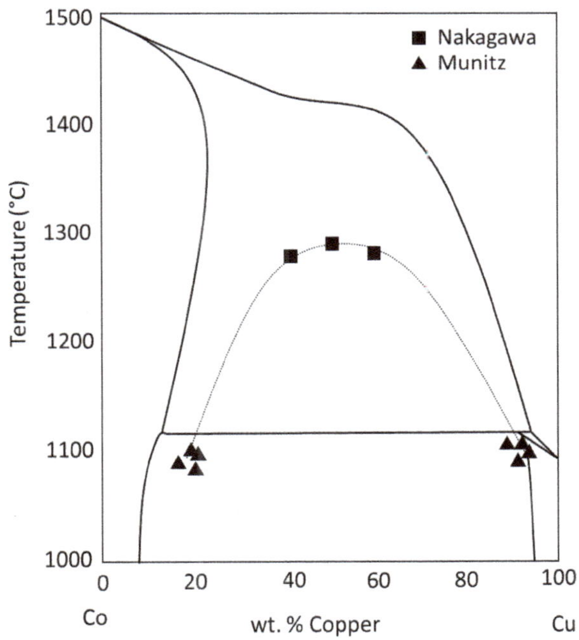

Figure 3. The Co-Cu phase diagram with the dashed line indicating the metastable liquid miscibility gap beneath the liquidus curves.

1.4. High-Entropy Alloys

The discovery of the high-entropy alloys (HEAs) [18,44–48] has inspired an enormous amount of research into multicomponent alloy design. Since their inception, there have been numerous reviews [49–60] and several books [61–66] that summarize the state-of-the-art for the materials community. These reviews compile and assess the microstructural developments, mechanical properties, crystallography, and thermodynamics of the high-entropy, complex concentrated, and multi-principal element alloy systems.

Many of the ternary alloys that exhibit liquid phase separation contain Cu, as it has a low affinity for mixing with other elements; however, there are a number of other non-Cu-containing ternary alloys with liquid miscibility gaps as well. Table 2 is a non-exhaustive list of the studied ternary alloys with liquid phase miscibility gaps, many of which contain Cu. As was the case with many binary alloys, the ΔH_{mix} of these systems are typically positive. One can think of the HEAs/CCAs/MPEAs as the addition of alloying elements to preexisting ternary alloys, some of which may actually contain a stable miscibility gap in the liquid, which is the case with CoCrCu [67] and many of the HEAs that contain Co, Cr, and Cu in equal parts with respect to the other alloying elements in the HEA. It would appear that the increase in the entropy of mixing ΔS_{mix} with additional alloying elements stabilizes the solution; however, this strongly depends on the enthalpy of mixing ΔH_{mix}, as well as other factors that determine miscibility [58].

Table 2. Ternary alloy systems that contain a stable liquid miscibility gap.

Ag-Al-Pb	[6]	Al-Mg-Mn	[6]
Ag-Co-Pd	[6]	Au-Cu-Pb	[6]
Ag-Cu-Fe	[68]	Au-In-Pb	[6]
Ag-Cu-Mn	[68]	Au-In-Pb	[6]
Ag-Cu-Ni	[68]	B-Cu-Fe	[13,69]
Ag-Cu-Pb	[68]	Bi-Ga-Zn	[6]
Ag-Cu-Se	[68]	Co-Cr-Cu	[67,70–72]
Ag-Cu-Ti	[68]	Co-Cr-Nb	[67]
Ag-Fe-Mn	[6]	Co-Cu-Fe	[7–12,14–17,73]
Ag-Fe-Ni	[6]	Cr-Cu-Fe	[11,74]
Ag-Nb-Ti	[75]	Cu-Fe-Mo	[76]
Ag-Ni-Sn	[6]	Cu-Fe-Nb	[76]
Al-Bi-Cu	[77]	Cu-Fe -Si	[11,13]
Al-Bi-Sb	[78]	Cu-Fe-Sn	[79]
Al-Bi-Sn	[80–83]	Cu-Fe-V	[11]
Al-Cu-Sn	[84]	Cu-Ni-Pb	[6]
Al-Ga-In	[6]	Fe-Si-Zn	[6]
Al-Ga-Sn	[6]	Pb-Pd-Sn	[6]

One of the first HEAs contained equal parts Co, Cr, Fe, Mn, and Ni, often written alphabetically as CoCrFeMnNi and referred to as the "Cantor alloy" after the alloy's inventor Brian Cantor [44]. Subsequent studies of the Cantor alloy involved substitution of various elements in place of the original five equiatomic elements in the alloy. One of the most popular substitutions is often Cu for Mn [18], as well as the addition of Al to create the widely-studied AlCoCrCuFeNi HEAs [85–87]. Since their inception, the majority of the high entropy alloys that are synthesized consist mostly of the 3D transition metals with other elements substituted intothe well-studied systems in the search for new stable phases, mechanical properties, and reproducible microstructures [58]. There have been several research papers that prescribe methodologies for computationally screening HEAs for those that are likely to create single-phase solid solutions [88–96]. A useful criteria for HEA/CCA/MPEA research is to know whether or not the combination of elements in a proposed HEA system will even mix in the liquid phase, as LPS traditionally is an unwanted phenomena when designing new materials with the goal of enhancing mechanical properties. If a single-phase liquid can be obtained from the proposed combination of elements, then the solidification microstructures will result in more uniformity.

There is much debate as to what exactly constitutes a high-entropy alloy (HEA), complex concentrated alloy (CCA), or multi-principal element alloy (MPEA). In essence, the core idea is very similar behind each definition: multicomponent, "baseless" alloys greater than three elements designed with the goal of surpassing the mechanical properties of traditional alloys. Whether or not which composition will solidify into a single, duplex, or multiple phases is not what this review is concerned with, but rather the observance of liquid phase separation in these multicomponent alloy systems.

Approximately 85% of HEAs in the literature to date are made up of predominantly 3D transition metals, many of which contain Al [58]. As it is impossible to visualize the phase diagram space of multicomponent alloy systems that contain 5+ elements, it can be difficult to know whether these alloy systems will phase separate in the liquid. Many of the well-studied HEA systems, such as CoCrCuFeNi for example [45], contain equiatomic CoCrCu, which has a very large liquid miscibility gap [67,70].

2. Solidification Microstructures

2.1. Dendritic Microstructure

Alloy solidification morphology and as-cast microstructure can have many forms depending on the solidification process. The most common microstructures of a solidified alloy can vary from plane front solidification, dendritic morphology, and eutectic microstructures, among others. Many of the solidification microstructures present in HEA literature consist of dendritic growth, which is typically indicative of crystal growth from a liquid with an imposed thermal gradient, as is the case with most arc-melting processes. Dendrites are solid tree-like, branching cellular structures that grow from a liquid phase. The conglomeration of atoms during solidification typically forms a nucleus of spherical shape, which then becomes unstable due to perturbations. The solid shape then begins to express the preferred growth directions of the underlying crystal and consumes atoms from the overall liquid to form a stable solidifying phase [97]. The liquid that is leftover after dendritic solidification is referred to as the interdendritic liquid, which solidifies last, and is referred to as the interdendritic region or interdendrite. The preferred dendritic growth directions for most cubic systems (FCC/BCC) are in the <100> directions, which leads the secondary dendrite arms to grow perpendicular from the primary arm. This is often an easy way to differentiate between the cubic and noncubic crystal structure of the dendritic phase.

The typical dendritic microstructure of an electromagnetically-levitated and solidified alloy is presented in Figure 4. The dendritic morphology usually indicates that the microstructure evolved from a single-phase liquid if the dendritic morphology is uniform; however, if the alloy has a small volume fraction of LPS, the remaining L2 globules may be pushed to the edge of the sample by the growing dendrites. There have been cases where liquid phase separation has occurred in the interdendritic liquid after the growth of primary dendrites [98,99]; however, the general morphology of dendritic microstructures indicates that there was no large-scale liquid-liquid immiscibility between the alloying elements.

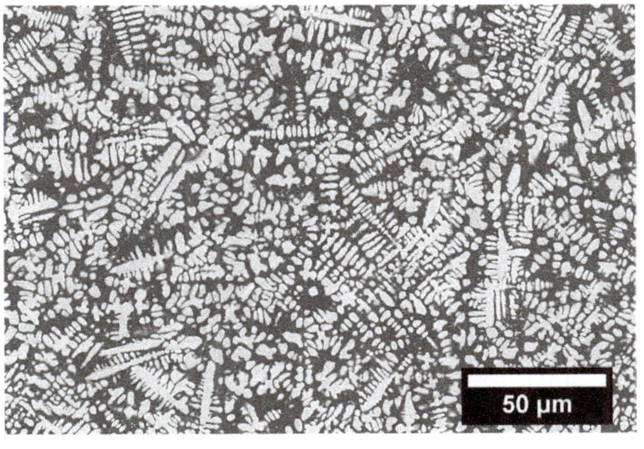

Figure 4. Backscattered electron image of an as-cast AlMoNi alloy with a dendritic microstructure.

Many HEAs/CCAs/MPEAs solidify with a duplex microstructure, where the dendritic and interdendritic regions have large compositional and crystallographic differences [53,58–60,64]. These have been shown to have interesting mechanical properties; however, they have yet to surpass the mechanical properties of commercial alloys.

2.2. Microstructures Resulting from Liquid Phase Separation

Alloys that undergo either stable or metastable LPS also have very distinct microstructures that can vary based on the solidification process. Slow cooling rates paired with a static environment can lead to the liquids separating. If the system is a little more dynamic, such as in the case of casting, the process can lead to trapping of the primary liquids in one another, referred to as emulsion (Figure 5). The separated liquids tend to be trapped as spherical globules inside the other liquid, and solidify as such, as is the case of the equiatomic CoCrCu alloy presented in Figure 6. As these liquids can be slightly different in composition than the primary liquid, they are referred to as secondary liquids. Based on morphology alone, one can distinguish the first phase to solidify from the interface between the the two liquids, as the higher melting point liquid will solidify at a higher temperature and will most of the time solidify with protrusions into the other liquid, as is the case with the solidification of CoCrCu presented in Figure 7. The backscattered electron images in Figures 6 and 7 display emulsion of the lighter (Cu-rich) and darker (CoCr-rich) liquids, as well as small protrusions coming from the CoCr-rich secondary liquid in Figure 7.

There have been several efforts to create uniform microstructures of immiscible liquid melts, such that the LPS is evenly distributed. These techniques include free directional and directional solidification [100], rheomixing [101], microgravity experiments [28,102,103], electromagnetic levitation processing [40], and rapid solidification [104]. Much of this work is aimed at the tailoring of the immiscible liquid droplets such that the microstructure has a uniform spread of the immiscible phase [3]. There have also been experiments aimed at the suppression of LPS with additional alloying elements [105].

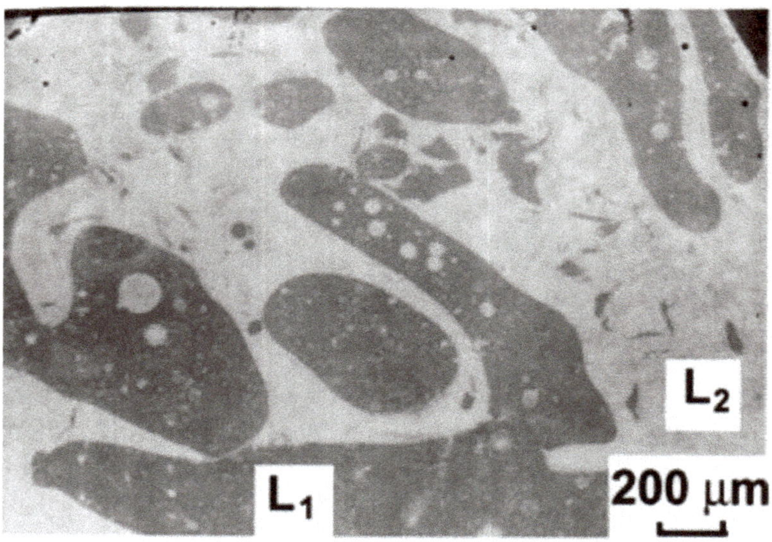

Figure 5. SEM image of liquid phase separation and emulsion of two immiscible liquids L1 and L2 in an undercooled CoCuFe alloy. Image presented with permission from the authors in [16].

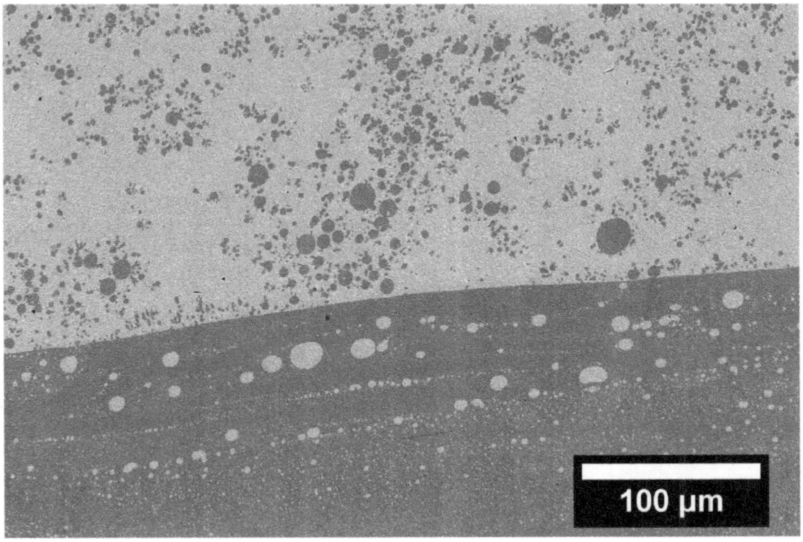

Figure 6. Backscattered electron image displaying emulsion of CoCr-rich (darker) and Cu-rich (lighter) liquids in an as-cast alloy of CoCrCu.

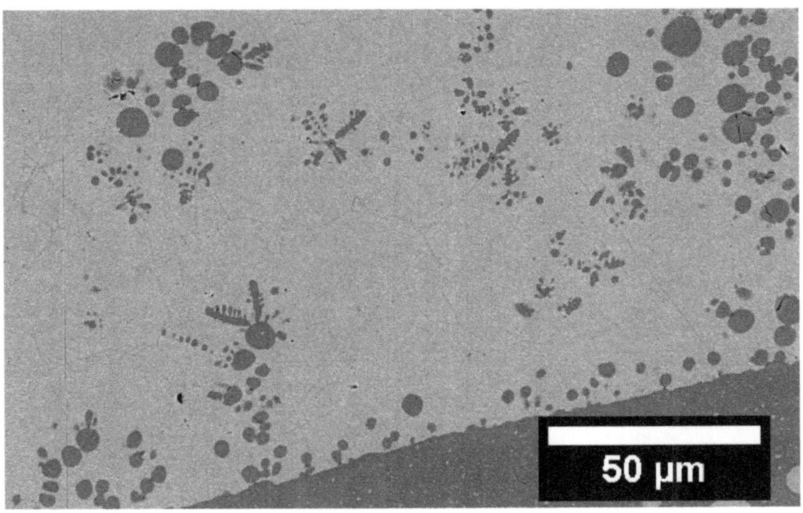

Figure 7. Backscattered electron image displaying emulsion and protrusions of the CoCr-rich (darker) phase into the Cu-rich (lighter) phase in an as-cast alloy of CoCrCu.

Traditionally, observing LPS in metallic systems was done via post-mortem analysis via metallography and microscopy, as metals are not transparent to light and have a very small transparency for X-rays. Recently, through the use of neutron transmission imaging techniques, the direct observation of liquid phase separation in metals was made possible via neutron radiographs taken during heating and cooling of immiscible CoCrCu alloys [72]. These experiments show for the first time an in situ observation of macroscopic LPS in metals and can be applied to any metallic system such that the neutron transmission through the each phase can provide enough contrast between them. Figure 8 displays neutron radiographs of two stacked CoCrCu arc-melted buttons in a small Al_2O_3 crucible with an inner diameter of 8 mm. Prior to the in situ testing, the arc-melted CoCrCu buttons underwent stable LPS and solidified with very heterogeneous Cu-rich and Cu-depleted regions, as presented in Figure 9. The resulting as-cast button consists of a non-uniform mix of the two solidified Cu-rich and Cu-depleted phases and can be seen as the lighter (Cu-rich) and darker (Cu-depleted) regions in Figure 8a. During melting (Figure 8b–d), the two liquid phases separate and stack according to density (Cu being the lighter contrast, more dense liquid phase). Note that the solid Cu-rich and CoCr-rich phases were already separated in the arc-melted buttons, but at a finer scale. The stable liquid phases then agglomerated and separated at the macro-scale. A full sequence of images in the form of a movie can be found in [72].

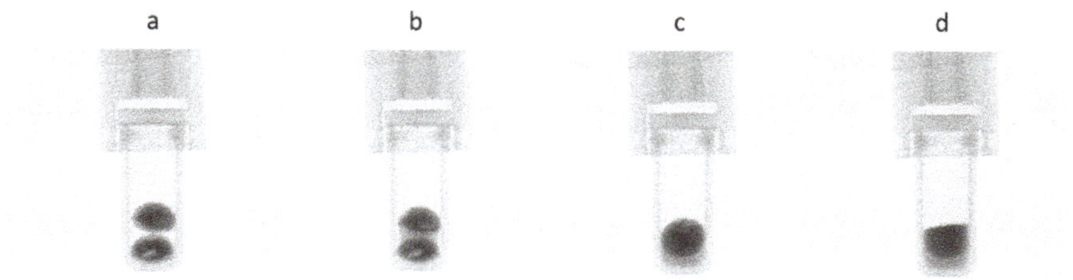

Figure 8. Melting and liquid phase separation of stacked CoCrCu samples. (**a**) During initial heating, the two as-cast buttons are intact; (**b**) the Cu-rich phase melts first between 1075 and 1100 °C and (**c**) pools at the bottom of the crucible; (**d**) the Cu-lean phase fully melts upon heating to 1500 °C and stacks based on density due to the influence of gravity. Images displayed with permission from the authors in [72].

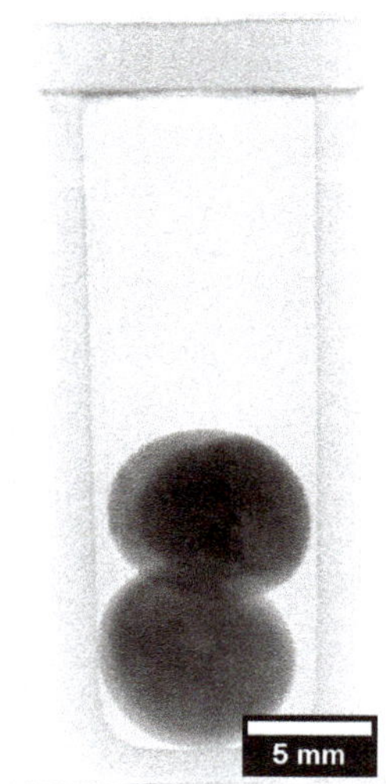

Figure 9. Neutron radiograph of two stacked arc-melted CoCrCu buttons in an alumina crucible with brighter regions corresponding to Cu-rich phases and darker regions corresponding to CoCr-rich phases. Image displayed with permission from the authors in [72].

3. High-Entropy Alloys Exhibiting Liquid Phase Separation

3.1. HEAs Containing Cu

The first occurrence of LPS in HEAs was observed by Hsu et al. in 2007 with a study of the alloying behavior of AlCoCrCuNi-based HEAs with additions of Fe, Ag, and Au [106]. The addition of Ag to the AlCoCrCuNi HEA to create AgAlCoCrCuNi was found to phase separate in the liquid, which resulted in the solidification microstructure consisting of Cu-rich globules embedded in Cu-depleted phases, contrary to the typical dendritic solidification microstructures observed for AlCoCrCuNi. Hsu suggested that in order to achieve effective mixing in the liquid, the ΔH_{mix} for atom pairs should not exceed 10 kJ/mol and that "mutual interaction between elements, based on their mixing enthalpies, should be taken into account when designing high-entropy alloys" [106].

These alloys were then revisited by Munitz et al. in 2013 where AgAlCoCrCuNi and AgAlCoCrCuFeNi were synthesized to study the melt separation behavior [107]. It was observed that the Cu-rich immiscible liquid tended to flow to the bottom of the buttons during arc-melting, as well as residual Cu-rich liquid being trapped in the interdendritic region of the Cu-depleted dendritic phase. Undercooling experiments were also carried out for a similar alloy of $Al_{1.8}CoCrCu_{3.5}FeNi$; however, no metastable liquid miscibility gap was found at the undercoolings obtained in the study (~150 K) [107].

A similar alloy composition of $AlCoCrCuFeNiSi_{0.5}$ doped with Y_2O_3 was synthesized via laser cladding with the intention to form a core-shell structure in HEAs, inspired by the LPS observed in HEAs and binary monotectics. The undoped $AlCoCrCuFeNiSi_{0.5}$ did not undergo LPS, while the addition of 1 wt.% nanosized Y_2O_3 caused the liquids to separate into egg-like globules of Cu-rich liquids inside the Cu-depleted liquid [108].

Recent studies into Co-free $Al_{2.2}CrCuFeNi_2$ revealed what the authors referred to as anomalous "sunflower-like" solidification microstructures [109], where it was suggested that LPS occurs in the no longer stable-depleted interdendritic liquid, occurring due to changes in the composition. Munitz et al. suggested that the liquid phase separation is due to constitutional changes and not temperature changes, where the authors referred to this phenomena as "constitutional LPS" (CLPS) [99]. For the $Al_{2.2}CrCuFeNi_2$ alloy, constitutional LPS occurred in the interdendritic liquid, where the interdendritic liquid decomposed into a CrFe-rich L_1 and a Cu-rich L_2. As the dendritic skeleton was already formed, the heavier Cu-rich liquid accumulated in the interdendritic region and the cast bottom, while the CrFe-rich spheres underwent solidification emulsified in the Cu-rich liquid.

A large study of several HEAs by Munitz et al. in 2017 was undertaken to explore the effects of Al, Co, Cr, Ni, Ti, and V on the miscibility gap temperature of several HEA systems. It was shown that Al, Co, Ni, and Ti lowered the miscibility gap temperature, while Cr, V, and Nb raised the miscibility gap temperature and increased LPS in these systems, the alloys of which are found in Table 3. Many of the HEAs studied by Munitz et al. contained equiatomic CoCrCu, which was experimentally determined to have a large liquid miscibility gap [67]. It is peculiar that systems such as CoCrCuFeNi will solidify dendritically [18], while similar alloys of CoCrCu [67,71,72] and CoCuFe [7–12,14–17,73] have been shown to have large liquid miscibility gaps. A recent study by Derimow et al. investigated the solidification microstructures of equiatomic CoCrCu with added Fe, Mn, Ni, V, FeMn, FeNi, FeV, MnNi, MnV, and NiV to the composition. It was found that only three of the alloys solidified dendritically (CoCrCuNi, CoCrCuFeNi, and CoCrCuMnNi), while the remaining combinations underwent stable LPS [71]. Derimow et al. also suggested that the positive mixing enthalpy of each of the systems was responsible for the LPS and presented a tree diagram for approximating the likelihood of which elements will cluster together in the melt, presented in Figure 10.

The tree diagram in Figure 10 indicates that AB atoms are more likely to cluster than ABC, AC, or BC, thereby rejecting the C element. This can be seen in case of the ternary CoCrCu, where the LPS consists of CoCr-rich and Cu-rich liquids [67,71] and solidifies with the microstructure shown in Figure 6.

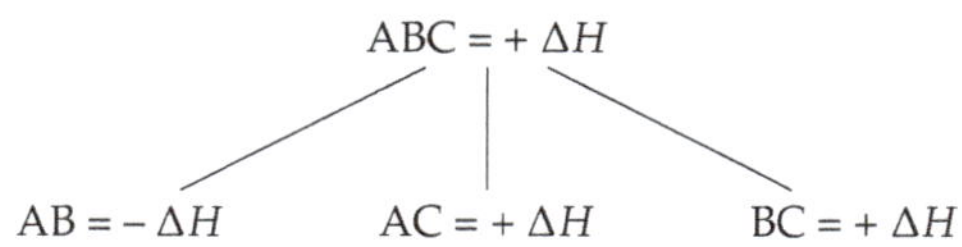

Figure 10. Tree diagram representing the probability of clustering based on the mixing enthalpies of binary combinations of elements ABC.

Table 3 compiles the multicomponent alloys found in the literature that have been shown to undergo LPS. Each of these systems have a solidification morphology similar to the microstructure presented in Figure 6. From this table, the common component in all but one of these systems is equiatomic Cu. This is in part due to the positive mixing enthalpy Cu has with many of the alloying elements in the system. It should be noted that although Ni appears to promote miscibility in Cu-containing HEAs, this may be ineffective if the repulsion between Cu and the majority of the alloying elements is too great.

Table 3. Multicomponent alloy systems with reported liquid miscibility gaps.

System	Classification	Type of LPS	Ref.
Ag-Al-Co-Cr-Cu-Fe-Ni	HEA	Stable	[107]
Ag-Al-Co-Cr-Cu-Ni	HEA	Stable	[106,107]
Al-Co-Ce-La-Zr	Bulk Metallic Glass	Stable/Metastable	[110]
$Al_{0.5}$-Co-Cr-Cu-Fe-V	HEA	Stable	[111]
Al-Cr-Cu-Fe-Ni	HEA	Stable	[99]
Al-Cu-La-Ni-Zr	Bulk Metallic Glass	Stable	[112]
B-Cu-Fe-P-Si	Fe-Cu-alloy	Stable	[113]
Co-Cr-Cu-Fe	HEA	Stable	[71,111]
Co-Cr-Cu-Fe-Mn	HEA	Stable	[71]
Co-Cr-Cu-Fe-Mo-Ni	HEA	Stable	[114]
Co-Cr-Cu-Fe-Ni	HEA	Metastable	[115–118]
Co-Cr-Cu-Fe-Ni-Nb	HEA	Stable	[111]
Co-Cr-Cu-Fe-Ti-V	HEA	Stable	[111]
Co-Cr-Cu-Fe-V	HEA	Stable	[71,111]
Co-Cr-Cu-Mn	HEA	Stable	[71]
Co-Cr-Cu-Mn-V	HEA	Stable	[71]
Co-Cr-Cu-Ni-V	HEA	Stable	[71]
Co-Cr-Cu-V	HEA	Stable	[71]
Cr-Cu-Fe-Mn-V	HEA	Stable	[111]
Cr-Cu-Fe-Mo-Ni	HEA	Stable	[119]
Cr-Cu-Fe -Ni	Cu-alloy	Stable/Metastable	[7]

3.2. *CoCrCuFeNi*

One of the seminal HEAs synthesized by Yeh et al. was the CoCrCuFeNi HEA [18]. Along with the Cantor alloy (CoCrFeMnNi) [44], these alloys served as many starting points for the addition and subtraction of alloying elements. The first study to note the liquid phase separation in HEAs in a similar composition of AgAlCoCrCuNi attributed the presence of LPS to the positive mixing enthalpies between Ag and the rest of the alloying elements [106].

An induction melting study by Wu et al. involving CoCrCuFeNi demonstrated the first occurrences of LPS in this alloy [116]. In Wu's study, several combinations of the CoCrCuFeNi alloy with varied Fe and Ni were studied to investigate the effects on the microstructure and crystallography of the system. When Fe and Ni were varied to create $CoCrCuFe_{0.5}Ni$ and $CoCrCuFeNi_{0.5}$, spherical Cu-rich separations were observed by post-mortem analysis of the solidified samples [116]. The authors rationalized that the two alloy compositions with positive mixing enthalpies were too great to be overcome by the entropy of the system, which would thereby lower the overall Gibbs energy of the system in the molten state.

Previous studies of supercooling and rapid solidification via electromagnetic levitation melting by Elder et al. characterized large undercoolings of 150 K for Cu-Fe and 75 K for Co-Cu that produce the LPS microstructures for these metastable alloys [120]. Elder et al. listed several techniques to achieve high undercoolings such as melt emulsification, melting in molten slag or fused silica (glass fluxing), free fall in a drop tube, and electromagnetic levitation techniques to achieve undercooled temperatures [120]. Using the molten fused silica technique, rapid solidification studies of CoCrCuFeNi by Liu et al. were carried out to investigate rapid solidification effects on the microstructure and phase stability of $CoCrCuFe_xNi$ HEAs (where x = 1, 1.5, and 2.0) [115]. It was found that LPS occurred in all three compositions below the critical undercooling temperature, ΔT_{crit}, where $\Delta T_{crit}^{x=1.0}$ = 160 K, $\Delta T_{crit}^{x=1.5}$ = 190 K, and $\Delta T_{crit}^{x=2.0}$ = 293 K. When $\Delta T > \Delta T_{crit}$, the microstructure is consistent with that of LPS, such that there were Cu-rich spheres present throughout the material. Due to the dendritic solidification behavior of CoCrCuFeNi through regular solidification routes [18], this alloy is classified as having a metastable liquid phase miscibility gap that is present when undercooled past ΔT_{crit} [115]. The metastable liquid miscibility gap was also confirmed by Wang et al. for CoCrCuFeNi, where the authors also achieved an exceptionally high degree of undercooling of 381 K ($0.23T_m$) for the alloy [117]. The ΔT_{crit} for equiatomic CoCrCuFeNi was also studied by Guo et al., and it was found that metastable LPS occurs when $\Delta T > \Delta T_{crit}$ = 100 K [118]. The authors also show that the yield strength and elongation of equiatomic CoCrCuFeNi significantly decrease when the alloy undergoes liquid phase separation due to the non-uniformity of the resultant microstructure [118].

Recently, Wang et al. showed that with the addition of 3 at.% Sn to CoCrCuFeNi, the alloy undergoes the same characteristic liquid phase separation when undercooled past ΔT_{crit} = 100 K [121]. The study showed that the LPS produced an increase in hardness of the Cu-depleted phases due to the separation of Cu and Sn in the liquid [121].

Previous studies on similar alloys have shown that Mo improves the strength in the AlCoCrFeNi and AlCoCrCuFeNi HEAs [122–124]. However, it has been shown that with varied Cu concentrations in $CoCrCu_xFeMoNi$ where x $\geq$ 0.5, LPS occurs in a similar fashion to the other Cu-containing HEAs [114]. The addition of Mo to the CoCrCuFeNi alloy was investigated by Wu et al. to elucidate the solidification process in these alloys, as there had been a lack of thorough studies on the solidification microstructures of these alloys [114]. Due to the Cu-rich sphere emulsion in Cu-depleted phases, likely due to the positive mixing enthalpy between Cu and the remaining alloying elements, Wu et al. suggested that the ΔH_{mix} criteria for the prediction of single phase formation be amended to include the possibility of liquid phase separation in the liquid when $\Delta H_{mix} > 0$. In order to further study the Mo-containing HEAs, Peng et al. synthesized a Co-depleted $CrCu_xFeMo_yNi$ HEA to further elucidate the effects of the large positive ΔH_{mix} between Cu and Mo on the solidification process and microstructure. It was found that Cu-rich and Cu-depleted LPS occurs in the $CrCu_xFeMo_yNi$ when x and y = 0.5 and 1, attributed to $\Delta H_{mix} > 0$ for these alloy combinations [119].

4. Closing

The field of high-entropy alloys, complex concentrated alloys, and multi-principal element alloys continues to grow, with new studies producing valuable insights for the materials community with the overarching goal of creating new alloys that exceed the properties of conventional materials. This relatively new class of material is not much different from the conventional alloys, being that they are still subject to the same thermodynamic rules that are imposed on them. The main caveats are that with the increase of alloying elements, orthogonal element phase diagram visualization becomes impossible; therefore, creative ideas are warranted to help understand the nature of the solidification of these alloys. Positive mixing enthalpy, if not compensated by the entropy of mixing, will cause liquid phase separation. It appears that Co, Ni, and Ti promote miscibility in multicomponent alloys, while Cr, V, and Nb will raise the miscibility gap temperature and increase LPS. Moreover, for equiatomic

CoCrCu, which has a large liquid miscibility gap, the addition of appropriate amounts of Ni eliminates immiscibility. The indication of such an example is the CoCrCuFeNi alloy, which will solidify dendritically, while similar alloys of CoCrCu and CoCuFe show strong immiscibility. Moreover, when Fe, Mn, Ni, V, FeMn, FeNi, FeV, MnNi, MnV, and NiV are added to to equiatomic CoCrCu, only three of the alloys solidify dendritically (CoCrCuNi, CoCrCuFeNi, and CoCrCuMnNi), while the remaining combinations undergo stable LPS. In the case of CoCrCuNiV, it appears that the addition of Ni in equiatomic amounts was not enough to overcome the positive mixing enthalpy interaction between Cu and V, as the CoCrCuV alloy also exhibits stable LPS. From the table of listed multicomponent alloys that undergo LPS, Cu is found in all but one of these combinations, which indicates that Cu containing HEAs may contain a metastable liquid miscibility gap such as the case with CoCrCuFeNi.

Author Contributions: Conceptualization, N.D. and R.A.; investigation, N.D. and R.A.; rescurces, R.A.; writing—original draft preparation, N.D. and R.A.; writing—review and editing, N.D. and R.A.; supervision, R.A.; funding acquisition, R.A.

Funding: This research was funded by the 'WINSTON CHUNG Professor in Sustainability' endowment at UC Riverside.

Acknowledgments: The authors would like to acknowledge Dr. Abraham Munitz for his lifetime of work on liquid phase separation.

Conflicts of Interest: The authors declare no conflict of interest.

References

1. Mott, B.W. Liquid immiscibility in metal systems. *Philos. Mag.* **1957**, *2*, 259–283. [CrossRef]
2. Mott, B. Immiscibility in Liquid Metal Systems. *J. Mater. Sci.* **1968**, *3*, 424–435. [CrossRef]
3. Ratke, L.; Diefenbach, S. Liquid immiscible alloys. *Mater. Sci. Eng. R* **1995**, *15*, 263–347. [CrossRef]
4. Ratke, L.; Korekt, G.; Drees, S. Solidification of Immiscible Alloys. In Proceedings of the Space Station Utilisation Symposium, Darmstadt, Germany, 30 September–2 October 1996; European Space Agency: Darmstadt, Germany, 1996; pp. 247–251.
5. Singh, R.N.; Sommer, F. Segregation and Immiscibility in Liquid Binary Alloys. *Rep. Prog. Phys.* **1997**, *60*, 57–150. [CrossRef]
6. Bale, C.W.; Bélisle, E.; Chartrand, P.; Decterov, S.A.; Eriksson, G.; Gheribi, A.E.; Hack, K.; Jung, I.H.; Kang, Y.B.; Melançon, J.; et al. FactSage thermochemical software and databases, 2010–2016. *Calphad Comput. Coupling Phase Diagr. Thermochem.* **2016**, *55*, 1–19. [CrossRef]
7. Munitz, A.; Abbachian, R.; Cotler, C.; Shacham, C. Liquid Phase Separation in Cu-Co-Fe and Cu-Fe-Ni-Cr Alloys. *High Temp. Mater. Process.* **1996**, *15*, 187–194. [CrossRef]
8. Munitz, A.; Abbaschian, R. Liquid separation in Cu-Co and Cu-Co-Fe alloys solidified at high cooling rates. *J. Mater. Sci.* **1998**, *33*, 3639–3649. [CrossRef]
9. Kim, D.I.; Abbaschian, R. The metastable liquid miscibility gap in Cu-Co-Fe alloys. *J. Phase Equilibria* **2000**, *21*, 25–31. [CrossRef]
10. Bamberger, M.; Munitz, A.; Kaufman, L.; Abbaschian, R. Evaluation of the stable and metastable Cu-Co-Fe phase diagrams. *Calphad Comput. Coupling Phase Diagr. Thermochem.* **2002**, *26*, 375–384. [CrossRef]
11. Wang, C.P.; Liu, X.J.; Ohnuma, I.; Kainuma, R.; Ishida, K. Phase equilibria in Fe-Cu-X (X: Co, Cr, Si, V) ternary systems. *J. Phase Equilibria* **2002**, *23*, 236–245. [CrossRef]
12. Cao, C.D.; Görler, G.P. Direct Measurement of the Metastable Liquid Miscibility Gap in Fe–Co–Cu Ternary Alloy System. *Chin. Phys. Lett.* **2005**, *22*, 482–484.
13. Yamauchi, I.; Irie, T.; Sakaguchi, H. Metastable liquid separation in undercooled Fe-Cu and Fe-Cu-Si melts containing a small B concentration and their solidification structure. *J. Alloy. Compd.* **2005**, *403*, 211–216. [CrossRef]
14. Munitz, A.; Bamberger, A.M.; Wannaparhun, S.; Abbaschian, R. Effects of supercooling and cooling rate on the microstructure of Cu-Co-Fe alloys. *J. Mater. Sci.* **2006**, *41*, 2749–2759. [CrossRef]
15. Curiotto, S.; Greco, R.; Pryds, N.H.; Johnson, E.; Battezzati, L. The liquid metastable miscibility gap in Cu-based systems. *Fluid Phase Equilibria* **2007**, *256*, 132–136. [CrossRef]

16. Munitz, A.; Venkert, A.; Landau, P.; Kaufman, M.J.; Abbaschian, R. Microstructure and phase selection in supercooled copper alloys exhibiting metastable liquid miscibility gaps. *J. Mater. Sci.* **2012**, *47*, 7955–7970. [CrossRef]
17. Liu, N.; Liu, F.; Yang, W.; Chen, Z.; Yang, G.C. Movement of minor phase in undercooled immiscible Fe-Co-Cu alloys. *J. Alloy. Compd.* **2013**, *551*, 323–326. [CrossRef]
18. Yeh, J.W.; Lin, S.J.; Chin, T.S.; Gan, J.Y.; Chen, S.K.; Shun, T.T.; Tsau, C.H.; Chou, S.Y. Formation of simple crystal structures in Cu-Co-Ni-Cr-Al-Fe-Ti-V alloys with multiprincipal metallic elements. *Metall. Mater. Trans. A* **2004**, *35*, 2533–2536. [CrossRef]
19. Miedema, A.R. A simple model for alloys I. *Philips Tech. Rev.* **1973**, *33*, 149–160.
20. Takeuchi, A.; Inoue, A. Metallic Glasses By Atomic Size Difference, Heat of Mixing and Period of Constituent Elements and Its Application To Characterization of the Main Alloying Element. *Mater. Trans.* **2005**, *46*, 2817–2829. [CrossRef]
21. Takeuchi, A.; Inoue, A. Mixing enthalpy of liquid phase calculated by miedema's scheme and approximated with sub-regular solution model for assessing forming ability of amorphous and glassy alloys. *Intermetallics* **2010**, *18*, 1779–1789. [CrossRef]
22. Hardy, H. A "sub-regular" solution model and its application to some binary alloy systems. *Acta Metall.* **1953**, *1*, 202–209. [CrossRef]
23. Nakagawa, Y.; Systemes, D.L.E.S. Liquid immiscibility in copper-iron and copper-cobalt systems in the supercooled state. *Acta Metall.* **1958**, *6*, 704–711. [CrossRef]
24. Munitz, A.; Abbaschian, R. Two-melt separation in supercooled Cu-Co alloys solidifying in a drop-tube. *J. Mater. Sci.* **1991**, *26*, 6458–6466. [CrossRef]
25. Munitz, A.; Elder-Randall, S.P.; Abbaschian, R. Supercooling effects in Cu-10 Wt Pct Co alloys solidified at different cooling rates. *Metall. Trans. A* **1992**, *23*, 1817–1827. [CrossRef]
26. Munitz, A.; Abbaschian, R. Microstructure of Cu-Co Alloys Solidified at Various Supercoolings. *Metall. Mater. Trans. A* **1996**, *27*, 4049–4059. [CrossRef]
27. Robinson, M.B.; Li, D.; Rathz, T.J.; Williams, G. Undercooling, liquid separation and solidification of Cu-Co alloys. *J. Mater. Sci.* **1999**, *34*, 3747–3753. [CrossRef]
28. Kolbe, M.; Gao, J.R. Liquid phase separation of Co-Cu alloys in the metastable miscibility gap. *Mater. Sci. Eng. A* **2005**, *413–414*, 509–513. [CrossRef]
29. Curiotto, S.; Pryds, N.H.; Johnson, E.; Battezzati, L. Liquid-liquid phase separation and remixing in the Cu-Co system. *Metall. Mater. Trans. A Phys. Metall. Mater. Sci.* **2006**, *37*, 2361–2368. [CrossRef]
30. Yang, W.; Xu, Z.; Wang, Z.; Li, S.; Liu, F.; Yang, G. Microstructure morphology and solute segregation in nonequilibrium solidification of metastable immiscible Cu50Co50 alloy. *Procedia Eng.* **2012**, *27*, 1518–1523. [CrossRef]
31. Zhang, Y.K.; Gao, J.; Yang, C.; Kolbe, M.; Binder, S.; Herlach, D.M. Asynchronous crystallization behavior of Co-rich droplets in phase-separated Cu-Co alloys. *Mater. Lett.* **2012**, *73*, 56–58. [CrossRef]
32. Zhang, Y.; Gao, J.; Yasuda, H.; Kolbe, M.; Wilde, G. Particle size distribution and composition in phase-separated Cu75Co25 alloys under various magnetic fields. *Scr. Mater.* **2014**, *82*, 5–8. [CrossRef]
33. Si, S.H.; Zhang, H.; He, Y.Z.; Li, M.X.; Guo, S. Liquid Phase Separation and the Aging Effect on Mechanical and Electrical Properties of Laser Rapidly Solidified Cu100-xCrx Alloys. *Metals* **2015**, *5*, 2119–2127. [CrossRef]
34. Jegede, O.E.; Cochrane, R.F.; Mullis, A.M. Metastable monotectic phase separation in Co–Cu alloys. *J. Mater. Sci.* **2018**, *53*, 11749–11764. [CrossRef]
35. Chuang, Y.Y.; Schmid, R.; Chang, Y.A. Thermodynamic analysis of the iron-copper system I: The stable and metastable phase equilibria. *Metall. Trans. A* **1984**, *15*, 1921–1930. [CrossRef]
36. Munitz, A. Liquid separation effects in Fe-Cu alloys solidified under different cooling rates. *Metall. Trans. B* **1987**, *18*, 565–575. [CrossRef]
37. Munitz, A. Metastable liquid phase separation in tungsten inert gas and electron beam copper/stainless-steel welds. *J. Mater. Sci.* **1995**, *30*, 2901–2910. [CrossRef]
38. Wang, C.P.; Liu, X.J.; Takaku, Y.; Ohnuma, I.; Kainuma, R.; Ishida, K. Formation of core-type macroscopic morphologies in Cu-Fe base alloys with liquid miscibility gap. *Metall. Mater. Trans. A Phys. Metall. Mater. Sci.* **2004**, *35*, 1243–1253. [CrossRef]
39. Chen, Y.Z.; Liu, F.; Yang, G.C.; Xu, X.Q.; Zhou, Y.H. Rapid solidification of bulk undercooled hypoperitectic Fe-Cu alloy. *J. Alloy. Compd.* **2007**, *427*, 3–7. [CrossRef]

40. Zhang, Y.K.; Gao, J.; Nagamatsu, D.; Fukuda, T.; Yasuda, H.; Kolbe, M.; He, J.C. Reduced droplet coarsening in electromagnetically levitated and phase-separated Cu-Co alloys by imposition of a static magnetic field. *Scr. Mater.* **2008**, *59*, 1002–1005. [CrossRef]
41. Liu, N. Investigation on the phase separation in undercooled Cu-Fe melts. *J. Non-Cryst. Solids* **2012**, *358*, 196–199. [CrossRef]
42. Zhang, J.T.; Wang, Y.H.; Cui, X.C.; Lin, J.B. The growth and congregation of minor phase in immiscible Cu-Fe alloys. *Medziagotyra* **2013**, *19*, 373–376. [CrossRef]
43. Zhang, J.; Cui, X.; Wang, Y. Liquid phase separation in immiscible Cu–Fe alloys. *Int. J. Cast Met. Res.* **2018**, *31*, 87–92. [CrossRef]
44. Cantor, B.; Chang, I.T.H.; Knight, P.; Vincent, A.J.B. Microstructural development in equiatomic multicomponent alloys. *Mater. Sci. Eng. A* **2004**, *375-377*, 213–218. [CrossRef]
45. Yeh, J.W.; Chen, S.K.; Lin, S.J.; Gan, J.Y.; Chin, T.S.; Shun, T.T.; Tsau, C.H.; Chang, S.Y. Nanostructured high-entropy alloys with multiple principal elements: Novel alloy design concepts and outcomes. *Adv. Eng. Mater.* **2004**, *6*, 299–303. [CrossRef]
46. Hsu, C.Y.; Yeh, J.W.; Chen, S.K.; Shun, T.T. Wear resistance and high-temperature compression strength of Fcc CuCoNiCrAl0.5Fe alloy with boron addition. *Metall. Mater. Trans. A* **2004**, *35*, 1465–1469. [CrossRef]
47. Huang, P.; Yeh, J. Multi-Principal Element Alloys with Improved Oxidation and Wear Resistance for Thermal Spray Coating. *Adv. Eng. Mater.* **2004**, *6*, 74–78. [CrossRef]
48. Chen, T.K.; Wong, M.S.; Shun, T.T.; Yeh, J.W. Nanostructured nitride films of multi-element high-entropy alloys by reactive DC sputtering. *Surf. Coat. Technol.* **2004**, *200*, 193–200. [CrossRef]
49. Tsai, M.H. Physical properties of high entropy alloys. *Entropy* **2013**, *15*, 5338–5345. [CrossRef]
50. Gao, M.C. Progress in high-entropy alloys. *Jom* **2013**, *65*, 1749–1750. [CrossRef]
51. Gao, M.C. Progress in high-entropy alloys. *Jom* **2014**, *66*, 1964–1965. [CrossRef]
52. Tsai, M.H.; Yeh, J.W. High-entropy alloys: A critical review. *Mater. Res. Lett.* **2014**, *2*, 107–123. [CrossRef]
53. Zhang, Y.; Zuo, T.T.; Tang, Z.; Gao, M.C.; Dahmen, K.A.; Liaw, P.K.; Lu, Z.P. Microstructures and properties of high-entropy alloys. *Prog. Mater. Sci.* **2014**, *61*, 1–93. [CrossRef]
54. Kozak, R.; Sologubenko, A.; Steurer, W. Single-phase high-entropy alloys—An overview. *Z. Krist.* **2015**, *230*, 55–68. [CrossRef]
55. Gao, M.C. Progress in high entropy alloys. *Jom* **2015**, *67*, 2251–2253. [CrossRef]
56. Pickering, E.J.; Jones, N.G. High-entropy alloys: A critical assessment of their founding principles and future prospects. *Int. Mater. Rev.* **2016**, *61*, 183–202. [CrossRef]
57. Ye, Y.F.; Wang, Q.; Lu, J.; Liu, C.T.; Yang, Y. High-entropy alloy: Challenges and prospects. *Mater. Today* **2016**, *19*, 349–362. [CrossRef]
58. Miracle, D.B.; Senkov, O.N. A critical review of high entropy alloys and related concepts. *Acta Mater.* **2017**, *122*, 448–511. [CrossRef]
59. Manzoni, A.M.; Glatzel, U. New multiphase compositionally complex alloys driven by the high entropy alloy approach. *Mater. Charact.* **2018**. [CrossRef]
60. Praveen, S.; Kim, H.S. High-entropy alloys: Potential candidates for high-temperature applications—An overview. *Adv. Eng. Mater.* **2018**, *20*, 1–22. [CrossRef]
61. Murty, B.; Yeh, J.; Ranganathan, S. *High-Entropy Alloys*, 1st ed.; Butterworth-Heinemann: London, UK, 2014.
62. Fisher, D. *High-Entropy Alloys—Microstructures and Properties*; Trans Tech Publications: Zurich, Switzerland, 2015.
63. Maiti, S. *Local Structure and Properties of Refractory High-Entropy Alloys: High-Entropy Alloys Critically Analyzed from a More Fundamental Point of View*; Lambert Academic Publishing: Saarbrücken, Germany, 2016.
64. Zhang, C.; Gao, M.C. *High-Entropy Alloys: Fundamentals and Applications*; Springer International Publishing: Cham, Switzerland, 2016; pp. 399–444.
65. Tian, F.; Varga, L.; Levente, V. *Theoretical Design of Single Phase High-Entropy Alloys*, 1st ed.; Lambert Academic Publishing: Saarbrücken, Germany, 2017.
66. Raj, A. *HEA: High Entropy Alloys*; CreateSpace Independent Publishing Platform: Scotts Valley, CA, USA, 2017.
67. Liu, X.J.; Jiang, Z.P.; Wang, C.P.; Ishida, K. Experimental determination and thermodynamic calculation of the phase equilibria in the Cu-Cr-Nb and Cu-Cr-Co systems. *J. Alloy. Compd.* **2009**, *478*, 287–296. [CrossRef]
68. Chang, Y.A.; Goldberg, D.; Neumann, J.P. Phase diagrams and thermodynamic properties of ternary copper-silver systems. *J. Phys. Chem. Ref. Data* **1977**, *6*, 621–674. [CrossRef]

69. Taguchi, K.; Ono-Nakazato, H.; Usui, T. Liquid Immiscibility in Fe-Cu-B System. *ISIJ Int.* **2006**, *46*, 633–636. [CrossRef]
70. Guo, J.B.; Cao, C.D.; Gong, S.L.; Song, R.B.; Bai, X.J.; Wang, J.Y.; Zheng, J.B.; Wen, X.X.; Sun, Z.B. Rapid solidification of Cu60Co30Cr10alloy under different conditions. *Trans. Nonferrous Met. Soc. China (Engl. Ed.)* **2013**, *23*, 731–734. [CrossRef]
71. Derimow, N.; Abbaschian, R. Solidification microstructures and calculated mixing enthalpies in CoCrCu containing alloys. *Mater. Today Commun.* **2018**, *15*, 1–10. [CrossRef]
72. Derimow, N.; Santodonato, L.; Mills, R.; Abbaschian, R. In-Situ Imaging of Liquid Phase Separation in Molten Alloys Using Cold Neutrons. *J. Imaging* **2018**, *4*, 5. [CrossRef]
73. Liu, N.; Liu, F.; Chen, Z.; Yang, G.; Yang, C.; Zhou, Y. Liquid-phase Separation in Rapid Solidification of Undercooled Fe-Co-Cu Melts. *J. Mater. Sci. Technol.* **2012**, *28*, 622–625. [CrossRef]
74. Dreval, L.A.; Turchanin, M.A.; Abdulov, A.R.; Bondar, A.A. Thermodynamic assessment of the Cu-Fe-Cr phase diagram. *Chem. Met. Alloy.* **2010**, *3*, 132–139.
75. Nagase, T.; Matsumoto, M.; Fujii, Y. Microstructure of Ti-Nb-Ag immiscible alloys with liquid phase separation. *Microscopy* **2017**, *66*, i22. [CrossRef]
76. Wang, C.P.; Liu, X.J.; Ohnuma, I.; Kainuma, R.; Ishida, K.; Hao, S.M. Phase equilibria in the Cu-Fe-Mo and Cu-Fe-Nb systems. *J. Phase Equilibria* **2000**, *21*, 54–62. [CrossRef]
77. Bo, L.; Li, S.; Wang, L.; Wu, D.; Zuo, M.; Zhao, D. Liquid-liquid phase separation and solidification behavior of Al55Bi36Cu9monotectic alloy with different cooling rates. *Results Phys.* **2018**, *8*, 1086–1091. [CrossRef]
78. Zhao, D.; Liu, R.; Wu, D.; Bo, L.; Wang, L. Liquid-liquid phase separation and solidification behavior of Al-Bi-Sb immiscible alloys. *Results Phys.* **2017**, *7*, 3216–3221. [CrossRef]
79. Wang, W.; Zhang, X.; Li, L.; Wei, B. Dual solidification mechanisms of liquid ternary Fe-Cu-Sn alloy. *Sci. China Phys. Mech. Astron.* **2012**, *55*, 450–459. [CrossRef]
80. Liu, H.X.; Wang, C.P.; Yu, Y.; Liu, X.J.; Takaku, Y.; Ohnuma, I.; Kainuma, R.; Ishida, K. Experimental investigation and thermodynamic calculation of the phase equilibria in the Al-Bi-Sn ternary system. *J. Phase Equilibria Diffus.* **2012**, *33*, 9–19. [CrossRef]
81. Dai, R.; Zhang, J.F.; Zhang, S.G.; Li, J.G. Liquid immiscibility and core-shell morphology formation in ternary Al-Bi-Sn alloys. *Mater. Charact.* **2013**, *81*, 49–55. [CrossRef]
82. Sun, X.; Li, M.; Jia, P.; Geng, H. Liquid–liquid phase equilibrium in ternary immiscible Al-Bi-Sn melts. *Phys. Chem. Liq.* **2016**, *54*, 740–746. [CrossRef]
83. Wang, L.; Li, S.; Bo, L.; Wu, D.; Zhao, D. Liquid-liquid phase separation and solidification behavior of Al-Bi-Sn monotectic alloy. *J. Mol. Liq.* **2018**, *254*, 333–339. [CrossRef]
84. Zhai, W.; Liu, H.M.; Wei, B. Liquid phase separation and monotectic structure evolution of ternary Al62.6Sn28.5Cu8.9 immiscible alloy within ultrasonic field. *Mater. Lett.* **2015**, *141*, 221–224. [CrossRef]
85. Tong, C.J.; Chen, Y.L.; Yeh, J.W.; Lin, S.J.; Chen, S.K.; Shun, T.T.; Tsau, C.H.; Chang, S.Y. Mechanical performance of the Al_xCoCrCuFeNi high-entropy alloy system with multiprincipal elements. *Metall. Mater. Trans. A* **2005**, *36*, 1263–1271. [CrossRef]
86. Tong, C.J.; Chen, Y.L.; Yeh, J.W.; Lin, S.J.; Chen, S.K.; Shun, T.T.; Tsau, C.H.; Chang, S.Y. Microstructure characterization of Al_xCoCrCuFeNi high-entropy alloy system with multiprincipal elements. *Metall. Mater. Trans. A* **2005**, *36*, 881–893. [CrossRef]
87. Santodonato, L.J.; Zhang, Y.; Feygenson, M.; Parish, C.M.; Gao, M.C.; Weber, R.J.K.; Neuefeind, J.C.; Tang, Z.; Liaw, P.K. Deviation from high-entropy configurations in the atomic distributions of a multi-principal-element alloy. *Nat. Commun.* **2015**, *6*, 5964. [CrossRef] [PubMed]
88. Yang, X.; Zhang, Y. Prediction of high-entropy stabilized solid-solution in multi-component alloys. *Mater. Chem. Phys.* **2012**, *132*, 233–238. [CrossRef]
89. Miracle, D.; Miller, J.; Senkov, O.; Woodward, C.; Uchic, M.; Tiley, J. Exploration and development of high entropy alloys for structural applications. *Entropy* **2014**, *16*, 494–525. [CrossRef]
90. Senkov, O.N.; Miller, J.D.; Miracle, D.B.; Woodward, C. Accelerated exploration of multi-principal element alloys for structural applications. *Calphad Comput. Coupling Phase Diagr. Thermochem.* **2015**, *50*, 32–48. [CrossRef]
91. Troparevsky, M.C.; Morris, J.R.; Kent, P.R.C.; Lupini, A.R.; Stocks, G.M. Criteria for predicting the formation of single-phase high-entropy alloys. *Phys. Rev. X* **2015**, *5*, 1–6. [CrossRef]

92. Toda-Caraballo, I.; Rivera-Díaz-del Castillo, P. A criterion for the formation of high entropy alloys based on lattice distortion. *Intermetallics* **2016**, *71*, 76–87. [CrossRef]
93. Tancret, F.; Toda-Caraballo, I.; Menou, E.; Díaz-Del-Castillo, P.E.J.R. Designing high entropy alloys employing thermodynamics and Gaussian process statistical analysis. *Mater. Des.* **2017**, *115*, 486–497. [CrossRef]
94. Gurao, N.P.; Biswas, K. In the quest of single phase multi-component multiprincipal high entropy alloys. *J. Alloy. Compd.* **2017**, *697*, 434–442. [CrossRef]
95. Sun, W.; Huang, X.; Luo, A.A. Phase formations in low density high entropy alloys. *Calphad Comput. Coupling Phase Diagr. Thermochem.* **2017**, *56*, 19–28. [CrossRef]
96. Tsai, M.H.; Li, J.H.; Fan, A.C.; Tsai, P.H. Incorrect predictions of simple solid solution high entropy alloys: Cause and possible solution. *Scr. Mater.* **2017**, *127*, 6–9. [CrossRef]
97. Dantzig, J.; Rappaz, M. *Solidification*, 2nd ed.; EPFL Press: Lausanne, Switzerland, 2016.
98. Takamatsu, Y.; Esaka, H.; Shinozuka, K. Liquid-phase separation in the interdendritic region after growth of primary β-Sn in undercooled Sn-2.8Ag-0.3Cu melt. *J. Electron. Mater.* **2012**, *41*, 2035–2044. [CrossRef]
99. Munitz, A.; Samuha, S.; Brosh, E.; Salhov, S.; Derimow, N.; Abbaschian, R. Liquid phase separation phenomena in Al 2.2 CrCuFeNi 2 HEA. *Intermetallics* **2018**, *97*, 77–84. [CrossRef]
100. Kamio, A.; Kumai, S.; Tezuka, H. Solidification structure of monotectic alloys. *Mater. Sci. Eng. A* **1991**, *146*, 105–121. [CrossRef]
101. Fan, Z.; Ji, S.; Zhang, J.; Fan, Z.; Ji, S.; Zhang, J. Processing of immiscible metallic alloys by rheomixing process. *Mater. Sci. Technol.* **2001**, *17*, 837–842. [CrossRef]
102. Barry Andrews, J. Low Gravity Containerless Processing of Immiscible Gold Rhodium Alloy. Available online: https://ntrs.nasa.gov/search.jsp?R=19870007312 (accessed on 19 November 2018).
103. Ratke, L.; Korekt, G.; Drees, S. Phase separation and solidification of immiscible metallic alloys under low gravity. *Adv. Space Res.* **1998**, *22*, 1227–1236. [CrossRef]
104. Luo, B.C.; Liu, X.R.; Wei, B. Macroscopic liquid phase separation of Fe-Sn immiscible alloy investigated by both experiment and simulation. *J. Appl. Phys.* **2009**, *106*. [CrossRef]
105. Lu, W.Q.; Zhang, S.G.; Li, J.G. Depressing liquid phase separation and macrosegeregation of Fe-Sn immiscible alloys by Cu alloying. *Mater. Sci. Technol.* **2014**, *30*, 231–235. [CrossRef]
106. Hsu, U.S.; Hung, U.D.; Yeh, J.W.; Chen, S.K.; Huang, Y.S.; Yang, C.C. Alloying behavior of iron, gold and silver in AlCoCrCuNi-based equimolar high-entropy alloys. *Mater. Sci. Eng. A* **2007**, *460-461*, 403–408. [CrossRef]
107. Munitz, A.; Kaufman, M.; Chandler, J.; Kalaantari, H.; Abbaschian, R. Melt separation phenomena in CoNiCuAlCr high entropy alloy containing silver. *Mater. Sci. Eng. A* **2013**, *560*, 633–642. [CrossRef]
108. Zhang, H.; Wu, W.; He, Y.; Li, M.; Guo, S. Formation of core—Shell structure in high entropy alloy coating by laser cladding. *Appl. Surf. Sci.* **2016**, *363*, 543–547. [CrossRef]
109. Guo, S.; Ng, C.; Liu, C.T. Anomalous solidification microstructures in Co-free Al xCrCuFeNi2 high-entropy alloys. *J. Alloy. Compd.* **2013**, *557*, 77–81. [CrossRef]
110. He, J.; Mattern, N.; Tan, J.; Zhao, J.Z.; Kaban, I.; Wang, Z.; Ratke, L.; Kim, D.H.; Kim, W.T.; Eckert, J. A bridge from monotectic alloys to liquid-phase-separated bulk metallic glasses: Design, microstructure and phase evolution. *Acta Mater.* **2013**, *61*, 2102–2112. [CrossRef]
111. Munitz, A.; Kaufman, M.; Abbaschian, R. Liquid phase separation in transition element high entropy alloys. *Intermetallics* **2017**, *86*, 59–72. [CrossRef]
112. Kündig, A.A.; Ohnuma, M.; Ping, D.H.; Ohkubo, T.; Hono, K. In situ formed two-phase metallic glass with surface fractal microstructure. *Acta Mater.* **2004**, *52*, 2441–2448. [CrossRef]
113. Ziewiec, K.; Garze, G.; Czeppe, T.; Artur, B.; Ruebenbauer, K. Microstructure and phase transformations in a liquid immiscible Fe60Cu20P10Si5B5 alloy. *Intermetallics* **2016**, *69*, 47–53. [CrossRef]
114. Wu, P.H.; Liu, N.; Yang, W.; Zhu, Z.X.; Lu, Y.P.; Wang, X.J. Microstructure and solidification behavior of multicomponent CoCrCuxFeMoNi high-entropy alloys. *Mater. Sci. Eng. A* **2015**, *642*, 142–149. [CrossRef]
115. Liu, N.; Wu, P.; Zhou, P.; Peng, Z.; Wang, X.; Lu, Y. Rapid solidification and liquid-phase separation of undercooled CoCrCuFexNi high-entropy alloys. *Intermetallics* **2016**, *72*, 44–52. [CrossRef]
116. Wu, P.H.; Liu, N.; Zhou, P.J.; Peng, Z.; Du, W.D.; Wang, X.J.; Pan, Y. Microstructures and liquid phase separation in multicomponent CoCrCuFeNi high entropy alloys. *Mater. Sci. Technol.* **2016**, *32*, 576–580. [CrossRef]

117. Wang, W.; Hu, L.; Luo, S.; Meng, L.; Geng, D.; Wei, B. Liquid phase separation and rapid dendritic growth of high-entropy CoCrCuFeNi alloy. *Intermetallics* **2016**, *77*, 41–45. [CrossRef]
118. Guo, T.; Li, J.; Wang, J.; Wang, Y.; Kou, H.; Niu, S. Liquid-phase separation in undercooled CoCrCuFeNi high entropy alloy. *Intermetallics* **2017**, *86*, 110–115. [CrossRef]
119. Peng, Z.; Liu, N.; Zhang, S.Y.; Wu, P.H.; Wang, X.J. Liquid-phase separation of immiscible CrCuxFeMoyNi high-entropy alloys. *Mater. Sci. Technol.* **2017**, *33*, 1352–1359. [CrossRef]
120. Elder, S.; Abbaschian, G.J. Supercooling and Rapid Solidification using EM Levitation. *Trans. Tech. Publ.* **1991**, *10*, 299–330.
121. Wang, S.; Chen, Z.; Feng, L.C.; Liu, Y.Y.; Zhang, P.; He, Y.Z.; Meng, Q.Q.; Zhang, J.Y. Nano-phase formation accompanying phase separation in undercooled CoCrCuFeNi-3 at.% Sn high entropy alloy. *Mater. Charact.* **2018**, *144*, 516–521. [CrossRef]
122. Hsu, C.Y.; Sheu, T.S.; Yeh, J.W.; Chen, S.K. Effect of iron content on wear behavior of AlCoCrFexMo0.5Ni high-entropy alloys. *Wear* **2010**, *268*, 653–659. [CrossRef]
123. Hsu, C.Y.; Wang, W.R.; Tang, W.Y.; Chen, S.K.; Yeh, J.W. Microstructure and mechanical properties of new AlCoxCrFeMo 0.5Ni high-entropy alloys. *Adv. Eng. Mater.* **2010**, *12*, 44–49. [CrossRef]
124. Zhu, J.M.; Zhang, H.F.; Fu, H.M.; Wang, A.M.; Li, H.; Hu, Z.Q. Microstructures and compressive properties of multicomponent AlCoCrCuFeNiMox alloys. *J. Alloy. Compd.* **2010**, *497*, 52–56. [CrossRef]

entropy

MDPI

Article

Configurational Entropy in Multicomponent Alloys: Matrix Formulation from Ab Initio Based Hamiltonian and Application to the FCC Cr-Fe-Mn-Ni System

Antonio Fernández-Caballero [1,2], Mark Fedorov [3], Jan S. Wróbel [3], Paul M. Mummery [1] and Duc Nguyen-Manh [2,*]

1 School of Mechanical Aerospace and Civil Engineering, University of Manchester, Manchester M13 9PL, UK; antonio.fernandez-caballero@postgrad.manchester.ac.uk (A.F.-C.); paul.m.mummery@manchester.ac.uk (P.M.M.)
2 CCFE, United Kingdom Atomic Energy Authority, Abingdon OX14 3DB, UK
3 Faculty of Materials Science and Engineering, Warsaw University of Technology, Woloska 141, 02-507 Warsaw, Poland; markfedorov93@gmail.com (M.F.); jan.wrobel@pw.edu.pl (J.S.W.)
* Correspondence: Duc.Nguyen@ukaea.uk

Received: 7 November 2018; Accepted: 11 January 2019; Published: 15 January 2019

Abstract: Configuration entropy is believed to stabilize disordered solid solution phases in multicomponent systems at elevated temperatures over intermetallic compounds by lowering the Gibbs free energy. Traditionally, the increment of configuration entropy with temperature was computed by time-consuming thermodynamic integration methods. In this work, a new formalism based on a hybrid combination of the Cluster Expansion (CE) Hamiltonian and Monte Carlo simulations is developed to predict the configuration entropy as a function of temperature from multi-body cluster probability in a multi-component system with arbitrary average composition. The multi-body probabilities are worked out by explicit inversion and direct product of a matrix formulation within orthonomal sets of point functions in the clusters obtained from symmetry independent correlation functions. The matrix quantities are determined from semi canonical Monte Carlo simulations with Effective Cluster Interactions (ECIs) derived from Density Functional Theory (DFT) calculations. The formalism is applied to analyze the 4-body cluster probabilities for the quaternary system Cr-Fe-Mn-Ni as a function of temperature and alloy concentration. It is shown that, for two specific compositions ($Cr_{25}Fe_{25}Mn_{25}Ni_{25}$ and $Cr_{18}Fe_{27}Mn_{27}Ni_{28}$), the high value of probabilities for Cr-Fe-Fe-Fe and Mn-Mn-Ni-Ni are strongly correlated with the presence of the ordered phases $L1_2$-$CrFe_3$ and $L1_0$-MnNi, respectively. These results are in an excellent agreement with predictions of these ground state structures by ab initio calculations. The general formalism is used to investigate the configuration entropy as a function of temperature and for 285 different alloy compositions. It is found that our matrix formulation of cluster probabilities provides an efficient tool to compute configuration entropy in multi-component alloys in a comparison with the result obtained by the thermodynamic integration method. At high temperatures, it is shown that many-body cluster correlations still play an important role in understanding the configuration entropy before reaching the solid solution limit of high-entroy alloys (HEAs).

Keywords: multicomponent; ab initio; configuration entropy; matrix formulation; cluster expansion; cluster variation method; monte carlo; thermodynamic integration

1. Introduction

Multicomponent systems, called High Entropy Alloys (HEAs), are crystalline solids that form predominantly in a single phase. These systems were brought to wider attention through the work of Cantor et al. [1] and Yeh et al. [2]. At or near equiatomic ratio of compositions, the configuration entropy is maximized and given by the formula $S_{config}/R = lnK$ where R is the ideal gas constant, and K is the total number of different components. It has been proposed to define $K = 4$ as the minimum number of components at or near equiatomic composition for these systems to be called HEAs [3–5].

There are various sources of entropy in a system in addition to configuration entropy including vibration, electronic and magnetic contributions. The co-existence of multiple phases in the equilibrium state of a chemical system with a given overall composition at a specified temperature is found by minimizing the Gibbs free energy of the whole system. When the system undergoes a phase transition of either order-disorder or spinodal decomposition, the relative differences of configurational entropy are dominant in comparison to those contributions from vibration and magnetic terms in FeCoCrNi [6] where the relative difference of configuration entropy dominates or is of similar magnitude to vibration entropy. These multicomponent alloys have transition metals Co, Cr, Fe, and Ni that exhibit magnetic behavior and have magnetic phase transitions characterized by Curie temperatures sensitive to concentrations [7].

In general, the formulation of configuration entropy given by the expression $S_{config}/R = lnK$ only holds for disordered alloys with equiatomic composition at temperatures near their melting point. On the other hand, at low temperatures, the lowest free energy structures can be ordered intermetallic phases or partially ordered structures. In the intermediate temperature range, there exists short range order, which is related to the nature of the chemical environment of each atomic species, containing ordering or segregation preferences specially at low temperatures, because random disordering is favored at the high temperatures. The importance of short-range order (SRO) was manifested in diffuse X-ray scattering measurements of alloys containing ordered superstructure domains. The SRO effect is also seen to significantly affect electrical resistivity properties [8,9], and to influence the alloy strength by hindering dislocation motion [10] in concentrated alloys. Recently, it has been demonstrated how short-range order can be predicted from ab initio based Hamiltonian in combination with Monte Carlo simulations [11] in multi-component Mo-Nb-Ta-V-W systems. The study predicted that a strong SRO parameter may lead to the formation of Mo-Ta ordering in the B2 structure after quenching down from temperatures as high as 3000 K. HEAs such as $Al_{1.3}CoCrCuFeNi$ have been reported to be susceptible to complex phase transitions including segregation, precipitation, chemical ordering and spinodal decomposition into a complex microstructure containing regions of Body-Centered Cubic (BCC), Face-Centered Cubic (FCC) and B2 phases [12]. Physical models have been successfully applied to predict the formation of these phases on the basis of the average number of valence electrons [13].

The prediction of equilibrium thermodynamic properties (free energies, and phase diagrams) is one of the goals of computational materials science. Lattice statistical models involving an Ising-like Hamiltonian developed from ab initio enthalpies of mixing have become an important tool in the computation of the thermodynamic properties of alloy systems. In particular, the cluster expansion method [14–16] expands the Hamiltonian into contributions from an optimized set of clusters, each term weighted by Effective Cluster Interactions. The thermodynamic properties at temperature are obtained from the ECIs by computing the free energies from semi-canonical Monte Carlo simulations in combination with the thermodynamic integration technique [17].

The Cluster Variation Method (CVM) formalism [18,19] expresses the free energy in terms of enthalpies of mixing and configuration entropies as a function of the temperature dependence for a specific set of clusters by minimizing the free energy from variational principles. The correlation function parameters in terms of which configuration entropy and enthalpy of mixing are expressed grow exponentially with component number K and size of the maximal cluster ω. Due to this, the CVM has been commonly applied to clusters consisting of four sites in a regular tetrahedron or the so-called tetrahedron-octahedron in the FCC lattice [20,21], or the four sites in a irregular tetrahedron in the BCC lattice [22]. Within the tetrahedron approximation, the CVM calculations for the FCC involve the empty lattice, isolated points, first nearest neighbor pair, triangle and tetrahedron [20]. It is generally accepted that the integrated Monte Carlo method is more accurate, but the calculations are time consuming due to the need for many passes to obtain free energies at any given temperature from the disordered high temperature configuration. A hybrid approach taking the Monte Carlo calculated correlation functions at temperature and computing the configuration entropy value from analytic CVM expression has been used in a binary FCC lattice model [23]. It is shown that accurate free energies can be obtained for ordered and disordered phases at arbitrary chemical concentration and temperature without thermodynamic integration provided that use is made of high-order CVM entropy expressions. However, the clusters optimized from the Cluster Expansion (CE) method reported in the literature do not often match those used for the CVM. More importantly, calculations of the configuration entropy for multi-component systems represent a very serious challenge for computational materials science. In this work, we close this gap by developing a new methodology based on matrix formulation to calculate analytically the cluster probabilities for arbitrary K-component alloys from the correlation functions obtained by the hybrid Monte Carlo and CE Hamiltonian. We apply our method to the FCC four-component CrFeMnNi system for investigating configuration entropy as a function of temperature and alloy composition. Two specific compositions are chosen to illustrate the importance of many-body cluster probability functions in multi-component systems. The first one is equatomic composition $Cr_{25}Fe_{25}Mn_{25}Ni_{25}$, which is of a great interest for the HEA community. The second one is $Cr_{18}Fe_{27}Mn_{27}Ni_{28}$ which has been used recently to design radiation tolerant materials for advanced nuclear reactor systems. This alloy has been studied in preference to Cantor's one [1] due to the removal of Co, which can cause activation by transmutation of ^{60}Co isotope in nuclear reactors [24].

This paper is organized as follows. In Section 2, a new matrix formulation for a K-component system based on the orthonormal sets of cluster expansion initially introduced within the Alloy Theoretic Automated Toolkit (ATAT) program [15,16] is presented. This formulation provides a rigorous link of cluster correlation functions obtained from Monte Carlo simulations with multi-body probabilities. Notation is developed to treat as general as possible arbitrary cluster sizes, ω, number of components, K, and temperature. In Section 3, the hybrid method is illustrated to calculate 4-body cluster probabilities as a function of temperature for the two specific compositions of equiatomic $\mathrm{Cr_{25}Fe_{25}Mn_{25}Ni_{25}}$ and also the composition $\mathrm{Cr_{18}Fe_{27}Mn_{27}}Ni_{18}$ from [24]. Here, the connection with the tetrahedron approximation in CVM is discussed due to the presence of a first nearest-neighbor 4-body cluster interaction within the investigated CE Hamiltonian. In Section 4, configuration entropy is discussed at two temperatures 1000 and 3000 K for different cluster decorations obtained from the CE method and compared with the integrated Monte Carlo results. The changes in configuration entropy are attributed to the presence of ordered phases that are more stable at low temperatures and the complementary tendency towards disordered random solution of the alloy at the given average composition. The main conclusions of this work are given in Section 5.

2. Methods

2.1. Matrix Formulation of Cluster Expansion

Let us consider an alloy system with arbitrary number, K, of components and crystalline lattice symmetry, in which the disordered phase is described by space group, G. The CE Hamiltonian at $T = 0$ K of the alloy can be found from enthalpies of mixing, $\Delta H_{DFT}^{Mixing}[\vec{\sigma}]$, of derivative alloy structures [16,25] denoted by varying length arrays of spin-like variables $\vec{\sigma}_i$ taking values from 0 to $K-1$. In general, a derivative structure of the alloy has non-zero atomic concentrations for each of $p = 1, ..., K$ different chemical elements forming it i.e., $x[\vec{\sigma}]_p \neq 0$. For each member structure, the reference energy in the calculation of the enthalpy of mixing is the pure element total energy, $E_{total}[p]$, with the same crystallographic symmetry. The alloy enthalpy of mixing is defined as follows:

$$\Delta H_{DFT}^{Mixing}[\vec{\sigma} \equiv \{\sigma_1, \sigma_2, \cdots\}] = E_{total}[\vec{\sigma} \equiv \{\sigma_1, \sigma_2, \cdots\}] - \sum_{p=1,\ldots,K} x_p[\vec{\sigma} \equiv \{\sigma_1, \sigma_2, \cdots\}] E_{tctal}[p]. \quad (1)$$

The enthalpies of mixing are calculated from the total energies $E_{total}[\vec{\sigma}]$ and $E_{total}[p]$ of the alloy and the pure element systems, respectively, by density functional theory for the underlying lattice. In the CE, the $\Delta H_{DFT}^{mixing}[\vec{\sigma}]$ of an arbitrary structure $\vec{\sigma}$ is expanded into a sum of reference clusters, $\Delta H_{CE}^{Mixing}[\vec{\sigma}]$ and can be written in the following formula [15–17]:

$$\Delta H_{CE}^{Mixing}[\vec{\sigma} \equiv \{\sigma_1, \sigma_2, \cdots\}] = \sum_{\omega,n,(s)} m_{\omega,n}^{(s)} J_{\omega,n}^{(s)} \langle \Gamma_{\omega',n'}^{(s')}[\vec{\sigma} \equiv \{\sigma_1, \sigma_2, \cdots\}] \rangle_{\omega,n,(s)}. \quad (2)$$

In Equation (2), each reference cluster is characterized by three labels ω, n and (s). The label ω denotes the total number sites; n is an auxiliary label that refers to highest order coordination shell contained in the reference cluster; and finally the label $(s) = (j_1, j_2, \cdots, j_\omega)$ denotes decoration of the cluster by point functions with dimension equal to ω and j_i taking values $0, ..., K-1$. For each reference cluster, there is an associated effective cluster interaction $J_{\omega,n}^{(s)}$, and multiplicity per lattice site, $m_{\omega,n}^{(s)}$. The term $\langle \Gamma_{\omega',n'}^{(s')}[\vec{\sigma}] \rangle_{\omega,n,(s)}$ used in the definition of enthalpy of mixing in Equation (2) is the thermally averaged cluster correlation function over all clusters $\omega', n', (s')$ which are equivalent to a space group symmetry element of the disordered phase within the reference decorated cluster defined by $\omega, n, (s)$. Overall, the triple product of multiplicity, ECI and correlation function, $m_{\omega,n}^{(s)} J_{\omega,n}^{(s)} \langle \Gamma_{\omega',n'}^{(s')}[\vec{\sigma}] \rangle_{\omega,n,(s)}$ gives the energetic contribution per lattice site of the reference cluster ω, n and (s) to the enthalpy of mixing of the particular structure configuration given by $\vec{\sigma}$.

For multicomponent systems, the choice of a set of basis functions is important for the matrix formulation of the CE method [14]. The simplest set consists of the successive $K-1$ powers of the pseudo-spin configuration variable $\{\sigma\}$ as was originally suggested by Taggart [26]. In this work, the correlation function is defined as the product of point functions initially proposed in the ATAT program [15,16]. Configurational average of the correlation functions is then given by the following formula:

$$\begin{aligned} \langle \Gamma_{\omega',n'}^{(s')}[\vec{\sigma}] \rangle_{\omega,n,(s)} &= \langle \gamma_{j_1,K}[\sigma_1] \gamma_{j_2,K}[\sigma_2] \cdots \gamma_{j_\omega,K}[\sigma_\omega] \rangle \\ &= \frac{1}{\Omega[\omega,n]} \sum_{u=1}^{\Omega[\omega,n]} \gamma_{j_1,K}[\sigma_{\overrightarrow{(\omega,n)}_{u_1}}] \gamma_{j_2,K}[\sigma_{\overrightarrow{(\omega,n)}_{u_2}}] \cdots \gamma_{j_\omega,K}[\sigma_{\overrightarrow{(\omega,n)}_{u_\omega}}], \end{aligned} \quad (3)$$

where the point function is written as:

$$\gamma_{j,K}[\sigma_i] = \begin{cases} 1, & \text{if } j = 0\,, \\ -\cos\left(2\pi \lceil \frac{j}{2} \rceil \frac{\sigma_i}{K}\right), & \text{if } j > 0 \text{ and odd}, \\ -\sin\left(2\pi \lceil \frac{j}{2} \rceil \frac{\sigma_i}{K}\right), & \text{if } j > 0 \text{ and even}, \end{cases} \quad (4)$$

In Equation (3), the correlation function of the system is averaged over the arbitrary crystal structure of the alloy system with configuration $\vec{\sigma}$ over all the set of $\Omega[\omega,n]$ clusters, $\{\overleftrightarrow{(\omega,n)_u}\}_{u=1,2,\cdots,\Omega[\omega,n]}$ where each cluster labeled by u contains ω sites, each site denoted by the vector, $\overrightarrow{(\omega,n)}_{u_i}$. The i-union of the ω vectors, $\overrightarrow{(\omega,n)}_{u_i}$, composes the cluster of vectors, $\left(\overrightarrow{(\omega,n)}_{u_1}, \overrightarrow{(\omega,n)}_{u_2}, \cdots, \overrightarrow{(\omega,n)}_{u_\omega}\right) \equiv \overleftrightarrow{(\omega,n)}_u$. The $u_1, u_2, \cdots, u_\omega$ are numeric labels referring to the values taken by variable u in Equation (3) that serve to enumerate the vectors making up the u cluster of vectors $\{\overleftrightarrow{(\omega,n)_u}\}$. The clusters $\overleftrightarrow{(\omega,n)}_u$ are symmetrically equivalent to the reference cluster (ω,n) sites by a symmetry operation and are enumerated by an index taking values $1, 2, \cdots, \Omega[\omega,n]$. $\Omega[\omega,n]$

is defined as the number of times the reference cluster (ω, n) is contained in a structural configuration which can be obtained from Monte Carlo simulation. The sites in the cluster $\{\overleftrightarrow{(\omega,n)_u}\}$ can allocate all of the possible decoration values $\in (s)$ which we express as $(s) = (j_1, j_2, \cdots, j_\omega)$. These integer indexes corresponding to the decorated cluster are parameters for the point functions in Equation (4).

For an arbitrary ω-sites cluster and K components, the number of ω-tuples formed with integer entries running from $0 \cdots K-1$ can be calculated. For $\omega = 2$, the formula calculates the number of symmetrically unique decorations (s) for a two-point cluster $\langle \Gamma^{(s')}_{\omega',n'}[\vec{\sigma}] \rangle_{\omega=2,n,(s)}$ and K component alloy system reduces to $(K+1)K/2$ [11], where K is the number of components. For higher order clusters, the total number of decorations depend on the cluster coordinates and the space group symmetry G of the high temperature disordered phase i.e., FCC or BCC and can not be simply expressed in terms of K (in general there could be less than $\binom{\omega+K-1}{K-1} = \dfrac{(\omega+K-1)!}{(K-1)!(\omega)!}$ number of symmetrically unique correlation functions). ATAT [15,16] numerically works out all the number of symmetrically unique decorated clusters $\omega', n', (s')$ equivalent to $\omega, n, (s)$ and uses one correlation function per set of equivalent decorated clusters $\langle \Gamma^{(s')}_{\omega',n'}[\vec{\sigma}] \rangle_{\omega,n,(s)}$. For convenience in notation, from now on, we will use $\langle \Gamma^{(s)}_{\omega,n}[\vec{\sigma}] \rangle = \langle \Gamma^{(s')}_{\omega',n'}[\vec{\sigma}] \rangle_{\omega,n,(s)}$. In general, two decorations (s) and (s') are symmetrically equivalent if there is at least one element in the space group symmetry $g \in G$ that transforms the sites $\{\vec{\tau}_1, \vec{\tau}_2, \cdots, \vec{\tau}_\omega\} \subset (\omega, n)$ into $\{\vec{\tau}_{g[1]}, \vec{\tau}_{g[2]}, \cdots, \vec{\tau}_{g[\omega]}\} \subset (\omega, n)$ as the permutation connecting the decorations $(s) = (1, 2, \cdots, \omega) \subset \{(s)_{(\omega,n)}\}$ and $(s') = (g[1], g[2], \cdots g[\omega]) \subset \{(s)_{(\omega,n)}\}$. This is given by the following equation:

$$g\{\vec{\tau}_1, \vec{\tau}_2, \cdots, \vec{\tau}_\omega\} = \{\vec{\tau}_{g[1]}, \vec{\tau}_{g[2]}, \cdots, \vec{\tau}_{g[\omega]}\}. \tag{5}$$

With the symmetry relations between decorations (s) and (s') in any given cluster, we are able to retrieve all K^ω possible ω-tuples from the set of unique symmetries corresponding to an ω cluster. The point functions $\gamma_{j,K}[\sigma_i]$ used to define the general correlation functions are related to the multi-body cluster probabilities (see below in Equation (9)) by direct products of a linear transformation [27], τ_K

$$(\tau_K)_{ji} \equiv \gamma_{j,K}[\sigma_i], \tag{6}$$

where the new matrix τ is constructed from these point functions,$\gamma_{j,K}[\sigma_i]$. It can be trivially shown that the inverse of Van der Monde matrices with complex entries equal to roots of unity is its complex conjugate. The inverse of the τ_K matrix can be obtained from the complex conjugate matrix of τ_K by taking the real and imaginary part and riffling their rows. From Equation (4), the following expression results for a system with K components:

$$(\tau_K^{-1})_{ij} = \begin{cases} \frac{1}{K}, & \text{if } j = 0\,, \\ -\frac{2}{K}\cos\left(2\pi \lceil \frac{j}{2} \rceil \frac{\sigma_i}{K}\right), & \text{if } j > 0 \text{ and } j-1 < K \text{ and j odd}, \\ -\frac{2}{K}\sin\left(2\pi \lceil \frac{j}{2} \rceil \frac{\sigma_i}{K}\right), & \text{if } j > 0 \text{ and j even}, \\ -\frac{1}{K}\cos\left(2\pi \lceil \frac{j}{2} \rceil \frac{\sigma_i}{K}\right), & \text{if } j-1 = K \text{ and j odd}. \end{cases} \tag{7}$$

To the best of our knowledge, Equation (7) represents a new formulation for the inverse of τ_K matrix to ensure that the basis set defined by Equation (4) is rigorously orthonormal. The size of τ_K matrix is KxK, where K is the number of components. It is convenient to perform the matrix multiplications with all K^ω decorations formed from ω-tuples with integer elements running from 0 to $K-1$. In particular, for the case of four component $K = 4$, these matrices τ become by applying Equations (4) and (7) for the inverse τ_K^{-1}:

$$\tau_4 = \begin{pmatrix} 1 & 1 & 1 & 1 \\ -1 & 0 & 1 & 0 \\ 0 & -1 & 0 & 1 \\ -1 & 1 & -1 & 1 \end{pmatrix} \quad \tau_4^{-1} = \frac{1}{4}\begin{pmatrix} 1 & -2 & 0 & -1 \\ 1 & 0 & -2 & 1 \\ 1 & 2 & 0 & -1 \\ 1 & 0 & 2 & 1 \end{pmatrix}. \tag{8}$$

We duplicate the symmetrically unique decorations (*s*) whenever two decorations are connected by symmetry of the disordered structure. For relating two equivalent decorations, we find it convenient to use a permutation representation of the space group operator as permutations of ω-site tuples. We use the property of invariance of the cluster expansion to obtain probability distributions from correlation functions [28]. As a consequence of the compact formalism and by using Equation (4) and (7), the expression of correlation functions can be rewritten into a matrix form:

$$\langle \Gamma_{\omega,n}^{(ij\cdots)}[\vec{\sigma}] \rangle = \sum_{\forall[(pq\cdots)]}^{K^{\omega}} \gamma_{i,K}[\vec{\sigma}_p]\gamma_{j,K}[\vec{\sigma}_q]\cdots y_{\omega,n}^{(pq\cdots)}[\vec{\sigma}] \equiv \overbrace{(\tau_K \otimes \cdots \otimes \tau_K)}^{\omega}{}_{ij\cdots,pq\cdots} y_{\omega,n}^{(pq\cdots)}[\vec{\sigma}]. \tag{9}$$

With the aid of the matrix formulation from Equation (9) and the generalized form of the inverse matrix τ_K^{-1} in Equation (7), one can express the ω-cluster probabilities into a matrix form:

$$y_{\omega,n}^{(pq\cdots)}[\vec{\sigma}] = \overbrace{(\tau_K^{-1} \otimes \cdots \otimes \tau_K^{-1})}^{\omega}{}_{pq\cdots,ij\cdots} \langle \Gamma_{\omega,n}^{(ij\cdots)}[\vec{\sigma}] \rangle, \tag{10}$$

where we have used the notation for direct product of matrices $\overbrace{(\tau_K^{-1} \otimes \cdots \otimes \tau_K^{-1})}^{\omega}{}_{pq\cdots,ij\cdots} = (\tau_K)_{p,i}(\tau_K)_{q,j}\cdots$. In addition, we also implied summation over repeated indexes on the right-hand side of Equation (10). Note that the size of these matrices increases exponentially with cluster size ω, K^{ω}xK^{ω}; in particular, for $\omega = 4$, the matrices have 256×256 entries. The cluster probabilities are normalized as expressed in Equation (11)

$$\sum_{\forall[(pq\cdots)]}^{K^{\omega}} y_{\omega,n}^{(pq\cdots)}[\vec{\sigma}] = 1. \tag{11}$$

As a consequence of the normalization of the cluster probabilities, it is possible to separate probabilities of the decorated sub-clusters from the probabilities of the maximal cluster by partial summations over all possible decorations for the sites that belong to the maximal cluster (ω, n) but not the i-th sub-cluster (ω_i, n_i):

$$\sum_{(pq\cdots)\in\{[(pq\cdots)_{\omega_i,n_i}]\}} y_{\omega,n}^{(pq\cdots)}[\vec{\sigma}] = y_{\omega_i,n_i}^{(pq\cdots)}[\vec{\sigma}]. \tag{12}$$

From Equation (10) and for the case with ω = 2, it follows that the generalized expression for SRO of species p and q at the nth shell, $\alpha_{2,n}^{(pq)}[\vec{\sigma}](p \neq q)$, can be interpreted as the tendency to order or segregate species p and q with respect the disordered random probability given by the product of their elemental, p, bulk concentration $x_p[\vec{\sigma}]$. For the four component Cr-Fe-Mn-Ni system, there are six chemically distinct SRO parameters: $\alpha_{2,n}^{(01)}[\vec{\sigma}]$ for Cr-Fe; $\alpha_{2,n}^{(02)}[\vec{\sigma}]$ for Cr-Mn; $\alpha_{2,n}^{(03)}[\vec{\sigma}]$ for Cr-Ni; $\alpha_{2,n}^{(12)}[\vec{\sigma}]$ for Fe-Mn; $\alpha_{2,n}^{(13)}[\vec{\sigma}]$ for Fe-Ni; and $\alpha_{2,n}^{(23)}[\vec{\sigma}]$ for Mn-Ni. The SRO allows a quantitative description of the interactions between atoms as a function of temperature to predict order-disorder transition temperatures [11,29]. In particular, the matrix formalism from Equations (7) and (10) allows one to generalize the SRO treatment for an arbitrary number of components, K:

$$y_{2,n}^{(pq)}[\vec{\sigma}] = x_p[\vec{\sigma}]x_q[\vec{\sigma}](1 - \alpha_{2,n}^{(pq)}[\vec{\sigma}]). \tag{13}$$

2.2. Configuration Entropy in the Matrix Formulation

In general, a thermodynamical system in state $\vec{\sigma}$ and with enthalpy of mixing given by the CE Hamiltonian $\Delta H_{CE}^{Mixing}[\vec{\sigma}]$ is described by a set of symmetry unique probability distributions $y_{\omega,n}^{(pq\cdots)}[\vec{\sigma}]$ characterized by decorations of a chosen maximal (ω, n) cluster. In practice, the chosen maximal cluster (ω, n) contains few points. As a mean field approximation [27], the CVM can be rationalized as a factorization of the probability distributions of the (ω, n) maximal cluster into integer powers $\eta_{\omega_1,n_1}, \eta_{\omega_2,n_2}\cdots, \eta_{\omega_{s[\omega,n]},n_{s[\omega,n]}}$ of the probability distributions of the sub-clusters $(\omega_i, n_i) \subseteq (\omega, n); i = 1, \cdots, [\omega, n]$ [19] with decorations $(pq\cdots)_{\omega_i,n_i}$ corresponding to the components $(pq\cdots)_{\omega_i,n_i}$ of all decorations $(pq\cdots) \in \{(pq\cdots)_{(\omega,n)}\}$ occupied by sites of the (ω_i, n_i) sub-cluster

$$y_{\omega,n}^{(pq\cdots)}[\vec{\sigma}] = (y_{\omega_1,n_1}^{(pq\cdots)_{\omega_1,n_1}}[\vec{\sigma}])^{\eta_{\omega_1,n_1}}(y_{\omega_2,n_2}^{(pq\cdots)_{\omega_2,n_2}}[\vec{\sigma}])^{\eta_{\omega_2,n_2}}\cdots(y_{\omega_{s[\omega,n]},n_{s[\omega,n]}}^{(pq\cdots)_{\omega_{s[\omega,n]},n_{s[\omega,n]}}}[\vec{\sigma}])^{\eta_{\omega_{s[\omega,n]},n_{s[\omega,n]}}}. \quad (14)$$

The following expression, a consequence of the CVM factorization scheme, can be derived from [19] and is the reason why the disordered configuration at high temperature is reproduced from multi-body probabilities, $\sum_{i=1}^{[\omega,n]} \omega_i \eta_{\omega_i,n_i} = -1$. This occurs because, at the high temperature limit, the CVM probabilities tend to products of composition of the species involved in the decorations of the cluster:

$$y_{\omega_i,n_i}^{(pq\cdots)_{\omega_i,n_i}}[\vec{\sigma}] \underset{T\to\infty}{\to} \prod_{j=1}^{\omega_i} y_{1,1}^{((pq\cdots)_j)_{\omega_i,n_i}}[\vec{\sigma}] = \prod_{j=1}^{\omega_i} x_{((pq\cdots)_j)_{\omega_i,n_i}}[\vec{\sigma}], \quad (15)$$

where $y_{1,1}^{((pq\cdots)_j)_{\omega_i,n_i}}[\vec{\sigma}] = x_{(pq\cdots_j)_{\omega_i,n_i}}[\vec{\sigma}]$ is the concentration of the $(pq\cdots)_j$ equal to one of the integers $0, 1, \cdots, K-1$ in the disordered state of the alloy. A natural consequence of the factorization scheme chosen for the CVM multi-body probabilities is that the configuration entropy assumes the formulation

$$S_{\omega,n}[\vec{\sigma}] \equiv \sum_{i=1}^{s[\omega,n]} \eta_{\omega_i,n_i}\tilde{S}_{\omega_i,n_i}[\vec{\sigma}] = \sum_{i=1}^{s[\omega,n]} \eta_{\omega_i,n_i} \sum_{\forall[(pq\cdots)_{\omega_i,n_i}]}^{K^{\omega_i}} k_B y_{\omega_i,n_i}^{(pq\cdots)_{\omega_i,n_i}}[\vec{\sigma}] ln(y_{\omega_i,r_i}^{(pq\cdots)_{\omega_i,n_i}}[\vec{\sigma}]), \quad (16)$$

where $\tilde{S}_{\omega_i,n_i}$, is the entropy contribution to cluster (ω, n) from the sub-cluster (ω_i, n_i):

$$\tilde{S}_{\omega_i,n_i}[\vec{\sigma}] = \sum_{\forall(pq\cdots)_{\omega_i,n_i}}^{K^{\omega_i}} k_B y_{\omega_i,n_i}^{(pq\cdots)_{\omega_i,n_i}}[\vec{\sigma}] ln(y_{\omega_i,n_i}^{(pq\cdots)_{\omega_i,n_i}}[\vec{\sigma}]). \quad (17)$$

The set of integers $\eta_{\omega_1,n_1}, \eta_{\omega_2,n_2}\cdots, \eta_{\omega_{s[\omega,n]},n_{s[\omega,n]}}$ are the mean field integer coefficients associated with the partition function of the alloy system. In the theory of regular mixtures, the coefficients can be found from the recursive heuristic expression after Kikuchi, [30], Barker [31] and whose formulation was explicitly derived by using group theoretic methods by Gratias et al. [32]. It requires the determination of two quantities: the site multiplicity N_{ω_i,n_i} and the sub-cluster multiplicity N_{ω_i,n_i}^{β}. The site multiplicity can be determined by calculating the number of symmetry operators, $\mathcal{N}_{\omega_i,n_i}$, that stabilize the ω_i, n_i cluster i.e., $g \in G$ such that the application of g into the set of cluster positions $\{\vec{\tau}_1, \vec{\tau}_2, \cdots, \vec{\tau}_\omega\}$ results in a permutation of the set. Then, $N_{\omega_i,n_i} = \frac{|G|}{|\mathcal{N}_{\omega_i,n_i}|}$, where $|G|$ is the order of the point group associated with the space group G and $|\mathcal{N}_{\omega_i,n_i}|$ is the order of the group $\mathcal{N}_{\omega_i,n_i}$. The sub-cluster multiplicity N_{ω_i,n_i}^{β} is just the frequency of the cluster β that is contained in the cluster (ω_i, n_i)

$$\eta_{\omega_i,n_i} = -N_{\omega_i,n_i} - \sum_{(\omega_i,n_i)\subset\beta\subseteq(\omega,n)} N_{\omega_i,n_i}^{\beta}\eta_{\beta}. \quad (18)$$

In particular, the formulation applied to a point cluster retrieves the Bragg–Williams approximation for the maximal cluster $(\omega = 1, n = 1)$, i.e., a site cluster, giving the entropy weight of $\eta_{\omega_1=1_1,n_1=1} = -1$ and for a 2-body cluster $(\omega = 2, n)$ in the nth coordination shell, we get $\eta_{\omega_2,n_2} = -N_{2,n}$ and $\eta_{\omega_1,n_1} = 2N_{2,n} - 1$, where $N_{2,n}$ is site multiplicity of the cluster $(2, n)$ calculated from the sites in this cluster and the space group G.

In this work, the above matrix formulation is applied in the hybrid CE-Monte Carlo method which performs the free energy minimization from the CE Hamiltonian in a combination with Monte Carlo simulations. Within the process, the Monte Carlo method produces the correlation functions for the equilibrium configurations found at each of the temperatures investigated. The hybrid approach uses these correlation functions in the analytic expressions for configuration entropy.

There is an alternative to the hybrid approach for the entropy calculation where the thermodynamic integration method can be used. Here, entropy is calculated from the configuration contribution to the specific heat at constant volume, C_{conf} derived from the fluctuations of enthalpy of mixing at temperature values in a fine grid of temperature values:

$$S_{conf}[T] = \int_0^T \frac{C_{conf}}{T'} dT' = \int_0^T \frac{\langle \Delta H_{CE}^{Mixing^2} \rangle - \langle \Delta H_{CE}^{Mixing} \rangle^2}{T'^3} dT' \tag{19}$$

where $\langle \Delta H_{CE}^{Mixing} \rangle^2$ and $\langle \Delta H_{CE}^{Mixing^2} \rangle$ are the square of the mean and mean square enthalpies of mixing, respectively, calculated by averaging over all the MC steps at the accumulation stage for a given temperature. The accuracy of evaluation of configuration entropy depends on the size of temperature integration step and the number of MC steps performed at the accumulation stage [33]. The integration of specific heat is performed from 0 K to the temperature T. In order to calculate the configuration entropy at a given temperature, the value for specific heats at lower temperatures is thus required. For example, assuming that the chosen temperature step is equal to 5 K, to evaluate configuration entropy numerically at 3000 K would require computing the specific heat at 600 smaller temperatures. In contrast, using Equation (17), the configuration entropy can be computed analytically from the correlation functions at any given temperature and alloy composition. Our experiences show that the computational time using the hybrid method can be of two orders faster than those by the thermodynamic integration.

2.3. Computational Details

The DFT enthalpies of mixing given by Equation (1) were calculated from fully relaxed spin-polarized DFT total energy calculations performed using the projector augmented wave (PAW) method [34] implemented in Vienna Ab initio Simulation Package (VASP) [35–39]. Exchange and correlation were treated in the generalized gradient approximation (GGA) and the Perdew-Burke-Ernzerhof (PBE) functional [40]. The core configurations of Fe, Cr, Mn and Ni in PAW potentials are $[Ar]3d^7 4s^1$, $[Ar]3d^5 4s^1$, $[Ar]3d^6 4s^1$ and $[Ar]3d^9 4s^1$, respectively. The total energies were calculated using the Monkhorst–Pack [41] mesh of $12 \times 12 \times 12$ k-point for a four-atom FCC cubic cell. The plane wave cut-off energy used in the calculations was 400 eV. The total energy convergence criterion was set to 10^{-6} eV/cell, and force components were relaxed to 10^{-3} eV/nm.

In the semi canonical Monte Carlo calculations performed, the temperature range and temperature steps are important. The accuracy of the thermodynamic integration method to calculate configuration entropy generally requires a smaller temperature integration step, $\Delta T = 5$ K and also depends on the number of Monte Carlo passes [33]. In particular, we use a cell containing 2048 atoms distributed into a $8 \times 8 \times 8$ primitive unit cell, and average compositions for the ensemble given by $Cr_{18}Fe_{27}Mn_{27}Ni_{28}$ and equiatomic $Cr_{25}Fe_{25}Mn_{25}Ni_{25}$. The Monte Carlo simulations were performed from random configurations at high temperature (3000 K) where the configuration entropy at the high-temperature limit is given by $k_B ln(4)$ for equiatomic and $-k_B\{0.18 * ln(0.18) + 2 * 0.27 * ln(0.27) + 0.28 * ln(0.28)\}$ for $Cr_{18}Fe_{27}Mn_{27}Ni_{28}$ [42]. By quenching down systematically with the temperature step of Δ T = 5 K, various equilibrium configurations were obtained at lower temperatures. For thermodynamic integration calculation of configuration entropy, we integrated numerically the specific heat at constant volume using the theoretical formula (19) starting from the lowest temperature value 0 K to 3000 K.

3. Cluster Probability Functions in FCC Cr-Fe-Mn-Ni Alloys

We apply the matrix formulation of cluster expansion outlined in Section 2.1 to the FCC CrFeMnNi system for investigating the temperature and composition dependent cluster probability distribution functions and the configuration entropy.

3.1. Cluster Expansion Hamiltonian for FCC CrFeMnNi

The DFT enthalpies of mixing for the FCC CrFeMnNi system were used to map iteratively into the cluster expansion Hamiltonian given by Equation (2) using the ATAT package [16]. The mapping has been performed systematically from the six binary and four ternary constituent subsystems of the considered quaternary. The database of structures for the cluster expansion consisted of 835 structures categorized by the difference of local environments in binaries (structures with two chemical elements: 58 CrFe, 55 CrMn, 77 CrNi, 58 FeMn, 54 FeNi and 52 MnNi), ternaries (structures with three chemical elements: 89 CrFeMn, 85 CrFeNi, 46 FeMnNi, and 66 CrMnNi); and 191 quaternaries CrFeMnNi. More information about the type of binary and ternary structures used in our DFT database has been detailed in our previous work [17]. Structures are typically ordered ones with their composition ranging for each constituent element from 5% to 95%. It is important to stress here that, different to other studies of HEAs, our DFT database for constructing the CE Hamiltonian didn't include randomly distributed structures such as the so-called special quasi-random structures (SQSs). The latter, however, can be generated within the present approach from Monte-Carlo simulations after obtaining the reliable ECIs. The set of clusters which have minimized the cross-validation score of 12.95 meV/atom, consists of six 2-body, two 3-body and one 4-body ECIs. In the present work, the clusters with the same sizes and relative positions have been included consistently in the considered subsystems. The prediction of the corresponding ground-state intermetallic phases is in a good agreement with the experimental binary phase diagrams available (for example, the ferromagnetic $FeNi_3$ in $L1_2$ structure and anti-ferromagnetic $MnNi$ in $L1_0$ structure) as well as with the previous theoretical study for the ternary and ferrimagnetic $CrFe_2Ni$ phase in $NiCu_2Zn$ structure [17]. A new intermetallic phase $FeCr_2MnNi_4$ is predicted for the quaternary system and full results of magnetic properties in CrFeMnNi systems will be discussed in a separate work.

The CE Hamiltonian for the quaternary system CrFeMnNi consists, in total, of 83 different decorated clusters distributed among 10 different non-decorated clusters: four decorated clusters for the point cluster, ($\omega = 1, n = 1$); six decorated clusters for each pair cluster ($\omega = 2, n = 1, \cdots, 6$); 10 decorated clusters for the non-decorated cluster ($\omega = 3, n = 1$); 18 decorated clusters for the non-decorated cluster ($\omega = 3, n = 2$); and 15 decorated clusters for the non-decorated cluster ($\omega = 4, n = 1$). The decoration labels, (s) required to specify the clusters, are listed in Table 1. Each non-decorated cluster is defined by the coordinates (specified with respect to the standard Cartesian coordinate system in units of lattice spacing) of the lattice sites that it includes. The decorations define the chemical species allocated to each site in strictly the same order i.e., for ($\omega = 2, n = 1$) cluster the decoration $(2,3)$ means that species 2 is allocated for site with coordinates $(1,1,1)$ and species 3 is allocated for site with coordinates $(1/2, 3/2, 3/2)$.

3.2. Full Set of Cluster Decorations

In general, the set of temperature dependent decorated cluster correlation functions, $\langle\Gamma^{(s)}_{\omega,n}[\vec{\sigma}]\rangle$, constitute a set of K^{ω} quantities for a given maximal cluster (ω, n) in the CE. From each of the K^{ω} temperature dependent decorated cluster correlation functions, the matrix formalism described in Section 2.1 generates the set of temperature dependent multi-body probability functions describing the temperature dependent behavior associated with the maximal cluster (ω, n). The symmetry unique decorations for all of the clusters in the CE, $\forall(\omega, n)$, are reported in the ATAT *clusters* output file and here they are listed in Table 1. If one cluster is included into another, it becomes a sub-cluster $(\omega_i, n_i) \in (\omega, n)$; these sub-clusters have been classified according to their inclusion into maximal clusters, and are listed in Table 2. If the sub-cluster (ω_i, n_i) is included in the maximal cluster, $(\omega_i, n_i) \in (\omega, n)$, then the decorations from the sub-cluster $(s)_{(\omega_i,n_i)}$ can be transferred into the decorations of the maximal cluster, $(s)_{(\omega,n)}$, by using the intrinsic space group symmetry of the parent lattice. In general, any given sub-cluster (ω_i, n_i) can be found several times within the cluster (ω, n). For convenience, symmetry of the cluster decorations is implemented by means of permutation operators in Table 2 including X for the empty site. The full set of decorated cluster correlation functions, $\langle \Gamma^{(s)}_{\omega,n}[\vec{\sigma}] \rangle$, corresponding to the maximal cluster (ω, n), is generated by studying which of the decorations $(s)_{\omega_i,n_i}$ form the sub-clusters (ω_i, n_i) in (ω, n). The decorations can have empty cluster, i.e., at least one integer entry in $0 \subseteq (s)_{\omega_i,n_i}$.

For the case of the maximal cluster $(\omega = 2, n = 1)$ consisting of sites $\{(1,1,1),(1,3/2,3/2)\}$, see Table 1, we have a full set of $4^2 = 16$ 2-tuples of decorations, each associated with a temperature dependent decorated cluster correlation function. It can be noted that the cluster $\{(1,1,1),(1,3/2,3/2)\}$ contains two sub-clusters: the point cluster $\{(1,1,1)\}$ that has decorations $\{(0),(1),(2),(3)\}$; and the cluster itself which has decorations $\{(1,1),(2,1),(3,1),(2,2),(2,3),(3,3)\}$. For each of these decorations, there are corresponding decorated cluster correlation functions. In this case, the point cluster is contained twice: once in $(1,1,1) \subset \{(1,1,1),(1,3/2,3/2)\}$ and also in $(1,3/2,3/2) \subset \{(1,1,1),(1,3/2,3/2)\}$. Similarly the cluster itself $\{(1,1,1),(1,3/2,3/2)\}$ is contained once. Thus a point cluster decorated by $(s) = (3)$ is transfered to $\omega = 2$-tuples notation in the maximal cluster as $(3,0)$ or $(0,3)$. This can be captured by the permutation operators $\{(1,X),(X,1)\}$ (see Table 2) where X stands for an empty cluster where a 0 should be placed. Similarly, the decorations denoted by $(s) = (1,2)$ in the 2-tuples notation are equivalent to the symmetry equivalent decorations $(s) = (1,2)$ and $(s') = (2,1) \equiv (1,2)$. The symmetry effect can again be captured by permutation operators $\{(1,2),(2,1)\}$, where X no longer appears, since the $\omega_i = 2$-tuples have the same order $(\omega = \omega_i)$ as the $\omega = 2$-tuples from the maximal cluster $(\omega = 2, n = 1)$. The analysis outlined above for the 2-body cluster $\{(1,1,1),(1,3/2,3/2)\}$ with $K = 4$ components can be extended to the remaining $(\omega = 2, n > 1)$ maximal clusters. The permutation operators that carry out the appropriate decorations $(s)_{\omega_i,n_i}$ of sub-clusters (ω_i, n_i) into (ω, n) cluster ω-tuples notation are detailed in Table 2. It should be noted that, for any 2-body cluster, the point and corresponding pair sub-clusters add up, forming $4 + 6 = 10$ symmetrically unique correlation functions $\langle \Gamma^{(ij)}_{\omega,n}[\vec{\sigma}] \rangle$.

Referring to the maximal cluster $(\omega = 3, n = 1)$, from Table 2, there are three contained sub-clusters $(\omega_1 = 1, n_1 = 1)$, $(\omega_2 = 2, n_2 = 1)$ and $(\omega_3 = 3, n_3 = 1)$ within the maximal cluster $(\omega = 3, n = 1)$. The point sub-cluster $(\omega_1 = 1, n_1 = 1)$ has, according to Table 1, four decorations associated with it; the 2-body cluster $(\omega_2 = 2, n_2 = 1)$ has six decorations associated with itself; and finally the sub-cluster $(\omega_3 = 3, n_3 = 1)$ has 10 decorations associated with itself. The total number of symmetry unique decorated cluster correlation functions, which is smaller than $K^\omega = 4^3 = 64$, but can nevertheless fully describe cluster $(\omega = 3, n = 1)$ with $K = 4$ components, is therefore given by $4 + 6 + 10 = 20$ decorated cluster correlation functions. From these 20 symmetry unique decorated cluster correlations, it is possible to generate a total of 64 decorated cluster correlation functions by using the permutation operators in Table 2. The number of permutation operators to transfer decorations from sub-cluster $(\omega_1 = 1, n_1 = 1)$ to $(\omega = 3, n = 1)$, is 3; from $(\omega_2 = 2, n_2 = 1)$, six permutation operators, and, from the cluster itself $(\omega_3 = 3, n_3 = 1)$, there are three permutation operators. The total number of permutation operators is therefore $3 + 6 + 6 = 15$. The remaining decorations, which are not listed in Table 1, add up to $64 - 20 = 44$ and are obtained by using appropriately the 15 permutation operators corresponding to the maximal cluster $(\omega = 3, n = 1)$. Similarly, the maximal cluster $(\omega = 3, n = 2)$ has 34 symmetrically unique decorations and 12 permutation operators; and $(\omega = 4, n = 1)$ has 35 symmetry unique decorations with 64 permutation operators (all possible permutations for $K = 4$).

Table 1. ω, highest coordination shell n, decoration (s) and coordinates of points in the relevant clusters on the Face-Centered Cubic (FCC) lattice. The coordinates are referred to the simple cubic Bravais lattice. Index (s) is the same as the sequence of points in the relevant cluster. The canonical order for decoration indexes, $(s)_i$, is 0, 1, 2 and 3 is Cr, Fe, Mn and Ni. All values of the Effective Cluster Interactions (ECIs) obtained from the present CE study are shown in the last column.

ω	n	(s)	Coordinates	ECI (meV/atom)	ω	n	(s)	Coordinates	ECI (meV/atom)
1	1	(0)	(1,1,1)	+0.00	3	1	(1,1,1)	(1,1,1)	+0.20
		(1)		+0.11			(2,1,1)	(3/2,1,1/2)	+0.90
		(2)		−0.04			(3,1,1)	(1,3/2,1/2)	+1.60
		(3)		−0.01			(2,2,1)		−3.40
2	1	(1,1)	(1,1,1)	+9.40			(3,2,1)		−0.50
		(2,1)	(1,3/2,3/2)	−0.10			(3,3,1)		+1.20
		(3,1)		+3.40			(2,2,2)		+0.20
		(2,2)		+0.40			(3,2,2)		+2.00
		(3,2)		+1.30			(3,3,2)		−0.50
		(3,3)		+6.00			(3,3,3)		+0.00
2	2	(1,1)	(1,1,1)	−9.20	3	2	(1,1,1)	(1,1,1)	−0.60
		(2,1)	(1,1,0)	+0.40			(2,1,1)	(1,3/2,1/2)	+1.00
		(3,1)		−4.40			(3,1,1)	(1,1,0)	+0.90
		(2,2)		−11.60			(1,2,1)		−1.80
		(3,2)		−3.50			(2,2,1)		+2.30
		(3,3)		−8.80			(3,2,1)		−0.70
2	3	(1,1)	(1,1,1)	0.90			(1,3,1)		−2.10
		(2,1)	(2,3/2,3/2)	2.60			(2,3,1)		−0.30
		(3,1)		3.50			(3,3,1)		−0.60
		(2,2)		1.60			(2,1,2)		−6.30
		(3,2)		−0.30			(3,1,2)		−1.20
		(3,3)		0.10			(2,2,2)		-0.60
2	4	(1,1)	(1,1,1)	−0.40			(3,2,2)		−0.10
		(2,1)	(2,1,2)	2.40			(2,3,2)		+1.60
		(3,1)		1.20			(3,3,2)		+0.60
		(2,2)		0.50			(3,1,3)		−2.00
		(3,2)		0.60			(3,2,3)		+0.30
		(3,3)		−0.80			(3,3,3)		−1.40
2	5	(1,1)	(1,1,1)	−1.00	4	1	(1,1,1,1)	(1,1,1)	−3.30
		(2,1)	(1,3/2,−1/2)	−3.50			(2,1,1,1)	(3/2,3/2,1)	+2.00
		(3,1)		−2.00			(3,1,1,1)	(3/2,1,1/2)	+0.70
		(2,2)		0.20			(2,2,1,1)	(1,3/2,1/2)	−2.90
		(3,2)		0.90			(3,2,1,1)		+0.60
		(3,3)		0.40			(3,3,1,1)		+0.70
2	6	(1,1)	(1,1,1)	0.80			(2,2,2,1)		−0.60
		(2,1)	(2,2,0)	2.10			(3,2,2,1)		−1.10
		(3,1)		1.00			(3,3,2,1)		+1.30
		(2,2)		−2.70			(3,3,3,1)		+2.60
		(3,2)		−1.30			(2,2,2,2)		−0.50
		(3,3)		0.50			(3,2,2,2)		+4.90
							(3,3,2,2)		+1.00
							(3,3,3,2)		−1.60
							(3,3,3,3)		−1.70

3.3. Four-Body Probability Functions from Monte Carlo Simulations

After performing semi-canonical Monte Carlo simulations for alloy compositions $Cr_{18}Fe_{27}Mn_{27}Ni_{28}$ and equiatomic $Cr_{25}Fe_{25}Mn_{25}Ni_{25}$, we use the correlation functions calculated at each of the temperatures to study the temperature dependent probabilities of decorated clusters corresponding to the equilibrium configuration $[\vec{\sigma}]$ of the Monte Carlo super-cell following the formalism described in the Methods section. For each non-decorated cluster ω, n and temperature, we generate the corresponding τ^{-1} matrices with dimension K^{ω}xK^{ω} and a vector formed by all the decorated correlation cluster functions, $\langle \Gamma_{\omega,n}^{(ij\cdots)}[\vec{\sigma}] \rangle$ with dimension K^{ω}. The Monte Carlo simulations output only the symmetrically unique correlation functions (see Table 1), which are less than the K^ω required for matrix operation. In order to generate the full set of K^ω from the symmetrically unique correlation functions, we devise a set of permutation operators g_i (see Table 2) that indicate how an arbitrary decoration ω-tuple of integers is obtained from the symmetrically unique decorations belonging to the non-decorated cluster or one of its sub-clusters (see the third column of Table 2 for identifying the cluster-sub-cluster relation). Table 2 contains all the operators necessary to generate cluster probabilities, $y^{(s)}_{\omega,n}$, for any possible decoration, (s), of 10 non-decorated clusters employed in the cluster expansion Hamiltonian.

As an important case of study, we chose the maximal cluster given by (ω = 4, n = 1) to obtain the 4-body configuration probabilities. Figure 1a,b show the plots of 35 symmetrically unique probabilities for the 4-body maximal cluster defined by the set of lattice sites $((1,1,1),(3/2,3/2,1),(3/2,1,1/2),(1,3/2,1/2))$ as a function of temperature.

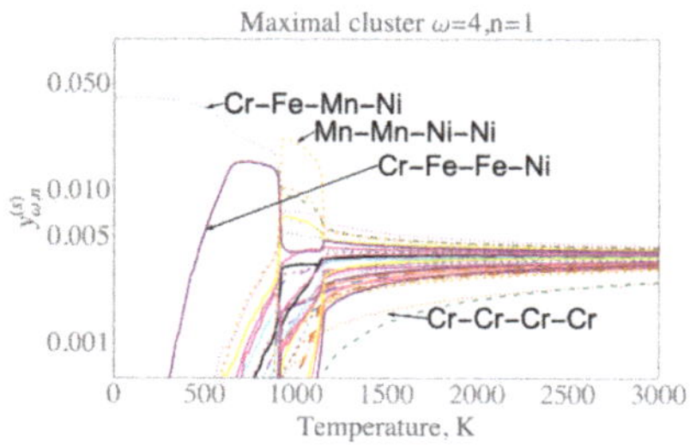

(a)

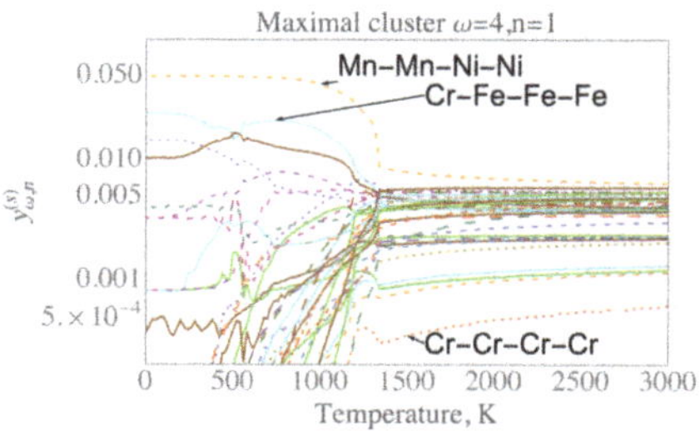

(b)

Figure 1. 4-body probabilities obtained from the hybrid Cluster Expansion (CE)-Monte Carlo calculations. (**a**) all the 4-body probabilities for the equiatomic composition $Cr_{25}Fe_{25}Mn_{25}Ni_{25}$ as a function of temperature; (**b**) the same as in (**a**) but for the composition $Cr_{18}Fe_{27}Mn_{27}Ni_{28}$.

The most important 4-body bonding configurations in the temperature range from 0 to 3000 K for the two alloy systems (the equiatomic $Cr_{25}Fe_{25}Mn_{25}Ni_{25}$ and $Cr_{18}Fe_{27}Mn_{27}Ni_{28}$) are highlighted in Figure 1a,b. It should be noted that, for both compositions, the cluster configuration referred to as Cr-Cr-Cr-Cr appears as the least probable of all the cluster configurations. This finding is consistent with the fact that Cr has the BCC ground-state and therefore the probability of finding a Cr cluster in the FCC lattice is negligible, in particular in the low temperature region.

For the equiatomic composition, the probability of Cr-Fe-Mn-Ni is particularly very high at temperatures between 0–900 K. The presence of 4-body Cr-Fe-Mn-Ni clusters at low temperatures demonstrates the relationship with our DFT/CE prediction of the new ordered phase $FeCr_2MnNi_4$ in the quaternary system. Furthermore, in the temperature range below 900 K, the cluster configuration Cr-Fe-Fe-Ni shows that it is the second most probable configuration. This configuration is directly correlated with the ordered Fe_2CrNi structure predicted in our earlier study of phase stability in the ternary Fe-Cr-Ni system [17]. In the temperature region between 900 K and 1200 K, an increase of probability of Mn-Mn-Ni-Ni cluster is significantly important. Beyond 1200 K, all of the cluster configuration probabilities tend to the solid solution or random configuration.

For the system with average composition $Cr_{18}Fe_{27}Mn_{27}Ni_{28}$, the cluster probability of decorations given by Mn-Mn-Ni-Ni appears to dominate until 1300 K, where the solid solution or random configuration begins. Furthermore, the second most probable cluster configuration in the temperature range 500–1200 K appears to be Cr-Fe-Fe-Fe. Again, these findings are remarkable and the origin of these clusters would need a more detailed discussion.

Table 2. List of permutation operators for generating the full set of decorations represented by K^{ω} dimensional integer arrays with entries taking values from 0 to $K-1$. The symmetry operators (see Equation (5)) represented here in permutation form in column 2 act on the set of 83 symmetry unique decorations indicated under the (s) column in Table 1 by permuting the entries in (s) or by introducing the empty cluster X for sub-clusters belong to a given cluster; examples with discussion are provided in Section 3.2. The space group of the disordered FCC structure, $g \in O_h^5$, is implicitly assumed in order to convolute the symmetry unique into the full set of decorations. The last four columns represent (ω_i, n_i) ith sub-cluster of the maximal cluster (ω, n); N_{ω_i,n_i} site multiplicity of (ω_i, n_i); $N_{\omega,n}^{\omega_i,n_i}$ sub-cluster multiplicity of the cluster $(\omega_i, n_i) \in (\omega, n)$; and η_{ω_i,n_i}, the sub-cluster (ω_i, n_i) contribution to configuration entropy expression corresponding to the maximal cluster (ω, n)

Maximal Cluster	Permutation Operators $\{g_1, g_2, \cdots\}$	Sub-Cluster $\{\omega_i, n_i\}$	N_{ω_i,n_i}	$N_{\omega,n}^{\omega_i,n_i}$	η_{ω_i,n_i}
$\omega=1, n=1$	$\{(1)\}$	$\omega_1=1, n_1=1$	1	1	−1
$\omega=2, n=1$	$\{(1,X),(X,1)\}$	$\omega_1=1, n_1=1$	1	2	11
	$\{(1,2),(2,1)\}$	$\omega_2=2, n_2=1$	6	1	−6
$\omega=2, n=2$	$\{(1,X),(X,1)\}$	$\omega_1=1, n_1=1$	1	2	5
	$\{(1,2),(2,1)\}$	$\omega_2=2, n_2=2$	3	1	−3
$\omega=2, n=3$	$\{(1,X),(X,1)\}$	$\omega_1=1, n_1=1$	1	2	23
	$\{(1,2),(2,1)\}$	$\omega_2=2, n_2=3$	12	1	−12
$\omega=2, n=4$	$\{(1,X),(X,1)\}$	$\omega_1=1, n_1=1$	1	2	11
	$\{(1,2),(2,1)\}$	$\omega_2=2, n_2=4$	6	1	−6
$\omega=2, n=5$	$\{(1,X),(X,1)\}$	$\omega_1=1, n_1=1$	1	2	23
	$\{(1,2),(2,1)\}$	$\omega_2=2, n_2=5$	12	1	−12
$\omega=2, n=6$	$\{(1,X),(X,1)\}$	$\omega_1=1, n_1=1$	1	2	7
	$\{(1,2),(2,1)\}$	$\omega_2=2, n_2=6$	4	1	−4
$\omega=3, n=1$	$\{(1,X,X),(X,X,1),(X,1,X)\}$	$\omega_1=1, n_1=1$	1	3	−13
	$\{(1,2,X),(2,1,X),(2,X,1),(1,X,2),(X,1,2),(X,2,1)\}$	$\omega_2=2, n_2=1$	6	3	18
	$\{(1,3,2),(3,2,1),(2,1,3),(3,1,2),(2,3,1),(1,2,3)\}$	$\omega_3=3, n_3=1$	8	1	−8
$\omega=3, n=2$	$\{(1,X,X),(X,X,1),(X,1,X)\}$	$\omega_1=1, n_1=1$	1	3	−19
	$\{(2,1,X),(1,2,X)\}$	$\omega_2=2, n_2=2$	1	1	9
	$\{(1,X,2),(2,X,1),(X,1,2),(X,2,1)\}$	$\omega_3=2, n_3=1$	6	2	18
	$\{(3,1,2),(1,3,2),(1,2,3)\}$	$\omega_4=3, n_4=2$	12	1	−12
$\omega=4, n=1$	$\{(1,X,X,X),(X,X,X,1),(X,X,1,X),(X,1,X,X)\}$	$\omega_1=1, n_1=1$	1	4	−5
	$\{(X,X,1,2),(X,X,2,1),(1,2,X,X),(2,1,X,X)$	$\omega_2=2, n_2=1$	6	6	6
	$(X,1,X,2),(X,2,X,1),(2,X,1,X),(1,X,2,X)$	$\omega_2=2, n_2=1$			
	$(1,X,X,2),(2,X,X,1),(X,1,2,X),(X,2,1,X)\}$	$\omega_2=2, n_2=1$			
	$\{(1,3,2,X),(3,1,X,2),(2,X,1,3),(X,2,3,1)$	$\omega_3=3, n_3=1$	8	4	0
	$(3,2,1,X),(1,X,3,2),(X,1,2,3),(2,3,X,1)$	$\omega_3=3, n_3=1$			
	$(2,1,3,X),(X,3,1,2),(1,2,X,3),(3,X,2,1)$	$\omega_3=3, n_3=1$			
	$(2,X,3,1),(X,2,1,3),(1,3,X,2),(3,1,2,X)$	$\omega_3=3, n_3=1$			
	$(1,X,2,3),(3,2,X,1),(2,3,1,X),(X,1,3,2)$	$\omega_3=3, n_3=1$			
	$(3,X,1,2),(1,2,3,X),(X,3,2,1),(2,1,X,3)\}$	$\omega_3=3, n_3=1$			
	$\{(1,4,3,2),(4,1,2,3),(3,2,1,4),(2,3,4,1)$	$\omega_4=4, n_4=1$	2	1	−2
	$(4,3,1,2),(1,2,4,3),(2,1,3,4),(3,4,2,1)$	$\omega_4=4, n_4=1$			
	$(3,1,4,2),(2,4,1,3),(1,3,2,4),(4,2,3,1)$	$\omega_4=4, n_4=1$			
	$(3,2,4,1),(2,3,1,4),(1,4,2,3),(4,1,3,2)$	$\omega_4=4, n_4=1$			
	$(1,2,3,4),(4,3,2,1),(3,4,1,2),(2,1,4,3)$	$\omega_4=4, n_4=1$			
	$(4,2,1,3),(1,3,4,2),(2,4,3,1),(3,1,2,4)\}$	$\omega_4=4, n_4=1$			

The prediction of the high probabilities for Mn-Mn-Ni-Ni and Cr-Fe-Fe-Fe clusters in $Cr_{18}Fe_{27}Mn_{27}Ni_{28}$ alloy composition shown in Figure 1b can be explained by the decoration of the first nearest neighbor four-atom cluster interaction obtained from the CE Hamiltonian. It is important to stress that, in this case, the CE method reproduces the well-known result from the tetrahedron approximation in the CVM [18,19]. The configuration Mn-Mn-Ni-Ni is understood to be related to $L1_0$ structure in Strukturbericht notation with Mn and Ni atoms, whereas the composition Cr-Fe-Fe-Fe is related the $L1_2$ structure in Strukturbericht notation with Cr and Fe atoms. Both of these structures are depicted in Figure 2 and they are in a full agreement with our first-principles investigations. Indeed, the *MnNi-L*1_0 structure is predicted by both the DFT and CE Hamiltonian to be one of the ground-state structures not only for the binary Mn-Ni system but also for the quaternary Fe-Cr-Mn-Ni one. The *CrFe*$_3$*-L*1_2 structure has also been found as the lowest-energy binary structure in the previous study of the phase stability of Fe-Cr binary in the FCC lattice (see Figure 2a from [17]). In particular, the latter structure has also been found to be stable due to the strong anti-ferromagnetic interaction between Cr and Fe from the magnetic cluster expansion (MCE) performed for the ternary CrFeNi system [43]. In that work, Monte Carlo simulations using MCE indicated that ordered magnetic structures in the Ni-rich corner (ferromagnetic) of the *FeCrNi* system persists until temperatures of 600 K, and that for the atomic compositions between $CrFe_2$ and $CrFe_3$ the anti-ferromagnetic order was retained beyond 500 K.

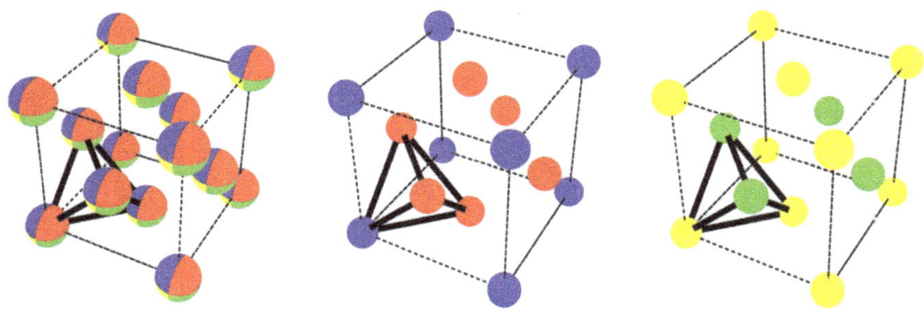

(a) (b) (c)

Figure 2. (**a**) most probable phase at high temperature (disordered structure); (**b**,**c**): two most probable ordered phases at low temperature in the equiatomic $Cr_{25}Fe_{25}Mn_{25}Ni_{25}$ and $Cr_{18}Fe_{27}Mn_{27}Ni_{28}$ HEAs compositions. Cr, Fe, Mn and Ni are illustrated in blue, red, yellow and green respectively. (**a**) A1 phase, sites are occupied by Cr, Mn, Fe, and Ni in probabilities determined by their average concentration in the system; (**b**) $L1_2$ phase corresponding to $CrFe_3$ with Cr and Fe; (**c**) $L1_0$ phase corresponding to MnNi with Mn and Ni.

4. Configuration Entropy in a Cr-Fe-Mn-Ni System

For each temperature value in the range 0–3000 K, the configuration probabilities corresponding to cluster ($\omega = 4, n = 1$) are used in Equation (17) to obtain the quantities $\tilde{S}_{\omega_i,n_i}[\vec{\sigma}]$. There is one quantity $\tilde{S}_{\omega_i,n_i}[\vec{\sigma}]$ and a corresponding weight η_{ω_i,n_i} for each of the (ω_i, n_i) sub-clusters : ($\omega = 1, n = 1$), ($\omega = 2, n = 1$); ($\omega = 3, n = 1$); and ($\omega = 4, n = 1$) included in the maximal cluster ($\omega = 4, n = 1$). The total configuration entropy, $S_{\omega,n}[\vec{\sigma}]$, is calculated by adding the η_{ω_i,n_i} weighted quantities $\tilde{S}_{\omega_i,n_i}[\vec{\sigma}]$. By applying the same arguments developed for the cluster ($\omega = 4, n = 1$), the total configuration entropy for each of the 10 different maximal clusters can be calculated. The sub-clusters and the permutation operators contained in each of the 10 maximal clusters appearing in the cluster expansion are detailed in Table 2.

Composition dependent entropies at fixed temperatures 1000 K and 3000 K are shown in Figure 3a,b. At each of these temperature values, any of the configuration entropy expressions from Table 2 provides approximately the same value. At 3000 K, the entropy is maximized in the center of the tetrahedron, namely at the equiatomic composition to 1.37 k_B, and it is decreased upon lowering the temperature at all of the composition points. The variation of configuration entropy is too complex to be captured in the fixed temperature plots, therefore we use specifically the equiatomic composition and calculate the configuration entropy as a function of temperature in Figure 4. Regarding the high temperature limit value of entropy in Figure 4, we use as maximal cluster that from the 4-body, 3-body and 2-body in the nearest neighbor cluster probabilities. Above 1200 K, the cluster approximation for the configuration entropy approaches the high temperature limit of disordered solute solution ($-k_B \sum_p^4 x_p[\vec{\sigma}] ln(x_p[\vec{\sigma}])$) much more rapidly than the thermodynamic integration. The reason the high temperature limit is preserved can be clearly seen from the factorization of the CE expression that resulted from analytical derivation in Equation (15). It is important to stress that, within the high-temperature limit, our theoretical prediction recovers the ideal configuration entropy of mixing for solution phase as it has been originally proposed by the phenomenological guidelines in designing high-entropy alloys using thermodynamic and topological parameters of the constituent elements [44].

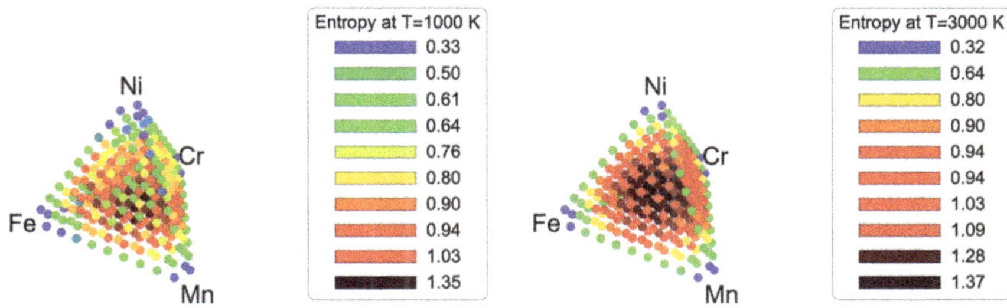

(**a**) (**b**)

Figure 3. Composition dependent entropies obtained from Monte Carlo simulations in CE. (**a**) Composition dependent entropy at fixed temperature 1000 K; (**b**) Composition dependent entropy at fixed temperature 3000 K.

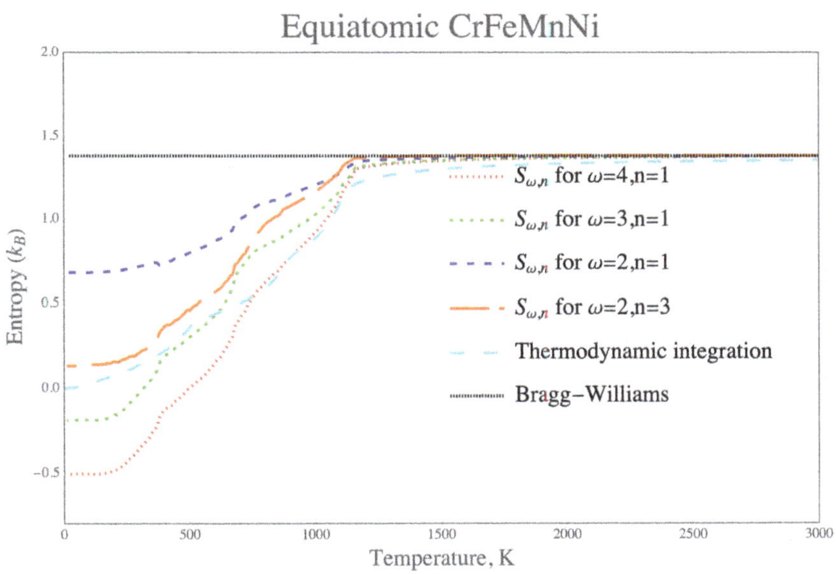

Figure 4. Temperature dependence of configuration entropy evaluated at various levels of cluster approximations and compared with the thermodynamic integration result at the equiatomic composition $Cr_{25}Fe_{25}Mn_{25}Ni_{25}$.

It is worth mentioning that the entropy increase (Figure 4) as a function of temperature for the equiatomic alloy composition can thus be understood by the corresponding decrease in probabilities (Figure 1) of the ordered state of the structures $L1_0$ towards the disorder configuration *A*1 in Figure 2. Similarly, in the alloy with composition $Cr_{18}Fe_{27}Mn_{27}Ni_{28}$, the increase of configuration entropy is correlated with the decrease in high probability of the configurations of Cr-Fe-Fe-Fe and Mn-Mn-Ni-Ni clusters originally related to the ordered FCC-like $L1_2$-$CrFe_3$ and $L1_0$-MnNi structure, respectively.

For the cluster ($\omega = 4, n = 1$), the calculated configuration entropy in Figure 4 appears to be negative at low temperature with respect to the positive values correctly predicted by the thermodynamic integration method. This demonstrates clearly that this maximum cluster within the tetrahedron approximation conventionally adopted within the CVM is only valid to describe the configuration entropy for the four-component Cr-Fe-Mn-Ni system in the high-temperature limit. It is shown from Figure 4 that the first nearest-neighbor triangle cluster ($\omega = 3, n = 1$), which plays an important role within the tetrahedron approximation, also gives physically incorrect and negative entropy contribution at low temperature. As it has been discussed in the introduction, in difference from the CVM tetrahedron approximation, besides the first nearest-neighbor cluster contributions, the present CE results also include other pair clusters up to the sixth nearest neighbors and the second nearest-neighbor triangle cluster contributions. For example, in Figure 4, the third nearest-neighbor pair cluster ($\omega = 2, n = 3$) gives the significantly positive contribution to configuration entropy in the entire range of temperature. These additional cluster contributions, in turn, ensure the correct behavior of the configuration entropy obtained from the thermodynamic integration. Therefore, from a statistical physics point of view, the hybrid technique combining Monte Carlo simulations with the CE method can be considered to be more advanced than the mean-field approach advocated within the CVM.

5. Conclusions

In this work, we develop a matrix formalism to study multi-body ordering probabilities beyond pair approximation previously used for investigating the SRO and configuration entropy in multi-component alloys by using a hybrid combination of CE and Monte Carlo methods. The cluster probabilities are worked out by explicit inversion within the orthonormal sets of the point functions adopted in the ATAT package and a direct product of a matrix formulation obtained from symmetrically independent correlation functions. The correlation functions are determined from semi-canonical Monte Carlo simulations and ECIs derived from DFT calculations. We apply our method to the quaternary FCC Cr-Fe-Mn-Ni system by considering 285 different alloy compositions covering the compositional space of the quaternary alloy as a function of temperature. To further assess our formulated expressions for configuration entropy, focus is put on two alloy compositions, equiatomic $Cr_{25}Fe_{25}Mn_{25}Ni_{25}$ and $Cr_{18}Fe_{27}Mn_{27}Ni_{28}$, to obtain cluster probabilities and understand the variation in configuration entropy with temperature due to the lowering of the ordering probability corresponding to the ordered configuration. The cluster probability plots against temperature show that, for the composition $Cr_{18}Fe_{27}Mn_{27}Ni_{28}$, there is a high probability of formation of the $L1_0$ MnNi phase at low temperatures below 1300 K and for the $L1_2$ $CrFe_3$ phase in the temperature range 500–1200 K. Similarly, for the equiatomic composition, the $L1_0$ MnNi phase appears stable in the temperature range 900–1200 K, while the lower temperature region is preferred for the configuration Cr-Fe-Mn-Ni. The configuration Cr-Cr-Cr-Cr was found to be the least probable configuration at all temperature ranges for both compositions of the Cr-Fe-Mn-Ni system. Furthermore, the configuration entropy as a function of temperature was derived from these probabilities: the high-temperature limit is in accordance with random solid solution approximation, but, at low temperatures, the entropy is seen to be reduced due to ordering or segregation tendencies which are in turn determined by multi-body probability functions including chemical short-range order.

We believe that the present study will help to promote further understanding of derivative phases as a function of temperature and multi-component alloy composition from disordered solid solutions in the phenomenological description of HEAs. By applying the formalism to the Cr-Fe-Mn-Ni system, it will also serve as a benchmarking example in designing radiation tolerant materials for advanced nuclear reactor systems by using the ab initio based CE method. The composition $Cr_{18}Fe_{27}Mn_{27}Ni_{28}$ multicomponent system has been studied for advanced nuclear applications due to promising irradiation resistance with regards to void swelling.

Author Contributions: A.F.-C. and D.N.-M. developed theoretical formalism and applied the formulation to the CrFeMnNi system. A.F.-C., D.N.-M. and J.S.W. wrote the manuscript and P.M.M. checked grammar J.S.W. and M.F. provided the results of DFT-based semi canonical Monte Carlo simulations for the CrFeMnNi system. All the authors made comments on the final version before its submission.

Funding: This research was funded by the Euratom research and training programme 2014–2018 and 2019–2020 under Grant Agreement No. 633053 and by the Research Council UK (RCUK) Energy Programme (Grant Number EP/P012450/1). The views and opinions expressed herein do not necessarily reflect those of the European Commission. A.F.-C. was funded by the EPSRC grant (EP/L01680X/1) through the Materials for Demanding Environments Center for Doctoral Training. M.F. and J.S.W. were funded by the Foundation for Polish Science grant HOMING (no. Homing/2016-1/12). The HOMING programme is co-financed by the European Union under the European Regional Development Fund.

Acknowledgments: The simulations were partially carried out by M.F. with the support of the Interdisciplinary Centre for Mathematical and Computational Modelling (ICM), University of Warsaw, under Grant No. GA69-30. D.N.-M. and J.S.W. acknowledges the support from high-performance computing facility MARCONI (Bologna, Italy) provided by EUROfusion. D.N.-M. also acknowledges the support from the Institute of Materials Science (IMS) at Los Alamos (NM, USA) for the IMS Rapid Response 2018 visit to the Los Alamos National Laboratory (LANL).

Conflicts of Interest: The authors declare no conflict of interest.

Abbreviations

The following abbreviations are used in this manuscript:

CE	Cluster Expansion
CVM	Cluster Variation Method
HEA	High-Entropy Alloy
K	Number of components in alloy
DFT	Density Functional Theory
GS	Ground States
G	Space Group of the disordered high temperature structure
(ω_i, n_i)	i-th sub-cluster of a maximal cluster (ω, n)
$\mathcal{N}_{\omega_i,n_i}$	operators $g \in G$ acting on the sites of cluster (ω_i, n_i) to rearrange the sites i.e., a permutation
N_{ω_i,n_i}	site multiplicity of the cluster ω_i
$N^{\beta}_{\omega,n}$	sub-cluster multiplicity of the cluster $\beta \in (\omega, n)$
$\Omega[\omega, n]$	Number of times a cluster ω, n is contained in a supercell structure, generally used for Monte Carlo simulations, for obtaining thermodynamic quantities.

References

1. Cantor, B.; Chang, I.; Knight, P.; Vincent, A. Microstructural development in equiatomic multicomponent alloys. *Mater. Sci. Eng. A* **2004**, *375–377*, 213–218. [CrossRef]
2. Yeh, J.W.; Chen, S.K.; Lin, S.J.; Gan, J.Y.; Chin, T.S.; Shun, T.T.; Tsau, C.H.; Chang, S.Y. Nanostructured High-Entropy Alloys with Multiple Principal Elements: Novel Alloy Design Concepts and Outcomes. *Adv. Eng. Mater.* **2004**, *6*, 299–303. [CrossRef]
3. Gali, A.; George, E.P. Tensile properties of high- and medium-entropy alloys. *Intermetallics* **2013**, *39*, 74–78. [CrossRef]
4. Carroll, R.; Lee, C.; Tsai, C.W.; Yeh, J.W.; Antonaglia, J.; Brinkman, B.A.W.; Leblanc, M.; Xie, X.; Chen, S.; Liaw, P.K.; et al. Experiments and Model for Serration Statistics in Low-Entropy, Medium-Entropy, and High-Entropy Alloys. *Sci. Rep.* **2015**, *5*, 16997. [CrossRef] [PubMed]
5. Xia, S.; Gao, M.C.; Yang, T.; Liaw, P.K.; Zhang, Y. Phase stability and microstructures of high entropy alloys ion irradiated to high doses. *J. Nucl. Mater.* **2016**, *480*, 100–108. [CrossRef]
6. Lucas, M.S.; Mauger, L.; Muñoz, J.A.; Xiao, Y.; Sheets, A.O.; Semiatin, S.L.; Horwath, J.; Turgut, Z. Magnetic and vibrational properties of high-entropy alloys. *J. Appl. Phys.* **2011**, *109*, 07E307. [CrossRef]
7. Calvo-Dahlborg, M.; Cornide, J.; Tobola, J.; Nguyen-Manh, D.; Wróbel, J.S.; Juraszek, J.; Jouen, S.; Dahlborg, U. Interplay of electronic, structural and magnetic properties as the driving feature of high-entropy CoCrFeNiPd alloys. *J. Phys. D Appl. Phys.* **2017**, *50*, 185002. [CrossRef]
8. Rossiter, P.L.; Wells, P. The dependence of the electrical resistivity on short-range order. *J. Phys. C Solid State Phys.* **1971**, *4*, 354–363. [CrossRef]
9. Spruiell, J.E.; Stansbury, E.E. X-Ray study of short-range order in nickel alloys containing 10.7 and 20.0 at. % Molybdenum. *J. Phys. Chem. Solids* **1965**, *26*, 811–822. [CrossRef]

10. Fisher, J.C. On the strength of solid solution alloys. *Acta Metall.* **1954**, *2*, 9–10. [CrossRef]
11. Fernández-Caballero, A.; Wróbel, J.S.; Mummery, P.M.; Nguyen-Manh, D. Short-Range Order in High Entropy Alloys: Theoretical Formulation and Application to Mo-Nb-Ta-V-W System. *J. Phase Equilib. Diffus.* **2017**, *38*, 391–403. [CrossRef]
12. Santodonato, L.J.; Zhang, Y.; Feygenson, M.; Parish, C.M.; Gao, M.C.; Weber, R.J.K.; Neuefeind, J.C.; Tang, Z.; Liaw, P.K. Deviation from high-entropy configurations in the atomic distributions of a multi-principal-element alloy. *Nat. Commun.* **2015**, *6*, 5964. [CrossRef] [PubMed]
13. Leong, Z.; Wróbel, J.S.; Dudarev, S.L.; Goodall, R.; Todd, I.; Nguyen-Manh, D. The Effect of Electronic Structure on the Phases Present in High Entropy Alloys. *Sci. Rep.* **2017**, *7*, 39803. [CrossRef] [PubMed]
14. De Fontaine, D. Cluster Approach to Order-Disorder Transformations in Alloys. *Solid State Phys.* **1994**, *47*, 33–176.
15. Van de Walle, A.; Asta, M.; Ceder, G. The Alloy Theoretic Automated Toolkit: A User Guide. *Calphad Comput. Coupling Phase Diagr. Thermochem.* **2002**, *26*, 539–553. [CrossRef]
16. Van de Walle, A. Multicomponent multisublattice alloys, nonconfigurational entropy and other additions to the Alloy Theoretic Automated Toolkit. *Calphad Comput. Coupling Phase Diagr. Thermochem.* **2009**, *33*, 266–278. [CrossRef]
17. Wróbel, J.S.; Nguyen-Manh, D.; Lavrentiev, M.Y.; Muzyk, M.; Dudarev, S.L. Phase stability of ternary fcc and bcc Fe-Cr-Ni alloys. *Phys. Rev. B* **2015**, *91*, 024108. [CrossRef]
18. Sanchez, J.M.; de Fontaine, D. The fcc Ising model in the cluster variation approximation. *Phys. Rev. B* **1978**, *17*, 2926–2936. [CrossRef]
19. Finel, A. The cluster variation method and some applications. In *Proceedings of NATO Advanced Study Institute on Statics and Dynamics of Alloy Phase Transformations*; Turchi, P.E.A., Gonis, A., Eds.; Springer: Rhodes, Greece, 1992; pp. 495–540.
20. Lu, Z.W.; Wei, S.H.; Zunger, A.; Frota-Pessoa, S.; Ferreira, L.G. First-principles statistical mechanics of structural stability of intermetallic compounds. *Phys. Rev. B* **1991**, *44*, 512–544. [CrossRef]
21. Lechermann, F.; Fähnle, M.; Sanchez, J.M. First-principles investigation of the Ni-Fe-Al system. *Intermetallics* **2005**, *13*, 1096–1109. [CrossRef]
22. Ackermann, H.; Inden, G.; Kikuchi, R. Tetrahedron approximation of the cluster variation method for b.c.c. alloys. *Acta Metall.* **1989**, *37*, 1–7. [CrossRef]
23. Tepesch, P.D.; Asta, M.; Ceder, G. Computation of configurational entropy using Monte Carlo probabilities in cluster-variation method entropy expressions. *Model. Simul. Mater. Sci. Eng.* **1998**, *6*, 787–797. [CrossRef]
24. Kumar, N.K.; Li, C.; Leonard, K.; Bei, H.; Zinkle, S. Microstructural stability and mechanical behavior of FeNiMnCr high entropy alloy under ion irradiation. *Acta Mater.* **2016**, *113*, 230–244. [CrossRef]
25. Hart, G.L.W.; Forcade, R.W. Algorithm for generating derivative structures. *Phys. Rev. B* **2008**, *77*, 224115. [CrossRef]
26. Taggart, G.B. Generalized Flinn operators with applications to multi-component alloys. *J. Phys. Chem. Solids* **1973**, *34*, 1917–1925. [CrossRef]
27. Cenedese, P.; Gratias, D. Multicomponent formalism in the mean-field approximation: A geometric interpretation of Chebychev polynomials. *Phys. A Stat. Mech. Appl.* **1991**, *179*, 277–287. [CrossRef]
28. Sluiter, M.H.; Kawazoe, Y. Invariance of truncated cluster expansions for first-principles alloy thermodynamics. *Phys. Rev. B* **2005**, *71*, 1–2. [CrossRef]
29. Blum, V.; Zunger, A. Prediction of ordered structures in the bcc binary systems of Mo, Nb, Ta, and W from first-principles search of approximately 3,000,000 possible configurations. *Phys. Rev. B Condens. Matter Mater. Phys.* **2005**, *72*, 3–6. [CrossRef]
30. Kikuchi, R. A Theory of Cooperative Phenomena. *Phys. Rev.* **1951**, *81*, 988–1003. [CrossRef]
31. Barker, J.A. Methods of Approximation in the Theory of Regular Mixtures. *Proc. R. Soc. A Math. Phys. Eng. Sci.* **1953**, *216*, 45–56.
32. Gratias, D.; Sanchez, J.; De Fontaine, D. Application of group theory to the calculation of the configurational entropy in the cluster variation method. *Phys. A Stat. Mech. Appl.* **1982**, *113*, 315–337. [CrossRef]
33. Wróbel, J.S.; Nguyen-Manh, D.; Kurzydłowski, K.J.; Dudarev, S.L. A first-principles model for anomalous segregation in dilute ternary tungsten-rhenium-vacancy alloys. *J. Phys. Condens. Matter* **2017**, *29*, 145403. [CrossRef] [PubMed]
34. Blöchl, P.E. Projector augmented-wave method. *Phys. Rev. B* **1994**, *50*, 17953–17979. [CrossRef]

35. Kresse, G.; Hafner, J. Ab initio molecular-dynamics simulation of the liquid-metal-amorphous-semiconductor transition in germanium. *Phys. Rev. B* **1994**, *49*, 14251–14269. [CrossRef]
36. Kresse, G.; Joubert, D. From ultrasoft pseudopotentials to the projector augmented-wave method. *Phys. Rev. B* **1999**, *59*, 1758–1775. [CrossRef]
37. Kresse, G.; Hafner, J. Ab initio molecular dynamics for liquid metals. *Phys. Rev. B* **1993**, *47*, 558–561. [CrossRef]
38. Kresse, G.; Furthmüller, J. Efficient iterative schemes for ab initio total-energy calculations using a plane-wave basis set. *Phys. Rev. B* **1996**, *54*, 11169–11186. [CrossRef]
39. Kresse, G.; Furthmüller, J. Efficiency of ab initio total energy calculations for metals and semiconductors using a plane-wave basis set. *Comput. Mater. Sci.* **1996**, *6*, 15–50. [CrossRef]
40. Perdew, J.P.; Burke, K.; Ernzerhof, M. Generalized Gradient Approximation Made Simple. *Phys. Rev. Lett.* **1996**, *77*, 3865–3868. [CrossRef]
41. Monkhorst, H.J.; Pack, J.D. Special points for Brillouin-zone integrations. *Phys. Rev. B* **1976**, *13*, 5188–5192. [CrossRef]
42. Van de Walle, A.; Asta, M. Self-driven lattice-model Monte Carlo simulations of alloy thermodynamic properties and phase diagrams. *Model. Simul. Mater. Sci. Eng.* **2002**, *10*, 521–538. [CrossRef]
43. Lavrentiev, M.Y.; Wróbel, J.S.; Nguyen-Manh, D.; Dudarev, S.L.; Ganchenkova, M.G. Magnetic cluster expansion model for random and ordered magnetic face-centered cubic Fe-Ni-Cr alloys. *J. Appl. Phys.* **2016**, *120*, 043902. [CrossRef]
44. Zhang, Y.; Lu, Z.; Ma, S.; Liaw, P.; Tang, Z.; Cheng, Y.; Gao, M. Guidelines in predicting phase formation of high-entropy alloys. *MRS Commun.* **2014**, *4*, 57–62. [CrossRef]

entropy

MDPI

Article

Entropy Determination of Single-Phase High Entropy Alloys with Different Crystal Structures over a Wide Temperature Range

Sebastian Haas [1], Mike Mosbacher [1], Oleg N. Senkov [2], Michael Feuerbacher [3], Jens Freudenberger [4,5], Senol Gezgin [6], Rainer Völkl [1] and Uwe Glatzel [1,*]

1. Metals and Alloys, University Bayreuth, 95447 Bayreuth, Germany; sebastian-haas@uni-bayreuth.de (S.H.); mike.mosbacher@uni-bayreuth.de (M.M.); rainer.voelkl@uni-bayreuth.de (R.V.)
2. UES, Inc., 4401 Dayton-Xenia Rd., Dayton, OH 45432, USA; oleg.senkov.ctr@us.af.mil
3. Institut für Mikrostrukturforschung, Forschungszentrum Jülich, 52425 Jülich, Germany; m.feuerbacher@fz-juelich.de
4. Leibniz Institute for Solid State and Materials Research Dresden (IFW Dresden), 01069 Dresden, Germany; j.freudenberger@ifw-dresden.de
5. TU Bergakademie Freiberg, Institut für Werkstoffwissenschaft, 09599 Freiberg, Germany
6. NETZSCH Group, Analyzing & Testing, 95100 Selb, Germany; senol.gezgin@netzsch.com

* Correspondence: uwe.glatzel@uni-bayreuth.de; Tel.: +49-921-55-5555

Received: 31 July 2018; Accepted: 20 August 2018; Published: 30 August 2018

Abstract: We determined the entropy of high entropy alloys by investigating single-crystalline nickel and five high entropy alloys: two fcc-alloys, two bcc-alloys and one hcp-alloy. Since the configurational entropy of these single-phase alloys differs from alloys using a base element, it is important to quantify the entropy. Using differential scanning calorimetry, c_p-measurements are carried out from −170 °C to the materials' solidus temperatures T_S. From these experiments, we determined the thermal entropy and compared it to the configurational entropy for each of the studied alloys. We applied the rule of mixture to predict molar heat capacities of the alloys at room temperature, which were in good agreement with the Dulong-Petit law. The molar heat capacity of the studied alloys was about three times the universal gas constant, hence the thermal entropy was the major contribution to total entropy. The configurational entropy, due to the chemical composition and number of components, contributes less on the absolute scale. Thermal entropy has approximately equal values for all alloys tested by DSC, while the crystal structure shows a small effect in their order. Finally, the contributions of entropy and enthalpy to the Gibbs free energy was calculated and examined and it was found that the stabilization of the solid solution phase in high entropy alloys was mostly caused by increased configurational entropy.

Keywords: HEA; entropy; multicomponent; differential scanning calorimetry (DSC); specific heat

1. Introduction

The conventional strategy of alloy design is based on the selection of one base element for the primary properties of a material, e.g., iron in steels or nickel in nickel-based superalloys. This base element dominates the chemical composition, usually representing more than 80 at. %, and relatively small amounts of other elements are added to modify the alloys' properties [1–3]. Thus the regions next to a single element of multicomponent phase diagrams have been well investigated in the past.

A novel way of alloy design concentrates on the unexplored centers of phase diagrams, where the alloys consist of elements in near equiatomic ratios. Such an alloy was pointed out by Cantor et al. [1] in 2004, when the equiatomic system of Cr, Mn, Fe, Co and Ni was found to result in a single-phase material. This simple face-centered cubic microstructure is free of any precipitates and stable over a wide temperature range [4]. Merely after long time annealing for 500 h at intermediate temperatures precipitations may segregate, e.g., a Cr-rich phase at 700 °C and three different phases (NiMn-rich, FeCo-rich, Cr-rich) at 500 °C [5]. However, a single-phase microstructure is the key factor for a high configurational entropy. In the same year Yeh et al. [3] pointed out the concept of high entropy alloys, independent of Cantor's work. His definition of high entropy alloys is to consist of at least five or more elements with concentrations of each element between 5 and 35 at. % [6].

Many elements, and therefore a high value of configurational entropy, can lead to a more stable solid-solution phase with randomly distributed atoms [3]. A solid-solution phase with statistically distributed atoms in the crystal lattice is claimed to lead to interesting and outstanding properties, e.g., high hardness, wear resistance, high temperature strength and stability, sluggish diffusion, oxidation and corrosion resistance [6,7].

Yeh et al. [8] named four core effects, which are characteristic for microstructures and properties of high entropy alloys: the formation of one random solid-solution phase to reach a high entropy effect [8]; severe lattice distortion in the random solid-solution [9]; sluggish diffusion kinetics [6,10]; the so called "cocktail effect" [8].

Nevertheless the major part of investigated alloys with compositional requirements of high entropy alloys do not form single solid-solutions, but consist of several, mostly intermetallic phases, which can be brittle, difficult to process. This observation particularly disagrees with the crucial issue of a single-phase microstructure.

Numerous examinations deal with the prediction of the conditions when a solid-solution phase is stable or additional intermetallic compounds are forming [7,10–12], but no reliable approaches have yet been proposed. Thus we follow the idea to calculate the total Gibbs free energy of an alloy system in single-phase state to compare it with formation enthalpies of several intermetallic compounds. Therefore the determination of thermal enthalpy and especially of entropy over a wide temperature range is necessary. A short insight in thermodynamics and in the way of calculating different parts of the entropy is given in the following part of the introduction:

The terms and definitions of the entropy theory, as well as the basic approach of determinations refer to Gaskell [13]. Changes in Gibbs free energy ΔG_{total} of a system at any state and temperature depends on the entropy S_{total} and on the enthalpy $H_{thermal}$ (see Equation (1)). The consideration of both thermochemical parameters leads to a description of the equilibrium state of an alloy system by minimization of the Gibbs free energy at a fixed temperature:

$$\Delta G_{total} = \Delta H_{thermal} - T\Delta S_{total}, \quad (1)$$

The entropy of an alloy consists of the configurational entropy S_{conf} [7] and the thermal entropy $S_{thermal}$ [13]:

$$S_{total} = S_{conf} + S_{thermal}, \quad (2)$$

$$\Delta S_{conf} = -R \cdot \sum_{i=1}^{n} x_i \cdot \ln x_i, \quad (3)$$

The calculation of configurational entropy, also called mixing entropy, is given in Equation (3) with n as the number of elements, x_i the concentration of each element i and R as the universal gas constant. The equation is derived from the mixing of noble gases and is adopted to fully disordered solid solution, which are assumed in our work. If elements are distributed non-equally between possible sub-lattices, then a more general equation should be used [14]. In case of an equiatomic alloy, Equation (3) is reduced to $\Delta S_{conf} = R \ln n$. Thus the configurational entropy of 5-component equiatomic alloy is ~1.6·R. At 0 K the resulting total entropy may not be zero. This does not violate the third law of thermodynamics ($\left.\frac{dS}{dT}\right|_{T=0} = 0$) which is often misinterpreted as $S_{(T=0)} = 0$. Crystalline solids may exhibit a non-zero entropy at the absolute zero point due to a randomly crystallographic orientation. A change of entropy in this point is not possible, because there is no ability of motion or diffusion.

The thermal entropy $S_{thermal}$ can be directly determined by measuring the temperature dependent heat capacity at constant pressure $c_p(T)$ by differential scanning calorimetry (DSC):

$$\Delta S_{thermal}(T) = \int_0^T c_p(T)\frac{1}{T}dT, \tag{4}$$

Next to the entropy, there is an enthalpy-contribution to the Gibbs free energy, mentioned in Equation (1). This thermal enthalpy can be calculated using Equation (5):

$$\Delta H_{thermal}(T) = \int_0^T c_p(T)dT, \tag{5}$$

Therefore, the total change in Gibbs free energy at a certain temperature is presented in the following Equation (6). Setting the value of configurational entropy to zero, just the thermal entropy is considered and the result is the change in thermal Gibbs free energy:

$$\Delta G_{total} = \Delta H_{thermal} - T\Delta S_{total} = \int_0^T c_p(T)dT - T\left(\Delta S_{conf} + \int_0^T c_p(T)\frac{1}{T}dT\right), \tag{6}$$

A schematic drawing of c_p from 0 K to temperatures in the range of incipient melting after reaching the solidus temperature T_S is given in Figure 1a. For lower temperatures $c_p(T)$ can be extrapolated down to 0 K following the Debye T^3-law [13]. For temperatures close to room temperature (RT) the heat capacity is close to 3R if there is no change in magnetic behavior and no phase transformation. This is known as the Dulong-Petit law, which states that every solid that consists of N atoms has 3N modes of vibration (corresponding to the freedom of motion in three dimensions). Energetic considerations using the equipartition theorem lead to $c_p = 3{\cdot}R$ for sufficiently high temperatures (RT). DSC measurements are carried out until the solidus temperature is reached and the c_p-value rises dramatically. In Figure 1b, the calculated thermal entropy and total entropy are schematically illustrated. The thermal entropy is shifted vertically by the configurational entropy over the whole temperature range, in case there are no contributions by phase changes. This is due to the low influence of various alloys on the values of thermal entropy, while the configurational entropy has a high impact. The vertical offset between thermal and total entropy, caused by the alloys' configuration, is still holding on after melting in the liquid state.

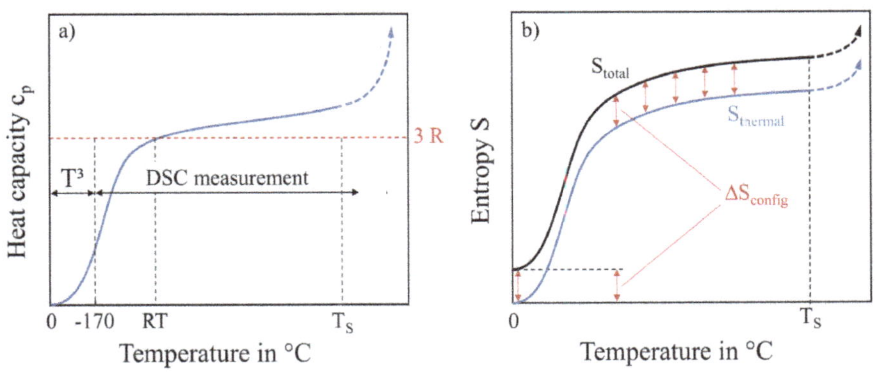

Figure 1. Schematic drawing (**a**) of the heat capacity of an ideal metal from 0 Kelvin to the liquid state with no change in magnetic behavior and no phase transition over the whole temperature range and (**b**) of the calculated thermal entropy and total entropy from 0 K to the liquid state.

2. Materials and Methods

The temperature dependent molar heat capacity $c_P(T)$ of the specimen was experimentally determined in alumina crucibles using differential scanning calorimetry (DSC 204, Netzsch, Selb, Germany) in the temperature range from −170 °C to 600 °C under a flushing flow of nitrogen. In this temperature range oxidation of the samples is not critical. For T = 20 °C − T_S (if T_S < 1600 °C) samples were measured using a Netzsch DSC 404 F1 Pegasus under an Ar 5.0 gas flush with a rate of 70 ml/min in crucibles composed of 80% Pt and 20% Rh. The crucibles are lined with ceramic inlays to prevent interaction of the metallic specimens with the Pt-Rh crucible. The calibration of temperature and enthalpy of the two Netzsch devices was performed using the calibration set 6.239.2–91.3 under the conditions of 10 K/min and a nitrogen flow of 40 mL/min. The two measurements were evaluated for each material and the curves were connected in the common temperature range from room temperature to 600 °C.

Single-crystalline (SX) nickel and SX-Cantor alloy (favored equiatomic composition of Cr, Mn, Fe, Co, Ni) were cast using a proprietary Bridgman investment casting furnace. Cylindrical specimens (Ø 5 mm, height 1 mm) were then cut out of the rods by electrical discharge machining (EDM). All other alloys were in poly-crystalline state. The bcc-alloys were provided by Senkov et al. [15], the noble metal alloy by Freudenberger et al. [16] and the hcp-alloy by Feuerbacher et al. [17]. The recast layer of the EDM samples was removed by etching and the base of all samples was finely ground with SiC paper up to 2000 grit to ensure good thermal contact to the DSC-sensor. All materials tested several times showed good reproducibility in their DSC-signal, even at different heating rates of 10 K/min and 20 K/min respectively.

Except for pure nickel all materials belong to the high entropy alloys group. They form a single-phase microstructure and contain at least four elements in desired, near equiatomic composition. The single-phase solid-solution has been confirmed by the authors using X-ray diffraction experiments. Homogenization of the samples was reached by very slow cooling rates after melting, except of fcc-noble metal that was annealed for 24 h at 1000 °C and bcc 5-component was annealed for 24 h at 1200 °C.

Although the ROM may only represent a very rough estimate of the melting temperature of the alloys, it has been applied for the bcc-5 component and hcp-alloys. The reasons are twofold: (i) the melting temperature of the bcc-5 component alloy cannot be experimentally determined with our set-up and (ii) we are facing significant reactions of the metals with the crucible during investigation and, therefore, the determined values would not reflect the samples under investigation.

The Mn content of 11 at. % in the Cantor alloy is due to the single-crystal investment casting process. Mn evaporates from the melt in the vacuum of 10^{-2} Pa of the Bridgman furnace during the slow withdrawal of the single-crystal.

3. Results and Discussion

The specific heat capacity in units of J/(g·K) has been converted into the molar heat capacity in units of J/(mol·K) by multiplication with the molar mass of the alloys corresponding to their chemical composition. Figure 2 shows the molar heat capacity of all materials as a function of temperature. For temperatures below −170 °C, the curves were extrapolated using the fit-function $a \cdot T^3$ (parameter a is listed in Table 1) and $c_P(0\ K) = 0$.

Table 1. Values of the parameter a in $J/mol \cdot K^4$ in the fit-function $a \cdot T^3$.

hcp	bcc-4 Comp.	bcc-5 Comp.	fcc-Cantor	fcc-Nickel	fcc-Noble
1.8×10^{-5}	7.6×10^{-6}	8.2×10^{-6}	9.8×10^{-6}	9.0×10^{-6}	1.0×10^{-5}

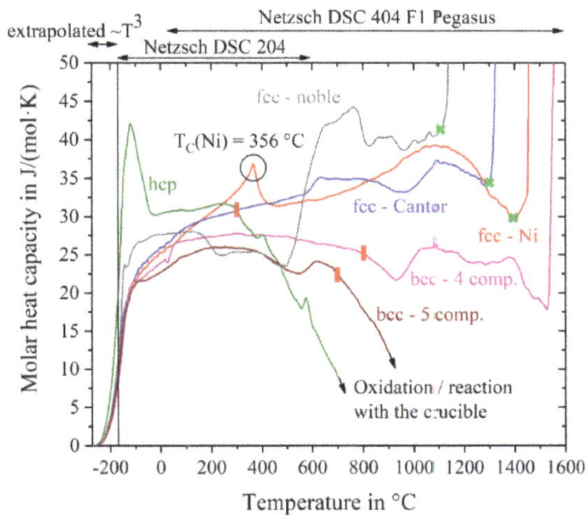

Figure 2. Molar heat capacities of pure nickel and five high entropy alloys. Green crosses indicate T_S, red lines show the beginning of oxidation or reactions with the crucible that make the data not evaluable.

Table 2. Sample labelling, crystal specification, chemical composition (measured by μ-XRF) and solidus temperature. In case of pure Ni, T_S is equal to the melting temperature T_m. The rule of mixture (ROM) is used for calculation, if no literature data is available. All alloys are in a single-phase state over the entire temperature region.

Alloy Name Used in This Paper	Crystal Type and Structure	Chemical Composition (at. %)	S_{conf} in J/(mol·K)	Solidus Temperature T_S
fcc-Ni	SX-fcc	Pure Ni (>99.99%)	0	1455 °C [18]
fcc-Cantor	SX-fcc	$Co_{22}Cr_{24}Fe_{22}Mn_{11}Ni_{21}$	1.58·R	1280 °C [19]
fcc-noble metal	PX-fcc	$Au_{16}Cu_{17}Ni_{17}Pd_{34}Pt_{16}$	1.56·R	1196 °C
bcc-4 component	PX-bcc	$Mo_{27}Nb_{27}Ta_{23}W_{23}$	1.59·R	2904 °C [15]
bcc-5 component	PX-bcc	$Hf_{21}Mo_{20}Nb_{21}Ti_{17}Zr_{21}$	1.38·R	2170 °C (ROM)
hcp	PX-hcp	$Ho_{18}Dy_{16}Y_{28}Gd_{20}Tb_{18}$	1.61·R	1416 °C (ROM)

The two single-crystal fcc-materials, Nickel and Cantor alloy, show a steady increase in c_p until shortly before their melting points. Both curves are very close to each other with the exception of the Curie-peak in Ni at around 356 °C, which lies in excellent agreement with literature data (e.g., 354 °C [20]). The Cantor alloy shows a broad plateau-like peak between 600 °C and 800 °C. This peak is most likely caused by a change in magnetic behavior. Jin et al. [21] have shown that Co and Fe shift the Curie temperature of Ni to higher temperatures and their investigations on Cantor alloy show the exact same plateau-like peak starting at around 600 °C. Incipient melting at the solidus temperatures is indicated by excessive jumps in the curves. The solidus temperatures of fcc-nickel, fcc-Cantor and fcc-noble do not correspond to the values listed in Table 2, but differ by about 30 K in case of fcc-Cantor and fcc-nickel and by about 70 K in case of fcc-noble. The other three alloys show decreasing specific heat capacities at elevated temperatures, which indicates reactions with the atmosphere and/or the crucible. High chemical interaction and oxidation with the ceramic liners of the crucibles occur with the alloys Hf-Mo-Nb-Ti-Zr and Ho-Dy-Y-Gd-Tb, so their curves are finally cut-off at 700 °C and 900 °C respectively. Melting intervals, starting from the solidus temperatures, are not evaluable for the bcc-5 component and hcp alloy. In case of the bcc-4 component alloy the curve mistakenly suggests the solidus temperature to be at about 1520 °C. However, Senkov et al. [15] expect a melting temperature of about 2904 °C, using the rule of mixture. This issue is likely to be due to chemical reactions with the Pt-Rh crucible and therefore we have to regard the values of the heat capacity of bcc-4 component for higher temperatures with caution. For further calculations concerning entropy, enthalpy and Gibbs free energy, the data of bcc- and hcp-alloys respectively are just used in the temperature range until the red lines in Figure 2.

Table 2 gives an overview of all tested samples, their crystal structures, chemical compositions, configurational entropies and solidus temperatures. Note that in case of bcc-5 component and the hcp alloy, there was no literature data concerning solidus temperature available. Therefore, the possible solidus temperatures were calculated using the rule of mixture (ROM).

Obvious shifts in the trend of c_p indicate phase transitions or changes in the magnetic behavior, e.g., the Curie point of nickel at 356 °C as mentioned above. The hcp-alloy Ho-Dy-Y-Gd-Tb shows a very pronounced peak at −120 °C, probably due to a not investigated magnetic phase transformation, because oxidation seems to be very unlikely in such a low temperature range.

Of particular interest are the heat capacities at room temperature, where all curves seem to approach a c_p value close to 25 J/(mol·K) = 3R (see Figure 3a–d), according to the Dulong-Petit-law [13]. The bcc-5 component equiatomic alloy, Hf-Mo-Nb-Ti-Zr, shows the lowest molar heat capacity at room temperature with about 23.5 J/(mol·K) (Figure 3b), the hcp-alloy Ho-Dy-Y-Gd-Tb shows the highest value at about 30.6 J/(mol·K) (Figure 3d). The high value of the molar heat capacity of the hcp-alloy might originate from magnetic ordering of the 4f electrons. This is likely the case as pure Gd shows ferromagnetic ordering below 19.9 °C. However, this suggestion needs to be verified in future studies. The other four alloys show RT c_p-values pretty close to 25 J/(mol·K). We applied the rule of mixture (ROM) to predict molar heat capacities of the alloys and their average molar mass (see Equation (7)). The calculated data is in very good agreement with the experimental results and shows a maximum deviation of 6%:

$$c_{p,alloy} = \frac{1}{n} \cdot \sum_{i=1}^{n} c_{pi}, \quad u_{alloy} = \frac{1}{n} \cdot \sum_{i=1}^{n} u_i, \tag{7}$$

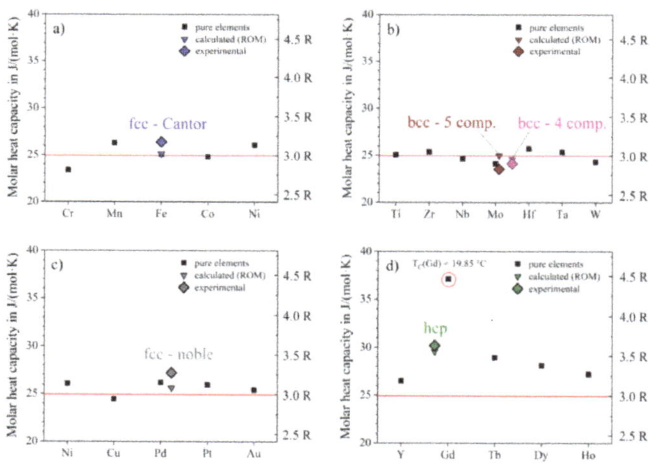

Figure 3. Molar heat capacities at room temperature of the investigated alloys, both experimental and rule of mixture (ROM) data for (**a**) fcc-Cantor, (**b**) both bcc-alloys, (**c**) fcc-noble and (**d**) the hcp-alloy. Values for pure elements were taken from [20]. The red line indicates the Dulong-Petit law with $c_p(RT) \approx 3 \cdot R$.

Figure 4 shows the thermal entropy $S_{thermal}$ determined with Equation (4). $S_{thermal}$ is equial zero at −273 °C andincreases continuously with increasing temperature having different slopes for different alloys. The curves are drawn till shortly before the solidus temperature of the alloys, except for Ho-Dy-Y-Gd-Tb, Hf-Mo-Nb-Ti-Zr and Mo-Nb-Ta-W where chemical reactions with the crucible/environment make the high temperature regions inaccessible. The highest thermal entropy appears for the hcp-alloy, caused by the highest heat capacity at low temperatures and additionally the early peak in the c_p-curve at about −120 °C. The fcc-nickel, fcc-Cantor alloy and fcc noble metal alloy have almost the same temperature dependence of $S_{thermal}$, which is noticeably weaker than that of the hcp-alloy. These fcc materials have similar $S_{thermal}$ values at any given temperature. The two bcc-alloys exhibit the lowest values of thermal entropy.

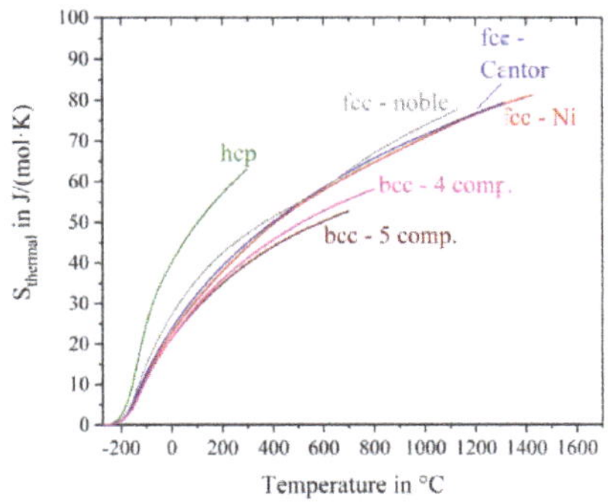

Figure 4. Thermal entropy $S_{thermal}$ of nickel and the five high entropy alloys.

The total entropy $S_{total} = S_{conf} + S_{thermal}$ over the whole temperature range starting from absolute zero is plotted in Figure 5. While the curve of nickel starts at 0 J/(mol·K) at absolute zero, the bcc-4 component alloy Mo-Nb-Ta-W exhibits a configurational entropy of R ln(4) = 1.38·R = 11.5 J/(mol·K), because of its four components. All other high entropy alloys, with five different elements show an offset of R ln(5) = 1.6·R = 13.4 J/(mol·K). In theory the configurational entropy is equal for all alloys with the same number and concentrations of elements. No considerations are done with respect tosimilarities between participating atoms, like differences in the atomic radius or crystal structures of the pure metals. In our case, all elements in the hcp-alloy promote a hexagonal close packed crystal structure and exhibit quite equal atomic radii. The bcc-5 component alloy however consists of elements of different crystal structures with larger atomic size differences. Nevertheless, the same value of S_{conf} is assumed, as stated in the Gibbs paradox [22]. For dislocation movement, however, different atom sizes in solid-solution crystal structures indeed play an important role. Consequently, this mixing paradox will be investigated thoroughly in a future work. It can be seen from Figure 5 that at any given temperature S_{total} is the smallest for fcc-Ni, followed by the bcc alloys, then the fcc alloys and being the highest for the hcp alloy.

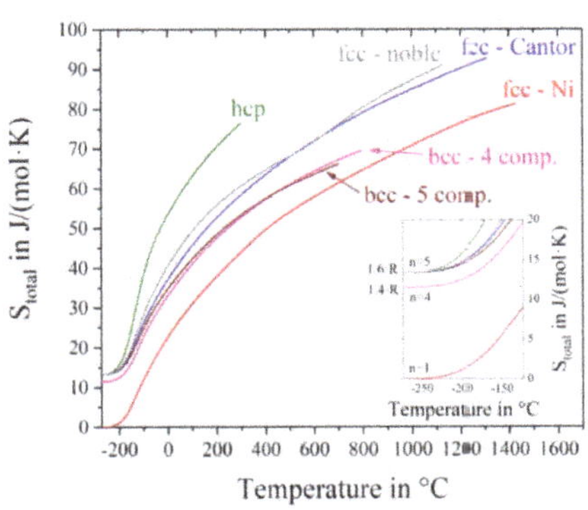

Figure 5. Total entropy as a sum of temperature dependent thermal entropy $S_{thermal}$ and constant composition dependent configurational entropy S_{conf}.

The contribution to the Gibbs free energy is the product of entropy and temperature. Using thermal and configurational entropy and their sum, we can examine their contributions to the Gibbs free energy separately. Figure 6a shows the energy contribution of thermal entropy. We can see a clear order with the alloy crystal structure: the highest contribution by thermal entropy is given by the hcp-alloy, mainly because of the high value of molar heat capacity at low temperatures (−120 °C) and also at room temperature. Pure nickel, noble and Cantor, all fcc-structures, overlie each other over a wide temperature range. The lowest heat capacities and therefore thermal entropy contributions are given by the bcc (4 and 5 component) alloys. Using the sum of thermal and configurational entropy, the total entropy times the temperature is shown in Figure 6b. Alloys with the same crystal structure and configurational entropy are close to each other. For example, the total entropy input to Gibbs free energy for fcc-Ni is smaller than for fcc Cantor and fcc nobel metal alloys. To explore the difference, two materials with fcc-structure, pure nickel and Cantor alloy, are drawn in Figure 7a in a temperature range from 600 °C to 1600 °C. Dashed lines show the configurational entropy contribution to Gibbs free energy of both materials, calculated by Equation (3). While the contribution of configurational entropy rises with higher temperatures for fcc-Cantor alloy, fcc-nickel does not exhibit any configurational entropy and, therefore, has no contribution to Gibbs free energy. Continuous lines show the product of the temperature and the total entropy. Thus the influence of configuration and thermal input can be quantified at certain temperatures. At 1000 °C the gap between continuous and dashed lines is almost similar, meaning that at this temperature there is a negligible difference in the influence of thermal entropy, but just the chemical composition of the solid-solutions affects the varying energy contributions. The energy level of total entropy contribution is about 90 kJ/mol for fcc-nickel and 108 kJ/mol for fcc-Cantor. The difference of 18.0 kJ/mol is very close to fcc-Cantor configurational entropy contribution of 1273 K·R·ln(5) = 17 kJ/mol ($T{\cdot}S_{conf}$) at this energy level, resulting that thermal entropy has an equal impact for materials with the same crystal structures (in this case fcc) and thus cannot play role in stabilizing a solid solution phase in multicomponent alloys. Figure 7b shows a bigger section of the area near the melting interval of both materials. The two values of melting enthalpy 13.3 kJ/mol and 15.0 kJ/mol have been detected by DSC and it is obvious that differences in entropy contribution to Gibbs free energy still remains dependent from the configurational entropy until the end of the solid state. Investigations on the fcc-noble alloy instead of fcc-Cantor yield to similar results and are, therefore, not shown in Figure 7 in detail.

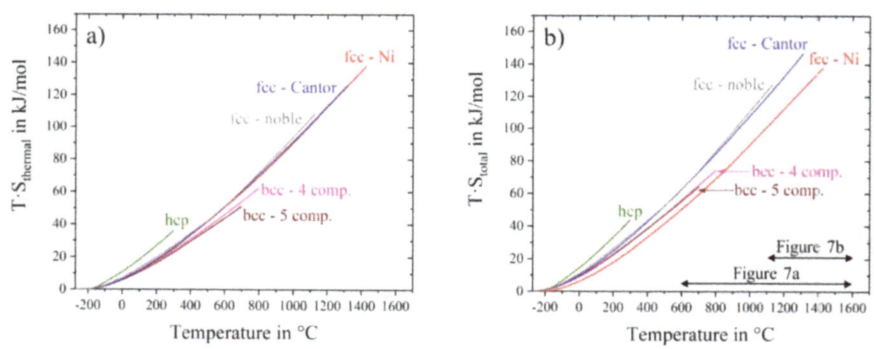

Figure 6. Contribution of thermal entropy (**a**) and total entropy (**b**) of nickel and all high entropy alloys to Gibbs free energy. The horizontal arrows refer to Figure 7 with magnified details of a special temperature range.

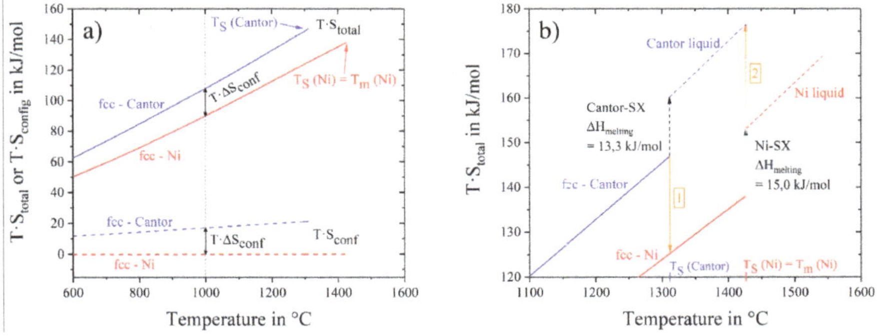

Figure 7. Contribution of total and configurational entropy of fcc-nickel and fcc-Cantor to Gibbs free energy in the range from (**a**) 600 °C to 1600 °C and from (**b**) 1100 °C to 1600 °C.

Gibbs energy has another input besides entropy, namely thermal enthalpy, $H_{thermal}$. The temperature dependence of the enthalpy, calculated with Equation (5), is displayed in Figure 8a, while Figure 8b,c show thermal Gibbs free energy and total Gibbs free energy, respectively. Similar to $S_{thermal}$, $H_{thermal}$ slightly depends on the type of the crystal structure having smallest values for the bcc alloys and highest values for the hcp alloy and shows almost no dependence on the number of components and alloy composition (Figure 8a). As a result of such behavior, the thermal part of the Gibbs free energy, $\Delta G_{thermal}$, also does not depend on the number and concentration of the alloying elements, but its temperature dependence is slightly stronger for the hcp alloy and weaker for the bcc alloys relative to the fcc materials (Figure 8b). On the other hand, the total Gibbs free energy, ΔG_{total}, which additionally includes the configurational term, has a stronger temperature dependence and thus becomes noticeably smaller at higher temperatures for the alloys with larger configurational entropy (Figure 8c). These observations indicate that, although the contribution of $S_{thermal}$ to the Gibbs free energy is much higher than that of than S_{config}, at any given temperature $\Delta G_{thermal}$ is nearly the same for the simple and complex alloys of the same type of crystal structure, i.e., $S_{thermal}$ does not play any role in stabilizing a solid solution phase in complex, multicomponent alloys: On the other hand, S_{config} increases with the number of constituents, which noticeably decreases ΔG_{total} of a multicomponent solid-solution relative to that of pure metals, especially at high temperatures. Moreover, in some specific cases ΔG_{total} of a multicomponent solid-solution with high S_{config} can become smaller than ΔG_{total} of competing intermetallic phases, resulting in a single-phase solid solution alloy.

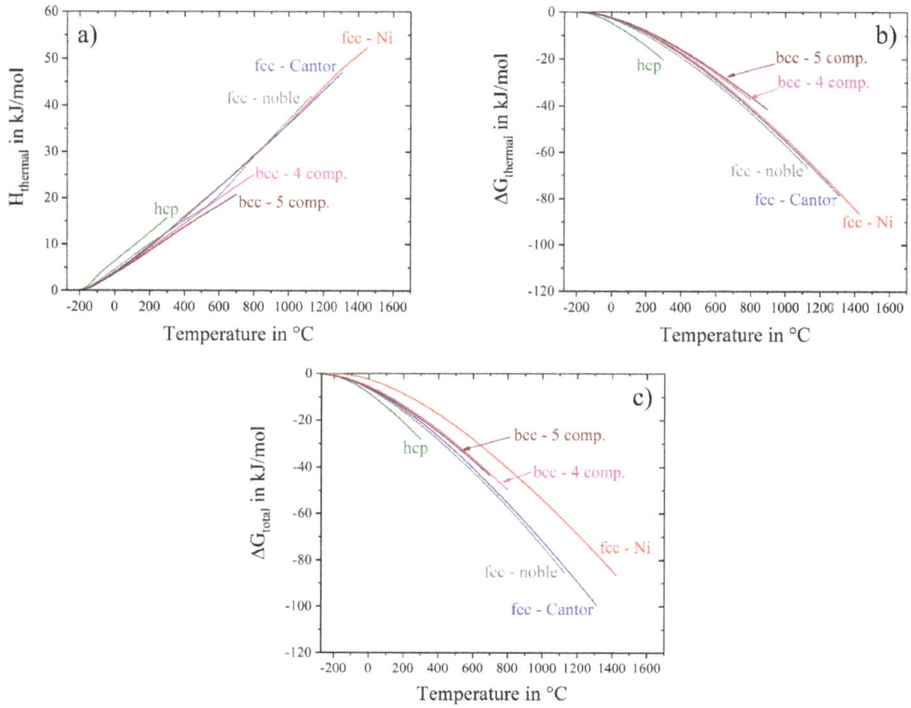

Figure 8. Temperature dependence of (**a**) thermal enthalpy, (**b**) thermal part of Gibbs free energy and (**c**) total Gibbs free energy for the studied materials.

4. Conclusions

In this work the heat capacity, entropy, enthalpy and Gibbs free energy of six different single-phase solid solution alloys (three fcc-alloys, two bcc-alloys and one hcp-alloy) were experimentally determined and investigated over a wide temperature range.

- At room temperature (RT), molar heat capacities of the studied alloys are close to 3R, in accordance with the Dulong-Petit law.
- The measured RT heat capacities of the studied alloys are in good agreement with the heat capacities calculated using the rule of mixture of pure elements.
- Thermal entropy and thermal enthalpy increase while the thermal part of the Gibbs free energy decreases with an increase in temperature. The temperature dependence of these quantities is the strongest for the hcp alloy and weakest for the bcc alloys, with the fcc alloys showing intermediate behavior. For the alloys with the same crystal structure, the thermal contributions to the Gibbs free energy do not depend on the number and concentration of the alloying elements. Therefore, in spite the thermal entropy, $S_{thermal}$, is much higher than the configurational entropy, S_{conf}, at T > 20 °C, $S_{thermal}$ does not increase thermal stability of a solid solution phase in complex, multicomponent alloys relative to simple alloys or pure metals.
- The compositional dependence of the Gibbs free energy of the solid solution alloys is totally due to the configurational entropy. Of the same crystal structure, solid solution alloys with higher S_{conf} have stronger temperature dependence of the Gibbs free energy and smaller ΔG_{total} values at a given temperature. Thus, although being smaller than $S_{thermal}$, only S_{conf} contributes to the thermal stability of the complex, multicomponent solid solution alloys.

Author Contributions: Conceptualization, U.G., R.V., S.H., M.M.; Methodology, S.H., M.M., S.G.; Software, S.G.; Validation, S.H., M.M., O.N.S.; Formal Analysis, S.H., O.N.S.; Investigation, S.H., M.M., O.N.S.; Resources, O.N.S., J.F., M.F., M.M.; Data Curation, S.H., O.N.S.; Writing-Original Draft Preparation, S.H.; Writing-Review & Editing, S.H., M.M., O.N.S., J.F., M.F.; Visualization, S.H., M.M.; Supervision, U.G.

Funding: This research was funded by German Research foundation (DFG), i.e., during the projects GL 181/50, GL 181/56, FR 1714/7 and FE 571/4, the three last within the priority programme SPP2006 "Compositionally Complex Alloys—High Entropy Alloys (CCA-HEA)". O.N. Senkov acknowledges financial support through the Air Force on-site Contract FA8650-15-D-5230 managed by UES, Inc., Dayton, OH, USA.

Conflicts of Interest: The authors declare no conflict of interest.

References

1. Cantor, B.; Chang, I.T.H.; Knight, P.; Vincent, A.J.B. Microstructural development in equiatomic multicomponent alloys. *Mater. Sci. Eng. A* **2004**, *375–377*, 213–218. [CrossRef]
2. Hsu, C.; Yeh, J.; Chen, S.; Shun, T. Wear resistance and high-temperature compression strength of Fcc $CuCoNiCrAl_{0.5}Fe$ alloy with boron addition. *Metall. Mater. Trans. A* **2004**, *35*, 1465–1469. [CrossRef]
3. Yeh, J.W.; Chen, S.K.; Lin, S.J.; Gan, J.Y.; Chin, T.S.; Shun, T.T.; Tsau, C.H.; Chang, S.J. Nanostructured high-entropy alloys with multiple principal elements: Novel alloy design concepts and outcomes. *Adv. Eng. Mater.* **2004**, *6*, 299–303. [CrossRef]
4. Brocq, L.M.; Akhatova, A.; Perrière, L.; Chebini, S.; Sauvage, X.; Leroy, E.; Champion, Y. Insights into the phase diagram of the CrMnFeCoNi high entropy alloy. *Acta Mater.* **2015**, *88*, 355–365. [CrossRef]
5. Otto, F.; Dlouhý, A.; Pradeep, K.G.; Kuběnová, M.; Raabe, D.; Eggeler, G.; George, E.P. Decomposition of the single-phase high-entropy alloy CrMnFeCoNi after prolonged anneals at intermediate temperatures. *Acta Mater.* **2016**, *112*, 40–52. [CrossRef]
6. Tsai, K.Y.; Tsai, M.H.; Yeh, J.W. Sluggish diffusion in Co-Cr-Fe-Mn-Ni high-entropy alloys. *Acta Mater.* **2013**, *61*, 4887–4897. [CrossRef]
7. Zhang, W.; Liaw, P.K.; Zhang, Y. Science and technology in high-entropy alloys. *Sci. China Mater.* **2017**, *61*, 2–22. [CrossRef]
8. Yeh, J.W. Recent Progress in High-Entropy Alloys. *Ann. Chim. Sci. Mater.* **2006**, *31*, 633–648. [CrossRef]
9. Yeh, J.W.; Chang, S.Y.; Der Hong, Y.; Chen, S.K.; Lin, S.J. Anomalous decrease in X-ray diffraction intensities of Cu-Ni-Al-Co-Cr-Fe-Si alloy systems with multi-principal elements. *Mater. Chem. Phys.* **2007**, *103*, 41–46. [CrossRef]
10. Vaidya, M.; Trubel, S.; Murty, B.S.; Wilde, G.; Divinski, S.V. Ni tracer diffusion in CoCrFeNi and CoCrFeMnNi high entropy alloys. *J. Alloys Compd.* **2016**, *688*, 994–1001. [CrossRef]
11. Yang, X.; Zhang, Y. Prediction of high-entropy stabilized solid-solution in multi-component alloys. *Mater. Chem. Phys.* **2012**, *132*, 233–238. [CrossRef]
12. Senkov, O.N.; Miracle, D.B. A new thermodynamic parameter to predict formation of solid solution or intermetallic phases in high entropy alloys. *J. Alloys Compd.* **2016**, *658*, 603–607. [CrossRef]
13. Gaskell, D. *Introduction to the Thermodynamics of Materials*; Taylor & Francis: Washington, DC, USA, 1995.
14. Miracle, D.B.; Senkov, O.N. A critical review of high entropy alloys and related concepts. *Acta Mater.* **2017**, *122*, 448–511. [CrossRef]
15. Senkov, O.N.; Wilks, G.B.; Miracle, D.B.; Chuang, C.P.; Liaw, P.K. Refractory high-entropy alloys. *Intermetallics* **2010**, *18*, 1758–1765. [CrossRef]
16. Freudenberger, J.; Rafaja, D.; Geissler, D.; Giebeler, L.; Ullrich, A.; Kauffmann, A.; Heilmaier, M.; Nielsch, K. Face Centred Cubic Multi-Component Equiatomic Solid Solutions in the Au-Cu-Ni-Pd-Pt System. *Metals* **2017**, *7*, 135. [CrossRef]
17. Feuerbacher, M.; Heidelmann, M.; Thomas, C. Hexagonal High-entropy Alloys. *Mater. Res. Lett.* **2014**, *3*, 1–6. [CrossRef]
18. Greenwood, N.N.; Earnshaw, A. *Chemistry of the Elements*, 2nd ed.; Elsevier: New York, NY, USA, 1997.
19. Gali, A.; George, E.P. Tensile properties of high- and medium-entropy alloys. *Intermetallics* **2013**, *39*, 74–78. [CrossRef]
20. Lide, D.R. *Handbook of Chemistry and Physics*; CRC Press: Boca Raton, FL, USA, 2013.

21. Jin, K.; Mu, S.; An, K.; Porter, W.D.; Samolyuk, G.D.; Stocks, G.M.; Bei, H. Thermophysical properties of Ni-containing single-phase concentrated solid solution alloys. *Mater. Des.* **2017**, *117*, 185–192. [CrossRef]
22. Gibbs, J.W. On the Equilibrium of Heterogeneous Substances. *Am. J. Sci. Arts* **1878**, *16*, 1820–1879. [CrossRef]

entropy

MDPI

Article

About the Reliability of CALPHAD Predictions in Multicomponent Systems

Stéphane Gorsse [1,2,*] and Oleg N. Senkov [3]

1 CNRS, Université de Bordeaux, ICMCB, UMR 5026, F-33600 Pessac, France
2 Bordeaux INP, ENSCBP, F-33600 Pessac, France
3 Air Force Research Laboratory, Materials and Manufacturing Directorate, Wright-Patterson AFB, OH 45433, USA; oleg.senkov.ctr@us.af.mil
* Correspondence: stephane.gorsse@icmcb.cnrs.fr

Received: 19 October 2018; Accepted: 22 November 2018; Published: 24 November 2018

Abstract: This study examines one of the limitations of CALPHAD databases when applied to high entropy alloys and complex concentrated alloys. We estimate the level of the thermodynamic description, which is still sufficient to correctly predict thermodynamic properties of quaternary alloy systems, by comparing the results of CALPHAD calculations where quaternary phase space is extrapolated from binary descriptions to those resulting from complete binary and ternary interaction descriptions. Our analysis has shown that the thermodynamic properties of a quaternary alloy can be correctly predicted by direct extrapolation from the respective fully assessed binary systems (i.e., without ternary descriptions) only when (i) the binary miscibility gaps are not present, (ii) binary intermetallic phases are not present or present in a few quantities (i.e., when the system has low density of phase boundaries), and (iii) ternary intermetallic phases are not present. Because the locations of the phase boundaries and possibility of formation of ternary phases are not known when evaluating novel composition space, a higher credibility database is still preferable, while the calculations using lower credibility databases may be questionable and require additional experimental verification. We estimate the level of the thermodynamic description which would be still sufficient to correctly predict thermodynamic properties of quaternary alloy systems. The main factors affecting the accuracy of the thermodynamic predictions in quaternary alloys are identified by comparing the results of CALPHAD calculations where quaternary phase space is extrapolated from binary descriptions to those resulting from ternary system descriptions.

Keywords: alloy design; structural metals; multi-principal element alloys; CALPHAD

---

1. Introduction

Computational approaches are now a vital aspect of materials science, which enable the prediction of unknown data, such as materials properties and phase stability, and the more rapid exploration of design space. CALPHAD (Calculation of Phase Diagrams) is an efficient computational thermodynamic technique which permits the prediction of the phase equilibria and thermodynamic properties of multicomponent systems from those of the respective binary and ternary subsystems. CALPHAD practices also include the simulation of solidification, mapping of onset driving forces and nucleation barriers for precipitation [1], modeling of phase transformations [2], and estimation of stacking fault energy [3], for example. The development of multicomponent thermodynamic databases caused a growing interest in the calculations of phase diagrams for high entropy alloys (HEAs) and complex concentrated alloys (CCAs) [3–7] allowing the exploration of the composition space to be accelerated [8–10]. In this context, it is of paramount importance to analyze the reliability of the predictions made using commercial dedicated HEA databases.

The reliability of the predictions depends upon the quality of the assessment of the Gibbs energy parameters of the phases stored in the databases. The CALPHAD approach faces new challenges when applied to the development of multi-principal element alloys (HEAs and CCAs) due to large extrapolations from the thermodynamic models which may lead to inaccurate evaluations of the Gibbs energies. Many CALPHAD databases have been developed for traditional alloys so the full thermodynamic assessment of the Gibbs energies is usually given only for compositions enriched with one main element. In contrast, HEAs and CCAs lie in the central regions of compositional space, far from the known boundaries delimited by the binary and ternary subsystems. Consequently, the calculations in these vast composition spaces rely on the availability and quality of the descriptions of the lower order constituent systems and require large extrapolations from these lower order systems by assigning higher order interaction parameters, which are refined using available experimental data. Reliability and accuracy of the predictions increases when using more complete, experimentally verified, thermodynamic descriptions specially designed for CCA compositions (e.g., TCHEA and PanHEA special HEA databases developed by Thermo-Calc Software AB and CompuTherm LLC, respectively). A complete thermodynamic description (full assessment) of binary and ternary systems is generally sufficient to correctly describe thermodynamic properties of higher order (quaternary, quinary, etc.) systems because the probability of the occurrence of quaternary or higher order intermetallic phases is low and decreases rapidly with an increase in the number of components. Therefore, complete thermodynamic description of binary and ternary systems is considered to be sufficient for the correct prediction of thermodynamic properties of higher order systems by extrapolation from the lower order systems. Unfortunately, even the most developed thermodynamic databases lack in the complete thermodynamic description (full assessment) of all binary and, especially, ternary systems based on the elementary components present in the database. For example, the TCHEA3 database includes 26 elements that can form 325 binary and 2600 ternary systems, but only 294 binary and 136 ternary systems are assessed in the full range of composition and temperature. Therefore, complete thermodynamic description is currently unavailable for a huge number of quaternary (or higher order) alloy systems, which may have a potential practical interest. For those reasons, it is very important to know what the minimum level of the thermodynamic assessment of binary and ternary systems is to be sufficient to qualitatively predict equilibrium phases and phase transformations in higher order CCAs.

The goal of this paper is to evaluate the quality of prediction of phases and phase compositions in a quaternary system at different levels of the assessment (thermodynamic description) of the respective ternary systems and a complete description of the respective binary systems. To achieve this goal, we used the current capabilities of a commercially specialized CALPHAD database and identified quaternary alloys systems that have a complete description of their binary and ternary sub-systems. We then calculated ternary and quaternary phase diagrams for these quaternary systems and considered them as "reference diagrams", assuming that they are 100% accurate relative to known experimental data due to a complete thermodynamic description. A parallel coordinate plots method, which is briefly described in the following section, was used to display three-dimensional data related to the extent of phase regions in the quaternary compositional space. Finally, we artificially reduced the levels of the thermodynamic description of the ternary systems, calculated the ternary and quaternary phase diagrams using this incomplete description, and compared the calculations with the reference diagrams. This analysis allowed us to identify conditions when the databases with incomplete thermodynamic descriptions can still be used to reliably predict thermodynamic properties of quaternary systems.

2. Methods

Senkov et al. [8,9] defined credibility criteria for CALPHAD calculations based on the fraction of fully thermodynamically assessed binary systems (FAB) and the fraction of fully assessed ternary systems (FAT) included in the database. By definition, a fully assessed system is one that has a complete thermodynamic description (of the respective elements, binary systems and ternary systems) within a given thermodynamic database allowing CALPHAD software to calculate thermodynamic properties for this system, which agree with the experimentally available data within the whole composition and temperature range. Using these theoretical credibility criteria, Wertz et al. [11] have recently demonstrated the importance of (a) the number of assessed constituent binary systems and (b) the compositional distance from them, on the accuracy of the predictions for ternary systems. Their study was carried out by comparing equilibrium calculations of the same ternary systems using eight different databases developed by CompuTherm LLC (Madison, WI, USA). Each of the studied ternary systems had complete thermodynamic descriptions (FAB = 1, FAT = 1) in at least one database, while other databases calculated the ternary space by extrapolation from binary descriptions with different levels of FAB and FAT = 0 (Figure 1a).

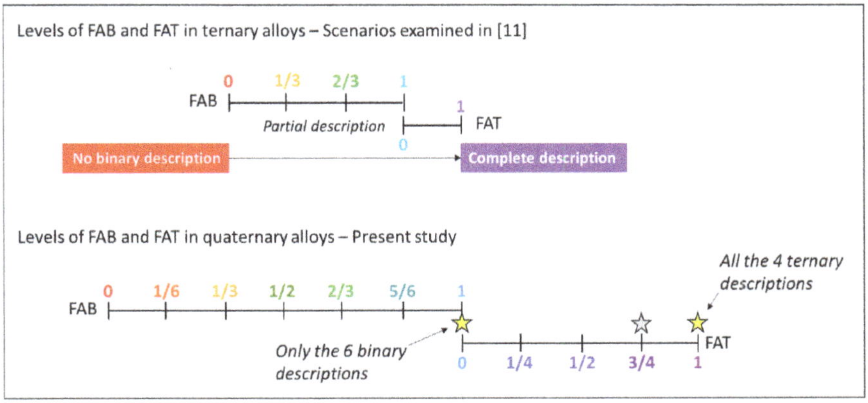

Figure 1. The possible values of the fraction of assessed binaries (FAB) and fraction of assessed ternaries (FAT) quantifying the fraction of binary/ternary descriptions assessed within a database for ternary and quaternary alloys. FAB = 1 and FAT = 0 represents partially assessed conditions, while FAB = 1 and FAT = 1 represents fully assessed conditions.

Here we expand a similar approach to higher order systems to evaluate the ability to predict quaternary phase diagrams. A quaternary system has 6 constituent binaries and 4 ternaries, so the possible FAB levels are 0, 1/6, 1/3, 1/2, 2/3, 5/6 or 1 for databases that include full thermodynamic descriptions for 0, 1, 2, 3, 4, 5 or 6 of the constituent binary systems, and the FAT values are 0, 1/4, 1/2, 3/4, or 1 for databases that include full thermodynamic descriptions for 0, 1, 2, 3, or 4 of the constituent ternary system (Figure 1b). The commercial database TCHEA3 from Thermo-Calc Software AB (Stockholm, Sweden) used in the present work includes 26 elements, 294 fully assessed binary and 136 fully-assessed ternary descriptions, providing few quaternary systems with FAB = 1 and FAT = 1 (e.g., Co-Cr-Fe-Ni and Co-Cu-Fe-Ni), and FAB = 1 and FAT = 3/4 (e.g., Al-Co-Ni-Ti, Al-Cu-Fe-Ni, Al-Cu-Fe-Ti, Al-Cu-Ni-Ti, Al-Mn-Ni-Ti, Al-Fe-Ni-Ti, Al-Ni-Si-Ti). Among them, we have selected three phase diagram templates for this study:

- Co-Cr-Fe-Ni (FAB = 1, FAT = 1), which consists of 3 disordered solid solutions (*fcc*, *bcc*, *hcp*) and one binary intermetallic having ternary solubility (σ);
- Co-Cu-Fe-Ni (FAB = 1, FAT = 1), which has only disordered solid solutions (*fcc* and *bcc* phases), with miscibility gap for *fcc*;
- Al-Co-Ni-Ti (FAB = 1, FAT = 3/4), which includes ordered and disordered solid solutions (*fcc*/L12 and *bcc*/B2) and several binary and ternary intermetallics.

TCHEA3 is an encrypted database which does not permit access to the values of the Gibbs free energy parameters of the phases. However, it is possible to fix the value of any parameter using the commands available in the Gibbs Energy System (GES) module implemented in Thermo-Calc® software. Using this command for all possible relevant parameters allows a comparison of the predictions for the reference systems, referred to as FAT=1 for Co-Cr-Fe-Ni and Co-Cu-Fe-Ni, and FAT = 3/4 for Al-Co-Ni-Ti (the Co-Ni-Ti ternary subcomponent is only partially assessed in TCHEA3), with the results obtained when (1) setting to zero all the ternary interaction parameters for all the relevant phases (solid solutions and a binary intermetallic having ternary solubility) but retaining ternary intermetallics and (2) additionally suspending all the ternary intermetallics. The scenarios (1) and (2) are referred to as FAT = 0 since they are equivalent to a database in which the ternary systems are not assessed for the three selected quaternary systems (the description of the quaternary relies only on the descriptions of the 6 constituent binaries). We have selected one comparison metric which is the extent of the compositional range of each phase throughout the 3D composition space. These data were retrieved from the phase point coordinates calculated at a fixed temperature for 12,341 equally spaced system points (2.5 at.% compositional steps) throughout the quaternary diagram. It is worth mentioning that our methodology uses the same commercial database for calculation of thermodynamic properties of a quaternary alloy at different levels of assessment of the respective ternary systems, which is achieved by artificial removal of the ternary interaction parameters. At the same time, the binary interaction parameters in the database containing the description of ternary systems can slightly differ from the published binary assessment to better reproduce the ternary phase diagram. However, these modifications are made in such a way that the calculated thermodynamic properties of the respective binary systems still agree with the accepted experimental data. Therefore, setting all ternary terms to zero (FAT=0) can still be treated as a condition with FAB = 1.

As previously discussed by Miracle and Senkov [12], and Wertz et al. [11], visualizing the extension and topology of binary and ternary terminal solid solution phase fields into a higher order dimensional composition space remains difficult because the properties of high dimensional objects diverge our two or three-dimensional intuition. An alternative approach to get an insight into the phase points (representative points of a phase, i.e., single-phase region and its boundaries) in high dimension is to use parallel coordinate plots [3,13]. In such a plot, the axes for the variables (e.g., composition of each alloying element and temperature) lie parallel to each other and span from the minimum to maximum value of each variable. The coordinates of a phase (i.e., phase points) or a phase field with two or more phases in equilibrium (i.e., system points) are visualized by segmented lines connecting each variable (molar composition and temperature), so the number of lines connecting the composition coordinates of a selected phase reflects the extent of that phase in the compositional space. It is then possible to evaluate the extent of a phase (or a phase field) in the n-dimensional space of a phase diagram as the ratio of the number of phase points over the number of system points. The ratio represents the (hyper)-volume fraction occupied by a phase in the n-dimensional composition space.

3. Results and Discussion

Figure 2a shows an unfolded tetrahedral quaternary phase diagram for the Co-Cr-Fe-Ni system showing the four isothermal sections of the ternary subsystems calculated at 800 °C under the condition FAT = 1. The single-phase fields are highlighted by red (*fcc*) green (*bcc*) gray (*hcp*) and blue (sigma) color codes, while the adjacent two- or three-phase regions are left un-colored (white). Figure 2b shows the difference between the extents of the phase fields in the subcomponent ternaries calculated for FAT = 1 and FAT = 0. The phase field regions where the calculations do not match are shown in white. It can be seen that the disagreements in the calculations appear mainly in the Co-Cr-Ni and Cr-Fe-Ni, while perfect match occurs in Co-Cr-Fe and Co-Fe-Ni. The disagreements of the FAT=1 and FAT=0 calculations for Co-Cr-Ni mainly appear in the regions associated with *hcp* and δ phases and at phase-field boundaries. In Cr-Fe-Ni, these entirely occur at the phase-field boundaries. In Figure 2c–d, the parallel coordinate plots show the *fcc* phase points in the Co-Cr-Fe-Ni quaternary system for FAT = 1 and FAT = 0, respectively. There is no significant difference between both predictions. As summarized in Figure 3, the phase points within the *fcc* and *bcc* phase fields calculated using FAT = 0 match 98%

of the FAT = 1 predictions. These results suggest that the phase equilibria and locations of the phase boundaries (solubility limits) delimiting the single solid solution regions in the Co-Cr-Fe-Ni quaternary diagram are weakly sensitive to the number of assessed ternary subsystems. In other words, ternary contributions are negligible, and the accuracy remains when extrapolating the Co-Cr-Fe-Ni quaternary system exclusively from the six fully-assessed binary subsystems. Correct prediction of miscibility gap and/or intermetallic phases may however require more complete description of ternary interactions.

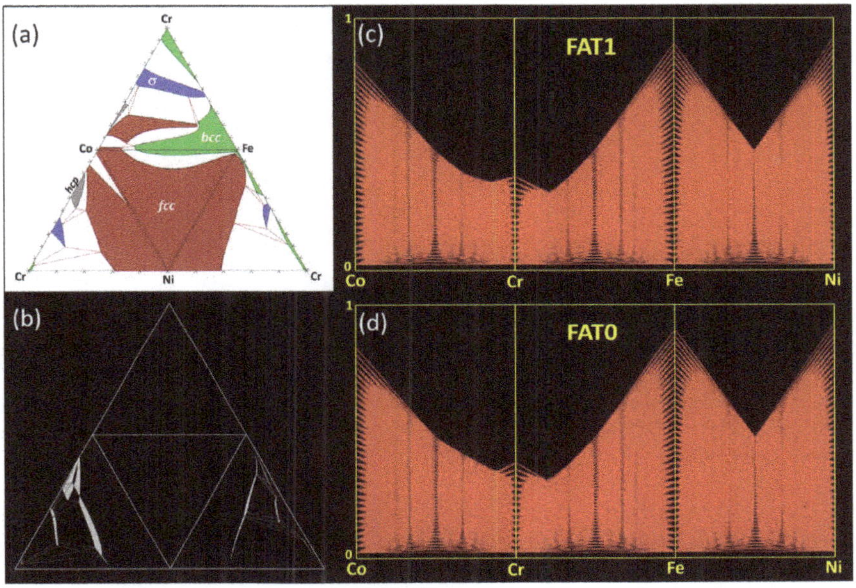

Figure 2. Co-Cr-Fe-Ni system: (**a**) Unfolded tetrahedral quaternary phase diagram showing the four isothermal sections of the ternary subsystems calculated at 800°C for FAT = 1, highlighting the single-phase fields. (**b**) Difference (appearing in white) between the extent of the various phase fields calculated for FAT = 1 and FAT = 0. (**c**,**d**) Parallel coordinate plots showing the predicted range of existence (4D compositional coordinates) of the *fcc* phase in the quaternary Co-Cu-Fe-Ni phase diagram at 800 °C for (c) FAT = 1 and (d) FAT = 0.

The effects from canceling ternary interactions are slightly different for the Co-Cu-Fe-Ni (Figure 4). This system contains disordered *fcc* and *bcc* phases and no intermetallic phases (Figure 4a). The calculations with FAT = 0 correctly predict the presence and location of these phases. Perfect match of FAT = 0 and FAT = 1 calculations occurs for the Co-Fe-Ni ternary, similar to the previous case with the Co-Cr-Fe-Ni system. However, the main disagreement between the FAT = 0 and FAT = 1 calculations arises from incorrect predictions of the miscibility gap boundaries for the high-temperature fcc phase in the Co-Cu-Ni and Cu-Fe-Ni ternary subcomponents (Figure 4b) and in the quaternary volume (Figure 4c,d). The absence of ternary interactions incorrectly shifts the miscibility gap boundaries toward higher concentrations of Ni, while the presence of these Ni-associated ternary interactions expands the high-temperature *fcc* phase to higher concentrations of Cu and Fe or Co and Cu (Figure 4b–d). Due to this, only 63% of the *fcc* phase points calculated from FAT = 0 match the FAT = 1 calculations, while almost 100% agreement occurs for the bcc phase fields (Figure 3).

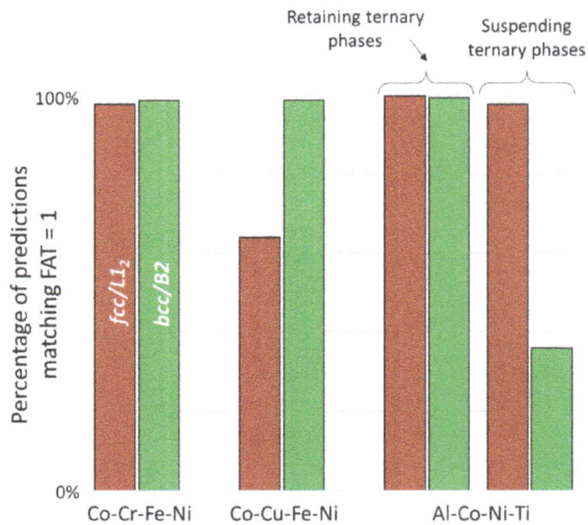

Figure 3. Percentage of predictions matching FAT = 1 for the Co-Cr-Fe-Ni and Co-Cu-Fe-Ni quaternary systems, for the *fcc*/$L1_2$ and *bcc*/B2 phase points, respectively, when neglecting ternary interactions (FAT = 0). For the Al-Co-Ni-Ti system, the first condition corresponds to a special case of FAT = 0, when all ternary interactions are canceled but the ternary phases are artificially retained, while the second condition corresponds to the actual FAT = 0 condition, when both ternary interactions and ternary phases are suspended.

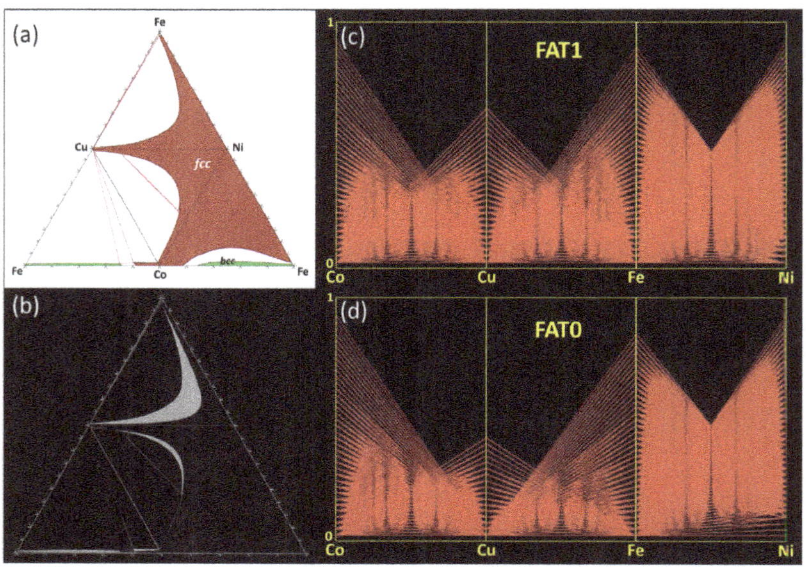

Figure 4. Co-Cu-Fe-Ni system: (**a**) Unfolded tetrahedral quaternary phase diagram showing the four isothermal sections of the ternary subsystems calculated at 800°C for FAT = 1, highlighting the single-phase fields. (**b**) Difference (appearing in white) between the extent of the various phase fields calculated for FAT = 1 and FAT = 0. Parallel coordinate plots showing the predicted range of existence (4D compositional coordinates) of the *fcc* phase in the quaternary Co-Cu-Fe-Ni phase diagram at 800 °C for (**c**) FAT = 1 and (**d**) FAT = 0.

The discrepancies between predictions performed based on ternary descriptions and those without are even more severe for the Al-Co-Ni-Ti system (Figure 5). In addition to the disordered fcc and bcc solid solutions, this system also contains several ordered ($L1_2$, B2) binary intermetallic (Al_5Co_2, Al_3Co, $Al_{13}Co_4$, Al_9Co_2, Al_3Ni, Al_3Ni_2, BCT, Al_2Ti, AlTi, $AlTi_3$, Ni_3Ti, (Co,Ni)Ti_2, Laves-C15 and Laves-C16) and several ternary intermetallic (H_$L2_1$, Laves_C14) phases (Figure 5a). The presence of many different phase field regions in this quaternary system results in high density of boundaries between these regions, where the main disagreements between FAT = 0 and FAT = 1 calculations occur. In addition, FAT = 0 calculations cannot predict composition regions for ternary intermetallic phases. Instead, they fill these regions with available binary phases (mainly the bcc and B2 phases). This can be well visualized by comparing the bcc and B2 phase points represented in parallel coordinate plots for FAT = 3/4 (Figure 5c) to FAB = 0 (Figure 5d).

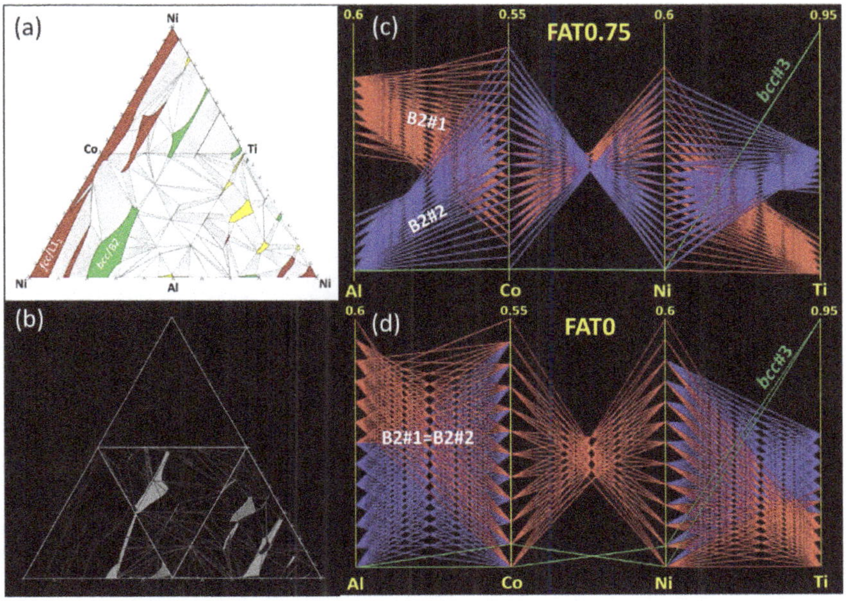

Figure 5. Al-Co-Ni-Ti system: (**a**) Unfolded tetrahedral quaternary phase diagram showing the four isothermal sections of the ternary subsystems calculated at 800°C (FAT = 3/4), highlighting the *single*-phase fields. In addition to the solid solution phases, the quaternary system includes several ordered solid solutions ($L1_2$, B2), binary intermetallics (Al_5Co_2, Al_3Co, $Al_{13}Co_4$, Al_9Co_2, Al_3Ni, Al_3Ni_2, BCT, Al_2Ti, AlTi, $AlTi_3$, Ni_3Ti, $NiTi_2$, Laves-C15 and Laves-C16) and several ternary intermetallics (H_$L2_1$, Laves_C14). (**b**) Difference (appearing in white) between the extent of the various phase fields calculated for FAT = 3/4 and FAT = 0 (the case when all ternary intermetallics are suspended, see Figure 3). Parallel coordinate plots showing the predicted range of existence (4D compositional coordinates) of the *bcc* phase in the quaternary Al-Co-Ni-Ti phase diagram at 800 °C for (**c**) FAT = 3/4 and (**d**) FAT = 0.

The different sensitivities to FAT of the calculated phase diagrams for Co-Cr-Fe-Ni, Co-Cu-Fe-Ni and Al-Co-Ni-Ti systems result from the different thermodynamic properties and topologies of these systems. In all the cases, the number of solid solution phases and solid-solution phase identifications match perfectly between FAT = 1 (or $\frac{3}{4}$) and FAT = 0 calculations. The main differences in the calculations occur at or near the boundaries between different phase fields. Therefore, higher density of these boundaries should result in lower accuracy of the FAT = 0 calculations. This observation is in agreement with earlier results by Wertz et al. [11]. The presence of ternary phases in the ternary systems, which the evaluated higher order system contains, will result in incorrect FAT = 0 calculations, as the extrapolations from the binary interactions cannot predict formation of ternary phases. Unfortunately, the locations of the phase boundaries and possible presence of ternary phases are not known when evaluating novel composition space. Therefore, a higher credibility database is still preferable, while the calculations using lower credibility databases may be questionable and require experimental verification.

The Co-Cr-Fe, Co-Cr-Ni, Co-Fe-Ni and Cr-Fe-Ni behave like near-ideal mixtures of the constituent elements so the Co-Cr-Fe-Ni quaternary does too, which results in the stabilization of the disordered solid solutions and the absence of ternary intermetallic phases. In such cases, neglecting ternary interaction parameters when considering FAT = 0 predictions gives approximatively the same results as for FAT = 1 because their contributions to the Gibbs free energy are also negligible. The Co-Cu-Fe-Ni quaternary system deviates to the near-ideal behavior due to a tendency for the demixing of the Cu-Co and Cu-Fe constituent binaries. In this case, the extent of the miscibility gap is sensitive to the ternary interactions and neglecting them causes incorrect predictions of the phase boundaries involving the solid solutions with the miscibility gap.

In contrast, the Al-Co-Ni-Ti system exhibits several ordered solid solutions and binary and ternary intermetallics. In this case, the most significant factor controlling the location of the phase boundaries arises from the phase assemblage, i.e., the presence and proximity of ternary intermetallics, whereas the removal of the ternary interaction parameters does not change significantly the number of phase points matching the reference predictions when the ternary phases are maintained (Figure 3). This sensitivity depends upon the neighboring phase regions. In particular, the assemblage around the *fcc* and $L1_2$ regions consists of various binary and ternary phase fields involving B2, Ni_3Ti and not containing ternary intermetallics, so the extent of the *fcc* region is not constrained when ternary intermetallics are removed. In consequence, using an incomplete database (FAT = 0) does not substantially affect the composition and boundaries of the *fcc* and $L1_2$ phases. On the other hand, *bcc* and B2 regions are surrounded by several ternary intermetallics (γ and H_L2_1), so neglecting them (FAT = 0) shifts significantly the *bcc* and B2 composition and boundaries, thus providing significant disagreement between FAT = 3/4 and FAT = 0 calculations (Figure 3).

4. Conclusions

The ability of a commercial CALPHAD database for high entropy alloys to predict quaternary phase diagrams has been evaluated through comparisons of predictions made from full binary system descriptions with the results from full ternary system descriptions. This point is of prime interest because even the most comprehensive databases (such as TCHEA3 from Thermo-Calc) include a limited number (about 5%) of fully assessed ternary systems. This entails that quaternary and higher order predictions are systematically obtained from incomplete thermodynamic descriptions and rely almost entirely on extrapolations from binary subsystem descriptions. The analysis has led to the following conclusions.

(1) The quality of CALPHAD calculations for 4-component alloys mainly depends on the ability to correctly describe (a) all the phases present in binary and ternary sub-systems, including novel ternary phases, and on (b) positions (composition and temperature) of boundaries between different phase-field regions.
(2) Direct extrapolation from binaries to quaternary (i.e., without ternary descriptions) systems may be acceptable only when binary miscibility gaps are not present, binary intermetallic phases are not present or present in a few quantities (i.e., when the system has low density of phase boundaries), and ternary intermetallic phases are not present.
(3) When ternary phases are present but not included in the thermodynamic descriptions, the prediction accuracy of the FAT = 0 calculations will be strongly affected only for neighboring phase regions.

Considering that quaternary and higher order interactions are negligibly weak, and the formation of novel quaternary intermetallics very unlikely, the present conclusions for quaternary alloys can be extended to higher order systems. Because the locations of the phase boundaries and possibility of formation of ternary phases are not known when evaluating novel composition space, a higher credibility database is still preferable, while the calculations using lower credibility databases may be questionable and require additional experimental verification.

Author Contributions: Conceptualization, S.G. and O.N.S.; Methodology, S.G. and O.N.S.; Formal Analysis, S.G. and O.N.S.; Writing-Original Draft Preparation, S.G.; Writing-Review & Editing, S.G. and O.N.S.; Visualization, S.G.

Funding: This research received no external funding.

Acknowledgments: Work by O.N.S. was supported through the Air Force on-site contract FA8650-15-D-5230 managed by UES, Inc., Dayton, Ohio.

Conflicts of Interest: The authors declare no conflicts of interest.

References

1. Gwalani, B.; Gorsse, S.; Choudhuri, D.; Styles, M.; Zheng, Y.; Mishra, R.S.; Banerjee, R. Modifying transformation pathways in high entropy alloys or complex concentrated alloys via thermo-mechanical processing. *Acta Mater.* **2018**, *153*, 169–185. [CrossRef]
2. Hutchinson, C.R.; Nie, J.-F.; Gorsse, S. Modeling the precipitation processes and strengthening mechanisms in a Mg-Al-(Zn) AZ91 alloy. *Metall. Mater. Trans. A* **2005**, *36*, 2093–2105. [CrossRef]
3. Gorsse, S.; Tancret, F. Current and emerging practices of CALPHAD toward the development of high entropy alloys and complex concentrated alloys. *J. Mater. Res.* **2018**, *33*, 2899–2923. [CrossRef]
4. Choudhuri, D.; Gwalani, B.; Gorsse, S.; Mikler, C.V.; Ramanujan, R.V.; Gibson, M.A.; Banerjee, R. Change in the primary solidification phase from fcc to bcc-based B2 in high entropy or complex concentrated alloys. *Scr. Mater.* **2017**, *127*, 186–190. [CrossRef]
5. Bracq, G.; Laurent-Brocq, M.; Perrière, L.; Pirès, R.; Joubert, J.-M.; Guillot, I. The fcc solid solution stability in the Co-Cr-Fe-Mn-Ni multi-component system. *Acta Mater.* **2017**, *128*, 327–336. [CrossRef]
6. Zhang, W.; Liaw, P.K.; Zhang, Y. Science and technology in high-entropy alloys. *Sci. China Mater.* **2018**, *61*, 2–22. [CrossRef]
7. Gorsse, S.; Couzinié, J.-P.; Miracle, D.B. From high-entropy alloys to complex concentrated alloys. Available online: https://doi.org/10.1016/j.crhy.2018.09.004 (accessed on 21 November 2018).
8. Senkov, O.N.; Miller, J.D.; Miracle, D.B.; Woodward, C. Accelerated exploration of multi-principal element alloys for structural applications. *CALPHAD* **2015**, *50*, 32–48. [CrossRef]
9. Senkov, O.N.; Miller, J.D.; Miracle, D.B.; Woodward, C. Accelerated exploration of multi-principal element alloys with solid solution phases. *Nat. Commun.* **2015**, *6*, 6529. [CrossRef] [PubMed]
10. Tancret, F.; Toda-Caraballo, I.; Menou, E.; Rivera-Díaz-Del-Castillo, P.E.J. Designing high entropy alloys employing thermodynamics and Gaussian process statistical analysis. *Mater. Des.* **2017**, *115*, 486–497. [CrossRef]
11. Wertz, K.N.; Miller, J.D.; Senkov, O.N. Toward multi-principal component alloy discovery: Assessment of CALPHAD thermodynamic databases for prediction of novel ternary alloy systems. *J. Mater. Res.* **2018**, *33*, 3204–3217. [CrossRef]
12. Miracle, D.B.; Senkov, O.N. A critical review of high entropy alloys and related concepts. *Acta Mater.* **2017**, *122*, 448–511. [CrossRef]
13. Inselberg, A. *Parallel Coordinates*; Springer: New York, NY, USA, 2009; pp. 1–50. [CrossRef]

entropy

MDPI

Article

Brute Force Composition Scanning with a CALPHAD Database to Find Low Temperature Body Centered Cubic High Entropy Alloys

T. P. C. Klaver *, D. Simonovic and M. H. F. Sluiter

Department of Materials Science and Engineering, Delft University of Technology, 2628 CD Delft, The Netherlands; darko.simonovic@gmail.com (D.S.); M.H.F.Sluiter@tudelft.nl (M.H.F.S.)

* Correspondence: klaver2@gmail.com; Tel.: +31-15-2784-345

Received: 11 October 2018; Accepted: 26 November 2018; Published: 29 November 2018

Abstract: We used the Thermo-Calc High Entropy Alloy CALPHAD database to determine the stable phases of AlCrMnNbTiV, AlCrMoNbTiV, AlCrFeTiV and AlCrMnMoTi alloys from 800 to 2800 K. The concentrations of elements were varied from 1–49 atom%. A five- or six-dimensional grid is constructed, with stable phases calculated at each grid point. Thermo-Calc was used as a massive parallel tool and three million compositions were calculated, resulting in tens of thousands of compositions for which the alloys formed a single disordered body centered cubic (bcc) phase at 800 K. By filtering out alloy compositions for which a disordered single phase persists down to 800 K, composition 'islands' of high entropy alloys are determined in composition space. The sizes and shapes of such islands provide information about which element combinations have good high entropy alloy forming qualities as well as about the role of individual elements within an alloy. In most cases disordered single phases are formed most readily at low temperature when several elements are almost entirely excluded, resulting in essentially ternary alloys We determined which compositions lie near the centers of the high entropy alloy islands and therefore remain high entropy islands under small composition changes. These island center compositions are predicted to be high entropy alloys with the greatest certainty and make good candidates for experimental verification. The search for high entropy islands can be conducted subject to constraints, e.g., requiring a minimum amount of Al and/or Cr to promote oxidation resistance. Imposing such constraints rapidly diminishes the number of high entropy alloy compositions, in some cases to zero. We find that AlCrMnNbTiV and AlCrMoNbTiV are relatively good high entropy alloy formers, AlCrFeTiV is a poor high entropy alloy former, while AlCrMnMoTi is a poor high entropy alloy former at 800 K but quickly becomes a better high entropy alloy former with increasing temperature.

Keywords: high entropy alloy; bcc; phase stability; CALPHAD; composition scanning

1. Introduction

High entropy alloys (HEAs) are at present a very active field of research within metallurgy. The vast number of possible compositions promises a very broad range of properties. While the vast majority of (near) equi-atomic combinations of alloying elements lead to alloys with poor properties, the small fraction of combinations with good properties still provides very promising prospects, spurring very active research in this area.

Originally, HEAs were defined as alloys with five or more principal elements in (near) equi-atomic amounts, which form a single disordered phase on a simple crystal lattice. Configurational entropy was thought to be the main stabilizing factor, though it was soon shown that other factors can be more important, see e.g., [1]. More recently, the focus of attention has widened. More alloys that are not (near) equi-atomic have been investigated [2]. Carbon and/or nitrogen have been deliberately introduced to steer ferritic/austenic stability and to form finely dispersed carbides and/or nitrides to improve mechanical properties, see e.g., [3]. Compositions are chosen to deliberately create multi-phase materials that have better mechanical properties [4,5]. Stacking fault energies and relative phase stabilities in multi-phase materials are engineered to induce TRIP and/or TWIP deformation mechanisms [6–12]. Despite the extensive research effort on HEAs, the number of true HEAs found is still rather limited [13]. The vast majority of compositions lead to the formation of alloys with very brittle phases, like Laves and sigma phases [14]. Even many of the compositions that lead to alloys with good properties for applications are not truly HEAs at lower temperature. These alloys (sometimes referred to as compositionally complex alloys) may be HEAs just below the solidification temperature, but at lower temperature their equilibrium state includes additional phases [15,16]. They often have good low temperature properties thanks to the sluggish formation of additional phases, which allows the disordered single phase to persist as a meta-stable state at lower temperature.

In this work we focus on finding HEAs that retain their single disordered phase down to relatively low temperature, consisting in part of elements that promote oxidation resistance (Al up to high temperature, Cr up to intermediate temperature in environments free of water vapour). The number of non-equi-atomic composition variations with five or more elements is so large that experimental testing, even with modern high-throughput screening using samples with composition gradients, is no longer feasible. Computationally however, using CALPHAD databases to determine the stable phases as a function of temperature on a fine grid in the composition space is possible. For the six element alloys AlCrMnNbTiV and AlCrMoNbTiV, and five element alloys AlCrFeTiV and AlCrMnMoTi, and their constituent alloys, we determined in a five (four) dimensional composition space where the 'islands' of low temperature HEA stability are located, i.e., for which compositions a single disordered phase remains stable down to low temperature. Apart from determining islands of low temperature HEA stability we also determine where the 'centers' of the islands are, i.e., which compositions remain HEAs under small compositional changes. These compositions are also likely to have some margin against the inevitable error inherent in the CALPHAD method, see e.g., the mismatches in the comparison between CALPHAD predicitons and experimental results drawn up by Saal et al. [15]. The island centre compositions are predicted to be low temperature HEAs with the greatest certainty and are good candidates for experimental verification. Apart from selecting compositions corresponding to the centers of islands of HEA stability, constraints can be imposed. For example, minimum amounts of Al and/or Cr can be required to promote oxidation resistance. Also, alloys can be selected for a narrow solidification temperature range to limit segregation during solidification.

The outline of this paper is as follows: in Section 2 we provide details on our computational approach. In Section 3 we first explain our choice of the five and six element alloys we investigated and present results of a simple composition optimization for these alloys. We then present results of convergence testing of the concentration step size used in brute force scanning of the composition space for these alloys. After that, we look at the overall HEA forming qualities of the alloys and the roles that individual elements play in them through binary element projections. Finally, we present results about the islands of HEA stability for our alloys, without and with constraints for minimum concentrations of certain elements. Conclusions are reported in Section 4.

2. Computational Details

The Thermo-Calc (TC) implementation of the CALPHAD method was used to calculate stable phases. The TC high entropy alloy v2.1 database (TCHEA2.1 [17,18]) was used within TC v2017b or 2018a, run under linux. The TCHEA2.1 database contains data for the elements Al, C, Co, Cr, Cu, Fe, Hf, Mn, Mo, N, Nb, Ni, Re, Ru, Si, Ta, Ti, V, W and Zr. For these elements, full information on all binary systems and 135 ternary systems is included, as well as partial information from another 308 ternary systems. Equilibrium data for some of the elements (including Fe) is available only for ~500 °C and above. To avoid the hazards of extrapolation, our calculations apply to the temperature range 800–2800 K. Below 800 K diffusion is exceedingly sluggish in transition metal HEAs, so that equilibrium calculations are in any case more applicable to the higher temperature ranges. We found that calculations over a continuous temperature range with TCHEA2.1 enter into infinite loops every few dozen compositions, making automated high-throughput calculations ineffective. Also, results are at times calculated over incomplete temperature ranges. Calculations did not go into infinite loops when calculated with a different TC database (SSOL) or when data was calculated at discrete temperatures rather than continuously over a temperature range. Hence, we calculated data with TCHEA2.1 every 50 K in the 800–2800 K range (41 temperatures).

We employed a high throughput approach that is in some ways similar to the high-throughput method used by Senkov et al. [19,20]. In their extensive study, the Pandat implementation of the CALPHAD method was used to calculate the equilibrium phases for over 100,000 equi-atomic alloys. Here we determine equilibrium phases for a large number of non-equi-atomic compositions for four alloys. We used the Console.sh command line interface within TC to run typically ~100 calculations in parallel on single cpu cores of a computing cluster. The calculation of the stable phases and their fractions at 41 temperatures takes less than a minute on one cpu core, allowing throughput of a few thousand compositions per core per day. For this work we calculated 3 million compositions in total. While e.g., using a genetic algorithm to find HEA compositions [21], possibly in combination with a constraint satisfaction algorithm [22] or performing a targeted search that optimizes an objective function (e.g., narrow solidification temperature range or single disordered phase stability down to low temperature) under constraints [23] are approaches that are all far less computationally demanding, using TC as a high throughput tool is not much limited by the required cpu time or the disk space required to store input and output files. Analysis can be time consuming if it is done post hoc in serial over hundreds of thousands of output files. Analysis should ideally be included right after each TC calculation so that it is carried out in parallel, either using external tools or the TC_Python module.

3. Results and Discussion

3.1. Selection of Alloys, Extending the HEA Temperature Range

HEAs containing Al, Cr and Ti are rather likely to have a bcc crystal structure. Pure Cr has a bcc crystal structure and while Ti has an hcp structure at room temperature, it assumes a bcc structure above 1155 K. While pure Al has an fcc crystal lattice, it is known to promote the bcc structure in transition metal based HEAs [24]. The work by Senkov et al. [19,20] reported both five element and six element bcc HEAs (Tables 14 and 15 in [19]). TCHEA2.1 did not confirm all bcc HEAs predicted in [22], but several six element HEAs containing Al and Cr, including AlCrMnNbTiV and AlCrMoNbTiV, were confirmed to be HEAs. AlCrMnNbTiV is predicted to be a single disordered bcc phase from ~1550–1750 K, AlCrMoNbTiV from ~1200–2100 K. According to TCHEA2.1 the five element HEA AlCrTaTiV starts to form a sigma phase just before solidification is complete. AlCrFeTiV is predicted to be a disordered single bcc phase from ~1050–1800 K, AlCrMnMoTi from ~1150–1800 K. We focused our work on AlCrMnNbTiV, AlCrMoNbTiV, AlCrFeTiV and AlCrMnMoTi.

For practical applications of these four series of alloys, it is preferable that the temperature at which other (brittle) phases appear is decreased and the amount of other phases formed is reduced. A simple way to achieve this is to determine what the composition of the disordered bcc phase and alternate phases is at a lower temperature, where multiple phases have formed. If the alternate phases were removed, the remaining bcc phase then forms a HEA at the lower temperature. This was tried for multiple iterations for AlCrMnNbTiV, see Figure 1.

Obviously, altering the concentrations of the individual elements within a HEA can be very effective in maintaining the HEA to a lower temperature and reducing the amount of alternate phases once they start to form. However, this way of strengthening the HEA character of an alloy produces a HEA that at low temperature is on the boundary of the HEA single phase region and the two or more phase region containing undesirable secondary phases.

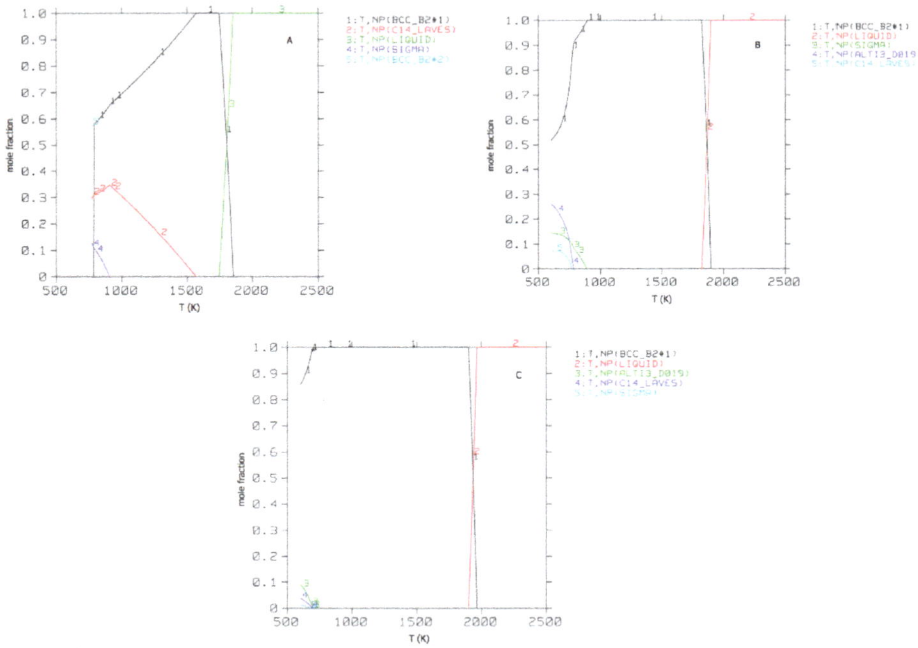

Figure 1. Phase fractions as a function of temperature for (**A**) equi-atomic AlCrMnNbTiV (**B**) $Al_{18}Cr_{10}Mn_{13}Nb_{12}Ti_{21}V_{26}$ (**C**) $Al_{15}Cr_{12}Mn_{17}Nb_3Ti_{10}V_{43}$.

The smallest change in composition in some directions already leads to the formation of secondary phases. In order to find an alloy that is a HEA 'with margin to spare', we want to find the compositions that remain HEAs under all small composition changes.

3.2. Convergence Testing for Scanning Part of the Composition Space

Scanning all possible five and six element alloy compositions at fine 1% increments requires going through more compositions than is feasible. In order to limit the number of compositions required, we limit the portion of the composition space that we cover and for that limited part of the composition space, we conduct convergence tests of the concentration increment, to see how fine a mesh is required. We limit the part of the composition space by requiring that no element in a HEA should be a majority constituent, i.e., the concentration of any element should be <50 atom%. Within the selected part of the composition space, atom percentages are varied from 1 to 49% for all but one element and the concentration of the last element is set to reach 100% in total. If the concentration of the last element has to be negative or larger than 50%, the composition is rejected. For five/six element alloys, each element has the role of 'filler-up' once and that of 'independent variable' four/five times. It should be noted that while the independent and filler-up elements have the same concentration increment, the possible concentrations of the filler-up element are shifted compared to those of the other elements. For example, the composition closest to a binary alloy has 49% of one element, 1% for four elements, leaving 47% for the filler-up element. Thus with a 4% concentration increment, the independently varied elements have concentrations of 49, 45, ... 5, 1% while the filler-up element has concentrations of 47, 43, ... , 7, 3%. Thus the possible element concentrations of the independently varied and filler-up elements are on sub-grids that have the same spacing but are shifted from each other. Hence a 4% concentration increment will result in some element concentrations being only 2% apart. Following the scheme outlined above, the numbers of compositions for five and six element alloys are as shown in Table 1.

Table 1. Numbers of compositions for five and six element alloys for different concentration spacings.

Concentration Spacing (%)	Compositions	
	Five elements	Six elements
6	13,530	74,412
4	60,905	473,382
2	862,750	2,114,580 *

* not calculated with TC in our study.

The results we are most interested in are the shapes of low temperature islands of HEA stability. The convergence tests should therefore determine how much these vary with the concentration spacing. We show a number of two-dimensional projections for $Al_aCr_bFe_cTi_dV_{1\text{-}a\text{-}b\text{-}c\text{-}d}$ in Figure 2 and for $Al_aCr_bMn_cNb_dTi_eV_{1\text{-}a\text{-}b\text{-}c\text{-}d\text{-}e}$ in Figure 3.

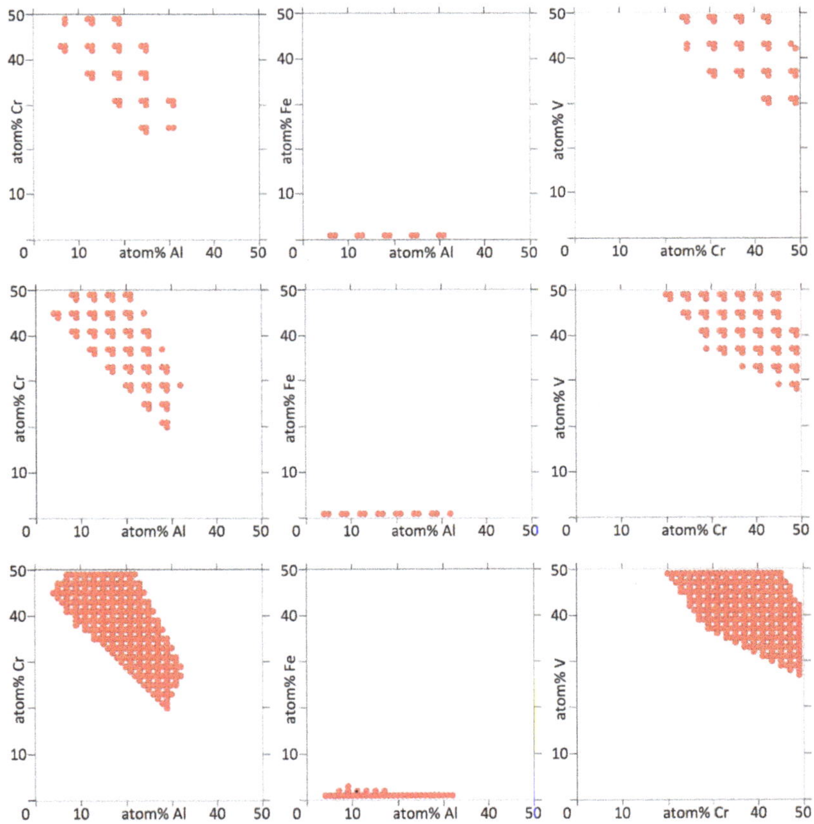

Figure 2. Two-dimensional projections for $Al_aCr_bFe_cTi_dV_{1\text{-}a\text{-}b\text{-}c\text{-}d}$, showing at which concentrations for two elements the alloy forms a bcc HEA at 800 K. Three concentration dimensions are flattened out to arrive at the two-dimensional projection. The concentrations of the three elements not shown can be any one or multiple combinations, i.e., a circle indicates that for the corresponding concentration of the two elements shown, there is at least one and in most cases there are many combinations of concentrations of the other three elements for which the alloy forms a HEA at 800 K. The concentration increments in the top, middle and bottom figures are 6, 4 and 2%, respectively.

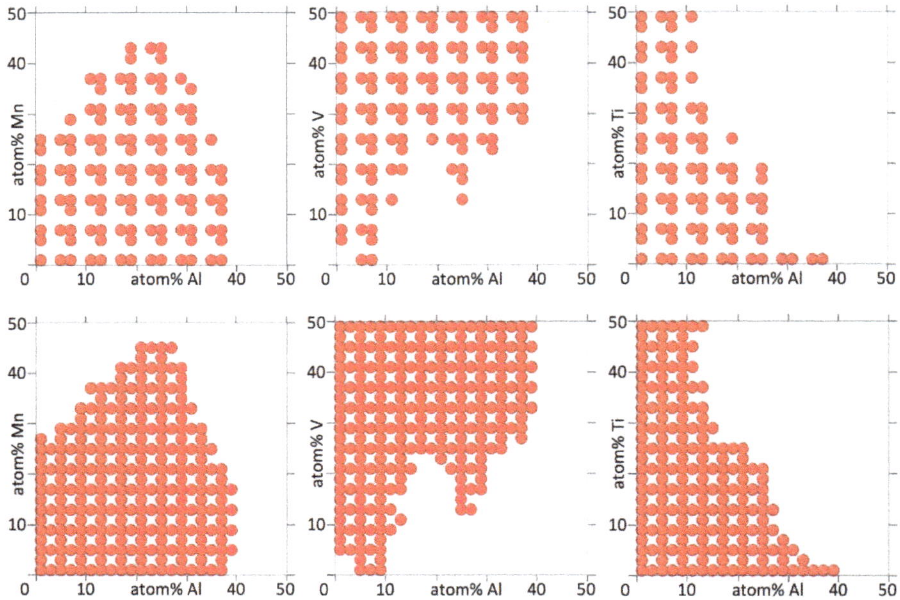

Figure 3. Two-dimensional projections for $Al_aCr_bMn_cNb_dTi_eV_{1-a-b-c-d-e}$, showing at which concentrations for two elements the alloy forms a bcc HEA at 800 K. Four concentration dimensions are flattened out to arrive at the two-dimensional projection. The concentrations of the four elements not shown are as explained in the caption of Figure 2. The concentration increments in the top and bottom figures are 6 and 4%, respectively.

In the small sampling of projections in Figures 2 and 3 there are only single islands of HEA stability, there are no small separate islands. Also, the islands are solid without holes in them. Generally the size of the islands is many times larger than the concentration spacing. The concentration spacing therefore only influences the outer edges of the islands. At a coarser spacing, some detail of the shapes of outer edges of the islands is lost, but the overall shapes of the islands are preserved. This means that for the cases shown, a relatively modest number of compositions on a coarse grid in composition space already provide most information about islands of low temperature HEA stability.

3.3. The Different Roles of Alloying Elements

Figures 4–7 show binary projections as in Figures 2 and 3 for all possible binary combinations in our alloys.

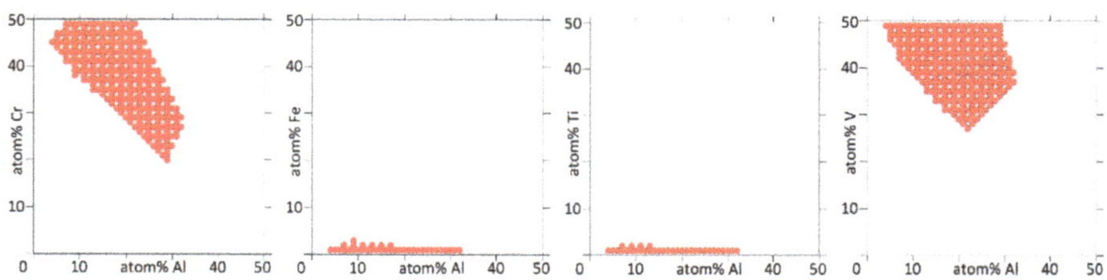

Figure 4. *Cont.*

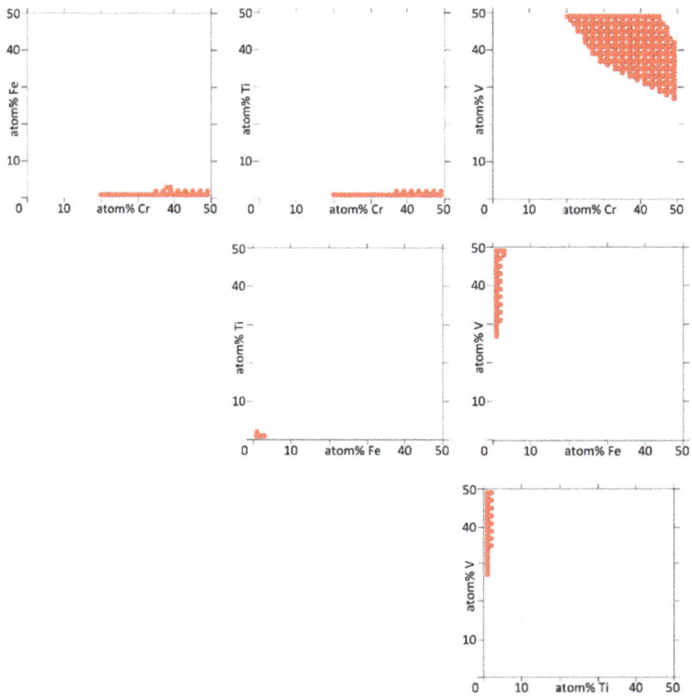

Figure 4. Two-dimensional projections for $Al_aCr_bFe_cTi_dV_{1\text{-}a\text{-}b\text{-}c\text{-}d}$, showing at which concentrations for two elements the alloy forms a bcc HEA at 800 K. Three concentration dimensions are flattened out to arrive at the two-dimensional projection. The concentrations of the three elements not shown are as explained in the caption of Figure 2. The concentration increments are 2%.

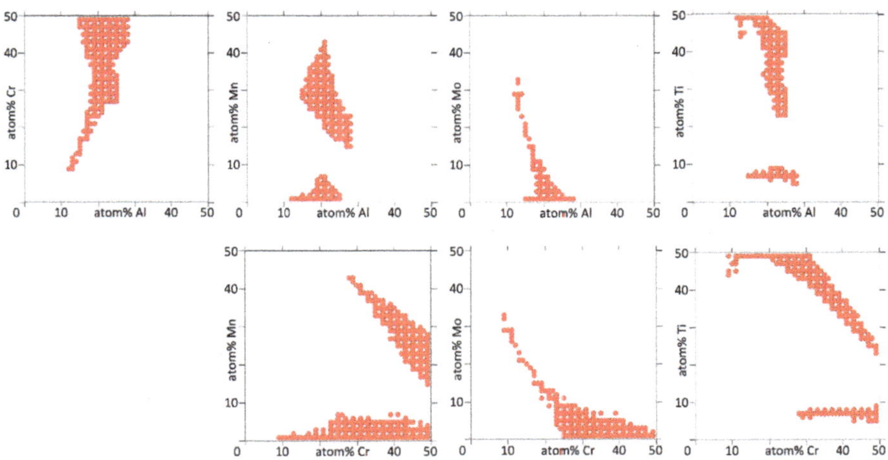

Figure 5. *Cont.*

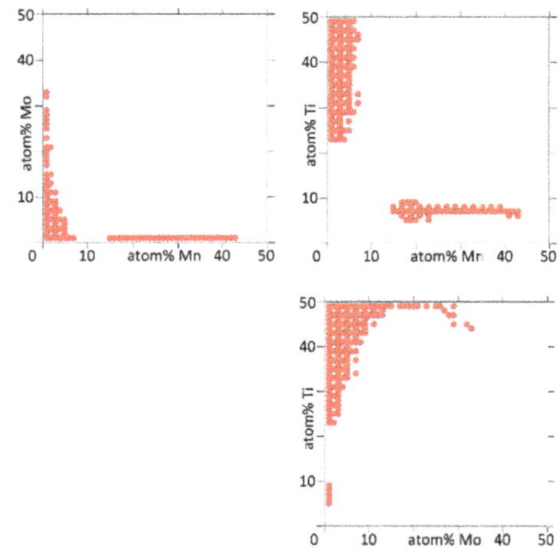

Figure 5. Two-dimensional projections for $Al_aCr_bMn_cMo_dTi_{1\text{-}a\text{-}b\text{-}c\text{-}d}$, showing at which concentrations for two elements the alloy forms a bcc HEA at 800 K. Three concentration dimensions are flattened out to arrive at the two-dimensional projection. The concentrations of the three elements not shown are as explained in the caption of Figure 2. The concentration increments are 2%.

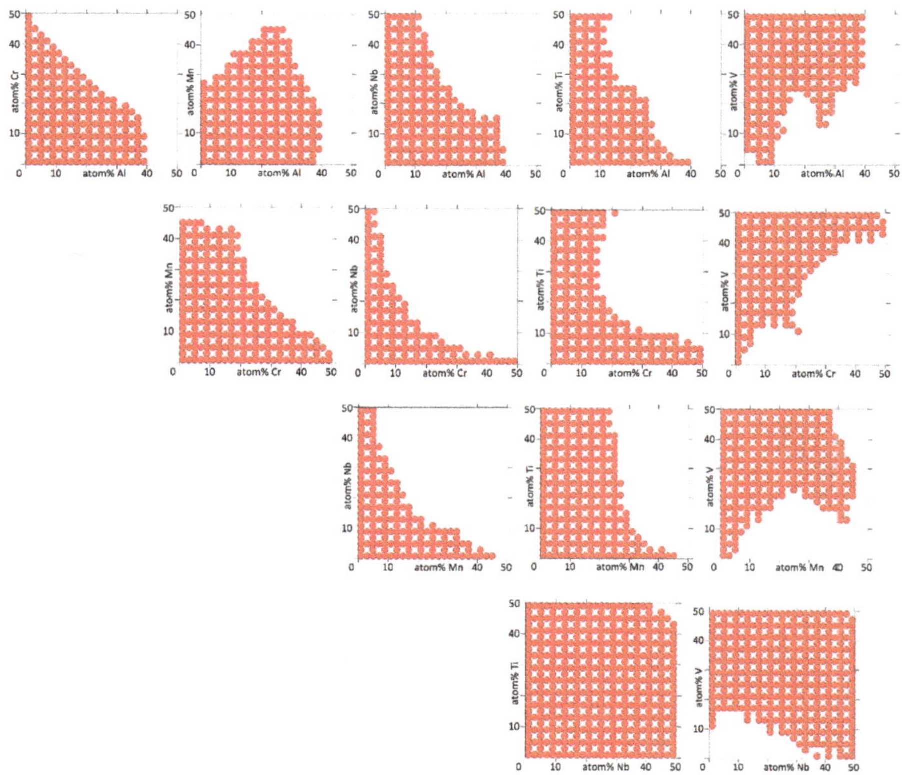

Figure 6. *Cont.*

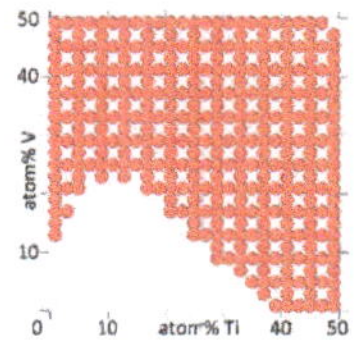

Figure 6. Two-dimensional projections for $Al_aCr_bMn_cNb_dTi_eV_{1-a-b-c-d-e}$, showing at which concentrations for two elements the alloy forms a bcc HEA at 800 K. Four concentration dimensions are flattened out to arrive at the two-dimensional projection. The concentrations of the four elements not shown are as explained in the caption of Figure 2. The concentration increments are 4%.

Figure 7. Two-dimensional projections for $Al_aCr_bMo_cNb_dTi_eV_{1-a-b-c-d-e}$, showing at which concentrations for two elements the alloy forms a bcc HEA at 800 K. Four concentration dimensions are flattened out to arrive at the two-dimensional projection. The concentrations of the four elements not shown are as explained in the caption of Figure 2. The concentration increments are 4%.

In interpreting Figures 4–7, it is worth pointing out that a lot of information is left out of the two-dimensional projections. What appears to be a single island may in fact consists of several separate islands in the dimension perpendicular to the projection (which contains all the information of the other elements than the two being shown), that overlap into a single island when shown as a two-dimensional projection.

Figures 4–7 show that the various elements in the four alloys play distinct roles. On the one hand, Fe and Ti in AlCrFeTiV hardly participate in forming a HEA. Single disordered bcc phases in AlCrFeTiV can form, but they are essentially ternary alloys, without Fe or Ti. On the other hand, Mo and Ti in AlCrMoNbTiV can form HEAs with the other elements at any combination of concentrations. In between these two extremes, a variety of other behaviors can be observed. HEA islands that cover part of the two-element projections may extend mutually over the full 0–50% range for both elements or over the full range for one element but part of the range for the other, or over part of the range for both elements. A minimum concentration of the two elements can be required, indicated by a lack of circles around the origin, such as for TiV in AlCrMnNbTiV, see Figure 6. The HEA island may be formed under an inversely proportional line, such as for CrNb in AlCrMnNbTiV, see Figure 6. The maximum percentage of one element as a function of the other may not follow a monotonous line, there may be minima and maxima such as for AlMn and AlV in AlCrMnNbTiV, see Figure 6. V in AlCrMnNbTiV in particular gives many minima and maxima in the two-dimensional projections in Figure 6. There may even be an archipelago of separate islands of stability, as is the case with the thin, stretched-out islands for AlCrMnMoTi, see Figure 5. Islands are seen to feature a great variety of shapes, including bays, peninsular outcroppings and satellite islands, see AlCr, AlNb and CrV in AlCrMoNbTiV, Figure 7. Contrary to the results of convergence testing in Section 3.2, some of these features would be lost if the calculations were carried out on a coarser grid.

Overall, the two six element alloys appear to be more promising candidates for forming low temperature HEAs than the two five element alloys. For AlCrMnNbTiV and AlCrMoNbTiV 17,830 (3.8%) and 17,289 (3.7%) out of 473,382 compositions sampled were single phase HEAs at 800 K. For AlCrFeTiV and AlCrMnMoTi only 356 (0.041%) and 785 (0.091%) out of 862,750 compositions sampled were single phase HEAs at 800 K. In Figures 6 and 7 on average 64% and 67% of the grid points of the two-dimensional projections for AlCrMnNbTiV and AlCrMoNbTiV have circles on them, while in Figures 4 and 5 these percentages are only 6.8% and 9.4% for AlCrFeTiV and AlCrMnMoTi. The supplementary material contains the list of compositions calculated and for each composition, whether that composition is a HEA at 800 K or not and what the phases and phase fractions are at 800 K for the four alloy systems.

3.4. Temperature Dependence of HEA Stability

Figure 8 shows the fraction of alloy compositions for which a HEA is formed as a function of temperature.

Figure 8 shows that both six element alloys are strong HEA formers, with 4% of compositions being HEAs at 800 K and the HEA fraction of fully solid alloys reaching over 90% at 2000 K. The much lower melting temperature of Mn (1519 K) compared to Mo (2896 K) increases the fraction of (partly) molten alloys at 2000 K but it does not greatly increase the onset of melting, since melting is likely to occur first for compositions rich in low-melting metals like Al. Also, alloys with little Mn or Mo are almost the same. In contrast to the six element alloys, AlCrFeTiV obviously has poor HEA forming qualities. AlCrMnMoTi is in between the six element alloys and AlCrFeTiV, with a very low fraction of HEAs at 800 K, but the fraction rapidly increases with temperature, surpassing that of the six element alloys and reaching 100% at 1750 K. Figure 9 shows the average concentrations of individual elements in HEAs as a function of temperature.

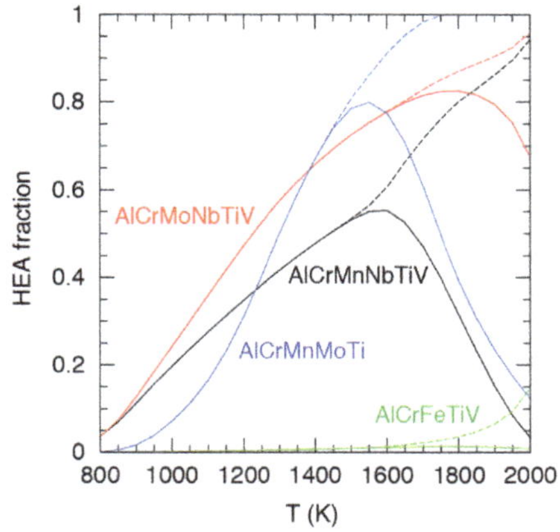

Figure 8. Fractions of alloy compositions for which a HEA is formed, as a function of temperature. Solid curves represent the number of HEAs as a fraction of all compositions, including those that are (partly) molten. Dashed curves represent the number of HEAs as a fraction of compositions for which the alloys are still completely solid.

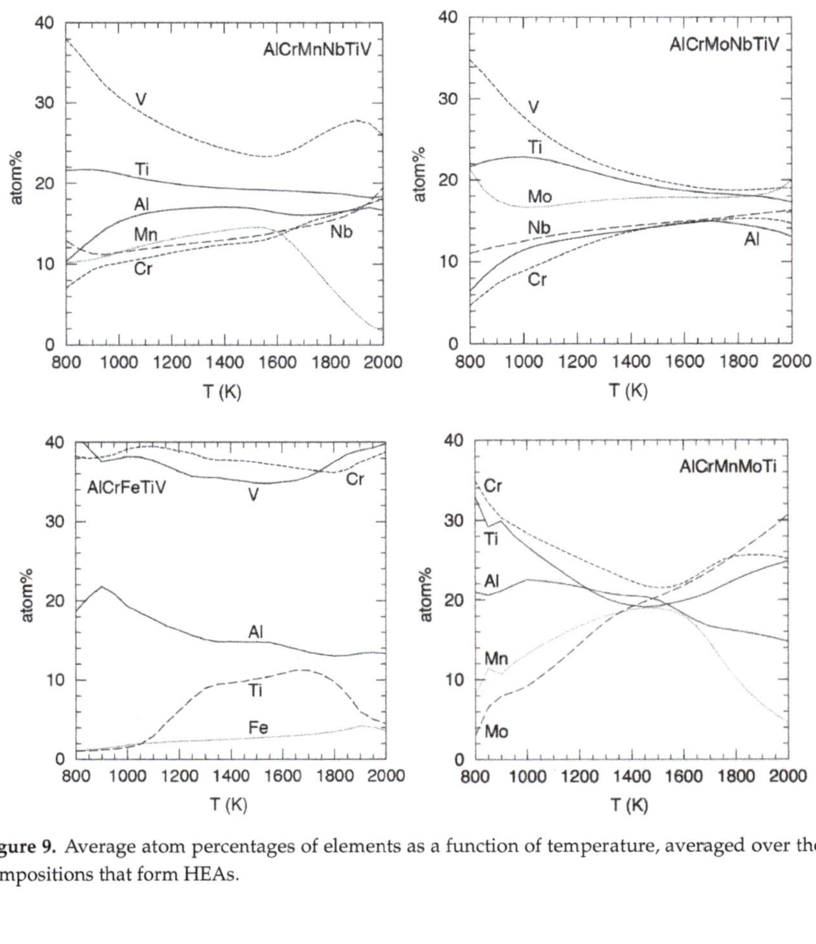

Figure 9. Average atom percentages of elements as a function of temperature, averaged over those compositions that form HEAs.

It is perhaps surprising to observe that in AlCrMnNbTiV and AlCrMnMoTi, Mn is the element that is the most reduced in concentration at higher temperatures, while pure Mn has a far higher melting temperature (1519 K) than pure Al (933 K). At present we are not able to explain this. For AlCrFeTiV the average composition shown in Figure 9 does not actually lie inside a HEA island for most of the temperature range. For the other three alloys the average concentrations shown in Figure 9 do lie inside HEA islands for all but a few of the lowest temperatures.

It should be noted that our calculations assume thermodynamic equilibrium and therefore homogenous phases. During solidification usually concentration gradients in the solid state develop so that our results may deviate from experimentally prepared materials.

3.5. HEA Island Centers

The center of a HEA island is here defined as the HEA composition that is furthest removed from any composition that is not a HEA. The distance between the island center and the closest non-HEA composition defines a body around the island center that contains a subset of the compositions that form the island. The size of the body indicates how much the concentrations of any element(s) can be varied from the island center while the alloy still remains a HEA. The island center—closest non-HEA distance can be calculated as the Euclidian distance (in which case the body is a high-dimensional spheroid) or Manhattan distance (in which case the body is a high-dimensional polyhedron, with a larger volume than the spheroid). Since we allow concentrations up to 50%, an island center may be close to 50% for one or two elements. Therefore it needs to be decided what to do with compositions on grid points on or outside the 50% boundary, for which there is no data. On the one hand, since the vast majority of compositions are not HEAs, it could be assumed that any composition on or outside the 50% boundary is not a HEA. This means that the sphere or polyhedron around the center must lie entirely within the 50% boundaries. On the other hand, if there is a part of an island of HEA stability bordering the 50% boundary, it is reasonable to assume that the island would not end abruptly at the 50% boundary but extend some distance beyond it as well. Therefore it could be argued that the center of the island needs to lie within the 50% boundary, but that part of the sphere or polyhedron may lie outside it. These two scenarios represent extremes for the smallest and biggest possible spheres/polyhedra and specific cases will usually lie somewhere in between. We shall present results for both scenarios, where all compositions at or beyond 50% are assumed to be non-HEAs ('boundary_on') or where compositions at or beyond 50% are assumed to be HEAs ('boundary_off'). For the former scenario, non-HEA composition data points are added (i.e., defined, not calculated with TC) for all compositions where one or two elements have a 50% concentration. As an example, Table 2 shows the island center(s) composition for AlCrMnNbTiV.

The alternative, equally valid island center compositions indicated by the asterisks in Table 2 are compositions like 3, 1, 1, 25, 21, 49% or 1, 1, 3, 21, 25, 49% Al, Cr, Mn, Nb, Ti, V.

Under boundary_on condition, the Euclidian distance between the island center and the nearest non-HEA compositions is $\sqrt{102}$ = 10.1%. This is only a few times the concentration increment, hence the figure of 10.1% is not very precise. However, it does mean that the alloy will remain a HEA under limited composition changes. For example, if any one element is changed 9% in one direction and four other elements are changed 2% in the opposite direction and one element is changed 1% in the opposite direction, the resulting alloy should still be a HEA. Table 3 shows the island center compositions for all four of our HEAs and the distances to the closest non-HEA compositions.

As in Figures 4–7, the compositions in Table 3 show that the six element alloys are much better HEA formers than the five element alloys at 800 K. The island radii for the five element alloys are so small that there are not really any HEA islands, just a few isolated HEA compositions, possibly with a very small number of their closest neighbor compositions. Finally in this section we show how HEA islands grow with temperature. Figure 10 shows the radii as a function of temperature for AlCrMnNbTiV and AlCrFeTiV.

Table 2. HEA island center composition for AlCrMnNbTiV, determined with different distance and boundary criteria at 800 K. Also shown are the five non-HEA compositions closest to the island center. The elements concentration spacing is 4%. An asterix behind an island center composition indicates there are other island center compositions nearby that have an equally long distance to a nearest non-HEA composition.

	Distance and Boundary Criteria			
	Euclidean Distance		Manhattan distance	
	Boundary_on	Boundary_off	Boundary_on	Boundary_off
	% Al Cr Mn Nb Ti V			
island centre(s)	1 1 1 25 31 41	1 1 1 23 25 49	1 3 1 21 37 37 * 1 1 13 5 39 41 *	1 3 1 21 25 49 *
1st nearest non-HEA	0 0 0 22 28 50 10.1% from center	11 1 1 17 25 45 12.3% from center	1 11 1 21 29 37 7 1 5 5 41 41 16% from center	1 11 1 25 13 49 24% from center
2nd nearest non-HEA	0 0 0 21 29 50 10.2% from center	9 1 1 17 29 43 12.3% from center	1 11 1 21 33 33 7 1 9 1 41 41 16% from center	1 11 1 25 17 45 24% from center
3rd nearest non-HEA	0 0 0 23 27 50 10.2% from center	1 9 1 27 17 45 12.6% from center	1 11 1 21 37 29 7 1 13 1 41 37 16% from center	1 11 1 25 21 41 24% from center
4th nearest non-HEA	0 0 1 21 28 50 10.4% from center	1 9 1 27 21 41 12.6% from center	9 3 1 13 37 37 9 1 11 1 37 41 16% from center	1 1 11 25 21 41 24% from center
5th nearest non-HEA	0 0 1 22 27 50 10.4% from center	9 1 1 15 29 45 12.6% from center	9 3 1 17 33 37 9 1 11 5 33 41 16% from center	1 9 1 27 13 49 24% from center

Unsurprisingly, the island radius pattern for AlCrMnNbTiV in Figure 10 is rather similar to the pattern of the AlCrMnNbTiV HEA fraction shown for AlCrMnNbTiV in Figure 8.

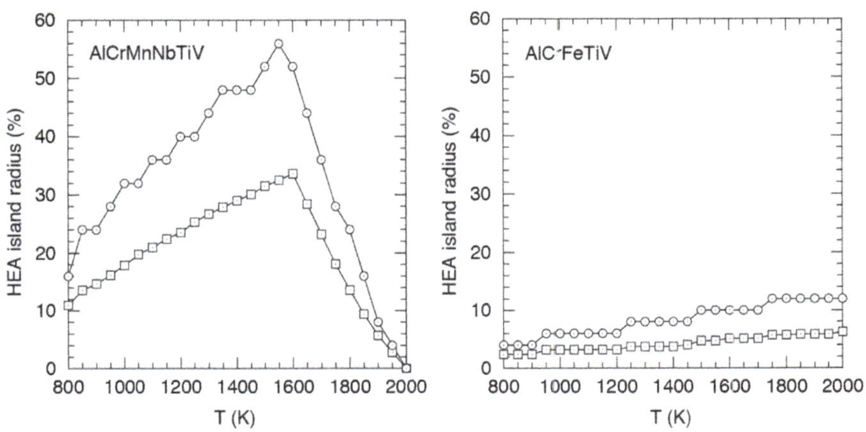

Figure 10. Euclidian (squares) and Manhattan (circles) radii of HEA islands around $Al_1Cr_1Mn_1Nb_{25}Ti_{31}V_{41}$ and $Al_{11}Cr_{43}Fe_1Ti_1V_{44}$ under boundary_off condition. Radii are here defined as the distances between the island center and the nearest non-HEA composition. Only compositions *less* than a radius away from the island center are guaranteed to be HEAs.

Table 3. HEA island center(s) compositions, determined with different distance and boundary criteria at 800 K. Below each composition is the distance to the nearest non-HEA composition. The elements concentration spacing is 4% for the six element alloys and 2% for the five element alloys. An asterix behind an island center composition indicates there are other island center compositions nearby that have an equally long distance to a nearest non-HEA composition.

	Island Center(s) and Distance to Nearest Non-HEA Composition			
	Euclidean Distance		Manhattan Distance	
	Boundary_on	Boundary_off	Boundary_on	Boundary_off
Alloy	% Elements			
AlCrMnNbTiV	1 1 1 25 31 41 10.1%	1 1 1 23 25 49 12.3%	1 3 1 21 37 37 * 1 1 13 5 39 41 * 16%	1 3 1 21 25 49 * 24%
AlCrMoNbTiV	1 1 37 1 21 39 * 10.2%	1 1 37 1 13 47 * 11.3%	1 1 39 1 21 37 * 20%	1 1 35 1 13 49 * 24%
AlCrFeTiV	11 43 1 1 44 * 2.4%	11 43 1 1 44 * 2.4%	11 43 1 1 44 * 4%	11 43 1 1 44 * 4%
AlCrMnMoTi	22 33 1 1 43 * 3.7%	22 33 1 1 43 * 3.7%	22 33 1 1 43 * 6%	22 33 1 1 43 * 6%

3.6. HEA Compositions with Minimum Concentration Constraints

The compositions in Table 3 are all essentially ternary alloys, meaning they are not really conventional HEAs. For the six element alloys, the elements mostly absent from the island center compositions include Al and Cr. While oxidation resistance depends on more than just having significant amounts of Al and/or Cr present in alloys, their presence is an important enabling factor for oxidation resistance. We repeated our search for HEA islands of maximum size, but now under the condition that minimum amounts of Al and/or Cr are present in the alloys or that four or more elements must be present in a concentration equal or greater than 10%. Tables 4 and 5 show HEA island center compositions and sizes determined under these constraints.

Table 4. HEA island center(s) compositions, determined with different distance and boundary criteria at 800 K, under the constraint of having minimum amounts of Al and/or Cr present. Below each composition is the distance to the nearest non-HEA composition. The elements concentration spacing is 4% for the six element alloys and 2% for the five element alloys. An asterix behind an island center composition indicates there are other island center compositions nearby that have an equally long distance to a nearest non-HEA composition.

	Island Center(s) and Distance to Nearest Non-HEA Composition			
	Euclidean Distance		Manhattan Distance	
	Boundary_on	Boundary_off	Boundary_on	Boundary_off
Alloy, Constraint	**% Al Cr Mn Nb Ti V**			
AlCrMnNbTiV Al ≥ 15%	25 7 25 1 1 41 * 21 7 21 1 9 41 * 7.5%	25 1 23 1 1 49 9.8%	25 7 21 1 5 41 * 12%	21 5 21 1 5 47 * 16%
AlCrMnNbTiV Cr ≥ 15%	17 15 17 1 9 41 5.7%	11 17 13 1 9 49 * 6.3%	5 17 5 11 17 45 * 17 17 13 1 9 43 * 8%	11 17 13 1 9 49 * 12%
AlCrMnNbTiV Al + Cr ≥ 15%	25 7 25 1 1 41 * 21 7 21 1 9 41 * 7.5%	25 1 23 1 1 49 9.8%	25 7 21 1 5 41 * 7 9 5 17 21 41 * 12%	21 5 21 1 5 47 * 7 9 5 13 17 49 16%
AlCrMoNbTiV Al ≥ 15%	17 1 9 21 9 43 * 17 5 21 1 11 45 * 4.9%	15 1 5 21 9 49 * 6.3%	17 1 9 21 9 43 * 17 5 21 1 11 45 * 8%	15 1 5 21 9 49 * 12%

Table 4. *Cont.*

	Island Center(s) and Distance to Nearest Non-HEA Composition			
	Euclidean Distance		Manhattan Distance	
	Boundary_on	Boundary_off	Boundary_on	Boundary_off
Alloy, Constraint	**% Al Cr Mn Nb Ti V**			
AlCrMoNbTiV Cr ≥ 15%	9 15 21 1 13 41 * 4.9%	15 17 1 5 13 49 13 17 11 1 9 49 * 4.9%	9 15 21 1 13 41 * 8%	15 17 1 5 13 49 13 17 11 1 9 49 * 8%
AlCrMoNbTiV Al + Cr ≥ 15%	9 7 29 1 13 41 * 6.3%	11 5 25 1 9 49 * 7.5%	9 7 29 1 13 41 * 12%	11 5 25 1 9 49 * 16%
AlCrFeTiV Al ≥ 15%	23 36 1 1 39 * 1.4%	23 36 1 1 39 * 1.4%	23 36 1 1 39 * 2%	23 36 1 1 39 * 2%
AlCrFeTiV Cr ≥ 15%	11 43 1 1 44 * 2.4%	11 43 1 1 44 * 2.4%	11 43 1 1 44 * 4%	11 43 1 1 44 * 4%
AlCrFeTiV Al + Cr ≥ 15%	11 43 1 1 44 * 2.4%	11 43 1 1 44 * 2.4%	11 43 1 1 44 * 4%	11 43 1 1 44 * 4%
AlCrMnMoTi Al ≥ 15%	22 33 1 1 43 * 3.7%	22 33 1 1 43 * 3.7%	22 33 1 1 43 * 6%	22 33 1 1 43 * 6%
AlCrMnMoTi Cr ≥ 15%	22 33 1 1 43 * 3.7%	22 33 1 1 43 * 3.7%	22 33 1 1 43 * 6%	22 33 1 1 43 * 6%
AlCrMnMoTi Al + Cr ≥ 15%	22 33 1 1 43 * 3.7%	22 33 1 1 43 * 3.7%	22 33 1 1 43 * 6%	22 33 1 1 43 * 6%

Table 5. HEA island center(s) compositions, determined with different distance and boundary criteria at 800 K, under the constraint of having four or more elements present in a ≥10% concentration. Below each composition is the distance to the nearest non-HEA composition. The elements concentration spacing is 4% for the six element alloys and 2% for the five element alloys. An asterix behind an island center composition indicates there are other island center compositions nearby that have an equally long distance to a nearest non-HEA composition.

	Distance and Boundary Criteria			
	Euclidean Distance		Manhattan Distance	
	Boundary_on	Boundary_off	Boundary_on	Boundary_off
Alloy, Constraint	**% Al Cr Mn Nb Ti V**			
AlCrMnNbTiV 4 elem. ≥ 10%	17 13 17 1 9 43 * 6.3%	11 1 9 17 13 49 17 11 17 1 9 45 * 5 11 1 13 21 49 * 5 1 11 13 25 45 * 7.5%	17 13 17 1 9 43 * 12%	17 11 17 1 9 45 * 16%
AlCrMnNbTiV 5 elem. ≥ 10%	17 13 17 1 11 41 6.3%	13 13 13 1 11 49 7.5%	17 13 17 1 11 41 12%	11 13 13 1 13 49 * 13 1 11 13 13 49 * 12%
AlCrMnNbTiV 6 elem. ≥ 10%	-	-	-	-
AlCrMoNbTiV 4 elem. ≥ 10%	11 5 29 1 13 41 * 6.3%	11 1 1 25 13 49 * 11 5 25 1 13 45 * 6.3%	11 1 29 5 13 41 * 12%	11 1 1 25 13 49 * 11 1 25 1 13 49 * 12%
AlCrMoNbTiV 5 elem. ≥ 10%	13 11 17 1 13 45 5.5%	13 11 17 1 13 45 5.7%	11 1 17 13 13 45 * 13 13 19 1 13 41 * 13 1 13 17 11 45 * 8%	11 1 13 13 13 49 * 13 13 17 1 13 43 * 13 11 29 1 17 29 8%
AlCrMoNbTiV 6 elem. ≥ 10%	-	-	-	-
AlCrFeTiV 4 elem. ≥ 10%	-	-	-	-
AlCrFeTiV 5 elem. ≥ 10%	-	-	-	-
AlCrMnMoTi 4 elem. ≥ 10%	16 17 1 17 49 * 1.4%	16 17 1 17 49 * 1.4%	16 17 1 17 49 * 2%	16 17 1 17 49 * 2%
AlCrMnMoTi 5 elem. ≥ 10%	-	-	-	-

Tables 4 and 5 show that the options for selecting HEAs with minimum Al and/or Cr content or four or more elements present in 10% or higher concentration are limited. Imposing such constraints decreases HEA island sizes, down to 0 when requiring that all elements have a 10% or higher concentration. AlCrMnNbTiV has the largest islands of HEA compositions that contain a high enough percentage of Al to promote oxidation resistance.

3.7. Melting Temperature Ranges

From a production point of view it is preferable to select alloys with narrow solidification temperature ranges in order to achieve solidification with minimal unmixing, or with unmixing on the smallest possible length scales. Since we only determine data at temperature intervals of 50 K, we can only roughly estimate solidification temperature ranges. Therefore, in Table 6 we list the number of 50 K spaced temperatures that fall within the solidification ranges.

Table 6. Number of 50 K spaced data points in the solidification temperature ranges of alloys.

Alloy	Average, All Compos.	Average Over HEA Compos. at 800 K	Average Over Non-HEA Compos. at 800 K	Max. of All Compos.	Max. of HEA Compos.
AlCrMnNbTiV	2.42	1.27	2.46	15	4
AlCrMoNbTiV	3.20	3.28	3.19	15	6
AlCrFeTiV	2.42	0.43	2.42	9	2
AlCrMnMoTi	4.21	1.68	4.21	15	4

Table 6 shows that alloy compositions that are a single phase HEAs at 800 K generally have rather narrow solidification ranges, in all cases six intervals of 50 K or less. On average the most promising HEA alloys from the AlCrMnNbTiV and AlCrMoNbTiV type have solidification ranges of just 1.27 and 3.28 50 K intervals. So it appears as if selecting compositions that are single phase HEAs at 800 K simultaneously also selects alloys that have desirable solidification behavior. However, the alloy with by far the narrowest solidification range, AlCrFeTiV, is also the poorest HEA former in our study. As in Section 3.4, the higher temperature results in this section have an extra deviation from experimental observations due to the artificial homogeneity and lack of any concentration gradients in our calculations.

4. Conclusions

We used the Thermo-Calc CALPHAD database to computationally investigate the HEA forming qualities between 800 and 2800 K of four alloys, AlCrMnNbTiV, AlCrMoNbTiV, AlCrFeTiV and AlCrMnMoTi. These alloys contain elements that provide oxidation resistance and were previously predicted to be HEAs at high temperature at equi-atomic compositions. Simple variations of the element concentrations away from being equi-atomic can already greatly extend the temperature range over which the alloys are HEAs. However, with a brute force compositions scanning approach, alloy compositions could be found that remain HEAs down to 800 K. By calculating the stable phases for these alloys on grids in five- or six-dimensional composition spaces, we were able to determine islands of low temperature HEA stability. Making binary alloy projections of these high-dimensional islands gives information about the overall HEA forming qualities of the alloys as well as about the roles of individual elements within the alloys. The HEA forming qualities of a combination of elements can also be gleaned from the percentage of compositions that form HEAs as a function of temperature. The compositions of the centers of the HEA islands remain HEAs under small composition changes and thus have some margin of error against inaccuracies in the TC HEA database. Applying our methodology to four alloys, we find that AlCrMnNbTiV and AlCrMoNbTiV are good HEA formers that have HEA islands of non-negligible size at 800 K and that these islands grow rapidly with increasing temperature. AlCrMnMoTi has very few HEA compositions at 800 K but rapidly develops them with increasing temperature. AlCrFeTiV is a poor HEA former at any temperature. For all alloys that have HEA islands, the island centre compositions correspond to what are essentially ternary alloys. Therefore these most interesting compositions are at best medium entropy alloys rather than high entropy alloys. The elements mostly absent from island centre compositions include Al and Cr for AlCrMnNbTiV and AlCrMoNbTiV. Alloys with these compositions thus lack elements that are important for oxidation resistance. Imposing constraints for minimal amounts of Al and/or Cr or four or more alloying elements with >10% concentration rapidly diminishes the number of available HEA compositions, though there are compositions that meet both the requirements of forming HEAs at 800 K and containing substantial amounts of Al and/or Cr. These requirements can be combined with the additional requirement of having a narrow solidification range. Alloy compositions around $Al_{25}Cr_7Mn_{25}Nb_1Ti_1V_{41}$ or $Al_{21}Cr_7Mn_{21}Nb_1Ti_9V_{41}$ offer the best compromise between these three different criteria, according to our CALPHAD predictions. Since CALPHAD predictions are sometimes at odds with experimental results [15], we propose these two compositions for experimental verification.

Supplementary Materials: The following are available online at http://www.mdpi.com/1099-4300/20/12/911/s1. The supplementary material contains the list of compositions calculated and for each composition whether that composition is a HEA at 800 K or not and what the phases and phase fractions are at 800 K for the four alloy systems.

Author Contributions: T.P.C.K. carried out the TC calculations, the processing of the TC output and prepared the manuscript. D.S. carried out the analysis to determine HEA island locations and sizes in composition spaces. M.H.F.S. contributed the main idea of determining HEA islands with a CALPHAD database, added further ideas during discussions throughout the research and did several rounds of critical reading and making suggestions during the preparation of the manuscript.

Funding: This research received funding through the ERA-NET Integrated computational materials engineering (ICME) program under project 4316 "HEAMODELL" as financed by NWO "domein Exacte en Natuurwetenschappen".

Acknowledgments: The authors thank Bengt Hallstedt, James Saal and Ake Jansson and others from Thermo-Calc support for their kind help with TC questions. The authors also thank Richard Huizinga for help with the TC license server and Fritz Körmann for general HEA discussions.

Conflicts of Interest: The authors declare no conflict of interest. The founding sponsor had no role in the design of the study; in the collection, analyses, or interpretation of data; in the writing of the manuscript, and in the decision to publish the results.

References

1. Maa, D.; Grabowski, B.; Körmann, F.; Neugebauer, J.; Raabe, D. Ab initio thermodynamics of the CoCrFeMnNi high entropy alloy: Importance of entropy contributions beyond the configurational one. *Acta Mater.* **2015**, *100*, 90–97. [CrossRef]
2. Raabe, D.; Tasan, C.C.; Springer, H.; Bausch, M. From High-Entropy Alloys to High-Entropy Steels. *Steel Res. Int.* **2015**, *86*, 1127–1138. [CrossRef]
3. Kivy, M.B.; Kriewall, C.S.; Zaeem, M.A. Formation of chromium-iron carbide by carbon diffusion in AlXCoCrFeNiCu high-entropy alloys. *Mater. Res. Lett.* **2018**, *6*, 321–326. [CrossRef]
4. Li, Z.; Raabe, D. Strong and Ductile Non-equiatomic High-Entropy Alloys: Design, Processing, Microstructure, and Mechanical Properties. *JOM* **2017**, *69*, 2099–2106. [CrossRef]
5. Gorsse, S.; Miracle, D.B.; Senkov, O.N. Mapping the world of complex concentrated alloys. *Acta Mater.* **2017**, *135*, 177–187. [CrossRef]
6. Li, Z.; Körmann, F.; Grabowski, B.; Neugebauer, J.; Raabe, D. Ab initio assisted design of quinary dual-phase high-entropy alloys with transformation-induced plasticity. *Acta Mater.* **2017**, *136*, 262–270. [CrossRef]
7. Deng, Y.; Tasan, C.C.; Pradeep, K.G.; Springer, H.; Kostka, A.; Raabe, D. Design of a twinning-induced plasticity high entropy alloy. *Acta Mater.* **2015**, *94*, 124–133. [CrossRef]
8. Uzera, B.; Picak, S.; Liu, J.; Jozaghi, T.; Canadinc, D.; Karaman, I.; Chumlyakov, Y.I.; Kireeva, I. On the mechanical response and microstructure evolution of NiCoCr single crystalline medium entropy alloys. *Mater. Res. Lett.* **2018**, *6*, 442–449. [CrossRef]
9. Nene, S.S.; Frank, M.; Liu, K.; Mishra, R.S.; McWilliams, B.A.; Cho, K.C. Extremely high strength and work hardening ability in a metastable high entropy alloy. *Sci. Rep.* **2018**, *8*, 9920. [CrossRef] [PubMed]

10. Huang, H.; Wu, Y.; He, J.; Wang, H.; Liu, X.; An, K.; Wu, W.; Lu, Z. Phase-Transformation Ductilization of Brittle High-Entropy Alloys via Metastability Engineering. *Adv. Mater.* **2017**, *29*, 1701678. [CrossRef] [PubMed]
11. Lilensten, L.; Couzinié, J.; Bourgon, J.; Perrière, L.; Dirras, G.; Prima, F.; Guillot, I. Design and tensile properties of a bcc Ti-rich high-entropy alloy with transformation-induced plasticity. *Mater. Res. Lett.* **2017**, *5*, 110–116. [CrossRef]
12. Luo, H.; Li, Z.; Raabe, D. Hydrogen enhances strength and ductility of an equiatomic high-entropy alloy. *Sci. Rep.* **2017**, *7*, 9892. [CrossRef] [PubMed]
13. Gao, M.C.; Alman, D.E. Searching for Next Single-Phase High-Entropy Alloy Compositions. *Entropy* **2013**, *15*, 4504–4519. [CrossRef]
14. Guruvidyathri, K.; Kumar, K.C.H.; Yeh, J.W.; Murty, B.S. Topologically Close-packed Phase Formation in High Entropy Alloys: A Review of Calphad and Experimental Results. *JOM* **2017**, *69*, 2113–2124. [CrossRef]
15. Saal, J.E.; Berglund, I.S.; Sebastian, J.T.; Liaw, P.K.; Olson, G.B. Equilibrium high entropy alloy phase stability from experiments and thermodynamic modeling. *Scripta Mater.* **2018**, *146*, 5–8. [CrossRef]
16. Otto, F.; Dlouhý, A.; Pradeep, K.G.; Kuběnová, M.; Raabe, D.; Eggeler, G.; George, E.P. Decomposition of the single-phase high-entropy alloy CrMnFeCoNi after prolonged anneals at intermediate temperatures. *Acta Mater.* **2016**, *112*, 40–52. [CrossRef]
17. Mao, H.; Chen, H.; Chen, Q. TCHEA1: A Thermodynamic Database Not Limited for "High Entropy" Alloys. *J. Phase Equilib. Diffus.* **2017**, *38*, 353–368. [CrossRef]
18. Chen, H.; Mao, H.; Chen, Q. Database development and Calphad calculations for high entropy alloys: Challenges, strategies, and tips. *Mater. Chem. Phys.* **2018**, *210*, 279–290. [CrossRef]
19. Senkov, O.N.; Miller, J.D.; Miracle, D.B.; Woodward, C. Accelerated exploration of multi-principal element alloys for structural applications. *CALPHAD* **2015**, *50*, 32–48. [CrossRef]
20. Senkov, O.N.; Miller, J.D.; Miracle, D.B.; Woodward, C. Accelerated exploration of multi-principal element alloys with solid solution phases. *Nat. Commun.* **2015**, *6*, 6529. [CrossRef] [PubMed]
21. Menou, E.; Toda-Caraballo, I.; Pineau, C.; Bertrand, E.; Ramstein, G.; Tancret, F. Evolutionary design of strong and stable high entropy alloys using multi-objective optimisation based on physical models, statistics and thermodynamics. *Mater. Des.* **2018**, *143*, 185–195. [CrossRef]
22. Abu-Odeh, A.; Galvan, E.; Kirk, T.; Mao, H.; Chen, Q.; Mason, P.; Malak, R.; Arróyave, R. Efficient exploration of the High Entropy Alloy composition-phase space. *Acta Mater.* **2018**, *152*, 41–57. [CrossRef]
23. Gheribi, A.E.; Pelton, A.D.; Bélisle, E.; le Digabel, S.; Harvey, J.-P. On the prediction of low-cost high entropy alloys using new thermodynamic multi-objective criteria. *Acta Mater.* **2018**, *161*, 73–82. [CrossRef]
24. Tang, Z.; Gao, M.C.; Diao, H.; Yang, T.; Liu, J.; Zuo, T.; Zhang, Y.; Lu, Z.; Cheng, Y.; Zhang, Y.; et al. Aluminum Alloying Effects on Lattice Types, Microstructures, and Mechanical Behavior of High-Entropy Alloys Systems. *JOM* **2013**, *65*, 1848–1858. [CrossRef]

entropy

MDPI

Article

First-Principles Design of Refractory High Entropy Alloy VMoNbTaW

Shumin Zheng [1,*] and Shaoqing Wang [2]

[1] School of Materials Science and Engineering, University of Science and Technology of China, Hefei 110016, China

[2] Shenyang National Laboratory for Materials Science, Institute of Metal Research, Chinese Academy of Sciences, Shenyang 110016, China; sqwang@imr.ac.cn

* Correspondence: smzheng15b@imr.ac.cn

Received: 28 November 2018; Accepted: 11 December 2018; Published: 13 December 2018

Abstract: The elastic properties of seventy different compositions were calculated to optimize the composition of a V–Mo–Nb–Ta–W system. A new model called maximum entropy approach (MaxEnt) was adopted. The influence of each element was discussed. Molybdenum (Mo) and tungsten (W) are key elements for the maintenance of elastic properties. The V–Mo–Nb–Ta–W system has relatively high values of C_{44}, bulk modulus (B), shear modulus (G), and Young's modulus (E), with high concentrations of Mo + W. Element W is brittle and has high density. Thus, low-density Mo can substitute part of W. Vanadium (V) has low density and plays an important role in decreasing the brittleness of the V–Mo–Nb–Ta–W system. Niobium (Nb) and tantalum (Ta) have relatively small influence on elastic properties. Furthermore, the calculated results can be used as a general guidance for the selection of a V–Mo–Nb–Ta–W system.

Keywords: high-entropy alloys; first-principles calculation; maximum entropy; elastic property

1. Introduction

In recent years, high entropy alloys (HEAs) have emerged as an interesting area of research [1]. HEAs have superior properties compared to conventional alloys [2]. Refractory high entropy alloys (RHEAs) were developed for high temperature use. RHEAs are mainly composed of Ti, V, Zr, Nb, Mo, Cr, Ta, W, and Hf. According to the literature, most RHEAs exceed the high use temperature of currently used refractory alloys Haynes@230@, MAR-M247@, INCONEL@718 [3], and conventional Ni-based superalloys [4]. This property makes RHEAs a promising candidate for the next generation of high-temperature applications. The VMoNbTaW alloy has received the most attention because of its characteristics, such as its good strength under extreme high temperature, but it is brittle between room temperature and 600 °C [5]. In a VMoNbTaW system, element W is a brittle element and has high density. Density is an important factor for transportation, especially for aircraft and aerospace. A high-temperature resistance is needed for turbine disks and blades, because the efficiency of gas turbines increases with working temperature [6].

Recently, many reports have shown that the best properties of RHEAs may generally be displaced from equilibrium compositions; thus, the studied compositions become complicated [7,8]. Some research focuses on the influence of elements on alloy properties, but most studies are often carried out for elements such as Al [9], Ti [10], Mo [11], and V [12]. It is feasible to study the influence of a single element, though single element optimization fails to meet application requirements most of the time. HEAs must have at least four elements in order to exhibit a high entropy effect [13]. Studying the influence of more than one element can enormously increase experimental efforts. First-principles calculation is an effective method for developing new RHEAs. Most data obtained in previous studies for RHEAs provide information for the hardness and compression of elements [3], but little is known about their elastic properties. Sufficiently large, homogeneous, and defect-free crystals are required to measure experimental elastic constants, so information on elastic properties is only available for a small portion of materials. Special quasi-random structure (SQS) [14] and coherent potential approximation adopted exact muffin-tin orbital (EMTO-CPA) are often used to predict the elastic properties of HEAs [15]. Elasticity is one of the fundamental properties to screen alloys and it directly relates to mechanical properties.

The present study reports a first-principles design of a VMoNbTaW alloy. The aims are to decrease the brittleness and density of a V–Mo–Nb–Ta–W system. The elastic properties of seventy different compositions were calculated. The influence of each element was discussed.

2. Methodology

CP2K was introduced for first-principles calculation and it is efficient for larger systems. CP2K is a quantum chemistry and solid-state physics software package [16]. QUICKSTEP was introduced to deal with the electronic structure. The Gaussian and plane wave (GPW) was used for the calculation of forces and energies [14]. Single-zeta valence Gaussian (SZV-MOLOPT-SR-GTH) was used as the basis set, while a 500Ry plane wave cutoff was used for the auxiliary grid. Fermi–Dirac smearing was used to accelerate the convergence to self-consistency with an electronic temperature of 300 K. In each self-consistent field (SCF) iteration step, the diagonalized Kohn–Sham matrix was introduced for solving eigenvalue issues. Additionally, Broyden mixing was used to accelerate the convergence to the total energy threshold. The value of the total energy threshold is 10^{-7} Hartree. A Broyden–Fletcher–Goldfarb–Shanno (BFGS) minimization algorithm was introduced to deal with the geometry optimization problems. The convergence criteria for the maximum geometry change and force were 1×10^{-3} Bohr and 1×10^{-3} Hartree/Bohr, respectively.

3. Maximum Entropy (MaxEnt) Model

MaxEnt structures were generated by a Monte Carlo simulation code in python. A repeat loop was written in the code to make sure all the elements were distributed homogeneously in the model [17]. In order to obtain a relatively homogeneous MaxEnt model, hundreds of structures were generated for selection. The screen criterion is the shortest distance between the same elements should locate in a narrow range—the narrower the better [18]. The most important advantage of the MaxEnt model is that it can demonstrate lattice distortion after relaxation. MaxEnt is a supercell model, while a $4 \times 4 \times 4$ face-centered cubic (FCC) model contains 256 atoms and a $4 \times 4 \times 4$ body-centered cubic (BCC) contains 128 atoms, so the MaxEnt model can present HEAs with complicated element concentrations. In order to test the accuracy and consistency of the MaxEnt model, ten MaxEnt models of BCC (TiZrNbMoV) were generated. Bulk moduli B and C_{44} were also calculated. All bulk moduli fluctuated around 143.3 ($\pm$2) GPa and all C_{44} fluctuated around 36.2 ($\pm$3) GPa. The scattered diagram is shown in Figure 1. Thus, the MaxEnt approach demonstrates a good consistence for each model. The MaxEnt approach has been elaborated in Reference [16]. The elastic properties of TaNbHfZrTi and CoCrFeNiMn were predicted based on the MaxEnt approach [18,19]. The accuracy of the predicted data was proven by experimental results [20,21]. Thus, the MaxEnt approach is accurate, believable, and suitable for the study of HEAs.

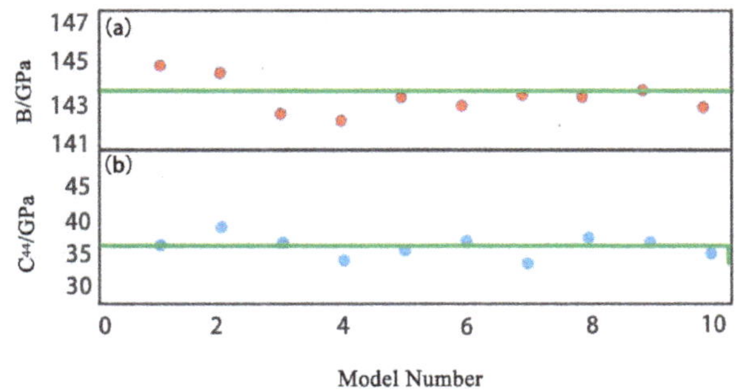

Figure 1. Bulk modulus B and C_{44} of ten maximum entropy (MaxEnt) models of a 4 × 4 × 4 BCC TiZrNbMoV alloy. (**a**) B, (**b**) C_{44}.

All components of the VMoNbTaW alloy have a BCC lattice and, thus, the formation of BCC substitution solutions was the most probable. This was confirmed by diffraction analysis of these alloys [3]. The 4 × 4 × 4 MaxEnt model of $V_{0.1}Mo_{0.2}Nb_{0.1}Ta_{0.4}W_{0.2}$ is shown as an example in Figure 2.

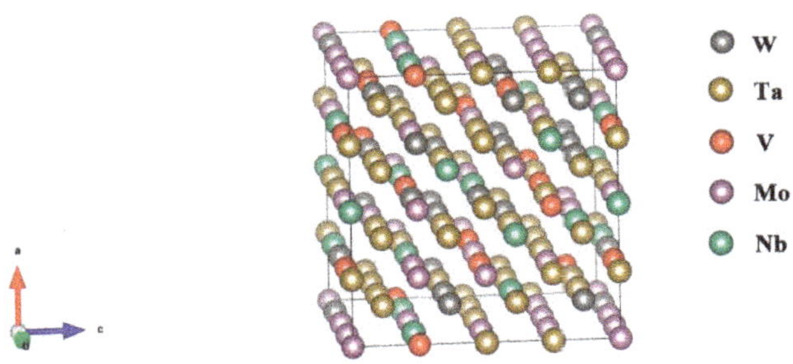

Figure 2. MaxEnt model of $V_{0.1}Mo_{0.2}Nb_{0.1}Ta_{0.4}W_{0.2}$.

4. Elastic Properties

The calculated bulk modulus B and equilibrium lattice configuration were determined from the minima of the curves according to the Birch–Murnaghan equation of state (B–M EOS), as presented in Equation (1). V, V_0, B, E, and E_0 are volume, equilibrium volume, bulk modulus, total energy, and equilibrium energy, respectively. In order not to exceed the elastic limit, the changes in V should be kept within 3%.

$$E(V) = E_0 + \frac{9V_0B}{16}\left\{\left[\left(\frac{V_0}{V}\right)^{\frac{2}{3}} - 1\right]^3 B' + \left[\left(\frac{V_0}{V}\right)^{\frac{2}{3}} - 1\right]^2 \left[6 - 4\left(\frac{V_0}{V}\right)^{\frac{2}{3}}\right]\right\} \quad (1)$$

The cubic crystal has three independent elastic constants: C_{11}, C_{12}, and C_{44}. They can be calculated by applying small strains to the equilibrium lattice configuration, which transforms the lattice vector **a** according to the rule [22] shown in Equations (2) and (3).

$$a' = a \cdot (I + \varepsilon) \quad (2)$$

$$\varepsilon = \begin{pmatrix} e_1 & e_6/2 & e_5/2 \\ e_6/2 & e_2 & e_4/2 \\ e_5/2 & e_4/2 & e_3 \end{pmatrix} \quad (3)$$

$e = (e_1, e_2, e_3, e_4, e_5, e_6)$ is the strain vector. The different values of e were applied to the equilibrium lattice configuration according to Table 1. The value of σ should keep within the range of $(-0.03, 0.03)$.

Table 1. Vector strain and the corresponding energy.

Strain Vector e	The Corresponding Energy for Distorted Structure
$(\sigma,0,0,0,0,0)$	$E(\sigma) = E(0) + \frac{1}{2}C_{11}V_0\sigma^2$
$(\sigma,\sigma,0,0,0,0)$	$E(\sigma) = E(0) + (C_{11} + C_{12})V_0\sigma^2$
$(0,0,0,\sigma,\sigma,\sigma)$	$E(\sigma) = E(0) + \frac{3}{2}C_{44}V_0\sigma^2$

The following equations were used to calculate Shear modulus G, Young's modulus E, and Poisson's ratio v.

$$G = \frac{3C_{44} + C_{11} - C_{12}}{5} \tag{4}$$

$$E = \frac{9BG}{3B + G} \tag{5}$$

$$v = \frac{3B - 2G}{2(3B + G)} \tag{6}$$

5. Results and Discussion

Due to the lack of experimental data of VMoNbTaW, the elastic properties of pure V, Mo, Nb, Ta, and W were calculated to prove the accuracy of the calculated data. Table 2 shows the elastic constants and moduli of V, Mo, Nb, Ta, and W. A comparison of calculated elastic properties with experimental data was made, and the agreement was found to be quite good. The accuracy of the calculated data in the present work was also proven by comparing with the calculated results in other studies.

Table 2. Elastic constants and moduli of V, Mo, Nb, Ta, and W.

Element		C_{11} (GPa)	C_{12} (GPa)	C_{44} (GPa)	B (GPa)	E (GPa)	G (GPa)	v	ρ (g/cm³)	Tm (°C)
V	expt.	232.0 [c]	119.0 [c]	46.0 [c]	155 [b]	127.6 [a]	46.7 [a]	0.365 [a]		
	present	261.0	109.5	45.5	159.7	154.0	57.6	0.340	6.11	1890
	other	205.0 [c]	130.0 [c]	5.0 [c]	194.0 [b]					
Mo	expt.	460.0 [a]/450.0 [c]	176.0 [a]/173.0 [c]	110.0 [a]/125.0 [c]	261.0 [b]					
	present	407.8	150.0	135.5	236.0	335.3	132.7	0.263	10.39	2622
	other	406.0 [c]	150.0 [c]	107.0 [c]	275.0 [b]					
Nb	expt.	242 [b]/253.0 [c]	129 [b]/133.0 [c]	31.0 [c]	169.0 [b]					
	present	270.0	135.0	39.4	180.0	138.9	50.64	0.370	8.57	2468
	other	267.0 [c]	147.0 [c]	27.0 [c]	171.0 [b]					
Ta	expt.	267.0 [a]/266.0 [c]	161.0 [a]/158.0 [c]	82.5 [a]/87.0 [c]	191.0 [b]	185.7 [a]	69.2 [a]	0.342 [a]		
	present	287.0	151.0	72.7	195.6	189.6	70.82	0.338	16.65	2996
	other	291.0 [c]	162.0 [c]	84.0 [c]	183.0 [b]					
W	expt.	501.0 [a]/533.0 [c]	198.0 [a]/205.0 [c]	151.4 [a]/163.0 [c]	308.0 [b]					
	present	558.0	181.5	173.8	307.9	451.0	179.6	0.256	19.35	3410
	other	553.0 [c]	207.0 [c]	178.0 [c]	306.0 [b]					

[a] Reference [23]. [b] Reference [24]. [c] Reference [25].

Seventy different compositions of the V–Mo–Nb–Ta–W system were calculated. The results are shown in Table 3. W is brittle and has high density, so three concentrations (0.1, 0.2, and 0.3) of W were studied, while four concentrations each of V, Mo, Nb, and Ta (0.1, 0.2, 0.3, and 0.4) were studied.

All the structures were found to fulfill the mechanical stability criteria. The mechanical stability criterion of the cubic structure is $C_{11} + 2C_{12} > 0$, $C_{11} > C_{12}$, $C_{44} > 0$. C_{44}, B, G, E, B/G and v are presented in scatter-plots in Figures 3–6, respectively. The correspondence between the numbers in Table 3 and the X-axis is shown in Table 4.

Table 3. Elastic constants and moduli of RHEAs. *B* (GPa), *E* (GPa), *G* (GPa) and *υ* represent the bulk modulus, Young's modulus, shear modulus, and Poisson's ratio and *a* (Å) stands for equilibrium lattice constants.

System	C_{11}	C_{12}	C_{44}	B	G	E	B/G	ν	a
(1)$V_{0.1}Mo_{0.1}Nb_{0.1}Ta_{0.4}W_{0.3}$	424.5	159.4	60.1	247.7	89.0	238.6	2.78	0.339	3.199
(2)$V_{0.1}Mo_{0.1}Nb_{0.2}Ta_{0.3}W_{0.3}$	421.8	158.7	89.8	246.4	106.5	279.2	2.31	0.311	3.200
(3)$V_{0.1}Mo_{0.1}Nb_{0.3}Ta_{0.2}W_{0.3}$	418.1	160.5	59.8	246.3	87.4	234.5	2.82	0.341	3.200
(4)$V_{0.1}Mo_{0.1}Nb_{0.3}Ta_{0.3}W_{0.2}$	407.4	155.2	31.5	239.3	69.3	189.5	3.45	0.368	3.212
(5)$V_{0.1}Mo_{0.1}Nb_{0.4}Ta_{0.1}W_{0.3}$	417.9	154.6	57.6	242.4	87.2	233.7	2.78	0.339	3.202
(6)$V_{0.1}Mo_{0.2}Nb_{0.1}Ta_{0.3}W_{0.3}$	407.0	176.1	85.1	253.1	97.4	259.0	2.60	0.329	3.187
(7)$V_{0.1}Mo_{0.2}Nb_{0.1}Ta_{0.4}W_{0.2}$	419.5	158.8	59.3	245.7	87.7	235.2	2.80	0.340	3.199
(8)$V_{0.1}Mo_{0.2}Nb_{0.1}Ta_{0.5}W_{0.1}$	370.2	172.9	17.6	238.6	50.0	140.2	4.77	0.402	3.212
(9)$V_{0.1}Mo_{0.2}Nb_{0.2}Ta_{0.2}W_{0.3}$	409.8	168.5	82.6	248.9	97.8	259.5	2.54	0.326	3.190
(10)$V_{0.1}Mo_{0.2}Nb_{0.2}Ta_{0.3}W_{0.2}$	397.4	168.5	58.2	244.5	80.8	218.4	3.02	0.351	3.200
(11)$V_{0.1}Mo_{0.2}Nb_{0.2}Ta_{0.4}W_{0.1}$	382.5	164.3	21.9	237.1	56.7	157.7	4.18	0.389	3.212
(12)$V_{0.1}Mo_{0.2}Nb_{0.3}Ta_{0.1}W_{0.3}$	401.7	167.9	81.9	245.8	95.9	254.7	2.56	0.327	3.190
(13)$V_{0.1}Mo_{0.2}Nb_{0.3}Ta_{0.2}W_{0.2}$	393.9	167.3	54.9	242.9	78.2	212.0	3.10	0.354	3.201
(14)$V_{0.1}Mo_{0.2}Nb_{0.3}Ta_{0.3}W_{0.1}$	389.9	158.5	23.6	235.6	60.4	167.0	3.90	0.382	3.212
(15)$V_{0.1}Mo_{0.2}Nb_{0.4}Ta_{0.2}W_{0.2}$	411.4	154.7	55.9	240.3	84.8	227.8	2.83	0.342	3.201
(16)$V_{0.1}Mo_{0.2}Nb_{0.4}Ta_{0.2}W_{0.1}$	399.9	153.5	18.4	235.4	60.3	166.8	3.90	0.382	3.214
(17)$V_{0.1}Mo_{0.2}Nb_{0.5}Ta_{0.1}W_{0.1}$	411.9	145.9	26.3	234.5	69.0	188.5	3.40	0.366	3.214
(18)$V_{0.1}Mo_{0.3}Nb_{0.1}Ta_{0.2}W_{0.3}$	348.1	207.5	92.0	254.3	83.3	225.4	3.05	0.352	3.176
(19)$V_{0.1}Mo_{0.3}Nb_{0.1}Ta_{0.3}W_{0.2}$	380.1	181.6	82.5	247.7	82.5	239.0	3.00	0.339	3.188
(20)$V_{0.1}Mo_{0.3}Nb_{0.2}Ta_{0.1}W_{0.3}$	355.6	199.7	92.6	251.7	86.7	233.4	2.90	0.345	3.177
(21)$V_{0.1}Mo_{0.3}Nb_{0.2}Ta_{0.2}W_{0.2}$	403.1	170.1	85.0	247.8	97.6	258.8	2.54	0326	3.189
(22)$V_{0.1}Mo_{0.3}Nb_{0.3}Ta_{0.1}W_{0.2}$	381.8	173.3	80.5	242.8	90.0	240.3	2.70	0.335	3.190
(23)$V_{0.1}Mo_{0.3}Nb_{0.3}Ta_{0.2}W_{0.1}$	394.7	169.0	79.7	244.2	93.0	247.6	2.63	0.331	3.190
(24)$V_{0.1}Mo_{0.4}Nb_{0.1}Ta_{0.1}W_{0.3}$	391.7	184.8	111.3	253.8	108.2	284.1	2.35	0.313	3.167
(25)$V_{0.1}Mo_{0.4}Nb_{0.1}Ta_{0.2}W_{0.2}$	352.2	200.2	97.0	250.9	88.6	237.9	2.83	0.342	3.176
(26)$V_{0.1}Mo_{0.4}Nb_{0.1}Ta_{0.3}W_{0.1}$	397.8	169.6	77.5	245.7	92.1	245.7	2.67	0.333	3.188
(27)$V_{0.1}Mo_{0.4}Nb_{0.2}Ta_{0.1}W_{0.2}$	356.7	191.4	93.2	246.5	88.9	238.2	2.77	0.339	3.178
(28)$V_{0.1}Mo_{0.4}Nb_{0.2}Ta_{0.2}W_{0.1}$	409.4	160.2	77.8	243.3	96.5	255.8	2.51	0.325	3.189
(29)$V_{0.2}Mo_{0.1}Nb_{0.1}Ta_{0.3}W_{0.3}$	444.7	144.2	61.8	244.3	97.1	257.4	2.51	0.324	3.174
(30)$V_{0.2}Mo_{0.1}Nb_{0.2}Ta_{0.2}W_{0.3}$	439.0	136.8	58.7	237.6	95.	253.1	2.48	0.322	3.177
(31)$V_{0.2}Mo_{0.1}Nb_{0.2}Ta_{0.3}W_{0.2}$	391.6	155.3	23.5	234.1	61.3	169.3	3.81	0.379	3.188
(32)$V_{0.2}Mo_{0.1}Nb_{0.3}Ta_{0.1}W_{0.3}$	406.0	147.2	51.2	233.4	82.4	221.4	2.83	0.342	3.179
(33)$V_{0.2}Mo_{0.1}Nb_{0.2}Ta_{0.2}W_{0.3}$	374.8	175.4	76.4	241.9	85.7	230.1	2.82	0.341	3.165
(34)$V_{0.2}Mo_{0.2}Nb_{0.1}Ta_{0.3}W_{0.2}$	395.2	161.6	57.1	239.5	80.9	218.3	2.96	0.348	3.175
(35)$V_{0.2}Mo_{0.2}Nb_{0.1}Ta_{0.4}W_{0.1}$	375.4	160.9	15.4	231.8	52.3	145.9	4.43	0.395	3.188
(36)$V_{0.2}Mo_{0.2}Nb_{0.2}Ta_{0.1}W_{0.3}$	377.5	169.1	76.8	238.6	87.8	234.6	2.72	0.336	3.167
(37)$V_{0.2}Mo_{0.2}Nb_{0.2}Ta_{0.2}W_{0.2}$	392.7	160.0	57.6	234.0	81.1	218.0	2.89	0.345	3.189
(38)$V_{0.2}Mo_{0.2}Nb_{0.2}Ta_{0.3}W_{0.1}$	402.0	142.4	28.9	228.9	69.2	188.8	3.30	0.363	3.189
(39)$V_{0.2}Mo_{0.2}Nb_{0.3}Ta_{0.1}W_{0.2}$	405.2	145.7	57.4	232.2	86.3	230.5	2.69	0.335	3.177
(40)$V_{0.2}Mo_{0.2}Nb_{0.3}Ta_{0.2}W_{0.1}$	395.1	141.0	24.7	225.7	65.7	179.7	3.43	0.367	3.190
(41)$V_{0.2}Mo_{0.2}Nb_{0.4}Ta_{0.1}W_{0.1}$	394.2	138.2	20.6	223.8	63.4	173.9	3.53	0.370	3.191
(42)$V_{0.2}Mo_{0.3}Nb_{0.1}Ta_{0.1}W_{0.3}$	340.0	196.9	85.6	244.6	79.9	216.3	3.06	0.353	3.153
(43)$V_{0.2}Mo_{0.3}Nb_{0.1}Ta_{0.2}W_{0.2}$	389.4	164.2	77.7	239.3	91.5	243.8	2.61	0.330	3.164
(44)$V_{0.2}Mo_{0.3}Nb_{0.2}Ta_{0.1}W_{0.2}$	380.0	164.3	75.3	236.2	88.3	235.7	2.67	0.334	3.165
(45)$V_{0.2}Mo_{0.3}Nb_{0.2}Ta_{0.2}W_{0.1}$	436.4	128.1	56.1	230.9	95.3	251.4	2.42	0.319	3.177
(46)$V_{0.2}Mo_{0.4}Nb_{0.1}Ta_{0.1}W_{0.2}$	309.4	206.7	84.6	241.0	71.3	194.7	3.38	0.365	3.152
(47)$V_{0.2}Mo_{0.4}Nb_{0.1}Ta_{0.2}W_{0.1}$	369.7	168.1	68.7	235.3	81.5	219.3	2.89	0.345	3.164
(48)$V_{0.2}Mo_{0.4}Nb_{0.2}Ta_{0.1}W_{0.1}$	361.0	172.5	65.9	235.3	77.2	208.8	3.05	0.352	3.165
(49)$V_{0.3}Mo_{0.1}Nb_{0.1}Ta_{0.2}W_{0.3}$	425.1	140.1	60.0	235.1	93.0	246.6	2.53	0.325	3.152
(50)$V_{0.3}Mo_{0.1}Nb_{0.1}Ta_{0.4}W_{0.1}$	315.8	174.6	12.8	221.7	35.9	102.4	6.16	0.423	3.177
(51)$V_{0.3}Mo_{0.1}Nb_{0.2}Ta_{0.1}W_{0.3}$	419.9	136.7	58.6	231.1	91.8	243.3	2.52	0.325	3.153
(52)$V_{0.3}Mo_{0.1}Nb_{0.2}Ta_{0.2}W_{0.2}$	384.7	147.7	15.5	226.7	56.7	157.1	4.00	0.384	3.164
(53)$V_{0.3}Mo_{0.1}Nb_{0.2}Ta_{0.3}W_{0.1}$	343.1	156.7	31.2	218.9	56.0	154.8	3.91	0.382	3.177
(54)$V_{0.3}Mo_{0.1}Nb_{0.3}Ta_{0.2}W_{0.1}$	303.9	172.1	28.8	216.0	43.6	122.6	4.95	0.405	3.178
(55)$V_{0.3}Mo_{0.1}Nb_{0.4}Ta_{0.1}W_{0.1}$	323.5	153.4	11.9	210.1	41.1	115.9	5.10	0.408	3.180
(56)$V_{0.3}Mo_{0.2}Nb_{0.1}Ta_{0.1}W_{0.3}$	307.4	201.1	68.8	236.6	62.5	172.4	3.78	0.379	3.139
(57)$V_{0.3}Mo_{0.2}Nb_{0.1}Ta_{0.2}W_{0.2}$	404.2	143.4	56.0	230.3	85.7	228.8	2.69	0.334	3.155
(58)$V_{0.3}Mo_{0.2}Nb_{0.1}Ta_{0.3}W_{0.1}$	374.1	140.7	21.2	225.5	59.6	164.0	3.66	0.375	3.164
(59)$V_{0.3}Mo_{0.2}Nb_{0.2}Ta_{0.1}W_{0.2}$	394.2	138.9	53.5	224.0	83.1	222.0	2.69	0.335	3.156
(60)$V_{0.3}Mo_{0.2}Nb_{0.2}Ta_{0.2}W_{0.1}$	367.6	145.6	18.4	219.6	55.4	153.4	3.96	0.384	3.169
(61)$V_{0.3}Mo_{0.2}Nb_{0.3}Ta_{0.1}W_{0.1}$	388.6	135.4	22.3	219.8	64.0	175.0	3.43	0.367	3.165
(62)$V_{0.3}Mo_{0.3}Nb_{0.1}Ta_{0.1}W_{0.2}$	337.5	178.0	67.7	231.2	72.5	197.0	3.19	0.358	3.141
(63)$V_{0.3}Mo_{0.3}Nb_{0.1}Ta_{0.2}W_{0.1}$	382.2	146.2	50.2	224.9	77.3	208.2	2.91	0.346	3.151
(64)$V_{0.3}Mo_{0.3}Nb_{0.2}Ta_{0.1}W_{0.1}$	375.9	174.2	47.6	221.1	68.9	188.0	3.50	0.360	3.156
(65)$V_{0.3}Mo_{0.4}Nb_{0.1}Ta_{0.1}W_{0.1}$	310.4	181.4	57.2	224.4	60.1	165.7	3.73	0.377	3.140
(66)$V_{0.4}Mo_{0.1}Nb_{0.1}Ta_{0.1}W_{0.3}$	390.5	146.4	53.3	227.8	80.7	216.7	2.82	0.341	3.127
(67)$V_{0.4}Mo_{0.2}Nb_{0.1}Ta_{0.1}W_{0.2}$	375.5	149.6	47.8	224.9	73.9	199.8	3.04	0.352	3.128
(68)$V_{0.4}Mo_{0.2}Nb_{0.1}Ta_{0.2}W_{0.1}$	382.1	139.9	18.3	220.6	59.4	163.6	3.72	0.376	3.140
(69)$V_{0.4}Mo_{0.2}Nb_{0.2}Ta_{0.1}W_{0.1}$	378.7	136.1	17.7	217.7	59.1	162.6	3.68	0.375	3.141
(70)$V_{0.5}Mo_{0.2}Nb_{0.1}Ta_{0.1}W_{0.1}$	363.3	139.1	13.1	213.8	52.7	146.1	4.06	0.386	3.115

Table 4. The correspondence between the numbers in Table 3 and the X-axis. The first point in the X-axis involved with V is the data of number 1 in Table 3. The first point in the X-axis involved with Ta is the data of number 5 in Table 3.

Table 2	V	Mo	Nb	Ta	W	Table 2	V	Mo	Nb	Ta	W	Table 2	V	Mo	Nb	Ta	W
1	1	1	1	65	51	25	25	63	8	36	37	49	49	10	18	45	67
2	2	2	30	52	52	26	26	64	9	58	7	50	50	11	19	69	16
3	3	3	53	29	53	27	27	65	36	8	38	51	51	12	46	17	68
4	4	4	54	53	29	28	28	66	37	37	8	52	52	13	47	46	46
5	5	5	65	1	54	29	29	6	10	59	61	53	53	14	48	63	17
6	6	18	2	54	55	30	30	7	38	38	62	54	54	15	63	47	18
7	7	19	3	66	30	31	31	8	39	60	39	55	55	16	69	18	19
8	8	20	4	70	1	32	32	9	60	9	63	56	56	39	20	19	69
9	9	21	34	30	56	33	33	30	11	39	64	57	57	40	21	48	47
10	10	22	32	55	31	34	34	31	12	61	40	58	58	41	22	64	20
11	11	23	33	67	2	35	35	32	13	68	9	59	59	42	49	20	48
12	12	24	55	2	57	36	36	33	40	10	65	60	60	43	50	49	21
13	13	25	56	31	32	37	37	34	41	40	41	61	61	44	64	21	22
14	14	26	57	56	3	38	38	35	42	62	10	62	62	59	23	22	49
15	15	27	66	3	33	39	39	36	61	11	42	63	63	60	24	50	23
16	16	28	67	32	4	40	40	37	62	41	11	64	64	61	51	23	24
17	17	29	70	4	5	41	41	38	68	12	12	65	65	70	25	24	25
18	18	49	5	33	58	42	42	55	14	13	66	66	66	17	26	25	70
19	19	50	6	57	34	43	43	56	15	42	43	67	67	45	27	26	50
20	20	51	34	5	59	44	44	57	43	14	44	68	68	46	28	51	26
21	21	52	35	34	35	45	45	58	44	43	13	69	69	47	52	27	27
22	22	53	58	6	36	46	46	67	16	15	45	70	70	48	29	28	28
23	23	54	59	35	6	47	47	68	17	44	14						
24	24	62	7	7	60	48	48	69	45	16	15						

5.1. C_{44}

According to Reference [26], there is a monotonous relation between hardness and C_{44}. In Figure 3, there is a regular distribution of all the points. They are distributed in two areas. The data points have the concentration of W + Mo $\geq$ 0.4 distributed at the top area. It is obvious that the values of C_{44} are bigger than the area below. There is also a data blank area between them. C_{44} increases with the increase of the W + Mo concentration. Thus, W and Mo show significant influence on C_{44}. This may be due to the fact that the C_{44} of Mo (125 Gpa) and W (163 Gpa) are higher than the C_{44} of V (46 Gpa), Nb (31 Gpa), and Ta (82 Gpa). The densities of W and Mo are 19.350 g/cm^3 and 10.390 g/cm^3. In order to decrease the density and keep the high hardness of the V–Mo–Nb–Ta–W system, increasing Mo concentration and decreasing W concentration may be a feasible method.

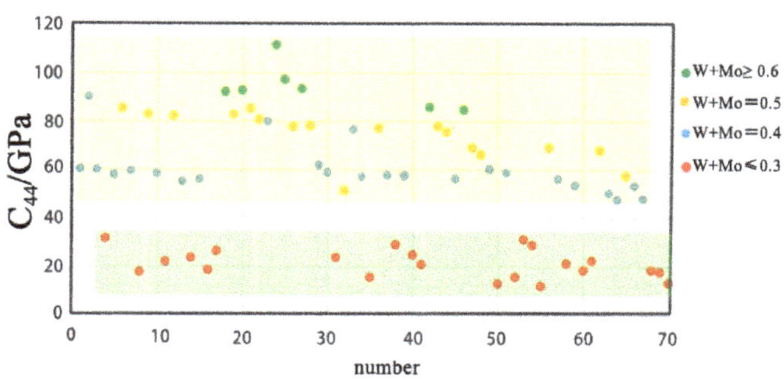

Figure 3. Scatter-plots of C_{44} of all seventy compositions.

5.2. Bulk Modulus

According to Reference [27], bulk modulus B can be used to describe the average atomic bond strength. The overall trend of the influence of alloying elements on B is shown in Figure 4. Figure 4a indicates that with the increase of V concentration, B decreases, while Figure 4b indicates B increases with the increase of Mo concentration. Additionally, Figure 4c indicates B decreases slightly with the increase of Nb concentration. Figure 4d shows that the concentration of Ta has no obvious influence on *B*. *B* increases with the increase of W concentration, as shown in Figure 4e. Furthermore, data points ran periodically with the changes of element concentrations. The trend in each period is the same as the overall trend of each element. For example, in Figure 4c, *B* decreases in the Nb = 0.1 area. This can be attributed to the increase in the concentration of element V. Arrow a in Figure 4e shows *B* decreases with the increase of Nb. Arrow c in Figure 4a shows *B* decreases with the decrease of W. A sharp variation in some points (1, 2, and 3) can be seen in Figure 4. For example, point 1 in Figure 4b shows that the initial concentration of W is 0.1, while the final concentration of W is 0.3. Thus, *B* increases sharply. In summary, Mo and W can help to increase *B*. Elements V and Nb have a negative effect on *B*. *B* has a high value in each period with W + Mo $\geq$ 0.4.

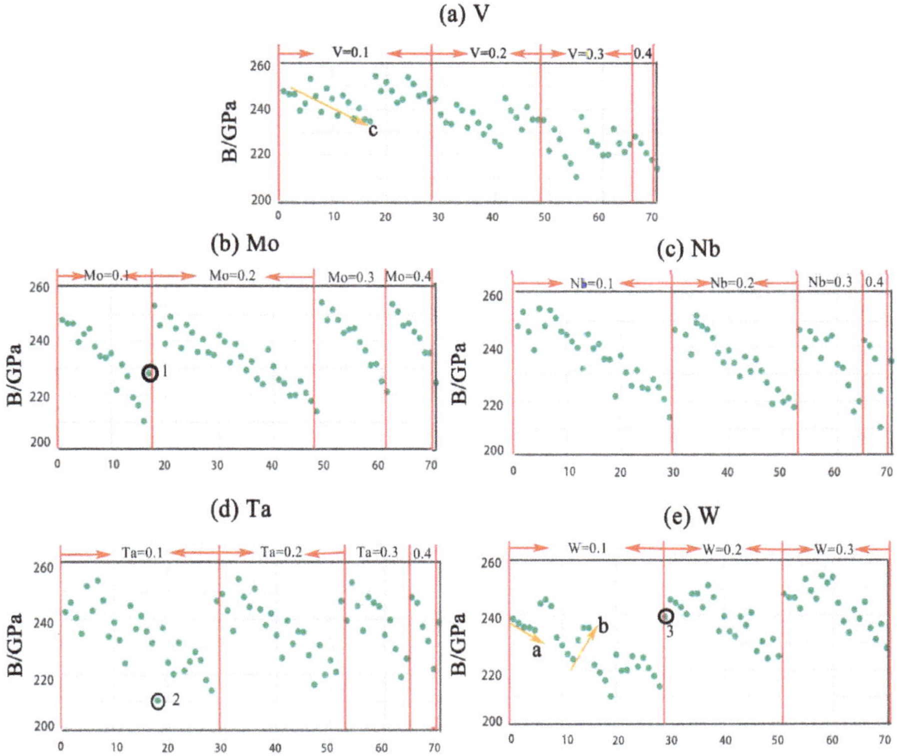

Figure 4. The trend of *B* along with element concentrations: (**a**) V, (**b**) Mo, (**c**) Nb, (**d**) Ta, and (**e**) W.

5.3. *G*, *E*, *B/G*, *v*

The hardness of materials can be related to Young's modulus *E* and the shear modulus *G* [28]. The general trend is that the larger these two moduli are, the harder the material. According to the Pugh criteria [29], materials with $B/G < 2$ are associated with brittleness; otherwise, the materials may behave as ductile. Materials with $v > 0.31$ have good ductility. Otherwise, the materials are considered brittle.

Figure 5 shows *G* and *E* have the same trend, while B/G and *v* also have the same trend. It can also be seen that there is an inverse relationship between them. Element V has a negative effect on *G* and *E*, and a positive effect on B/G and *v*. Thus, the trend of *E*, B/G, and *v* can be predicted from the trend of *G*. Figure 6 shows the trends of Mo, Nb, Ta, and W. It is obvious that W has a positive effect on *G* and *E* and exhibits a negative effect on B/G and *v*, while elements Nb and Ta have no obvious effect.

In summary, element V can help to increase the ductility of the V–Mo–Nb–Ta–W system. *G* and *E* have a relatively high value with W + Mo ≥ 0.4.

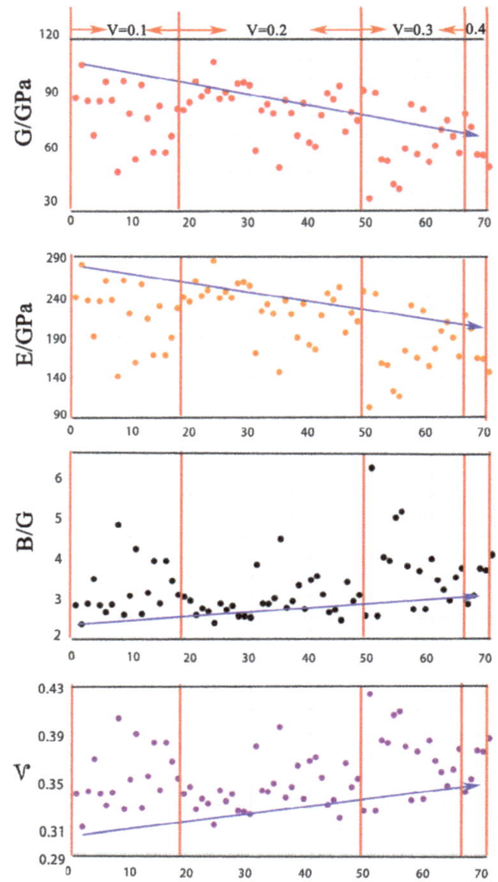

Figure 5. The trends of *G*, *E*, *B*/*G*, and *ν* along with V concentrations.

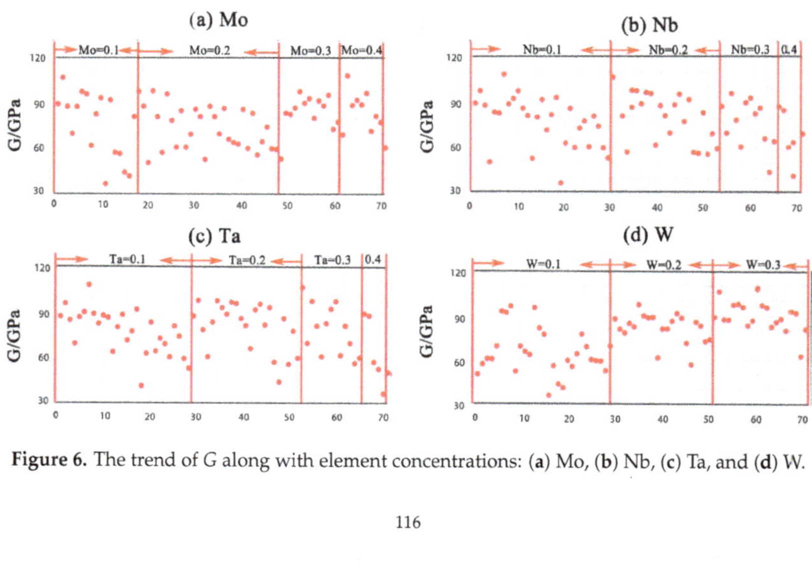

Figure 6. The trend of *G* along with element concentrations: (**a**) Mo, (**b**) Nb, (**c**) Ta, and (**d**) W.

6. Conclusions

In order to improve the ductility and decrease the density of the V–Mo–Nb–Ta–W system, the elastic properties of seventy different compositions were studied. This work concludes as follows:

1. Mo and W are key elements in the V–Mo–Nb–Ta–W system. The V–Mo–Nb–Ta–W system has relatively high values of C_{44}, B, E, and G, with high concentrations of Mo + W. The concentration of Mo + W shows the most prominent effect on C_{44}. Due to the high density of W, element Mo can be used to substitute part of W. In this case, the concentration of W should be reduced.
2. V has low density (6.110 g/cm^3) and plays an important in decreasing the brittleness of the V–Mo–Nb–Ta–W system.
3. In comparison, Nb and Ta have relatively small influence on elastic properties.

Author Contributions: All authors contributed extensively to this study. S.Z. did the calculation work and wrote the paper; S.Z. and S.W. conceived the idea and analyzed the data. All authors have read and approved the final manuscript.

Funding: This work was supported by the National Natural Science Foundation of China (No.51471164) and the National Key R&D Program of China (No. 2016YFB0701302).

Acknowledgments: The authors thank om the Informalization Construction Project of Chinese Academy of Sciences for their computational support during the 11th Five-Year Plan Period (No.INFO-115-B01). The Special Program for Applied Research on Super Computation of the NSFC-Guangdong Joint Fund (the second phase) is also highly acknowledged. Some of the calculations in this study were done on a Tianhe-II high performance computer system in the National Supercomputer Center in Guangzhou, China.

Conflicts of Interest: The authors declare no conflict of interest.

References

1. Yeh, J.W.; Chen, S.K.; Lin, S.J.; Gan, J.Y.; Chin, T.S.; Shun, T.T.; Tsau, C.H.; Chang, S.Y. Nanostructured High-Entropy Alloys with Multiple Principal Elements; Noval Alloy Design Concepts and Outcomes. *Adv. Eng. Mater.* **2004**, *6*, 299–303. [CrossRef]
2. Gludovatz, B.; Hohenwarter, A.; Catoor, D.; Chang, E.H.; George, E.P.; Ritchie, R.O. A Fracture-resistant High-entrop Alloy for Cryogenic Applications. *Science* **2014**, *345*, 1153–1158. [CrossRef] [PubMed]
3. Miracle, D.B.; Senkov, O.N. A Critical Review of High Entropy Alloys and Related. *Acta. Mater.* **2017**, *122*, 448–511. [CrossRef]
4. Senkov, O.N.; Woodward, C.; Miracle, D.B. Microstructure and Properties of Aluminum-Containing Refractory High-Entropy Alloys. *JOM* **2014**, *66*, 2030–2042. [CrossRef]
5. Senkov, O.N.; Wilks, G.B.; Scott, J.M.; Miracle, D.B. Mechanical Properties of $Nb_{25}Mo_{25}Ta_{25}W_{25}$ and $V_{20}Nb_{20}Mo_{20}Ta_{20}W_{20}$ Refractory High Entropy Elloys. *Intermetallics* **2011**, *19*, 698–706. [CrossRef]
6. Han, Z.D.; Luan, H.W.; Liu, X.; Chen, N.; Li, X.Y.; Shao, Y.; Yao, K.F. Microstructures and Mechanical Properties of TixNbMoTaW Refractory High Entropy Alloys. *Mater. Sci. Eng. A* **2018**, *712*, 380–385. [CrossRef]
7. Yuan, L.; Yan, Z.; Heng, Z.; Naijuan, W.; Xiang, C.; Huawei, Z.; Yanxiang, L. Microstructure and mechanical properties of refractory HfMo0.5NbTiV0.5Six high-entropy composites. *J. Alloy Compd.* **2017**, *694*, 869–876.
8. Senkov, O.N.; Isheim, D.; Seidman, D.N.; Pilchak, A.L. Development of a Refractory High Entropy Superalloy. *Entropy* **2016**, *18*, 102. [CrossRef]
9. Senkov, O.N.; Şenkova, S.V.; Woodward, C. Effect of Aluminum on The Microstructure and Properties of Two Refractory High-Entropy Alloys. *Acta Mater.* **2014**, *68*, 214–228.
10. Mina, Z.; Xianglin, Z.; Jinghao, L. Microstructure and Mechanical Properties of a Refractory CoCrMoNbTi High-Entropy Alloy. *J. Mater. Eng. Perform.* **2017**, *26*, 3657–3665.
11. Shao-Ping, W.; Jian, X. (TiZrNbTa)-Mo High-entropy Alloys: Dependence of Microstructure and Mechanical Properties on Mo Concentration and Modeling of Solid Solution Strengthening. *Intermetallic* **2018**, *95*, 59–72.
12. Zhang, Y.; Yang, X.; Liaw, P.K. Alloy Design and Properties Optimization of High-Entropy Alloys. *JOM* **2012**, *64*, 830–838. [CrossRef]
13. Liam, X.Z. Metal Mixology. *Nature* **2016**, *533*, 306–307.

14. Zaddach, A.J.; Niu, C.; Koch, C.C.; Irving, D.L. Mechanical Properties and Stacking Fault Energies of NiFeCrCoMn High-Entropy Alloy. *JOM* **2013**, *65*, 1780–1789. [CrossRef]
15. Tian, F.Y.; Varga, L.K.; Chen, N.; Delczeg, L.; Vitos, L. Ab Initio Investigation of High-entropy Alloys of 3d Elements. *Phys. Rev. B* **2013**, *87*, 075144. [CrossRef]
16. VandeVondele, J.; Krack, M.; Mohamed, F.; Parrinello, M.; Chassaing, T.; Hutter, J. QUICKSTEP; Fast and Accurate Density Functional Calculations Using a Mixed Gaussian and Plane Waves Approach. *Comput. Phys. Commun.* **2005**, *167*, 103. [CrossRef]
17. Wang, S.Q. Atomic Structure Modeling of Multi-Principal-Element Alloys by the Principle of Maximum Entropy. *Entropy* **2013**, *15*, 5536–5548. [CrossRef]
18. Zheng, S.M.; Feng, W.Q.; Wang, S.Q. Elastic Properties of High Entropy Alloys by MaxEnt Approach. *Comp. Mater. Sci.* **2018**, *142*, 332–337. [CrossRef]
19. Zheng, S.M.; Wang, S.Q. Elastic Properties of Face-centered Cubic, Body-centered Cubic and Hexagonal High Entropy Alloys by MaxEnt Approach. *Mater. Res. Express.* **2018**, *5*, 076503. [CrossRef]
20. Dirras, G.; Lilensten, L.; Djemia, P.; Laurent-Brocq, M.; Tingaud, D.; Couzinié, J.P.; Perrière, L.; Chauveau, T.; Guillot, I. Elastic and Plastic Properties of As-cast Equimolar TiHfZrTaNb High-entropy Alloy. *Mater. Sci. Eng. A* **2016**, *654*, 30–38. [CrossRef]
21. Laplanche, G.; Gadaud, P.; Horst, O.; Otto, F.; Eggeler, G.; George, E.P. Temperature Dependencies of The Elastic Moduli and Thermal Expansion Coefficient of An Equiatomic, Single-phase CoCrFeMnNi High-entropy Alloy. *J. Alloy Compd.* **2015**, *623*, 348–353. [CrossRef]
22. Wang, S.Q.; Ye, H.Q. First-principles Study on Elastic Properties and Phase Stability of III–V Compounds. *J. Phys. Condensed Matter* **2003**, *240*, 45–54. [CrossRef]
23. Meyers, M.A.; Chawla, K.K. *Mechanical Behavior of Materials*, 2nd ed.; Cambridge University Press: Cambridge, UK, 2008; pp. 112–114.
24. Vitos, L. *Computational Quantum Mechanics for Materials Engineers: The EMTO method and applications*; Springer: Berlin, Germany, 2007; p. 129.
25. Soderlind, P. Theory of Elastic Constants of Cubic Transition Metals and Alloys. *Phys. Rev. B* **1993**, *48*, 5844–5851. [CrossRef]
26. Jhi, S.H.; Ihm, J.; Louie, S.G.; Cohen, M.L. Electronic Mechanism of Hardness Enhancement in Transition-metal Carbonitrides. *Nature* **1999**, *399*, 132–134. [CrossRef]
27. Clerc, D.G.; Ledbetter, H.M. Mechanical Hardness; A Semiempirical Theory Based on Screened Eelctroatatics and Elasticshear. *J. Phys. Chem. Solids* **1998**, *59*, 1071–1095. [CrossRef]
28. Chen, K.Y.; Zhao, L.R.; Rodgers, J.; Tse, J.S. Alloying Effects on Elastic Properties of TiN-based Nitrides. *J. Phys. D Appl. Phys.* **2003**, *36*, 2725–2729. [CrossRef]
29. Pugh, S.F. XCII. Relations Between the elastic moduli and the plastic properties of polycrystalline pure metals. *Philos. Mag.* **1954**, *45*, 823–843. [CrossRef]

entropy

Review

Additive Manufacturing of High-Entropy Alloys: A Review

Shuying Chen *, Yang Tong and Peter K. Liaw *

Department of Materials Science and Engineering, The University of Tennessee, Knoxville, TN 37909, USA; ytong1@vols.utk.edu

* Correspondence: schen38@vols.utk.edu (S.C.); pliaw@utk.edu (P.K.L.); Tel.: +1-865-974-6356 (P.K.L.)

Received: 15 November 2018; Accepted: 3 December 2018; Published: 5 December 2018

Abstract: Owing to the reduced defects, low cost, and high efficiency, the additive manufacturing (AM) technique has attracted increasingly attention and has been applied in high-entropy alloys (HEAs) in recent years. It was found that AM-processed HEAs possess an optimized microstructure and improved mechanical properties. However, no report has been proposed to review the application of the AM method in preparing bulk HEAs. Hence, it is necessary to introduce AM-processed HEAs in terms of applications, microstructures, mechanical properties, and challenges to provide readers with fundamental understanding. Specifically, we reviewed (1) the application of AM methods in the fabrication of HEAs and (2) the post-heat treatment effect on the microstructural evolution and mechanical properties. Compared with the casting counterparts, AM-HEAs were found to have a superior yield strength and ductility as a consequence of the fine microstructure formed during the rapid solidification in the fabrication process. The post-treatment, such as high isostatic pressing (HIP), can further enhance their properties by removing the existing fabrication defects and residual stress in the AM-HEAs. Furthermore, the mechanical properties can be tuned by either reducing the pre-heating temperature to hinder the phase partitioning or modifying the composition of the HEA to stabilize the solid-solution phase or ductile intermetallic phase in AM materials. Moreover, the processing parameters, fabrication orientation, and scanning method can be optimized to further improve the mechanical performance of the as-built-HEAs.

Keywords: high-entropy alloys; additive manufacturing; microstructure; mechanical properties

---

1. Introduction

High-entropy alloys (HEAs) and multi-principal-element (MPE) alloys were proposed by Yeh [1] and Cantor [2] in the 2000s, respectively, attracting increasing interest all over the world. It was originally defined as an alloy composed of five or more multi-principal elements, with equi- or near equi-atomic percentages [1]. However, recently, researchers extended the concept of HEAs, which now include alloys with three or four principal elements as well [3]. Usually, they have a single crystal structure, such as a body-centered-cubic (BCC) [4–6], face-centered cubic (FCC) [7–11], or hexagonal-closed packed (HCP) structure [12–15]. HEAs present superior properties, such as a combination of high yield strength and ductility [16], good microstructural stability and retained mechanical strength at elevated temperatures [17–23], strong resistance to wear [24,25], fatigue [26–31], corrosion, and oxidation [32–36].

Most previous studies have focused on cast materials and their microstructure tuning through different post-processing methods, such as cold rolling, forging, or annealing treatment [37]. However, casting defects, such as shrinkages and pores, exist in as-cast materials, thereby requiring further processing to remove these defects. Compared with the conventional up–down fabrication method, additive manufacturing (AM), a flexible processing technique, has been applied to the fabrication of

HEAs to produce materials with a complex geometry. AM, also known as three-dimensional (3D) printing, enables the fabrication of 3D objects based upon computer-aided design (CAD) models, which have been accepted as a transformative technology across multiple industries [38,39]. The AM technique has been becoming increasingly important in the materials science field and has been broadly applied in the industry for manufacturing products with complex shapes. Currently, several engineering materials can be produced by AM, such as aircraft components made of Ti alloys [40,41], Al alloys [42], stainless steel [43], and Polyamide 12 [44], which can significantly increase production efficiency and decrease the production cost due to the combined advantage of the net-shaping capability and design freedom [45]. Especially, several methods based on the AM concept were developed and commonly used in manufacturing products. They are classified as the laser metal deposition (LMD), selective laser melting (SLM, also called laser beam melting, LBM), laser metal fusion (LMF), direct metal laser sintering (DMLS) [46], or selective electron beam melting (SEBM) [39]. Direct metal deposition (DMD) and direct laser fabrication (DLF) are two representative LMD processes.

LMD is characterized by the part being cladded layer by layer [47], as shown in Figure 1. The powders are melted after being carried by the inert gas into a laser beam and are then fed onto the workpiece to fuse with the thin layer deposited previously [47]. LMD can produce the 3D product with the ultrafine microstructure and highly complex geometry based on the layer-by-layer incremental shaping and consolidation of the feedstock to a wide range of configurations. The composition of the feeding powder could be modified by in-situ alloying during LMD, e.g., alloying varied amounts of Al to a CoCrFeMnNi powder blend, which will lead to even higher flexibility and throughput production [48,49].

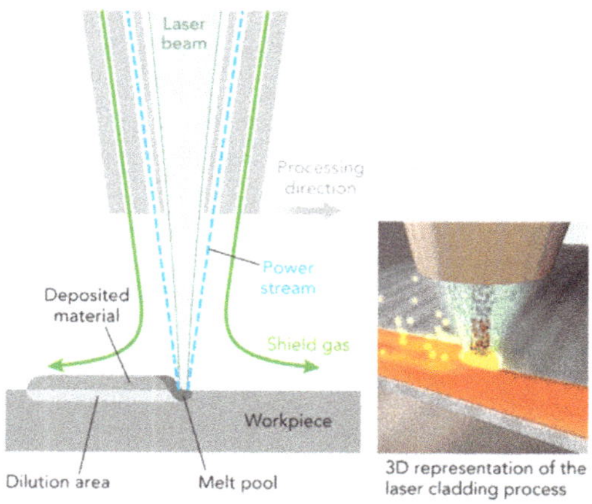

Figure 1. Schematic image of LMD. Image displayed with the permission from the authors in [50].

SEBM is a melting fabrication process by a high-power electron beam in vacuum. Metal powders are normally fed from a hopper and then distributed by a rake across a build plate. A powder layer, which is uniformly supplied on the base plate, is pre-heated by the electron beam raster scanning, as shown in Figure 2. The points built in the slice data obtained from 3D-CAD file are sequentially melted by the focused electron beam. The process of powder feeding, pre-heating, and melting will be repeated until the bulk metallic parts is completed [51]. SEBM has attracted increasingly attention in the past decade due to its unique advantages of the high energy density of the high scan speed, incident electron beam, and reduced operation cost, which make it a suitable method to produce materials used the harsh environment, such as titanium alloys, aluminum alloys, cobalt chromium alloys, and nickel-based alloys [52].

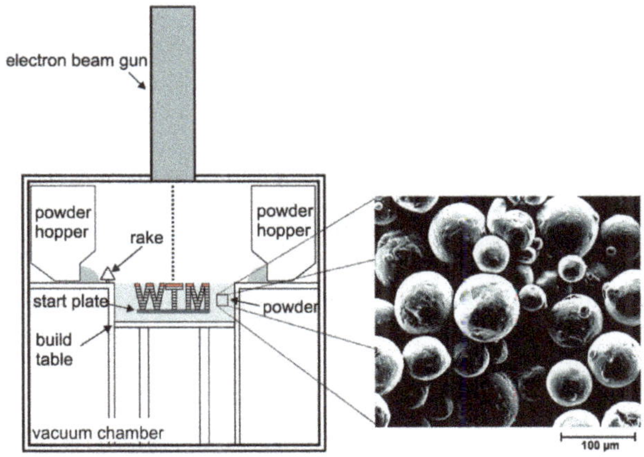

Figure 2. Schematic image of the SEBM process. Image displayed with permission from the authors in [53].

SLM is an AM technology applying a high-energy laser beam, by which the part is fabricated in a layer-by-layer mode through the selective melting and consolidation of the metal powder [54], as shown in Figure 3. The layer thicknesses vary in a range of 20 and 100 μm. Compared with the traditional casting and forging method, SLM attracts increasing attention due to its impressive features, such as the ability to net-shape manufacture without the dies and high geometry complexity [55].

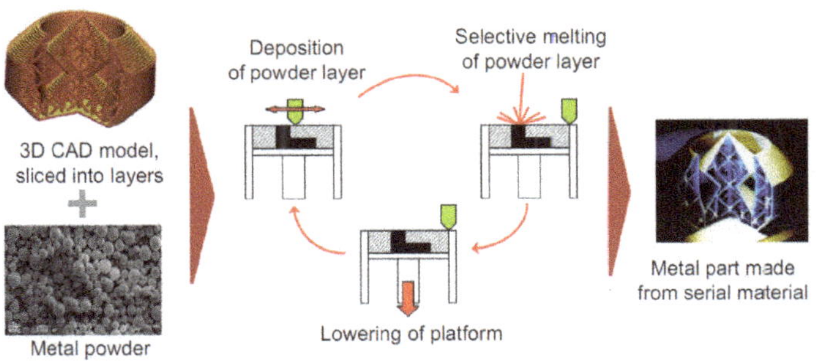

Figure 3. Schematic image of the SLM process. Image displayed with permission from the authors in [56].

AM has gained increasing attention ascribed to its ability of producing parts with complex shapes. The laser-melting method has been used to HEAs to obtain coatings and bulk materials [57,58]. Brif et al. [55] studied the FeCoCrNi HEA alloy by the powder-bed technology, suggesting that the alloy remained in its single-phase solid-solution state and displayed excellent strength and ductility. Similarly, the phase evolution and mechanical properties of AlCoCrFeNi [52] and CoCrFeMnNi [59] HEAs were reported. In the current study, we review the application of AM in HEAs in terms of their microstructures, mechanical properties, and deformation mechanisms and compare them with conventional cast-HEAs.

2. Microstructure Evolution during the AM Process

2.1. CoCrFeNi HEAs

Most of the previous studies are the HEA coatings [60,61]. Here, we highlight the performance of AM applications on bulk HEAs. The CoCrFeNi alloy system has been studied most, and more compositions or systems have been developed based on this alloy. For instance, $Al_xCoCrFeNi$ and CoCrFeNiMn HEAs have been commonly investigated in recent years. Brif et al. [55] studied FeCoCrNi HEAs fabricated by SLM, followed by annealing treatment. X-ray diffraction (XRD) results showed a single BCC solid solution with a uniform chemical composition. No segregation could be found. Later, Karthik et al. [62] proposed a new concept of the metal–metal composite, which consists of an aluminum–magnesium alloy, the AA5083 matrix, and the nanocrystalline CoCrFeNi HEA reinforcement precipitates in 12 vol%. Scanning electron microscope (SEM) images presented a very uniform distribution of HEA particles across the layers in a multi-layer composite. No intermetallic compounds, severe deformation, or accumulation of the HEA particles could be found at the interfaces, which was further confirmed by transmission electron microscope (TEM) examination. Furthermore, the TEM results presented a dynamic recrystallized aluminum matrix with fine equiaxed grains, which illustrated various dislocation densities, and some of the grains exhibited cells and subgrains in their formative processes.

In order to improve their mechanical performance, Zhou et al. [63] obtained the C-containing FeCoCrNi HEAs fabricated by SLM with varying processing parameters. A single-FCC solid solution without the carbide phase was detected, and a uniform distribution of carbon in the matrix was found as well. Equiaxed grains existed mostly in the middle and bottom parts of the sample, and most grains in the top part were columnar. The geometry of grains in the SLM samples was found to be related to the cooling rate and thermal gradient. The electron-backscattered diffraction (EBSD) results revealed that the grain size and microstructure of the SLM specimens were highly dependent upon the laser power and scanning speed. For example, the specimen with a low power presented a larger number of equiaxed grains, and grains became irregular by increasing the scanning speed.

2.2. $Al_xCoCrFeNi$ HEAs

Based on the CoCrFeNi composition, more AM investigations were performed on both composites and simple HEAs. For instance, with the addition of Al to CorCrFeNi, various microstructures were obtained by different processes. A single FCC phase in DLF $Al_{0.3}CoCrFeNi$ alloys was observed by Joseph et al. [64]. A large grain structure was found to be parallel and transverse to the build direction, with a strong texture of <001>, which could be due to the extensional growth of the material along the orientation of the deposition caused by the quick cooling rate and large thermal gradient in the molten pool. It should be noted that a small amount of very fine Ni- and Al-rich particles could be detected at grain boundaries.

By increasing the content of Al to CoCrFeNi, Fujieda et al. [51] reported the BCC and FCC structures in AlCoCrFeNi after SEBM. Further EBSD results showed that the BCC crystal grew along the direction of <100>, which is a preferred orientation for crystal growth in the BCC alloy, which is along the build direction, coinciding with the heat-flux direction. Similarly, Shiratori et al. [65] prepared the AlCoCrFeNi by SEBM as well. More complex solid solutions of B2/BCC and FCC phases were obtained, which is different from the SEBM-specimen with BCC and FCC phases and the casting specimen with B2/BCC phases. Moreover, the confirmed Al–Ni-rich B2/Cr–Fe-rich BCC phases were found to be oriented along the build direction. The detailed elemental distribution in each phase is shown in Figure 4. An AlCoCrFeNi HEA with BCC, B2, and FCC phases along the grain boundaries was fabricated by Kuwabara et al. [52] by applying the SEBM method. Similarly, the SEBM samples presented fine and columnar grains along the building direction with a texture of <100>.

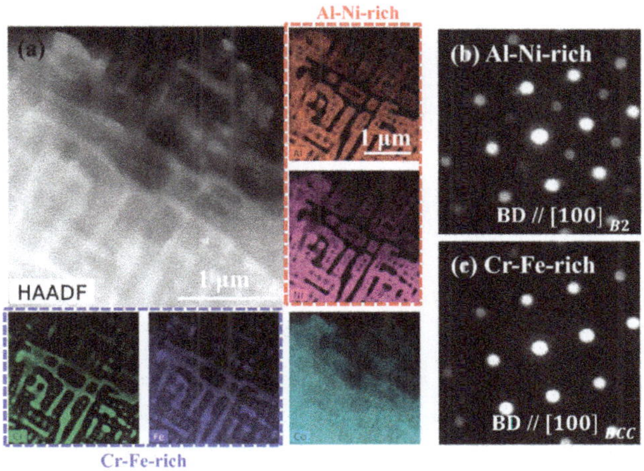

Figure 4. (**a**) The high-angle annular dark-field scanning transmission electron microscopy (HAADF–STEM) image and elemental maps obtained from the energy-dispersive X-ray spectroscopy (EDX). (**b**,**c**) Nano-beam diffraction (NBD) patterns obtained from the Al–Ni-rich and Cr–Fe-rich regions, respectively. Image displayed with permission from the authors in [65].

Li et al. [49] further investigated the properties of the ultrafine nanocrystals (UNs)-modified FeCoCrAlCu high-entropy alloy composites (HEACs) fabricated by the LMD, which presented the fine microstructure without the micro-crack with many $AlCu_2Zr$ UNs attached to the HEAC matrix. The lattice distortion was observed as well, due to the ultrafine microstructure and high diffusion of UNs destroying the atomic-equilibrium state, which increased the potential/free energy, thus leading to the formation of the lattice distortion. On the one hand, compression stress resulting from the alloying elements of the small atomic radius was generated. On the other hand, tensile stress resulting from the alloying elements of the large atomic radius was also achieved. The reciprocal interaction of these two kinds of stress fields led to decreasing stress and the formation of a relatively stable atomic group, favoring the formation of UNs.

Niu et al. [66] investigated the phase evolution with varying the volumetric-energy density (VED), as shown in Figure 5. During the SLM process, the phase is mainly composed of A2 and B2 phases. Specifically, the B2 phase was mostly found to be distributed on the boundary of the molten pool, indicating that the B2 phase was the original structure due to the strongest combination between Al and Ni elements. Thus, the other elements were homogeneously dispersed around the Al–Ni B2 phase and formed the A2 phase. As the VED increased, the B2 phase increased, while the A2 phase tended to decrease, which was mainly due to the larger VED, inducing a faster cooling rate and thus leading to more B2 phases [66].

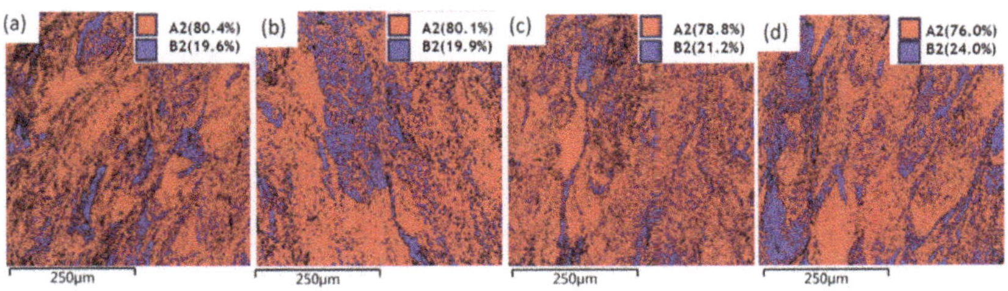

Figure 5. Phase dispersion of different VEDs in SLM samples: (**a**) 68.4 J/mm^3; (**b**) 83.3 J/mm^3; (**c**) 97.2 J/mm^3; and (**d**) 111.1 J/mm^3. Image displayed with permission from the authors in [66].

Sistla et al. [67] studied the Al/Ni ratio effect on the microstructure in $Al_xFeCoCrNi_{2-x}$ (x = 0.3 and 1) HEAs fabricated by the LMD method. The XRD and SEM results suggest that the studied solid solution transforms from BCC to FCC structures, which is consistent with the casting HEAs [67]. Joseph et al. [68] firstly proposed the HIP effect on the microstructure and mechanical behaviors in DLF $Al_xCoCrFeNi$ HEAs. The FCC, duplex FCC + BCC, and BCC solid solution were found in $Al_{0.3}CoCrFeNi$, $Al_{0.6}CoCrFeNi$, and $Al_{0.85}CoCrFeNi$ HEAs, respectively, which was similar to the casting specimens in previous reports [69]. It turned out that the HIP process removed all second-phase grain-boundary phases and segregation of the $Al_{0.3}CoCrFeNi$ alloy, indicating that the isothermal holding at 1100 °C for 2 h during HIP leads to the chemical homogenization of the material and the effectively dissolving of the grain-boundary phases, even though the HIP may induce microstructural coarsening.

2.3. CoCrFeMnNi HEAs

With the development of the AM research on HEAs, Cantor's alloys have been employed in 3D printing as well. Haase et al. [48] investigated the 3D printing of the elemental powder blend in the CoCrFeMnNi system, producing elongated features with cellular dendrite structures. In fact, the fast cooling rates during the fabrication process could contribute to the formation of a fine cellular dendrite structure. The EDX-elemental mapping suggested a more homogeneous distribution of five principal elements than that in the casting specimens. Li et al. [70] firstly studied the processability of the non-equilibrium microstructure in the SLM CoCrFeMnNi. By increasing the VED, the microstructure became much denser. They showed that the HIP process removed the microcracks, and most micropores were closed, leading to an increased density, large grains, and a more homogeneous elemental distribution. The non-equilibrium processing of SLM resulted in a greater residual-stress difference than that in the HIP specimens. The TEM examination revealed the existence of an original FCC phase and a tetragonal-precipitation phase, which were due to the ultrafine grains and a large amount of dislocations induced in the SLM process. Moreover, nanotwins were found during the SLM process without plastic deformation, which may have been due to the low stacking fault energy caused by rapid solidification.

Guo et al. [71] firstly proposed the post-machining on the SLM CoCrFeMnNi HEA specimens to investigate the machinability. SEM illustrated that the mechanical polishing led to a uniform elemental distribution and smooth surface without clear waviness even though some microcracks and very small pores were present. Piglione et al. [59] proposed the printability of a CoCrFeMnNi HEA with single-layer and multi-layer builds fabricated by the laser-powder-bed fusion. A homogeneous distribution of composition was found in the bulk of the HEA with a single FCC crystal structure and a high degree of consolidation, without the apparent elemental segregation. In the single layer, the high cooling rate resulted in much finer cells, compared with previous reports [72] with the same composition. The high hardness values may be due to the restricted moving dislocation by cell boundaries during plastic deformation. Moreover, the cells were found to be aligned along the <001> orientation because the preferred growth direction is aligned with maximum heat flux. Similar cells with a reduced number could be found in multi-layers. The alternating sequence of columnar grains dominated by two crystallographic orientations was reported, which was caused by the coupling of the extensional growth and grain selection after remelting.

In order to explore the AM materials application in a harsh environment, Qiu et al. [57] firstly reported the deformation mechanism at cryogenic temperature in the CrMnFeCoNi HEA fabricated by the LAM process. A single FCC crystal structure with a typical dendrite structure and a lattice constant of 3.598 Å were observed. The growth direction of the dendrite was found to be perpendicular to the laser-scanning direction due to the quick directional solidification. Zhu et al. [45] investigated the hierarchical microstructure in the CoCrFeNiMn HEA by SLM. The as-built specimens were obtained by controlling the processing parameters of the laser energy density, laser power, and scanning speed. The SEM images revealed that the columnar grains formed along the building direction (BD), with a <001> texture, implying epitaxial growth, which indicates that the formation of the cellular structure was correlated solidification conditions. A sub-micron cellular structure was detected under TEM, which was displayed with a large amount of dislocations and with clean interiors. No obvious segregation could be found from EDS, which was due to a fast cooling rate kinetically suppressing the segregation. After a heat treatment (HT), the cellular structure vanished with a much lower dislocation density left, suggesting their thermodynamically metastable characteristic [45]. Later, Fujieda et al. [73] investigated the SEBM method to prepare the CoCrFeNiTi-based HEA. The XRD results presented Ni_3Ti intermetallic compounds with a uniform distribution in the matrix, different from the segregation in casting specimens, as shown in Figure 6. The needle-like NiTi precipitates with a basket-weave morphology contributed to the high yield strength in this SEBM specimen.

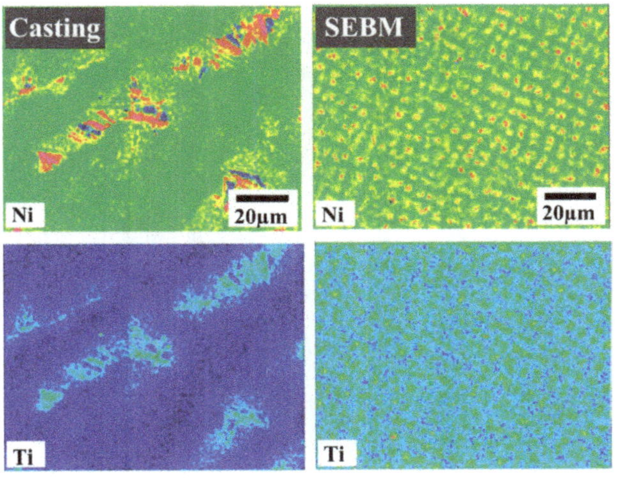

Figure 6. EPMA elemental maps of Ni and Ti in the as-cast sample and SEBM sample. Image displayed with the permission from the authors in [73].

2.4. $Ti_{25}Zr_{50}Nb_0Ta_{25}$ HEAs

The $Ti_{25}Zr_{50}Nb_0Ta_{25}$ refractory HEA fabricated by laser-melt deposition has been recently reported. Phase separation was found in the specimen, as shown in Figure 7a. The segregation near the grain boundary was found to be enriched in Ta and depleted in other elements, which was confirmed as a BCC crystal structure, similar to the matrix but with different compositions, shown in Figure 7b. Moreover, it was suggested that the grain size during the fabrication process was more likely influenced by the chemical partitioning of Ta in Zr-rich areas rather than the cooling rate [74]. The other MoNbTaW refractory alloy was prepared by the DMD method, but an additional remelting procedure between the powder deposition was required due to the different melting points of elements. The resultant compositional homogeneity of alloys was improved by preheating, laser parameters, geometries, and heat conduction [75].

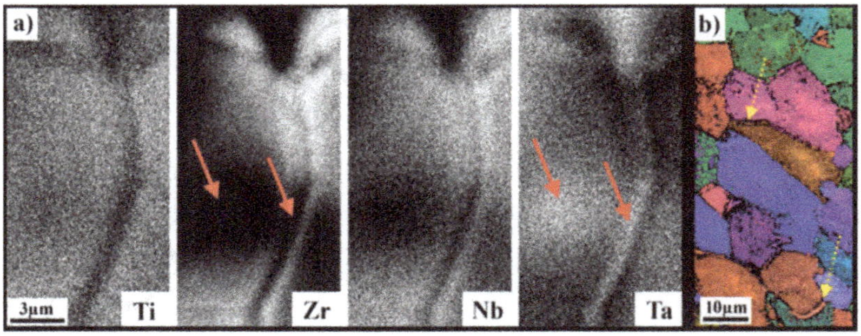

Figure 7. (**a**) Ta-enriched in the grain boundary and Ta-rich dendritic core inside the grain (illustrated by red arrows). Nb, Zr, and Ti concentrations showing an opposite trend (illustrated by red arrows). (**b**) EBSD presenting the grain-orientation map. Image displayed with permission from the authors in [74].

3. Mechanical Properties of AM-Processed HEAs

3.1. CoCrFeNi HEAs

Brif et al. [55] studied the mechanical performance of FeCoCrNi prepared by SLM and compared it with a more traditional processing route of arc melting, as shown in Figure 8. It was apparent that the AM specimen presented a much higher yield strength than that in the arc-melt specimen, while still retaining a significant portion of ductility. The enhanced mechanical properties of AM specimens were ascribed to a fine microstructure, due to the large temperature gradients and rapid solidification in SLM.

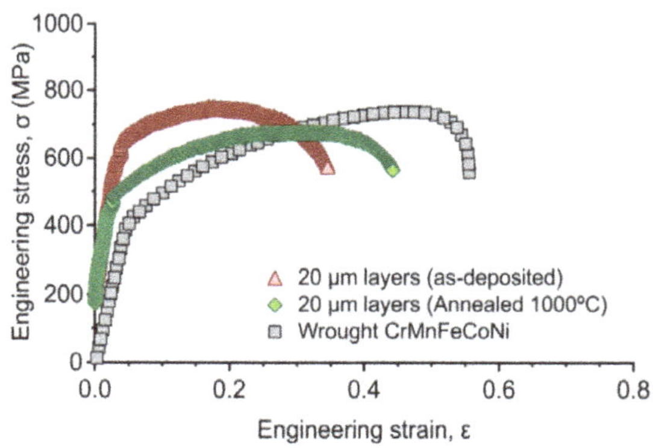

Figure 8. Representative tensile stress–strain curves of AM specimens with a 20-μm-layer thickness. Image displayed with permission from the authors in [55].

Karthik et al. [62] investigated the mechanical properties of the aluminum-magnesium alloy, an AA5083 matrix reinforced with 12 vol.% of CoCrFeNi HEA nanoparticles, i.e., the HEAp/5083 composite, fabricated by friction deposition. The composite demonstrated a high strength, resulting from the reinforcement precipitates of nanocrystalline CoCrFeNi, with a uniform distribution in the ultra-fine-grained aluminum matrix. It was suggested that the strengthening mechanism in the single-layer/HEAp/5083 composite could be the existing load transfer from the Al matrix to the reinforcement precipitation. The fracture morphology also revealed fine HEA reinforcement particles inside many of the dimples. The encouraging mechanical properties in the HEAp/5083 composite were due to the deficiency of brittle intermetallics at the particle/matrix and layer interfaces. The influence of SLM-processing parameters on mechanical behavior was studied in other C-containing FeCoCrNi alloys [63]. It was found that the yield strength was not significantly affected by the scanning speed and laser power. By applying the highest scanning speed of 1200 mm/s and a laser power of 400 W, the highest yield strength with a comparable ductility was obtained. It was apparent that the addition of carbon resulted in enhanced strength for FeCoCrNi, due to solid-solution strengthening. Work hardening was suggested to be caused by the interaction of dislocation–dislocation and dislocation–cellular walls since there was no deformation twinning.

3.2. Al_xCoCrFeNi HEAs

The mechanical properties of the addition of Al to CoCrFeNi alloys were extensively investigated in the AM HEAs. Joseph et al. [64] studied the asymmetry of tension/compression in the $Al_{0.3}$CoCrFeNi alloy with a strong preferred orientation, fabricated by the DLF method. A significant difference in work hardening after yielding could be found in compression and tension curves, even though they had similar yield strengths. The limited work hardening in the tensile experiment was caused by cracks propagating along the grain boundaries, which may have been due to the precipitates enriched in nickel and Al, developing along the grain boundaries, which was observed in the laser-fabricated Rene88DT superalloy as well. Interestingly, deformation twins could be found in compression specimens, rather than tension specimens, because the stress required for the deformation twin could not be reached in the tensile sample before the final fracture. It was suggested that the loading axis in the tensile and compressive specimens were parallel to the depositing direction, which had a significant regulation with the texture of <001>, and was well arranged for deformation twins in the compressive experiment, but failed to orient for twinning in the tension experiment, due to the fact that only in certain directions can the polar deformation twinning accommodate the shape change. Thus, a high working-hardening rate was found in the preferred direction that favors twinning in the compression sample, revealing that the strong texture in the primary material integrated with the activation of twinning contributed to the difference in work-hardening behavior in the tensile and compressive experiments.

With the increase in Al content, the mechanical properties in the AlCoCrFeNi alloy was investigated by the SEBM method [65]. The compressive stress–strain curve presented much better plasticity in SEBM specimens than that in the cast specimens [51,52], due to their finer grains obtained at a higher cooling rate. The cooling rate in the solidification process during the SEBM varied in the range of 10^{-3}–10^{-5}/s, which was much greater than the conventional casting case even with a water-cooled copper mold. This trend could be a reason for the improved deformability in SEBM alloys. Moreover, the generation of the FCC structure could contribute to the good ductility in SEBM materials. It was concluded that the cast specimen of AlCoCrFeNi almost solely consisted of the B2 and BCC phase mixture, which lacks the slip system and leads to a brittle fracture. However, in SEBM specimens, a substantial fraction of the ductile FCC phase was found, which will favor the multi-slip system and thus enhance ductility during the compression experiment, as shown in Figure 9.

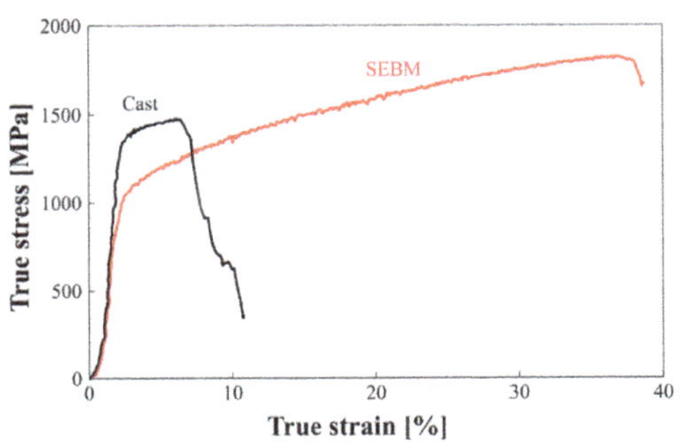

Figure 9. True stress–strain curves in the cast and SEBM samples. Image displayed with permission from the authors in [65].

Similar work and results were obtained in AlCoCrFeNi alloys [51,52] as well. Moreover, the anisotropy of the compressive specimen was observed, which was ascribed to a large number of grain boundaries [51]. The SEBM sample along the BDs exhibited improved properties, but the reduced yield strength and plasticity were found in samples perpendicular to the build direction (BD). Moreover, their compressive properties were much higher than the tensile properties, which was consistent with previous studies [68]. As previously discussed, FCC and B2 ordered phases existed in the SEBM materials. The fracture of the tensile specimens was ascribed to the preferential formation and propagation of the cracks along BCC/FCC boundaries induced by the significant difference in the elastic properties of these adjacent phases. The micro-voids found in the failed specimen were attributed to the shrinkage induced by the denser FCC phases than BCC or B2 structures [45]. Li et al. [49] proposed the mechanical performance of the composite of the FeCoCrAlCu alloy deposited on the titanium alloy via the LMD method. The excellent wear resistance of materials was primarily attributed to their multiple structure, such as the quasi-crystalline/nanocrystalline phases, and the free micro-crack microstructure. The enhanced strength and ductility induced by adding the proper amount of Y_2O_3 content could lead to the improvement of the wear resistance. Moreover, the spreading nanoscale particles were capable of withstanding the external normal load, leading to increased wear performance.

3.3. CoCrFeMnNi HEAs

The mechanical properties and formability of CoCrFeMnNi HEAs were widely investigated by various AM methods [41–51]. Haase et al. [48] revealed the high formability and significant high hardness of CoCrFeMnNi alloys fabricated by the LMD method, producing excellent yield strength and ductility. It was clear that the yield strength of LMD-produced materials presented a much higher value than that in cast materials, which may have been due to the pronounced texture, lowering the mean Schmid factor, thus requiring higher stress to initiate dislocation motion. Furthermore, the initial high density of dislocation caused by the fast solidification and cooling in the LMD process contributed to the increased yield strength as well.

Li et al. [70] studied the mechanical properties of the CoCrFeMnNi HEA by the SLM method. Holes and cracks were reduced by increasing the VED, leading to higher ultimate tensile strength. The tensile strength was improved by HIP, which was ascribed to the closure of micro-pores and micro-cracks, accompanied by the reduced elongation. Overall, the SLM specimen presented a higher tensile strength than that in the slowly solidified HEA, which could be explained in terms of the Hall–Petch theory, which shows that decreasing the grain size leads to high strength. Moreover, the precipitation of the σ phase, preserving from high temperatures owing to rapid solidification, contributed to the improvement of the tensile strength. Furthermore, a similar investigation was performed on the strengthening mechanism in an SLM CoCrFeNiMn HEA with hierarchical microstructures [45]. All the SLM specimens presented much better mechanical properties than that in the as-cast specimens. The postmortem STEM indicated a significant dislocation arrest and retainment mechanism within the deformation cells, resulting in a distinct increase in the dislocation density within the cell walls, which is consistent with Piglione's [59] work. Moreover, the planar sliding interact with the cellular structure substantially for forming a 3D dislocation network, which dominates the deformation process in the as-built HEAs, accompanied by the additional contribution from deformation twinning. Based on the calculation of the yield strength dependent on the dislocation density, they induced that the improved strength was mostly controlled not only by grain-boundary strengthening and friction stress but also by dislocation hardening. Furthermore, the excellent uniform elongation of the as-built HEA was due to the capability of steady strain-hardening at large stress levels. Specifically, dislocations being significantly arrested and retained in the cell interior, and substantial interactions between slip bands and cellular structures, resulted in the generation of a complicated dislocation configuration, which led to a comparable strain-hardening ability.

Another way of studying LAM-fabricated HEAs was investigated to explore the deformation mechanisms at cryogenic temperatures [57]. Their yield strength and ductility were enhanced when the temperature decreased from 298 to 77 K, as shown in Figure 10. The strength in the LAM specimen was found to be higher than that in the as-cast sample, which was due to the suppressed elemental segregation and possible high dislocation density induced by the fast solidification rate in the LAM sample. Deformation twinning was found in the deformed sample. When increasing the plastic strain, the deformation twinning was more obvious. The increased deformation twinning was more dispersed as the strain reached the highest level. However, the deformation mechanism could not be dominated by deformation twinning even at high strain levels due to the limited quantity and its heterogeneous distribution. The local misorientation map indicated that the dislocation is the dominant deformation mechanism, especially with a low strain value. The density of dislocations increased rapidly and uniformly with increasing strain, which revealed that dislocations played an essential role during deformation. It was noticeable that the initial dislocation density in the LAM specimen was substantial, larger than that in the reported materials, which may have been due to rapid solidification and cooling in the LAM process. Such microstructures will undoubtedly produce an enhanced yield strength. It should be noted that, under the cryogenic condition, the deformation twinning at high strain levels could be an obstacle to moving dislocation, resulting in a continuous accumulation of dislocations, thus leading to a high work-hardening rate as well as an increased strength.

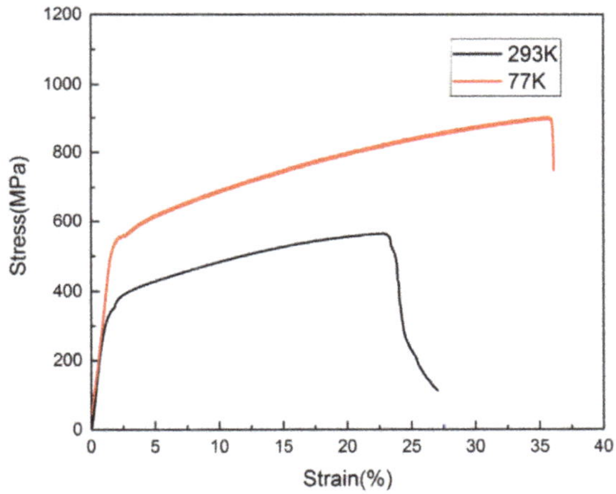

Figure 10. Tensile engineering stress–strain curves in the CoCrFeMnNi HEA tested at 293 K and 77 K, respectively. Image displayed with permission from the authors in [57].

Fujieda et al. [73] investigated the tensile properties of the CoCrFeNiTi-based HEA, achieved using SEBM and the subsequent solution treatment. The untreated SEBM sample presented a much higher tensile strength than that in the as-cast specimen, due to the uniform distribution of the needle-like Ni_3Ti in the matrix, while a reduced ductility was found as well, which should be attributed to the cracks progressing along the boundaries between excessive Ni_3Ti intermetallic compounds and the matrix.

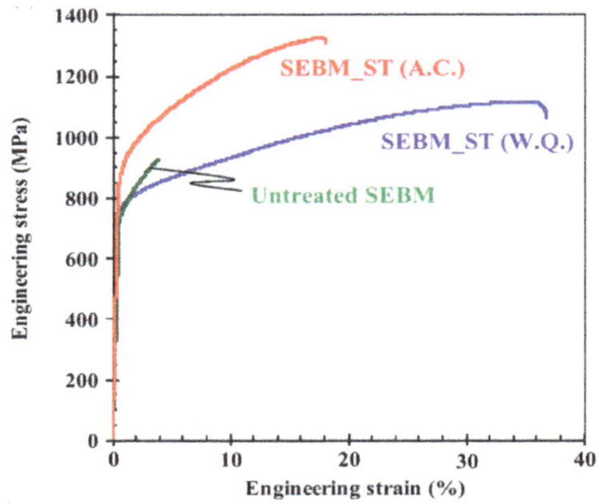

Figure 11. Engineering stress-engineering strain curves of deposited, and solution-treated SEBM specimens tested at room temperature. Image displayed with permission from the authors in [73].

3.4. Post-Treatment Effect

It has been proposed that post-treatment has a significant effect on the mechanical properties of AM-processed HEAs. It is believed that defects such as residual stress may exist in a specimen during deposition, which may affect the mechanical properties. For example, after annealing at 750 °C, a CoCrFeNi specimen presented a reduced yield and tensile strength with a minor reduction in ductility. When annealing temperature increased to 1000 °C, this CoCrFeNi specimen presented a lower yield strength and tensile strength, but improved ductility, which was due to the stress relief and grain growth (shown in Figure 8) [55].

The HIP effects on mechanical properties were studied in the DLF $Al_xCoCrFeNi$ alloy (x = 0.3, 0.6, and 0.85) [68]. The DLF/HIP $Al_{0.3}CoCrFeNi$ presented an enhanced ductility and work-hardening rate, without losing its yield strength. However, the DLF/HIP $Al^{0.3}$ showed an improved yield strength but reduced ductility. A small volume fraction of deformation twinnings was found in the tension-failed specimens of the DLF/HIP $Al_{0.3}CoCrFeNi$, which was due to a lower stress compared with the critical stress to activate the twinning. The formation of the hard B2 grain-boundary phase by the HIP process was detrimental to tension, which led to reduced ductility. This result was proved by the fact that the B2 phase in $Al_xCoCrFeNi$ failed to accommodate shape changes. Sistla et al. [67] studied the heat-treatment effect on the $Al_xFeCoCrNi_{2-x}$ prepared by DMD. It was found that the Fe–Cr matrix in the as-deposited, annealed, and as-quenched specimen was enhanced by the precipitates of the Ni- and Al-rich phases. The Al–Ni composition will transform to a two-phase domain with NiAl + Ni_3Al at higher temperatures, resulting in further improvment in the strength of materials.

A SEBM–CoCrFeNiTi-based HEA exhibited good tensile properties [73]. In order to further improve the mechanical properties of SEBM materials, a solution treatment was performed, leading to the disappearance of Ni_3Ti. Only small particles composed of Ni and Ti elements were present in the specimen. It is obvious that the ductility was improved without lowering their strengths after solution treatment (ST), as shown in Figure 11. The excellent ductility induced by the delay at the beginning of the plastic instability was attributed to the dynamic recovery induced by the screw-dislocations cross slips or the dislocations coalescence and annihilation during the plastic deformation. Specifically, the air-cooled (A.C.) specimen illustrated a higher yield strength but a lower ductility than that in the water-quenched (W.Q.) specimen, which was ascribed to the different precipitate size of the ordering phases that can act as a weak obstacle to the moving dislocation. The particle size of the sample obtained by the A.C. method was about 40 nm, higher than that of the W.C. sample, with a particle size of 10 nm. It was concluded that the moving dislocation was able to cut through the small particles with a diameter of several tens of nanometers. Moreover, the critical resolved shear stress increased proportionally to the square root of the precipitate diameter.

3.5. Comparison of Mechanical Properties

Figure 12 presents the yield strength and ductility obtained from both compression and tension experiments in HEAs, conventional Al, TiAl, and Cu alloys, and steels alloys fabricated by AM [60,76]. It is apparent that HEAs present a much higher yield strength and impressive plasticity than that in Al, TiAl, and Cu alloys and in steels. Even though the AlCoCrFeNi alloy fabricated by casting or by SEBM was found to have a high yield strength in the tensile tests, limited ductility was observed as well, which was due to the brittle phase appearance in the matrix. The modified $Co_{1.5}CeFeNi_{1.5}Mo_{0.1}$ HEA fabricated by SEBM showed an excellent combination of yield strength and ductility in the tensile experiments due to the uniform dispersion of small Ni_3Ti particles in the matrix. The CoCrFeNi alloys prepared by SLM demonstrated a higher yield strength than that in the LMD CoCrFeNi and CoCrFeNi HEA/5083 composite, but comparable to the SLM C-containing FeCoCrNi HEAs, which was ascribed to their fine microstructure in SLM materials. The LAM CoCrFeMnNi HEA suggested that the HIP-post process enhanced yield strength, but slightly lowered elongation. The SLM CoCrFeMnNi HEAs depending on various scanning speeds and heat treatments displayed a wide range of ductility and comparable yield strength. The DLF $Al_{0.3}CoCrFeNi$ alloys revealed a much lower yield strength, compared with other methods, as apparent in Figure 12. We can conclude that different AM methods induce substantial differences in both yield strength and plasticity, and the post-heat-treatment enhances their properties by removing various defects and releasing the residual stress present in the materials.

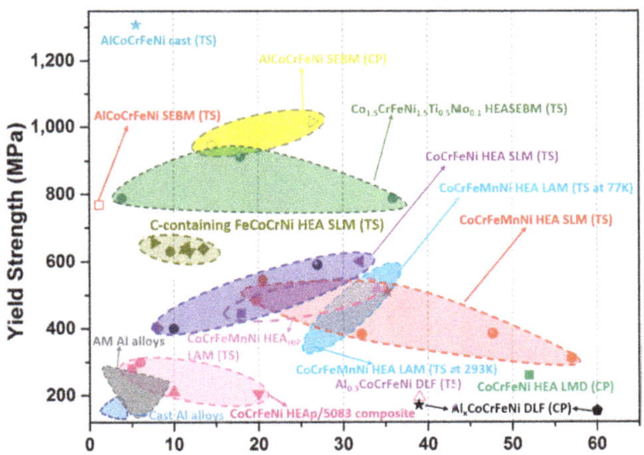

Figure 12. Yield strength vs. elongation using various AM methods. (TS: Tensile; CP: Compression).

4. Future Work

HEAs are potential materials that can act as functional and structural materials due to their excellent properties. The complex geometries of HEAs cannot be realized by the conventional method. Therefore, AM paves the way to the fabrication of HEAs with complicated shapes. However, defects are present in AM materials, which degrade mechanical properties. Thus, it is essential to explore ways to optimize the microstructure and enhance the mechanical properties of AM-processed HEAs. Suggestions are as follows:

1. The residual stress can occur in as-built components during the high cooling process, which may cause crack growth. It is necessary to further optimize the processing parameters and heat treatment regime to improve the microstructures and mechanical properties.
2. Most work highlights tension or compression experiments. Investigations on their fracture toughness, fatigue properties, oxidation, irradiation, or corrosion properties are rare. It is essential to study AM HEAs in a wide range of applications.
3. Moreover, most reported alloys are Al_xCoCrFeNi or CoCrFeMnNi systems. Little work has been performed on light-weight HEAs or refractory HEAs, which is another challenge for AM investigation in the future.
4. It is essential to develop a broader compositional range and set of mechanical properties, which could be a guideline to properly choosing an appropriate set of reinforcement compositions and designing an AM product suitable for a complex service environment.
5. Investigating materials with large differences in melting temperature is another challenge.
6. The starting powder feature should have an important influence on the fabricated materials during the AM process. Hence, it is essential to investigate an effective way of blending the powder directly or indirectly to reduce agglomeration and increase homogeneity.

5. Conclusions

The application of AM in the fabrication of HEAs and the post-heat-treatment were reviewed. The improved mechanical behavior of AM samples, compared with casting materials, is ascribed to the refinement of microstructures caused by large temperature gradients, rapid solidification, and cooling in the fabrication process, which is of significant importance in highly alloyed systems. The post-heat-treatment enhances their properties by removing various defects and releasing the residual stress formed in AM materials. Furthermore, reducing the pre-heating temperature to prevent phase segregation or modifying the composition of the HEA to stabilize one of the phases of BCC, FCC, and B2 structures were considered to improve the mechanical properties in AM materials. It is feasible to further improve the mechanical performance of as-built materials by optimizing the processing parameters, the scanning method, and the fabrication orientation. Promising results pave the way to investigate HEAs as engineering materials by AM methods.

Author Contributions: Conceptualization and Writing—Original Draft preparation, S.C. editing: Y.T. and P.K.L.; supervision, P.K.L.

Acknowledgments: The authors very much appreciate the support from the Department of Energy (DOE) Office of Fossil Energy, National Energy Technology Laboratory (NETL) (DE-FE-0011194) and the National Science Foundation (DMR-1611180 and 1809640), with J. Mullen, V. Cedro, R. Dunst, S. Markovich, G. Shiflet, and D. Farkas as program managers. P.K.L. very much appreciate the support from the U.S. Army Office Project (W911NF-13-1-0438) with the program managers, M. P. Bakas, S. N. Mathaudhu, and D. M. Stepp. S.C. and P.K.L. would like to acknowledge the financial support of the Center for Materials Processing (CMP), at The University of Tennessee, with the director of Claudia J. Rawn. S.C. and P.K.L. also thanks the support from the University of Tennessee's Open Publishing Support Fund.

Conflicts of Interest: The authors declare no conflict of interest.

References

1. Yeh, J.-W.; Chen, S.-K.; Lin, S.-J.; Gan, J.-Y.; Chin, T.-S.; Shun, T.-T.; Tsau, C.-H.; Chang, S.-Y. Nanostructured high-entropy alloys with multiple principal elements: Novel alloy design concepts and outcomes. *Adv. Eng. Mater.* **2004**, *6*, 299–303. [CrossRef]
2. Cantor, B.; Chang, I.T.H.; Knight, P.; Vincent, A.J.B. Microstructural development in equiatomic multicomponent alloys. *Mater. Sci. Eng. A* **2004**, *375–377*, 213–218. [CrossRef]
3. Wu, Z.; Bei, H.; Pharr, G.M.; George, E.P. Temperature dependence of the mechanical properties of equiatomic solid solution alloys with face-centered cubic crystal structures. *Acta Mater.* **2014**, *81*, 428–441. [CrossRef]
4. Senkov, O.N.; Wilks, G.B.; Miracle, D.B.; Chuang, C.P.; Liaw, P.K. Refractory high-entropy. *Intermetallics* **2010**, *18*, 1758–1765. [CrossRef]

5. Senkov, O.N.; Scott, J.M.; Senkov, S.V.; Miracle, D.B.; Woodward, C.F. Microstructure and room temperature properties of a high-entropy TaNbHfZrTi alloy. *J. Alloys Compd.* **2011**, *509*, 6043–6048. [CrossRef]
6. Chen, S.Y.; Tong, Y.; Tseng, K.-K.; Yeh, J.-W.; Poplawsky, D.; Wen, J.G.; Gao, M.C.; Kim, G.; Chen, W.; Ren, Y.; et al. Phase transformations of HfNbTaTiZr high-entropy alloy at intermediate temperatures. *Scr. Mater.* **2019**, *158*, 50–56. [CrossRef]
7. Chen, S.Y.; Xie, X.; Chen, B.L.; Qiao, J.W.; Zhang, Y.; Ren, Y.; Dahmen, K.A.; Liaw, P.K. Effects of temperature on serrated flows of $Al_{0.5}CoCrCuFeNi$ high-entropy alloy. *JOM* **2015**, *67*, 2314–2320. [CrossRef]
8. Zhang, Y.; Zuo, T.T.; Tang, Z.; Gao, M.C.; Dahmen, K.A.; Liaw, P.K.; Lu, Z.P. Microstructures and properties of high-entropy. *Prog. Mater. Sci.* **2014**, *61*, 1–93. [CrossRef]
9. Tang, W.-Y.; Yeh, J.-W. Effect of aluminum content on plasma-nitrided $Al_xCoCrCuFeNi$ high-entropy alloys. *Metall. Mater. Trans. A* **2009**, *40*, 1479–1486. [CrossRef]
10. Thurston, K.V.S.; Gludovatz, B.; Hohenwarter, A.; Laplanche, G.; George, E.P.; Ritchie, R.O. Effect of temperature on the fatigue-crack growth behavior of the high-entropy alloy CrMnFeCoNi. *Intermetallics* **2017**, *88*, 65–72. [CrossRef]
11. Brechtl, J.; Chen, S.Y.; Xie, X.; Ren, Y.; Qiao, J.W.; Liaw, P.K.; Zinkle. S.J. Towards a Greater Understanding of Serrated Flows in an Al-Containing High-Entropy-Based. *Int. J. Plast.* **2018**, in press. [CrossRef]
12. Feuerbacher, M.; Heidelmann, M.; Thomas, C. Hexagonal High-entropy Alloys. *Mater. Res. Lett.* **2014**, *3*, 1–6. [CrossRef]
13. Lilensten, L.; Couzinié, J.P.; Perrière, L.; Bourgon, J.; Emery, N.; Guillot, I. TaNbHfZrTi New structure in refractory high-entropy alloys. *Mater. Lett.* **2014**, *132*, 123–125. [CrossRef]
14. Youssef, K.M.; Zaddach, A.J.; Niu, C.; Irving, D.L.; Koch, C.C. A Novel Low-Density, High-Hardness, High-entropy Alloy with Close-packed Single-phase Nanocrystalline Structures. *Mater. Res. Lett.* **2014**, *3*, 95–99. [CrossRef]
15. Gao, M.C.; Zhang, B.; Guo, S.M.; Qiao, J.W.; Hawk, J.A. High-Entropy Alloys in Hexagonal Close-Packed Structure. *Metall. Mater. Trans. A* **2015**, *47*, 3322–3332. [CrossRef]
16. Zhou, Y.J.; Zhang, Y.; Wang, Y.L.; Chen, G.L. Microstructure and compressive properties of multicomponent $Al_x(TiVCrMnFeCoNiCu)_{100-x}$ high-entropy alloys. *Mater. Sci. Eng. A* **2007**, *454–455*, 260–265. [CrossRef]
17. Senkov, O.N.; Scott, J.M.; Meisenkothen, F.; Senkova, S.V.; Miracle, D.B.; Woodward, C.F. Microstructure and elevated temperature properties of a refractory TaNbHfZrTi alloy. *J. Mater. Sci.* **2012**, *47*, 4062–4074. [CrossRef]
18. Senkov, O.N.; Wilks, G.B.; Scott, J.M.; Miracle, D.B. Mechanical properties of $Nb_{25}Mo_{25}Ta_{25}W_{25}$ and $V_{20}Nb_{20}Mo_{20}Ta_{20}W_{20}$ refractory high entropy alloys. *Intermetallics* **2012**, *19*, 698–706. [CrossRef]
19. Lee, C.; Song, G.; Gao, M.C.; Feng, R.; Chen, P.; Brechtl, J.; Chen, Y.; An, K.; Guo, W.; Poplawsky, J.D.; et al. Lattice Distortion in a Strong and Ductile Refractory High-entropy Alloy. *Acta Mater.* **2018**, *160*, 158–172. [CrossRef]
20. Feng, W.; Qi, Y.; Wang, S. Effects of Short-Range Order on the Magnetic and Mechanical Properties of $FeCoNi(AlSi)_x$ High Entropy Alloys. *Metals* **2017**, *7*, 482. [CrossRef]
21. Klimova, M.; Stepanov, N.; Shaysultanov, D.; Chernichenko, R.; Yurchenko, N.; Sanin, V.; Zherebtsov, S. Microstructure and Mechanical Properties Evolution of the Al, C-Containing CoCrFeNiMn-Type High-Entropy Alloy during Cold Rolling. *Materials* **2017**, *11*, 53. [CrossRef] [PubMed]
22. Manzoni, A.; Haas, S.; Daoud, H.; Glatzel, U.; Förster, C.; Wanderka, N. Tensile Behavior and Evolution of the Phases in the $Al_{10}Co_{25}Cr_8Fe_{15}Ni_{36}Ti_6$ Compositionally Complex/High Entropy Alloy. *Entropy* **2018**, *20*, 646. [CrossRef]
23. Liu, X.; Yin, H.; Xu, Y. Microstructure, Mechanical and Tribological Properties of Oxide Dispersion Strengthened High-Entropy Alloys. *Materials* **2017**, *10*, 1312. [CrossRef] [PubMed]
24. Chuang, M.H.; Tsai, M.H.; Wang, W.R.; Lin, S.J.; Yeh, J.W. Microstructure and wear behavior of $Al_xCo_{1.5}CrFeNi_{1.5}Ti_y$ high-entropy alloys. *Acta Mater.* **2011**, *59*, 6308–6325. [CrossRef]
25. Chen, M.; Shi, X.H.; Yang, H.; Liaw, P.K.; Gao, M.C.; Hawk, J.A ; Qiao, J. Wear behavior of $Al_{0.6}CoCrFeNi$ high-entropy alloys: Effect of environments. *J. Mater. Res.* **2018**, *33*, 3310–3320. [CrossRef]
26. Chen, P.; Lee, C.; Wang, S.-Y.; Seifi, M.; Lewandowski, J.J.; Dahmen, K.A.; Jia, H.; Xie, X.; Chen, B.; Yeh, J.-W.; et al. Fatigue behavior of high-entropy alloys: A review. *Sci. China Technol. Sci.* **2017**, *61*, 168–178. [CrossRef]

27. Hemphill, M.A.; Yuan, T.; Wang, G.Y.; Yeh, J.W.; Tsai, C.W.; Chuang, A.; Liaw, P.K. Fatigue behavior of $Al_{0.5}$CoCrCuFeNi high entropy alloys. *Acta Mater.* **2012**, *60*, 5723–5734. [CrossRef]
28. Tang, Z.; Yuan, T.; Tsai, C.-W.; Yeh, J.-W.; Lundin, C.D.; Liaw, P.K. Fatigue behavior of a wrought $Al_{0.5}$CoCrCuFeNi two-phase high-entropy alloy. *Acta Mater.* **2015**, *99*, 247–258. [CrossRef]
29. Seifi, M.; Li, D.; Yong, Z.; Liaw, P.K.; Lewandowski, J.J. Fracture Toughness and Fatigue Crack Growth Behavior of As-Cast High-Entropy Alloys. *JOM* **2015**, *67*, 2288–2295. [CrossRef]
30. Shukla, S.; Wang, T.; Cotton, S.; Mishra, R.S. Hierarchical microstructure for improved fatigue properties in a eutectic high entropy alloy. *Scr. Mater.* **2018**, *156*, 105–109. [CrossRef]
31. Liu, K.; Komarasamy, M.; Gwalani, B.; Shukla, S.; Mishra, R.S. Fatigue behavior of ultrafine grained triplex $Al_{0.3}$CoCrFeNi high entropy alloy. *Scr. Mater.* **2019**, *158*, 116–120. [CrossRef]
32. Shi, Y.; Yang, B.; Liaw, P. Corrosion-Resistant High-Entropy Alloys: A Review. *Metals* **2017**, *7*, 43. [CrossRef]
33. Shi, Y.; Yang, B.; Xie, X.; Brechtl, J.; Dahmen, K.A.; Liaw, P.K. Corrosion of Al_xCoCrFeNi high-entropy alloys: Al-content and potential scan-rate dependent pitting behavior. *Corros. Sci.* **2017**, *119*, 33–45. [CrossRef]
34. Kai, W.; Cheng, F.P.; Liao, C.Y.; Li, C.C.; Huang, R.T.; Kai, J.J. The oxidation behavior of the quinary $FeCoNiCrSi_x$ high-entropy alloys. *Mater. Chem. Phys.* **2017**, *210*, 362–369. [CrossRef]
35. Chen, H.; Kauffmann, A.; Gorr, B.; Schliephake, D.; Seemüller, C.; Wagner, J.N.; Christ, H.J.; Heilmaier, M. Microstructure and mechanical properties at elevated temperatures of a new Al-containing refractory high-entropy alloy Nb-Mo-Cr-Ti-Al. *J. Alloys Compd.* **2016**, *661*, 206–215. [CrossRef]
36. Gorr, B.; Azim, M.; Christ, H.J.; Mueller, T.; Schliephake, D.; Heilmaier, M. Phase equilibria, microstructure, and high temperature oxidation resistance of novel refractory high-entropy alloys. *J. Alloys Compd.* **2015**, *624*, 270–278. [CrossRef]
37. Lyu, Z.; Lee, C.; Wang, S.-Y.; Fan, X.; Yeh, J.-W.; Liaw, P.K. Effects of Constituent Elements and Fabrication Methods on Mechanical Behavior of High-Entropy Alloys: A Review. *Metall. Mater. Trans. A* **2018**, 1–28. [CrossRef]
38. Williams, S.T.; Todd, I. Design for additive manufacturing with site-specific properties in metals and alloys. *Scr. Mater.* **2017**, *135*, 105–110. [CrossRef]
39. Seifi, M.; Salem, A.; Beuth, J.; Harrysson, O.; Lewandowski, J.J. Overview of Materials Qualification Needs for Metal Additive Manufacturing. *JOM* **2016**, *68*, 747–764. [CrossRef]
40. Zhu, Y.Y.; Li, J.; Tian, X.J.; Wang, H.M.; Liu, D. Microstructure and mechanical properties of hybrid fabricated Ti-6.5Al-3.5Mo-1.5Zr-0.3Si titanium alloy by laser additive manufacturing. *Mater. Sci. Eng. A* **2014**, *607*, 427–434. [CrossRef]
41. Chern, A.H.; Nandwana, P.; Yuan, T.; Kirka, M.M.; Dehoff, R.R.; Liaw, P.K.; Duty, C.E. A review on the fatigue behavior of Ti-6Al-4V fabricated by electron beam melting additive manufacturing. *Int. J. Fatigue* **2019**, *119*, 173–184. [CrossRef]
42. Li, X.P.; Wang, X.J.; Saunders, M.; Suvorova, A.; Zhang, L.C.; Liu, Y.J.; Fang, M.H.; Huang, Z.H.; Sercombe, T.B. A selective laser melting and solution heat treatment refined Al–12Si alloy with a controllable ultrafine eutectic microstructure and 25% tensile ductility. *Acta Mater.* **2015**, *95*, 74–82. [CrossRef]
43. Wang, Y.M.; Voisin, T.; McKeown, J.T.; Ye, J.; Calta, N.P.; Li, Z.; Zeng, Z.; Zhang, Y.; Chen, W.; Roehling, T.T.; et al. Additively manufactured hierarchical stainless steels with high strength and ductility. *Nat. Mater.* **2018**, *17*, 63–71. [CrossRef]
44. Guo, J.; Bai, J.; Liu, K.; Wei, J. Surface quality improvement of selective laser sintered polyamide 12 by precision grinding and magnetic field-assisted finishing. *Mater. Des.* **2017**, *138*, 1–10. [CrossRef]
45. Zhu, Z.G.; Nguyen, Q.B.; Ng, F.L.; An, X.H.; Liao, X.Z.; Liaw, P.K.; Nai, S.M.L.; Wei, J. Hierarchical microstructure and strengthening mechanisms of a CoCrFeNiMn high entropy alloy additively manufactured by selective laser melting. *Scr. Mater.* **2018**, *154*, 20–24. [CrossRef]
46. Herzog, D.; Seyda, V.; Wycisk, E.; Emmelmann, C. Additive manufacturing of metals. *Acta Mater.* **2016**, *117*, 371–392. [CrossRef]
47. Lewandowski, J.J.; Seifi, M. Metal additive manufacturing: A review of mechanical properties. *Annu. Rev. Mater. Res.* **2016**, *46*, 151–186. [CrossRef]
48. Haase, C.; Tang, F.; Wilms, M.B.; Weisheit, A.; Hallstedt, B. Combining thermodynamic modeling and 3D printing of elemental powder blends for high-throughput investigation of high-entropy alloys—Towards rapid alloy screening and design. *Mater. Sci. Eng. A* **2017**, *688*, 180–189. [CrossRef]

49. Li, J.; Craeghs, W.; Jing, C.; Gong, S.; Shan, F. Microstructure and physical performance of laser-induction nanocrystals modified high-entropy alloy composites on titanium alloy. *Mater. Des.* **2017**, *117*, 363–370. [CrossRef]
50. Laser metal deposition defined. Available online: https://www.industrial-lasers.com/articles/print/volume-250/issue-6/features/laser-metal-deposition.html (accessed on 1 November 2010).
51. Fujieda, T.; Shiratori, H.; Kuwabara, K.; Kato, T.; Yamanaka, K.; Koizumi, Y.; Chiba, A. First demonstration of promising selective electron beam melting method for utilizing high-entropy alloys as engineering materials. *Mater. Lett.* **2015**, *159*, 12–15. [CrossRef]
52. Kuwabara, K.; Shiratori, H.; Fujieda, T.; Yamanaka, K.; Koizumi, Y.; Chiba, A. Mechanical and corrosion properties of AlCoCrFeNi high-entropy alloy fabricated with selective electron beam melting. *Addit. Manuf.* **2018**, *23*, 264–271. [CrossRef]
53. Juechter, V.; Scharowsky, T.; Singer, R.F.; Körner, C. Processing window and evaporation phenomena for Ti–6Al–4V produced by selective electron beam melting. *Acta Mater.* **2014**, *76*, 252–258. [CrossRef]
54. Gibson, I.; Rosen, D.W.; Stucker, B. *Additive Manufacturing Technologies: Rapid Prototyping to Direct Digital Manufacturing*; Springer: New York, NY, USA, 2009.
55. Brif, Y.; Thomas, M.; Todd, I. The use of high-entropy alloys in additive manufacturing. *Scr. Mater.* **2015**, *99*, 93–96. [CrossRef]
56. Brandt, M. *Laser Additive Manufacturing. Materials, Design, Technologies, and Applications*; Woodhead Publishing: Duxford, UK, 2017.
57. Qiu, Z.; Yao, C.; Feng, K.; Li, Z.; Chu, P.K. Cryogenic deformation mechanism of CrMnFeCoNi high-entropy alloy fabricated by laser additive manufacturing process. *Int. J. Lightweight Mater. Manuf.* **2018**, *1*, 33–39. [CrossRef]
58. Huang, C.; Zhang, Y.; Vilar, R.; Shen, J. Dry sliding wear behavior of laser clad TiVCrAlSi high entropy alloy coatings on Ti–6Al–4V substrate. *Mater. Des.* **2012**, *41*, 338–343. [CrossRef]
59. Piglione, A.; Dovgyy, B.; Liu, C.; Gourlay, C.M.; Hooper, P.A.; Pham, M.S. Printability and microstructure of the CoCrFeMnNi high-entropy alloy fabricated by laser powder bed fusion. *Mater. Lett.* **2018**, *224*, 22–25. [CrossRef]
60. Gorsse, S.; Hutchinson, C.; Goune, M.; Banerjee, R. Additive manufacturing of metals: A brief review of the characteristic microstructures and properties of steels, Ti-6Al-4V and high-entropy alloys. *Sci. Technol. Adv. Mater.* **2017**, *18*, 584–610. [CrossRef] [PubMed]
61. Li, W.; Liu, P.; Liaw, P.K. Microstructures and properties of high-entropy alloy films and coatings: A review. *Mater. Res. Lett.* **2018**, *6*, 199–229. [CrossRef]
62. Karthik, G.M.; Panikar, S.; Ram, G.D.J.; Kottada, R.S. Additive manufacturing of an aluminum matrix composite reinforced with nanocrystalline high-entropy alloy particles. *Mater. Sci. Eng. A* **2017**, *679*, 193–203. [CrossRef]
63. Zhou, R.; Liu, Y.; Zhou, C.; Li, S.; Wu, W.; Song, M.; Liu, B.; Liang, X.; Liaw, P.K. Microstructures and mechanical properties of C-containing FeCoCrNi high-entropy alloy fabricated by selective laser melting. *Intermetallics* **2018**, *94*, 165–171. [CrossRef]
64. Joseph, J.; Stanford, N.; Hodgson, P.; Fabijanic, D.M. Tension/compression asymmetry in additive manufactured face centered cubic high entropy alloy. *Scr. Mater.* **2017**, *129*, 30–34. [CrossRef]
65. Shiratori, H.; Fujieda, T.; Yamanaka, K.; Koizumi, Y.; Kuwabara, K.; Kato, T.; Chiba, A. Relationship between the microstructure and mechanical properties of an equiatomic AlCoCrFeNi high-entropy alloy fabricated by selective electron beam melting. *Mater. Sci. Eng. A* **2016**, *656*, 39–46. [CrossRef]
66. Niu, P.D.; Li, R.D.; Yuan, T.C.; Zhu, S.Y.; Chen, C.; Wang, M.B.; Huang, L. Microstructures and properties of an equimolar AlCoCrFeNi high entropy alloy printed by selective laser melting. *Intermetallics* **2019**, *104*, 24–32. [CrossRef]
67. Sistla, H.R.; Newkirk, J.W.; Liou, F.F. Effect of Al/Ni ratio, heat treatment on phase transformations and microstructure of $Al_xFeCoCrNi_{2-x}$ ($x = 0.3, 1$) high entropy alloys. *Mater. Des.* **2015**, *81*, 113–121. [CrossRef]
68. Joseph, J.; Hodgson, P.; Jarvis, T.; Wu, X.; Stanford, N.; Fabijanic, D.M. Effect of hot isostatic pressing on the microstructure and mechanical properties of additive manufactured $Al_xCoCrFeNi$ high entropy alloys. *Mater. Sci. Eng. A* **2018**, *733*, 59–70. [CrossRef]
69. Wang, W.-R.; Wang, W.-L.; Yeh, J.-W. Phases, microstructure and mechanical properties of $Al_xCoCrFeNi$ high-entropy alloys at elevated temperatures. *J. Alloys Compd.* **2014**, *589*, 143–152. [CrossRef]

70. Li, R.; Niu, P.; Yuan, T.; Cao, P.; Chen, C.; Zhou, K. Selective laser melting of an equiatomic CoCrFeMnNi high-entropy alloy: Processability, non-equilibrium microstructure and mechanical property. *J. Alloys Compd.* **2018**, *746*, 125–134. [CrossRef]
71. Guo, J.; Goh, M.; Zhu, Z.; Lee, X.; Nai, M.L.S.; Wei, J. On the machining of selective laser melting CoCrFeMnNi high-entropy alloy. *Mater. Des.* **2018**, *153*, 211–220. [CrossRef]
72. Liu, W.H.; Wu, Y.; He, J.Y.; Nieh, T.G.; Lu, Z.P. Grain growth and the Hall–Petch relationship in a high-entropy FeCrNiCoMn alloy. *Scr. Mater.* **2013**, *68*, 526–529. [CrossRef]
73. Fujieda, T.; Shiratori, H.; Kuwabara, K.; Hirota, M.; Kato, T.; Yamanaka, K.; Koizumi, Y.; Chiba, A.; Watanabe, S. CoCrFeNiTi-based high-entropy alloy with superior tensile strength and corrosion resistance achieved by a combination of additive manufacturing using selective electron beam melting and solution treatment. *Mater. Lett.* **2017**, *189*, 148–151. [CrossRef]
74. Dobbelstein, H.; Gurevich, E.L.; George, E.P.; Ostendorf, A.; Laplanche, G. Laser metal deposition of compositionally graded TiZrNbTa refractory high-entropy alloys using elemental powder blends. *Addit. Manuf.* **2018**, in press. [CrossRef]
75. Dobbelstein, H.; Thiele, M.; Gurevich, E.L.; George, E.P.; Ostendorf, A. Direct Metal Deposition of Refractory High Entropy Alloy MoNbTaW. *Phys. Procedia* **2016**, *83*, 624–633. [CrossRef]
76. Scudino, S.; Unterdörfer, C.; Prashanth, K.G.; Attar, H.; Ellendt, N.; Uhlenwinkel, V.; Eckert, J. Additive manufacturing of Cu–10Sn bronze. *Mater. Lett.* **2015**, *156*, 202–204. [CrossRef]

entropy

MDPI

Article

Effect of Annealing on Microstructure and Mechanical Properties of $Al_{0.5}CoCrFeMo_xNi$ High-Entropy Alloys

Yan-Xin Zhuang *, Xiu-Lan Zhang and Xian-Yu Gu

Key Laboratory of Electromagnetic Processing of Materials, Ministry of Education, Northeastern University, Shenyang 110819, China; 15204001761@163.com (X.-L.Z.); xianyu062@126.com (X.-Y.G.)

* Correspondence: yxzhuang@epm.neu.edu.cn; Tel.: +86-024-8368-0136

Received: 11 September 2018; Accepted: 19 October 2018; Published: 23 October 2018

Abstract: The effect of annealing temperature on the microstructure, phase constituents and mechanical properties of $Al_{0.5}CoCrFeMo_xNi$ high-entropy complex alloys has been investigated at a fixed annealing time (10 h). The 600 °C-annealing has no obvious effect on their microstructures, while the annealing at 800–1200 °C enhances the precipitation of (Al,Ni)-rich ordered BCC phase or/and (Cr,Mo)-rich σ phase, and thereby greatly affects the microstructure and mechanical properties of the alloys. All the annealed $Al_{0.5}CoCrFeNi$ alloys are composed of FCC and (Al,Ni)-rich ordered BCC phases; the phase constituent of the $Al_{0.5}CoCrFeMo_{0.1}Ni$ alloy changes from FCC + BCC (600 °C) to FCC + BCC + σ (800 °C) and then to FCC + BCC (1100 °C); the phase constituents of the $Al_{0.5}CoCrFeMo_{0.2}Ni$ and $Al_{0.5}CoCrFeMo_{0.3}Ni$ alloys change from FCC + BCC + σ to FCC + BCC with the annealing temperature rising from 600 to 1200 °C; while all the annealed $Al_{0.5}CoCrFeMo_{0.4}Ni$ and $Al_{0.5}CoCrFeMo_{0.5}Ni$ alloys consist of FCC, BCC and σ phases. The phase constituents of most of the alloys investigated are in good agreement with the calculated results from Thermo-Calc program. The alloys annealed at 800 °C under current investigation conditionshave relative fine precipitations and microstructure, and thereby higher hardness and yield stress.

Keywords: high-entropy alloys; annealing; microstructure; mechanical properties; phase constituent

1. Introduction

The recently developed high-entropy alloys (HEAs), also known as multi-principal elements alloys or complex concentrated solid solution alloys, have attracted increasing attention due to their unique microstructures and adjustable properties [1–6]. The HEAs, which contains more than five principal elements with concentrations from 5 to 35 at.% for each principal element [2], tend to form simple solution structures (FCC, BCC, HCP or mixed) rather than many complex phases. To date, many efforts have been made to understand and control the structure and properties within as-cast and/or homogenized HEAs [7–15]. The original concept of HEAs and the strict restriction on the HEA design strategy proposed by Yeh [2] have also been relaxed. Depending on their composition, the as-cast and/or homogenized HEAs can have many interesting mechanical and physical properties, and wide potential applications.

It has been realized that the solid solutions in as-cast HEAs are the firstly formed solid phases upon cooling from the molten liquids, and are generally metastable phases at room temperature [16]. The room-temperature-metastable solid solutions could transform to other phases at an appropriate annealing temperature, and the annealing process can modify the microstructure and properties of the HEAs. Aging the AlCoCrCuFeNi HEA at elevated temperatures causes the structure gradually to transform from stabilized BCC to FCC, decreases the yield strength of the alloy, and increases the plastic strain of the material [17]. Aging the $Cu_{0.5}CoCrFeNi$ alloy at 1100–1350 °C induces the precipitation of a

Cr-rich phase in the FCC matrix of the alloy, and improves the anti-corrosion properties of the alloy [18]. The AlNi-based B2 phase and Cr-rich σ phase formed in the $Al_{0.5}$CrFeCoNiCu HEAs annealed at 700–900 °C [19]. A needle-like Cu-rich FCC phase has been precipitated from the BCC dendrite region at the annealing temperatures higher than 973 K in the AlCoCuFeNi HEA [20]. The microstructure and properties of the Al_xCoCrFeNi HEAs have a strong dependence on the Al content and annealing temperature [21]. The phase stability of the metastable solid solutions is actually becoming a critical issue for HEAs.

Over the last decades, the single-FCC or BCC HEAs have been intensively investigated and developed. In fact, it is very hard to reach a balance between high strength and high ductility for the single-phase HEAs due to the fact that the single-FCC HEAs normally have good ductility and poor strength, while the single-BCC HEAs generally have poor ductility and good strength. By adjusting the composition of the alloys, a eutectic high-entropy alloy $AlCoCrFeNi_{2.1}$ with a regular FCC/BCC lamellar structure has been developed, and the alloy has an excellent combination of high strength and high ductility [22], which was attributed to the coupling between the ductile FCC and brittle BCC phases during tension deformation [23]. A transformation-induced plasticity-assisted dual-phase (TRIP-DP) HEA has also been developed, in which two high-entropy phases (FCC γ matrix and laminate HCP ε phase) present [24]. The TRIP-DP HEA combines the solid-solution strengthening effect in HEAs with the TRIP effect, exhibits multiple deformation mechanisms and dynamic strain partitioning behavior [25], and has improved strength and ductility compared to its corresponding single-phase HEA. This combined increases in strength and ductility distinguishes the TRIP-DP-HEA alloy from other structural materials [24]. All these results have showed that the dual- or multi-phase HEAs can display an excellent combination between strength and ductility, and are becoming the future direction for designing advanced HEAs.

The $Al_{0.5}$CoCrFeNi alloy has a mixed FCC + B2 dual-phase structure with FCC as the dominant structure [7,21]. Our previous results have shown that the addition of Mo into this alloy system could enhance the formation of the σ phase, and tune the mechanical properties of the as-cast $Al_{0.5}CoCrFeMo_xNi$ [26]. Although we have evaluated the high-temperature equilibrium phases existing in $Al_{0.5}CoCrFeMo_xNi$ alloy using thermodynamic calculation, the results need to be assured by experiments. On the other hand, an appropriate annealing process can further modify the microstructure and improve its mechanical properties. In order to understand the temperature effect on this HEA system, the aged-and-quenched microstructure, phase transformation and mechanical properties have been investigated. The pseudo binary phase diagram derived from Thermo-Calc program has been compared with the experimental results.

2. Materials and Methods

Ingots with nominal compositions of $Al_{0.5}CoCrFeMo_xNi$ (x = 0, 0.1, 0.2, 0.3, 0.4 and 0.5, denoted by Mo_0, $Mo_{0.1}$, $Mo_{0.2}$, $Mo_{0.3}$, $Mo_{0.4}$ and $Mo_{0.5}$, respectively) were prepared by arc-melting the mixture of constituent elements with purity better than 99.9 wt.% in a water-cooled copper hearth under a titanium-gettered high-purity argon atmosphere. The ingots were remelted at least four times to assure their chemical homogeneity. Samples for microstructural observation and annealing processes were cut from the ingots, mechanically ground and polished through standard routines. Some samples were sealed in quartz tubes under a vacuum better than 1×10^{-2} Pa, and then annealed at given temperatures for 10 h. After the annealing process, the samples were quenched in water. The phase constitutions of the alloys were characterized using X-ray diffraction (XRD) with Cu K_α radiation (X'Pert Pro, PANalytical B.V., Almelo, The Netherlands). The microstructure and chemical compositions were examined on the polished samples using scanning electron microscopy (SEM, SSX-550, SHIMADZU corp., Kyoto, Japan) equipped with energy dispersive spectrometry (EDS). Hardness was characterized using a Vickers hardness tester (Wolpert 452SVD, Wolpert Wilson instrument, Shanghai, China) under a load of 5 Kgf for 15 s. The hardness measurements were made on at least seven points to yield an average value for each sample. Room temperature compressive tests were performed on cylindrical specimens with 5 mm in diameter and 10 mm in length using a SHIMADZU precision universal tester (AG-X, SHIMADZU corp., Kyoto, Japan) with a strain rate of 8.3×10^{-4} s^{-1}.

Thermodynamic calculations were conducted using the Thermo-Calc TCCS program in conjunction with a commercial TTNI7 database, which is developed for Ni-based alloys and contains 22 elements including Fe, Co, Ni, Al, Cr, Cu, Nb and Mo. The pseudo binary phase diagram, fraction and composition of each equilibrium phase at different temperatures were derived using the program.

3. Results and Discussion

3.1. Phase Transformation of Annealed $Al_{0.5}CoCrFeMo_xNi$ Alloys

Figure 1 summaries the XRD patterns of the $Al_{0.5}CoCrFeMo_xNi$ multicomponent high-entropy alloys annealed at different temperatures for 10 h followed by water quenching. The XRD patterns of the as-cast alloys are also given in the figure. The addition of Mo and the annealing process affect the phase constituents in the alloys. The addition of Mo enhances the formation of σ and ordered BCC phases in the as-cast $Al_{0.5}CoCrFeMo_xNi$ alloys as reported in our previous work [26]. The Bragg peaks in the XRD patterns of the as-cast $Al_{0.5}CoCrFeNi$ alloy can be indexed to a simple FCC phase. When the $Al_{0.5}CoCrFeNi$ alloy was annealed at the temperatures between 600–1200 °C for 10 h, the Bragg peaks corresponding to a BCC phase can found on the XRD patterns even though the peak at 2θ of about 44 is very weak for the alloy annealed at 1200 °C, meaning that the annealing induces the formation of the BCC phase. Similarly, the as-cast and 600 °C-annealed $Mo_{0.1}$ alloys have a dominant FCC phase and a minor BCC phase, while a σ phase has been found in the $Mo_{0.1}$ alloys annealed at 800 °C and disappears again in the alloys annealed at 1000–1200 °C. The as-cast $Mo_{0.2}$ alloy and the alloys annealed at 600–1100 °C consist of three phases, namely FCC, BCC and σ phases, while the $Mo_{0.2}$ alloys annealed at 1200 °C only have FCC and BCC phases. The as-cast $Mo_{0.3}$ alloy and the alloys annealed at 600–1100 °C are composed of the FCC, BCC and σ phases, while only FCC and BCC phases can be found in the alloy annealed at 1200 °C. The as-cast and annealed $Mo_{0.4}$ alloys have the FCC, BCC and σ phases even though the peaks for BCC phase are very weak in the alloy annealed at 1200 °C. The as-cast $Mo_{0.5}$ alloy and the alloys annealed at 600–1200 °C have FCC, BCC and σ phases even though the peaks for the BCC phase are very weak in the alloy annealed at 1200 °C. It can be concluded that both the Mo content and the annealing temperature affect the phase constituents of the $Al_{0.5}CoCrFeMo_xNi$ alloys. On the other hand, some kind of preferred orientation can also be observed in some of the alloys.

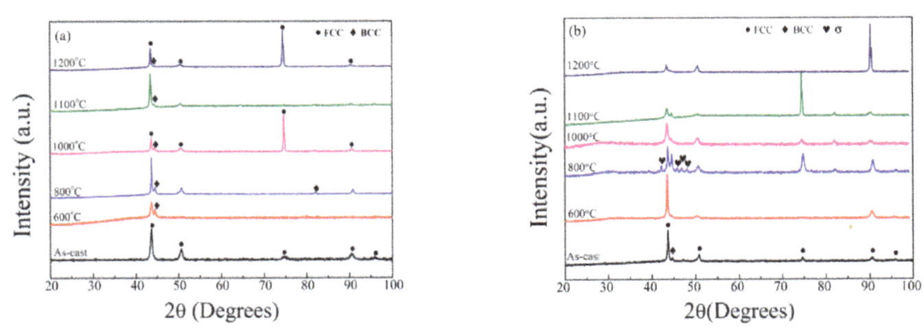

Figure 1. *Cont.*

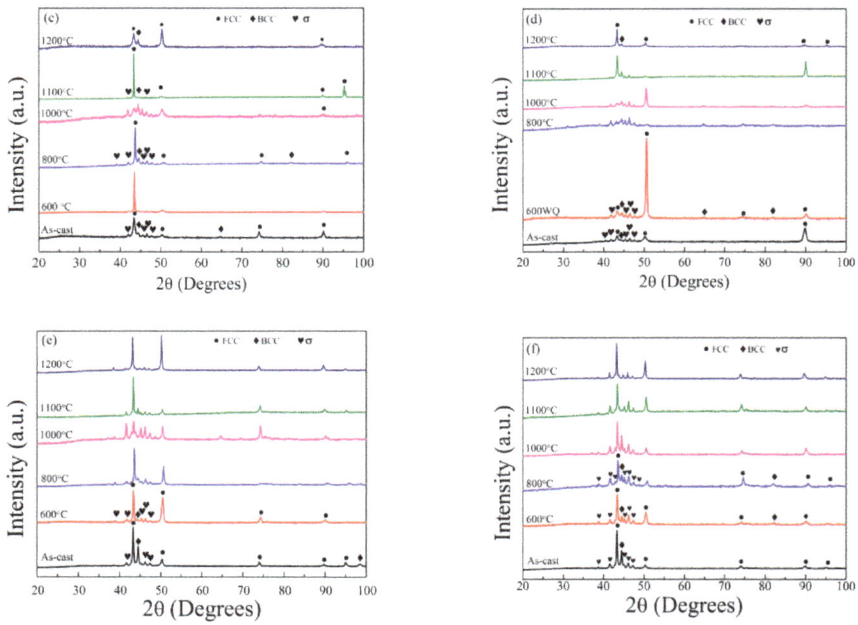

Figure 1. X-ray diffraction (XRD) patterns of the as-cast $Al_{0.5}CoCrFeMo_xNi$ alloys and the alloys annealed at different temperatures for 10 h followed by a water quenching. (**a**) x = 0; (**b**) x = 0.1; (**c**) x = 0.2; (**d**) x = 0.3; (**e**) x = 0.4; (**f**) x = 0.5.

3.2. Microstructures of Annealed $Al_{0.5}CoCrFeMo_xNi$ Alloys

Figure 2 presents the microstructure of the as-cast Mo_0, $Mo_{0.1}$ and $Mo_{0.2}$ high entropy alloys and the alloys annealed at different temperatures for 10 h. Table 1 lists distribution of element in different phases of the alloys annealed at 1000 °C. The as-cast Mo_0, $Mo_{0.1}$ and $Mo_{0.2}$ alloys have typical dendrite microstructures. The dendrite (DR) phase is FCC phase, while the interdendrite (ID) region consists of (Al,Ni)-rich phase, FCC phase and/or (Cr,Mo)-rich phase [26]. The 600 °C annealing has no obvious effect on the microstructure of the three alloys. However, the annealing at temperatures between 800 and 1200 °C has a distinct influence on the microstructure of the three alloys. The needle-like precipitations have been found in the Mo_0 alloys annealed at 800–1100 °C, and the EDS shows that both the needle-like precipitation (marked as NLP in the figure) and the dark matrix phase in the ID region have high Al and Ni, can be regarded as a (Al,Ni)-rich phase (refer to the data in Table 1). The (Al,Ni)-rich phase have an ordered BCC (B2) structure [21,26]. When increasing annealing temperature, the number of needle-like precipitation decreases, while its size increases. When the annealing temperature is up to 1200 °C, the needle-like (Al,Ni)-rich precipitation disappears. It is clear that the two phases (FCC and (Al,Ni)-rich ordered BCC) exist in the Mo_0 alloy annealed at temperatures of 600–1200 °C, which is in good agreement with the observation from XRD patterns. Similarly, many obvious precipitations have been found in the $Mo_{0.1}$ and $Mo_{0.2}$ alloys annealed at 800–1100 °C. By contrast with the Mo_0 alloys, a new phase enriched with Cr and Mo (bright phase marked as WP) has been observed in the $Mo_{0.1}$ and $Mo_{0.2}$ alloys. The (Cr,Mo)-rich phase is the σ phase [26]. There are three phases, namely FCC matrix, needle-like (Al,Ni)-rich ordered BCC phase and round-like (Cr,Mo)-rich σ phases, existing in the $Mo_{0.1}$ alloys annealed at 800–1000 °C and the $Mo_{0.2}$ alloys annealed at 600–1100 °C, while there are two phases (FCC and (Al,Ni)-rich ordered BCC phases) in the $Mo_{0.1}$ alloys annealed at 1100 and 1200 °C and the $Mo_{0.2}$ alloy annealed at 1200 °C. It can also be observed that the (Cr,Mo)-rich σ phase exists in the DR and ID regions in the $Mo_{0.1}$ and

$Mo_{0.2}$ alloys annealed at 800 °C, but only appears together with the (Al,Ni)-rich ordered BCC phase in the alloys annealed at higher temperatures.

The alloys annealed at 800 °C have fine microstructures, and are supposed to have better mechanical properties. The fine microstructure can be attributed to solid-state decomposition of the B2 phase and FCC matrix. The phenomena can be often found in $Al_xCoCrFeNi$ alloys [27–30]. Heat treatment at 620 °C for the $Al_{0.3}CoCrFeNi$ alloy caused the transformation of FCC to FCC + $L1_2$ or FCC + B2 + σ phases depending on the processing routes, and the microstructural variation realized by different process pathways was attributed to the competition between the thermodynamic driving force and activation barrier for the second-phase nucleation [27]. An aging process lead to the spinodal decomposition reactions of the FCC matrix and the interdendrite (Al,Ni)-rich phase in $Al_{0.5}CoCrFeNi$ alloy, and the phase segregation effect was explained based on the mixing enthalpy between different atom pairs [28]. Recently, Rao et al. found that a (Al,Ni)-rich $L1_2$ phase was precipitated from the FCC matrix of $Al_{0.5}CoCrFeNi$ alloy, a Cr-rich BCC nanoprecipitate was observed in its B2 phase, and the Cr-rich BCC nanoprecipitate was the origin of the σ phase [29]. Banerjee et al. also reported that the solid-state decomposition resulted in the formation of FCC + $L1_2$ and BCC + B2, accompanied by a compositional partitioning [30]. Cleary, the solid-state decomposition of the FCC matrix and (Al,Ni)-rich ordered BCC in the Mo-free $Al_{0.5}CoCrFeNi$ alloy is responsible for its fine structure at 800 °C. In the Mo-containing alloys, the addition of Mo enhances the formation of σ phase. Both Cr and Mo are strong σ-forming elements [31]. The σ phase forms either in a three-step formation from B2 region or directly from the FCC matrix [29].

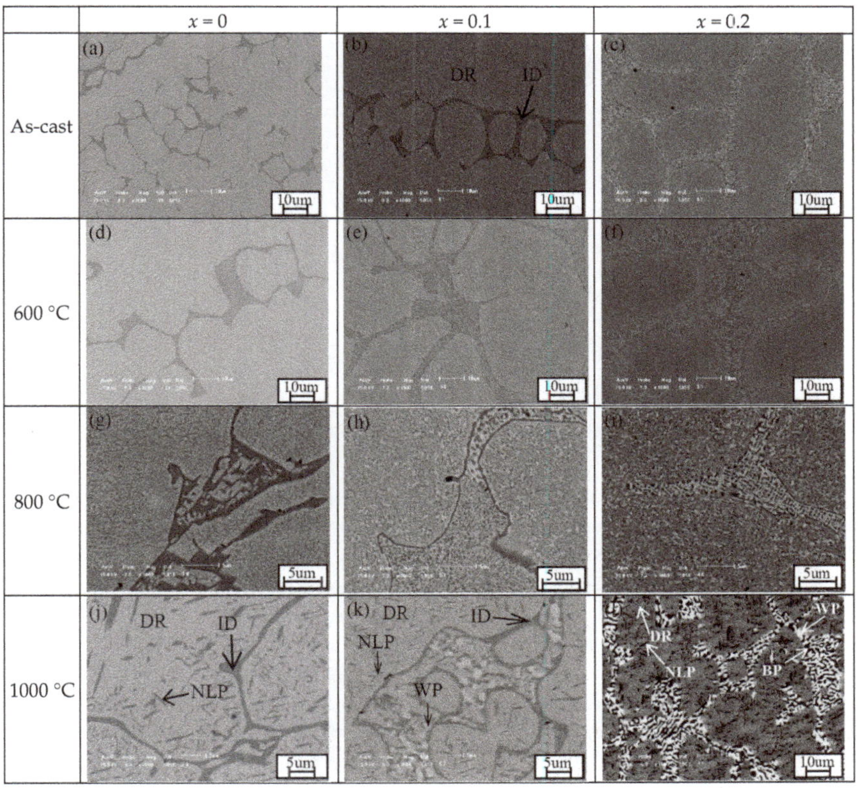

Figure 2. *Cont.*

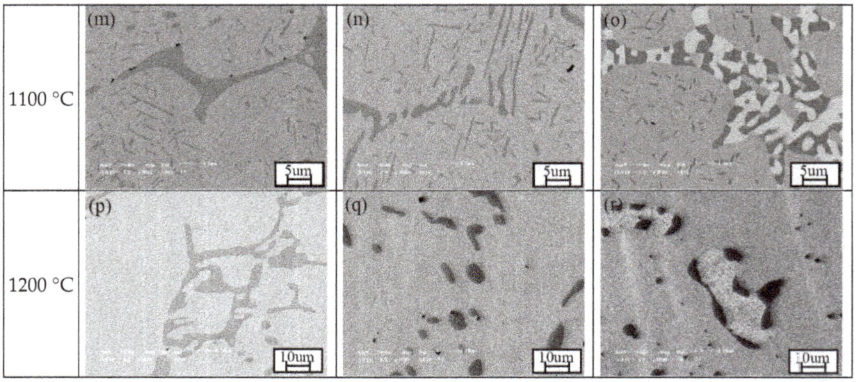

Figure 2. Back-scattering scanning electron microscope (SEM) images of the as-cast Mo_0, $Mo_{0.1}$ and $Mo_{0.2}$ high entropy alloys and the alloys annealed at different temperatures for 10 h. DR is for FCC dendrite phase, ID is for (Al,Ni)-rich interdendrite phase, NLP is for needle-like (Al,Ni)-rich recipitate, WP is for white (Cr,Mo)-rich precipitate, and BP is for black (Al,Ni)-rich precipitate. (**a**) as-cast Mo_0 alloy; (**b**) as-cast $Mo_{0.1}$ alloy; (**c**) as-cast $Mo_{0.2}$ alloy; (**d**) Mo_0 alloy annealed at 600 °C; (**e**) $Mo_{0.1}$ alloy annealed at 600 °C; (**f**) $Mo_{0.2}$ alloy annealed at 600 °C; (**g**) Mo_0 alloy annealed at 800 °C; (**h**) $Mo_{0.1}$ alloy annealed at 800 °C; (**i**) $Mo_{0.2}$ alloy annealed at 800 °C; (**j**) Mo_0 alloy annealed at 1000 °C; (**k**) $Mo_{0.1}$ alloy annealed at 1000 °C; (**l**) $Mo_{0.2}$ alloy annealed at 1000 °C; (**m**) Mo_0 alloy annealed at 1100 °C; (**n**) $Mo_{0.1}$ alloy annealed at 1100 °C; (**o**) $Mo_{0.2}$ alloy annealed at 1100 °C; (**p**) Mo_0 alloy annealed at 1200 °C; (**q**) $Mo_{0.1}$ alloy annealed at 1200 °C; (r) $Mo_{0.2}$ alloy annealed at 1200 °C.

Table 1. Distribution of elements (at.%) in different regions of the alloys annealed at 1000 °C.

Alloys	Phases	Fe	Co	Ni	Al	Cr	Mo
Mo_0-1000	DR: FCC	22.3	23.8	21.5	7.2	25.2	-
	ID: (Al,Ni)-rich BCC	11.0	15.5	35.5	30.6	7.4	-
	NLP: (Al,Ni)-rich BCC	16.0	18.6	22.4	31.2	11.8	-
$Mo_{0.1}$-1000	DR: FCC	24.3	23.5	16.4	7.6	25.5	2.7
	ID: (Al,Ni)-rich BCC	14.8	19.8	22.6	30.9	10.8	1.1
	NLP: (Al,Ni)-rich BCC	16.8	16.8	28.5	20.7	15.7	1.6
	WP: (Cr,Mo)-rich σ	19.3	18.5	6.7	4.4	40.5	10.6
$Mo_{0.2}$-1000	DR: FCC	23.7	21.0	20.6	8.2	21.2	5.3
	BP: (Al,Ni)-rich BCC	13.7	15.2	28.2	20.6	15.9	6.4
	WP: (Cr,Mo)-rich σ	20.9	20.5	14.9	6.0	28.1	9.6
	NLP: (Al,Ni)-rich BCC	20.2	22.9	21.9	11.0	19.6	4.5

Figure 3 summarizes the microstructure of the $Mo_{0.3}$, $Mo_{0.4}$ and $Mo_{0.5}$ alloys at various annealing temperature. Table 2 lists the elemental distribution in each phase for the three alloys annealed at 1000 °C. All the three as-cast alloys are composed of FCC, (Al,Ni)-rich ordered BCC, and (Cr,Mo)-rich σ phases as stated in our previous work [26]. The 600 °C-annealing has also no obvious effect on the microstructure of the three alloys, while 800 °C-annealing induces the precipitation of a large amount of fine (Al,Ni)-rich BCC and (Cr,Mo)-rich σ phases in the alloys even though their typical dendrite microstructure remains. Obvious coarsening occurred when the alloys were annealed at 1000–1100 °C, and the morphology of BCC and σ phases changes greatly. The BCC and σ phases become larger, and most of them appear together, implying that certain crystallographic orientation exists between the ordered BCC and σ phases. A few precipitations with even higher Mo content can also be found in the $Mo_{0.5}$ alloy annealed at 1000 °C (marked as WP2 in the corresponding image). When the alloys annealed at 1200 °C, the amount of ordered BCC and σ phases obviously decrease.

The wide composition range of each phase has been observed in the annealed alloys, as shown in Tables 1 and 2. The addition of Mo into the $Al_{0.5}$CoCrFeNi alloys can change the liquidus temperature or solvus temperature, the equilibrium phase composition, and even the formation sequence of the equilibrium phases [26]. The variable mixing enthalpy between different atoms and competition among the elements could be possible reasons for the variable composition of each phase. On the other hand, the needle-like (Al,Ni)-rich precipitate has a relatively small size, which might cause the inaccurately measured composition. Further work is going to clarify the reason for the wide composition. However, the wide composition range of each phase has no effect on the phase structure.

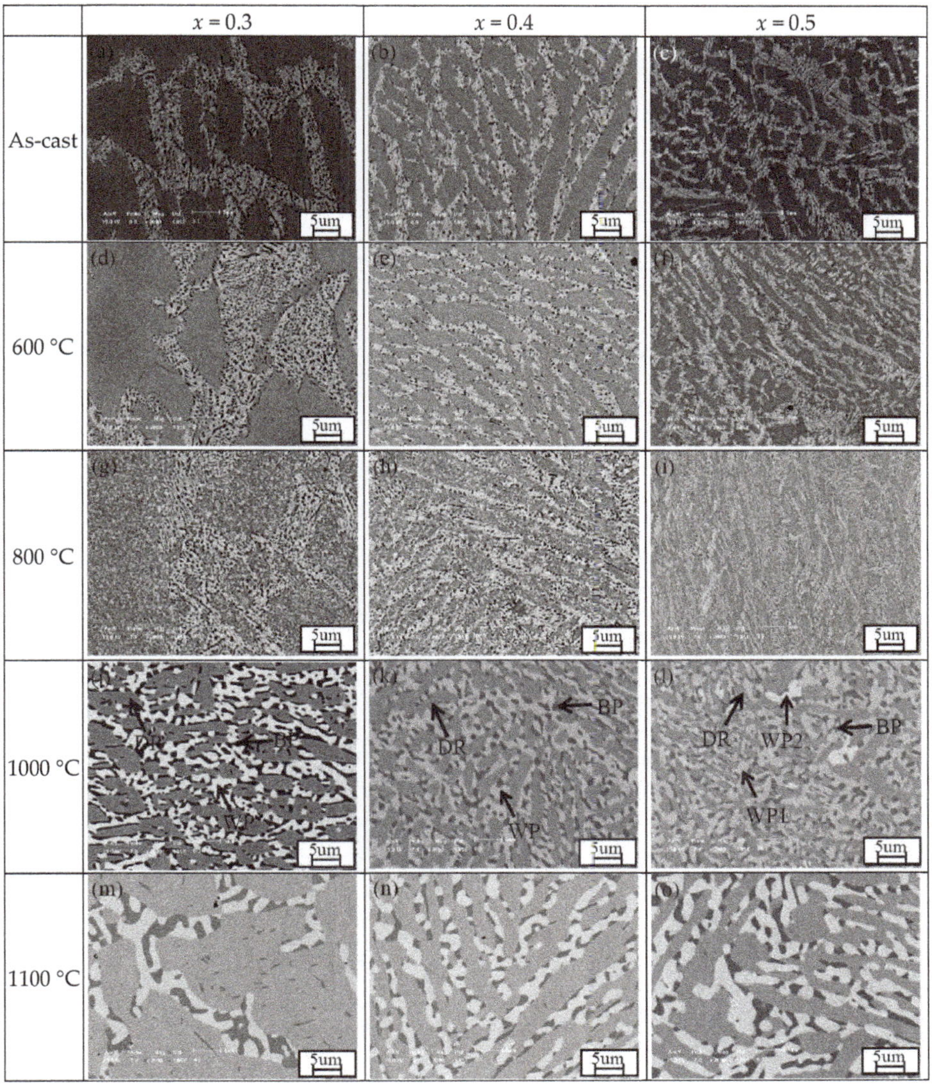

Figure 3. *Cont.*

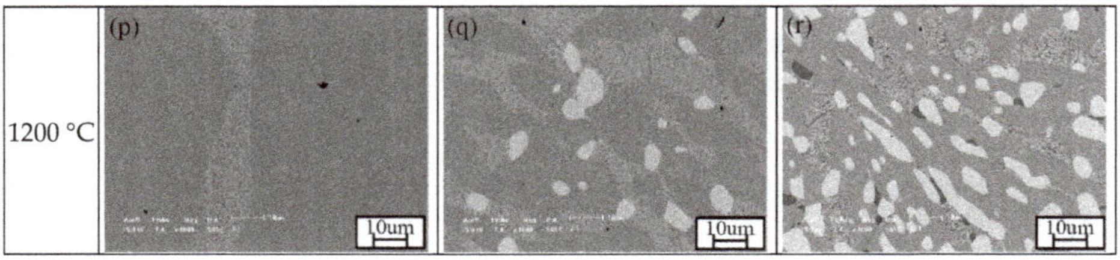

Figure 3. Back-scattering SEM images of the as-cast $Mo_{0.3}$, $Mo_{0.4}$, and $Mo_{0.5}$ high entropy alloys and the alloys annealed at different temperatures for 10 h. DR is for dendrite FCC, WP1 is for white (Cr,Mo)-rich precipitate with relative low Mo content, WP2 is the whit (Cr,Mo)-rich precipitate with relative high Mo content, and BP is for black (Al,Ni)-rich precipitate. (**a**) as-cast $Mo_{0.3}$ alloy; (**b**) as-cast $Mo_{0.4}$ alloy; (**c**) as-cast $Mo_{0.5}$ alloy; (**d**) $Mo_{0.3}$ alloy annealed at 600 °C; (**e**) $Mo_{0.4}$ alloy annealed at 600 °C; (**f**) $Mo_{0.5}$ alloy annealed at 600 °C; (**g**) $Mo_{0.3}$ alloy annealed at 800 °C; (**h**) $Mo_{0.4}$ alloy annealed at 800 °C; (**i**) $Mo_{0.5}$ alloy annealed at 800 °C; (**j**) $Mo_{0.3}$ alloy annealed at 1000 °C; (**k**) $Mo_{0.4}$ alloy annealed at 1000 °C; (**l**) $Mo_{0.5}$ alloy annealed at 1000 °C; (**m**) $Mo_{0.3}$ alloy annealed at 1100 °C; (**n**) $Mo_{0.4}$ alloy annealed at 1100 °C; (**o**) $Mo_{0.5}$ alloy annealed at 1100 °C; (**p**) $Mo_{0.3}$ alloy annealed at 1200 °C; (**q**) $Mo_{0.4}$ alloy annealed at 1200 °C; (r) $Mo_{0.5}$ alloy annealed at 1200 °C.

Table 2. Distribution of elements (at.%) in different regions of the alloys annealed at 1000 °C.

Alloys	Regions	Fe	Co	Ni	Al	Cr	Mo
$Mo_{0.3}$-1000	DR: FCC	22.8	18.8	21.3	10.1	20.7	6.3
	BP: (Al,Ni)-rich BCC	10.7	15.7	33.1	30.2	7.6	2.8
	WP: (Cr,Mo)-rich σ	18.1	18.6	7.7	3.6	30.3	21.6
$Mo_{0.4}$-1000	DR: FCC	25.7	20.7	18.8	7.6	21.2	6.0
	BP: (Al,Ni)-rich BCC	9.3	17.5	30.6	25.0	11.5	6.1
	WP: (Cr,Mo)-rich σ	19.0	18.7	9.8	3.0	25.1	24.4
$Mo_{0.5}$-1000	DR: FCC	23.3	21.3	21.1	7.0	17.7	9.5
	BP: (Al,Ni)-rich BCC	12.4	10.5	35.5	29.4	6.3	6.0
	WP1: (Cr,Mo)-rich σ	16.4	17.2	9.9	2.8	25.0	28.7
	WP2: (Cr,Mo)-rich	14.0	15.9	8.4	2.0	17.3	42.3

3.3. Calculated Pseudo Binary Phase Diagram and Its Comparison with Experiments

It is clear that the phase constituents in the $Al_{0.5}$CoCrFeMoxNi alloys depend on the content of Mo and the annealing temperature. Figure 4 is the calculated pseudo-binary phase diagram of $Al_{0.5}CoCrFeMo_xNi$ using the Thermo-Calc program, where the calculated equilibrium phase constituents in each region have been labeled. The alloys investigated in this work have also been marked in the diagram as the dash lines. The phase constituents in most of the alloys, which have been derived based on the XRD and microstructures above, are in good agreement with the calculated results. Only 6 samples among the 30 samples have a little difference with the calculated phase constituents from Thermo-Calc program. Careful examination found that the difference happened either near the boundary between different regions or at the low temperatures. For example, the Mo_0 alloy at 800 °C, the $Mo_{0.1}$ alloy at 1000 °C, the $Mo_{0.2}$ alloy at 1100 °C and the $Mo_{0.4}$ alloy at 1200 °C locate near the boundary between two regions. The difference might be attributed to the unspecified database for the alloy system, which makes the boundary line shift a little. Another reason could be the sluggish diffusion of elements in the high-entropy alloys, whereby the equilibrium state might not be reached in the annealing time used. The examples are the Mo_0 and $Mo_{0.1}$ alloys at 600 °C, where the calculated equilibrium phases in these two alloys are FCC, NiAl and σ phases, but only FCC and (Al,Ni)-rich ordered BCC phases have been identified from our experiments. Gwalani et al. reported that a different processing pathway resulted in different phases and the equilibrium phases in $Al_{0.3}$CoCrFeNi alloy were FCC + B2 + σ phases [27]. Rao et al. also reported that the stable phases in $Al_{0.5}CoCrFeNi$ alloy are BCC + $L1_2$ + σ phases [29]. Anyway, the calculated phase diagram could give us potential help in designing alloys and their corresponding annealing processes.

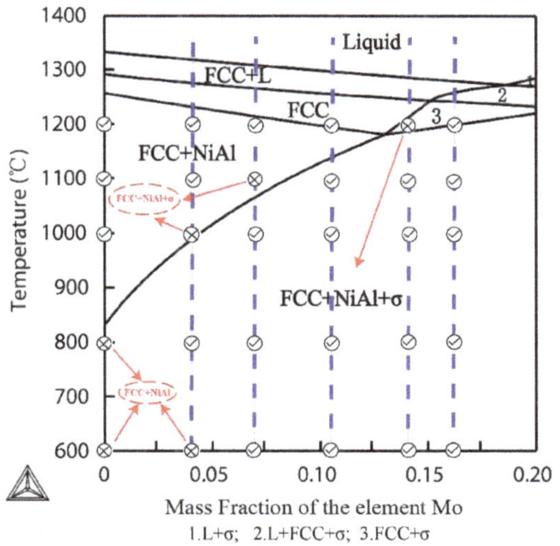

Figure 4. Pseudo binary phase diagram of $Al_{0.5}CoCrFeMo_xNi$ HEAs derived from Thermo-Calc. The dashed line represents the compositions of the alloys experimentally investigated in this work. The symbol ⊗ represents that the phase constituents of the alloys are different from the calculated results, and the text in red is the phase constituents of the alloys identified from experiments. The alloys marked with symbol ⊘ have the same phase constituents as calculated results.

3.4. Mechanical Properties of Annealed $Al_{0.5}CoCrFeMo_xNi$ Alloys

Figure 5 depicts the room temperature hardness of the $Al_{0.5}CoCrFeMo_xNi$ high entropy alloys annealed at various annealing temperatures. The data at 25 °C represent the values of as-cast alloys. The hardness of the alloys increases with the increasing Mo, and this can be attributed to the lattice distortion and formation of BCC and σ phases. On the other hand, with annealing temperature rising, the hardness of the alloys first increases, reaches its maximum at 800 °C, and then decreases afterwards. The alloys annealed at 800 °C have relatively fine precipitation and microstructure, which could be responsible for their higher hardness. Afterwards, the fine precipitations become larger and the effect of the interface strengthening become smaller.

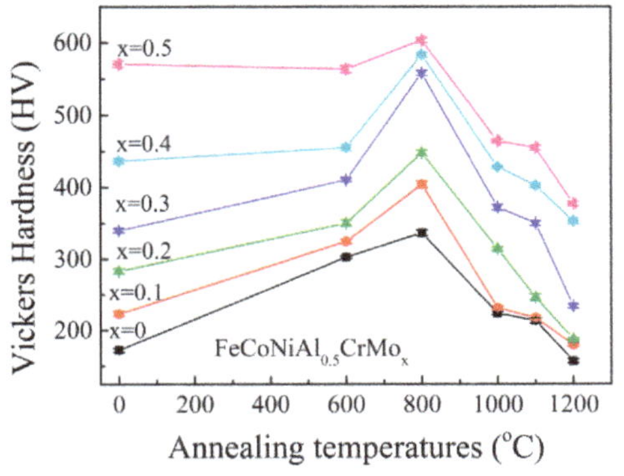

Figure 5. Hardness of $Al_{0.5}CoCrFeMo_xNi$ high-entropy alloys annealed at different temperatures for 10 h followed by water quenching.

Figure 6 presents the fracture strain and yield strength of the as-cast $Al_{0.5}CoCrFeMo_xNi$ high-entropy alloys and the alloys annealed at 800 °C for 10 h. It is clear that the fracture strain decreases, and the yield strength increases with Mo content increasing from 0 to 0.5. The alloys annealed at 800 °C have higher yield compressive strength and smaller compressive fracture strain than those of the as-cast alloys, which is attributed to the formation of fine (Al,Ni)-rich ordered BCC and (Cr,Mo)-rich σ phases. Both the Mo content and annealing processes have a great influence on the microstructure and mechanical properties of $Al_{0.5}CoCrFeMo_xNi$ high-entropy alloys. For the as-cast alloys, the $Mo_{0.3}$ and $Mo_{0.4}$ alloys can have balanced properties of compressive strength and ductility, while an appropriate annealing process can provide more chances to modify the mechanical properties of the alloys. For example, the $Mo_{0.5}$ alloy annealed at 800 °C can have a high compressive yield strength up to 2.1 GPa (ultimate fracture strength of 2.6 GPa) with an accepted fracture strain of 13%. The $Mo_{0.1}$ alloy annealed at 800 °C has a compressive yield strength of 1.2 GPa, an ultimate fracture strength of 2.2 GPa, and an ultimate fracture strain of 17%. Both the Mo content and annealing process can be used to tune the mechanical properties of the alloys.

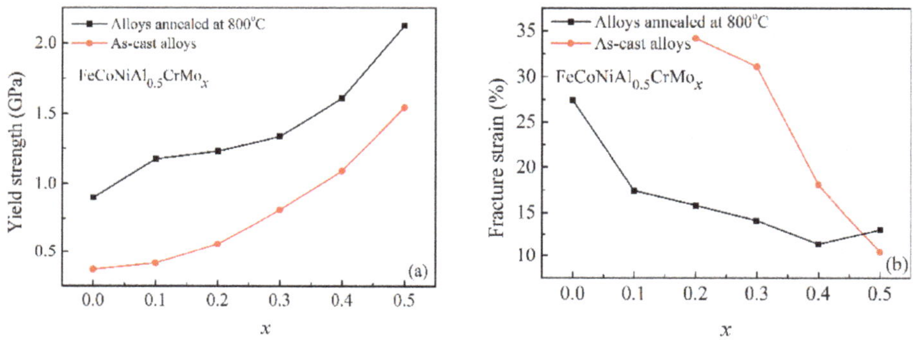

Figure 6. Yield strength (**a**) and fracture strain (**b**) of the as-cast alloys and the alloys annealed at 800 °C for 10 h.

4. Conclusions

The evolution of microstructure, phase and mechanical properties of the $Al_{0.5}CoCrFeMo_xNi$ high-entropy alloys have been investigated. Both the Mo content and annealing process can be used to tune the mechanical properties of the alloys. The following conclusions can be established from the current work.

(1) The annealing process at 600 °C for 10 h has no obvious effect on the microstructures of the $Al_{0.5}CoCrFeMo_xNi$, while it can increase its hardness to some extent. The annealing at 800–1200 °C for 10 h causes the precipitation of (Al,Ni)-rich ordered BCC phase or/and (Cr,Mo)-rich σ phase, and greatly affects the microstructure and mechanical properties of the alloys.

(2) The evolution of structure with temperature can be classified into four types:

   (a) Mo_0 alloy: mixed structure (FCC + BCC/B2).
   (b) $Mo_{0.1}$ alloy: mixed structure (FCC + BCC/B2) below 600 °C → FCC + BCC/B2+ σ (800–1000 °C) → FCC + BCC/B2 (1100–1200 °C).
   (c) $Mo_{0.2}$–$Mo_{0.3}$ alloys: mixed structure (FCC + BCC/B2 + σ) below 1100 °C → FCC + BCC/B2 (1200 °C).
   (d) $Mo_{0.4}$–$Mo_{0.5}$ alloys: mixed structure (FCC + BCC/B2 + σ).

(3) The alloys annealed at 800 °C for 10 h have relatively finer microstructure and higher hardness and higher yield stress than the as-cast alloys, which can be attributed to the solid-state decomposition of FCC and B2 phases. The precipitations become larger at 1000–1100 °C.

(4) The morphology and amount of each phase in the alloys vary with the Mo content and annealing temperatures. The $Mo_{0.5}$ alloy annealed at 800 °C have a high compressive yield strength up to 2.1 GPa with an accepted fracture strain of 13%.

Author Contributions: Y.-X.Z. and X.-L.Z. conceived and designed the experiments. X.-L.Z. prepared the $Al_{0.5}CoCrFeMo_xNi$ high entropy alloys and conducted the annealing processes. X.-L.Z. and X.-Y.G. performed the microstructural characterization and mechanical testing. Y.-X.Z. and X.-Y.G. analyzed the data. Y.-X.Z. wrote the paper. All authors have read and approved the final manuscript.

Funding: This research was funded by the Fundamental Research Funds for the Central University (Grant No. N150902001), Foundation of Liaoning Educational Committee for key laboratory (Grant No. LZ2015042), and NSF of China (Grant No. 51171041).

Conflicts of Interest: The authors declare no conflict of interest.

References

1. Cantor, B.; Chang, I.T.H.; Knight, P.; Vincent, A.J.B. Microstructural development in equiatomic multicomponent alloys. *Mater. Sci. Eng. A* **2004**, *375–377*, 213–218. [CrossRef]
2. Yeh, J.W.; Chen, S.K.; Lin, S.J.; Gan, J.Y.; Chin, T.S.; Shun, T.T.; Tsau, C.H.; Chang, S.Y. Nanostructured high-entropy alloys with multiple principal elements: novel alloy design concepts and outcomes. *Adv. Eng. Mater.* **2004**, *6*, 299–303. [CrossRef]
3. Zhang, Y.; Zuo, T.T.; Tang, Z.; Gao, M.C.; Dahmen, K.A.; Liaw, P.K.; Lu, Z.P. Microstructures and properties of high-entropy alloys. *Prog. Mater. Sci.* **2014**, *61*, 1–93. [CrossRef]
4. Tsai, M.H. Three Strategies for the design of advanced high-entropy alloys. *Entropy* **2016**, *18*, 252. [CrossRef]
5. Lu, Z.P.; Wang, H.; Chen, M.W.; Baker, I.; Jeh, J.W.; Liu, C.T.; Nieh, T.G. An assessment on the future development of high-entropy alloys: Summary from a recent workshop. *Intermetallics* **2015**, *65*, 67–76. [CrossRef]
6. Miracle, D.B.; Senkov, O.N. A critical review of high entropy alloys and related concepts. *Acta Mater.* **2017**, *122*, 448–511. [CrossRef]
7. Wang, W.R.; Wang, W.L.; Wang, S.C.; Tsai, Y.C.; Lai, C.H.; Yeh, J.W. Effects of Al addition on the microstructure and mechanical property of $Al_xCoCrFeNi$ high-entropy alloys. *Intermetallics* **2012**, *26*, 44–51. [CrossRef]
8. He, J.Y.; Liu, W.H.; Wang, H.; Wu, Y.; Liu, X.J.; Nieh, T.G.; Lu, Z.P. Effects of Al addition on structural evolution and tensile properties of the FeCoNiCrMn high-entropy alloy system. *Acta Mater.* **2014**, *62*, 105–133. [CrossRef]
9. Feng, R.; Gao, M.C.; Lee, C.; Mathes, M.; Zuo, T.; Chen, S.; Hawk, J.A.; Zhang, Y.; Liaw, P.K. Design of Light-Weight High-Entropy Alloys. *Entropy* **2016**, *18*, 333. [CrossRef]
10. Salishchev, G.A.; Tikhonovsky, M.A.; Shaysultanov, D.G.; Stepanov, N.D.; Kuznetsov, A.V.; Kolodiy, I.V.; Tortika, A.S.; Senkov, O.N. Effect of Mn and V on structure and mechanical properties of high-entropy alloys based on CoCrFeNi system. *J. Alloys Compd.* **2014**, *591*, 11–21. [CrossRef]
11. Qiu, Y.; Hu, Y.J.; Taylor, A.; Styles, M.J.; Marceau, R.K.W.; Ceguerra, A.V.; Gibson, M.G.; Liu, Z.K.; Fraser, H.L.; Birbilis, N. A lightweight single-phase AlTiVCr compositionally complex alloy. *Acta Mater.* **2017**, *123*, 115–124. [CrossRef]
12. Wang, Z.; Baker, I.; Guo, W.; Poplawsky, J.D. The effect of carbon on the microstructures, mechanical properties, and deformation mechanisms of thermo-mechanically treated $Fe_{40.4}Ni_{11.3}Mn_{34.8}Al_{7.5}Cr_6$ high entropy alloys. *Acta Mater.* **2017**, *126*, 346–360. [CrossRef]
13. Zhuang, Y.X.; Liu, W.J.; Chen, Z.Y.; Xue, H.D.; He, J.C. Effect of elemental interaction on microstructure and mechanical properties of FeCoNiCuAl alloys. *Mater. Sci. Eng. A* **2012**, *556*, 395–399. [CrossRef]
14. Zhou, Y.J.; Zhang, Y.; Wang, Y.L.; Chen, G.L. Solid solution alloys of $AlCoCrFeNiTi_x$ with excellent room-temperature mechanical properties. *Appl. Phys. Lett.* **2007**, *90*, 181904. [CrossRef]

15. Tracy, C.L.; Park, S.; Rittman, D.; Zinkle, S.J.; Bei, H.; Lang, M.; Ewing, R.C.; Mao, W.L. High pressure synthesis of a hexagonal close-packed phase of the high-entropy alloy CrMnFeCoNi. *Nat. Commun.* **2017**, *8*, 15634. [CrossRef] [PubMed]
16. Ng, C.; Guo, S.; Luan, J.H.; Wang, Q.; Lu, J.; Shi, S.Q.; Liu, C.T. Phase stability and tensile properties of Co-free $Al_{0.5}CrCuFeNi_2$ high entropy alloys. *J. Alloys Compd.* **2014**, *584*, 530–537. [CrossRef]
17. Wen, L.H.; Kou, H.C.; Li, J.S.; Chang, H.; Xue, X.Y.; Zhou, L. Effect of aging temperature on microstructure and properties of AlCoCrCuFeNi high-entropy alloy. *Intermetallics* **2009**, *17*, 266–269. [CrossRef]
18. Lin, C.M.; Tsai, H.L.; Bor, H.Y. Effect of aging treatment on microstructure and properties of high-entropy $Cu_{0.5}CoCrFeNi$ alloy. *Intermetallics* **2010**, *18*, 1244–1250. [CrossRef]
19. Jones, N.G.; Izzo, R.; Mignanelli, P.M.; Christofidou, K.A.; Tone, H.J. Phase evolution in an Al0.5CrFeCoNiCu high entropy alloys. *Intermetallics* **2016**, *71*, 43–50. [CrossRef]
20. Zhuang, Y.X.; Xue, H.D.; Chen, Z.Y.; Hu, Z.Y.; He, J.C. Effect of annealing treatment on microstructures and mechanical properties of FeCoNiCuAl high entropy alloys. *Mater. Sci. Eng. A* **2013**, *572*, 30–35. [CrossRef]
21. Wang, W.R.; Wang, W.L.; Yeh, J.W. Phases, microstructure and mechanical properties of $Al_xCoCrFeNi$ high entropy alloys at elevated temperatures. *J. Alloys Compd.* **2014**, *589*, 143–152. [CrossRef]
22. Lu, Y.; Dong, Y.; Guo, S.; Liang, L.; Kang, H.; Wang, T.; Wen, B.; Wang, Z.; Jie, J.; Cao, Z.; et al. A promising new class of high-temperature alloys: Eutectic high-entropy alloy. *Sci. Rep.* **2014**, *4*, 6200. [CrossRef] [PubMed]
23. Gao, X.; Lu, Y.; Zhang, B.; Liang, N.; Wu, G.; Sha, G.; Liu, J.; Zhao, Y. Microstructural origins of high strength and high ductility in an $AlCoCrFeNi_{2.1}$ eutectic high-entropy alloy. *Acta Mater.* **2017**, *141*, 59–66. [CrossRef]
24. Li, Z.; Pradeep, K.G.; Deng, Y.; Raabe, D.; Tasan, C.C. Metastable high-entropy dual-phase alloys overcome the strength-ductility trade-off. *Nature* **2016**, *534*, 227–230. [CrossRef] [PubMed]
25. Li, Z.; Tasan, C.C.; Pradeep, K.G.; Raabe, D. A TRIP-assisted dual-phase high-entropy alloy: Grain size and phase fraction effects on deformation behavior. *Acta Mater.* **2017**, *131*, 323–335. [CrossRef]
26. Zhuang, Y.X.; Zhang, X.L.; Gu, X.Y. Effect of molybdenum on phases, microstructure and mechanical properties of $Al_{0.5}CoCrFeMo_xNi$ high entropy alloys. *J. Alloys Compd.* **2018**, *743*, 514–522. [CrossRef]
27. Gwalani, B.; Gorsse, S.; Choudhuri, D.; Styles, M.; Zheng, Y.; Mishra, R.S.; Banerjee, R. Modifying transformation pathways in high entropy alloys or complex concentrated alloys via thermo-mechanical processing. *Acta Mater.* **2018**, *153*, 169–185. [CrossRef]
28. Lin, C.M.; Tsai, H.L. Evolution of microstructure, hardness, and corrosion properties of high-entropy $Al_{0.5}CoCrFeNi$ alloy. *Intermetallics* **2011**, *19*, 288–294. [CrossRef]
29. Rao, J.C.; Diao, H.Y.; Ocelik, V.; Vainchtein, D.; Zhang, C.; Kuo, C.; Tang, Z.; Guo, W.; Poplawsky, J.D.; Zhou, Y.; et al. Secondary phases in $Al_xCoCrFeNi$ high-entropy alloys: An in-situ TEM heating study and thermodynamic appraisal. *Acta Mater.* **2017**, *131*, 206–220. [CrossRef]
30. Choudhuri, D.; Gwalani, B.; Gorsse, S.; Mikler, C.V.; Ramanujan, R.V.; Gibson, M.A.; Banerjee, R. Change in the primary solidification phase from fcc to bcc-based B2 in high entropy or complex concentrated alloys. *Scr. Mater.* **2017**, *127*, 186–190. [CrossRef]
31. Tsai, M.H.; Chang, K.C.; Li, J.H.; Tsai, R.C.; Cheng, A.H. A second criterion for sigma phase formation in high entropy alloys. *Mater. Res. Lett.* **2016**, *4*, 90–95. [CrossRef]

entropy

MDPI

Article

Effect of Annealing on Microstructure and Tensile Behavior of CoCrNi Medium Entropy Alloy Processed by High-Pressure Torsion

Praveen Sathiyamoorthi [1,2], Jae Wung Bae [1,2], Peyman Asghari-Rad [1,2], Jeong Min Park [1,2], Jung Gi Kim [1] and Hyoung Seop Kim [1,2,3,*]

1 Department of Material Science and Engineering, Pohang University of Science and Technology (POSTECH), Pohang 37673, Korea; spravin88@gmail.com (P.S.); j.w.bae900@gmail.com (J.W.B.); peyman.asghari.rad@gmail.com (P.A.-R.); ack0401@gmail.com (J.M.P.); junggi91@gmail.com (J.G.K.)

2 Center for High Entropy Alloys, Pohang University of Science and Technology (POSTECH), Pohang 37673, Korea

3 Graduate Institute of Ferrous Technology (GIFT), Pohang University of Science and Technology (POSTECH), Pohang 37673, Korea

* Correspondence: hskim@postech.ac.kr; Tel.: +82-54-279-2150

Received: 19 September 2018; Accepted: 2 November 2018; Published: 6 November 2018

Abstract: Annealing of severely plastic deformed materials is expected to produce a good combination of strength and ductility, which has been widely demonstrated in conventional materials. In the present study, high-pressure torsion processed CoCrNi medium entropy alloy consisting of a single face-centered cubic (FCC) phase with a grain size of ~50 nm was subjected to different annealing conditions, and its effect on microstructure and mechanical behavior was investigated. The annealing of high-pressure torsion processed CoCrNi alloy exhibits partial recrystallization and near full recrystallization based on the annealing temperature and time. The samples annealed at 700 °C for 2 min exhibit very fine grain size, a high fraction of low angle grain boundaries, and high kernel average misorientation value, indicating partially recrystallized microstructure. The samples annealed for a longer duration (>2 min) exhibit relatively larger grain size, a low fraction of low angle grain boundaries, and low kernel average misorientation value, indicating nearly full recrystallized microstructure. The annealed samples with different microstructures significantly influence the uniform elongation, tensile strength, and work hardening rate. The sample annealed at 700 °C for 15 min exhibits a remarkable combination of tensile strength (~1090 MPa) and strain to failure (~41%).

Keywords: medium entropy alloy; high-pressure torsion; partial recrystallization; tensile strength

1. Introduction

High entropy alloys (HEAs), also known as compositionally complex alloys, baseless alloys, and concentrated solid solution alloys, exhibit unique and remarkable properties as compared with the conventional alloys [1–3]. The unique properties of HEAs are mainly attributed to the distinct alloy design concept based on multi-principal elements [4,5]. The alloys with multi-principal elements are widely classified into medium entropy alloys (MEAs) and HEAs based on the configurational entropy [5,6]. Among the HEAs, alloys based on the Co–Cr–Fe–Mn–Ni system, Al–Co–Cr–Cu–Fe–Ni system, and refractory elements are widely reported in the literature [1,2,5].

Recently, CoCrNi MEA has received widespread attention from researchers because of its unique properties such as exceptional fracture toughness at cryogenic temperature, high friction stress, superior dynamic shear properties, and annealing-induced hardening at intermediate temperatures (500–600 °C) [7–10]. It is widely reported that HEAs with a single face-centered cubic (FCC) phase exhibit low yield strength [11–13]. Several strategies have been reported to increase the strength of a single FCC phase HEAs by solid solution strengthening, precipitation strengthening, grain boundary strengthening, and partial recrystallization through thermo-mechanical treatments [11,14–16]. Among the several strategies to enhance the mechanical properties, thermo-mechanical treatment is an effective technique to achieve a good combination of strength and ductility by achieving microstructure with different grain size and/or different volume fraction of secondary phases [11,17,18]. The CoCrNi MEA with a single FCC phase, like any other HEA with a single FCC phase, also exhibits lower yield strength [14,19]. Recently, it has been reported that the addition of 3 at% W resulted in a 20% increase in intrinsic strength [14]. In another study, cold rolling followed by annealing of CoCrNi alloy is reported to have an ultra-fine-grained size (~750 nm) with a good combination of tensile strength (964 MPa) and strain to failure (66%) [20]. In this connection, high-pressure torsion (HPT) followed by annealing has been effectively used in conventional alloys and HEAs to improve their mechanical properties [21–23]. Generally, HPT is an effective process in fabricating materials with nano-grains or ultra-fine grained structure by simultaneous application of torsion and high hydrostatic pressure [22]. The HPT processed CoCrNi MEA is reported to have a grain size of ~50 nm with an exceptional ultimate tensile strength of ~2.2 GPa and a considerable strain to failure of 9% [24]. It is also reported that annealing of the HPT processed CoCrNi MEA in the temperature range of 500–600 °C has resulted in an increase in hardness due to a decrease in dislocation density and grain boundary relaxation [7]. However, the systematic investigation of post HPT annealing of CoCrNi MEA with respect to microstructure evolution and strength enhancement is limited.

Thus, the present study is aimed at achieving a better combination of strength and ductility in the HPT processed CoCrNi MEA by post-HPT annealing. Consequently, in the present study, the post-HPT annealing effect on the microstructural and mechanical behaviors of CoCrNi MEA is investigated.

2. Experimental Methods

The CoCrNi alloy for the present study was fabricated using a vacuum induction melting of pure metals followed by homogenization, cold rolling, recrystallization, HPT, and annealing of HPT samples. The detailed procedure about the fabrication and secondary processing route can be found in our earlier publications [7,24]. The microstructure of starting material for HPT consists of a single FCC phase with an average grain size of ~19 μm and profuse annealing twins [24]. After the HPT process, the microstructure consists of a very fine grain size of ~50 nm with a high density of nano-twins and stacking faults [24]. In this study, the HPT processed sample with a single FCC phase was subjected to different heat treatment conditions in the Ar atmosphere to investigate the effect of post-HPT annealing. The post-HPT annealed samples were characterized using X-ray diffraction (XRD, Rigaku D/MAX-2500), field emission scanning electron microscopy (FE-SEM, FEI XL30S) coupled with electron backscatter diffraction (EBSD), Vickers hardness, and tensile testing equipment (Instron 1361). The deformed microstructures of post tensile samples were characterized using a focused ion beam (FIB) coupled with transmission Kikuchi diffraction analysis (TKD, FEI-FIB/TKD).

Tensile samples (dimension: 1.5 mm gauge length and 1 mm gauge width) were cut from the annealed HPT disc using electric discharge machining. Tensile tests were carried out at a quasi-static strain rate of 10^{-3} s^{-1} and the strain on the sample was measured using digital image correlation (DIC, ARAMIS v6.1, GOM Optical Measuring Techniques). The Vickers hardness measurement was performed using a load of 300 gf and a dwell time of 15 s. For EBSD experiment, the samples were mechanically polished using SiC carbide papers followed by diamond and colloidal polishing. For TKD experiment, a transparent sample near to the fracture surface of deformed tensile samples was lifted out by FIB.

3. Results and Discussion

Figure 1 shows the XRD patterns of CoCrNi alloy under different processing conditions: HPT and post-HPT annealing. It can be seen that in all the processing conditions, the alloy exhibits a single FCC phase.

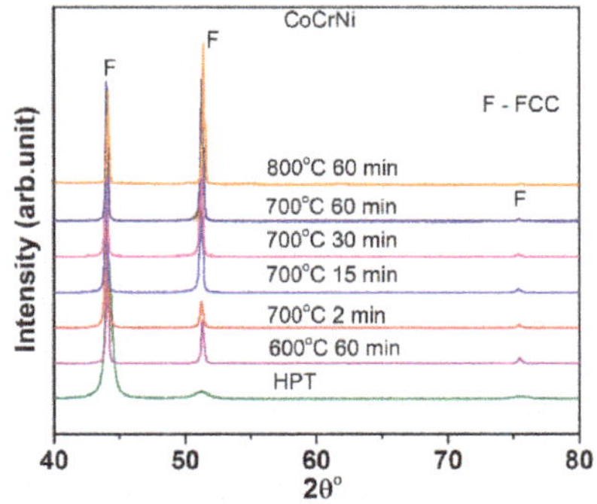

Figure 1. X-ray diffraction (XRD) patterns of CoCrNi alloy in different processing conditions illustrating the presence of a single face-centered cubic (FCC) phase.

The formation of sigma phase/secondary phases is widely reported for HEAs processed by HPT and annealing. In $Al_{0.5}$CoCrFeMnNi HEA, the formation of sigma phase is reported when the HPT processed sample is annealed at 800 °C for 1 h [25]. Similarly, for the HPT processed $CrFe_2NiMnV_{0.25}C_{0.125}$ HEA, sigma phase has been observed during annealing at 550 °C [26]. In CoCrFeMnNi HEA, the formation of sigma and secondary nano-phases have been reported after annealing of the HPT processed sample [27]. However, in the present alloy, there is no evidence of formation of sigma phase or secondary phase in the temperature range of 600–800 °C. The formation of sigma phase in this alloy was also not observed when annealed in the temperature range of 500–600 °C [7].

Figure 2 shows the inverse pole figure (IPF) map of the post-HPT annealed samples, and the corresponding kernel average misorientation (KAM) map and misorientation angle chart are shown in Figures 3 and 4, respectively.

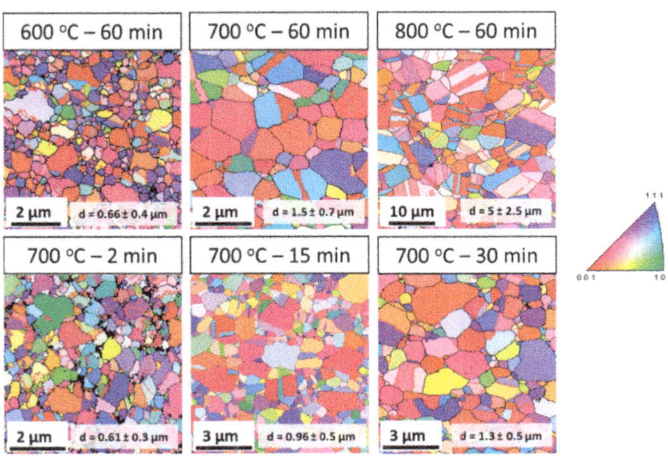

Figure 2. Electron backscatter diffraction (EBSD) inverse pole figure map of post-high-pressure torsion (HPT) annealing samples. The color scale on the right side corresponds to [001] inverse pole figure.

The IPF map indicates an increase in grain size with increasing annealing temperature and time. The samples annealed at 600 °C for 60 min and 700 °C for 2 min consist of a microstructure with almost similar average grain size (~600–670 nm), while the samples annealed at 700 °C over 2 min and 800 °C for 60 min show that the average grain size is in the range of approximately 960 nm to 5 μm.

The KAM map is generally used to represent the strain distribution based on the dislocation density. The KAM maps in Figure 3 were estimated up to the third nearest neighbor kernel with a maximum misorientation of 5°. The KAM map (Figure 3) shows that the strain distribution in samples annealed at 600 °C for 60 min and 700 °C for 2 min is relatively high as compared with samples at other annealing conditions. This clearly indicates that the strain induced into the material by HPT processing is not completely recovered in samples annealed at 600 °C for 60 min and 700 °C for 2 min. In general, deformed samples exhibit a larger KAM value (>1°), while recrystallized samples exhibit a lower KAM value (<1°) [28]. An almost similar average KAM value of approximately 0.83–0.84 is observed for samples annealed at 600 °C for 60 min and 700 °C for 2 min, while the average KAM value is in the range of 0.37°–0.45° for samples annealed at 700 °C over 2 min and 800 °C for 60 min.

The misorientation chart (Figure 4) indicates that the fraction of low angle grain boundaries is higher in samples annealed at 600 °C for 60 min and 700 °C for 2 min, but decreases significantly with increasing annealing time and temperature. However, a high fraction of 60° misorientation angle can be seen in all the conditions, with the fraction increasing with temperature and time. The presence of a high fraction of low angle grain boundaries indicates that the samples are not fully recrystallized, and the high fraction of 60° misorientation angle indicates the presence of abundant annealing twins.

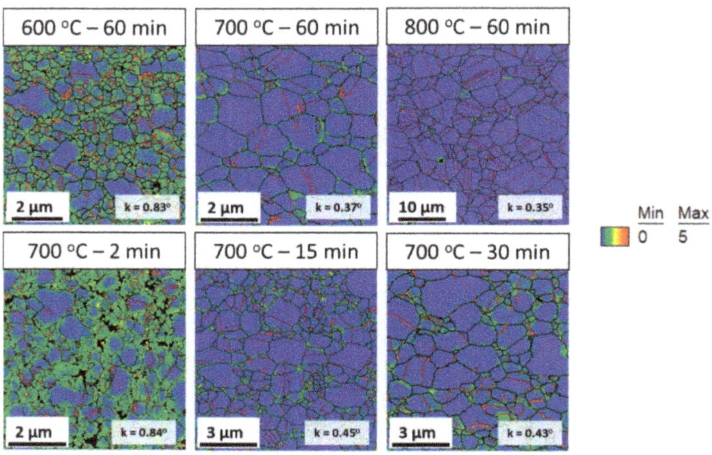

Figure 3. Kernel average misorientation (KAM) map of post-HPT annealing samples, and the corresponding color scale (right side).

Based on the IPF map, KAM map, and misorientation angle chart, the annealed samples can be divided into two categories: (a) microstructure with fine grain size, high KAM value, and high fraction of low angle grain boundaries (partially recrystallized); and (b) microstructure with relatively coarser grain size, low KAM value, and low fraction of low angle grain boundaries (nearly fully recrystallized). Accordingly, the sample annealed at 600 °C for 60 min, and 700 °C for 2 min will be referred to as samples with partially recrystallized microstructures. Similarly, the samples annealed at 700 °C for 15 min, 700 °C for 30 min, 700 °C for 60 min, and 800 °C for 60 min will be referred to as samples with recrystallized microstructures.

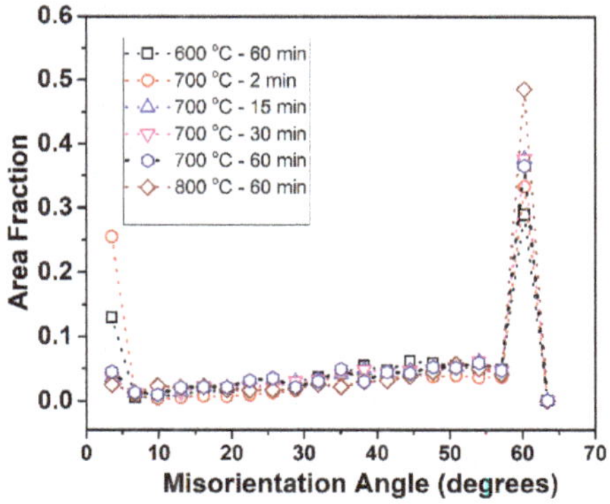

Figure 4. Misorientation angle chart of HPT processed CoCrNi alloy after annealing at different temperatures and times.

The mechanical properties of CoCrNi alloy at different processing conditions were assessed by Vickers hardness measurement and uniaxial tensile test. The hardness values under different processing conditions are shown in Figure 5a. The hardness of the HPT processed sample decreases with increasing annealing temperature and time. The annealed HPT samples with partially recrystallized microstructure show the smallest decrease in hardness as compared with the samples with near fully recrystallized microstructure. This can be attributed to the fine grain size and high KAM value in the samples with partially recrystallized microstructure.

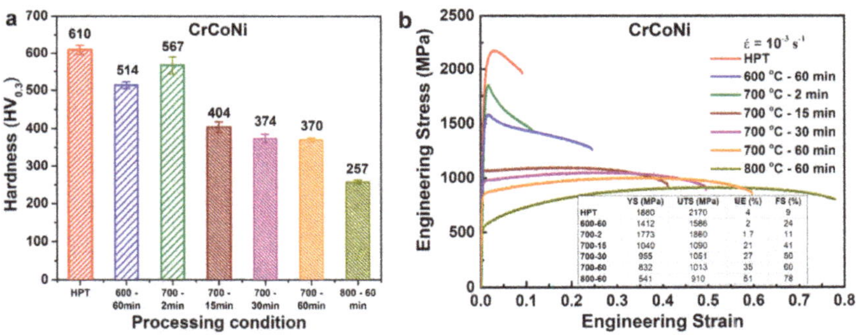

Figure 5. Mechanical properties of CoCrNi alloy in different processing conditions. (**a**) Hardness and (**b**) engineering stress–strain curve.

The engineering stress–strain curves of CoCrNi alloy at different processing conditions are shown in Figure 5b. The tensile curves clearly indicate that the annealing of the HPT processed sample leads to a reasonably good combination of strength and ductility. The tensile strength of the annealed samples follows a similar trend to that of hardness value. The annealed samples with partially recrystallized microstructure show the lowest yield strength reduction compared with the samples with recrystallized microstructure. In steels, it has been reported that the strength increases with an increase in pre-strain [29]. As KAM indicates the strain distribution based on the dislocation density, the annealed samples with high KAM can be regarded as a material with pre-strain. Thus, high strength in the samples with partially recrystallized microstructure can be attributed to the fine grain size and high KAM value.

The post-HPT annealing has significantly improved the uniform elongation and strain to fracture at the cost of a reduction in strength for samples with recrystallized microstructure. It is interesting to note that only the strain to fracture has been improved significantly and not the uniform elongation for the samples with partially recrystallized microstructure. The uniform elongations in the samples annealed at 600 °C for 60 min and 700 °C for 2 min are almost similar to each other and are also similar to the HPT processed sample. However, the strain to fracture is significantly higher in the samples annealed at 600 °C for 60 min and 700 °C for 2 min than the HPT processed sample, with the sample annealed at 600 °C for 60 min showing higher elongation to fracture.

In high-grade pipeline steels, it has been reported that the samples with a high percentage of low angle grain boundaries resulted in the smaller difference between yield strength and tensile strength [30]. In the present study, the samples with a high fraction of low angle grain boundaries exhibit a smaller difference between the yield strength and tensile strength. In addition, it has been shown that defects such as dislocation and vacancies can pass through the low angle grain boundaries, unlike the pile-up of defects at high angle grain boundaries [31]. This could be a possible reason for reduced work hardenability, thereby resulting in reduced uniform elongation in materials with a high fraction of low-angle grain boundaries. In addition, low angle grain boundaries are considered to be resistant to intergranular fracture in metallic materials [32]. This could be a possible reason for reasonable post-necking deformation in samples with partially recrystallized microstructure. Thus, the presence of low angle grain boundaries resulted in an increase in strain to fracture without enhancement in uniform elongation.

In conventional alloys, enhancement of uniform elongation in post-deformed annealed samples is observed when the fraction of recrystallized grains exceeds a certain volume fraction. In Cu–Al alloys, a reasonable improvement in uniform elongation is observed when the recrystallized grains are over 80% of the volume fraction [33]. In nanostructured Cu processed by rolling at liquid nitrogen, an increase in uniform elongation is reported when the fraction of secondary recrystallized grains is about 25% [34]. In the present study, an increase in uniform elongation is observed when the fraction of low-angle grain boundaries and the average KAM value is lower than 5% and 0.5°, respectively.

The shape of the flow curves (Figure 5) is distinct between the samples with partially recrystallized and recrystallized microstructures, clearly indicating the influence of microstructure on the flow curve. In samples with the partially recrystallized microstructure, the flow curve shows softening behavior. The rate of softening in the sample annealed at 600 °C for 60 min is different from that of the sample annealed at 700 °C for 2 min, with the one at 700 °C for 2 min showing pronounced softening. As the average KAM value and the average grain size are almost similar for 700 °C for 2 min and 600 °C for 60 min samples, the pronounced softening in the one at 700 °C for 2 min clearly indicates that the fraction of low angle grain boundaries significantly influences the softening behavior. Thus, the sample annealed at 700 °C for 2 min with a higher fraction of low-angle grain boundaries shows more pronounced softening than the 600 °C for 2 min sample. In the samples with recrystallized microstructure, the flow curves show strain hardening behavior, and the strain hardening ability is observed to increase with increasing annealing temperature and time.

In order to further understand the flow curve behavior, the work hardening rate (WHR) curves are plotted as a function of true strain for different processing conditions and are shown in Figure 6. In general, the WHR is not discussed with respect to plastic deformation after the plastic instability. However, the WHR after plastic instability is also shown in Figure 6a to understand the enhancement in strain to fracture without a significant enhancement of uniform elongation in samples with partial recrystallization. For the HPT sample, the WHR decreases rapidly and enters into the softening after yielding, and decreases continuously till fracture. This is commonly observed in severely deformed or nano-grained materials because of the limited dislocation activity and ineffectiveness in dislocation generation and accumulation during deformation. For the samples with a partially recrystallized microstructure, the WHR decreases initially and enters the softening region, and with further strain, the WHR curve increases and slowly approaches a peak, and then decreases gradually. The amount of increase in WHR from softening is higher for the sample annealed at 700 °C for 2 min than that at 600 °C for 60 min. Interestingly, the WHR curve of the 600 °C for 60 min sample increases after softening and reaches hardening region (i.e., from a negative value to positive value of WHR) and then decreases again as the strain approaches the fracture strain. As previously mentioned, the difference softening behavior in the partially recrystallized microstructure samples can be attributed to the difference in the fraction of low angle grain boundaries. The increase in WHR curve in the softening zone can be a possible reason for the increase in strain to failure without an increase in uniform elongation in the annealed samples with partially recrystallized microstructure.

The WHR curves of samples with recrystallized microstructure are shown in Figure 6b. The WHR curves show three distinct stages of WHR. In the first stage (Stage A), the WHR decreases rapidly at the initial stage of plastic deformation (elastic to plastic), and is observed in all the annealed samples. In 700 °C for 60 min and 800 °C for 60 min samples, the second stage (Stage B) shows a gradual decrease in WHR followed by a slow decrease (Stage C) with the increase in true strain. In the 700 °C for 30 min sample, the second stage of WHR is very narrow, and is not significant as in 700 °C for 60 min and 800 °C for 60 min samples. In the 700 °C for 15 min annealed sample, Stage B shows an increase in WHR followed by the slow decrease in WHR (Stage C).

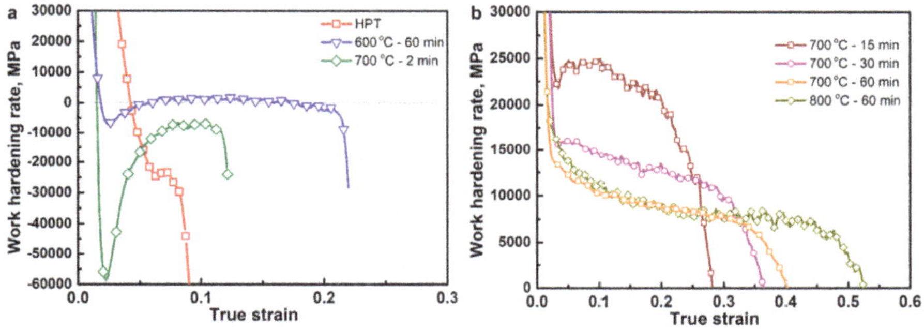

Figure 6. Work hardening rate as a function of true strain for samples under different processing conditions.

The difference in WHR stage, especially Stage B, of the annealed samples can be attributed to the difference in grain size and the amount of strain present in the sample before tensile testing [15,35]. The increase in WHR in Stage B has been reported for ultra-fine grained CoCrMnFeNi HEAs, and it is attributed to the yield drop phenomenon observed in ultra-fine grained materials [15]. In commercial pure Ti, it is demonstrated that the increase in WHR in Stage B is observed when the pre-strain reaches 3.5% and is attributed to the presence of high density of dislocations and deformation twinning in pre-strained samples. In the present study, the increase in WHR in Stage B of the 700 °C for 15 min sample could be attributed to the fine grain size (~960 nm) and the presence of strain (KAM value ~0.45°) (Figure 3). As the grain size and KAM value of the 700 °C for 30 min samples fall in between the those of the 700 °C for 15 min sample and 700 °C for 60 min sample, Stage B is observed only for a narrow strain as there could be a balance between increase and gradual decrease in WHR.

Figure 7 shows the deformed microstructure (transmission Kikuchi diffraction inverse pole figure map) near the fracture surfaces of the samples with partially recrystallized microstructure (600 °C for 60 min) and recrystallized microstructure (700 °C for 30 min). As a result of the large strain developed in the material, the confidence index is low, and there are some unindexed regions in Figure 7. Both the microstructures show the presence of fine annealing twins. It is to be noted that the formation of deformation twins is difficult to form in ultra-fine grained materials as the critical twining stress increases with decreasing grain size [36,37]. Indeed, the formation of deformation twinning is not observed in ultra-fine grained CoCrMnFeNi HEA [15]. The presence of very fine twins in Figure 7 could be annealing twins, and the deformation twins may not have formed as a result of the fine grain size of the annealed samples in the present study. Besides, the WHR curve also did not show a plateau region indicating the formation of twins during deformation.

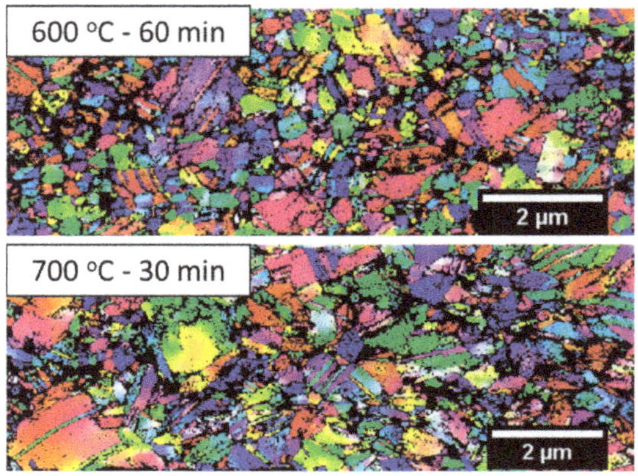

Figure 7. Transmission Kikuchi diffraction inverse pole figure map of post-deformed samples of 600 °C for 60 min and 700 °C for 30 min annealed samples.

Figure 8 shows the comparison of tensile properties of the present study with the conventional alloys and other high strength HEAs. (Figure 8 is adopted and modified from the literature [38]. The values for HEA wire with nano-twins are taken from the literature [39]). It is clear that the strengths of the CoCrNi in HPT processed state and after annealing conditions are quite high as compared with the other HEAs and other high strength conventional alloys.

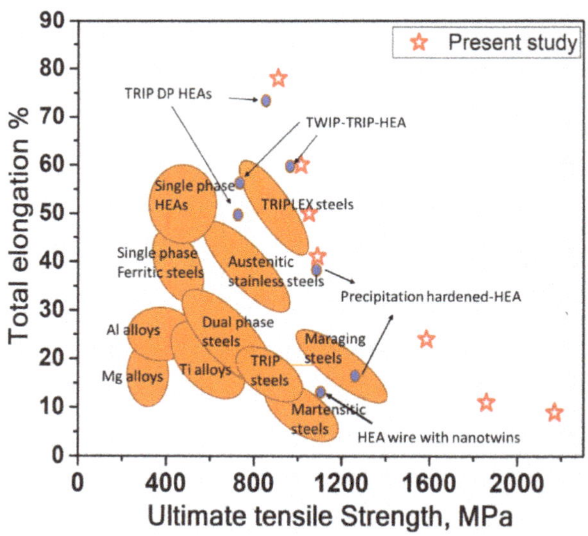

Figure 8. Comparison of tensile properties of HPT processed and post-HPT annealed CoCrNi alloy with other high strength high entropy alloys (HEAs) and conventional alloys (This figure is adopted and modified from the literature [38]. The values for HEA wire with nano-twins are taken from the literature [39]).

The superior mechanical properties achieved in the present alloy can be primarily attributed to the presence of fine grains, the formation of fine annealing twins, high shear stress, and high friction stress. It is well known that the strength increases with the decrease in grain size based on the Hall–Petch relation [9,40]. The presence of very fine twins can enhance the mechanical properties by reducing the dislocation mean free path and resisting the dislocation slip, a mechanism very similar to the

Hall–Petch relation [33]. The CoCrNi alloy is reported to have very low stacking fault energy and high shear stress [19,41]. In materials with high shear stress and low stacking fault energy, the cross-slip during deformation becomes difficult and thereby enhances the mechanical properties [36,42]. It is reported that the friction stress of CoCrNi alloy is higher than that of many conventional FCC metals because of the fluctuation of Peierls potential for dislocation motion, and it can enhance the strength of the material [9]. Thus, the presence of fine grains, fine annealing twins, high shear stress, high friction stress, and low-angle grain boundaries resulted in superior mechanical properties in the post-HPT annealed CoCrNi samples.

4. Summary

In the present study, the effect of annealing on the microstructural and mechanical properties of HPT processed CoCrNi HEA was investigated. The results clearly indicate that the initial microstructure plays an important role in the mechanical properties of CoCrNi alloy. The initial microstructure with fine grain size of ~660 nm, high fraction of low angle grain boundaries, and high KAM value shows strain softening followed by strain hardening, whereas only strain hardening is observed for the microstructure with grain size >900 nm, a low fraction of low-angle grain boundaries, and low KAM value. The sample annealed at 700 °C for 15 min shows the optimized room temperature mechanical properties with a remarkable combination of strength (1090 MPa) and strain to failure (41%). Thus, the present study demonstrates that the severe plastic deformation followed by annealing is one of the effective ways of enhancing the mechanical properties in CoCrNi MEA.

Author Contributions: Conceptualization, P.S.M.; Data curation, P.S.M.; Formal analysis, P.S.M. and J.G.K.; Investigation, P.S.M., J.W.B., P.A.-R., and J.M.P.; Methodology, P.S.M., J.W.B., P.A.-R., and J.M.P.; Supervision, H.S.K.; Writing—original draft, P.S.M.; Writing—review & editing, P.S.M., J.G.K., and H.S.K.

Funding: This research was supported by Future Materials Discovery Project through the National Research Foundation of Korea (NRF) funded by the Ministry of Science and ICT (NRF-2016M3D1A1023383). Praveen Sathiyamoorthi is supported by Korea Research Fellowship program through the National Research Foundation of Korea (NRF) funded by the Ministry of Science and ICT (2017H1D3A1A01013666).

Conflicts of Interest: The authors declare no conflict of interest.

References

1. Praveen, S.; Kim, H.S. High-Entropy Alloys: Potential Candidates for High-Temperature Applications—An Overview. *Adv. Eng. Mater.* **2018**, *20*, 1700645. [CrossRef]
2. Miracle, D.B.; Senkov, O.N. A critical review of high entropy alloys and related concepts. *Acta Mater.* **2017**, *122*, 448–511. [CrossRef]
3. Jo, Y.H.; Jung, S.; Choi, W.M.; Sohn, S.S.; Kim, H.S.; Lee, B.J.; Kim, N.J.; Lee, S. Cryogenic strength improvement by utilizing room-temperature deformation twinning in a partially recrystallized VCrMnFeCoNi high-entropy alloy. *Nat. Commun.* **2017**, *8*, 15719. [CrossRef] [PubMed]
4. Yeh, J.W.; Chen, S.K.; Lin, S.J.; Gan, J.Y.; Chin, T.S.; Shun, T.T.; Tsau, C.H.; Chang, S.Y. Nanostructured high-entropy alloys with multiple principal elements: novel alloy design concepts and outcomes. *Adv. Eng. Mater.* **2004**, *6*, 299–303. [CrossRef]
5. Murty, B.S.; Yeh, J.-W.; Ranganathan, S. *High-Entropy Alloys*; Butterworth-Heinemann: Oxford, UK, 2014.
6. Yeh, J.-W. Physical Metallurgy of High-Entropy Alloys. *JOM* **2015**, *67*, 2254–2261. [CrossRef]
7. Praveen, S.; Bae, J.W.; Asghari-Rad, P.; Park, J.M.; Kim, H.S. Annealing-induced hardening in high-pressure torsion processed CoCrNi medium entropy alloy. *Mater. Sci. Eng. A* **2018**, *734*, 338–340. [CrossRef]
8. Gludovatz, B.; Hohenwarter, A.; Thurston, K.V.; Bei, H.; Wu, Z.; George, E.P.; Ritchie, R.O. Exceptional damage-tolerance of a medium-entropy alloy CrCoNi at cryogenic temperatures. *Nat. Commun.* **2016**, *7*, 10602. [CrossRef] [PubMed]
9. Yoshida, S.; Bhattacharjee, T.; Bai, Y.; Tsuji, N. Friction stress and Hall-Petch relationship in CoCrNi equi-atomic medium entropy alloy processed by severe plastic deformation and subsequent annealing. *Scr. Mater.* **2017**, *134*, 33–36. [CrossRef]

10. Ma, Y.; Yuan, F.; Yang, M.; Jiang, P.; Ma, E.; Wu, X. Dynamic shear deformation of a CrCoNi medium-entropy alloy with heterogeneous grain structures. *Acta Mater.* **2018**, *148*, 407–418. [CrossRef]
11. Bae, J.W.; Moon, J.; Jang, M.J.; Yim, D.; Kim, D.; Lee, S.; Kim, H.S. Trade-off between tensile property and formability by partial recrystallization of CrMnFeCoNi high-entropy alloy. *Mater. Sci. Eng. A* **2017**, *703*, 324–330. [CrossRef]
12. Otto, F.; Dlouhý, A.; Somsen, C.; Bei, H.; Eggeler, G.; George, E.P. The influences of temperature and microstructure on the tensile properties of a CoCrFeMnNi high-entropy alloy. *Acta Mater.* **2013**, *61*, 5743–5755. [CrossRef]
13. Gali, A.; George, E.P. Tensile properties of high- and medium-entropy alloys. *Intermetallics* **2013**, *39*, 74–78. [CrossRef]
14. Wu, Z.; Guo, W.; Jin, K.; Poplawsky, J.D.; Gao, Y.; Bei, H. Enhanced strength and ductility of a tungsten-doped CoCrNi medium-entropy alloy. *J. Mater. Res.* **2018**, 1–9. [CrossRef]
15. Sun, S.J.; Tian, Y.Z.; Lin, H.R.; Dong, X.G.; Wang, Y.H.; Zhang, Z.J.; Zhang, Z.F. Enhanced strength and ductility of bulk CoCrFeMnNi high entropy alloy having fully recrystallized ultrafine-grained structure. *Mater. Des.* **2017**, *133*, 122–127. [CrossRef]
16. Seol, J.B.; Bae, J.W.; Li, Z.; Chan Han, J.; Kim, J.G.; Raabe, D.; Kim, H.S. Boron doped ultrastrong and ductile high-entropy alloys. *Acta Mater.* **2018**, *151*, 366–376. [CrossRef]
17. Ming, K.; Bi, X.; Wang, J. Precipitation strengthening of ductile Cr 15 Fe 20 Co 35 Ni 20 Mo 10 alloys. *Scr. Mater.* **2017**, *137*, 88–93. [CrossRef]
18. Otto, F.; Hanold, N.L.; George, E.P. Microstructural evolution after thermomechanical processing in an equiatomic, single-phase CoCrFeMnNi high-entropy alloy with special focus on twin boundaries. *Intermetallics* **2014**, *54*, 39–48. [CrossRef]
19. Wu, Z.; Bei, H.; Pharr, G.M.; George, E.P. Temperature dependence of the mechanical properties of equiatomic solid solution alloys with face-centered cubic crystal structures. *Acta Mater.* **2014**, *81*, 428–441. [CrossRef]
20. Dan Sathiaraj, G.; Skrotzki, W.; Pukenas, A.; Schaarschuch, R.; Jose Immanuel, R.; Panigrahi, S.K.; Arout Chelvane, J.; Satheesh Kumar, S.S. Effect of annealing on the microstructure and texture of cold rolled CrCoNi medium-entropy alloy. *Intermetallics* **2018**, *101*, 87–98. [CrossRef]
21. Shahmir, H.; Langdon, T.G. Using heat treatments, high-pressure torsion and post-deformation annealing to optimize the properties of Ti-6Al-4V alloys. *Acta Mater.* **2017**, *141*, 419–426. [CrossRef]
22. Zhilyaev, A.; Langdon, T. Using high-pressure torsion for metal processing: Fundamentals and applications. *Prog. Mater. Sci.* **2008**, *53*, 893–979. [CrossRef]
23. Torbati-Sarraf, S.A.; Sabbaghianrad, S.; Langdon, T.G. Using Post-Deformation Annealing to Optimize the Properties of a ZK60 Magnesium Alloy Processed by High-Pressure Torsion. *Adv. Eng. Mater.* **2018**, *20*. [CrossRef]
24. Praveen, S.; Bae, J.W.; Asghari-Rad, P.; Park, J.M.; Kim, H.S. Ultra-high tensile strength nanocrystalline CoCrNi equi-atomic medium entropy alloy processed by high-pressure torsion. *Mater. Sci. Eng. A* **2018**, *735*, 394–397. [CrossRef]
25. Reddy, T.S.; Wani, I.S.; Bhattacharjee, T.; Reddy, S.R.; Saha, R.; Bhattacharjee, P.P. Severe plastic deformation driven nanostructure and phase evolution in a Al 0.5 CoCrFeMnNi dual phase high entropy alloy. *Intermetallics* **2017**, *91*, 150–157. [CrossRef]
26. Shahmir, H.; Tabachnikova, E.; Podolskiy, A.; Tikhonovsky, M.; Langdon, T.G. Effect of carbon content and annealing on structure and hardness of CrFe2NiMnV0.25 high-entropy alloys processed by high-pressure torsion. *J. Mater. Sci.* **2018**, *53*, 11813–11822. [CrossRef]
27. Schuh, B.; Mendez-Martin, F.; Völker, B.; George, E.P.; Clemens, H.; Pippan, R.; Hohenwarter, A. Mechanical properties, microstructure and thermal stability of a nanocrystalline CoCrFeMnNi high-entropy alloy after severe plastic deformation. *Acta Mater.* **2015**, *96*, 258–268. [CrossRef]
28. Li, H.; Hsu, E.; Szpunar, J.; Utsunomiya, H.; Sakai, T. Deformation mechanism and texture and microstructure evolution during high-speed rolling of AZ31B Mg sheets. *J. Mater. Sci.* **2008**, *43*, 7148–7156. [CrossRef]
29. De, P.S.; Kundu, A.; Chakraborti, P.C. Effect of prestrain on tensile properties and ratcheting behaviour of Ti-stabilised interstitial free steel. *Mater. Des.* **2014**, *57*, 87–97. [CrossRef]
30. Xu, D.Y.; Yu, H. Effects of Low-Angle Grain Boundaries on the Yield-Strength Ratio of High Grade Pipeline Steels. *Adv. Mater. Res.* **2010**, *168–170*, 1581–1585. [CrossRef]

31. Zhang, Z.F.; Wang, Z.G. Comparison of fatigue cracking possibility along large- and low-angle grain boundaries. *Mater. Sci. Eng. A* **2000**, *284*, 285–291. [CrossRef]
32. Watanabe, T.; Tsurekawa, S. The control of brittleness and development of desirable mechanical properties in polycrystalline systems by grain boundary engineering. *Acta Mater.* **1999**, *47*, 4171–4185. [CrossRef]
33. An, X.H.; Wu, S.D.; Zhang, Z.F.; Figueiredo, R.B.; Gao, N.; Langdon, T.G. Enhanced strength–ductility synergy in nanostructured Cu and Cu–Al alloys processed by high-pressure torsion and subsequent annealing. *Scr. Mater.* **2012**, *66*, 227–230. [CrossRef]
34. Wang, Y.; Chen, M.; Zhou, F.; Ma, E. High tensile ductility in a nanostructured metal. *Nature* **2002**, *419*, 912–915. [CrossRef] [PubMed]
35. Chang, L.; Zhou, C.-Y.; He, X.-H. The Effects of Prestrain and Subsequent Annealing on Tensile Properties of CP-Ti. *Metals* **2017**, *7*. [CrossRef]
36. Tian, Y.Z.; Zhao, L.J.; Chen, S.; Shibata, A.; Zhang, Z.F.; Tsuji, N. Significant contribution of stacking faults to the strain hardening behavior of Cu-15%Al alloy with different grain sizes. *Sci. Rep.* **2015**, *5*, 16707. [CrossRef] [PubMed]
37. Komarasamy, M.; Kumar, N.; Tang, Z.; Mishra, R.S.; Liaw, P.K. Effect of Microstructure on the Deformation Mechanism of Friction Stir-Processed Al0.1CoCrFeNi High Entropy Alloy. *Mater. Res. Lett.* **2014**, *3*, 30–34. [CrossRef]
38. Li, Z.; Tasan, C.C.; Springer, H.; Gault, B.; Raabe, D. Interstitial atoms enable joint twinning and transformation induced plasticity in strong and ductile high-entropy alloys. *Sci. Rep.* **2017**, *7*, 40704. [CrossRef] [PubMed]
39. Huo, W.; Fang, F.; Zhou, H.; Xie, Z.; Shang, J.; Jiang, J. Remarkable strength of CoCrFeNi high-entropy alloy wires at cryogenic and elevated temperatures. *Scr. Mater.* **2017**, *141*, 125–128. [CrossRef]
40. Hansen, N. Hall–Petch relation and boundary strengthening. *Scr. Mater.* **2004**, *51*, 801–806. [CrossRef]
41. Liu, S.F.; Wu, Y.; Wang, H.T.; He, J.Y.; Liu, J.B.; Chen, C.X.; Liu, X.J.; Wang, H.; Lu, Z.P. Stacking fault energy of face-centered-cubic high entropy alloys. *Intermetallics* **2018**, *93*, 269–273. [CrossRef]
42. Okamoto, N.L.; Fujimoto, S.; Kambara, Y.; Kawamura, M.; Chen, Z.M.; Matsunoshita, H.; Tanaka, K.; Inui, H.; George, E.P. Size effect, critical resolved shear stress, stacking fault energy, and solid solution strengthening in the CrMnFeCoNi high-entropy alloy. *Sci. Rep.* **2016**, *6*, 35863. [CrossRef] [PubMed]

entropy

MDPI

Review

High-Pressure Induced Phase Transitions in High-Entropy Alloys: A Review

Fei Zhang [1,2], Hongbo Lou [1], Benyuan Cheng [1,3], Zhidan Zeng [1] and Qiaoshi Zeng [1,4,*]

1 Center for High Pressure Science and Technology Advanced Research, Pudong, Shanghai 201203, China; fei.zhang@hpstar.ac.cn (F.Z.); hongbo.lou@hpstar.ac.cn (H.L.); benyuan.cheng@hpstar.ac.cn (B.C.); zengzd@hpstar.ac.cn (Z.Z.)

2 State Key Laboratory for Advanced Metals and Materials, University of Science and Technology Beijing, Beijing 100083, China

3 China Academy of Engineering Physics, Mianyang 621900, China

4 Jiangsu Key Laboratory of Advanced Metallic Materials, School of Materials Science and Engineering, Southeast University, Nanjing 211189, China

* Correspondence: zengqs@hpstar.ac.cn; Tel.: +86-021-8017-7102

Received: 24 January 2019; Accepted: 26 February 2019; Published: 2 March 2019

Abstract: High-entropy alloys (HEAs) as a new class of alloy have been at the cutting edge of advanced metallic materials research in the last decade. With unique chemical and topological structures at the atomic level, HEAs own a combination of extraordinary properties and show potential in widespread applications. However, their phase stability/transition, which is of great scientific and technical importance for materials, has been mainly explored by varying temperature. Recently, pressure as another fundamental and powerful parameter has been introduced to the experimental study of HEAs. Many interesting reversible/irreversible phase transitions that were not expected or otherwise invisible before have been observed by applying high pressure. These recent findings bring new insight into the stability of HEAs, deepens our understanding of HEAs, and open up new avenues towards developing new HEAs. In this paper, we review recent results in various HEAs obtained using in situ static high-pressure synchrotron radiation x-ray techniques and provide some perspectives for future research.

Keywords: high pressure; polymorphic transition; high-entropy alloy

1. Introduction

Developing multicomponent metallic alloys with superior properties has played a vital role in the advancement of human civilizations since the Bronze Age. Conventional metallic alloys are usually based on one or two principle elements, such as Fe-, Al-, Mg-, and TiAl- based alloys. Adding alloying elements into the host lattice of the principle elements forming solid solutions has been the major strategy to optimize the microstructures and properties of alloys. However, in the traditional metallurgy, the development of alloys was restricted by the limited solubility of the alloying element in the solvent lattice, which leaves the areas besides the corner of their multicomponent phase diagrams unexplored. Deviation from the phase diagram corner was believed to readily result in the formation of useless, brittle intermetallic compounds. In 2004, Yeh et al. and Cantor et al.'s discoveries of single solid-solution phase alloys formed with multi-principal elements challenged the traditional metallurgy experience and established an exciting new concept for alloy design. By mixing five or more elements with equimolar or near-equimolar ratios, the system could be stabilized in a single phase solid-solution by their maximized configurational entropy [1,2]. Since then, this new class of so-called high-entropy alloys (HEAs) has attracted considerable attention and research interests in the advanced metallic materials community [3–5]. Numerous new HEAs have been developed with simple crystal structures such as face-centered cubic (*fcc*) [1,2], body-centered cubic (*bcc*) [1,6,7], and hexagonal close-packing (*hcp*) [8–12]. With unique compositions and atomic structures, HEAs show many interesting properties for potential applications, such as high ductility and strength over a wide temperature range and excellent resistance to both corrosion and wear [13–21]. On the other hand, due to their complex composition with a high chemical disorder, many fundamental questions of the HEAs remains challenging to address [3,22,23].

One of the critical unsolved questions about HEAs is their phase stability. Many pure elements in the periodic table show rich polymorphic phase transitions [24]. By mapping out the phase diagram (transition paths) with varying temperature and pressure, the phase stability of each structure can be clarified. A famous example is iron. Phase transitions between three different prototype polymorphs with *fcc*, *hcp*, and *bcc* structures were extensively studied in iron [25–27]. The *bcc* phase of iron was confirmed to be stable at ambient conditions, while the *fcc* phase is stable at high temperatures and the *hcp* phase is favorable at high pressures. As a combination of multiple elements, HEAs do not simply inherit the structure and properties of their constituent elements as expected with a "linear effect" [28]. Regardless of the various compositions and structures, HEAs are reported to be surprisingly stable over a large temperature range in previous experiments. It is believed that HEAs are thermodynamically stabilized by their high configurational entropy which can largely lower down the Gibbs free energy. Also, the high chemical complexity and packing disorder cause severe local lattice distortion and extremely sluggish atomic diffusion in HEAs, which could further stabilize HEAs kinetically. For instance, the equiatomic CoCrFeMnNi alloy (also named as Cantor's alloy) [2], is a prototype *fcc*-structured HEA. Extensive studies demonstrate that the Cantor's alloy can maintain its *fcc* structure from cryogenic temperatures up to its melting temperature without any polymorphic phase transition [13,29–31].

It has been empirically established that the competition between configuration entropy and enthalpy, the difference between the atomic radius and electronegativity of constituent elements [32–34], and also the overall valence electron concentration [35] are a few key thermodynamic parameters for the formation of HEAs. All these parameters are very susceptible to pressure tuning. Actually, pressure is a very powerful tool to tune the atomic/electronic structure of various materials and has been employed to understand materials and to search for novel materials through rich pressure-induced phase transitions in diverse systems, such as pure elements [24], alloys [36–42], oxides [43–46], and metallic glass [47–49]. Among them, the Ce_3Al system is of particular interest, where a pressure-induced intermetallic compound and metallic glass to *fcc* solid solution transitions were discovered due to the significant reduction of the difference between both the atomic radii and electronegativity of Ce and Al during compression [42,49]. For HEAs, their synthesis process is closely associated with the competition between intermetallic compounds, metallic glasses, and simple crystalline solid solutions. Therefore, in contrast to the seeming "ultra-stability" during heating or cooling, HEAs might exhibit rich tunable behavior under high pressure.

Very recently, the structural stability of various HEA systems has been explored using in situ high-pressure synchrotron radiation-based x-ray diffraction techniques. Many interesting polymorphic transitions have been discovered. These results are summarized in Table 1 and are subsequently reviewed in detail. The microstructural and compositional metastability of HEAs was nicely reviewed by Wei et al. [50]. In this paper, we focus on very recent results about phase stability and transitions under high pressure and provide a brief review of the relevant experimental methods, issues, and perspectives for future study.

Table 1. A summary of the pressure-induced polymorphic transitions in high-entropy alloys (HEAs) investigated by in situ high-pressure XRD (ME: methanol:ethanol = 4:1 (volume ratio) mixture; MEW: methanol:ethanol:water = 16:3:1 (volume ratio) mixture).

Composition	Initial Structure	Synthesis Method	Grain Size (μm)	Pressure Medium	Max. P (GPa)	Transition P (GPa)	Phase Transition	Ref.
CoCrFeMnNi	*fcc*	Homogenization	/	Silicone oil	54.1	14.7	*fcc-hcp*	[51]
CoCrFeMnNi	*fcc*	Gas-atomization	~ 5	Helium	22.1	41.1	*fcc-hcp*	[52]
CoCrFeMnNi	*fcc*	Gas-atomization	~ 5	Silicone oil	6.9	31.4	*fcc-hcp*	[53]
CoCrFeMnNi	*fcc*	Gas-atomization	~ 5	Amorphous boron	2.2–6.6	36.6	*fcc-hcp*	[53]
CoCrFeMnNi	*fcc*	High pressure tortion	~ 0.01	Silicone oil	12.3	31.4	*fcc-hcp*	[53]
CoCrFeMnNi	*fcc*	Cold rolled	~ 100	Neon	7.1	20	*fcc-hcp*	[54]
CoCrFeMnNi	*fcc*	/	/	Neon	/	48.9	*no*	[55]
CoCrFeMnNi	*fcc*	High pressure sintered	~ 0.1	Silicone oil	/	31	*no*	[56]
CoCrFeCuNi	*fcc*	High pressure sintered	~ 0.1	Silicone oil	/	31	*no*	[56]
NiCoCrFe	*fcc*	Homogenization	/	ME	13.5	39	*fcc-hcp*	[57]
NiCoCr	*fcc*	Homogenization	/	ME	45	45	*fcc-hcp*	[57]
NiCoCrFePd	*fcc*	Homogenization	/	Neon	/	74	*no*	[57]
CoCrCuFeNiPr	*dual fcc*	Milled powder	/	Silicone oil	/	106.4	*disordered-ordered fcc*	[58]
$Al_{0.3}CoCrFeNi$	*fcc*	Annealed	~ 100	Neon	/	61	*no*	[59]
AlCoCrCuFeNi	*fcc+bcc*	As-cast	/	MEW	/	24	*no*	[60]
AlCoCrFeNi	*bcc*	Melt-spun ribbon	/	Silicone oil	/17.6	42	*σ2–distorted bcc*	[61]
$Al_{0.6}CoCrFeNi$	*bcc*	Gas-atomization	~ 10	Silicone oil/Helium	10.6	40	*bcc-orthorhombic-bct*	[62]
$Al_{0.6}CoCrFeNi$	*fcc*	Gas-atomization + Annealed	~ 5	Silicone oil/Helium	17.5	40	*fcc-hcp*	[62]
TiZrHfNb	*bcc*	/	/	Neon	/	50.8	*no*	[55]
$Al_2CoCrFeNi$	*bcc*	Annealed	~ 100	Neon	/	61	*no*	[59]
$(TaNb)_{0.67}(HfZrTi)_{0.33}$	*bcc*	As-cast	/	/	/	96	*no*	[63]
ReRuCoFe	*hcp*	/	/	Neon	/	80.4	*no*	[55]
$Ir_{0.19}Os_{0.22}Re_{0.21}Rh_{0.20}Ru_{0.19}$	*hcp*	/	/	/	/	45	*no*	[64]
HoDyYGdTb	*hcp*	As-cast	/	Silicone oil	/	60.1	*hcp*→Sm-type→d*hcp*→d*fcc*	[65]

2. Experimental Methods

Diamond anvil cells (DACs) are the most commonly used device to generate high pressure for the in situ studies of materials. DACs are versatile devices for generating pressures up to hundreds of GPa and for combining a full range of in situ measurements [66]. DACs are composed of two opposing diamond anvils that squeeze materials in between them to generate hydrostatic/non-hydrostatic pressure. Since diamonds are transparent to almost the entire electromagnetic spectrum, various in situ electromagnetic radiation detection approaches can be employed to study the structure and properties of samples inside DACs, such as in situ x-ray and neutron diffraction techniques, x-ray emission/absorption spectroscopy, Raman spectroscopy, and Brillouin scattering. Among them, in situ angular dispersive x-ray diffraction (XRD) based on an intense synchrotron radiation x-ray source is the primary technique in the high-pressure structural study of various materials [67].

The size of the DAC anvil culets is typically small with a diameter of approx. 500 μm to 20 μm depending on the target maximum pressure. The sample chamber is a small hole (with a diameter of approx. 1/3 of the culet size and a height of approx. 50 μm to 30 μm) drilled in the gasket indent which is pre-indented by the two anvils. Therefore, only tiny samples with a typical maximum dimension less than tens of microns can be accommodated in the sample chamber in a DAC. Particularly, the sample thickness should be less than the height of the gasket hole during the entire compression process to avoid bridging the anvils (otherwise, it will be uniaxially compressed with large shear stress).

Metals with high shear strength such as T301 stainless steel, Re, and W are usually used as the gasket materials for typical high-pressure experiments. Different pressure transmitting mediums (including solid, liquid, and gas) can be selected based on the target pressure range and the required degree of hydrostaticity or operational convenience in the experiment. The pressure in the sample chamber can be determined either by the pressure–volume (*P*–*V*) equation of state (EOS) of standard materials (e.g., MgO, NaCl, Au, and Pt) using in situ high-pressure XRD [68] or by the ruby fluorescence peak shift excited by an optical laser or x-ray [69]. The standard material or ruby balls should be loaded close to the sample in the sample chamber to minimize the pressure difference. Due to the small sample volume, a high-brightness synchrotron radiation X-ray beam focused down to tens of microns (<20 μm) is required. To go through the two thick diamond anvils (3–5 mm in total) and to cover a large enough range in *d* space with the limited two-theta opening of DACs, a high-energy x-ray is usually required (>20 keV). Using in situ high-pressure synchrotron radiation XRD diffraction coupled with a DAC (a typical experimental setup is shown in Figure 1), researchers have been able to obtain detailed structural information of samples as a function of pressure, including their unit cell parameters, atomic positions, thermal parameters, and even electron density distributions.

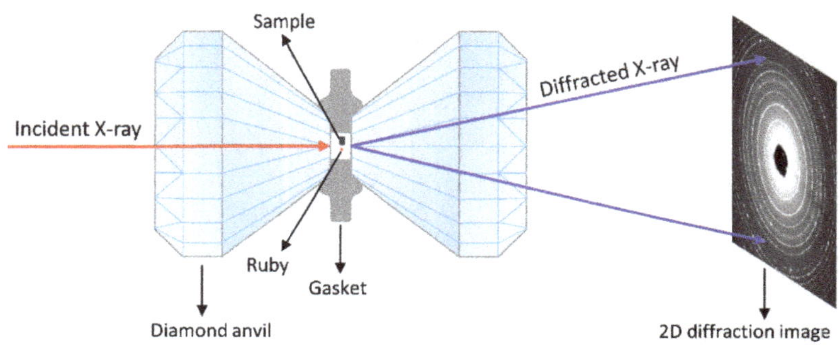

Figure 1. A schematic illustration of the in situ high-pressure synchrotron radiation XRD setup using a diamond anvil cell (DAC).

3. Structural Stability and Evolution of HEAs under High Pressure

Over the last fourteen years, intense effort has been devoted to developing numerous HEAs, which provides hundreds of new alloys for fundamental and applied studies. Recently, the structural stability of some typical HEAs has been investigated using in situ high-pressure synchrotron radiation XRD. Herein, we briefly review these results in separate groups according to their initial crystal structures.

3.1. *Fcc*-Structured HEAs

Among the various HEAs, single-phase *fcc*-structured alloy systems tend to show a relatively low yield strength but an excellent ductility and strain hardening capability. The CoCrFeMnNi (Cantor's alloy) as a prototype *fcc*-structured HEA has attracted the most extensive investigation. Cantor's alloy shows high structural stability over a broad temperature range at ambient pressures (from extreme low temperature (~3 K) up to its melting temperature) [13,29–31]. In contrast, Cantor's alloy undergoes unexpected polymorphic transformations from *fcc* to *hcp* phases under applied high pressure [51–54]. Under the best hydrostatic pressure conditions provided by the pressure-medium helium, the *fcc* to *hcp* phase transition starts at approx. 22 GPa but does not fully complete even up to approx. 41 GPa (Figure 2a) [52]. The transition is sluggish and irreversible. The *hcp* phase can be retained during the decompression down to ambient pressure, as shown in Figure 2b. The fabricated *hcp* CoCrFeMnNi HEA can almost maintain its volume fraction during decompression.

Therefore, *hcp-fcc* dual-phase composites with tunable volume fractions can be readily synthesized by decompression from different maximum pressures between approx. 22 and 41 GPa (Figure 2a) [52].

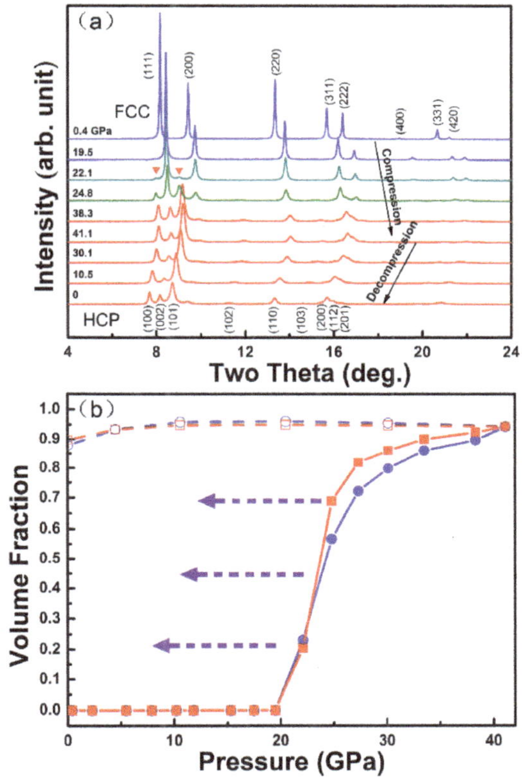

Figure 2. (**a**) The in situ high-pressure XRD patterns of the CoCrFeMnNi HEA under high pressure in a DAC at room temperature [52]: The x-ray wavelength is 0.2952 Å and (**b**) the change of the *hcp* phase volume fraction as a function of pressure during compression (solid symbols) and decompression (open symbols). The volume fractions of the *hcp* phase were calculated based on the peak area changes of the *fcc*-(200) (blue circles) and *hcp*-(101) peaks (red squares) in Figure 2a, which yield consistent results of the volume fractions. During decompression, the volume fraction of the *hcp* almost remains constant, which makes the synthesis of the *hcp-fcc* dual-phase composite possible by following a different decompression path, as shown by the dashed arrows [52].

According to previous theoretical simulations, the Gibbs free energy of the *hcp* phase of the CoCrFeMnNi HEA may be smaller than its well-known *fcc* phase at room temperature [70,71]. However, there was no clear experimental evidence to confirm the simulation results. For example, the CoCrFeMnNi HEA samples synthesized by various melt-quenching methods always only show an *fcc* structure, and the irreversibility of the *fcc* to *hcp* phase transition under high pressure also questions the relative stability of the *fcc* and *hcp* phases [51,53,54]. To clarify the phase stability, the synthesized *hcp* phase was further examined using in situ high-temperature XRD measurements at different pressures [52]. During heating at constant pressures, the *hcp* phase transforms back to the *fcc* phase and the critical transition temperatures increase with increasing pressure. Therefore, these results demonstrate that the well-known *fcc* phase of the CoCrFeMnNi HEA is thermodynamically favorable at high temperatures. In contrast, the *hcp* phase is indeed more stable at relatively lower temperatures and higher pressures [52].

The pressure-induced *fcc* to *hcp* polymorphic transition has been observed in Cantor's alloy by independent research groups. However, quite different onset pressures were reported ranging from approx. 7 GPa to even above 49 GPa [51,53–55]. Tracy et al. loaded a CoCrFeMnNi HEA sample (annealed at 1200 °C for 24 h) in a DAC with silicone oil as the pressure medium and compressed it up to 54.1 GPa. A sluggish martensitic transformation from the *fcc* to an *hcp* phase was also observed starting at approx. 14 GPa [51]; Huang et al. reported that the *fcc* to *hcp* phase transition occurred at approx. 7.1 GPa in a CoCrFeMnNi HEA sample. The initial sample was processed by a series of heating, cold rolling, and milling and then loaded into a DAC with neon as the pressure medium [54]; Yu et al. prepared a nanograined (approx. 100 nm) CoCrFeMnNi HEA sample by mechanical alloying and a high-pressure sintering process. They compressed the sample with silicone oil as the pressure medium in a DAC up to 31GPa, but no phase transition was observed [56]. Ahmad et al. investigated the structure of Cantor's alloy up to approx. 49 GPa with neon as the pressure-transmitting medium in a DAC. Surprisingly, no obvious phase transition was observed as well [55]. This experimental inconsistency suggests that the polymorphic phase transition in Cantor's alloy may be susceptible to the sample and experimental conditions, such as the different sample grain sizes and different pressure mediums used in the experiments above.

To clarify this speculation, Zhang et al. [52] systematically investigated the effect of the non-hydrostaticity of the pressure environment and the grain size of the samples on their pressure-induced phase transitions. The experiments were carefully designed to study only one factor at each time. To address the effect of the pressure environment, they loaded the same sample with three distinct pressure mediums with different degrees of hydrostaticity, such as helium (the most hydrostatic), amorphous boron (the most non-hydrostatic), and silicone oil (quasi-hydrostatic in-between). According to the in situ high-pressure XRD results (as shown in Figure 3), the onset pressures for the *fcc* to *hcp* transition were estimated to be approx. 22 GPa in helium, approx. 2 to 6 GPa in amorphous boron, and approx. 7 GPa in silicone oil. These results demonstrate that the degree of the pressure medium's hydrostaticity has a positive effect on the onset pressure of the *fcc*-to-*hcp* phase transition in Cantor's alloy [53].

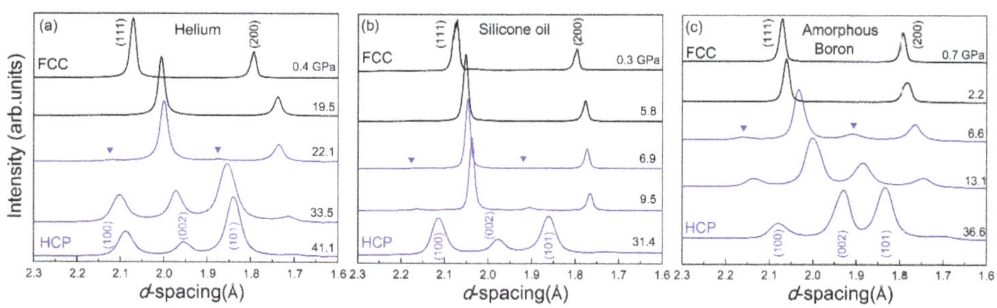

Figure 3. The in situ high-pressure XRD patterns of the CoCrFeMnNi HEA sample with helium (**a**), silicone oil (**b**), and amorphous boron (**c**) as the pressure mediums [53].

To study the effect of grain size (an important internal factor) on the pressure-induced phase transition in the CoCrFeMnNi HEA, Zhang et al. [53] loaded two distinct samples into one sample chamber in a symmetric DAC. The two samples were carefully located with equivalent positions to the chamber center to ensure they had identical pressure environments. To highlight the grain size effect, the two selected samples had a huge difference in grain size; one was synthesized by gas-atomization (GA) (approx. 5 μm), and the other was obtained by high-pressure torsion (HPT) (approx. 10 nm). When the two samples were compressed from 0.3 GPa up to 31.4 GPa, both of them showed an *fcc*-to-*hcp* phase transition but with quite different onset pressures (as shown in Figure 4). For the HPT sample with nano-sized grains, the phase transition was observed from approx. 12.3 GPa, while the GA sample with a bigger grain size had a much lower onset pressure of approx. 6.9 GPa. The underlying mechanism is still not clear and calls for further investigation [53].

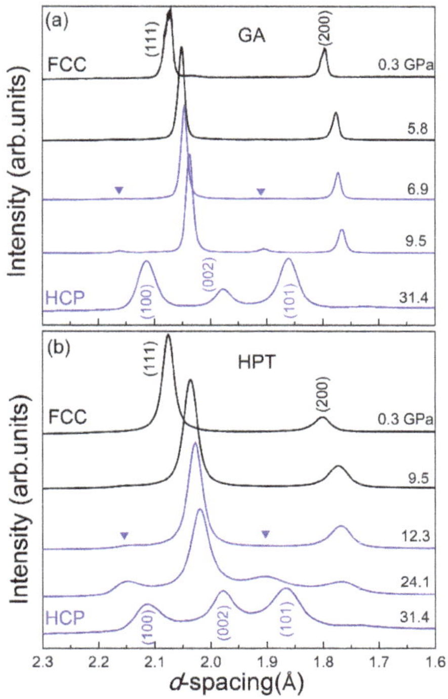

Figure 4. The in situ high-pressure XRD patterns of the gas-atomization (GA) and high-pressure torsion (HPT) CoCrFeMnNi HEA samples loaded in the same DAC with silicone oil as their pressure medium [53].

The effect of different alloying elements on the pressure-induced polymorphic phase transition was investigated by Zhang et al. [57] in three *fcc*-structured medium-entropy alloys and HEAs (NiCoCr, NiCoCrFe, and NiCoCrFePd). They observed a similar martensitic phase transition from *fcc* to *hcp* in the CoCrFeNi alloy starting at approx. 13.5 GPa, as shown in Figure 5a. This phase transformation was also sluggish and irreversible, as reported in Cantor's alloy. The volume fraction of the *hcp* phase was only about 36% when the pressure reached 39 GPa. However, with different alloying elements, the NiCoCr and NiCoCrFePd alloys exhibited distinct compression behaviors under high pressure, as shown in Figure 5b,c, respectively. Only a small amount (<5 wt.%) of the *hcp* phase emerged at 34.4 GPa in the NiCoCr alloy, and its amount barely changes with increasing pressure. In the NiCoCrFePd system, as the element Mn in the Cantor's alloy is replaced by the Pd, no obvious *hcp* phase emerges up to 74 GPa [57].

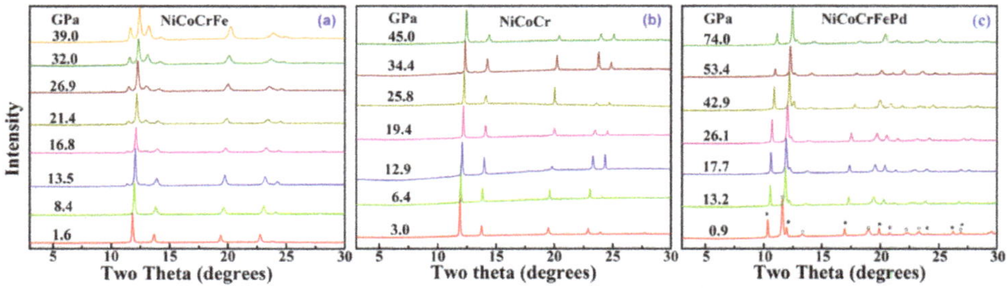

Figure 5. The XRD profiles of CoCrFeNi (**a**), CoCrNi (**b**), and CoCrFeNiPd (**c**) at high pressures: The diffraction peaks marked by symbol * are from the Au pressure standard [57].

Many of the HEAs have a minor second phase. Ma et al. [58] investigated an equiatomic CoCrCuFeNiPr HEA sample with dual phases (major disordered-*fcc* and minor ordered-*fcc* phases) using in situ synchrotron radiation high-pressure energy-dispersive XRD (EDXRD). They observed a pressure-induced fast ordering transition from approx. 8 GPa to 16.0 GPa followed by a slow transition up to 106.4 GPa. The initially ordered domain in the CoCrCuFeNiPr HEA was believed to act as embryos. With increasing pressure, the embryos grow into the ordered phase [58].

3.2. Bcc-Structured HEAs

In addition to the *fcc*-structured HEAs, *bcc*-structured alloys are another major member of the HEA family, which include both the chemically disordered A2 and ordered B2 phases. *Bcc*-structured HEAs often exhibit high yield strength in a very high-temperature regime. Therefore, the stability of the *bcc*-structured HEAs is an exciting topic that has been explored extensively at various conditions, recently also by high pressure.

Ahmad et al. investigated the structural stability of a TiZrHfNb alloy with a disordered *bcc* structure during compression up to 50.8 GPa; no phase transition was found [55]. Yusenko et al. explored another *bcc*-structured $Al_2CoCrFeNi$ HEA; it had no phase transition up to 60 GPa as well [59]. Guo et al. studied the structural evolution of a superconducting $(TaNb)_{0.67}(HfZrTi)_{0.33}$ HEA during compression up to approx. 100 GPa; its *bcc* structure seemed very robust without any detectable structural transition [63]. No phase transition has ever observed in the *bcc*-structured HEAs. Therefore, it seems that the *bcc*-structured HEAs are incredibly stable, and much higher pressure may be needed to induce phase transitions (compared to the *fcc* family).

To lower down the transition pressure of possible polymorphic phase transitions, Cheng et al. [61] employed a creative strategy to focus on relatively less stable compositions. They chose an equiatomic AlCoCrFeNi HEA and monitored its structural evolution during compression up to 42 GPa. The AlCoCrFeNi alloy had an ordered *bcc*-structure (B2 phase) and was reported to sit in the transition zone between the *fcc* and *bcc* phases, as x varies in the $Al_xCoCrFeNi$ HEA system ($0 < x < 2$) [72]. Indeed, they discovered a phase transition from the initial B2 phase to a highly distorted form starting at relatively low pressure of approx. 17.6 GPa, by combining ex situ high-resolution transmission electron microscope (HRTEM) with in situ high-pressure synchrotron radiation XRD data, as shown in Figure 6. Besides the XRD peak splitting, severe peak weakening and broadening occurred during compression, which may have been caused by the significant lattice distortion developed in the sample. Therefore, their work was unable to resolve the atomic structure of the high-pressure phase. Nevertheless, it is the first time that a pressure-induced polymorphism was suggested in a *bcc*-structured HEA [61].

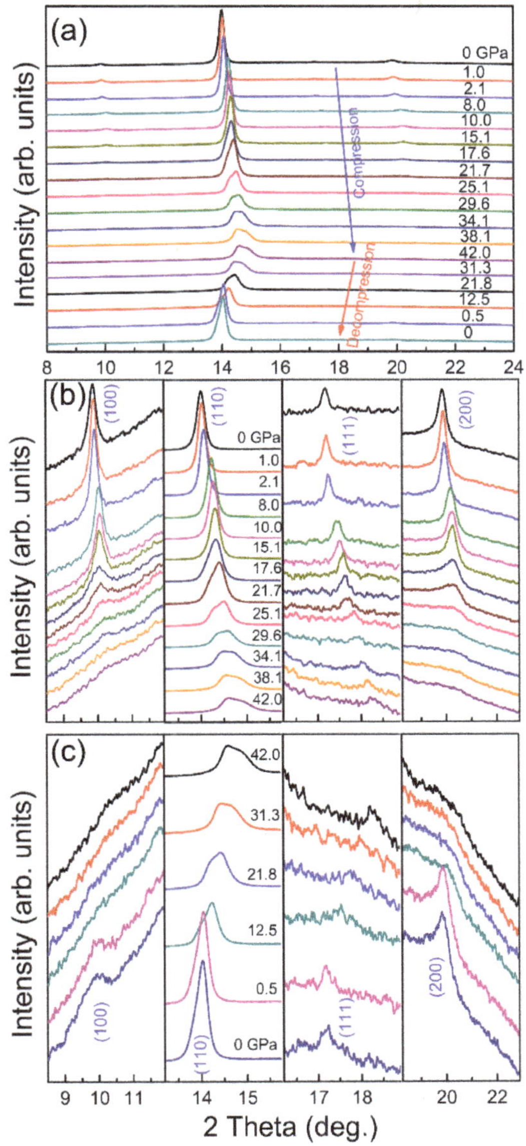

Figure 6. The structural evolution of the AlCoCrFeNi HEA as a function of pressure monitored by in situ high-pressure XRD patterns at room temperature (**a**) and the locally enlarged plot of the XRD patterns for each peak upon compression (**b**) and decompression (**c**) to show more details of the peak shape and width: The x-ray wavelength is 0.4959 Å [61].

With slightly lower Al content but still located in the *bcc-fcc* transition zone, the $Al_{0.6}CoCrFeNi$ HEA was studied by Wang et al. using in situ synchrotron radiation XRD in a DAC with both silicone oil and helium as the pressure-transmitting medium up to approx. 40 GPa [62]. The $Al_{0.6}CoCrFeNi$ HEA powders were prepared by the GA method. A single *bcc* phase was obtained with a high quenching rate in the GA process. They revealed a *bcc*-to-orthorhombic phase transition, which started at approx. 10.6 GPa and completed at approx. 21.4 GPa. Interestingly, another body-center-tetragonal (*bct*) phase emerged and coexisted with the high-pressure synthesized orthorhombic phase when the pressure was released. These results indicate that the orthorhombic phase may be metastable at ambient conditions but could be partially maintained due to the possible large energy barrier [62]. Moreover, after annealing at 1000 °C for 2 h, the initial GA *bcc* $Al_{0.6}$CoCrFeNi HEA can transform into a more stable *fcc* phase. During compression, the *fcc* phase of the $Al_{0.6}$CoCrFeNi HEA could completely transform into an *hcp* phase similar to Cantor's alloy. The samples recovered from high-pressure compression were characterized by transmission electron microscopy (TEM) and further confirmed that all of the five polymorphs could stably/metastably exist at ambient conditions (transition path between them is shown in Figure 7). Severe lattice distortion, which is tunable by high pressure or temperature was suggested to play a crucial role in the formation of various polymorphs and the transition between them in the $Al_{0.6}$CoCrFeNi HEA. These findings suggest that HEAs could behave quite differently from the expectation of a linear combination of its constituent element; they may also exhibit structural flexibility/tunability far beyond that of their solution components [62].

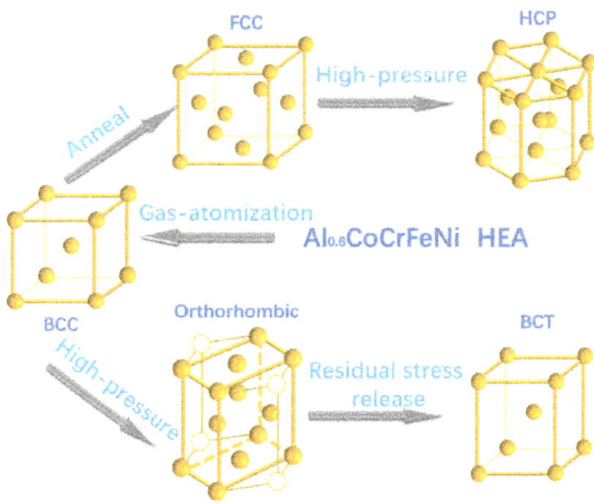

Figure 7. A schematic illustration of the atomic structure for five polymorphs observed in the $Al_{0.6}$CoCrFeNi HEA and the transition paths between them [62].

3.3. *Hcp*-Structured HEAs

HEAs commonly form with *fcc* or *bcc* structures. Recently, *hcp*-structured HEAs were observed in high-pressure experiments via an *fcc* to *hcp* polymorphic transition and was obtained in the melt-quenched alloys mainly consisting of heavy *hcp* metals, e.g., the CoOsReRu, CoFeReRu and CoReRuV, $Ir_{0.19}Os_{0.22}Re_{0.21}Rh_{0.20}Ru_{0.19}$, or rare earth *hcp* elements [8–12]. Ahmad et al. [55] investigated the structural stability of the quarternary equiatomic ReRuCoFe alloy under high pressure. They compressed the sample from 0.9 GPa to 80.4 GPa in a DAC. The unit cell parameters of *a* and *c*, its ratio *a*/*c*, and the sample volume all continuously decreased with increasing pressure, which indicates the *hcp* structure of the ReRuCoFe alloy is stable under compression up to approx. 80 GPa. Yusenko et al. investigated the structural stability of the *hcp* $Ir_{0.19}Os_{0.22}Re_{0.21}Rh_{0.20}Ru_{0.19}$ HEA at room temperature during compression up to 45 GPa but also observed no phase transition [64].

For the *hcp*-structured rare earth HEAs, Yu et al. reported a series of pressure-induced phase transitions in the HoDyYGdTb HEA by in situ XRD measurements in a DAC using synchrotron radiation x-ray. Four polymorphs were observed following a transition sequence of *hcp*→Sm-type→d*hcp*→d*fcc* during compression up to 60.1 GPa (Figure 8), which resembles the rich pressure-induced polymorphic transitions in its constituent elements [65]. The Sm-type phase firstly appeared when the pressure reached 4.4 GPa. At 13.6 GPa, the *hcp* (102) diffraction peak disappeared, indicating the completion of the *hcp* to Sm-type phase transition. With further increasing pressure to

26.7 GPa, the d*hcp* phase emerged and persisted to 38.3 GPa, and then, the d*hcp*-d*fcc* phase transition occurred at 40.2 GPa.

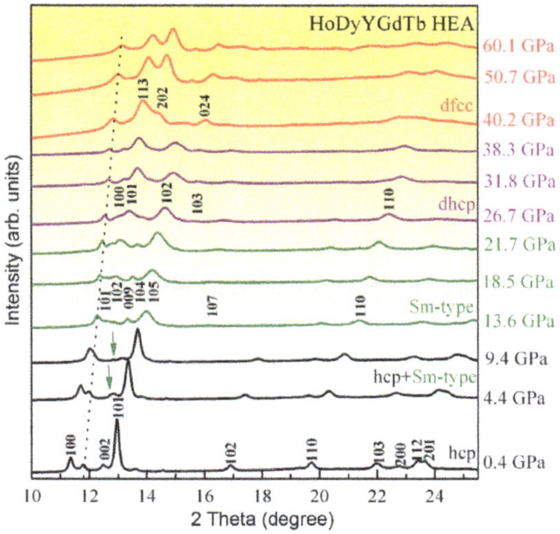

Figure 8. The in situ high-pressure XRD patterns of the HoDyYGdTb HEA during compression [65].

4. Conclusions and Outlooks

HEAs are the focus of advanced metallic alloy research and have been attracting more and more attention over the last decade. Recent studies on HEAs under high pressure have added another dimension to the exploration of HEAs. The exciting findings in HEAs during compression under high pressure deepen our understanding of HEAs, providing a new avenue towards new HEAs development and helpful guidance for applications at extreme conditions. As a new research direction, growing interest in the structure and properties of HEAs under high pressure is expected to continue. Future outlooks are briefly summarized below:

(1) Synergic effect of pressure–composition–temperature. HEAs open up an almost infinite composition space for alloy design. Hundreds of different HEAs have been developed, but so far, only a few of them have been studied under high pressure. Inspired by the existing high-pressure work, more exciting novel phenomena and new structures are expected with broader exploration in more HEAs. Meanwhile, the elusive composition effect on the phase transitions of HEAs remains to be addressed. Besides, combining high pressure with temperature (from cryogenic temperatures up to the melting temperatures) can further clarify the stability of various HEAs and is worth more effort in future research.

(2) Combing multiple high-pressure techniques for better understanding of HEAs. Over the last few years, the major high-pressure research of HEAs has focused on the crystal structure evolution during compression and decompression. However, we still lack an in-depth understanding of the transformation mechanism. The atomic size ratio, electronegativity difference, valence electron concentration, magnetic states, etc., which are all critical for the formation and transformation of HEAs, have not been systematically explored under high pressure yet. To get more detailed information of the atomic and electronic structure of the multicomponent HEAs under high pressure, besides XRD measurements, more experiments combining other powerful in situ element-sensitive techniques are required, such as in situ high-pressure extended x-ray absorption fine structure (EXAFS), in situ high-pressure x-ray emission spectroscopy (XES), and in situ high-pressure x-ray magnetic circular dichroism (XMCD).

(3) Involving more variables for high-pressure studies of HEAs. Existing studies have shown that the phase transitions of HEAs are sensitive to shear stress. Therefore, high-pressure torsion (HPT) which generates extreme shear deformation under high pressure could be another powerful technique for HEAs structure tuning with well-controlled shear stress and deformation. In addition, the results reviewed in this paper focus on the static compression of HEAs using DACs. The strain rate effect on the high-pressure behaviors of HEAs has not been extensively investigated. The dynamic compression of HEAs with another dimension of an extremely high strain rate is also worth more exploration.

(4) Properties studies of HEAs using large-volume press. With unique compositions and disordered atomic structures, HEAs show many unusual properties. Under high-pressure compression, HEAs with new structure could be synthesized. Meanwhile, the grain size and defects could also be considerably changed, which could affect their properties as well. Guo et al. measured the superconducting behavior of the $(TaNb)_{0.67}(HfZrTi)_{0.33}$ HEA under high pressure. They surprisingly observed extraordinarily robust superconductivity even up to 190.6 GPa [63]. Although the properties of HEAs under high pressure may be interesting, besides the equation of states (EOS) which can be readily measured by in situ XRD in DACs, the vast other properties have not been well-studied. One critical issue of the DAC samples is the requisite tiny sample size. Fortunately, since the critical pressures reported for the polymorphic transitions in HEAs are mostly around 20 GPa or below, a large-volume press (LVP) with approx. 1000 times larger sample volumes than DACs under similar pressure conditions (typically <25 GPa) [73] could be used to synthesize millimeter or centimeter-sized HEAs readily for various properties characterization. Very recently, Yu et al. used a 10-MN double-stage LVP and compressed Cantor's alloy with a diameter of 1.5 mm and a height of 2 mm to 20 GPa [74]. They synthesized a bulk equiatomic CoCrFeMnNi HEA containing a mixture of *fcc* and *hcp* phases for property characterization. Cantor's alloy recovered from high-pressure treatment (20 GPa) showed a doubled hardness of the as-cast *fcc* samples because of enhanced dislocations, twins, stacking faults, and the *hcp* laths [74].

(5) Theoretical calculations. Recent progress in the experimental discovery of the polymorphic phase transitions in HEAs was first inspired by the finite-temperature ab initio calculation work done by Ma et al., which predicted that the hcp phase in certain magnetic states would be more stable than the *fcc* phase of Cantor's alloy at room temperature [70]. HEAs with complex compositions are a challenge for theoretical simulations, but it is quite encouraging that much exciting work has been successfully done on HEAs [23,75]. In the high-pressure community, many calculation methods have also been successfully established to handle materials under high pressure [76–78]. Theoretical simulations definitely will continue to play a vital part in predicting new phenomena and in interpreting elusive experimental results of HEAs. So far, there is still limited computational calculation works on the high-pressure behaviors of HEAs. However, we believe more exciting works can be expected in HEAs under high pressure by combing advanced experimental tools with simulation methods closely.

The concept of high entropy has been extended into many material systems including high-entropy nitrides, carbides, oxides, and metallic glasses. Therefore, the proposed research above is suitable for other new high-entropy materials as well.

Funding: This work was funded by the National Thousand Youth Talents Program in China and the National Natural Science Foundation of China, grant numbers 51871054 and U1530402, and the Fundamental Research Funds for the Central Universities.

Acknowledgments: The authors would like to thank Peter K. Liaw for the useful discussions and suggestions and Freyja O'Toole for editing the manuscript.

Conflicts of Interest: The authors declare no conflict of interest.

References

1. Yeh, J.W.; Chen, S.K.; Lin, S.J.; Gan, J.Y.; Chin, T.S.; Shun, T.T.; Tsau, C.H.; Chang, S.Y. Nanostructured High-Entropy Alloys with Multiple Principal Elements: Novel Alloy Design Concepts and Outcomes. *Adv. Eng. Mater.* **2004**, *6*, 299–303. [CrossRef]
2. Cantor, B.; Chang, I.T.H.; Knight, P.; Vincent, A.J.B. Microstructural development in equiatomic multicomponent alloys. *Mater. Sci. Eng. A* **2004**, *375–377*, 213–218. [CrossRef]
3. Zhang, Y.; Zuo, T.T.; Tang, Z.; Gao, M.C.; Dahmen, K.A.; Liaw, P.K.; Lu, Z.P. Microstructures and properties of high-entropy alloys. *Prog. Mater. Sci.* **2014**, *61*, 1–93. [CrossRef]
4. Lu, Z.P.; Wang, H.; Chen, M.W.; Baker, I.; Yeh, J.W.; Liu, C.T.; Nieh, T.G. An assessment on the future development of high-entropy alloys: Summary from a recent workshop. *Intermetallics* **2015**, *66*, 67–76. [CrossRef]
5. Zhang, W.R.; Liaw, P.K.; Zhang, Y. Science and technology in high-entropy alloys. *Sci. China Mater.* **2018**, *61*, 2–22. [CrossRef]
6. Senkov, O.N.; Wilks, G.B.; Miracle, D.B.; Chuang, C.P.; Liaw, P.K. Refractory high-entropy alloys. *Intermetallics* **2010**, *18*, 1758–1765. [CrossRef]
7. Senkov, O.N.; Wilks, G.B.; Scott, J.M.; Miracle, D.B. Mechanical properties of Nb25Mo25Ta25W25 and V20Nb20Mo20Ta20W20 refractory high entropy alloys. *Intermetallics* **2011**, *19*, 698–706. [CrossRef]
8. Gao, M.C.; Alman, D.E. Searching for Next Single-Phase High-Entropy Alloy Compositions. *Entropy* **2013**, *15*, 4504–4519. [CrossRef]
9. Takeuchi, A.; Amiya, K.; Wada, T.; Yubuta, K.; Zhang, W. High-Entropy Alloys with a Hexagonal Close-Packed Structure Designed by Equi-Atomic Alloy Strategy and Binary Phase Diagrams. *JOM* **2014**, *66*, 1984–1992. [CrossRef]
10. Feuerbacher, M.; Heidelmann, M.; Thomas, C. Hexagonal High-entropy Alloys. *Mater. Res. Lett.* **2015**, *3*, 1–6. [CrossRef]
11. Gao, M.C.; Zhang, B.; Guo, S.M.; Qiao, J.W.; Hawk, J.A. High-Entropy Alloys in Hexagonal Close-Packed Structure. *Metall. Mater. Trans. A* **2016**, *47*, 3322–3332. [CrossRef]
12. Qiao, J.W.; Bao, M.L.; Zhao, Y.J.; Yang, H.J.; Wu, Y.C.; Zhang, Y.; Hawk, J.A.; Gao, M.C. Rare-earth high entropy alloys with hexagonal close-packed structure. *J. Appl. Phys.* **2018**, *124*, 195101. [CrossRef]
13. Gludovatz, B.; Hohenwarter, A.; Catoor, D.; Chang, E.H.; George, E.P.; Ritchie, R.O. A fracture-resistant high-entropy alloy for cryogenic applications. *Science* **2014**, *345*, 1153–1158. [CrossRef] [PubMed]
14. Li, Z.; Pradeep, K.G.; Deng, Y.; Raabe, D.; Tasan, C.C. Metastable high-entropy dual-phase alloys overcome the strength–ductility trade-off. *Nature* **2016**, *534*, 227–230. [CrossRef] [PubMed]
15. He, J.Y.; Wang, H.; Huang, H.L.; Xu, X.D.; Chen, M.W.; Wu, Y.; Liu, X.J.; Nieh, T.G.; An, K.; Lu, Z.P. A precipitation-hardened high-entropy alloy with outstanding tensile properties. *Acta Mater.* **2016**, *102*, 187–196. [CrossRef]
16. Huang, H.; Wu, Y.; He, J.; Wang, H.; Liu, X.; An, K.; Wu, W.; Lu, Z. Phase-Transformation Ductilization of Brittle High-Entropy Alloys via Metastability Engineering. *Adv. Mater.* **2017**, *29*, 1701678. [CrossRef] [PubMed]
17. Praveen, S.; Kim, H.S. High-Entropy Alloys: Potential Candidates for High-Temperature Applications—An Overview. *Adv. Eng. Mater.* **2017**, *20*, 1700645. [CrossRef]
18. Liang, Y.-J.; Wang, L.; Wen, Y.; Cheng, B.; Wu, Q.; Cao, T.; Xiao, Q.; Xue, Y.; Sha, G.; Wang, Y.; et al. High-content ductile coherent nanoprecipitates achieve ultrastrong high-entropy alloys. *Nat. Commun.* **2018**, *9*, 4063. [CrossRef] [PubMed]
19. Lu, C.; Niu, L.; Chen, N.; Jin, K.; Yang, T.; Xiu, P.; Zhang, Y.; Gao, F.; Bei, H.; Shi, S.; et al. Enhancing radiation tolerance by controlling defect mobility and migration pathways in multicomponent single-phase alloys. *Nat. Commun.* **2016**, *7*, 13564. [CrossRef] [PubMed]
20. Zou, Y.; Ma, H.; Spolenak, R. Ultrastrong ductile and stable high-entropy alloys at small scales. *Nat. Commun.* **2015**, *6*, 7748. [CrossRef] [PubMed]
21. Chuang, M.-H.; Tsai, M.-H.; Wang, W.-R.; Lin, S.-J.; Yeh, J.-W. Microstructure and wear behavior of $Al_xCo_{1.5}CrFeNi_{1.5}Ti_y$ high-entropy alloys. *Acta Mater.* **2011**, *59*, 6308–6317. [CrossRef]
22. Tsai, M.-H.; Yeh, J.-W. High-Entropy Alloys: A Critical Review. *Mater. Res. Lett.* **2014**, *2*, 107–123. [CrossRef]

23. Miracle, D.B.; Senkov, O.N. A critical review of high entropy alloys and related concepts. *Acta Mater.* **2017**, *122*, 448–511. [CrossRef]
24. Tonkov, E.Y.; Ponyatovsky, E.G. *Phase Transformations of Elements Under High Pressure;* CRC Press: Boca Raton, FL, USA, 2004.
25. Mao, H.K.; Bassett, W.A.; Takahashi, T. Effect of Pressure on Crystal Structure and Lattice Parameters of Iron up to 300 kbar. *J. Appl. Phys.* **1967**, *38*, 272–276. [CrossRef]
26. Bassett, W.A.; Huang, E. Mechanism of the Body-Centered Cubic—Hexagonal Close-Packed Phase Transition in Iron. *Science* **1987**, *238*, 780–783. [CrossRef] [PubMed]
27. Yoo, C.S.; Akella, J.; Campbell, A.J.; Mao, H.K.; Hemley, R.J. Phase Diagram of Iron by in Situ X-ray Diffraction: Implications for Earth's Core. *Science* **1995**, *270*, 1473–1475. [CrossRef]
28. Miracle, D.B. High-Entropy Alloys: A Current Evaluation of Founding Ideas and Core Effects and Exploring "Nonlinear Alloys". *JOM* **2017**, *69*, 2130–2136. [CrossRef]
29. Otto, F.; Dlouhy, A.; Somsen, C.; Bei, H.; Eggeler, G.; George, E.P. The influences of temperature and microstructure on the tensile properties of a CoCrFeMnNi high-entropy alloy. *Acta Mater.* **2013**, *61*, 5743–5755. [CrossRef]
30. Schneeweiss, O.; Friák, M.; Dudová, M.; Holec, D.; Šob, M.; Kriegner, D.; Holý, V.; Beran, P.; George, E.P.; Neugebauer, J.; et al. Magnetic properties of the CrMnFeCoNi high-entropy alloy. *Phys. Rev. B* **2017**, *96*, 014437. [CrossRef]
31. Wu, Z.; Bei, H.; Pharr, G.M.; George, E.P. Temperature dependence of the mechanical properties of equiatomic solid solution alloys with face-centered cubic crystal structures. *Acta Mater.* **2014**, *81*, 428–441. [CrossRef]
32. Zhang, Y.; Zhou, Y.J.; Lin, J.P.; Chen, G.L.; Liaw, P.K. Solid-Solution Phase Formation Rules for Multi-component Alloys. *Adv. Eng. Mater.* **2008**, *10*, 534–538. [CrossRef]
33. Yang, X.; Zhang, Y. Prediction of high-entropy stabilized solid-solution in multi-component alloys. *Mater. Chem. Phys.* **2012**, *132*, 233–238. [CrossRef]
34. Otto, F.; Yang, Y.; Bei, H.; George, E.P. Relative effects of enthalpy and entropy on the phase stability of equiatomic high-entropy alloys. *Acta Mater.* **2013**, *61*, 2628–2638. [CrossRef]
35. Guo, S.; Ng, C.; Lu, J.; Liu, C.T. Effect of valence electron concentration on stability of fcc or bcc phase in high entropy alloys. *J. Appl. Phys.* **2011**, *109*, 103505. [CrossRef]
36. Ming, L.C.; Manghnani, M.H.; Katahara, K.W. Investigation of a→w transformation in the Zr-Hf system to 42 GPa. *J. Appl. Phys.* **1981**, *52*, 1332–1335. [CrossRef]
37. Smith, D.; Joris, O.P.J.; Sankaran, A.; Weekes, H.E.; Bull, D.J.; Prior, T.J.; Dye, D.; Errandonea, D.; Proctor, J.E. On the high-pressure phase stability and elastic properties of β -titanium alloys. *J. Phys. Condens. Matter* **2017**, *29*, 155401. [CrossRef] [PubMed]
38. Mao, W.L.; Campbell, A.J.; Heinz, D.L.; Shen, G. Phase relations of Fe–Ni alloys at high pressure and temperature. *Phys. Earth Planet. Interiors* **2006**, *155*, 146–151. [CrossRef]
39. Velisavljevic, N.; Chesnut, G.N. Direct hcp→bcc structural phase transition observed in titanium alloy at high pressure. *Appl. Phys. Lett.* **2007**, *91*, 101906. [CrossRef]
40. Ahart, M.; DeVreugd, C.; Li, J.; Viehland, D.; Gehring, P.M.; Hemley, R.J. X-ray diffraction study of the pressure-induced bcc-to-hcp phase transition in the highly magnetostrictive Fe0.81Ga0.19 alloy. *Phys. Rev. B* **2013**, *88*, 184102. [CrossRef]
41. Sakai, T.; Takahashi, S.; Nishitani, N.; Mashino, I.; Ohtani, E.; Hirao, N. Equation of state of pure iron and $Fe_{0.9}Ni_{0.1}$ alloy up to 3Mbar. *Phys. Earth Planet. Interiors* **2014**, *228*, 114–126. [CrossRef]
42. Zeng, Q.S.; Ding, Y.; Mao, W.L.; Luo, W.; Blomqvist, A.; Ahuja, R.; Yang, W.; Shu, J.; Sinogeikin, S.V.; Meng, Y.; et al. Substitutional alloy of Ce and Al. *Proc. Nat. Acad. Sci. USA* **2009**, *106*, 2515–2518. [CrossRef] [PubMed]
43. Dubrovinsky, L.S.; Saxena, S.K.; Lazor, P.; Ahuja, R.; Eriksson, O.; Wills, J.M.; Johansson, B. Experimental and theoretical identification of a new high-pressure phase of silica. *Nature* **1997**, *388*, 362. [CrossRef]
44. Hemley, R.J.; Jephcoat, A.P.; Mao, H.K.; Ming, L.C.; Manghnani, M.H. Pressure-induced amorphization of crystalline silica. *Nature* **1988**, *334*, 52. [CrossRef]
45. Bai, L.; Li, Q.; Corr, S.A.; Meng, Y.; Park, C.; Sinogeikin, S.V.; Ko, C.; Wu, J.; Shen, G. Pressure-induced phase transitions and metallization in VO_2. *Phys. Rev. B* **2015**, *91*, 104110. [CrossRef]
46. Cheng, B.; Li, Q.; Zhang, H.; Liu, R.; Liu, B.; Yao, Z.; Cui, T.; Liu, J.; Liu, Z.; Sundqvist, B.; et al. Pressure-induced metallization and amorphization in VO_2 nanorods. *Phys. Rev. B* **2016**, *93*, 184109. [CrossRef]

47. Sheng, H.W.; Liu, H.Z.; Cheng, Y.Q.; Wen, J.; Lee, P.L.; Luo, W.K.; Shastri, S.D.; Ma, E. Polyamorphism in a metallic glass. *Nat. Mater.* **2007**, *6*, 192–197. [CrossRef] [PubMed]
48. Zeng, Q.-S.; Ding, Y.; Mao, W.L.; Yang, W.; Sinogeikin, S.V.; Shu, J.; Mao, H.-K.; Jiang, J.Z. Origin of Pressure-Induced Polyamorphism in $Ce_{75}Al_{25}$ Metallic Glass. *Phys. Rev. Lett.* **2010**, *104*, 105702. [CrossRef] [PubMed]
49. Zeng, Q.; Sheng, H.; Ding, Y.; Wang, L.; Yang, W.; Jiang, J.-Z.; Mao, W.L.; Mao, H.-K. Long-Range Topological Order in Metallic Glass. *Science* **2011**, *332*, 1404–1406. [CrossRef] [PubMed]
50. Wei, S.; He, F.; Tasan, C.C. Metastability in high-entropy alloys: A review. *J. Mater. Res.* **2018**, *33*, 2924–2937. [CrossRef]
51. Tracy, C.L.; Park, S.; Rittman, D.R.; Zinkle, S.J.; Bei, H.B.; Lang, M.; Ewing, R.C.; Mao, W.L. High pressure synthesis of a hexagonal close-packed phase of the high-entropy alloy CrMnFeCoNi. *Nat. Commun.* **2017**, *8*, 15634. [CrossRef] [PubMed]
52. Zhang, F.; Wu, Y.; Lou, H.B.; Zeng, Z.D.; Prakapenka, V.B.; Greenberg, E.; Ren, Y.; Yan, J.Y.; Okasinski, J.S.; Liu, X.J.; et al. Polymorphism in a high-entropy alloy. *Nat. Commun.* **2017**, *8*, 15687. [CrossRef] [PubMed]
53. Zhang, F.; Lou, H.; Chen, S.; Chen, X.; Zeng, Z.; Yan, J.; Zhao, W.; Wu, Y.; Lu, Z.; Zeng, Q. Effects of non-hydrostaticity and grain size on the pressure-induced phase transition of the CoCrFeMnNi high-entropy alloy. *J. Appl. Phys.* **2018**, *124*, 115901. [CrossRef]
54. Huang, E.W.; Lin, C.M.; Jain, J.; Shieh, S.R.; Wang, C.P.; Chuang, Y.C.; Liao, Y.-F.; Zhang, D.Z.; Huang, T.; Lam, T.N.; et al. Irreversible phase transformation in a CoCrFeMnNi high entropy alloy under hydrostatic compression. *Mater. Today Commun.* **2018**, *14*, 10–14. [CrossRef]
55. Ahmad, A.S.; Su, Y.; Liu, S.Y.; Ståhl, K.; Wu, Y.D.; Hui, X.D.; Ruett, U.; Gutowski, O.; Glazyrin, K.; Liermann, H.P.; et al. Structural stability of high entropy alloys under pressure and temperature. *J. Appl. Phys.* **2017**, *121*, 235901. [CrossRef]
56. Yu, P.F.; Zhang, L.J.; Cheng, H.; Zhang, H.; Ma, M.Z.; Li, Y.C.; Li, G.; Liaw, P.K.; Liu, R.P. The high-entropy alloys with high hardness and soft magnetic property prepared by mechanical alloying and high-pressure sintering. *Intermetallics* **2016**, *70*, 82–87. [CrossRef]
57. Zhang, F.X.; Zhao, S.J.; Jin, K.; Bei, H.B.; Popov, D.; Park, C.Y.; Neuefeind, J.C.; Weber, W.J.; Zhang, Y.W. Pressure-induced fcc to hcp phase transition in Ni-based high entropy solid solution alloys. *Appl. Phys. Lett.* **2017**, *110*, 011902. [CrossRef]
58. Ma, Y.; Fan, J.; Zhang, L.; Zhang, M.; Cui, P.; Dong, W.; Yu, P.; Li, Y.; Liaw, P.K.; Li, G. Pressure-induced ordering phase transition in high-entropy alloy. *Intermetallics* **2018**, *103*, 63–66. [CrossRef]
59. Yusenko, K.V.; Riva, S.; Crichton, W.A.; Spektor, K.; Bykova, E.; Pakhomova, A.; Tudball, A.; Kupenko, I.; Rohrbach, A.; Klemme, S.; et al. High-pressure high-temperature tailoring of High Entropy Alloys for extreme environments. *J. Alloys Compd.* **2018**, *738*, 491–500. [CrossRef]
60. Li, G.; Xiao, D.; Yu, P.; Zhang, L.; Liaw, P.K.; Li, Y.; Liu, R. Equation of State of an AlCoCrCuFeNi High-Entropy Alloy. *JOM* **2015**, *67*, 2310–2313. [CrossRef]
61. Cheng, B.; Zhang, F.; Lou, H.; Chen, X.; Liaw, P.K.; Yan, J.; Zeng, Z.; Ding, Y.; Zeng, Q. Pressure-induced phase transition in the AlCoCrFeNi high-entropy alloy. *Scr. Mater.* **2019**, *161*, 88–92. [CrossRef]
62. Wang, L.; Zhang, F.; Nie, Z.; Wang, L.; Wang, F.; Wang, B.; Zhou, S.; Xue, Y.; Cheng, B.; Lou, H.; et al. Abundant polymorphic transitions in the $Al_{0.6}$CoCrFeNi high-entropy alloy. *Mater. Today Phys.* **2019**, *8*, 1–9. [CrossRef]
63. Guo, J.; Wang, H.; von Rohr, F.; Wang, Z.; Cai, S.; Zhou, Y.; Yang, K.; Li, A.; Jiang, S.; Wu, Q.; et al. Robust zero resistance in a superconducting high-entropy alloy at pressures up to 190 GPa. *Proc. Natl. Acad. Sci. USA* **2017**, *114*, 13144–13147. [CrossRef] [PubMed]
64. Yusenko, K.V.; Riva, S.; Carvalho, P.A.; Yusenko, M.V.; Arnaboldi, S.; Sukhikh, A.S.; Hanfland, M.; Gromilov, S.A. First hexagonal close packed high-entropy alloy with outstanding stability under extreme conditions and electrocatalytic activity for methanol oxidation. *Scr. Mater.* **2017**, *138*, 22–27. [CrossRef]
65. Yu, P.F.; Zhang, L.J.; Ning, J.L.; Ma, M.Z.; Zhang, X.Y.; Li, Y.C.; Liaw, P.K.; Li, G.; Liu, R.P. Pressure-induced phase transitions in HoDyYGdTb high-entropy alloy. *Mater. Lett.* **2017**, *196*, 137–140. [CrossRef]
66. Mao, H.-K.; Chen, B.; Chen, J.; Li, K.; Lin, J.-F.; Yang, W.; Zheng, H. Recent advances in high-pressure science and technology. *Matter Radiat. Extremes* **2016**, *1*, 59–75. [CrossRef]
67. Shen, G.Y.; Mao, H.-K. High-pressure studies with x-rays using diamond anvil cells. *Rep. Prog. Phys.* **2017**, *80*, 016101. [CrossRef] [PubMed]

68. Mao, H.-K.; Chen, X.-J.; Ding, Y.; Li, B.; Wang, L. Solids, liquids, and gases under high pressure. *Rev. Mod. Phys.* **2018**, *90*, 015007. [CrossRef]
69. Mao, H.K.; Xu, J.; Bell, P.M. Calibration of the ruby pressure gauge to 800 kbar under quasi-hydrostatic conditions. *J. Geophys. Res.* **1986**, *91*, 4673. [CrossRef]
70. Ma, D.; Grabowski, B.; Körmann, F.; Neugebauer, J.; Raabe, D. Ab initio thermodynamics of the CoCrFeMnNi high entropy alloy: Importance of entropy contributions beyond the configurational one. *Acta Mater.* **2015**, *100*, 90–97. [CrossRef]
71. Tian, F.; Varga, L.K.; Shen, J.; Vitos, L. Calculating elastic constants in high-entropy alloys using the coherent potential approximation: Current issues and errors. *Comput. Mater. Sci.* **2016**, *111*, 350–358. [CrossRef]
72. Wang, W.R.; Wang, W.L.; Wang, S.C.; Tsai, Y.C.; Lai, C.H.; Yeh, J.W. Effects of Al addition on the microstructure and mechanical property of AlxCoCrFeNi high-entropy alloys. *Intermetallics* **2012**, *26*, 44–51. [CrossRef]
73. Yamazaki, D.; Ito, E.; Yoshino, T.; Tsujino, N.; Yoneda, A.; Guo, X.; Xu, F.; Higo, Y.; Funakoshi, K. Over 1Mbar generation in the Kawai-type multianvil apparatus and its application to compression of $(Mg_{0.92}Fe_{0.08})SiO_3$ perovskite and stishovite. *Phys. Earth Planet. Interiors* **2014**, *228*, 262–267. [CrossRef]
74. Yu, P.; Zhang, L.; Cheng, H.; Tang, H.; Fan, J.; Liaw, P.K.; Li, G.; Liu, R. Formation, reverse transformation, and properties of ε-martensite phase in the CoCrFeMnNi high-entropy alloy under high-pressure. *J. Alloys Compd.* **2019**, *779*, 1–6. [CrossRef]
75. Tian, F. A Review of Solid-Solution Models of High-Entropy Alloys Based on Ab Initio Calculations. *Front. Mater.* **2017**, *4*, 36. [CrossRef]
76. Winkler, B.; Milman, V. Density functional theory based calculations for high pressure research. *Cryst. Mater.* **2014**, *229*, 112–122. [CrossRef]
77. Wang, Y.; Ma, Y. Perspective: Crystal structure prediction at high pressures. *J. Chem. Phys.* **2014**, *140*, 040901. [CrossRef] [PubMed]
78. Zhang, L.; Wang, Y.; Lv, J.; Ma, Y. Materials discovery at high pressures. *Nat. Rev. Mater.* **2017**, *2*, 17005. [CrossRef]

entropy

MDPI

Review

Welding of High Entropy Alloys—A Review

Jing Guo [1,*], Cong Tang [1], Glynn Rothwell [1], Lisa Li [1], Yun-Che Wang [2], Qingxiang Yang [3] and Xuejun Ren [1,*]

1 Department of Mechanical and Maritime Engineering, Faculty of Engineering and Technology, Liverpool John Moores University, Liverpool L3 5UG, UK; C.Tang@2015.ljmu.ac.uk (C.T.); G.Rothwell@ljmu.ac.uk (G.R.); B.Li@ljmu.ac.uk (L.L.)
2 Department of Civil Engineering, National Cheng Kung University, Tainan 701, Taiwan; yunche@mail.ncku.edu.tw
3 State Key Laboratory of Metastable Materials Science and Technology, College of Materials Science and Engineering, Yanshan University, Qinhuangdao 066004, China; qxyang@ysu.edu.cn
* Correspondence: j.guo@ljmu.ac.uk (J.G.); x.j.ren@ljmu.ac.uk (X.R.)

Received: 22 February 2019; Accepted: 15 April 2019; Published: 24 April 2019

Abstract: High-entropy alloy (HEA) offers great flexibility in materials design with 3–5 principal elements and a range of unique advantages such as good microstructure stability, mechanical strength over a broad range of temperatures and corrosion resistance, etc. Welding of high entropy alloy, as a key joining method, is an important emerging area with significant potential impact to future application-oriented research and technological developments in HEAs. The selection of feasible welding processes with optimized parameters is essential to enhance the applications of HEAs. However, the structure of the welded joints varies with material systems, welding methods and parameters. A systemic understanding of the structures and properties of the weldment is directly relevant to the application of HEAs as well as managing the effect of welding on situations such as corrosion that are known to be a service life limiting factor of welded structures in conditions such as marine environments. In this paper, key recent work on welding of HEAs is reviewed in detail focusing on the research of main HEA systems when applying different welding techniques. The experimental details including sample preparation, sample size (thickness) and welding conditions reflecting energy input are summarized and key issues are highlighted. The microstructures and properties of different welding zones, in particular the fusion zone (FZ) and the heat affected zones (HAZ), formed with different welding methods are compared and presented in details and the structure-property relationships are discussed. The work shows that the weldability of HEAs varies with the HEA composition groups and the welding method employed. Arc and laser welding of AlCoCrFeNi HEAs results in lower hardness in the FZ and HAZ and reduced overall strength. Friction stir welding results in higher hardness in the FZ and achieves comparable/higher strength of the welded joints in tensile tests. The welded HEAs are capable of maintaining a reasonable proportion of the ductility. The key structure changes including element distribution, the volume fraction of face centered cubic (FCC) and body centered cubic (BCC) phase as well as reported changes in the lattice constants are summarized and analyzed. Detailed mechanisms governing the mechanical properties including the grain size-property/hardness relationship in the form of Hall–Petch (H–P) effect for both bulk and welded structure of HEAs are compared. Finally, future challenges and main areas to research are highlighted.

Keywords: high-entropy alloys; welding; Hall–Petch (H–P) effect; lattice constants

1. Introduction

High-entropy alloys (HEAs) are based on the promising alloy design ideas of configurational entropy maximization, which offer great flexibility in developing different material systems [1–6].

Compared with the classic alloys, HEAs possess better performance in areas such as hardness, wear resistance, fatigue and resistance to corrosion and oxidation [3–6] as well as novel physical properties [2,3]. In general, HEA has high intrinsic strength and ductility, and the low diffusion rate of HEAs at high and low temperatures makes their applications in harsh condition possible [1,7,8]. There are many different material systems and processing methods developed in the past decades, which offer great flexibility in materials and processing methods selection for different conditions with a suitable welding process. In the meantime, the complexity of the material systems requires development of detailed understanding of the material behavior in different manufacturing processes including both hot and cold working operations. Joining of HEAs in similar or dissimilar material systems is increasingly important for expanding the applications of HEAs [9], among which welding process of HEAs is challenging due to the complex chemical, physical, and mechanical nature of welding and the need for tailoring the properties and structures of the weldment for complex loading conditions in similar and dissimilar welding.

HEA can be produced by many different routes such as liquid processing approaches (e.g., arc melting, Bridgman solidification, atomization and laser cladding), additive manufacturing, mechanical alloying (e.g., powder metallurgy), mixing elements of the vapor state including sputter deposition, atomic layer deposition and vapor phase deposition [1,10–18]. Welding is a complex process due to the fact that many conditions or factors affect the welding process. The quality of the welded joints can be assessed by the microstructure of the welding zones formed and mechanical structural integrity of the weldment as well as corrosion and fatigue [12]. There are many different welding processes such as arc welding, laser welding, friction stir welding, electron beam welding, electrical resistance spot welding as well as other processes such as electroslag welding, vacuum diffusion welding, etc. [9,12–16]. These welding processes differ significantly in the welding mechanism, working principles of the equipment and efficiency, but each has been found to have wide applications in different industries and application conditions. The energy input and the maximum temperature could be many orders of magnitude different from each of the welding methods, which directly influence the elemental and material behavior such as evaporation, melting and cooling. For example, the heat density ranges from 10^5–10^6 W/cm^2 for gas metal arc welding or plasma arc welding to 10^7–10^8 W/cm^2 for laser or electron beam welding [15,16]. The structural zones formed (Figure 1) after different welding processes could be significantly different in geometry, size, microstructure and compositions, which is dependent on many factors such as the behavior of material systems (elements, evaporation, melting, flow, phase changes, cooling, etc.), the original structures of the workpiece, size of the specimen and the welding parameters, etc. The structure and properties of the welded joints (such as size/shape of the welding zones, the hardness distribution, residual stress, defects, etc.) also affects the performance of the overall joints under different modes of loading and service conditions [17]. These directly influence the suitability of each welding process on joining different HEAs, the application of HEAs and future developments in related areas [1,5,6,18].

In this paper key recent work on welding of HEAs is systematically reviewed. The main works on welding of different HEAs with different welding techniques are presented. The experimental details including samples preparation, sample size (thickness) and welding conditions reflecting the speeds and energy input are summarized. The different welding zones formed with different welding methods and the microstructures are compared. The hardness change of the fusion zone (FZ) and heat affected zones (HAZ) in different welding processes is compared and linked to the material systems. The mechanism of microstructure and property changes in the FZ and HAZ and their influence on the mechanical properties is discussed. Data on the Hall–Petch (H–P) effect (grain size-property/hardness relationship) for both bulk and welded structure of HEAs is compared. The future challenges and main areas to research are highlighted on the weldability of HEAs, welding process selection and potential effects of the microstructure and properties in different welding zones of HEAs on their applications.

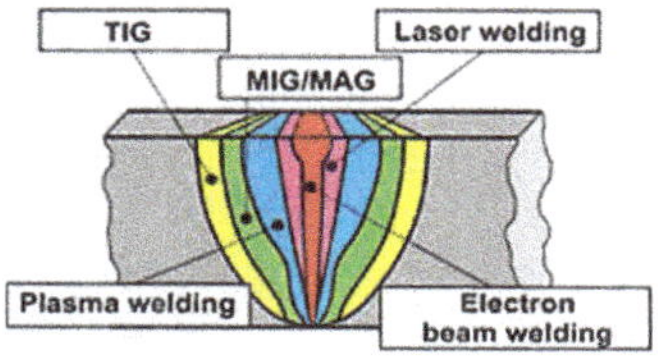

Figure 1. Welded zones formed by different welding methods. [http://is.gliwice.pl/en/strona-cms/electron-beam-welding-laboratory].

2. Welding of HEAs in Different Welding Processes

2.1. Arc Welding of HEAs

Arc welding is one of the most common welding methods, in which the energy from an arc of electric current between the material and a consumable/non-consumable electrode stick (either) is employed to melt the workpiece(s) with/without a protective atmosphere. Typical arc welding includes shielded metal arc welding (SMAW), gas tungsten arc welding (GTAW) (also known as TIG (Tungsten inert gas)), gas metal arc welding (GMAW), flux-cored arc welding (FCAW), submerged arc welding (SAW) [15,16]. Most of the work on welding HEAs has been focused on gas tungsten arc welding (GTAW) [19,20]. In the welding process, a non-consumable tungsten electrode is used to produce the weld by melting the base metal (Figure 2a), and the weld area is protected from atmospheric contamination by an inert shielding gas such as argon or helium. In the work by Sokkalingam et al [19], two Al0.5CoCrFeNi–HEA plates (homogenized at 1423 K for 24 hours followed by furnace cooling) of a thickness of 2.5 mm were butt-welded with a current of 40 A and voltage of 12 V at a welding speed of 80 mm/min. The FZ and HAZ formed exhibited distinctively different structures from the base metal (BM) (as shown in Figure 2b). The dominant structure in the FZ mainly consisted of fine elongated grains with diameter of 8–12 μm and lengths of 80–120 μm. The HAZ had a distinctive coarse grain structure with grain boundaries delineated by the body centered cubic (BCC) phase. Vickers micro-hardness tests across the three zones revealed an inferior hardness of the FZ and HAZ compared to the BM, particularly at the center of the welding pool (Figure 2c). Tensile tests of welded sample perpendicular to the welding seam showed that both the yield strength (YS) and the ultimate tensile strength (UTS) were reduced after welding, but the welded sample has maintained a high proportion of the ductility (Figure 2d), which is better than samples with other welding processes (detailed in the following sections).

CoCrFeMnNi system is another major HEAs system with significant potential in structural applications [20–23]. Wu et al [20] investigated the welding of CoCrFeMnNi HEAs with low-energy-density, high-heat-input gas tungsten arc (GTA). The raw material of the workpiece was arc-melted and drop-cast into an ingot of 25.4 mm × 12.7 mm × 127 mm, a rolling operation along the longitudinal direction was then performed in air at room temperature to reach a final thickness of 1.6 mm. The sheets were annealed at 900 °C for 1h to obtain an equiaxed microstructure. Buttwelding of two pieces of the sheet alloys were performed at a power level of 8.4 V and 75 A with a welding speed of 25.4 mm/min. The FZ of the welded joints exhibited a centerline grain structure of large columnar grains grown from the fusion lines to the centerlines (Figure 3). Tensile tests of the specimen at 293 K and 77 K showed that the effective YS of the GTA weld was close the BM, but the UTS was significantly lower, in particular for the low temperature test. The ductility of the GTA sample was maintained over 50% (Figure 3). The difference in the microstructures of the FZ between the two material systems (i.e., Al0.5CoCrFeNi and CoCrFeMnNi) is apparent, both showed columnar grain structure, but the effect of the welding on the mechanical response is different. The FZ of Al0.5CoCrFeNi–HEA (Figure 2d) showed a drop in YS and UTS, and maintained high proportion of the ductility against the BM; while the drop in YS for CoCrFeMnNi (Figure 3) was relatively limited but there was a clear drop of the UTS, but it had only maintained about half of the ductility. The work also revealed limited depletion behavior of Mn in the weld zone, which is different from welding of high-Mn stainless steels [24]. Nano-twins (Σ3) common to CoCrFeMnNi HEA and other face-centered cubic (FCC) metals and alloys with low stacking fault energy were also observed [25]. The nano-twins and twin bundles formed during deformation had significantly beneficial effect on mechanical properties such as tensile strength.

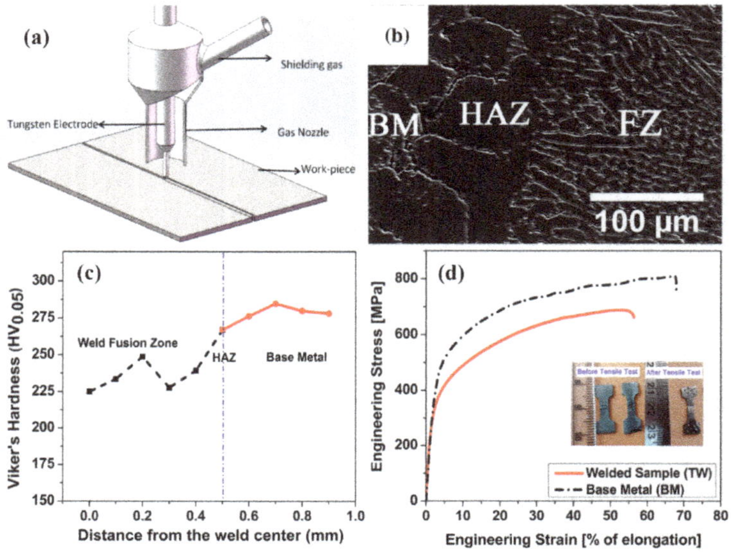

Figure 2. Structure and hardness of the welded zones and engineering stress-stain curves of the welded samples of Al0.5CoCrFeNi high entropy alloy (HEA) [19]: (**a**) Schematic of gas tungsten arc (GTA) welding process, (**b**) Scanning electron micrographs of base metal (BM)–heat affected zone (HAZ)–fusion zone (FZ) interfaces, (**c**) Microhardness profile on the surface of the welded sample and (**d**) Engineering stress–engineering strain curve for the BM and welded sample.

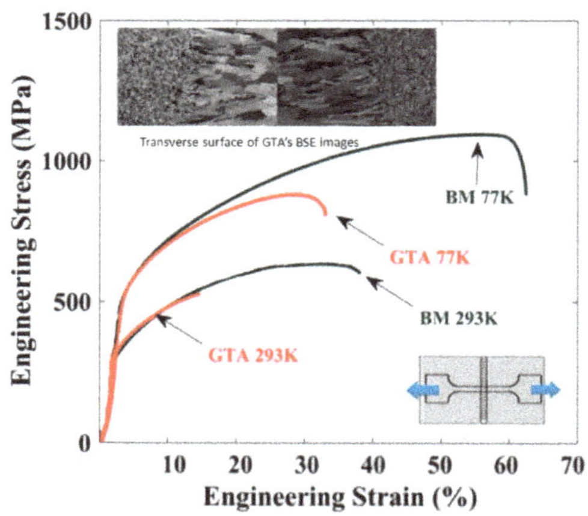

Figure 3. Structure of the welding zones and tensile test results of welded joint of CoCrFeMnNi alloy by GTA [20].

2.2. *Laser Welding/Laser Beam Welding*

Research on laser beam welding (LBW) of HEAs has been reported in several recent publications [26–30]. In a laser welding process, pieces of metal are melted and joined through a concentrated heat source provided by the laser beam, forming a narrow, deep weld (Figure 4a). LBW has much higher welding rates than arc welding and is more suitable for high volume applications. LBM is based on keyhole or penetration mode welding [17]. Compared to arc welding, the heat density for laser is much higher and the welding point/zones is narrower and the cooling could be much faster [17,18]. Kashaev et al [26] studied LBW of a CoCrFeNiMn-type HEA produced by self-propagating high-temperature synthesis (SHS). The synthesis of the initial CoCrFeNiMn-type alloy was carried out with the use of thermite-type SHS powders containing oxides of the target elements (NiO, Cr_2O_3, Co_3O_4, Fe_2O_3, MnO_2 and high purity Al as the metal reducer). The SHS-fabricated alloy was characterized by ~2 times reduced Mn content in comparison with that of the other principal components and the presence of impurities including Al, C, S, and Si. In the experiments, welded butt-joints were produced using an 8.0kW fiber laser with a fiber optic (300 μm core diameter) and a 300 mm focal length with a welding speed in the range between 3.0 m/min and 6.0 m/min. Detailed metallurgical analysis revealed that the difference in microstructure and grain orientation distribution between the BM, HAZ and the FZ was not significant. As shown in Figure 4b, the FZ was not fully symmetric. Detailed Vickers hardness tests at the top, middle and bottom of cross-section showed a pronounced increase in microhardness from (153±3) HV 0.5 (BM) to (208±6) HV 0.5 (FZ) (Figure 4b). Nam et al [28] recently also reported the change of hardness on LBW of equiatomic Co0.2Cr0.2Fe0.2Mn0.2Ni0.2. The specimen in the work was prepared *via* vacuum induction melting, casting and homogenisation at 1100 °C for 24 h followed by air cooling. The tensile test data (Figure 4c) showed that the rolled specimen was much stronger with significant work hardening than the cast specimen. Systematic tests also showed that the stress-strain curves were not very sensitive to the change of welding speed, further details can be found in the paper [28]. Tests on the cross section of the weld of rolled specimen and cast specimen revealed hardness increase in the FZ and HAZ over the BM, but to a distinctively different extent. The hardness increase in the FZ of the rolled HEA specimen was limited while the hardness increase in the FZ of the cast specimen was over 30% (~130 to ~170 HV 0.5) (Figure 4d). Similar levels of hardness increase were also reported in the work on welding of CrMnFeCoNi plates with thickness of 1 or 2 mm [29], in which metallurgical analysis showed the FZ formed a dendritic structure with the dendritic and interdendritic regions relatively rich in Fe and Mn, respectively. The hardness of the FZ was much higher than the BM (~185 $HV_{0.1}$ *vs.* ~143 $HV_{0.1}$). Despite the difference in material systems and the condition of the BM, the thickness and welding powers and speeds [26,28,29], all these works showed an increase in the hardnesses of the FZ and HAZ.

The trend of the other reported works on LBW showed a difference in terms of the hardness changes in the FZ [19,27]. In the work by Sokkalingam et al [19], 1 mm thick sheets of Al0.5CoCrFeNi were welded with a power of 1.5 kW and a traverse speed of 600 mm/min. The FZ that was exposed to rapid heating and cooling rate showed clear grain refinement (Figure 5a) with two kinds of dendritic structures: few longer columnar dendrites at the region next to fusion line, at the interloop boundaries and weld center with an average dendritic spacing of 4.8 μm and smaller equiaxed dendrites at in-between areas. Different from the results shown in Figure 4, Vickers microhardness test showed that the FZ became softer than the BM (Figure 5b). Similar hardness drop in LBW has also been reported by Nam et al [27] on Co0.2Cr0.2Fe0.2Mn0.2Ni0.2 (Figure 5c,d). In the work, the HEA slab used in laser butt-welding was homogenized at 1100 °C for 24 h and hot-rolled to 3 mm, followed by air cooling, then cold-rolled to 1.5 mm at room temperature (25 °C). As shown in Figure 5d the hardness of the FZ is much lower than the BM. The data also clearly showed that the width of the HAZ was affected by the welding speed but the bound of hardness of the FZ was not sensitive to the welding speed.

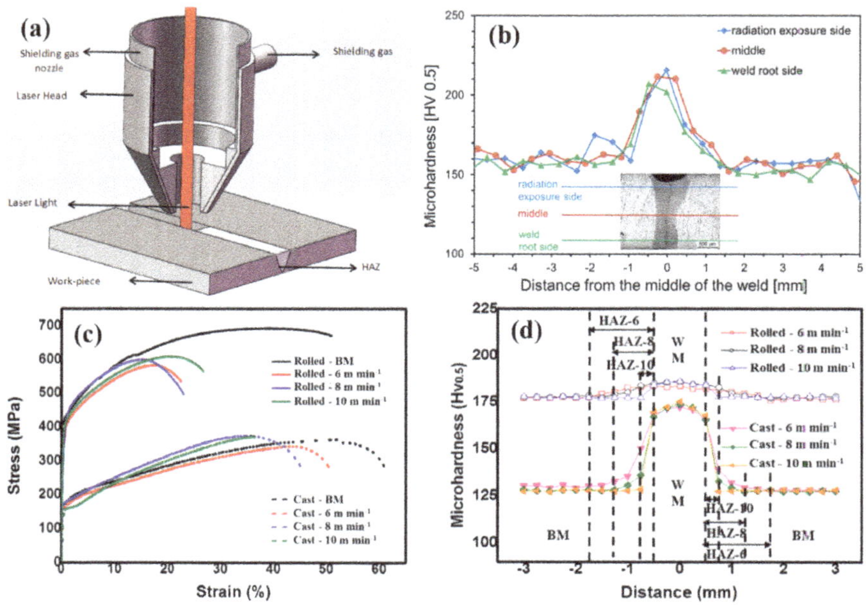

Figure 4. Hardness distribution in laser welded joints of HEAs: (**a**) Schematic diagram showing the laser welding process, (**b**) Microhardness profile of a butt–joint [26], (**c**) Stress-strain curves of the cast and rolled BM at various welding velocities [28] and (**d**) Hardness distribution in the transverse welds of the cast and rolled HEAs at various welding velocities: 6–10 m min^{-1} [28].

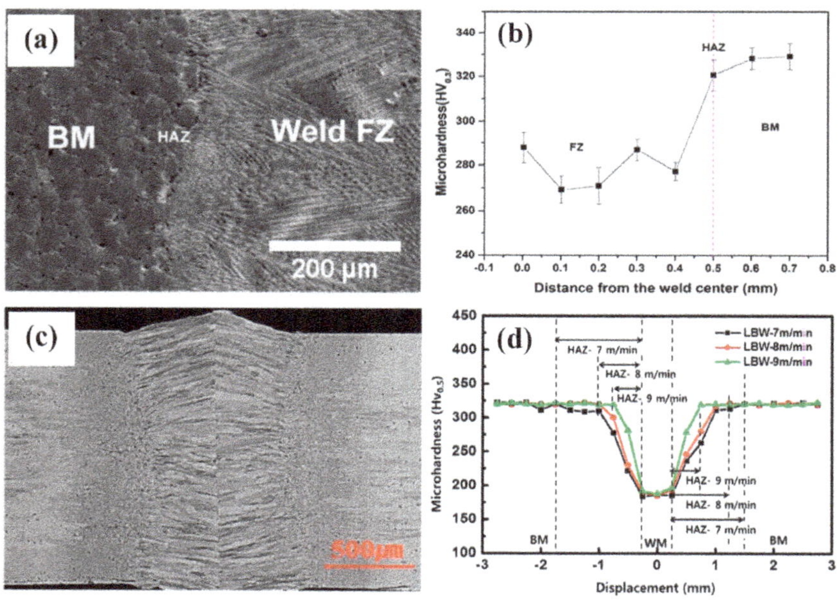

Figure 5. Structure and hardness of laser beam welded Al0.5CoCrFeNi HEAs: (**a**) Welding zones and boundary [30], (**b**) Microhardness profile on the surface [30], (**c**) Structure of the welded joints [27] and (**d**) Hardness distributions in the transverse weld for various welding velocities [27].

2.3. Electron Beam Welding

In an electron beam welding (EBW) process, a beam of high-velocity electrons is applied to two materials to be melted and joined. The heat density is high and the melted zone is normally thinner than that in laser and arc welding. In the work by Wu et al [31] on weldability of a high entropy CrMnFeCoNi alloy using a controlled test, a face centered cubic (FCC) CrMnFeCoNi alloy was selected. Welds produced by EBW showed no cracking. Tensile tests data showed that the welded joints possessed mechanical properties comparable to those of the BM at both room and cryogenic temperatures. Compared with the BM, deformation twinning was more pronounced in the FZ of the tested alloy. In another work, Wu et al [20] investigated low-heat-input EBW of CoCrFeMnNi HEA. The EB welds were made at a power level of 125 kV/2.2mA, and at a welding speed of 38 mm min^{-1}. The work also showed that the welded specimen has maintained the strength and ductility of the BM indicating good weldability of the HEA in this condition (Figure 6). In the work by Nahmany et al [32], electron beam surface re-melting was employed to modify the surface properties of two five-component AlxCrFeCoNi HEAs (x-0.6 and 0.8) prepared by vacuum arc-melting. The effects of electron beam heat on the structure and mechanical properties of deep penetration welding of HEAs were investigated. The quality of weld was found to be dependent on the welding condition (Figure 7), when the welding heat input was increased from 72 J/mm (P2-1 and P3-2) to 108J/mm (P2-3, P3-4), cracks were observed, which is thought to be due to the residual stresses. In addition, the hardness of the FZ also increased relative to the BM, which is different from EBW. The reason for the hardness difference requires further studies but this highlighted the effect of depth and energy input on the quality of the welding process.

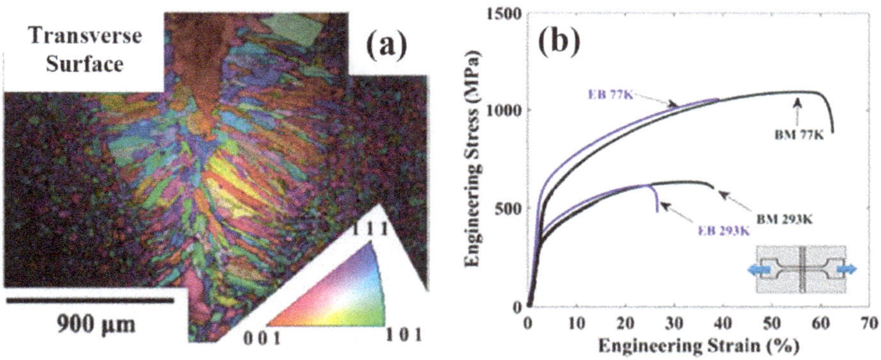

Figure 6. Structures and properties of electron beam butt-welded joints of CrMnFeCoNi alloy: (**a**) Electron backscattered diffraction (EBSD) maps showing the grain structure of the welded joints on transverse surface and (**b**) Tensile test results of welded joint [20].

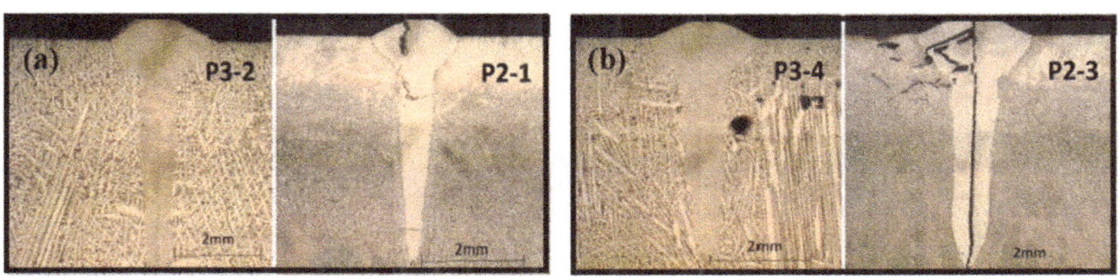

Figure 7. Structures of electron beam deep penetration welding of AlCrFeCoNi HEAs with different heat input: (**a**) 72 J/mm [32] and (**b**) 108 J/mm [32].

2.4. Friction Stir Welding

Several works have studied the friction stir welding (FSW) of HEA systems [29,33–37]. Different from arc, laser and electron beam welding, FSW (Figure 8) is a form of solid-state joining, in which two facing workpieces are jointed through the heat generated by friction between the rotating of a non-consumable tool and the workpiece material [38]. The volume of the material affected in the welding process is much wider than other fusion based welding processes. Zhu et al [34] studied the FSW of a typical FCC CoCrFeNiAl0.3 HEA. The sample was made by arc-melting and cast into ingot plates with a dimension of 2 mm × 10 mm × 30 mm. The welding process was performed with speeds of 30 and 50 mm/min while the rotation rate and load force were kept at 400 rpm and 1500 kg, respectively. The tool has a shoulder diameter of 12 mm, probe diameter of 4 mm and probe length of 1.8 mm. The FSW joint consisted of four different regions: the stir zone (SZ), thermomechanically affected zone (TMAZ), HAZ and BM (Figure 8). The SZ exhibited refined grain size arising from recrystallization and it exhibited higher hardness due to grain size refinement. The TMAZ exhibited a mixed microstructure comprising coarse and fine grains due to partial recrystallization. The XRD results indicated that the HEA remained an FCC structure after FSW. The SZ showed a refined equiaxed microstructure due to recrystallization, and the hardness of the FZ was found to be much higher (220 HV) than the BM (180 HV). Another study [35] reported the work on FSW of a ductile Co16Fe28Ni28Cr28 HEA of a low content of Co with a particular focus on microstructural evolution and weld strength in comparison to typical FCC HEAs. The work revealed a similar trend of hardness increase in the SZ compared to the hardness of the BM (~250 HV vs. 150 HV). The work also showed that the grain size decreased slightly while the hardness increased with increasing the welding speed.

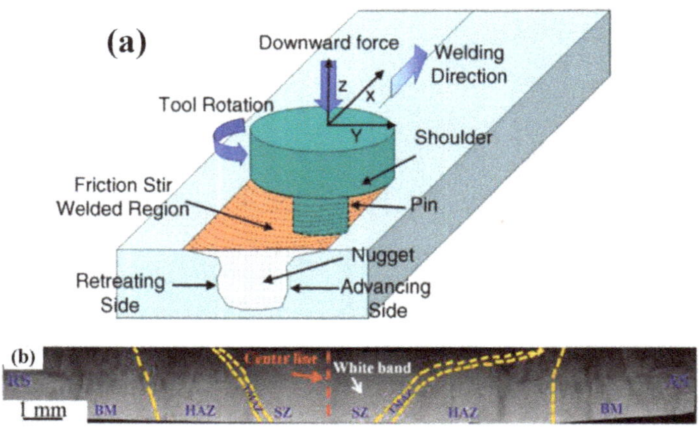

Figure 8. (**a**) Schematic diagram showing the friction stir welding (FSW) process and (**b**) Microstructure formed in FSW of Co16Fe28Ni28Cr28 alloys [34]: stir zone (SZ), thermomechanically affected zone (TMAZ), heat affected zone (HAZ) and BM.

Jo et al [29] investigated the microstructure and mechanical properties of friction stir welded CrMnFeCoNi HEA. The material was prepared by vacuum induction melting and hot rolling at 1100 °C. The dimension of the FSW plates was 55 mm × 60 mm × 2 mm. The pin diameter was 4–5.76 mm, the pin length was 1.85 mm, the shoulder diameter was 12 mm and the tilt angle was 3°. The FSW was carried out at a welding speed of 150 mm/min and tool rotation speeds of 600 and 70 RPM. Detailed metallurgical analysis showed that FSW refined the grain size in the weld region by a factor of ∼14 when compared with the BM (Figure 9a). The hardness in the weld region was much higher than the BM (~215 HV vs.144 HV) (Figure 9b). The tensile strength and ductility of FSW CrMnFeCoNi were comparable to that of annealed CrMnFeCoNi. This is probably associated with considerable microstructure refinement in the SZ and further recrystallization in FSW.

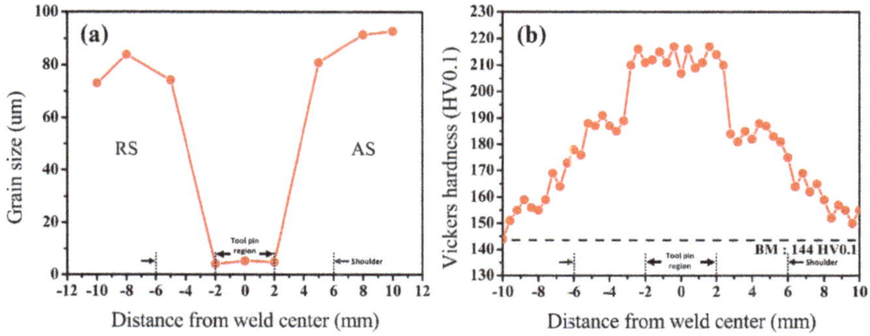

Figure 9. (**a**) Grain size at different distance from weld center positions and (**b**) Vickers hardness in the cross-section of FSW CrMnFeCoNi HEA [29].

Shaysultanov et al [36] studied FSW of a carbon-doped CoCrFeNiMn HEA in butt joints. Along with the principal elements, a small amount (0.9 at.%) of C was added to the alloy produced by SHS. The CoCrFeNiMn alloy was produced using thermite-type SHS, in which a mixture of powders (oxides of the target elements NiO, Cr_2O_3, Co_3O_4, Fe_2O_3, MnO_2, pure carbon C, and Al as the metal reducer) was used as the starting material. The as-cast alloy was cold rolled and annealed at 900 °C to produce a refined microstructure. The microhardness measurement showed an approximately 40 HV increase in the area of the SZ in comparison with the BM. Tensile tests were performed on samples perpendicular to the welded seam as well as on the sample along the welding direction (Figure 10). The data for both sample conditions showed a clear noticeable rise in strength in comparison with the BM. This can be associated with the microstructure refinement and some increase in the volume fraction of $M_{23}C_6$ carbides [36]. The sample with the loading axis perpendicular to the weld maintained around 50% of the ductility, while the sample fully taken from the seam showed comparable ductility to the BM.

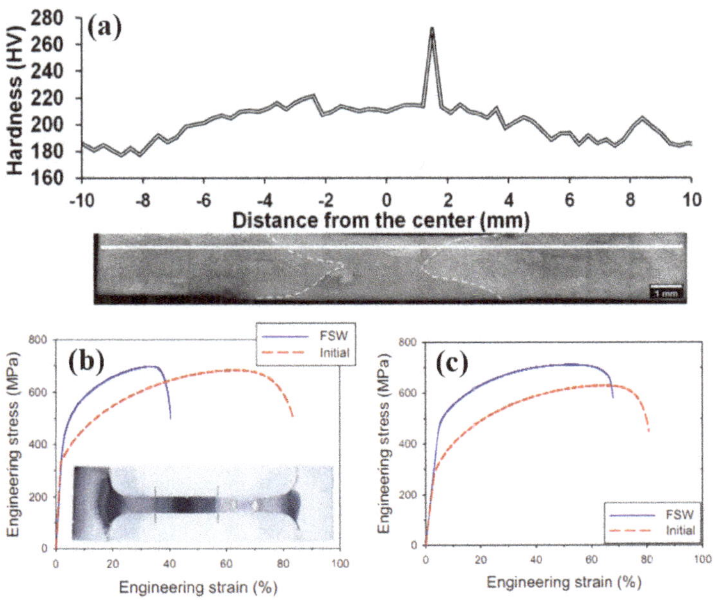

Figure 10. Microhardness distribution across the weld seam (**a**) and tensile stress-strain curves of the FSW specimens cut across (**b**) or along (**c**) the weld seam [36].

3. Grain Structure, Element Distribution and Precipitations in Welded Structures of HEAs

The structures at different levels (grain, precipitation and lattice) are important to the integrity of the weldment. The structure, precipitation and element segregation/redistribution in a welding process are also important for some service conditions such as corrosion, creep etc. [1,12,15,16]. Apart from the difference in the general structure and hardness profiles of welding zones detailed in Section 2, research work on welding of HEAs also revealed significant difference in secondary phase and precipitation associated with different welding processes and material systems. In arc welding processes, such as GTAW, which is like a miniature casting process, the weld metal experiences a rapid cooling rate. The main grain structure is elongated columnar dendrites nucleated from the fusion line and equiaxed grains near the weld centerline [19,20]. Sokkalingham et al [19] has performed detailed XRD analysis on GTAW for Al0.5CoCrFeNi and found that both the volumetric fraction and lattice constant for both FCC and BCC phases change. The volume fraction of FCC increases (from 76 to 97.8%) but the lattice constant of FCC decreased from 0.3568 to 0.2424, and the lattice constant of BBC increased significantly, from 0.2132 to 0.3020. This is an interesting finding, and further research is required on how this is linked to the element segregation and how this change may affect the mechanical behaviour including shear and corrosion [39,40].

Composition distribution and element segregation across the welded joint at macro and micro scales are important to the understanding of the mechanical and corrosion resistance of the welded joint, which may be associated with many different mechanisms, such as evaporation, time and temperature for element diffusion, oxidation and precipitation, etc. [4,19,20,30,41–43]. For CoCrFeMnNi alloy [20], the GTA weld exhibited microsegregation behaviour in the Mn- and Ni-rich interdendritic region and Co-, Cr-, and Fe-rich dendrite cores, but the depletion of Mn was not observed, while for EBW weld, there was clear Mn depletion in the FZ (Figure 11) [20]. As shown in Figure 11b, the average value of the Mn is lower than the nominal value (20%). This was probably caused by evaporation of Mn due to the high power density associated with the EB welding process. For Al0.5CoCrFeNi, the elemental analysis in FZ revealed that the dentrites were rich in Fe and Co, depleted of Al and Ni, and had an even distribution of the Cr element [19]. Similar results have been observed in LBE of Al0.5CoCrFeNi [30]. It is not fully clear if evaporation of Al had played role influencing the structure and properties. However, elemental analysis showed that the concentration of Al element in dentrites and interdentrites of the as-welded (AW) sample is lower than the grains and the grain boundaries of the BM. The dendrites of the AW have a Al concentration of 4.78 %, while the grain of the BM has a concentration of 5.27%; The data also shows that the interdentrites of the AW region consist of 10.69% Al, while the Al concentration in the BM-grain boundary is about14.43%. So, it is possible that evaporation of Al may have occurred during the LBE process, which requires further quantitative studies together with other mechanisms proposed by the authors, such as precipitation of BBC phase (Al-Ni rich phase). In both material systems (i.e., AlCoCrFeNi and CoCrFeMnNi), the primary dendrite arm spacing and the extent of elemental segregation were less in the welds than in the cast ingot as the cooling in welding is much higher than in casting. In LBW, the heat density and the cooling rate is much higher than arc welding, the FZ gives the appearance of a columnar grained microstructure with random crystallographic orientation, and no significant change in microstructure near the FZ-BM boundary was observed [29]. In another work [26], the differences in microstructure and grain orientation distributions between the BM, HAZ and the FZ were not significant for CoCrFeNiMn-type HEA in LBW. For CrMnFeCoNi, the interdendritic and dendritic regions are enriched in Mn and Fe, respectively, similar to that at the FZ-BM boundary. LBW also resulted in precipitation of the nanoscale B_2 phase particles for CoCrFeNiMn-type HEA. In the weldment of Al0.5CoCrFeNi, the equiaxed polygonal grains with grain size of 60 μm in homogenized state transformed to longer columnar dendrites at the region nearer to fusion boundary and inter-loop boundaries with an average dendritic spacing of 4.8 μm and fine equiaxed dendrites at the weld center on welding [19]. The work also found that, in the LB weldment, the hardening factor (Al-Ni rich BCC phase) was less in the weld metal than that in the BM, therefore the weld metal exhibited a lower hardness than that of BM.

For FSW, the SZ comprised of refined grains arising from dynamic recrystallization. The HAZ exhibited a columnar structure [34,35]. In another work [29], EDS-line scan showed that in general the dendrites were enriched in Fe and depleted in Ni and Mn. The fluctuation in composition was significant: between ~5 and 15 at%. The cause for this compositional fluctuation in the FZ, which is expected to stay hot longer than the FZ-BM boundary and thereby allow for greater homogenization, is not clear and needs further study. In the work, relatively dark spherical particles were also detected in the High-angle annular dark-field image (HAADF) and all the element maps were found to be rich in Mn, S and O indicating that these were Mn sulfides/oxides. In the work on FSW of carbon-doped CoCrFeNiMn HEA [36], the microstructure of the SZ was shown to contain $M_{23}C_6$ carbides. More importantly, the increase in the volume fraction and size of the carbides (to 7% and 150 ± 62 nm, respectively) after FSW in comparison with that in the BM can be clearly noted. All these may have contributed to the strength and hardness increase of the welded zone in FSW as illustrated in Figures 9 and 10. This suggests that, for FSW, there is no specific post welding processing needed to obtain a hardened joint, i.e., the alloy hardens due to the "natural" heat treatment during welding [36]. This is an advantage of interstitial alloying of HEAs. This indicates that higher strength of the FSW joints can be achieved in similar alloys by tailoring microstructure *via* adjusting the chemical composition and processing parameters even [26,33,44,45]. A typical example is shown in Figure 12, which shows that the FSW can be used for processing HEAs with enhanced strength and ductility in a friction stir processing engineered dual phase HEA [33].

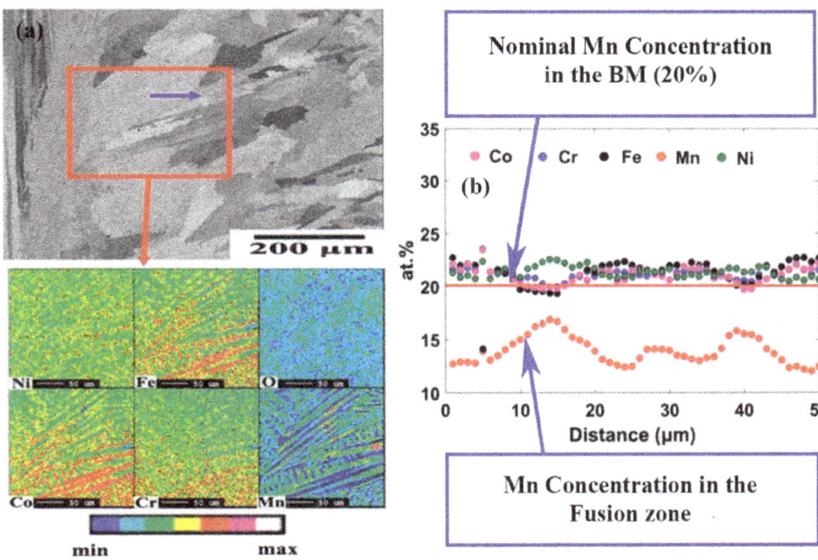

Figure 11. Elemental mapping of fusion zone (FZ) for electron beam welding of CoCrFeMnNi: (**a**) Microstructure of the electron beam (EB) weld zone area and electron microprobe analyzer (EMPA) compositional mapping of the marked area in upper figure and (**b**) Compositional profile along the arrow shown in (**a**) [20].

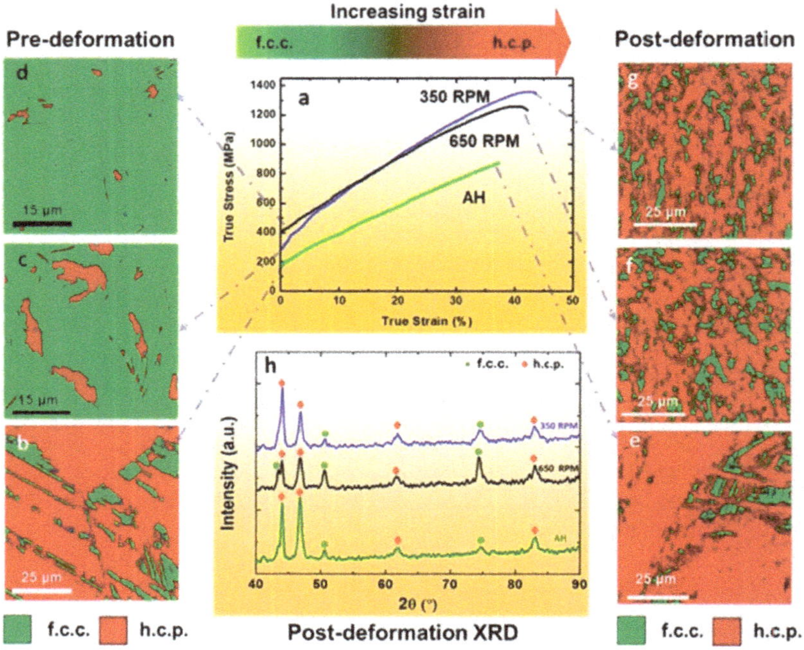

Figure 12. Structure and stress strain curves of high entropy alloys (HEAs) with enhanced properties through friction stir welding [45].

4. Discussion and Future Works

The recent work summarized in Sections 2 and 3 showed that HEAs have good welding characteristics. A broad range of welding methods were found to be applicable to the welding of HEAs and most cases showed no significant defects or cracking. The size of different welding zones varies with the welding method, such as electron beam, laser and arc welding. FSW resulted in a much wider FZ due to the involvement of the rotating pin. In most of the cases, the shape of the welded beam is symmetrical, which represents good weldability for structure and quality control. However, there are reported cases in which some butt-joints showed a non-symmetrical weld shape (Figure 4b), this is most likely due to coarse columnar grains of the base material [26]. Most of the reported works were based on butt-welding, which is one of the simplest forms of welding configuration. More complex setup such as tensile, shear, bending, torsion, and impact, in which the stress condition is more complex and close to real service conditions, need to be considered.The residual stresses also need to be studied systematically. Residual stress is influenced by many factors such as localized heating and cooling, differential volumetric change occurs both at macroscopic and microscopic level [46]. Limited work has been reported on the measurement of residual stresses in the work reviewed, which were mostly based on relatively simple, less constrained samples. In the case of EBM deep penetration welding of AlxCrFeCoNi HEAs, it was shown that residual stresses could cause cracking (Figure 7b). HEAs have greater freedom in designing composition, mechanical/physical properties, such as yield strength, stiffness, thermal conductivity and expansion, all these may offer opportunity to optimize the material design to manage the residual stresses in welding processes.

As detailed in Sections 2 and 3, the works published clearly show that the change of the strength/hardness of the FZ and HAZ varies between the composition systems and the welding process employed. This is a significant difference from other engineering materials such as carbon or stainless steels, for which, the FZs normally have a hardness increase [47,48]. For nonferrous metals such as magnesium, the hardness for the FZ normally becomes higher but the hardness of the HAZ may drop [49]. For HEAs, hardness drop in the FZ was observed for Al0.5CoCrFeNi in laser and arc welding, while the hardness drop for CoCrFeMnNi is relatively limited or slightly increased. Sokkalingam et al [19,30] suggested that the hardness drop for the Al0.5CoCrFeNi in the LB weldment was due to the reduction of the hardening factor (Al-Ni rich BCC phase) in the weld metal than in the base metal. The processing method of the sample also influences the hardness changes. For example, as shown in Figure 4, the hardness change for the cast specimen after welding was much more profound than that for the rolled specimen. Other parameters such as welding speed and power also showed some influence on the hardness change but not as significant as the material groups and welding methods. Another issue, which requires further work, lies in the question how the hardness/strength change of the FZ and HAZ may affect the overall ductility and toughness of the welded joints. Hardness/strength drop is not an ideal situation in terms of strength following general engineering principles, however, this may potentially offer a way to enhance the ductility and toughness of the welded joints. This can be beneficial to some service condition where toughness is more critical than strength or hardness. Preliminary numerical modelling work by the authors (in-process, result not shown) indicates that soft FZs in HEAs may help to spread the load more evenly and reduce stress concentration, which commonly exists in welded joints with a hardened FZ [17], thus improve the overall ductility or toughness of the welded structure. In the work by Wu et al [31] on EBW of CrMnFeCoNi, tensile test results showed that, compared with the BM, deformation twinning was more pronounced in the FZ of the tested alloy. Further combined experimental and numerical modelling work is required to systematically investigate the structure and stress in the FZ, HAZ and the HAZ-BM or HAZ-FZ interface within welded joints of HEAs under controlled material, welding and testing conditions.

For HEAs and welded structures, many factors may contribute to the strength of the materials (such as grain refinement, precipitation, residual stress, volume ratio of different phases) [15,16,22–25,41–43]. Grain size is still an important but not necessarily the dominating factor depending on the HEA system and the welding processes utilized. For FSW, the SZ becomes much harder [19,20,26] than the BM. One major mechanism contributing to the hardness increase is the refinement of the grain, but the trend of the grain size effect does not fully follow the theories or data established for conventional bulk HEAs. For bulk HEA alloys, the Hall–Petch (H–P) trend between grain size and strength is applicable for a range of HEAs based on either tensile tests or hardness data including CoCrFeNi [50–53], CoCrFeNiMn [7,54–56] and AlxCoCrFeNi [57–61]. While for welding, the suitability of H–P relationship varies with the materials and welding conditions. For example, the data for FSW of Co16Cr28Fe28Ni28 [35] and Al0.1CoCrFeNi [34,62] followed the H–P trend well, but for arc welding and laser welding, the hardness and grain size relationship is opposite to the H–P trend [19,30]. These differences reflect that the controlling operative strengthening mechanisms are more complicated than pure grain refinement, and other factors such as phase change, precipitation all may affect the hardness to different extents. A more quantitative data driven methodology is required to clarify or establish the dominating mechanism(s). For example, the combined contribution of grain refinement and precipitation was quantified in [36] through combined Hall-Petch coefficient analysis. The work showed that a decrease in grain size by a factor of two (from 9.2 μm in the initial condition to 4.6 μm after FSW) resulted in an increase in strength by ~55 MPa. The rest of the increase in the yield strength (~145 MPa) was attributed to the precipitation strengthening associated with some increase in the volume fraction of the $M_{23}C_6$ carbides. Such quantitative analysis is not only important for establishing the strengthening mechanism, it is also essential to provide guidelines for developing effective pre- or post-treatments of welded structures. The published research work has successfully highlighted the key issues influencing the mechanical properties of the welded joints, which is important for expanding the applications of HEAs and welding processes. However, the data available is still limited to draw direct comprehensive conclusion on the property differences between different welding methods. Even though the general heat input is different between various welding techniques, but more systematic data on the cooling curves of comparable materials systemsare required to be conclusive when quantifying the main strengthening mechanisms. Physical simulation with controlled cooling (such as Gleeble^{TM}) will help to quantify the effect cooling history on the microstructure and properties of different welding zones [62]. Establishment of the mechanisms of phase change, grain refinement and precipitation is also directly beneficial to the development of new materials and interruptive control technologies to maintain or improve the beneficial properties of HEAs such as high temperature stability and corrosion resistance of welded structures of HEAs [63–65].

5. Summary

Recent works on welding of HEAs with various welding methods of different setup and heat input were reviewed in detail focusing on the research on main HEA systems when applying different welding techniques. The structures and properties of the welding zones in particular the FZ and the HAZ formed with different welding methods were compared and presented in details and the structure properties relationships were discussed. The works showed that weldability of HEAs varies with the composition groups and the welding methods employed. Arc and laser welding of AlCoCrFeNi HEAs resulted in lower hardness in the FZ and HAZ and reduced strength. FSW resulted in higher hardness in the FZ and maintained the strength of the welded joints under tensile load. The welded of HEAs are capable to maintaining reasonable proportion of strength and the ductility. The key structural changes including element distribution, the volume fraction of FCC and BCC as well as some reported lattice were summarized and analyzed. The effects of evaporation for high energy welding (such as LBM and EBM) on composition and structure requires further quantitative study. Detailed mechanism(s) governing the mechanical properties including the contribution of the grain size-properties/hardness relationship in the form of Hall–Petch (H–P) effect for both bulk and welded structure of HEAs were discussed. Future research is required to establish the strengthening mechanisms of the welded joints and the effect of the observed hardness changes in the FZ on the strength and toughness of welded structures of HEAs. Such quantitative analysis will provide guidelines for developing effective pre- or post-treatments of welded structures. Studies of residual stress for different welding processes and pre-post weld treatments, in particular for under complex loading conditions, is required for enhancing the applications of HEAs. It is also beneficial to the development of new materials and interruptive control technologies to maintain or improve the beneficial properties of HEAs such as high temperature stability and corrosion resistance of welded structures of HEAs.

Author Contributions: All the authors contributed to the review. Original Draft preparation: J.G., C.T.; Analysis and editing: L.L., G.R., Y.-C.W. and Q.Y.; supervision: G.R., X.R.

Funding: This project has received funding from the European Union's Horizon 2020 research and innovation programme under the Marie Skłodowska-Curie grant agreement No: 793114".

Acknowledgments: The authors would like to acknowledge the Royal Society for the support through an International Exchange Grant.

Conflicts of Interest: The authors declare no conflict of interest.

References

1. Zhang, Y.; Zuo, T.T.; Tang, Z.; Gao, M.C.; Dahmen, K.A.; Liaw, P.K.; Lu, Z.P.; Zhang, A. Microstructures and properties of high-entropy alloys. *Prog. Mater. Sci.* **2014**, *61*, 1–93. [CrossRef]
2. Ye, Y.; Wang, Q.; Lu, J.; Liu, C.; Yang, Y. High-entropy alloy: Challenges and prospects. *Mater. Today* **2016**, *19*, 349–362. [CrossRef]
3. Tsai, M.-H. Physical Properties of High Entropy Alloys. *Entropy* **2013**, *15*, 5338–5345. [CrossRef]
4. Dominguez, L. Combinatorial High Throughput Synthesis of High Entropy Alloys. Ph.D. Thesis, University of Sheffield, Sheffield, UK, 2016.
5. Kumar, A.; Gupta, M. An Insight into Evolution of Light Weight High Entropy Alloys: A Review. *Metals* **2016**, *6*, 199. [CrossRef]

6. Diao, H.; Feng, R.; Dahmen, K.; Liaw, P. Fundamental deformation behavior in high-entropy alloys: An overview. *Curr. Opin. Mater. Sci.* **2017**, *21*, 252–266. [CrossRef]
7. Otto, F.; Dlouhy, A.; Somsen, C.; Bei, H.; Eggeler, G.; George, E. The influences of temperature and microstructure on the tensile properties of a CoCrFeMnNi high-entropy alloy. *Acta Mater.* **2013**, *61*, 5743–5755. [CrossRef]
8. Gludovatz, B.; Hohenwarter, A.; Catoor, D.; Chang, E.H.; George, E.P.; Ritchie, R.O. A fracture-resistant high-entropy alloy for cryogenic applications. *Science* **2014**, *345*, 1153–1158. [CrossRef] [PubMed]
9. Zherebtsov, S.; Stepanov, N.; Shaysultanov, D.; Malopheyev, S.; Vysotskiy, I.; Sanin, V.; Kashaev, N.; Kaibyshev, R. Use of Novel Welding Technologies for High-Entropy Alloys Joining. *Mater. Sci.* **2018**, *941*, 919–924. [CrossRef]
10. Alaneme, K.K.; Bodunrin, M.O.; Oke, S.R. Processing, alloy composition and phase transition effect on the mechanical and corrosion properties of high entropy alloys: A review. *J. Mater. Res. Technol.* **2016**, *5*, 384–393. [CrossRef]
11. Yim, D.; Kim, H. Fabrication of the High-Entropy Alloys and Recent Research Trends: A Review. *Korean J. Met. Mater.* **2017**, *55*, 671–683.
12. Ma, C.; Peng, Q.; Mei, J.; Han, E.-H.; Ke, W. Microstructure and corrosion behavior of the heat affected zone of a stainless steel 308L-316L weld joint. *J. Mater. Sci. Technol.* **2018**, *34*, 1823–1834. [CrossRef]
13. Liu, Y. Interfacial Behavior and Joint Performance of High-entropy Alloy CoCrFeMnNi and Pure Cu Joints Obtained by Vacuum Diffusion Welding. *J. Mech. Eng.* **2017**, *53*, 84–91. [CrossRef]
14. Cui, L.; Ma, B.; Feng, S.Q.; Wang, X.L. Microstructure and Mechanical Properties of High-Entropy Alloys CoCrFeNiAl by Welding. *Adv. Mater. Res.* **2014**, *936*, 1635–1640. [CrossRef]
15. Lippold, J.; Kiser, S.; DuPont, J. *Welding Metallurgy and Weldability of Nickel-Base Alloys*; Wiley: Hoboken, NJ, USA, 2013.
16. Vendan, S.; Gao, L.; Garg, A.; Kavitha, P.; Dhivyasri, G.; SG, R. *Interdisciplinary Treatment to ARC Welding Power Sources*; Springer: Singapore, 2018.
17. Kong, X.; Yang, Q.; Li, B.; Rothwell, G.; English, R.; Ren, X. Numerical study of strengths of spot-welded joints of steel. *Mater. Des.* **2008**, *29*, 1554–1561. [CrossRef]
18. Chen, S.; Tong, Y.; Liaw, P. Additive Manufacturing of High-Entropy Alloys: A Review. *Entropy* **2018**, *20*, 937. [CrossRef]
19. Sokkalingam, R.; Mishra, S.; Cheethirala, S.R.; Muthupandi, V.; Sivaprasad, K. Enhanced Relative Slip Distance in Gas-Tungsten-Arc-Welded Al0.5CoCrFeNi High-Entropy Alloy. *Met. Mater. Trans. A* **2017**, *48*, 3630–3634. [CrossRef]
20. Wu, Z.; David, S.A.; Leonard, D.N.; Feng, Z.; Bei, H. Microstructures and mechanical properties of a welded CoCrFeMnNi high-entropy alloy. *Sci. Technol. Weld. Join.* **2018**, *23*, 585–595. [CrossRef]
21. Laplanche, G.; Horst, O.; Otto, F.; Eggeler, G.; George, E. Microstructural evolution of a CoCrFeMnNi high-entropy alloy after swaging and annealing. *J. Alloy Compd.* **2015**, *647*, 548–557. [CrossRef]
22. Zhang, Z.; Mao, M.M.; Wang, J.; Gludovatz, B.; Zhang, Z.; Mao, S.X.; George, E.P.; Yu, Q.; Ritchie, R.O. Nanoscale origins of the damage tolerance of the high-entropy alloy CrMnFeCoNi. *Nat. Commun.* **2015**, *6*, 10143. [CrossRef]
23. Miracle, D.; Senkov, O. A critical review of high entropy alloys and related concepts. *Acta Mater.* **2017**, *122*, 448–511. [CrossRef]
24. Shigesato, G.; Sugiyama, M. Development of in situ observation technique using scanning ion microscopy and demonstration of Mn depletion effect on intragranular ferrite transformation in low-alloy steel. *QJM Int. J. Med.* **2002**, *51*, 359–367. [CrossRef]
25. Joo, S.-H.; Kato, H.; Jang, M.; Moon, J.; Tsai, C.; Yeh, J.; Kim, H. Tensile deformation behavior and deformation twinning of an equimolar CoCrFeMnNi high-entropy alloy. *Mater. Sci. Eng. A* **2017**, *689*, 122–133. [CrossRef]
26. Kashaev, N.; Ventzke, V.; Stepanov, N.; Shaysultanov, D.; Sanin, V.; Zherebtsov, S. Laser beam welding of a CoCrFeNiMn-type high entropy alloy produced by self-propagating high-temperature synthesis. *Intermetallics* **2018**, *96*, 63–71. [CrossRef]
27. Nam, H.; Park, C.; Kim, C.; Kim, H.; Kang, N. Effect of post weld heat treatment on weldability of high entropy alloy welds. *Sci. Technol. Weld. Join.* **2018**, *23*, 420–427. [CrossRef]
28. Nam, H.; Parka, C.; Moon, J.; Na, Y.; Kim, H.; Kang, N. Laser weldability of cast and rolled high-entropy alloys for cryogenic applications. *Mater. Sci. Eng. A* **2019**, *742*, 224–230. [CrossRef]

29. Jo, M.-G.; Kim, H.-J.; Kang, M.; Madakashira, P.P.; Park, E.S.; Suh, J.-Y.; Kim, D.-I.; Hong, S.-T.; Han, H.N. Microstructure and mechanical properties of friction stir welded and laser welded high entropy alloy CrMnFeCoNi. *Met. Mater. Int.* **2018**, *24*, 73–83. [CrossRef]
30. Sokkalingam, R.; Sivaprasad, K.; Muthupandi, V.; Duraiselvam, M. Characterization of Laser Beam Welded Al0.5CoCrFeNi High-Entropy Alloy. *Key Eng. Mater.* **2018**, *775*, 448–453.
31. Wu, Z.; David, S.; Feng, Z.; Bei, H. Weldability of a high entropy CrMnFeCoNi alloy. *Scr. Mater.* **2016**, *124*, 81–85. [CrossRef]
32. Nahmany, M.; Hooper, Z.; Stern, A.; Geanta, V.; Voiculescu, I. AlxCrFeCoNi High-Entropy Alloys: Surface Modification by Electron Beam Bead-on-Plate Melting. *Metallogr. Microst. Anal.* **2016**, *5*, 229–240. [CrossRef]
33. Nene, S.S.; Liu, K.; Frank, M.; Mishra, R.S.; Brennan, R.E.; Cho, K.C.; Li, Z.; Raabe, D. Enhanced strength and ductility in a friction stir processing engineered dual phase high entropy alloy. *Sci. Rep.* **2017**, *7*, 16167. [CrossRef]
34. Zhu, Z.; Sun, Y.; Goh, M.; Ng, F.; Nguyen, Q.; Fujii, H.; Nai, S.; Wei, J.; Shek, C. Friction stir welding of a CoCrFeNiAl 0.3 high entropy alloy. *Mater. Lett.* **2017**, *205*, 142–144. [CrossRef]
35. Zhu, Z.; Sun, Y.; Ng, F.; Goh, M.; Liaw, P.; Fujii, H.; Nguyen, Q.; Xu, Y.; Shek, C.; Nai, S.; et al. Friction-stir welding of a ductile high entropy alloy: Microstructural evolution and weld strength. *Mater. Sci. Eng. A* **2018**, *711*, 524–532. [CrossRef]
36. Shaysultanov, D.; Stepanov, N.; Malopheyev, S.; Vysotskiy, I.; Sanin, V.; Mironov, S.; Kaibyshev, R.; Salishchev, G.; Zherebtsov, S. Friction stir welding of a carbon-doped CoCrFeNiMn high-entropy alloy. *Mater. Charact.* **2018**, *145*, 353–361. [CrossRef]
37. Park, S.; Park, C.; Na, Y.; Kim, H.-S.; Kang, N. Effects of (W, Cr) carbide on grain refinement and mechanical properties for CoCrFeMnNi high entropy alloys. *J. Alloy Compd.* **2019**, *770*, 222–228. [CrossRef]
38. Zhang, S.; Chen, G.; Liu, Q.; Li, H.; Zhang, G.; Wang, G.; Shi, Q. Numerical analysis and analytical modeling of the spatial distribution of heat flux during friction stir welding. *J. Manuf. Process.* **2018**, *33*, 245–255. [CrossRef]
39. Feng, R.; Gao, M.C.; Lee, C.; Mathes, M.; Zuo, T.; Chen, S.; Hawk, J.A.; Zhang, Y.; Liaw, P.K.; Zhang, A. Design of Light-Weight High-Entropy Alloys. *Entropy* **2016**, *18*, 333. [CrossRef]
40. Lee, C.; Song, G.; Gao, M.C.; Feng, R.; Chen, P.; Brechtl, J.; Chen, Y.; An, K.; Guo, W.; Poplawsky, J.D.; et al. Lattice distortion in a strong and ductile refractory high-entropy alloy. *Acta Mater.* **2018**, *160*, 158–172. [CrossRef]
41. Moravcik, I.; Gouvea, L.; Hornik, V.; Kovacova, Z.; Kitzmantel, M.; Neubauer, E.; Dlouhy, I. Synergic strengthening by oxide and coherent precipitate dispersions in high-entropy alloy prepared by powder metallurgy. *Scr. Mater.* **2018**, *157*, 24–29. [CrossRef]
42. Lin, C.-M.; Tsai, H.-L. Evolution of microstructure, hardness, and corrosion properties of high-entropy Al0.5CoCrFeNi alloy. *Intermetallics* **2011**, *19*, 288–294. [CrossRef]
43. Tang, Z.; Senkov, O.N.; Parish, C.M.; Zhang, C.; Zhang, F.; Santodonato, L.J.; Wang, G.; Zhao, G.; Yang, F.; Liaw, P.K. Tensile ductility of an AlCoCrFeNi multi-phase high-entropy alloy through hot isostatic pressing (HIP) and homogenization. *Mater. Sci. Eng. A* **2015**, *647*, 229–240. [CrossRef]
44. Stepanov, N.; Shaysultanov, D.; Chernichenko, R.; Ikornikov, D.; Sanin, V.; Zherebtsov, S. Mechanical properties of a new high entropy alloy with a duplex ultra-fine grained structure. *Mater. Sci. Eng. A* **2018**, *728*, 54–62. [CrossRef]
45. Nene, S.S.; Frank, M.; Liu, K.; Mishra, R.S.; McWilliams, B.A.; Cho, K.C. Extremely high strength and work hardening ability in a metastable high entropy alloy. *Sci. Rep.* **2018**, *8*, 9920. [CrossRef]
46. De, A.; DebRoy, T. A perspective on residual stresses in welding. *Sci. Technol. Weld. Join.* **2011**, *16*, 204–208. [CrossRef]
47. Tavares, S.; Almeida, B.; Correa, D.; Pardal, J. Failure of super 13Cr stainless steel due to excessive hardness in the welded joint. *Eng. Fail. Anal.* **2018**, *91*, 92–98. [CrossRef]
48. Soltani, H.M.; Tayebi, M. Comparative study of AISI 304L to AISI 316L stainless steels joints by TIG and Nd:YAG laser welding. *J. Alloy. Compd.* **2018**, *767*, 112–121. [CrossRef]
49. Peter, I.; Rosso, M. Investigations on Tungsten Inert Gas Welded Magnesium Alloy. *Iop Conf. Ser. Mater. Sci. Eng.* **2018**, *416*, 012030. [CrossRef]

50. Wu, Z.; Bei, H.; Otto, F.; Pharr, G.; George, E. Recovery, recrystallization, grain growth and phase stability of a family of FCC-structured multi-component equiatomic solid solution alloys. *Intermetallics* **2014**, *46*, 131–140. [CrossRef]
51. Liu, B.; Wang, J.; Liu, Y.; Fang, Q.; Wu, Y.; Chen, S.; Liu, C. Microstructure and mechanical properties of equimolar FeCoCrNi high entropy alloy prepared via powder extrusion. *Intermetallics* **2016**, *75*, 25–30. [CrossRef]
52. Sathiyamoorthi, P.; Basu, J.; Kashyap, S.; Pradeep, K.; Kottada, R.S. Thermal stability and grain boundary strengthening in ultrafine-grained CoCrFeNi high entropy alloy composite. *Mater. Des.* **2017**, *134*, 426–433. [CrossRef]
53. Wang, J.; Guo, T.; Li, J.; Jia, W.; Kou, H. Microstructure and mechanical properties of non-equilibrium solidified CoCrFeNi high entropy alloy. *Mater. Chem. Phys.* **2018**, *210*, 192–196. [CrossRef]
54. Liu, W.; Wu, Y.; He, J.; Nieh, T.; Lu, Z. Grain growth and the Hall–Petch relationship in a high-entropy FeCrNiCoMn alloy. *Scr. Mater.* **2013**, *68*, 526–529. [CrossRef]
55. Okamoto, N.L.; Fujimoto, S.; Kambara, Y.; Kawamura, M.; Chen, Z.M.T.; Matsunoshita, H.; Tanaka, K.; Inui, H.; George, E.P. Size effect, critical resolved shear stress, stacking fault energy, and solid solution strengthening in the CrMnFeCoNi high-entropy alloy. *Sci. Rep.* **2016**, *6*, 35863. [CrossRef]
56. Sun, S.; Tian, Y.; Lin, H.; Dong, X.; Wang, Y.; Zhang, Z.; Zhang, Z. Enhanced strength and ductility of bulk CoCrFeMnNi high entropy alloy having fully recrystallized ultrafine-grained structure. *Mater. Des.* **2017**, *133*, 122–127. [CrossRef]
57. Gwalani, B.; Soni, V.; Lee, M.; Mantri, S.; Ren, Y.; Banerjee, R. Optimizing the coupled effects of Hall-Petch and precipitation strengthening in a Al 0.3 CoCrFeNi high entropy alloy. *Mater. Des.* **2017**, *121*, 254–260. [CrossRef]
58. Wu, S.; Wang, G.; Yi, J.; Jia, Y.; Hussain, I.; Zhai, Q.; Liaw, P. Strong grain-size effect on deformation twinning of an Al0.1CoCrFeNi high-entropy alloy. *Mater. Res. Lett.* **2017**, *5*, 276–283. [CrossRef]
59. Gangireddy, S.; Gwalani, B.; Mishra, R.S. Grain size dependence of strain rate sensitivity in a single phase FCC high entropy alloy Al0.3CoCrFeNi. *Mater. Sci. Eng. A* **2018**, *736*, 344–348. [CrossRef]
60. Liu, G.; Liu, L.; Liu, X.; Wang, Z.; Han, Z.; Zhang, G.; Kostka, A. Microstructure and mechanical properties of Al0.7CoCrFeNi high-entropy-alloy prepared by directional solidification. *Intermetallics* **2018**, *93*, 93–100. [CrossRef]
61. Chen, G.; Li, L.; Qiao, J.; Jiao, Z.; Ma, S.; Ng, F.L.; Zhu, Z.; Zhao, D.; Wang, Z. Gradient hierarchical grain structures of Al0.1CoCrFeNi high-entropy alloys through dynamic torsion. *Mater. Lett.* **2019**, *238*, 163–166. [CrossRef]
62. Liu, W.; Lu, F.; Yang, R.; Tang, X.; Cui, H. Gleeble simulation of the HAZ in Inconel 617 welding. *J. Mater. Process. Technol.* **2015**, *225*, 221–228. [CrossRef]
63. Qiu, Y.; Gibson, M.A.; Fraser, H.L.; Birbilis, N. Corrosion characteristics of high entropy alloys (HEAs). *Mater. Sci. Technol.* **2015**, *31*, 1235–1243. [CrossRef]
64. Komarasamy, M.; Kumar, N.; Tang, Z.; Mishra, R.; Liaw, P. Effect of Microstructure on the Deformation Mechanism of Friction Stir-Processed Al 0.1 CoCrFeNi High Entropy Alloy. *Mater. Res. Lett.* **2014**, *3*, 30–34. [CrossRef]
65. Lu, P.; Saal, J.E.; Olson, G.B.; Li, T.; Swanson, O.J.; Frankel, G.; Gerard, A.Y.; Quiambao, K.F.; Scully, J.R. Computational materials design of a corrosion resistant high entropy alloy for harsh environments. *Scr. Mater.* **2018**, *153*, 19–22. [CrossRef]

entropy

MDPI

Article

Intermediate-Temperature Creep Deformation and Microstructural Evolution of an Equiatomic FCC-Structured CoCrFeNiMn High-Entropy Alloy

Chengming Cao [1], Jianxin Fu [1], Tongwei Tong [1], Yuxiao Hao [1], Ping Gu [1], Hai Hao [2] and Liangming Peng [1,*]

1 CAS Key Laboratory of Mechanical Behavior and Design of Materials, Department of Modern Mechanics, School of Engineering Science, University of Science and Technology of China, Hefei 230027, China; ccming@mail.ustc.edu.cn (C.C.); fjianxin@mail.ustc.edu.cn (J.F.); tw199309@mail.ustc.edu.cn (T.T.); haozai@mail.ustc.edu.cn (Y.H.); guping@ustc.edu.cn (P.G.)

2 School of Materials Science and Engineering, Dalian University of Technology, Dalian 116024, China; haohai@dlut.edu.cn

* Correspondence: penglm@ustc.edu.cn; Tel.: +86-551-6360-6964

Received: 30 November 2018; Accepted: 9 December 2018; Published: 12 December 2018

Abstract: The tensile creep behavior of an equiatomic CoCrFeNiMn high-entropy alloy was systematically investigated over an intermediate temperature range (500–600 °C) and applied stress (140–400 MPa). The alloy exhibited a stress-dependent transition from a low-stress region (LSR-region I) to a high-stress region (HSR-region II). The LSR was characterized by a stress exponent of 5 to 6 and an average activation energy of 268 kJ mol^{-1}, whereas the HSR showed much higher corresponding values of 8.9–14 and 380 kJ mol^{-1}. Microstructural examinations on the deformed samples revealed remarkable dynamic recrystallization at higher stress levels. Dislocation jogging and tangling configurations were frequently observed in LSR and HSR at 550 and 600 °C, respectively. Moreover, dynamic precipitates identified as $M_{23}C_6$ or a Cr-rich σ phase were formed along grain boundaries in HSR. The diffusion-compensated strain rate versus modulus-compensated stress data analysis implied that the creep deformation in both stress regions was dominated by stress-assisted dislocation climb controlled by lattice diffusion. Nevertheless, the abnormally high stress exponents in HSR were ascribed to the coordinative contributions of dynamic recrystallization and dynamic precipitation. Simultaneously, the barriers imposed by these precipitates and severe initial deformation were referred to so as to increase the activation energy for creep deformation.

Keywords: high entropy alloy; tensile creep behavior; microstructural evolution; creep mechanism

1. Introduction

High-entropy alloys (HEAs), as a novel type of material generally consisting of five or more principle elements, have attracted extensive attention in recent years because of their attractive crystallographic and mechanical properties [1–7]. Although it is anticipated that simple solid solution phases are easier to form rather than the intermetallic compounds or complex-ordered phases due to their high configurational entropy [8,9], only a few HEAs have exactly single and stable solid solution structures, among which CoCrFeNiMn alloy is widely concerned [10–14]. It possesses a single face-center-cubic (fcc) solid solution phase even after a series of thermomechanical processing [15,16] and its constituent elements have more sluggish diffusion compared to other conventional alloys [17]. These characterizations contribute to excellent mechanical properties with a satisfactory combination between tensile strength and ductility [18,19]. Apart from previous investigations mainly focusing on its temperature dependence of strength [20,21], which provided a fundamental understanding of the strengthening mechanism in fcc solid solution HEAs, recent reports have elucidated the thermally activated process, where dislocation lines overcome nanoscale clusters or short-range orders for controlling the deformation rate [22,23]. This intrinsic mechanism indicated effective heat-resistance during high-temperature plastic deformation processes. Owing to the aforementioned sluggish elemental diffusion, it is optimistic that the CoCrFeNiMn alloy will exhibit a promising perspective in high-temperature applications.

When referring to high-temperature performance for most alloy systems, creep resistance is one of the most important standards for service life and safety reliability of engineering structures. In particular, it is necessary to evaluate the stress exponents and activation energies to demonstrate the mechanism operative in the power-law creep deformation, as these two parameters are usually employed to predict the steady-state creep strain. However, only quite limited reports on tensile creep properties of HEAs at elevated temperatures have been available. The investigation on creep behavior of AlxCoCrFeNi (x = 0.15, 0.60) alloys using the stress relaxation method [24] demonstrated a lower creep resistance for higher Al content alloy, which was attributed to the easier dislocation cross-slip due to its higher stacking fault energy. The high-stress exponent of 8.8 and activation energy of 334 kJ mol^{-1} were simply explained by the increasing grain boundary in the initial body-center-cubic (bcc) phase. The dislocation configurations and microstructural evolution were not examined to confirm the validity of the proposed creep mechanism. Recently, He et al. [25,26] investigated the high-temperature plastic flow behavior of single-phase FeCoNiMn and precipitation-hardened $(FeCoNiCr)_{94}Ti_2Al_4$ HEAs using strain-rate-jump/stress increment tests at different temperatures (750–900 °C). In general, the deformation in these two alloys was divided into two regimes, dependent on the applied strain rate or testing temperatures. Specifically, the obtained higher stress exponent (6–9) and activation energy (>600 kJ mol^{-1}) in the latter alloy were proposed to be associated with the strong interactions between dislocations and coherent precipitates, resulting in a threshold stress term in the flow constitutive equation [26]. It should be pointed out that the testing temperatures were up to 750–900 °C and far beyond the softening temperature (~600 °C) of the investigated alloys [19]. From a scientific view, the strain-rate-jump or stress increment test method is inadequate using a single specimen at respective temperatures, as the microstructure during elevated-temperature plastic flow is generally not invariant. Moreover, such plastic-flow behavior is, to some extent, obviously different from creep deformation under a constant applied stress. Consequently, it is still necessary to systematically investigate the intermediate temperature (not higher than 600 °C) creep deformation and examine the microstructural evolution for exploring the intrinsic deformation mechanism of HEAs.

In the present study, the tensile creep tests under constant stress were performed on multiple fine-grained CoCrFeNiMn HEA specimens in the temperature range of 500–600 °C. Specifically, some tests were interrupted at the steady-state stages and cooled quickly to reserve the crept microstructures for examining the microstructural evolution. The objective was to reveal the inherent relationship between microstructural evolution and creep properties of single-phase CoCrFeNiMn high-entropy alloy.

2. Experimental Procedures

Ingots with a nominal composition of $Co_{20}Cr_{20}Fe_{20}Ni_{20}Mn_{20}$ (in atomic percentage) were prepared by vacuum induction melting the constituent elements with at least 99.9 mass % purity under argon atmosphere. To ensure chemical homogeneity, the ingots were re-melted at least five times and then drop-cast into a rectangular steel mold with a dimension of 120 × 65 × 10 mm^3. The ingots were homogenized at 1100 °C for 24 h in a vacuum, followed by furnace cooling. Cold rolling was subsequently conducted on these ingots with a reduction of 40% in thickness, with several cross-rolling steps to ensure flatness. The rolled sheets were eventually annealed at 900 °C for 1 h.

Flat dog-boned specimens with a dimension of 25 mm in gauge length and 5.6 mm × 1.5 mm in cross-section were electric-discharge machined for creep tests. The specimens were mechanically ground to 2000-grid sand paper to remove surface asperities. Tensile creep tests were conducted at temperatures of 500, 550, and 600 °C under applied stresses of 140–400 MPa on a CSS-3905 multi-functional testing machine. The temperature was measured with a temperature accuracy of ±1 °C by three thermocouples closely attached to the upper, middle, and lower sections of the specimen, respectively. The creep strain was continuously measured using a Linear Variable Differential Transducer (LVDT) extensometer with a strain resolution of ± 0.1 μm. The acquisition of strain-time data was accomplished by a computer, and data processing was conducted through a computer program. At least seven applied stress levels were chosen to obtain a wide range of creep rates for each testing temperature. Several specimens were subjected to interrupted tests at a steady-state stage and cooled quickly under the load using liquid nitrogen to freeze the structures produced during creep deformation.

The microstructural observation was conducted using optical microscopy (OM) (AxioImager.A1m, Jena, Germany) after etching in an aqueous solution of $HCl+H_2O_2+Cu(NO_3)_2$. The phase constitution was identified by X-ray diffraction (XRD) with Cu-Ka radiation (PANalytical X'pert PRO, Almelo, Netherlands). Thin foils from the interrupted specimens for transmission electron microscope (TEM) observations were twin-jet electropolished in an ethanol solution containing 5% perchloric acid at −25 °C and an applied voltage of 35 V. TEM investigations were conducted on a JEOL JEM2100F microscope operated at 200 kV. Fracture surfaces were examined by a scanning electron microscope (SEM) (XL30 ESEM, Philip, Netherlands) equipped with an energy-dispersive spectrometer (EDS)).

3. Results

3.1. Initial Microstructures

Figure 1 shows the initial microstructures of the alloy in as-cast and annealed states prior to high-temperature creep deformation. It is evident that the cast alloy exhibits typical dendritic and interdendritic structures, caused by the segregation during the quick freezing process in the mold. After being cold-rolled and recrystallized, equiaxed and homogenous grain structures are obtained with many annealing twins visible inside the grains. The average grain size is evaluated to be approximately 25 μm. Energy-dispersive spectroscopy (EDS) analyses demonstrate that different grains have almost identical elemental compositions as Cr = 19.9, Co = 19.9, Fe = 19.9, Ni = 19.8, and Mn=19.5 with a little loss of Mn due to its slight evaporation at an elevated temperature. As depicted by XRD patterns in Figure 1c, the alloy consists of a single fcc solid solution phase without precipitates or intermetallic compounds in both cast and recrystallized states. Furthermore, almost no peak shift is observed in the two states, with a cell parameter of a = 0.361 nm for the fcc phase. The present results are different from those observed in the $Al_{0.3}$CoCrFeNi HEA where precipitation of nanometer scale-ordered L12, B2, and sigma phases occurred in the case of different thermo-mechanical processing routes [27].

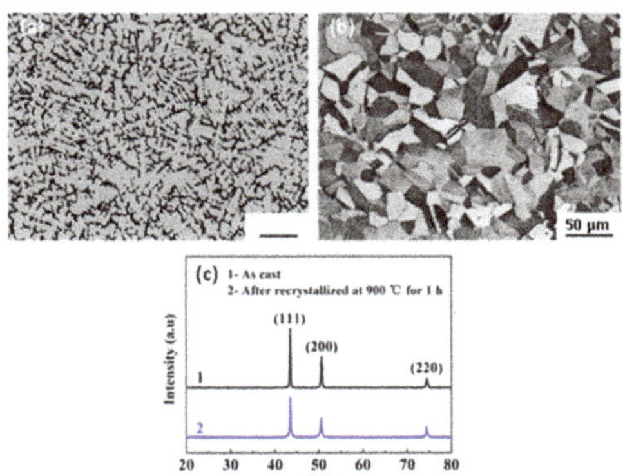

Figure 1. Microstructures of the alloy in different states. (**a**) As-cast, (**b**) thermo-mechanical treatment (cold-rolled and annealed at 900 °C for 1 h), and (**c**) X-ray diffraction (XRD) patterns.

3.2. Steady-State Creep Deformation Behavior

Figure 2 shows the selected creep curves of the alloy at 500–600 °C under different stress levels. It can be found that each of the individual curves exhibits a rather short primary creep stage and a relatively long steady-state region where the creep strain increases linearly with time, especially under lower stress levels. Necking and fracture of the specimens do occur, and hence, a tertiary stage of the curves is recorded in cases of high stress, where the creep rate accelerates with time until the final fracture occurs.

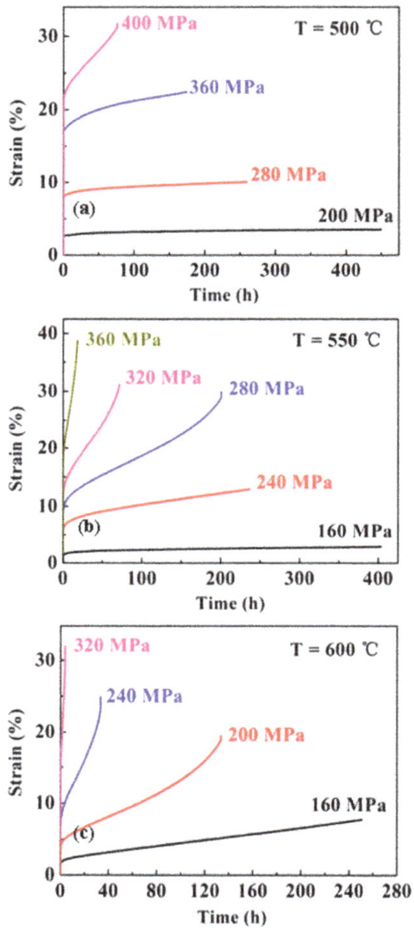

Figure 2. Selected creep strain versus time at (**a**) 500 °C, (**b**) 550 °C, and (**c**) 600 °C.

It is generally accepted that the steady-state creep rate $\dot{\varepsilon}$ of metallic materials can be correlated with the applied stress σ using the well-known power-law equation as follows [28]:

$$\dot{\varepsilon} = \frac{AD_0Gb}{kT}\left(\frac{b}{d}\right)^p\left(\frac{\sigma}{G}\right)^n \exp\left(-\frac{Q}{RT}\right) \quad (1)$$

where A is a material-dependent constant, G the shear modulus, b the Burgers vector, D_0 the frequency factor, k the Boltzmann's constant, T the absolute temperature, d the grain size of polycrystalline materials, Q the activation energy for creep deformation, and p and n the grain size and stress exponents, respectively. The stress exponent n and activation energy value Q are determined according to Equation (1) by plotting the steady-state creep rate against applied stress on a double logarithmic scale and the reciprocal of the absolute temperatures $1/T$ at constant stress levels on a semi-logarithmic scale, as shown in Figures 3 and 4, respectively. It is evident that the data exhibit two distinct regions. In the low-stress region (subsequently denoted as LSR-region I), the stress exponent values vary in the range of 5 to 6, and the average activation energy Q takes a value of 268 kJ mol^{-1}. On the contrary, in the high-stress region (subsequently denoted as HSR-region II), the stress exponents are in the range of 8.9–14, and the activation energy is Q = 359–410 kJ/mol with an average value of 380 kJ mol^{-1}. The applied stress at which the transition occurred increased from 200 to 350 MPa, with the temperature decreasing from 600 to 500 °C.

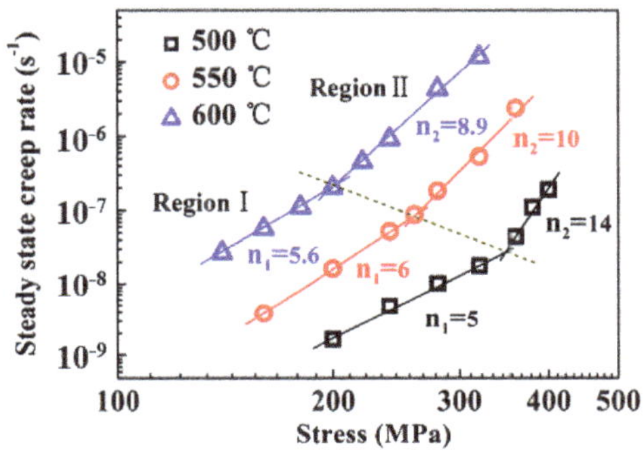

Figure 3. Dependence of creep rates on applied stress showing the transition in stress exponents.

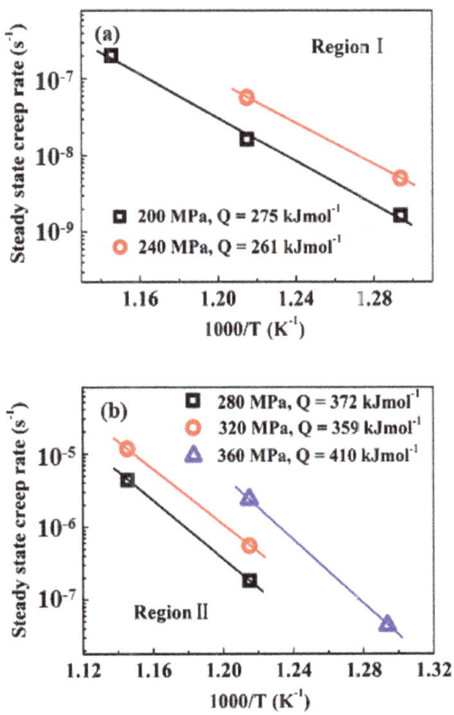

Figure 4. Arrhenius plot of steady-state creep rate versus temperature to determine the activation energy for (**a**) low-stress region I and (**b**) high-stress region II.

It should be noted that the average activation energy of 268 kJ mol^{-1} in LSR-region I is only slightly lower than the range of activation energies for lattice diffusion of the constituent elements in the alloy (288–317 kJ mol^{-1} [17]), and thus can be considered to be comparable to the lattice diffusion of the constituent elements in the alloy. In addition, this value is also quite close to those (284–333 kJ mol^{-1}) reported for the steady-state flow behavior of this alloy at higher temperatures and lower stress levels [26]. The two characteristic parameters suggest a dislocation-climb mechanism operative in LSR. Nevertheless, the abnormally high stress exponent in HSR-region II is obviously beyond the values of 3 to 5 responsible for the dislocation glide or climb mechanism [29,30] reported in some Mg [31], Al [32], and Ti [33,34] alloys. The underlying operative mechanism for HSR creep deformation in the present alloy will be addressed in the following section. In order to eliminate the influence of temperature on creep rate, the normalized creep rates $\dot{\varepsilon}kT\exp(Q/RT)/G$ versus the shear modulus-compensated stresses σ/G were plotted in Figure 5, where activation energies Q were taken to be 268 kJ mol^{-1} and 380 kJ mol^{-1} for region I and region II, respectively, and the shear modulus $G = 85 - 16/(e^{448/T} - 1)$ [35]. It is evident that almost all data points in the two stress regions at different temperatures can be represented by a single line with a respective slope of 5.5 and 10.6 for LSR and HSR. This strongly implies that creep deformation in both regions may be controlled by lattice diffusion of constituent elements in the alloy. However, it should be noted that the correlation between $\dot{\varepsilon}kT\exp(Q/RT)/G$ and σ/G in HSR still yields an abnormally high stress exponent.

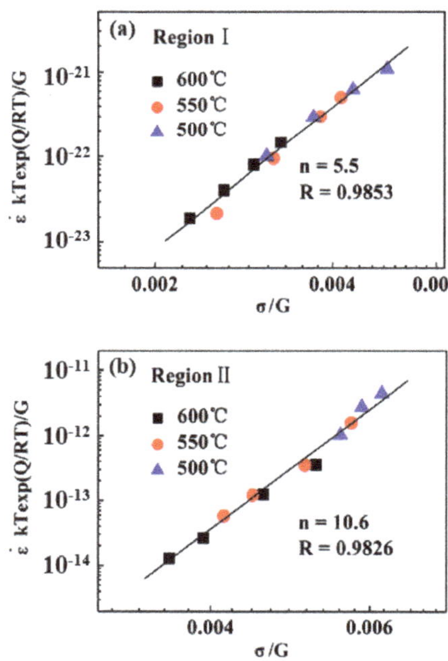

Figure 5. Normalized creep rate versus shear modulus-compensated stress for (**a**) low-stress region I and (**b**) high-stress region II.

3.3. *Crept Microstructures*

Figure 6 shows the microstructures of the interrupted samples at different temperatures and stress levels to illustrate the evolution of grain morphologies. Compared to the initial microstructure of the alloy, obvious coarsening can be observed when the specimens are subjected to low stress levels (500 °C/200MPa and 600 °C/140MPa), as depicted in Figure 6a,d. Similar grain coarsening phenomena are also visible at intermediate stress levels (500 °C/320MPa and 600 °C/200MPa) near the transition stress indicated in Figure 3. However, the volume fraction of small equiaxed grains increases at the coarse grain boundaries (indicated by arrows in Figure 6b,e), revealing strong evidence of the occurrence of dynamic recovery and recrystallization during creep deformation. Moreover, higher stress levels (500 °C/400MPa and 600 °C/320MPa) lead to the formation of completely recrystallized and refined microstructures (Figure 6c,f). The resulting average grain sizes are statistically evaluated to be approximately 56, 45, and 22 μm at 500 °C, whereas 48, 43, and 31 μm at 600 °C under different chosen stress levels.

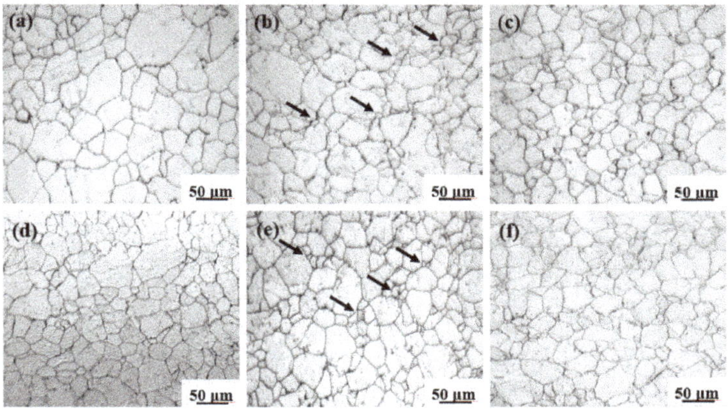

Figure 6. Optical microstructures showing the evolution of grain morphology after being crept at (**a**) 500 °C/200MPa, (**b**) 500 °C/320MPa, (**c**) 500 °C/400MPa, (**d**) 600 °C/140MPa, (**e**) 600 °C/200MPa, and (**f**) 600 °C/320MPa.

The dislocation substructures are shown in Figures 7–9, with selected area electron diffraction (SAED) patterns inserted to identify the precipitations. In general, the dislocation densities in HSR were distinctly higher than those in LSR at three testing temperatures. After being deformed at 500 °C/200 MPa (LSR), only a few short straight dislocation segments were displayed inside the grains (Figure 7a), and in particular, several lath-shaped areas were formed with closely-spaced dislocations arranged (Figure 7b). However, the substructures reveal increasing dislocation activity in LSR with increasing temperature, as the dislocations are extensively curved or even looped (Figures 8a and 9a). Apart from the limited number of alignments of pile-ups in Figure 8b, it is noted that cusped configurations are frequently observed inside grains with the bowed segments on either side. Such cusped configurations may be attributed to intrinsic pinning due to the occurrence of jogs along the screw dislocations [34]. In contrast, the dislocations in HSR exhibit a higher density of configurations. Numerous short straight segments are tangled and tend to form cell substructures at 500 °C/400 MPa (Figure 7c). However, the dislocations at 550 °C/360 MPa show a relatively homogeneous distribution still with high densities of curved dislocations tangling with each other (Figure 8c). The tangled dislocations at 600 °C/320 MPa are severely curved, with a large number of loops formed in a wide range of sizes (Figure 9c).

Another striking aspect of microstructural evolution relies on the precipitation of dispersoids during a long-term deformation process at elevated temperatures. On the whole, considerable precipitates of irregular shape were predominantly observed at the grain boundaries in HSR. The sizes of these precipitates were within a range of 50–200 nm and increased with testing temperature. In cases of 500 °C/400MPa and 600 °C/320MPa (Figures 7d and 9d), quantitative micro-EDS analysis indicates that the chemical compositions (at. %) of the precipitates take almost identical values of Cr = 26.6, Mn = 20.2, Fe = 18.9, Co = 17.5, Ni =16.6 (in at. %) and a minor amount of carbon. The SAED pattern along the [011] zone axis exhibits intense diffraction spots from the fcc matrix, accompanied by weak spots from the precipitates. The crystal structure of these precipitates is identified as fcc $M_{23}C_6$ carbide with a lattice parameter of 1.06 nm, which is consistent with both the general identifications of grain-boundary precipitates in austenitic steel [36] and the second phase observed in coarse-grained CoCrFeNiMn alloy after being subjected to prolonged exposures at 700 °C [37]. The tiny presence of carbon in the alloy may have originated from either the potential contamination of starting materials, or the melting system. Nevertheless, as depicted in Figures 8d and 9b for 550 °C/360MPa and 600 °C/140MPa, the micro-EDS analysis demonstrates chemical compositions (at %) of the precipitates with Cr = 48.8, Mn = 15.5, Fe = 14.0, Co = 11.5, and Ni = 10.2. The present results are quite consistent with previous reports for the prolonged annealing CrMnFeCoNi system [38,39], where the precipitates were identified as a quinary variant of the binary Cr-Fe σ phase. The interplanar spacings are calculated to be $d_1/d_2/d_3$ = 0.648/0.402/0.337 nm and show a satisfactory agreement with the values of $d_1/d_2/d_3$ = 0.648/0.383/0.330 nm for $(1\bar{1}0)/(11\bar{1})/(20\bar{1})$ lattice planes in the [112] zone axes for the σ-FeCrMo phase [40]. Accordingly, the precipitates in the 50 °C/360MPa crept alloy are concluded to be Cr-rich tetragonal σ phase with lattice parameters of a/c = 0.916/0.509 nm. Similarly, the weak spots along the [221] zone axes in the SAED pattern in Figure 9b are also indicated to be very close to the Cr-rich σ phase, with lattice parameters of a/c = 0.895/0.473 nm.

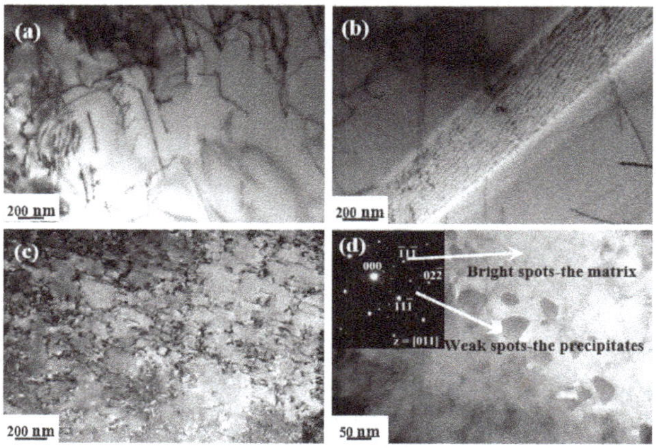

Figure 7. Transmission electron microscope (TEM) images of dislocation substructures in the interrupted specimens after being crept at 500 °C under (**a**) and (**b**) 200 MPa (low-stress region (LSR)); (**c**) and (**d**) 400 MPa (high-stress region (HSR)).

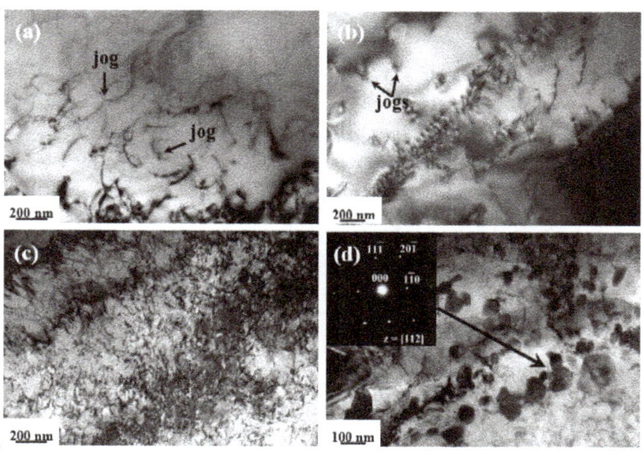

Figure 8. TEM images of dislocation substructures in the interrupted specimens after being crept at 550 °C under (**a**) and (**b**) 160 MPa (LSR); (**c**) and (**d**) 360 MPa (HSR).

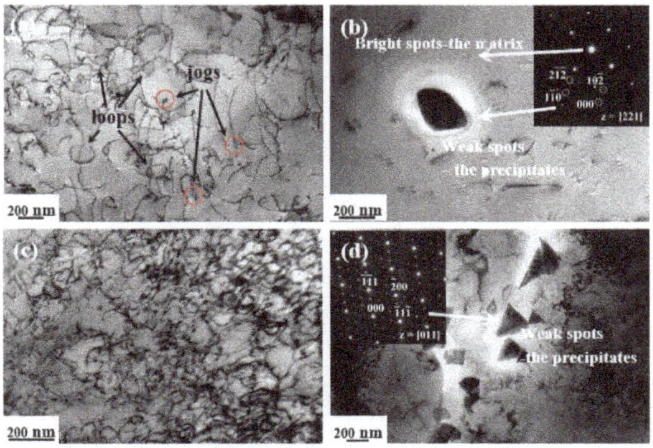

Figure 9. TEM images of dislocation substructures in the interrupted specimens after being crept at 600 °C under (**a**) and (**b**) 140 MPa (LSR); (**c**) and (**d**) 320 MPa (HSR).

3.4. Fractographs

All the fractographs in Figure 10 for HSR from 500 °C to 600 °C exhibit typical ductile rupture, where both the width and depth of dimples increase with temperature. However, slip striations around the dimples become notable at 600 °C. Meanwhile, a small number of particles identified as Mn-containing oxides by EDS are visible on the fracture surfaces, which is attributed to an in-situ oxidation during fracture process [26]. It is worth noting that the crept samples fracture in a ductile transgranular manner at high stress, which is contrary to common knowledge, as large amounts of brittle are generally considered to be destructive to the stability of grain boundaries during the tertiary creep stage [41,42]. The ductile creep failure may be associated with the dislocation interactions during high-temperature deformation, which contribute to the progressive generation of vacancies and nucleation of voids.

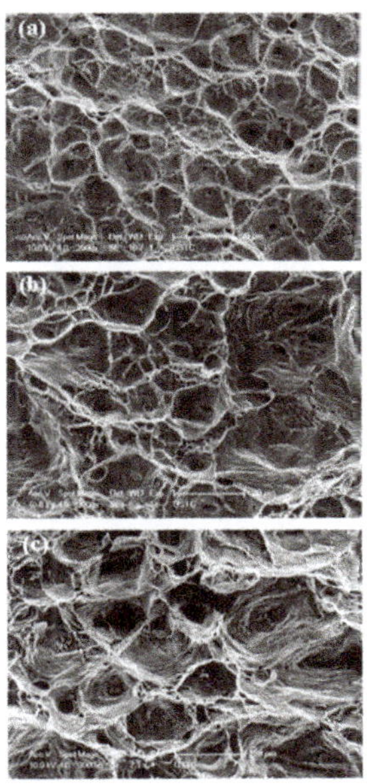

Figure 10. Microstructural features of the fracture surfaces after being deformed at high stress levels. (**a**) 500 °C/400MPa, (**b**) 550 °C/360MPa, and (**c**) 600 °C/320MPa.

4. Discussion

The observed jog configurations are indicative of a dislocation climb process, and thus the stress-assisted dislocation climb controlled by lattice diffusion may be the operative creep mechanism of the present high-entropy alloy in LSR. Contrary to LSR, creep deformation in HSR exhibits abnormally high stress exponents of 8.9–14 and an average activation energy of 380 kJ mol^{-1}. Combined with the dislocation configurations in Figures 8c and 9c, the same creep mechanism may also be considered rate-controlling. The unusually high stress exponents and activation energy in several traditional alloys have been frequently described using the power-law breakdown equation [43–46]. However, no deviation from the linear relationship on double-logarithmic curves of $\log \dot{\varepsilon}$~$\log \sigma$ were observed, and therefore, the power-law breakdown assumption may be inappropriate for the present alloy.

It is documented that dynamic recrystallization during the creep process could cause an increase in the stress exponent [47–49]. Recrystallization is generally accompanied by intense diffusion of solutes, resulting in a refined-grain microstructure and a larger opportunity for grain boundary sliding. Under such circumstances, the creep rate would increase with decreasing grain size [50]. As shown in Figure 6, obvious dynamic recrystallization occurred in the present CoCrFeNiMn alloy in HSR. In fact, low stacking-fault energy of alloys is conducive to promoting recrystallization [51]. Based on Equation (1) where p is generally equal to 2 for a dislocation-controlled creep, an increase in the stress exponent Δn due to dynamic recrystallization is expressed as [47]:

$$\Delta n = n_{app} - n = \frac{2\ln(d_1/d_2)}{\ln(\sigma_2/\sigma_1)} \quad (2)$$

where n_{app} is the experimentally measured apparent stress exponent, n is the stress exponent prior to the occurrence of recrystallization, d_1 and d_2 denote the average grains size before and after dynamic recrystallization, and σ_1 and σ_2 are the corresponding stress levels. Using the relevant evaluated grain sizes, the Δn values were calculated to be 6.5 and 1.4 for 500 °C and 600 °C, respectively. The corresponding modified stress exponents n then fall down to ~7.5. However, this value is still higher than those for traditional solid solution alloys, which implies that the sole recrystallization is insufficient and additional mechanisms should be involved to be responsible for the entire increase in stress exponents.

TEM observations provide clear evidence of dynamic precipitation in HSR. As a result, precipitation hardening may have contributed to the high stress exponents [25]. In general, the precipitates are predominantly formed along the grain boundaries and would exert a boundary obstacle stress σ_{bo}, which prevent the annihilation of moving dislocations at the boundaries. The stress controlling the dislocation velocity is reduced to an effective value obtained by subtracting the stress necessary for overcoming precipitates from the applied stress. Equation (1) is then replaced by [52,53]:

$$\dot{\varepsilon} = \frac{AD_0Gb}{kT}\left(\frac{b}{d}\right)^p\left(\frac{\sigma - \sigma_{b0}}{G}\right)^n \exp\left(-\frac{Q}{RT}\right) \quad (3)$$

By re-plotting the double-logarithm relationship between $\dot{\varepsilon}/\exp(-Q/RT)$ and $(\sigma - \sigma_{bo})/G$ in HSR, the stress exponents can then be decreased by subtracting the precipitation-hardening portion. Unfortunately, it is difficult to precisely estimate the values of boundary obstacle stress and the decrease in the stress exponent, due to the ambiguous interaction between dynamic precipitation and recrystallization. Nevertheless, combinative effects of precipitation hardening and recrystallization can provide reasonable insights into the intrinsic reason for the abnormal stress exponents and activation energy for creep in CoCrFeNiMn alloy. Moreover, the estimated abnormal activation energy in HSR might also stem partially from precipitation hardening [22]. The nano-sized dispersoids could effectively fasten the grain boundaries, and thus extra activation energy is needed to overcome the barriers for recrystallization and further growth of grains. In addition, the samples experience drastic plastic deformation in a short time to establish a constant stress state for subsequent creep deformation. As a result, some grains are severely elongated, which, in turn, exerts an extra influence on increasing the barriers for dynamic recovery and recrystallization.

5. Conclusions

The intermediate-temperature creep behavior and microstructural evolution of CoCrFeNiMn HEA have been studied. Optical micrographs and TEM images of the dislocation substructure after creep deformation revealed the effects of dynamic recrystallization and dynamic precipitation on the deformation mechanisms. The following conclusions can be drawn:

1. An obvious transition in stress exponent and average activation energy was observed: At low stresses, the values were approximately 5 to 6 and 268 kJ/mol, whereas at high-stress regions, the corresponding values were 8.9–14 and 380 kJ/mol. Stress-assisted dislocation climb controlled by lattice diffusion is suggested as a possible rate-dominating mechanism at low stresses.
2. At high stresses, obvious dynamic recrystallization occurred, leading to the refinement of average grain size. Simultaneously, nano-sized $M_{23}C_6$ carbides and the Cr-rich tetragonal σ phase were dynamically precipitated. High density of tangled and curved dislocation substructures was observed.
3. The anomalously high stress exponent and activation energy in HSR were attributed to the combinative effects of dynamic recrystallization and precipitation. In particular, the precipitates act as barriers for dislocation motion via producing boundary obstacle stress, resulting in high activation energy for creep deformation. However, lattice-diffusion controlled dislocation climb is still responsible for the deformation mechanism in HSR.

Author Contributions: L.P. initiated this research project. C.C. performed the creep tests. T.T. and Y.H. prepared the entropy alloy. J.F. performed the microstructural characterization and thermal-mechanical processing under the supervision of P.G., H.H. and L.P. All authors discussed the results and approved the final manuscript.

Funding: This research was funded by the National Natural Science Foundation of China (11572306) and the Fundamental Research Funds for Central Universities (WK2090050040).

Conflicts of Interest: The authors declare no conflict of interest.

References

1. Zhang, Y.; Zuo, T.T.; Tang, Z.; Gao, M.C.; Dahmen, K.A.; Liaw, P.K.; Lu, Z.P. Microstructures and properties of high-entropy alloys. *Prog. Mater. Sci.* **2014**, *61*, 1–93. [CrossRef]
2. Sheng, H.F.; Gong, M.; Peng, L.M. Microstructural characterization and mechanical properties of an $Al_{0.5}$CoCrFeCuNi high-entropy alloy in as-cast and heat-treated/quenched conditions. *Mater. Sci. Eng. A* **2013**, *567*, 14–20. [CrossRef]
3. Borkar, T.; Gwalani, B.; Choudhuri, D.; Mikler, C.V.; Yannetta, C.J.; Chen, X.; Ramanujan, R.V.; Styles, M.J.; Gibson, M.A.; Banerjee, R. A combinatorial assessment of $Al_xCrCuFeNi_2$ ($0 < x < 1.5$) complex concentrated alloys: Microstructure, microhardness, and magnetic properties. *Acta Mater.* **2016**, *116*, 63–76.
4. Manzoni, A.M.; Haas, S.; Daoud, H.; Glatzel, U.; Förster, C.; Wanderka, N. Tensile behavior and evolution of the phases in the $Al_{10}Co_{25}Cr_8Fe_{15}Ni_{36}Ti_6$ compositionally complex/high entropy alloy. *Entropy* **2018**, *20*, 646. [CrossRef]
5. Miracle, D.B.; Senkov, O.N. A critical review of high entropy alloys and related concepts. *Acta Mater.* **2017**, *122*, 448–511. [CrossRef]
6. Chen, R.R.; Qin, G.; Zheng, H.T.; Wang, L.; Su, Y.Q.; Chiu, Y.L.; Ding, H.S.; Guo, J.J.; Fu, H.Z. Composition design of high entropy alloys using the valence electron concentration to balance strength and ductility. *Acta Mater.* **2018**, *144*, 129–137. [CrossRef]
7. Bu, Y.Q.; Peng, S.Y.; Wu, S.W.; Wei, Y.J.; Wang, G.; Liu, J.B.; Wang, H.T. Unconventional deformation behaviours of nanoscaled high-entropy alloys. *Entropy* **2018**, *20*, 778. [CrossRef]
8. Otto, F.; Yang, Y.; Bei, H.; George, E.P. Relative effects of enthalpy and entropy on the phase stability of equiatomic high-entropy alloys. *Acta Mater.* **2013**, *61*, 2628–2638. [CrossRef]

9. Zhang, Y.; Zhou, Y.J.; Lin, J.P.; Chen, G.L.; Liaw, P.K. Solid-solution phase formation rules for multi-component alloys. *Adv. Eng. Mater.* **2008**, *10*, 534–538. [CrossRef]
10. Ji, W.; Wang, W.M.; Wang, H.; Zhang, J.Y.; Wang, Y.C.; Zhang, F.; Fu, Z.Y. Alloying behavior and novel properties of CoCrFeNiMn high-entropy alloy fabricated by mechanical alloying and spark plasma sintering. *Intermetallics* **2015**, *56*, 24–27. [CrossRef]
11. Sathiaraj, G.D.; Bhattacharjee, P.P. Analysis of microstructure and microtexture during grain growth in low stacking fault energy equiatomic CoCrFeMnNi high entropy and Ni–60 wt. % Co alloys. *J. Alloys Compd.* **2015**, *637*, 267–276. [CrossRef]
12. Tasan, C.C.; Deng, Y.; Pradeep, K.G.; Yao, M.J.; Springer, H.; Raabe, D. Composition dependence of phase stability, deformation mechanisms, and mechanical properties of the CoCrFeMnNi high-entropy alloy system. *JOM* **2014**, *66*, 1993–2001. [CrossRef]
13. Varvenne, C.; Luque, A.; Curtin, W.A. Theory of strengthening in fcc high entropy alloys. *Acta Mater.* **2016**, *118*, 164–176. [CrossRef]
14. Wu, Z.; Bei, H.; Otto, F.; Pharr, G.M.; George, E.P. Recovery, recrystallization, grain growth and phase stability of a family of FCC-structured multi-component equiatomic solid solution alloys. *Intermetallics* **2014**, *46*, 131–140. [CrossRef]
15. Liu, W.H.; Wu, Y.; He, J.Y.; Nieh, T.G.; Lu, Z.P. Grain growth and the Hall–Petch relationship in a high-entropy FeCrNiCoMn alloy. *Scr. Mater.* **2013**, *68*, 526–529. [CrossRef]
16. Otto, F.; Hanold, N.L.; George, E.P. Microstructural evolution after thermomechanical processing in an equiatomic, single-phase CoCrFeMnNi high-entropy alloy with special focus on twin boundaries. *Intermetallics* **2014**, *54*, 39–48. [CrossRef]
17. Tsai, K.Y.; Tsai, M.H.; Yeh, J.W. Sluggish diffusion in Co–Cr–Fe–Mn–Ni high-entropy alloys. *Acta Mater.* **2013**, *61*, 4887–4897. [CrossRef]
18. Gludovatz, B.; Hohenwarter, A.; Catoor, D.; Chang, E.H.; George, E.P.; Ritchie, R.O. A fracture-resistant high-entropy alloy for cryogenic applications. *Science* **2014**, *345*, 1153–1158. [CrossRef]
19. Otto, F.; Dlouhý, A.; Somsen, C.; Bei, H.; Eggeler, G.; George, E.P. The influences of temperature and microstructure on the tensile properties of a CoCrFeMnNi high-entropy alloy. *Acta Mater.* **2013**, *61*, 5743–5755. [CrossRef]
20. Gali, A.; George, E.P. Tensile properties of high- and medium-entropy alloys. *Intermetallics* **2013**, *39*, 74–78. [CrossRef]
21. Wu, Z.; Bei, H.; Pharr, G.M.; George, E.P. Temperature dependence of the mechanical properties of equiatomic solid solution alloys with face-centered cubic crystal structures. *Acta Mater.* **2014**, *81*, 428–441. [CrossRef]
22. Hong, S.I.; Moon, J.S.; Hong, K.; Kim, H.S. Thermally activated deformation and the rate controlling mechanism in CoCrFeMnNi high entropy alloy. *Mater. Sci. Eng. A* **2017**, *682*, 569–576. [CrossRef]
23. Wu, Z.; Gao, Y.; Bei, H. Thermal activation mechanisms and Labusch-type strengthening analysis for a family of high-entropy and equiatomic solid-solution alloys. *Acta Mater.* **2016**, *120*, 108–119. [CrossRef]
24. Cao, T.; Shang, J.; Zhao, J.; Cheng, C.; Wang, R.; Wang, H. The influence of Al elements on the structure and the creep behavior of Al_xCoCrFeNi high entropy alloys. *Mater. Lett.* **2016**, *164*, 344–347. [CrossRef]
25. He, J.Y.; Zhu, C.; Zhou, D.Q.; Liu, W.H.; Nieh, T.G.; Lu, Z.P. Steady state flow of the FeCoNiCrMn high entropy alloy at elevated temperatures. *Intermetallics* **2014**, *55*, 9–14. [CrossRef]
26. He, J.Y.; Wang, H.; Wu, Y.; Liu, X.J.; Nieh, T.G.; Lu, Z.P. High-temperature plastic flow of a precipitation-hardened FeCoNiCr high entropy alloy. *Mater. Sci. Eng. A* **2017**, *686*, 34–40. [CrossRef]
27. Gwalani, B.; Gorsse, S.; Choudhuri, D.; Styles, M.; Zheng, Y.F.; Mishra, R.S.; Banerjee, R. Modifying transformation pathways in high entropy alloys or complex concentrated alloys via thermo-mechanical processing. *Acta Mater.* **2018**, *153*, 169–185. [CrossRef]
28. Mukherjee, A.K.; Bird, J.E.; Dorn, J.E. *Experimental Correlations for High-Temperature Creep*; Technical Report; California Univ., Lawrence Radiation Lab.: Berkeley, CA, USA, 1968.
29. Kassner, M.E.; Pérez-Prado, M.T. Five-power-law creep in single phase metals and alloys. *Prog. Mater. Sci.* **2000**, *45*, 1–102. [CrossRef]
30. Tobolová, Z.; Čadek, J. An interpretation of steady state creep. *Philos. Mag.* **1972**, *26*, 1419–1428. [CrossRef]
31. Naghdi, F.; Mahmudi, R. The microstructure and creep characteristics of cast Mg–4Zn–0.5Ca and Mg–4Zn–0.5Ca–2RE alloys. *Mater. Sci. Eng. A* **2014**, *610*, 315–325. [CrossRef]

32. Chaudhury, P.K.; Farghalli, A.M. Creep and ductility in an Al-Cu solid-solution alloy. *Metall. Trans. A* **1987**, *18*, 2105–2113. [CrossRef]
33. Nie, X.; Liu, H.Q.; Zhou, X.Z.; Yi, D.Q.; Huang, B.Y.; Hu, Z.; Xu, Y.F.; Yang, Q.; Wang, D.C.; Gao, Q. Creep of Ti–5Al–5Mo–5V–1Fe–1Cr alloy with equiaxed and lamellar microstructures. *Mater. Sci. Eng. A* **2016**, *651*, 37–44. [CrossRef]
34. Viswanathan, G.B.; Karthikeyan, S.; Hayes, R.W.; Mills, M.J. Creep behaviour of Ti-6Al-2Sn-4Zr-2Mo: II. Mechanisms of deformation. *Acta Mater.* **2002**, *50*, 4965–4980. [CrossRef]
35. Laplanche, G.; Gadaud, P.; Horst, O.; Otto, F.; Eggeler, G.; George, E.P. Temperature dependencies of the elastic moduli and thermal expansion coefficient of an equiatomic, single-phase CoCrFeMnNi high-entropy alloy. *J. Alloys Compd.* **2015**, *623*, 348–353. [CrossRef]
36. Heczko, M.; Polák, J.; Kruml, T. Microstructure and dislocation arrangements in Sanicro 25 steel fatigued at ambient and elevated temperatures. *Mater. Sci. Eng. A* **2017**, *680*, 168–181. [CrossRef]
37. Pickering, E.J.; Muñoz-Moreno, R.; Stone, H.J.; Jones, N.G. Precipitation in the equiatomic high-entropy alloy CrMnFeCoNi. *Scr. Mater.* **2016**, *113*, 106–109. [CrossRef]
38. Otto, F.; Dlouhý, A.; Pradeep, K.G.; Kuběnová, M.; Raabe, D.; Eggeler, G.; George, E.P. Decomposition of the single-phase high-entropy alloy CrMnFeCoNi after prolonged anneals at intermediate temperatures. *Acta Mater.* **2016**, *112*, 40–52. [CrossRef]
39. Zhu, Z.G.; Ma, K.H.; Yang, X.; Shek, C.H. Annealing effect on the phase stability and mechanical properties of $(FeNiCrMn)_{(100-x)}Co_x$ high entropy alloys. *J. Alloys Compd.* **2017**, *695*, 2945–2950. [CrossRef]
40. Shen, Y.Z.; Zhou, X.L.; Shi, T.T.; Huang, X.; Shang, Z.X.; Liu, W.W.; Ji, B.; Xu, Z.Q. Sigma phases in an 11%Cr ferritic/martensitic steel with the normalized and tempered condition. *Mater. Charact.* **2016**, *122*, 113–123. [CrossRef]
41. Han, W.Y.; Yang, G.Y.; Xiao, L.; Li, J.H.; Jie, W.Q. Creep properties and creep microstructure evolution of Mg-2.49Nd-1.82Gd-0.19Zn-0.4Zr alloy. *Mater. Sci. Eng. A* **2017**, *684*, 90–100. [CrossRef]
42. Zhang, Z.; Hu, Z.F.; Tu, H.Y.; Schmauder, S.; Wu, G.X. Microstructure evolution in HR3C austenitic steel during long-term creep at 650°C. *Mater. Sci. Eng. A* **2017**, *681*, 74–84. [CrossRef]
43. Sherby, O.D.; Burke, P.M. Mechanical behavior of crystalline solids at elevated temperature. *Prog. Mater Sci.* **1968**, *13*, 323–390. [CrossRef]
44. Srikant, G.; Marple, B.; Charit, I.; Murty, K. Characterization of stress rupture behavior of commercial-purity-Ti via burst testing. *Mater. Sci. Eng. A* **2007**, *463*, 203–207. [CrossRef]
45. Tanhaee, Z.; Mahmudi, R. The microstructure and creep characteristics of cast Mg–3Si and Mg–3Si–1Gd alloys. *Mater. Sci. Eng. A* **2016**, *673*, 148–157. [CrossRef]
46. Raj, S.V. Power-law and exponential creep in class M materials: Discrepancies in experimental observations and implications for creep modeling. *Mater. Sci. Eng. A* **2002**, *322*, 132–147. [CrossRef]
47. Allameh, S.M. High creep exponents in a nearly-lamellar γ-based titanium aluminide intermetallic. *J. Mater. Sci.* **2001**, *36*, 3539–3547. [CrossRef]
48. Luan, Q.D.; Duan, Q.Q.; Wang, X.G.; Liu, J.; Peng, L.M. Tensile properties and high temperature creep behavior of microalloyed Ti–Ti3Al–Nb alloys by directional solidification. *Mater. Sci. Eng. A* **2010**, *527*, 4484–4496. [CrossRef]
49. Rettberg, L.H.; Pollock, T.M. Localized recrystallization during creep in nickel-based superalloys GTD444 and René N5. *Acta Mater.* **2014**, *73*, 287–297. [CrossRef]
50. Ning, Z.L.; Liu, H.H.; Cao, F.Y.; Wang, S.T.; Sun, J.F.; Qian, M. The effect of grain size on the tensile and creep properties of Mg–2.6Nd–0.35Zn–xZr alloys at 250°C. *Mater. Sci. Eng. A* **2013**, *560*, 163–169. [CrossRef]
51. Stepanov, N.D.; Shaysultanov, D.G.; Yurchenko, N.Y.; Zherebtsov, S.V.; Ladygin, A.N.; Salishchev, G.A.; Tikhonovsky, M.A. High temperature deformation behavior and dynamic recrystallization in CoCrFeNiMn high entropy alloy. *Mater. Sci. Eng. A* **2015**, *636*, 188–195. [CrossRef]
52. Zhang, J.S.; Li, P.E.; Chen, W.X.; Jin, J.Z. Grain boundary precipitation strengthening in high temperature creep of Fe-15Cr-25Ni alloys. *Scr. Metall.* **1989**, *23*, 547–551. [CrossRef]
53. Li, P.E.; Zhang, J.S.; Wang, F.G.; Jin, J.Z. Influence of Intergranular carbide density and grain size on creep of Fe-15Cr-25Ni alloys. *Metall. Trans. A* **1991**, *23*, 1379–1381. [CrossRef]

entropy

MDPI

Article

Mechanical Properties and Microstructure of a NiCrFeCoMn High-Entropy Alloy Deformed at High Strain Rates

Bingfeng Wang [1,2], Xianrui Yao [2], Chu Wang [2], Xiaoyong Zhang [1,*] and Xiaoxia Huang [2]

1 State Key Laboratory for Powder Metallurgy, Central South University, Changsha 410083, China; wangbingfeng@csu.edu.cn

2 School of Materials Science and Engineering, Central South University, Changsha 410083, China; yaoxianrui@csu.edu.cn (X.Y.); wangchu@csu.edu.cn (C.W.); huangxiaoxiacsu@outlook.com (X.H.)

* Correspondence: zhangxiaoyong@csu.edu.cn; Tel.: +86-731-88876244

Received: 1 October 2018; Accepted: 17 November 2018; Published: 21 November 2018

Abstract: The equiatomic NiCrFeCoMn high-entropy alloy prepared by arc melting has a single crystallographic structure. Mechanical properties and microstructure of the NiCrFeCoMn high-entropy alloy deformed at high strain rates (900 s^{-1} to 4600 s^{-1}) were investigated. The yield strength of the NiCrFeCoMn high-entropy alloy is sensitive to the change of high strain rates. Serration behaviors were also observed on the flow stress curves of the alloy deformed at the strain rates ranging from 900 s^{-1} to 4600 s^{-1}. The Zerilli–Armstrong constitutive equation can be used to predict the flow stress curves of the NiCrFeCoMn high-entropy alloy. Large amounts of deformation bands led to obvious serration behaviors of the NiCrFeCoMn high-entropy alloy under dynamic loading.

Keywords: high-entropy alloy; electron microscopy; plasticity methods; plasticity; serration behavior

1. Introduction

The concept of multi-component high-entropy alloys (HEAs) has been presented in the beginning of this century. Most HEAs usually contain five elements with nearly equal atomic ratios [1]. HEAs have excellent mechanical properties, such as low strength and high plasticity [2–12]. From recent literatures, the hardness of FeCoNiCrMn high-entropy alloy has even increased to 6700 MPa [13]. The $Al_{0.6}CoCrFeNi$ high-entropy alloy also displayed the excellent strength–ductility combination for nanoscale deformation twins induced by dynamic loading and high-density dislocation substructure [14]. However, the NiCrFeCoMn high-entropy alloy has still been regarded as a typical case for its single-phase face-centered cubic (FCC) [15] and relative promising mechanical properties. For example, its yield strength increases with the angle of rotation at high-pressure torsion (HPT) at room temperature at the cost of reduced ductility [16] and can reach up to 350 MPa when reducing its grain size [17,18]. Therefore, the NiCrFeCoMn high-entropy alloy prepared by arc melting can be used to fabricate big parts of industry devices or as the transitional layer between the two types of alloys, e.g., the HEA solder used for welding pure titanium and chromium–nickel–titanium stainless steel [19]. At the present time, dynamic impacts are widely found in aeronautical engineering, the automotive industry, and marine engineering. The as-cast NiCrFeCoMn high-entropy alloy could be applied to many dynamic deformation processes such as penetration, impact cyclic loading, and shock loading. Hence, it is vital to comprehend the dynamic behavior of as-cast NiCrFeCoMn high-entropy alloy under low/high-speed loading and expand its applications.

Many researchers were devoted to investigating the deformation behavior of the high-entropy alloy at different strain rate levels (1×10^{-8} s^{-1} to 10 s^{-1}) [20]. From the recent research, serration behavior was found in the high-entropy alloy CoCrFeMnNi prepared by powder metallurgy at a low strain rate of 1×10^{-3} s^{-1} and the high strain rates (1×10^{3} s^{-1} to 3×10^{3} s^{-1}) [21]. Serration behavior was supposed to be associated with Cottrell atmosphere interaction with moving dislocations, slip bands, and dynamic strain aging, i.e., the dynamic breakaway/locking of dislocations from/by mobile solute atoms at intermediate temperatures. Until now, only several researches on dynamic behavior of high-entropy alloys under high strain rates (beyond 1×10^{3} s^{-1}) have been published. Kumar et al. [22] had investigated the strain-rate sensitivity of yield strength for the $Al_{0.1}CrFeCoNi$ high-entropy alloy under high strain rates. Li et al. [23] had also suggested that the $Al_{0.3}CoCrFeNi$ high-entropy alloy exhibits high strain-rate sensitivity. Dirras et al. [24] had found that the $Ti_{20}Hf_{20}Zr_{20}Ta_{20}Nb_{20}$ high-entropy alloy would be strongly localized under deformation at high strain rates. He et al. [25] had discussed the strain-rate sensitivity effect for the FeCoNiCrMn high-entropy alloy and found a higher strain-rate sensitivity of 0.022 than that of traditional FCC metals. Park et al. [26] had even found that the strain-rate dependency of the yield strength under dynamic conditions would be much higher than that under quasi-static conditions. However, the mechanism of the special phenomenon for the high-entropy alloy deformed at high strain rates (beyond 1×10^{3} s^{-1}) is not clear yet. Serration characteristics are ubiquitous in many structural and functional materials such as high-entropy alloy ($AlCoCr_{1.5}Fe_{1.5}NiTi_{0.5}$, $Al_{0.3}CoCrFeNi$ etc.). It would lead to the instability of mechanical properties. By studying the serration behavior, it may be possible to have an early warning signal for oncoming epileptic seizures and perhaps economic trends.

In the present work, we used the split-Hopkinson pressure bar to investigate the dynamic mechanical behavior of NiCrFeCoMn high-entropy alloy prepared by arc melting. The aims are: (1) To report the mechanical properties and microstructure under dynamic loadings, (2) to obtain the plastic model, and (3) to discuss the microstructural mechanism for the serration behavior when the alloy deforms at high strain rates.

2. Experiments and Procedures

The material was prepared by arc melting and the chemical composition is given in Table 1. Figure 1 shows the optical micrograph of the initial microstructure of the sample. The obtained specimen exhibits an as-cast dendrite structure and the columnar crystals are distributed evenly. Elemental scanning results in Figure 2 show the distribution of each element evenly in the alloy. From the XRD pattern shown in Figure 3, it can be found that the NiCrFeCoMn high-entropy alloy is composed of a simple FCC solid solution.

Table 1. Chemical composition of the NiCrFeCoMn high-entropy alloy.

Elements	Cr	Mn	Fe	Co	Ni
wt.%	16.42	18.06	21.64	21.71	22.18
at.%	17.76	18.49	21.79	20.72	21.24

Figure 1. Optical micrograph of the as-cast NiCrFeCoMn alloy.

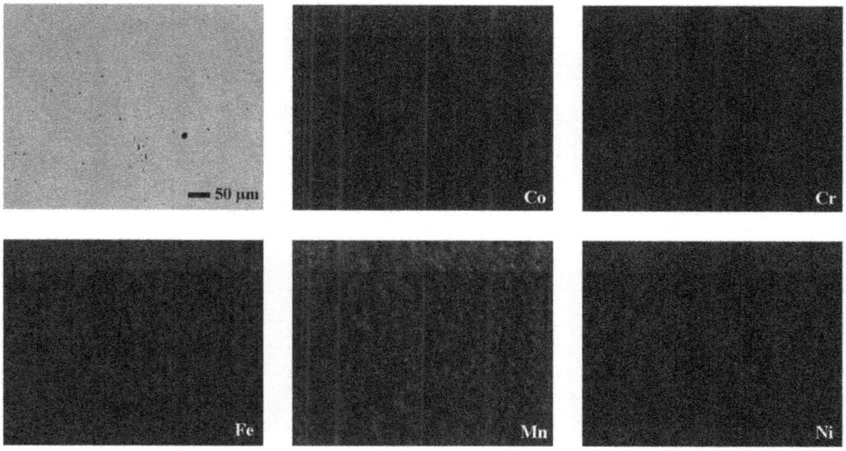

Figure 2. Element planar pattern of the as-cast NiCrFeCoMn alloy.

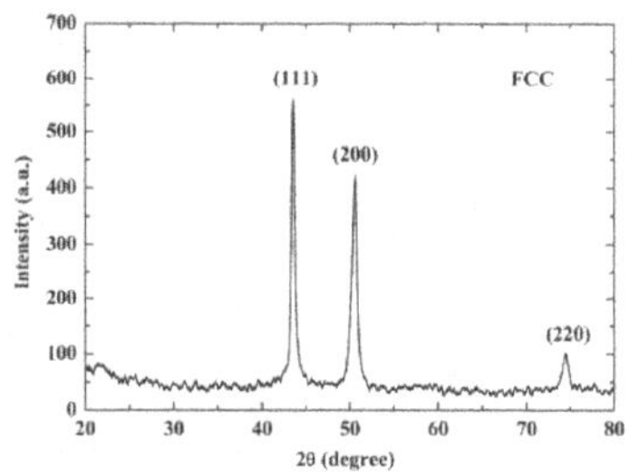

Figure 3. The XRD pattern of the as-cast NiCrFeCoMn alloy.

In this investigation, cylindrical specimens were used for mechanical testing. To ensure uniaxial compressive condition, the end faces of the compressive specimens were ground on each side with SiC paper. During the process of the impact loading and the electrical signal collection, we had adopted advanced waving plastic and anti-jamming techniques. Two kinds of compressive tests were adopted to do mechanical testing at an ambient temperature (298 K) as follows: (1) Quasi-static compressive tests were performed with an INSTRON 3369 machine at the strain rate of 1×10^{-3} s^{-1}, (2) dynamic compressive tests were performed with a split-Hopkinson pressure bar at the strain rates of 900 s^{-1}, 1700 s^{-1}, and 4600 s^{-1}. The cylindrical specimens had three diameters. Specimens for the strain rate of about 900 s^{-1} had a height of 8.4 mm and diameter of 6 mm, for the strain rate of about 1700 s^{-1} they had a height of 5.6 mm and diameter of 4 mm, and for the strain rate of about 4600 s^{-1} they had a height of 2.8 mm and diameter of 2 mm. The strain rate, the true strain, and the true stress of cylindrical specimens can be obtained by the following equations:

$$\dot{\varepsilon} = -\frac{2C_0}{L_S}\varepsilon_r(t), \tag{1}$$

$$\varepsilon = -\frac{2C_0}{L_S}\int_0^t \varepsilon_r(t)dt, \tag{2}$$

$$\varepsilon = -\frac{2C_0}{L_S}\int_0^t \varepsilon_r(t)dt, \tag{3}$$

where E_0 and C_0 are elastic modulus and elastic wave speed in a split-Hopkinson pressure bar, A_0 is the cross-sectional area of the bar, A_s and L_s are the cross-sectional area and the length of the cylindrical specimens, and $\varepsilon_r(t)$ and $\varepsilon_t(t)$ are the experimentally measured strain of incident and transmitted stress pulse on the split-Hopkinson pressure bars respectively.

Cylindrical specimens were sectioned to two halves along the impacting axis by line cutting. Afterwards, the half sections were polished with the etchant of 50 mL hydrochloric acid, 50 mL water, and 10 g $CuSO_4 \cdot 5H_2O$. Samples were cured and ground using SiC papers. Polishing steps were employed by using diamond pastes and polishing cloths. A mixture solution of 90% acetic acid and 10% perchloric acid at room temperature and at an applied voltage of 27 V for 15 s was then used for electro-polishing. Optical microscopy was performed with a POLYVAR-MET microscope. The crystallographic structure was identified by X-ray diffraction (Rigaku D/MAX-2500 X-ray diffractometer, Rigaku Corporation, Tokyo, Japan) using a Cu target at an operating voltage of 40 kV and current of 250 mA. Electron backscatter diffraction (EBSD) patterns were collected using a ZEISS EVOMA10 scanning electron microscope (SEM, Carl Zeiss SMT Ltd., Cambridge, UK) equipped with a detector and operated at an accelerating voltage of 20 kV. The working distance for SEM is about 10 mm. Transmission electron microscopy (TEM, Royal Philips, Amsterdam, Netherlands) observations were carried out with a Tecnai G2 T20 ST transmission electron microscope operated at 200 kV.

3. Results

Figure 4 presents the true stress–strain curves of the NiCrFeCoMn high-entropy alloy. It can be seen that the yield strength of the NiCrFeCoMn high-entropy alloy increased from 490 MPa to 800 MPa, with the strain rates varying from 900 s^{-1} to 4600 s^{-1}. The flow stress curves are likely smoothed when the specimens are deformed at the strain rate of 0.001 s^{-1} at an ambient temperature (298 K). However, the serrations appeared on the flow stress curves of the specimens deformed at high strain rates (e.g., 900 s^{-1}, 1700 s^{-1}, and 4600 s^{-1}). Furthermore, with the increasing of the strain rates, the serrations on the flow stress curves became more serious.

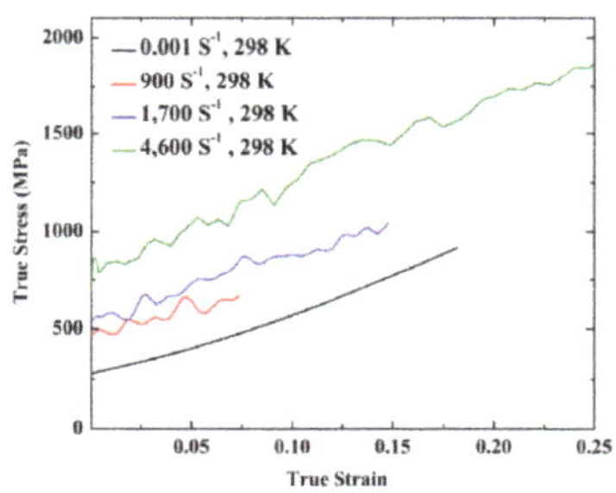

Figure 4. Compressive true stress–strain curves of the NiCrFeCoMn high-entropy alloy at different strain rates.

The EBSD technique was used to investigate the microstructure and the micro-orientation of a NiCrFeCoMn high-entropy alloy. Figure 5a–c are the electron backscattered diffraction images of the as-received specimen, the specimen deformed at the strain rate of 0.001 s^{-1}, and the specimen deformed at the strain rate of 4600 s^{-1}. First, the noise of the images was reduced. Then, to Figure 5b,c were added a layer of full Euler angles after adding a layer of band contrast. The size of the Kuwahara filter was 3-pixel points $\times$ 3-pixel points and the smoothing angle was 5°. Among them, high-angle boundaries (60°) were marked with black lines. It is evident that the alloy in the as-cast condition consisted of grains on the order of micrometers in size. A visual comparison of Figure 5b,c suggests a higher density of high-angle boundaries in the specimen deformed at a high strain rate of 4600 s^{-1} compared to the specimen deformed at a low strain rate of 0.001 s^{-1}. Further, the intensive deformation bands were generated in the specimen deformed at a high strain rate of 4600 s^{-1}, as shown in Figure 5c. Therefore, the obvious serrations in flow stress curves of the specimen deformed at high strain rates are caused by the intensive deformation bands.

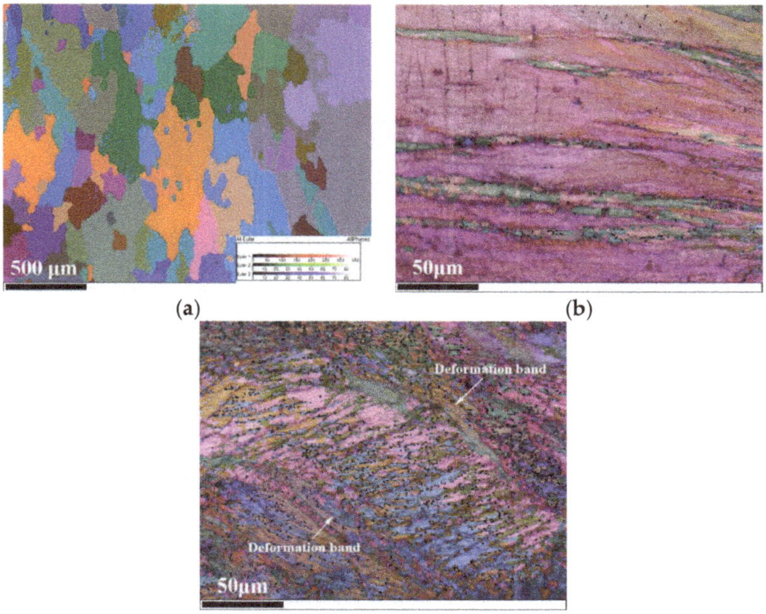

Figure 5. Electron backscattered diffraction images. (**a**) the Euler image of the as-received specimen; (**b**) the Euler+BC image of the deformed specimen at the strain rate 1×10^{-3} s^{-1}; (**c**) the Euler+BC image of the deformed specimen at the strain rate 4600 s^{-1}.

Bright field electron images taken for the specimens deformed at strain rates of 1700 s^{-1} and 4600 s^{-1} are shown in Figure 6. Figure 6a,c show the high-density dislocations and deformation bands that were formed in the NiCrFeCoMn high-entropy alloy. These deformation bands were distributed in parallel. The deformation band consisted of nanograins, as shown in Figure 6b,d. Comparing Figure 6a,b and Figure 6c,d, the parallel deformation bands were more clear when the specimen was deformed at relative lower strain rates as shown in Figure 6a,c, and the sizes of the nanograins in the specimens under a strain rate of 1700 s^{-1} were larger than those in the specimens deformed at a strain rate of 4600 s^{-1}.

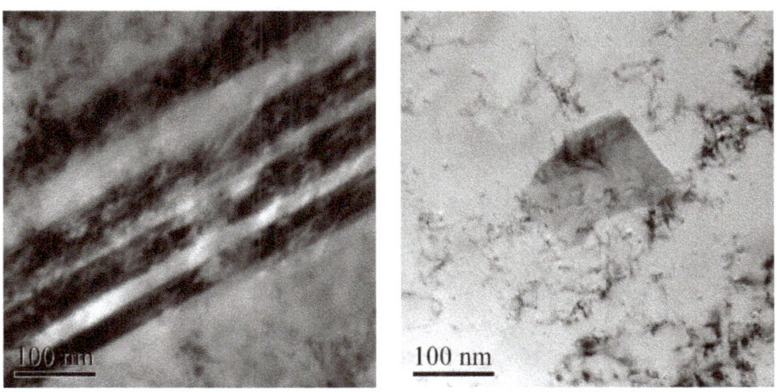

(a) (b)

Figure 6. *Cont.*

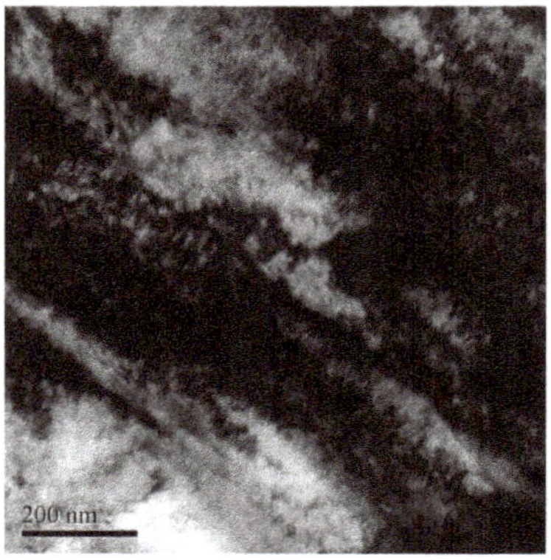

(c)

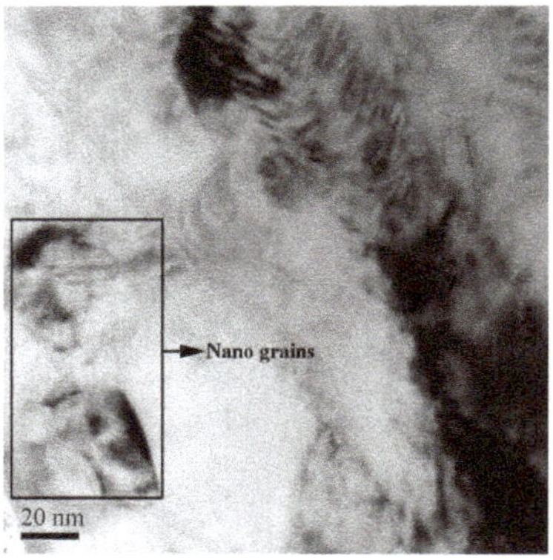

(d)

Figure 6. Bright field electron images showing microstructure in the specimens deformed at high strain rates. (**a**) and (**b**) are for the specimen deformed at the strain rate of about 1700 s^{-1} (**c**) and (**d**) are for the specimen deformed at the strain rate of about 4600 s^{-1}.

4. Discussion

4.1. The Constitutive Model and Strain-Rate Sensitivity

The main dynamic constitutive equations are the Johnson–Cook and Zerilli–Armstrong plastic models. Among them, the Johnson–Cook model is the most widely used for its relative simple expression. It can be represented as follows [27,28]:

$$\sigma = (\mathrm{A} + \mathrm{B}\varepsilon^{n})\left[1 + \mathrm{C}\ln\left(\frac{\dot{\varepsilon}}{\dot{\varepsilon_0}}\right)\right]\left[1 - \left(\frac{T - T_r}{T_m - T_r}\right)^{m}\right], \quad (4)$$

where A, B, and C are material constants, σ and ε are the flow stress and the equivalent plastic strain respectively, $\dot{\varepsilon}$ and T are the equivalent plastic strain rate and the experimental temperature respectively, and T_r and T_m are the reference temperature (usually room temperature) and the melting point, respectively.

The dislocation mechanism is important for studying the plastic deformation of metallic materials under dynamic deformation. Therefore, the Zerilli–Armstrong plastic model improves the Johnson–Cook model on the basis of dislocation mechanism [29]. In this work, the Zerilli–Armstrong plastic model was used for predicting the strain rate flow behavior of the as-cast NiCrFeCoMn high-entropy alloy. It can be represented as follows:

$$\sigma = \mathrm{C}_0 + \mathrm{C}_1 \times \varepsilon^{\mathrm{P}} \times exp\left[-\mathrm{C}_2 T + \mathrm{C}_3 T\ln\left(\frac{\dot{\varepsilon}}{\dot{\varepsilon_0}}\right)\right], \quad (5)$$

where C_0, C_1, C_2, C_3, and P are material constants. Note that T is 298 K.

Taking initial values of the Johnson–Cook model as follows: A = 1, B = 4, C = 100, n = 1, then the parameter values and the constitutive relation of stress-strain with strain rate could be obtained by using the MATLAB program (Version 7.0). Therefore, the constitutive equation based on the Johnson–Cook plastic model can be obtained as follows:

$$\sigma = \left(0.6039 + 4.81\varepsilon^{1}\right)\left[1 + 121.6\ln\left(\frac{\dot{\varepsilon}}{\dot{\varepsilon_0}}\right)\right]. \quad (6)$$

As above, taking initial values of the Zerilli–Armstrong model as follows: $C_0 = 400$, $C_1 = 9 \times 10^5$, $C_2 = 0.01$, $C_3 = 0.001$, P = 0.5, then the constitutive equation based on the Zerilli-Armstrong plastic model can be obtained as follows:

$$\sigma = 473.3 + 9.3 \times 10^5 \times \varepsilon^{0.5} \times exp\left[-0.0415T + 0.0026T \ln\left(\frac{\dot{\varepsilon}}{\dot{\varepsilon}_0}\right)\right]. \quad (7)$$

Figure 7 shows the comparison of the calculated results obtained from Equations (6) and (7) and the experimental data of the NiCrFeCoMn high-entropy alloy specimens. It can be seen that the results predicted by the Zerilli–Armstrong plastic model are in better agreement with the experimental results.

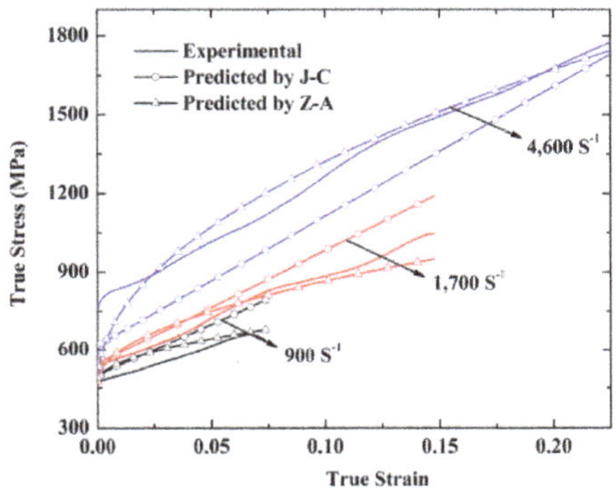

Figure 7. Comparison between the experimental results and the stress calculated by the Johnson–Cook and the Zerilli–Armstrong plastic models at 298K.

Figure 8 shows the yield strength vs. strain rate curves of the as-cast NiCrFeCoMn high-entropy alloy as a function of strain rate at ambient temperature. The strain-rate sensitivity is defined as follows:

$$\mathrm{m} = \frac{d(\log \sigma)}{d(\log \dot{\varepsilon})}. \quad (8)$$

Notice that the yield strength distributes from 200 MPa to 800 MPa and the slope of the tangent for the curve is increasing with the increase of strain rates. Therefore, the as-cast NiCrFeCoMn high-entropy alloy has distinguished strain-rate sensitivity at high strain rates.

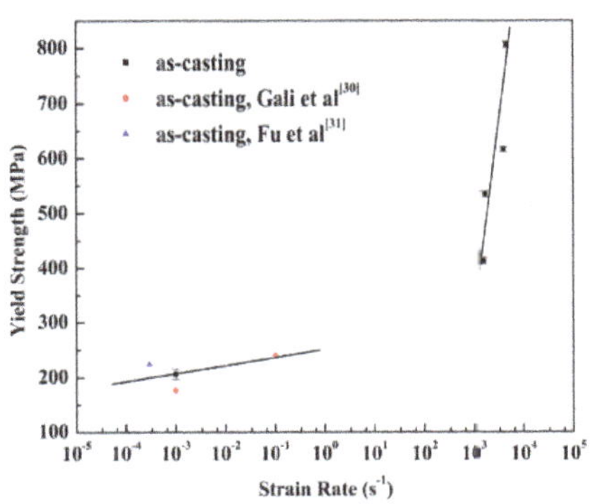

Figure 8. The yield strength vs. strain rate curves of the NiCrFeCoMn high-entropy alloy [30,31].

He et al. [32] studied the stress exponent (the reciprocal of strain rate sensitivity) of the NiCrFeCoMn high-entropy alloy deformed under strain rates less than 10^{-2} s^{-1}. He found that the stress exponent was in a positive relationship with the strain rates. When the strain rate ranged from 3.205×10^{-5} to 8.013×10^{-4} s^{-1}, the stress exponent increased simultaneously. Moon et al. [33] also studied the strain rate sensitivity at the elevated temperature and the cryogenic temperature, respectively. The results showed that the flow stress at 77 K was higher than that at room temperature. Therefore, the strain rate sensitivity of the flow stress at RT was higher than that at 77 K under strain rates less than 1×10^{-2} s^{-1}. The following formula can explain this phenomenon:

$$V^* = \sqrt{3}\mathrm{k}T\frac{\partial \ln \dot{\varepsilon}}{\partial \sigma}, \tag{9}$$

where k is the Boltzmann constant and V^* is activation volume.

The NiCrFeCoMn high-entropy alloy had positive activation volume at strain rates less than 1×10^{-2} s^{-1}. However, when the specimen deformed under high strain rates (beyond 1×10^{3} s^{-1}), the deformation time was very short. There was not enough time available for thermal energy to help dislocations overcome the barriers. The NiCrFeCoMn high-entropy alloy had a different deformation mechanism at dynamic loading. According to the literature [22], the large jump in yield strength at high strain rates is probably due to the phonon drag effect on the motion of dislocations. The phonon drag phenomenon becomes very effective during plastic deformation at high strain rates [22]. A phonon is an elastic lattice vibration propagating in a crystal [26]. The viscous drag generating from the interaction of the dislocations has negative impacts on the deformation progress. The drag effects by the phonons can be ignored under quasi-static conditions for low dislocation velocities. However, high dislocation velocities under high strain rates would enhance the phonon drag effects greatly and lead to phonons scattering, which also hinders dislocation movement. Therefore, the dynamic deformations lead to a much higher strain rate dependence of the flow stress than those under quasi-static conditions.

4.2. Mechanism for the Serration Behavior

High-entropy alloys present serrations on the flow stress curves during the plastic deformation, often at a normal strain rate around 1×10^{-4} s^{-1} or at low temperatures. Serration behavior exhibits in the stress-strain curves of the $Al_{0.5}CoCrCuFeNi$ high-entropy alloy at 7 K, 7.5 K, and 9 K at a strain rate of 4×10^{-4} s^{-1} [19]. The $Al_5Cr_{12}Fe_{35}Mn_{28}Ni_{20}$ high-entropy alloy exhibits typical serration behaviors at the elevated temperatures of 573 K and 673 K, with a strain rate of 1×10^{-4} s^{-1} [34]. Several microstructure mechanisms are proposed to explain the serration behavior, e.g., the Portevin-Le Chatelier (PLC) effect. Serration behavior was supposed to be associated with Cottrell atmosphere interaction with moving dislocations, slip bands, and dynamic strain aging. In Figure 4, it can be seen that the stress-strain curves of the as-cast NiCrFeCoMn high-entropy alloy deformed at a strain rate of 1×10^{-3} s^{-1} show no serrations, and marked serrations are present on the stress-strain curves of specimens deformed at strain rates above 900 s^{-1}. With increasing strain rates, amplitudes of the serrations on the stress-strain curves become much larger. Investigations on the microstructure in the as-cast NiCrFeCoMn high-entropy alloy show that the high density dislocations and the deformation bands are generated in the specimens deformed at high strain rates, shown in Figures 5 and 6. The Portevin-Le Chatelier effect, i.e., Cottrell atmosphere interaction with moving a simple dislocation structure, may not be the main reason for the serration behavior of the as-cast NiCrFeCoMn high-entropy alloy becoming deformed at dynamic loadings. On the other hand, if the dynamic deformation becomes more serious and the value of the strain rate increases, deformation bands are generated in the as-cast NiCrFeCoMn high-entropy alloy and the amplitudes of the serrations on the stress-strain curves become much larger. Therefore, large amounts of the deformation bands, leading to the serration behaviors, play an important role for the mechanical properties of the high-entropy alloy.

5. Conclusions

The as-cast equiatomic NiCrFeCoMn high-entropy alloy has a simple FCC crystallographic structure. Serration behavior is observed in the stress- strain curves of the as-cast NiCrFeCoMn high-entropy alloy deformed at dynamic loadings. The yield strength of the high-entropy alloy, which distributes from 490 to 800 MPa, presents a positive relationship with the strain rates. The Johnson–Cook plastic model and the Zerilli–Armstrong plastic model of the NiCrFeCoMn high-entropy alloy are obtained. However, the results predicted by Zerilli–Armstrong correspond better with the experimental results. The serration behavior of the NiCrFeCoMn high-entropy alloy at a high strain rate is sensitive to the strain rates. The high density of deformation bands plays an important role in the deformation behavior and mechanical properties of the as-cast NiCrFeCoMn high-entropy alloy deformed at dynamic loadings.

Author Contributions: Conceptualization, B.W. and X.Z.; methodology, X.Y. and X.H.; validation, B.W. and Y.Z.; formal analysis, B.W., X.H., C.W. and X.Y.; investigation, B.W., X.H. and X.Y.; resources, B.W.; data curation, B.W., X.H., C.W. and X.Y.; writing—original draft preparation, B.W. and X.Y.; writing—review and editing, B.W. and Y.Z.; visualization, C.W.; supervision, B.W. and Y.Z..; project administration, B.W. and Y.Z.; funding acquisition, B.W. and Y.Z.

Funding: This research was funded by National Natural Science of China, grant number 51771231, and by State Key Laboratory of Powder Metallurgy, Central South University, Changsha, China, grant number 20181106.

Acknowledgments: Authors wish to express their most sincere gratitude to M.A. Meyers at University of California, San Diego, and Yong Liu at Central South University for their advice and help. Authors would like to express their sincere thanks to Yang Wang and Yu Wang at University of Science and Technology of China, and Xiang Zan at Hefei University of Technology for dynamic testing.

Conflicts of Interest: The authors declare no conflict of interest.

References

1. Yeh, J.W.; Chen, S.K.; Lin, S.J.; Gan, J.Y.; Chin, T.S.; Shun, T.T.; Tsai, C.H.; Chang, S.Y. Nanostructured high-entropy alloys with multiple principal elements: Novel alloy design concepts and outcomes. *Adv. Eng. Mater.* **2004**, *6*, 299–303. [CrossRef]
2. Yeh, J.W. Alloy design strategies and future trends in high-entropy alloys. *JOM* **2013**, *65*, 1759–1771. [CrossRef]
3. Zhang, Y.; Yan, X.; Liao, W.; Zhao, K. Effects of nitrogen content on the structure and mechanical properties of ($Al_{0.5}CrFeNiTi_{0.25}$)Nx high-entropy films by reactive sputtering. *Entropy* **2018**, *20*, 624. [CrossRef]
4. Salishchev, G.A.; Tikhonovsky, M.A.; Shaysultanov, D.G.; Stepanov, N.D.; Kuznetsov, A.V.; Kolodiy, I.V.; Tortikab, A.S.; Senkov, O.N. Effect of Mn and V on structure and mechanical properties of high-entropy alloys based on CoCrFeNi system. *J. Alloy. Compd.* **2014**, *591*, 11–21. [CrossRef]
5. Stepanov, N.; Tikhonovsky, M.; Yurchenko, N.; Zyabkin, D.; Klimova, M.; Zherebtsov, S.; Efimov, A.; Salishchev, G. Effect of cryo-deformation on structure and properties of CoCrFeNiMn high-entropy alloy. *Intermetallics* **2015**, *59*, 8–17. [CrossRef]
6. Schuh, B.; Mendez-Martin, F.; Völker, B.; George, E.P.; Clemens, H.; Pippan, R.; Hohenwarter, A. Mechanical properties, microstructure and thermal stability of a nanocrystalline CoCrFeMnNi high-entropy alloy after severe plastic deformation. *Acta Mater.* **2015**, *96*, 258–268. [CrossRef]
7. Otto, F.; Dlouhý, A.; Somsen, C.; Bei, H.; Eggeler, G.; George, E.P. The influences of temperature and microstructure on the tensile properties of a cocrfemnni high-entropy alloy. *Acta Mater.* **2013**, *15*, 5743–5755. [CrossRef]
8. Zhai, S.; Wang, W.; Xu, J.; Xu, S.; Zhang, Z.; Wang, Y. Effect of Co and Gd additions on microstructures and properties of FeSiBAlNi high entropy alloys. *Entropy* **2018**, *20*, 487. [CrossRef]
9. Feng, W.; Qi, Y.; Wang, S. Effects of short-range order on the magnetic and mechanical properties of FeCoNi(AlSi)x high entropy alloys. *Metals* **2017**, *7*, 482. [CrossRef]
10. Cicala, G.; Giordano, D.; Tosto, C.; Filippone, G.; Recca, A.; Blanco, I. Polylactide (PLA) filaments a biobased solution for additive manufacturing: Correlating rheology and thermomechanical properties with printing quality. *Materials* **2018**, *7*, 1191. [CrossRef] [PubMed]
11. Cicco, D.D.; Asaee, Z.; Taheri, F. Use of nanoparticles for enhancing the interlaminar properties of fiber-reinforced composites and adhesively bonded joints—A review. *Nanomaterials* **2017**, *11*, 360. [CrossRef] [PubMed]

12. Laplanche, G.; Kostka, A.; Horst, O.M.; Eggeler, G.; George, E.P. Microstructure evolution and critical stress for twinning in the CrMnFeCoNi high-entropy alloy. *Acta Mater.* **2016**, *118*, 152–163. [CrossRef]
13. Kilmametov, A.; Kulagin, R.; Mazilkin, A.; Seils, S.; Boll, T.; Heilmaier, M.; Hahn, H. High-pressure torsion driven mechanical alloying of CoCrFeMnNi high entropy alloy. *Scr. Mater.* **2019**, *158*, 29–33. [CrossRef]
14. Wang, L.; Qiao, J.W.; Ma, S.G.; Jiao, Z.M.; Zhang, T.W.; Chen, G.; Zhao, D.; Zhang, Y.; Wang, Z.H. Mechanical response and deformation behavior of $Al_{0.6}$CoCrFeNi high-entropy alloys upon dynamic loading. Mater. *Sci. Eng. A* **2018**, *727*, 208–213. [CrossRef]
15. Klimova, M.; Stepanov, N.; Shaysultanov, D.; Chernichenkoet, R.; Yurchenko, N.; Sanin, V.; Zherebtsov, S. Microstructure and mechanical properties evolution of the Al, C-containing CoCrFeNiMn-type high-entropy alloy during cold rolling. *Materials* **2018**, *11*, 53. [CrossRef] [PubMed]
16. Zherebtsov, S.; Stepanov, N.; Ivanisenko, Y.; Shaysultanov, D.; Yurchenko, N.; Klimova, M.; Salishchev, G. Evolution of microstructure and mechanical properties of a CoCrFeMnNi High-Entropy Alloy during high-pressure torsion at room and cryogenic temperatures. *Metals* **2018**, *8*, 123. [CrossRef]
17. He, J.Y.; Wang, H.; Huang, H.L.; Xu, X.D.; Chen, M.W.; Wu, Y.; Liu, X.J.; Nieh, T.G.; An, K.; Lu, Z.P. A precipitation-hardened high-entropy alloy with outstanding tensile properties. *Acta Mater.* **2016**, *102*, 187–196. [CrossRef]
18. Stepanov, N.D.; Shaysultanov, D.G.; Yurchenko, N.Y.; Zherebtsov, S.V.; Ladygin, A.N.; Salishchev, G.A.; Tikhonovsky, M.A. High temperature deformation behavior and dynamic recrystallization in CoCrFeNiMn high entropy alloy, Mater. *Sci. Eng. A* **2015**, *636*, 188–195. [CrossRef]
19. Ke, H.B.; Wen, P.; Peng, H.L.; Wang, W.H.; Greer, A.L. Homogeneous deformation of metallic glass at room temperature reveals large dilatation. *Scr. Mater.* **2011**, *64*, 966–969. [CrossRef]
20. Zhang, Y.; Liu, J.P.; Chen, S.Y.; Xie, X.; Liaw, P.K.; Dahmen, K.A.; Qiao, J.W.; Wang, Y.L. Serration and noise behaviors in materials. *Prog. Mater. Sci.* **2017**, *90*, 358–460. [CrossRef]
21. Wang, B.; Huang, X.; Fu, A.; Liu, Y.; Liu, B. Serration behavior and microstructure of high entropy alloy CoCrFeMnNi prepared by powder metallurgy. *Mater. Sci. Eng. A* **2018**, *726*, 37–44. [CrossRef]
22. Kumar, N.; Ying, Q.; Nie, X.; Mishra, R.S.; Tang, Z.; Liaw, P.K.; Brennan, R.E.; Doherty, K.J.; Cho, K.C. High strain-rate compressive deformation behavior of the $Al_{0.1}$CrFeCoNi high entropy alloy. *Mater. Des.* **2015**, *86*, 598–602. [CrossRef]
23. Li, Z.; Zhao, S.; Diao, H.; Liaw, P.K.; Meyers, M.A. High-velocity deformation of $Al_{0.3}$CoCrFeNi high-entropy alloy: Remarkable resistance to shear failure. *Sci. Rep.* **2017**, *7*, 42742. [CrossRef] [PubMed]
24. Dirras, G.; Couque, H.; Lilensten, L.; Heczel, A.; Tingaud, D.; Couzinié, J.-P.; Perrière, L.; Gubicza, J.; Guillot, I. Mechanical behavior and microstructure of $Ti_{20}Hf_{20}Zr_{20}Ta_{20}Nb_{20}$ high-entropy alloy loaded under quasi-static and dynamic compression conditions. *Mater. Charact.* **2016**, *111*, 106–113. [CrossRef]
25. He, J.; Wang, Q.; Zhang, H.; Dai, L.; Mukai, T.; Wu, Y.; Liu, X.; Wang, H.; Nieh, T.G.; Lu, Z. Dynamic deformation behavior of a face-centered cubic FeCoNiCrMn high-entropy alloy. *Sci. Bull.* **2018**, *63*, 362–368. [CrossRef]
26. Park, J.M.; Moon, J.; Bae, J.W.; Jang, M.J.; Park, J.; Lee, S.; Kim, H.S. Strain rate effects of dynamic compressive deformation on mechanical properties and microstructure of CoCrFeMnNi high-entropy alloy. *Mater. Sci. Eng. A* **2018**, *719*, 155–163. [CrossRef]
27. Wang, F.J.; Zhang, Y.; Chen, G.L. Atomic packing efficiency and phase transition in a high entropy alloy. *J. Alloy. Compd.* **2009**, *478*, 321–324. [CrossRef]
28. Zhou, Y.J.; Zhang, Y.; Wang, Y.L.; Chen, G.L. Microstructure and compressive properties of multicomponent $Al_X(TiVCrMnFeCoNiCu)_{(100-X)}$ high-entropy alloys. *Mater. Sci. Eng. A* **2007**, *454*, 260–265. [CrossRef]
29. Zerilli, F.J.; Armstrong, R.W. Dislocation-mechanics-based constitutive relations for material dynamics calculations. *Appl. Phys.* **1987**, *61*, 1816–1825. [CrossRef]
30. Gali, A.; George, E.P. Tensile properties of high- and medium-entropy alloys. *Intermetallics* **2013**, *39*, 74–78. [CrossRef]
31. Fu, J.X.; Cao, C.M.; Tong, W.; Hao, Y.X.; Peng, L.M. The tensile properties and serrated flow behavior of a thermomechanically treated CoCrFeNiMn high-entropy alloy. *Mater. Sci. Eng. A* **2017**, *690*, 418–426. [CrossRef]
32. He, J.Y.; Zhu, C.; Zhou, D.Q.; Liu, W.H.; Nieh, T.G.; Lu, Z.P. Steady state flow of the FeCoNiCrMn high entropy alloy at elevated temperatures. *Intermetallics* **2014**, *55*, 9–14. [CrossRef]

33. Moon, J.; Hong, S.I.; Bae, J.W.; Jang, M.J.; Yim, D.; Kim, H.S. On the strain rate-dependent deformation mechanism of CoCrFeMnNi high entropy alloy at liquid nitrogen temperature. *Mater. Res. Lett.* **2017**, *5*, 472–477. [CrossRef]
34. Antonaglia, J.; Xie, X.; Tang, Z.; Tsai, C.W.; Qiao, J.W.; Zhang, Y.; Laktionova, M.O.; Tabachnikova, E.D.; Yeh, J.W.; Senkov, O.N.; et al. Temperature effects on deformation and serration behavior of high-entropy alloys (HEAs). *JOM* **2014**, *66*, 2002–2008. [CrossRef]

entropy

MDPI

Article

Unconventional Deformation Behaviours of Nanoscaled High-Entropy Alloys

Yeqiang Bu [1], Shenyou Peng [2], Shiwei Wu [3], Yujie Wei [2], Gang Wang [3], Jiabin Liu [1,*] and Hongtao Wang [4,*]

1 School of Materials Science and Engineering, Zhejiang University, Hangzhou 310027, China; yeqiangbu@zju.edu.cn
2 LNM, Institute of Mechanics, Chinese Academy of Sciences, Beijing 100190, China; pengshenyou@lnm.imech.ac.cn (S.P.); yujie_wei@lnm.imech.ac.cn (Y.W.)
3 Laboratory for Microstructures, Shanghai University, Shanghai 200444, China; wushiwei@shu.edu.cn (S.W.); g.wang@shu.edu.cn (G.W.)
4 Institute of Applied Mechanics, Zhejiang University, Hangzhou 310027, China
* Correspondence: liujiabin@zju.edu.cn (J.L.); htw@zju.edu.cn (H.W.)

Received: 7 September 2018; Accepted: 28 September 2018; Published: 11 October 2018

Abstract: The bulk high-entropy alloys (HEAs) exhibit similar deformation behaviours as traditional metals. These bulk behaviours are likely an averaging of the behaviours exhibited at the nanoscale. Herein, in situ atomic-scale observation of deformation behaviours in nanoscaled CoCrCuFeNi face-centred cubic (FCC) HEA was performed. The deformation behaviours of this nanoscaled FCC HEA (i.e., nanodisturbances and phase transformations) were distinct from those of nanoscaled traditional FCC metals and corresponding bulk HEA. First-principles calculations revealed an obvious fluctuation of the stacking fault energy and stability difference at the atomic scale in the HEA. The stability difference was highlighted only in the nanoscaled HEA and induced unconventional deformation behaviours. Our work suggests that the nanoscaled HEA may provide more chances to discover the long-expected essential distinction between the HEAs and traditional metals.

Keywords: nanoscaled high-entropy alloys; nanodisturbances; phase transformations; atomic-scale unstable

1. Introduction

Traditional alloys, such as steels and copper alloys, are fabricated based on one or two principle constituent elements. Yeh et al. [1] proposed the concept of high-entropy alloys (HEAs) that provides a novel basis to design new alloys. These HEAs are composed of multi-principle elements at equiatomic or near-equiatomic ratios, distinguishing them from traditional alloys. Consequently, the deformation behaviours of HEAs are believed to be different from traditional alloys [2], but no convincing experiments have yet been reported to show an essential distinction of the plastic deformation behaviours between HEAs and traditional metals. On the contrary, most previous results in bulk HEAs have shown similar scenarios with traditional metals, in which the plastic deformation is primarily carried by dislocations or twins [3–5]. The underlying mechanisms of macroscopic mechanical responses are essentially the collective behaviours of atomic-scale configurations. Therefore, the physical processes during alloy deformation exhibit no remarkable features if the salient atomic configuration details are essentially blurred and only the average effect can be measured and observed [6,7]. As expected, such macroscopic deformation behaviours are generally controlled by only a few key parameters such as the elastic moduli, the stable and unstable stacking fault energies (SFEs), the microstructure parameters and the temperature.

Close observation at the nanoscaled regime where discrete plasticity dominates may uncover the essential features that distinguish the HEAs from traditional alloys. Extensive investigations using in situ high-resolution transmission electron microscopy (HRTEM) have revealed some interesting deformation behaviours of nanoscaled pure metals [8–17]. Size-dependent behaviours have thus been uncovered such as the reversible deformation twinning and detwinning processes found in nanoscaled W samples [12] and the dislocation-originated stacking fault tetrahedra in nanoscaled Au samples [13]. Meanwhile, surface-mediated plasticity deformation behaviours have frequently been observed such as the partial dislocations emitted from the surface in sub-10 nm-sized Au [14] and the liquid-like deformations in sub-10 nm-sized Ag nanoparticles [15]. It would be interesting to determine whether nanoscaled HEAs behave similarly in the discrete plasticity regime, in contrast to their collective behaviour. The face-centred cubic (FCC) HEA CoCrCuFeNi is a typical HEA which was proposed in the earliest paper about the concept of HEA [1]. Numerous research papers related to HEAs were applied in CoCrCuFeNi from then. Therefore, CoCrCuFeNi was taken as a model HEA to reveal the obscured potential high-entropy effect on plastic deformation behaviours at the nanoscale in this work.

2. Materials and Methods

The CoCrCuFeNi button ingots were prepared by melting high-purity Co, Cr, Cu, Fe and Ni at equiatomic ratios in a vacuum arc furnace. During specimen preparation, the ingot was first sliced into a rod of $0.25 \times 0.25 \times 10$ mm^3 in dimension. Firstly, we employed a tungsten carbide cutter to make a pair of gaps on the surface of the rod. Then, we used a plier to pull this rod apart in the long dimension and then led to the formation of fresh triangular nano-tips on the fractured surface, which served as the specimens for the in situ TEM experiments (Figure S1 in supplementary). A JEM-2100F field emission TEM equipped with a Nanofactory TEM-scanning tunnelling microscope (STM) sample holder was used in the in situ TEM experiments. Two fractured ends exhibiting nano-tips were mounted on the sample holder, with one at the fixed end of the holder and the other at the piezo-manipulator end (Figure S2 in Supplementary Materials). One nano-tip could thus be driven to touch the other nano-tip on the counter fractured surface, guided by the piezo-actuated nanoscaled manipulator. A nanoscaled welding joint 10–30 nm in size could then be formed instantly via pulsed joule heating. Uniaxial tensile stress could be applied by step-by-step retraction via the nano-manipulator. This method essentially furnished the fabrication, mechanical testing and easily atomic-resolution observation of the nanoscaled HEA. The structure of the nanoscaled HEA fabricated by this method is shown in Figure S3. The crystalline interplanar spacing of the $(\bar{1}1\bar{1})$ plane is 2.07 Å in the obtained nanoscaled HEA, i.e., its lattice parameter is 3.59 Å. This is consistent with the reported lattice parameter (3.579 Å) of the bulk CoCrCuFeNi [1,18]. This shows that the structure of the nanoscaled HEA is not affected by the process of preparation. Meanwhile, there is no oxide layer on the surface of the nano-tips before the fabrication of the nanoscaled samples, as shown in Figure S4, which indicates that there is no impact of the oxide layer on the deformation behaviours (These nano-tips were exposed to air for only a few minutes.). The energy dispersion spectrum (EDS) mapping results are shown in Figure S5, which indicates that the constituent elements distribute uniformly in the nao-tip (EDS hardly displays the details of composition in several-atoms scale, even the atomic resolution EDS only gives the statistical chemical composition of each atom column in planar view).

The first-principles calculations were performed using the Vienna ab initio simulation package based on density functional theory, wherein all of the HEA samples were relaxed to the energy precision of 0.01 meV. The SFE was calculated with $E_{sf} = (E_{\text{fault}} - E_{\text{perfect}})/A$, where E_{perfect} and E_{fault} are the free energy of perfect and faulted structures, respectively; and A is the area of each layer. The structure with one layer fault was obtained by a rigid displacement between two adjacent layers, where the magnitude was equal to that of the Burgers vector, $\boldsymbol{b}_p = 1/6$<112>.

3. Results

3.1. *Nanodisturbances*

Figure 1a displays a nanoscaled HEA during in situ straining in the TEM (Vedio S1 in Supplementary Materials), where the loading direction (LD) is around $[3\bar{1}1]$ and the beam zone is [011]. The deformed nanoscaled HEA contains several dislocations. The deformation of this nanoscaled HEA is closely related to the behaviours of dislocations. Dislocation cores are labelled in Figure 1 by "⊥", and an enlarged image of a dislocation core is shown in Figure 1b. According to the analysis of the Burgers circuit, these are full dislocations with $\boldsymbol{b} = 1/2[101]$. Some of the dislocations (circled in Figure 1a) appear in pairs and are thus dislocation dipoles, seen in an enlarged image in Figure 1c. Using in situ HRTEM, we could dynamically observe the evolution of these dislocation dipoles. Figure 1d–f show inverse fast Fourier transform (IFFT) images of the one-dimensional {111} plane fringes in the area around a dislocation dipole. At t = 20 s, the circled area exhibited a distorted lattice which was induced by high stress in the nanoscaled HEA. At t = 20.5 s, a dislocation dipole pair was produced in the distorted area. Under stress driving, the dislocation dipole was observed to expand along the $(\bar{1}1\bar{1})$ slip plane at t = 21 s. This deformation mode could be called nanodisturbance, which has been proposed on body-centred cubic (BCC) "gum metal" by Gutkin et al. [19]. This kind of dislocation dipole nucleation and expansion could act as a mechanism of dislocation multiplication during the deformation process of the nanoscaled HEA.

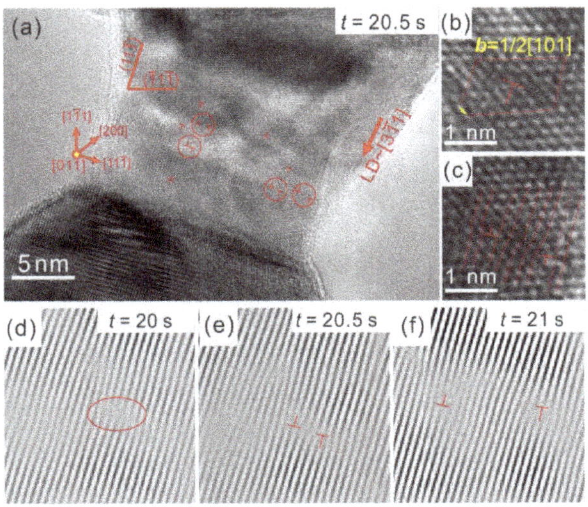

Figure 1. The nanodisturbances in the nanoscaled high-entropy alloy (HEA). (**a**) deformed nanoscaled HEA sample containing several dislocations (Beam // [011], loading direction (LD) $\approx$ $[3\bar{1}1]$); (**b**) analysis of the Burgers vector; (**c**) enlarged image of the dislocation dipole; (**d**–**f**) Inverse fast Fourier transform (IFFT) images of the one-dimensional {111} plane fringes showing the formation and expansion of the dislocation dipole.

3.2. *Phase Transformation*

Figure 2a–c shows the tensile process of a nanoscaled HEA sample (Video S2 in Supplementary Materials), where the LD is around the $[3\bar{1}1]$ and beam zone is [011]. Figure 2e–g show enlarged images corresponding to the areas in red squares in Figure 2a–c, respectively. As shown in Figure 2e, the crystal lattice exhibits clear characteristics of a FCC structure; i.e., the angle between two close-packed planes in the {111} family is 70.5°. As the nanoscaled HEA continues to be stretched, the angles between two {111} planes reduce to 64° at t = 104 s (Figure 2f), and then to 60° at t = 384 s (Figure 2g). The angle of 64° indicates that the lattice structure deviates from the original FCC structure notably. The angle of 60° represents a typical BCC structure with a [111] zone. Figure 2d plots the angles between the close-packed planes in the red square areas of Figure 2a–c as a function of time under stress, where the angle exhibits a slow and successive transition from ~70° to ~60°. The change of the angle corresponds to the transition that the initial FCC lattice transforms to BCC lattice during in situ tension. The resultant orientation relationship between the FCC and BCC agrees with the K–S relationship, i.e., $[011]_{FCC}$ // $[111]_{BCC}$, $(\bar{1}1\bar{1})_{FCC}$ // $(\bar{1}10)_{BCC}$. The transformation process could be explained by K–S model. The FCC lattice sheared 19.5° along the <112>$_{FCC}$ direction on the $\{111\}_{FCC}$ plane and sheared 10.5° along the <110>$_{FCC}$ direction on the $\{112\}_{FCC}$ plane. Therefore, the stress-induced FCC→BCC transition realized and the BCC lattice formed. The slow and successive transformation was believed to relate with the high lattice friction in HEAs [20–22], the transformation dislocations slip and lattice shear more slowly in HEAs compared to the traditional abrupt Martensitic transformation. Such stress-induced FCC→BCC transition has a good reproducibility, and Figure S6 shows another example.

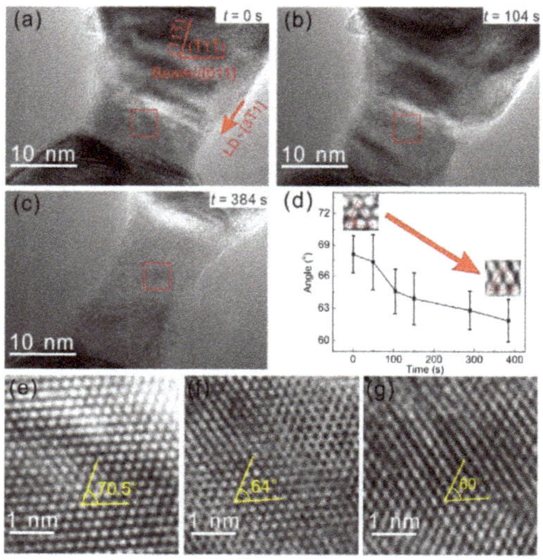

Figure 2. The phase transformation from face-centred cubic (FCC) to body-centred cubic (BCC) in the nanoscaled HEA. (**a**–**c**) elongation process of the nanoscaled sample (Beam // [011], LD ≈ $[3\bar{1}1]$); (**d**) variation of the angle between two close-packed planes in the red square area during in situ tension; (**e**–**g**) High-resolution transmission electron microscopy (HRTEM) images of the red square zones in (**a**–**c**), respectively.

3.3. Fluctuation of Stacking Fault Energy

The SFE is one of the most significant parameters determining the deformation behaviours of alloys, and is closely related to phase transformation and structural stability. Herein, we employed first-principles calculations to determine the SFE of this HEA, and the results of 52 independent SFE calculations are shown in Figure 3a. These results exhibit a fluctuant distribution of the SFE in HEA, where the SFE value covers a wide range and even is negative. The negative SFE values indicate that some of the HEA atomic configurations are unstable. The instability seems to be strongly correlated with the non-uniform distribution of atoms. As an example, we analyze a typical calculation sample illustrated in Figure 3b and find that the atoms in the dashed boxes (also seen in Figure 3c,d) are not a uniform distribution of all kinds of elements in CoCrFeNiCu. Some areas have more Co and Cr atoms and less Cu atoms (Figure 3c), which leads to a negative SFE of −24 mJ/m^2. Some areas have more Cu atoms and less Co and Cr atoms (Figure 3d), which leads to a SFE as high as 109 mJ/m^2. We conclude that the elemental inhomogeneity at the atomic scale leads to SFE difference in local, and the fluctuant distribution of the SFE induces a stability difference at atomic-scale. However, such fluctuant SFE is averaged in the bulk HEAs, the atomic-scale stability difference is also blurred. The small-scaled sample size may highlight such atomic details.

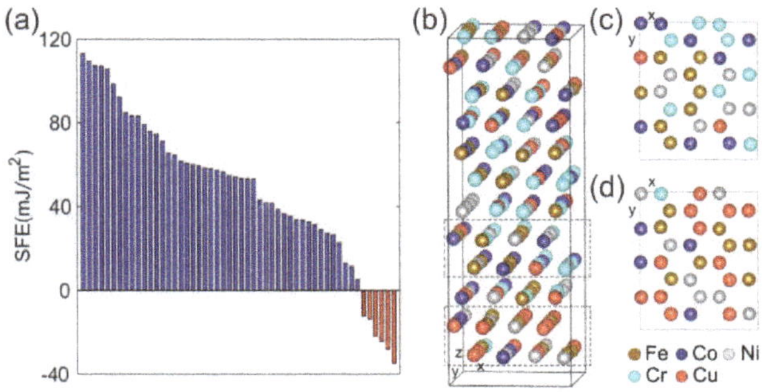

Figure 3. The fluctuant distribution of the stacking fault energies (SFEs). (**a**) the SFEs of 52 independent calculations; (**b**) typical structure of a calculation sample; (**c**) atomic configuration producing a negative SFE of -24 mJ/m^2 from the upper dashed box in (**b**); (**d**) atomic configuration producing a high SFE of 109 mJ/m^2 from the lower dashed box in (**b**).

4. Discussion

In such in situ HRTEM experiments, the effects of electron beam on the nanoscaled samples should be verified and eliminated. Although the CoCrCuFeNi HEA was verified to be stable under the severe electron irradiation [23], all in situ experiments are still performed under the weak electron beam for minimizing the effect of electron irradiation. Meanwhile, our verification experiments (see Supplementary Figure S6) and theoretical estimation (see Appendix A) confirm the negligible influence of electron irradiation on the in situ experiments.

Nanodisturbances could be an effective mechanism for dislocation multiplication, and the process of nanodisturbances evolving into dislocation dipoles has been observed in the BCC gum metal [19,24]. However, there has not yet been any experimental observation of this novel dislocation-generating mechanism for FCC structure metals. Some theoretical calculations have investigated the nanodisturbance phenomenon in Au and Cu nanowires [25,26], indicating that the nanodisturbance deformation mode could dominate over traditional dislocation generation at high stresses and 0 K. Obviously, the temperature condition of 0 K was not satisfied in this work. However, we still observed the nanodisturbance deformation mode in FCC HEAs for the first time. The high-level stress in nanoscaled samples and the intrinsic features of HEAs both play significant roles on the occurrence of the nanodisturbances. On the one hand, we observed nanodisturbance in the nanoscaled FCC HEA at the relative loose condition (at room tempreture). On the other hand, nanodisturbance is hardly observed in the bulk HEAs because of the relative low stress level in the bulk HEA. Furthermore, we believe that it is the nanoscaled size that triggers the emergence of the intrinsic features of the HEA.

The previous works showed that there are no phase transformation during the deformation process [27]. However, herein we observed the FCC$\rightarrow$BCC transformation in nanoscaled CoCrCuFeNi HEA. A similar transformation from FCC to body-centered tetragonal (BCT) has been observed in nanoscaled fractured Au [14,28], where the phase transformation therein was considered to be stimulated by the relaxation of surface stress. The surface stress has an inverse relationship with the sample size, and thus surface stress of a nanoscaled sample is sufficiently high to stimulate a transformation. However, in this study, the phase transformation in the nanoscaled HEA occurred before fracturing, and thus the surface stress had not been completely released. Therefore, the cause of phase transformation in the nanoscaled HEA is not solely surface stress, but an important role may also be played by the unstable nature of nanoscaled HEA. The combined effect of high surface stress and an unstable nature, therefore, stimulates the occurrence of the phase transformation.

Both nanodisturbances and phase transformations are not regular deformation behaviours in bulk CoCrCuFeNi HEA. These unconventional deformation behaviours observed in nanoscaled HEA are believed to be related to atomic configuration details present at such a small scale. Our first-principles calculations could well illuminate the unconventional deformation behaviours in such nanscaled HEAs. The first-principles results show that the elemental inhomogeneity at the atomic scale leads to SFE difference in local, and the fluctuant distribution of the SFE induces a stability difference of atomic-scale HEA. However, the entire structure averaged in the bulk HEA possesses a constant SFE. The stability difference can be outlined and plays a dominant role only when the sample dimensions reach the nanoscale. At that time, nanoscaled HEA exhibits deformation behaviours different with bulk counterparts. We further speculate that the nanoscaled HEA provides more of a chance to discover the long-expected essential distinction between the HEAs and traditional metals.

Besides the FCC HEAs, the BCC and hexagonal close-packed (HCP) HEAs may also possess some distinct characteristic physical properties but are blurred in bulk. It is worth investigating the other nanoscaled HEAs with various structures in future research to reveal the essential distinction between the HEAs and traditional metals.

5. Conclusions

In summary, we employed in situ HRTEM to investigate the deformation behaviours of nanoscaled HEA. Unconventional deformation behaviours (i.e., nanodisturbances and phase transformations) were observed in the nanoscaled HEA. The first-principles calculations revealed obvious fluctuant distribution of the SFE at atomic scale, which was resulted from the elemental inhomogeneity. The SFE fluctuation leaded to stability difference at the atomic scale, which plays a dominant role in the deformation of the nanoscale sample but tiny roles in bulk counterparts. The nanoscaled HEA provided a chance to highlight the stability difference and therefore exhibited unconventional deformation behaviours. Our investigations reveal some HEA features and are significant for understanding the nature of HEA.

Supplementary Materials: The following are available online at http://www.mdpi.com/1099-4300/20/10/778/s1, Figure S1: The nano-tips on the fractured surface, Figure S2: Two teared parts with nano-tips are mounted on a Nanofactory transmission electron microscope (TEM)-scanning tunnelling microscope (STM) TEM holder, Figure S3: The structure keep unchanged during the preparation process, Figure S4: The high-resolution TEM image of the nano-tip, Figure S5: The verification experiments about the effects of electron irradiation, Video S1: Nanodisturbances deformation mode in the nanoscaled HEA, Video S2: Phase transformation in the nanoscaled HEA.

Author Contributions: H.W. and J.L. initiated this research project. Y.B. performed the TEM experiments. S.P. and Y.W. performed the first-principles calculations. S.W. and G.W. synthesized the initial samples. H.W., J.L. and Y.B. analyzed the data and wrote the manuscript.

Funding: This research was funded by the National Natural Science Foundation of China (Nos. 11572281, 11725210 and 11672355), and the Fundamental Research Funds for the Central Universities (No. 2018XZZX001-05), and the Zhejiang Provincial Science & Technology Program of China (2017C31076).

Conflicts of Interest: The authors declare no conflict of interest.

Appendix A. Theoretical Estimation of the Effects of Electron Irradiation

The effects of electron irradiation could be divided into two aspects, i.e., knock-on displacement and heating effect. We could not observe any surface diffusion or atoms hopping occurs in the absence of external force. Therefore, the knock-on displacement driven by electron irradiation has no influence on the nanoscaled samples. For the heating effect, the estimation was based on the Fisher's model [29] and the Bethe–Bloch equation [30]. Herein, the nanoscaled samples are all with a certain thickness

(~10 nm), and exposed at the weak beam intensity. Fisher has proposed a model to estimate the temperature increase induced by electron irradiation [29]:

$$\Delta T = \frac{I}{4\pi Ke}(\frac{\Delta E}{d})(1 + 2\ln\frac{b}{r_0}), \quad (1)$$

where ΔT is the maximum temperature rise heated by electron irradiation, K is the thermal conductivity of the sample, I is the beam current, ΔE is the total energy loss per electron in a sample of thickness d, b is the radius of the heat sink, and r_0 is the beam radius. The term $\frac{\Delta E}{d}$ could be approximately equal to $\frac{\mathrm{d}E}{\mathrm{d}x}$ (where x is the position in thickness), which could be calculated by the Bethe–Bloch equation [30], as follows:

$$-\frac{\mathrm{d}E}{\mathrm{d}x} = \frac{2\pi Z\rho(e^2/4\pi\varepsilon_0)^2}{mv^2}\left\{\ln\left[\frac{E(E+mc^2)^2\beta^2}{2I_e^2mc^2}\right] + (1-\beta^2) - (1-\sqrt{1-\beta^2}+\beta^2)\ln 2 + \frac{1}{8}(1-\sqrt{1-\beta^2})^2\right\}, \quad (2)$$

where Z is the atomic number of the samples, ρ is the atomic density, ε_0 is the vacuum dielectric constant, m is the electron rest mass, v is the electron velocity, c is the speed of light, E is the electron energy, I_e is the average excitation energy of electrons in the target, and $\beta = v/c$.

We take the $Z_{ave} = (Z_{Co}+ Z_{Cr}+ Z_{Cu}+ Z_{Fe}+ Z_{Ni})/5 = 26.8$ as the atomic number of this HEA samples containing Co, Cr, Cu, Fe and Ni five principle elements. $\rho = \rho_{mass}/((m_{Co/atom}+ m_{Cr/atom}+ m_{Cu/atom} + m_{Fe/atom}+ m_{Ni/atom})/5) = 7.231 \times 1027\ m^{-3}$. The thermal conductivity of this HEA is estimated to be K= 16.2 $W{\cdot}m^{-1}{\cdot}K^{-1}$. Given I = 4.8 nA, $e = 1.6 \times 10^{-19}$ C, b = 1.5 mm, r_0 = 200 nm, $\varepsilon_0 = 8.85 \times 10^{-12}\ F{\cdot}m^{-1}$, $m = 9.3 \times 10^{-31}$ kg, $v = 2.0837{\times}10^8\ m{\cdot}s^{-1}$, $c = 3.0 \times 10^8\ m{\cdot}s^{-1}$, E = 200 KeV, $I_e = 8.8Z$ = 235.8 eV, the estimated results of the maximum temperature increase of the HEA sample is 0.061 K. Therefore, the heating effect induced by electron irradiation could be neglected.

References

1. Yeh, J.W.; Chen, S.K.; Lin, S.J.; Gan, J.Y.; Chin, T.S.; Shun, T.T.; Tsau, C.H.; Chang, S.Y. Nanostructured high-entropy alloys with multiple principal elements: Novel alloy design concepts and outcomes. *Adv. Eng. Mater.* **2004**, *6*, 299–303. [CrossRef]
2. Zhang, Y.; Zuo, T.T.; Tang, Z.; Gao, M.C.; Dahmen, K.A.; Liaw, P.K.; Lu, Z.P. Microstructures and properties of high-entropy alloys. *Prog. Mater. Sci.* **2014**, *61*, 1–93. [CrossRef]
3. Gludovatz, B.; Hohenwarter, A.; Catoor, D.; Chang, E.H.; George, E.P.; Ritchie, R.O. A fracture-resistant high-entropy alloy for cryogenic applications. *Science* **2014**, *345*, 1153–1158. [CrossRef] [PubMed]
4. He, J.Y.; Wang, H.; Wu, Y.; Liu, X.J.; Nieh, T.G.; Lu, Z.P. High-temperature plastic flow of a precipitation-hardened FeCoNiCr high entropy alloy. *Mater. Sci. Eng. A* **2017**, *686*, 34–40. [CrossRef]
5. Liu, J.B.; Chen, C.X.; Xu, Y.Q.; Wu, S.W.; Wang, G.; Wang, H.T.; Fang, Y.T.; Meng, L. Deformation twinning behaviors of the low stacking fault energy high-entropy alloy: An in-situ TEM study. *Scr. Mater.* **2017**, *137*, 9–12. [CrossRef]
6. Wang, P.; Xu, S.; Liu, J.; Li, X.; Wei, Y.; Wang, H.; Gao, H.; Yang, W. Atomistic simulation for deforming complex alloys with application toward TWIP steel and associated physical insights. *J. Mech. Phys. Solids* **2017**, *98*, 290–308. [CrossRef]
7. Wang, P.; Wu, Y.; Liu, J.; Wang, H. Impacts of atomic scale lattice distortion on dislocation activity in high-entropy alloys. *Extrem. Mech. Lett.* **2017**, *17*, 38–42. [CrossRef]
8. Shan, Z.W.; Mishra, R.K.; Asif, S.A.S.; Warren, O.L.; Minor, A.M. Mechanical annealing and source-limited deformation in submicrometre-diameter Ni crystals. *Nat. Mater.* **2008**, *7*, 115–119. [CrossRef] [PubMed]
9. Kiener, D.; Minor, A.M. Source truncation and exhaustion: Insights from quantitative in situ tem tensile testing. *Nano Lett.* **2011**, *11*, 3816–3820. [CrossRef] [PubMed]
10. Uchic, M.D.; Dimiduk, D.M.; Florando, J.N.; Nix, W.D. Sample dimensions influence strength and crystal plasticity. *Science* **2004**, *305*, 986–989. [CrossRef] [PubMed]
11. Zhu, T.; Li, J. Ultra-strength materials. *Prog. Mater. Sci.* **2010**, *55*, 710–757. [CrossRef]

12. Wang, J.W.; Zeng, Z.; Weinberger, C.R.; Zhang, Z.; Zhu, T.; Mao, S.X. In situ atomic-scale observation of twinning-dominated deformation in nanoscale body-centred cubic tungsten. *Nat. Mater.* **2015**, *14*, 594–600. [CrossRef] [PubMed]
13. Wang, J.W.; Narayanan, S.; Huang, J.Y.; Zhang, Z.; Zhu, T.; Mao, S.X. Atomic-scale dynamic process of deformation-induced stacking fault tetrahedra in gold nanocrystals. *Nat. Commun.* **2013**, *4*, 2340. [CrossRef] [PubMed]
14. Zheng, H.; Cao, A.J.; Weinberger, C.R.; Huang, J.Y.; Du, K.; Wang, J.B.; Ma, Y.Y.; Xia, Y.N.; Mao, S.X. Discrete plasticity in sub-10-nm-sized gold crystals. *Nat. Commun.* **2010**, *1*, 144. [CrossRef] [PubMed]
15. Sun, J.; He, L.B.; Lo, Y.C.; Xu, T.; Bi, H.C.; Sun, L.T.; Zhang, Z.; Mao, S.X.; Li, J. Liquid-like pseudoelasticity of sub-10-nm crystalline silver particles. *Nat. Mater.* **2014**, *13*, 1007–1012. [CrossRef] [PubMed]
16. Lu, Y.; Song, J.; Huang, J.Y.; Lou, J. Surface dislocation nucleation mediated deformation and ultrahigh strength in sub-10-nm gold nanowires. *Nano Res.* **2011**, *4*, 1261–1267. [CrossRef]
17. Wang, J.W.; Sansoz, F.; Huang, J.Y.; Liu, Y.; Sun, S.H.; Zhang, Z.; Mao, S.X. Near-ideal theoretical strength in gold nanowires containing angstrom scale twins. *Nat. Commun.* **2013**, *4*, 1742. [CrossRef] [PubMed]
18. Guo, S.; Ng, C.; Wang, Z.; Liu, C.T. Solid solutioning in equiatomic alloys: Limit set by topological instability. *J. Alloys Compd.* **2014**, *583*, 410–413. [CrossRef]
19. Gutkin, M.Y.; Ishizaki, T.; Kuramoto, S.; Ovid'ko, I.A. Nanodisturbances in deformed Gum Metal. *Acta Mater.* **2006**, *54*, 2489–2499. [CrossRef]
20. Wu, Z.; Bei, H.; Pharr, G.M.; George, E.P. Temperature dependence of the mechanical properties of equiatomic solid solution alloys with face-centered cubic crystal structures. *Acta Mater.* **2014**, *81*, 428–441. [CrossRef]
21. Wang, Z.; Baker, I.; Cai, Z.; Chen, S.; Poplawsky, J.D.; Guo, W. The effect of interstitial carbon on the mechanical properties and dislocation substructure evolution in $Fe_{40.4}Ni_{11.3}Mn_{34.8}Al_{7.5}Cr_6$ high entropy alloys. *Acta Mater.* **2016**, *120*, 228–239. [CrossRef]
22. Liu, S.Y.; Wei, Y.J. The Gaussian distribution of lattice size and atomic level heterogeneity in high entropy alloys. *Extrem. Mech. Lett.* **2017**, *11*, 84–88. [CrossRef]
23. Nagase, T.; Rack, P.D.; Noh, J.H.; Egami, T. In-situ TEM observation of structural changes in nano-crystalline CoCrCuFeNi multicomponent high-entropy alloy (HEA) under fast electron irradiation by high voltage electron microscopy (HVEM). *Intermetallics* **2015**, *59*, 32–42. [CrossRef]
24. Cui, J.P.; Hao, Y.L.; Li, S.J.; Sui, M.L.; Li, D.X.; Yang, R. Reversible movement of homogenously nucleated dislocations in a β-titanium alloy. *Phys. Rev. Lett.* **2009**, *102*, 045503. [CrossRef] [PubMed]
25. Bobylev, S.V.; Ovid'ko, I.A. Nanodisturbances in deformed nanowires. *Phys. Rev. Lett.* **2009**, *103*, 135501. [CrossRef] [PubMed]
26. Bobylev, S.V.; Ovid'ko, I.A. Nanodisturbances and nanoscale deformation twins in fcc nanowires. *Phys. Rev. B* **2011**, *83*, 054111. [CrossRef]
27. Zhang, L.J.; Fan, J.T.; Liu, D.J.; Zhang, M.D.; Yu, P.F.; Jing, Q.; Ma, M.Z.; Liaw, P.K.; Li, G.; Liu, R.P. The microstructural evolution and hardness of the equiatomic CoCrCuFeNi high-entropy alloy in the semi-solid state. *J. Alloys Compd.* **2018**, *745*, 75–83. [CrossRef]
28. Nie, A.; Wang, H. Deformation-mediated phase transformation in gold nano-junction. *Mater. Lett.* **2011**, *65*, 3380–3383. [CrossRef]
29. Fisher, S.B. On the temperature rise in electron irradiated foils. *Radiat. Eff.* **1970**, *5*, 239–243. [CrossRef]
30. Jenčič, I.; Bench, M.W.; Robertson, I.M.; Kirk, M.A. Electron-beam-induced crystallization of isolated amorphous regions in Si, Ge, GaP, and GaAs. *J. Appl. Phys.* **1995**, *78*, 974–982. [CrossRef]

entropy

MDPI

Article

Small-Scale Plastic Deformation of Nanocrystalline High Entropy Alloy

Sanghita Mridha [1], Mageshwari Komarasamy [1], Sanjit Bhowmick [2], Rajiv S. Mishra [1] and Sundeep Mukherjee [1,*]

[1] Department of Materials Science and Engineering, University of North Texas, Denton, TX 76203, USA; sanghita.mridha@gmail.com (S.M.); Mageshwari.Komarasamy@unt.edu (M.K.); Rajiv.Mishra@unt.edu (R.S.M.)

[2] Bruker Nano Surfaces, Minneapolis, MN 55344, USA; Sanjit.Bhowmick@bruker.com

* Correspondence: sundeep.mukherjee@unt.edu; Tel.: +1-940-565-4170

Received: 22 October 2018; Accepted: 16 November 2018; Published: 20 November 2018

Abstract: High entropy alloys (HEAs) have attracted widespread interest due to their unique properties at many different length-scales. Here, we report the fabrication of nanocrystalline (NC) $Al_{0.1}CoCrFeNi$ high entropy alloy and subsequent small-scale plastic deformation behavior via nano-pillar compression tests. Exceptional strength was realized for the NC HEA compared to pure Ni of similar grain sizes. Grain boundary mediated deformation mechanisms led to high strain rate sensitivity of flow stress in the nanocrystalline HEA.

Keywords: nanocrystalline materials; high entropy alloy; sputtering; deformation and fracture; strain rate sensitivity

1. Introduction

High entropy alloys (HEAs) represent an alloy design paradigm of combining five or more elements in equiatomic or near-equiatomic proportions [1,2]. In certain compositions, high configurational entropy suppresses intermetallic compound formation and leads to single-phase solid solution [3]. HEAs have attracted widespread interest due to their intriguing physical and mechanical properties [3]. Some of the appealing properties include exceptional ductility [3], outstanding thermal stability [4], and cryogenic fracture toughness [5]. Bulk of the research on HEAs have focused on alloy development [3], phase stability [6], and mechanical behavior of coarse grained (CG) and fine-grained systems [3,7,8]. But there are limited reports on nanocrystalline (NC) HEAs and their small-scale deformation behavior [9,10]. NC metals typically show very high strength [11] and good fatigue resistance [12]. Body-centered cubic NbMoTaW refractory NC HEA exhibited exceptional strength at small scales and ductility [9]. Furthermore, NC HEA retained yield strength (YS) of over 5 GPa up to 600 °C denoting an exceptional nano-structural stability [10]. Similar studies for face-centered cubic (FCC) systems could provide insights into their deformation mechanisms at reduced length-scale and pave the way for new application domains towards low-cost, durable, ductile, and strong FCC HEAs.

$Al_{0.1}CoCrFeNi$ HEA is a canonical example of a FCC single phase multi-principal element alloy whose mechanical properties have been widely reported. Komarasamy et al. [13] examined the work hardening mechanisms in coarse-grained (CG) (a few mm) and fine-grained (FG) (~3–14 μm) $Al_{0.1}CoCrFeNi$ HEA and concluded that both the conditions exhibited deformation twinning mediated plasticity. Following that, Choudhuri et al. [14] investigated the plastic deformation mechanisms of the same alloy in CG and FG conditions using transmission electron microscopy. After quasi-static tensile testing, both the microstructures exhibited nanoscale (~2 nm) twins signifying twinning assisted plastic deformation. Wu et al. [15] also noted a similar behavior with deformation twins in both CG and FG materials. Furthermore, an increase in average twin spacing and a reduction in twin thickness was observed in FG condition as compared with CG material. Kumar et al. [16] examined the high strain rate compression behavior of CG $Al_{0.1}$CoCrFeNi HEA and observed a similar work hardening behavior as quasi-static compression. Yu et al. [17] investigated the deformation mechanism in $Al_{0.1}$CoCrFeNi HEA subjected to high pressure torsion (HPT). Deformation via dislocation slip was noted at low strains while deformation twinning was activated at large plastic strains. Furthermore, a large strain rate sensitivity of 0.035 was obtained signaling grain boundary related deformation mechanism in the HPT condition. In $Al_{0.1}$CoCrFeNi HEA, Feng et al. [18] introduced high density of nano-twins and stacking faults, and investigated the mechanical behavior at small length scales. The pillars exhibited a compressive strength of 4.0 GPa with 15% compressive ductility. The exceptional mechanical properties were attributed to the stability of stacking faults and its effective hindrance to dislocation motion.

In this paper, we report on the nano-mechanical behavior of NC $Al_{0.1}$CoCrFeNi HEA (average grain size ~40 nm) synthesized using magnetron sputtering technique. Stress–strain response was obtained by nano-pillar compression, concurrent with direct observation of their deformation behavior inside a scanning electron microscope (SEM). Different strain rates were used for strain rate sensitivity (m) analysis.

2. Materials and Methods

Target alloy of composition $Al_{0.1}$CoCrFeNi was prepared using high purity elements (99.99%) and arc melted in an Ar atmosphere. A thin film of the alloy was deposited by magnetron sputtering technique (AJA International, Scituate, MA, USA) on a silicon substrate at room temperature (RT), with the base and process pressure maintained at ~3×10^{-6} torr, and ~5×10^{-3} torr, respectively. An Ar atmosphere (flow rate 10 sccm) was used and applied power was 75 W. Nano-pillars were synthesized by milling the thin film using Focused Ion Beam (FIB) (FEI). A concentric circular pattern was used to mill out the nano-pillar. Gallium ion beam with current of 10 pA, and operating voltage of 30 kV was used for milling and the pillars had an average diameter of ~450 nm. ImageJ software was used to determine the grain size. In situ compression tests were performed on the nano-pillars with a SEM equipped with PicoIndenter PI 85 (Bruker Nano Surfaces, Minneapolis, MN, USA) using a 2 μm diameter diamond flat punch. Uniaxial compression tests were conducted at strain rates of 1.2×10^{-1}, 1.9×10^{-2}, and 7.5×10^{-3} s^{-1}.

3. Results and Discussion

Figure 1a,b shows the X-ray diffraction (XRD) patterns for $Al_{0.1}$CoCrFeNi HEA thin film and target, respectively, demonstrating a single phase FCC structure. SEM image of the thin film (Figure 1c) shows uniformly distributed nano-sized grains with an average grain size of ~40 ± 5 nm. SEM images of NC nano-pillar HEA before and after compression tests are shown in Figure 2. The pillar diameter was around 450 nm (Figure 2a). Figure 2b shows the nano-pillar with the diamond flat punch just before the compression test. Uniform plastic deformation of the nano-pillar with no evidence of buckling was observed (Figure 2c). Furthermore, crack propagation parallel to the loading direction can be observed in Figure 2c.

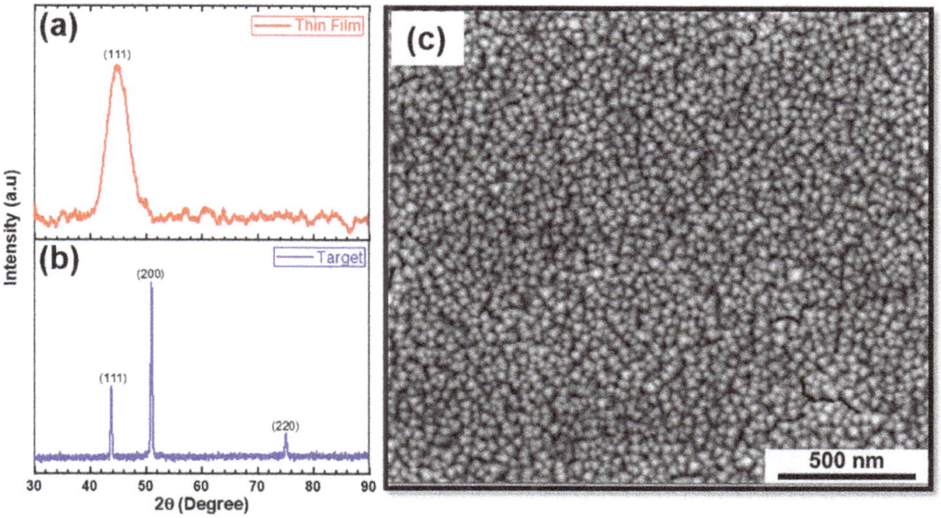

Figure 1. X-Ray diffraction (XRD) patterns of (**a**) $Al_{0.1}$CoCrFeNi thin film and (**b**) $Al_{0.1}$CoCrFeNi target, showing peaks corresponding to the face-centered cubic (FCC) phase; (**c**) high-magnification scanning electron microscope (SEM) image of the thin film showing nano-sized grains with an average grain size of ~40 ± 5 nm.

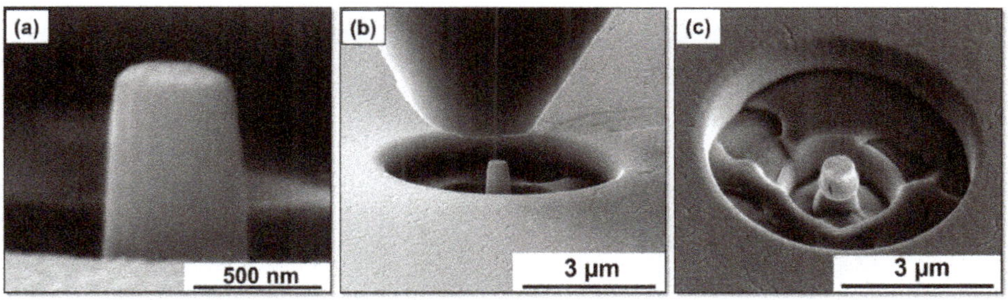

Figure 2. SEM images of the (**a**) nano-pillar with diameter 450 nm (**b**) diamond punch and the nano-pillar, and (**c**) nano-pillar after the compression test.

Compressive engineering stress-strain curve at a strain rate of 7.5×10^{-3} s^{-1} is shown in Figure 3a. Inset images of the pillar at various deformation intervals clearly show that the plastic deformation was uniform without buckling of the nano-pillar. YS of nano-pillar $Al_{0.1}$CoCrFeNi HEA at 7.5×10^{-3} s^{-1} strain rate was ~3829 MPa. YS of the CG material of the same composition was ~190 MPa indicating a 20 fold increase in strength for the NC HEA. $Al_{0.1}$CoCrFeNi forms a single phase alloy with FCC crystal structure without any secondary phases. Therefore, only grain size strengthening contribution was investigated. To that end, YS versus $d^{-1/2}$ correlation (where d represents grain size) for various HEAs, pure Ni, and the current study are shown in Figure 3b. Hall–Petch relation for various HEAs and Ni are shown in Figure 3b, based on the following equation:

$$\sigma_{YS} = \sigma_o + kd^{-1/2} \tag{1}$$

where, σ_o is the strength of the material for infinitely large grain size also called lattice friction stress, k is the Hall-Petch coefficient, and d is the grain size. The Hall–Petch equation with coefficients, σ_o and k, for all the conditions are given in the bottom inset of Figure 3b. Hall–Petch coefficients based on two independent HEA investigations were used to calculate the expected grain size strengthening contribution for the current condition with ~50 nm grain size. The calculated YS based on Otto et al.'s [19] and Nilesh et al.'s [8] Hall–Petch coefficients were 2255 and 1849 MPa, respectively. Remarkably, the obtained YS in the current investigation was ~1500 MPa higher than the predicted strength values. This may be attributed to the shift in deformation mode from dislocation-controlled to grain boundary mediated plastic deformation. For the same grain size, HEAs exhibited strength values two fold higher than that of pure Ni [20]. In addition to Hall–Petch and grain boundary mediated deformation mechanism, size effect may also dominate the YS of NC $Al_{0.1}CoCrFeNi$ HEA, which is out of the scope of the current paper. In CoCrCuFeNi HEA, Zhang et al. [21] investigated size-dependent YS based on micro-/nano-pillar uniaxial compression tests and noted that the HEA did exhibit size-dependent mechanical properties.

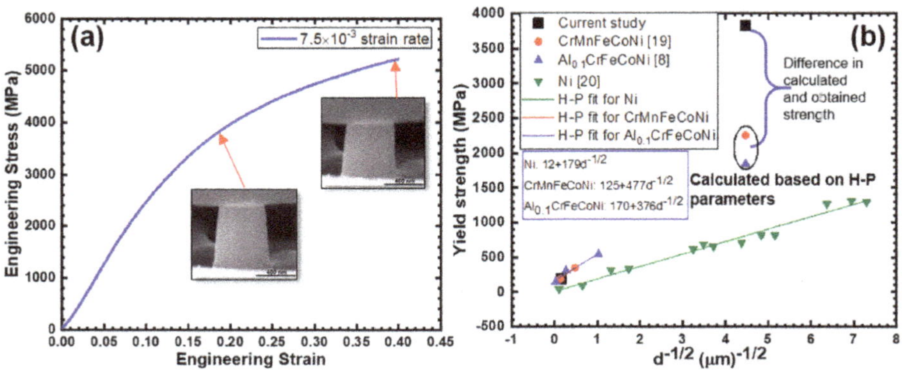

Figure 3. (**a**) Compressive engineering stress–strain curve at strain rate of 7.5×10^{-3} s^{-1} showing YS of 3829 MPa. Insets show the *in situ* image of pillar at various stages; (**b**) YS vs. $d^{-1/2}$ plot with Hall–Petch equation fit for Ni and various HEAs.

Effect of strain rate on yield strength and strain rate sensitivity calculation are shown in Figure 4a,b, respectively. Stress-strain curves for 1.9×10^{-2} s^{-1} and 1.2×10^{-1} s^{-1} strain rates were shifted along the strain axis for a clear representation. With the increase in strain rate, yield strength of the NC nano-pillar $Al_{0.1}CoCrFeNi$ HEA increased. This indicates a positive strain rate sensitivity of flow stress at room temperature which was observed in other investigations as well [22]. Strain rate sensitivity (m) is defined as:

$$m = \frac{\partial \ln \sigma}{\partial \ln \varepsilon} \tag{2}$$

where, σ is the flow stress and ε is the strain rate. Following this equation, slope of 1% flow stress and strain rate in logarithmic scale yielded a strain rate sensitivity of 0.08 as presented in Figure 4b. Comparison with m values reported in literature for other materials is given as inset in Figure 4b. As can be clearly seen, m value increases with the reduction in grain size. For example, CG and NC copper exhibited m values of 0.009 and 0.06, respectively [11]. Furthermore, m value of CG HEA was higher than CG conventional metals/alloys, which was attributed to fluctuating lattice energy controlled deformation mechanism. In the current investigation, due to the nano-scale grain size, there is high possibility that dislocation controlled processes were suppressed and grain-boundary mediated processes were activated. Furthermore, m value of 0.08 for NC HEA would translate into lower apparent activation volume of dislocation as compared to NC copper with m value of 0.06. Therefore, in addition to differences in CG material due to lattice distortion controlled dislocation activity, current results suggest that grain boundary mediated plastic deformation was influenced by the inherent lattice distortion of HEA.

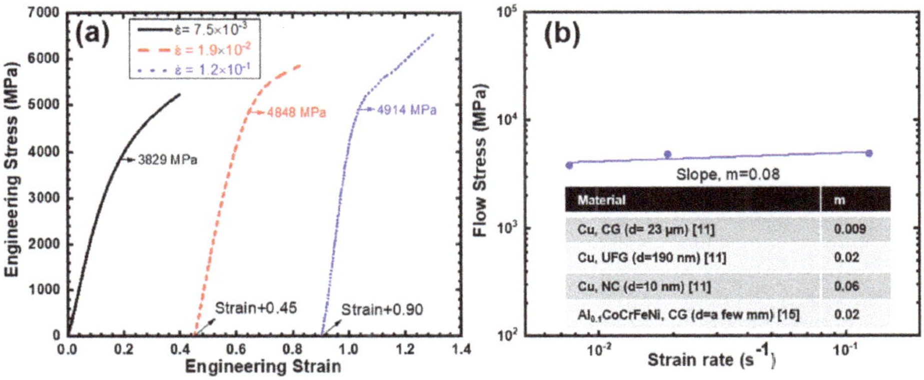

Figure 4. (**a**) Compressive engineering stress-strain plot at different strain rates showing the increase in tensile strength with increase in strain rate; (**b**) flow stress at 1% offset strain versus strain rate plot in logarithmic scale to calculate strain rate sensitivity (*m*).

An important implication of the current investigation is the viability of nano-crystalline HEAs for use in high-strength applications without expensive refractory elements such as Mo, Ta, W, and Nb. In fact, body centered cubic NbMoTaW HEA pillar of 1 μm diameter exhibited yield strength similar to the current $Al_{0.1}CoCrFeNi$ face centered cubic nano-pillar [9]. Therefore, the current nanostructured alloy is a cost-effective alternative towards achieving ultra-strong and ductile wires for small-scale applications.

4. Conclusions

In conclusion, exceptional strength was seen for nano-crystalline $Al_{0.1}CoCrFeNi$ HEA similar to refractory HEAs of comparable length scales. This was attributed to grain boundary mediated plastic deformation processes. The strain rate sensitivity was higher than conventional NC material implying an even lower activation volume of dislocations.

Author Contributions: Conceptualization, S.M.; Formal analysis, S.B. and R.S.M.; Investigation, S.M. and M.K.; Supervision, S.M.; Writing—original draft, S.M., M.K. and S.M.

Funding: This research received no external funding.

Conflicts of Interest: The authors declare no conflict of interest.

References

1. Yeh, J.W.; Chen, S.K.; Lin, S.J.; Gan, J.Y.; Chin, T.S.; Shun, T.T.; Tsau, C.H.; Chang, S.Y. Nanostructured high-entropy alloys with multiple principal elements: Novel alloy design concepts and outcomes. *Adv. Eng. Mater.* **2004**, *6*, 299–303. [CrossRef]
2. Cantor, B.; Chang, I.T.H.; Knight, P.; Vincent, A.J.B. Microstructural development in equiatomic multicomponent alloys. *Mater. Sci. Eng. A.* **2004**, *375–377*, 213–218. [CrossRef]
3. Miracle, D.B.; Senkov, O.N. A critical review of high entropy alloys and related concepts. *Acta Mater.* **2017**, *122*, 448–511. [CrossRef]
4. Dolique, V.; Thomann, A.L.; Brault, P.; Tessier, Y.; Gillon, P. Thermal stability of AlCoCrCuFeNi high entropy alloy thin films studied by in-situ XRD analysis. *Surf. Coat. Technol.* **2010**, *204*, 1989–1992. [CrossRef]
5. Gludovatz, B.; Hohenwarter, A.; Catoor, D.; Chang, E.H.; George, E.P.; Ritchie, R.O. A fracture-resistant high-entropy alloy for cryogenic applications. *Science* **2014**, *345*, 1153–1158. [CrossRef] [PubMed]
6. Otto, F.; Yang, Y.; Bei, H.; George, E.P. Relative effects of enthalpy and entropy on the phase stability of equiatomic high-entropy alloys. *Acta Mater.* **2013**, *61*, 2628–2638. [CrossRef]
7. Zhang, Y.; Zuo, T.T.; Tang, Z.; Gao, M.C.; Dahmen, K.A.; Liaw, P.K.; Lu, Z.P. Microstructures and properties of high-entropy alloys. *Prog. Mater. Sci.* **2014**, *61*, 1–93. [CrossRef]

8. Kumar, N.; Komarasamy, M.; Nelaturu, P.; Tang, Z.; Liaw, P.K.; Mishra, R.S. Friction stir processing of a high entropy alloy $Al_{0.1}$CoCrFeNi. *JOM* **2015**, *67*, 1007–1013. [CrossRef]
9. Zou, Y.; Ma, H.; Spolenak, R. Ultrastrong ductile and stable high-entropy alloys at small scales. *Nat. Commun.* **2015**, *6*, 1–8. [CrossRef] [PubMed]
10. Zou, Y.; Wheeler, J.M.; Ma, H.; Okle, P.; Spolenak, R. Nanocrystalline High-Entropy Alloys: A New Paradigm in High-Temperature Strength and Stability. *Nano Lett.* **2017**, *17*, 1569–1574. [CrossRef] [PubMed]
11. Chen, J.; Lu, L.; Lu, K. Hardness and strain rate sensitivity of nanocrystalline Cu. *Scr. Mater.* **2006**, *54*, 1913–1918. [CrossRef]
12. Padilla, H.A.; Boyce, B.L. A Review of fatigue behavior in nanocrystalline metals. *Exp. Mech.* **2010**, *50*, 5–23. [CrossRef]
13. Komarasamy, M.; Kumar, N.; Tang, Z.; Mishra, R.S.; Liaw, P.K. Effect of microstructure on the deformation mechanism of friction stir-processed $Al_{0.1}$CoCrFeNi high entropy alloy. *Mater. Res. Lett.* **2015**, *3*, 30–34. [CrossRef]
14. Choudhuri, D.; Komarasamy, M.; Ageh, V.; Mishra, R.S. Investigation of plastic deformation modes in $Al_{0.1}$CoCrFeNi high entropy alloy. *Mater. Chem. Phys.* **2018**, *217*, 308–314. [CrossRef]
15. Wu, S.W.; Wang, G.; Yi, J.; Jia, Y.D.; Hussain, I.; Zhai, Q.J.; Liaw, P.K. Strong grain-size effect on deformation twinning of an $Al_{0.1}$CoCrFeNi high-entropy alloy. *Mater. Res. Lett.* **2017**, *5*, 276–283. [CrossRef]
16. Kumar, N.; Ying, Q.; Nie, X.; Mishra, R.S.; Tang, Z.; Liaw, P.K.; Brennan, R.E.; Doherty, K.J.; Cho, K.C. High strain-rate compressive deformation behavior of the $Al_{0.1}$CrFeCoNi high entropy alloy. *Mater. Des.* **2015**, *86*, 598–602. [CrossRef]
17. Yu, P.F.; Cheng, H.; Zhang, L.J.; Zhang, H.; Jing, Q.; Ma, M.Z.; Liaw, P.K.; Li, G.; Liu, R.P. Effects of high pressure torsionon microstructures and properties of an $Al_{0.1}$CoCrFeNi high-entropy alloy. *Mater. Sci. Eng.* **2016**, *655*, 283–291. [CrossRef]
18. Feng, X.; Zhang, J.; Wu, K.; Liang, X.; Liu, G.; Sun, J. Ultrastrong Al0.1CoCrFeNi high-entropy alloys at small scales: effects of stacking faults vs. nanotwins. *Nanoscale* **2018**, *10*, 13329–13334. [CrossRef] [PubMed]
19. Otto, F.; Dlouhý, A.; Somsen, C.; Bei, H.; Eggeler, G.; George, E.P. The influences of temperature and microstructure on the tensile properties of a CoCrFeMnNi high-entropy alloy. *Acta Mater.* **2013**, *61*, 5743–5755. [CrossRef]
20. Xiao, C.; Mirshams, R.A.; Whang, S.H.; Yin, W.M. Tensile behavior and fracture in nickel and carbon doped nanocrystalline nickel. *Mater. Sci. Eng. A.* **2001**, *301*, 35–43. [CrossRef]
21. Zhang, H.; Siu, K.W.; Liao, W.; Wang, Q.; Yang, Y.; Lu, Y. In situ mechanical characterization of CoCrCuFeNi high-entropy alloy micro/nano-pillars for their size-dependent mechanical behavior. *Mater. Res. Express* **2016**, *3*, 94002. [CrossRef]
22. Komarasamy, M.; Kumar, N.; Mishra, R.S.; Liaw, P.K. Anomalies in the deformation mechanism and kinetics of coarse-grained high entropy alloy. *Mater. Sci. Eng. A* **2016**, *654*, 256–263. [CrossRef]

entropy

MDPI

Article

Impact of Chemical Fluctuations on Stacking Fault Energies of CrCoNi and CrMnFeCoNi High Entropy Alloys from First Principles

Yuji Ikeda [1,2,*], Fritz Körmann [1,3], Isao Tanaka [2,4,5,6] and Jörg Neugebauer [1]

1 Computational Materials Design, Max-Planck-Institut für Eisenforschung GmbH, 40237 Düsseldorf, Germany; koermann@mpie.de (F.K.); neugebauer@mpie.de (J.N.)
2 Materials Science and Engineering, Kyoto University, Kyoto 606-8501, Japan; tanaka@cms.mtl.kyoto-u.ac.jp
3 Materials Science and Engineering, Delft University of Technology, 2628 CD Delft, The Netherlands
4 Center for Elements Strategy Initiative for Structure Materials (ESISM), Kyoto University, Kyoto 606-8501, Japan
5 Center for Materials Research by Information Integration, National Institute for Materials Science (NIMS), Tsukuba 305-0047, Japan
6 Nanostructures Research Laboratory, Japan Fine Ceramics Center, Nagoya 456-8587, Japan
* Correspondence: y.ikeda@mpie.de

Received: 13 August 2018; Accepted: 29 August 2018; Published: 30 August 2018

Abstract: Medium and high entropy alloys (MEAs and HEAs) based on 3d transition metals, such as face-centered cubic (fcc) CrCoNi and CrMnFeCoNi alloys, reveal remarkable mechanical properties. The stacking fault energy (SFE) is one of the key ingredients that controls the underlying deformation mechanism and hence the mechanical performance of materials. Previous experiments and simulations have therefore been devoted to determining the SFEs of various MEAs and HEAs. The impact of local chemical environment in the vicinity of the stacking faults is, however, still not fully understood. In this work, we investigate the impact of the compositional fluctuations in the vicinity of stacking faults for two prototype fcc MEAs and HEAs, namely CrCoNi and CrMnFeCoNi by employing first-principles calculations. Depending on the chemical composition close to the stacking fault, the intrinsic SFEs vary in the range of more than 150 mJ/m^2 for both the alloys, which indicates the presence of a strong driving force to promote particular types of chemical segregations towards the intrinsic stacking faults in MEAs and HEAs. Furthermore, the dependence of the intrinsic SFEs on local chemical fluctuations reveals a highly non-linear behavior, resulting in a non-trivial interplay of local chemical fluctuations and SFEs. This sheds new light on the importance of controlling chemical fluctuations via tuning, e.g., the annealing condition to obtain the desired mechanical properties for MEAs and HEAs.

Keywords: high-entropy alloy; stacking-fault energy; density functional theory

1. Introduction

High entropy alloys (HEAs) or complex concentrated alloys (CCAs) based on 3d transition metals have attracted enormous attention recently, in particular due to their outstanding mechanical properties. The equiatomic CrMnFeCoNi HEA, also often termed the Cantor alloy [1], has an excellent combination of strength and ductility [2–5]. Different strategies have been proposed to further improve the mechanical properties, e.g., by tuning the chemical compositions of CrMnFeCo and CrMnFeCoNi towards nonequiatomic alloys [6–10] or by resorting to so-called medium entropy alloys (MEAs), such as CrCoNi alloys [11–13].

A key factor in controlling the underlying deformation mechanism and therewith tuning the mechanical properties is the stacking fault energy (SFE). Low SFEs can induce, e.g., transformation-induced plasticity (TRIP) or twinning-induced plasticity (TWIP) [14–16], and for this reason, SFEs of HEAs and CCAs have been investigated previously in numerous experimental [13,17–19] as well as theoretical studies [20–31]. Interestingly, in a recent experimental work [17], the measured SFEs for equiatomic CrMnFeCoNi revealed large fluctuations, and it was proposed that the SFEs of CrMnFeCoNi may sensitively depend on the local chemical environments in the vicinity of the stacking faults (SFs). Also, in computational works [24,26,29,30], large fluctuations of SFEs have been found for MEAs and HEAs based on 3d transition metals, like CrCoNi and CrMnFeCoNi, which indicates a strong dependence of the SFEs on local chemical fluctuations close to the SFs. A recent experimental work [32] revealed that chemical inhomogeneity in a $Cr_{10}Mn_{30}Fe_{50}Co_{10}$-based alloy also caused a large deterioration of mechanical properties. Although these results suggest an important role of chemical fluctuations in these alloys, the impact of such fluctuations on the SFEs of MEAs and HEAs has not been intensively investigated yet. In a recent study employing first-principles calculations [24], it was proposed that the SFEs of CrCoNi and CrCoFeNi may depend on the valence electron concentration (VEC) of the elements near the SF. In that study, however, only a very limited number of configurations and local compositions near the SF were evaluated, prohibiting a further quantitative analysis.

In the present study, we comprehensively investigate the impact of compositional fluctuations near the intrinsic SF (ISF) on the intrinsic SFE (ISFE) for the face-centered cubic (fcc) equiatomic CrCoNi and CrMnFeCoNi alloys based on first-principles calculations. The ISFEs were calculated using supercells with and without an ISF, while the chemical disorder was modeled using the coherent potential approximation (CPA) [33–35]. The combination of supercells and the CPA makes it possible to systematically investigate arbitrary local composition ratios in the vicinity of the ISFs and to elucidate the impact of compositional fluctuations for individual elements using relatively small, and thus computationally efficient, supercells.

2. Computational Details

We investigate the impact of the local chemical environment in the vicinity of ISFs on the ISFEs, as described in Figure 1a. In principle, the impact of the local chemical environment can be investigated by employing large supercells. As this approach turns out to be computationally too demanding for systematically screening a large number of different compositional fluctuations near the ISFE, we resort to an alternative approach. To reduce the computational cost while keeping the key physical ingredients, we combined the CPA with the supercell approach. Specifically, the impact of the local chemical environment near the ISFs on the ISFEs was investigated using six-layer supercells with and without an ISF as shown in Figure 1b, and the compositional fluctuation in the vicinity of the ISF was introduced based on the CPA by modifying the mixing ratios of the chemical elements in the L1 layers close to the ISF (see Figure 1b) from the equiatomic while keeping the equiatomic ratios in the subsequent L2 and L3 layers. The ISF was introduced by tilting the $\langle 111 \rangle$ axis of the perfect-fcc simulation cell by $6\langle 11\bar{2} \rangle / a$, where a is the fcc lattice constant [36]. The ISFEs, γ_{ISF}, were computed with

$$\gamma_{\text{ISF}} = \frac{E^{\text{fcc + ISF}} - E^{\text{fcc}}}{A}, \tag{1}$$

where $E^{\text{fcc + ISF}}$ and E^{fcc} are the energies of the simulation cells with and without ISFs, respectively, and A denotes the area of the ISFs. As we focused on the impacts of the chemical fluctuations near the ISFs, the computed ISFEs are shown as the differences from the ones without compositional fluctuations. The impacts of lattice vibrations [26,29], magnetic excitations [21], chemical short-range order (SRO) [30], and volume changes [9] on the ISFEs have been investigated previously and were therefore not included in the present study. We also note that local lattice distortions, which cannot be considered in the CPA, possibly affect the ISFEs.

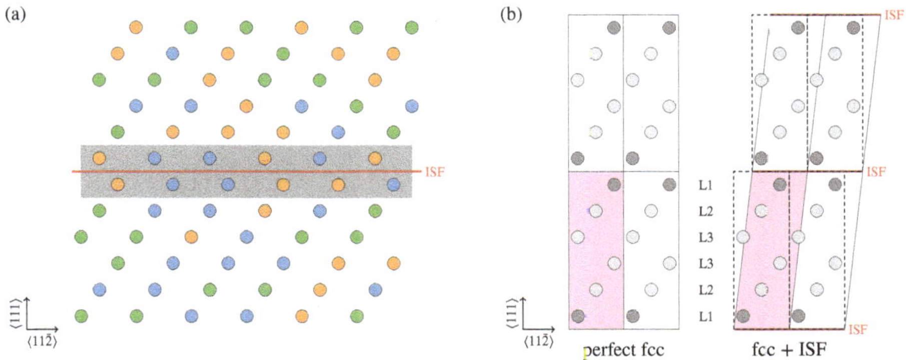

Figure 1. (**a**) Schematic of the intrinsic stacking faults (ISF) in a three-component face-centered cubic (fcc) equiatomic disordered alloy. The red line indicates the ISF. The circles represent atoms, which are colored differently according to the chemical elements. Notice that local chemical fluctuations can exist close to the ISF (gray background region) even if, on average, the alloy has an equiatomic composition; (**b**) Projected atomic positions of the simulation cells with and without ISFs. The red background regions represent the six-layer simulation cells, while the dashed boxes indicate the cell shape of the perfect fcc structure. The red lines indicate the ISFs. The circles represent atoms which are colored differently according to the given mixing ratios of the constitutive chemical elements in the CPA to model the compositional fluctuations. The atomic layers are labeled L1, L2, and L3 according to the distance from the ISFs.

SFEs are also often computed based on the axial Ising model [37], in which the SFE is derived from the energy differences among the fcc, hexagonal close-packed (hcp), and sometimes double hcp phases. By construction, however, the axial Ising model cannot capture the impact of local chemical fluctuations close to the SFs, because the perfect fcc and hcp structures do not include the SFs explicitly. The presently employed supercell approach, in contrast, enabled us to explicitly investigate the impacts of local chemical fluctuations close to the SFs.

The electronic structure calculations were performed with the exact-muffin-tin-orbital (EMTO) method [38–42] in combination with the full-charge-density (FCD) method [43,44] within the density functional theory (DFT) framework. The DFT energies were calculated within the generalized gradient approximation (GGA) of the Perdew–Burke–Ernzerhof (PBE) form [45] in the following perturbative manner [46]. The electronic densities were first calculated using the local-density approximation (LDA), and then the total energies were calculated within the GGA-PBE via the FCD method based on the obtained electronic densities. This approach makes the calculations faster and often more stable while keeping the accuracy of the GGA [46], and it has therefore been employed in numerous studies using the EMTO approach [47–56]. The Brillouin zones were sampled by $22 \times 22 \times 4$ *k*-point meshes per six-atom computational unit cell, where the ⟨111⟩ direction in Figure 1b was set to be the *z*-axis. As experiments [57] and first-principles calculations [52,53] have revealed that both CrCoNi and CrMnFeCoNi are paramagnetic (PM) at room temperature, we simulated the magnetic disorder with random magnetic moments by employing the disordered local moment (DLM) model [42,58,59] in combination with the CPA. The lattice constant was fixed to 3.56 Å for CrCoNi and 3.6 Å for CrMnFeCoNi, which are close to the experimental values [1,60–64], and the atomic positions were fixed to keep the rigid-sphere packing.

3. Results and Discussion

We first discuss the impact of compositional fluctuations in the vicinity of the ISF for CrCoNi with increasing or decreasing individual elemental concentrations. The results are shown in Figure 2a.

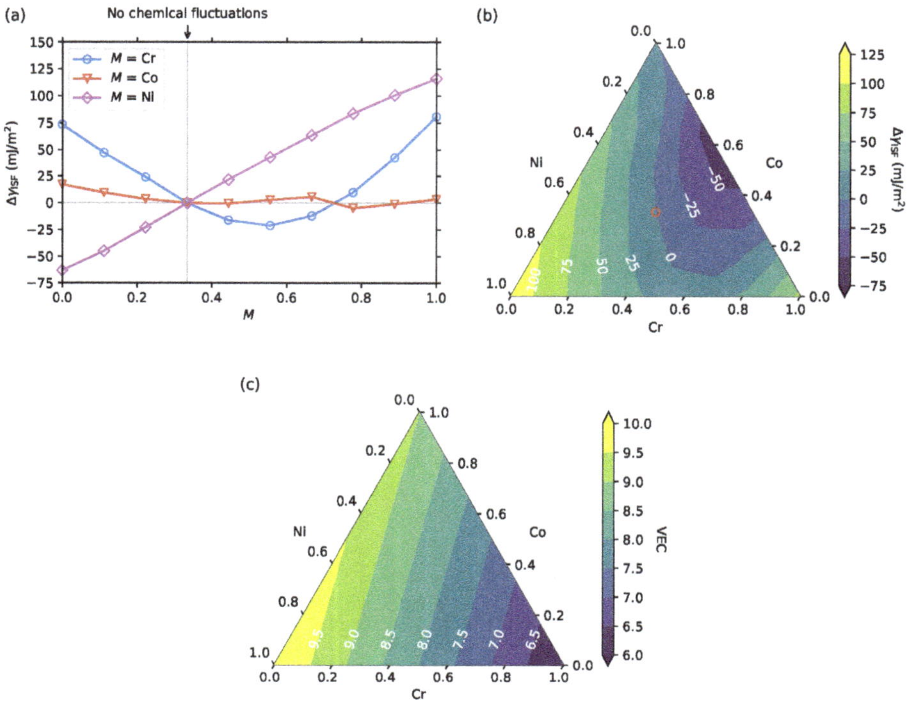

Figure 2. (**a**,**b**) Differences between the computed ISFEs of CrCoNi from those where no compositional fluctuation exists in the L1 layers. (**a**) Result as a function of the local concentration of the element M (M = Cr, Co, Ni) in the L1 layers. The other elements were kept equiatomic in the L1 layers. The vertical gray line indicates the point of the ideal solid solution, i.e., without chemical fluctuations near the SF; (**b**) Result as a function of the local composition ratio in the L1 layers. The red circle indicates the equiatomic composition ratio; (**c**) The valence electron concentration (VEC) as a function of the composition ratio.

The ISFE was found to increase monotonically with the increase of local Ni concentration close to the ISFs. This suggests that Ni segregation towards the ISF is thermodynamically limited and that there is a strong driving force to deplete the Ni concentration in the vicinity of the ISFs. This can be intuitively understood by the fact that pure Ni energetically prefers the fcc phase (groundstate of Ni), whereas in the vicinity of the ISF, the stacking order of the close-packed planes is similar to that of the hcp structure. If the layers close to the ISF are fully occupied by Ni, the ISFE increases by more than 100 $\mathrm{mJ/m^2}$, and when Ni is fully suppressed from the ISF, the ISFE decreases by more than 50 $\mathrm{mJ/m^2}$. This is consistent with previous computational results based on supercell models [24] in which the ISF with the local composition ratio of $Cr_8Co_{10}Ni_{14}$ was 59 $\mathrm{mJ/m^2}$ higher than the ISF with the local composition ratio of $Cr_{12}Co_{10}Ni_{10}$.. It should be noted that the results in Ref. [24] were obtained based on non-spin-polarized calculations and another type of supercell approach (slab + vaccum layer), which prohibits a quantitative comparison.

The impact of Cr turned out to be somewhat more complex. The minimum ISFE was found at a Cr concentration of about 0.5, whereas the ISFE increased when the L1 layers were either highly occupied by Cr or mostly free from it. When Cr fully occupied the layers close to the ISF, the ISFE increased by more than 80 $\mathrm{mJ/m^2}$. A similar non-linear behavior of the ISFE was also found for the ferromagnetic (FM) fcc Cr_xCo_{1-x} binary alloys by first-principles calculations [65]. This suggests that such a complex impact of Cr on the ISFE may be also found in other random 3d transition metal alloys.

When Co fully occupied the L1 layers, the ISFE did not change largely from when no chemical fluctuations existed, which suggests that Co in CrCoNi reveals a small fcc–hcp energy difference. This could be reasoned as follows. The groundstate of pure Co is the FM hcp phase, while with increasing temperature, it experiences both a structural transition to the fcc phase at around 700 K [66] and a magnetic transition at around 1400 K [66]. As found in previous works [67,68], magnetic fluctuations in Co contribute to the stability of the fcc phase at elevated temperatures. For example, in Ref. [68], the fcc–hcp energy difference of pure Co was found to be significantly reduced in the DLM state compared to the FM state. From the above considerations on pure Co, it can be intuited that the small ISFE change with respect to the local Co concentration close to the ISF is caused by the paramagnetic state of CrCoNi.

To further elucidate the impact of the non-linear trend of the compositional fluctuations for the ISFEs, we further explored the full compositional region for the L1 layers, as shown in Figure 2b. Overall, the ISFE varied in the range of approximately 180 mJ/m^2, depending on the local composition ratio of the L1 layers. The strongest decrease in the ISFE of more than 60 mJ/m^2 was found for nearly equiatomic Cr and Co occupying the L1 layers.

Large variation in the ISFE for CrCoNi was also found in previous computational works using supercell models with and without ISFs, where the computed ISFEs of CrCoNi were distributed in the range of 59 mJ/m^2 [24], approximately 230 mJ/m^2 [29], or 205 mJ/m^2 [30], or with a standard deviation of approximately 110 mJ/m^2 [26]. Note that the ISFE variation range in Ref. [24] was relatively small compared with that in Refs. [26,29,30]. This might be related to the limited set of considered configurations, or the usage of non-spin-polarized calculations and the suppression of energy fluctuations due to the magnetic degrees of freedom. This is consistent with the finding in Ref. [29], where the fluctuations of the ISFEs in CrMnFeCoNi were much smaller in the nonmagnetic state compared to the spin-polarized one.

As found in Figure 2a,b, the ISFE of CrCoNi varied non-linearly with respect to the local chemical composition close to the ISFs. This became even clearer by considering the VEC dependence on the ISFE, which is often employed to predict the phase stability of HEAs and CCAs [69,70]. The relations between the ISFE and the VEC near the ISFs for CrCoNi and CrFeCoNi have been discussed previously [24]. Figure 2c shows the VEC values for the same composition region as that in Figure 2b. By construction, VEC depends linearly on the chemical concentrations, while the trend does not match that of the actually computed ISFE, except for under Ni-rich conditions. This indicates that the VEC is, in general, not a sufficient quantitative descriptor for estimating the variation of ISFEs.

We next considered the five-component fcc CrMnFeCoNi HEA, the so-called Cantor alloy. The impact of compositional fluctuations in the L1 layers on the computed ISFE are shown in Figure 3a. The ISFE depended on the local concentrations of Cr, Co, and Ni similarly, as found for CrCoNi. The ISFE monotonically increased with an increase in the local Ni concentration in the L1 layers. The ISFE also increased both when Cr nearly fully occupied the L1 layers and when it was nearly fully excluded in the L1 layers, while the ISFE decreased when the local Cr concentration in the L1 layers was around 0.5. Co showed a relatively small impact on the ISFE compared to the other elements also for CrMnFeCoNi with a slight decrease in the ISFE when fully occupying the L1 layers. The impact of Mn on the ISFE is clearly non-linear, as also found for Cr. When the local Mn concentration in the L1 layers was less than 0.4, the ISFE was hardly affected. In contrast, when the local Mn concentration in the L1 layers was larger than 0.4, the ISFE drastically decreased down to approximately 150 mJ/m^2 with Mn fully occupying the L1 layers.

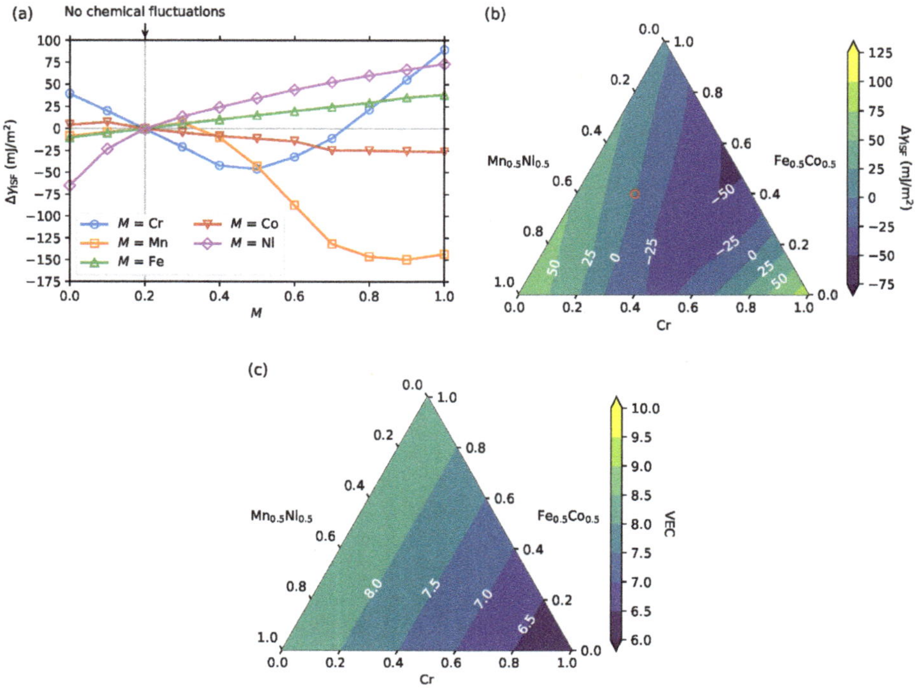

Figure 3. (**a**,**b**) Differences between the computed intrinsic SFE (ISFEs) of CrMnFeCoNi from those where no compositional fluctuation exists in the L1 layers. (**a**) Result as a function of the local concentration of the element M (M = Cr, Mn, Fe, Co, Ni) in the L1 layers. The other elements were kept equiatomic in the L1 layers. The vertical gray line indicates the point of the ideal solid solution, i.e., without chemical fluctuations near the SF; (**b**) Result as a function of the local composition ratio in the L1 layers in the pseudoternary region of Cr, $Fe_{0.5}Co_{0.5}$, and $Mn_{0.5}Ni_{0.5}$. The red circle indicates the equiatomic composition ratio; (**c**) The valence electron concentration (VEC) as a function of the composition ratio in the same pseudoternary region as that of (**b**).

Recent experiments [63,71] have reported the phase decomposition of the CrMnFeCoNi alloy into body-centered cubic Cr, $L1_0$ MnNi, and B2 FeCo after annealing at 450–500 °C. To resolve the possible relationship between the precursor of the phase decomposition and the corresponding local chemical fluctuations near the ISFs, the ISFEs were also computed for the local composition ratios in the L1 layers in the pseudoternary region of Cr, $Fe_{0.5}Co_{0.5}$, and $Mn_{0.5}Ni_{0.5}$. The results are shown in Figure 3b. The ISFE increased with an increasing amount of $Mn_{0.5}Ni_{0.5}$, while the lowest ISFE was found when $Mn_{0.5}Ni_{0.5}$ was fully excluded from the L1 layers and replaced by a mixture of about 0.5 Cr and 0.5 $Fe_{0.5}Co_{0.5}$. This suggests that ISFs are suppressed in the presence of MnNi clusters. The VEC-derived linear dependencies are shown in Figure 3c and reveal once more that the VEC is, in general, not a good predictor of the nonlinear dependence of ISFE on local chemical fluctuations.

4. Conclusions

We investigated the impact of compositional fluctuations on the ISFEs for the fcc equiatomic CrCoNi and CrMnFeCoNi alloys by employing first-principles calculations by combining the supercell and CPA approaches. For both alloys, the ISFEs were found to vary within a range of more than 150 mJ/m^2 depending on the local chemical environment in the vicinity of the ISFs. The chemical dependencies were shown to be highly non-linear and therewith, strongly deviated from linear

Vegard's law-like behavior. Cr caused non-linear behavior in CrCoNi. In the CrMnFeCoNi alloy, the presence of MnNi clusters could suppress ISFs.

The strong SFE dependence on the local chemical compositions in the vicinity of the SFs indicated, on the one hand, that the SFs in HEAs and CCAs may promote particular types of chemical segregations towards the SFs. On the other hand, if chemical fluctuations exist in HEAs and CCAs on a large enough scale, SFs are likely to occur in the local chemical environments with low SFEs. These potential behaviors further complicate the prediction of physical descriptors to determine deformation mechanisms in HEAs and CCAs compared to, e.g., unary metals or ordered alloys. The complexity might be further enhanced in the presence of chemical SRO, which impacts the probabilities of local chemical fluctuations compared to the ideal mixing state. At the same time, the revealed dependence of SFEs on local chemical fluctuations in the vicinity of SFs opens the route towards tuning alloy properties of HEAs and CCAs via controlling chemical fluctuations by tuning, e.g., the annealing conditions in the alloy processing route. Our results encourage further experimental analyses employing, e.g., transmission electron microscopy/energy-dispersive X-ray spectroscopy or atomic probe tomography to explore the role of chemical fluctuations in the vicinity of SFs in more detail.

Author Contributions: Y.I. performed the DFT calculations which were analyzed together with F.K. All authors have equally contributed to the overall discussion and preparation of the manuscript as well as read and approved the final manuscript version.

Funding: Funding from the Deutsche Forschungsgemeinschaft (SPP 2006), from the Ministry of Education, Culture, Sports, Science and Technology (MEXT), Japan, through Elements Strategy Initiative for Structural Materials (ESISM) of Kyoto University, from the Grant-in-Aid for Scientific Research on Innovative Areas "Nano Informatics" (Grant No. 25106005) from the Japan Society for the Promotion of Science (JSPS), and from NWO/STW (VIDI grant 15707) are gratefully acknowledged.

Acknowledgments: Discussions with Blazej Grabowski and Zhiming Li (Max-Planck Institut für Eisenforschung GmbH) are gratefully acknowledged.

Conflicts of Interest: The authors declare no conflict of interest. The founding sponsors had no role in the design of the study; in the collection, analyses, or interpretation of data; in the writing of the manuscript, and in the decision to publish the results.

References

1. Cantor, B.; Chang, I.; Knight, P.; Vincent, A. Microstructural development in equiatomic multicomponent alloys. *Mater. Sci. Eng. A* **2004**, *375*, 213–218. [CrossRef]
2. Gali, A.; George, E. Tensile properties of high- and medium-entropy alloys. *Intermetallics* **2013**, *39*, 74–78. [CrossRef]
3. Otto, F.; Dlouhý, A.; Somsen, C.; Bei, H.; Eggeler, G.; George, E. The influences of temperature and microstructure on the tensile properties of a CoCrFeMnNi high-entropy alloy. *Acta Mater.* **2013**, *61*, 5743–5755. [CrossRef]
4. Gludovatz, B.; Hohenwarter, A.; Catoor, D.; Chang, E.H.; George, E.P.; Ritchie, R.O. A fracture-resistant high-entropy alloy for cryogenic applications. *Science* **2014**, *345*, 1153–1158. [CrossRef] [PubMed]
5. Laplanche, G.; Kostka, A.; Horst, O.; Eggeler, G.; George, E. Microstructure evolution and critical stress for twinning in the CrMnFeCoNi high-entropy alloy. *Acta Mater.* **2016**, *118*, 152–163. [CrossRef]
6. Li, Z.; Pradeep, K.G.; Deng, Y.; Raabe, D.; Tasan, C.C. Metastable high-entropy dual-phase alloys overcome the strength–ductility trade-off. *Nature* **2016**, *534*, 227–230. [CrossRef] [PubMed]
7. Li, Z.; Tasan, C.C.; Springer, H.; Gault, B.; Raabe, D. Interstitial atoms enable joint twinning and transformation induced plasticity in strong and ductile high-entropy alloys. *Sci. Rep.* **2017**, *7*, 40704. [CrossRef] [PubMed]
8. Li, Z.; Tasan, C.C.; Pradeep, K.G.; Raabe, D. A TRIP-assisted dual-phase high-entropy alloy: Grain size and phase fraction effects on deformation behavior. *Acta Mater.* **2017**, *131*, 323–335. [CrossRef]
9. Li, Z.; Körmann, F.; Grabowski, B.; Neugebauer, J.; Raabe, D. *Ab initio* assisted design of quinary dual-phase high-entropy alloys with transformation-induced plasticity. *Acta Mater.* **2017**, *136*, 262–270. [CrossRef]
10. Li, Z.; Raabe, D. Strong and Ductile Non-equiatomic High-Entropy Alloys: Design, Processing, Microstructure, and Mechanical Properties. *JOM* **2017**, *69*, 2099–2106. [CrossRef]

11. Wu, Z.; Bei, H.; Pharr, G.; George, E. Temperature dependence of the mechanical properties of equiatomic solid solution alloys with face-centered cubic crystal structures. *Acta Mater.* **2014**, *81*, 428–441. [CrossRef]
12. Gludovatz, B.; Hohenwarter, A.; Thurston, K.V.S.; Bei, H.; Wu, Z.; George, E.P.; Ritchie, R.O. Exceptional damage-tolerance of a medium-entropy alloy CrCoNi at cryogenic temperatures. *Nat. Commun.* **2016**, *7*, 10602. [CrossRef] [PubMed]
13. Laplanche, G.; Kostka, A.; Reinhart, C.; Hunfeld, J.; Eggeler, G.; George, E. Reasons for the superior mechanical properties of medium-entropy CrCoNi compared to high-entropy CrMnFeCoNi. *Acta Mater.* **2017**, *128*, 292–303. [CrossRef]
14. Grässel, O.; Krüger, L.; Frommeyer, G.; Meyer, L. High strength Fe–Mn–(Al, Si) TRIP/TWIP steels development—properties—application. *Int. J. Plast.* **2000**, *16*, 1391–1409. [CrossRef]
15. Frommeyer, G.; Brüx, U.; Neumann, P. Supra-Ductile and High-Strength Manganese-TRIP/TWIP Steels for High Energy Absorption Purposes. *ISIJ Int.* **2003**, *43*, 438–446. [CrossRef]
16. Pierce, D.; Jiménez, J.; Bentley, J.; Raabe, D.; Oskay, C.; Wittig, J. The influence of manganese content on the stacking fault and austenite/ϵ-martensite interfacial energies in Fe–Mn–(Al–Si) steels investigated by experiment and theory. *Acta Mater.* **2014**, *68*, 238–253. [CrossRef]
17. Smith, T.; Hooshmand, M.; Esser, B.; Otto, F.; McComb, D.; George, E.; Ghazisaeidi, M.; Mills, M. Atomic-scale characterization and modeling of 60° dislocations in a high-entropy alloy. *Acta Mater.* **2016**, *110*, 352–363, [CrossRef]
18. Okamoto, N.L.; Fujimoto, S.; Kambara, Y.; Kawamura, M.; Chen, Z.M.T.; Matsunoshita, H.; Tanaka, K.; Inui, H.; George, E.P. Size effect, critical resolved shear stress, stacking fault energy, and solid solution strengthening in the CrMnFeCoNi high-entropy alloy. *Sci. Rep.* **2016**, *6*, 35863. [CrossRef] [PubMed]
19. Liu, S.; Wu, Y.; Wang, H.; He, J.; Liu, J.; Chen, C.; Liu, X.; Wang, H.; Lu, Z. Stacking fault energy of face-centered-cubic high entropy alloys. *Intermetallics* **2018**, *93*, 269–273. [CrossRef]
20. Zaddach, A.J.; Niu, C.; Koch, C.C.; Irving, D.L. Mechanical Properties and Stacking Fault Energies of NiFeCrCoMn High-Entropy Alloy. *JOM* **2013**, *65*, 1780–1789. [CrossRef]
21. Huang, S.; Li, W.; Lu, S.; Tian, F.; Shen, J.; Holmström, E.; Vitos, L. Temperature dependent stacking fault energy of FeCrCoNiMn high entropy alloy. *Scr. Mater.* **2015**, *108*, 44–47. [CrossRef]
22. Patriarca, L.; Ojha, A.; Sehitoglu, H.; Chumlyakov, Y. Slip nucleation in single crystal FeNiCoCrMn high entropy alloy. *Scr. Mater.* **2016**, *112*, 54–57. [CrossRef]
23. Beyramali Kivy, M.; Asle Zaeem, M. Generalized stacking fault energies, ductilities, and twinnabilities of CoCrFeNi-based face-centered cubic high entropy alloys. *Scr. Mater.* **2017**, *139*, 83–86. [CrossRef]
24. Zhang, Y.; Zhuang, Y.; Hu, A.; Kai, J.; Liu, C. The origin of negative stacking fault energies and nano-twin formation in face-centered cubic high entropy alloys. *Scr. Mater.* **2017**, *130*, 96–99. [CrossRef]
25. Zhang, Z.; Sheng, H.; Wang, Z.; Gludovatz, B.; Zhang, Z.; George, E.P.; Yu, Q.; Mao, S.X.; Ritchie, R.O. Dislocation mechanisms and 3D twin architectures generate exceptional strength-ductility-toughness combination in CrCoNi medium-entropy alloy. *Nat. Commun.* **2017**, *8*, 14390. [CrossRef] [PubMed]
26. Zhao, S.; Stocks, G.M.; Zhang, Y. Stacking fault energies of face-centered cubic concentrated solid solution alloys. *Acta Mater.* **2017**, *134*, 334–345. [CrossRef]
27. Huang, H.; Li, X.; Dong, Z.; Li, W.; Huang, S.; Meng, D.; Lai, X.; Liu, T.; Zhu, S.; Vitos, L. Critical stress for twinning nucleation in CrCoNi-based medium and high entropy alloys. *Acta Mater.* **2018**, *149*, 388–396. [CrossRef]
28. Alkan, S.; Ojha, A.; Sehitoglu, H. Determination of latent hardening response for FeNiCoCrMn for twin-twin interactions. *Acta Mater.* **2018**, *147*, 149–164. [CrossRef]
29. Niu, C.; LaRosa, C.R.; Miao, J.; Mills, M.J.; Ghazisaeidi, M. Magnetically-driven phase transformation strengthening in high entropy alloys. *Nat. Commun.* **2018**, *9*, 1363. [CrossRef] [PubMed]
30. Ding, J.; Yu, Q.; Asta, M.; Ritchie, R.O. Tunable stacking fault energies by tailoring local chemical order in CrCoNi medium-entropy alloys. *arXiv* **2018**, arXiv:1806.00718.
31. Ikeda, Y.; Grabowski, B.; Körmann, F. *Ab initio* phase stabilities and mechanical properties of multicomponent alloys: A comprehensive review for high entropy alloys and compositionally complex alloys. *Mater. Charact.* **2018**, in press, [CrossRef]
32. Li, Z.; Raabe, D. Influence of compositional inhomogeneity on mechanical behavior of an interstitial dual-phase high-entropy alloy. *Mater. Chem. Phys.* **2018**, *210*, 29–36. [CrossRef]

33. Soven, P. Coherent-Potential Model of Substitutional Disordered Alloys. *Phys. Rev.* **1967**, *156*, 809–813. [CrossRef]
34. Gyorffy, B.L. Coherent-Potential Approximation for a Nonoverlapping-Muffin-Tin-Potential Model of Random Substitutional Alloys. *Phys. Rev. B* **1972**, *5*, 2382–2384. [CrossRef]
35. Vitos, L.; Abrikosov, I.A.; Johansson, B. Anisotropic Lattice Distortions in Random Alloys from First-Principles Theory. *Phys. Rev. Lett.* **2001**, *87*, 156401. [CrossRef] [PubMed]
36. Gholizadeh, H.; Draxl, C.; Puschnig, P. The influence of interstitial carbon on the γ-surface in austenite. *Acta Mater.* **2013**, *61*, 341–349. [CrossRef]
37. Denteneer, P.J.H.; van Haeringen, W. Stacking-fault energies in semiconductors from first-principles calculations. *J. Phys. C* **1987**, *20*, L883. [CrossRef]
38. Vitos, L.; Skriver, H.; Johansson, B.; Kollár, J. Application of the exact muffin-tin orbitals theory: the spherical cell approximation. *Comput. Mater. Sci.* **2000**, *18*, 24–38. [CrossRef]
39. Vitos, L. Total-energy method based on the exact muffin-tin orbitals theory. *Phys. Rev. B* **2001**, *64*, 014107. [CrossRef]
40. Ruban, A.V.; Skriver, H.L. Screened Coulomb interactions in metallic alloys. I. Universal screening in the atomic-sphere approximation. *Phys. Rev. B* **2002**, *66*, 024201. [CrossRef]
41. Ruban, A.V.; Simak, S.I.; Korzhavyi, P.A.; Skriver, H.L. Screened Coulomb interactions in metallic alloys. II. Screening beyond the single-site and atomic-sphere approximations. *Phys. Rev. B* **2002**, *66*, 024202. [CrossRef]
42. Vitos, L. *Computational Quantum Mechanics for Materials Engineers: The EMTO Method and Applications*; Springer Science & Business Media: New York, NY, USA, 2007. [CrossRef]
43. Vitos, L.; Kollár, J.; Skriver, H.L. Full charge-density calculation of the surface energy of metals. *Phys. Rev. B* **1994**, *49*, 16694–16701. [CrossRef]
44. Vitos, L.; Kollár, J.; Skriver, H.L. Full charge-density scheme with a kinetic-energy correction: Application to ground-state properties of the 4d metals. *Phys. Rev. B* **1997**, *55*, 13521–13527. [CrossRef]
45. Perdew, J.P.; Burke, K.; Ernzerhof, M. Generalized Gradient Approximation Made Simple. *Phys. Rev. Lett.* **1996**, *77*, 3865–3868. [CrossRef] [PubMed]
46. Asato, M.; Settels, A.; Hoshino, T.; Asada, T.; Blügel, S.; Zeller, R.; Dederichs, P.H. Full-potential KKR calculations for metals and semiconductors. *Phys. Rev. B* **1999**, *60*, 5202–5210. [CrossRef]
47. Olsson, P.; Abrikosov, I.; Vitos, L.; Wallenius, J. Ab initio formation energies of Fe–Cr alloys. *J. Nucl. Mater.* **2003**, *321*, 84–90. [CrossRef]
48. Ruban, A.V.; Khmelevskyi, S.; Mohn, P.; Johansson, B. Magnetic state, magnetovolume effects, and atomic order in $Fe_{65}Ni_{35}$ Invar alloy: A first principles study. *Phys. Rev. B* **2007**, *76*, 014420. [CrossRef]
49. Ropo, M.; Kokko, K.; Vitos, L. Assessing the Perdew-Burke-Ernzerhof exchange-correlation density functional revised for metallic bulk and surface systems. *Phys. Rev. B* **2008**, *77*, 195445. [CrossRef]
50. Delczeg, L.; Delczeg-Czirjak, E.K.; Johansson, B.; Vitos, L. Assessing common density functional approximations for the ab initio description of monovacancies in metals. *Phys. Rev. B* **2009**, *80*, 205121. [CrossRef]
51. Delczeg, L.; Johansson, B.; Vitos, L. *Ab initio* description of monovacancies in paramagnetic austenitic Fe-Cr-Ni alloys. *Phys. Rev. B* **2012**, *85*, 174101. [CrossRef]
52. Körmann, F.; Ma, D.; Belyea, D.D.; Lucas, M.S.; Miller, C.W.; Grabowski, B.; Sluiter, M.H.F. "Treasure maps" for magnetic high-entropy-alloys from theory and experiment. *Appl. Phys. Lett.* **2015**, *107*, 142404, [CrossRef]
53. Ma, D.; Grabowski, B.; Körmann, F.; Neugebauer, J.; Raabe, D. *Ab initio* thermodynamics of the CoCrFeMnNi high entropy alloy: Importance of entropy contributions beyond the configurational one. *Acta Mater.* **2015**, *100*, 90–97. [CrossRef]
54. Heidelmann, M.; Feuerbacher, M.; Ma, D.; Grabowski, B. Structural anomaly in the high-entropy alloy ZrNbTiTaHf. *Intermetallics* **2016**, *68*, 11–15. [CrossRef]
55. Ruban, A.V.; Dehghani, M. Atomic configuration and properties of austenitic steels at finite temperature: Effect of longitudinal spin fluctuations. *Phys. Rev. B* **2016**, *94*, 104111. [CrossRef]
56. Ruban, A.V. First-principles modeling of the Invar effect in $Fe_{65}Ni_{35}$ by the spin-wave method. *Phys. Rev. B* **2017**, *95*, 174432. [CrossRef]

57. Jin, K.; Sales, B.C.; Stocks, G.M.; Samolyuk, G.D.; Daene, M.; Weber, W.J.; Zhang, Y.; Bei, H. Tailoring the physical properties of Ni-based single-phase equiatomic alloys by modifying the chemical complexity. *Sci. Rep.* **2016**, *6*, 20159. [CrossRef] [PubMed]
58. Staunton, J.; Gyorffy, B.; Pindor, A.; Stocks, G.; Winter, H. The "disordered local moment" picture of itinerant magnetism at finite temperatures. *J. Magn. Magn. Mater.* **1984**, *45*, 15–22. [CrossRef]
59. Gyorffy, B.L.; Pindor, A.J.; Staunton, J.; Stocks, G.M.; Winter, H. A first-principles theory of ferromagnetic phase transitions in metals. *J. Phys. F* **1985**, *15*, 1337–1386. [CrossRef]
60. Bhattacharjee, P.; Sathiaraj, G.; Zaid, M.; Gatti, J.; Lee, C.; Tsai, C.W.; Yeh, J.W. Microstructure and texture evolution during annealing of equiatomic CoCrFeMnNi high-entropy alloy. *J. Alloys Compd.* **2014**, *587*, 544–552. [CrossRef]
61. Wu, Z. Temperature and Alloying Effects on the Mechanical Properties of Equiatomic FCC Solid Solution Alloys. Ph.D. Thesis, The University of Tennessee, Knoxville, TN, USA, 2014.
62. Laurent-Brocq, M.; Akhatova, A.; Perrière, L.; Chebini, S.; Sauvage, X.; Leroy, E.; Champion, Y. Insights into the phase diagram of the CrMnFeCoNi high entropy alloy. *Acta Mater.* **2015**, *88*, 355–365. [CrossRef]
63. Schuh, B.; Mendez-Martin, F.; Völker, B.; George, E.; Clemens, H.; Pippan, R.; Hohenwarter, A. Mechanical properties, microstructure and thermal stability of a nanocrystalline CoCrFeMnNi high-entropy alloy after severe plastic deformation. *Acta Mater.* **2015**, *96*, 258–268. [CrossRef]
64. Laplanche, G.; Gadaud, P.; Bärsch, C.; Demtröder, K.; Reinhart, C.; Schreuer, J.; George, E. Elastic moduli and thermal expansion coefficients of medium-entropy subsystems of the CrMnFeCoNi high-entropy alloy. *J. Alloys Compd.* **2018**, *746*, 244–255. [CrossRef]
65. Tian, L.Y.; Lizárraga, R.; Larsson, H.; Holmström, E.; Vitos, L. A first principles study of the stacking fault energies for fcc Co-based binary alloys. *Acta Mater.* **2017**, *136*, 215–223. [CrossRef]
66. Nishizawa, T.; Ishida, K. The Co (Cobalt) system. *Bull. Alloy Phase Diagr.* **1983**, *4*, 387–390.
67. Uhl, M.; Kübler, J. Exchange-Coupled Spin-Fluctuation Theory: Application to Fe, Co, and Ni. *Phys. Rev. Lett.* **1996**, *77*, 334–337. [CrossRef] [PubMed]
68. Lizárraga, R.; Pan, F.; Bergqvist, L.; Holmström, E.; Gercsi, Z.; Vitos, L. First Principles Theory of the hcp-fcc Phase Transition in Cobalt. *Sci. Rep.* **2017**, *7*, 3778. [CrossRef] [PubMed]
69. Guo, S.; Ng, C.; Lu, J.; Liu, C.T. Effect of valence electron concentration on stability of fcc or bcc phase in high entropy alloys. *J. Appl. Phys.* **2011**, *109*, 1035051. [CrossRef]
70. Tsai, M.H.; Tsai, K.Y.; Tsai, C.W.; Lee, C.; Juan, C.C.; Yeh, J.W. Criterion for Sigma Phase Formation in Cr- and V-Containing High-Entropy Alloys. *Mater. Res. Lett.* **2013**, *1*, 207–212. [CrossRef]
71. Otto, F.; Dlouhý, A.; Pradeep, K.; Kuběnová, M.; Raabe, D.; Eggeler, G.; George, E. Decomposition of the single-phase high-entropy alloy CrMnFeCoNi after prolonged anneals at intermediate temperatures. *Acta Mater.* **2016**, *112*, 40–52. [CrossRef]

entropy

MDPI

Article

Elemental Phase Partitioning in the γ-γ'' $Ni_2CoFeCrNb_{0.15}$ High Entropy Alloy

Bin Han [1,*], Jie Wei [1], Feng He [1,2], Da Chen [1], Zhi Jun Wang [2], Alice Hu [1,3,4], Wenzhong Zhou [1,3] and Ji Jung Kai [1,3,*]

1 Department of Mechanical Engineering, City University of Hong Kong, Hong Kong, China; jiewei@cityu.edu.hk (J.W.); fenghe@mail.nwpu.edu.cn (F.H.); dachen5-c@my.cityu.edu.hk (D.C.); alicehu@cityu.edu.hk (A.H.); wenzzhou@cityu.edu.hk (W.Z.)

2 State Key Laboratory of Solidification Processing, Northwestern Polytechnical University, Xi'an 710072, China; zhjwang@nwpu.edu.cn

3 Centre for Advanced Nuclear Safety and Sustainable Development, City University of Hong Kong, Hong Kong, China

4 City University of Hong Kong, Shenzhen Research Institute, Shenzhen 518057, China

* Correspondence: binhan@cityu.edu.hk (B.H.); jijkai@cityu.edu.hk (J.J.K.); Tel.: +86-852-3442-2176 (B.H.)

Received: 28 October 2018; Accepted: 16 November 2018; Published: 28 November 2018

Abstract: The partitioning of the alloying elements into the γ'' nanoparticles in a $Ni_2CoFeCrNb_{0.15}$ high entropy alloy was studied by the combination of atom probe tomography and first-principles calculations. The atom probe tomography results show that the Co, Fe, and Cr atoms incorporated into the Ni_3Nb-type γ'' nanoparticles but their partitioning behaviors are significantly different. The Co element is much easier to partition into the γ'' nanoparticles than Fe and Cr elements. The first-principles calculations demonstrated that the different partitioning behaviors of Co, Fe and Cr elements into the γ'' nanoparticles resulted from the differences of their specific chemical potentials and bonding states in the γ'' phase.

Keywords: high entropy alloy; gamma double prime nanoparticles; elemental partitioning; atom probe tomography; first-principles calculations

1. Introduction

Recently, a new class of structural materials, known as high entropy alloys (HEAs), have attracted considerable attention due to their excellent properties and potential applications in the aerospace and energy industries [1–11]. Compared with the conventional alloys, the face-centered cubic (FCC) HEAs exhibit unique properties such as outstanding ductility [7], exceptional fracture toughness [12] as well as excellent corrosion resistance [13]. However, the single-phase FCC HEAs are insufficiently strong, which limits their engineering applications.

The strategy of introducing the dispersed hard $D0_{22}$-structured gamma double prime (γ'') or $L1_2$-structured gamma prime (γ') nanoparticles into the FCC matrix (γ phase) has been proved to be one of the most effective approaches to enhance the strength of the FCC HEAs, as it is the case in many superalloys [10,14–16]. It is known that the alloying elements in the γ'' or γ' phase plays an important role on the stability and the mechanical properties of the nano-precipitated alloys [17–20]. Therefore, it is critical to clarify the partitioning of the alloying elements into the nanoparticles of the FCC HEAs. However, this issue still lacks research because the observation of alloying elements in the nanoparticles which embedded in the FCC matrix is still a challenge. Although the energy dispersive X-ray spectroscopy (EDS) equipped on scan electron microscope (SEM) or on transmission electron microscope (TEM) has been widely used to determine the material composition, it is difficult to distinguish the composition of the nanoparticles from that of the surrounding matrix. Atom probe tomography (APT), the only technique which can generate the three-dimensional (3D) atom maps of materials in the real space with nearly atomic-scale resolution, has been proved to be a powerful method of characterizing the composition of different kinds of nanoparticles [15,18,21,22]. Recently, we have successfully clarified the partitioning of the alloying elements into the γ' nanoparticles in the $NiFeCoCrTi_{0.2}$ HEA by APT [22]. In this work, the γ'' nanoparticles in a $Ni_2CoFeCrNb_{0.15}$ HEA was investigated by APT. Different partitioning behaviors of the alloying elements into the γ'' nanoparticles were observed. It was found that the Co element tends to partition into the γ'' nanoparticles but Fe and Cr elements are largely depleted from the γ'' nanoparticles. The APT results were confirmed by the first-principles calculations from the perspective of the electronic states.

2. Materials and Methods

An ingot with a composition $Ni_2CoFeCrNb_{0.15}$ was produced by arc melting Fe, Co, Ni, Cr, and Nb metals with high purity (>99.9%) in an argon atmosphere. After repeatedly melted five times, the ingot was then drop-casted into a copper mold to make a slab with a dimension of 5 mm × 10 mm × 50 mm. Afterwards, the slab was solution-treated at 1473 K for 2 h, followed by water quenching. Then, the homogenized slab was cold rolled with a total thickness reduction of 70% and subsequently recrystallized at 1473 K for 4 minutes (min) and water-quenched. At last, aging was performed at 923 K for 40 h and 100 h, respectively, followed by water quenching.

The TEM specimen was prepared by mechanically grinding and followed by ion-milling using a precision ion polishing system (PIPS, Model 695, Gatan, Pleasanton, CA, USA). The TEM (TEM, JEOL 2100F, Tokyo, JAPAN) was operated under 200 keV. Needle specimens for APT analysis were prepared by gallium focused-ion-beam (FIB), with a FIB-SEM dual-beam system (Scios, FEI, Hillsboro, OR, USA), using a conventional lift-out technique [23]. The APT analysis was performed using a local electrode atom probe (LEAP5000 XR, CAMECA, Madison, WI, USA). The samples were run in the voltage mode at a specimen temperature of 50 K, with 200 kHz pulses at a pulse fraction of 20%. An Integrated Visualization and Analysis Software (IVAS, Version 3.8.2) protocol was employed to reconstruct the 3D atomic maps [24].

3. Results and Discussion

To confirm the formation of $D0_{22}$-structured γ'' nanoparticles, the TEM analysis was performed before the APT measurement. Figure 1 shows a bright-field (BF) and a dark-field (DF) TEM images of the sample aged for 40 h. The DF-TEM image recorded from the spot marked with a yellow circle in the inset selected area diffraction pattern (SADP). The nanoparticles can be clearly observed in both the BF-TEM and the DF-TEM images. From the SADP it can be confirmed that the matrix has an FCC structure (γ phase), whilst the nanoparticles have a $D0_{22}$ structure (γ'' phase) which is revealed by the additional faint spots. Figure 1c shows the size distribution of the nanoparticles with an average size of 13.5 ± 2.9 nm. The size of the nanoparticles was measured from the length of the nanoparticle along their long axis.

Figure 2 shows the APT results of the sample aged for 40 h. In the three-dimensional (3D) atom map, the γ'' nanoparticles are delineated by 50 at.% Ni iso-concentration surfaces in red. It can be observed that the γ'' nanoparticles are disk-like. From the sliced atom maps, it can be found that the γ'' nanoparticles mainly consist of Ni and Nb. In addition, Co element shows a strong tendency to partition into the γ'' nanoparticles, but Fe and Cr elements are largely depleted from the γ'' nanoparticles. To clarify the accurate composition of these γ'' nanoparticles, the proximity histogram, which is calculated over the iso-concentration surfaces, is plotted (Figure 2b). Therein, the chemical elements are displayed as a function of the distance from the iso-concentration surfaces. The proximity histogram shows that the average concentration of Co is up to 8.2 ± 0.3 at.%, but the average concentrations of Fe and Cr are only 1.3 ± 0.4 and 1.6 ± 0.1 at.% in the γ'' nanoparticles, respectively. To investigate the composition stability of γ'' nanoparticles, the sample aged for 100 h was also analyzed by APT. The composition of the γ'' nanoparticles are summarized in Table 1. It was found that the composition of γ'' nanoparticles in the sample aged for 100 h is almost the same as that in the sample aged for 40 h, which indicates that the composition of γ'' nanoparticles have reached the steady state after 40 h aging. The APT results demonstrate that the Co element is much easier to partition into the γ'' nanoparticles than Fe and Cr elements. In addition, Co, Fe, and Cr prefer Ni sublattice sites in the Ni_3Nb-type γ'' nanoparticles because the Ni composition in the γ'' nanoparticles are only 65 at.%. It should be noted that a small part of Co, Fe, or Cr atoms may also occupy Nb sublattice sites as the Nb composition in the γ'' nanoparticles is about 24%. However, it is difficult to determine which kind of elements occupied Nb sublattice sites only from APT results. Some similar results are also found in the Ni-based superalloys [25–27]. For example, Lawitzki et al. reported that the alloying elements such as Cr occupied both the Ni and Nb sublattice sites of γ'' phase in the 718 alloy [25].

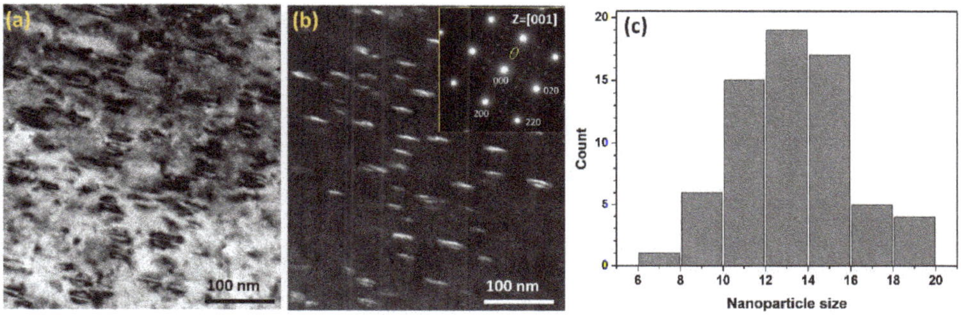

Figure 1. The BF-TEM (**a**); and the DF-TEM (**b**) images of the sample aged for 40 h. The inset in (**b**) is the SADP along the zone-axis z = [001]. (**c**) The size distribution of the γ'' nanoparticles.

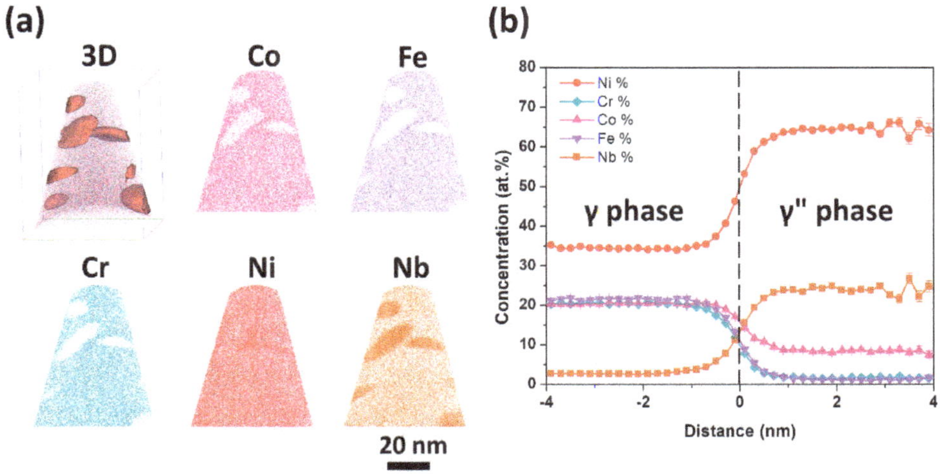

Figure 2. (**a**) The 3D atom map ($62 \times 64 \times 80$ nm^3) and the 4 nm-thick sliced atom maps of Co, Fe, Cr, Ni, and Nb of the 40 h aged sample. In the 3D map, the nanoparticles are delineated by 50 at.% Ni iso-concentration surfaces in red for better illustration. (**b**) The proximity histogram of the iso-concentration surfaces illustrated in the 3D atom map. The alloying elements are shown as a function of the distance from the iso-concentration surface (vertical dashed line).

Table 1. Chemical composition of γ'' nanoparticles (at.%).

Aging Time	Co	Fe	Cr	Ni	Nb
40 h	8.2 ± 0.3	1.3 ± 0.1	1.6 ± 0.1	64.3 ± 0.4	24.6 ± 0.2
100 h	8.1 ± 0.1	1.2 ± 0.1	1.4 ± 0.1	65.2 ± 0.3	24.0 ± 0.2

First-principles calculations were performed to confirm the site substitution preferences in the γ'' nanoparticles and to investigate the origin of the different partitioning behaviors of the Co, Fe, and Cr elements into the γ'' nanoparticles. The calculations employed the plane-wave pseudopotential approximations with the generalized gradient approximations, as implemented in the Vienna ab initio simulation package (VASP) [28]. A plane wave cutoff energy of 500 eV and 9 × 9 × 9 Monkhorst–Pack k-point grids were used in the calculation. A 3D periodic supercell with $D0_{22}$-structured $Ni_{24}Nb_8$ was employed to determine the total energies of the cells. The $D0_{22}$-structured Ni_3Nb was fully relaxed, and the lattice parameters were determined to be a = b = 3.643 Å and c = 7.484 Å, which is in good agreement with both the previous reported experimental and theoretical results [27,29].

The formation energies for an element X (X = Co, Fe, and Cr) to substitute a Ni site and a Nb site of the $D0_{22}$-structured Ni_3Nb were defined as [30]

$$E_{X\to Ni} = \left(E^{tot}_{Ni_{23}XNb_8} + \mu_{Ni}\right) - \left(E^{tot}_{Ni_{24}Nb_8} + \mu_X\right) \tag{1}$$

$$E_{X\to Nb} = \left(E^{tot}_{Ni_{24}Nb_7X} + \mu_{Nb}\right) - \left(E^{tot}_{Ni_{24}Nb_8} + \mu_X\right) \tag{2}$$

where E^{tot} is the total energy and μ is the chemical potential. The chemical potential is defined as the energy per atom of the element in its stable pure phase. Our calculations show that the total energy of $Ni_{24}Nb_8$ ($E^{tot}_{Ni_{24}Nb_8}$) is −222.84 eV and the chemical potentials of Ni and Nb are −5.47 and −10.20 eV, respectively. Table 2 summarized the total energies, the chemical potentials, and the formation energies. The calculation results demonstrate that Co and Fe atoms prefer to occupy the Ni sublattice sites rather than Nb sublattice sites, as the formation energies, $E_{Co/Fe\to Ni}$, are significantly lower than $E_{Co/Fe\to Nb}$. However, the formation energies for Cr to occupy Ni and Nb sublattice sites are almost same, which indicates that Cr atoms occupy both the Ni and Nb sublattice sites. In addition, the formation energy of Co that occupies the Ni sublattice site is nearly zero, which is much lower than that of Fe and Cr, indicating that Co is more stable in the $D0_{22}$-structured Ni_3Nb than Fe and Cr. Similar results were also reported in the $L1_2$-structured Ni_3Ti phase [31]. The calculation results confirmed the APT observation that the concentration of Co in the γ'' nanoparticles are much higher than that of Fe and Cr.

Table 2. The Calculated chemical potentials, total energies and formation energies with the unit of eV.

	μ	$E^{tot}_{Ni_{23}XNb_8}$	$E^{tot}_{Ni_{24}Nb_7X}$	$E_{X\to Ni}$	$E_{X\to Nb}$
Co	−7.01	−224.16	−217.74	0.08	1.71
Fe	−8.23	−225.00	−218.79	0.19	1.61
Cr	−9.50	−225.55	220.65	1.38	1.50

To further clarify the origin of the formation energy differences of Co, Fe, and Cr in the $D0_{22}$-structured Ni_3Nb, the total energy, and the chemical potential of the solute atoms are carefully checked, as the formation energy is determined by these two parts. It is found that the formation energy difference between Fe and Cr is mainly caused by the difference of their chemical potentials because the total energies of $E^{tot}_{Ni_{23}FeNb_8}$ and $E^{tot}_{Ni_{23}CrNb_8}$ are almost the same. However, both the chemical potential and the total energy of Co ($E^{tot}_{Ni_{23}CoNb_8}$) are higher than that of Fe and Cr, which indicates that the formation energy differences between Co and Fe/Cr not only result from their chemical potential differences, but also come from their total energy differences. As the total energy originates from the charge distribution of the system, we calculated the charge density difference of $Ni_{23}CoNb_8$,

$Ni_{23}FeNb_8$, and $Ni_{23}CrNb_8$ systems with reference to $Ni_{24}Nb_8$ system, respectively. From the charge density difference, we can find that the charge accumulation appears between Co and Nb atoms (Figure 3a) but does not appear between Fe and Nb (Figure 3b), or Cr and Nb (Figure 3c). The charge distribution results indicate that the bonding for Co and Nb is much stronger than that of Fe and Cr. The strong Co-Nb bond stabilizes Co atoms in the $D0_{22}$-structured Ni_3Nb, which demonstrates that the bonding state of Co plays an important role in lowering the formation energy compared with that of Fe and Cr.

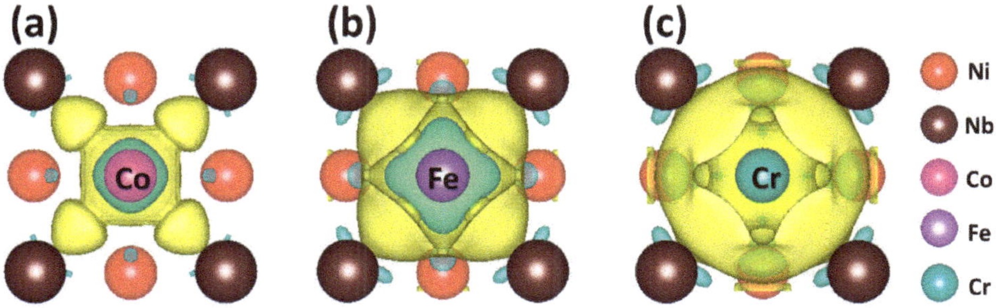

Figure 3. The charge density difference on the (001) plane of the $Ni_{23}CoNb_8$ (**a**), $Ni_{23}FeNb_8$ (**b**), and $Ni_{23}CrNb_8$ (**c**) systems with reference to the $Ni_{24}Nb_8$ system. The yellow regions and blue regions correspond to the increased and decreased charge density (0.001 eV/Bohr^{-3}), respectively.

4. Conclusions

In summary, the partitioning of alloying elements into the γ'' nanoparticles in a $Ni_2CoFeCrNb_{0.15}$ HEA was studied by APT and first-principles calculations. It was found that the composition of Co in the γ'' nanoparticles is up to 8.2 at.% but the composition of Fe and Cr are only 1.3 at.% and 1.6 at.%, respectively. This indicates that the Co element is much easier to partition into the γ'' nanoparticles than Fe and Cr elements. In addition, Co and Fe atoms prefer to occupy the Ni sublattice sites but Cr occupies both Ni and Nb sublattice sites. The first-principles calculations demonstrated that the different partitioning behaviors of Co, Fe, and Cr elements into the γ'' nanoparticles are attributed to the differences of their specific chemical potentials and the bonding states in the γ'' phase. This research paves the way for the composition control of the γ'' nanoparticles in the HEAs.

Author Contributions: B.H., J.W., and F.H. conceived and designed the experiments; B.H. performed the APT experiment under the supervision of J.J.K.; J.W. performed the first-principles calculations under the supervision of W.Z. and A.H.; F.H. designed and prepared the bulk samples under the supervision of Z.J.W.; and D.C. performed the TEM analysis. All authors discussed the results and approved the final manuscript.

Funding: This work was supported by the Hong Kong Research Grant Council for financial support (Grant Nos. CityU 21202517 and 11212915), National Natural Science Foundation of China (Grant Nos. 11605148 and 51771149), and the Collaborative Research Funds (CityU 11205515 and 11209314) from the Research Grant Council, Hong Kong.

Acknowledgments: We would like to express our thanks to Junhua Luan for technical support.

Conflicts of Interest: The authors declare no conflict of interest.

References

1. Gludovatz, B.; Hohenwarter, A.; Catoor, D.; Chang, E.H.; George, E.P.; Ritchie, R.O. A fracture-resistant high-entropy alloy for cryogenic applications. *Science* **2014**, *345*, 1153–1158. [CrossRef] [PubMed]
2. Li, Z.; Pradeep, K.G.; Deng, Y.; Raabe, D.; Tasan, C.C. Metastable high-entropy dual-phase alloys overcome the strength-ductility trade-off. *Nature* **2016**, *534*, 227–230. [CrossRef] [PubMed]
3. Zhang, Z.; Sheng, H.; Wang, Z.; Gludovatz, B.; Zhang, Z.; George, E.P.; Yu, Q.; Mao, S.X.; Ritchie, R.O. Dislocation mechanisms and 3D twin architectures generate exceptional strength-ductility-toughness combination in CrCoNi medium-entropy alloy. *Nat. Commun.* **2017**, *8*, 14390. [CrossRef] [PubMed]

4. Tracy, C.L.; Park, S.; Rittman, D.R.; Zinkle, S.J.; Bei, H.; Lang, M.; Ewing, R.C.; Mao, W.L. High pressure synthesis of a hexagonal close-packed phase of the high-entropy alloy CrMnFeCoNi. *Nat. Commun.* **2017**, *8*, 15634. [CrossRef] [PubMed]
5. Huang, H.; Wu, Y.; He, J.; Wang, H.; Liu, X.; An, K.; Wu, W.; Lu, Z. Phase-Transformation Ductilization of Brittle High-Entropy Alloys via Metastability Engineering. *Adv. Mater.* **2017**, *29*, 1701678. [CrossRef] [PubMed]
6. Jo, Y.H.; Jung, S.; Choi, W.M.; Sohn, S.S.; Kim, H.S.; Lee, B.J.; Kim, N.J.; Lee, S. Cryogenic strength improvement by utilizing room-temperature deformation twinning in a partially recrystallized VCrMnFeCoNi high-entropy alloy. *Nat. Commun.* **2017**, *8*, 15719. [CrossRef] [PubMed]
7. Otto, F.; Dlouhý, A.; Somsen, C.; Bei, H.; Eggeler, G.; George, E.P. The influences of temperature and microstructure on the tensile properties of a CoCrFeMnNi high-entropy alloy. *Acta Mater.* **2013**, *61*, 5743–5755. [CrossRef]
8. He, F.; Wang, Z.; Wu, Q.; Li, J.; Wang, J.; Liu, C.T. Phase separation of metastable CoCrFeNi high entropy alloy at intermediate temperatures. *Scr. Mater.* **2017**, *126*, 15–19. [CrossRef]
9. Rao, J.C.; Diao, H.Y.; Ocelík, V.; Vainchtein, D.; Zhang, C.; Kuo, C.; Tang, Z.; Guo, W.; Poplawsky, J.D.; Zhou, Y.; et al. Secondary phases in AlxCoCrFeNi high-entropy alloys: An in-situ TEM heating study and thermodynamic appraisal. *Acta Mater.* **2017**, *131*, 206–220. [CrossRef]
10. Liang, Y.-J.; Wang, L.; Wen, Y.; Cheng, B.; Wu, Q.; Cao, T.; Xiao, Q.; Xue, Y.; Sha, G.; Wang, Y.; et al. High-content ductile coherent nanoprecipitates achieve ultrastrong high-entropy alloys. *Nat. Commun.* **2018**, *9*, 4063. [CrossRef] [PubMed]
11. Zhang, W.; Liaw, P.K.; Zhang, Y. Science and technology in high-entropy alloys. *Sci. China Mater.* **2018**, *61*, 2–22. [CrossRef]
12. Gludovatz, B.; Hohenwarter, A.; Thurston, K.V.; Bei, H.; Wu, Z.; George, E.P.; Ritchie, R.O. Exceptional damage-tolerance of a medium-entropy alloy CrCoNi at cryogenic temperatures. *Nat. Commun.* **2016**, *7*, 10602. [CrossRef] [PubMed]
13. Lee, C.P.; Chang, C.C.; Chen, Y.Y.; Yeh, J.W.; Shih, H.C. Effect of the aluminium content of AlxCrFe1.5MnNi0.5 high-entropy alloys on the corrosion behaviour in aqueous environments. *Corros. Sci.* **2008**, *50*, 2053–2060. [CrossRef]
14. Yang, T.; Zhao, Y.L.; Liu, W.H.; Zhu, J.H.; Kai, J.J.; Liu, C.T. Ductilizing brittle high-entropy alloys via tailoring valence electron concentrations of precipitates by controlled elemental partitioning. *Mater. Res. Lett.* **2018**, *6*, 600–606. [CrossRef]
15. Zhao, Y.L.; Yang, T.; Tong, Y.; Wang, J.; Luan, J.H.; Jiao, Z.B.; Chen, D.; Yang, Y.; Hu, A.; Liu, C.T.; et al. Heterogeneous precipitation behavior and stacking-fault-mediated deformation in a CoCrNi-based medium-entropy alloy. *Acta Mater.* **2017**, *138*, 72–82. [CrossRef]
16. Thomas, A.; El-Wahabi, M.; Cabrera, J.M.; Prado, J.M. High temperature deformation of Inconel 718. *J. Mater. Process. Technol.* **2006**, *177*, 469–472. [CrossRef]
17. Daoud, H.M.; Manzoni, A.M.; Wanderka, N.; Glatzel, U. High-Temperature Tensile Strength of Al10Co25Cr8Fe15Ni36Ti6 Compositionally Complex Alloy (High-Entropy Alloy). *JOM* **2015**, *67*, 2271–2277. [CrossRef]
18. Jiao, Z.B.; Luan, J.H.; Miller, M.K.; Yu, C.Y.; Liu, Y.; Liu, C.T. Precipitate transformation from NiAl-type to Ni 2 AlMn-type and its influence on the mechanical properties of high-strength steels. *Acta Mater.* **2016**, *110*, 31–43. [CrossRef]
19. Chen, Y.Q.; Francis, E.; Robson, J.; Preuss, M.; Haigh, S.J. Compositional variations for small-scale gamma prime (γ') precipitates formed at different cooling rates in an advanced Ni-based superalloy. *Acta Mater.* **2015**, *85*, 199–206. [CrossRef]
20. Wang, W.Z.; Jin, T.; Jia, J.H.; Liu, J.L.; Hu, Z.Q. Effects of cobalt on creep rupture properties and dislocation structures in nickel base superalloys. *Mater. Sci. Eng. A* **2015**, *624*, 220–228. [CrossRef]
21. Han, B.; Shimizu, Y.; Seguini, G.; Arduca, E.; Castro, C.; Ben Assayag, G.; Inoue, K.; Nagai, Y.; Schamm-Chardon, S.; Perego, M. Evolution of shape, size, and areal density of a single plane of Si nanocrystals embedded in SiO_2 matrix studied by atom probe tomography. *RSC Adv.* **2016**, *6*, 3617–3622. [CrossRef]

22. Han, B.; Wei, J.; Tong, Y.; Chen, D.; Zhao, Y.; Wang, J.; He, F.; Yang, T.; Zhao, C.; Shimizu, Y.; et al. Composition evolution of gamma prime nanoparticles in the Ti-doped CoFeCrNi high entropy alloy. *Scr. Mater.* **2018**, *148*, 42–46. [CrossRef]
23. Miller, M.; Russell, K.; Thompson, G. Strategies for fabricating atom probe specimens with a dual beam FIB. *Ultramicroscopy* **2005**, *102*, 287–298. [CrossRef] [PubMed]
24. Vurpillot, F.; Gault, B.; Geiser, B.P.; Larson, D. Reconstructing atom probe data: A review. *Ultramicroscopy* **2013**, *132*, 19–30. [CrossRef] [PubMed]
25. Lawitzki, R.; Hassan, S.; Karge, L.; Wagner, J.; Wang, D.; von Kobylinski, J.; Krempaszky, C.; Hofmann, M.; Gilles, R.; Schmitz, G. Differentiation of γ'- and γ''-precipitates in Inconel 718 by a complementary study with small-angle neutron scattering and analytical microscopy. *Acta Mater.* **2019**, *163*, 28–39. [CrossRef]
26. Theska, F.; Stanojevic, A.; Oberwinkler, B.; Ringer, S.P.; Primig, S. On conventional versus direct ageing of Alloy 718. *Acta Mater.* **2018**, *156*, 116–124. [CrossRef]
27. Kusabiraki, K.; Komatsu, H.; Ikeuchi, S. Lattice constants and compositions of the metastable Ni3Nb phase precipitated in a Ni-15Cr-8Fe-6Nb alloy. *Metall. Mater. Trans. A* **1998**, *29*, 1169–1174. [CrossRef]
28. Kresse, G.; Hafner, J. Ab initio molecular dynamics for liquid metals. *Phys. Rev. B* **1993**, *47*, 558. [CrossRef]
29. Connétable, D.; Galliano, F.; Odemer, G.; Blanc, C.; Andrieu, É. DFT study of the solubility of hydrogen and carbon in Ni3Nb-D0a and Ni3Nb-D022 systems. *J. Alloys Compd.* **2014**, *610*, 347–351. [CrossRef]
30. Booth-Morrison, C.; Mao, Z.; Noebe, R.D.; Seidman, D.N. Chromium and tantalum site substitution patterns in Ni3Al(L12) γ'-precipitates. *Appl. Phys. Lett.* **2008**, *93*, 033103. [CrossRef]
31. Xu, J.-H.; Lin, W.; Freeman, A.J. Electronic structure and phase stability of A3Ti (A = Fe, Co, Ni, and Cu). *Phys. Rev. B* **1993**, *48*, 4276–4286. [CrossRef]

entropy

MDPI

Article

The Effects of Mo and Nb on the Microstructures and Properties of CrFeCoNi(Nb,Mo) Alloys

Chun-Huei Tsau * **and Meng-Chi Tsai**

Institute of Nanomaterials, Chinese Culture University, Taipei 111, Taiwan; asd99586@yahoo.com.tw
* Correspondence: chtsau@staff.pccu.edu.tw

Received: 29 July 2018; Accepted: 27 August 2018; Published: 29 August 2018

Abstract: The effects of niobium and molybdenum additions on the microstructures, hardness and corrosion behaviors of CrFeCoNi(Nb,Mo) alloys were investigated. All of the CrFeCoNi(Nb,Mo) alloys displayed dendritic microstructures. The dendrites of CrFeCoNiNb and $CrFeCoNiNb_{0.5}Mo_{0.5}$ alloys were a hexagonal close packing (HCP) phase and the interdendrites were a eutectic structure of HCP and face-centered cubic (FCC) phases. Additionally, the dendrites of CrFeCoNiMo alloys were a simple cubic (SC) phase and the interdendrites were a eutectic structure of SC and FCC phases. The volume fraction of dendrites and interdendrites in these alloys were calculated. The influences of the volume fraction of dendrite in the alloys on the overall hardness were also discussed. The CrFeCoNiNb alloy had the larger volume fraction of dendrite and thus had the highest hardness among these alloys. The CrFeCoNi(Nb,Mo) alloys also showed better corrosion resistances in 1 M H_2SO_4 and 1 M NaCl solutions by comparing with commercial 304 stainless steel. The $CrFeCoNiNb_{0.5}Mo_{0.5}$ alloy possessed the best corrosion resistances in these solutions among the CrFeCoNi(Nb,Mo) alloys.

Keywords: CrFeCoNi(Nb,Mo); microstructure; hardness; corrosion; sulfuric acid; sodium chloride

1. Introduction

High-entropy alloys (HEA) has been announced for more than ten years [1–3]. The concept of HEA provides a new field for alloys design and thus becomes a very important field of materials development. This high-entropy alloy concept is now widely used to develop the high-performance alloys [4,5] and refractory alloys [6,7], also it is also applied in the thin film processes [8–10]. All of these researches are focus on the unique properties of the high-entropy alloys. Corrosion resistance is an important property of high-entropy alloys for structural applications; and many high-entropy alloys possess good corrosion resistances in different solutions are reported, such as $FeCoNiCrCu_x$ high-entropy alloys in 3.5% sodium chloride solution [11], $Al_{7.5}Cr_{22.5}Fe_{35}Mn_{20}Ni_{15}$ high-entropy alloy in different solutions [12] and $Al_{0.5}CoCrFeNi$ alloy in a 3.5% NaCl solution [13]. That is, the high-entropy alloy concept is used to develop structural alloys with good corrosion resistance.

CrFeCoNi alloy has a very good corrosion resistance property is reported in our previous study [14]. It has a granular FCC structure and some HCP precipitates. However, the hardness of CrFeCoNi alloy is too low (HV144) to limit its structural application. Molybdenum has a benefit on the corrosion resistance of stainless steels is well known [15,16]. Niobium also can improve the corrosion resistance of FeCuNbSiB amorphous alloys [17]. Therefore, this study adds Mo and/or Nb into CrFeCoNi alloy and tests their properties. The microstructures revolution, hardness and polarization behaviors of corrosion of the CrFeCoNi(Nb,Mo) alloys in H_2SO_4 and NaCl solutions are all tested to evaluate their commercial application.

2. Experimental

The CrFeCoNiNb, $CrFeCoNiNb_{0.5}Mo_{0.5}$ and CrFeCoNiMo alloys were prepared by arc melting using appropriate amounts of the elements with purities above 99.9%. The alloys were made under a partial pressure of argon atmosphere (400 torrs). The bottoms were remelted at least 4 times to ensure homogeneity. Table 1 lists the chemical compositions of the alloys, the maximum deviation of each element in the alloys was less than 1 atomic percent. The microstructural evolution of the as-cast alloys was observed using a field emission scanning electron microscope with an energy dispersive spectrometer (SEM/EDS, JEOL JSM-6335, JEOL Ltd., Tokyo, Japan), which was operated at 15 kV. The structures were characterized by X-ray diffraction (XRD) using a Rigaku ME510-FM2 (Rigaku Ltd., Tokyo, Japan) with Cu-K (with a wavelength of 1.5406 Å) radiation operated at 30 kV at a scanning rate of 0.04 degree/s. The microstructures and lattice images of the alloys were obtained using a high-resolution transmission electron microscope (HREM, JEOL JEM-3000F, JEOL Ltd., Tokyo, Japan), which was operated at 300 kV. The corresponding diffraction patterns (DP) were obtained from the high-resolution lattice images by fast Fourier transformation (FFT) in Gatan digital micrograph software. The hardness of the alloys was measured using both a Mitutoyo Akashi MVK-G1500 microhardness tester (Mitutoyo Co., Kanagawa, Japan) under a load of 10 gf and a Matsuzawa Seiki MV1 Vicker's hardness tester (Matsuzawa Co., Akita, Japan) under a load of 30 kgf.

Table 1. The average chemical compositions of the as-cast CrFeCoNi(Nb,Mo) alloys analyzed by SEM/EDS.

Alloys	Compositions (Atomic Percent)					
	Cr	Fe	Co	Ni	Nb	Mo
CrFeCoNiMo	20.1	20.2	19.5	20.5	N/A	19.7
$CrFeCoNiNb_{0.5}Mo_{0.5}$	21.4	19.4	19.8	17.6	11.0	10.8
CrFeCoNiNb	19.8	19.8	19.0	19.5	22.0	N/A

Polarization curves of the as-cast alloys were obtained in a potentiostat/galvanostat (Autolab PGSTAT302N, Metrohm Autolab B.V., Utrecht, The Netherlands) using a three-electrode system at a scanning rate of 1 mV/s. The CrFeCoNi(Nb,Mo) alloys for polarization testing were mounted in epoxy resin and the exposed surface area of each was fixed at 19.64 mm^2 (with a diameter of 5 mm). The reference electrode was a saturated silver chloride electrode (Ag/AgCl) and the counter electrode was a smooth Pt sheet. All the potentials that are below a saturated silver chloride electrode (SSE), whose reduction potential is 222 mV higher than that of the standard hydrogen electrode (SHE) at 25 °C [18]. The specimens whose polarization curves were obtained were all mechanically wet-polished using 1200 SiC grit paper. Test solutions with a concentration of 1 M were prepared from reagent-grade sulfuric acid (H_2SO_4) and sodium chloride (NaCl) that were dissolved in distilled water. To eliminate any effect of dissolved oxygen, the solutions were deaerated by bubbling nitrogen gas through them before and during the polarization experiments. The polarization test started after the specimen, counter electrode and reference electrode were placed in the bubbling solution for 900 s.

3. Results and Discussion

The microstructures of as-cast CrFeCoNi(Nb,Mo) alloys are displayed in Figure 1. They indicated that the Nb and Mo additions could change the granular microstructures of CrFeCoNi alloy to a dendritic microstructure of CeFeCoNi(Nb,Mo) alloys; and their interdendrites all showed a eutectic structure. Also, some precipitates were only observed in the as-cast CrFeCoNiNb alloy, shown in Figure 1a; the precipitates were not found in the other two alloys under the as-cast state. The chemical compositions of the phases in these CrFeCoNi(Nb,Mo) alloys are listed in Table 2. The deviation of each element-content in the phases were less than 1 atomic percent. The dendrites of these three alloys had higher niobium and/or molybdenum contents. Additionally, the precipitates in CrFeCoNiNb alloy had similar compositions with the HCP dendrites but the lattice constants of these two phases were quite different.

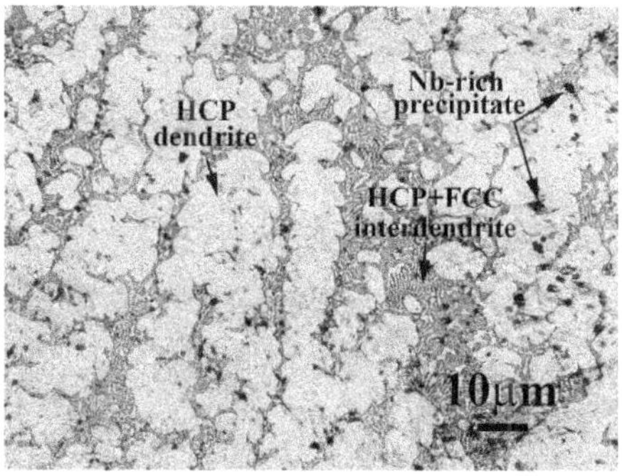

(a)

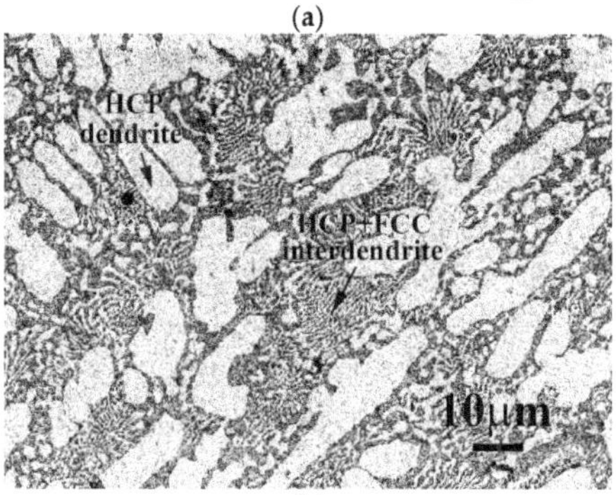

(b)

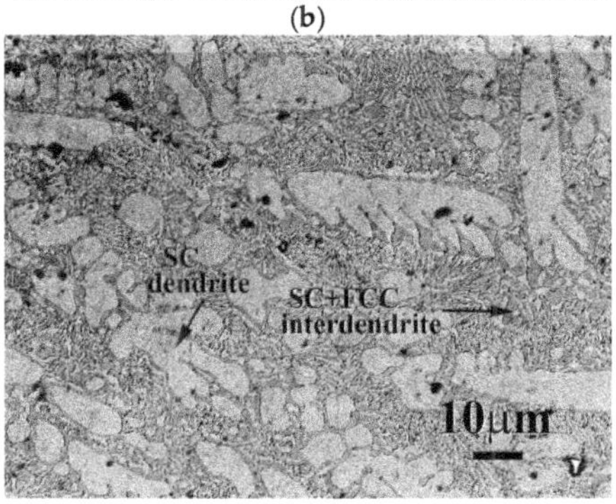

(c)

Figure 1. The SEM micrographs of as-cast (**a**) CrFeCoNiNb; (**b**) $CrFeCoNiNb_{0.5}Mo_{0.5}$; and (**c**) CrFeCoNiMo alloys.

Figure 2 shows the XRD patterns of the as-cast CrFeCoNi(Nb,Mo) alloys. The lattice constants of the phases are also marked in the figure. It indicated that every alloy had two major phases. The major phases in the as-cast CrFeCoNiNb and $CrFeCoNiNb_{0.5}Mo_{0.5}$ alloys were the FCC and HCP phases, the lattice constants of both the FCC and HCP phases in these two alloys were also very close. No peak of the precipitates was found in the as-cast CrFeCoNiNb alloy because its volume fraction was too small. However, the phases in the as-cast CrFeCoNiMo alloy were an FCC phase and a simple cubic

(SC) phase, the SC phase (isometric cubic) had a large lattice constant of 8.398 Å which indicated that a unit cell had many atoms.

Table 2. The average chemical compositions of the phases in the CrFeCoNi(Nb,Mo) alloys analyzed by SEM/EDS.

Alloys	Compositions (Atomic Percent)					
	Co	Cr	Fe	Ni	Mo	Nb
CrFeCoNiNb						
FCC	18.4	26.7	24.3	25.5	N/A	5.1
HCP	19.7	17.1	18.1	14.8	N/A	30.3
precipitate	19.1	16.7	16.7	15.7	N/A	31.8
$CrFeCoNiNb_{0.5}Mo_{0.5}$						
FCC	18.4	22.1	20.4	22.8	9.2	7.1
HCP	19.1	17.3	23.6	14.2	15.6	16.9
CrFeCoNiMo						
FCC	21.3	19.2	22.0	24.0	13.5	N/A
SC	17.2	21.9	17.7	14.4	28.9	N/A

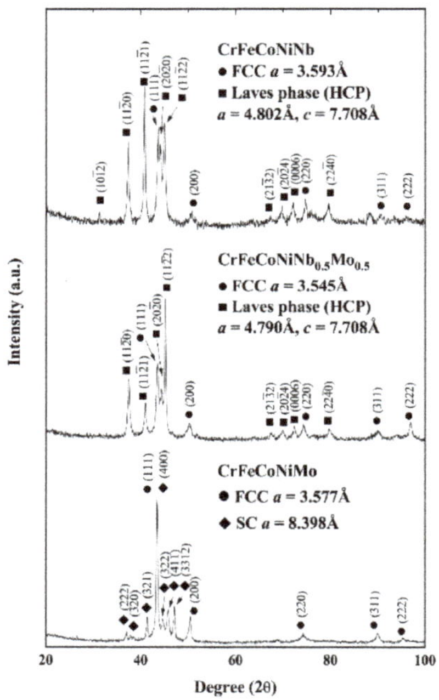

Figure 2. X-ray diffraction (XRD) patterns of the as-cast CrFeCoNi(Nb,Mo) alloys.

The transmission electron microscopy (TEM) bright field (BF) images of the dendrite, the matrix of the interdendrite and the precipitate of as-cast CrFeCoNiNb alloy are shown in Figure 3a–c, respectively. Figure 3a displays the image of the dendrite and inserts are the corresponding lattice image and the FFT DP, which were taken from the zone axis of [01$\bar{1}$0]; and the FFT DP indicates that the dendrite was a single HCP phase. Figure 3b is the TEM image of the matrix of interdendrite in the alloy, the inserts are the corresponding lattice image and FFT DP, which were taken from the zone zxis of [011]; and this FFT DP indicated that it was an FCC structure. Figure 3c shows the TEM image of the precipitate and its corresponding lattice image and FFT DP, which were taken from the zone axis of

[01$\bar{1}$1]. The precipitates showed a HCP structure and its lattice constants of a- and c-axes were 2.88 and 4.70 Å, respectively.

Figure 3. Transmission electron microscopy (TEM) (bright field) BF images of the phases in as-cast CrFeCoNiNb alloy: (**a**) an image of dendrite, inserts are the corresponding lattice image and fast Fourier transformation (FFT) diffraction patterns (DP) taken from the zone axis of [01$\bar{1}$0] which shows a hexagonal close packing (HCP) structure; (**b**) an image of interdendrite, inserts are the corresponding lattice image and FFT DP taken from the zone axis of [011] which shows an FCC structure; and (**c**) an image of precipitate, inserts are the corresponding lattice image and FFT DP taken from the zone axis of [01$\bar{1}$1] which shows a HCP structure.

Figure 4a shows the TEM BF images of the dendrite in $CrFeCoNiNb_{0.5}Mo_{0.5}$ alloy. inserts are the corresponding lattice image and FFT DP, which were taken from the zone axis of $[\bar{2}4\bar{2}3]$; and this FFT DP indicated that this phase was a HCP structure. Figure 4b show the TEM BF image of the interdendrite in $CrFeCoNiNb_{0.5}Mo_{0.5}$ alloy, inserts are the corresponding lattice image and FFT DP, which were taken from the zone axis of [011]; and this FFT DP indicated an FCC structure.

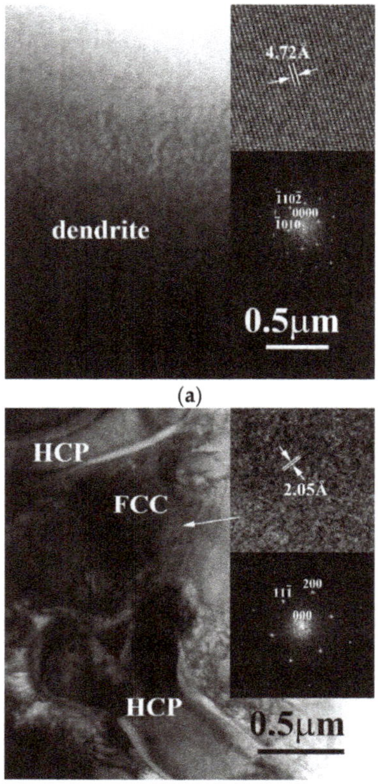

(b)

Figure 4. TEM BF images of the phases in as-cast $CrFeCoNiNb_{0.5}Mo_{0.5}$ alloy: (**a**) an image of dendrite, inserts are the corresponding lattice image and FFT DP taken from the zone axis of $[\bar{2}4\bar{2}3]$ which shows a HCP structure; (**b**) an image of interdendrite, inserts are the corresponding lattice image and FFT DP taken from the zone axis of [011] which shows an FCC structure.

Figure 5a shows a TEM BF image of the dendrite of as-cast CrFeCoNiMo alloy, inserts are the corresponding lattice image and FFT DP, which were taken from the zone axis of [001]. The image indicated that the dendrite was a single phase. However, the lattice image indicated that the unit cell of this phase had a large lattice constant of 8.398 Å. Therefore, the diffraction spots of the FFT DP were very close which indicated that this phase had a large lattice constant. Additionally, the lattice points from the lattice image showed only 1-fold symmetry and thus the unit cell had a SC structure (i.e., an isometric cubic structure) with a large lattice constant. This also meant the SC structure had a complex structure, this complex structure needs further investigation. Figure 5b displays the image of the eutectic interdendrite in the as-cast CrFeCoNiMo alloy. Both of the corresponding lattice image and FFT DP taken from the zone axis of $[\bar{1}12]$ confirmed that the matrix of interdendrite was an FCC structure.

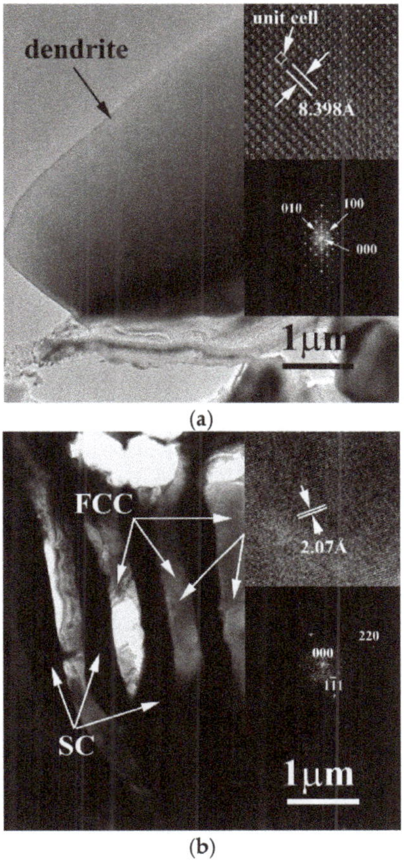

Figure 5. TEM BF images of the phases in as-cast CrFeCoNiMo alloy: (**a**) an image of dendrite, inserts are the corresponding lattice image and FFT DP taken from the zone axis of [001] which shows a SC structure; (**b**) an image of interdendrite, inserts are the corresponding lattice image and FFT DP taken from the zone axis of $[\bar{1}12]$ which shows an FCC structure.

The influence of niobium and molybdenum additions on the volume fractions of the dendrites and interdendrites of these alloys were different. Table 3 lists the volume fraction of the dendrites of as-cast CrFeCoNi(Nb,Mo) alloys. To determine of the volume fraction of dendrites was by drawing arbitrary lines in photos and measuring the intercept lengths of intercepted dendrites. From this, the volume fraction of the dendrites was calculated by the equation [19]:

$$V_d = L_d = \frac{\sum L_a}{L_T} \tag{1}$$

where V_d is the volume fraction of the dendrites, L_d is the linear fraction of the dendrites, L_a is the intercepted length of each dendrite and L_T is the total length. The results indicated that the CrFeCoNiNb alloy had more volume fraction of dendrites; and molybdenum addition would decrease the volume fraction of dendrites.

Table 3. The volume fraction of the dendrites in the as-cast CrFeCoNi(Nb,Mo) alloys.

Alloys	Volume Fraction of the Dendrites (vol.%)
CrFeCoNiMo	42 ± 6
$CrFeCoNiNb_{0.5}Mo_{0.5}$	36 ± 6
CrFeCoNiNb	68 ± 4

The overall hardness of as-cast CrFeCoNi(Nb,Mo) alloys and the microhardness of the dendrites and interdendrites of the alloys are list in Table 4. The hardness of as-cast CrFeCoNi alloy only had HV 144. However, the hardness increased sharply after additions of niobium and/or molybdenum. Because of niobium and molybdenum had larger atomic radiuses. The atomic radiuses of Cr, Fe, Co, Ni, Nb and Mo are 0.128, 0.124, 0.125, 0.125, 0.143 and 0.140 nm, respectively [20]. Therefore, the hardness increased significantly because of the larger lattice distortion and forming the different phases after additions of niobium and molybdenum. The interdendrites of the alloys had almost the same hardness (about HV 400). The dendrites of the alloys were harder than the interdendrites (about HV 700). The overall hardness of an alloy was contributed by the volume fractions of the dendrite (the hard part) and the interdendrite (the soft part) in this alloy. Therefore, decreasing the volume fraction of the hard part, that is, the dendrites, would result in decreasing the overall hardness of the alloy. Both the dendrites and interdendrites of $CrFeCoNiNb_{0.5}Mo_{0.5}$ alloy had the highest hardness but the overall hardness of this alloy was lowest among these alloys. This was contributed by the lowest volume fraction of dendrite (the hard part) in the $CrFeCoNiNb_{0.5}Mo_{0.5}$ alloy. On the contrary, the CrFeCoNiNb alloy had the highest volume fraction of dendrite and thus had the highest overall hardness among these alloys.

Table 4. The hardness of the CrFeCoNi(Nb,Mo) alloys.

Alloys	Hardness		
	Overall	Dendrite	Interdendrite
CrFeCoNiMo	604 ± 8	692 ± 18	405 ± 9
$CrFeCoNiNb_{0.5}Mo_{0.5}$	533 ± 6	745 ± 10	412 ± 7
CrFeCoNiNb	652 ± 8	693 ± 23	398 ± 24

The polarization behaviors of the as-cast CrFeCoNi(Nb,Mo) alloys and 304 stainless steel in 1M deaerated H_2SO_4 solution at 30 °C are shown in Figure 6. The polarization data were also compared with those of commercial 304 stainless steel (304SS) whose composition was by weight 71.61% Fe, 18.11% Cr, 8.24% Ni, 1.12% Mn, 0.75% Si, 0.05% Co, 0.02% Mo, 0.05% C, 0.03% P and 0.02% S. The important data of these polarization curves are listed in Table 5. The corrosion potential (E_{corr}) of CrFeCoNi(Nb,Mo) alloys were very close and all nobler than that of 304SS. Table 6 lists the standard electrode potential of selected elements [21]. The standard electrode potential of niobium is lower than that of molybdenum, which means that the niobium is more active than molybdenum. Therefore, the CrFeCoNiMo alloy had the highest E_{corr} and the CrFeCoNiNb alloy had the lowest E_{corr} among these three alloys. The polarization curve below E_{corr} was the cathodic polarization curve; and the curve above E_{corr} was the anodic polarization curve. The corrosion current densities (i_{corr}) of CrFeCoNi(Nb,Mo) alloys were less or equal to the i_{corr} of 304SS. The polarization curve of 304 stainless steel displayed a large anodic peak; and the anodic peaks of CrFeCoNi(Nb,Mo) alloys were significantly less than that of 304SS and thus CrFeCoNi(Nb,Mo) alloys had lower passivation potential (E_{pp}) and anodic critical current density (i_{crit}). This meant that the CrFeCoNi(Nb,Mo) alloys were easy to enter passivation regions and form passive films during corrosion in H_2SO_4 solution by comparing with 304 stainless steel. The large anodic peak of 304 stainless steel in H_2SO_4 solution was caused by formation of iron hydroxide and higher oxides of iron and chromium [22]. The lowest current densities of the passivation regions (i_{pass}) of these alloys were around 10–20 A/cm^2. Additionally, the main passivation region of CrFeCoNi(Nb,Mo) alloys were broader than that of 304SS. All of these alloys had a similar breakdown potential (E_b) of about 1 V (SSE).

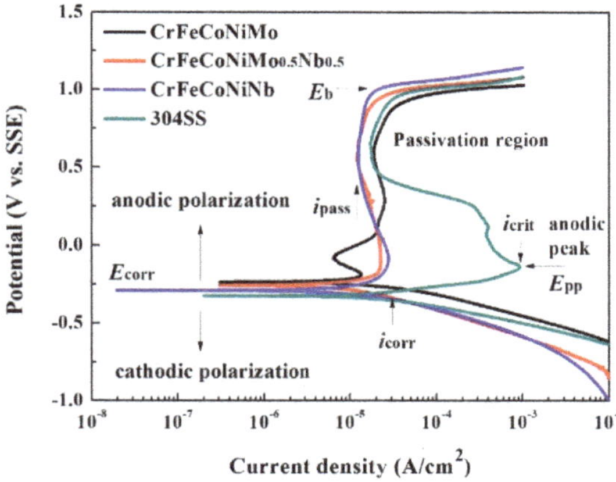

Figure 6. Polarization curves of CrFeCoNi(Nb,Mo) alloys and 304 stainless steel in 1 M deaerated H_2SO_4 solution at 30 °C.

Table 5. Polarization data of CrFeCoNi(Nb,Mo) alloys and 304 stainless steel in 1 M deaerated H_2SO_4 solution at 30 °C.

Alloys	i_{corr} μA/cm^2	E_{corr} V vs. SSE	E_{pp} V vs. SSE	i_{crit} mA/cm^2	i_{pass} μA/cm^2
CrFeCoNiNb	22.3	−0.290	−0.090	0.028	12.4
$CrFeCoNiNb_{0.5}Mo_{0.5}$	12.9	−0.256	−0.165	0.022	12.2
CrFeCoNiMo	30.0	−0.236	−0.174	0.013	18.9
304SS	30.0	−0.320	−0.140	0.930	17.2

Table 6. Standard electrode potential at 25 °C [21].

Reaction	Electrode Potential ($E°$ vs. SSE)
Cr, Cr^{3+}	−0.962
Fe, Fe^{2+}	−0.662
Co, Co^{2+}	−0.449
Ni, Ni^{2+}	−0.472
Nb, Nb^{3+}	−1.322
Mo, Mo^{3+}	−0.422

The micrographs of CrFeCoNi(Nb,Mo) alloys after the polarization test in 1 M deaerated H_2SO_4 solution at 30 °C are shown in Figure 7. Both of the dendrites and interdendrites of CrFeCoNiNb alloy were significantly corroded after test, as shown in Figure 7a; but the FCC phase (the matrix of interdendrite) was severely corroded than the HCP phase. On the contrary, only the FCC phase (the matrix of interdendrite) of $CrFeCoNiNb_{0.5}Mo_{0.5}$ alloy was slightly corroded, the HCP phase almost maintained its original shape, as shown in Figure 7b. This also proved that the $CrFeCoNiNb_{0.5}Mo_{0.5}$ alloy had the minimum i_{corr} among these CrFeCoNi(Nb,Mo) alloys, as listed in Table 5. The micrograph of CrFeCoNiMo alloy also displayed a severely corroded surface after polarization test, as shown in Figure 7c. Also, the FCC phase (the matrix of interdendrite) was more corroded than the SC phase. Therefore, in the local cells of the CrFeCoNi(Nb,Mo) alloys, the FCC phase in the interdendrites behaved as an anode and another phase (e.g., the HCP phase of CrFeCoNiNb and $CrFeCoNiNb_{0.5}Mo_{0.5}$ alloys and the SC phase of CrFeCoNiMo alloy) behaved as a cathode.

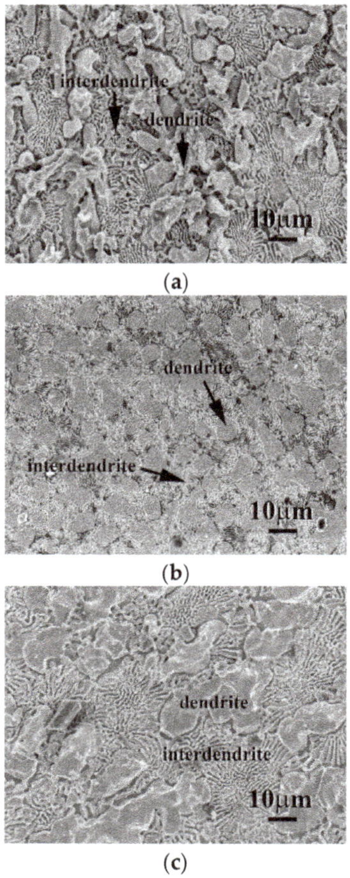

Figure 7. SEM micrographs of the alloys after polarization test in 1 M deaerated H_2SO_4 solution at 30 °C, (**a**) CrFeCoNiNb alloy; (**b**) $CrFeCoNiNb_{0.5}Mo_{0.5}$ alloy; and (**c**) CrFeCoNiMo alloy.

The polarization curves of the as-cast CrFeCoNi(Nb,Mo) alloys in 1 M deaerated NaCl solution at 30 °C are shown in Figure 8. The values of E_{corr} and i_{corr} of these alloys are listed in Table 7. All of these data are also compared with commercial 304 stainless steel. The i_{corr} of these four alloys were also close. In addition, the E_{corr} of the CrFeCoNi(Nb,Mo) alloys were very close and much nobler than 304 stainless steel. However, the passivation regions of CrFeCoNi(Nb,Mo) alloys were much broader than that of 304 stainless steel. Both of the CrFeCoNiMo and $CrFeCoNiNb_{0.5}Mo_{0.5}$ alloys had significantly anodic peaks and passivation regions; but CrFeCoNiNb and 304 stainless steel did not display anodic peaks. Adding Mo can reportedly increase the corrosion resistance of the alloy in a solution that contains chloride ions because molybdenum can increase the stability of the passivation films of steels [15,23]. The polarization curves of CrFeCoNi(Nb,Mo) alloys indicated that the increasing of Mo-content resulted in forming anodic peak and passivation regions. In addition, the cathodic limiting current densities (i_L) were observed in the polarization curves of CrFeCoNiNb and CrFeCoNiMo alloys. The cathodic limiting current density (i_L) related to the maximum reaction rate, which was limited by the diffusion rate of hydroxyl ions (OH^-) in solution [18].

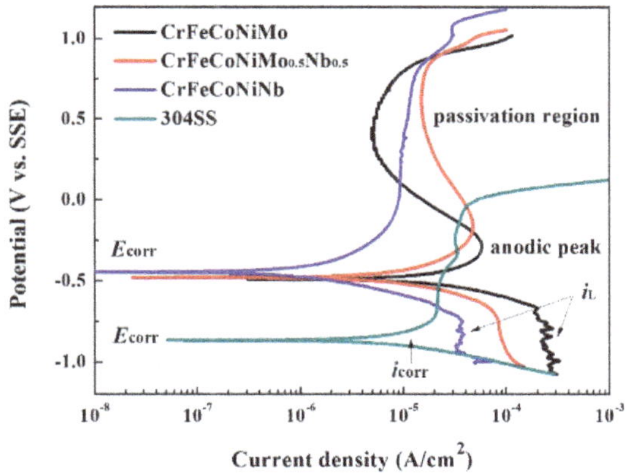

Figure 8. Polarization curves of CrFeCoNi(Nb,Mo) alloys and 304 stainless steel in 1 M deaerated NaCl solution at 30 °C.

Table 7. Polarization data (i_{corr} and E_{corr}) of CrFeCoNi(Nb,Mo) alloys and 304 stainless steel in 1 M deaerated NaCl solution at 30 °C.

Alloys	i_{corr}	E_{corr}
	A/cm^2	V vs. SSE
CrFeCoNiNb	1.2	−0.443
CrFeCoNiNb$_{0.5}$Mo$_{0.5}$	6.7	−0.477
CrFeCoNiMo	13.0	−0.489
304SS	12.9	−0.860

The micrographs of the as-cast CrFeCoNi(Nb,Mo) alloys after polarization test in 1 M deaerated NaCl solution at 30 °C are shown in Figure 9. Similar to the results of these alloys tested in 1 M deaerated H_2SO_4 solution at 30 °C, the major corroded areas of these alloys were the matrixes of the interdendrites (FCC phase) of CrFeCoNi(Nb,Mo) alloys. On the contrary, almost no corrosion occurred on the dendrites of these CrFeCoNi(Nb,Mo) alloys. Furthermore, no deep-type pitting was observed indicated that these alloys had a good corrosion resistance in NaCl solution. This also proved that molybdenum could improve the localized corrosion resistance.

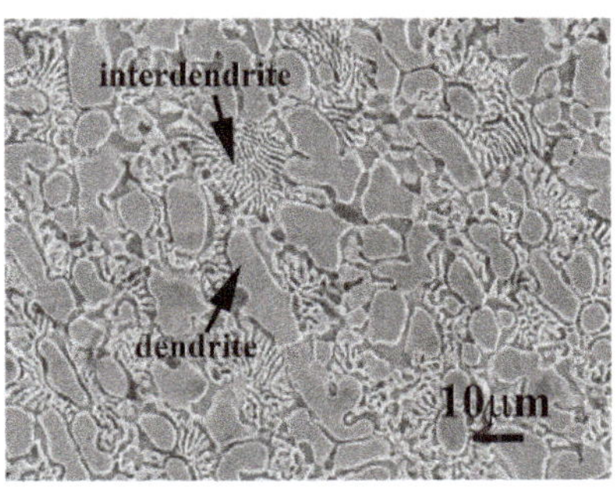

(a)

Figure 9. *Cont.*

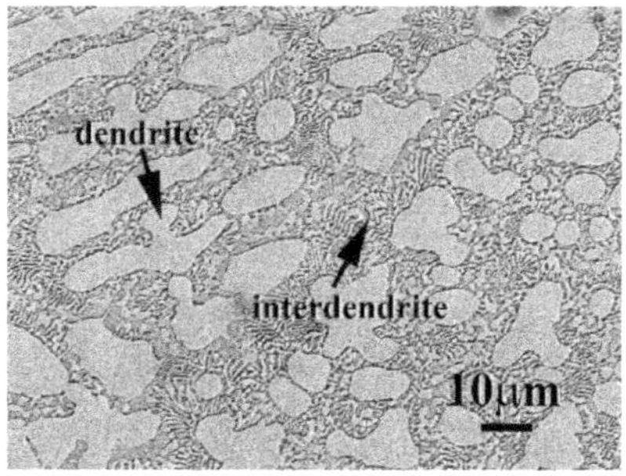

(b)

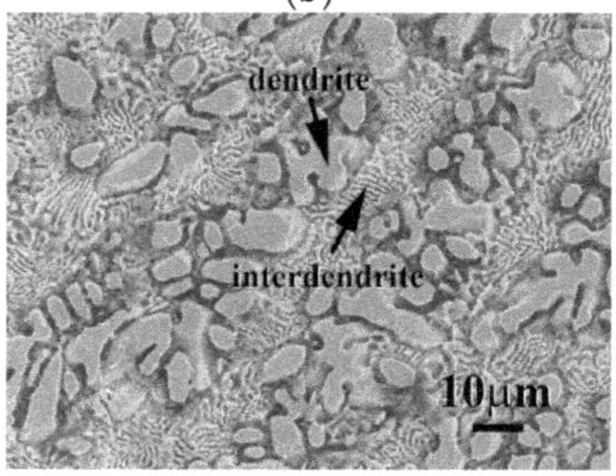

(c)

Figure 9. SEM micrographs of the alloys after polarization test in 1 M deaerated NaCl solution at 30 °C, (**a**) CrFeCoNiNb alloy; (**b**) $CrFeCoNiNb_{0.5}Mo_{0.5}$ alloy; and (**c**) CrFeCoNiMo alloy.

4. Conclusions

All of the CrFeCoNi(Nb,Mo) alloys displayed dendritic microstructures. The major two phases of CrFeCoNiNb and $CrFeCoNiNb_{0.5}Mo_{0.5}$ alloys were the HCP and FCC phases, where the dendrites were a single HCP phase. The major two phases of CrFeCoNiMo alloys were the SC and FCC phases, where the dendrites were a single SC phase. All of the interdendrites in the CrFeCoNi(Nb,Mo) alloys were eutectic structures.

The microhardness and overall hardness of CrFeCoNi(Nb,Mo) alloys increased by comparing with CrFeCoNi alloy because the elements of niobium and molybdenum had larger atomic radiuses. The microstructures significantly influenced the overall hardness of these alloys. The highest overall hardness of CrFeCoNiNb alloy was caused by its larger volume fraction of the dendrites. On the contrary, the lowest overall hardness of $CrFeCoNiNb_{0.5}Mo_{0.5}$ alloy was caused by its less volume fraction of the dendrites.

The corrosion resistances of CrFeCoNi(Nb,Mo) alloys in 1M deaerated H_2SO_4 and NaCl solutions were better than commercial 304 stainless steel. Additionally, the $CrFeCoNiNb_{0.5}Mo_{0.5}$ alloy had the best corrosion resistances in these solutions form the polarization curves and the micrographs after corrosion test. In these CrFeCoNi(Nb,Mo) alloys, the FCC phase behaved as an anode of the local cell in the alloy and was thus severely corroded than another phase in 1M deaerated H_2SO_4 and NaCl solutions.

Author Contributions: C.-H.T. conceived and designed the experiments; M.-C.T. performed the experiments; M.-C.T. and C.-H.T. analyzed the data; C.-H.T. contributed reagents/materials/analysis tools; C.-H.T. wrote the paper. Both authors have read and approved the final manuscript.

Acknowledgments: We are grateful to the Ministry of Science and Technology of Republic of China for its financial support under the project MOST 106-2221-E-034-008.

Conflicts of Interest: The authors declare no conflict of interest.

References

1. Yeh, J.W.; Chen, S.K.; Lin, S.J.; Gan, J.Y.; Chin, T.S.; Shun, T.T.; Tsau, C.H.; Chang, S.Y. Nanostructured high-entropy alloys with multiple principal elements: Novel alloy design concepts and outcomes. *Adv. Eng. Mater.* **2004**, *6*, 299–303. [CrossRef]
2. Murty, B.S.; Yeh, J.W.; Ranganathan, S. *High-Entropy Alloys*; Butterworth-Heinemann: London, UK, 2014; pp. 13–36.
3. Zhang, W.; Liaw, P.K.; Zhang, Y. Science and technology in high-entropy alloys. *Sci. China Mater.* **2018**, *61*, 2–22. [CrossRef]
4. Wu, J.M.; Lin, S.J.; Yeh, J.W.; Chen, S.K.; Huang, Y.S.; Chen, H.C. Adhesive wear behavior of $Al_xCoCrCuFeNi$ high-entropy alloys as a function of aluminum content. *Wear* **2006**, *261*, 513–519. [CrossRef]
5. Huo, W.Y.; Shi, H.F.; Ren, X.; Zhang, J.Y. Microstructure and wear behavior of CoCrFeMnNbNi High-Entropy Alloy Coating by TIG Cladding. *Adv. Mater. Sci. Eng.* **2015**, *2015*, 647351. [CrossRef]
6. Senkov, O.N.; Wilks, G.B.; Scott, J.M.; Miracle, D.B. Mechanical properties of $Nb_{25}Mo_{25}Ta_{25}W_{25}$ and $V_{20}Nb_{20}Mo_{20}Ta_{20}W_{20}$ refractory high entropy alloys. *Intermetallics* **2011**, *19*, 698–706. [CrossRef]
7. Gao, M.C.; Carney, C.S.; Doğan, Ő.N.; Jablonksi, P.D.; Hawk, J.A.; Alman, D.E. Design of refractory high-entropy alloys. *JOM* **2015**, *67*, 2653–2669. [CrossRef]
8. Antonaglia, J.; Xie, X.; Tang, Z.; Tsai, C.W.; Qiao, J.W.; Zhang, Y.; Laktionova, M.O.; Tabachnikova, E.D.; Yeh, J.W.; Senkov, O.N.; et al. Temperature effects on deformation and serration behavior of high-entropy alloys (HEAs). *JOM* **2014**, *66*, 2002–2008.
9. Sheng, W.J.; Yang, X.; Wang, C.; Zhang, Y. Nano-Crystallization of high-entropy amorphous $NbTiAlSiW_xN_y$ films prepared by magnetron sputtering. *Entropy* **2016**, *18*, 226. [CrossRef]
10. Zhang, Y.; Yan, X.-H.; Liao, W.-B.; Zhao, K. Effects of nitrogen content on the structure and mechanical properties of $(Al_{0.5}CrFeNiTi_{0.25})Nx$ high-entropy films by reactive sputtering. *Entropy* **2018**, *20*, 624. [CrossRef]
11. Hsu, Y.J.; Chiang, W.C.; Wu, J.K. Corrosion behavior of $FeCoNiCrCu_x$ high-entropy alloys in 3.5% sodium chloride solution. *Mater. Chem. Phys.* **2005**, *92*, 112–117. [CrossRef]
12. Tsau, C.H.; Lee, P.Y. Microstructures of $Al_{7.5}Cr_{22.5}Fe_{35}Mn_{20}Ni_{15}$ high-entropy alloy and its polarization behaviors in sulfuric acid, nitric acid and hydrochloric acid solutions. *Entropy* **2016**, *18*, 288. [CrossRef]
13. Lin, C.M.; Tsai, H.L. Evolution of microstructure, hardness, and corrosion properties of high-entropy $Al_{0.5}CoCrFeNi$ alloy. *Intermetallics* **2011**, *19*, 288–294. [CrossRef]
14. Tsau, C.H.; Lin, S.X.; Fang, C.H. Microstructures and corrosion behaviors of FeCoNi and CrFeCoNi equimolar alloys. *Mater. Chem. Phys.* **2017**, *186*, 534–540. [CrossRef]
15. Hashimoto, K.; Asami, K.; Teramoto, K. An X-ray photo-electron spectroscopic study on the role of molybdenum in increasing the corrosion resistance of ferritic stainless steels in HCl. *Corros. Sci.* **1979**, *19*, 3–14. [CrossRef]
16. Pardo, A.; Merino, M.C.; Coy, A.E.; Viejo, F.; Arrabal, R.; Matykina, E. Effect of Mo and Mn additions on the corrosion behavior of AISI 304 and 316 stainless steel in H_2SO_4. *Cossos. Sci.* **2008**, *50*, 780–794.
17. Mariano, N.A.; Souza, C.A.C.; May, J.E.; Kuri, S.E. Influence of Nb content on the corrosion resistance and saturation magnetic density of FeCuNbSiB alloys. *Mater. Sci. Eng. A* **2003**, *354*, 1–5. [CrossRef]
18. Revie, R.W.; Uhlig, H.H. *Corrosion and Corrosion Control: An Introduction to Corrosion Science and Engineering*, 4th ed.; John Wiley & Sons: Hoboken, NJ, USA, 2008; pp. 31–63.
19. Voort, G.F.V. *Metallography-Principles and Practice*; ASM International: Materials Park, OH, USA, 1999; pp. 425–426.
20. Smith, W.F. *Foundations of Materials Science and Engineering, 3rd ed*; McGraw-Hill: New York, NY, USA, 2004; pp. 877–878.
21. Chawla, S.L. *Materials Selection for Corrosion Control*; ASM International: Materials Park, OH, USA, 1993; p. 18.

22. Abdallah, M. Corrosion behavior of 304 stainless steel in sulphuric acid solutions and its inhibition by some substituted pyrazolones. *Mater. Chem. Phys.* **2003**, *81*, 786–792. [CrossRef]
23. Tomio, A.; Sagara, M.; Doi, T.; Amaya, H.; Otsuka, N.; Kudo, T. Role of alloyed molybdenum on corrosion resistance of austenitic Ni-Cr-Mo-Fe alloys in H_2S-Cl^- environments. *Corros. Sci.* 2015; 98, 391–398.

entropy

MDPI

Article

Effect of Molybdenum Additives on Corrosion Behavior of $(CoCrFeNi)_{100-x}Mo_x$ High-Entropy Alloys

Wenrui Wang [1,*], Jieqian Wang [1], Honggang Yi [1], Wu Qi [1] and Qing Peng [2]

1 School of Mechanical Engineering, University of Science and Technology Beijing, Beijing 100083, China; jieqian.wang@xs.ustb.edu.cn (J.W.); hg.yi@ustb.edu.cn (H.Y.); wu.qi@xs.ustb.edu.cn (W.Q.)
2 Nuclear Engineering and Radiological Sciences, University of Michigan, Ann Arbor, MI 48109, USA; qpeng@umich.edu
* Correspondence: gmbitwrw@ustb.edu.cn; Tel.: +86 010-6233-4845

Received: 6 October 2018; Accepted: 26 November 2018; Published: 28 November 2018

Abstract: The present work investigates the influence of micro-alloyed Mo on the corrosion behavior of $(CoCrFeNi)_{100-x}Mo_x$ high-entropy alloys. All of the $(CoCrFeNi)_{100-x}Mo_x$ alloys exhibit a single face-centered cubic (FCC) solid solution. However, the $(CoCrFeNi)_{97}Mo_3$ alloy exhibits an ordered sigma (σ) phase enriched in Cr and Mo. With the increase of x (the Mo content) from 1 to 3, the hardness of the $(CoCrFeNi)_{100-x}Mo_x$ alloys increases from 124.8 to 133.6 Vickers hardness (HV), and the compressive yield strength increases from 113.6 MPa to 141.1 MPa, without fracture under about a 60% compressive strain. The potentiodynamic polarization curve in a 3.5% NaCl solution indicates that the addition of Mo has a beneficial effect on the corrosion resistance to some certain extent, opposed to the σ phase. Furthermore, the alloys tend to form a passivation film in the 0.5 M H_2SO_4 solution in order to inhibit the progress of the corrosion reaction as the Mo content increases.

Keywords: $(CoCrFeNi)_{100-x}Mo_x$ alloys; high entropy alloy; microstructure; mechanical properties; corrosion behavior

1. Introduction

Traditional alloys only have one major element as a matrix [1]. With the increase of the amount of alloying elements and the concentration of minor elements, the alloy forms a fragile intermetallic phase, which not only increases the difficulty of the microstructure, but also may result in a reduction in the mechanical performance [2–4]. To overcome these difficulties, high entropy alloys (HEAs) are invented with extensive research interests; HEAs usually compose of five or more kinds of major elements, with the concentration of each principal element being between 5 and 35 at % [5–7]. HEAs tend to generate a face-centered cubic (FCC), body-centered cubic (BCC), or hexagonal closed-packed (HCP) multicomponent solid solution phase [8,9]. Some HEAs have been confirmed to achieve a series of excellent properties, such as high strength, high hardness, and glorious corrosion resistance [10–12].

The CoCrFeNi alloy has received extensive attention for its outstanding corrosion resistance, ductility, and structure stability [13,14]. However, because of the poor mechanical strength, the application of the CoCrFeNi alloy in engineering has been limited. It has been shown that an ordered σ strengthening phase can be formed by a certain amount of Mo additives, resulting in precipitation strengthening [15–17]. However, the excessive addition of Mo causes a large amount of the coarse σ strengthening phase, which may lead to a rapid increase of the alloy brittleness. Furthermore, because of the low electronic potential of Mo, the excessive content of Mo may reduce the corrosion resistance [18–20]. Referring to the chemical composition of austenitic stainless steel, the content of Mo in stainless steel is generally less than 3 wt %. Therefore, we also supply a small amount of Mo in addition to the CoCrFeNi alloy, in order to study the corrosion resistance of the $(CoCrFeNi)_{100-x}Mo_x$ alloy for the development of an HEA system, with a good performance of both strength and corrosion resistance.

In this paper, the as-cast $(CoCrFeNi)_{100-x}Mo_x$ (x =1, 2, and 3 wt %) HEAs have been prepared mainly by vacuum arc melting. The excellent corrosion behavior was investigated by the electrochemical experiments.

2. Materials and Methods

2.1. Samples Fabrication

Elements Co, Cr, Fe, Ni, and Mo with purities of over 99.9 wt % were prepared as raw materials previously. The as-cast $(CoCrFeNi)_{100-x}Mo_x$ (x = 1, 2, and 3, represented by Mo_1, Mo_2, and Mo_3, respectively) were prepared by vacuum arc melting and were fabricated under the WK-∏ vacuum arc melting furnace at least five times in the crucible, so as to ensure the chemical homogeneity. The size of the ingot was approximately Φ 35 × 10 mm, and the ingot was annealed for 3 h under 500 °C and was cooled in the air so as to release the residual stress caused by rapid cooling during casting.

2.2. Microstructure of the $(CoCrFeNi)_{100-x}Mo_x$ Alloys

The crystalline phases of the $(CoCrFeNi)_{100-x}Mo_x$ alloys were identified by X-ray diffraction (XRD), using an Ultima IV X-ray diffractometer with Cu K_α radiation. The X-ray diffractometer has an operating voltage of 30 kV and an operating current of 20 mA with the diffraction angle (2θ) from 20 to 90°, at a scanning rate of 4°/min.

Etching the $(CoCrFeNi)_{100-x}Mo_x$ alloys with aqua regia (HNO_3: HCl = 1:3, volume fraction) and using FE-SEM JEOL JEM-7600F scanning-electron microscopy (SEM) (JEOL Ltd., Tokyo, Japan) equipped with an energy dispersive spectrometer (EDS) allowed for an analysis of the microstructure and composition.

2.3. Mechanical Properties

The microhardness was measured using a Wolpert-401MVD Vickers hardness tester (WOLPERT Co., Norwood, MA, USA) with loads of 500 g and a duration of 10 s. The measurements were performed at 10 different locations on each sample, and the average value of the 10 measurements was calculated. Compressive tests were carried out on the Φ 3 × 6 mm samples, using the universal testing machine (CMT 4305) (MTS Co., Eden Prairie, MN, USA) with a strain rate of 10^{-3} s^{-1}.

2.4. Electrochemical Corrosion Test

The electrochemical experiments were performed on $(CoCrFeNi)_{100-x}Mo_x$ alloys, using the Potentiostat Workstation Versa STAT MC (PARSTAT 4000, AMETEK Co., Princeton, NJ, USA). A three-electrode electrochemical cell using a saturated calomel electrode (SCE) as a reference electrode, a platinum plate as an auxiliary electrode, and a sample as a working electrode were tested. The electrochemical experiments of the $(CoCrFeNi)_{100-x}Mo_x$ alloys were conducted in a 3.5% NaCl and 0.5 M H_2SO_4 solution at room temperature, respectively. The potentiodynamic polarization measurements were taken at a scan rate of 1 mV/s from a potential scanning range of −0.5 V to 1.1 V.

2.5. Corroded Microstructure

The electrochemically tested alloys were cleaned using an ultrasonic cleaner, and then dried in nitrogen. SEM and EDS were used to study the morphology of the corrosion surface of the high-entropy alloys.

3. Results and Discussion

3.1. Microstructure of the $(CoCrFeNi)_{100-x}Mo_x$ Alloys

Figure 1 represents the XRD pattern of the $(CoCrFeNi)_{100-x}Mo_x$ (x = 1, 2, 3) alloys. All of the samples of alloys have a single FCC solid solution structure, which was confirmed by the predecessors [21]. As the Mo content increases, the peak intensity changes, but the FCC phase is kept. As the Mo content becomes 3 wt % (Mo_3 structure), the small peak on the left of the matrix FCC phase in the XRD patterns is identified as Cr and Mo rich σ phase, which agrees with the previous remarks [22].

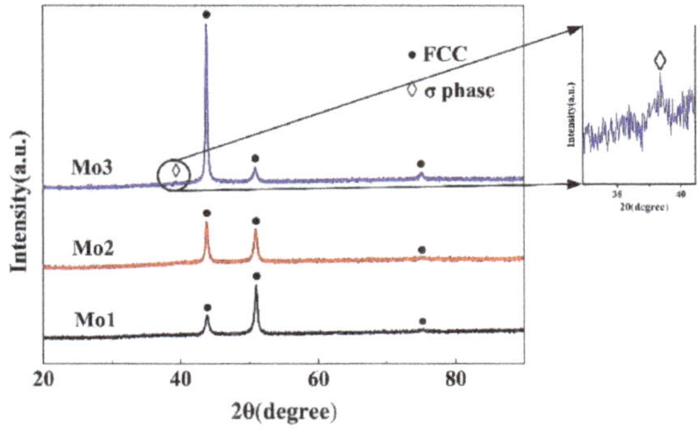

Figure 1. XRD (X-ray diffraction) patterns of the $(CoCrFeNi)_{100-x}Mo_x$ (x = 1, 2, 3) high entropy alloys.

The SEM images of the $(CoCrFeNi)_{100-x}Mo_x$ (x = 1, 2, and 3) alloys presented in Figure 2 shows that the alloys are composed of typical dendritic structures. Region A is interdendrites and region B is dendrites. Figure 2c presents the SEM image of the Mo_3 alloy, and region D is the grain boundary. Figure 2d is a part of the SEM of the Mo_3 alloy. The EDS results of the $(CoCrFeNi)_{100-x}Mo_x$ alloys are shown in Table 1. According to the EDS, the dendrite is the Co and Fe rich phase, and the interdendrite is the Cr and Mo rich phase. Combined with the XRD and EDS results, the Mo_3 alloy exhibits Cr and Mo rich σ phase in the interdendrite. When the content of Mo is 1 and 2 wt %, there is no formation of a precipitate phase in the alloy because of the high entropy effect. However, HEAs undergo spinodal decomposition inside the crystal grains during cooling, leading to the formation of microstructures with the same structure but different compositions.

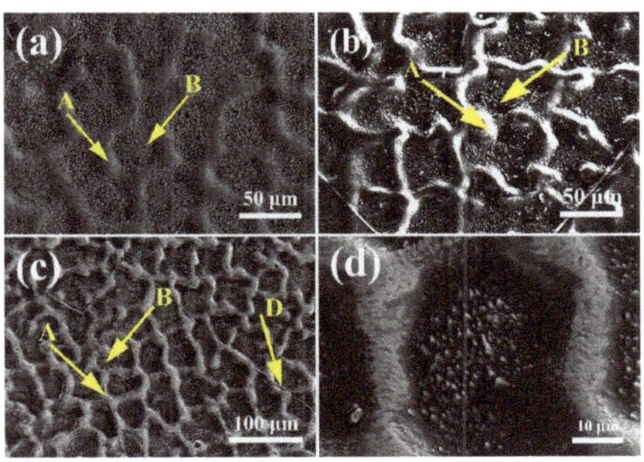

Figure 2. SEM (scanning-electron microscopy) images of the $(CoCrFeNi)_{100-x}Mo_x$ high-entropy alloys: (**a**) x = 1; (**b**) x = 2; (**c**) x = 3; (**d**) 10 × magnification.

Table 1. Element concentration determined using the energy dispersive spectrometer (EDS) of the three samples of the $(CoCrFeNi)_{100-x}Mo_x$ alloys (at %).

Element		Cr	Fe	Co	Ni	Mo
Mo_1	Dendrite	25.25	23.54	26.07	24.53	0.61
	interdendrite	27.69	22.31	24.03	24.94	1.04
Mo_2	Dendrite	26.02	25.17	24.55	23.56	0.69
	interdendrite	27.20	23.20	23.33	23.87	2.40
Mo_3	Dendrite	24.99	23.47	25.28	24.96	1.31
	interdendrite	26.82	22.54	24.81	24.35	1.50

3.2. Mechanical Properties

Figure 3 shows the Vickers hardness (HV) of the $(CoCrFeNi)_{100-x}Mo_x$ alloys as a function of the Mo content. The alloy hardness increases from 124.83 to 133.60 HV, with the Mo content increasing from 1 to 3 wt %. When the content of the Mo element is 3 wt %, the presence of the σ phase results in a remarkable increase in the hardness.

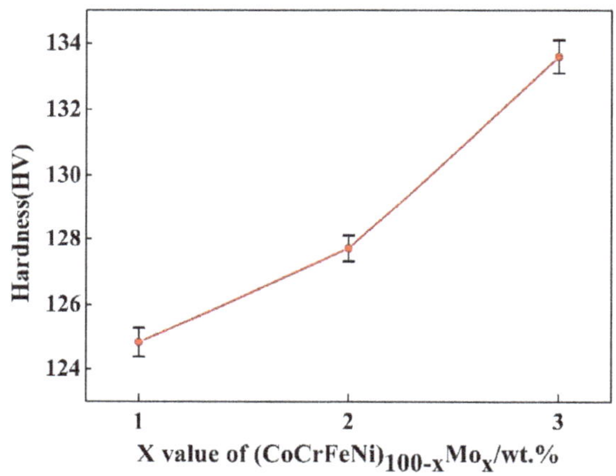

Figure 3. Vickers hardness of the $(CoCrFeNi)_{100-x}Mo_x$ high-entropy alloys as a function of the Mo content.

Besides the hardness, we have examined the stress–strain relationships. The compressive stress–strain curves and the inner longitudinal-section SEM images of the $(CoCrFeNi)_{100-x}Mo_x$ (x = 1, 2, and 3) alloys are shown in Figure 4. After yielding, the strength of the alloys increases continuously. All of the three samples do not break under about a 60% compressive strain, indicating that the alloys possess a good ductility, flexibility, and fracture strain. As shown in Figure 4b–d, the deformation of the Mo_3 subgrain boundaries is more prominent in the angle of 45°, probably due to the resolved shear stress.

Table 2 lists the mechanical properties of the $(CoCrFeNi)_{100-x}Mo_x$ alloys. The Mo_1 and Mo_2 alloys exhibit a similar behavior. The yield stress of the Mo_3 alloy increases significantly because of the second-phase hardening by the σ phase, as reported in the literature [23,24].

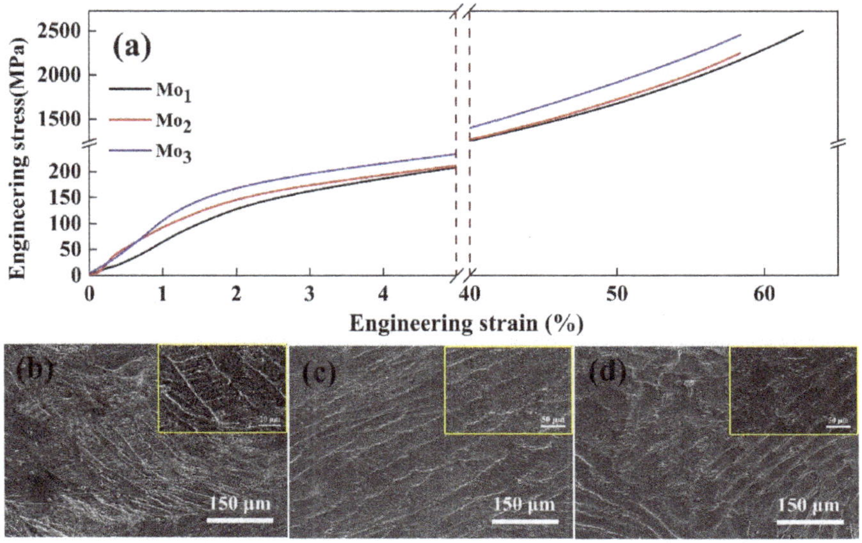

Figure 4. (**a**) Compressive stress–strain curves and the inner longitudinal-section SEM images of the $(CoCrFeNi)_{100-x}Mo_x$ alloys after compression deformation: (**b**) $x = 1$; (**c**) $x = 2$; (**d**) $x = 3$.

Table 2. Mechanical properties (yield stress, compressive strength, and fracture strain) of the $(CoCrFeNi)_{100-x}Mo_x$ alloys for $x = 1$, 2, and 3 as Mo_1, Mo_2, and Mo_3, respectively.

Alloy	Yield Stress σ_y (MPa)	Compressive Strength σ_{max} (MPa)	Fracture Strain ε_p (%)
Mo_1	113.6	Not fractured	>60
Mo_2	119.7	Not fractured	>60
Mo_3	141.1	Not fractured	>60

3.3. *Environmental Effect on Corrosion Behavior*

3.3.1. Corrosion Behavior in Chloride-Containing Solutions

Figure 5 lists the polarization curve of the $(CoCrFeNi)_{100-x}Mo_x$ ($x = 1$, 2, and 3) alloys in a 3.5% NaCl solution. The corrosion potential of the alloys gradually shifts to more positive potentials with a decreasing Mo content.

Table 3 presents the electrochemical parameters of the $(CoCrFeNi)_{100-x}Mo_x$ ($x = 1$, 2, and 3) alloys in a 3.5% NaCl solution. The corrosion current densities (i_{corr}) of the Mo_1, Mo_2, and Mo_3 alloys were 0.4, 0.24, and 6.6 $\mu A/cm^2$, and the corrosion potentials (E_{corr}) were −199, −277, and −493 mV, respectively. The breakdown potential (E_b) gradually shifts to more positive potentials. E_{pit} is a primary passivation potential. ΔE is the passive region width, defined as the difference between the E_b and E_{pit}. The i_{corr} value of the Mo_3 alloy is an order of magnitude higher than the other two. The corrosion resistance of Mo_3 was dropped. Therefore, the presence of the Cr and Mo rich σ phase in the Mo_3 alloy leads to the diminution of the corrosion resistance.

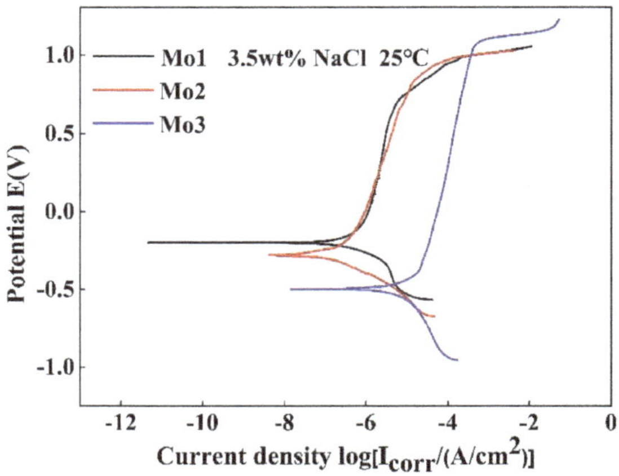

Figure 5. Polarization curves of the $(CoCrFeNi)_{100-x}Mo_x$ (x = 1, 2, and 3) alloys in a 3.5% NaCl solution.

Table 3. Electrochemical parameters of the $(CoCrFeNi)_{100-x}Mo_x$ alloys in a 3.5% NaCl solution.

Alloy	E_{corr} (mV)	I_{corr} (μA/cm^2)	E_b (mV)	E_{pit} (mV)	ΔE (mV)
Mo_1	−199	0.402	992	−5	997
Mo_2	−277	0.235	968	−108	1076
Mo_3	−493	6.610	1053	−358	1411

Figure 6 presents the microstructure of the $(CoCrFeNi)_{100-x}Mo_x$ alloys after potentiodynamic polarization in a 3.5% NaCl solution. DR is dendrite and IR is interdendrite. According to the SEM images, the majority of types of corrosion were mainly the pitting corrosion. The Mo_3 alloy is more susceptible to pitting corrosion, which is consistent with the polarization curve results. Region A is a Cr-rich phase and region B is a Mo-rich phase. There is an element of segregation that causes the corrosion to occur. The interdendritic phase of Mo_3 is the Cr- and Mo-rich phase, and the dendrite is a Cr- and Mo-depleted phase. Therefore, galvanic corrosion occurred at the junctions of dendrites. The XRD and SEM result show that the Mo and Cr rich σ phase appeared inside the interdendrite, the corrosion occurred at the interfaces around σ phase, as shown in Figure 6c,d. The results of the EDS of the alloy after corrosion are summarized in Table 4.

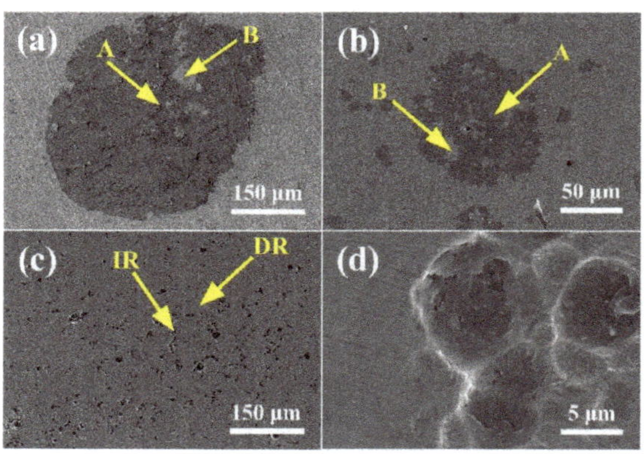

Figure 6. SEM images of $(CoCrFeNi)_{100-x}Mo_x$ after potentiodynamic polarization in a 3.5% NaCl solution: (**a**) x = 1; (**b**) x = 2; (**c**) x = 3; (**d**) the interdendrite morphology of the Mo_3 alloy.

Table 4. EDS results for the Mo_3 alloy after the potentiodynamic polarization in a 3.5% NaCl solution.

Element	Region	Cr (%)	Fe (%)	Co (%)	Ni (%)	Mo (%)
Mo_1	A	26.17	24.56	23.32	25.01	0.93
	B	30.79	23.79	22.87	22.51	0.04
Mo_2	A	26.44	23.72	24.32	23.83	1.70
	B	35.81	23.79	19.78	20.11	0.52
Mo_3	DR	24.96	25.21	24.72	24.88	0.23
	IR	29.06	23.21	22.43	22.76	2.54

3.3.2. Corrosion Behavior in Acid Solutions

Figure 7 shows the polarization curve of the $(CoCrFeNi)_{100-x}Mo_x$ (x = 1, 2, and 3) alloys in 0.5 M H_2SO_4. Table 5 presents the electrochemical parameters of the $(CoCrFeNi)_{100-x}Mo_x$ (x = 1, 2, and 3) alloys in 0.5 M H_2SO_4. The i_{pp} is the lunt current density. It can be seen from the electrochemical parameters that the i_{corr} was 34.1, 28.0, and 15.4 $\mu A/cm^2$, respectively. The E_{corr} shifted to more positive potentials, and the i_{corr} value dropped as the Mo content increased. This suggests that the alloys tend to form a passivation film to inhibit the progress of the corrosion reaction.

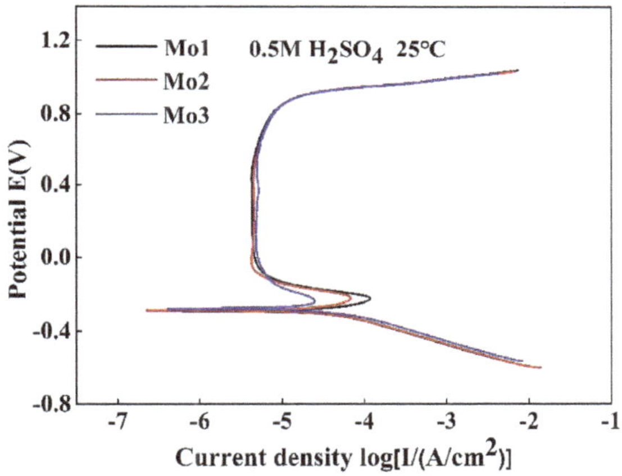

Figure 7. Polarization curves of the $(CoCrFeNi)_{100-x}Mo_x$ (x = 1, 2, and 3) alloys in a 0.5 M H_2SO_4 solution

Table 5. Electrochemical parameters of the $(CoCrFeNi)_{100-x}Mo_x$ alloys in a 0.5 M H_2SO_4 solution.

Alloy	E_{corr} (mV)	I_{corr} ($\mu A/cm^2$)	E_{pp} (mV)	i_{pp} ($\mu A/cm^2$)	ΔE (mV)
Mo_1	−294	34.1	−225	117	655
Mo_1	−290	28.0	−225	68	681
Mo_1	−279	15.4	−239	25	751

Figure 8 shows the SEM microstructure of the $(CoCrFeNi)_{100-x}Mo_x$ alloys after potentiodynamic polarization in a 0.5 M H_2SO_4 solution, combined with EDS, because the low potential of Mo is enriched and the Cr_2O_3 is insufficient in region A. Consequently, region A is more susceptible to corrosion in the H_2SO_4 solution. When the content of Mo is 3 wt %, as Figure 8c indicates, the results of the EDS analysis show that the concentration of element Cr in region A is reduced, and the content of Mo in region B is significantly higher than that in region A, the effect of Mo was to form Mo (VI) oxyhydroxide or molybdate ($MoO_4{}^{2-}$), decreasing the rate of dissolution in active zones. Hence, the corrosion of the Mo_3 alloy is concentrated in region A.

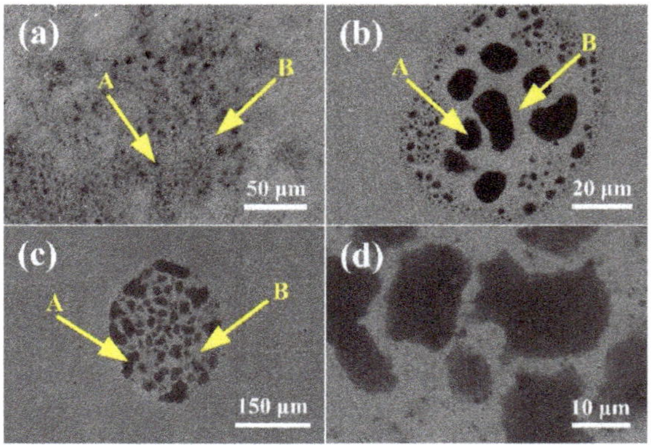

Figure 8. SEM images of $(CoCrFeNi)_{100-x}Mo_x$ after potentiodynamic polarization in a 0.5 M H_2SO_4 solution: (**a**) $x = 1$; (**b**) $x = 2$; (**c**) $x = 3$; (**d**) partial view of the Mo_3 alloy.

Figure 9a shows the comparison of the corrosion behavior between the HEAs and the conventional corrosion resistant alloys in a 3.5% NaCl solution, compared with those of the conventional corrosion resistant alloys [25,26]. The HEAs are located in the upper part of Figure 9a, the E_{corr} of the HEAs are more positive than those of the Mn alloys, Ni alloys, and some of the Ti alloys. On the other hand, the i_{corr} of the HEAs are much lower than some of Mn alloys and are comparable with the Ti alloys, which indicates that the corrosion resistance of the HEAs is comparable or even better than those of the conventional alloys. However, the partial i_{corr} of the HEAs is higher than that of the total, because the presence of the σ phase is catastrophic for HEAs. Figure 9b presents the comparison of the corrosion behavior between the HEAs and the conventional corrosion resistant alloys in the 0.5 M H_2SO_4 solution. Compared with the conventional alloys [27,28], the E_{corr} of the HEAs are much more positive than those of the Ti alloys and Ni alloys. The i_{corr} of the HEAs are much lower than the Ti alloys, Ni alloys, and some of the Cu alloys. As a general trend, the corrosion resistance of the HEAs in the 0.5 M H_2SO_4 solution is better than those of the conventional alloys.

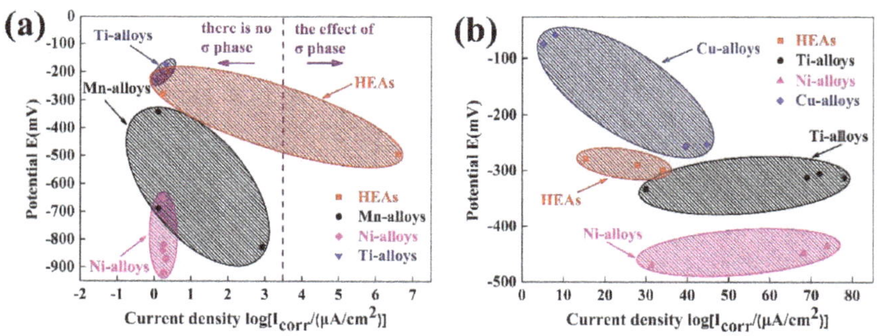

Figure 9. Comparison of the i_{corr} and E_{corr} between high entropy alloys (HEAs) and conventional alloys: (**a**) in a 0.5 M NaCl solution; (**b**) in a 0.5 M H_2SO_4 solution.

4. Conclusions

The $(CoCrFeNi)_{100-x}Mo_x$ ($x = 1$, 2, and 3) alloys have been synthesized. Their microstructures, mechanical properties, and corrosion behaviors have been experimentally investigated. The microstructures of the $(CoCrFeNi)_{100-x}Mo_x$ ($x = 1$, 2, and 3) alloys belong to a single FCC structure. The increase in Mo promotes the formation of the Cr- and Mo-rich σ phase. The hardness and compressive yield strength increase obviously with an increase of the Mo content from 1 to 3 wt %. Regarding the potentiodynamic polarization curves of the $(CoCrFeNi)_{100-x}Mo_x$ alloys in a 3.5% NaCl solution, the curves of the Mo_1 and Mo_2 alloys indicated that the increase of the Mo content increases the corrosion resistance of the chloride environment to some extent. However, the Cr and Mo rich σ phase is present at the grain boundary of the Mo_3 alloy, resulting in a decreasing in corrosion resistance in the 3.5% NaCl solution. Furthermore, the potentiodynamic polarization curves of the $(CoCrFeNi)_{100-x}Mo_x$ alloys in the 0.5 M H_2SO_4 solution yielded an extensive passive region, and as the content of the Mo increased, the corrosion current density gradually decreased. Therefore, in an acidic solution, the addition of Mo has a positive effect on the corrosion resistance of the $(CoCrFeNi)_{100-x}Mo_x$ alloys.

Author Contributions: Conceptualization, W.W.; formal analysis, J.W.; methodology, J.W. and W.Q.; resources, W.W.; writing (original draft), J.W.; writing (review and editing), J.W., H.Y. and Q.P.

Funding: The authors are grateful for the financial support provided by the Fundamental Research Funds for the Central Universities (FRF-TP-16-044A1 and FRF-GF-17-B18) and the National Natural Science Foundation of China (21703007).

Acknowledgments: The authors would like to thank the project team assistance with funding. Many thanks to the tutor for his thoughtful and thorough guidance.

Conflicts of Interest: The authors declare no conflict of interest.

References

1. Peng, Q.; Meng, F.; Yang, Y.; Lu, C.; Deng, H.; Wang, L.; De, S.; Gao, F. Shockwave generates <100> dislocation loops in bcc iron. *Nat. Commun.* **2018**, *9*, 4880. [CrossRef] [PubMed]
2. Ye, Q.; Feng, K.; Li, Z.; Lu, F.; Li, R.; Huang, J.; Wu, Y. Microstructure and corrosion properties of CrMnFeCoNi high entropy alloy coating. *Appl. Surface Sci.* **2017**, *396*, 1420–1426. [CrossRef]
3. Shun, T.T.; Chang, L.Y.; Shiu, M.H. Age-hardening of the $CoCrFeNiMo_{0.85}$ high-entropy alloy. *Mater. Charact.* **2013**, *81*, 92–96. [CrossRef]
4. Senkov, O.N.; Zhang, F.; Miller, J.D. Phase composition of a $CrMo_{0.5}NbTa_{0.5}TiZr$ high entropy alloy: Comparison of experimental and simulated data. *Entropy* **2013**, *15*, 3796–3809. [CrossRef]
5. Qiu, X.W.; Zhang, Y.P.; He, L.; Liu, C.-G. Microstructure and corrosion resistance of AlCrFeCuCo high entropy alloy. *J. Alloys Compd.* **2013**, *549*, 195–199. [CrossRef]
6. Zhang, Y.; Tang, T.T.Z.Z.; Gao, M.C.; Dahmen, K.A.; Liaw, P.K.; Lu, Z.P. Microstructures and properties of high-entropy alloys. *Prog. Mater. Sci.* **2014**, *61*, 1–93. [CrossRef]
7. Tung, C.C.; Yeh, J.W.; Shun, T.-T.; Chen, S.K.; Huang, Y.S.; Chen, H.C. On the elemental effect of AlCoCrCuFeNi high-entropy alloy system. *Mater. Lett.* **2007**, *61*, 1–5. [CrossRef]
8. Yeh, J.W. Recent progress in high entropy alloys. *Ann. Chim. Sci. Mat.* **2006**, *31*, 633–648. [CrossRef]
9. Zhang, H.; Pan, Y.; He, Y.Z. Synthesis and characterization of FeCoNiCrCu high-entropy alloy coating by laser cladding. *Mater. Des.* **2011**, *32*, 1910–1915. [CrossRef]
10. Chen, Y.; Duval, T.; Hung, U.; Yeh, J.; Shih, H. Microstructure and electrochemical properties of high entropy alloys—A comparison with type-304 stainless steel. *Corros. Sci.* **2005**, *47*, 2257–2279. [CrossRef]
11. Qiu, X.-W.; Wu, M.J.; Liu, C.G.; Zhang, Y.P.; Huang, C.-X. Corrosion performance of $Al_2CrFeCoxCuNiTi$ high-entropy alloy coatings in acid liquids. *J. Alloys Compd.* **2017**, *708*, 353–357. [CrossRef]
12. Zhang, Y.; Yan, X.H.; Liao, W.B.; Zhao, K. Effects of nitrogen content on the structure and mechanical properties of $(Al_{0.5}CrFeNiTi_{0.25})$ N_x high-entropy films by reactive sputtering. *Entropy* **2018**, *20*, 624. [CrossRef]
13. Hsu, Y.J.; Chiang, W.C.; Wu, J.K. Corrosion behavior of $FeCoNiCrCu_x$ high-entropy alloys in 3.5% sodium chloride solution. *Mater. Chem. Phys.* **2005**, *92*, 112–117. [CrossRef]
14. Shi, Y.; Yang, B.; Liaw, P.K. Corrosion-resistant high-entropy alloys: A review. *Metals* **2017**, *7*, 43. [CrossRef]
15. Zhu, J.; Zhang, H.; Fu, H.; Wang, A.; Li, H.; Hu, Z. Microstructures and compressive properties of multicomponent $AlCoCrCuFeNiMo_x$ alloys. *J. Alloys Compd.* **2010**, *497*, 52–56. [CrossRef]
16. Zhu, J.; Fu, H.; Zhang, H.; Wang, A.; Li, H.; Hu, Z. Microstructures and compressive properties of multicomponent $AlCoCrFeNiMo_x$ alloys. *Mater. Sci. Eng. A* **2010**, *527*, 6975–6979. [CrossRef]
17. Miao, J.; Guo, T.; Ren, J.; Zhang, A.; Su, B.; Meng, J. Optimization of mechanical and tribological properties of FCC CrCoNi multi-principal element alloy with Mo addition. *Vacuum* **2018**, *149*, 324–330. [CrossRef]

18. Nilsson, J.O.; Kangas, P.; Wilson, A.; Karlsson, T. Mechanical properties, microstructural stability and kinetics of σ-phase formation in 29Cr-6Ni-2Mo-0.38 N superduplex stainless steel. *Metall. Mater. Trans. A* **2000**, *31*, 35–45. [CrossRef]
19. Lopez, N.; Cid, M.; Puiggali, M. Influence of o-phase on mechanical properties and corrosion resistance of duplex stainless steels. *Corros. Sci.* **1999**, *41*, 1615–1631. [CrossRef]
20. Schwind, M.; Källqvist, J.; Nilsson, J.O.; Ågren, J.; Andrén, H.O. σ-phase precipitation in stabilized austenitic stainless steels. *Acta Mater.* **2000**, *48*, 2473–2481. [CrossRef]
21. Tsau, C.H.; Tsai, M.C. The effects of Mo and Nb on the microstructures and properties of CrFeCoNi (Nb, Mo) alloys. *Entropy* **2018**, *20*, 648. [CrossRef]
22. Chou, Y.; Yeh, J.; Shih, H. The effect of molybdenum on the corrosion behaviour of the high-entropy alloys $Co_{1.5}CrFeNi_{1.5}Ti_{0.5}Mo_x$ in aqueous environments. *Corros. Sci.* **2010**, *52*, 2571–2581. [CrossRef]
23. Zhang, L.; Yu, P.; Zhang, M.; Liu, D.; Zhou, Z.; Ma, M.; Liaw, P.; Li, G.; Liu, R. Microstructure and mechanical behaviors of GdxCoCrCuFeNi high-entropy alloys. *Mater. Sci. Eng. A* **2017**, *707*, 708–716. [CrossRef]
24. Li, T.; Liu, B.; Liu, Y.; Guo, W.; Fu, A.; Li, L.; Yan, N.; Fang, Q. Microstructure and mechanical properties of particulate reinforced NbMoCrTiAl High Entropy based composite. *Entropy* **2018**, *20*, 517. [CrossRef]
25. Wu, C.; Zhang, S.; Zhang, C.; Zhang, H.; Dong, S. Phase evolution and cavitation erosion-corrosion behavior of $FeCoCrAlNiTi_x$ high entropy alloy coatings on 304 stainless steel by laser surface alloying. *J. Alloys Compd.* **2017**, *698*, 761–770. [CrossRef]
26. Qiu, X.W.; Liu, C.G. Microstructure and properties of $A_{l2}CrFeCoCuTiNi_x$ high-entropy alloys prepared by laser cladding. *J. Alloys Compd.* **2013**, *553*, 216–220. [CrossRef]
27. Lee, C.; Chang, C.; Chen, Y.; Yeh, J.; Shih, H. Effect of the aluminium content of $Al_xCrFe_{1.5}MnNi_{0.5}$ high-entropy alloys on the corrosion behaviour in aqueous environments. *Corros. Sci.* **2008**, *50*, 2053–2060. [CrossRef]
28. Xiao, D.; Zhou, P.; Wu, W.; Diao, H.; Gao, M.; Song, M.; Liaw, P. Microstructure, mechanical and corrosion behaviors of AlCoCuFeNi-(Cr, Ti) high entropy alloys. *Mater. Des.* **2017**, *116*, 438–447. [CrossRef]

entropy

MDPI

Article

Effect of Zr Addition on the Microstructure and Mechanical Properties of CoCrFeNiMn High-Entropy Alloy Synthesized by Spark Plasma Sintering

Hongling Zhang [1], Lei Zhang [1], Xinyu Liu [2], Qiang Chen [1] and Yi Xu [1,*]

1 School of materials Science & Engineering, Southwest Jiaotong University, Chengdu 610031, China; zhanghl@my.swjtu.edu.cn (H.Z.); zhanglei_5367@163.com (L.Z.); 18684016703@163.com (Q.C.)

2 Chengdu Advanced Metal Materials Industry Technology Research Institute Co., Ltd, Chengdu 610031, China; cgvermouth@gmail.com

* Correspondence: xybwbj@swjtu.cn; Tel.: +86-28-87600782

Received: 12 September 2018; Accepted: 16 October 2018; Published: 23 October 2018

Abstract: As a classic high-entropy alloy system, CoCrFeNiMn is widely investigated. In the present work, we used ZrH_2 powders and atomized CoCrFeNiMn powders as raw materials to prepare $CoCrFeNiMnZr_x$ (x = 0, 0.2, 0.5, 0.8, 1.0) alloys by mechanical alloying (MA), followed by spark plasma sintering (SPS). During the MA process, a small amount of Zr ($x \leq 0.5$) can be completely dissolved into CoCrFeNiMn matrix, when the Zr content is above 0.5, the ZrH_2 is excessive. After SPS, CoCrFeNiMn alloy is still as single face-centered cubic (FCC) solid solution, and $CoCrFeNiMnZr_x$ ($x \geq 0.2$) alloys have two distinct microstructural domains, one is a single FCC phase without Zr, the other is a Zr-rich microstructure composed of FCC phase, B2 phase, Zr_2Ni_7, and σ phase. The multi-phase microstructures can be attributed to the large lattice strain and negative enthalpy of mixing, caused by the addition of Zr. It is worth noting that two types of nanoprecipitates (body-centered cubic (BCC) phase and Zr_2Ni_7) are precipitated in the Zr-rich region. These can significantly increase the yield strength of the alloys.

Keywords: high-entropy alloy; mechanical alloying; spark plasma sintering; nanoprecipitates; mechanical properties

1. Introduction

During the past few years, high-entropy alloys (HEAs), a type of multi-principal-element alloy, have drawn widespread attention form worldwide material scientists [1–3]. Compared to conventional alloys, which are composed of one or two major elements, HEAs usually contain 5–13 principal elements, and the content of each component is between 5 and 35 at.% [4]. Due to the high mixing entropy in the multi-component alloy systems, HEAs generally form a simple face-centered cubic (FCC) or body-centered cubic (BCC) structure solid solution and exhibit many novel properties, for example, superb mechanical properties [5,6], outstanding resistances to wear [7], oxidation [8] and corrosion [9], as well as a good temperature stability [10]. Therefore, the concept of HEAs provides a new approach to design alloys with excellent properties to meet different environmental requirements.

Casting [11] and powder metallurgy [12] are two common methods for preparing HEAs. For casting, there are two kinds of defects which are the vaporization of low melting-point elements and the elements segregation in the ingot [13]. For powder metallurgy, the HEAs can be prepared by MA and then consolidated. The HEAs powders prepared by MA are uniform and fine, and even form nanocrystalline. Hence a rapid sintering is needed to maintain this microstructure [14]. SPS is a new field-assisted sintering technique which can effectively suppress the grain coarsening by rapid heating and efficient densification in a few minutes [15–17]. Compared with MA, gas atomization can produce more homogeneous powders. The preparation of HEAs powders by gas atomization is a rapid solidification process [18], in which grain growth and element segregation are inhibited, and the high cooling rate even leads to the formation of an amorphous phase [14].

In the previous studies, two classical HEAs, CoCrFeNiMn with a single FCC phase and AlCoCrFeNi with a single BCC phase, have been extensively studied. Among these studies, there are lots of experiments that have explored the influence of alloying elements such as Mo [5], Nb [6], Ti [19], Si [20], and Zr [21] on the microstructure and properties of AlCoCrFeNi. The addition of these alloying elements leads to the formation of intermetallic compounds and significantly improves the yield strength of AlCoCrFeNi. For example, after the addition of Zr, Laves phase appears in the BCC-phase matrix, the yield strength of AlCoCrFeNi increases from 1320 MPa to 1560 MPa, and the plastic strain increases from 22.5% to 29.5%. It is feasible to change the phase composition and improve the mechanical properties of AlCoCrFeNi by adding an alloying element. However, there are little similar data on the CoCrFeNiMn alloy [22,23]. In the present work, the CoCrFeNiMn alloy was selected as the matrix and the effect of Zr addition was investigated. There are several reasons for choosing Zr as an alloying element. Firstly, both Zr and Ti are located in the same family in the periodic table of elements, so they have some similar physico-chemical properties. In previous research, the addition of Ti can result in precipitates and improve the superplasticity of CoCrFeNiMn [23]. Secondly, the large atomic size difference and negative enthalpies of mixing [24] between Zr and other constituent elements can produce a strong lattice distortion and even change the phase composition, which may improve the mechanical properties of the CoCrFeMnNi.

In this paper, we focused on the novel high-entropy alloy system of $CoCrFeNiMnZr_x$ (x = 0, 0.2, 0.5, 0.8, 1.0). First of all, CoCrFeNiMn metallic powders were prepared by gas atomization, and then, Zr derived from the in-situ dehydrogenation of ZrH_2 powders was solid-dissolved into CoCrFeNiMn by MA. The $CoCrFeNiMnZr_x$ alloy powders were subsequently sintered by SPS. The alloying behavior, microstructures, and mechanical properties of both powders and sintered alloys were investigated.

2. Experimental

Equimolar ratio CoCrFeNiMn powders were prepared by gas atomization with high purity Ar. The atomization pressure was 4 MPa. Then, using CoCrFeNiMn powders and ZrH_2 powders as raw materials, the $CoCrFeNiMnZr_x$ (x values in molar ratio, x = 0, 0.2, 0.5, 0.8 and 1.0, denoted by Zr_0, $Zr_{0.2}$, $Zr_{0.5}$, $Zr_{0.8}$, and $Zr_{1.0}$, respectively) alloys in nominal compositions were prepared by high-energy ball milling (referred as MA) and SPS. The reaction (ZrH_2 = Zr(s) + H_2(g)) occurs during MA and SPS. In this experiment, the mixed powders and zirconia balls (5 mm and 10 mm in diameter with the mass ration of 1:1) were put into stainless-steel vials at a mass ratio of 10:1, then mounted on a planetary ball miller (QM-S3P4, Nanjing NanDa Instrument Plant, Nanjing, China) and milled at a rate of 300 rpm under the protection of argon gas. All the mixed powders were milled for 30 h and stopped for 10 min every 20 min to prevent the powders from overheating. Beforehand, in order to study the alloying behavior, mixed $Zr_{0.5}$ alloy powders were milled for 50 h, and a small amount of powder was taken every 10 h for X-ray diffraction tests. Finally, the gas-atomized CoCrFeNiMn powders without ball milling (Zr_0w) and the alloy powders with 30 h of ball milling were sintered by SPS (Dr. Sinter-3.20 MKII, SCM, Japan) at 900 °C for 8 min under a uniaxial pressure of 40 MPa with a vacuum of 1×10^{-3} Pa. The sintered columnar samples have a diameter of 30 mm and a height of 10 mm.

The oxygen content of the CoCrFeNiMn powder was analyzed by the fusion method on an O/N analyzer (736 series, LECO, Saint Joseph, MI, USA). X-ray diffraction with Cu Kα radiation (XRD, XPertPowder, PANalytical, Almelo, Netherlands) was used to analyze the phase composition of the alloy powder and bulk at a speed of 4°/min, and the 2 thetas ranging from 20° to 100°. The PDF-2 2004 database was used for the phase assignment. The morphology and chemical composition were characterized by scanning electron microscopy (SEM, Quanta, FEI, Hillsboro, TX, USA) equipped with energy dispersive spectrometry (EDS, Inca X-Max, Oxford instruments, Oxford, UK) operating at 20 kV. Cuboidal specimens with a side length of 10 mm and a height of 5 mm were cut from sintered samples and then ground and polished for SEM and EDS analyses. The crystal structure was identified by transmission electron microscopy (TEM, JEM2100F, JEOL, Tokyo, Japan) operating at 200 kV. The disc-shaped TEM samples with a diameter of 3 mm were electropolished (TenuPol-5, Struers, Ballerup, Denmark) in an electrolyte composed of 20 vol.% perchloric acid and 80 vol.% methanol at −15 °C, and the applied voltage and current were 60 mV and 80 mA, respectively. Archimedes' method was used to measure the density of samples. Cylindrical specimens (7 mm in height and 3.5 mm in diameter) were used for the compressive tests (810 series, MTS, Minneapolis, MN, USA) at room temperature. Samples of each composition were tested for 3 times.

3. Results and Discussion

3.1. Phase and Microstructure of the Gas Atomized HEA Powders

The chemical analysis of the atomized HEA powder is shown in Table 1. As can be seen, the composition of the CoCrFeNiMn powder prepared by gas atomization was consistent with the nominal composition. The low oxygen content of the powder means that very little oxidation occurred during the atomization. The XRD pattern indicates the formation of the CoCrFeNiMn HEA powders with a single FCC phase (Figure 1a). The morphology and microstructure of the gas-atomized powders are depicted in Figure 1b,c. The atomized powders are spherical or quasi-spherical in diameter, ranging from 5 to 150 μm, and a small amount of fine powders adhere to the surface of large powders to form satellite structures. During the process of gas atomization, solidified fine droplets can easily adhere to the surface of large droplets that remain molten. The zoom-in image indicates that the microstructure consists of cellular and dendritic structures at a submicron scale.

Table 1. The chemical compositions of the gas atomized CoCrFeNiMn high-entropy alloy (HEA) powder.

Elements	Co	Cr	Fe	Ni	Mn	O
at.%	20.12	18.85	19.45	20.05	19.71	0.0365

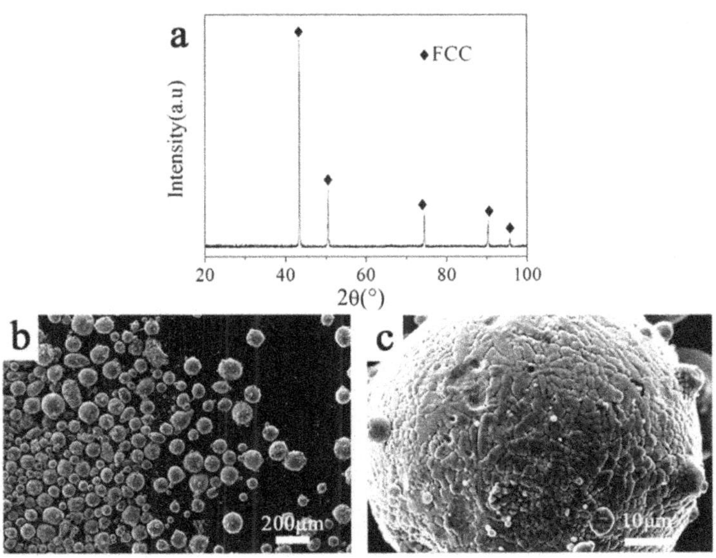

Figure 1. The X-ray diffraction (XRD) pattern (**a**), morphology (**b**) and microstructure (**c**) of the gas-atomized Zr_0 HEA powders.

3.2. Microstructure and Phase Evolution during MA

To investigate the alloying behavior in the MA process, X-ray diffraction (XRD) was performed on the milled $Zr_{0.5}$ powders at 10 h intervals until the milling time reached 50 h (Figure 2). It can be seen that the XRD pattern of primary blending powders includes diffraction peaks of CoCrFeNiMn alloy with FCC structure and ZrH_2 (PDF 073-2076). After 10 h of ball milling, the diffraction peaks of ZrH_2 could still be detected, but their intensity was significantly reduced. This indicates that ZrH_2 decomposed and the Zr dissolved into the matrix. After 20 h of ball milling, some diffraction peaks of ZrH_2 disappeared due to the alloying. As the milling time increasing to 30 h, all the diffraction peaks of ZrH_2 became absent and the XRD pattern became similar to that of Figure 1, which suggests that the Zr was fully dissolved into the FCC matrix to form a supersaturated solid solution. Additionally, as the ball milling time was extended, the diffraction peaks appeared to broaden. Throughout the milling process, the lattice strain increase and grain refinement are the main reasons for the peak broadening and intensity reduction [25]. On the other hand, ZrO_2 diffraction peaks were observed after 40 h of ball milling, and the intensity of the ZrO_2 diffraction peaks increase with the extending of ball milling time. The raw materials are usually contaminated by milling media in the process of ball milling [26,27].

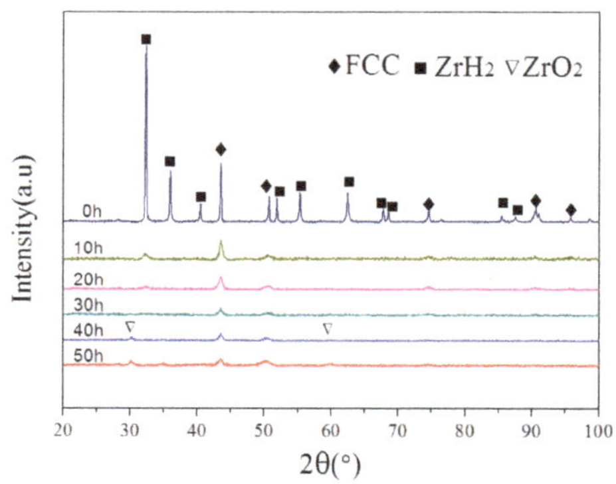

Figure 2. The XRD patterns of the $Zr_{0.5}$ alloy powders with different milling times.

Table 2 shows the average crystal size and lattice strain of the $Zr_{0.5}$ alloy powders with different milling times that have been calculated using the Williamson–Hall equation method. As the milling time was extended from 10 h to 50 h, the crystal size decreased from 14.8 nm to 8.1 nm, and the lattice strain increased from 0.237% to 0.746%. Thus, the high lattice strain and the formation of nanocrystalline are the main reasons for the mentioned diffraction peaks broadening and the diffraction intensity reduction.

Table 2. The crystal size and lattice strain of the $Zr_{0.5}$ HEA powders with different milling times.

Milling Time (h)	Crystal Size (nm)	Lattice Strain (%)
10	14.8	0.237
20	13.4	0.385
30	12.5	0.473
40	11.2	0.564
50	8.1	0.746

Figure 3 shows the morphologies of the $Zr_{0.5}$ alloy powders that have undergone different milling times. The initial powder is a mixture of spherical atomized CoCrFeNiMn powders and irregularly shaped ZrH_2 powders, and the size of the ZrH_2 powders are smaller than the CoCrFeNiMn powders (Figure 3a). The difference in size between the two powders will affect the alloying process. After 10 h of ball milling, the relatively large-sized CoCrFeNiMn powders became elliptical and fine sheet powders were deposited on their surfaces. This indicates that the alloy powders were deformed, broken, and welded during ball milling and, at the same time, the fine Zr (or ZrH_2) powders and broken CoCrFeNiMn powders were agglomerated on the surface of the large powder particles. As the ball milling time extended, the CoCrFeNiMn powders were wrapped by Zr-rich alloy powders and the outer layer of the CoCrFeNiMn powders was alloyed with Zr under the impact of the high-energy small balls. Meanwhile, the cycle of crushing and agglomeration of the outer layer continued, which causes the powders to gradually refine and promotes the alloying and diffusion among the different alloy elements [28,29]. After 30 h of ball milling, the particle size hardly changed, which means that the crushing and agglomeration reached a dynamic balance in the MA process.

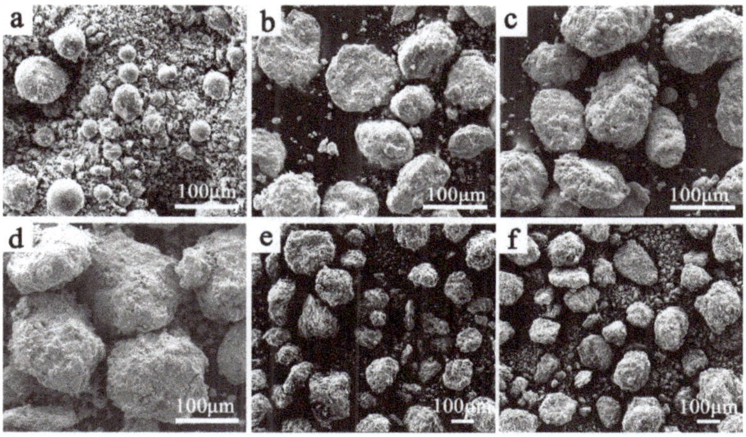

Figure 3. The scanning electron microscopy (SEM) images of the $Zr_{0.5}$ alloy powders with different milling times: (**a**) 0 h; (**b**) 10 h; (**c**) 20 h; (**d**) 30 h; (**e**) 40 h; (**f**) 50 h.

Figure 4a displays the XRD patterns of HEA powders with different contents of Zr after ball milling of 30 h. Compared with Figure 2, no extra peaks can be found in the XRD patterns of the Zr_0, $Zr_{0.2}$, and $Zr_{0.5}$ alloys, which suggests that the ZrH_2 was completely decomposed and the Zr was effectively dissolved into the CoCrFeNiMn HEA matrix when the Zr content is not higher than 0.5. Nevertheless, the diffraction peak of ZrH_2 can be found in the XRD patterns when the Zr content is up to 0.8 and 1.0, and the intensity of the peak increases with the increase of the ZrH_2 content.

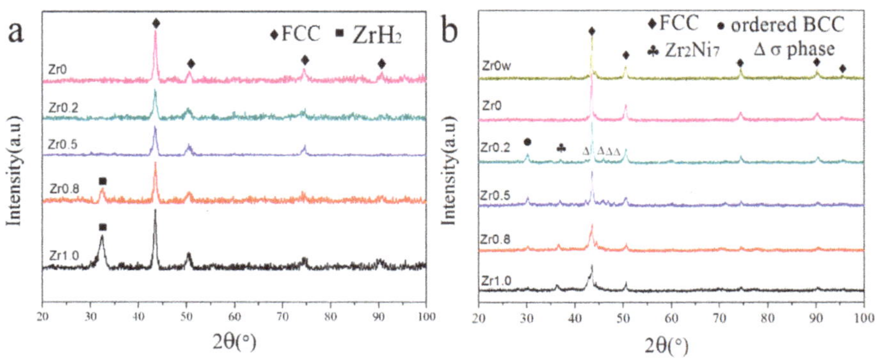

Figure 4. The XRD patterns of HEAs: (**a**) CoCrFeNiMnZr_x (x = 0, 0.2, 0.5, 0.8, 1.0) alloy powders after milling for 30 h, (**b**) sintered CoCrFeNiMnZr_x (x = 0w, 0, 0.2, 0.5, 0.8, 1.0) HEAs.

3.3. Phase Evolution and Microstructure after SPS

XRD patterns of the sintered $CoCrFeNiMnZr_x$ (x = 0w, 0, 0.2, 0.5, 0.8, 1.0) alloys are shown in Figure 4b. Obviously, the ZrH_2 diffraction peaks in the $Zr_{0.8}$ and $Zr_{1.0}$ powders disappear in the sintered samples. This suggests that the excessive ZrH_2 was decomposed and that the Zr remainder was fully alloyed during SPS. The FCC phase is still the dominant phase, meanwhile, some weak diffraction peaks which are identified as Zr_2Ni_7 (PDF 071-0543), σ phase, and ordered BCC (B2) phase appear. Compared with the XRD patterns of the powders after ball milling, the broad diffraction peaks became narrow due to internal energy releases during the SPS process.

The Zr_0w and Zr_0 alloys have the same diffraction peaks and consist of an FCC phase, indicating that the MA and SPS processes do not alter the FCC phase of the alloys. The diffraction peaks of the Zr_2Ni_7, the B2 phase, and the σ phase can be found in the $Zr_{0.2}$, $Zr_{0.5}$, $Zr_{0.8}$, and $Zr_{1.0}$ alloys. The σ phase was widely reported in previous research [30,31] and it can be identified as the NiCoCr (PDF 021-1271) σ phase, which has a tetragonal structure (a = 8.85 Å and c = 4.59 Å) in the present experiment. The B2 phase has been often found in HEAs containing CoCrFeNi [4,21,32]. These indicate that during the SPS process, some of the alloy atoms diffuse under thermal activation conditions to form ordered solid solutions and intermetallic compounds.

With the increase of the Zr content, the diffraction intensity of the B2 phase decreases while the diffraction intensity of the Zr_2Ni_7 compound increases. This indicates that when the Zr content is low, the free energy is reduced by the formation of the ordered phase. However, as the Zr content increases, the formation of the Zr_2Ni_7 is more effective to reduce the free energy, and the formation of the B2 phase is suppressed. On the other hand, the diffraction peaks of the σ phase almost disappear in the $Zr_{1.0}$ alloy, which also means that the formation of the Zr_2Ni_7 suppresses the formation of the σ phase. These can be attributed to the lattice strain and a lager negative enthalpy of mixing, caused by the addition of Zr.

Figure 5 displays the SEM (operating at back-scattered electron (BSE) mode) images of the $CoCrFeNiMnZr_x$ alloys synthesized by MA and SPS. As can be seen from Figure 5a,b, both the Zr_0w and Zr_0 alloys exhibit a single phase feature, which is consistent with the XRD results. After the addition of Zr, there are two distinct microstructural domains: the bright and gray regions (Figure 5c–f).

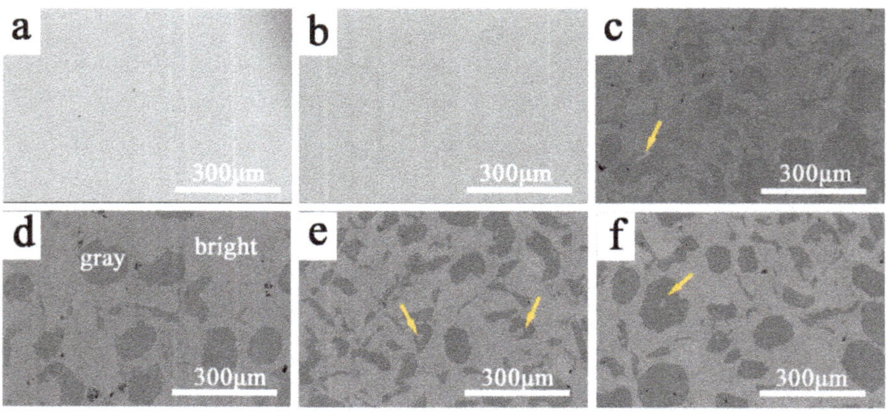

Figure 5. The SEM (operating at back-scattered electron (BSE) mode) images of the $CoCrFeNiMnZr_x$ alloys synthesized by mechanical alloying (MA) and spark plasma sintering (SPS). (**a**) x = 0w; (**b**) x = 0; (**c**) x = 0.2; (**d**) x = 0.5; (**e**) x = 0.8; (**f**) x = 1.0. Any Zr-rich particles were wrapped by a plastic face-centered cubic (FCC) phase and indicated by yellow arrows.

EDS was performed to investigate the two regions using the $Zr_{0.2}$ alloy and the results are shown in Figure 6 and Table 3. On the one hand, the bright region contains a higher amount of Zr than the nominal composition whereas Zr is barely found in the gray regions. On the other hand, the content of Co, Cr, Fe, Ni, and Mn are relatively lower in the bright region, and they are almost equal in the atomic ratio in both regions. Obviously, some CoCrFeNiMn HEA particles are not completely broken by the high-energy ball milling due to their high toughness [14,33], and Zr is not uniformly distributed in the matrix. The gray regions can be identified as a pure CoCrFeNiMn FCC phase and the bright region consists of FCC phase, B2 phase, Zr_2Ni_7, and σ phase. For the gray region, the shapes present are ovals and strips, which were inherited from the MA process. Furthermore, the hard and brittle Zr-rich particles were wrapped by a plastic FCC phase during the ball milling process, a few small bright microstructures are indicated by yellow arrows (Figure 5). Table 4 shows the density of the sintered $CoCrFeNiMnZr_x$ alloys. It can be seen that the addition of a relatively low-density Zr element reduces the density of the alloy system (the density of Co, Cr, Fe, Ni, Mn, and Zr are 8.9 $g{\cdot}cm^{-3}$, 7.19 $g{\cdot}cm^{-3}$, 7.87 $g{\cdot}cm^{-3}$, 8.9 $g{\cdot}cm^{-3}$, 7.44 $g{\cdot}cm^{-3}$, and 6.49 $g{\cdot}cm^{-3}$, respectively).

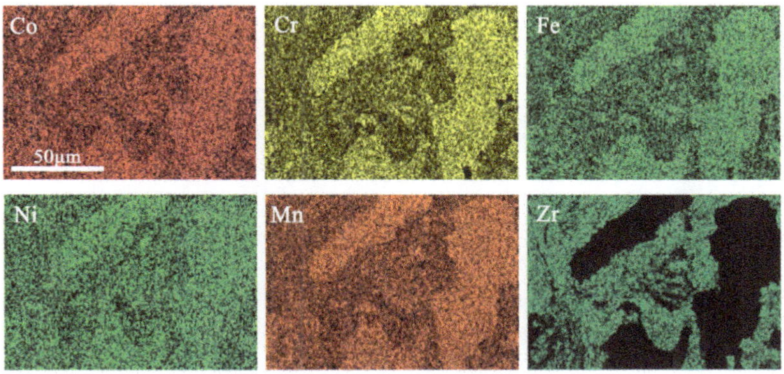

Figure 6. The energy dispersive spectrometry (EDS) mappings of the $Zr_{0.2}$ alloy synthesized by MA and SPS.

Table 3. The chemical compositions of the $Zr_{0.2}$ alloy synthesized by MA and SPS.

Elements	Co	Cr	Fe	Ni	Mn	Zr
Gray region (at.%)	19.2	21.0	20.0	18.9	20.9	0
Bright Region (at.%)	14.9	16.3	15.0	14.9	16.1	22.8

Table 4. The density of the $CoCrFeNiMnZr_x$ alloys synthesized by MA and SPS.

Samples	Zr_0w	Zr_0	$Zr_{0.2}$	$Zr_{0.5}$	$Zr_{0.8}$	$Zr_{1.0}$
Density (g/cm^3)	7.91 ± 0.03	7.92 ± 0.02	6.86 ± 0.04	6.87±0.05	6.91 ± 0.03	6.93 ± 0.02

Figure 7a shows the grain structure of Zr_0 and the selected area electron diffraction (SAED) pattern is related to the red circle marking area. It indicates that the sintered Zr_0 alloy has a micron grain size and a single FCC structure which is consistent with XRD results. Meanwhile, some twins are found. After the Zr addition, there are two obviously different areas distinguished by grain size (Figure 7b). The SAED pattern of red circle marking area demonstrates that the micro-crystal has an FCC structure. EDS results (not shown here) revealed that these nanocrystals are Zr-rich while the micro-crystals do not contain Zr. This confirms the SEM results: the Zr and the outer layer of the atomized powder are alloyed during the ball milling process, and as time passes, the fine Zr-rich powders agglomerated on the outer layer is continuously refined. Furthermore, many spherical nanoprecipitates are found in the nanocrystal area (Figure 7c). Figure 7d–g show the high-resolution TEM (HRTEM) images with corresponding fast Fourier transformations (FFT) of nanoprecipitates. These indicate that there are two kinds of nanoprecipitates which are BCC structure precipitates and Zr_2Ni_7 precipitates. In the previous studies [3,34], nanoprecipitates have been considered to be beneficial to the mechanical properties of the material due to the ability of pinning dislocations.

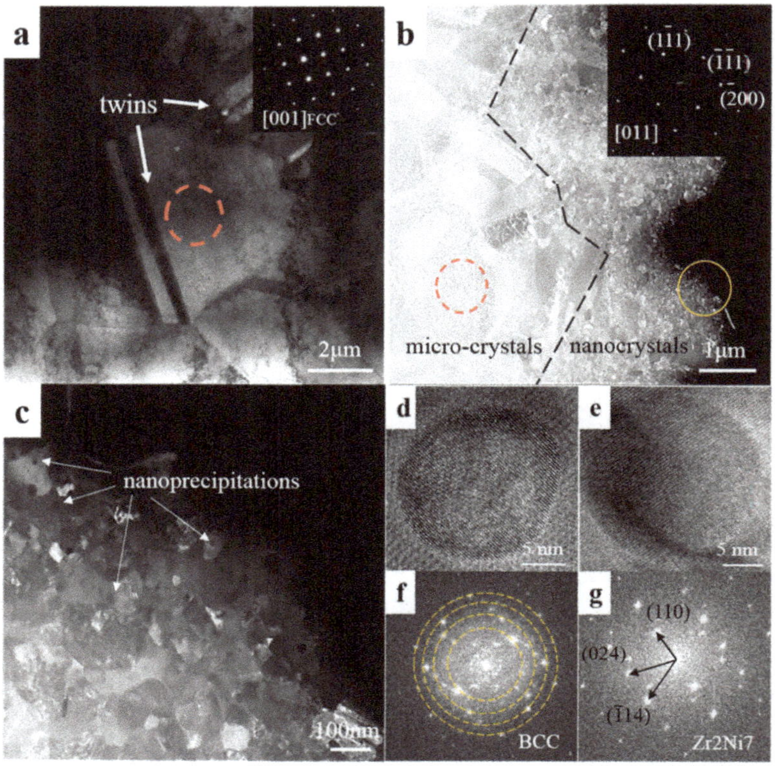

Figure 7. The transmission electron microscopy (TEM) images of sintered (**a**) Zr_0 and (**b**) $Zr_{0.2}$ HEAs. The insets in (**a**,**b**) are the selected area electron diffraction (SAED) patterns of the red circle marking area. The dashed line in (**b**) distinguishes the two regions with different grain sizes: the micro-crystals on the left side of the dashed line and the nanocrystals on the right side of the dashed line; (**c**) is the magnifying image of the yellow circle marking area in (**b**); (**d**,**e**) are the high-resolution TEM (HRTEM) of nanoprecipitation; (**f**,**g**) are the fast Fourier transformations (FFT) of (**d**,**e**), respectively.

3.4. Mechanical Properties

Figure 8 shows the compressive stress-strain curves of the CoCrFeNiMnZr$_x$ alloys at room temperature. The values of yield strength (YS) σ_y, compressive strength (CS) σ_{max}, and plastic strain limit (PS) ε_p of alloys are listed in Table 5. The maximum yield strength reaches 820 MPa when the Zr content is up to 0.8, but the compressive strength of the Zr$_x$ (x = 0.2, 0.5, 0.8, 1.0) alloys are lower than that of alloys without Zr. Because of the addition of Zr, the yield strength evidently improved while the ductility declined simultaneously. Zr_0 and Zr_0w alloys, exhibit excellent plasticity, and both of the two alloys are not broken in the pressure range of the compressor. The yield strength of Zr_0 is higher than that of Zr_0w, which suggests that MA can improve the strength of the alloy by refining the grain [14]. Furthermore, the yield strength of the CoCrFeNiMnZr$_x$ (x = 0.2, 0.5, 0.8, 1.0) alloys are significantly higher than that of the Zr_0 alloy, but the plasticity strongly reduced at the same time. This can be attributed to solid solution strengthening and second phase strengthening caused by the addition of Zr, as well as fine grain strengthening obtained by mechanical ball milling.

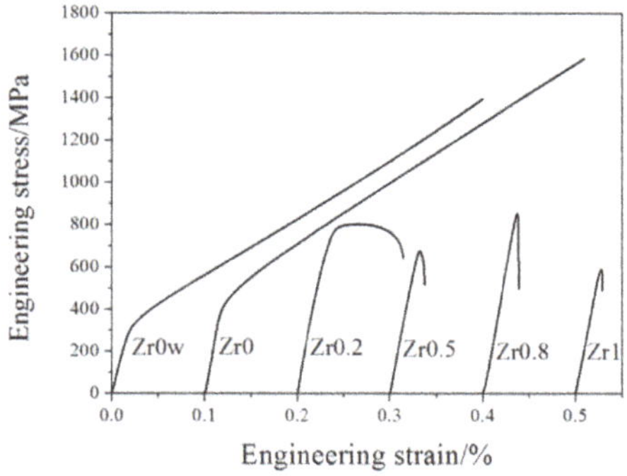

Figure 8. The compressive engineering stress–strain curves of the CoCrFeNiMnZr$_x$ alloys.

Table 5. The mechanical properties of the CoCrFeNiMnZr$_x$ alloys.

x	YS σ_y (MPa)	CS σ_{max} (MPa)	PS ε_p (%)
0w	350	>1400	>40
0	420	>1600	>40
0.2	800	880	12
0.5	800	810	4
0.8	820	850	3
1.0	600	600	2.5

The fractographs of the CoCrFeNiMnZr$_x$ (x = 0.2, 0.5, 0.8, 1.0) alloys after compressive tests are depicted in Figure 9. The fracture surface of the Zr$_{0.2}$ alloy presents a step-like pattern with quasi-cleavage characteristics. A small radiation zone appears on the right side of Figure 9a, indicating the localized ductile fracture. Complete cleavage fracture patterns can be seen in Figure 9b–d, which thus deteriorates the alloy plasticity. For the Zr$_{0.5}$ alloy, the fracture surface consists of many small cleavage planes, with both inter-granular fractures and trans-granular fractures having occurred. Large bright and flat cleavage planes appear in the fractures of Zr$_{0.5}$ and Zr$_{1.0}$ alloys, which is in accord with the totally brittle fracture of the alloys. The above results indicate that the CoCrFeNiMnZr$_x$ (x = 0.2, 0.5, 0.8, 1.0) alloys exhibit brittle fractures, which can be attributed to a large lattice distortion and the brittle intermetallic compounds present in the Zr-rich region.

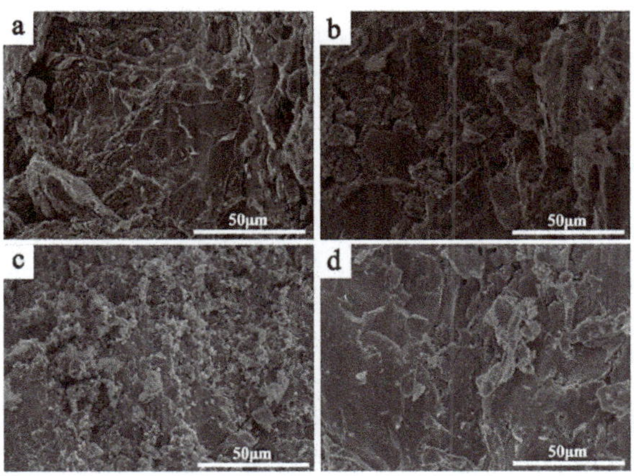

Figure 9. SEM micrographs of the fracture surface of CoCrFeNiMnZr$_x$ alloys. (**a**) x = 0.2; (**b**) x = 0.5; (**c**) x = 0.8; (**d**) x = 1.0.

4. Conclusions

The alloying behavior, microstructures, and mechanical properties of both powders and bulk CoCrFeNiMnZr$_x$ alloys were investigated. The main conclusions of the present work are given below:

(1) Metastable single FCC phase CoCrFeNiMnZr$_x$ alloy powders were prepared after 30 h of ball milling, which means that MA could enhance the solid solubility of large size elements in the matrix. However, toughness powder is difficult to break during ball milling, which would affect the alloying process and lead to the element segregation. Toughness powders can retain their original properties during ball milling, which means that MA can be used to prepare HEAs composites.

(2) The phase composition of HEAs can be changed by adjusting the content of the alloying element. After SPS, CoCrFeNiMn maintains a single FCC phase. However, two distinct microstructural domains, one composed of a micro-sized CoCrFeNiMn alloy without Zr and the other one composed of Zr-rich multi-phase microstructures consisting of a nano-sized FCC phase, B2 phase, Zr_2Ni_7, and σ phase, appear in the CoCrFeNiMnZr$_x$ ($x \geq 0.2$) alloys. When the elements with different physicochemical properties are dissolved into the HEA matrix, the local structural stability will be destroyed, and the local atomic arrangement will change from disorder to order.

(3) The addition of Zr significantly increases the yield strength of the CoCrFeNiMn alloy. The $Zr_{0.8}$ alloy exhibits the highest yield strength up to 820 MPa which is almost twice as much as the Zr_0 alloy (420 MPa). The high yield strength makes it a promising candidate for precision equipment manufacturing.

Author Contributions: Conceptualization, Y.X.; data curation, L.Z.; formal analysis, H.Z.; investigation, H.Z. and L.Z.; methodology, Y.X; project administration, Y.X; resources, Y.X; writing—original draft, H.Z. and L.Z.; writing—review & editing, H.Z., X.L., Q.C. and Y.X.

Acknowledgments: The authors thank the National Key Project of Research and Development Program of China (2016YFB1100202) for financial support.

Conflicts of Interest: The authors declare no conflict of interest.

References

1. Chuang, M.H.; Tsai, M.H.; Wang, W.R.; Lin, S.J.; Yeh, J.W. Microstructure and wear behavior of Al$_x$Co$_{1.5}$CrFeNi$_{1.5}$Ti$_y$ high-entropy alloys. *Acta Mater.* **2011**, *59*, 6308–6317. [CrossRef]
2. Löbel, M.; Lindner, T.; Mehner, T.; Lampke, T. Influence of titanium on microstructure, phase formation and wear behaviour of AlCoCrFeNiTi$_x$ high-entropy alloy. *Entropy* **2018**, *20*, 505. [CrossRef]
3. He, J.Y.; Wang, H.; Huang, H.L.; Xu, X.D.; Chen, M.W.; Wu, Y.; Liu, X.J.; Nieh, T.G.; An, K.; Lu, Z.P. A precipitation-hardened high-entropy alloy with outstanding tensile properties. *Acta Mater.* **2016**, *102*, 187–196. [CrossRef]
4. Yeh, J.W.; Chen, S.K.; Lin, S.J.; Gan, J.Y.; Chin, T.S.; Shun, T.T.; Tsau, C.H.; Chang, S.Y. Nanostructured high-entropy alloys with multiple principal elements: Novel alloy design concepts and outcomes. *Adv. Eng. Mater.* **2004**, *6*, 299–303. [CrossRef]
5. Zhu, J.M.; Fu, H.M.; Zhang, H.F.; Wang, A.M.; Li, H.; Hu, Z.Q. Microstructures and compressive properties of multicomponent AlCoCrFeNiMo$_x$ alloys. *Mater. Sci. Eng. A Struct. Mater.* **2010**, *527*, 6975–6979. [CrossRef]
6. Ma, S.G.; Zhang, Y. Effect of Nb addition on the microstructure and properties of AlCoCrFeNi high-entropy alloy. *Mater. Sci. Eng. A Struct. Mater.* **2012**, *532*, 480–486. [CrossRef]
7. Hsu, C.Y.; Sheu, T.S.; Yeh, J.W.; Chen, S.K. Effect of iron content on wear behavior of AlCoCrFe$_x$Mo$_{0.5}$Ni high-entropy alloys. *Wear* **2010**, *268*, 653–659. [CrossRef]
8. Liu, C.M.; Wang, H.M.; Zhang, S.Q.; Tang, H.B.; Zhang, A.L. Microstructure and oxidation behavior of new refractory high entropy alloys. *J. Alloy. Compd.* **2014**, *583*, 162–169. [CrossRef]
9. Shang, C.; Axinte, E.; Ge, W.; Zhang, Z.; Wang, Y. High-entropy alloy coatings with excellent mechanical, corrosion resistance and magnetic properties prepared by mechanical alloying and hot pressing sintering. *Surf. Interfaces* **2017**, *9*, 36–43. [CrossRef]

10. Schuh, B.; Mendez-Martin, F.; Völker, B.; George, E.P.; Clemens, H.; Pippan, R.; Hohenwarter, A. Mechanical properties, microstructure and thermal stability of a nanocrystalline CoCrFeMnNi high-entropy alloy after severe plastic deformation. *Acta Mater.* **2015**, *96*, 258–268. [CrossRef]
11. Singh, S.; Wanderka, N.; Murty, B.S.; Glatzel, U.; Banhart, J. Decomposition in multi-component AlCoCrCuFeNi high-entropy alloy. *Acta Mater.* **2011**, *59*, 182–190. [CrossRef]
12. Ji, W.; Wang, W.; Wang, H.; Zhang, J.; Wang, Y.; Zhang, F.; Fu, Z. Alloying behavior and novel properties of CoCrFeNiMn high-entropy alloy fabricated by mechanical alloying and spark plasma sintering. *Intermetallics* **2015**, *56*, 24–27. [CrossRef]
13. Pradeep, K.G.; Wanderka, N.; Choi, P.; Banhart, J.; Murty, B.S.; Raabe, D. Atomic-scale compositional characterization of a nanocrystalline AlCrCuFeNiZn high-entropy alloy using atom probe tomography. *Acta Mater.* **2013**, *61*, 4696–4706. [CrossRef]
14. Liu, Y.; Wang, J.; Fang, Q.; Liu, B.; Wu, Y.; Chen, S. Preparation of superfine-grained high entropy alloy by spark plasma sintering gas atomized powder. *Intermetallics* **2016**, *68*, 16–22. [CrossRef]
15. Tokita, M. Mechanism of spark plasma sintering and its application to ceramics. *Nyu Seramikkusu* **1997**, *10*, 43–53.
16. Zhou, Y.; Hirao, K.; Yamauchi, Y.; Kanzaki, S. Densification and grain growth in pulse electric current sintering of alumina. *J. Eur. Ceram. Soc.* **2004**, *24*, 3465–3470. [CrossRef]
17. Okuni, T.; Miyamoto, Y.; Abe, H.; Naito, M. Joining of silicon carbide and graphite by spark plasma sintering. *Ceram. Int.* **2014**, *40*, 1359–1363. [CrossRef]
18. Yang, M.; Dai, Y.; Song, C.; Zhai, Q. Microstructure evolution of grey cast iron powder by high pressure gas atomization. *J. Mater. Process. Technol.* **2010**, *210*, 351–355. [CrossRef]
19. Zhou, Y.J.; Zhang, Y.; Wang, Y.L.; Chen, G.L. Solid solution alloys of $AlCoCrFeNiTi_x$ with excellent room-temperature mechanical properties. *Appl. Phys. Lett.* **2007**, *90*, 181904. [CrossRef]
20. Zhu, J.M.; Fu, H.M.; Zhang, H.F.; Chen, G.L. Synthesis and properties of multiprincipal component $AlCoCrFeNiSi_x$ alloys. *Mater. Sci. Eng. A Struct. Mater. Prop. Microstruct. Process.* **2010**, *527*, 7210–7214. [CrossRef]
21. Chen, J.; Niu, P.; Liu, Y.; Lu, Y.; Wang, X.; Peng, Y.; Liu, J. Effect of Zr content on microstructure and mechanical properties of AlCoCrFeNi high entropy alloy. *Mater. Des.* **2016**, *94*, 39–44. [CrossRef]
22. Stepanov, N.D.; Yurchenko, N.Y.; Tikhonovsky, M.A.; Salishchev, G.A. Effect of carbon content and annealing on structure and hardness of the CoCrFeNiMn-based high entropy alloys. *J. Alloy. Compd.* **2016**, *687*, 59–71. [CrossRef]
23. Shahmir, H.; Nili-Ahmadabadi, M.; Shafie, A.; Langdon, T.G. Hardening and thermal stability of a nanocrystalline $CoCrFeNiMnTi_{0.1}$ high-entropy alloy processed by high-pressure torsion. *IOP Conf. Ser. Mater. Sci. Eng.* **2017**, *194*, 012017. [CrossRef]
24. Takeuchi, A.; Inoue, A. Classification of bulk metallic glasses by atomic size difference, heat of mixing and period of constituent elements and its application to characterization of the main alloying element. *Mater. Trans.* **2005**, *46*, 2817–2829. [CrossRef]
25. Yeh, J.W.; Chang, S.Y.; Hong, Y.D.; Chen, S.K.; Lin, S.J. Anomalous decrease in X-ray diffraction intensities of Cu–Ni–Al–Co–Cr–Fe–Si alloy systems with multi-principal elements. *Mater. Chem. Phys* **2007**, *103*, 41–46. [CrossRef]
26. Joo, S.H.; Kato, H.; Jang, M.J.; Moon, J.; Kim, E.B.; Hong, S.J.; Kim, H.S. Structure and properties of ultrafine-grained CoCrFeMnNi high-entropy alloys produced by mechanical alloying and spark plasma sintering. *J. Alloy. Compd.* **2017**, *698*, 591–604. [CrossRef]
27. Fu, Z.; Chen, W.; Wen, H.; Zhang, D.; Chen, Z.; Zheng, B.; Zhou, Y.; Lavernia, E.J. Microstructure and strengthening mechanisms in an FCC structured single-phase nanocrystalline $Co_{25}Ni_{25}Fe_{25}Al_{7.5}Cu_{17.5}$ high-entropy alloy. *Acta Mater.* **2016**, *107*, 59–71. [CrossRef]
28. Fang, S.; Chen, W.; Fu, Z. Microstructure and mechanical properties of twinned $Al_{0.5}CrFeNiCo_{0.3}C_{0.2}$ high entropy alloy processed by mechanical alloying and spark plasma sintering. *Mater. Des.* **2014**, *54*, 973–979. [CrossRef]
29. Zhang, K.B.; Fu, Z.Y.; Zhang, J.Y.; Shi, J.; Wang, W.M.; Wang, H.; Wang, Y.C.; Zhang, Q.J. Nanocrystalline CoCrFeNiCuAl high-entropy solid solution synthesized by mechanical alloying. *J. Alloy. Compd.* **2009**, *485*, L31–L34. [CrossRef]

30. Zhou, W.; Fu, L.M.; Liu, P.; Xu, X.D.; Chen, B.; Zhu, G.Z.; Wang, X.D.; Shan, A.D.; Chen, M.W. Deformation stimulated precipitation of a single-phase CoCrFeMnNi high entropy alloy. *Intermetallics* **2017**, *85*, 90–97. [CrossRef]
31. Lim, K.R.; Lee, K.S.; Lee, J.S.; Kim, J.Y.; Chang, H.J.; Na, Y.S. Dual-phase high-entropy alloys for high-temperature structural applications. *J. Alloy. Compd.* **2017**, *728*, 1235–1238. [CrossRef]
32. Zhang, K.B.; Fu, Z.Y.; Zhang, J.Y.; Wang, W.M.; Wang, H.; Wang, Y.C.; Zhang, Q.J.; Shi, J. Microstructure and mechanical properties of $CoCrFeNiTiAl_x$ high-entropy alloys. *Mater. Sci. Eng. A Struct. Mater.* **2009**, *508*, 214–219. [CrossRef]
33. Löbel, M.; Lindner, T.; Kohrt, C.; Lampke, T. Processing of AlCoCrFeNiTi high entropy alloy by atmospheric plasma spraying. *IOP Conf. Ser. Mater. Sci. Eng.* **2017**, *181*, 012015. [CrossRef]
34. Jiang, S.; Wang, H.; Wu, Y.; Liu, X.; Chen, H.; Yao, M.; Gault, B.; Ponge, D.; Raabe, D.; Hirata, A.; et al. Ultrastrong steel via minimal lattice misfit and high-density nanoprecipitation. *Nature* **2017**, *544*, 460–464. [CrossRef] [PubMed]

entropy

MDPI

Article

Phases, Microstructures and Mechanical Properties of CoCrNiCuZn High-Entropy Alloy Prepared by Mechanical Alloying and Spark Plasma Sintering

Yuchen Sun, Boren Ke, Yulin Li, Kai Yang, Mingqi Yang, Wei Ji * and Zhengyi Fu *

State Key Laboratory of Advanced Technology for Materials Synthesis and Processing, Wuhan University of Technology, Wuhan 430070, China; 15370786151231@whut.edu.cn (Y.S.); ke@whut.edu.cn (B.K.); 1422915370@whut.edu.cn (Y.L.); yangletian@whut.edu.cn (K.Y.); mingqi.yang@whut.edu.cn (M.Y.)

* Correspondence: jiwei@whut.edu.cn (W.J.); zyfu@whut.edu.cn (Z.F.)

Received: 14 November 2018; Accepted: 26 January 2019; Published: 29 January 2019

Abstract: In the study, an equiatomic CoCrNiCuZn high-entropy alloy (HEA) was prepared by mechanical alloying (MA) and the phases, microstructures, and thermal properties of the alloy powder were explored. The results suggest that a solid solution with body-centered cubic (BCC) phase and a crystalline size of 10 nm formed after 60 h of milling. Subsequently, the alloy powder was consolidated by spark plasma sintering (SPS) at different temperatures (600 °C, 700 °C, 800 °C, and 900 °C). Two kinds of face-centered cubic (FCC) phases co-existed in the as-sintered samples. Besides, Vickers hardness and compressive strength of the consolidated alloy sintered at 900 °C were respectively 615 HV and 2121 MPa, indicating excellent mechanical properties.

Keywords: high-entropy alloy; spark plasma sintering; mechanical alloying; mechanical property; microstructure

1. Introduction

Conventional alloy is generally composed of one or two main elements and a small amount of other elements, to enhance its mechanical properties, such as steel and NiAl intermetallics [1,2]. The emergence of high-entropy alloys (HEAs) [3] has broken this traditional notion. A HEA is loosely defined as alloy composed of more than five principal elements with an equimolar ratio (5–35 at.%). High-entropy alloy has high entropy effect, lattice distortion effect, sluggish cooperative diffusion effect, and cocktail effect. It often has simple solid-solutions or amorphous structure [4]. Well-designed HEAs have good mechanical properties including high hardness, high strength, good corrosion, and wear resistance [5].

HEAs can be prepared by various routes, such as vacuum arc-melting and casting [6,7]. However, these routes are not suitable for HEA systems which contain elements with very different melting points. For example, the melting temperature of Cr is 1000 °C above the atmospheric boiling point of Zn, so some systems such as CoCrNiCuZn high-entropy alloy cannot be synthesized by arc-melting route. Besides, arc-melting is not suitable for industrial manufacturing and final products have some limitations in shape and size [8]. Mechanical alloying (MA) is a convenient route to synthesize nanocrystalline HEAs materials. MA can reduce the preparation cost of nanocrystalline materials [9,10]. In addition, HEAs can be easily consolidated from the as-milled powders with spark plasma sintering (SPS) technique [11–13].

In this study, we synthesized the CoCrNiCuZn high-entropy alloy by MA and SPS. The phases, microstructures and mechanical properties of the consolidated alloys were also explored.

2. Experimental

Metal powders (Co, Cr, Ni, Cu, and Zn with a purity of more than 99.5 wt.% and a particle size of ~45 μm) were mixed according to the equiatomic composition and milled in a planetary ball-miller (300 rpm for 60 h, argon atmosphere) with stainless steel vials and balls as milling media (a ball-to-powder mass ratio of 20:1). N-heptane was used as the processing controlling agent (PCA) to avoid cold welding and oxidation. The MA process was monitored with an interval of 6 h. After 60 h of ball milling, the powder was consolidated by SPS (Dr. Sinter-3.20 MKII, Sumitomo, Osaka, Japan) at different temperatures (600 °C, 700 °C, 800 °C, and 900 °C) under 30 MPa with dwell time of 10 min in argon atmosphere.

The phases of ball milled (QM-BP, Nanjing Nanda Instrument Plant, Nanjing, China) alloys were characterized by X-ray diffractometer (XRD, Rigaku Ultima III, Tokyo, Japan) with a Cu Kα radiation to investigate the crystal structure. The microstructure was analyzed by a scanning electron microscope (SEM, Hitachi 3400, Tokyo, Japan) and a transmission electron microscope (TEM, JEOL JEM-2010HT, Tokyo, Japan). The thermal analysis of as-milled powder was conducted by a differential scanning calorimeter (DSC, NETZSCH 449C, Selb, Germany) heating the alloy to 1500 °C (5 °C/min) in flowing argon atmosphere. According to the Archimedes principle, the density of HEA was determined. The hardness of sectioned and polished specimens was determined by vickers hardness tester (Wolpert-430SV, Aachen, Germany). The compressive properties at room temperature were determined by a MTS810 testing machine (MTS 810, MTS Systems Corporation, Eden Prairie, MN, USA) with a loading rate of 1 mm/min. The dimensions of sample is 2 mm × 2 mm × 5 mm. The fracture surface was analyzed by SEM. A thin foil of sintered material obtained by mechanical thinning and ion milling was analyzed by TEM. At least 5 measurements were performed to calculate the means of vickers hardness and compressive strength.

3. Results and Discussion

3.1. Mechanical Alloying of CoCrNiCuZn HEAs

3.1.1. X-Ray Analysis

The XRD patterns of the CoCrNiCuZn high-entropy alloy (Figure 1) indicated that a major peak formed after 30-h milling. The diffraction patterns of all alloying elements can be observed in the XRD patterns of primitive blending powder. After 6-h MA, the diffraction peaks of the principle elements were still observed, but the intensity was dramatically decreased. With the increase in milling time to 18 h, some peaks were significantly broadened and some peaks were invisible. After 30-h milling, only 3 peaks of a BCC structure ((1 1 0), (2 0 0), (2 1 1)) could be identified, indicating the formation of a simple solid solution. The BCC solid-solution had a lattice parameter of 2.8831 Å. After 60 h MA, the XRD patterns showed no obvious change. In the milling process, the decreased intensity, broadened or disappeared peak might be caused by high lattice strain, refined crystallite size, and decreased crystallinity [14,15].

The crystallite size (CS) and lattice strain (LS) of CoCrNiCuZn HEA obtained after milling for different time were calculated by Scherrer's formula after eliminating the interferences of instruments and strain [16,17]. The CS of the BCC phase was significantly refined to 19 nm after 18-h MA and then decreased to 13 nm after 30-h milling (Table 1). Further increasing of milling time had no significant influence on the crystallite size. The equilibrium between crystalline refinement and cold welding of BCC phase might be reached after 30-h milling. The lattice strain of milled powders increased with milling time and reached 0.70% after 60 h milling [18].

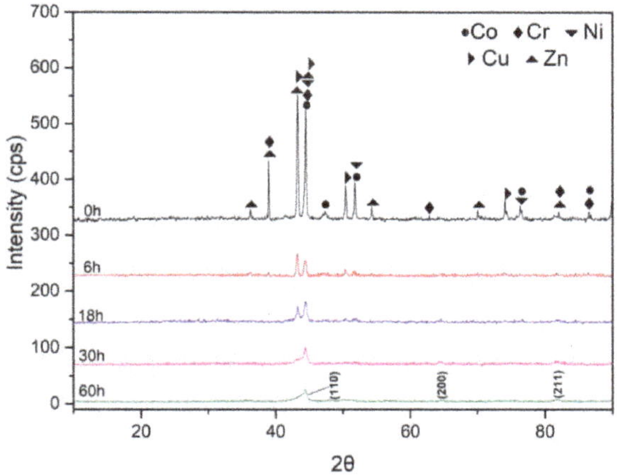

Figure 1. The change of XRD patterns of CoCrNiCuZn high-entropy alloy (HEA) powder obtained after milling for different time (from 0 h to 60 h).

Table 1. Crystallite size (CS), lattice strain (LS), lattice parameter (LP) of CoCrNiCuZn HEA obtained after milling for different time (0 h to 60 h).

Milling Time (h)	CS (nm)	LS (%)
0	–	–
6	22	0.64
18	19	0.65
30	13	0.67
60	13	0.70

3.1.2. Microstructure and Composition

Figure 2 shows the microstructure of CoCrNiCuZn HEA powder obtained after ball milling for different time (0 h, 6 h, 18 h, 30 h, and 60 h). The non-milled powder has a different particle size. Through the MA process, milled HEA powder agglomerated into an elliptical shape with the size of ~3 μm and the elliptical particles evolved into ~1 μm thick sheets. The nanocrystalline nature of CoCrNiCuZn HEA obtained after 60 h MA was characterized by the selected area electron diffraction (SAED) pattern and TEM bright field image (Figure 3). The crystal size measured from bright field TEM image was approximately 10 nm, which was consistent with the calculation results by the Scherrer's formula. The existence of nanoscaled crystallite indicated that the microsized alloy particles in SEM images were the aggregates of nanosized grains.

The rings in the SAED pattern (Figure 3) indicated that the nanocrystalline HEA powder after 60 h milling only consisted of a BCC phase. The result was consistent with XRD analysis results. The results confirmed that the CoCrNiCuZn high-entropy alloy with a structure of simple BCC solid solution had been successfully fabricated by mechanical alloying.

Zhang and Guo proposed the criteria for the formation of solid solution and phase stability of HEA prepared by casting [19–23]. According to the results, the as-calculated values of ΔS_{mix} (J·K^{-1} mol^{-1}), ΔH_{mix} (kJ·mol^{-1}), and δ for CoCrNiCuZn HEA were respectively 1.61R, 0.96 and 4.4%, which were consistent with the formation criteria of HEAs. Table 2 shows mixing enthalpies of atomic pairs in the CoCrNiCuZn alloy system [24,25]. The main advantage of MA is the extension of solid solubility. Therefore, the simple solid solution is more likely formed in the as-milled HEA than that in the as-cast HEA. The calculated values of ΔS_{mix}, ΔH_{mix} and δ for CoCrNiCuZn HEA indicated that the simple solid solution should be formed in the MA process.

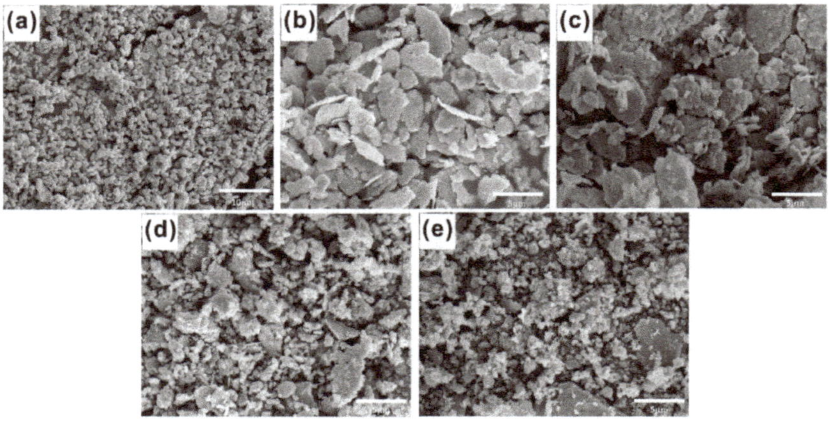

Figure 2. SEM images of CoCrNiCuZn HEA powder obtained after milling for different time: (**a**) 0 h, (**b**) 6 h, (**c**) 18 h, (**d**) 30 h, and (**e**) 60 h.

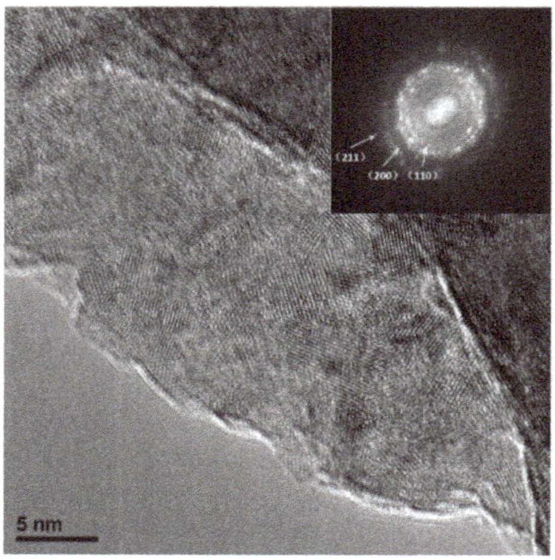

Figure 3. TEM image and selected area electron diffraction (SAED) pattern of CoCrNiCuZn HEA powder obtained after 60 h milling.

Table 2. Enthalpies (kJ·mol^{-1}) between every two elements in CoCrNiCuZn HEA.

Elements	Co	Cr	Ni	Cu	Zn
Co	0	−4	0	6	−5
Cr	-	0	−7	12	5
Ni	-	-	0	4	−9
Cu	-	-	-	0	1
Zn	-	-	-	-	0

3.1.3. Thermal Analysis

Figure 4 shows the DSC results of the CoCrNiCuZn high-entropy alloy powder obtained after 60 h milling. The first endothermic peak at around 100 °C is related to the energy absorption of the PCA evaporation [15]. Then the evaporated matter was eliminated by the flowing argon during testing. In the temperature range of 200~400 °C, the curve was relatively stable. When the temperature was above 600 °C, an endothermic line is observed, indicating that phase changes started at around this temperature. Two endothermic peaks at 1244.8 °C and 1321.8 °C were considered as the melting points of different phases [26], proving that there were two phases after the phase change occurred.

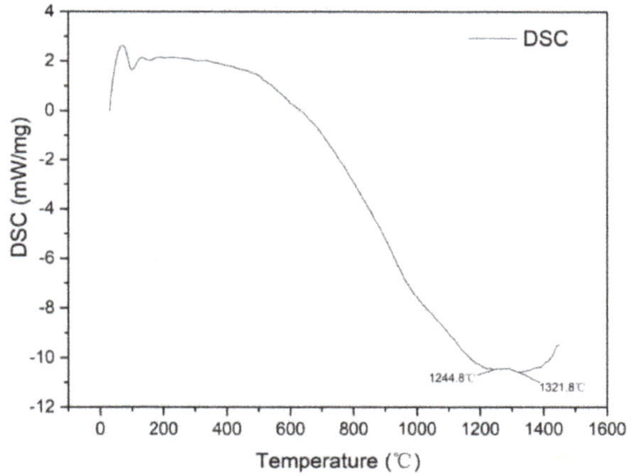

Figure 4. The trend and peaks of the thermal analysis curves (DSC, Mass) of CoCrNiCuZn HEA powder after 60 h ball milling.

3.2. Consolidation by SPS

3.2.1. X-Ray Analysis

Figure 5 shows the XRD patterns of the HEA powder after 60 h ball milling and the samples sintered at 600 °C, 700 °C, 800 °C, and 900 °C, respectively. Two FCC phases were formed at 900 °C and respectively recorded as FCC1 and FCC2. This is consistent with thermal analysis results.

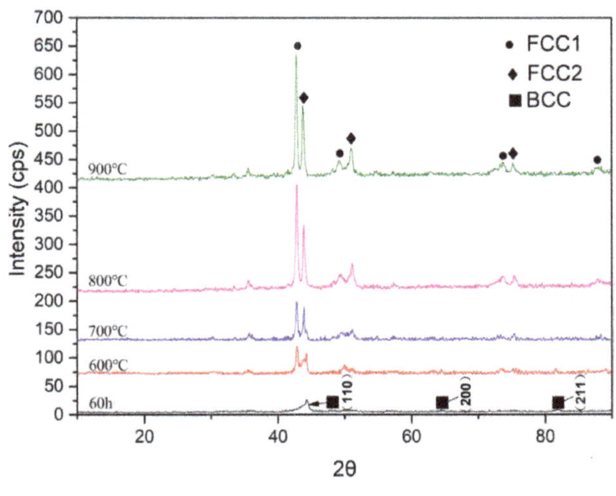

Figure 5. XRD patterns of CoCrNiCuZn HEA powder after 60 h ball milling and CoCrNiCuZn HEA samples fabricated by SPS at different sintering temperatures (600–1000 °C).

The above results indicated that both the as-milled CoCrNiCuZn powders and the as-sintered CoCrNiCuZn samples mainly had simple solid solution structures. This phenomenon can be explained by the Gibbs free energy of mixing defined as:

$$G_{mix} = H_{mix} - TS_{mix}, \quad (1)$$

where H_{mix} is the mixing entropy; G_{mix} is the Gibbs free energy of the mixture; S_{mix} is the mixing entropy and T is absolute temperature. The entropies of solid solution phases were much higher than those of the intermetallics. The increase in the mixing entropy largely decreased Gibbs free energy. Therefore, especially at high temperatures, the solid solution phases were preferentially formed rather than intermetallics and other complex phases [27].

3.2.2. Microstructure

The densities of alloys sintered at 600 °C, 700 °C, 800 °C, and 900 °C are respectively 5.26 g/cm^3, 6.26 g/cm^3, 7.84 g/cm^3, and 7.89 g/cm^3 measured by Archimedes principle. Figure 6 shows TEM bright field image and corresponding SAED patterns of CoCrNiCuZn HEA obtained after SPS at 900 °C. In the TEM image, two different morphologies were observed. Corresponding SAED patterns in Figure 6b,c indicated that the larger particles had a FCC1 structure, whereas the smaller ones had an FCC2 structure. The result was consistent with the XRD results.

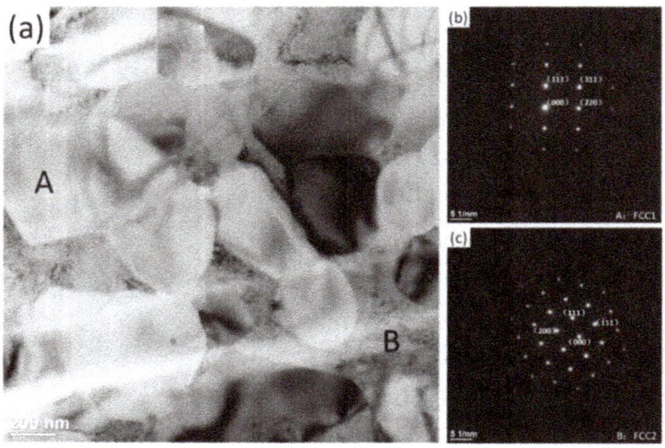

Figure 6. TEM image and SAED patterns of the CoCrNiCuZn HEA bulk obtained after SPS at 900 °C: (**a**) TEM bright field image of bulk CoCrNiCuZn HEA after SPS, (**b**) and (**c**) corresponding SAED patterns respectively indicate Region A with a FCC1 phase and Region B with an FCC2 phase.

Figure 7 shows the corresponding fractographic feature of the alloys sintered at 700 °C, 800 °C, and 900 °C, respectively. Section structure and stepped structure can be respectively observed in Figure 7a,b. The bulk alloys sintered at 900 °C showed a significant plasticity trend because the FCC phase exhibited a higher plasticity than BCC phase [15].

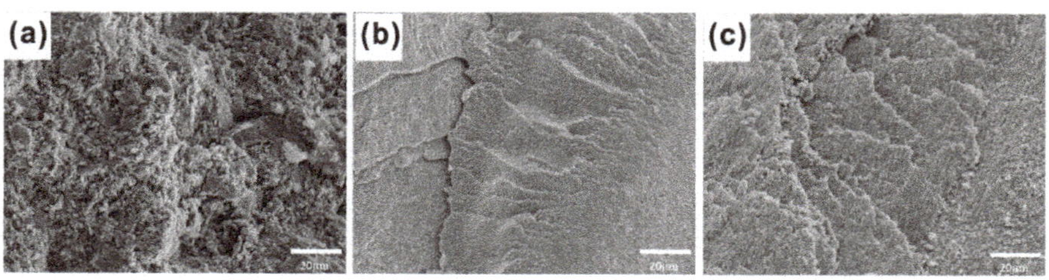

Figure 7. Slip fracture morphology of CoCrNiCuZn HEA samples fabricated by SPS at different sintering temperatures: (**a**) 700 °C, (**b**) 800 °C, and (**c**) 900 °C.

3.2.3. Mechanical Properties

Figure 8 shows the room-temperature compressive properties of the CoCrNiCuZn HEA consolidated at different temperatures. The strength increases with increasing of sintering temperature. The compressive strength of the sample sintered at 900 °C reached 2121 MPa, which was higher than that of most previously reported HEAs [6,27]. The Vickers hardness of HEA bulk sintered at 900 °C reached 615 HV, which was also superior to most commercial hard facing alloys [28]. The high compressive strength and high hardness are ascribed to the ultrafine grains (as shown in Figure 6a) and solid solution strengthening.

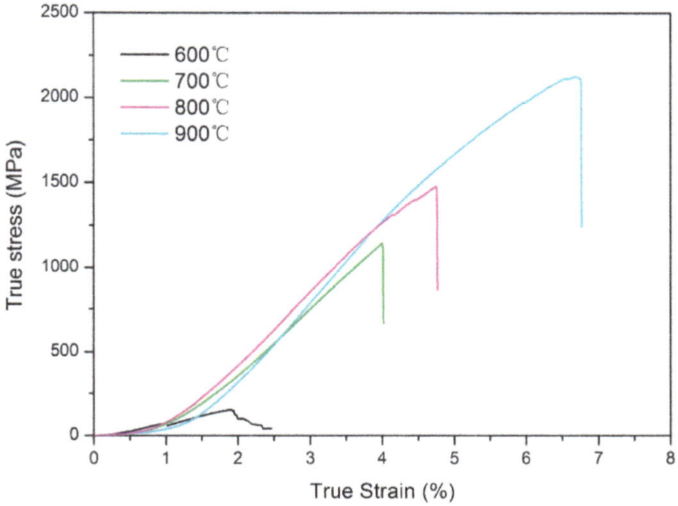

Figure 8. Compressive strain–stress curves at room temperature of CoCrNiCuZn HEA samples fabricated by SPS at different sintering temperatures (600 °C–900 °C).

4. Conclusions

The equiatomic CoCrNiCuZn HEA powder was successfully synthesized by MA. After 30-h ball milling, a BCC phase structure with a grain size of 10 nm was formed. The thermal analysis curve proved that the BCC phase structure gradually converted into FCC phase at above 600 °C. The XRD and TEM results demonstrated that the high-entropy alloy obtained after sintering had two FCC phases. The sample sintered at 900 °C had a Vickers hardness of 615 HV and a compressive strength of 2121 MPa. The combination of mechanical properties is superior to most of reported HEA systems and commercial hard facing alloys.

Author Contributions: Conceptualization, W.J. and Z.F.; methodology, Y.S. and B.K.; investigation, Y.S. and Y.L.; analysis, K.Y. and M.Y.; writing—original draft preparation, Y.S. and W.J.; writing—review and editing, W.J. and Z.F.; supervision, W.J. and Z.F.; project administration, W.J.; funding acquisition, Z.F.

Funding: This research was funded by the National Natural Science Foundation of China, grant numbers 51521001 and 51832003, the Students Innovation and Entrepreneurship Training Program of WHUT, grant numbers 2018CLA127 and 20181049701037, and the Self-Determined and Innovative Research Funds of WHUT, grant numbers 2018III020 and 2018IVA094.

Conflicts of Interest: The authors declare no conflict of interest.

References

1. Greer, A.L. Confusion by design. *Nature* **1993**, *366*, 303–304. [CrossRef]

2. Inoue, A.; Wang, X. Bulk amorphous FC20 (Fe-C-Si) alloys with small amounts of B and their crystallized structure and mechanical properties. *Acta. Mater.* **2000**, *48*, 1383–1395. [CrossRef]
3. Yeh, J.; Chen, S.; Lin, J.; Gan, T.; Chin, T.; Shun, C.; Tsau, S.; Chang, S. Nanostructured high-entropy alloys with multiple principal elements: Novel alloy design concepts and outcomes. *Adv. Eng. Mater.* **2004**, *6*, 299–303. [CrossRef]
4. Yeh, J. Recent progress in high-entropy alloys. *Ann. Chim. Sci. Mat.* **2006**, *31*, 633–648. [CrossRef]
5. Chen, Y.; Duval, T.; Hung, U.; Yeh, J.; Shih, H. Microstructure and electrochemical properties of high entropy alloys-a comparison with type-304 stainless steel. *Corros. Sci.* **2005**, *47*, 2257–2279. [CrossRef]
6. Wang, Y.; Li, B.; Ren, M.; Yang, C.; Fu, H. Microstructure and compressive properties of AlCrFeCoNi high entropy alloy. *Mater. Sci. Eng. A* **2008**, *491*, 154–158. [CrossRef]
7. Joseph, J.; Jarvis, T.; Wu, X.; Stanford, N.; Hodgson, P.; Fabijanic, D. Comparative study of the microstructures and mechanical properties of direct laser fabricated and arc-melted Al_xCoCrFeNi high entropy alloys. *Mater. Sci. Eng. A* **2015**, *633*, 184–193. [CrossRef]
8. Suryanarayana, C.; Ivanov, E.; Boldyrev, V. The science and technology of mechanical alloying. *Mater. Sci. Eng. A* **2001**, *304*, 151–158. [CrossRef]
9. Zhang, K.; Fu, Z.; Zhang, J.; Shi, J.; Wang, W.; Wang, H.; Wang, Y.; Zhang, Q. Nanocrystalline CoCrFeNiCuAl high-entropy solid solution synthesized by mechanical alloying. *J. Alloy. Comp.* **2009**, *485*, L31–L34. [CrossRef]
10. Chen, W.; Fu, Z.; Fang, S.; Xiao, H.; Zhu, D. Alloying behavior, microstructure and mechanical properties in a $FeNiCrCo_{0.3}Al_{0.7}$ high entropy alloy. *Mater. Des.* **2013**, *51*, 854–860. [CrossRef]
11. Li, Q.; Wang, G.; Song, X.; Fan, L.; Hu, W.; Xiao, F.; Yang, Q.; Ma, M.; Zhang, J.; Liu, R. $Ti_{50}Cu_{23}Ni_{20}Sn_7$ bulk metallic glasses prepared by mechanical alloying and spark-plasma sintering. *J. Mater. Process. Tech.* **2009**, *209*, 3285–3288. [CrossRef]
12. Fu, Z.; Chen, W.; Xiao, H.; Zhou, L.; Zhu, D.; Yang, S. Fabrication and properties of nanocrystalline $Co_{0.5}FeNiCrTi_{0.5}$ high entropy alloy by MA-SPS technique. *Mater. Des.* **2013**, *44*, 535–539. [CrossRef]
13. Bouad, N.; Marin-Ayral, R.; Tedenac, J. Mechanical alloying and sintering of lead telluride. *J. Alloy. Comp.* **2000**, *297*, 312–318. [CrossRef]
14. Yeh, J.; Chang, S.; Hong, Y.; Chen, S.; Lin, S. Anomalous decrease in X-ray diffraction intensities of Cu-Ni-Al-Co-Cr-Fe-Si alloy systems with multi-principal elements. *Mater. Chem. Phos.* **2007**, *103*, 41–46. [CrossRef]
15. Ji, W.; Fu, Z.; Wang, W.; Wang, H.; Zhang, J.; Wang, Y.; Zhang, F. Mechanical alloying synthesis and spark plasma sintering consolidation of CoCrFeNiAl high-entropy alloy. *J. Alloy. Comp.* **2014**, *589*, 61–66. [CrossRef]
16. Varalakshmi, S.; Kamaraj, M.; Murty, B. Synthesis and characterization of nanocrystalline AlFeTiCrZnCu high entropy solid solution by mechanical alloying. *J. Alloy. Comp.* **2008**, *460*, 253–257. [CrossRef]
17. Monshi, A.; Foroughi, M.; Monshi, M. Modified Scherrer Equation to Estimate More Accurately Nano-Crystallite Size Using XRD. *World J. Nano Sci. Eng.* **2012**, *2*, 154–160. [CrossRef]
18. De Keijser, T.H.; Langford, J.I.; Mittemeijer, E.J.; Vogels, A.B.P. Use of the Voigt function in a single-line method for the analysis of X-ray diffraction line broadening. *J. Appl. Cryst.* **1982**, *15*, 308–314. [CrossRef]
19. Zhang, Y.; Zhou, Y.; Lin, J.; Chen, G.; Liaw, P. Solid-solution phase formation rules for multi-component alloys. *Adv. Eng. Mater.* **2008**, *10*, 534–538. [CrossRef]
20. Inoue, A. Stabilization of metallic supercooled liquid and bulk amorphous alloys. *Acta. Mater.* **2000**, *48*, 279–306. [CrossRef]
21. Fang, S.; Xiao, X.; Lei, X.; Li, W.; Dong, Y. Relationship between the widths of supercooled liquid regions and bond parameters of Mg-based bulk metallic glasses. *J. Non-Cryst. Solids.* **2003**, *321*, 120–125. [CrossRef]
22. Zhu, J.; Liaw, P.; Liu, C. Effect of electron concentration on the phase stability of $NbCr_2$-based Laves phase alloys. *Mater. Sci. Eng. A* **1997**, *239*, 260–264. [CrossRef]
23. Guo, S.; Ng, C.; Lu, J.; Liu, C. Effect of valence electron concentration on stability of fcc or bcc phase in high entropy alloys. *J. Appl. Phys.* **2011**, *109*, 103505. [CrossRef]
24. Takeuchi, A.; Inoue, A. Calculations of mixing enthalpy and mismatch entropy for ternary amorphous alloys. *Mater. Trans. JIM.* **2000**, *41*, 1372–1378. [CrossRef]
25. Takeuchi, A.; Inoue, A. Classification of bulk metallic glasses by atomic size difference, heat of mixing and period of constituent elements and its application to characterization of the main alloying element. *Mater. Trans.* **2005**, *46*, 2817–2829. [CrossRef]

26. Zhang, K.; Fu, Z. Effects of annealing treatment on phase composition and microstructure of $CoCrFeNiTiAl_x$ high-entropy alloys. *Intermetallics* **2012**, *22*, 24–32. [CrossRef]
27. Pi, J.; Pan, Y.; Zhang, L.; Zhang, H. Microstructure and property of AlTiCrFeNiCu high-entropy alloy. *J. Alloy. Comp.* **2011**, *509*, 5641–5645. [CrossRef]
28. Kapoor, S.; Liu, R.; Wu, X.; Yao, M. Temperature-dependence of hardness and wear resistance of stellite alloys. *Eng. Technol.* **2012**, *67*, 964–973.

entropy

MDPI

Article

Lattice Distortion and Phase Stability of Pd-Doped NiCoFeCr Solid-Solution Alloys

Fuxiang Zhang [1,*], Yang Tong [1], Ke Jin [1], Hongbin Bei [1], William J. Weber [1,2] and Yanwen Zhang [1]

[1] Materials Science and Technology Division, Oak Ridge National Laboratory, 1 Bethel Valley Rd, Oak Ridge, TN 37831, USA; tongy1@ornl.gov (Y.T.); pkuphyjk@gmail.com (K.J.); hbei1@utk.edu (H.B.); weberwj@ornl.gov (W.J.W.); zhangy1@ornl.gov (Y.Z.)

[2] Department of Materials Science and Engineering, University of Tennessee, Knoxville, TN 37996, USA

* Correspondence: zhangf@ornl.gov; Tel.: +01-865-574-0835

Received: 12 November 2018; Accepted: 21 November 2018; Published: 25 November 2018

Abstract: In the present study, we have revealed that $(NiCoFeCr)_{100-x}Pd_x$ (x= 1, 3, 5, 20 atom%) high-entropy alloys (HEAs) have both local- and long-range lattice distortions by utilizing X-ray total scattering, X-ray diffraction, and extended X-ray absorption fine structure methods. The local lattice distortion determined by the lattice constant difference between the local and average structures was found to be proportional to the Pd content. A small amount of Pd-doping (1 atom%) yields long-range lattice distortion, which is demonstrated by a larger (200) lattice plane spacing than the expected value from an average structure, however, the degree of long-range lattice distortion is not sensitive to the Pd concentration. The structural stability of these distorted HEAs under high-pressure was also examined. The experimental results indicate that doping with a small amount of Pd significantly enhances the stability of the fcc phase by increasing the fcc-to-hcp transformation pressure from ~13.0 GPa in NiCoFeCr to 20–26 GPa in the Pd-doped HEAs and NiCoFeCrPd maintains its fcc lattice up to 74 GPa, the maximum pressure that the current experiments have reached.

Keywords: solid-solution alloys; lattice distortion; phase transformation

1. Introduction

High-entropy alloys (HEAs) are usually a single-phase solid-solution with multi principle elements randomly distributed in the lattice [1,2]. Due to the size difference of individual atoms, lattice distortion is believed to be one of the core effects, which greatly affects the mechanical and physical properties [3–8]. The distorted local lattice provides pinning sites to slow down dislocation motion and therefore improve the mechanical performance of high-entropy alloys [9–12]. The intrinsic lattice distortion in HEAs can shorten the free-electron migration paths and reduce the electrical and thermal conductivities [1,13,14], which can enhance the recombination of radiation defects due to a strong localized heating effect. In addition, distorted local lattice sites can retard the motion of radiation defects to delay their growth. Therefore, HEAs are strong candidates for nuclear materials by showing excellent radiation resistance [13,15]. However, a quantitative description of the lattice distortion in HEAs is a challenge and previous experimental results are controversial [1,16,17]; especially with respect to alloys with the fcc structure. For example, no obvious lattice distortion was reported previously for a NiCoFeMnCr alloy [17]. Recently, we developed a new analytical method based on atomic pair distribution function (PDF) measurement that can quantitatively describe the local lattice distortion in some high- and medium-entropy alloys [18–20]. PDF analysis has shown that the local lattice distortion in the NiCoFeCr is negligible (< 0.1%), while the NiCoFeCrPd HEA has a very large local lattice distortion (0.79%). Since there is a large mismatch of atomic size between Pd and other atoms, it is interesting to investigate the effect of Pd content on the lattice distortion in $(NiCoFeCr)_{100-x}Pd_x$ solid-solution alloys.

Besides atomic size mismatch, different atomic configurations can also affect local lattice distortion in HEAs. For instance, the local bonding environment of individual atoms varies in solid-solution alloys, leading to the fluctuation of nearest atomic pair distances and short-range order [21]. Extended X-ray absorption fine structure (EXAFS) is atomic mass sensitive and is a powerful tool to measure the distance of different atomic pairs in solid-solution alloys. However, some approximations need to be made for those alloys with components that are neighbors in the periodic table. With EXAFS measurement, we have successfully revealed the short-range order in NiPd [22] and NiCoCr [21] solid-solution alloys.

The uncertainty for atoms being located exactly on the lattice sites is another type of structural disorder that contributes to the excess configurational entropy [1,23]. Previous TEM analysis indicated that the lattice distortion destabilized the structure, and phase segregation was observed in solid-solution alloys, such as NiCoFeCrMn [24,25], during annealing and ion irradiation. Under high pressure, the fcc lattice of some alloys can transform to another close-packed hcp structure. However, the phase transformation behavior in the HEAs is not simple. Experiments demonstrated that NiCoFeCr [26] and NiCoFeCrMn [27,28] started to transform to an hcp structure at ~13 GPa, whereas phase transition was not found in NiCoFeCrPd alloy even up to 74 GPa [26]. Moreover, it recognized that magnetic contributions to the free energy may play a critical role. In this paper, the long/short-range lattice distortion and structural stability of Pd-doped $(NiCoFeCr)_{100-x}Pd_x$ HEAs were experimentally studied with total X-ray scattering, X-ray diffraction methods and the local bonding environment of atoms in the solid-solution is derived by EXAFS measurements.

2. Materials and Methods

Elemental metals Ni, Co, Fe, Cr, and Pd (> 99.9% pure) in the designed atomic ratios with the formula of $(NiCoFeCr)_{100-x}Pd_x$ (x = 1, 3, 5 and 20) were carefully weighed and mixed by arc melting. The arc-melted buttons were flipped and re-melted at least five times before drop casting to ensure the homogeneity. The ambient total scattering measurements were performed at synchrotron beamline F2 of CHESS (Cornell High Energy Synchrotron Source), Cornell University, with an X-ray energy of E = 61.332 KeV and beam size of 500 × 500′μm^2. A two-dimensional stationary detector with 200 × 200′μm^2 pixel size was placed ~20 cm behind the sample to collect data. Fit2D software [29] was used to correct for a beam polarization and a dark current. In order to obtain real-space PDF, the measured patterns were Fourier transformed by PDFgetX3 [30] and then normalized reciprocal-space structure function in a Q range of 30 Å^{-1}. Using PDFGui software [31], the measured PDFs were refined with the fcc structure models. The in situ high-pressure XRD (X-ray powder diffraction) was conducted with the diamond anvil cell technique in transmission mode at beamline 16-BM-D of the APS (Advanced Photon Source), Argonne National Laboratory. For high-pressure XRD experiments, a methanol/ethanol (4/1) mixture was used as the pressure transition medium. The wavelength of the X-ray was 0.4989 Å and 0.3103 Å for ambient and high-pressure measurements, respectively. For all of the synchrotron XRD experiments, the instrument parameters were calibrated with CeO_2 as the standard, and XRD profiles were analyzed with the Rietveld refinement method using the program Fullprof [32]. The EXAFS spectra at the K-edge of elements Ni, Co, Fe, Cr were conducted in a fluorescence mode with a grazing exit configuration (grazing angle of 2–3°) at beam 13-ID-E of APS, Argonne National Laboratory. Athena program [33] was used for the reduction and analysis of the EXAFS data. The fitting of the EXAFS spectra was conducted with Artemis in Demeter software package [33] in a fixed k-range (3.0–12.0 Å^{-1}) and an fcc structural model was used to simulate the structure of the solid solutions.

3. Results and Discussion

3.1. Local Lattice Distortion

The local lattice distortion induced by atomic size mismatch in HEAs has been estimated by a hard sphere model [34]. However, an experimental investigation has revealed that the hard-sphere model considerably overestimated the local lattice distortion in the HEAs. The measurement of the Bragg peak width in XRD or neutron diffraction profiles contains information of both static and dynamic displacements, which, however, cannot be resolved [17]. PDF analysis based on total scattering measurements can effectively reveal the local lattice distortion in terms of variation of local bond distance. Figure 1a shows the observed PDF profile of the NiCoFeCrPd HEA and the calculated one based on a random solid-solution model. Except for the first peak, the calculated pattern matches the observed one very well. The mismatch in the first atomic shell is an indication of local lattice distortion in NiCoFeCrPd. In order to quantitively describe the local lattice distortion in solid-solution alloys, we introduced a local lattice distortion parameter ε.

$$\varepsilon = \left(a_{var} - a_{avg}\right)/a_{avg} \tag{1}$$

Where a_{avg} is the lattice parameter obtained from fitting the PDF profile over the whole r-range and a_{var} is the lattice parameter obtained by fitting the PDF profile from r_{min} = 1.5 Å (data below this value was excluded because of large oscillations) to the varied r_{max} value. The local lattice distortion in the fcc HEAs is strongly localized in the first atomic shell, as shown in Figure 1a. Therefore, we only focused on the lattice strain in the first atomic shell, ε_{1st}. Our results show that NiCoFeCr has a negligible ε_{1st} (< 0.1%), whereas the NiCoCr and FeCoNiCrMn possess a small positive ε_{1st}, suggesting that the local bond distances are larger than the expected value from their average structures. NiCoFeCrPd has the largest ε_{1st} (0.79%) (Figure 1b) reported so far. The large lattice strain in NoCoFeCrPd is caused by the large size mismatch between Pd and other elements. With similar analysis of the Pd-doped $(NiCoFeCr)_{100-x}Pd_x$ HEAs, the local lattice distortion is found to be proportional to the content of Pd (Figure 1b).

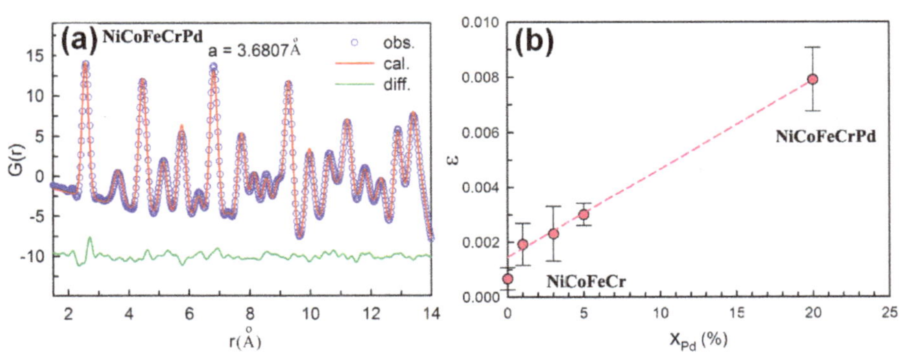

Figure 1. (**a**) Pair distribution function of NiCoFeCrPd HEA (high-entropy alloys). The blue symbols are experimental data and the red line is fit to the data using a random solid-solution model. The slight shift of the measured PDF (pair distribution function) from the fitted data (difference shown at *r* from 2 to 3 Å indicates the local lattice distortion; (**b**) Local lattice distortion in the first atomic shell as a function of the Pd concentration in the alloys.

3.2. Long-Range Lattice Distortion

XRD experimental results have shown that the Bragg peaks are broadened and the intensities are reduced in HEAs [35] because of the larger uncertainty for atoms being exactly on the crystalline lattice sites. No obvious long-range lattice distortion has been observed previously from X-ray or neutron diffraction measurements. For most HEAs, their lattice remains the fcc, bcc or hcp structure. However, long-range lattice distortion is found in Pd-doped $(NiCoFeCr)_{100-x}Pd_x$ HEAs from XRD measurement. Figure 2 shows the XRD patterns of $(NiCoFeCr)_{100-x}Pd_x$ (x = 1, 3, 5 and 20 atom%) HEAs under ambient conditions. The red dots are observed patterns and the green line are calculated patterns based on Rietveld refinement. As shown in the enlarged patterns (Figure 2b), there is a clear deviation at the (200) Bragg peak. The observed (200) peak exhibits a larger d-spacing than the average, whereas all of the other Bragg peaks match the average positions very well. The deviation of the (200) lattice planes is ~0.004 Å. Since XRD reveals the long-range order, the mismatch of the (200) peak indicates that the Pd-doped $NiCoFeCrPd_x$ HEAs have a distorted lattice from the ideal fcc structure, though no peak splitting was observed. Experimental analysis also suggests that the deviation of the (200) peak in the alloys with different concentrations of Pd is nearly the same. Due to the substitution of larger Pd atoms into the structure, it is not difficult to understand the change of lattice constant (Table 1) and local lattice distortion with the Pd content in the solid-solution alloys. It is surprising that even 1atom% Pd-doping in $(NiCoFeCr)_{99}Pd_1$ HEAs can cause a long-range structural distortion on the (200) lattice planes. The systematic larger (200) lattice plane spacing suggests that the large Pd atoms may be not randomly distributed in the lattice. To obtain the short-range order information for these HEAs, a method capable of characterizing the local bonding environment is strongly needed.

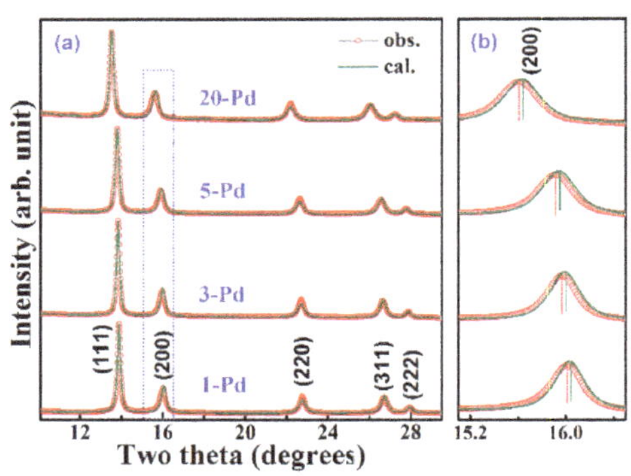

Figure 2. (**a**) The XRD (X-ray powder diffraction) profiles measured with synchrotron X-rays (λ = 0.4989Å). The red symbols are measured data and the green lines are calculated profiles based on Rietveld refinement; (**b**) the enlarged part of the XRD profiles and the observed (200) Bragg peak in all the samples obviously shifted to lower two theta angles with larger d-values.

Table 1. The lattice constant and the nearest atomic pair distance in the solid-solution alloys measured with XRD and EXAFS (extended X-ray absorption fine structure).

Sample	Lattice Constant (Å)	Nearest Atomic Pair Distance (Å)	
		XRD	EXAFS
$(NiCoFeCr)_{99}Pd_1$	3.5767(1)	2.5291	2.53(1)
$(NiCoFeCr)_{97}Pd_3$	3.5860(2)	2.5357	2.54(1)
$(NiCoFeCr)_{95}Pd_5$	3.5975(2)	2.5438	2.54(1)
NiCoFeCrPd	3.6679(4)	2.5936	2.55(4)

3.3. Local Bonding Environment

Neither XRD nor total scattering measurements can give the atomic bonding information in the solid-solution alloys. In order to detect the bond distance in the $(NiCoFeCr)_{100-x}Pd_x$ (x = 1, 3, 5 and 20 atom%) HEAs, we measured the K-edge X-ray absorption spectrums of Ni, Co, Fe, and Cr elements. Since these four elements have similar X-ray scattering ability, it is difficult to distinguish the individual elements in the solid-solution alloys. As an approximation, we assumed that, except for Pd, the atoms around the core have the same X-ray scattering ability. Figure 3 is the k^3-weighted FTs of the Fe K-edge EXAFS for the $(NiCoFeCr)_{100-x}Pd_x$ (x = 1, 3, 5 and 20 atom%) HEAs. We assumed that Pd and Fe atoms are randomly distributed in the fcc lattice and the red dash lines in Figure 3 are the fittings within the first shell in the radial distance of 1–3.5 Å. The first shell peak is generally well fitted with this approximation. The derived average distance between the nearest atomic pairs is shown in Table 1. In general, for the lower Pd-doped alloys, the distance measured with EXAFS is in good agreement with that measured with XRD but for the equiatomic NiCoFeCrPd alloy, the distance measured with EXAFS is obviously smaller (1.7%) than that from the XRD measurement, which suggests that there is a larger lattice strain in the NiCoFeCrPd HEA than the lower Pd-doped alloys. This is in agreement with the total scattering measurements. From the PDF analysis, we have confirmed that the local lattice distortion in NiCoFeCrPd is 3–4 times larger than that in the lower Pd-doped solid-solution alloys [20].

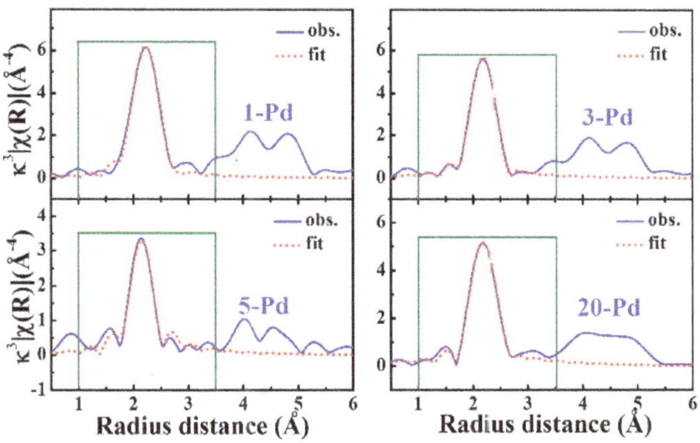

Figure 3. The κ^3-weighted FTs (Fourier Transforms) of the Fe K-edge EXAFS (extended X-ray absorption fine structure) in $(NiCoFeCr)_{100-x}Pd_x$ solid-solution alloys. The solid blue line is observed and the red dash line is the fitting with the nearest neighbors.

3.4. Structural Stability at High Pressures

Multi-component concentrated solid-solution alloys can have an fcc, bcc or hcp structure. Theoretical calculations suggested that some fcc HEAs are metastable because their hcp counterparts have similar Gibbs free energies at ambient conditions. Previous experiments have revealed that some of the fcc alloys can transform to the hcp structure under high-pressure conditions, such as NiCoFeCr [26] and NiCoFeCrMn [27,28] alloys that transformed to an hcp structure at ~13 GPa. The hcp structure is quenchable to ambient conditions, though the phase transition is very sluggish. The phase stability is composition sensitive. For FeMnCoCr alloys [36], the hcp structure can coexist with the fcc structure from the sample preparation process. By properly tuning the chemical composition, a dual-phase alloy can possess excellent mechanical properties. For the five-element system of NiCoFeCrPd, the fcc structure is stable up to 74 GPa [26]. The larger size of Pd atoms plays a key role in structural stability. It is thus interesting to study the effect of Pd content on the phase transition. We pressurized the Pd-doped $(NiCoFeCr)_{100-x}Pd_x$ HEAs with diamond-anvil cell techniques, and the experimental results indicate that a small amount of Pd greatly affects the structural stability. The critical pressure for the fcc to hcp phase transition is strongly increased to more than 20 GPa in 1 atom%, 3 atom% and 5 atom% doped $(NiCoFeCr)_{100-x}Pd_x$ solid-solution alloys. Figure 4 shows the XRD profiles of 3 atom% Pd doped $(NiCoFeCr)_{97}Pd_3$ alloy at different pressures, and the hcp structure starts to appear at 20.9 GPa. The transition is very sluggish, and the amount of the hcp structure is only ~33% at 34.1 GPa. The hcp structure is stable once it is formed, and the quenched alloy has a mixed structure of fcc and hcp. The critical transition pressure is not sensitive to the amount of Pd doped and is observed at 26.0, 20.9, and 21.0 GPa for the 1%, 3% and 5% Pd-doped $(NiCoFeCr)_{100-x}Pd_x$ HEAs, respectively. However, no hcp structure was found in the equiatomic solid-solution alloy NiCoFeCrPd up to 74 GPa [26]. Since Pd has a much larger atomic size than other elements in these alloys, the substitution of Pd for other atoms increases the lattice parameter and atomic-pair distances. When the Pd content in the alloys is sufficient, all the smaller atoms will have more free space to move, which may allow adaption to the lattice distortion at high pressures. This may be the main reason why Pd can cause changes in the critical transition pressure for Pd-doped NiCoFeCrPd solid-solution alloys. Therefore, the high local lattice distortion greatly in the equiatomic HEA can enhance the stability of the fcc lattice.

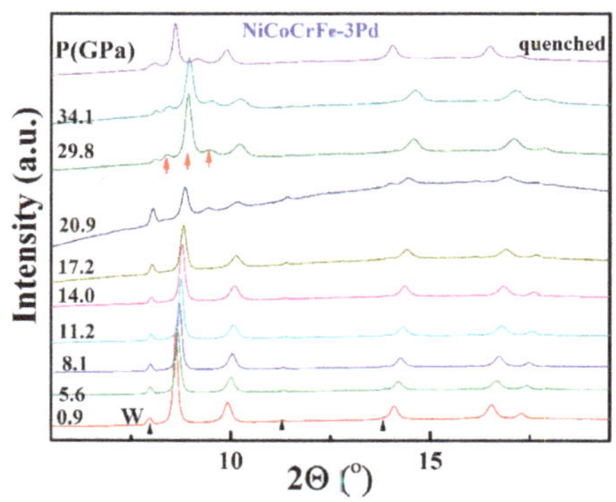

Figure 4. The XRD profiles of $(NiCoFeCr)_{97}Pd_3$ measured at different pressures. The fcc lattice starts to transform to hcp structure at 20.9 GPa. The weak diffraction peaks marked with small black arrows are from the W gasket during measurement.

We further analyzed the long-range lattice distortion of $(NiCoFeCr)_{100-x}Pd_x$ solid-solution alloys under high pressure, i.e., the deviation at the (200) Bragg peak. Figure 5 shows the deviation of each observed Bragg peaks at different pressures for the 1% Pd-doped NiCoFeCr. The deviation of the (200) Bragg peak obviously increased with pressure. Before the hcp structure starts to form, the deviation of $(NiCoFeCr)_{99}Pd_1$ has reached 0.01 Å. In a strict sense, the structure of the $(NiCoFeCr)_{99}Pd_1$ alloy is not fcc anymore. A similar behavior is also observed in the $(NiCoFeCr)_{97}Pd_3$ and $(NiCoFeCr)_{95}Pd_5$ alloys. Accordingly, the external high pressure can enhance the long-range lattice distortion.

Figure 6 shows the *P-V* curves of Pd-doped NiCoFeCr alloys. As a comparison, the *P-V* curves for NiCoFeCr and NiCoFeCrPd are also shown. When fitted with a 3rd Birch-Murnaghan equation of state, the bulk modulus is 190(5), 171(8) and 186(4) GPa for the 1%, 3% and 5% Pd-doped NiCoFeCr solid-solution alloys, respectively. The bulk modulus of the Pd-doped solid-solution alloys is smaller than that of NiCoFeCr (206 GPa) but larger than that of NiCoFeCrPd (168 GPa). The addition for larger Pd atoms makes the alloys more compressible with smaller bulk modulus because the addition of Pd atoms increased the lattice parameters and the smaller atoms, Ni, Co, Fe, and Cr, may have more "free" space to move in order to adapt the structure during pressurization.

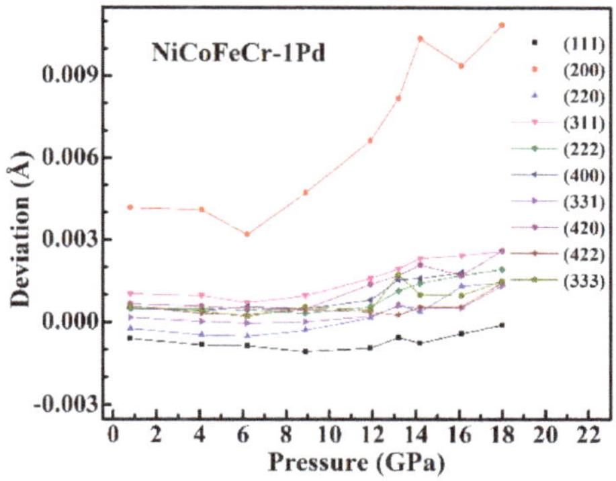

Figure 5. The deviation of the observed Bragg peaks from the ideal fcc structure at different pressures. The deviation of (200) peak is obvious and it increases with pressure.

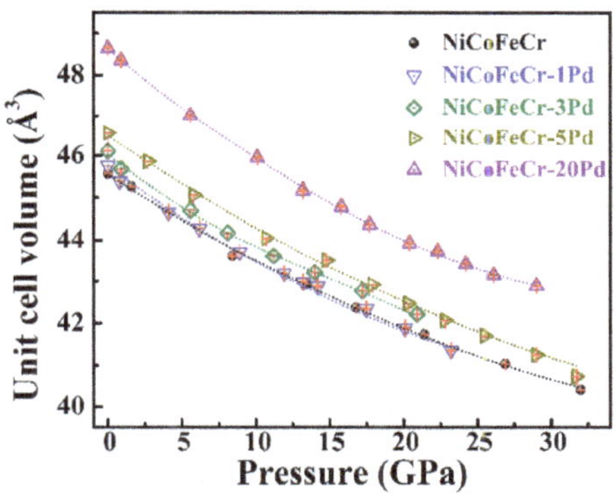

Figure 6. The pressure dependence of the unit cell volume. The dashed lines *P-V* curves which are fitted with 3-rd-order Birch-Murnaghan equation of state

4. Conclusions

We have systematically studied the effects of Pd doping in $(NiCoFeCr)_{100-x}Pd_x$ solid-solution alloys on lattice distortion and phase stabilities. The short-range order is strongly dependent on the Pd content in the alloys. Both PDF and EXAFS measurements suggest that the lattice is locally strained in the highly Pd doped solid-solution alloys. Even a small amount of Pd addition in the alloys can cause a long-range lattice distortion by showing a larger (200) lattice plan spacing than the expected from the average structure. High-pressure studies revealed that a small amount of Pd in the solid-solution alloys greatly enhanced the phase stability of the fcc structure, and the critical pressure for the fcc to hcp transition increased from ~13 GPa to more than 20 GPa in 1%, 3% and 5% Pd-doped alloys, while the fcc is stable up to 74 GPa in the equiatomic NiCoFeCrPd HEA.

Author Contributions: F.Z. and Y.T. performed the experiments, K.J. and H.B. prepared the samples, W.J.W. and Y.Z. contributed to discussion and writing.

Funding: This research was funded by US DOE grant number DE-AC05-00OR22725.

Acknowledgments: This work was supported as part of the Energy Dissipation to Defect Evolution (EDDE), an Energy Frontier Research Center funded by the U.S. Department of Energy, Office of Science, Basic Energy of Sciences under contract number DE-AC05-00OR22725. The X-ray diffraction and total scattering measurement were conducted at the Cornell High Energy Synchrotron Source (CHESS) which is supported by the National Science Foundation and the National Institutes of Health/National Institute of General Medical Sciences under NSF award DMR-1332208. The EXAFS measurement was performed at GeoSoilEnviroCARS (The University of Chicago, Sector 13), Advanced Photon Source (APS), Argonne National Laboratory. GeoSoilEnviroCARS is supported by the National Science Foundation - Earth Sciences (EAR - 1634415) and Department of Energy-GeoSciences (DE-FG02-94ER14466). This research used resources of the Advanced Photon Source, a U.S. Department of Energy (DOE) Office of Science User Facility operated for the DOE Office of Science by Argonne National Laboratory under Contract No. DE-AC02-06CH11357.

Conflicts of Interest: The authors declare no conflict of interest.

References

1. Miracle, D.B.; Senkov, O.N. A critical review of high entropy alloys and related concepts. *Acta Mater.* **2017**, *122*, 448–511. [CrossRef]
2. Cantor, B.; Chang, I.T.H.; Knight, P.; Vincent, A.J.B. Microstructural development in equiatomic multicomponent alloys. *Mater. Sci. Eng. A* **2004**, *375–377*, 213–218. [CrossRef]
3. Zhang, Y.; Zuo, T.T.; Tang, Z.; Gao, M.C.; Dahmen, K.A.; Liaw, P.K.; Lu, Z.P. Microstructures and properties of high-entropy alloys. *Prog. Mater. Sci.* **2014**, *61*, 1–93. [CrossRef]
4. Yeh, J.W. Alloy design strategies and future trends in high-entropy alloys. *Jom* **2013**, *65*, 1759–1771. [CrossRef]
5. Zhu, J.M.; Fu, H.M.; Zhang, H.F.; Wang, A.M.; Li, H.; Hu, Z.Q. Microstructures and compressive properties of multicomponent AlCoCrFeNiMox alloys. *Mater. Sci. Eng. A* **2010**, *527*, 6975–6979. [CrossRef]
6. Miracle, D.B.; Miller, J.D.; Senkov, O.N.; Woodward, C.; Uchic, M.D.; Tiley, J. Exploration and development of high entropy alloys for structural applications. *Entropy* **2014**, *16*, 494–525. [CrossRef]
7. Tsai, K.Y.; Tsai, M.H.; Yeh, J.W. Sluggish diffusion in Co-Cr-Fe-Mn-Ni high-entropy alloys. *Acta Mater.* **2013**, *61*, 4887–4897. [CrossRef]
8. Chou, H.P.; Chang, Y.S.; Chen, S.K.; Yeh, J.W. Microstructure, thermophysical and electrical properties in AlxCoCrFeNi ($0 \leq x \leq 2$) high-entropy alloys. *Mater. Sci. Eng. B Solid-State Mater. Adv. Technol.* **2009**, *163*, 184–189. [CrossRef]
9. Wu, Z.; Bei, H.; Otto, F.; Pharr, G.M.; George, E.P. Recovery, recrystallization, grain growth and phase stability of a family of FCC-structured multi-component equiatomic solid solution alloys. *Intermetallics* **2014**, *46*, 131–140. [CrossRef]
10. Shun, T.T.; Du, Y.C. Microstructure and tensile behaviors of FCC Al0.3CoCrFeNi high entropy alloy. *J. Alloys Compd.* **2009**, *479*, 157–160. [CrossRef]
11. Kuznetsov, A.V.; Shaysultanov, D.G.; Stepanov, N.D.; Salishchev, G.A.; Senkov, O.N. Tensile properties of an AlCrCuNiFeCo high-entropy alloy in as-cast and wrought conditions. *Mater. Sci. Eng. A* **2012**, *533*, 107–118. [CrossRef]
12. Toda-Caraballo, I.; Rivera-Díaz-Del-Castillo, P.E.J. A criterion for the formation of high entropy alloys based on lattice distortion. *Intermetallics* **2016**, *71*, 76–87. [CrossRef]
13. Zhang, Y.; Stocks, G.M.; Jin, K.; Lu, C.; Bei, H.; Sales, B.C.; Wang, L.; Beland, L.K.; Stoller, R.E.; Samolyuk, G.D.; et al. Influence of chemical disorder on energy dissipation and defect evolution in advanced alloys. *Nat. Commun.* **2015**, *6*, 8736. [CrossRef]
14. Lu, C.; Yang, T.; Jin, K.; Velisa, G.; Xiu, P.; Song, M.; Peng, Q.; Gao, F.; Zhang, Y.; Bei, H.; et al. Enhanced void swelling in NiCoFeCrPd high-entropy alloy by indentation-induced dislocations. *Mater. Res. Lett.* **2018**, *6*, 584–591. [CrossRef]
15. Lu, C.; Niu, L.; Chen, N.; Jin, K.; Yang, T.; Xiu, P.; Zhang, Y.; Gao, F.; Bei, H.; Shi, S.; et al. Enhancing radiation tolerance by controlling defect mobility and migration pathways in multicomponent single-phase alloys. *Nat. Commun.* **2016**, *7*, 13564. [CrossRef]
16. Yao, M.J.; Pradeep, K.G.; Tasan, C.C.; Raabe, D. A novel, single phase, non-equiatomic FeMnNiCoCr high-entropy alloy with exceptional phase stability and tensile ductility. *Scr. Mater.* **2014**, *72–73*, 5–8. [CrossRef]
17. Owen, L.R.; Pickering, E.J.; Playford, H.Y.; Stone, H.J.; Tucker, M.G.; Jones, N.G. An assessment of the lattice strain in the CrMnFeCoNi high-entropy alloy. *Acta Mater.* **2017**, *122*, 11–18. [CrossRef]
18. Tong, Y.; Jin, K.; Bei, H.; Ko, J.Y.P.; Pagan, D.C.; Zhang, Y.; Zhang, F.X. Local lattice distortion in NiCoCr, FeCoNiCr and FeCoNiCrMn concentrated alloys investigated by synchrotron X-ray diffraction. *Mater. Des.* **2018**, *155*, 1–7. [CrossRef]

19. Tong, Y.; Velisa, G.; Zhao, S.; Guo, W.; Yang, T.; Jin, K.; Lu, C.; Bei, H.; Ko, J.Y.P.; Pagan, D.C.; et al. Evolution of local lattice distortion under irradiation in medium- and high-entropy alloys. *Materialia* **2018**, *2*, 73–81. [CrossRef]
20. Tong, Y.; Zhao, S.; Jin, K.; Bei, H.; Ko, J.Y.P.; Zhang, Y.; Zhang, F.X. A comparison study of local lattice distortion in Ni80Pd20 binary alloy and FeCoNiCrPd high-entropy alloy. *Scr. Mater.* **2018**, *156*, 14–18. [CrossRef]
21. Zhang, F.X.; Zhao, S.; Jin, K.; Xue, H.; Velisa, G.; Bei, H.; Huang, R.; Ko, J.Y.P.; Pagan, D.C.; Neuefeind, J.C.; et al. Local Structure and Short-Range Order in a NiCoCr Solid Solution Alloy. *Phys. Rev. Lett.* **2017**, *118*, 205501. [CrossRef]
22. Zhang, F.X.; Ullah, M.W.; Zhao, S.; Jin, K.; Tong, Y.; Velisa, G.; Xue, H.; Bei, H.; Huang, R.; Park, C.; et al. Local structure of NiPd solid solution alloys and its response to ion irradiation. *J. Alloys Compd.* **2018**, *755*, 242–250. [CrossRef]
23. Zhang, F.; Tong, Y.; Jin, K.; Bei, H.; Weber, W.J.; Huq, A.; Lanzirotti, A.; Newville, M.; Pagan, D.C.; Ko, J.Y.P.; et al. Chemical complexity induced local structural distortion in NiCoFeMnCr high-entropy alloy. *Mater. Res. Lett.* **2018**, *6*, 450–455. [CrossRef]
24. Otto, F.; Dlouhý, A.; Pradeep, K.G.; Kuběnová, M.; Raabe, D.; Eggeler, G.; George, E.P. Decomposition of the single-phase high-entropy alloy CrMnFeCoNi after prolonged anneals at intermediate temperatures. *Acta Mater.* **2016**, *112*, 40–52. [CrossRef]
25. Ma, D.; Yao, M.; Pradeep, K.G.; Tasan, C.C.; Springer, H.; Raabe, D. Phase stability of non-equiatomic CoCrFeMnNi high entropy alloys. *Acta Mater.* **2015**, *98*, 288–296. [CrossRef]
26. Zhang, F.X.; Zhao, S.; Jin, K.; Bei, H.; Popov, D.; Park, C.; Neuefeind, J.C.; Weber, W.J.; Zhang, Y. Pressure-induced fcc to hcp phase transition in Ni-based high entropy solid solution alloys. *Appl. Phys. Lett.* **2017**, *110*, 011902. [CrossRef]
27. Zhang, F.; Wu, Y.; Lou, H.; Zeng, Z.; Prakapenka, V.B.; Greenberg, E.; Ren, Y.; Yan, J.; Okasinski, J.S.; Liu, X.; et al. Polymorphism in a high-entropy alloy. *Nat. Commun.* **2017**, *8*, 15687. [CrossRef]
28. Tracy, C.L.; Park, S.; Rittman, D.R.; Zinkle, S.J.; Bei, H.; Lang, M.; Ewing, R.C.; Mao, W.L. High pressure synthesis of a hexagonal close-packed phase of the high-entropy alloy CrMnFeCoNi. *Nat. Commun.* **2017**, *8*, 15634. [CrossRef]
29. Hammersley, A.P.; Svensson, S.O.; Thompson, A.; Graafsma, H.; Kvick, Å.; Moy, J.P. Calibration and correction of distortions in two-dimensional detector systems. *Rev. Sci. Instrum.* **1995**, *66*, 2729–2733. [CrossRef]
30. Juhás, P.; Davis, T.; Farrow, C.L.; Billinge, S.J.L. PDFgetX3: A rapid and highly automatable program for processing powder diffraction data into total scattering pair distribution functions. *J. Appl. Crystallogr.* **2013**, *46*, 560–566. [CrossRef]
31. Farrow, C.L.; Juhas, P.; Liu, J.W.; Bryndin, D.; Boin, E.S.; Bloch, J.; Proffen, T.; Billinge, S.J.L. PDFfit2 and PDFgui: Computer programs for studying nanostructure in crystals. *J. Phys. Condens. Matter* **2007**, *19*, 335219. [CrossRef]
32. Rodríguez-Carvajal, J. Recent advances in magnetic structure determination by neutron powder diffraction. *Phys. B Phys. Condens. Matter* **1993**, *192*, 55–69. [CrossRef]
33. Ravel, B.; Newville, M. ATHENA, ARTEMIS, HEPHAESTUS: Data analysis for X-ray absorption spectroscopy using IFEFFIT. *J. Synchrotron Radiat.* **2005**, *12*, 537–541. [CrossRef] [PubMed]
34. Ye, Y.F.; Liu, C.T.; Yang, Y. A geometric model for intrinsic residual strain and phase stability in high entropy alloys. *Acta Mater.* **2015**, *94*, 152–161. [CrossRef]
35. Yeh, J.; Chen, S.; Gan, J.; Lin, S. Communications: Formation of Simple Crystal Structures in Cu-Co-Ni-Cr-Al-Fe-Ti-V alloys with multiprincipal metallic elements. *Metall. Mater. Trans.* **2004**, *35*, 2533–2536. [CrossRef]
36. Li, Z.; Pradeep, K.G.; Deng, Y.; Raabe, D.; Tasan, C.C. Metastable high-entropy dual-phase alloys overcome the strength-ductility trade-off. *Nature* **2016**, *534*, 227–230. [CrossRef] [PubMed]

entropy

MDPI

Article

Effect of Mn Addition on the Microstructures and Mechanical Properties of CoCrFeNiPd High Entropy Alloy

Yiming Tan, Jinshan Li *, Jun Wang and Hongchao Kou

State Key Laboratory of Solidification Processing, Northwestern Polytechnical University, Xi'an 710072, China; tanym2015@mail.nwpu.edu.cn (Y.T.); nwpuwj@nwpu.edu.cn (J.W.); hchkou@nwpu.edu.cn (H.K.)

* Correspondence: ljsh@nwpu.edu.cn

Received: 16 February 2019; Accepted: 12 March 2019; Published: 16 March 2019

Abstract: $CoCrFeNiPdMn_x$ (x = 0, 0.2, 0.4, 0.6, 0.8) high entropy alloys (HEAs) were prepared and characterized. With an increase in Mn addition, the microstructures changed from dendrites (CoCrFeNiPd with a single face-centered-cubic (FCC) phase) to divorced eutectics ($CoCrFeNiPdMn_{0.2}$ and $CoCrFeNiPdMn_{0.4}$), to hypoeutectic microstructures ($CoCrFeNiPdMn_{0.6}$), and finally to seaweed eutectic dendrites ($CoCrFeNiPdMn_{0.8}$). The addition of Mn might change the interface energy anisotropy of both the FCC/liquid and MnPd-rich intermetallic compound/liquid interfaces, thus forming the seaweed eutectic dendrites. The hardness of the FCC phase was found to be highly related to the solute strengthening effect, the formation of nanotwins and the transition from CoCrFeNiPd-rich to CoCrFeNi-rich FCC phase. Hierarchical nanotwins were found in the MnPd-rich intermetallic compound and a decrease in either the spacing of primary twins or secondary twins led to an increase in hardness. The designing rules of EHEAs were discussed and the pseudo binary method was revised accordingly.

Keywords: high entropy alloys; solidification; alloy design; eutectic dendrites; hierarchical nanotwins

1. Introduction

High entropy alloys (HEAs) [1] or multi-principal element alloys [2] are now attracting more and more attention [3–10]. In contrast to the traditional alloys with one principal element or two, HEAs have at least four principal elements and usher in an expansive alloy space for exploring potential new materials with brilliant properties [11–26]. Initially, studies of HEAs concentrated to a greater extent on the solid-solution phases, e.g., the HEAs with a single face-centered-cubic (FCC) phase, with a single body-centered-cubic (BCC) phase or with dual FCC and BCC phases. Lots of studies suggested that the high configurational entropy would be able to stabilize thermodynamically the solid-solution phases [1,4,5,27,28]. As the researches move forward, more and more studies suggested that the high configurational entropy alone could not determine completely the constituent phases, because most of the HEAs consisted of multi-phases [29–34].

Although the HEAs with a single solid-solution phase have some advantages (e.g., higher melting points than the HEAs with multi-phases, higher strength for the HEAs with a single BCC phase, better ductility for the HEAs with a single FCC phase etc.), their good properties are usually accompanied by some disadvantages, which are fatal for technological applications. One is that the HEAs with a single solid-solution phase usually have inadequate liquidity, poor castability and hence considerable chemical inhomogeneity [21,35]. The other is that the HEAs with a single solid-solution phase could not achieve a balance between high strength and good ductility (e.g., the HEAs with a single FCC phase were ductile but not strong enough while the HEAs with a single BCC phase were adequately strong but at risk of brittleness [21,35–37]).

To tackle the aforementioned problems, eutectic HEAs (EHEAs) [21] were proposed. On the one hand, EHEAs should have the general character of traditional eutectic alloys. In this sense, EHEAs should have better fluidity and thus better castability and less casting defects [15,35]. On the other hand, EHEAs as one kind of in-situ composites with lamellar or rod-like eutectic microstructures might reach the balance between strength and ductility via mixing the soft FCC phase with the hard BCC phase or intermetallic compound [15,35,38–44]. Some EHEAs indeed have outstanding properties. Lu et al. [15] reported the $AlCoCrFeNi_{2.1}$ EHEA with simultaneous high strength (944 MPa) and good ductility (25.6%). The excellent mechanical properties do not depend significantly on derivation of eutectic compositions [35]. After cold-rolling and annealing, its strength reached up to 1.2 GPa and its elongation could remain at about 12% [40]. After cryo-rolling and annealing, its strength could reach up to 1.47 GPa while its ductility could even increase to 14% [41]. He et al. [42,43] designed the $CoCrFeNiNb_x$ EHEAs and found that the microstructures were stable from 600 °C to 900 °C.

The current work aims to report a new EHEA. From Ref. [45], CoCrFeNiPd is a single FCC solid solution HEA. From the Mn-Pd phase-diagram [46], Mn_xPd_y is a relative stable intermetallic compound. We hence chose CoCrFeNiPd as a FCC solid solution phase and Mn_xPd_y as an intermetallic compound (IMC) phase to design pseudo binary EHEAs via adjusting the content of IMC forming element Mn to finally get the eutectic structure. The effect of Mn addition on the microstructures was investigated and a seaweed eutectic dendrite solidification microstructure was found in the $CoCrFeNiPdMn_{0.8}$ EHEA. The effect of Mn addition on the mechanical properties was studied by nano-indentation and compression tests. The size effects of primary and secondary twins on the hardness of Mn_xPd_y phase were shown. The designing rules of EHEAs were improved.

2. Materials and Methods

2.1. Material Preparation

The ingots were prepared by arc melting under a Ti-gettered, high-purity argon atmosphere. Elements of Co, Cr, Fe, Ni, Mn and Pd with purities better than 99.95 wt.% were chosen as the raw materials. To prevent the mass loss due to evaporation of Mn, a high purity Fe-68.7at.%Mn intermediate alloy was prepared in advance and the total mass loss of each ingot was less than 0.3 wt.%. In order to ensure the chemical homogeneity, electromagnetic stirring was used during the melting process; each ingot was re-melted at least five times in the water-chilled copper crucible, held at a liquid state for at least 5 min and flipped before each melting process. The prepared button-shaped ingots were approximately 20 mm in diameter and 10 mm in thickness.

2.2. Material Characterization

The crystal structures were analyzed by X-ray diffraction (XRD, DX2700, Fang Yuan Company, Dandong, China) using Co kα radiation and a 2θ scattering range of 20°–120°. The microstructures were characterized by the field emission scanning electron microscopy (SEM, Zeiss SUPRA 55, Zeiss Inc., Jena, Germany) operated at 15 kV. The SEM samples were first polished and then etched for a few seconds within the solution of hydrochloric acid, sulfuric acid and supersaturated copper sulfuric (30 mL, 10 mL, 1 g). After the SEM observations, the samples for transmission electron microscopy (TEM) analysis were cut from the center of the SEM samples, prepared by mechanically polishing to a thickness of 45 µm, punched into disks with a diameter of 3 mm and then thinned by ion milling (GATAN 691, Gatan Inc., Warrendale, PA, United State). The chemical components and element distributions in different phases were measured by an electron probe micro-analyzer (EPMA, Shimadzu 1720, Shimadzu Inc., Kyoto, Japan) and an energy dispersive spectrometer (EDS) attached to TEM ($TecnaiFG^2$).

The hardness and elastic modulus of constituent phases in the as-cast alloys were investigated by the Nano-indenter XP® system (MTS Inc., Eden Prairie, MN, United State) at room temperature with a diamond Berkovich indenter at a peak load of 20 mN and a load rate of 0.1 mN·s^{-1}. The peak load was held for about 5 s to eliminate the instrument noise and five different points were measured for each phase. The samples for nano-indentation were mechanically polished to 1 mm thickness and then electro-polished in an electrolyte of 90 vol.% ethanol and 10 vol.% perchloric acid, with a voltage of 30 V and a polishing time of about 20 s in Struers LectroPol-5. The compression tests were conducted at room temperature in an electronic testing machine (INSTRON 3382, Instron Inc., Norwood, MA, United State) with a strain rate of 1×10^{-3} s^{-1}. Cuboid specimens were produced by electric-discharged machining from the cast buttons. The samples were 6 mm in height and 3 mm in length and width, giving an aspect ratio of 2. In order to show the solidification path, the thermal histories of as-cast alloys were measured by a differential scanning calorimetry (DSC, Netzsch 449 C, Netzsch Inc., Selb, Germany) under a flow of purified argon for protection and with a rate of 20 K min^{-1}. The mass of samples was about 15 mg.

3. Results

3.1. Crystal Structures and Microstructures

Figure 1 shows the XRD patterns of as-cast CoCrFeNiPdMn$_x$ (x = 0–0.8) HEAs. It should be noted that the CoCrFeNiPdMn$_x$ HEA in what follows was denoted as Mn$_x$ for short (e.g., Mn$_{0.2}$ stands for the CoCrFeNiPdMn$_{0.2}$ alloy). The Mn$_0$ HEA was of a single FCC phase with a lattice parameter of a = 3.669 Å. The Mn$_{0.2}$, Mn$_{0.4}$, Mn$_{0.6}$ and Mn$_{0.8}$ HEAs had a dual FCC phase and Mn$_x$Pd$_y$ intermetallic compound. Because the diffraction peaks of Mn$_x$Pd$_y$ intermetallic compound are intensified with increasing Mn addition, one could draw a conclusion that the Mn addition promotes the formation of Mn$_x$Pd$_y$ intermetallic compound. However, the diffraction peaks of MnPd, Mn$_2$Pd$_3$ and Mn$_3$Pd$_5$ as well as those of Mn$_7$Pd$_9$ and Mn$_{11}$Pd$_{21}$ were quite similar. The XRD results alone were therefore not able to distinguish the crystal structure of the Mn$_x$Pd$_y$ intermetallic compound.

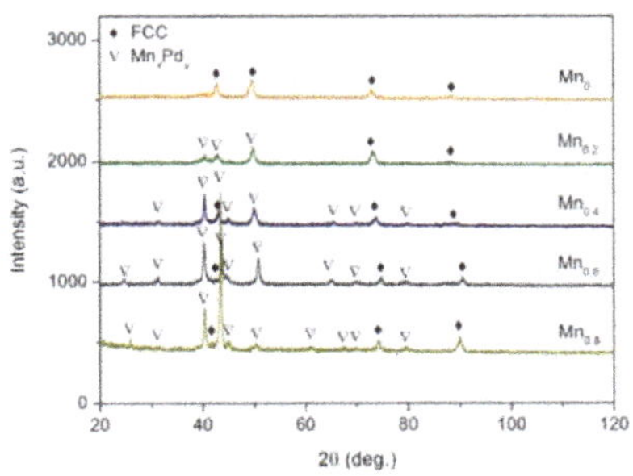

Figure 1. XRD patterns of as-cast CoCrFeNiPdMnx (x = 0, 0.2, 0.4, 0.6, 0.8) HEAs.

Typical microstructures of as-cast Mn$_x$ (x = 0.2–0.8) HEAs are shown in Figure 2. The Mn$_0$ HEA exhibited a single solid-solution phase and the coarse dendrites were of several hundred or even a thousand microns; see Figure 2a,a$_1$ in different magnifications. For the Mn$_{0.2}$ EHEA, the microstructure consisted of a main FCC solid-solution phase in the dendrite and a sporadic distributed granular Mn$_x$Pd$_y$ intermetallic compound; see Figure 2b. Because the Mn$_x$Pd$_y$ intermetallic compound distributed within the inter-dendrites, it could be reasonable to conclude that the microstructure belonged to divorced eutectics; see Figure 2b$_1$ in which the FCC phase and the Mn$_x$Pd$_y$ intermetallic compound are in dark grey and light grey, respectively. It should be pointed out that at the inter-dendrites, a eutectic microstructure could be found but its volume fraction was very small. The microstructure of Mn$_{0.4}$ EHEA was quite similar to the Mn$_{0.2}$ EHEA, except that both the volume fractions of eutectics and granular Mn$_x$Pd$_y$ intermetallic compound were much larger; see Figure 2c,c$_1$ in different magnifications. The microstructure changes from a hypoeutectic microstructure for the Mn$_{0.6}$ EHEA (e.g., a primary FCC dendrite around which were the lamellar eutectics) to a fully eutectic microstructure for the Mn$_{0.8}$ EHEA (e.g., a eutectic dendrite with a fine lamellar spacing around which were the coarse granular eutectics); see Figure 2d–e_1 in different magnifications. In order to show the characteristics of the eutectic dendrite pattern in the $Mn_{0.8}$ EHEA, two additional figures with different amplifications are shown in Figure 2e_2,e_3. Figure 2e_2 shows an overall view of eutectic dendrites and Figure 2e_3 presents some details for tip splitting of eutectic dendrites. Because the tips repeatedly split into several parts and grew on themselves, the microstructure of $Mn_{0.8}$ HEA belonged to seaweed eutectic dendrites [47,48].

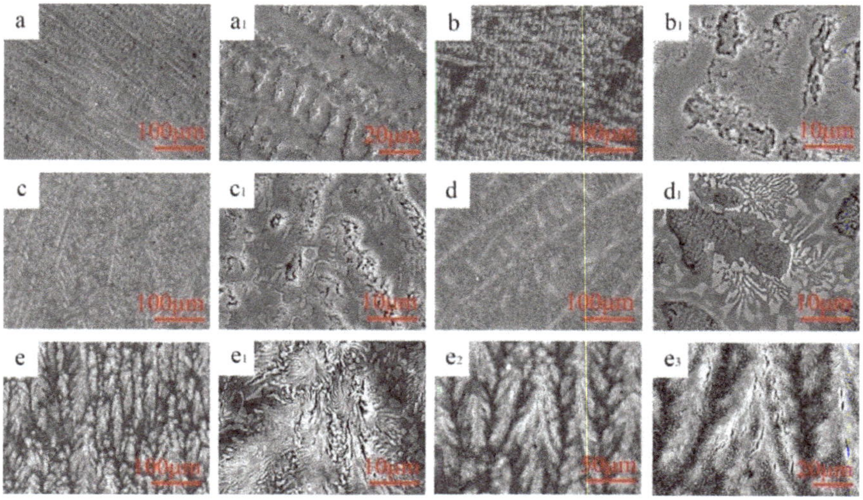

Figure 2. SEM images of as-cast CoCrFeNiPdMn_x (x = 0, 0.2, 0.4, 0.6, 0.8): Mn_0 (**a**,**a$_1$**), $Mn_{0.2}$ (**b**,**b$_1$**), $Mn_{0.4}$ (**c**,**c$_1$**), $Mn_{0.6}$ (**d**,**d$_1$**) and $Mn_{0.8}$ (**e**,**e$_1$**,**e$_2$**,**e$_3$**).

3.2. Phase Identification

The TEM results of as-cast Mn_x (x = 0.2, 0.4, 0.6, 0.8) HEAs are shown in Figures 3–6. In each figure, the bright-field TEM images (a, d), the selected area electron diffraction (SAED) pattern of FCC (c) and Mn_xPd_y (d) phases, the EDS mapping of Co (e), Cr (f), Fe (g), Ni (h), Pd (i) and Mn (j) elements are shown. To show the effect of Mn addition on the phase transition in the Mn_x HEAs, the chemical compositions of FCC and Mn_xPd_y phases were measured by EDS attached to TEM and EMPA. In the current work, four points were randomly selected for each phase in the fine lamellar region by EDS and five points were measured randomly for each phase in the surrounding coarse granular eutectic region by EPMA. Because the average compositions measured by EDS and EPMA were quite close, only the EPMA results for the FCC solid-solution phase and Mn_xPd_y intermetallic compound are summarized in Tables 1 and 2, respectively.

Table 1. EPMA results of the FCC phase in the CoCrFeNiPdMn_x (x = 0–0.8) HEAs (in atomic fraction).

HEA	Co	Cr	Fe	Ni	Pd	Mn	FCC Phase
$Mn_{0.2}$	22.32 ± 0.91	21.23 ± 0.48	19.30 ± 0.41	19.79 ± 0.28	13.49 ± 0.66	2.09 ± 0.36	CoCrFeNiPd-rich
$Mn_{0.4}$	22.61 ± 0.53	21.92 ± 0.39	20.41 ± 0.90	20.02 ± 0.60	11.75 ± 0.67	3.56 ± 0.25	CoCrFeNiPd-rich
$Mn_{0.6}$	22.29 ± 0.51	22.12 ± 0.76	21.32 ± 0.36	21.22 ± 0.97	8.09 ± 0.89	4.22 ± 0.45	CoCrFeNi-rich
$Mn_{0.8}$	20.83 ± 0.36	21.44 ± 0.45	20.84 ± 0.53	24.46 ± 0.59	6.67 ± 0.70	5.12 ± 0.30	CoCrFeNi-rich

Table 2. EPMA results of the Mn_xPd_y phase in the CoCrFeNiPdMn_x (x = 0-0.8) HEAs (in atomic fraction).

HEA	Co	Cr	Fe	Ni	Pd	Mn	MnxPdy
$Mn_{0.2}$	3.87 ± 0.85	7.48 ± 0.51	8.55 ± 0.17	5.49 ± 0.41	47.32 ± 0.65	27.29 ± 0.65	Mn_3Pd_5
$Mn_{0.4}$	4.49 ± 0.83	7.32 ± 0.21	8.71 ± 0.21	5.53 ± 0.31	46.01 ± 0.76	29.44 ± 0.57	Mn_3Pd_5
$Mn_{0.6}$	1.60 ± 0.09	4.78 ± 0.36	4.27 ± 0.45	3.39 ± 0.34	43.66 ± 0.93	42.31 ± 0.17	Mn_7Pd_9
$Mn_{0.8}$	2.56 ± 0.38	6.28 ± 0.11	4.69 ± 0.79	3.81 ± 0.44	40.35 ± 0.67	41.71 ± 0.23	Mn_7Pd_9

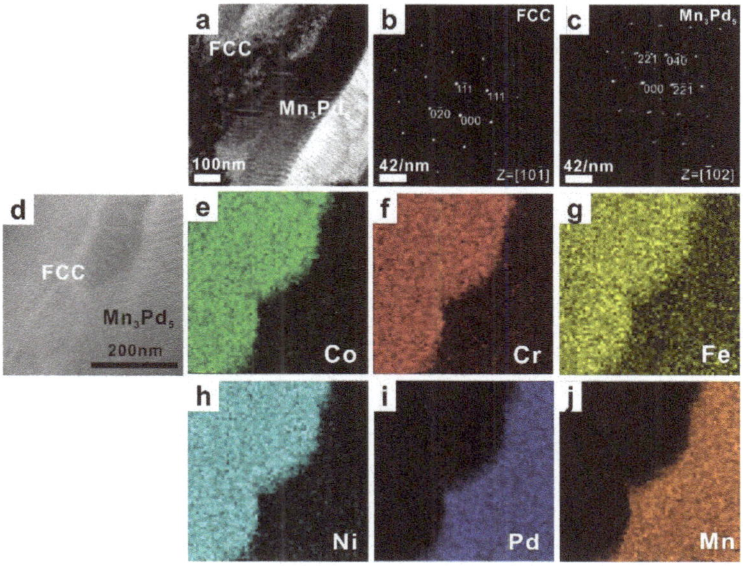

Figure 3. TEM images (**a**) and (**d**), the corresponding SAED patterns of FCC (**b**) and Mn_3Pd_5 (**c**) phases, and the EDS mapping of Co (**e**), Cr (**f**), Fe (**g**), Ni (**h**), Pd (**i**), Mn (**j**) for the as-cast $CoCrFeNiPdMn_{0.2}$ HEA.

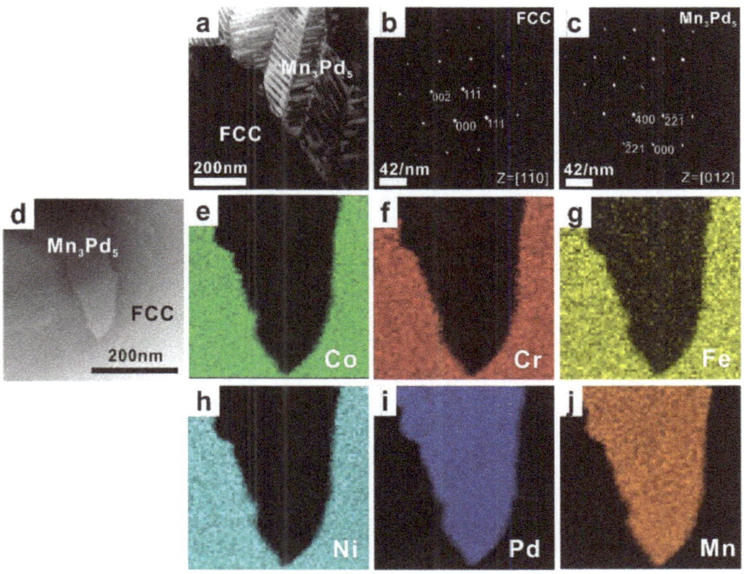

Figure 4. TEM images (**a**) and (**d**), the corresponding SAED patterns of FCC (**b**) and Mn_3Pd_5 (**c**) phases, and the EDS mapping of Co (**e**), Cr (**f**), Fe (**g**), Ni (**h**), Pd (**i**), Mn (**j**) for the as-cast $CoCrFeNiPdMn_{0.4}$ HEA.

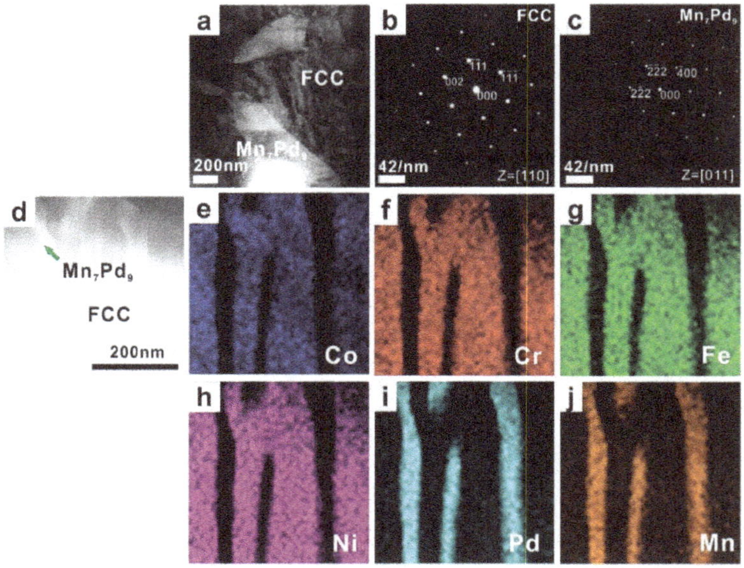

Figure 5. TEM images (**a**) and (**d**), the corresponding SAED patterns of FCC (**b**) and Mn_7Pd_9 (**c**) phases, and the EDS mapping of Co (**e**), Cr (**f**), Fe (**g**), Ni (**h**), Pd (**i**), Mn (**j**) for the as-cast $CoCrFeNiPdMn_{0.6}$ EHEA.

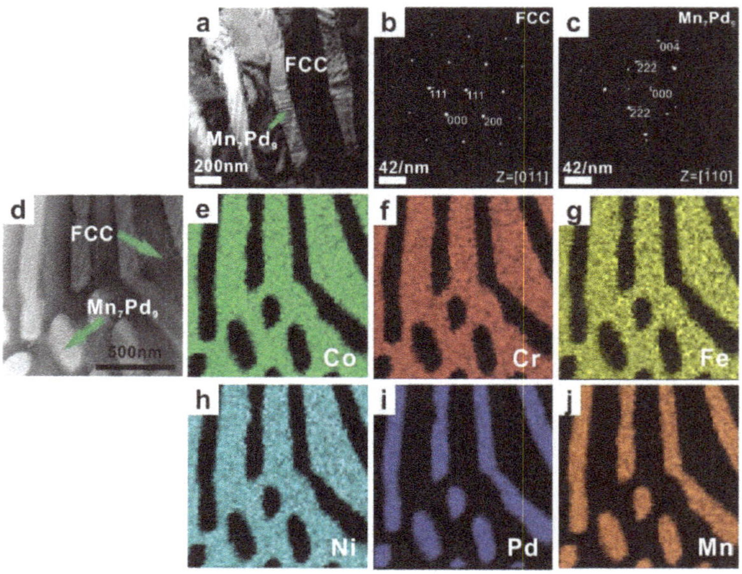

Figure 6. TEM images (**a**) and (**d**), the corresponding SAED patterns of FCC (**b**) and Mn_7Pd_9 (**c**) phases, and the EDS mapping of Co (**e**), Cr (**f**), Fe (**g**), Ni (**h**), Pd (**i**), Mn (**j**) for the as-cast $CoCrFeNiPdMn_{0.8}$ RHEA.

For the $Mn_{0.2}$ EHEA, the FCC phase was rich in Co, Cr, Fe, Ni and Pd but depleted of Mn, whereas for the Mn_xPd_y phase, the compositions of Co, Cr, Fe and Ni were negligible; see Figure 3 and Table 1. According to the SAED patterns taken from the FCC-region and Mn_xPd_y-region, the matrix was of a FCC structure while the Mn_xPd_y phase was a Mn_3Pd_5 intermetallic compound with lattice parameters of a = 0.2285 nm, b = 0.1998 nm and c = 0.2278 nm, being consistent with the XRD results in Figure 1. It should be pointed out that even though the composition of Pd in the FCC phase (≈13.5%) was much larger than that of Mn (≈2%), it was still considerably smaller than that in Mn_3Pd_5 intermetallic compound (≈47%). Therefore, the fact that the FCC phase was rich in Pd cannot be shown by the EDS mapping; see Figure 3i. For the $Mn_{0.4}$ EHEA, the same result could be found from the SAED patterns, i.e., the matrix was the FCC phase and the Mn_xPd_y phase was the Mn_3Pd_5 intermetallic compound. The FCC phase was still a (CoCrFeNiPd)-rich one and similar EDS mappings could be found; see Figure 4e–j.

With the further addition of Mn element, the FCC phases became rich in Co, Cr, Fe and Ni for the $Mn_{0.6}$ and $Mn_{0.8}$ EHEAs, while the compositions of Mn (~42.3% and 41.7%) and Pd (~43.7% and ~40.4%) were comparable in the Mn_xPd_y phases; see Figures 5e–j and 6e–j, Tables 1 and 2. According to the SAED patterns in Figures 5c and 6c, the Mn_xPd_y phase could be the Mn_7Pd_9 or the $Mn_{11}Pd_{21}$ intermetallic compound with lattice parameters of $a = b$ = 0.2267 nm, c = 0.203 nm or $a = b$ = 0.2235 nm, c = 0.1816 nm. Because the $Mn_{11}Pd_{21}$ phase was neither confirmed experimentally nor theoretically [46], the Mn_xPd_y phase in the $Mn_{0.6}$ and $Mn_{0.8}$ EHEAs was ultimately determined to be the Mn_7Pd_9 intermetallic compound.

3.3. Solidification Path

To confirm further the effect of Mn addition on solidification microstructures, the cooling histories were measured by DSC; see Figure 7. For the $Mn_{0.2}$ (the solid line) and $Mn_{0.4}$ (the dashed line) EHEAs, two completely separated exothermal peaks could be found during the solidification process. The first and the second peak should correspond to the primary solidification of the FCC phase and following growth of the Mn_3Pd_5 intermetallic compound or eutectic growth. For the $Mn_{0.6}$ EHEA (the dotted line), two exothermal peaks still exited during the solidification process but they overlapped with each other. As shown in Figures 1 and 2, an increase of the Mn content promoted the formation of Mn_xPd_y phase, thus intensifying the second exothermal peak during solidification and narrowed the distance between the two peaks as shown in Figure 7. For the $Mn_{0.8}$ EHEA, only one solidification peak could be found; see the dashed-dotted line in Figure 7. The peak should correspond to eutectic solidification.

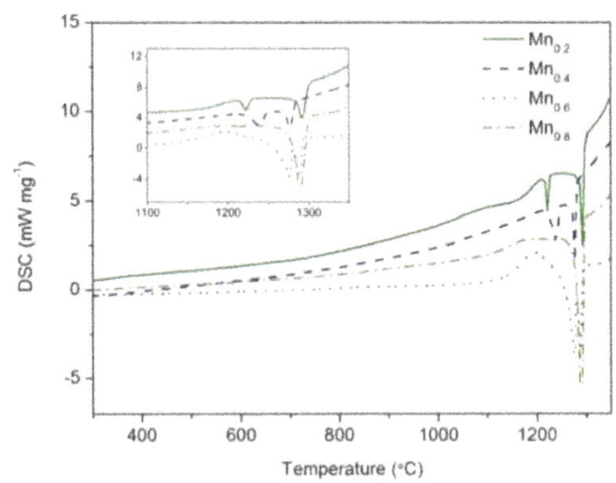

Figure 7. DSC solidification curves of as-cast CoCrFeNiPdMn$_x$ (x = 0.2–0.8) HEAs. Insert shows the magnified exothermic peaks during solidification.

3.4. Mechanical Properties

The nanoindentor was used to measure the hardness and elastic modulus of FCC and Mn_xPd_y phases; see Figure 8a,b. It should be noted that the lamellar spacing of lamellar eutectics in the $Mn_{0.6}$ and $Mn_{0.8}$ EHEAs was so fine that it was beyond the measurability of the nanoindentor. In this case, the coarse granular eutectics were measured. From Figure 8a, it can be seen that with an increase in the Mn addition, the hardness and elastic modulus of FCC phase first increased and then decreased. From Table 1, the FCC phase was CoCrFeNiPd-rich for the $Mn_{0.2}$ and $Mn_{0.4}$ EHEAs while it was CoCrFeNi-rich for the $Mn_{0.6}$ and $Mn_{0.8}$ EHEAs. From Ref. [49], the measured hardness of CoCrFeNiPd HEA 3.16 GPa was nearly twice of CoCrFeNi HEA 1.47 GPa. This was the reason why the hardness of Mn_0, $Mn_{0.2}$ and $Mn_{0.4}$ HEAs was much larger than that of $Mn_{0.6}$ and $Mn_{0.8}$ HEAs; see Figure 8a. For the Mn_xPd_y phase, the hardness of the Mn_3Pd_5 intermetallic compound in the $Mn_{0.2}$ (4.9 GPa) and $Mn_{0.4}$ (5.3 GPa) EHEAs was much larger than that of the Mn_7Pd_9 intermetallic compound in the $Mn_{0.6}$ (3.1 GPa) and $Mn_{0.8}$ (3.4 GPa) EHEAs. For both the FCC and Mn_xPd_y phases, the evolution tendencies of hardness were the same as those of the elastic modulus.

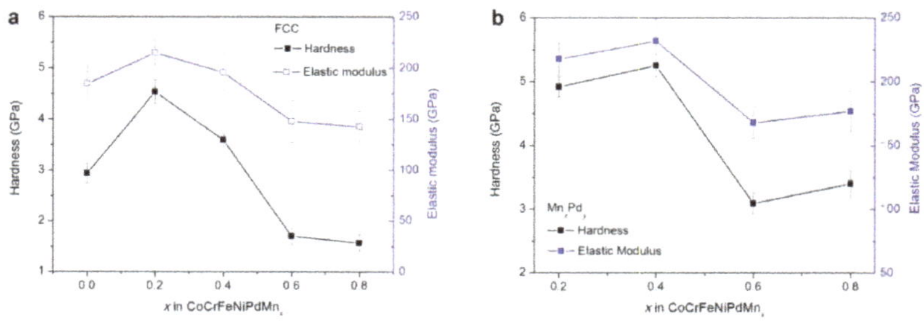

Figure 8. Hardness and elastic modulus of FCC phase (**a**) Mn_xPd_y phase (**b**) in the $CoCrFeNiPdMn_x$ (x = 0.2–0.8) HEAs.

To show further the effect of Mn addition on the mechanical properties, compression tests were conducted for the as-cast Mn_x HEAs; see Figure 9 One can see that with the increase of Mn addition, the yielding strength held constantly at about 650 MPa. The fracture strain (strength) decreased from about 50% (2.4 GPa) for the $Mn_{0.2}$ HEA to about 35% (1.9 GPa) for the $Mn_{0.8}$ HEA. The current EHEAs had good strength and ductility.

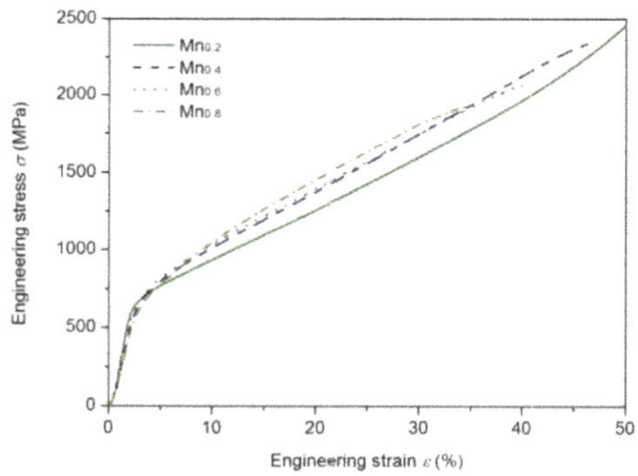

Figure 9. Compressive engineering stress-strain curves of as-cast $CoCrFeNiPdMn_x$ (x = 0.2–0.8) HEAs.

4. Discussion

4.1. *Effect of Mn Addition on Microstructures*

With an increase in the Mn content, the microstructures of Mn_x HEAs changed from dendrites for the Mn_0 HEA to divorced eutectics for the $Mn_{0.2}$ and $Mn_{0.4}$ EHEAs, to hypoeutectic microstructures for the $Mn_{0.6}$ EHEA and finally to eutectic dendrites for the $Mn_{0.8}$ EHEA. The eutectic dendrite solidification pattern in the $Mn_{0.8}$ EHEA was formed by cooperative growth of the FCC phase and Mn_7Pd_9 intermetallic compound. From Tables 1 and 2, the FCC phase was lacking Mn while the Mn_7Pd_9 intermetallic compound was lacking Co, Cr, Fe and Ni. Therefore, lateral solute diffusion of

Co, Cr, Fe, Ni and Mn formed the eutectic pattern while longitudinal solute diffusion of Pd made the eutectic interface unstable to a eutectic dendrite.

From Figure 2e_2,e_3, seaweed eutectic dendrites were found for the $Mn_{0.8}$ EHEA. Unlike the normal dendrite pattern where the structure branches with pronounced orientation order, the seaweed pattern is characterized by tip-splitting and the key factor for its formation is weak interface energy anisotropy [50]. Generally, the formation of seaweed dendrites is highly related to alloy compositions and solidification conditions [51–54]. For example, the effect of Zn content on the microstructures of directional solidification of Al-Zn alloys was studied by X-ray tomographic microscopy and phase-field simulation [51]. Accordingly, an increase in the Zn content modified the interface energy anisotropy, thus leading to the transition from <100> dendrites at low Zn content to <110> dendrites at high Zn content, between which were the <320> seaweed dendrites. For both the undercooled Cu-8.9 wt.% Ni and Cu-3.98 wt.% Ni alloys [53,54], a transition from <100> dendrites to mixed <100> and <111> seaweed dendrites and then to <111> dendrites was reported.

For eutectic solidification that consisted of at least two solid phases, its morphology was determined by a combination effect of eutectic phases and the formation mechanism became more complex. Eutectic seaweed dendrites were reported in the undercooled Co-24.0at.%Sn eutectic alloy, in which the weak interface energy anisotropy ascribed to an alternate arrangement of lamellae and alloy physical properties [55]. For the current Mn_x HEAs, primary FCC dendrites were found in the divorced eutectics (e.g., $Mn_{0.2}$ and $Mn_{0.4}$) and the hypoeutectic microstructures (e.g., $Mn_{0.6}$), indicated that its interface energy anisotropy was not weak. From Tables 1 and 2, the addition of Mn changed not only the compositions of FCC phase but also those of the Mn_7Pd_9 intermetallic compound. Therefore, it was quite possible that the addition of Mn influenced the interface energy anisotropy of both the FCC/liquid and Mn_xPd_y/liquid interfaces, thus forming the seaweed eutectic dendrites in the $Mn_{0.8}$ EHEA.

4.2. *Effect of Mn Addition on Mechanical Properties*

Because an increase in Mn addition results in a transition from the CoCrFeNiPd-rich to the CoCrFeNi-rich FCC phase in the Mn_x HEAs (Table 1) and the hardness of CoCrFeNiPd HEA is much higher than that of CoCrFeNi HEA [49], the hardness of the FCC phase should decrease with increasing Mn addition. This was however, not the case, e.g., the hardness increased first and then decreased; see Figure 8a. The larger hardness of the FCC phase in the $Mn_{0.2}$ and $Mn_{0.4}$ EHEAs than that in the Mn_0 HEA could be ascribed to the solute strengthening effect. But this effect alone cannot explain the fact that the hardness of the FCC phase in the $Mn_{0.2}$ EHEA was larger than that in the $Mn_{0.4}$ EHEA. The TEM results showed that a small amount of Mn addition might promote but a large amount would suppress the formation of nanotwins in the FCC phase; see Figure 10. Abundant nanotwins of about 50 nm could be found in the $Mn_{0.2}$ EHEA, whereas for the $Mn_{0.4}$ EHEA, it was almost free of nanotwins and so were the $Mn_{0.6}$ and $Mn_{0.8}$ EHEAs (not shown here). Therefore, the solute strengthening effect and the formation of nanotwins made the hardness increase first with increased Mn addition, the suppression of nanotwins then decreased the hardness and finally the transition from the CoCrFeNiPd-rich to the CoCrFeNi-rich FCC phase made the hardness decrease considerably.

Besides, hierarchical nanotwins were found in the Mn_xPd_y intermetallic compounds of $Mn_{0.2}$-$Mn_{0.8}$ EHEAs; see Figure 10. With the help of Image-Pro Plus software, the spacing of the primary twins (λ_1) and secondary twins (λ_2) in the Mn_xPd_y intermetallic compound were measured for the $Mn_{0.2}$-$Mn_{0.8}$ EHEAs; see Table 3. With an increase in the Mn addition, λ_1 decreased but λ_2 remained unchanged for the Mn_3Pd_5 intermetallic compound. For the Mn_7Pd_9 intermetallic compound, λ_1 did not change significantly but λ_2 decreased. The measured spacing of primary twins (242.1 nm, 180.3 nm) and secondary twins (10.0 nm, 9.98 nm) in the Mn_3Pd_5 intermetallic compound were much larger than those in the Mn_7Pd_9 intermetallic compound (15.0 nm, 15.0 nm for λ_1, 2.22 nm and 1.46 nm for λ_2) but the hardness of the former was much larger than that of the latter. However, for the same phase, a decrease of either λ_1 or λ_2 would increase the hardness of the intermetallic compound, being consistent with Yuan and Wu [56] who studied the size effects of primary and secondary twins on the atomistic deformation mechanisms in the hierarchically nanotwinned metals.

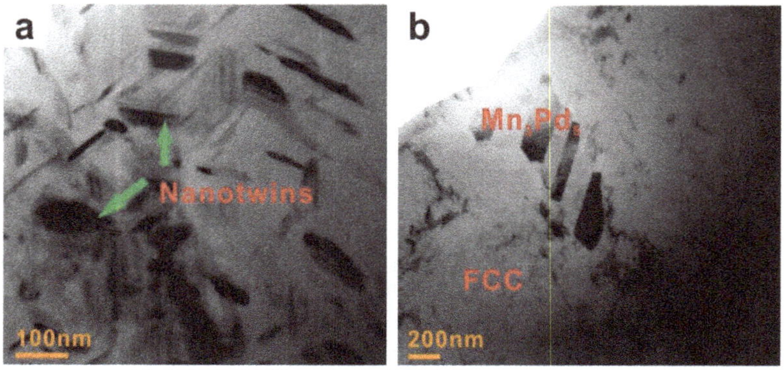

Figure 10. TEM images of FCC phase in the $Mn_{0.2}$ (**a**) and $Mn_{0.4}$ (**b**) EHEAs.

Table 3. The measured spacing of primary twins (λ_1) and secondary twins (λ_2) for the Mn_xPd_y phase in the $CoCrFeNiPdMn_x$ (x = 0.2–0.8) EHEAs.

EHEA (Mn_xPd_y)	Spacing of Primary Twins λ_1 (nm)	Spacing of Secondary Twins λ_2 (nm)
Mn0.2 (Mn_3Pd_5)	242.10 ± 26.63	10.02 ± 1.10
Mn0.4 (Mn_3Pd_5)	180.33 ± 19.84	9.99 ± 1.22
Mn0.6 (Mn_7Pd_9)	14.96 ± 16.46	2.22 ± 0.24
Mn0.8 (Mn_7Pd_9)	15.02 ± 1.65	1.46 ± 0.16

4.3. Designing Rules for EHEAs

Even though the EHEAs have good processing and mechanical properties, most of the reported EHEAs were found by the trial and error method. Up to now, several studies were carried out for designing EHEAs [42,57–59]. Lu et al. [57] started from their representative $AlCoCrFeNi_{2.1}$ EHEA. They divided the constituent elements into two different groups, i.e., Al and Ni with very high negative mixing of enthalpy (-22 kJ·mol^{-1}), and Co, Cr and Fe with similar atomic size and very small negative mixing of enthalpy; see Table 4. Their method was to substitute Al by Zr, Nb, Hf and Ta that had very high negative mixing of enthalpy with Ni. After using the enthalpy mixing of equimolar binary alloys to obtain the eutectic points, four new EHEAs, i.e., $Zr_{0.6}CoCrFeNi_{2.1}$, $Nb_{0.74}CoCrFeNi_{2.1}$, $Hf_{0.55}CoCrFeNi_{2.1}$ and $Ta_{0.65}CoCrFeNi_{2.1}$, were reported. In their subsequent work [58], the eutectic composition containing (Ni, Co, Cr, Fe)-rich solid-solution phase in the (Co, Cr, Fe, Ni)-(Nb, Ta, Zr, Hf) binary systems were averaged to obtain the eutectic compositions of pseudo binary alloy $CoCrFeNiM_x$ (M = Nb, Ta, Zr and Hf). Consequently, four new EHEAs, i.e., $Zr_{0.51}CoCrFeNi$, $Nb_{0.6}CoCrFeNi$, $Hf_{0.49}CoCrFeNi$ and $Ta_{0.47}CoCrFeNi$, were found. Even though the actual eutectic compositions were very close to the predicted ones using the above simple methods, the former method was based on a known EHEA, which might limit its application [59] and for the latter, there should be a eutectic reaction between the added element and any element in the base alloy which is not always the case for EHEAs. For example, for the $CoCrFeNiMnPd_x$ EHEAs, eutectic reactions happen only in the Mn-Pd and Cr-Pd binary alloys while for the $CoCrFeNiPdMn_x$ EHEAs, eutectic reactions can be found only in the Pd-Mn binary alloy.

He et al. [42] designed a pseudo binary alloy, i.e., the CoCrFeNi HEA with a single FCC solid-solution phase as the base alloy and Nb as the additional element. Such simple pseudo binary method was followed by Jin et al. [59]. First, they chose one HEA with a single solid-solution phase and one stable binary intermetallic compound. After that, they obtained the HEA with dual phase by mixing the two phases. To ensure the formation of an eutectic structure, three conditions were proposed: (1) The single solid-solution phase should be stable enough without any segregation and precipitation; (2) the binary intermetallic compound should be stable from room temperature to its melting point; (3) the intermetallic compound must have the most negative mixing of enthalpy among all the binary combinations in the alloy. With $CoCrFeNi_2$, $Co_2CrFeNi$ and $CoCrFe_2Ni$ as the HEAs with a single FCC solid-solution phase and NiAl as the binary intermetallic compound, they found three new EHEAs.

Table 4. The mixing enthalpy ΔH_{mix} (kJ·mol^{-1}) of atom pairs in the current $CoCrFeNiPdMn_x$ (x = 0.2–0.8) and some other EHEAs.

	Co	Cr	Ni	Mn	Pd	Al	Nb	Ta	Zr	Hf
Fe	−1	−1	−2	0	−4	−11	−16	−15	−25	−21
Co		−4	0	−5	−1	−19	−25	−24	−41	−35
Cr			−7	2	−15	−10	−9	−7	−12	−9
Ni				−8	0	−22	−30	−29	−49	−42
Mn					−23	−19	−4	−4	−15	−12

For the $CoCrFeNiMnPd_x$ and $CoCrFeNiPdMn_x$ EHEAs, CoCrFeNi can be taken to be the HEA with a single FCC solid-solution phase and MnPd can be taken to be the binary intermetallic compound; their mixing led to the CoCrFeNiMnPd EHEA [47]. Even though the mixing enthalpy between Mn and Pd was the most negative one (Table 4), the MnPd intermetallic compound was not stable enough from room temperature to its melting point [46]. As a result, the Mn_xPd_y intermetallic compound in the eutectics depending on the compositions could be Mn_2Pd_3, Mn_3Pd_5 or Mn_7Pd_9 [46,47]. In one word, the pseudo binary method could be a simple way for designing EHEAs but the designing rules still need to be studied further to achieve general and effective rules. According to our study, the consistent elements in the EHEAs with a solid-solution phase and an intermetallic compound can be divided into two groups, i.e., two of them with very high mixing of enthalpy forms the intermetallic compound and the rest of them with very small mixing of enthalpy forms the solid-solution phase. There should be a eutectic reaction in the binary alloy system for the two elements in the first group. One of the eutectic phases is the solid-solution phase which should have a good solubility for all the elements in the second group. The other one is the intermetallic compound which might have negligible solubility for all the elements in the second group.

5. Conclusions

In the current work, the Mn_x (x = 0, 0.2, 0.4, 0.6, 0.8) HEAs were prepared and characterized. Our main conclusions were as follows:

(1) With an increase in Mn addition, the microstructures of $CoCrFeNiPdMn_x$ HEAs changed from dendrites to divorced eutectics, to hypoeutectic microstructures and finally to eutectic dendrites. For the $Mn_{0.2}$ and $Mn_{0.4}$ ($Mn_{0.6}$ and $Mn_{0.8}$) EHEA, the FCC phase was a CoCrFeNiPd-rich (CoCrFeNi-rich) phase and the Mn_xPd_y intermetallic compound was Mn_3Pd_5 (Mn_7Pd_9). Addition of Mn might influence the interface energy anisotropy of both the FCC/liquid and Mn_xPd_y/liquid interfaces, thus forming the seaweed eutectic dendrites in the $Mn_{0.8}$ EHEA.

(2) With an increase in Mn addition, the hardness of FCC phase increased first and then decreased. The solute strengthening effect of Mn and the formation of nanotwins made the hardness increase firstly, the suppression of nanotwins then decreased the hardness and finally the transition from the CoCrFeNiPd-rich to the CoCrFeNi-rich FCC phase made the hardness decrease considerably. For the Mn_3Pd_5 and Mn_7Pd_9 intermetallic compounds, a decrease of either λ_1 or λ_2 would increase the hardness.

(3) The current EHEA system violates to some extent all the designing rules for EHEAs. The pseudo binary method was improved accordingly and the current work might be helpful for accelerating designing of potential EHEAs.

Author Contributions: Y.T. and J.L. conceived and designed the experiments; Y.T. performed the experiments; Y.T., J.W., H.K. performed the data analysis and drafted the manuscript. Y.T., J.W., H.K. participated in the data analysis, discussion, and interpretation. Y.T. and J.L. completed the paper.

Funding: This research was funded by the National Nature Science Foundation of China, grant number 51571161, 51774240 and 51690163, the Natural Science Basic Research Plan in Shaanxi Province of China, grant number 2016JQ5003 and the Program of Introducing Talents of Discipline to Universities, grant number B08040.

Conflicts of Interest: The authors declare no conflicts of interest.

References

1. Yeh, J.W.; Chen, S.K.; Lin, S.J.; Gan, J.Y.; Chin, T.S.; Shun, T.T.; Tsau, C.H.; Chang, S.Y. Nanostructures high-entropy alloys with multiple principal elements: Novel alloy design concepts and outcomes. *Adv. Eng. Mater.* **2004**, *6*, 299–303. [CrossRef]
2. Cantor, B.; Chang, I.T.H.; Knight, P.; Vincent, A.J.B. Microstructural development in equiatomic multicomponent alloys. *Mater. Sci. Eng. A* **2004**, *375–377*, 213–218. [CrossRef]
3. Zhang, Y.; Zuo, T.T.; Tang, Z.; Gao, M.C.; Dahmen, K.A.; Liaw, P.K.; Lu, Z.P. Microstructures and properties of high-entropy alloys. *Prog. Mater. Sci.* **2014**, *61*, 1–93. [CrossRef]
4. Tsai, M.H.; Yeh, J.W. High-entropy alloys: A critical review. *Mater. Res. Lett.* **2014**, *2*, 107–123. [CrossRef]
5. Tsai, M.H. Three Strategies for the Design of Advanced High-Entropy Alloys. *Entropy* **2016**, *18*, 252. [CrossRef]
6. Yeh, J.W. Physical metallurgy of high-entropy alloys. *JOM* **2015**, *67*, S499–S503. [CrossRef]
7. Pickering, E.J.; Jones, N.G. High-entropy alloys: A critical assessment of their founding principles and future prospects. *Int. Mater. Rev.* **2016**, *61*, 183–202. [CrossRef]
8. Ye, Y.F.; Wang, Q.; Lu, J.; Liu, C.T.; Yang, Y. High entropy alloy: Challenges and prospects. *Mater. Today* **2016**, *19*, 349–362. [CrossRef]
9. Miracle, D.B.; Senkov, O.N. A critical review of high entropy alloys and related concepts. *Acta Mater.* **2017**, *122*, 448–511. [CrossRef]
10. Gorsse, S.; Miracle, D.B.; Senkov, O.N. Mapping the world of complex concentrated alloys. *Acta Mater.* **2017**, *135*, 177–187. [CrossRef]
11. Gludovatz, B.; Hohenwarter, A.; Catoor, D.; Chang, E.H.; George, E.P.; Rirtchie, R.O. A fracture-resistant high-entropy alloy for cryogenic applications. *Science* **2014**, *345*, 1153. [CrossRef] [PubMed]
12. Li, Z.M.; Pradeep, K.G.; Deng, Y.; Raabe, D.; Tasan, C.C. Metastable high-entropy dual-phase alloys overcome the strength–ductility trade-off. *Nature* **2016**, *534*, 227–230. [CrossRef]
13. Lei, Z.F.; Liu, X.J.; Wu, Y.; Wang, H.; Jiang, S.; Wang, S.; Hui, X.; Wu, Y.; Gault, B.; Kontis, P. Enhanced strength and ductility in a high-entropy alloy via ordered oxygen complexes. *Nature* **2018**, *563*, 546–550. [CrossRef] [PubMed]
14. Yang, T.; Zhao, Y.L.; Tong, Y.; Jiao, Z.B.; Wei, J.; Cai, J.X.; Han, X.D.; Chen, D.; Hu, A.; Kai, J.J. Multicomponent intermetallic nanoparticles and superb mechanical behaviors of complex alloys. *Science* **2018**, *362*, 933–937. [CrossRef] [PubMed]
15. Senkov, O.N.; Wilks, G.B.; Scott, J.M.; Miracle, D.B. Mechanical properties of Nb25Mo25Ta25W25 and V20Nb20Mo20Ta20W20 refractory high entropy alloys. *Intermetallics* **2011**, *19*, 698–706. [CrossRef]
16. Chen, S.Y.; Yang, X.; Dahmen, K.A.; Liaw, P.K.; Zhang, Y. Microstructures and Crackling Noise of AlxNbTiMoV High Entropy Alloys. *Entropy* **2014**, *16*, 870–884. [CrossRef]
17. Yao, H.W.; Qiao, J.W.; Gao, M.C.; Hawk, J.A.; Ma, S.G.; Zhou, H.F. MoNbTaV Medium-Entropy Alloy. *Entropy* **2016**, *18*, 189. [CrossRef]
18. Ye, Y.X.; Liu, C.Z.; Wang, H.; Nieh, T.G. Friction and wear behavior of a single-phase equiatomic TiZrHfNb high-entropy alloy studied using a nanoscratch technique. *Acta Mater.* **2018**, *147*, 78–89. [CrossRef]
19. Tang, Z.; Yuan, T.; Tsai, C.W.; Yeh, J.W.; Lundin, C.D.; Liaw, P.K. Fatigue behavior of a wrought Al0.5CoCrCuFeNi two-phase high-entropy alloy. *Acta Mater.* **2015**, *99*, 247–258. [CrossRef]

20. Zou, Y.; Ma, H.; Spolenak, R. Ultrastrong ductile and stable high-entropy alloys at small scales. *Nat. Commun.* **2015**, *6*, 7748. [CrossRef] [PubMed]
21. Lu, Y.P.; Dong, Y.; Guo, S.; Jiang, L.; Kang, H.J.; Wang, T.M.; Wen, B.; Wang, Z.J.; Jie, J.C.; Cao, Z.Q.; et al. A promising new class of high-temperature alloys: Eutectic high entropy alloys. *Sci. Rep.* **2014**, *4*, 6200. [CrossRef] [PubMed]
22. Yu, Y.; He, F.; Qiao, Z.H.; Wang, Z.J.; Liu, W.M.; Yang, J. Effects of temperature and microstructure on the triblogical properties of CoCrFeNiNbx eutectic high entropy alloys. *J. Alloys Compd.* **2019**, *775*, 1376–1385. [CrossRef]
23. Li, D.Y.; Li, C.X.; Feng, T.; Zhang, Y.D.; Sha, G.; Lewandowski, J.J.; Liaw, P.K.; Zhang, Y. High-entropy AlCoCrFeNi alloy fibers with high tensile strength and ductility at ambient and cryogenic temperatures. *Acta Mater.* **2017**, *123*, 285–294. [CrossRef]
24. Zhang, Y.; Zuo, T.T.; Cheng, Y.Q.; Liaw, P.K. High-entropy alloys with high saturation magnetization, electrical resistivity, and malleability. *Sci. Rep.* **2013**, *3*, 1455. [CrossRef] [PubMed]
25. Yuan, Y.; Wu, Y.; Tong, X.; Zhang, H.; Wang, H.; Liu, X.J.; Ma, L.; Suo, H.L.; Lu, Z.P. Rare-earth high-entropy alloys with giant magnetocaloric effect. *Acta. Mater.* **2017**, *125*, 481–489. [CrossRef]
26. Zhou, Q.; Du, Y.; Han, W.C.; Ren, Y.; Zhai, H.M.; Wang, H.F. Identifying the origin of strain rate sensitivity in a high entropy bulk metallic glass. *Scr. Mater.* **2019**, *164*, 121–125. [CrossRef]
27. Zhang, Y.; Zhou, Y.J.; Lin, J.P.; Chen, G.L.; Liaw, P.K. Solid-solution phase-formation rules for multi-component alloys. *Adv. Eng. Mater.* **2008**, *10*, 534–538. [CrossRef]
28. Yang, X.; Zhang, Y. Prediction of high entropy stabilized solid solution in multi-component alloys. *Mater. Chem. Phys.* **2012**, *132*, 233–238. [CrossRef]
29. Otto, F.; Yang, Y.; Bei, H.; George, E.P. Relative effects of enthalpy and entropy on the phase stability of equiatomic high entropy alloys. *Acta Mater.* **2013**, *61*, 2628–2638. [CrossRef]
30. Tsai, M.H.; Li, J.H.; Fan, A.C.; Tsai, P.H. Incorrect predictions of simple solid solution high entropy alloys: Cause and possible solution. *Scr. Mater.* **2017**, *127*, 6–9. [CrossRef]
31. King, D.J.M.; Middle burgh, S.C.; Mcgregor, A.G.; Cortie, M.B. Predicting the formation and stability of single phase high entropy alloys. *Acta. Mater.* **2016**, *104*, 172–179. [CrossRef]
32. Sohn, S.; Liu, Y.H.; Liu, J.B.; Gong, P.; Prades-Rodel, S.; Blatter, A.; Scanley, B.E.; Broadbridge, C.C.; Schroers, J. Noble metal high entropy alloys. *Scr. Mater.* **2017**, *126*, 29–32. [CrossRef]
33. Stepanov, N.D.; Shaysultanov, D.G.; Salishchev, G.A.; Tikhonovsky, M.A.; Oleynik, E.E.; Tortika, A.S.; Senkov, O.N. Effect of V content on microstructure and mechanical properties of the CoCrFeMnNiVx high entropy alloys. *J. Alloys Compd.* **2015**, *628*, 170–185. [CrossRef]
34. Xu, X.D.; Liu, P.; Guo, S.; Hirata, A.; Fujita, T.; Nieh, T.G. Nanoscale phase separation in a fcc-based CoCrCuFeNiAl high entropy alloy. *Acta. Mater.* **2015**, *84*, 145–152. [CrossRef]
35. Lu, Y.P.; Gao, X.Z.; Jiang, L.; Chen, Z.M.; Wang, T.; Jie, J.; Kang, H.; Zhang, Y.; Guo, S.; Ruan, H. Directly cast bulk eutectic and near-eutectic high entropy alloys with balanced strength and ductility in a wide temperature range. *Acta Mater.* **2017**, *124*, 143–150. [CrossRef]
36. Guo, N.N.; Wang, L.; Luo, L.S.; Li, X.Z.; Su, Y.Q.; Guo, J.J.; Fu, H.Z. Microstructure and mechanical properties of refractory MoNbHfZrTi high-entropy alloy. *Mater. Des.* **2015**, *81*, 87–94. [CrossRef]
37. Chen, R.; Qin, G.; Zheng, H.T.; Wang, L.; Su, Y.Q.; Chiu, Y.L.; Ding, H.S.; Guo, J.J.; Fu, H.Z. Composition design of high entropy alloy using the valence electron concentration to balance strength and ductility. *Acta Mater.* **2018**, *144*, 129–137. [CrossRef]
38. Lu, Y.P.; Gao, X.X.; Dong, Y.; Wang, T.M.; Chen, H.; Maob, H.; Zhao, Y.; Jiang, H.; Cao, Z.; Li, T. Preparing bulk ultrafine-microstructure high entropy alloys via direct solidification. *Nanoscale* **2017**, *10*, 1039. [CrossRef]
39. Ai, C.; He, F.; Guo, M.; Zhou, J.; Wang, Z.J.; Yuan, Z.W.; Guo, Y.J.; Liu, Y.L.; Liu, L. Alloy design, mechanical and macromechanical properties of CoCrFeNiTax eutectic high entropy alloys. *J. Alloys Compd.* **2018**, *735*, 2653–2662. [CrossRef]
40. Wani, I.S.; Bhattacharjee, T.; Sheikh, S.; Lu, Y.P.; Chatterjee, S.; Bhattacharjee, P.P.; Guo, S.; Tsuji, N. Ultrafine-grained AlCoCrFeNi eutectic high entropy alloy. *Mater. Res. Lett.* **2016**, *4*, 174–179. [CrossRef]
41. Bhattacharjee, T.; Wani, I.S.; Sheikh, S.; Clark, I.T.; Okawa, T.; Guo, S.; Bhattacharjee, P.P.; Tsuji, N. Simultaneous strength-ductility enhancement of a nano-lamellar AlCoCrFeNi2.1 eutectic high entropy alloy by cryo-rolling and annealing. *Sci. Rep.* **2018**, *8*, 3276. [CrossRef]

42. He, F.; Wang, Z.J.; Cheng, P.; Wang, Q.; Li, J.J.; Dang, Y.Y.; Wang, J.C.; Liu, C.T. Designing eutectic high entropy alloys of CoCrFeNiNbx. *J. Alloys Compd.* **2016**, *656*, 284–289. [CrossRef]
43. He, F.; Wang, Z.J.; Shang, X.L.; Leng, C.; Li, J.J.; Wang, J.C. Stability of lamellar structures in CoCrFeNiNbx eutectic high entropy alloys at elevated temperatures. *Mater. Des.* **2016**, *104*, 259–264. [CrossRef]
44. Huo, W.Y.; Zhou, H.; Fang, F.; Xie, Z.H.; Jiang, J.Q. Microstructure and mechanical properties of CoCrFeNiZrx eutectic high entropy alloys. *Mater. Des.* **2017**, *134*, 226–233. [CrossRef]
45. Lucas, M.S.; Mauger, L.; Muoz, J.A.; Xiao, Y.M.; Sheets, A.O.; Semiatin, S.L.; Horwath, J.; Turgut, Z. Magnetic and vibrational properties of high-entropy alloys. *J. Appl. Phys.* **2011**, *109*, 07E307. [CrossRef]
46. Karadeniz, E.P.; Lang, P.; Moszner, F.; Pogatscher, S.; Ruban, A.V.; Uggowitzer, P.J.; Kozeschnik, E. Thermodynamics of Pd-Mn phases and extension to the Fe-Mn-Pd system. *Calphad* **2015**, *51*, 314–333. [CrossRef]
47. Tan, Y.M.; Li, J.S.; Wang, J.; Kou, H.C. Seaweed eutectic-dendritic solidification pattern in a CoCrFeNiMnPd eutectic high-entropy alloy. *Intermetallics* **2017**, *85*, 74–79. [CrossRef]
48. Tan, Y.M.; Li, J.S.; Wang, J.; Kolbe, M.; Kou, H.C. Microstructure characterization of CoCrFeNiMnPdx eutectic high entropy alloys. *J. Alloys Compd.* **2018**, *731*, 600–611. [CrossRef]
49. Wang, J.; Guo, T.; Li, J.S.; Jia, W.J.; Kou, H.C. Microstructure and mechanical properties of non-equilibrium solidified CoCrFeNi high entropy alloy. *Mater. Chem. Phys.* **2018**, *200*, 192–196. [CrossRef]
50. Ben-Jacob, E.; Garik, P. The formation of patterns in non-equilibrium growth. *Nature* **1990**, *343*, 523–530. [CrossRef]
51. Friedli, J.; Fife, J.L.; Di Napoli, P.; Rappaz, M. Dendritic Growth Morphologies in Al-Zn Alloys—Part I: X-ray Tomographic Microscopy. *Metall. Mater. Trans. A* **2013**, *44*, 5522–5531. [CrossRef]
52. Dantzig, J.A.; Di Napoli, P.; Friedli, J.; Rappaz, M. Dendritic Growth Morphologies in Al-Zn Alloys—Part II: Phase-Field Computations. *Metall. Mater. Trans. A* **2013**, *44*, 5532–5543. [CrossRef]
53. Castle, E.G.; Mullis, A.M.; Cochrane, R.F. Evidence for an extensive, undercooling-mediated transition in growth orientation, and novel dendritic seaweed microstructures in Cu–8.9 wt.% Ni. *Acta Mater.* **2014**, *66*, 378–387. [CrossRef]
54. Castle, E.G.; Mullis, A.M.; Cochrane, R.F. Mechanism selection for spontaneous grain refinement in undercooled metallic melts. *Acta Mater.* **2014**, *77*, 76–84. [CrossRef]
55. Liu, L.; Li, J.F.; Zhou, Y.H. Solidification interface morphology pattern in the undercooled Co-24.0 at.% Sn eutectic melt. *Acta Mater.* **2011**, *59*, 5558–5567. [CrossRef]
56. Yuan, F.P.; Wu, X.L. Size effects of primary/secondary twins on the atomistic deformation mechanisms in hierarchically nanotwinned metals. *J. Appl. Phys.* **2013**, *113*, 203516. [CrossRef]
57. Lu, Y.P.; Jiang, H.; Guo, S.; Wang, T.M.; Cao, Z.Q.; Li, T.J. A new strategy to design eutectic high entropy alloys using mixing enthalpy. *Intermetallics* **2017**, *91*, 124–128. [CrossRef]
58. Jiang, H.; Han, K.M.; Gao, X.X.; Lu, Y.P.; Cao, Z.Q.; Gao, M.C.; Hawk, J.A.; Li, T.J. A new strategy to design eutectic high-entropy alloys using simple mixture method. *Mater. Des.* **2018**, *142*, 101–105. [CrossRef]
59. Jin, X.; Zhou, Y.; Zhang, L.; Du, X.Y.; Li, B.S. A new pseudo binary strategy to design eutectic high entropy alloys using mixing enthalpy and valence electron concentration. *Mater. Des.* **2018**, *143*, 49–55. [CrossRef]

entropy

MDPI

Article

Effect of Atomic Size Difference on the Microstructure and Mechanical Properties of High-Entropy Alloys

Chan-Sheng Wu [1], Ping-Hsiu Tsai [1], Chia-Ming Kuo [1,2] and Che-Wei Tsai [2,*]

[1] Department of Materials Science and Engineering, National Tsing Hua University, Hsinchu 30013, Taiwan; wu.cs@nsrrc.org.tw (C.-S.W.); lansunma@gapp.nthu.edu.tw (P.-H.T.); cmkuo@mx.nthu.edu.tw (C.-M.K.)

[2] High Entropy Materials Center, National Tsing Hua University, Hsinchu 30013, Taiwan

* Correspondence: chewei@mx.nthu.edu.tw

Received: 29 October 2018; Accepted: 10 December 2018; Published: 14 December 2018

Abstract: The effects of atomic size difference on the microstructure and mechanical properties of single face-centered cubic (FCC) phase high-entropy alloys are studied. Single FCC phase high-entropy alloys, namely, CoCrFeMnNi, $Al_{0.2}$CoCrFeMnNi, and $Al_{0.3}CoCrCu_{0.3}$FeNi, display good workability. The recrystallization and grain growth rates are compared during annealing. Adding Al with 0.2 molar ratio into CoCrFeMnNi retains the single FCC phase. Its atomic size difference increases from 1.18% to 2.77%, and the activation energy of grain growth becomes larger than that of CoCrFeMnNi. The as-homogenized state of $Al_{0.3}CoCrCu_{0.3}$FeNi high-entropy alloy becomes a single FCC structure. Its atomic size difference is 3.65%, and the grain growth activation energy is the largest among these three kinds of single-phase high-entropy alloys. At ambient temperature, the mechanical properties of $Al_{0.3}CoCrCu_{0.3}$FeNi are better than those of CoCrFeMnNi because of high lattice distortion and high solid solution hardening.

Keywords: high-entropy alloys; mechanical property; recrystallization

1. Introduction

A multi-principal-element alloying system, developed by Yeh et al. in 2004 [1,2], is called high-entropy alloy (HEA). HEAs are defined as equiatomic or near-equiatomic alloys containing at least five elements, whose atomic concentration ranges from 5% to 35%. The multiprincipal-elemental mixtures of HEAs result in high entropy, lattice distortion, sluggish diffusion, and cocktail effects [3]. High entropy causes single-phase structures to become stable; as such, HEAs usually consist of simple solid solution phases with face-centered cubic (FCC) and body-centered cubic (BCC) structures rather than other intermetallic compounds [2]. A lattice is highly distorted because all atoms are solutes that can disorderly fill in a FCC or BCC lattice, and atomic sizes differ among elements in this alloying system, thereby possessing strengthening effect in high-solid-soluted HEAs [4]. Lattice distortion impedes atomic movement and slows down the diffusion rate of atoms in HEAs; these conditions lead to higher recrystallization temperatures, that is, the activation energies of grain growth in deformed HEAs are higher than those in conventional alloys [1,5].

In FCC single-phase HEAs, CoCrFeMnNi has been widely studied [2,6–11]. The atomic radius of each component in CoCrFeMnNi is close to each other, and the atomic size difference (δ) is approximately 1.18% only. δ is defined as follows:

$$\delta = 100\sqrt{\sum_{i=1}^{n} c_i \left(1 - r_i/\bar{r}\right)^2}, \quad (1)$$

where $\bar{r} = \sum_{i=1}^{n} c_i r_i$ is the average radius, c_i and r_i are the atomic percentage and atomic radius of the i element, respectively [12–14]. Other reports have also suggested that δ should consider the shear modulus, adjacency, and differences in the modulus of another element in single BCC phase HEA [15]. Moreover, researchers clarified that δ might be dependent on chemical composition of the alloys and the local lattice distortion in HEAs would differ from place to place [16,17].

Grain growth kinetics is usually described by the following equation:

$$d^n - d_0^n = kt, \tag{2}$$

where d is the grain size after annealing time t, d_0 is the initial grain size (this term approaches to zero for the as-rolled state), k is a temperature-dependent constant, and n is the grain growth exponent. k can be described by Equation (3):

$$k = k_0 e^{-Q/RT}, \tag{3}$$

where k_0 is a constant, R is Boltzmann's constant, T is the temperature, and Q is the activation energy of grain growth.

Otto et al. [6] observed that the activation energy of the grain growth of CoCrFeMnNi is 325 kJ/mol. However, the relation between lattice distortion and grain growth activation energy remains unknown. To clarify this phenomenon, other researchers [9,18] added aluminum, whose atomic radius is larger than that of other elements, to CoCrFeMnNi. Al is a BCC stabilizer in HEAs [9,19]. Thus, Al addition is limited to a molar ratio of 0.2 to prevent the formation of dual-phases in CoCrFeMnNi. With a moderate amount of Al addition, a FCC single-phase structure and a high atomic size difference should be obtained in the designed alloy.

$Al_{0.5}CoCrCuFeNi$ high-entropy alloy is composed of matrix and Cu-rich phases, which become a FCC phase after they are treated with a solution [19–23]. A Ni–Al-rich phase precipitates in the matrix phase, and a Cu-rich phase precipitates at midtemperature, leading to the brittleness of the alloy at intermediate temperature. Thus, to design single-phase structures, decreasing the amounts of Al and Cu in $Al_{0.5}CoCrCuFeNi$ is needed.

According to the Hume–Rothery rule, solid solutions can be obtained until the atomic size difference is larger than 15% [24]. In addition, factor Ω can be utilized to predict whether an alloy is solid soluted:

$$\Omega = T_m \frac{\Delta S_{mix}}{|\Delta H_{mix}|}, \tag{4}$$

where T_m is the melting point (Kelvin) of the alloy, and ΔS_{mix} and ΔH_{mix} are the mixing entropy and enthalpy of the alloy, respectively [13]. For solid solution formation, Ω should be larger than 1. ΔH_{mix} ranging from −15 kJ/mol to 5 kJ/mol is beneficial to obtaining single-phase structures [12].

For high-entropy alloys, an empirical statistical pattern is observed by calculating the valence electron concentration (VEC) to predict the formation of stable phases [25,26]. A phase consists of BCC and FCC when the VEC is below 6.8 and above 8.0, respectively. Otherwise, BCC and FCC phases are formed in HEAs.

In this study, $Al_{0.3}CoCrCu_{0.3}FeNi$ (δ = 3.65%; ΔH_{mix} = −4.38 kJ/mol; Ω = 5.67; VEC = 8.35) is designed to be the FCC single-phase high-entropy alloy. The lattice distortion and activation energy of grain growth in $Al_{0.3}CoCrCu_{0.3}FeNi$ are compared with those of other FCC single-phase HEAs to determine the effect of atomic size difference on microstructure and mechanical properties.

2. Materials and Experimental Methods

High-entropy CoCrFeMnNi, $Al_{0.2}CoCrFeMnNi$, and $Al_{0.3}CoCrCu_{0.3}FeNi$ alloys were prepared by arc melting in a vacuum chamber at a pressure of 0.01 Torr, and its constituent elements had at least 99.99 wt.% purity. Melting was performed at a current of 500 A in a water-cooled copper hearth and repeated at least four times to confirm chemical homogeneity. The dimensions of the final solidified ingot were cuboid and had a width of 20 mm, a length of 40 mm, and a height of 10 mm. The ingots were homogenized at 1100 °C for 6 h, quenched with water quenching or cooled in a furnace, and cold rolled at a thickness reduction of 70%. The specimens were annealed between 900 °C and 1100 °C for various times and finally quenched with water.

The specimens were prepared by cutting, grinding, and polishing in a sequence. The crystalline structure of the present alloys was characterized using an X-ray diffractometer SHIMADZU-XRD6000 equipped with Cu-target radiation (K_α = 1.54 Å) at 30 kV and 20 mA. The sample was scanned at 2θ angle from 20° to 100° at a scanning rate of 2°/min. The microstructures were observed under a scanning electron microscope (JEOL-5410) at an acceleration voltage of 20 kV for a working distance of 24 mm. All of the specimens were etched with 0.5 g of copper (II) chloride, 10 mL of hydrochloric acid, and 10 mL of ethanol mixing liquid solution to observe the grains. The grain size was calculated and statistically measured using ImageJ. The grain size was also circled at least 400 grains to yield an average grain size for each specimen. For the tensile test, all of the specimens were tested at ambient temperature by Instron 4468 at a stain rate of 10^{-3} s^{-1}. Figure 1 presents the dimensions of the samples used in the tensile test.

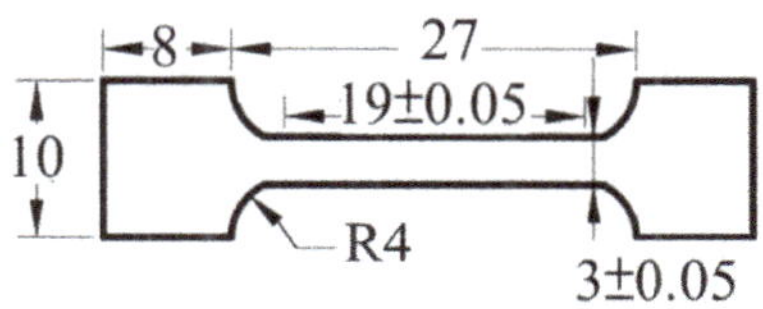

Figure 1. Dimensions of tensile specimens (unit: mm).

3. Results

3.1. Microstructure and Crystalline Structure

$Al_{0.2}$CoCrFeMnNi is designed from CoCrFeMnNi by adding 0.2 molar ratio of Al to contribute increased lattice mismatch. Figure 2 shows the X-ray diffraction (XRD) analysis of $Al_{0.2}$CoCrFeMnNi HEAs in its homogenized state quenched with water and cooled in a furnace. The lattice constant of $Al_{0.2}$CoCrFeMnNi is 3.582 Å. The atomic size difference of $Al_{0.2}$CoCrFeMnNi can be calculated by Equation (1), and the value is 2.77%. Figure 3 shows the XRD analysis of the designed $Al_{0.3}$CoCr$Cu_{0.3}$FeNi HEAs in its homogenized state after water quenching and furnace cooling. The diffraction pattern also clearly reveals the appearance of peaks of the FCC structure only. The lattice constant of $Al_{0.3}$CoCr$Cu_{0.3}$FeNi is 3.585 Å. The atomic size difference of $Al_{0.3}$CoCr$Cu_{0.3}$FeNi is 3.65%, which is calculated by Equation (1).

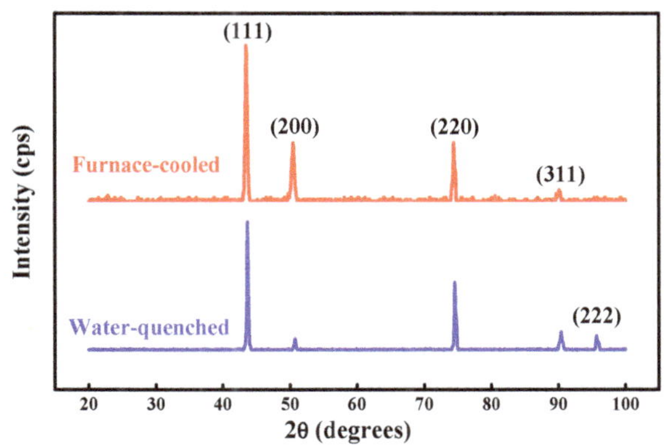

Figure 2. X-ray diffraction (XRD) patterns of homogenized $Al_{0.2}$CoCrFeMnNi alloys after water quenching and furnace cooling.

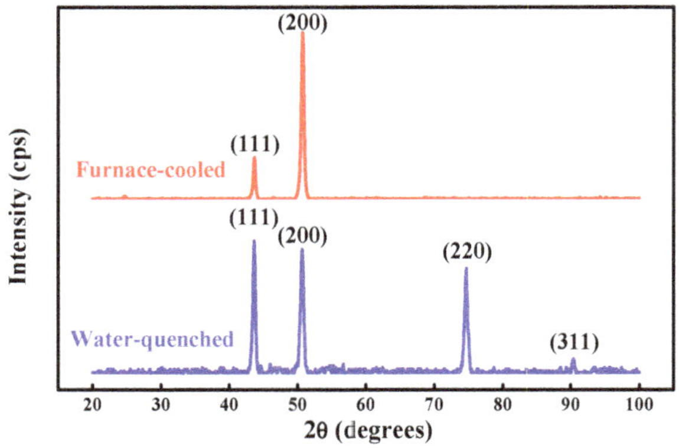

Figure 3. XRD patterns of homogenized $Al_{0.3}CoCrCu_{0.3}FeNi$ alloys after water quenching and furnace cooling.

3.2. Grain Growth Activation Energy of $Al_{0.2}CoCrFeMnNi$ and $Al_{0.3}CoCrCu_{0.3}FeNi$ Alloys

All of the present alloys exhibit only single FCC phases without any other precipitations and show good workability at room temperature. The reduction of thickness can reach 70% without any cracks neither on the rims nor inside the as-rolled specimens. Recrystallization occurs after annealing and is performed above 900 °C for various times. The movement of each solute atom is more difficult than that in traditional alloying systems because of sluggish effect in high-entropy alloys. Under this condition, the recrystallization temperature in HEAs becomes higher than that in conventional alloys. Recrystallization takes place in high-lattice-strain energy regions, such as slip band, deformation twin intersections, and grain boundaries, which are the preferred nucleation sites for new strain-free grains. The average grain size of $Al_{0.2}CoCrFeMnNi$ is 42.6 μm after it is annealed at 1000 °C for 120 min, water quenching is subsequently performed Figure 4a. The grain sizes at different annealing temperatures for various times are shown in Table 1 The grain growth activation of $Al_{0.2}CoCrFeMnNi$ is calculated with Equations (2) and (3), and the linear fitting result is shown in Figure 5. The slope of the fitting line represents the activation energy (Q) for grain growth, which is 434.4 kJ/mol in $Al_{0.2}CoCrFeMnNi$.

The grain growth is observed at 900 °C in $Al_{0.3}CoCrCu_{0.3}FeNi$, and the evolution of the microstructure in $Al_{0.3}CoCrCu_{0.3}FeNi$ is shown in Figure 6. High-resolution scanning electron microscopy is utilized after the etching condition is optimized for observation because of the small grain size. The grain sizes at different temperatures in various times are shown in Table 2. The grain growth activation is calculated by Equations (2) and (3), and the fitting result is shown in Figure 7. The activation energy of grain growth is 761.3 kJ/mol.

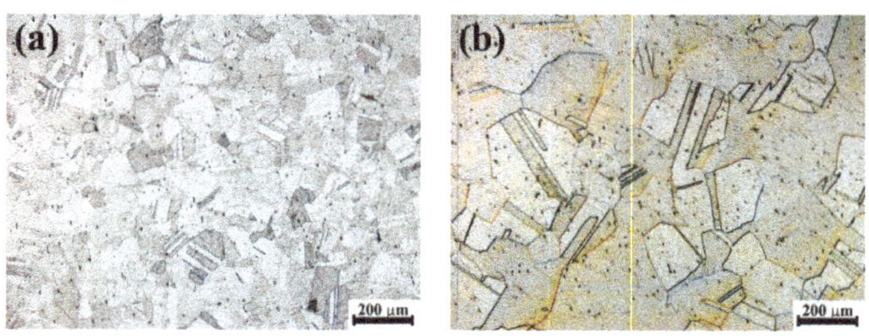

Figure 4. Microstructure of $Al_{0.2}CoCrFeMnNi$ with cold rolling and annealing at (**a**) 1000 °C and (**b**) 1100 °C for 120 min.

Table 1. Grain size of $Al_{0.2}CoCrFeMnNi$ annealed at 1000 °C, 1050 °C, and 1100 °C with different times (unit: μm).

Grain Size (μm)		Annealing Temperature (°C)		
		1000	1050	1100
Annealing time (min)	10	27.0	42.6	80.2
	30	40.1	65.7	118.1
	60	48.1	80.3	152.8
	120	63.2	98.4	201.2

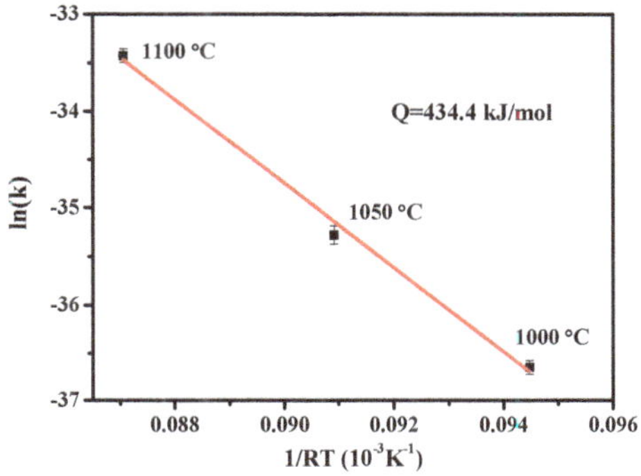

Figure 5. Grain growth activation energy of $Al_{0.2}CoCrFeMnNi$.

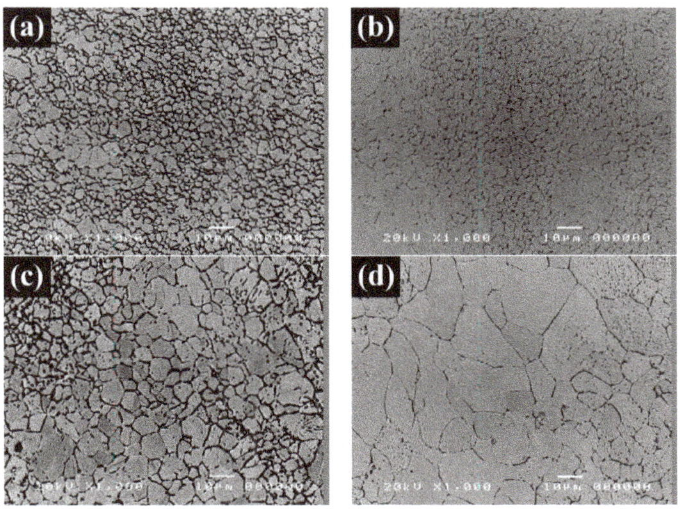

Figure 6. Microstructure of $Al_{0.3}CoCrCu_{0.3}FeNi$ with cold rolling and annealing at 900 °C for (**a**) 120; (**b**) 300; (**c**) 600 and (**d**) 1200 min.

Table 2. Grain size of $Al_{0.3}CoCrCu_{0.3}FeNi$ annealed at 900 °C, 950 °C, and 1000 °C at different times (unit: μm).

Grain Size (μm)		Annealing Temperature (°C)		
		900	950	1000
Annealing time (min)	120	3.35	12.9	50.8
	300	5.13	16.9	70.6
	600	8.21	21.7	103.3
	1200	10.50	30.3	141.0

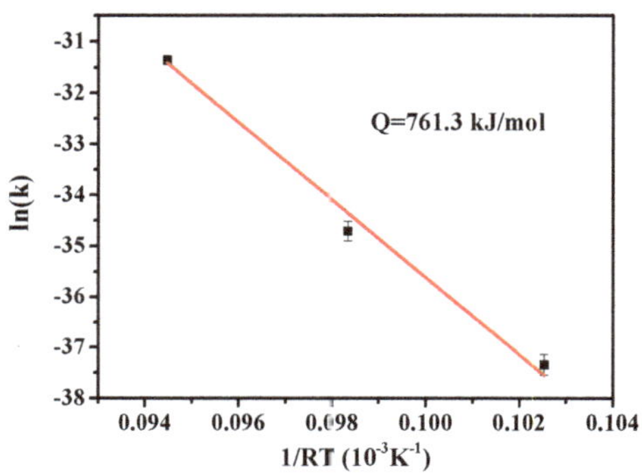

Figure 7. Grain growth activation energy of $Al_{0.3}CoCrCu_{0.3}FeNi$.

3.3. Relationship between Atomic Size Difference and Grain Growth Activation Energy

Three high-entropy alloy systems with a single FCC phase are observed: CoCrFeMnNi, $Al_{0.2}CoCrFeMnNi$, and $Al_{0.3}CoCrCu_{0.3}FeNi$. These alloys are selected and compared with others. The outstanding phase stability of these HEAs can avoid from the formation of second phases that can influence grain growth at increased temperature.

The calculated results of atomic size difference and grain growth activation is shown in Table 3. The atomic size difference of $Al_{0.3}CoCrCu_{0.3}FeNi$ is 3.65% as calculated by Equation (1). This value is the highest in these three kinds of HEAs, and 2.77% and 1.18% belong to $Al_{0.2}CoCrFeMnNi$ and in CoCrFeMnNi, respectively. The activation energy of $Al_{0.3}CoCrCu_{0.3}FeNi$ for grain growth is also the highest among others.

Table 3. Value of atomic size difference and grain growth activation energy.

	Atomic Size Difference (%)	Grain Growth Activation Energy (kJ/mol)	ΔH_{mix} (kJ/mol)
$Al_{0.3}CoCrCu_{0.3}FeNi$	3.65	761.3	−4.38
$Al_{0.2}CoCrFeMnNi$	2.77	434.4	−6.24
CoCrFeMnNi	1.18	325.0	−4.16

The mixing enthalpy between each solute element is also shown in Table 4 to determine how the mixing enthalpy affects the activation energy of grain growth. However, the mixing enthalpy of these three HEAs is slightly related to the atomic size difference or the activation energy of grain growth.

Table 4. ΔH_{mix} for the element pairs (unit: kJ/mol).

	Al	Co	Cr	Cu	Fe	Mn	Ni
Al		−19	−10	−1	−11	−19	−22
Co			−4	6	−1	−5	0
Cr				12	−1	2	−7
Cu					13	4	4
Fe						0	−2
Mn							8
Ni							

4. Discussion

4.1. Effect of Atomic Size Difference on Microstructures

The comparison result of the microstructure between $Al_{0.2}CoCrFeMnNi$ and $Al_{0.3}CoCrCu_{0.3}FeNi$ single FCC-type high-entropy alloys according to XRD analysis in Figures 2 and 3 shows that the as-homogenized state and the furnace-cooled state both have outstanding phase stability without any detrimental non-FCC phases regardless of the cooling condition. Haas, Sebastian, et al. [27] reported that the Gibbs free energy of solid solution alloys is completely due to configurational entropy and contributes to the thermal stability of solid solution alloys. The consequent single FCC phase is gained with a significant sluggish diffusion effect in HEAs [3], and this parameter is beneficial to the following heavily cold-rolling procedure without showing the undesired brittleness and ensuring great workability.

The atomic size difference of $Al_{0.3}CoCrCu_{0.3}FeNi$ is larger than that of $Al_{0.2}CoCrFeMnNi$. Different grain sizes are found (Tables 2 and 3) when the samples are treated with the same thermomechanical process, that is, 70% cold rolled, annealed at 1000 °C for 120 min, and quenched with water. This result reveals that the atomic size difference likely affects the ability of dislocation movement, causing a different grain growth behavior in these two HEAs.

4.2. Effect of Atomic Size Difference on Mechanical Properties

Large atomic size difference (δ) corresponds to the great amount of activation energy needed for grain growth. Large δ is associated with a high degree of lattice distortion in the single-phase solid solution and cause atoms to spontaneously move at the most stable state, thereby decreasing the potential energy of the existing defects in a low level; that is, defects are found in stable sites. The energy of grain growth comes from the different energy levels between lattice distortion and defects [28]. When the number of alloying elements is low, or the alloy is low entropy, the energy of grain growth in such an alloy, which is nearly pure metal, is attributed to undistorted grain and perfect dislocation, and the energy level difference between the former and the latter can be regarded as the driving force for recrystallization then to make a perfect dislocation being released by a new grain.

In other cases, if the difference between these two energies is insufficient, the driving force for recrystallization is too low to induce dislocation rearrangement. In other words, higher annealing temperature or longer annealing time is needed so that it can enable the recrystallization to start at dislocation site and new grain to grow subsequently.

Table 4 shows that aluminum has the highest binding energy to each element, possibly leading to the solute-pinning effect and making each solute atom suitable to their lattice site. In other words, atoms become self-accommodating to the stable sites. Thus, when a great amount of aluminum is added, the grain is distorted remarkably because of large δ, causing energy level of a grain to be higher than undistorted one. Also, solute atoms are pinned in their preferred lattice site, and each solute atom can act as an obstacle of the movement of dislocations, thus obtaining the lower energy level of pinned dislocations, that is, dislocations become more stable or immovable.

4.3. Comparison of Tensile Properties with Different HEAs

Otto et al. [6] reported that the tensile properties of CoCrFeMnNi at different temperatures with different grain sizes are yield strength of 350 MPa, ultimate tensile strength of 650 MPa, and recrystallized grain size of 4.4 μm. With cold rolling and annealing, the grain size of CoCrFeMnNi can be small. The compatible small scale of the grain size of $Al_{0.3}CoCrCu_{0.3}FeNi$ was designed to compare its mechanical properties with those of CoCrFeMnNi. After cold rolling and annealing were performed at 900 °C for 5 h and water quenching was conducted, the grain size is approximately 5.13 μm under the optimized thermomechanical treatment. The final microstructure is shown in Figure 6b.

Figure 8 illustrates the tensile stress–strain curves of the designed $Al_{0.3}CoCrCu_{0.3}FeNi$ in the as-homogenized states and 900 °C/5 h annealing state. Table 5 shows the comparison of the mechanical properties between CoCrFeMnNi and $Al_{0.3}CoCrCu_{0.3}FeNi$ with different grain sizes. The yield strength and the ultimate tensile strength are 500 and 717 MPa in the annealed $Al_{0.3}CoCrCu_{0.3}FeNi$, respectively. The elongation of $Al_{0.3}CoCrCu_{0.3}FeNi$ is smaller than that of CoCrFeMnNi, suggesting that the plastic deformation in CoCrFeMnNi is involved in one-to-one atom-vacancy exchange mechanism [29,30]. This finding can be accounted for this high ductility.

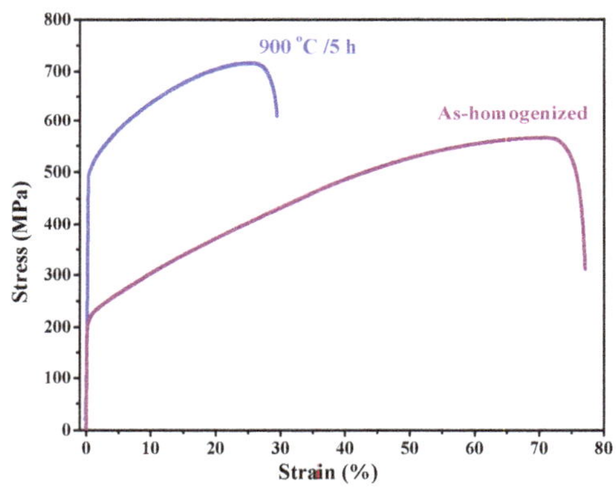

Figure 8. Mechanical properties of $Al_{0.3}CoCrCu_{0.3}FeNi$ alloy annealed at 900 °C for 5 h and the as-homogenized state.

Table 5. Comparisons of mechanical property between CoCrFeMnNi [6] and $Al_{0.3}CoCrCu_{0.3}FeNi$.

	CoCrFeMnNi [6]		**$Al_{0.3}CoCrCu_{0.3}FeNi$**	
Annealing condition temperature/time	1150 °C/1 h	800 °C/1 h	As-homogenized	900 °C/5 h
Grain size (μm)	155	4.4	516	5.13
YS (MPa)	190	350	217	500
UTS (MPa)	560	650	566	717
Elongation (%)	78	60	77	29

The lattice distortion of $Al_{0.3}CoCrCu_{0.3}FeNi$ is higher than that of CoCrFeMnNi because the former has a larger atomic size difference (δ) than the latter. Different δ can be related to the mechanical behavior, for an example, higher yield and tensile strength in $Al_{0.3}CoCrCu_{0.3}FeNi$ with the same FCC structure and the close value of grain sizes compared with CoCrFeMnNi. The high lattice distortion can introduce the concentrated strain field around the lattice, causing the movement of dislocations to be more difficult. The large amount of the added aluminum can introduce a high degree of interaction to each alloying element, causing the pinning effect on dislocations. Finally, in high-entropy alloys with high lattice distortion, although the lattice distortion effect on properties of HEAs is yet an open question still await to be solved [31], once a large atomic size difference is obtained, the increment of tensile strength can be predicted.

5. Conclusions

$Al_{0.2}CoCrFeMnNi$ is a single FCC high-entropy alloy structure composed of Al added to CoCrFeMnNi at a molar ratio of up to 0.2. The microstructure of $Al_{0.3}CoCrCu_{0.3}FeNi$ is also a single FCC phase. The recrystallization and grain growth behavior of $Al_{0.3}CoCrCu_{0.3}FeNi$ and $Al_{0.2}CoCrFeMnNi$ are observed. The calculation of the grain sizes under different annealing conditions reveals that the activation energy of the grain growth of $Al_{0.2}CoCrFeMnNi$ is 434.4 kJ/mol with an atomic size difference of 2.77%. The activation energy of the grain growth of $Al_{0.3}CoCrCu_{0.3}FeNi$ is 761.3 kJ/mol with an atomic size difference of 3.65%. A large atomic size difference indicates that a high activation energy is needed for grain growth. The lattice distortion of $Al_{0.3}CoCrCu_{0.3}FeNi$ is much higher than that of single FCC-phase CoCrFeMnNi. The mechanical properties of $Al_{0.3}CoCrCu_{0.3}FeNi$ are superior to those of CoCrFeMnNi under similar conditions. This result is attributed to high lattice distortion and pinning effect on dislocation because of the large atomic size difference in high-entropy $Al_{0.3}CoCrCu_{0.3}FeNi$ alloy.

Author Contributions: Conceptualization, C.-W.T.; formal analysis, C.-S.W. and P.-H.T.; data curation, C.-S.W. and P.-H.T.; writing—original draft preparation, C.-S.W.; writing—review and editing, C.-M.K.; supervision, C.-W.T.; project administration, C.-W.T.

Funding: This research received no external funding.

Acknowledgments: We are pleased to acknowledge the financial support for this research by Ministry of Science and Technology, R.O.C (MOST 104-2218-E-007-017 and 106-2218-E-007-019). The support provided by the High Entropy Materials Center from The Featured Areas Research Center Program within the framework of the Higher Education Sprout Project by the Ministry of Education (MOE) in Taiwan is greatly appreciated.

Conflicts of Interest: The authors declare no conflict of interest.

References

1. Yeh, J.W.; Chen, S.K.; Lin, S.J.; Gan, J.Y.; Chin, T.S.; Shun, T.T.; Tsau, C.H.; Chang, S.Y. Nanostructured high-entropy alloys with multiple principal elements: Novel alloy design concepts and outcomes. *Adv. Eng. Mater.* **2004**, *6*, 299–303. [CrossRef]
2. Cantor, B.; Chang, I.T.H.; Knight, P.; Vincent, A.J.B. Microstructural development in equiatomic multicomponent alloys. *Mater. Sci. Eng. A-Struct. Mater. Prop. Microstruct. Process.* **2004**, *375*, 213–218. [CrossRef]
3. Yeh, J.-W. Recent progress in high-entropy alloys. *Ann. Chim. Sci. Matér.* **2006**, *31*, 633–648. [CrossRef]
4. Yeh, J.W.; Chen, S.K.; Gan, J.Y.; Lin, S.J.; Chin, T.S.; Shun, T.T.; Tsau, C.H.; Chang, S.Y. Formation of simple crystal structures in Cu-Co-Ni-Cr-Al-Fe-Ti-V alloys with multiprincipal metallic elements. *Metall. Mater. Trans. A-Phys. Metall. Mater. Sci.* **2004**, *35A*, 2533–2536. [CrossRef]
5. Tsai, K.Y.; Tsai, M.H.; Yeh, J.W. Sluggish diffusion in Co-Cr-Fe-Mn-Ni high-entropy alloys. *Acta Mater.* **2013**, *61*, 4887–4897. [CrossRef]
6. Otto, F.; Hanold, N.L.; George, E.P. Microstructural evolution after thermomechanical processing in an equiatomic, single-phase cocrfemnni high-entropy alloy with special focus on twin boundaries. *Intermetallics* **2014**, *54*, 39–48. [CrossRef]
7. Otto, F.; Dlouhý, A.; Somsen, C.; Bei, H.; Eggeler, G.; George, E.P. The influences of temperature and microstructure on the tensile properties of a cocrfemnni high-entropy alloy. *Acta Mater.* **2013**, *61*, 5743–5755. [CrossRef]
8. Gludovatz, B.; Hohenwarter, A.; Catoor, D.; Chang, E.H.; George, E.P.; Ritchie, R.O. A fracture-resistant high-entropy alloy for cryogenic applications. *Science* **2014**, *345*, 1153–1158. [CrossRef]

9. He, J.Y.; Liu, W.H.; Wang, H.; Wu, Y.; Liu, X.J.; Nieh, T.G.; Lu, Z.P. Effects of al addition on structural evolution and tensile properties of the feconicrmn high-entropy alloy system. *Acta Mater.* **2014**, *62*, 105–113. [CrossRef]
10. Wu, Z.; Bei, H.; Pharr, G.M.; George, E.P. Temperature dependence of the mechanical properties of equiatomic solid solution alloys with face-centered cubic crystal structures. *Acta Mater.* **2014**, *81*, 428–441. [CrossRef]
11. Zaddach, A.J.; Scattergood, R.O.; Koch, C.C. Tensile properties of low-stacking fault energy high-entropy alloys. *Mater. Sci. Eng. A* **2015**, *636*, 373–378. [CrossRef]
12. Zhang, Y.; Zhou, Y.J.; Lin, J.P.; Chen, G.L.; Liaw, P.K. Solid-solution phase formation rules for multi-component alloys. *Adv. Eng. Mater.* **2008**, *10*, 534–538. [CrossRef]
13. Yang, X.; Zhang, Y. Prediction of high-entropy stabilized solid-solution in multi-component alloys. *Mater. Chem. Phys.* **2012**, *132*, 233–238. [CrossRef]
14. Guo, S.; Liu, C.T. Phase stability in high entropy alloys: Formation of solid-solution phase or amorphous phase. *Prog. Nat. Sci.* **2011**, *21*, 433–446. [CrossRef]
15. Lee, C.; Song, G.; Gao, M.C.; Feng, R.; Chen, P.; Brechtl, J.; Chen, Y.; An, K.; Guo, W.; Poplawsky, J.D. Lattice distortion in a strong and ductile refractory high-entropy alloy. *Acta Mater.* **2018**, *160*, 158–172. [CrossRef]
16. Lubarda, V. On the effective lattice parameter of binary alloys. *Mech. Mater.* **2003**, *35*, 53–68. [CrossRef]
17. Owen, L.R.; Jones, N.G. Lattice distortions in high-entropy alloys. *J. Mater. Res.* **2018**, *33*, 2954–2969. [CrossRef]
18. Tang, Z.; Gao, M.C.; Diao, H.; Yang, T.; Liu, J.; Zuo, T.; Zhang, Y.; Lu, Z.; Cheng, Y.; Zhang, Y.; et al. Aluminum alloying effects on lattice types, microstructures, and mechanical behavior of high-entropy alloys systems. *Jom* **2013**, *65*, 1848–1858. [CrossRef]
19. Tung, C.-C.; Yeh, J.-W.; Shun, T.-T.; Chen, S.-K.; Huang, Y.-S.; Chen, H.-C. On the elemental effect of alcocrcufeni high-entropy alloy system. *Mater. Lett.* **2007**, *61*, 1–5. [CrossRef]
20. Tsai, C.-W.; Chen, Y.-L.; Tsai, M.-H.; Yeh, J.-W.; Shun, T.-T.; Chen, S.-K. Deformation and annealing behaviors of high-entropy alloy Al0.5CoCrCuFeNi. *J. Alloys Compd.* **2009**, *486*, 427–435. [CrossRef]
21. Tsai, C.-W.; Tsai, M.-H.; Yeh, J.-W.; Yang, C.-C. Effect of temperature on mechanical properties of al0.5cocrcufeni wrought alloy. *J. Alloys Compd.* **2010**, *490*, 160–165. [CrossRef]
22. Sheng, H.F.; Gong, M.; Peng, L.M. Microstructural characterization and mechanical properties of an al0.5cocrfecuni high-entropy alloy in as-cast and heat-treated/quenched conditions. *Mater. Sci. Eng. A* **2013**, *567*, 14–20. [CrossRef]
23. Tong, C.J.; Chen, M.R.; Chen, S.K.; Yeh, J.W.; Shun, T.T.; Lin, S.J.; Chang, S.Y. Mechanical performance of the alxcocrcufeni high-entropy alloy system with multiprincipal elements. *Metall. Mater. Trans. A-Phys. Metall. Mater. Sci.* **2005**, *36A*, 1263–1271. [CrossRef]
24. Hume-Rothery, W. *Atomic Theory for Students of Metallurgy*; Institute of Metals: London, UK, 1952.
25. Guo, S.; Ng, C.; Lu, J.; Liu, C. Effect of valence electron concentration on stability of fcc or bcc phase in high entropy alloys. *J. Appl. Phys.* **2011**, *109*, 103505. [CrossRef]
26. Tian, F.; Varga, L.K.; Chen, N.; Shen, J.; Vitos, L. Empirical design of single phase high-entropy alloys with high hardness. *Intermetallics* **2015**, *58*, 1–6. [CrossRef]
27. Haas, S.; Mosbacher, M.; Senkov, O.; Feuerbacher, M.; Freudenberger, J.; Gezgin, S.; Völkl, R.; Glatzel, U. Entropy determination of single-phase high entropy alloys with different crystal structures over a wide temperature range. *Entropy* **2018**, *20*, 654. [CrossRef]
28. Huang, P.K.; Yeh, J.W. Inhibition of grain coarsening up to 1000 degrees C in (AlCrNbSiTiV)N superhard coatings. *Scr. Mater.* **2010**, *62*, 105–108. [CrossRef]
29. Zhu, C.; Lu, Z.; Nieh, T. Incipient plasticity and dislocation nucleation of fecocrnimn high-entropy alloy. *Acta Mater.* **2013**, *61*, 2993–3001. [CrossRef]
30. Basu, I.; Ocelík, V.; De Hosson, J.T.M. Size effects on plasticity in high-entropy alloys. *J. Mater. Res.* **2018**, *33*, 3055–3076. [CrossRef]
31. Song, H.; Tian, F.; Hu, Q.-M.; Vitos, L.; Wang, Y.; Shen, J.; Chen, N. Local lattice distortion in high-entropy alloys. *Phys. Rev. Mater.* **2017**, *1*, 023404. [CrossRef]

entropy

MDPI

Article

Influence of Titanium on Microstructure, Phase Formation and Wear Behaviour of $AlCoCrFeNiTi_x$ High-Entropy Alloy

Martin Löbel *, Thomas Lindner, Thomas Mehner and Thomas Lampke

Materials and Surface Engineering Group, Institute of Materials Science and Engineering, Chemnitz University of Technology, D-09125 Chemnitz, Germany; th.lindner@mb.tu-chemnitz.de (T.Li.); thomas.mehner@mb.tu-chemnitz.de (T.M.); thomas.lampke@mb.tu-chemnitz.de (T.La.)

* Correspondence: martin.loebel@mb.tu-chemnitz.de; Tel.: +49-371-531-31865

Received: 31 May 2018; Accepted: 29 June 2018; Published: 2 July 2018

Abstract: The novel alloying concept of high-entropy alloys (HEAs) has been the focus of many recent investigations revealing an interesting combination of properties. Alloying with aluminium and titanium showed strong influence on microstructure and phase composition. However, detailed investigations on the influence of titanium are lacking. In this study, the influence of titanium in the alloy system $AlCoCrFeNiTi_x$ was studied in a wide range (molar ratios x = 0.0; 0.2; 0.5; 0.8; 1.0; 1.5). Detailed studies investigating the microstructure, chemical composition, phase composition, solidification behaviour, and wear behaviour were carried out. Alloying with titanium showed strong influence on the resulting microstructure and lead to an increase of microstructural heterogeneity. Phase analyses revealed the formation of one body-centred cubic (bcc) phase for the alloy without titanium, whereas alloying with titanium caused the formation of two different bcc phases as main phases. Additional phases were detected for alloys with increased titanium content. For $x \geq 0.5$, a minor phase with face-centred cubic (fcc) structure was formed. Further addition of titanium led to the formation of complex phases. Investigation of wear behaviour revealed a superior wear resistance of the alloy $AlCoCrFeNiTi_{0.5}$ as compared to a bearing steel sample.

Keywords: HEA; high-entropy alloy; CCA; compositionally complex alloy; phase composition; microstructure; wear behaviour

1. Introduction

High-entropy alloys (HEAs) are an emerging class of new materials. Their alloying concept differs from conventional alloys, which are composed of one main element, and an improvement of properties is achieved by adding minor amounts of other elements. In contrast, HEAs are multicomponent alloys comprising at least five elements with approximately equimolar composition [1]. Despite their complex composition, only simple solid solutions with fcc or bcc structure were formed for several alloy systems. The formation of brittle and intermetallic phases can be successfully suppressed. One of the first alloys with a single fcc phase is the equimolar alloy CoCrFeMnNi investigated by Cantor et al. [2]. Due to their unique structure, HEAs exhibit an interesting combination of properties e.g., high hardness and strength in combination with adequate ductility. Furthermore, a high wear and corrosion resistance can be obtained [3–6]. Two main groups can be distinguished: HEAs forming (i) cubic or (ii) hexagonal phases [4].

Detailed structural investigations of HEAs revealed that only a few alloys form a single phase microstructure comprising fcc or bcc phases. Most alloys are composed of more than one phase, partly including complex or intermetallic phases. For these alloys, the term compositionally complex alloys (CCA) was introduced [7].

One of the most intensely investigated HEA systems, primarily forming cubic phases is AlCoCrCuFeNi [8,9]. Due to the positive enthalpy of mixing ΔH_{mix} among Cu and the elements Fe and Cr, segregation was observed, deteriorating mechanical and corrosion properties. Therefore, subsequent investigations focused on the copper-free derivative AlCoCrFeNi [3,10–13]. Early studies concentrated on the equimolar composition. However, investigations showed that optimum properties are usually achieved when choosing a differing chemical composition [14].

Investigating the influence of different alloying elements showed a strong influence of aluminium on the microstructure, phase composition, and properties [15,16]. For a low aluminium content, fcc phases are stabilised, whereas high contents of aluminium act as a strong bcc phase stabiliser [17,18]. In addition, the formation of complex phases can be suppressed for a high aluminium content [19].

Furthermore, the influence of additional alloying elements has been investigated. One element which shows distinct effects on microstructure, phase composition, and mechanical properties is titanium. Due to its large atomic radius, titanium leads to solid solution strengthening, increasing its hardness and strength. However, for a high titanium content, the formation of intermetallic phases and compounds was determined, leading to embrittlement. The alloy system AlCoCrFeNiTi shows high wear resistance also in comparison to conventional steels [20–24].

Detailed investigations on the influence of the alloying element titanium are required for the development of lightweight HEAs. The aim of the present study is the determination of the influence of this alloying element in the alloy system $AlCoCrFeNiTi_x$ regarding its influence on microstructure, phase composition and wear behaviour.

2. Materials and Methods

Bulk samples of the alloy system $AlCoCrFeNiTi_x$ with the molar ratios of x = 0.0; 0.2; 0.5; 0.8; 1.0; 1.5 were produced by arc-melting. Elemental granules with a purity of $\geq$99.9% were used as raw materials. The elemental granules were weighed and mixed according to the intended molar ratios. Each sample had a total weight of 10 g. Arc-melting of the samples was conducted in a water-cooled copper crucible.

After evacuating and reaching a pressure of 2×10^{-4} mbar, the furnace chamber was filled with argon to a pressure of 1.1 bar. A tungsten electrode was used to ignite an arc. All samples were remoulded three times and turned after each step to achieve chemical homogeneity. The resulting samples had a diameter of approximately 20 mm. For the arc furnace, a low cooling rate of <50 K/s was determined in preliminary studies.

Metallographic cross-sections were prepared according to standard metallographic procedures. Investigations of the microstructure were carried out by scanning electron microscopy (SEM) in a LEO 1455VP (Zeiss, Jena, Germany) with an acceleration voltage of 25 kV. For the visualisation of material contrast, a backscattered-electron detector (BSD) was used. The analyses of the overall chemical composition was carried out by energy dispersive X-ray spectroscopy (EDS) with a GENESIS spectrometer (EDAX, Mahwah, NJ, USA) at a magnification of 500× within an analysis area of approximately 43,500 μm^2. Three measurements were carried out for every sample. Microhardness measurements (Vickers hardness HV0.5) were conducted with a Wilson Tukon 1102 device (Buehler, Uzwil, Switzerland) in metallographic cross-sections. The average microhardness and standard deviation was calculated from ten single indents. Phase analyses was conducted by X-ray diffraction (XRD) with a D8 Discover diffractometer (Bruker AXS, Billerica, MA, USA) using Co Kα radiation (tube parameters: 40 kV; 40 mA). The diffractograms were measured in a diffraction angle range (2θ) of 20° to 130° with a step size of 0.01° and 3.4 s/step, which corresponds to 653 s/step due to the utilisation of a 1D Lynxeye XE detector. For phase identification, the powder diffraction file (PDF) database 2014 (International Centre for Diffraction Data) was used. The solidification behaviour was investigated by differential scanning calorimetry (DSC) with a STA 409 C device (Netzsch, Selb, Germany) under argon atmosphere in the temperature range from 1800 K to room temperature with a cooling rate of 20 K/min.

To investigate the tribological behaviour under adhesive, oscillating, and abrasive wear conditions, ball-on-disk, oscillating wear and scratch tests have been carried out. For the ball-on-disk tests a Tetra Basalt Tester (Tetra, Ilmenau, Germany) were used. The oscillating wear tests were carried out with a Wazau SVT 40 device (Wazau, Berlin, Germany) and a CSM Revetest-RST device (CSM Instruments SA, Peseux, Switzerland) has been used for the scratch tests. The applied parameters are summarised in Table 1.

Table 1. Wear test parameters.

Ball-on-Disk Test		Oscillating Wear Test		Scratch Test	
Force	20 N	Force	26 N	Mode	progressive
Radius	5 mm	Frequency	40 Hz	Force	1–200 N
Speed	96 RPM	Time	900 s	Speed	2.5 mm/min
Cycles	15916	Amplitude	0.5 mm	Length	5 mm
Counter body	Al_2O_3	Counter body	Al_2O_3	Tip	Rockwell C
Diameter	6 mm	Diameter	10 mm	Radius	200 µm

The measurements of the resulting wear depths after the ball-on-disk test were conducted by a contact stylus instrument with a Hommel Etamic T8000 device (Jenoptik, Villingen-Schwenningen, Germany). Resulting wear marks of the oscillating wear and scratch tests were analysed by laser scanning microscopy (LSM) with a Keyence VK-X200 device (Keyence, Osaka, Japan) to determine the resulting wear depth. The reference material bearing steel EN 1.3505 (100Cr6) was investigated under identical conditions. Wear marks of the scratch test were investigated with the digital microscope Keyence VHX-500 (Keyence, Osaka, Japan).

3. Results and Discussion

3.1. Chemical Composition

The mean chemical composition of all samples was measured. Results are summarised in Table 2.

Table 2. Mean chemical composition (in at.%) of $AlCoCrFeNiTi_x$ samples, measured by EDS and nominal values. The measured chemical compositions are in good agreement with the target compositions.

x		Al	Co	Cr	Fe	Ni	Ti
0.0	nominal	20.0	20.0	20.0	20.0	20.0	20.0
	measured	21.3	19.4	19.9	19.9	19.5	0.0
0.2	nominal	19.2	19.2	19.2	19.2	19.2	3.9
	measured	20.3	18.9	19.0	19.2	18.7	3.9
0.5	nominal	18.2	18.2	18.2	18.2	18.2	9.1
	measured	20.0	17.8	17.4	17.6	18.2	9.0
0.8	nominal	17.2	17.2	17.2	17.2	17.2	13.8
	measured	19.0	16.7	16.8	16.7	17.3	13.5
1.0	nominal	16.7	16.7	16.7	16.7	16.7	16.7
	measured	17.9	16.2	16.7	16.6	16.4	16.3
1.5	nominal	15.4	15.4	15.4	15.4	15.4	23.1
	measured	16.7	15.2	15.4	15.3	15.6	22.0

The chemical composition of the arc-melted samples was in good agreement with the nominal values. Only for the aluminium content did a distinct deviation of >1 at.% occur. The titanium content was in good agreement with the nominal values for all samples.

3.2. Microstructure

SEM images of the microstructure using a BSD to visualise material contrast with different magnifications are shown in Figure 1.

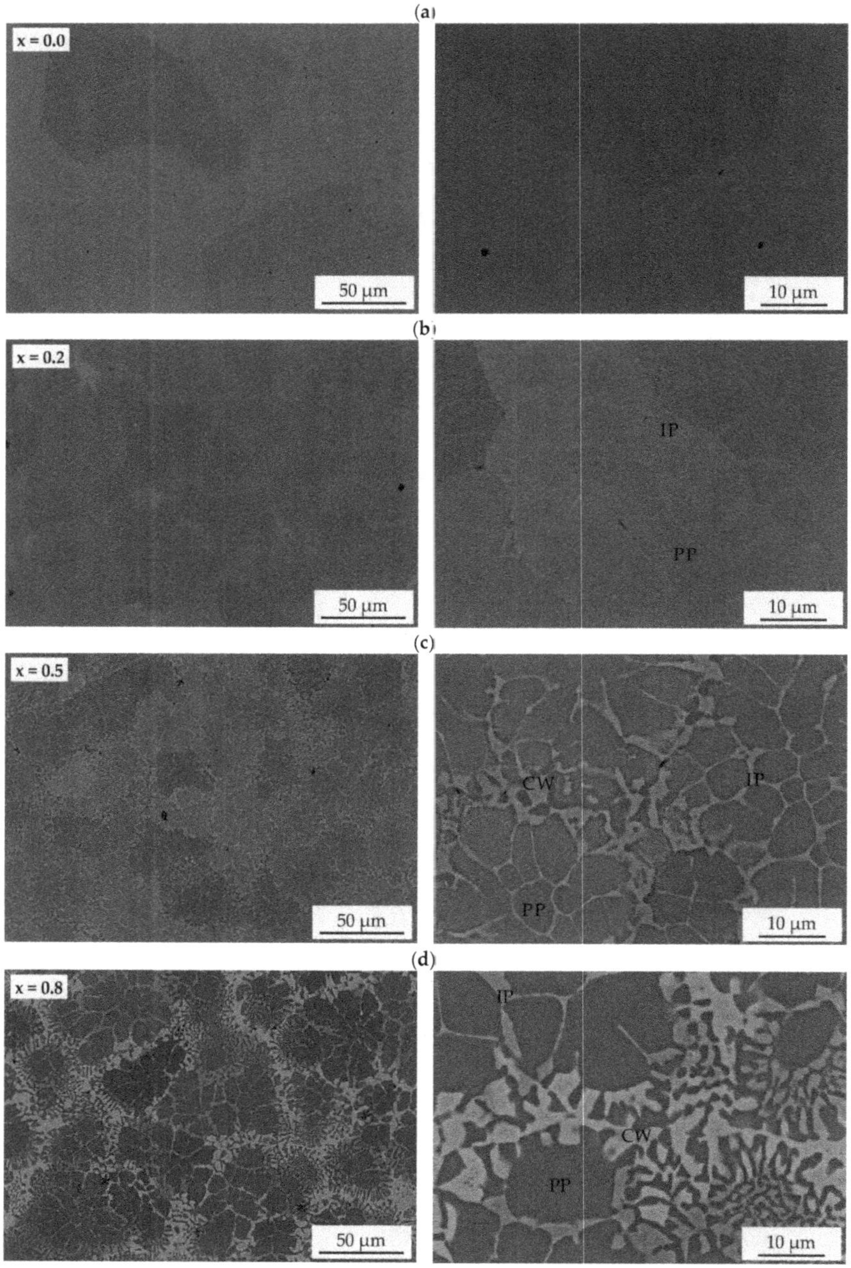

Figure 1. *Cont.*

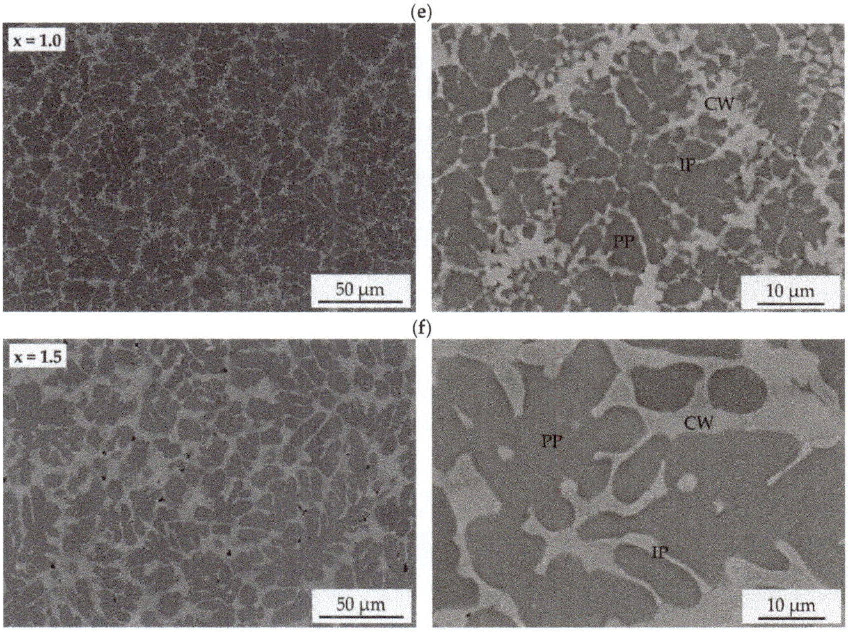

Figure 1. SEM micrographs (BSD detector) of arc-melted $AlCoCrFeNiTi_x$ samples: (**a**) x = 0.0; (**b**) x = 0.2; (**c**) x = 0.5; (**d**) x = 0.8; (**e**) x = 1.0; (**f**) x = 1.5 with phase declaration (PP: primary phase; CW: cell wall; IP: interdendritic phase). The formation of additional phases and an increase of heterogeneity in the microstructure can be observed for an increased titanium content.

For the alloy without titanium (x = 0.0), a homogeneous microstructure occurred. The grains solidified with no preferred direction. Low differences of the chemical composition were confirmed by a minor BSD contrast. Alloying with titanium causes the formation of a more heterogeneous microstructure. Within the grains, material contrast was observed for the alloy with x = 0.2, indicating differences in the chemical composition. Between both of these areas, no distinct boundaries occurred, indicating a directional solidification with minor change in orientation. During the solidification, the precipitation of the primary phase caused a depletion of alloy elements in the residual liquid phase, causing the precipitation of a secondary phase with a different chemical composition. With further increased titanium content (x = 0.5), a dendritic structure appeared. A bright-appearing phase (interdendritic phase) was observed at the grain boundaries of the primary phase, indicating that this area is rich in elements with a high atomic number. A third phase (cell wall) solidifies as a remainder between the dendritic structures. For the alloy with higher titanium content (x = 0.8), the content of this third phase increased. Also, for the samples with further increased titanium content (x = 1.0 and x = 1.5), a dendritic structure comprising three distinguishable phases appeared. The cross-sections of all arc-melted samples exhibited cavities and shrinkage porosity caused by a different contraction of the present phases, and small amounts of breakouts due to the metallographic preparation or sample production (black areas). The presence of these defects was most distinct for the alloy with the highest titanium content (x = 1.5).

3.3. Solidification Behaviour and Phase Analyses

The solidification behaviour was determined by DSC measurements. The resulting cooling curves in a temperature range of 1800 K to 800 K are shown in Figure 2.

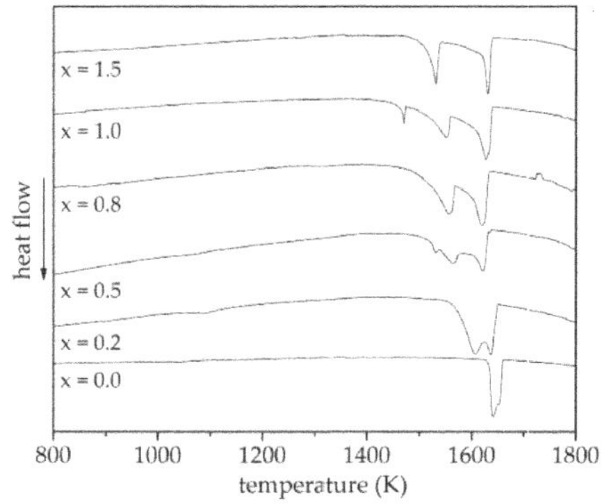

Figure 2. DSC cooling curves of arc-melted AlCoCrFeNiTi$_x$ samples. A single exothermic reaction can be observed for the alloy without titanium, whereas several exothermic reactions occur for all alloys containing titanium, revealing the formation of additional phases.

The DSC cooling curve of the alloy without titanium (x = 0.0) exhibited one peak at a temperature of 1640 K. All investigated alloys showed a major peak at a temperature above 1600 K, corresponding to the major primarily formed phase. For the alloy with (x = 0.2), two distinct exothermic reactions occurred, showing that two major phases were formed. The similar temperature range indicated a continuous solidification. For the alloys with increased titanium content (x = 0.5 and 0.8), the DSC cooling curves also showed two major exothermic reactions. However, the increasing temperature shift of the peaks indicated a changed solidification behaviour and chemical composition of the second major phases, while the primary phase precipitated at the same temperature. For the equimolar alloy (x = 1.0), a further peak appeared at a temperature of 1470 K, showing that an additional phase was formed. The DSC cooling curve of the alloy with the highest titanium content (x = 1.5) only displayed two exothermic reactions, which can be ascribed to the two major phases also detected for the alloys with lower titanium content. No further peak of another phase visible in the SEM images (Figure 1f) appeared, indicating a small phase content or a similar solidification temperature.

For the assignment of phases, the resulting diffractograms of the XRD phase analyses are shown in Figure 3.

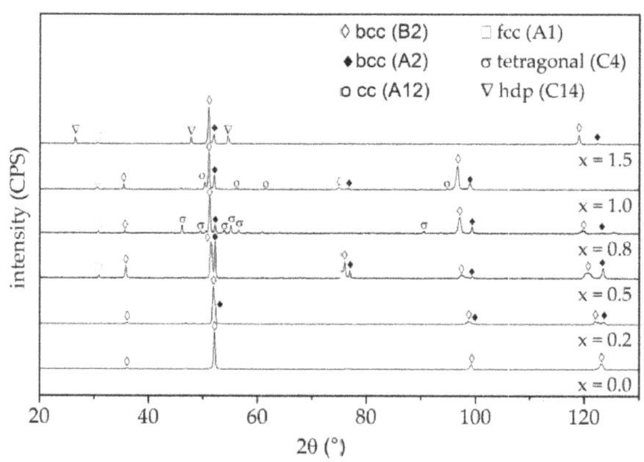

Figure 3. XRD diffractograms of arc-melted AlCoCrFeNiTi$_x$ samples. A single phase is formed for the alloy without titanium, whereas diffraction peaks of additional phases occur for an increased titanium content. Complex phases are formed for x $\geq$ 0.8.

The diffraction diagram of the alloy without titanium (x = 0.0) exhibited high intensity diffraction peaks which could be ascribed to a chemically ordered bcc phase with B2 structure. Diffraction peaks of this phase appeared for all investigated alloys, which was in accordance with DSC results, revealing that this phase is the primary phase (PP). For all titanium containing alloys, the diffractograms showed an additional phase with a bcc structure: a chemically disordered bcc phase with A2 structure. This phase corresponded to the second exothermic reaction in the DSC measurements, and hence to the interdendritic phase (IP) in the alloys $x \geq 0.2$. The stabilisation of bcc phases due to a high aluminium content has been reported elsewhere in detail [19]. For the alloy with the lowest titanium content (x = 0.2), no additional phases were be detected. However, with increased titanium content further diffraction peaks occurred. For all alloys with a titanium content of $x \geq 0.5$, an additional peak at a diffraction angle of 31.0° appeared. This diffraction angle could be ascribed to an fcc phase. Further diffraction peaks of this phase overlapped with the bcc (B2) phase. Previous investigations of the alloy system $Al_xCoCrFeNiTi$ revealed the solidification of the fcc phase as a remainder in the cell walls [19]. The diffraction diagram of the alloy with a titanium content of x = 0.8 exhibited several additional diffraction peaks, which can be assigned to a tetragonal σ phase. This phase has also been detected by Moravcik et al. for the alloy $AlCoCrFeNiTi_{0.5}$ produced by spark plasma sintering (SPS). Microstructural investigations revealed the formation of this phase embedded in a mixture of other phases around a primary phase. However, subsequent heat treatment resulted in the dissolution of the σ phase, showing that the formation of this phase sensitively depends on manufacturing conditions [25]. For the equimolar alloy (x = 1.0), no diffraction peaks of the σ phase appeared. Additional peaks occurred, which can be ascribed to a centred cluster (cc) with A12 structure type. DSC results also revealed the formation of an additional phase, which solidifies as a remainder after the two major bcc phases and forms cell walls in the microstructure. This phase has been detected in the same alloy system in preliminary studies by Lindner et al. and for a high aluminium content ($AlCoCrFeNiTi_{1.5}$), the formation of this phase could be suppressed [19]. The diffractogram of the alloy with the highest titanium content (x = 1.5) exhibited diffraction peaks which can be ascribed to the bcc (A2), bcc (B2) and an fcc phase. However, the diffraction peaks at 2θ = 35.4°; 96.7° and 98.9° of these phases did not appear. This might be caused by the relatively coarse microstructure and texture of the samples due to the comparatively slow cooling conditions. The diffraction diagram did not exhibit peaks of the σ or cc (A12) phase. Additional diffraction peaks can be ascribed to a hexagonal Laves phase (C14/$MgZn_2$ type). A similar phase with slightly changed lattice parameters has been detected by Zhou et al. [20]. The present phases and corresponding major crystallographic information are summarised in Table 3.

Table 3. Summary of phases detected by XRD analyses for arc-melted $AlCoCrFeNiTi_x$ samples.

x	Phase	Struktur-Bericht	Lattice	Structure	Pearson Symbol	Space Group	Lattice Parameter (Å)
0.0	PP	B2	bcc	CsCl	cP2	$Pm\bar{3}m$ (221)	2.88
0.2	PP	B2	bcc	CsCl	cP2	$Pm\bar{3}m$ (221)	2.89
	IP	A2	bcc	W	cI2	$Im\bar{3}m$ (229)	2.87
0.5	PP	B2	bcc	CsCl	cP2	$Pm\bar{3}m$ (221)	2.92
	IP	A2	bcc	W	cI2	$Im\bar{3}m$ (229)	2.87
	CW	A1	fcc	Cu	cF4	$Fm\bar{3}m$ (225)	5.82
0.8	PP	B2	bcc	CsCl	cP2	$Pm\bar{3}m$ (221)	2.92
	IP	A2	bcc	W	cI2	$Im\bar{3}m$ (229)	2.87
	CW	A1	fcc	Cu	cF4	$Fm\bar{3}m$ (225)	5.85
	CW	C4	tetragonal	TiO_2	tP6	$P4_2/mnm$ (136)	a = 8.80 c = 4.56
1.0	PP	B2	bcc	CsCl	cP2	$Pm\bar{3}m$ (221)	2.93
	IP	A2	bcc	W	cI2	$Im\bar{3}m$ (229)	2.88
	CW	A1	fcc	Cu	cF4	$Fm\bar{3}m$ (225)	5.87
	CW	A12	cc	α-Mn	cI58	$I\bar{4}3m$ (217)	8.92
1.5	PP	B2	bcc	CsCl	cP2	$Pm\bar{3}m$ (221)	2.94
	IP	A2	bcc	W	cI2	$Im\bar{3}m$ (229)	2.88
	CW	A1	fcc	Cu	cF4	$Fm\bar{3}m$ (225)	5.87
	CW	C14	hexagonal	$MgZn_2$	hP12	$P6_3/mmc$ (194)	a = 4.80 c = 7.81

In addition to the formation of further phases, a shift of lattice parameters was observed with increasing titanium content. The lattice parameters of the bcc (B2) phase and the fcc (A1) phase were slightly increased. This behaviour indicated that more titanium with a large atomic radius was resolved in these phases.

3.4. Hardness and Wear Behaviour

The influence of the titanium content on the average microhardness HV0.5 was investigated. The results are summarised in Figure 4.

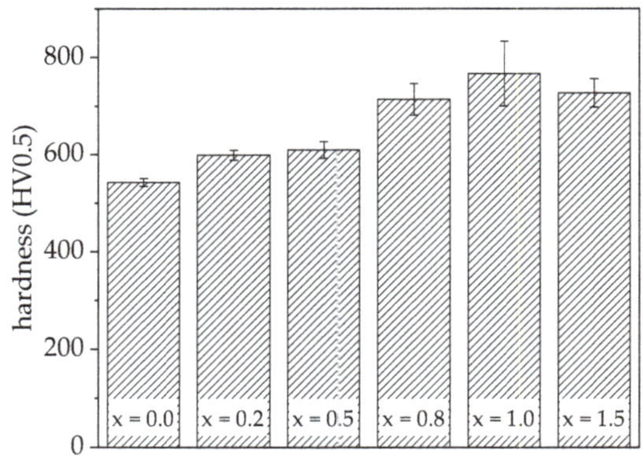

Figure 4. Microhardness of arc-melted $AlCoCrFeNiTi_x$ samples. Microhardness increases with titanium content, reaching a maximum of 770 HV0.5 for the equimolar alloy.

For the alloy without titanium (x = 0.0), an average microhardness of 550 HV0.5 was measured. With increasing titanium content, an increase of microhardness was observed, reaching a maximum for the equimolar composition with a microhardness of 770 HV0.5. The increasing hardness indicated the formation of additional phases, which was in accordance with microstructural investigations and phase analyses. Furthermore, solid solution strengthening contributes to an increase of hardness—an increase of lattice parameters could be proven for the bcc (B2) and fcc (A1) phases. However, for the alloy with the highest titanium content (x = 1.5), a reduced microhardness of 730 HV0.5 was measured, which might be a result of the increased presence of cavities and shrinkage porosity in that alloy.

The wear behaviour was investigated under adhesive, oscillating, and abrasive wear conditions in ball-on-disk, oscillating wear, and scratch tests. The results are summarised in Figure 5.

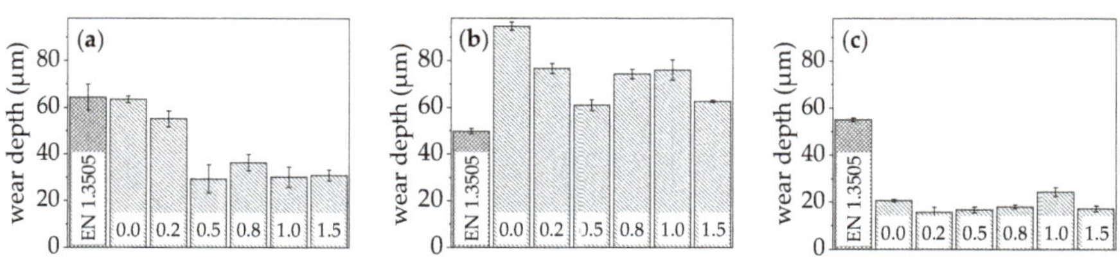

Figure 5. Wear depths of bearing steel EN 1.3505 and $AlCoCrFeNiTi_x$ in: (**a**) Ball-on-disk; (**b**) Oscillating wear and (**c**) Scratch tests. The wear depths of $AlCoCrFeNiTi_x$ are decreased in comparison to EN 1.3505, except under oscillating wear conditions. Overall, the best results are obtained for x = 0.5.

The investigation of the wear behaviour in the ball-on-disk test revealed a high wear depth for the alloy without titanium (x = 0.0). With the addition of titanium (x = 0.2), a slight decrease of wear depth could be achieved. Further increase to x = 0.5 resulted in a distinct decrease of wear depth. The alloy x = 0.5 exhibited a multiphase character, only comprising cubic phases. In comparison to the alloys with lower titanium content, the microhardness was increased, which enhanced wear resistance in ball-on-disk tests. However, further increase of titanium content did not cause a reduction of wear depth or an improvement of the wear resistance. Phase analyses revealed the formation of additional complex phases reducing wear resistance. All samples containing titanium exhibited a lower wear depth in comparison with the bearing steel EN 1.3505, and hence a higher wear resistance in the ball-on-disk test.

In the oscillating wear tests, the highest wear depth was measured for the alloy without titanium (x = 0.0). Adding minor amounts of titanium resulted in a decrease of the wear depth. For the sample x = 0.5, the lowest wear depth was measured, showing that a multiphase character only comprising cubic phases was advantageous under oscillating wear condition. The further addition of titanium led to an increase of the wear depth. Additional tetragonal or cc phases did not contribute to an improvement of wear resistance. For the alloy with the highest titanium content (x = 1.5), a low wear depth was measured, which was in the range of the alloy with x = 0.5. In comparison with bearing steel EN 1.3505, all investigated samples exhibited a higher wear depth, and thereby lower wear resistance under oscillating wear conditions.

Under abrasive tribological conditions in the scratch tests, the lowest wear depth was measured for the sample with a titanium content of x = 0.2. With an increase in titanium content, the wear depth slightly increased. The highest wear depth was measured for the equimolar alloy (x = 1.0). In comparison to the bearing steel EN 1.3505, all investigated samples exhibit a distinctly lower wear depth and thereby higher wear resistance. The wear tracks of all samples were investigated by optical microscopy. In Figure 6 images of the sample surface where the highest load was applied in the progressive mode scratch test are shown.

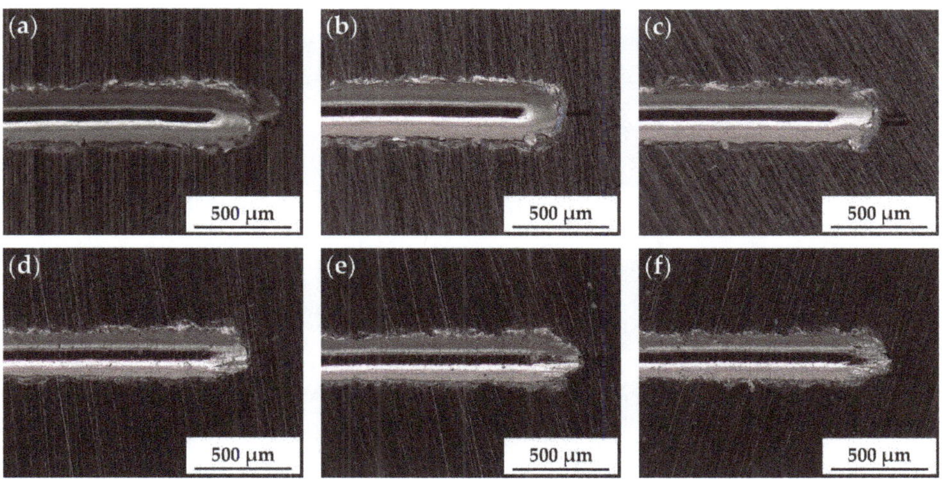

Figure 6. Surface of the arc-melted AlCoCrFeNiTix samples: (**a**) x = 0.0; (**b**) x = 0.2; (**c**) x = 0.5; (**d**) x = 0.8; (**e**) x = 1.0; (**f**) x = 1.5 after progressive mode scratch test. Ductile behaviour occurs for $x \leq 0.5$, whereas cracks or spalling of material are visible for $x \geq 0.8$.

The investigation of the surface after scratch test under abrasive conditions reveals no cracks or spalling of material along the main scratch for the sample without titanium (x = 0.0) and the two samples with the lowest titanium content (x = 0.2; x = 0.5). However, spalling of material and distinct secondary cracks perpendicular to the main scratch can be observed for all samples with a titanium content exceeding x = 0.5. The formation of secondary cracks and spalling of material indicated the brittle behaviour of these alloys, which is caused by the formation of additional, complex phases—tetragonal, cc and a hexagonal Laves phase for a titanium content $x \geq 0.8$. These phases possess low numbers of slip systems, which results in reduced ductility.

4. Summary and Conclusions

The influence of the alloying element titanium was studied in detail in the alloy system $AlCoCrFeNiTi_x$. For the alloy without titanium (x = 0.0), a single chemically ordered bcc phase with B2 structure is formed. In contrast, a multiphase microstructure was revealed for all titanium-containing alloys. With an increase of titanium content, an increase of heterogeneity of the microstructure is observed. Furthermore, the hardness can be distinctly increased, whereas the maximum hardness is achieved for the equimolar composition. Phase analyses prove the formation of two major bcc phases for all titanium-containing samples. One chemically ordered bcc phase with B2 structure and one chemically disordered bcc phase with A2 structure is formed. With an increase in titanium content, additional phases occur. For $x \geq 0.5$ a minor fraction of an fcc phase was detected. A further increase in titanium content results in additional, more complex phases. This could also be proved by analysing the solidification behaviour. Analyses of the lattice parameters revealed a shift to bigger values with increasing titanium content, especially for the bcc (B2) and the fcc (A1) phase. The alloy system $AlCoCrFeNiTi_x$ exhibits an increased wear resistance in comparison with the bearing steel EN 1.3505, except under oscillating wear conditions.

Correlations between phase composition, microstructure, and wear resistance can be concluded. Microstructure design for high wear resistance requires cubic phases. Hereby, a multiple bcc/fcc phase character exhibits an advantageous behaviour. Complex phases (cc and tetragonal) increase the hardness, but should be avoided in order to achieve a high wear resistance, as the presence of these phases causes increased brittleness. $AlCoCrFeNiTi_{0.5}$ is a promising candidate for wear protection applications in both bulk and coating materials.

Author Contributions: M.L. and T.L. (Thomas Lindner) conceived and designed the experiments. M.L., T.L. (Thomas Lindner) and T.M. performed the experiments, analysed the data, and wrote the paper. T.L. (Thomas Lampke) directed the research, and contributed to the discussion and interpretation of the results.

Acknowledgments: The publication costs of this article were funded by the German Research Foundation/DFG and the Chemnitz University of Technology in the funding programme Open Access Publishing. The authors thank Benjamin Sattler for sample manufacturing, Marc Pügner for conducting the XRD measurements and Thomas Uhlig for conducting the DSC measurements.

Conflicts of Interest: The authors declare no conflict of interest.

References

1. Yeh, J.-W.; Lin, S.-J.; Chin, T.-S.; Gan, J.-Y.; Chen, S.-K.; Shun, T.-T.; Tsau, C.-H.; Chou, S.-Y. Formation of simple crystal structures in Cu-Co-Ni-Cr-Al-Fe-Ti-V alloys with multiprincipal metallic elements. *Metall. Mater. Trans. A* **2004**, *35*, 2533–2536. [CrossRef]
2. Cantor, B.; Chang, I.; Knight, P.; Vincent, A. Microstructural development in equiatomic multicomponent alloys. *Mater. Sci. Eng. A* **2004**, *375–377*, 213–218. [CrossRef]
3. Murty, B.S.; Yeg, J.-W.; Ranganathan, S. *High-Entropy Alloys*; Elsevier: Amsterdam, The Netherlands, 2014.
4. Gao, M.C.; Yeh, J.-W.; Liaw, P.K.; Zhang, Y. *High-Entropy Alloys: Fundamentals and Applications*, 1st ed.; Springer International Publishing: Cham, Switzerland, 2016.
5. Miracle, D.B.; Senkov, O.N. A critical review of high entropy alloys and related concepts. *Acta Mater.* **2017**, *122*, 448–511. [CrossRef]
6. Zhang, W.; Liaw, P.K.; Zhang, Y. Science and technology in high-entropy alloys. *Sci. China Mater.* **2018**, *61*, 2–22. [CrossRef]
7. Miracle, D.B.; Miller, J.D.; Senkov, O.N.; Woodward, C.; Uchic, M.D.; Tiley, J. Exploration and Development of High Entropy Alloys for Structural Applications. *Entropy* **2014**, *16*, 494–525. [CrossRef]
8. Tsai, M.-H.; Yeh, J.-W. High-Entropy Alloys: A Critical Review. *Mater. Res. Lett.* **2014**, *2*, 107–123. [CrossRef]

9. Singh, S.; Wanderka, N.; Murty, B.S.; Glatzel, U.; Banhart, J. Decomposition in multi-component AlCoCrCuFeNi high-entropy alloy. *Acta Mater.* **2011**, *59*, 182–190. [CrossRef]
10. Soare, V.; Mitrica, D.; Constantin, I.; Badilita, V.; Stoiciu, F.; Popescu, A.-M.; Carcea, I. Influence of remelting on microstructure, hardness and corrosion behaviour of AlCoCrFeNiTi high entropy alloy. *Mater. Sci. Technol.* **2015**, *10*, 1194–2200. [CrossRef]
11. Löbel, M.; Lindner, T.; Kohrt, C.; Lampke, T. Processing of AlCoCrFeNiTi high entropy alloy by atmospheric plasma spraying. *IOP Conf. Ser. Mater. Sci. Eng.* **2017**, *181*, 12015. [CrossRef]
12. Wang, R.; Zhang, K.; Davies, C.; Wu, X. Evolution of microstructure, mechanical and corrosion properties of AlCoCrFeNi high-entropy alloy prepared by direct laser fabrication. *J. Alloys Compd.* **2017**, *694*, 971–981. [CrossRef]
13. Wang, Y.P.; Li, B.S.; Ren, M.X.; Yang, C.; Fu, H.Z. Microstructure and compressive properties of AlCrFeCoNi high entropy alloy. *Mater. Sci. Eng. A* **2008**, *491*, 154–158. [CrossRef]
14. Kumar, A.; Gupta, M. An Insight into Evolution of Light Weight High Entropy Alloys: A Review. *Metals* **2016**, *6*, 199. [CrossRef]
15. Cao, T.; Shang, J.; Zhao, J.; Cheng, C.; Wang, R.; Wang, H. The influence of Al elements on the structure and the creep behavior of Al_xCoCrFeNi high entropy alloys. *Mater. Let.* **2016**, *164*, 344–347. [CrossRef]
16. Li, C.; Li, J.C.; Zhao, M.; Jiang, Q. Effect of aluminium contents on microstructure and properties of Al_xCoCrFeNi alloys: 16th International Symposium on Metastable, Amorphous and Nanostructured Materials. *J. Alloys Compd.* **2010**, *504*, 515–518. [CrossRef]
17. Chou, H.-P.; Chang, Y.-S.; Chen, S.-K.; Yeh, J.-W. Microstructure, thermophysical and electrical properties in Al_xCoCrFeNi ($0 \leq x \leq 2$) high-entropy alloys. *Mater. Sci. Eng. B Adv.* **2009**, *163*, 184–189. [CrossRef]
18. Tung, C.-C.; Yeh, J.-W.; Shun, T.-T.; Chen, S.-K.; Huang, Y.-S.; Chen, H.-C. On the elemental effect of AlCoCrCuFeNi high-entropy alloy system. *Mater. Lett.* **2007**, *61*, 1–5. [CrossRef]
19. Lindner, T.; Löbel, M.; Mehner, T.; Dietrich, D.; Lampke, T. The Phase Composition and Microstructure of Al_xCoCrFeNiTi Alloys for the Development of High-Entropy Alloy Systems. *Metals* **2017**, *7*, 162. [CrossRef]
20. Zhou, Y.J.; Zhang, Y.; Wang, Y.L.; Chen, G.L. Solid solution alloys of AlCoCrFeNiTi_x with excellent room-temperature mechanical properties. *Appl. Phys. Lett.* **2007**, *90*, 181904. [CrossRef]
21. Chen, M.R.; Lin, S.J.; Yeh, J.W.; Chen, S.K.; Huang, Y.S.; Tu, C.P. Microstructure and Properties of $Al_{0.5}$CoCrCuFeNiTi_x (x = 0–2.0) High-Entropy Alloys. *Mater. Trans.* **2006**, *47*, 1395–1401. [CrossRef]
22. Chuang, M.-H.; Tsai, M.-H.; Wang, W.-R.; Lin, S.-J.; Yeh, J.-W. Microstructure and wear behavior of $Al_xCo_{1.5}CrFeNi_{1.5}Ti_y$ high-entropy alloys. *Acta Mater.* **2011**, *59*, 6308–6317. [CrossRef]
23. Yu, Y.; Wang, J.; Li, J.; Kou, H.; Liu, W. Characterization of BCC phases in AlCoCrFeNiTi_x high entropy alloys. *Mater. Lett.* **2015**, *138*, 78–80. [CrossRef]
24. Wang, Y.; Ma, S.; Chen, X.; Shi, J.; Zhang, Y.; Qiao, J. Optimizing Mechanical Properties of AlCoCrFeNiTi_x High-Entropy Alloys by Tailoring Microstructures. *Acta Metall. Sin.* **2013**, *26*, 277–284. [CrossRef]
25. Moravcik, I.; Cizek, J.; Gavendova, P.; Sheikh, S.; Guo, S.; Dlouhy, I. Effect of heat treatment on microstructure and mechanical properties of spark plasma sintered AlCoCrFeNi$Ti_{0.5}$ high entropy alloy. *Mater. Lett.* **2016**, *174*, 53–56. [CrossRef]

entropy

MDPI

Article

Prediction of Strength and Ductility in Partially Recrystallized $CoCrFeNiTi_{0.2}$ High-Entropy Alloy

Hanwen Zhang [1], Peizhi Liu [2], Jinxiong Hou [1], Junwei Qiao [1,2,*] and Yucheng Wu [2,*]

[1] College of Materials Science and Engineering, Taiyuan University of Technology, Taiyuan 030024, China; zhanghanwen1234@hotmail.com (H.Z.); houginxiong@hotmail.com (J.H.)

[2] Key Laboratory of Interface Science and Engineering in Advanced Materials, Ministry of Education, Taiyuan University of Technology, Taiyuan 030024, China; liupeizhi@tyut.edu.cn

* Correspondence: qiaojunwei@gmail.com (J.Q.); wyc@tyut.edu.cn (Y.W.)

Received: 24 February 2019; Accepted: 28 March 2019; Published: 11 April 2019

Abstract: The mechanical behavior of a partially recrystallized fcc-$CoCrFeNiTi_{0.2}$ high entropy alloys (HEA) is investigated. Temporal evolutions of the morphology, size, and volume fraction of the nanoscaled $L1_2$-$(Ni,Co)_3Ti$ precipitates at 800 °C with various aging time were quantitatively evaluated. The ultimate tensile strength can be greatly improved to ~1200 MPa, accompanied with a tensile elongation of ~20% after precipitation. The temporal exponents for the average size and number density of precipitates reasonably conform the predictions by the PV model. A composite model was proposed to describe the plastic strain of the current HEA. As a consequence, the tensile strength and tensile elongation are well predicted, which is in accord with the experimental results. The present experiment provides a theoretical reference for the strengthening of partially recrystallized single-phase HEAs in the future.

Keywords: high entropy alloys; precipitation kinetics; strengthening mechanisms; elongation prediction

1. Introduction

High entropy alloys (HEAs), a new class of structural materials, have attracted a great deal of attention in recent years on account of their special intrinsic characteristics [1–9], such as high configuration entropy [10], sluggish atomic diffusion [11], and large lattice distortion [12]. Nevertheless, recent studies indicate that single-phase HEAs, especially single-phase fcc HEAs, the strength is insufficiently for structural applications [13,14]. In other words, strengthening is badly needed so that satisfactory mechanical properties can be achieved. Klimova et al. [15] reported that in the Al- and C-containing CoCrFeNiMn-type high-entropy alloy, substructure strengthening was found to be dominated at low rolling reductions (<40%), while grain (twin) boundary strengthening prevailed at higher strains. Chemical short-range order also has an important influence on the mechanical properties of $FeCoNi(AlSi)_x$ high entropy alloys [16]. Among the numerous strengthening mechanisms, Precipitation hardening is an effective technique widely used for strengthening high entropy alloys [9,17–22]. For example, He et al. [17,23] reported nano-sized $L1_2$ coherent precipitates in a face centered-cubic (fcc) NiCoFe alloy with minor additions of Al and Ti, specifically $(NiCoFeCr)_{94}Ti_2Al_4$, which exhibits a strength increment of about 327 MPa. It is widely studied that these HEAs are all fully recrystallized, and the grain/grain boundary is thermally stable. In contrast, the partial recrystallized HEAs are rarely studied, the microstructure-property relationship and consequent strengthening mechanism are lacking. In this study, we pay attention to the effect of the partially crystallized microstructure on the mechanical properties. The grain growth mechanisms for the nanoscale precipitation and strengthening mechanisms are fully investigated. In addition, a quantitative model for estimation of tensile ductility is established.

2. Materials and Methods

The mixture of Co, Cr, Fe, Ni, and Ti with purity of at least 99.9% (weight percent, wt.%) was prepared by arc-melting. Under a Ti-gettered argon atmosphere, the sample was cast into an 85 × 10 × 2 mm^3 copper mould and $CoCrFeNiTi_{0.2}$ (atomic percent, at.%) alloy ingots were prepared. A stable uniform structure was obtained at 1200 °C homogenized for 5 h. The sliced samples were cold-rolled to 75% of the total reduction ratio, and then aged at 800 °C for 3 h, 5 h, 8 h, 10 h, 24 h and 48 h respectively, followed by water quenching. The phase identification was carried out by X-ray diffraction (XRD) using Cu Kα radiation. Then, optical microscopy (OM), scanning electron microscopy (ZEISS SUPRA55 SEM) operated at 20 kV, with the working distance of 9.1 mm, energy dispersive spectrometer (EDS), and JEM-2010 transmission electron microscopy (TEM) were used to observe the surface microstructures. Dog-bone-like tensile specimens with a gauge length of 10 mm, a gauge width of 2 mm and a thickness of 0.5 mm were prepared from aged specimens by electrical discharge machining. Instron 5969 universal testing machine was used to carry out quasi-static tensile tests at room temperature at a constant strain rate of 1×10^{-3} s^{-1} greater than or equal to five times.

3. Results and Discussion

3.1. Precipitation Kinetics

As can be seen from Figure 1a, the XRD diagrams of homogenized and aged alloys are displayed. In each alloy, a series of fcc diffraction peaks can be found, which indicates that the matrix of the four alloys is composed of the same fcc phase. In aged samples, an extra series of minor peaks, named $L1_2$-Ni_3Ti, can be detected, which indicates that there is a great probability of the existence of a secondary phase. Figure 2a–e,g–h shows the corresponding SEM micrographs of these HEAs in different statuses. It can be concluded that in Figure 2a, the homogenized alloy demonstrates a single-phase structure, in accordance with the XRD pattern. There are only a few etch pits on the surface probably introduced during electropolishing. After the homogenization process been carried out, the grains are all transformed into equiaxed grains with an average grain size of ~198 μm. Figure 2b exhibits the microstructural feature of the cold-rolled $CoCrFeNiTi_{0.2}$ HEAs. Because of the large CR reduction, there are serious deformation and elongation of grains in the rolling direction, and a high-density of lamellar deformation bands can be dramatically observed after electro-polishing. Similar microstructures have been reported in $Fe_{34.95}Ni_{27.5}Co_{17.5}Al_{11.5}Cr_{8.5}B_{0.05}$ HEAs [24]. Since the recrystallization temperature of the present HEA is above 800 °C, a large number of slip lines (deformation bands) remain in the matrix. As a consequence, in the recrystallization region, the precipitates spread along the grain boundaries, as shown in Figure 2c. Besides, some plate-shaped precipitates are formed. Generally, upon aging, precipitates often preferentially form at grain boundaries, which is the microstructural heterogeneous site. Moreover, when the aging temperature is low, for instance 800 °C here, the diffusion along the grain boundary is much faster than that in the lattice [23]. This is the main reason why the plate-shaped precipitates grow at grain boundaries. Whereas in the area where the grains are not recrystallized, a great many precipitates are uniformly distributed throughout the matrix, as evidenced in Figure 2d. The measured average compositions of the precipitates and fcc matrix in the 800 °C/5 h aged alloy, as well as the composition variation across a single precipitate is, as shown in Figure 1b. It indicates that the precipitates are enriched with Ni and Ti, but depleted in Co, Cr, and Fe. Meanwhile a part of Ni is substituted by Co. With the above EDS analysis, it is finally identified that the precipitates in the 800 °C/5 h aged alloy is the $(Ni,Co)_3Ti$ phase. Figure 2i shows the bright field TEM images of aged alloys at 800 °C/5 h. Spheroidal and plate-shaped particles with an average size of 145 nm are found to disperse in a matrix. From the illustration on the right side of Figure 2i, the selected area electron diffraction (SAED) pattern along Z = [011] taken from the precipitates are indexed. The main diffraction points can effectively prove that the matrix does has an fcc structure, whilst additional weak spots observed in the image can also scientifically prove the existence of precipitates with superlattice $L1_2$ structure. Han et al. have identified that $L1_2$ phase as $(Ni,Co)_3Ti$ type γ' phase [25].

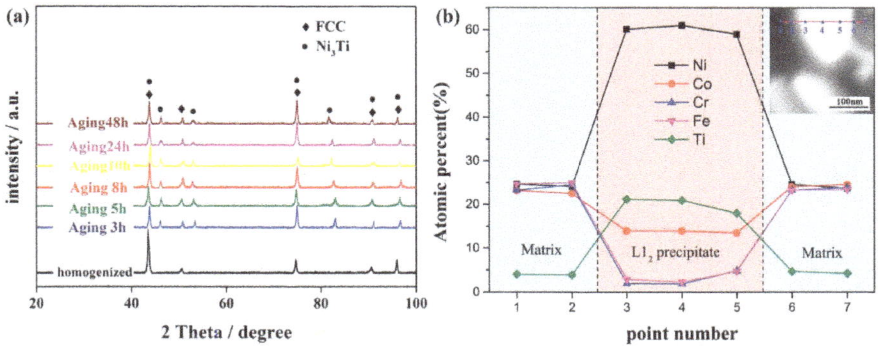

Figure 1. The XRD patterns of the homogenized and the aged alloys (**a**), and EDS-measured composition variation across a single precipitate in the 800 °C/5 h aged alloy(**b**).

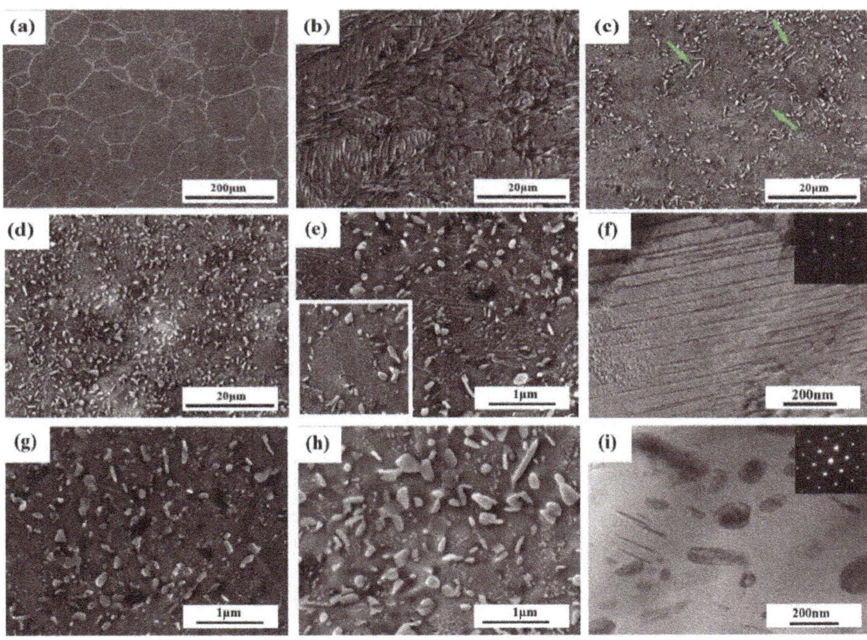

Figure 2. SEM images of the homogenized (**a**), the cold rolled (**b**), the 800 °C/3 h aged (**c**–**d**) alloys and the aged alloy of different aging time: 5 h (**e**), 10 h (**g**), and 48 h (**h**). The left insert in (**e**) is the corresponding enlarged views of the slip lines in 800 °C/5 h aged alloy. (**f**) and (**i**) show the TEM image of the $L1_2$ precipitates and the slip lines in 800 °C/5 h aged alloy, respectively.

As mentioned above, many slip lines remain in the matrix on account of partial recrystallization, which is revealed in Figure 2e. In order to further examine the nature of these slip lines, an enlarged area is selected as exhibited in the left inset of Figure 2e. A significant amount of slip lines along different directions are clearly observed inside the grains. Although the directions of these slip lines are inconsistent, the value of the intersection angle between them is virtually constant as to be about 60°. Figure 2f is the corresponding TEM image of these slip lines in 800 °C/5 h aged alloy. The slip lines are distributed along different directions, and the intersection angle between one another is about 60° as well, which are consistent with SEM images.

To unveil the evolution of precipitation with the time, SEM images for 800 °C/5, 10, and 48 h aged alloys are carefully taken. The precipitates in Figure 2e are basically spheroidal and plate-shaped, uniformly distributed throughout the fcc matrix, the volume fraction is about 15.6 %. As time goes on, the shape of the precipitates turns to be droplet-like and plate-shaped. The average size of precipitates is closely related to the aging time. The longer the aging time is, the larger the average size of precipitates will be. For example, after three hours of aging, the average size of particles is 120 nm, but the average value rises to 385 nm after aging for 48h. The size distributions of the two aged alloys are plotted in Figure 3. Moreover, the precipitate number density n_v is negatively correlated with the aging time. A similar variation of the precipitate with aging time has been reported in many HEAs, such as $(NiCoFeCr)_{94}Ti_2Al_4$ [26] and $(FeCoNiCr)_{100\text{-}x\text{-}y}Ti_xAl_y$ [23] HEAs.

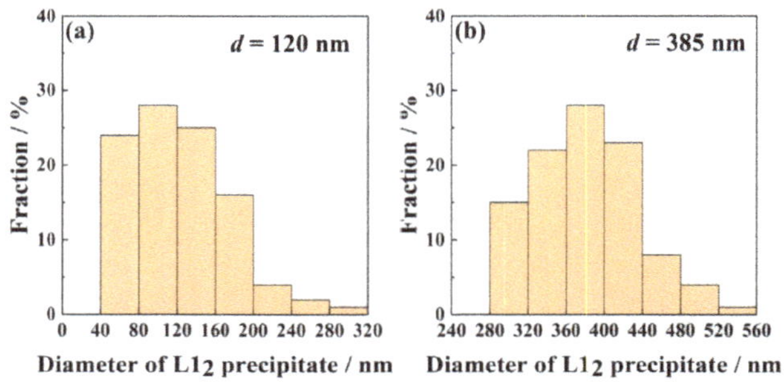

Figure 3. The size distributions of the 800 °C/3 h aged alloy(**a**) and 800 °C/48 h aged alloy(**b**).

The precipitate number density n_v, was determined from $n_v = n_a/d$, where n_a is the areal density of precipitates measured from SEM micrographs. The precipitate size was defined using an area-equivalent diameter (i.e., diameter = $2\sqrt{(\text{area}\pi)}$) measured from SEM micrographs. The average precipitate size d, was calculated according to precipitate size distributions. Dividing the areal density by the precipitate diameter to get the number density in three-dimensional space. It should be noted that many slip lines are clearly revealed in the matrix in 800 °C/3 h aged alloy, as shown in Figure 2e. With the aging time, the amount of slip lines is gradually decreased. When the aging time is reached to 48 h, the slip lines do not basically exist in the matrix. This indicates that although the alloy does not fully recrystallize at 800 °C, the recrystallization has been going on with the extension of time.

It is very important to study the precipitates growth kinetics in order to further reveal the microstructure-property relationship. The volume fraction ($\varphi(t)$), number density ($n_v(t)$), and average size ($d(t)$) are all varies with time, and the relationship is shown in Figure 4a,b, which is corresponding to $L1_2$ precipitates in the $CoCrFeNiTi_{0.2}$ HEAs aged at different times in 800 °C. It can be seen from Figure 4a that the value of $\varphi(t)$ remains constant when the aging time changed from 3 to 48 h. It is pointed out that the nucleation stage of the precipitation process has been bypassed and entered the coarsening stage directly, therefore $\varphi(t)$ remain stable on its equilibrium value (φ_{eq}) [26]. The stored energy is negatively correlated with the aging time, which is consistent with the theory. With the increase of aging time, the stress concentration is caused by the growth of precipitates, leading to the fracture of the alloy more easily, which is corresponding to the reduction of stored energy. The precipitation number density $n_v(t)$ is a function of aging time, as demonstrated in Figure 4b, which indicates the power-law relationship between the two variables at a given temperature. When the aging time is increased from 3 to 48 h, the number density is accordingly decreased from $(5.6 \pm 0.3) \times 10^{20}$ to $(6.4 \pm 0.2) \times 10^{17}$ m^{-3}. Definitely, the power-law exponent for $n_v(t)$, namely the slope of the linear fitted curve is -0.78. Similar temporal exponents also appear in some Ni-Al-based ternary alloys [27–30].

The size of precipitates also depends on the aging time. It can be concluded from Figure 4b that the average precipitation size *d(t)* also follows the power law relationship. The power law index is 0.41.

A model can be used to explain the power-law relationship among the number density, average precipitation size and aging time scientifically and effectively. Philip and Voorhees established the PV model for Ostwald ripening in multi-component systems [31]. According to the model, the coarsening process of precipitates is in accordance with a similar power-law relationship. In other words, the size of precipitates raised to a 1/3 power, due to the increasing aging time. To apply the PV model reasonably to the $CoCrFeNiTi_{0.2}$ HEA, the first step is to concentrate the chemical composition. The baseline of the alloy is fcc-CoCrFeNi. In addition, Ni is the only element with an fcc structure in the alloy, so it can be treated substantially as a Ni-based alloy. Besides that, the lattice constant of CoCrFeNi (0.3572 nm) is highly close to that of Ni (0.3517 nm) [30]. Hence, it seems logical to treat the $CoCrFeNiTi_{0.2}$ as a Ni-based pseudo binary Ni-Ti alloy. Therefore, the PV model of stable coarsening reaction can be used to calculate the coarsening behavior of $L1_2$ precipitation in $CoCrFeNiTi_{0.2}$ HEAs [32–35]:

$$d^3(t) - d^3(t_0) = K(t - t_0), \tag{1}$$

$$n_v(t)^{-1} - n_v(t_0)^{-1} = 4.74\frac{K}{\varphi_{eq}}(t - t_0), \tag{2}$$

where K represents the coarsening rate constant, *d(t)*, and $n_v(t)$ refer to the average size and number density of precipitates at time *t*. Obviously, in Figure 3a,b, the time index for $n_v(t)$ is -0.78, and that for *d(t)* is 0.41, which is basically consistent with the predicted values of -1 (Equation (3)) and $1/3$ (Equation(2)) corresponding to the PV model. The main reason for the slight deviation is probably caused by the fact that the aging time is within a limited duration, but in principle, the coarsening stage of precipitates cannot reach a stable state [36–39].

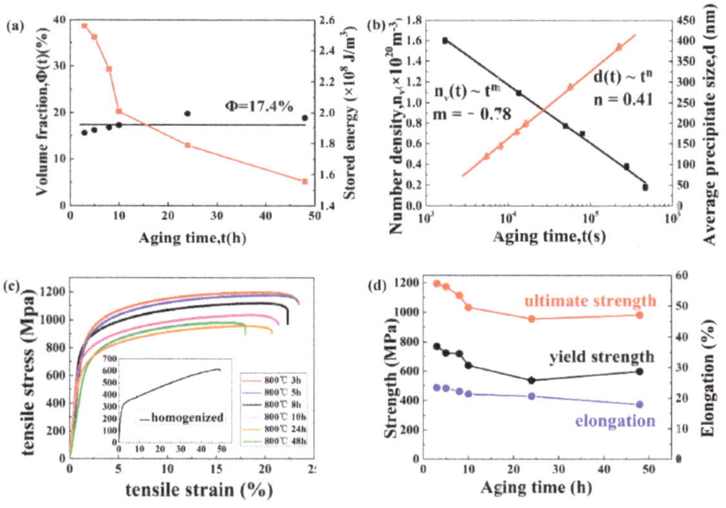

Figure 4. Temporal evolution of (**a**) volume fraction ($\Phi(t)$) and stored energy ($E(t)$), (**b**) number density ($n_v(t)$), and average size (d(t)) of $L1_2$ precipitates in the $CoCrFeNiTi_{0.2}$ high entropy alloys (HEA). Tensile stress-strain curves of the homogenized and the annealed alloys(**c**). Detailed tensile strength and the elongation value are plotted in (**d**).

3.2. Strengthening Mechanisms

In order to study the effect of microstructure on mechanical properties, uniaxial tensile tests were carried out. As can be seen from Figure 4c, the tensile engineering stress-strain curves of the current seven HEAs under room temperature are measured. The homogenized alloy exhibits high ductility, which reaches 48%, but the yield and ultimate strengths of the alloy are low, which are only 315 and 609 MPa, respectively. A similar value of strength is obtained in CoCrFeMnNi HEAs [1]. Compared to the homogenized alloy, the aged alloys (3 h) exhibit a striking improvement in the mechanical performances, the yield and ultimate strength can reach about ~800 and ~1200 MPa respectively, accompanied by ~20% homogeneous elongation. Detailed tensile strength and the elongation of the aged alloys are plotted in Figure 4d. It is noteworthy that both the tensile strength and elongation are gradually decreased with the aging time. The results can be ascribed to the growing up of the second-phase precipitates with the increasing aging time, which can pin the dislocation movement and accordingly promote the dislocation accumulation, leading to stress concentration, which in turn reduce both the strength and ductility [23].

In precipitation hardening alloys, the particles morphological distribution is the main factor of strength, which covers particle size, particle shape and spacing between particles. The classical Orowan bowing/looping and particles shearing are the main models used to describe the precipitation hardening, which mainly includes. Orowan looping often occurs when the radius of the coherent particles exceeds the critical value or when the particles are incoherent with the matrix. Whereas when the precipitates are small and coherent, the shear mechanism takes place. For the current HEA, the sizes of the precipitates are all more than 100 nm, and the particles are not easily plastically deformed. Therefore, it is concluded that $L1_2$ particles strengthen the alloy via Orowan mechanism. The critical stress σ or can be expressed in the following [40]:

$$\sigma_{or} = M \cdot \frac{0.4Gb}{\pi\sqrt{1-\upsilon}} \cdot \frac{ln\left(\frac{\bar{d}}{b}\right)}{\lambda}, \tag{3}$$

where M = 3.06 is the Taylor factor, G = 87.5 GPa is the shear modulus, υ = 0.31 is the Poisson ratio, and b = $\sqrt{2}/2 \times a_{CoCrFeNiTi_{0.2}}$ = 0.255 nm represents the burger vector for an fcc structure. $\bar{d} = \sqrt{2/3} \cdot d$ represents the average precipitate diameter on the slip planes. $\lambda = \bar{d}(\sqrt{\pi/(4f)} - 1)$ represents the average interparticle spacing. Orowan stresses for precipitates in the six aged HEA with different volume fractions and sizes are calculated respectively.

As mentioned above, a large number of slip lines exist in the alloys along various directions, due to partial recrystallization. These slip lines can be regarded as sub-grain boundaries. Generally speaking, grain refinement can greatly improve the strength of the alloy. The volume fraction of the grain-boundaries is negatively correlated with the grain size. Smaller grain size offers a higher volume fraction of grain-boundaries, which will hinder the movement of dislocations. So here the extensive slip lines can improve the strength by grain-boundary strengthening. To apply the mechanism to the current $CoCrFeNiTi_{0.2}$ HEA, let us consider its equivalent grain size. Figure 5a–b is the corresponding schematic illustration. The slip lines within a grain are selected as the statistical target. Draw twelve lines passing through the center of the grain at 30-degree intervals and measure the length of each line, as Figure 5a shows. Then, it is counted how many segments the 12 lines are divided by the slip lines within a grain and how many segments each slip line within a grain are divided by the 12 lines. Accordingly, it is obtained the length of each line being cut off and the average length. From the two dimensions, the equivalent length of the spacing can be obtained between the slip lines, that is, the average grain diameter. In the statistics at least 50 grains were used for counting in different aging times. In this way, the distance between slip lines within the grains is transformed into the average grain diameter of sub-grain boundaries (Figure 5b). Consequently, the strength increment provided by grain-size refinement can be theoretically predicted.

Hall-Petch equation can explain the relationship between yield strength and grain size scientifically [41]:

$$\sigma_y = \sigma_0 + k_y/d^{\frac{1}{2}}, \tag{4}$$

where σ_y represents the yield stress, σ_0 represents the lattice friction stress, k_y represents the strengthening coefficient, and d represents the average grain diameter. We can express the increase of yield strength caused by grain size difference ($\Delta\sigma_G$) as:

$$\Delta\sigma_G = k_y\left(d_p^{\frac{-1}{2}} - d_A^{-\frac{1}{2}}\right). \quad (5)$$

In the formula, d_P denotes the grain size of the thermo-mechanically processed alloys, and d_A represents the grain size of the homogenized alloy.

In order to show the conclusion more intuitively and clearly, the histogram, as shown in Figure 5c, is summarized to explain the strength contribution of the above two strengthening mechanisms. The black part represents the intrinsic strength of the alloy, or the so-called lattice friction strength, according to Moon et al [42], the calculated value is 156 MPa. It can be seen from the diagram that the contribution of precipitation strengthening to strength increment is much greater than that of grain-boundary strengthening. However, with the growth of precipitated particles size, the contribution of precipitation strengthening is gradually decreased, similar to the $(FeCoNiCr)_{100\text{-}x\text{-}y}Ti_xAl_y$ HEAs reported before [23]. Moreover, with the aging time, the number of slip lines is decreased, leading to the increased distance between slip lines, thus causing gradually weakened contribution of grain boundary strengthening, which is consistent with the experimental results in Figure 4c.

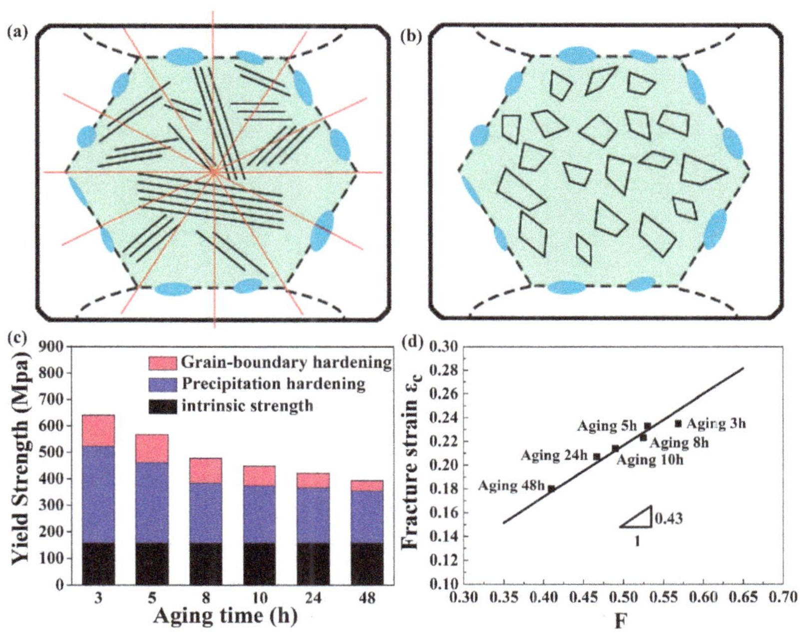

Figure 5. Schematic illustrations showing the slip line be equivalent to sub-grain boundary (**a**) and (**b**). Strengthening contributions from precipitation hardening and grain-boundary hardening in all the aged alloys (**c**). The elongation displayed as a function of the volume fraction and mean particle diameter of $L1_2$ phase in the current HEA (**d**).

3.3. Prediction of Elongation

Compared with the strength which is relatively easy to evaluate in the existing models, it is quite challenging to quantitatively evaluate the tensile ductility. Aging time is one of the important reasons for the variation of tensile ductility in current HEA. It is due to that with the increase of aging time, the $L1_2$ particles grow up gradually, leading to stress concentration, which will lead to a significant reduction in ductility. Here, the present HEAs containing $L1_2$ particles can be regarded as ceramic particles-reinforced metal-matrix composites (MMCs) [43]. In discontinuous reinforced metal matrix composites, when the size of the reinforcing phase is micron order and the volume fraction reaches about 20%, the most balanced strength-ductility properties are expected to be exhibited. Similar to metal-based composite alloys, the $L1_2$ particles in existing alloys are about 0.1–0.4 μm and the volume fraction is 15.6–19.8%, which appears to be consistent with the model. In order to estimate the elongation of HEAs accurately, a model based on MMCS phenomenologically is adopted [44]:

$$\varepsilon_c/\varepsilon_m = (1-f)(1+\varepsilon_{cav})(1-f_{con}), \tag{6}$$

where ε_c represents the failure strain of the composites, ε_m represents the failure strain of the unreinforced matrix, and f represents the volume fraction of particles. In the equation mentioned above, $\varepsilon_{cav} = f^{4/3}/s$ [44] is the contribution to the failure strain from the cavity formation, and s represents the aspect ratio of the particles. The parameter f_{con} in Equation (6) represents the ratio of the constrained to the matrix volume, and can be expressed as [44]:

$$f_{con} = 2s/\left[5\left(f^{-1}-1\right)\right]. \tag{7}$$

Combining Equations (6) and (7), it is readily obtained

$$\varepsilon_c/\varepsilon_m = (1-f)\left(1+\frac{f^{4/3}}{s}\right)\left(1-\frac{2s}{5(f^{-1}-1)}\right) = F. \tag{8}$$

At present, ε_c can be plotted as a function of F, and the result is demonstrated in Figure 4d. It is pointed out that there is a good linear relationship with the slope of the curve to be about 0.43. It can be concluded from the Equation (8) that the slope is predicted to be the fracture strain of the unreinforced alloy matrix, ε_m. Compared with the elongation value of homogenized alloy (~0.49), the value of 0.43 is close to the experimental value. The main reason for the slight offset in Figure 4f is the inhomogeneous distribution of $L1_2$ particles with non-random orientation. Nevertheless, it can be concluded from Figure 5d that the tensile ductility in HEA is determined by the volume fraction and morphology of $L1_2$ precipitates.

4. Conclusions

In this study, a systematic study of the aging behavior of the $CoCrFeNiTi_{0.2}$ HEA is conducted. $L1_2$ particles are readily precipitated after annealing at 800 °C for a different time. The temporal exponents for the average size and number density of precipitates are in reasonable accord with the predictions by the PV model for particle coarsening in the current alloys. The ultimate tensile strength can be greatly improved to ~1200 MPa, accompanied with a tensile elongation of ~20% after precipitation. Both grain boundary and precipitates can contribute to strengthening. The tensile strength and tensile elongation are well predicted, which is consistent with the experimental results. The present experiment provides a theoretical reference for the strengthening of partially recrystallized single-phase HEAs in the future.

Author Contributions: Conceptualization, J.Q.; Data curation, P.L.; Project administration, Y.W.; Software, J.H.; Writing-original draft, H.Z.; Writing-review and editing, H.Z.

Funding: This work was financially supported by the financial support from National Key Laboratory for Remanufacturing, Academy of Armored Forces Engineering (No. 61420050204).

Conflicts of Interest: The authors declare no conflict of interest.

References

1. Gludovatz, B.; Hohenwarter, A.; Catoor, D.; Chang, E.H.; George, E.P.; Ritchie, R.O. A fracture-resistant high-entropy alloy for cryogenic applications. *Science* **2014**, *345*, 1153–1158. [CrossRef] [PubMed]

2. Li, Z.; Pradeep, K.G.; Deng, Y.; Raabe, D.; Tasan, C.C. Metastable high-entropy dual-phase alloys overcome the strength-ductility trade-off. *Nature* **2016**, *534*, 227–230. [CrossRef] [PubMed]
3. Zhang, Z.; Sheng, H.; Wang, Z.; Gludovatz, B.; Zhang, Z.; George, E.P.; Yu, Q.; Mao, S.X.; Ritchie, R.O. Dislocation mechanisms and 3D twin architectures generate exceptional strength-ductility-toughness combination in CrCoNi medium-entropy alloy. *Nat. Commun.* **2017**, *8*, 14390. [CrossRef] [PubMed]
4. Tracy, C.L.; Park, S.; Rittman, D.R.; Zinkle, S.J.; Bei, H.; Lang, M.; Ewing, R.C.; Mao, W.L. High pressure synthesis of a hexagonal close-packed phase of the high-entropy alloy CrMnFeCoNi. *Nat. Commun.* **2017**, *8*, 15634. [CrossRef]
5. Jo, Y.H.; Jung, S.; Choi, W.M.; Sohn, S.S.; Kim, H.S.; Lee, B.J.; Kim, N.J.; Lee, S. Cryogenic strength improvement by utilizing room-temperature deformation twinning in a partially recrystallized VCrMnFeCoNi high-entropy alloy. *Nat. Commun.* **2017**, *8*, 15719. [CrossRef] [PubMed]
6. Huang, H.; Wu, Y.; He, J.; Wang, H.; Liu, X.; An, K.; Wu, W.; Lu, Z. Phase-Transformation Ductilization of Brittle High-Entropy Alloys via Metastability Engineering. *Adv. Mater.* **2017**, *29*, 1701678.
7. Otto, F.; Dlouhý, A.; Somsen, C.; Bei, H.; Eggeler, G.; George, E.P. The influences of temperature and microstructure on the tensile properties of a CoCrFeMnNi high-entropy alloy. *Acta Mater.* **2013**, *61*, 5743–5755. [CrossRef]
8. He, F.; Wang, Z.; Wu, Q.; Li, J.; Wang, J.; Liu, C.T. Phase separation of metastable CoCrFeNi high entropy alloy at intermediate temperatures. *Scr. Mater.* **2017**, *126*, 15–19. [CrossRef]
9. Rao, J.C.; Diao, H.Y.; Ocelík, V.; Vainchtein, D.; Zhang, C.; Kuo, C.; Tang, Z.; Guo, W.; Poplawsky, J.D.; Zhou, Y.; Liaw, P.K.; De Hosson, J.T.M. Secondary phases in Al x CoCrFeNi high-entropy alloys: An in-situ TEM heating study and thermodynamic appraisal. *Acta Mater.* **2017**, *131*, 206–220. [CrossRef]
10. Yeh, J.-W. Alloy Design Strategies and Future Trends in High-Entropy Alloys. *Jom* **2013**, *65*, 1759–1771. [CrossRef]
11. Tsai, K.Y.; Tsai, M.H.; Yeh, J.W. Sluggish diffusion in Co–Cr–Fe–Mn–Ni high-entropy alloys. *Acta Mater.* **2013**, *61*, 4887–4897. [CrossRef]
12. Yeh, J.-W.; Chang, S.-Y.; Hong, Y.-D.; Chen, S.-K.; Lin, S.-J. Anomalous decrease in X-ray diffraction intensities of Cu–Ni–Al–Co–Cr–Fe–Si alloy systems with multi-principal elements. *Mater. Chem. Phys.* **2007**, *103*, 41–46. [CrossRef]
13. Wang, W.-R.; Wang, W.-L.; Wang, S.-C.; Tsai, Y.-C.; Lai, C.-H.; Yeh, J.-W. Effects of Al addition on the microstructure and mechanical property of AlxCoCrFeNi high-entropy alloys. *Intermetallics* **2012**, *26*, 44–51. [CrossRef]
14. He, J.Y.; Liu, W.H.; Wang, H.; Wu, Y.; Liu, X.J.; Nieh, T.G.; Lu, Z.P. Effects of Al addition on structural evolution and tensile properties of the FeCoNiCrMn high-entropy alloy system. *Acta Mater.* **2014**, *62*, 105–113. [CrossRef]
15. Klimova, M.; Stepanov, N.; Shaysultanov, D.; Chernichenko, R.; Yurchenko, N.; Sanin, V.; Zherebtsov, S. Microstructure and Mechanical Properties Evolution of the Al, C-Containing CoCrFeNiMn-Type High-Entropy Alloy during Cold Rolling. *Materials* **2018**, *11*, 53. [CrossRef] [PubMed]
16. Feng, W.; Qi, Y.; Wang, S. Effects of Short-Range Order on the Magnetic and Mechanical Properties of FeCoNi(AlSi)x High Entropy Alloys. *Metals* **2017**, *7*, 482. [CrossRef]
17. He, J.Y.; Wang, H.; Huang, H.L.; Xu, X.D.; Chen, M.W.; Wu, Y.; Liu, X.J.; Nieh, T.G.; An, K.; Lu, Z.P. A precipitation-hardened high-entropy alloy with outstanding tensile properties. *Acta Mater.* **2016**, *102*, 187–196. [CrossRef]
18. Ma, Y.; Wang, Q.; Jiang, B.B.; Li, C.L.; Hao, J.M.; Li, X.N.; Dong, C.; Nieh, T.G. Controlled formation of coherent cuboidal nanoprecipitates in body-centered cubic high-entropy alloys based on Al 2 (Ni,Co,Fe,Cr) 14 compositions. *Acta Mater.* **2018**, *147*, 213–225. [CrossRef]
19. Zhao, Y.L.; Yang, T.; Tong, Y.; Wang, J.; Luan, J.H.; Jiao, Z.B.; Chen, D.; Yang, Y.; Hu, A.; Liu, C.T.; Kai, J.J. Heterogeneous precipitation behavior and stacking-fault-mediated deformation in a CoCrNi-based medium-entropy alloy. *Acta Mater.* **2017**, *138*, 72–82. [CrossRef]
20. Gwalani, B.; Soni, V.; Choudhuri, D.; Lee, M.; Hwang, J.Y.; Nam, S.J.; Ryu, H.; Hong, S.H.; Banerjee, R. Stability of ordered L1 2 and B 2 precipitates in face centered cubic based high entropy alloys—Al 0.3 CoFeCrNi and Al 0.3 CuFeCrNi 2. *Scr. Mater.* **2016**, *123*, 130–134. [CrossRef]

21. Gwalani, B.; Choudhuri, D.; Soni, V.; Ren, Y.; Styles, M.; Hwang, J.Y.; Nam, S.J.; Ryu, H.; Hong, S.H.; Banerjee, R. Cu assisted stabilization and nucleation of L1 2 precipitates in Al 0.3 CuFeCrNi 2 fcc-based high entropy alloy. *Acta Mater.* **2017**, *129*, 170–182. [CrossRef]
22. Tsao, T.K.; Chang, Y.J.; Chang, K.C.; Yeh, J.W.; Chiou, M.S.; Jian, S.R.; Kuo, C.M.; Wang, W.R.; Murakami, H. Developing New Type of High Temperature Alloys–High Entropy Superalloys. *Int. J. Metall. Mater. Eng.* **2015**, *1*. [CrossRef]
23. He, J.Y.; Wang, H.; Wu, Y.; Liu, X.J.; Mao, H.H.; Nieh, T.G.; Lu, Z.P. Precipitation behavior and its effects on tensile properties of FeCoNiCr high-entropy alloys. *Intermetallics* **2016**, *79*, 41–52. [CrossRef]
24. Zhang, C.; Zhu, C.; Harrington, T.; Vecchio, K. Design of non-equiatomic high entropy alloys with heterogeneous lamella structure towards strength-ductility synergy. *Scr. Mater.* **2018**, *154*, 78–82. [CrossRef]
25. Han, B.; Wei, J.; Tong, Y.; Chen, D.; Zhao, Y.; Wang, J.; He, F.; Yang, T.; Zhao, C.; Shimizu, Y.; Inoue, K.; Nagai, Y.; Hu, A.; Liu, C.T.; Kai, J.J. Composition evolution of gamma prime nanoparticles in the Ti-doped CoFeCrNi high entropy alloy. *Scr. Mater.* **2018**, *148*, 42–46. [CrossRef]
26. Zhao, Y.Y.; Chen, H.W.; Lu, Z.P.; Nieh, T.G. Thermal stability and coarsening of coherent particles in a precipitation-hardened (NiCoFeCr) 94 Ti 2 Al 4 high-entropy alloy. *Acta Mater.* **2018**, *147*, 184–194. [CrossRef]
27. Schmuck, C.; Caron, P.; Hauet, A.; Blavette, D. Ordering and precipitation of γ' phase in low supersaturated Ni-Cr-Al model alloy: An atomic scale investigation. *Philos. Mag. A* **1997**, *76*, 527–542. [CrossRef]
28. Davies, C.K.; Nash, P.; Stevens, R.N. Precipitation in Ni-Co-AI alloys. *J. Mater. Sci.* **1980**, *15*, 1521–1532. [CrossRef]
29. Hein, W. Nucleation, growth and coarsening of γ'-precipitates in Ni-5.0 at.% Al-5.8 at.% Ti. *Acta Metall.* **1989**, *37*, 2145–2152. [CrossRef]
30. Sudbrack, C.K.; Ziebell, T.D.; Noebe, R.D.; Seidman, D.N. Effects of a tungsten addition on the morphological evolution, spatial correlations and temporal evolution of a model Ni–Al–Cr superalloy. *Acta Mater.* **2008**, *56*, 448–463. [CrossRef]
31. Zhao, Y.Y.; Nieh, T.G. Correlation between lattice distortion and friction stress in Ni-based equiatomic alloys. *Intermetallics* **2017**, *86*, 45–50. [CrossRef]
32. Philippe, T.; Voorhees, P.W. Ostwald ripening in multicomponent alloys. *Acta Mater.* **2013**, *61*, 4237–4244. [CrossRef]
33. Booth-Morrison, C.; Weninger, J.; Sudbrack, C.K.; Mao, Z.; Noebe, R.D.; Seidman, D.N. Effects of solute concentrations on kinetic pathways in Ni–Al–Cr alloys. *Acta Mater.* **2008**, *56*, 3422–3438. [CrossRef]
34. Karnesky, R.A.; Dunand, D.C.; Seidman, D.N. Evolution of nanoscale precipitates in Al microalloyed with Sc and Er. *Acta Mater.* **2009**, *57*, 4022–4031. [CrossRef]
35. van Dalen, M.E.; Gyger, T.; Dunand, D.C.; Seidman, D.N. Effects of Yb and Zr microalloying additions on the microstructure and mechanical properties of dilute Al–Sc alloys. *Acta Mater.* **2011**, *59*, 7615–7626. [CrossRef]
36. Voorhees, P.W. The Theory of Ostwald Ripening. *J. Stat. Phys.* **1985**, *38*, 231–252. [CrossRef]
37. Snyder, V.A.; Alkemper, J.; Voorhees, P.W. Transient ostwald ripening and the disagreement between steady-state coarsening theory and experiment. *Acta Mater.* **2001**, *49*, 699–709. [CrossRef]
38. Hadjiapostolidou, D.; Shollock, B.A. Long term coarsening in Rene 80 Ni-base superalloy. *Superalloy* **2008**, *2008*, 733–739.
39. Chen, M.K.; Voorhees, P.W. The dynamics of transient Ostwald ripening. *Model. Simul. Mater. Sci. Eng.* **1993**, *1*, 591. [CrossRef]
40. Brown, L.M.; Ham, R.K. Dislocation-particle interactions. In *Strengthening Methods in Crystals*; Applied Science Publishers: London, UK, 1971; p. 9.
41. Hall, E.O. The Deformation and Ageing of Mild Steel: III Discussion of Results. *Proc. Phys. Soc. Sect. B* **1951**, *64*, 747. [CrossRef]
42. Moon, J.; Jang, M.J.; Bae, J.W.; Yim, D.; Park, J.M.; Lee, J.; Kim, H.S. Mechanical behavior and solid solution strengthening model for face-centered cubic single crystalline and polycrystalline high-entropy alloys. *Intermetallics* **2018**, *98*, 89–94. [CrossRef]
43. Nieh, T.G.; Chellman, D.J. Modulus measurements in discontinuous reinforced aluminum composites. *Scr. Metall.* **1984**, *18*, 925–928. [CrossRef]
44. Whitehouse, A.F.; Clyne, T.W. Cavity formation during tensile straining of particulate and short fibre metal matrix composites. *Acta Metall. Mater.* **1993**, *41*, 1701–1711. [CrossRef]

entropy

MDPI

Article

Tensile Behavior and Evolution of the Phases in the $Al_{10}Co_{25}Cr_8Fe_{15}Ni_{36}Ti_6$ Compositionally Complex/High Entropy Alloy

Anna Maria Manzoni [1,*], Sebastian Haas [2], Haneen Daoud [3], Uwe Glatzel [2], Christiane Förster [1] and Nelia Wanderka [1]

1 Helmholtz-Zentrum Berlin für Materialien und Energie GmbH, Hahn-Meitner-Platz 1, D-14109 Berlin, Germany; christiane.foerster@helmholtz-berlin.de (C.F.); wanderka@helmholtz-berlin.de (N.W.)

2 Metals and Alloys, University Bayreuth, Ludwig-Thoma-Str. 36b, D-95447 Bayreuth, Germany; sebastian-haas@uni-bayreuth.de (S.H.); Uwe.Glatzel@uni-bayreuth.de (U.G.)

3 Neue Materialien Bayreuth GmbH, Gottlieb-Keim-Str. 60, D-95448 Bayreuth, Germany; haneen.daoud@nmbgmbh.de

* Correspondence: anna_manzoni@gmx.net; Tel.: +49-30-8062-42326; Fax: +49-30-8062-43059

Received: 3 August 2018; Accepted: 24 August 2018; Published: 29 August 2018

Abstract: Compositionally complex alloys, or high entropy alloys, are good candidates for applications at higher temperatures in gas turbines. After their introduction, the equiatomic $Al_{17}Co_{17}Cr_{17}Cu_{17}Fe_{17}Ni_{17}$ (at.%) served as a starting material and a long optimization road finally led to the recently optimized $Al_{10}Co_{25}Cr_8Fe_{15}Ni_{36}Ti_6$ (at.%) alloy, which shows promising mechanical properties. Investigations of the as-cast state and after different heat treatments focus on the evolution of the microstructure and provide an overview of some mechanical properties. The dendritic solidification provides two phases in the dendritic cores and two different ones in the interdendritic regions. Three of the four phases remain after heat treatments. Homogenization and subsequent annealing produce a γ-γ' based microstructure, similar to Ni-based superalloys. The γ phase is Co-Cr-Fe rich and the γ' phase is Al-Ni-Ti rich. The understanding of the mechanical behavior of the investigated alloy is supported and enhanced by the study of the different phases and their nanohardness measurements. The observations are compared with mechanical and microstructural data from commercial Ni-based superalloys, Co-based alloys, and Co-Ni-based alloys at the desired application temperature of ~800 °C.

Keywords: high entropy alloys; microstructure; transmission electron microscopy; compositionally complex alloys

1. Introduction

High entropy alloys (single phase) and compositionally complex (more phases, based on the high entropy effect) alloys have shown promising mechanical properties since the beginning of their discovery. The current trend in the development of the new high entropy alloys (HEA) or compositionally complex alloys (CCA) is focused on their use in industrial applications as a replacement for Ni-based or Co-based alloys. The latter commonly find their applications in aircraft turbines. However, there is a high demand for increasingly efficient materials, not only for aircraft turbines, but also for the next generation of USC (ultra-supercritical) power plants. In this case, the efficiency must be increased, which means that the materials need to withstand temperatures of up to 750 °C [1]. Usually, Ni- and Co-based alloys are considered good candidates. In the temperature window of 600–800 °C, the commercially used alloys are restricted to a small group, especially alloy 800, Inconel 706, and Inconel 617, which are used as turbine disks.

In order to mirror the good performance of these alloys, several optimization steps have been tested in high entropy alloys and compositionally complex alloys. To make a new alloy compete with Co- and Ni-based superalloys, its tensile properties are of high importance, but unfortunately, until now, there have only been a few works on the tensile properties of HEA and CCA [2]. For the aforementioned application as turbine disks, thermomechanical strengthening mechanisms, such as cold-rolling, which can show good results [3], are not applicable. This is also the case for several potential strengthening phases, such as oxides [4] or carbides [5], which are usually obtained by powder metallurgy. Both these phases and the non-equilibrium method have been excluded because their performance in the present type of application is inferior to a coherent strengthening phase like γ' in γ in a cast alloy that has been carefully homogenized, as has been shown in a recent review paper [6]. This is also the strengthening mechanism that is dominant in the Ni- and Co-based superalloys.

One aspect of particular importance is the amount of γ' phase in the γ matrix [7]. It is responsible for a high ultimate tensile strength. An increase of the γ' volume fraction is therefore a most interesting path for material optimization, not only in superalloys, but also in compositionally complex alloys.

In 2011, we started a methodical search for the improvement of the equiatomic $Al_{17}Co_{17}Cr_{17}Cu_{17}Fe_{17}Ni_{17}$ compositionally complex alloy's mechanical properties [8,9]. This equiatomic alloy displays at least five phases and is very brittle [8–11]. First ideas, like the reduction of the amount of Cu [12–15], the increase of Al and Cr for better oxidation properties [14,16], or the reduction of the number of phases [13,14], had to be omitted because the associated mechanical properties did not meet the expectations. Especially the idea of a single phase, which makes the concept of high entropy alloys unique [17], was disappointing because of the material's low strength [14,15].

A change of path then had to be made, and it was decided to promote the formation of a coherent strengthening phase, i.e., the aformentioned γ-γ' morphology. It could be obtained in the $Al_8Co_{17}Cr_{17}Cu_8Fe_{17}Ni_{33}$ alloy, after the optimum heat treatment, but the γ' precipitates were very small (about 10 nm sized) and had a low volume fraction (about 20%); thus the alloy was extremely weak [12,14,15]. The addition of small amounts of W, Mo, and Ti in the $Al_8Co_{17}Cr_{14}Cu_8Fe_{17}Ni_{34.8}Mo_{0.1}Ti_{0.1}W_1$ alloy allowed a slight increase in size of the γ' precipitates (up to 50 nm) at a lower volume fraction (about 7%), but it had a similar weakness [15]. The search had to go on.

The systematic change in the composition of the alloy and the calculation of the equilibrium diagrams using the Thermo-Calc software (Thermo-Calc Software AB, Solna, Sweden) [18] and the TTNi7 database [19] finally led to the alloy $Al_{10}Co_{25}Cr_8Fe_{15}Ni_{36}Ti_6$, which has a wide temperature range of existence of a γ-γ' two phase region and a high proportion of the ordered γ' phase. The mechanical properties of this alloy are based on a stable microstructure in the temperature range up to 800 °C, as reported recently [16]. Temperature tensile strength and elongation to failure up to 800 °C [16] are even superior to those of Alloy 800 and Inconel 617 [20]. At this temperature, $Al_{10}Co_{25}Cr_8Fe_{15}Ni_{36}Ti_6$ also has better mechanical properties than other high entropy alloys and compositionally complex alloys like solid solution $Co_{20}Cr_{20}Fe_{20}Mn_{20}Ni_{20}$ (~100 MPa, [21]) or equiatomic $Al_{17}Co_{17}Cr_{17}Cu_{17}Fe_{17}Ni_{17}$ (~180 MPa, [22]).

In order to further optimize the alloy, a wider knowledge of the microstructure and its correlation with the mechanical properties is necessary. A major point is also the reproducibility of the tensile tests, which can be explained by microstructural features. Therefore, this study focuses, on the one hand, on the microstructural characterization of the as-cast and the heat treated $Al_{10}Co_{25}Cr_8Fe_{15}Ni_{36}Ti_6$ alloy, as well as on the evolution of its phases, and on the other hand, on tensile results at room temperature and at 800 °C, and the correlation between the two parts. A combination of complementary imaging techniques at several scale levels has been employed for the characterization of the phases in the alloy (optical microscopy, scanning electron microscopy (SEM), transmission electron microscopy (TEM)). Tensile tests have been performed at room temperature and 800 °C to investigate the mechanical properties concerning strength and ductility. Nanohardness measurements of each phase have been carried out to support the mechanical properties.

2. Materials and Methods

2.1. Alloy Preparation

The $Al_{10}Co_{25}Cr_8Fe_{15}Ni_{36}Ti_6$ compositionally complex alloy (CCA) was prepared from the constituent elements of 99.99% purity. It was melted in a vacuum induction furnace and solidified in a circular ceramic mould of a 140 mm length and a 10 mm diameter. Samples were solidified directionally with a preferred [001] direction using the Bridgman method. Homogenization was carried out at 1220 °C for 20 h with a subsequent cooling in the furnace. The homogenized samples were aged at 900 °C or at 950 °C for different times, i.e., 5 and 50 h (900 °C) and 50, 100, 200, 500, and 100 h (950 °C), and then cooled down in the furnace to room temperature to prevent cracking at the grain boundaries.

2.2. Microstructural Observations

Specimens for optical microscopy (OM), scanning electron microscopy (SEM), X-ray diffraction (XRD), and nanohardness measurements were mechanically polished down in a final polishing step with a 50 nm sized OP-U colloidal silica suspension. Specimens for SEM investigation were etched with a solution of 3 g Mo-acid in 100 mL H_2O, 100 mL HCl, and 100 mL HNO_3.

SEM investigations were carried out in two SEMS: the first is a Zeiss Leo LEO Gemini 1530, equipped with an EDS spectrometer (Noran System Six; Thermo Fisher Scientific Inc., Waltham, MA, USA); and the second SEM is a Zeiss 1540EsB Cross Beam. All images were recorded at 5 keV and at working distances between 7.5 and 8.7 mm.

XRD observations were carried out in a Bruker AXS (Bruker AXS GmbH, Karlsruhe, Germany), D8 diffractometer, using Cu Kα radiation in the θ–2θ configuration.

Specimens for TEM were electropolished with a solution of 83% ethanol, 10% perchloric acid, and 7% glycerin at a temperature of −7 °C and a voltage of 30 V. A TEM Philips CM30 (ThermoFischer Scientific, Waltham, MA, USA), operated at 300 kV and equipped with an energy-dispersive X-ray spectroscopy (EDS) detector (EDAX), was used in this study. The beam size for EDS measurements was 10 nm.

Phase volume fractions and precipitate sizes were determined in 2D using the training and classification of the Weka segmentation method [23] in the open source software Fiji [24] based on ImageJ [25,26]. In some cases, these measurements were accompanied by thresholds set on EDS mappings or OM, SEM, and TEM images taken along the [100] zone axis of the γ'-γ' morphology were used for the most varied view on particles and the reduction of systematic errors. An average of 1000 particles per state have been analyzed, except for the as-cast and the homogenized state, where the Weka method could not be applied due to the small size. An average of 100 particles have been observed here.

2.3. Mechanical Tests

Vickers microhardness measurements were carried out using a Polyvar met (Reichert-Jung/Leica Microsystems, Wetzlar, Germany) microscope and a MHT-10 (Anton Paar, Graz, Austria) microhardness tester under a load of ~0.5 N for 15 s. The average microhardness value was calculated from the minimum ten indentations. Nanohardness measurements were performed in a picodentor HM500 (Helmut Fischer GmbH, Sindelfingen, Germany) under a load of 25 mN for 20 s.

Flat specimens for tensile tests were manufactured by electrical discharge machining (EDM) and the recast layer was removed by fine grinding. The square cross section of the samples was $1 \times 1.9\ mm^2$ and the gauge length was 8 mm. The samples were attached form-closed to ceramic Al_2O_3-clampings. For tests at elevated temperatures, a radiant heated, fire-proof furnace was used and the tensile tests were performed with a deformation rate of 0.01 mm/s. The strain data was recorded optically by a high-resolution camera and the force for deformation by a load cell. About four stress-strain value couples were logged in one second. Young's modulus was determined from the stress-strain curves by determining the slope of the elastic region.

3. Results and Discussion

3.1. Morphological Evolution

In a previous study [20], it has been found that the as-cast alloy solidifies dendritically, as is the case for most high entropy and compositionally complex alloys. After homogenization at 1220 °C for 20 h, the alloy does not show any dendritic structure.

In the same study, it has also been found that there are two main morphological features in the microstructure of the different states, i.e., coarse Al-Ni rich needles (in the annealed states) and a fine γ' morphology (in all states). An overview of these three phases in the four states described in [20] (as-cast, homogenized, annealed 900 °C 5 h and annealed 900 °C 50 h) is given in Figure 1. The directional solidification in the [100] direction of the samples implies that the XRD spectrum records only the (100) type planes of the γ and γ' phase and the (220) type planes of the Al-Ni rich needles, oriented 45° towards the matrix. The latter show the strongest peak in the as-cast state, because of their high volume fraction of 10 ± 3%. The Al-Ni rich phase is absent in the homogenized state (1220 °C 20 h), and in the annealed states, its peak is visible again.

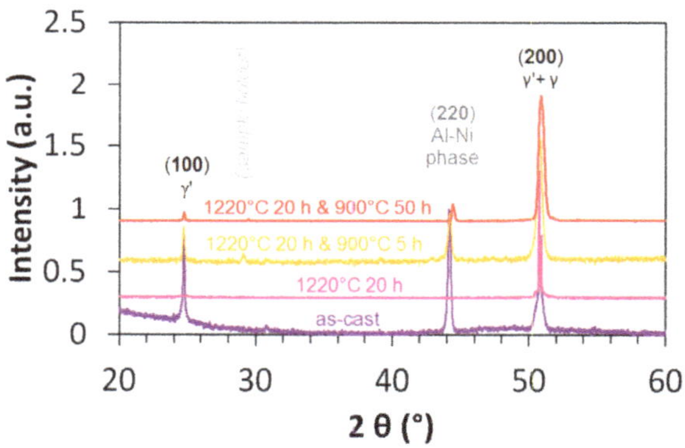

Figure 1. Comparison of the XRD spectra of four states of the $Al_{10}Co_{25}Cr_8Fe_{15}Ni_{36}Ti_6$ alloy.

In the following, the three phases will be presented separately.

3.1.1. Al-Ni Rich Needles

Figure 2 shows the optical micrographs (a,b) and SEM images (c,d) of four exemplary annealed states of the $Al_{10}Co_{25}Cr_8Fe_{15}Ni_{36}Ti_6$ alloy. The annealing treatments at 900 °C or 950 °C lead to the formation of Al-Ni rich needle-like precipitates, displayed in bright grey in OM and dark grey in the SEM. These can reach up to 15 µm in length after 900 °C 5 h annealing (Figure 2a), and up to 40 µm after 900 °C 50 h (Figure 2b), 950 °C 100 h (Figure 2c), and 950 °C 1000 h (Figure 2d). The needles can grow into plates after the heat treatments at 950 °C, especially at the grain boundaries (Figure 2d,e). In all states of the investigated alloy, some unwanted pores and Ti-nitrides can be identified.

These needles were not predicted by the calculation of the phase diagram in Reference [20]. They are non-existent in the homogenized state of the alloy, i.e., at 1220 °C (see [20]). They form at the lower temperatures of 900 and 950 °C.

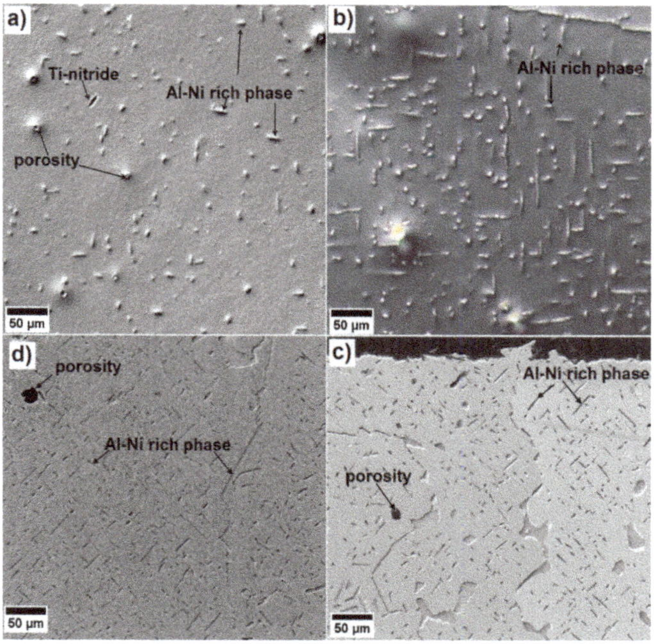

Figure 2. (**a**,**b**) Optical micrographs of the microstructure of the $Al_{10}Co_{25}Cr_8Fe_{15}Ni_{36}Ti_6$ alloy after annealing at (**a**) 900 °C 5 h and (**b**) 900 °C 50 h; (**c**,**d**) SEM images after annealing at (**c**) 950 °C 100 h and (**d**) 950 °C 1000 h.

The needles are oriented perpendicularly to each other. They are scarce (about 1–3 vol.%) in the 900 °C 5 h annealed alloy, but their volume fraction increases after the longer heat treatments and at a higher temperature. A comparison of the different volume fractions is shown in Figure 3. It is interesting to note that the annealing treatments at 950 °C increase the volume fraction to about 9 (100 h) or even 13% (200–1000 h). The volume fractions of needles in all states are shown in Figure 3. It can be assumed that an equilibrium is reached after 200 h at 950 °C because the volume fraction does not change after this time.

The needles' lengths remain about the same, even after the longer heat treatments at higher temperatures of 950 °C 100 h and 950 °C 1000 h, but in some cases, the needles become plate-like, especially at the grain boundaries (see Figure 2c,d).

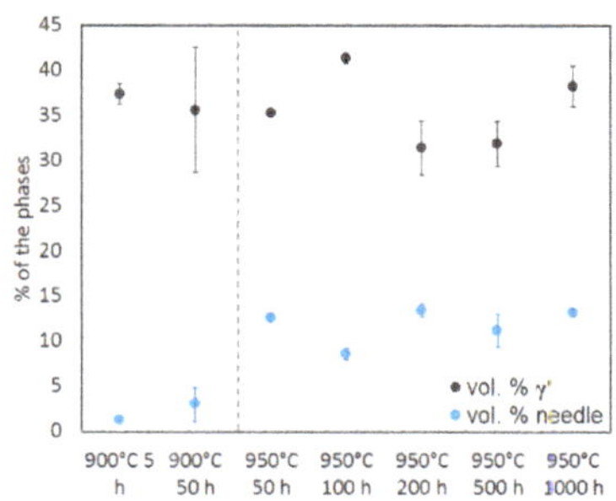

Figure 3. Volume fractions of the γ' and the needle phase in the different annealing states of the $Al_{10}Co_{25}Cr_8Fe_{15}Ni_{36}Ti_6$ alloy.

3.1.2. The γ-γ′ Morphology in Detail

Higher resolution observations show the second most important morphological feature in the $Al_{10}Co_{25}Cr_8Fe_{15}Ni_{36}Ti_6$ alloy, i.e., the already mentioned γ-γ′ morphology. As has been stated before, the microstructure of the samples aged at 900 °C and at 950 °C demonstrates that the two-phase microstructure (consisting of the γ and γ′ phases) obtained in the homogenized alloy and predicted by Thermo-Calc is unstable and decomposes to form the already described additional needle-like phase.

Figure 4a–d shows dark field (DF) TEM micrographs of four states of the $Al_{10}Co_{25}Cr_8Fe_{15}Ni_{36}Ti_6$ alloy, imaged using the (110) superlattice reflection. The two-phase microstructure with a γ-γ′ morphology is known from Ni-based superalloys. The brightly displayed particles are γ′ precipitates of an $L1_2$ structure embedded in a darkly displayed γ matrix of a disordered A1 structure. The corresponding diffraction pattern of the [001] zone axis is shown in the inset of every micrograph. Figure 4e,f shows two SEM micrographs of the γ-γ′ morphology in the etched specimens annealed at 950 °C. These show an additional feature, i.e., small secondary γ′ particles inside the γ matrix, made visible by the etching process. They will be treated in detail in a subsequent article and will not be mentioned here again.

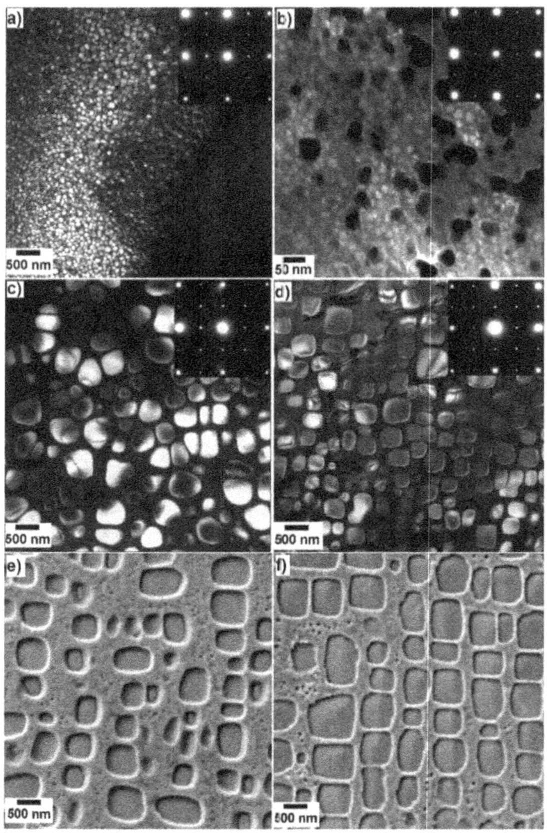

Figure 4. Evolution of the γ-γ′ morphology in six states of the $Al_{10}Co_{25}Cr_8Fe_{15}Ni_{36}Ti_6$ alloy: (**a**–**d**) DF TEM in (**a**) the as-cast; (**b**) homogenized at 1220 °C 20 h; subsequent annealing (**c**) at 900 °C 5 h and (**d**) at 900 °C 50 h. The upper right corner shows the corresponding SAED taken along the [001] zone axis. The DF images have been recorded with the (110) reflex; (**e**,**f**) SEM micrographs of the alloy after annealing at (**e**) 950 °C 200 h and (**f**) 950 °C 1000 h. Note the different scale bar in (**b**).

The primary γ' particles have different shapes and sizes in the different states. They are rather round and 40–90 nm in diameter in the as-cast state (Figure 4a). They have undefined shapes with diameters of 6–24 nm after homogenization at 1220 °C 20 h (Figure 4b). They grow intensely after the subsequent annealing and start to adopt a cubic shape with diameters of about 200–550 nm after 900 °C 5 h (Figure 4c). They display a clear cubic shape and a diameter of 200–530 nm after 50 h annealing (Figure 4d). The heat treatments at higher temperatures and conducted for longer times further increase the particles' sizes from 400–700 nm in the 950 °C 100 h to 400–800 nm in the 950 °C 1000 h annealed alloy.

The volume fractions of the γ' particles in all states have been summarized in Table 1 and displayed in Figure 3. They are between 30 and 55 vol.% at all states and no tendency can be determined.

Table 1. Concentration of the different phases in at.% in four states of the $Al_{10}Co_{25}Cr_8Fe_{15}Ni_{36}Ti_6$ alloy, as taken with TEM/EDS. The error bar is given by the standard deviation 2σ. (De: dendrite, ID: interdendritic region). The composition of the primary γ' precipitates in the homogenized alloy cannot be measured because of their small sizes.

Phase	Composition (at.%)					
	Al	Co	Cr	Fe	Ni	Ti
Nominal	10	25	8	15	36	6
As-cast						
De -γ matrix	8.8 ± 2.2	**29.5 ± 1.8**	7.7 ± 1.2	**20.0 ± 4.2**	30.7 ± 4.3	3.2 ± 1.7
De -γ′ precipitate	12.1 ± 1.4	22.3 ± 1.9	3.1 ± 1.1	7.7 ± 1.7	**45.5 ± 2.9**	9.3 ± 1.2
ID core	**20.6 ± 2.5**	23.3 ± 1.4	3.6 ± 0.9	9.1 ± 1.2	34.3 ± 2.2	9.1 ± 0.5
Homogenized 1220 °C 20 h						
γ Matrix γ′ precipitate	9.3 ± 1.6	26.1 ± 0.8	7.2 ± 0.5	14.8 ± 1.0	**37.1 ± 0.8**	5.5 ± 0.5
Annealed 1220 °C 20 h—900 °C 5 h						
γ matrix	8.9 ± 1.7	**28.1 ± 1.2**	**9.4 ± 0.7**	**18.8 ± 0.9**	31.1 ± 1.1	3.6 ± 0.4
Primary γ′ precipitate	14.0 ± 1.0	21.4 ± 0.8	3.6 ± 0.4	8.2 ± 0.6	**44.4 ± 1.1**	8.3 ± 0.8
Al-Ni needle	**24.7 ± 1.9**	20.8 ± 0.7	2.9 ± 0.2	9.4 ± 0.1	36.2 ± 1.1	6.0 ± 0.2
Annealed 1220 °C 20 h—900 °C 50 h						
γ matrix	6.9 ± 0.6	**29.5 ± 0.5**	**9.3 ± 0.4**	**20.4 ± 0.6**	30.4 ± 1.0	3.5 ± 0.4
Primary γ′ precipitate	11.4 ± 0.6	22.5 ± 0.6	3.5 ± 0.4	8.8 ± 0.7	**45.0± 1.5**	8.7 ± 0.5
Al-Ni needle	**24.4 ± 1.3**	21.9 ± 1.7	3.6 ± 0.2	10.7 ± 0.4	33.9 ± 0.6	5.6 ± 0.2

In superalloys, the size and morphology of the γ' precipitates determine the mechanical properties. It is supposed that the same correlation can be observed in the $Al_{10}Co_{25}Cr_8Fe_{15}Ni_{36}Ti_6$ alloy.

In order to find out the growth mechanism of the γ' precipitates, especially after ageing, we compare the two-dimensional γ' particle size distribution for all states. The quantification of the particle size distribution (PSD) was made by characterizing about 100 (in the as-cast and homogenized states) or 1000 (in the annealed states) primary γ' precipitates using the following method: the area was enclosed using the Weka segmentation method and, in rare cases, additionally by hand, and the area equivalent diameter $r = \sqrt{(\text{area drawn})}$ was obtained. The obtained values for r were then grouped into classes of 5 (as-cast), 2 (homogenized), or 50 nm (annealed). Their relative distribution in six selected states is shown in detail in Figure 5. The Gauss normal distribution (GND) is shown by a dotted line. Additionally, the PSD was calculated according to the LSW (Lifshitz-Slyozov-Wagner) theory of Ostwald ripening [27–29], following the equation [27]:

$$f(u) = \frac{4u^2}{9}\left(\frac{3}{3+u}\right)^{\frac{7}{3}}\left(\frac{-1.5}{u-1.5}\right)^{\frac{11}{3}}\exp\left(\frac{u}{u-1.5}\right) \tag{1}$$

where $u = r/r_{avg}$, r is the diameter of the analyzed precipitate, and r_{avg} is the average diameter of all precipitates. The LSW distribution is shown in a red continuous line, even for the as-cast and the homogenized sample, where no ripening is expected.

The PSD in the as-cast sample (Figure 5a) spreads from about 35 to about 90 nm in diameter. After homogenization, the distribution is much narrower, i.e., from about 6 to about 24 nm in diameter (Figure 4b). For the annealed states (Figure 5c–f), the PSD is widely spread and can range from 50 to 800 nm diameter.

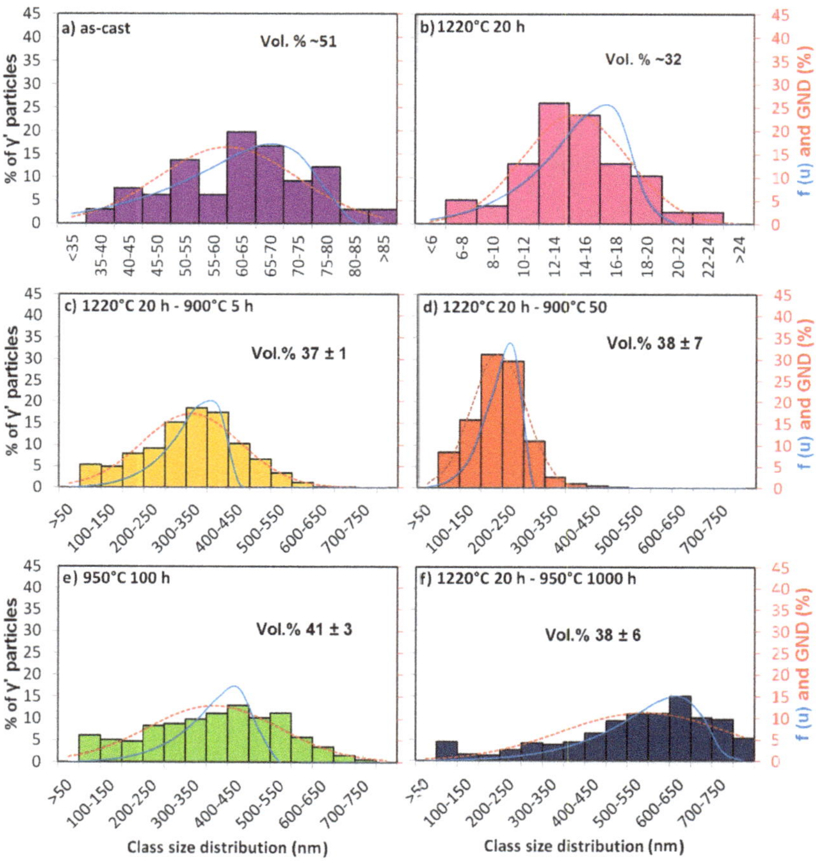

Figure 5. Relative γ' particle size distribution of the $Al_{10}Co_{25}Cr_8Fe_{15}Ni_{36}Ti_6$ alloy in (**a**) the as-cast state; (**b**) homogenized state at 1220 °C 20 h; and subsequent annealing (**c**) at 900 °C 5 h, (**d**) at 900 °C 50 h, (**e**) at 950 °C 100 h, and (**f**) at 950 °C 1000 h. The dashed line corresponds to the Gauss normal distribution (GND) and the continuous line to the LSW distribution for every case. Note the different class sizes and class size distributions in the as-cast and the homogenized states. Volume fraction shave been added in the graphs for comparison.

For all samples, the PSD fits better with the GND than with the LSW prediction. This is in accordance with research on Ostwald ripening in Ni-based alloys [30], which shows a PSD for several alloys that is broader than the one predicted by the LSW theory. A disagreement with the LSW theory has also been observed in a model Ni-14.2 Cr-5.2 Al (at.%) superalloy [31]. According to the investigations in the study from Ref. [31], the disagreement with the LSW theory can be caused by various reasons, two of which are the most plausible: (i) a transient coarsening, and (ii) the possible non-constant composition of the γ' precipitates as a function of ageing time until it reached the equilibrium values. The last argument is in agreement with the results in the present work, where the solute concentration of Al in the γ' precipitates of the annealed samples decreases with time from 14.0 at.% (after 900 °C 5 h ageing) to 11.4 at.% (after 900 °C 50 h), as determined by TEM-EDS, and shown in Table 1.

A comparison of the GND of the γ' sizes of all ten investigated states is shown in Figure 6. As has been shown in Figure 4a,b, the as-cast (purple) and the homogenized (pink) states stand out due to their small γ' sizes.

The three states after annealing at 900 °C show the most peculiar behavior. The treatment of 5 h annealing (orange dotted line) produces the largest γ' particles, but also the most wide spread of sizes. The treatment of 50 h annealing (red dotted line) shows the smallest annealing γ' precipitates, and the narrowest distribution. This is contradictory to the LSW theory, which predicts the growing of the precipitates' average radius with $t^{1/3}$. As this result is quite unexpected, several counting methods and many micrographs have been compared to a total number of over 6000 particles of these three states and the result has been confirmed by all methods. It can be concluded that the 900 °C 50 h annealing produces a unique behavior in this $Al_{10}Co_{25}Cr_8Fe_{15}Ni_{36}Ti_6$ alloy.

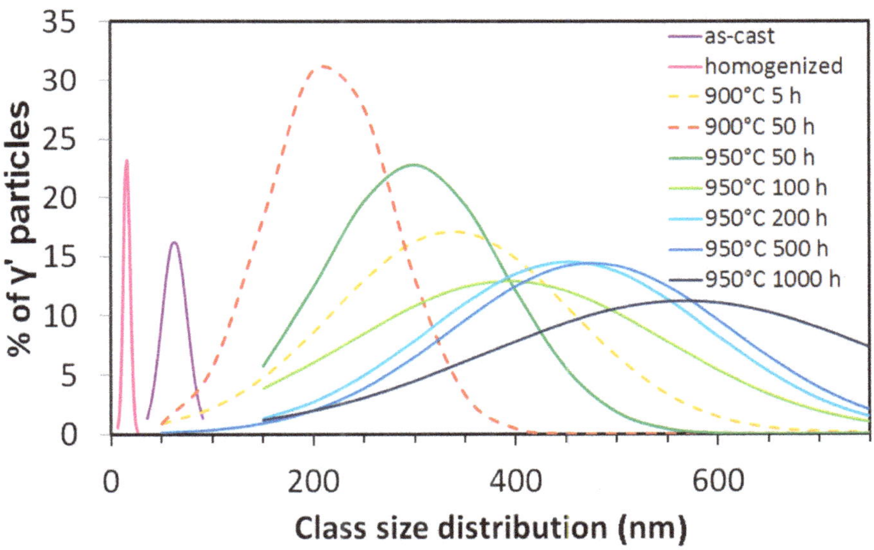

Figure 6. The Gauss Normal Distributions of the relative γ' particle size distribution of the $Al_{10}Co_{25}Cr_8Fe_{15}Ni_{36}Ti_6$ alloy in all investigated states. The 900 °C annealed states are shown in dashed lines for an easier comparison.

Unlike the annealing at 900 °C, the annealing at 950 °C produces a γ' particle growth that is in accordance with the LSW theory of particle growth (Equation (2)):

$$R^3 = R_0^{\,3} + Kt \tag{2}$$

where R is the average particle radius after time t, R_0 is the initial particle radius, 3 is the exponent for diffusion-controlled growth, and K is the rate constant [32]. In our case, R_0 corresponds to the average particle radius after homogenization, which is around 4 nm. It can be neglected compared to the radii of the particles after annealing, and thus the equation can be simplified to Equation (3):

$$R \propto t^{1/3} \tag{3}$$

This relationship of the average γ' particles radii versus $t^{1/3}$ and the corresponding error bars are displayed in Figure 7. It shows the linear growth of the average particle radius with the cubic root of time, with a constant of about K = 1.9.

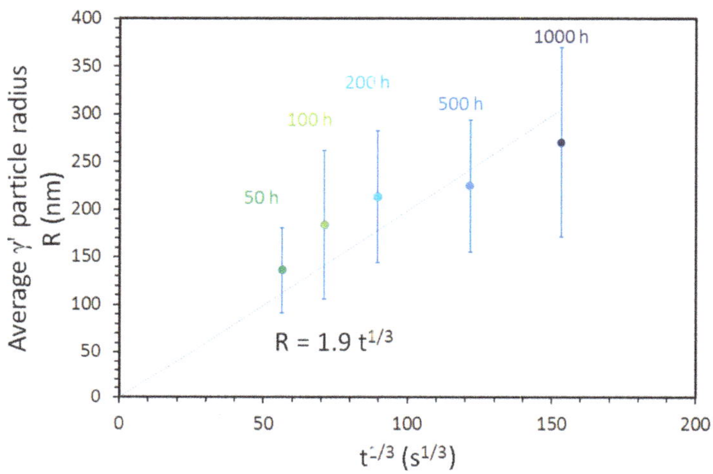

Figure 7. Growth of the γ' particles in the states annealed at 950 °C, according to the LSW theory. The error bars are given by the standard deviation 2σ.

3.2. *Mechanical Observations*

The microstructure of the samples were correlated to tensile tests at room temperature and at 800 °C, in addition to Vickers micro- and nanohardness measurements.

3.2.1. Tensile Tests

The as-cast and the homogenized state have not been submitted to tensile tests. The tensile data from the 900 °C 5 h and the 900 °C 50 h annealed states was taken from [20]. They are shown in Table 2, together with the new tensile data of the 900 °C 50 h and the 950 °C 100 h annealed states. The engineering tensile curves of the latter two are shown in Figure 8, both at room temperature and at 800 °C.

From [20], it is known that a higher RT ultimate tensile strength σ_{TS} can be observed for the 900 °C 50 h state (1039 MPa) than for the 900 °C 5 h state (786 MPa). At a 800 °C testing temperature, their tensile strengths are almost identical (~650 MPa).

The new experiments of the 900 °C 50 h state show slightly higher RT σ_{TS} values (~1200 MPa) and lower 800 °C σ_{TS} values (~580 MPa).

The most interesting to note in Figure 8 is the reproducibility of the tensile stress-strain curves. It is very high in the 900 °C 50 h state, but the results for the investigated 950 °C 100 h specimens are widely spread. It is almost impossible to determine an average σ_{TS} value for the investigated specimens of the 950 °C 100 h alloy. This is even worse at RT than at 800 °C.

Table 2. Mechanical and morphological data for the six investigated states of the $Al_{10}Co_{25}Cr_8Fe_{15}Ni_{36}Ti_6$ alloy (as-cast, 1220 °C 20 h, 1220 °C 20 h + 900 °C 5 h, 1220 °C 20 h + 900 °C 50 h, and 1220 °C 20 h + 950 °C 100 h). The error bars correspond to the standard deviation 2σ. DE: dendritic region; ID: interdendritic region.

	PropertyState	As-Cast	Homogenized 1220 °C-20 h	Annealed 1220 °C-20 h/900 °C-5 h	Annealed 1220 °C-20 h/900 °C-50 h	Annealed 1220 °C-20 h/950 °C-100 h
Morphological data	Size of primary γ′ precipitates (nm)	60 ± 12	14 ± 3	368 ± 92	315 ± 62	388 ± 132
	Volume fraction of primary γ′ precipitates	~57	~32	~45	~40	~45
	Volume fraction of the (needle-like) Al-Ni phase	~5	0	~1	~4	~9
Mechanical data	Micro HV50/20	343 ± 22 (DE)	341 ± 12	345 ± 14	345 ± 17	373 ± 11
	Nano HV0.025/20 γ-γ′	433 ± 32 (DE)	457 ± 5	448 ± 7	451 ± 5	454 ± 91
	Nano HV0.025/20 Al-Ni phase	736 ± 87 (ID)	-	569 ± 61	560 ± 148	647 ± 72
	Ultimate tensile strength σ_{TS} at RT (MPa)	-	-	786 from [20]	1197 ± 6	758 ± 40
	Yield strength at RT (MPa)	-	-	568 from [20]	648 ± 1	473 ± 38
	Elongation at RT (%)			12 from [20]	26 ± 1	5 ± 1
	Young's modulus at RT (GPa)	-	-	90 ± 18	99 ± 37	118 ± 32
	Ultimate tensile strength σ_{TS} at 800 °C (MPa)	-	-	672 from [20]	575 ± 7	493 ± 40
	Yield strength at 800 °C (MPa)	-	-	535 from [20]	450	310 ± 15
	Elongation to fracture at 800 °C (%)	-	-	27 from [20]	20 ± 4	14 ± 5
	Young's modulus at 800 °C (GPa)	-	-	52 ± 6	54 ± 4	51 ± 10

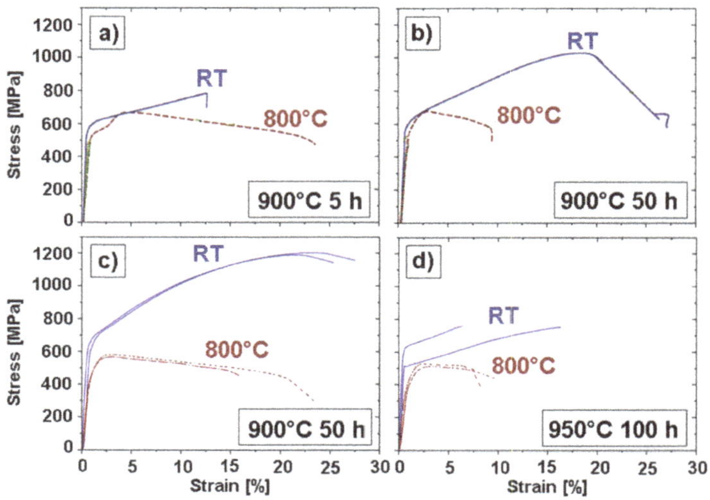

Figure 8. Engineering stress-strain curves of (**a**) the 900 °C 5 h (from [20]); (**b**) the 900 °C 50 h (from [20]); (**c**) the 900 °C 50 h; and (**d**) the 950 °C 100 h annealed states at room temperature and at 800 °C.

A reason for this difference in fracture behavior of the 900 °C and the 950 °C annealed states can be found in the microstructure of the deformed states (see Figure 9).

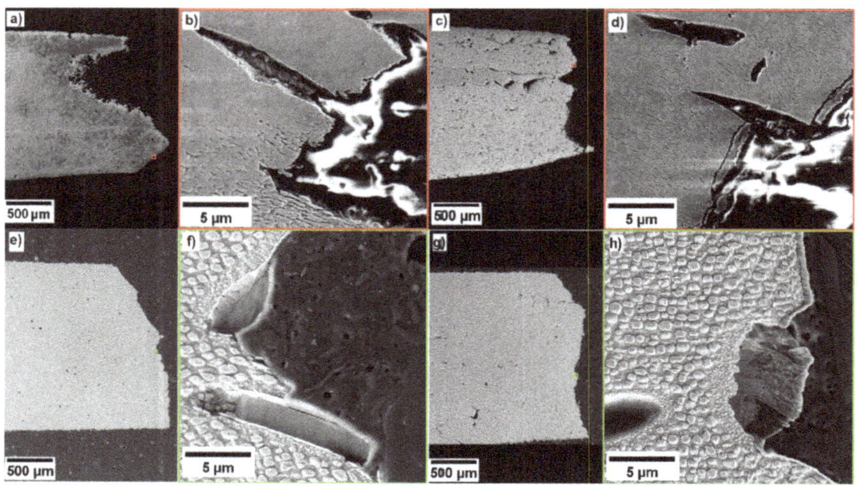

Figure 9. SEM images of fractures of the states (**a**–**d**) 900 °C 50 h tested at (**a**,**b**) RT and (**c**,**d**) 800 °C and (**e**–**h**) 950 °C 100 h tested at (**e**,**f**) RT and (**g**,**h**) at 800 °C. The colored boxes show the zoom to reveal the microstructure.

There are three main factors that play a role in the difference in fracture behavior:

1. The deformation temperature and the implied softening of the material
2. The abundance of the Al-Ni rich needles
3. The state of the γ' precipitates

They will be compared in detail in the following, by their order of magnitude.

1. The deformation temperature is the most important factor. The heating from RT to 800 °C implies a softening of the material and allows for a more ductile tensile behavior. The fracture micrographs of the samples deformed at RT, shown in Figure 9a,e, show little (900 °C 50 h state, Figure 9a) or no (950 °C 100 h state, Figure 9e) necking behavior and thus a quite brittle fracture. The samples deformed at 800 °C (Figure 9c,g) display an important necking and thus a ductile fracture.

This factor is no surprise and the investigated CCA behaves like many other alloys.

2. The amount of needles is lowest in the 900 °C 5 h annealed alloy (about 1%) and highest in the 950 °C 100 h annealed alloy (about 13%). Both the longer annealing times and the higher annealing temperature increase the amount of needles, but the temperature has the stronger influence. In all five states annealed at 950 °C, the volume fraction of needles is up to ten times higher than in the 900 °C annealed states.

When observing the fracture micrographs, it was hardly possible to find any needle in the 900 °C 5 h state, and none at the fracture line (see Figure 9d). The 900 °C 50 h specimen in Figure 9b does show some needles, but the fracture seems to be uninfluenced by them because the fracture line mainly passes through the γ-γ' region. The 950 °C states in Figure 9f,h, however, show a clear fracture preference along the needles all over the specimen.

Note that the aforementioned secondary γ' particles have a preference for the regions around the needles. This is not the case in the 900 °C annealed states. This observation is helpful for determining the broken needles at the fracture line—the secondary γ' ribbons are always visible, even if the needle has quarried out.

These observations lead to the conclusion that the amount of needles is a major factor in the fracture behavior of the $Al_{10}Co_{25}Cr_8Fe_{15}Ni_{36}Ti_6$ alloy, both concerning the earlier fracture and the reproducibility of the results. Thus, in the following, the higher temperature heat treatment at 950 °C will be omitted and the next observations will be focused on the specimens annealed at the more promising heat treatments at 900 °C.

It is interesting to note that the γ' particles in the 900 °C annealed alloys (Figure 9b,d) show a heavy deformation from their original cuboid shape. They are heavily elongated in the 900 °C 50 h annealed state after deformation at RT (Figure 9b) and they adopted a starry shape in the 900 °C 5 h annealed state after deformation at 800 °C. There is hardly any deformation of the γ' particles visible in the 950 °C annealed states (Figure 9f,g). It can thus be concluded that the γ' particles absorbed the plastic deformation in the 900 °C annealed alloys, while in the 950 °C annealed alloys, the important part of the deformation energy concentrated in the interface needle-matrix.

3. The γ-γ' morphology at room temperature has been analyzed in detail (see Section 3.1.2) and is responsible for the difference in behavior between the 900 °C 5 h and the 900 °C 50 h state. As stated above, after 900 °C 5 h annealing, the γ' precipitates have not yet established their optimum cuboidal shape. After 900 °C 50 h annealing, however, with the cuboids sides parallel to the tensile load direction and thus a higher geometrical ordering, the dislocation movement is more difficult. This effect is in accordance with the order hardening mechanism, which is often observed in Ni-based alloys [33,34].

To understand the morphological evolution of the γ' precipitates between 900 °C 5 h and 50 h ageing, the aspect ratios of the γ' precipitates were determined by using the precipitates' diameters r measured in the [100] and the [010] directions. The distribution of these aspect ratios is shown in Figure 10.

In the 900 °C 5 h annealed alloy, there is a widespread, bimodal distribution of the aspect ratios (see Figure 5c). About 37% of the analyzed precipitates have an aspect ratio below 0.75. At the same time, about 39% of them have an aspect ratio >0.9 (see Figure 10a). This result indicates that the precipitates still have no clearly defined cubic shape, as can also be seen in Figure 4c. The secondary precipitates have not been taken into account in the distribution of Figure 5c.

The 50 h annealed alloy has a narrower distribution of aspect ratios. The maximum of the distribution shown in Figure 10b is located at about 0.85–0.9. In this case, the precipitates are close to cubic.

The longer heat treatment allows the primary γ' precipitates to approach their energetically most favorable shape, i.e., cuboids, which are a compromise between the lowest surface-to-volume ratio (spheres) and the lowest elastic energy (perfect cubes).

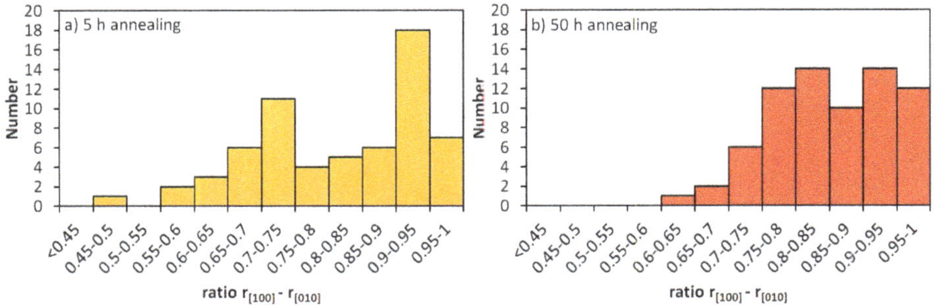

Figure 10. Distribution of the ratios of the γ' precipitate lengths in the $Al_{10}Co_{25}Cr_8Fe_{15}Ni_{36}Ti_6$ alloy along the [100] and [010] directions: (**a**) after 900 °C 5 h and (**b**) after 900 °C 50 h.

3.2.2. Micro- and Nanohardness

The microhardness measurements were performed in the as-cast alloy in the dendritic region and in the annealed specimens randomly, including the Al-Ni rich needles. No microhardness change can be observed through the states as-cast homogenized and annealed at 900 °C. It remains 341–345 HV50/20 independently of the heat treatment. The microhardness is thus not much influenced by the presence of the Al-Ni rich needles nor by the average size of the γ' precipitates, which varies from ~14 nm (homogenized) to 312 ± 55 nm (annealed at 900 °C 5 h). The volume fraction of γ' does not contribute to the microhardness, either. The reason for the similar value of the microhardness of the samples with different states is most probably the size of the indents, which is usually about 20 µm. Thus, the measured microhardness is an average over all phases in the alloy.

The microhardness in the state annealed at 950 °C is slightly higher (about 370 compared to about 345 HV50/20 in the other states). This can probably be linked to the significantly higher amount of Al-Ni rich needles (13 vol.% compared to 1–3 vol.%)

A nanoindentation hardness map confirms that there are differences in hardness between the phases. Figure 11 shows a set of 100 nanoindents on the surface of two exemplary states, i.e., as-cast and annealed at 900 °C 50 h. The small load of 25 mN has been chosen in order to make sure that the influence of the small needles has the highest impact in the indent. Note that another set with a load of 100 mN shows less difference between the needles and the γ-γ' region and is therefore not shown here.

The as-cast alloy (Figure 11a) shows the most important local hardness changes, due to the dendritic morphology. The interdendritic regions are much harder (>700 HV0.025/20, shown in red) than the dendritic regions (~430 HV0.025/20). In the 900 °C 50 h specimen (Figure 11b), it was possible to place the indents on some Al-Ni rich needle phases (shown in red). These are harder (>500 HV0.025/20) than the γ-γ' region (~450 HV0.025/20) around them. In order to obtain better statistics, several hardness indents were placed on various Al-Ni needles (images not shown here). The 1220 °C 20 h homogenized sample (not shown here) has a very homogeneous nanohardness of

457 ± 5 HV0.025/20. The nanohardness values of the samples aged at 900 °C 5 h and 950 °C 100 h are comparable to that of the sample aged at 900 °C 50 h. All hardness values are summarized in Table 2.

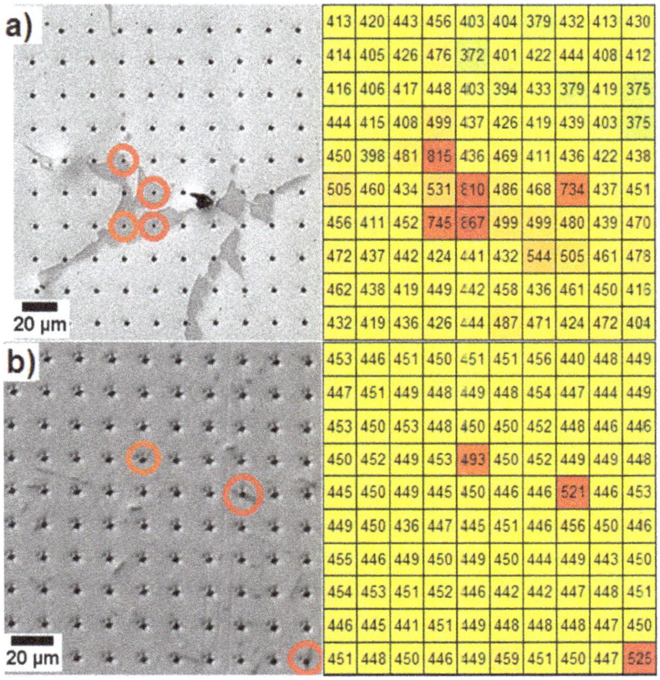

Figure 11. Picodentor hardness maps of two investigated states of the $Al_{10}Co_{25}Cr_8Fe_{15}Ni_{36}Ti_6$ alloy. Left column: the position of the nanoindentations; right column: the corresponding color-coded HV0.0025/20 Vickers hardness values (red = highest values). (**a**) The as-cast sample; (**b**) the sample aged at 900 °C 50 h.

Unlike in the tensile tests, the preferred cuboidal shape of γ' in the 900 °C 50 h annealed alloy does not have any effect. However, the dislocation movement created by an indentation is much more arbitrary than in a directional tensile test. The orientation of the dislocation lines could thus play a role in the alloy's mechanical properties, emphasizing its anisotropic behavior.

Table 2 summarizes the γ' precipitates' size and the γ' precipitates' volume fraction, the micro- and nanohardness, the ultimate tensile strength σ_{TS} at room temperature, and the ultimate tensile strength at 800 °C for the five mechanically investigated states.

3.3. Comparison with Other Alloys

The microscopic data obtained in this work allows an overview of different properties for the following alloys:

- High entropy alloys $Al_{10}Co_{25}Cr_8Fe_{15}Ni_{36}Ti_6$, $Co_{20}Cr_{20}Fe_{20}Mn_{20}Ni_{20}$, $Al_8Cr_{17}Co_{17}Cu_8Fe_{17}Ni_{33}$
- Ni-based Alloy 800H and Inconel 617, Alloy 800, and Inconel 706 single crystal Ni-based CMSX-4
- Co-based Co-9Al-9W-2Ta-0.02B
- Co-Ni based Co-30Ni-10Al-5Mo-2Ta and TMW-4M3-1.

γ' sizes and volume fractions are compared with the ultimate tensile strength at 800 °C and are summarized in Table 3, along with additional phases.

Table 3. Comparison of different microstructural and mechanical parameters of $Al_{10}Co_{25}Cr_8Fe_{15}Ni_{36}Ti_6$ and other (commercial) alloys.

	Alloy	Size of the Primary γ' Precipitates (after Optimum Heat Treatment)	Volume Fraction of the Primary γ' Precipitates (after Optimum Heat Treatment)	Additional Phases (except γ Matrix)	Ultimate Tensile Strength at 800 °C (MPa)	Ref.
HEAS	$Al_8Cr_{17}Co_{17}Cu_8Fe_{17}Ni_{33}$	20 nm	20%	$L1_2$ at grain boundaries	Not known	[14,15]
	$Al_{10}Co_{25}Cr_8Fe_{15}Ni_{36}Ti_6$	191 nm	36%	Al-Ni rich needles	565	this work
	$Co_{20}Cr_{20}Fe_{20}Mn_{20}Ni_{20}$	-	-	-	~170	[21]
	$Al_{17}Co_{17}Cr_{17}Cu_{17}Fe_{17}Ni_{17}$	-	-	A2, B2, fcc1, fcc2, $L1_2$	180	[22]
Ni-based alloys	CMSX-4	500 nm	70%	-	~1100	[33,35]
	Inconel 617	80 nm	5%	$Ni_2(Cr,Mo)$; MC; $M_{23}C_6$; M_6C	~500	[36,37]
	Inconel 706	10–100 nm	<25%	γ'', η, carbides, nitrides, (Ni_3Nb)	Not known	[38]
	Alloy 800	<10 nm	1%	$Cr_{23}C_6$, TiC	~180	[39–41]
	Alloy 800H	<100 nm	small	$Cr_{23}C_6$, TiC	~200	[39,42]
Co-based alloy	Co-9Al-9W-2Ta-0.02B	~200 nm	~70%	-	~580	[43,44]
Co-Ni based alloys	Co-30Ni-10-A-5Mo-2Ta	30–50 nm	~74%	-	>600	[45]
	TMW 4M3-1	<2.5 μm	17%	Secondary & tertiary γ', σ, η	below 1122	[46,47]

In the category "Size of γ' precipitates", the $Al_{10}Co_{25}Cr_8Fe_{15}Ni_{36}Ti_6$ alloy is comparable to Co-based Co-9Al-9W-2Ta-0.02B.

The volume fraction of the γ' precipitates of the compared alloys can be classified into three groups: the highest volume fractions of about 70% are obtained in the single crystal CMSX-4, Co-based Co-9Al-9W-2Ta-0.02B and Co-Ni based Co-30Ni-10Al-5Mo-2Ta. An average volume fraction of 25–40% is obtained in the $Al_{10}Co_{25}Cr_8Fe_{15}Ni_{36}Ti_6$ alloy and Inconel 706. All other compared alloys have very low γ' volume fractions, of 5% or below.

Alloys with many additional phases, like the equiatomic $Al_{17}Co_{17}Cr_{17}Cu_{17}Fe_{17}Ni_{17}$, Inconel 706, Alloy 800, and Alloy 800H, usually have a rather low ultimate tensile strength at 800 °C. The exceptions are Inconel 617 and TMW-4M3-1. Generally speaking, a high number of phases does not seem to be advisable for an ultimate tensile strength at 800 °C.

The ultimate tensile strength at 800 °C is highest in the single Crystal CMSX-4 and in Co-Ni-based TMW-4M3-1. The $Al_{10}Co_{25}Cr_8Fe_{15}Ni_{36}Ti_6$ alloy's ultimate tensile strength is slightly higher, but most similar to the Co-30Ni-10Al-5Mo-2Ta, Co-9Al-9W-2Ta-0.02B, and Inconel 617.

There seems to be a correlation between the γ' volume fraction and the ultimate tensile strength at 800 °C. The latter would probably benefit from an increase of the γ' volume fraction in the $Al_{10}Co_{25}Cr_8Fe_{15}Ni_{36}Ti_6$ alloy.

Of all the eleven compared alloys, the $Al_{10}Co_{25}Cr_8Fe_{15}Ni_{36}Ti_6$ alloy shows the most similarities to the single crystal CMSX-4 alloy and the Co-9Al-9W-2Ta-0.02B alloy. The size of the γ' precipitates, the volume fraction, and the ultimate tensile strength at 800 °C lie between those of CMSX-4 and Inconel 617. Further optimization will probably continue the improvement of the $Al_{10}Co_{25}Cr_8Fe_{15}Ni_{36}Ti_6$ alloy's properties. We particularly expect in the future to improve the ultimate tensile strength at 800 °C by increasing the volume fraction of γ' precipitates by an adequate heat treatment or through the addition of traces of γ' builders.

4. Summary

The $Al_{10}Co_{25}Cr_8Fe_{15}Ni_{36}Ti_6$ alloy solidifies dendritically and the dendrite cores are characterized by a two-fold morphology consisting of γ' precipitates inside a disordered fcc matrix. The dendritic structure is not present after the homogenization at 1220 °C/20 h. However, the formation of γ' precipitates could not be prevented in the homogenized sample.

After ageing at 900 °C or 950 °C, a third phase joins the γ-γ' morphology, i.e., the Al-Ni rich needles. The longer annealing times and higher annealing temperatures increase the volume fraction of the needles, from ~1% after 900 °C 5 h to ~13% after 950 °C 200 h. It stabilizes after 200 h at 950 °C.

The γ-γ' morphology can be observed in all states from the as-cast to the 1220 °C 20 h homogenized state and the subsequently annealed states at 900 °C and 950 °C. The size of the γ' particles is smallest after homogenization and increases drastically to over 800 nm in average diameter after the longest and hottest annealing at 950 °C 1000 h. After annealing at 950 °C, the γ' precipitates grow according to the LSW theory of Ostwald ripening. This is not the case after annealing at 900 °C. The softest annealing at 900 °C 5 h produces larger average precipitates than the annealing at 900 °C 50 h.

Tensile tests of the chosen states 900 °C 5 h, 900 °C 50 h, and 950 °C 100 h reveal that the amount of Al-Ni rich needles is the main factor responsible for an early and unpredictable fracture. The high amount in the 950 °C treated states is thus unwanted and this annealing temperature can be excluded in the future. In the 900 °C annealed samples, the longer annealing time of 50 h intensifies the cuboidal shape of the γ' precipitates, which implies better tensile results both at RT and at 800 °C. The small amount of needles does not seem to disturb the deformation mechanism.

Microhardness values are practically identical in all investigated states. At this scale, neither the γ-γ' morphology nor the Al-Ni rich needles influence the hardness. However, nanohardness maps show a difference in hardness between the softer γ-γ' region and the harder Al-Ni rich needles in the annealed alloys.

When comparing the $Al_{10}Co_{25}Cr_8Fe_{15}Ni_{36}Ti_6$ alloy to ten other alloys, i.e., Ni-based, Co-based, Co-Ni-based, and other high entropy alloys, it could be concluded that the data from the $Al_{10}Co_{25}Cr_8Fe_{15}Ni_{36}Ti_6$ alloy is amongst the best in the categories γ' particle size, γ' volume fraction, and ultimate tensile strength at the desired 800 °C. The properties are approaching those of single crystal CMSX-4 and Co-based Co-9Al-9W-2Ta-0.02B at much lower costs.

Author Contributions: Conceptualization, A.M.M.; Methodology, A.M.M., H.D, S.H., and C.F.; Software, A.M.M. and S.H.; Validation, A.M. and S.H.; Formal Analysis, A M.M. and S.H.; Investigation, A.M.M., H.D. and S.H.; Writing—Original Draft Preparation, A.M.M.; Writing—Review & Editing, S.H. and N.W.; Visualization, A.M.M. and S.H.; Supervision, N.W. and U.G.; Project Administration, A.M.M., N.W., and U.G.; Funding Acquisition, A.M.M., N.W., and U.G.

Funding: This work was financed by the German Research foundation (DFG), i.e., during the projects WA 1378/15-2, GL 181/25-2, MA 7004/1-1, and GL 181/50-1.

Acknowledgments: The authors would like to thank C. Leistner for help with casting. Many thanks to M. Osenberg for his most valuable help with image segmentation procedures.

Conflicts of Interest: The authors declare no conflict of interest.

References

1. Di Martino, S.F.; Faulkner, R.G.; Hogg, S.C.; Vujic, S.; Tassa, O. Characterisation of microstructure and creep properties of alloy 617 for high-temperature applications. *Mater. Sci. Eng. A Struct. Mater. Prop. Microstruct. Process.* **2014**, *619*, 77–86. [CrossRef]
2. Zhang, W.R.; Liaw, P.K.; Zhang, Y. Science and technology in high-entropy alloys. *Sci. China Mater.* **2018**, *61*, 2–22. [CrossRef]
3. Klimova, M.; Stepanov, N.; Shaysultanov, D.; Chernichenko, R.; Yurchenko, N.; Sanin, V.; Zherebtsov, S. Microstructure and mechanical properties evolution of the al, c-containing cocrfenimn-type high-entropy alloy during cold rolling. *Materials* **2018**, *11*, 53. [CrossRef] [PubMed]
4. Liu, X.Y.; Yin, H.; Xu, Y. Microstructure, Mechanical and tribological properties of oxide dispersion strengthened high-entropy alloys. *Materials* **2017**, *10*, 1312. [CrossRef] [PubMed]
5. Nam, S.; Kim, M.J.; Hwang, J.Y.; Choi, H. Strengthening of Al0.15CoCrCuFeNiTix–C (x = 0, 1, 2) high-entropy alloys by grain refinement and using nanoscale carbides via powder metallurgical route. *J. Alloys Compd.* **2018**, *762*, 29–37. [CrossRef]
6. Manzoni, A.M.; Glatzel, U. New multiphase compositionally complex alloys driven by the high entropy alloy approach. *Mater. Charact.* **2018**, in press. [CrossRef]
7. Durand-Charre, M. *The Microstructure of Superalloys*; CRC Press: Boca Raton, FL, USA; London, UK; New York, NY, USA; Washington, DC, USA, 1997; p. 124.
8. Singh, S.; Wanderka, N.; Kiefer, K.; Siemensmeyer, K.; Banhart, J. Effect of decomposition of the Cr–Fe–Co rich phase of AlCoCrCuFeNi high entropy alloy on magnetic properties. *Ultramicroscopy* **2011**, *111*, 619–622. [CrossRef] [PubMed]
9. Singh, S.; Wanderka, N.; Murty, B.S.; Glatzel, U.; Banhart, J. Decomposition in multi-component AlCoCrCuFeNi high-entropy alloy. *Acta Mater.* **2011**, *59*, 182–190. [CrossRef]
10. Tong, C.J.; Chen, M.R.; Chen, S.K.; Yeh, J.W.; Shun, T.T.; Lin, S.J.; Chang, S.Y. Mechanical performance of the Al x CoCrCuFeNi high-entropy alloy system with multiprincipal elements. *Metall. Mater. Trans. A* **2005**, *36*, 1263–1271. [CrossRef]
11. Tong, C.J.; Chen, Y.L.; Chen, S.K.; Yeh, J.W.; Shun, T.T.; Tsau, C.H.; Lin, S.J.; Chang, S.Y. Microstructure characterization of Al x CoCrCuFeNi high-entropy alloy system with multiprincipal elements. *Metall. Mater. Trans. A* **2005**, *36*, 881–893. [CrossRef]
12. Daoud, H.M.; Manzoni, A.; Völkl, R.; Wanderka, N.; Glatzel, U. Microstructure and tensile behavior of Al8Co17Cr17Cu8Fe17Ni33 (at.%) high-entropy alloy. *JOM* **2013**, *65*, 1805–1814. [CrossRef]
13. Manzoni, A.; Daoud, H.; Völkl, R.; Glatzel, U.; Wanderka, N. Phase separation in equiatomic AlCoCrFeNi high-entropy alloy. *Ultramicroscopy* **2013**, *132*, 212–215. [CrossRef] [PubMed]
14. Manzoni, A.; Daoud, H.; Mondal, S.; van Smaalen, S.; Völkl, R.; Glatzel, U.; Wanderka, N. Investigation of phases in $Al_{23}Co_{15}Cr_{23}Cu_8Fe_{15}Ni_{16}$ and $Al_8Co_{17}Cr_{17}Cu_8Fe_{17}Ni_{33}$ high entropy alloys and comparison with equilibrium phases predicted by Thermo-Calc. *J. Alloys Compd.* **2013**, *552*, 430–436. [CrossRef]

15. Manzoni, A.M.; Daoud, H.M.; Voelkl, R.; Glatzel, U.; Wanderka, N. Influence of W, Mo and Ti trace elements on the phase separation in $Al_8Co_{17}Cr_{17}Cu_8Fe_{17}Ni_{33}$ based high entropy alloy. *Ultramicroscopy* **2015**, *159*, 265–271. [CrossRef] [PubMed]
16. Daoud, H.M.; Manzoni, A.M.; Völkl, R.; Wanderka, N.; Glatzel, U. Oxidation Behavior of $Al_8Co_{17}Cr_{17}Cu_8Fe_{17}Ni_{33}$, $Al_{23}Co_{15}Cr_{23}Cu_8Fe_{15}Ni_{15}$, and $Al_{17}Co_{17}Cr_{17}Cu_{17}Fe_{17}Ni_{17}$ Compositionally Complex Alloys (High-Entropy Alloys) at Elevated Temperatures in Air. *Adv. Eng. Mater.* **2015**, *17*, 1134–1141. [CrossRef]
17. Jien-Wei, Y. Recent progress in high entropy alloys. *Ann. Chim. Sci. Mat.* **2006**, *31*, 633–648.
18. *The Version Tccr*; Thermocalc Software AB: Stockholm, Sweden. Available online: http://www.Thermocalc.Com (accessed on 15 June 2016).
19. *Thermotech Ni-Based Superalloys Database, TTNi7, 7.0*, Thermo-Calc Software AB: Stockholm, Sweden, 2006.
20. Daoud, H.M.; Manzoni, A.M.; Wanderka, N.; Glatzel, U. High-temperature tensile strength of $Al_{10}Co_{25}Cr_8Fe_{15}Ni_{36}Ti_6$ compositionally complex alloy (high-entropy alloy). *JOM* **2015**, *67*, 2271–2277. [CrossRef]
21. Otto, F.; Dlouhy, A.; Somsen, C.; Bei, H.; Eggeler, G.; George, E.P. The influences of temperature and microstructure on the tensile properties of a CoCrFeMnNi high-entropy alloy. *Acta Mater.* **2013**, *61*, 5743–5755. [CrossRef]
22. Kuznetsov, A.V.; Shaysultanov, D.G.; Stepanov, N.D.; Salishchev, G.A.; Senkov, O.N. Tensile properties of an AlCrCuNiFeCo high-entropy alloy in as-cast and wrought conditions. *Mater. Sci. Eng. A Struct. Mater. Prop. Microstruct. Process.* **2012**, *533*, 107–118. [CrossRef]
23. Arganda-Carreras, I.; Kaynig, V.; Rueden, C.; Eliceiri, K.W.; Schindelin, J.; Cardona, A.; Seung, H.S. Trainable weka segmentation: A machine learning tool for microscopy pixel classification. *Bioinformatics* **2017**, *33*, 2424–2426. [CrossRef] [PubMed]
24. Schindelin, J.; Arganda-Carreras, I.; Frise, E.; Kaynig, V.; Longair, M.; Pietzsch, T.; Preibisch, S.; Rueden, C.; Saalfeld, S.; Schmid, B.; et al. Fiji: An open-source platform for biological-image analysis. *Nat. Methods* **2012**, *9*, 676. [CrossRef] [PubMed]
25. Rueden, C.T.; Schindelin, J.; Hiner, M.C.; DeZonia, B.E.; Walter, A.E.; Arena, E.T.; Eliceiri, K.W. ImageJ2: ImageJ for the next generation of scientific image data. *BMC Bioinform.* **2017**, *18*, 529. [CrossRef] [PubMed]
26. Schneider, C.A.; Rasband, W.S.; Eliceiri, K.W. Nih image to imagej: 25 years of image analysis. *Nat. Methods* **2012**, *9*, 671. [CrossRef] [PubMed]
27. Wagner, C. Theorie der Alterung von Niederschlägen durch Umlösen (Ostwald-Reifung). *Z. Elektrochem Ber. Bunsenges. Phys. Chem.* **1961**, *65*, 581–591.
28. Lifshitz, I.M.; Slyozov, V.V. The kinetics of precipitation from supersaturated solid solutions. *J. Phys. Chem. Solids* **1961**, *19*, 35–50. [CrossRef]
29. Kahlweit, M. Ostwald ripening of precipitates. *Adv. Colloid Interface Sci.* **1975**, *5*, 1–35. [CrossRef]
30. Baldan, A. Review progress in Ostwald ripening theories and their applications to nickel-base superalloys Part I: Ostwald ripening theories. *J. Mater. Sci.* **2002**, *37*, 2171–2202. [CrossRef]
31. Sudbrack, C.K.; Yoon, K.E.; Mao, Z.; Noebe, R.D.; Isheim, D.; Seidman, D.N. *Temporal Evolution of Nanostructures in a Model Nickel-Base Superalloy: Experiments and Simulations*; Minerals, Metals & Materials Soc.: Warrendale, PA, USA, 2003; pp. 43–50.
32. Gleiter, H. Microstructure. In *Physical Metallurgy*; Cahn, R.W., Hassen, P., Eds.; North-Holland Physics Publishing: Amsterdam, The Netherlands; Oxford, UK; New York, NY, USA; Tokyo, Japan, 1983; Volume 1, p. 1973.
33. Sengupta, A.; Putatunda, S.K.; Bartosiewicz, L.; Hangas, J.; Nailos, P.J.; Peputapeck, M.; Alberts, F.E. Tensile behavior of a new single-crystal nickel-based superalloy (CMSX-4) at room and elevated temperatures. *J. Mater. Eng. Perform.* **1994**, *3*, 73–81. [CrossRef]
34. Müller, L.; Glatzel, U.; Feller-Kniepmeier, M. Calculation of the internal stresses and strains in the microstructure of a single crystal nickel-base superalloy during creep. *Acta Metall. Mater.* **1993**, *41*, 3401–3411. [CrossRef]
35. Reed, R.C. *The Superalloys. Fundamentals and Applications*; Cambridge University Press: Cambridge, UK, 2006.
36. Wu, Q.Y.; Song, H.J.; Swindeman, R.W.; Shingledecker, J.P.; Vasudevan, V.K. Microstructure of long-term aged in617 Ni-base superalloy. *Metal. Mater. Trans. A Phys. Met. Mater. Sci.* **2008**, *39*, 2569–2585. [CrossRef]

37. Inconel Alloy 617. Available online: http://www.Specialmetals.Com/documents/inconel%20alloy%20617.Pdf (accessed on 3 December 2015).
38. Wanderka, N.; Naundorf, V.; Banhart, J.; Mukherji, D.; Genovesse, D.D.; Rosler, J. Microstructural characterization of Inconel 706 alloy. *Surf. Interface Anal.* **2004**, *36*, 546–551. [CrossRef]
39. Alloy 800. Available online: http://www.Sandmeyersteel.Com/images/alloy-800-spec-sheet.Pdf (accessed on 3 December 2015).
40. Coppola, R.; Fiorentin, S.R. Study of γ'-precipitation kinetics in alloy 800 at 575 °C by small angle neutron scattering. *Nucl. Instr. Meth. Phys. Res. Sect. B Beam Interact. Mater. Atoms* **1987**, *22*, 564–572. [CrossRef]
41. Vittori, M. Gamma particle coarsening and yield in alloy 800. *J. Mater. Sci.* **1981**, *16*, 3461–3469. [CrossRef]
42. Nilsson, J.O.; Thorvaldsson, T. Low cycle fatigue behavior of alloy 800 h at 600–800 °C—Effect of grain size and gamma particle dispersion. *Fatigue Fracture Eng. Mater. Struct.* **1985**, *8*, 373–384. [CrossRef]
43. Zhong, F.; Li, S.S.; Sha, J.B. Tensile behaviour of Co–Al–W–Ta–B–Mo alloys with a coherent γ/γ' microstructure at room and high temperatures. *Mater. Sci. Eng. A Struct. Mater. Prop. Microstruct. Process.* **2015**, *637*, 175–182. [CrossRef]
44. Feng, G.; Li, H.; Li, S.S.; Sha, J.B. Effect of Mo additions on microstructure and tensile behavior of a Co–Al–W–Ta–B alloy at room temperature. *Scripta Mater.* **2012**, *67*, 499–502. [CrossRef]
45. Makineni, S.K.; Samanta, A.; Rojhirunsakool, T.; Alam, T.; Nithin, B.; Singh, A.K.; Banerjee, R.; Chattopadhyay, K. A new class of high strength high temperature cobalt based γ–γ' Co–Mo–Al alloys stabilized with Ta addition. *Acta. Mater.* **2015**, *97*, 29–40. [CrossRef]
46. Gu, Y.; Zhong, Z.; Yuan, Y.; Osada, T.; Cui, C.; Yokokawa, T.; Harada, H. An advanced cast-and-wrought superalloy (TMW-4M3) for turbine disk applications beyond 700 C. In Proceedings of the International Symposium on Superalloys, Champion, PA, USA, 10 September 2012; Volume 903, pp. 903–910.
47. Osada, T.; Gu, Y.F.; Nagashima, N.; Yuan, Y.; Yokokawa, T.; Harada, H. Optimum microstructure combination for maximizing tensile strength in a polycrystalline superalloy with a two-phase structure. *Acta Mater.* **2013**, *61*, 1820–1829. [CrossRef]

entropy

MDPI

Article

Microstructure and Mechanical Properties of Precipitate Strengthened High Entropy Alloy $Al_{10}Co_{25}Cr_8Fe_{15}Ni_{36}Ti_6$ with Additions of Hafnium and Molybdenum

Sebastian Haas [1], Anna M. Manzoni [2], Fabian Krieg [1] and Uwe Glatzel [1,*]

[1] Metals and Alloys, University Bayreuth, 95440 Bayreuth, Germany; sebastian-haas@uni-bayreuth.de (S.H.); fabian.krieg@uni-bayreuth.de (F.K.)

[2] Helmholtz-Zentrum Berlin für Materialien und Energie GmbH, 14109 Berlin, Germany; anna.manzoni@helmholtz-berlin.de

* Correspondence: uwe.glatzel@uni-bayreuth.de; Tel.: +49-921-55-5555

Academic Editor: Yong Zhang

Received: 28 January 2019; Accepted: 8 February 2019; Published: 12 February 2019

Abstract: High entropy or compositionally complex alloys provide opportunities for optimization towards new high-temperature materials. Improvements in the equiatomic alloy $Al_{17}Co_{17}Cr_{17}Cu_{17}Fe_{17}Ni_{17}$ (at.%) led to the base alloy for this work with the chemical composition $Al_{10}Co_{25}Cr_8Fe_{15}Ni_{36}Ti_6$ (at.%). Characterization of the beneficial particle-strengthened microstructure by scanning electron microscopy (SEM) and observation of good mechanical properties at elevated temperatures arose the need of accomplishing further optimization steps. For this purpose, the refractory metals hafnium and molybdenum were added in small amounts (0.5 and 1.0 at.% respectively) because of their well-known positive effects on mechanical properties of Ni-based superalloys. By correlation of microstructural examinations using SEM with tensile tests in the temperature range of room temperature up to 900 °C, conclusions could be drawn for further optimization steps.

Keywords: HEA; high entropy alloys; compositionally complex alloys; mechanical characterization

1. Introduction

For centuries, constructions with high mechanical requirements have been built using steel or iron-based alloys, while titanium or aluminum have been used for light-weight constructions. In gas turbines nickel- or cobalt-based superalloys are used because of their extremely good mechanical properties and oxidation resistance over long periods of time at high temperatures. The main properties of these alloys are given by their base element, while the addition of other elements in small amounts leads to fine adjustments of a specific behavior [1].

An unconventional approach of alloy design is based not on a main element, but on a chemical composition that exhibits a large number of elements with none of them dominating. In this group one differentiates between single-phase high entropy alloys (HEA) [2] and multiphase compositionally complex alloys (CCA). Multicomponent alloys were examined first by Yeh [2] and Cantor [3]. Investigating the microstructure of alloys with different numbers of elements in equiatomic mixtures, Cantor found the five-component, equiatomic alloy $Co_{20}Cr_{20}Fe_{20}Mn_{20}Ni_{20}$ to form a single-phase solid-solution. High configurational entropy, due to the large number of elements and similar concentrations, leads to a decrease in the Gibbs free energy and thus a stabilization of one solid-solution phase if no intermetallic phase with low enthalpy of formation exists [2]. Such a single-phase material was claimed to have outstanding properties, e.g., a good strength-ductility behavior over a wide temperature range, thermal stability, wear resistance and high resistance against oxidation of the material [4].

Most alloys containing at least five elements in near equiatomic composition, however, do not crystallize as a single-phase solid-solution, but rather form intermetallic compounds as secondary phases with even lower Gibbs free energy. This leads to the formation of compositionally complex alloys (CCA). They cover a wide range of chemical compositions, with a high number of occurring phases and therefore a wide perspective in terms of applications. We restricted ourselves to have at least five elements and no element should dominate the composition. In the domain of high-temperature materials a gap between steels (<650 °C) and nickel-based superalloys (>850 °C) might be filled by less cost-intensive compositionally complex alloys. The focus should be on mechanical strength, oxidation resistance, processability and, last but not least, the cost factor. We therefore compared our alloy with commercially available alloys. The Ni-Co-Cr-Mo alloy IN 617 shows high-temperature strength and oxidation resistance, while Alloy 800 H has good creep properties and a quite ductile behavior at temperatures below 600 °C. The base alloy of our work shows a similar melting range and is optimized with respect to the features mentioned above. Possible industrial applications could be: Chemical industry appliances and especially land-based steam turbines as well as parts for gas turbines in sections with lower temperatures (700–800 °C).

The development of CCA for mechanical applications at both room and higher temperature is based on different microstructural approaches (see the review article by Manzoni [5]): Interpenetrating phases lead to a high resistance of the material, and additionally interesting electrical or magnetic properties may appear [6]. One example to strengthen the equiatomic high entropy alloy CoCrFeMnNi is the addition of 2 at.% carbon. This leads to a fcc-microstructure with embedded $Cr_{23}C_6$ particles and higher strength than for the CoCrFeMnNi alloy. The best strength-ductility relation of this carbide-strengthened alloy can be reached by applying the correct thermomechanical treatment [7]. Another example of particle strengthened CCA is the alloy $Al_{12.35}Co_{17.5}Cr_{17.5}Fe_{35.15}Ni_{17.5}$ (in at.%). This microstructure exhibits a bcc-matrix with coherently embedded particles of B2 structure. The cuboidal nanoscale particles lead to high strength at room and high temperatures [8].

These three mentioned approaches show the diversity in designing CCA for mechanical applications, using strengthening mechanisms based on very different microscopic features. Singh et al. [9] examined the six-component, equiatomic alloy AlCoCrCuFeNi, with a focus on solidification behavior, using different techniques to analyze the microstructure. Even a high cooling rate by splat-quenching leads to a phase decomposition and the cast alloy shows a microstructure with several phases of different crystal structures, unlike expected solid-solution stabilization due to a high configurational entropy. Recognizing the difficulties of avoiding phase separation, a transition was made from developing a HEA to optimizing a CCA.

Much effort was put into optimizing alloys with near equiatomic composition of AlCoCrCuFeNi, investigating changes in composition and finding correct heat treatment parameters [10], resulting in the chemical composition $Al_{10}Co_{25}Cr_8Fe_{15}Ni_{36}Ti_6$ (in at.%). This alloy composition is the base alloy for this work, exhibiting $L1_2$-ordered, coherently embedded precipitates in a fcc-matrix, comparable to nickel-based superalloys. Additionally the base alloy shows another third phase, appearing in a needle-like shape up to a length of 50 µm. These needles are very rich in aluminum (28 at.%), with a reduction of all other elements, especially Fe, Co and Cr. TEM investigations identified this phase to have a Heusler type structure.

A way to improve the mechanical behavior is directional solidification, where grains are oriented in loading direction. As a material itself has quite good creep properties, for example in the case of superalloys, the grain boundaries play a strong role in weakening the material at higher temperatures. Avoiding grain boundaries inclined at 45° by the loading direction, shear stresses along these weaker boundaries can be eliminated and the strength is just defined by the microstructure inside of the longitudinal orientated grains [11].

Nickel-based superalloys provide further ideas to improve the high-temperature properties, due to similarities in microstructure. By the addition of Hf and/or Mo, different approaches for optimization are expected: Molybdenum is supposed to partition to the matrix, to have a solid-solution strengthening effect due to its bigger atomic size and therefore to lead to a distorted lattice, acting as an obstacle for dislocation movement [12]. The positive effect of hafnium is due to a strengthening of grain boundaries: A columnar-grained microstructure for example shows an enhancement of creep ductility and lifetime by the addition of hafnium that strengthens the vicinity of grain boundaries vertical to the load stress [13]. Doherty et al. [14] explain this by Hf participating more likely to the γ'-phase $Ni_3(Al, Hf)$, leading to a strengthening of it. The grain boundaries are therefore strengthened by the fracture-retarding effect of the interlocking γ'-configuration in this areas [15].

2. Materials and Methods

2.1. Alloy Preparation

In this work all chemical compositions are given in at.%. The composition of the base alloy $Al_{10}Co_{25}Cr_8Fe_{15}Ni_{36}Ti_6$ was changed by adding 0.5 at.% hafnium or 1.0 at.% molybdenum on the expense of aluminum. The added amounts and the reduction of aluminum were identified by simulations using the ThermoCalc software [16] with the database TTNi7 [17]. The base alloy in its original composition is also considered.

All constituents with a purity of 99.99% were cleaned in ethanol using an ultrasonic bath and were then melted in a vacuum induction furnace. The material was distributed randomly in a ceramic crucible in the middle of a water-cooled Cu-coil. After evacuating the chamber twice to a pressure of $5{\cdot}10^{-4}$ mbar it was flooded with argon to prevent the evaporation of elements, especially chromium. The ceramic mold was heated up to a temperature of 1400 °C by a second coil and a graphite receptor, thus the material remained in liquid state after casting. To achieve directionally solidified grains in the [001]-direction, the Bridgman process was used and the mold was withdrawn through a water cooled baffle with a speed of 3 mm/min. The cast rods, with a diameter of 20 mm and a length of about 110 mm, were homogenized for 20 h at 1220 °C in the case of the Mo-containing alloy and at 1140 °C in the case of the Hf-containing alloy to avoid eutectic formation determined by differential scanning calorimetry. Subsequent annealing was performed in two different ways for both alloys, 900 °C/50 h and 950 °C/100 h respectively. After heat treatment the rods cooled down to room temperature in the furnace. To remove the oxide layer the samples were initially sand-blasted and afterwards treated with aqua regia. The rods were cut by electrical discharge machining to obtain samples for microscopic and mechanical characterization.

2.2. Microstructural Observations

Flat disks were cut from the rods and cut again in the middle to examine the cross and longitudinal section of the microstructure. These surfaces were embedded in a conductive resin, ground, polished with 1 μm diamond slurry and finally polished chemically. The specimens were etched with a solution of 3 g Mo-acid in 100 mL H_2O, 100 mL HCl and 100 mL HNO_3 to achieve a better phase contrast by dissolving γ'- and Heusler type phase. For the examination we used a scanning electron microscope (SEM) Zeiss 1540EsB Cross Beam, operating under an accelerating voltage of 30 kV and using the SE2-detector for imaging. Precipitate size and volume fraction were determined using the classification of the Weka segmentation method [18] in the open source software Fiji [19], based on ImageJ [20,21]. More than 500 particles per state were analyzed.

2.3. Mechanical Tests

Electrical discharge machining was employed to obtain flat specimens for high temperature tensile tests. The square cross section of the samples was 1.0 × 1.9 mm^2 and the gauge length was 8 mm, while the entire length of the sample was 25 mm. The specimens were cut out in such a way that the tensile direction was parallel to the [001]-grain-orientation. Before assembling the specimen, their surfaces were ground and a type-S thermocouple was welded for the regulation of temperature on the lower end of the gauge length. The sample was then attached to the ceramic clamping, the radiant heated furnace was closed and heated up to the desired temperature. Tensile tests were performed with a deformation rate of 0.01 mm/s (corresponding to 1.3×10^{-3} 1/s). A load cell and a high-resolution camera were logging about four pairs of values for stress and strain in one second. This lead to engineering stress–strain curves for each test, providing mechanical parameters like ultimate tensile strength (UTS), yield strength (YS) and strain to failure (ε_f).

3. Results and Discussion

3.1. Chemical and Microstructural Analysis of $Al_{10}Co_{25}Cr_8Fe_{15}Ni_{36}Ti_6$

The alloy $Al_{10}Co_{25}Cr_8Fe_{15}Ni_{36}Ti_6$ exhibited a dendritic microstructure after the casting process. Dendrites could be dissolved by a homogenization heat-treatment at 1220 °C for 20 h. Subsequent annealing for 50 h at 900 °C lead to a three-phase microstructure: Figure 1a) shows the large (several 10 μm), randomly distributed Heusler type phase, with its characteristic needle-like shape and a volume-fraction of <5%. The γ'-microstructure is displayed in Figure 1b) with a higher magnification with $L1_2$-ordered cuboidal shaped γ'-precipitates and small matrix channels with round secondary γ'-particles. The matrix has a face-centered cubic structure [22].

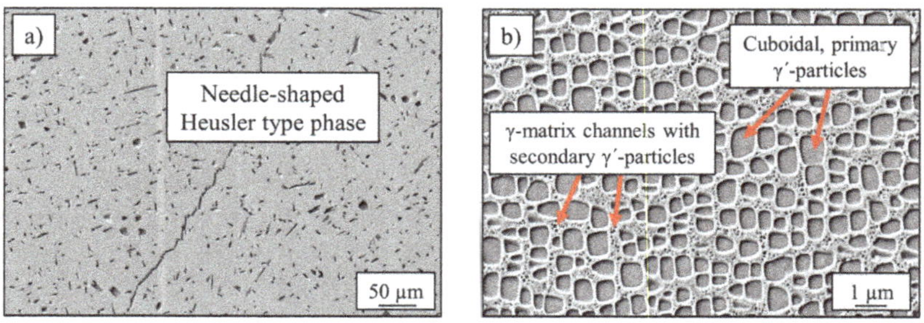

Figure 1. SEM-images of (**a**) Heusler type phase with a needle-like shape and a length up to 50 μm and (**b**) cuboidal γ'-particles with an edge-length up to 400 nm and matrix channels with small and spherical secondary γ'-particles (some 10 nm).

The chemical analyzation of all phases, determined by TEM/EDS, is listed in Table 1 and results in a Co-Fe-Cr rich fcc-matrix and Ni-Al-Ti rich $L1_2$-ordered precipitates [10]. The Heusler type phase is very rich in Al, while all other elements are depleted, especially Fe, Co and Cr.

Table 1. Chemical composition (in at.%) of all occurring phases in the annealed state (900 °C for 50 h) of the alloy $Al_{10}Co_{25}Cr_8Fe_{15}Ni_{36}Ti_6$ analyzed by TEM/EDS [10].

Element Content in at.%	γ-Matrix	γ′-Particles	Heusler Type Phase
Al	6.9 ± 0.6	11.4 ± 0.6	24.4 ± 1.3
Co	29.5 ± 0.5	22.5 ± 0.6	21.9 ± 1.7
Cr	9.3 ± 0.4	3.5 ± 0.4	3.6 ± 0.2
Fe	20.4 ± 0.6	8.8 ± 0.7	10.7 ± 0.4
Ni	30.4 ± 1.0	45.0 ± 1.5	33.9 ± 0.6
Ti	3.5 ± 0.4	8.7 ± 0.5	5.6 ± 0.2

3.2. Impact of Bridgman Process on Mechanical Properties

An increase of tensile strength was expected by the use of the Bridgman process, resulting in directional solidified samples. The base alloy was cast twice, conventionally cast with randomly orientated grains and directionally solidified, as described in Section 2 "Materials and Methods". The heat treatment was equal for both conditions (annealing at 900 °C for 50 h). The phase-characteristics (content and size) concerning Heusler type phase and γ'-phase were identical for both types of processing techniques and are displayed in Figure 1. Figure 2 shows the grain structures of both states, confirming elongated grains in the direction of load (σ) in the case of directional solidification, while the conventionally cast microstructure exhibits grain boundaries across the orientation of external stress.

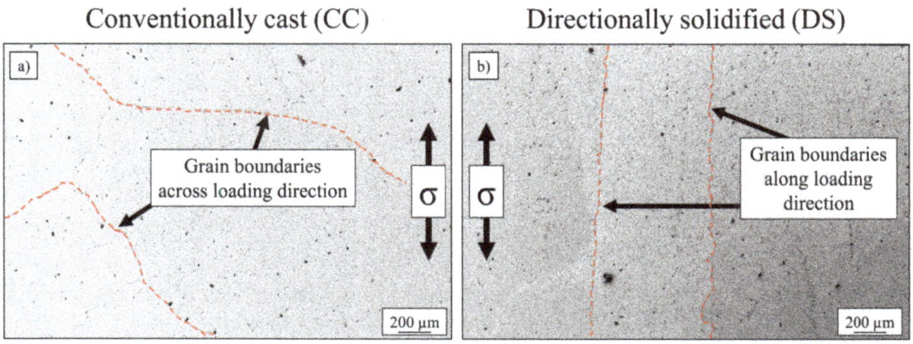

Figure 2. SEM-image of (**a**) conventionally cast, polycrystalline and (**b**) directionally solidified microstructures of the base alloy $Al_{10}Co_{25}Cr_8Fe_{15}Ni_{36}Ti_6$ after annealing at 900 °C for 50 h.

Results of tensile tests, carried out over a temperature range from room temperature to 900 °C, are displayed in Figure 3. Directional solidification (DS) showed an improvement in two ways as compared to conventional casting (CC). On one hand, DS-samples scattered much less, as standard deviation of particularly ultimate tensile strength (UTS) and also strain to failure (ε_f) were reduced strongly for each temperature level, see also Table 2. While the deviation is 50 MPa and 8% respectively in CC-state at the most, the maximum deviation in the DS-state is only 16 MPa and 5%. Furthermore, curve progressions for DS-samples seem to be much smoother and more like expected, as the strain to failure increases and the ultimate tensile strength decreases from 600 °C to 900 °C. All curves at room temperature exhibit high ductility with strain to failure levels about 25% for DS-samples and up to 50% for CC-samples.

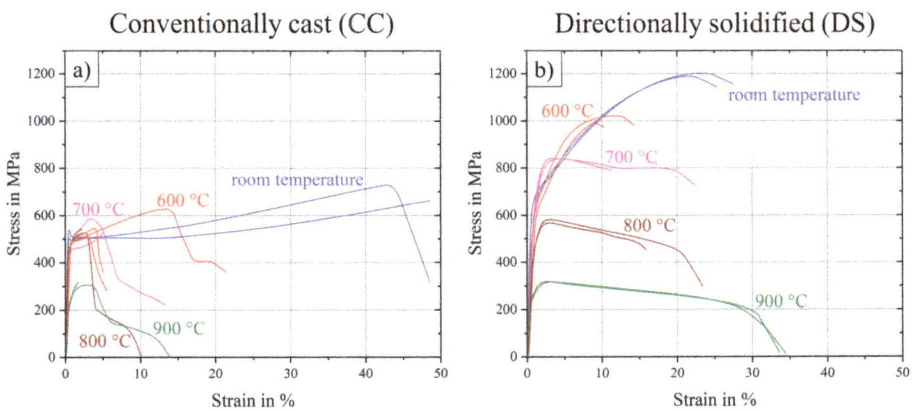

Figure 3. Stress–strain curves at room temperature up to 900 °C for (**a**) conventionally cast samples and (**b**) directionally solidified samples of the base alloy $Al_{10}Co_{25}Cr_8Fe_{15}Ni_{36}Ti_6$ in the annealed state (900 °C/50 h).

Table 2. Ultimate tensile strength (UTS) and strain to failure (ε_f) at different temperature levels for the conventionally cast (CC) state and the directionally solidified (DS) state.

	UTS in MPa		YS in MPa		ε_f in %	
	CC	DS	CC	DS	CC	DS
~23 °C	696 ± 33	1197 ± 6	520 ± 19	627 ± 9	49 ± 0	27 ± 1
600 °C	561 ± 50	1006 ± 16	470 ± 8	594 ± 3	11 ± 8	12 ± 2
700 °C	560 ± 27	840 ± 1	461 ± 10	547 ± 42	9 ± 4	17 ± 5
800 °C	536 ± 7	575 ± 7	445 ± 2	399 ± 1	6 ± 4	20 ± 4
900 °C	312 ± 6	319 ± 1	219 ± 6	243 ± 2	8 ± 6	34 ± 1

On the other hand, the most important improvement was the increase of UTS in case of the DS-samples by a factor of 1.5–1.8 in the range of room temperature up to 700 °C. Surprisingly, at higher temperatures of 800 and 900 °C the improvement by DS processing was only small (a factor of 1.1 and 1.0 respectively for 800 and 900 °C).

3.3. *Influence of Refractory Elements on Microstructural Characteristics*

The original heat treatment of the base alloy has been investigated and adapted for the alloys containing small amounts of hafnium and molybdenum: The homogenization treatment for the base alloy (1220 °C/20 h) was supposed to work for $Al_9Co_{25}Cr_8Fe_{15}Ni_{36}Ti_6Mo_1$ and the Hf-containing alloy as well. In the case of the Mo-alloy the Heusler type phase was completely dissolved after this treatment, while the Hf-alloy exhibited eutectic formations at the grain boundaries and unsolved Heusler type phase in spherical form, attached to the grain boundaries between eutectic regions. In this case the heat treatment temperature needed to be adapted and finally no eutectic formation occurred at 1140 °C, but the Heusler type phase still remained unsolved in the homogenized state. These two phenomena are shown in Figure 4.

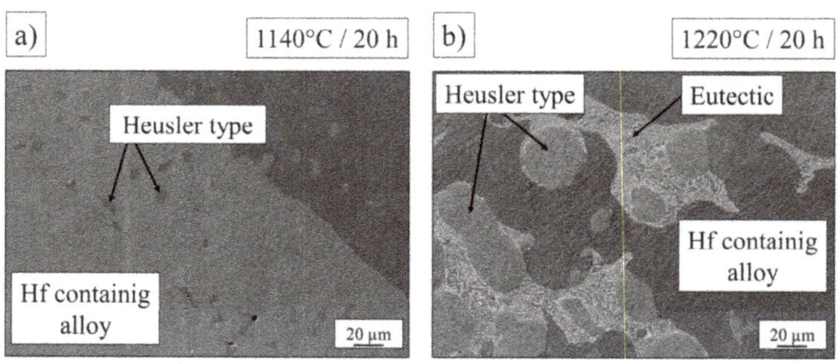

Figure 4. Microstructure of the Hf-containing alloy $Al_{9.5}Co_{25}Cr_8Fe_{15}Ni_{36}Ti_6Hf_{0.5}$ after homogenization at (**a**) 1140 °C/20 h and (**b**) 1220 °C/20 h.

The annealing step 900 °C/50 h was taken from the initial heat treatment, where γ'-particles precipitate in the fcc-matrix. The γ'-morphology of all alloys after standard treatment is shown in Figure 5 with clear changes: While the base alloy showed cuboidal particles with rounded corners, the Mo-alloy exhibited spherical, and the Hf-alloy showed cubic, sharp-cornered precipitates. These geometries were due to different values of misfit between the γ'-phase and the matrix. It was a result of the differences in lattice parameters of both phases [23]. Experiments for quantitative determination of lattice parameters and therefore misfit values have been carried out with synchrotron radiation at photon source BESSY II in Berlin, Germany over a wide temperature range and are in the process of being evaluated.

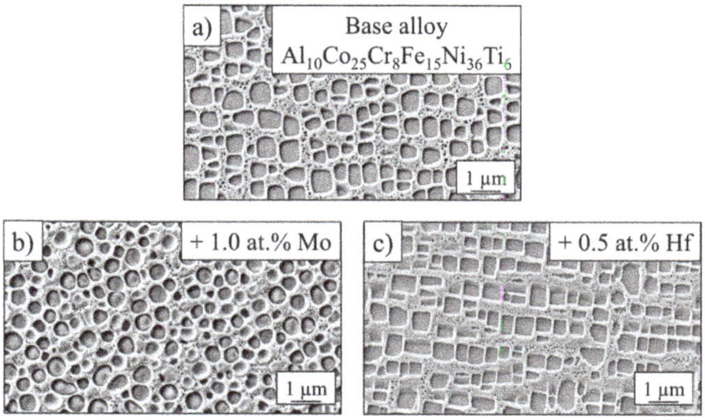

Figure 5. γ'-microstructure of the base alloy (**a**) and with the addition of molybdenum (**b**) and hafnium (**c**) after annealing at 900 °C/50 h.

Nickel-based superalloys with extraordinary strength- and creep-properties at high temperatures exhibited more cubic precipitates, a higher γ'-volume content (60–70%) and larger γ'-particles (up to 500 nm) [24] than the base alloy $Al_{10}Co_{25}Cr_8Fe_{15}Ni_{36}Ti_6$ ($V\gamma'$= 40%, $d\gamma'$ = 400 nm) after heat treatment at 900 °C/50 h. Therefore, several studies with variation of annealing time and temperature (±50 K; +500 h) have been conducted, resulting in an enhancement of both size and volume fraction after an annealing treatment at 950 °C for 100 h for the base alloy, already described in [25]. SEM-images of both conditions for the base alloy, as well as for $Al_{9.5}Co_{25}Cr_8Fe_{15}Ni_{36}Ti_6Hf_{0.5}$ and $Al_9Co_{25}Cr_8Fe_{15}Ni_{36}Ti_6Mo_1$ are shown in Figure 6, where larger cubic particles in (b), (e) and larger spherical particles in (h) can be detected compared to their original appearance in (a), (d) and (g) after shorter annealing at a lower temperature. Figure 6f shows the spherical Heusler type phase accumulation in the case of the Hf-containing alloy, while the needle-shaped Heusler type phase in the case of the base alloy and the Mo-containing alloy is represented in (c) and (i).

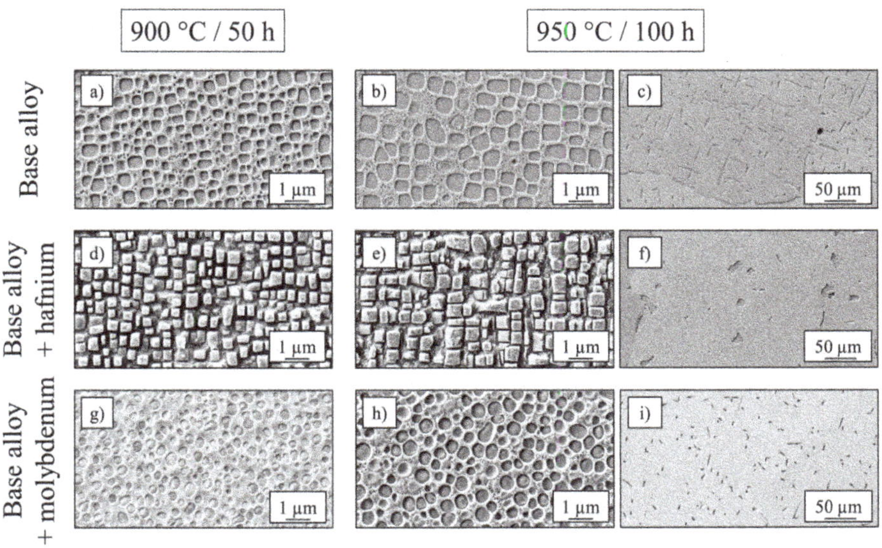

Figure 6. SEM-images showing the γ'- and Heusler type morphology for three different alloys and two different annealing treatments 900 °C/50 h (**a**,**d**,**g**) and 950 °C/100 h (**b**,**c**,**e**,**f**,**h**,**i**).

Since volume fraction and precipitate size play an important role for the mechanical behavior, all samples were investigated carefully after both annealing steps using SEM-images. The resulting volume fractions of the γ'- and Heusler type phase, as well as the size of γ'-particles are listed in Table 3. The size corresponds to the diameter in the case of round particles and to the edge-length in case of cuboidal particles.

Table 3. Volume fractions of γ'- and Heusler type phase, size and shape of γ'-precipitates after the annealing treatments 900 °C/50 h and 950 °C/100 h.

		Base Alloy	Base Alloy +Mo	Base Alloy +Hf
900 °C 50 h	$d_{\gamma'}$ in nm	200 ± 70	190 ± 70	210 ± 70
	$V_{\gamma'}$ in %	38 ± 7	44 ± 1	46 ± 2
	$V_{Heusler}$ in %	3 ± 2	3 ± 3	6 ± 1
950 °C 100 h	$d_{\gamma'}$ in nm	400 ± 100	360 ± 100	420 ± 100
	$V_{\gamma'}$ in %	41 ± 3	37 ± 1	38 ± 8
	$V_{Heusler}$ in %	9 ± 1	3 ± 1	5 ± 1
Shape of γ'-precipitates		Cuboidal (round corners)	Round	Cuboidal (sharp corners)

An increase of γ'-size and γ'-volume fraction, as well as an increase of Heusler type volume fraction is confirmed by Table 3 in the case of the base alloy for the annealing step 950 °C/100 h. While the size of precipitates increased with longer treatment at higher temperatures for the Mo- and Hf-containing alloys, too, the volume fractions of the Heusler type phase and γ'-phase were not enhanced, but decreased about 7–8% in the case of the γ'-phase and remained constant in the case of the Heusler type phase.

3.4. High-Temperature Tensile Tests of the Alloys after Annealing at 950 °C for 100 h

The presented stress–strain diagrams of the base alloy after standard heat-treatment (900 °C/50 h) in Figure 3 are completed by Figure 7, where samples were annealed at 950 °C for 100 h, resulting in larger γ'-precipitates. Samples tested at room temperature showed a brittle fracture behavior, as well as specimens tested at 600 and 700 °C. In general, large scattering occurred respective to strain to failure and curve progression. Ultimate tensile strength scattered less within one temperature series.

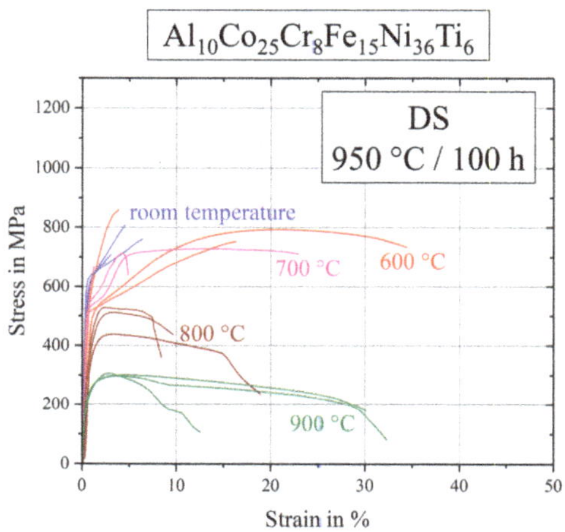

Figure 7. Stress–strain curves for the base alloy $Al_{10}Co_{25}Cr_8Fe_{15}Ni_{36}Ti_6$ after annealing at 950 °C for 100 h.

Figure 8 shows the stress–strain curves of $Al_9Co_{25}Cr_8Fe_{15}Ni_{36}Ti_6Mo_1$ after the annealing step 950 °C/100 h in detail. Scattering of the curve progressions was very high and particularly strain to failure values show large scatter at all temperatures tested. At room temperature and at 600 °C the alloy exhibited a ductile behavior, while the ultimate tensile strength at higher temperatures was reached very quickly after short elongations. In these cases, the first cracks appeared rapidly and fracture propagation was fast.

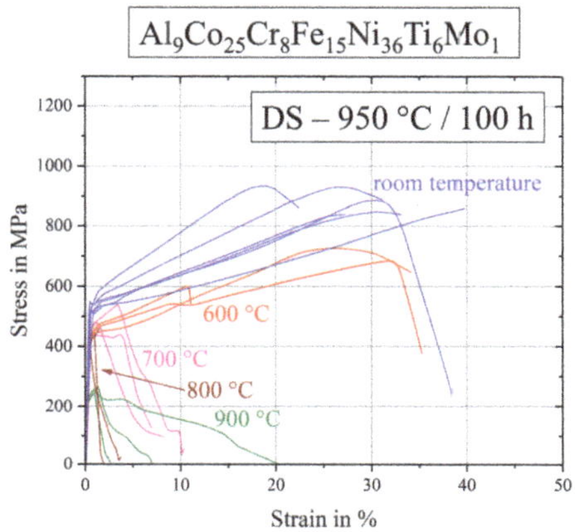

Figure 8. Stress–strain curves for the Mo-containing alloy $Al_9Co_{25}Cr_8Fe_{15}Ni_{36}Ti_6Mo_1$ after annealing at 950 °C for 100 h.

Stress-strain curves of the hafnium containing alloy are displayed in Figure 9. First of all, various tests at the same temperature were very reproducible with only a little scatter. Tests at room temperature reach the highest ultimate tensile strength with a strain to failure of about 20%. Samples deformed at 600 °C exhibited only half the strain to failure. In the range of 600–900 °C the samples showed the expected evolution of strength and plasticity, as ultimate tensile strength decreased and strain to failure increased with increasing temperature.

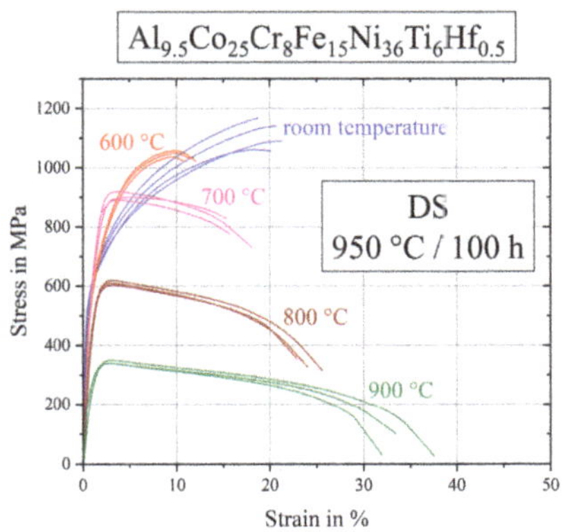

Figure 9. Stress–strain curves for the Hf-containing alloy $Al_{9.5}Co_{25}Cr_8Fe_{15}Ni_{36}Ti_6Hf_{0.5}$ after annealing at 950 °C for 100 h.

3.5. Discussion and Comparison of Mechanical Properties

In this work, compositionally complex alloys were improved with the goal to increase mechanical properties in the temperature range around 700–800 °C. Therefore, discussion will mainly depend on this temperature region.

Figure 10 shows the comparison between different variants of the base alloy respective ultimate tensile strength and strain to failure. The progress of this values from low to high temperatures is indicated by arrows and the significant temperature ranges are marked by differently colored areas. While the polycrystalline samples showed little strength and ductility in the marked red area, both parameters increased using directional solidification (green and blue areas). The two colored areas (green, blue) for directional solidified alloys, with different annealing treatments in contrast, revealed a clear difference: Annealing at 900 °C/50 h did in fact lead to smaller precipitates, but these samples exhibited a higher ultimate tensile strength, higher strain to failure and less scatter. The reason for this desirable behavior was due to the volume content of Heusler type phase that was drastically reduced from 9% to 3% after annealing at 900 °C/50 h, see Table 3.

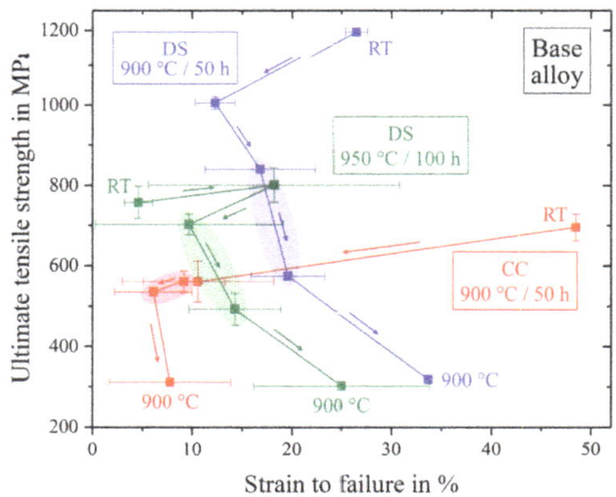

Figure 10. Evolution of ultimate tensile strength and strain to failure from room temperature (RT) to 900 °C for different manufactured and annealed types of the base alloy $Al_{10}Co_{25}Cr_8Fe_{15}Ni_{36}Ti_6$.

An overview about all tested tensile samples is shown in Figure 11, including the different types of the base alloy (blue), the Hf- and Mo-containing alloys in red and green respectively, as well as two conventionally used nickel-based alloys that are used in the temperature-range of 680–820 °C, highlighted by the vertical yellow stripe.

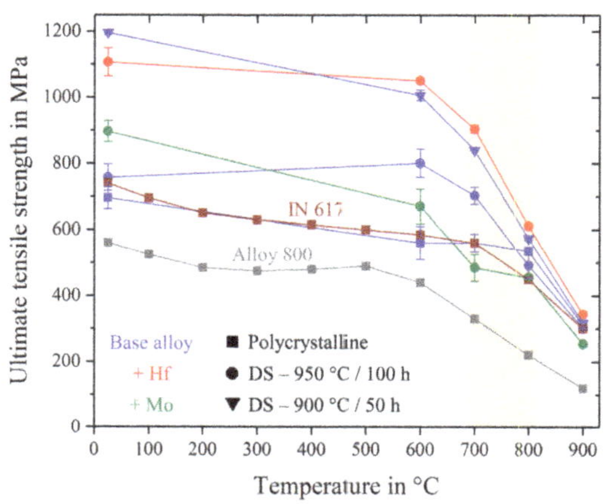

Figure 11. Ultimate tensile strength for all investigated alloys over the temperature range from room temperature to 900 °C. Inconel 617 and Alloy 800 H in a polycrystalline state taken from references [26,27].

Figure 12 shows more detailed views on two types of comparison and allows viewing of the effective aspects of mechanical behavior. Directionally solidified samples were produced to neglect the huge factor of grain-structure and grain-size in the mechanical behavior and to investigate the pure microstructure influence independently.

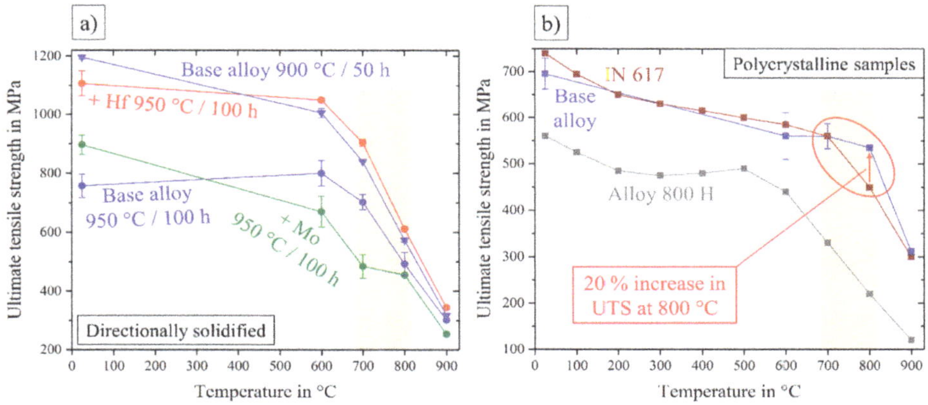

Figure 12. Ultimate tensile strength over the temperature range from room temperature to 900 °C for all directionally solidified alloys (**a**) and the conventionally cast, polycrystalline base alloy, compared with the two commercial nickel-based alloys Inconel 617 [26] and Alloy 800 H [27] (**b**).

Thus Figure 12a compares all DS-samples and the base alloy even after two different annealing treatments, leading to the following remarks: Differences in UTS of all tested specimen were getting smaller in the higher temperature range 800–900 °C. Between 700 and 800 °C, and also at 600 °C, there was a clear order: The worst behavior was exhibited by the Mo-containing alloy, the base alloy had a remarkably higher strength and the best mechanical properties could be observed in the case of the Hf-containing alloy. Next to the alloys, annealed under the same conditions, the base alloy after a treatment at 900 °C/50 h was observed in the range of the Hf-containing alloy. Since the base and the Mo-containing alloy (950 °C/50 h) showed the same characteristics of the Heusler type phase, round γ'-particles contributed in a bad way to the mechanical behavior. As the two strongest materials, the base alloy (900 °C/50 h) and the Hf-containing alloy (950 °C/100 h) exhibited almost identical stress–strain diagrams concerning strength, ductility and reproducibility, the small difference in γ'-morphology (sharp corners at the Hf-containing alloy), as well as the shape of Heusler type phase were not a reason for the worse fracture behavior. Consequently, good high-temperature tensile properties occurred in the case of cubic precipitates, assumed that the Heusler type phase content was kept quite low. If the Heusler type phase content rose three times higher in the case of the base alloy, the ultimate tensile strength fell down drastically.

For a comparison with commercially used alloys, the polycrystalline base alloy could be used, see Figure 12b: While the base alloy showed UTS exceeding that of Alloy 800 H over the whole temperature range, it was very similar or only a little worse than the UTS of the alloy Inconel 617 at temperatures up to 700 °C. In the important temperature range between 700 and 800 °C, UTS of the base alloy was equal in the beginning and even exceeded IN 617 at 800 °C by a factor of 1.2.

Figure 13a shows the yield strength of all tested alloys over the temperature range from room temperature to 900 °C, where the samples showed a remaining plastic deformation of 0.2%. An interesting and application oriented fact is shown in Figure 13b, where the polycrystalline base alloy was compared to the conventionally nickel-based alloys. Similar to the ultimate tensile strength in Figure 12 the yield strength of the base alloy exceeded Alloy 800 H quite significantly. The more competitive alloy IN 617, however, showed similar values of UTS, except at 800 °C, but the yield strength was not able to reach the levels of the base alloy. In the important temperature range the yield strength of the base alloy overran IN 617 by a factor of about 1.7 at 700 °C and 1.5 at 800 °C.

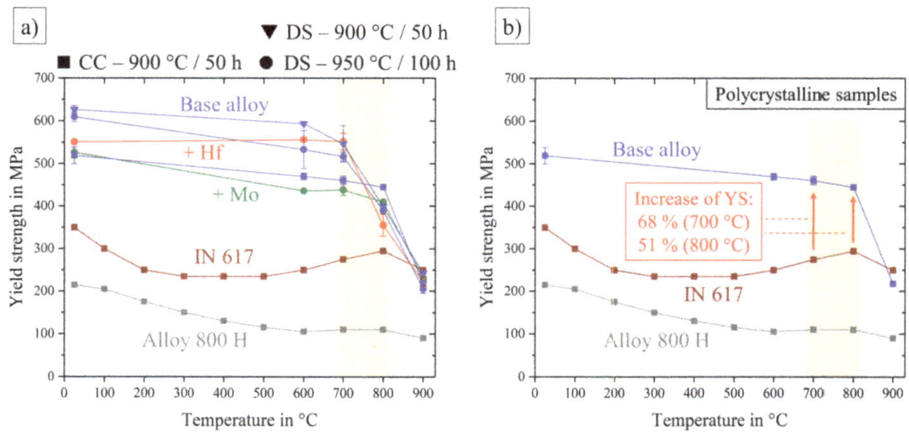

Figure 13. Yield strength over the temperature range from room temperature to 900 °C for all alloys (**a**) and the conventionally cast, polycrystalline base alloy, compared with two commercial nickel-based alloys Inconel 617 [26] and Alloy 800 H [27] (**b**).

4. Conclusions and Outlook

Summarizing the base alloy $Al_{10}Co_{25}Cr_8Fe_{15}Ni_{36}Ti_6$ and the influence of Hf (0.5 at.%) and Mo (1.0 at.%) additions on microstructural and mechanical properties:

1. All alloys showed a remarkably high UTS up to 1.26 GPa at room temperature.
2. Casting by the Bridgman process lead to a directionally solidified microstructure and an increase of strength in the direction of the grains by a factor of 1.7 at RT and 1.8 at 600 °C.
3. Addition of 1 at.% molybdenum lead to round γ'-particles, while the Heusler type phase remained in its needle-like shape.
4. An amount of 0.5 at.% hafnium sharpened the corners of the γ'-particles and lead to a spherical Heusler type phase.
5. The γ'-particle size increased for all alloys by annealing for longer times at higher temperature. Volume fractions of γ' and particularly Heusler type were only increasing in the case of the base alloy, while the Hf- and Mo-containing alloys showed a decrease of γ'-volume fraction and no obvious changes concerning the Heusler type phase.
6. Under the same heat treatment conditions, Mo addition lowered the ultimate tensile strength of the base alloy due to the more round morphology of γ'-particles. The Hf-containing alloy, however, showed an increase in UTS due to the lower content and/or the spherical shape of Heusler type phase.
7. The annealing treatment of 900 °C/50 h for the base alloy lead to a similar mechanical behavior to that of the Hf-containing alloy, with high strength, good strain to failure and high reproducibility. This is referred to a three times reduction of Heusler type volume fraction.
8. In comparison to the two commercially used nickel-based alloys IN 617 and Alloy 800 H, the polycrystalline base alloy showed better mechanical behavior, particularly in the temperature range 700–800 °C and especially in yield strength.

The top priority for future work is to reduce or maybe even avoid the Heusler type phase and to clarify the question about its role on mechanical behavior. We can state that a spherical shape and a low content of <3% is desirable for good mechanical properties. Thus it is necessary to gain knowledge about the chemical and thermodynamic stability resulting in an only two-phase microstructure.

Author Contributions: Conceptualization, U.G., S.H., A.M.M.; Methodology, S.H., A.M.M., F.K.; Software, F.K., A.M.M.; Validation, S.H., A.M.M.; Formal Analysis, S.H., A.M.M.; Investigation, S.H., A.M.M.; Resources, S.H., A.M.M.; Data Curation, S.H., A.M.M.; Writing-Original Draft Preparation, S.H.; Writing-Review & Editing, S.H., A.M.M., U.G.; Visualization, S.H., A.M.M.; Supervision, U.G.

Funding: This research work was funded by German Research foundation (DFG) projects GL 181/50-1 and MA 7004/1-1 and supported by the priority program SPP2006 "Compositionally Complex Alloys—High Entropy Alloys (CCA-HEA)".

Conflicts of Interest: The authors declare no conflict of interest.

References

1. Davis, J.R. *Metals Handbook*, 2nd ed.; CRC Press: Boca Raton, FL, USA, 1998.
2. Yeh, J.W.; Chen, S.K.; Lin, S.J.; Gan, J.Y.; Chin, T.S.; Shun, T.T.; Tsau, C.H.; Chang, S.J. Nanostructured high-entropy alloys with multiple principal elements: Novel alloy design concepts and outcomes. *Adv. Eng. Mater.* **2004**, *6*, 299–303. [CrossRef]
3. Cantor, B. Multicomponent and high entropy alloys. *Entropy* **2014**, *16*, 4749–4768. [CrossRef]
4. Tsai, K.Y.; Tsai, M.H.; Yeh, J.W. Sluggish diffusion in Co-Cr-Fe-Mn-Ni high-entropy alloys. *Acta Mater.* **2013**, *61*, 4887–4897. [CrossRef]
5. Manzoni, A.M.; Glatzel, U. New multiphase compositionally complex alloys driven by the high entropy alloy approach. *Mater. Charact.* **2018**, in press. [CrossRef]
6. Glicksman, M.E. *Principles of Solidification: An Introduction to Modern Casting and Crystal Growth Concepts*; Springer: New York, NY, USA, 2011.
7. Cheng, H.; Wang, H.Y.; Xie, Y.C.; Tang, Q.H.; Dai, P.Q. Controllable fabrication of a carbide-containing FeCoCrNiMn high-entropy alloy: Microstructure and mechanical properties. *Mater. Sci. Technol.* **2017**, *33*, 2032–2039. [CrossRef]
8. Wang, Q.; Ma, Y.; Jiang, B.; Li, X.; Shi, Y.; Dong, C.; Liaw, P.K. A cuboidal B2 nanoprecipitation-enhanced body-centered-cubic alloy $Al_{0.7}CoCrFe_2Ni$ with prominent tensile properties. *Scr. Mater.* **2017**, *20*, 85–89.
9. Singh, S.; Wanderka, N.; Murty, B.S.; Glatzel, U.; Banhart, J. Decomposition in multi-component AlCoCrCuFeNi high-entropy alloy. *Acta Mater.* **2011**, *59*, 182–190. [CrossRef]
10. Manzoni, A.M.; Singh, S.; Daoud, H.M.; Popp, R.; Völkl, R.; Glatzel, U.; Wanderka, N. On the path to optimizing the Al-Co-Cr-Cu-Fe-Ni-Ti high entropy alloy family for high temperature applications. *Entropy* **2016**, *18*, 104. [CrossRef]
11. Sherby, O.D.; Miller, A.K. Combining Phenomenology and Physics in Describing the High Temperature Mechanical Behavior of Crystalline Solids. *J. Eng. Mater. Technol.* **1979**, *101*, 387–395. [CrossRef]
12. Fleischmann, E.; Miller, M.K.; Affeldt, E.; Glatzel, U. Quantitative experimental determination of the solid solution hardening potential of rhenium, tungsten and molybdenum in single-crystal nickel-based superalloys. *Acta Mater* **2015**, *87*, 350–356. [CrossRef]
13. Duhl, D.N.; Sullivan, C.P. Some effects of hafnium additions on the mechanical properties of a columnar-grained nickel-base superalloy. *JOM* **1971**, *23*, 38–40. [CrossRef]
14. Doherty, J.E.; Kear, B.H.; Giamei, A.F. On the origin of the ductility enhancement in Hf-doped Mar-M200. *J. Met* **1971**, *23*, 59–62. [CrossRef]
15. Kotval, P.S.; Venables, J.D.; Calder, R.W. The role of hafnium in modifying the microstructure of cast nickel-base superalloys. *Metall. Trans.* **1972**, *3*, 457–462. [CrossRef]
16. The Version Tccr; Thermocalc Software AB: Stockholm, Sweden. Available online: https://www.thermocalc.Com (accessed on 15 June 2016).
17. *Thermotech Ni-Based Superalloys Database, TTNI7, Version 7.0*; ThermoCalc Software AB: Stockholm, Sweden, 2006.
18. Arganda-Carreras, I.; Kaynig, V.; Rueden, C.; Eliceiri, K.W.; Schindelin, J.; Cardona, A.; Seung, H.S. Trainable weka segmentation: A machine learning tool for microscopy pixel classification. *Bioinformatics* **2017**, *33*, 2424–2426. [CrossRef] [PubMed]
19. Schindelin, J.; Arganda-Carreras, I.; Frise, E.; Kaynig, V.; Longair, M.; Pietzsch, T.; Preibisch, S.; Rueden, C.; Saalfeld, S.; Schmid, B.; et al. Fiji: An open-source platform for biological-image analysis. *Nat. Methods* **2012**, *9*, 676–682. [CrossRef] [PubMed]

20. Rueden, C.T.; Schindelin, J.; Hiner, M.C.; DeZonia, B.E.; Walter, A.E.; Arena, E.T.; Eliceiri, K.W. ImageJ2: ImageJ for the next generation of scientific image data. *BMC Bioinform.* **2017**, *18*, 529. [CrossRef] [PubMed]
21. Schneider, C.A.; Rasband, W.S.; Eliceiri, K.W. Nih image to imagej: 25 years of image analysis. *Nat. Methods* **2012**, *9*, 671–675. [CrossRef] [PubMed]
22. Daoud, H.M.; Manzoni, A.M.; Wanderka, N.; Glatzel, U. High-temperature tensile strength of $Al_{10}Co_{25}Cr_8Fe_{15}Ni_{36}Ti_6$ compositionally complex alloy (high-entropy alloy). *JOM* **2015**, *67*, 2271–2277. [CrossRef]
23. Thompson, M.E.; Su, C.S.; Voorhees, P.W. The equilibrium shape of a misfitting precipitate. *Acta Met. mater.* **1994**, *42*, 2107–2122. [CrossRef]
24. Pyczak, F.; Devrient, B.; Mughrabi, H. The Effects of Different Alloying Elements on the Thermal Expansion Coefficients, Lattice Constants and Misfit of Nickel-Based Superalloys Investigated by X-Ray Diffraction. *Superalloys* **2004**, 827–836.
25. Manzoni, A.M.; Haas, S.; Daoud, H.; Glatzel, U.; Förster, C.; Wanderka, N. Tensile Behavior and Evolution of the Phases in the $Al_{10}Co_{25}Cr_8Fe_{15}Ni_{36}Ti_6$ Compositionally Complex/High Entropy Alloy. *Entropy* **2018**, *20*, 646. [CrossRef]
26. Special Metals. Available online: http://www.specialmetals.com/assets/smc/documents/alloys/inconel/inconel-alloy-617.pdf (accessed on 7 December 2018).
27. Specification Sheet: Alloy 800, 800H, and 800AT. Available online: https://www.sandmeyersteel.com/A800-A800H-A800AT.html (accessed on 7 December 2018).

entropy

MDPI

Article

The Effect of Scandium Ternary Intergrain Precipitates in Al-Containing High-Entropy Alloys

Sephira Riva [1], Shahin Mehraban [1], Nicholas P. Lavery [1], Stefan Schwarzmüller [2], Oliver Oeckler [2], Stephen G. R. Brown [1] and Kirill V. Yusenko [1,3,*]

1 College of Engineering, Swansea University, Swansea SA1 8EN, Wales, UK; sephira.riva@icloud.com (S.R.); S.Mehraban@Swansea.ac.uk (S.M.); N.P.Lavery@Swansea.ac.uk (N.P.L.); S.G.R.Brown@swansea.ac.uk (S.G.R.B.)

2 Faculty of Chemistry and Mineralogy, Institute for Mineralogy, Crystallography and Materials Science, Leipzig University, Scharnhorststr. 20, 04275 Leipzig, Germany; stefan.schwarzmueller@uni-leipzig.de (S.S.); oliver.oeckler@gmx.de (O.O.)

3 Institute of Solid State Chemistry, Pervomaiskaia str. 91, 620990 Ekaterinburg, Russia

* Correspondence: k.yusenko@swansea.ac.uk

Received: 4 June 2018; Accepted: 19 June 2018; Published: 22 June 2018

Abstract: We investigate the effect of alloying with scandium on microstructure, high-temperature phase stability, electron transport, and mechanical properties of the $Al_2CoCrFeNi$, $Al_{0.5}CoCrCuFeNi$, and $AlCoCrCu_{0.5}FeNi$ high-entropy alloys. Out of the three model alloys, $Al_2CoCrFeNi$ adopts a disordered CsCl structure type. Both of the six-component alloys contain a mixture of body-centered cubic (*bcc*) and face centered cubic (*fcc*) phases. The comparison between in situ high-temperature powder diffraction data and ex situ data from heat-treated samples highlights the presence of a reversible *bcc* to *fcc* transition. The precipitation of a $MgZn_2$-type intermetallic phase along grain boundaries following scandium addition affects all systems differently, but especially enhances the properties of $Al_2CoCrFeNi$. It causes grain refinement; hardness and electrical conductivity increases (up to 20% and 14% respectively) and affects the CsCl-type → *fcc* equilibrium by moving the transformation to sensibly higher temperatures. The maximum dimensionless thermoelectric figure of merit (*ZT*) of 0.014 is reached for $Al_2CoCrFeNi$ alloyed with 0.3 wt.% Sc at 650 °C.

Keywords: high-entropy alloys; in situ X-ray diffraction; grain refinement; thermoelectric properties; scandium effect

1. Introduction

High-Entropy Alloys (HEAs) are defined according to the atomic percentage of their principal elements, between 5 and 35 at %, or according to their configurational entropy at random state, $\Delta S > 1.5\ R$ ($R = 8.314\ J \cdot K^{-1} \cdot mol^{-1}$). Their high compositional complexity draws a hyper-dimensional space whose limits have yet to be explored, for its investigation has mostly focused on equiatomic or near-equiatomic compositions [1,2]. In fact, the study of HEAs is driven by the search of new single-phase systems, whose formation is opposed by the diverse mixing enthalpies, atomic size, and valence electron concentration of the constituent elements. Consequently, single-phase HEAs are rare, whereas systems consisting of multiple solid solutions and ordered intermetallic phases are more common [3].

Following the pioneering publications in 2004 by Yeh [4] and Cantor [5], HEAs have attracted growing research interest due to their outstanding mechanical and thermal properties; including high compression yield, fracture strength, ductility and toughness, as well as extreme corrosion, wear and fatigue resistance. HEAs have thus been proposed as candidates for applications in which high temperature stability is pivotal; e.g., as replacements for conventional binders after liquid-phase or spark-plasma sintering, for liquefied gas storage and for high-temperature thermoelectrics [1,6]. Thermoelectric properties can be tuned with respect to valence electron concentration (VEC) and a dimensionless thermoelectric figure of merit (*ZT*) of 0.012 at 505 °C was reached for $Al_2CoCrFeNi$. The phase evolution of HEAs has been extensively studied both ex situ and in situ [7–11].

Following a traditional trend in alloy development, HEAs have seen the addition of selected secondary phases to further tune their mechanical properties. This approach has led to the development of a new class of metal-matrix composites, containing oxides [12], silicon carbide [13] or nano-diamonds [14]. Alloying with elements in low concentrations, on the other hand, has resulted in the precipitation of intermetallic compounds in the matrix phase. However, while intermetallics deeply affect yield strength, hardness, tensile properties, and matrix stabilization—due to the competition between mixing enthalpy of atom pairs and mixing entropy—their proper distribution, size, shape, and volume fraction represent a cause for concern [1]. The synthesis of precipitation-hardened HEAs following the introduction of intermetallics has proven largely unsuccessful when binary compounds are concerned, and no studies have been performed on the formation of stable ternary inclusions as pinning centers in complex multi-principal component alloys [15–17]. This is mostly due to the difficulties in choosing appropriate alloying elements. Their selection should be guided by the following considerations: miscibility for most or all HEA constitutive elements in liquid state, low formation enthalpy for ternary compounds and crystallization in common structure types (e.g., σ-phase, Laves phase).

We recently highlighted the compound-forming ability of scandium and its outstanding effect on the mechanical properties of multicomponent alloys [18]. Scandium forms over three hundred binary and ternary phases with most elements of the periodic table, many of which crystallize in highly symmetrical structures (e.g., space groups $Pm\overline{3}m$, $Fm\overline{3}m$, $P6_3/mmc$). Scandium-based intermetallics (i.e., Al_3Sc, *V*- and *W*-phases) are responsible for the enhanced properties of several commercial aluminum alloys and newly developed multicomponent systems [19]. These features make scandium a perfect candidate to achieve precipitation-hardened HEAs. Moreover, its low density makes it ideal in combination with HEA based on 3*d*-elements, which make up to 85% of the known systems [1]. The $Al_xCoCrCu_yFeNi$ HEA can be considered a model alloy. The nature of the solid solution can be tuned by changing the aluminum and copper content, since Al and Cu act as *bcc*- and *fcc*-stabilizers respectively [20]. Thus, $Al_2CoCrFeNi$ and $AlCoCrCu_{0.5}FeNi$ have been widely reported as pure *bcc* phases, while $Al_{0.5}CoCrCuFeNi$ as purely *fcc*-structured [21–23].

We herein report on the effects of 0.3–5 wt.% scandium addition to the microstructure, mechanical and transport properties, thermal stability, and phase evolution upon temperature of the model $Al_2CoCrFeNi$, $AlCoCrCu_{0.5}FeNi$, and $Al_{0.5}CoCrCuFeNi$ HEAs. We show that scandium forms the same stable ternary intermetallic in all three alloys, but that the compound interacts differently with *bcc*- and *fcc*-structured alloys.

2. Materials and Methods

The target alloys were prepared using induction melting from pure metallic powders, in a BN crucible in an Ar-filled Customised DAB01 glove-box (Saffron Scientific Equipment Limited, Knaresborough, UK). Complete melting of the samples was achieved above 1300 °C. After 5 min at the melting temperature, the sample was cooled down naturally to room temperature. The samples were re-melted three times to assure homogeneity. A brief rationale of the synthesized specimens and their performed analysis is reported in Supplementary Materials Table S1.

Differential scanning calorimetry (DSC) measurements were performed on small pieces of sintered samples (50 mg) placed in an Al_2O_3 crucible and heated in a Netzsch STA 449 F1 Jupiter. Heating (Selb, Germany) and cooling were performed in flowing Ar gas with a temperature ramp of 10 $K\cdot min^{-1}$ from 35 to 1300 °C.

The transition temperatures highlighted by DSC were used to decide annealing conditions. Samples were heat-treated above their first reversible or irreversible transition temperature with the following specifics: $Al_2CoCrFeNi$, $Al_{0.5}CoCrCuFeNi$ and $AlCoCrCu_{0.5}FeNi$ at 850 °C, 12 h; $Al_2CoCrFeNi$ + 3 wt.% Sc at 900 °C, 12 h; $Al_{0.5}CoCrCuFeNi$ + 3 wt.% Sc and $AlCoCrCu_{0.5}FeNi$ + 3 wt.% Sc at 930 °C for 6 h. During annealing, pellets of each sample were sealed in a silica tube under vacuum (<10^{-7} Pa) and heated in a furnace. After annealing, the tubes were quenched in ice-cold water.

For microstructure and elemental analysis, all samples were mounted in carbonized resin, ground and polished using MetaDiTM Supreme Polycrystalline Diamond Suspension (1 μm) (Coventry, UK).

The morphology and elemental compositions were analyzed using a Hitachi S-4800 Field Emission scanning-electron microscope (SEM) equipped with energy dispersive X-ray (EDX) analyzer (Tokio, Japan). The average elemental composition was obtained from 2.5 mm maps (Table S2).

The Vickers hardness was measured on a WilsonR VH3100 Automatic Knoop/Vickers Hardness tester (Buehler, Lake Bluff, IL, USA); 25 individual points under a 9.81 N (1 kg) testing load were measured to get statistically significant results.

The density was measured according to Archimedes' principle in water, in the ATTENSION equipment (Biolin Scientific, Stockholm, Sweden). Six measurements were taken for each sample to obtain statistically relevant results

The small punch tests were performed on discs of diameter 12.5 mm and thickness circa 0.8 mm. Measurements were performed with a properly modified Tinius Olsen H25KS Benchtop Tester (Salfords, UK). The setup included a lower die (diameter of 8 mm) and a punch (4 mm diameter). Each experiment was reproduced twice for statistical significance; however, since each specimen had a slightly different thickness, the final results were normalized for the standard 0.5 mm thickness as described in [24].

For powder X-ray diffraction (PXRD), samples were powdered using a Fritsch mini-mill Pulverisette 23 (Idar-Oberschtein, Germany) (steel vial and ball, 10 min at 50 rpm). PXRD data for the annealed powdered samples were collected at ID06B-LVP beam-line at the European Synchrotron Research Facility, ESRF (room temperature, λ = 0.22542 Å) using position sensitive detector. LaB_6 (NIST SRM 660c) was used as external standard for calibration.

In situ high-temperature PXRD patterns were collected at the I-11 beam-line at the DIAMOND light source (λ = 0.494984 Å). LaB_6 (NIST SRM 660c) was used as external standard for wavelength and sample to detector distance calibration. A wide-angle Mythen-2 Si position sensitive detector. The detector was moved at constant angular speed with 10 s scan time at each temperature and 60 s waiting time to let the temperature stabilize. The powdered alloys were sealed in a 0.5 mm silica glass capillary in vacuum and heated in the capillary furnace from 25 to 1200 °C with axial rotation [25]. In all samples, oxidation was detected above 1000 °C, which can be due to the reaction of metallic alloy with silica at high temperature, resulting in capillary destruction. In situ low-temperature PXRD profiles were collected at the P02.1 beam-line at the PETRA III synchrotron (λ = 0.207150 Å). LaB_6 (NIST SRM 660c) was used as external standard for calibration. A wide-angle position sensitive detector based on Mythen-2 Si strip modules was used. The detector was moved at constant angular speed with 10 s scan time at each temperature and 60 s waiting time to let the temperature stabilize. The powdered alloys were sealed in 0.5 mm silica glass capillaries in vacuum and cooled in nitrogen flow from 300 to 100 K. Temperature was directly measured with a thermocouple during the experiment; error arising from the distance between sensor and capillary was estimated to be below 5%.

Temperature dependent PXRD patterns were analyzed using Powder3D software [26]. Phase composition has been verified using the Powder Diffraction File database [27]. Parametric sequential refinements were performed using the TOPAS 5.0 software [28]. Profile parameters for the Lorentzian function, cell parameters, and phase fractions were refined simultaneously for all phases.

The Seebeck coefficient S ($\mu V \cdot K^{-1}$) and electrical conductivity σ ($kS \cdot cm^{-1}$) were measured simultaneously under He atmosphere with a Linseis – Seebeck and Electric Resistivity Unit (LSR-3 1100, Linseis, Selb, Germany) four-point setup with PtRh/Pt and Pt contacts and a continuous reverse of the polarity of the thermocouples (bipolar setup, measurement current: 100 mA) using cuboid samples (ca. 8 × 2 × 3 mm). Three heating cycles up to 875 °C (10 K·min^{-1}, 3 data points per temperature) were performed. Thermal diffusivity was measured up to 875 °C (heating/cooling rate 10 K·min^{-1}) under He atmosphere with a Linseis LFA1000 (Selb, Germany) apparatus. Simultaneous heat loss and finite pulse corrections were applied using Dusza's model [29]. Values were averaged from five measurement points at each temperature. For calculation of thermal conductivity κ, they were multiplied with the Dulong–Petit heat capacity C_p and the density as derived by the weight and the volume determined by Archimedes' principle. The single values of each sample are given in Table S3. According to experimental C_p values of materials with similar compositions (e.g., 0.60 J·g^{-1}·K^{-1} for AlCoCrFeNi at 25 °C) [30], the room temperature heat capacity of these materials is about 20% higher than the Dulong–Petit value of 0.49 J·g^{-1}·K^{-1}; this probably adds this uncertainty to the values of κ and thus *ZT*.

3. Results and Discussions

$Al_2CoCrFeNi$, $Al_{0.5}CoCrCuFeNi$ and $AlCoCrCu_{0.5}FeNi$ were synthesized via induction melting in the atomic compositions reported in the Table S1. While all three systems are widely reported as single-phase [1], refinements of PXRD data obtained with synchrotron radiation highlight the presence of a secondary *fcc* phase in the *bcc*-structured $AlCoCrCu_{0.5}FeNi$ alloy (Figure 1c) and of a very small secondary *bcc* phase in the mainly *fcc*-structured $Al_{0.5}CoCrCuFeNi$ alloy (Figure 1b). Unlike $Al_{0.5}CoCrCuFeNi$ and $AlCoCrCu_{0.5}FeNi$, $Al_2CoCrFeNi$ appears to be a solid solution forming a disordered CsCl structure-type. (Figure 1a). This is consistent with a previously reported investigation on CoCrCuFeNi-based systems—which displayed a phase separation due to the positive mixing enthalpy between copper and other elements [31]—and with the Hume–Rotary classification maps in ref. [32].

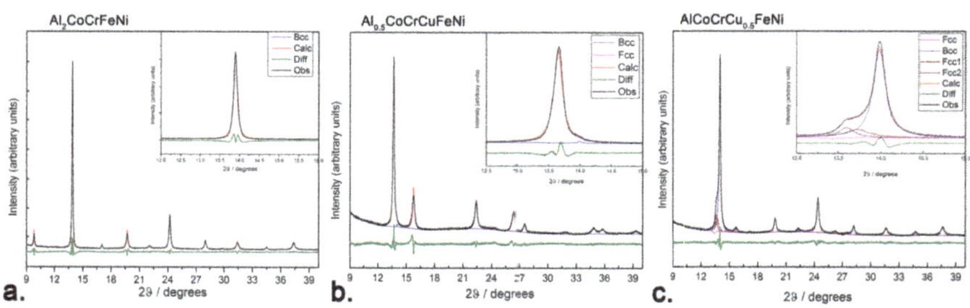

Figure 1. Powder X-ray diffraction (PXRD) Rietveld refinements performed from the DIAMOND light source (I11, λ = 0.494984 Å) of (**a**) $Al_2CoCrFeNi$. The section between 12 and 16 degrees 2θ is enlarged to show the symmetric shape of the first reflection. (**b**) $Al_{0.5}CoCrCuFeNi$. The section between 12 and 15 degrees 2θ is enlarged to show the asymmetry of the first reflection, which can only be fitted by taking into account a small amount of *bcc* phase. (**c**) $AlCoCrCu_{0.5}FeNi$. The section between 13 and 15 degrees 2θ is enlarged to show the presence of an *fcc* phase.

The presence of the (100) diffraction line in the *bcc*-structured PXRD pattern indicates an ordered superstructure, generally attributed to Al and Ni ordering [22,33,34]. Lattice parameters of the as-cast alloys are the following: a_{B2} = 2.877(2) Å for $Al_2CoCrFeNi$; a_{fcc} = 3.601(1) Å for $Al_{0.5}CoCrCuFeNi$ and a_{bcc} = 2.891(3) Å for $AlCoCrCu_{0.5}FeNi$. All values are consistent with the literature within the experimental error [20,22,35–37]. Small differences between the reported results arise from the high sensitivity of the HEA to synthetic pathway and to minor compositional variations. Therefore, nominally equivalent starting materials can in turn display profoundly different crystal structures, microstructures, and phase transitions.

The microstructures and elemental distributions of as-cast $Al_2CoCrFeNi$, $Al_{0.5}CoCrCuFeNi$ and $AlCoCrCu_{0.5}FeNi$ are reported by means of EDX element mapping in Figures S1–S3, respectively. Two phases are present in the as-cast $AlCoCrCu_{0.5}FeNi$ alloy, whose microstructure is characterized by a matrix and a circular secondary phase of darker color. Elemental distributions appear completely homogeneous (Figure S3). Annealing results in the growth of a darker secondary phase and the slight segregation of Al from the rest of the elements. The as-cast $Al_{0.5}CoCrCuFeNi$ HEA consists of dendritic-like and interdendritic-like regions, the first being richer in Co, Cr and Fe; and the latter Cu-rich (Figure S2). Both the dendrite-like and interdendrite-like matrix have been previously linked to a simple *fcc* phase, and copper segregation has been explained through its high mixing enthalpy with cobalt, chromium, iron, and nickel [20]. The microstructure changes drastically after annealing, as the two phases cannot be easily differentiated. Nevertheless, element segregation persists, with Cu and Al separating from Co, Cr, Fe, and Ni. The as cast $Al_2CoCrFeNi$ microstructure is dominated by large (~100 μm) unstructured grains which, unlike previous studies, show no trace of non-equiaxed dendrites [21,37]. With respect to other elements, chromium segregation is clearly visible, though it can be reduced by annealing (see Figure S1). On the other hand, annealing causes the formation of a homogeneously dispersed secondary phase, which appears as black dots.

The microstructure of all alloys is strongly affected by even a 3 wt.% addition of scandium. In the case of $Al_2CoCrFeNi$ (Figure 2a, Figure S4), the large grains of homogeneous compositions are refined and scandium segregates in the inter-grain volume. The scandium-based intermetallic forming in the inter-granular region appears to contain all elements in the same relative fractions as the main phase—with the notable exception of chromium (Table S1). In $Al_{0.5}CoCrCuFeN$ (Figure 2b), scandium is dispersed more homogeneously and aids the formation of a globular microstructure. Nevertheless, most of it segregates in the inter-granular region, with copper- and nickel-rich areas (Figure S5). Lastly, following scandium addition the original columnar cellular microstructure of the $AlCoCrCu_{0.5}FeNi$ HEA turns into equiaxed non-dendritic-like grains (Figure 2c). The chemical inhomogeneity due to scandium segregation appears surprising, considering the strongly negative mixing enthalpy of the metal with iron, nickel, cobalt and aluminum [38–40]. The driving force of the segregation is thus the formation of the secondary phase.

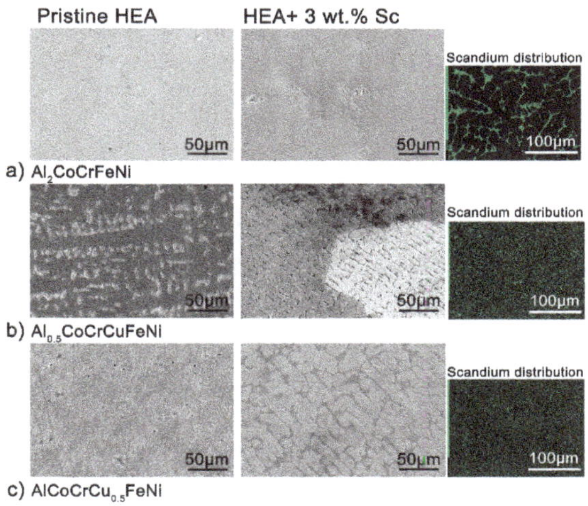

Figure 2. Overview of the effect of scandium addition on the microstructure of the three systems (scanning electron microscope-back scatter detector (SEM-BSE) images). **Left**: Microstructure of the as-cast (**a**) $Al_2CoCrFeNi$, (**b**) $Al_{0.5}CoCrCuFeNi$ and (**c**) $AlCoCrCu_{0.5}FeNi$ High-Entropy Alloys (HEAs) before and after 3 wt.% scandium addition. **Right**: scandium distribution of the areas displayed in (**b**) according to energy dispersive X-ray (EDX) maps.

The secondary phase is the same in all systems and can be indexed as a ternary intermetallic analogous of $MgZn_2$-type (Figure 3). Compounds of this structure type have been reported for scandium with several metals, in compositions such as AlCuSc, AlCoSc, $Al_{1.06}Cr_{0.94}Sc$, AlFeSc, and AlNiSc [18]. As confirmed by elemental composition maps (Figures S4–S6), the ternary phase is a highly disordered structure containing all five elements and scandium.

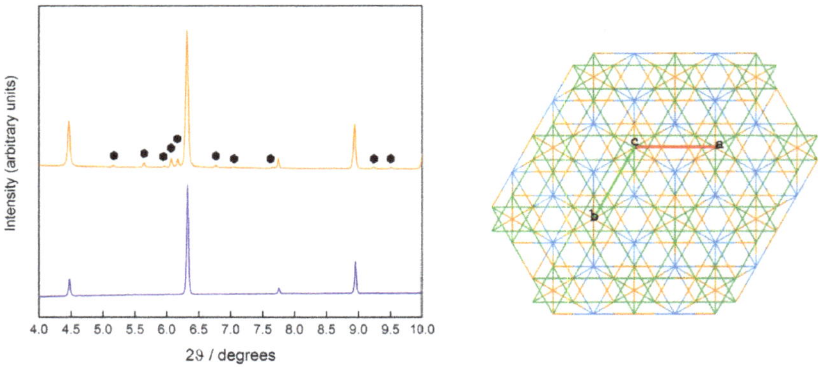

Figure 3. Left. PXRD profile (DIAMOND = 0.22542 Å) of the as-cast disordered CsCl-structured $Al_2CoCrFeNi$ alloy before (blue) and after (orange) a 3 wt.% Sc addition. The PXRD profile of the scandium phase is indexed with black hexagons. **Right**. Crystal structure of the hexagonal $MgZn_2$-type intermetallic along the c-axis, depicted as its AlCuSc analogue (Al in green, Cu in orange and Sc in blue; a, b and c are cell axis).

Knowledge about scandium-containing ternary compounds is still fragmentary. Out of all the cited $MgZn_2$-type intermetallic phases, only AlCuSc has been thoroughly investigated, due to its effect on the mechanical properties of Al-based alloys, as part of the so-called *W*-phase. In particular, it was shown that the microhardness of the *W*-phase is much higher than the one of the Al_3Sc phase (5150–5170 MPa against 3900–4300 MPa) [18].

The formation of a very hard phase in the HEA matrix affects its mechanical properties. As shown in Figure 4, single phase *bcc* alloys are harder than duplex-structured and *fcc* alloys. This is hardly surprising, considering the stronger interatomic forces involved in the *bcc* vs. *fcc* packing of alloys. On the other hand, increasing scandium content in $Al_{0.5}CoCrCuFeNi$ does not affect hardness. Indeed, it is even detrimental to the hardness of the $AlCoCrCu_{0.5}FeNi$ alloy and is not accompanied by an increase in ductility (as shown by the disk punch tests presented in Figure S7). Only in the originally hard $Al_2CoCrFeNi$ HEA the formation of the intermetallic results in an increase in hardness. The addition of 0.5 wt.% Sc causes a 20% hardness enhancement, as well as visible grain refinement. Disc punch tests performed on the HEA with 0, 0.5 and 2 wt.% scandium additions show a decisive increment in brittleness, proportional to the concentration of scandium (Figure S8).

Differential scanning calorimetry (DSC) was performed on alloys from room temperature to 1300 °C with a 10 K·min^{-1} heating rate. The second cycle of heating and cooling, which is less influenced by effects of the synthetic route on the specimens—i.e., internal stress-strain, magnetic ordering—is reported in Figure 5.

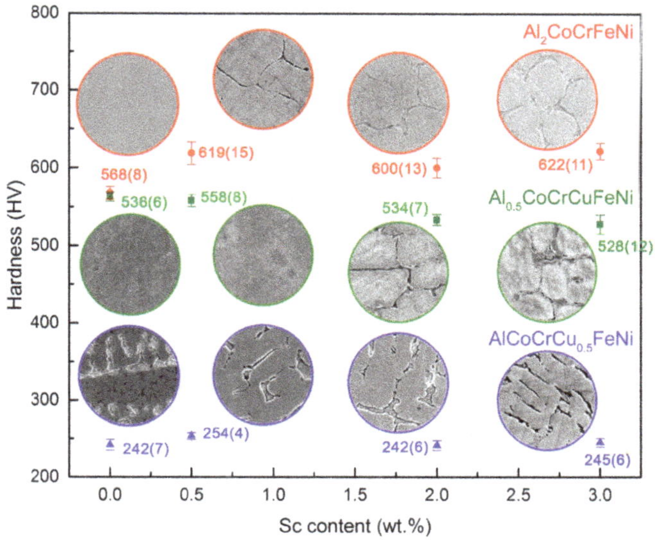

Figure 4. Vickers hardness values for $Al_2CoCrFeNi$ (red), $Al_{0.5}CoCrCuFeNi$ (green) and $AlCoCrCu_{0.5}FeNi$ (blue) HEAs with 0, 0.5, 2 and 3 wt.% Sc additions. Values are an average of 25 indentations at 1 HV. SEM images of the microstructures of all alloys are shown in circles of 30 μm diameter.

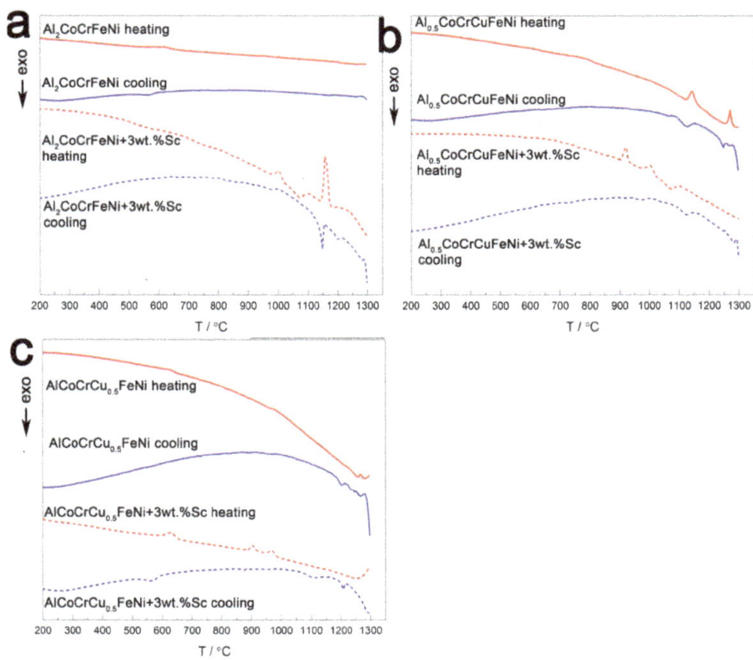

Figure 5. Differential scanning calorimetry (DSC) of (**a**) $Al_2CoCrFeNi$, (**b**) $Al_{0.5}CoCrCuFeNi$ and (**c**) $AlCoCrCu_{0.5}FeNi$ with (dotted line) and without (solid line) 3 wt.% scandium. The second heating/cooling cycle is reported for each specimen, in red and blue respectively.

The DSC profile of $Al_2CoCrFeNi$ (Figure 5a) has a sigmoid-like deviation between 600 and 700 °C. In a theoretical work, Gao associates this feature with the transition bcc_1 + bcc_2 + CsCl → *bcc*+CsCl, but this is incompatible with the crystal structure of the $Al_2CoCrFeNi$ alloy as per Figure 1a. The divergence between the results reported here and Gao's interpretation might arise from the profound differences in the crystal structures of the nominally equivalent starting material [41]. To unequivocally identify the cause of the transition, a more detailed investigation of the effect of temperature on phase stability is needed. The corresponding Sc-containing sample displays a sharp reversible peak at 1150 °C, probably corresponding to the melting and crystallization of the intermetallic phase, preceded by irreversible peaks. These two endothermic peaks, located at 906 °C and 966 °C, might correspond to phase transitions occurring in the scandium-phase. The sigmoid-like deviation clearly visible in the pristine alloy is hardly distinguishable from the background line in the Sc-containing specimen but is located at higher temperature (between 700 and 800 °C).

$Al_{0.5}CoCrCuFeNi$ shows a slight reversible peak centered at 760 °C, as well as two reversible peaks above 1150 °C (Figure 5b). In previously reported DSC curves, the two endothermic peaks of $Al_{0.5}CoCrCuFeNi$ at 1140 and 1270 °C have been linked to the melting of interdendritic-like and dendritic-like material, respectively. The results of Jones et al. highlight a third reversible peak, appearing at 850 °C and related to the dissolution and recrystallization of the $L1_2$ phase, which might form during prolonged heat treatment below 850 °C and is dependent from the sample cooling rate [42,43]. Its absence is indicative of the purity of the as-cast *fcc* sample, which is confirmed by high-resolution PXRD data (Figure 1b). The scandium containing specimen displays three irreversible signals upon heating (the first two, endothermic, at 921 and 1000 °C; the latter, exothermic, at 1073 °C). Upon cooling, the Sc-containing specimen behaves very similarly to its corresponding pristine alloy. The irreversible transitions occurring in the sample might thus relate solely to the intermetallic, and have little impact on the matrix.

Finally, $AlCoCrCu_{0.5}FeNi$ displays a reversible transition at circa 624 °C, which is maintained in the Sc-containing sample (Figure 5c). Irreversible phenomena occur in the pristine alloy above 1150 °C, as in the previous sample. The scandium-containing alloy presents two irreversible endothermic peaks (at 906 and 966 °C) upon heating and two exothermic peaks upon cooling (at 1102 and 1206 °C), which are too far from the heating ones to be considered part of the same phenomenon.

Reversible phenomena thus occur in all systems above 600 °C; whereas irreversible peaks appear in the scandium-containing alloys at ca. 900 °C. To investigate the nature of these transitions, the pristine alloys were annealed above their average first reversible transition temperature, and the scandium-containing specimens at the temperatures of their first irreversible transition. Annealing time was shortened for the samples at the highest temperature (930 °C) in order not to lose aluminum. Therefore, $Al_2CoCrFeNi$, $Al_{0.5}CoCrCuFeNi$, and $AlCoCrCu_{0.5}FeNi$ were annealed at 850 °C for 12 h. $Al_2CoCrFeNi$ + 3 wt.% Sc was treated at 900 °C for 12 h, while $Al_{0.5}CoCrCuFeNi$ + 3 wt.% Sc and $AlCoCrCu_{0.5}FeNi$ + 3 wt.% Sc at 930 °C for 6 h. The corresponding element distribution and microstructures are presented in the following pages. Figure 6 reports the microstructure and element distribution of the annealed $Al_2CoCrFeNi$ + 3 wt.% Sc alloy. With respect to Figure S1b (the annealed pristine alloy), microstructure is refined and the intermetallic scandium phase grows in a dendritic-like structure. The secondary phase which that have appeared in the pristine alloy (in the form of black dots) is not displayed by the matrix. A comparison with the scandium-containing alloy prior to annealing (Figure S4) shows a more homogeneous distribution of chromium, even though the metal still visibly segregates along the grain boundaries.

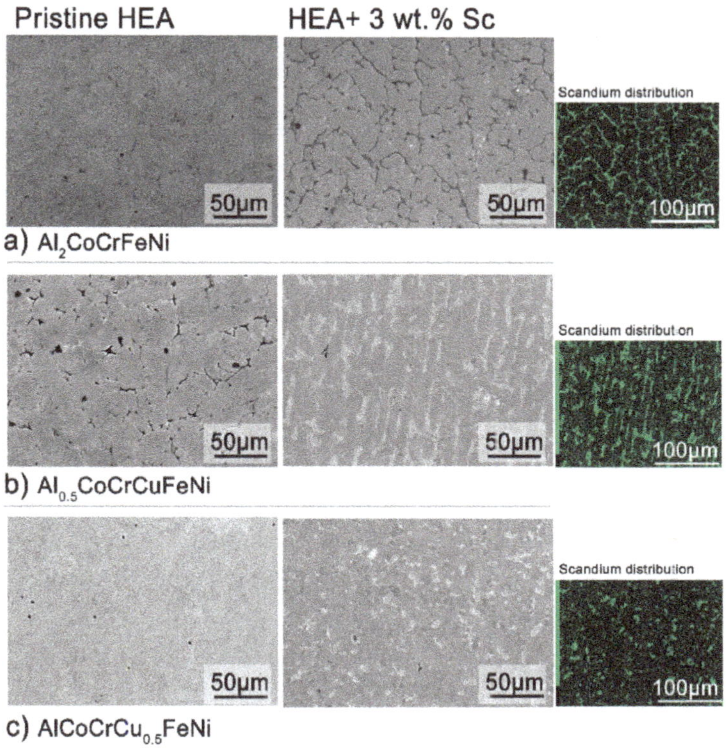

Figure 6. SEM images of the microstructure of annealed $Al_2CoCrFeNi$ (**a**), $Al_{0.5}CoCrCuFeNi$ (**b**) and $AlCoCrCu_{0.5}FeNi$ (**c**) before and after 3 wt.% Sc addition. Blank alloys were annealed at 850 °C for 12h; $Al_2CoCrFeNi$ + 3 wt.% Sc at 900 °C, 12 h; $Al_{0.5}CoCrCuFeNi$ and $AlCoCrCu_{0.5}FeNi$ at 930 °C, 6h. EDX maps show the scandium element distribution of the areas depicted on the right side (green).

The heat treatment of the pristine $Al_{0.5}CoCrCuFeNi$ (Figure 6b) causes the disruption of the original dendritic- and interdentritic-like microstructure. However, the microstructure of the scandium-containing alloy after annealing is quite similar to the as-cast pristine alloy: a columnar cellular structure displaying strong element segregation (Figure S2b). The light phase consists mostly of Cu and Sc, and within it regions rich in Al and Ni. The darker areas are rich in Co, Cr, and Fe. It is of interest to note that this element segregation is different from the one prior annealing (Figure S5), as Ni is depleted from the matrix.

The annealed $AlCoCrCu_{0.5}FeNi$ + 3 wt.% Sc has a complex microstructure (Figure 6c). Acicular particles grow in an unstructured matrix, the bigger of those being easily removed from the specimen during polishing (and leaving behind elongated valleys). Elemental segregation occurs to form several diverse regions: A Co-rich matrix, areas rich in Al and Ni, and Cu-Sc precipitates. Iron clearly segregates from Al and Ni, but does not follow the trend of other elements. Finally, Cr coalesces in circular well-shaped areas. Combining this information, we can note that the brittle phase is, as expected, rich in Cu and Sc. The annealed scandium-containing alloy is very different not only from the pristine alloy, but also from the original as cast specimen. Indeed, the annealed pristine alloy shows no clustering of Cr, Fe, and Cu (Figure S3); and the scandium-containing as cast alloy clearly highlights the coexistence of Co, Cr, and Fe (Figure S6).

To further investigate the nature of the transitions occurring in our HEA systems and how they are affected by the intermetallic phases, we performed in situ PXRD of disordered CsCl-type $Al_2CoCrFeNi$

and the *fcc*-structured $Al_{0.5}$CoCrFeNi with and without 3 wt.% scandium. Quantitative data are available only for the main phases, whose reflections had enough intensity to be detected and refined. The development of scandium-rich phases could not be followed directly.

Several works have highlighted that $Al_{0.5}$CoCrCuFeNi consists of a mixture of two *fcc* phases of similar cell parameters, corresponding to dendritic and interdendritic phase [44–46]. Chen reported a single *fcc* structure below 500 °C, and a duplex *fcc/bcc* structure above 600 °C [47]. According to Jones, on the other hand, two *fcc* phases of similar cell parameters co-exist and are stable up until their melting temperatures (1150 °C for the Cu based solid solution and 1350 °C for the multi-component phase) [44]. As shown in Figure 1b, our system consists of an *fcc* phase (a_{fcc} = 3.601(1) Å) and a small (<10%) amount of *bcc* phase (a_{bcc} = 2.867(4)). The second fcc_2 phase (a_{fcc_2} = 3.609(2) Å) reported in Figure 7a can be considered a minor admixture in an exsolution equilibrium with the *bcc* phase. Heating does not affect the primary phase: reflections become sharper, but their relative intensity remains similar. The addition of scandium (Figure 7b) has no impact on the *fcc* matrix, but influences the $bcc \rightarrow fcc_2$ equilibrium by shifting it towards the formation of *bcc*.

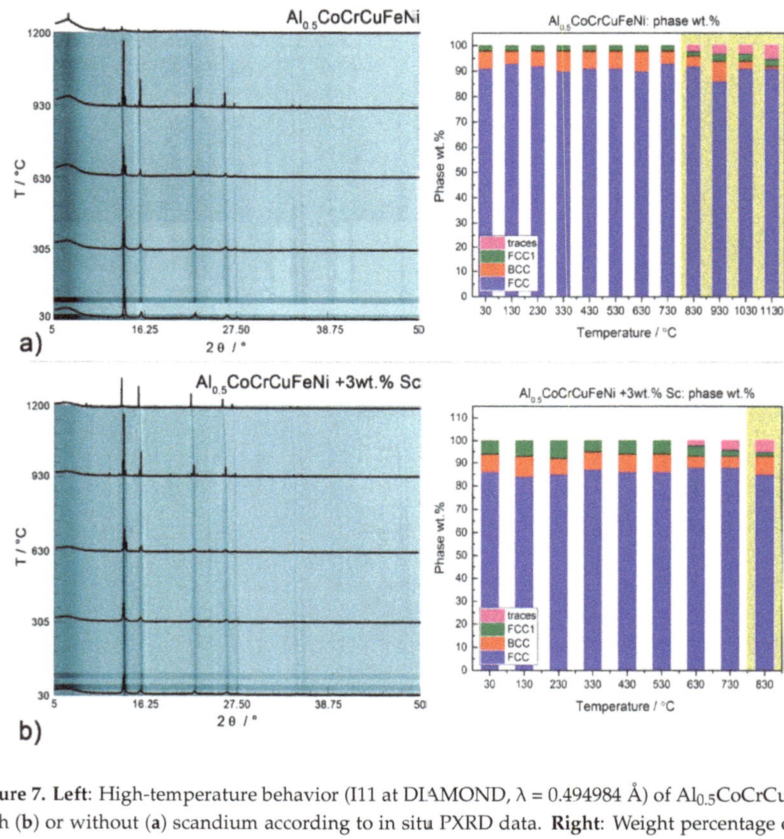

Figure 7. Left: High-temperature behavior (I11 at DIAMOND, λ = 0.494984 Å) of $Al_{0.5}$CoCrCuFeNi with (**b**) or without (**a**) scandium according to in situ PXRD data. **Right**: Weight percentage of the major phases in each sample: *fcc* (blue), *bcc* (red), fcc_1 (green), others (pink). The yellow areas mark the start of oxidation.

The thermal behavior of the CsCl-type Al_2CoCrFeNi sample is displayed in Figure 8a. We observe the decomposition of 40 wt.% of the CsCl-type phase above 620 °C and the exsolution of an *fcc* phase (a_{fcc_1} = 3.635(8) Å). At 750 °C a second fcc_2 phase of lattice parameter a_{fcc_2} = 3.631(9) Å develops from fcc_1 and grows almost linearly with temperature. The continuous presence of the (100) diffraction line in the PXRD pattern of the alloy suggests that the Al-rich sub-structure responsible for the CsCl-type crystal structure might not be involved in the exsolution process. Results for the Sc-containing $Al_2CoCrFeNi$ alloy are presented in Figure 8b. The exsolution of the *fcc* phase from the CsCl-type one follows a different pathway, with fcc_1 and fcc_2 forming at the same time. The transformation occurs at higher temperatures: a 3 wt.% scandium addition is enough to stabilize the main CsCl-type phase for about 150 °C.

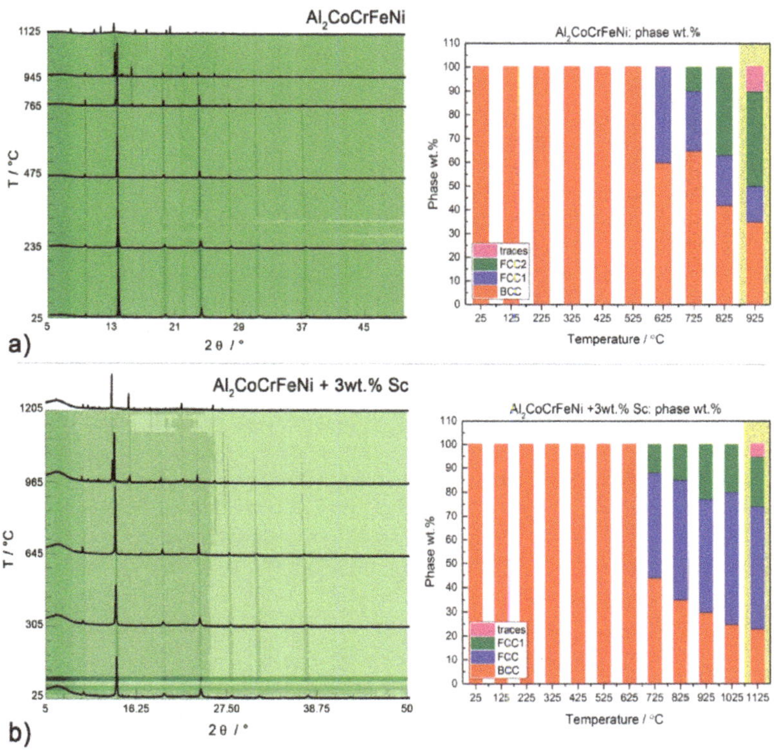

Figure 8. **Left**: High-temperature behavior (I11 at DIAMOND, λ = 0.494984 Å) of CsCl-type $Al_2CoCrFeNi$ with (**b**) or without (**a**) scandium according to in situ PXRD data. **Right**: Weight percentage of the major phases in each sample: *fcc* (blue), *bcc* (red), fcc_1 (green), others (pink). The yellow areas mark the start of oxidation.

The thermal expansion coefficient of $Al_2CoCrFeNi$ and $Al_2CoCrFeNi$ + 3 wt.% Sc can be evaluated by in situ high-temperature PXRD measurements by fitting the corresponding dataset to:

$$\alpha(T) = \alpha_{T_o} \cdot \exp\left[\alpha \cdot (T - T_o) + \frac{\beta}{2} \cdot \left(T^2 - T_o^2\right)\right] \quad (1)$$

where α_{T_o} is the cell parameter α at the reference temperature (T_o). Figure 9 reports the variation of the lattice parameter α_{CsCl} upon heating and cooling. A satisfactory fitting can be obtained for $Al_2CoCrFeNi$ for α and β equalling 3.6(2) $\times 10^{-6} \cdot K^{-1}$ and 1.69(3) $\times 10^{-8} \cdot K^{-2}$, respectively (Figure 9a); and for $Al_2CoCrFeNi$ + 3 wt.% Sc for α and β equaling 4.2(1) $\times 10^{-6} \cdot K^{-1}$ and 1.62(2) $\times 10^{-8} \cdot K^{-2}$, respectively (Figure 9b). As highlighted in Figure 9c, the scandium-containing CsCl-type phase has slightly larger cell parameters than the regular alloy and follows the same trend with temperature. This trend is consistent to the one of pure iron in the investigated temperature range, but differs strongly from the behavior of other cubic metals which constitute the alloy (in particular, from chromium and nickel) [48–51].

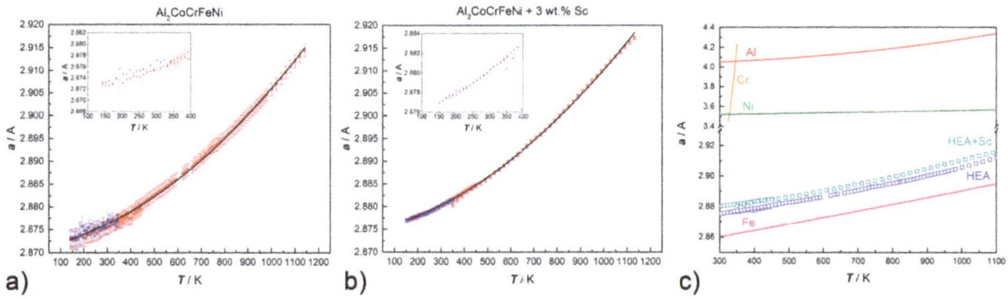

Figure 9. Thermal expansion data fitted with Equation (1) for (**a**) $Al_2CoCrFeNi$ and (**b**) $Al_2CoCrFeNi$ + 3 wt.% Sc. In red: data collected upon heating from 100–400 K (PETRAIII, λ = 0.207150 Å) and from 300–1100 K (DIAMOND, λ = 0.494984 Å); in blue: data collected upon cooling from 300–100 K (PETRAIII, λ = 0.207150 Å). Low temperature data are highlighted in the inset. (**c**) Thermal expansion curves for $Al_2CoCrFeNi$ and $Al_2CoCrFeNi$ + 3 wt.% Sc with respect to their constitutive cubic metals [48–51].

The addition of scandium to the Al-containing HEAs results in the precipitation of a ternary intermetallic of $MgZn_2$-type along the grain boundaries. Microstructural data as well as high-temperature PXRD suggest an extraordinary stability of Sc-based precipitates, which form in all alloys containing aluminum and first raw transition metals. The matrix coherence following the formation of the secondary phase is maintained and the Vickers hardness increases up to 20% due to precipitation hardening (for 0.5 wt.% Sc addition to $Al_2CoCrFeNi$). The intermetallic phase has high thermal stability and affects the *bcc* → *fcc* exsolution equilibrium by stabilizing the body-centered cubic phase with respect to the face-centered cubic one. This effect is likely related to the segregation of part of the *fcc*-stabilizing elements (i.e., Ni and Co) in the ternary compound. Out of *bcc*-stabilizers, aluminum segregates in the $MgZn_2$-type phase, but chromium has much lower affinity for the secondary phase and coalesces in the matrix (as highlighted by EDX maps). Lattice thermal expansion data of the CsCl-type alloy with or without scandium show that both systems follow the same trend upon heating. However, the cubic phase in the scandium-containing HEA has slightly larger cell parameters: this is a further indication of the compositional difference between the matrix of the two systems.

While the thermoelectric properties of $Al_xCoCrFeNi$ ($0 \leq x \leq 3$) [6,30,52] are discussed in detail in literature, the effect of Sc alloying on the transport properties of $Al_2CoCrFeNi$ is herein reported for the first time (Figure 10). The absolute Seebeck coefficient of *n*-type materials decreases slightly with increasing Sc content. The electrical conductivity for both compositions increases by about 1000 $S{\cdot}cm^{-1}$ compared to pure $Al_2CoCrFeNi$ [6,52]. In accordance with the higher electrical conductivity, the thermal conductivity shows similar, also slightly increased values compared to ref. [6]; whereas the thermal conductivity for pure $Al_2CrCoFeNi$ from ref. [52] is extraordinarily high.

Power factor and *ZT* value increase with temperature until a peak *ZT* value of 0.012 at 700 °C (for 5 wt.% Sc addition) and 0.014 at 650 °C (for 0.3 wt.% Sc) is reached. The subsequent decrease of *ZT* with temperature is mainly attributed to the decrease of the absolute value of *S* due to excitation of minority charge carriers. According to PXRD patterns of the samples after thermoelectric measurements, no degradation was observed (Figure S12). In addition, the samples show a good cyclability within three subsequent heating and cooling cycles (Figures S13 and S14).

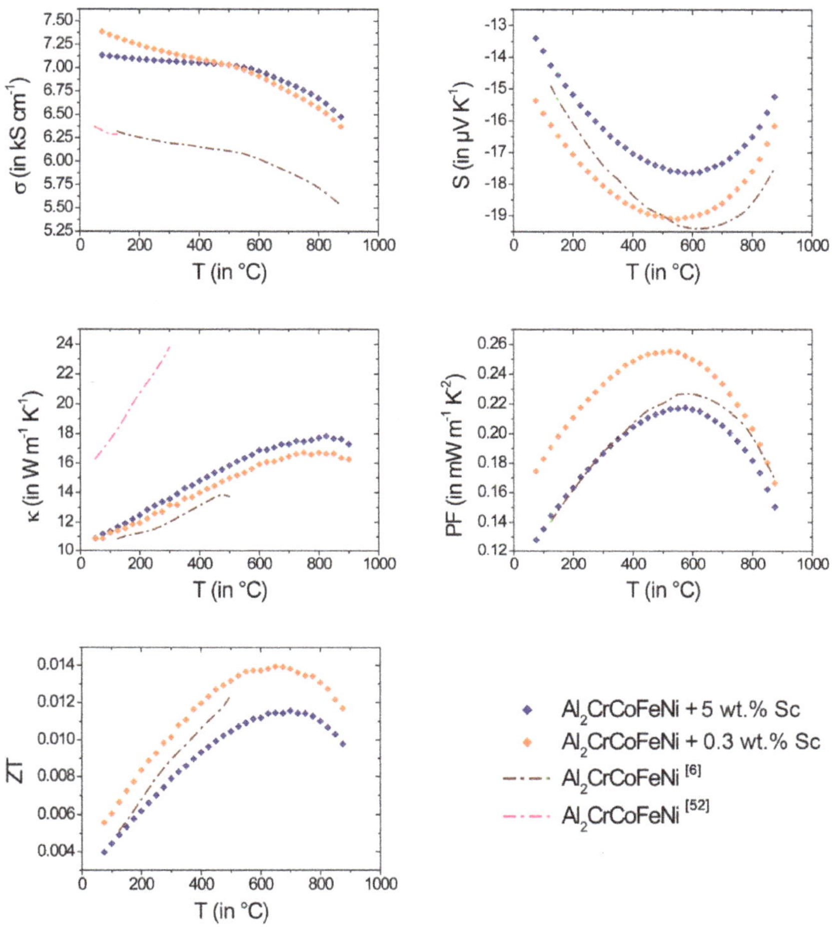

Figure 10. Thermoelectric properties of $Al_2CrCoFeNi$ + 5 wt.% Sc (blue) and $Al_2CrCoFeNi$ + 0.3 wt.% Sc (orange) averaged over heating and cooling cycles (without first heating): electrical conductivity σ (**top, left**), Seebeck coefficient S (**top, right**), thermal conductivity κ (**middle, left**), power factor PF (**middle, right**) and thermoelectric figure of merit *ZT* (**bottom, left**). For comparison, literature data for $Al_2CrCoFeNi$ (brown [6] and pink [52] dashed line) were added.

4. Conclusions

Ever since HEAs have been proposed as promising materials for high-temperature applications, their phase stability, and the effect of selected intermetallic compounds on their performances upon heating have seldom been studied. We report a thorough investigation on the effect of small scandium additions to the microstructure, phase stability, and mechanical and thermoelectric properties of the model HEAs $Al_2CoCrFeNi$, $Al_{0.5}CoCrCuFeNi$ and $AlCoCrCu_{0.5}FeNi$. High-temperature in situ PXRD investigations highlights the existence of secondary phases in these systems, traditionally though as consisting of a single solid solution. The addition of scandium to the studied HEAs causes the precipitation of a $MgZn_2$-type intermetallic phase. This hexagonal compound leads to chemical inhomogeneity along grain boundaries, causing grain refinement and a powerful increase in hardness (for $Al_2CoCrFeNi$, a 0.5 wt.% addition of scandium enhances hardness by 20%). Upon heating, *bcc* and CsCl-type phases turn into one or more *fcc* phases, whereas *fcc* phases appears very stable. Scandium acts as stabilizer of the *bcc* phase by affecting the exsolution equilibrium *bcc* → *fcc*. Regarding thermoelectric properties, the alloying of scandium in $Al_2CoCrFeNi$ increases the electrical conductivity by 14%. The maximum *ZT* value of 0 014 reached for 0.3 w% Sc addition at 650 °C is in the same order of magnitude as for other HEAs, but too low to make these materials promising for current thermoelectric applications. Still, HEAs are a young material class and optimized HEA will have the advantage of extreme thermal stability.

The results reported here open new possibilities in the design and further improvement of the properties of multicomponent alloys. They could be used as the basis for designing new stable and metastable multicomponent single-phase alloys with improved mechanical properties. The effect of minor elements additions to HEA should be thoroughly investigated to tackle specific technological needs.

Supplementary Materials: The following are available online at http://www.mdpi.com/1099-4300/20/7/488/s1.

Author Contributions: Conceptualization, K.V.Y.; Investigation, K.V.Y., S.R., S.S., S.M.; Resources, N.P.L.; Writing-Original Draft Preparation, S.R.; Writing-Review & Editing, S.R., S.S., O.O., K.V.Y., Supervision, O.O., S.G.R.B.

Funding: This research was funded by the National research network of Wales (Ser Cymru, project NRN046) and by the European Space Agency (contract number 4000111643/NL/PA).

Acknowledgments: The Authors gratefully acknowledge the financial support provided by the Welsh Government and Higher Education Funding Council for Wales through the Sêr Cymru National Research Network in Advanced Engineering and Materials and by the European Space Agency. The authors thank the Materials Advanced Characterization Centre (MACH1) at Swansea University, the DIAMOND light source (Oxfordshire, UK) and the PETRA III synchrotron (Hamburg, Germany) for providing measurement time and technical support. A PhD scholarship from the "Studienstiftung des deutschen Volkes" for S.S. is gratefully acknowledged.

Conflicts of Interest: The authors declare no conflict of interest.

References

1. Miracle, D.B.; Senkov, O.N. A critical review of high entropy alloy and related concepts. *Acta Mater.* **2017**, *122*, 448–511. [CrossRef]
2. Zhang, W.; Liaw, P.K.; Zhang, Y. Science and technology in high-entropy alloys. *Sci. China Mater.* **2018**, *61*, 2–22. [CrossRef]
3. Dahlborg, U.; Cornide, J.; Calvo-Dahlborg, M.; Hansen, T.C.; Fitch, A.; Leong, Z.; Chambreland, S.; Goodall, R. Structure of some CoCrFeNi and CoCrFeNiPd multicomponent HEA alloys by diffraction techniques. *J. Alloy Compd.* **2016**, *681*, 330–341. [CrossRef]
4. Yeh, J.W.; Chen, S.K.; Lin, S.J.; Gan, J.Y.; Chin, T.S.; Shun, T.T.; Tsau, C.H.; Chang, S.Y. Nanostructured high-entropy alloys with multiple principal elements novel alloy design concepts and outcomes. *Adv. Eng. Mater.* **2004**, *6*, 299–303. [CrossRef]
5. Cantor, B.; Chang, I.T.H.; Knight, P.; Vincent, A.J.B. Microstructural development in equiatomic multicomponent alloys. *Mater. Sci. Eng. A* **2004**, *375–377*, 213–218. [CrossRef]
6. Shafeie, S.; Guo, S.; Hu, Q.; Fahlquist, H.; Erhart, P.; Palmqvist, A. High-entropy alloys as high-temperature thermoelectric materials. *J. Appl. Phys.* **2015**, *118*, 184905. [CrossRef]
7. Samaei, A.T.; Mirsayar, M.M.; Aliha, M.R.M. The microstructure and mechanical behavior of modern high temperature alloys. *Eng. Solid Mech.* **2015**, *3*, 1–20. [CrossRef]
8. Antonaglia, J.; Xie, X.; Tang, Z.; Tsai., C.-W.; Qiao, J.W.; Zhang, Y.; Laktionova, M.O.; Tabachnikova, E.D.; Yeh, J.W.; Senkov, O.N.; et al. Temperature effect on deformation and serration behavior of high-entropy alloys (HEAs). *JOM* **2014**, *66*, 2002–2008. [CrossRef]
9. Manzoni, A.M.; Singh, S.; Daoud, H.M.; Popp, R.; Völkl, R.; Glatzel, U.; Wanderka, N. On the path to optimizing the Al-Co-Cr-Co-Fe-Ni-Ti high entropy alloy family for high temperature applications. *Entropy* **2016**, *18*, 104. [CrossRef]
10. Santodonato, L.J.; Zhang, Y.; Feygenson, M.; Parish, C.M.; Gao, M.C.; Weber, R.J.W.; Neuefeind, J.C.; Tang, Z.; Liaw, P.K. Deviation from high-entropy configurations in the atomic distributions of a multi-principal-element alloy. *Nat. Comm.* **2015**, *6*, 5964. [CrossRef] [PubMed]

11. Yusenko, K.V.; Riva, S.; Crichton, W.A.; Spektor, K.; Bykova, E.; Pakhomova, A.; Tudball, A.; Kupenko, I.; Rohrbach, A.; Klemme, S.; et al. High-pressure high-temperature tailoring of High-Entropy Alloys for extreme environments. *J. Alloy Compd.* **2018**, *738*, 491–500. [CrossRef]
12. Prasad, H.; Singh, S.; Panigrahi, B.B. Mechanical activated synthesis of alumina dispersed FeNiCoCrAlMn high entropy alloy. *J. Alloy Compd.* **2017**, *692*, 720–726. [CrossRef]
13. Rogal, L.; Kalita, D.; Tarasek, A.; Bobrowski, P.; Czerwinski, F. Effect of SiC nano-particles on microstructure and mechanical properties of the CoCrFeMnNi high entropy alloy. *J. Alloy Compd.* **2017**, *708*, 344–352. [CrossRef]
14. Riva, S.; Tudball, A.; Mehraban, S.; Lavery, N.P.; Brown, S.G.R.; Yusenko, K.V. A novel high-entropy alloy-based composite material. *J. Alloy Compd.* **2018**, *730*, 544–551. [CrossRef]
15. Hsu, U.S.; Hung, U.D.; Yeh, J.W.; Chen, S.K.; Huang, Y.S.; Yang, C.C. Alloying behavior of iron, gold and silver in AlCoCrCuNi-based equimolar high-entropy alloys. *Mater. Sci. Eng. A* **2007**, *460–461*, 403–408. [CrossRef]
16. Zhu, J.M.; Fu, H.M.; Zhang, H.F.; Wang, A.M.; Li, H.; Hu, Z.Q. Synthesis and properties of multiprincipal component $AlCoCrFeNiSi_x$ alloys. *Mater. Sci. Eng. A* **2010**, *527*, 7210–7214. [CrossRef]
17. Li, B.S.; Wang, Y.P.; Ren, M.X.; Yang, C.; Fu, H.Z. Effects of Mn, Ti and V on the microstructure and properties of AlCrFeCoNiCu high entropy alloys. *Mater. Sci. Eng. A* **2008**, *498*, 482–486. [CrossRef]
18. Riva, S.; Yusenko, K.V.; Lavery, N.P.; Jarvis, D.J.; Brown, S.G.R. The scandium effect in multicomponent alloys. *Int. Mater. Rev.* **2016**, *61*, 203–228. [CrossRef]
19. Riva, S.; Fung, C.M.; Searle, J.R.; Clark, R.N.; Lavery, P.N.; Brown, S.G.R.; Yusenko, K.V. Formation and disruption of W-phase in High-Entropy Alloys. *Metals* **2016**, *6*, 106. [CrossRef]
20. Tung, C.-C.; Yeh, J.-W.; Shun., T.-T.; Chen, S.-K.; Huang, Y.-S.; Chen, H.-C. On the elemental effect of AlCoCrCuFeNi high-entropy alloy system. *Mater. Lett.* **2007**, *61*, 1–5. [CrossRef]
21. Li, C.; Li, J.C.; Zhao, M.; Jiang, Q. Effect of aluminum contents on microstructure and properties of AlxCoCrFeNi alloys. *J. Alloy Compd.* **2010**, *504*, S515–S518. [CrossRef]
22. Kao, Y.F.; Chen, T.J.; Chen, S.K.; Yeh, J.W. Microstructure and mechanical property of as-cast, -homogenized and -deformed AlxCoCrFeNi ($0 \leq x \leq 2$) high-entropy alloys. *J. Alloy Compd.* **2009**, *488*, 57–64. [CrossRef]
23. Li, B.-Y.; Peng, K.; Hu, A.-P.; Zhou, L.-P.; Zhu, J.-J.; Li, D.-Y. Structure and properties of FeCoNiCrCo0.5Alx high-entropy alloys. *Trans. Nonferrous Met. Soc. China* **2013**, *23*, 735–741. [CrossRef]
24. Norris, S.D.; Parker, J.D. Deformation processes during disc bend loading. *Mater. Sci. Technol.* **1996**, *12*, 163–170. [CrossRef]
25. Thompson, S.P.; Parker, J.E.; Marchal, J.; Potter, J.; Birt, A.; Yuan, F.; Fearn, R.D.; Lennie, A.R.; Street, S.R.; Tang, C.C. Fast X-ray powder diffraction on I11 at Diamond. *J. Synchrotron Radiat.* **2011**, *18*, 637–648. [CrossRef] [PubMed]
26. Rajiv, P.; Dinnebier, R.; Jansen, M. Powder 3D Parametric: A program for automated sequential and parametric Rietveld refinement using Topas. *Mater. Sci. Forum* **2010**, *651*, 97–104. [CrossRef]
27. Kabekkodu, S. (Ed.) *PDF-2 Release 2012 (Database)*; International Centre for Diffraction Data: Newtown Square, PA, USA, 2012; Available online: http://www.icdd.com/index.php/pdf-2/ (accessed on 22 June 2018).
28. TOPAS v.4.0, Bruker-AXS 5465 East Cheryl Parkway–Bruker AXS–2009. Available online: https://www.bruker.com/products/x-ray-diffraction-and-elemental-analysis/x-ray-diffraction/xrd-software/topas.html (accessed on 22 June 2018).
29. Dusza, L. Combined Solution of the Simultaneous Heat Loss and Finite Pulse Corrections with the Laser Flash Method. *High Temp.–High Press.* **1995**, *27*, 467–473. [CrossRef]
30. Uporov, S.; Bykov, V.; Pryanichnikov, S.; Shubin, A.; Uporova, N. Effect of synthesis route on structure and properties of AlCoCrFeNi high-entropy alloy. *Intermetallics* **2017**, *83*, 1–8. [CrossRef]
31. Ye, Y.F.; Wang, Q.; Zhao, Y.L.; He, Q.F.; Lu, J.; Yang, Y. Elemental segregation in solid-solution high-entropy alloys: Experiments and modelling. *J. Alloy Compd.* **2016**, *681*, 167–174. [CrossRef]
32. Calvo-Dahlborg, M.; Brown, S.G.R. Hume-Rothery for HEA classification and self-organizing map for phases and properties prediction. *J. Alloy Compd.* **2017**, *724*, 353–364. [CrossRef]
33. Lucas, M.S.; Mauger, L.; Munoz, J.A.; Xiao, Y.M.; Sheets, A.O.; Semiatin, S.L.; Horwath, J.; Turgut, Z. Magnetic and vibrational properties of high-entropy alloys. *J. Appl. Phys.* **2011**, *109*, 07E307. [CrossRef]

34. Li, C.; Zhao, M.; Li, J.C.; Jiang, Q. B2 structure of high-entropy alloys with addition of Al. *J. Appl. Phys.* **2008**, *104*, 113504. [CrossRef]
35. Tang, Z.; Yuan, T.; Tsai, C.-W.; Yeh, J.-W.; Lundin, C.D.; Liaw, P.K. Fatigue behavior of a wrought Al0.5CoCrCuFeNi two-phase high-entropy alloy. *Acta Mater.* **2015**, *99*, 247–258. [CrossRef]
36. Novak, T.G.; Vora, H.D.; Mishra, R.S.; Young, M.L.; Dahotre, N.B. Synthesis of $Al_{0.5}$CoCrCuFeNi and $Al_{0.5}$CoCrFeMnNi high-entropy alloys by laser melting. *Metall. Mater. Trans. B* **2014**, *45*, 1603–1607. [CrossRef]
37. Wang, W.-R.; Wang, W.-L.; Wang, S.-C.; Tsai, Y.-C.; Lai, C.-H.; Yeh, J.-W. Effects of Al addition on the microstructure and mechanical property of Al_xCoCrFeNi high-entropy alloys. *Intermetallics* **2012**, *26*, 44–51. [CrossRef]
38. Sudavtsova, V.S.; Shevchenko, M.O.; Berezutskii, V.V.; Ivanov, M.I. Thermodynamic properties of liquid Fe-Sc alloys. *Powder Metall. Met. Ceram.* **2013**, *52*, 456–464. [CrossRef]
39. Shevchenko, M.A.; Ivanov, M.I.; Berezutskii, V.V.; Kudin, V.G.; Sudavtsova, V.S. Thermodynamic properties of alloys in the Ni-Sc and Ni-Y systems. *Russ. J. Phys. Chem. A* **2014**, *88*, 892–902. [CrossRef]
40. Shevchenko, M.A.; Berezutskii, V.V.; Ivanov, M.I.; Kudin, V.G.; Sudatsova, V.S. Thermodynamic properties of the alloys of the Al-Co and Al-Co-Sc systems. *Russ. J. Phys. Chem. A* **2014**, *88*, 729–734. [CrossRef]
41. Gao, M.C.; Yeh, J.-W.; Liaw, P.K.; Zhang, Y. *High-Entropy Alloys: Fundamentals and applications*, 1st ed.; Springer International Publishing: Cham, Switzerland, 2016.
42. Jones, N.G.; Frezza, A.; Stone, H.J. Phase equilibria in an $Al_{0.5}$CrFeCoNiCu high entropy alloy. *Mater. Sci. Eng. A* **2014**, *615*, 214–221. [CrossRef]
43. Jones, N.G.; Christofidou, K.A.; Stone, H.J. Rapid precipitation in an $Al_{0.5}$CrFeCoNiCu high entropy alloy. *Mater. Sci. Technol.* **2015**, *31*, 1171–1177. [CrossRef]
44. Jones, N.G.; Izzo, R.; Mignanelli, P.M.; Christofidou, K.A.; Stone, H.J. Phase evolution in an $Al_{0.5}$CrFeCoNiCu high entropy alloy. *Intermetallics* **2016**, *71*, 43–50. [CrossRef]
45. Jones, N.G.; Aveson, J.W.; Bhowmik, A.; Conduit, B.D.; Stone, H.J. On the entropic stabilization of an $Al_{0.5}$CrFeCoNiCu. *Intermetallics* **2014**, *54*, 148–153. [CrossRef]
46. Pickering, E.J.; Stone, H.J.; Jones, N.G. Fine-scale precipitation in the high-entropy alloy $Al_{0.5}$CrFeCoNiCu. *Mater. Sci. Eng. A* **2015**, *645*, 65–71. [CrossRef]
47. Chen, S.; Xie, X.; Chen, B.; Qiao, J.; Zhang, Y.; Ren, Y.: Dahmen, K.A.; Liaw, P.K. Effects of temperature on serrated flows of $Al_{0.5}$CoCrCuFeNi High-Entropy Alloy. *JOM* **2015**, *67*, 2314–2320. [CrossRef]
48. Figgins, B.F.; Jones, G.O.; Riley, D.P. LXXVII. The thermal expansion of aluminium at low temperatures as measured by an X-ray diffraction method. *Philos. Mag.* **1956**, *1*, 747–758. [CrossRef]
49. Basinski, Z.S.; Hume-Rothery, W.; Sutton, A.L. The lattice expansion of iron. *Proc. R. Soc. Lond. A* **1955**, *229*, 459–467. [CrossRef]
50. Owen, E.A.; Yates, E.L. LXVI. X-ray measurement of the thermal expansion of pure nickel. *Lond. Edinb. Dublin Philos. Mag. J. Sci.* **1936**, *21*, 809–819. [CrossRef]
51. Ross, R.G.; Hume-Rothery, W. High temperature X-ray metallography: I. A new Debye-Sherrer camera for use at very high temperatures II. A new parafocusing camera III. Applications to the study of chromium, hafnium, molybdenum, rhodium, ruthenium and tungsten. *J. Less Common Met.* **1963**, *5*, 258–270. [CrossRef]
52. Chou, H.-P.; Chang, Y.-S.; Chen, S.-K.; Yeh, J.-K. Microstructure, thermophysical and electrical properties in Al_xCoCrFeNi ($0 \leq x \leq 2$) high-entropy alloys. *Mater. Sci. Eng. B* **2009**, *163*, 184–189. [CrossRef]

entropy

MDPI

Article

Effect of Cold Rolling on the Phase Transformation Kinetics of an $Al_{0.5}CoCrFeNi$ High-Entropy Alloy

Jun Wang *, Haoxue Yang, Tong Guo, Jiaxiang Wang, William Yi Wang and Jinshan Li

State Key Laboratory of Solidification Processing, Northwestern Polytechnical University, Xi'an 710072, China; hxyang@mail.nwpu.edu.cn (H.Y.); believegt@163.com (T.G.); 2013301034@mail.nwpu.edu.cn (J.W.); wywang@nwpu.edu.cn (W.Y.W.); ljsh@nwpu.edu.cn (J.L.)

* Correspondence: nwpuwj@nwpu.edu.cn; Tel.: +86-29-88460568; Fax: +86-29-88460294

Received: 25 October 2018; Accepted: 29 November 2018; Published: 30 November 2018

Abstract: The solid state phase transformation kinetics of as-cast and cold rolling deformed $Al_{0.5}CoCrFeNi$ high-entropy alloys have been investigated by the thermal expansion method. The phase transformed volume fractions are determined from the thermal expansion curve using the lever rule method, and the deformed sample exhibits a much higher transformation rate. Two kinetic parameters, activation energy (E) and kinetic exponent (n) are determined using Kissinger–Akahira–Sunose (KAS) and Johnson–Mehl–Avrami (JMA) method, respectively. Results show that a pre-deformed sample shows a much lower activation energy and higher kinetic exponent compared with the as-cast sample, which are interpreted based on the deformation induced defects that can promote the nucleation and growth process during phase transformation.

Keywords: high-entropy alloy; phase transformation; kinetics; deformation; thermal expansion

1. Introduction

High-entropy alloys (HEAs) are a new class of alloys designed based on a unique alloy concept that have multi alloy components (normally five or more principal elements) with equal or near-equal atomic composition [1,2]. HEAs are mainly solid solution based alloys, which own very attractive properties, like high strength and hardness, high fracture toughness, excellent corrosion resistance and unique physical properties, making them the potential engineering materials in many industry areas [3–7].

Thermal induced phase transformations are one of the main transitions inside metallic materials which can be used to control the final phases of the material, and thus to tune the properties. In order to know the details of the phase transformation processes, one of the useful methods is to study the phase transformation kinetics, which can give a lot of useful information like phase transformation speed and transition mode [8,9]. However, until now, there have only been very few papers referring to the kinetic analysis of high-entropy alloys.

The as-cast HEAs always exhibit coarse grains especially for big ingots. In order to obtain fine equiaxed grains, methods like cold rolling or forging are incorporated during the treating process [10,11]. Whether this kind of treatment can affect the subsequent phase transformation process is still unclear for HEAs.

$Al_xCoCrFeNi$ ($0 \leq x \leq 2$) HEAs are one of the alloy systems that intrigue the research interest due to many aspects. First, this alloy system owns very good mechanical [12–15] and physical properties [16–18]. Second, with the increasing Al content, the main phase of the alloy moves from pure face-centered-cubic (FCC) to FCC + pure body-centered-cubic (BCC), and pure BCC phase, making the mechanical and physical properties adjustable [16–25]. Thus, the microstructure, phases and properties of this alloy system have been intensively investigated. In this study, a binary phase

(FCC + BCC) $Al_{0.5}CoCrFeNi$ HEA with a balanced strength and plasticity are chosen to investigate the effect of cold rolling on the subsequent solid state phase transformation kinetics.

2. Experimental Procedures

The ingots with a nominal composition of $Al_{0.5}CoCrFeNi$ were prepared by melting pure elements (at least 99.95 wt.%) in a vacuum induction-melting furnace. The furnace is firstly heated to 500 °C and held for 2 h to remove the water vapor, and vacuumed to below 10 Pa. Then pure argon is backfilled to expel the rest of air until the vacuum goes back to standard atmospheric pressure. This process is repeated three times in order to gain an oxygen-free environment as much as possible. Afterwards, melting and casting is performed with the protection of high purity argon and the alloy is heated at about 1550 °C for 15 min. Approximately 15 kg ingot is produced by casting the melt into a steel crucible with a height of 180 mm, upper inner diameter of 140 mm and bottom inner diameter of 130 mm.

Samples with the size of 8 × 20 × 70 mm^3 are taken from the center of the ingot. Then, the sample is cold rolled to a thickness reduction of 20%. Then, samples were machined to ϕ 6 × 25 mm for thermal expansion measurement both from as-cast and cold rolled plates. The thermal expansion curves were tested using a Netzsch® DIL-402C dilatometer (Selb, Germany) under the protection of argon with constant heating rate of 4 K/min, 6 K/min, 8 K/min and 10 K/min.

The microstructure and phases were characterized by scanning electron microscope (SEM, TESCAN MIRA3 XMU (Brno, Czech Republic), and the working distance and the energy of beam used are 15 mm and 20 kV, respectively) and X-ray diffractometer (XRD, DX 2700 (Dandong, China), and the voltage and current used during measurement are 40 kV and 30 mA, respectively), respectively. Uniaxial tensile tests are carried out with a MTS SANS CMT5105 (Shenzhen, China) mechanical tester at the strain rate of 10^{-3} s^{-1} and the specimens are prepared along the rolling direction (RD) direction.

3. Results and Discussion

3.1. Microstructure

The microstructure and XRD patterns of $Al_{0.5}CoCrFeNi$ high-entropy alloys at as-cast and 20% cold rolled (CR) condition are shown in Figures 1 and 2. Except for the grain after cold rolling showing a very small amount elongation along the rolling direction, the as-cast and cold rolled sample show quite similar microstructure (Figure 1). Both the samples are mainly FCC phase together with a small amount of BCC phase (Figure 2), indicating 20% cold rolling has no significant effect on the microstructure and phase transition of the $Al_{0.5}CoCrFeNi$ high-entropy alloy.

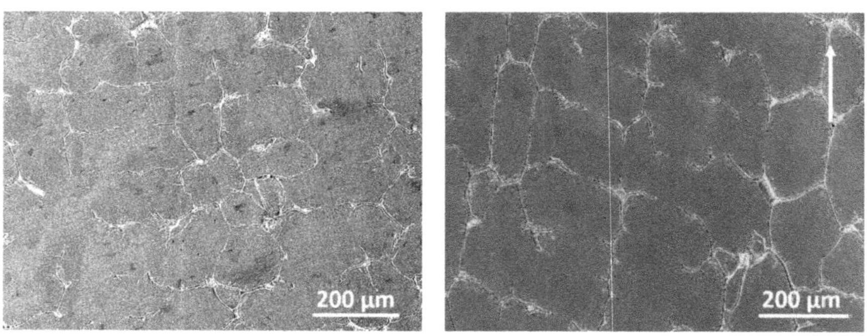

(a) (b)

Figure 1. The microstructure of the $Al_{0.5}CoCrFeNi$ high-entropy alloys at (**a**) as-cast and (**b**) 20% cold rolled condition. The arrow represents the rolling direction.

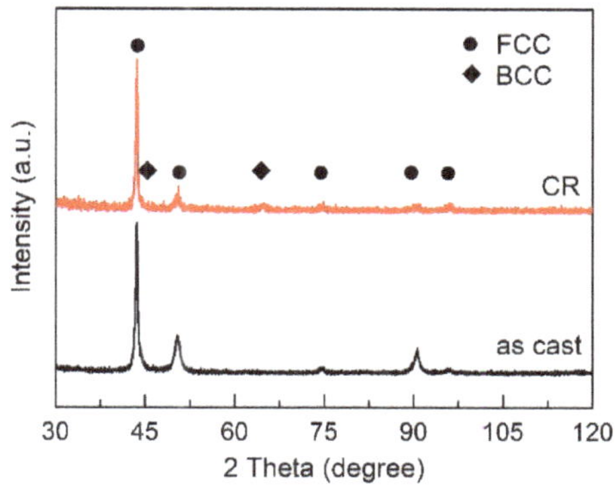

Figure 2. The XRD patterns of $Al_{0.5}$CoCrFeNi high-entropy alloy at as-cast and 20% cold rolled condition.

3.2. Transformed Volume Fraction

Figure 3a shows the thermal expansion curves of $Al_{0.5}$CoCrFeNi HEA at as-cast and 20% CR condition measured at the heating rate of 10 K/min. There is a peak around 1100 K for both curves indicating a phase transition exist, which is proved to be an FCC-BCC phase transition [8]. The first step for phase transition kinetic analysis is to determine the phase transformed volume fraction (f) as a function of temperature or time. Figure 3b shows how to calculate the phase transformed volume as a function of temperature with lever rule method using thermal expansion curves shown in Figure 3a. The lines AD and BE can be obtained by extending the linear expansion section of the thermal expansion curve. Then, the point C is the intersection of AC and thermal expansion curve (f = 0 when C is at point D and f = 100% when C is point E). Finally, the theoretical volume fraction of phase transition can be calculated using the simple lever rule method assuming the volume changes of phase transition is in proportion to the length variations, which is expressed in Equation (1) [26]:

$$f = \frac{|AC|}{|AB|}, \tag{1}$$

where f is the volume fraction, and $|AC|$ and $|AB|$ are the length of the dashed line shown in Figure 3b, indicating the instant thermal expansion length and overall length, respectively.

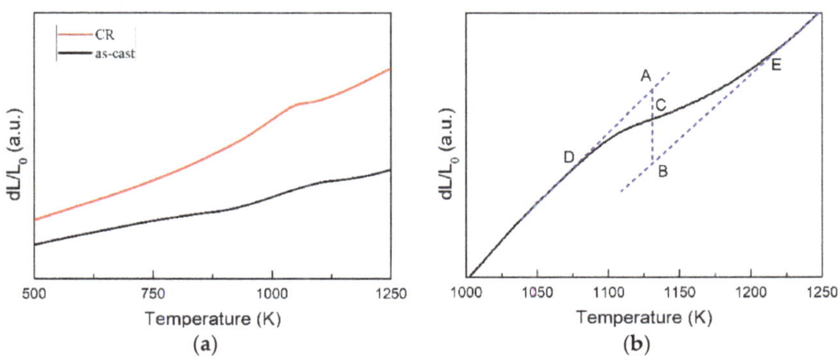

Figure 3. (**a**) thermal expansion curves of $Al_{0.5}$CoCrFeNi high-entropy alloy measured at as-cast and 20% cold rolled condition at heating rate of 10 K/min; (**b**) an example of as-cast sample shows how to calculate the phase transformed volume fraction as function of temperature.

Figure 4 shows the transformed volume fraction as a function of temperature at different heating rates for the as-cast (Figure 4a) and 20% CR (Figure 4b) samples. All of the f–T curves show the similar "S" type trend, indicating that the transformation is controlled by a typical nucleation-growth mechanism [27]. Compared with the as-cast sample, it can be seen from Figure 4 that the phase transition of cold rolled sample starts at a much higher temperature (1033 K and 1055 K at 4 K/min for as-cast and cold rolled samples, respectively) but ends at a much lower temperature (1195 K and 1151 K at 4 K/min for as-cast and cold rolled samples, respectively), indicating a much higher transformation rate when the sample is deformed.

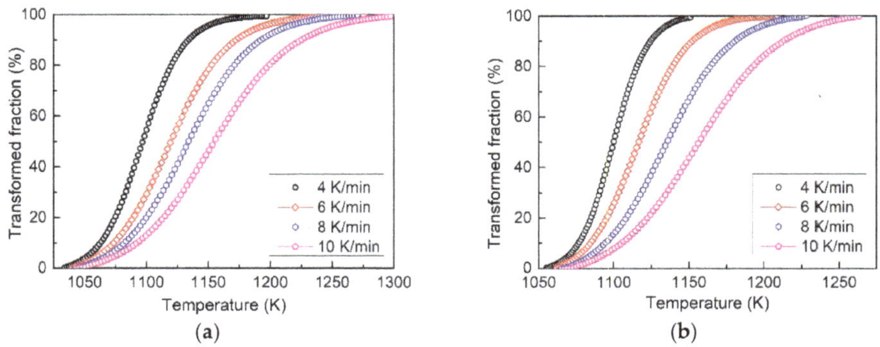

Figure 4. The phase transformed volume fraction as a function of temperature curves of (**a**) as-cast and (**b**) 20% cold rolled $Al_{0.5}CoCrFeNi$ high-entropy alloy.

3.3. Activation Energy

Activation energy (E) and kinetic exponent (n) are the two most important kinetic parameters. Activation energy is the parameter that can assess phase transition energy barrier. The Kissinger–Akahira–Sunose (KAS) method [28] and Friedman method [29] can be used for the determination of activation energies. Here, we use KAS method, which has the following form [28]:

$$\ln\left(\frac{T^2}{\varphi}\right) = -C + \frac{E}{RT'} \quad (2)$$

where T is temperature at certain transformed volume fraction, ϕ is the heating rate, E is the activation energy, C is a constant and R is the molar gas constant. Assuming the activation energy is the same at the same transformed fraction at different heating rates, then, by linear regression between $\ln(T^2/\phi)$ and $1/T$ using the data taken from Figure 4, the activation energy can be directly determined by the slope of the curve.

The determined activation energy E for as cast and 20% CR $Al_{0.5}CoCrFeNi$ HEA at four different transformed volume fractions, f = 0.2, 0.4, 0.6 and 0.8 are shown in Figure 5. The activation energies of as-cast and 20% CR $Al_{0.5}CoCrFeNi$ HEA show the same decreasing trend with the increasing transformed volume fraction, f, however, the value are quite different. For the as-cast sample, the activation energies are 262 kJ/mol and 126 kJ/mol when transformed volume fraction is 0.2 and 0.8, respectively. The mean value is 181 kJ/mol—while, for the 20% CR sample, E are 223 kJ/mol and 110 kJ/mol when transformed volume fraction is 0.2 and 0.8, respectively. The average activation energy is 159 kJ/mol, which is 10% lower than the as-cast sample, indicating that the phase transformation process of the deformed sample is much easier than the as-cast sample.

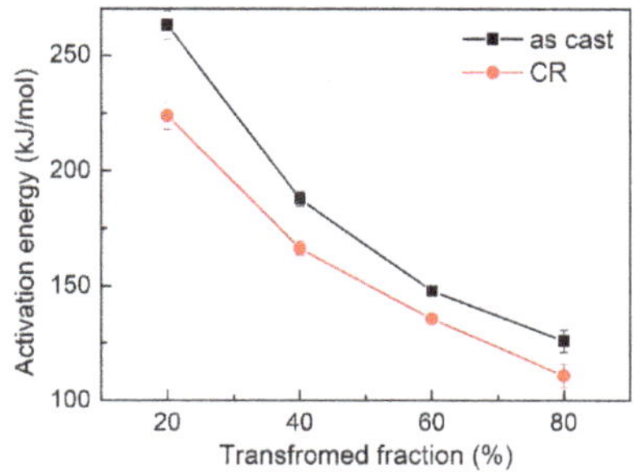

Figure 5. The activation energies at different phase transformed volume fractions for as-cast and cold rolled $Al_{0.5}$CoCrFeNi high-entropy alloys calculated using the KAS method.

3.4. Kinetic Exponent

The kinetic exponent, also called the Avrami exponent, is a key kinetic parameter that can directly obtain the phase transition mode. The kinetic exponent is generally determined based on the well-known Johnson–Mehl–Avrami (JMA) model [30], which has the following form:

$$f = 1 - exp(-Kt^n), \tag{3}$$

where n is the Avrami exponent, K is a rate constant and t is the time. In case of phase transformations that are carried out at constant heating rate, the time can be expressed as [31]:

$$t - t_0 = \frac{T - T_0}{\varnothing}, \tag{4}$$

where T_0 and t are the starting temperature and time of phase transition. Then, the kinetic exponent can be obtained by some mathematical treatment of Equations (3) and (4), which can be expressed as:

$$n = -\frac{\ln(-\ln(1-f))}{\frac{E}{RT}}. \tag{5}$$

Based on Equation (5), we can determine the kinetic expoent n from the slope of $\ln(-\ln(1-f))$ vs. $1/T$ curve.

For a real phase transformation, n will change with the transformed volume fraction. A simple derivative equation of Equation (5) can determine the Avrami exponent as variables, which has been used in many previous published papers [31–33]. In this case, n is called local Avrami exponent, which can be described as:

$$n = -\frac{\partial \ln(-\ln(1-f))}{\partial \frac{E}{RT}}. \tag{6}$$

By taken the activation energy at a different transformed volume fraction using Equation (2), Avrami exponent n can be calculated by the differential of the curve of $E/RT - \ln[-\ln(1-\mathrm{f})]$. Figure 6 represents the kinetic exponent at different transformed volume fractions at the heating rate 10 K/min for the as-cast and CR sample. According to Figure 6, two important pieces of information can be drawn. First, both of the as-cast and deformed samples show three typical stages of transformation: (I) rapid decrease of the kinetic exponent when $f < 5\%$, indicating that the nucleation rate of the sample is decreasing at the beginning; (II) slow decrease of the kinetic exponent from 4 to 1.5 during $9\% < f < 95\%$, indicating the decreasing nucleation rate, and the growth mode of the nuclei changes from interface controlled to diffusion controlled; (III) strong increase of the kinetic exponent when $f > 95\%$, an indication that some new inhomogeneous nucleation occurs inside the sample. Second the kinetic exponent, n, is higher for the sample pre-deformed than the as-cast sample. This means that the nucleation rate of the deformed sample is much higher and the growth of the nuclei is much easier than the as-cast sample.

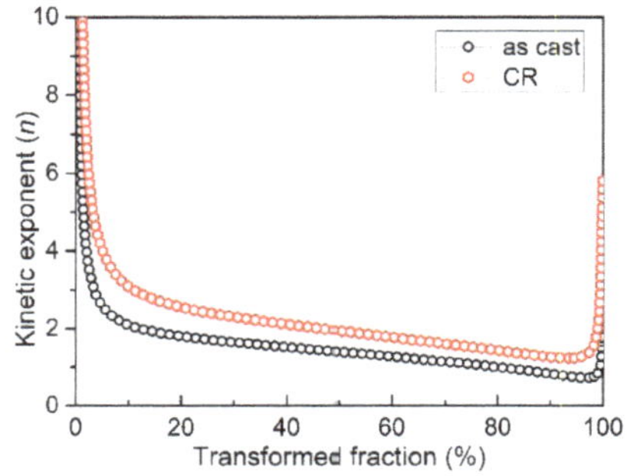

Figure 6. Variation of *n* with phase transformed fraction for as-cast and deformed $Al_{0.5}CoCrFeNi$ high-entropy alloy at the heating rate of 10 K/min.

3.5. Discussion

For the deformed sample like cold rolling, normally, there will be an anisotropic effect due to the stress state during deformation are different at different directions. For the cold rolled sample, the rolling direction (RD) and the transverse direction (TD) always owns different textures, stress and properties. For the above results, all of the analysis is based on the sample that is taken from the RD direction. Figure 7 shows the thermal expansion curves that are both taken from the RD and TD direction of a 20% CR sample. According to the figure, it can be seen that the rolling direction has quite a limited effect on the thermal expansion behavior. The reason may be due to the fact that a 20% reduction is not large enough to generate a big difference, and it can also be found from Figure 1 that the CR sample has a similar microstructure as the as-cast sample. Large differences can be made evident at much larger CR reduction thickness, and the grain will be severely deformed and elongated along the rolling direction [12]. In this case, the thermal expansion behavior could be different for different directions. Thus, it can be concluded that there is no typical anisotropic effect in the 20% CR sample, and, during the solid state phase transformation process, the nucleation and growth process can be treated as homogeneous.

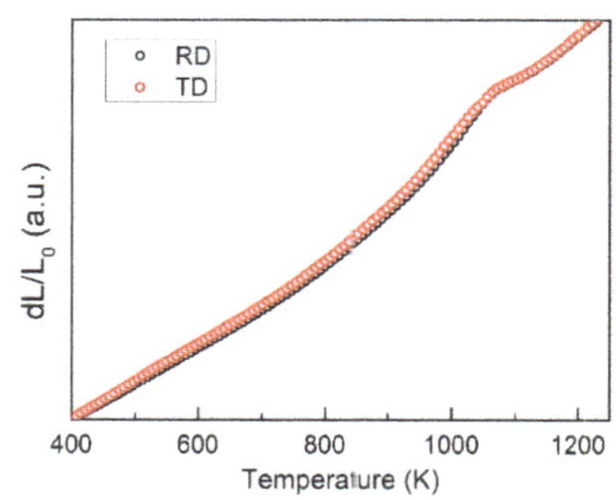

Figure 7. The thermal expansion curves of the samples taken from the rolling direction (RD) and transverse direction (TD) for 20% cold rolled $Al_{0.5}CoCrFeNi$ high-entropy alloy measured at the heating rate of 10 K/min.

There is not much difference for the microstructure (Figure 1) and phases (Figure 2) between the as-cast and deformed sample; however, according to the f–T curve, activation energy and kinetic exponent, there is a large difference. The microstructure of $Al_xCoCrFeNi$ ($0 \leq x \leq 2$) HEAs has been intensively investigated in different heat treatment conditions [17–25]. Different from the thermo-mechanical processing that can affect the transformation pathways in $Al_{0.3}$CoCrFeNi [34], 20% cold rolling cannot alter the phase transition product [12]. Thus, the microstructure and phases exhibit no difference as shown in Figures 1 and 2. However, the deformed sample shows a much faster transformation rate, much lower activation energy and much larger kinetic exponents, an indication of a large different for the transformation kinetics. Figure 8 exhibits the tensile stress–strain curves of $Al_{0.5}$CoCrFeNi HEAs at as-cast and 20% CR conditions. There exists a large difference: the deformed sample shows a much larger stress but lower plasticity. The yield stress increased from 402 MPa (as-cast sample) to 755 MPa (CR sample), while the strain decreased from 33.7% (as-cast sample) to 6.2% (CR sample), indicating a strong strain-hardening effect of the cold rolling. In the present study, due to the small reduction rate of the CR, there is not much difference in the macroscopic microstructure; however, 20% CR is enough to generate a large crystal lattice distortion, and many crystal defects, like vacancy, dislocation and substructure inside the grain. Then, the deformed sample stored a large quantity of energy, which is beneficial for the phase transition during the continuous heating. Thus, because of all the above factors that are generated by cold rolling, the deformed sample shows a much higher transformation rate since all the defects and stored energy inside the sample can promote the nucleation rate and grain growth speed.

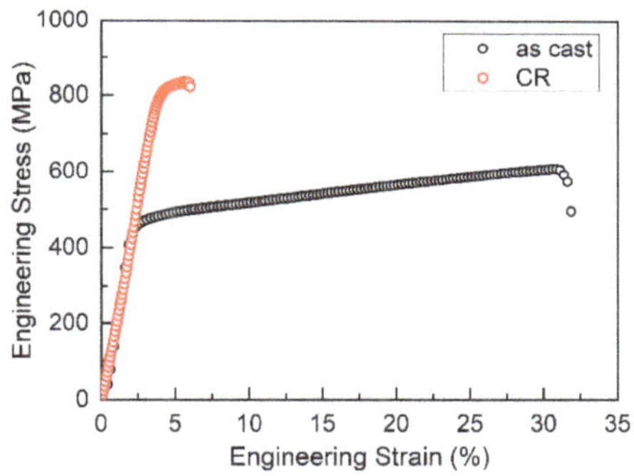

Figure 8. The engineering stress–strain curves of $Al_{0.5}$CoCrFeNi high-entropy alloys at as-cast and 20% CR conditions.

4. Conclusions

The effects of cold rolling on the phase transformation kinetics of an $Al_{0.5}$CoCrFeNi high-entropy alloy are studied by the thermal expansion method. The transformed volume fraction–temperature curves of as-cast and 20% cold rolling $Al_{0.5}$CoCrFeNi high-entropy alloys are determined by the lever rule method using thermal expansion curves. Using the KAS method, the mean activation energy for as-cast and deformed HEAs calculated are 181 kJ/mol and 159 kJ/mol, respectively. The kinetic exponent is determined using the local Avrami exponent method based on the JMA equation. Both as-cast and deformed samples show the three stages of kinetic exponent variations, and the deformed alloy shows a much larger value compared with the as-cast alloy, indicating a much higher nucleation and growth rate. The reason for the kinetic exponent variations are thought to be related with the deformation induced strain and defects inside the sample that can generate a much higher energy that favors the nucleation and growth process.

Author Contributions: Conceptualization, J.W, and J.S.L.; Investigation, J.W., H.X.Y., T.G., J.X.W.; Writing—original draft, J.W., H.X.Y., T.G.; Writing—review & editing, J.W., W.Y.W., J.S.L.

Funding: This research was funded by the Natural Science Foundation of China, grant number 51571161 and 51774240 and the fund of State Key Laboratory of Solidification Processing in NWPU, grant number 121-TZ-2015.

Conflicts of Interest: The authors declare no conflict of interest.

References

1. Cantor, B.; Chang, I.T.H.; Knight, P.; Vincent, A.J.B. Microstructural development in equiatomic multicomponent alloys. *Mater. Sci. Eng. A* **2004**, *375–377*, 213–218. [CrossRef]
2. Yeh, J.W.; Chen, S.K.; Lin, S.J.; Gan, J.Y.; Chin, T.S.; Shun, T.T.; Tsau, C.H.; Chang, S.Y. Nanostructured high-entropy alloys with multiple principal elements: Novel alloy design concepts and outcomes. *Adv. Eng. Mater.* **2004**, *6*, 299–303. [CrossRef]
3. Zhang, W.; Liaw, P.K.; Zhang, Y. Science and technology in high-entropy alloys. *Sci. China Mater.* **2018**, *61*, 2–22. [CrossRef]
4. Miracle, D.B.; Senkov, O.N. A critical review of high entropy alloys and related concepts. *Acta Mater.* **2017**, *122*, 448–511. [CrossRef]
5. Wang, J.; Li, J.; Wang, J.; Bu, F.; Kou, H.; Li, C.; Zhang, P.; Beaugnon, E. Effect of Solidification on Microstructure and Properties of FeCoNi(AlSi)0.2 High-Entropy Alloy Under Strong Static Magnetic Field. *Entropy* **2018**, *20*, 275. [CrossRef]
6. Tsai, M.H. Physical Properties of High Entropy Alloys. *Entropy* **2013**, *15*, 5338–5345. [CrossRef]
7. Klimova, M.; Stepanov, N.; Shaysultanov, D.; Chernichenko, R.; Yurchenko, N.; Sanin, V.; Zherebtsov, S. Microstructure and Mechanical Properties Evolution of the Al, C-Containing CoCrFeNiMn-Type High-Entropy Alloy during Cold Rolling. *Materials* **2018**, *11*, 53. [CrossRef] [PubMed]
8. Wang, J.; Niu, S.; Guo, T.; Kou, H.; Li, J. The FCC to BCC phase transformation kinetics in an $Al_{0.5}CoCrFeNi$ high entropy alloy. *J. Alloy. Compd.* **2017**, *710*, 144–150. [CrossRef]
9. Liu, F.; Sommer, F.; Bos, C.; Mittemeijer, E.J. Analysis of solid state phase transformation kinetics: Models and recipes. *Inter. Mater. Rev.* **2007**, *52*, 193–212. [CrossRef]
10. Zhang, Y.; Li, J.; Wang, J.; Niu, S.; Kou, H. Hot deformation behavior of as-cast and Homogenized $Al_{0.5}CoCrFeNi$ High Entropy Alloys. *Metals* **2016**, *6*, 277. [CrossRef]
11. Li, D.; Zhang, Y. The ultrahigh charpy impact toughness of forged $Al_xCoCrFeNi$ high entropy alloys at room and cryogenic temperatures. *Intermetallics* **2016**, *70*, 24–28. [CrossRef]
12. Guo, T.; Li, J.; Wang, J.; Wang, W.Y.; Liu, Y.; Luo, X.; Kou, H.; Beaugnon, E. Microstructure and properties of bulk $Al_{0.5}CoCrFeNi$ high-entropy alloy by cold rolling and subsequent annealing. *Mater. Sci. Eng. A* **2018**, *729*, 141–148. [CrossRef]
13. Zhang, Y.; Li, J.; Wang, J.; Wang, W.Y.; Kou, H.; Beaugnon, E. Temperature dependent deformation mechanisms of $Al_{0.3}CoCrFeNi$ high-entropy alloy, starting from serrated flow behavior. *J. Alloy. Compd.* **2018**, *757*, 39–43. [CrossRef]
14. Niu, S.; Kou, H.; Guo, T.; Zhang, Y.; Wang, J.; Li, J. Strengthening of nanoprecipitations in an annealed Al0.5CoCrFeNi high entropy alloy. *Mater. Sci. Eng. A* **2016**, *671*, 82–86. [CrossRef]
15. Li, D.; Li, C.; Feng, T.; Zhang, Y.; Sha, G.; Lewandowski, J.J.; Liaw, P.K.; Zhang, Y. High-entropy $Al_{0.3}CoCrFeNi$ alloy fibers with high tensile strength and ductility at ambient and cryogenic temperatures. *Acta Mater.* **2017**, *123*, 285–294. [CrossRef]
16. Yang, T.; Xia, S.; Liu, S.; Wang, C.; Liu, S.; Fang, Y.; Zhang, Y.; Xue, J.; Yan, S.; Wang, Y. Precipitation behavior of $Al_xCoCrFeNi$ high entropy alloys under ion irradiation. *Sci. Rep.* **2016**, *6*, 32146. [CrossRef] [PubMed]
17. Chou, H.P.; Chang, Y.S.; Chen, S.K.; Yeh, J.W. Microstructure, thermophysical and electrical properties in $Al_xCoCrFeNi$ ($0 \leq x \leq 2$) high-entropy alloys. *Mater. Sci. Eng. B* **2009**, *163*, 184–189. [CrossRef]
18. Kao, Y.F.; Chen, S.K.; Chen, T.J.; Chu, P.C.; Yeh, J.W.; Lin, S.J. Electrical, magnetic, and Hall properties of $Al_xCoCrFeNi$ high-entropy alloys. *J. Alloy. Compd.* **2011**, *509*, 1607–1614. [CrossRef]
19. Kao, Y.F.; Lee, T.D.; Chen, S.K.; Chang, Y.S. Electrochemical passive properties of $Al_xCoCrFeNi$ (x = 0, 0.25, 0.50, 1.00) alloys in sulfuric acids. *Corros. Sci.* **2010**, *52*, 1026–1034. [CrossRef]
20. Jasiewicz, K.; Cieslak, J.; Kaprzyk, S.; Tobola, J. Relative crystal stability of $Al_xFeNiCrCo$ high entropy alloys from XRD analysis and formation energy calculation. *J. Alloy. Compd.* **2015**, *648*, 307–312. [CrossRef]
21. Xia, S.Q.; Yang, X.; Yang, T.F.; Liu, S.; Zhang, Y. Irradiation Resistance in $Al_xCoCrFeNi$ High Entropy Alloys. *JOM* **2015**, *67*, 2340–2344. [CrossRef]
22. Nahmany, M.; Hooper, Z.; Stern, A.; Geanta, V.; Voiculescu, I. $Al_xCrFeCoNi$ high-entropy alloys: Surface modification by electron beam bead-on-plate melting. *Metallogr. Micros. Anal.* **2016**, *5*, 1–12.
23. Zhang, C.; Zhang, F.; Diao, H.; Gao, M.C.; Tang, Z.; Poplawsky, J.D.; Liaw, P.K. Understanding phase stability of Al-Co-Cr-Fe-Ni high entropy alloys. *Mater. Des.* **2016**, *109*, 425–433. [CrossRef]

24. Ma, Y.; Jiang, B.; Li, C.; Wang, Q.; Dong, C.; Liaw, P.K.; Xu, F.; Sun, L. The BCC/B2 morphologies in Al_xNiCoFeCr high-entropy alloys. *Metals* **2017**, *7*, 57. [CrossRef]
25. Rao, J.C.; Diao, H.Y.; Ocelík, V.; Vainchtein, D.; Zhang, C.; Kuo, C.; Tang, Z.; Guo, W.; Poplawsky, J.D.; Zhou, Y.; et al. Secondary phases in Al_xCoCrFeNi high-entropy alloys: An in-situ TEM heating study and thermodynamic appraisal. *Acta Mater.* **2017**, *131*, 206–220. [CrossRef]
26. Kou, H.C.; Wang, J.; Chang, H.; Tang, B.; Li, J.S.; Hu, R.; Zhou, L. Kinetic analysis of the isochronal crystallization of $Ti_{40}Zr_{25}Ni_8Cu_9Be_{18}$ metallic glass. *J. Non-Cryst. Solids* **2009**, *355*, 420–424. [CrossRef]
27. Christian, J.W. *Theory of Transformations in Metals and Alloys*, 3rd ed.; Pergamon Press: Oxford, UK, 2002.
28. Kissinger, H.E. Reaction Kinetics in Differential Thermal Analysis. *Anal. Chem.* **1957**, *29*, 1702–1706. [CrossRef]
29. Bertolino, V.; Cavallaro, G.; Lazzara, G. Effect of the biopolymer charge and the nanoclay morphology on nanocomposite materials. *Ind. Eng. Chem. Res.* **2016**, *55*, 7373–7380. [CrossRef]
30. Wang, J.; Kou, H.C.; Li, J.S.; Gu, X.F.; Zhong, H.; Chang, H.; Zhou, L An integral fitting method for analyzing the isochronal transformation kinetics: Application to the crystallization of a Ti-based amorphous alloy. *J. Phys. Chem. Solids* **2009**, *70*, 1448–1453. [CrossRef]
31. Majhi, K.; Varma, K.B.R. Crystallization kinetic studies of CaBi2B2O7 glasses by non-isothermal methods. *J. Mater. Sci.* **2009**, *44*, 385–391. [CrossRef]
32. Lu, W.; Niu, J.; Wang, T.; Xia, K.; Xiang, Z.; Song, Y.; Mi, Z.; Zhang, W.; Tian, W.; Yan, Y. Phase transformation kinetics and microstructural evolution of MnAl permanent magnet alloys. *J. Alloy. Compd.* **2016**, *685*, 992–996. [CrossRef]
33. Zhou, Z.; Lai, M.; Tang, B.; Kou, H.; Chang, H.; Zhu, Z.; Li, J.; Zhou. L. Non-isothermal phase transformation kinetics of ω phase in TB-13 titanium alloys. *Mater. Sci. Eng. A* **2010**, *527*, 5100–5104. [CrossRef]
34. Gwalani, B.; Gorsse, S.; Choudhuri, D.; Styles, M.; Zheng, Y.; Mishra, R.S.; Banerjee, R. Modifying transformation pathways in high entropy alloys or complex concentrated alloys via thermo-mechanical processing. *Acta Mater.* **2018**, *153*, 169–185. [CrossRef]

entropy

MDPI

Article

Al-Ti-Containing Lightweight High-Entropy Alloys for Intermediate Temperature Applications

Minju Kang, Ka Ram Lim, Jong Woo Won, Kwang Seok Lee and Young Sang Na *

Korea Institute of Materials Science, 797 Changwondae-ro, Seongsan-gu, Changwon, Gyeongnam 642-831, Korea; minju214@gmail.com (M.K.); krlim@kims.re.kr (K.R.L.); jwwon@kims.re.kr (J.W.W.); ksl1784@kims.re.kr (K.S.L.)

* Correspondence: nys1664@kims.re.kr; Tel.: +82-55-280-3377

Received: 19 April 2018; Accepted: 8 May 2018; Published: 9 May 2018

Abstract: In this study, new high-entropy alloys (HEAs), which contain lightweight elements, namely Al and Ti, have been designed for intermediate temperature applications. Cr, Mo, and V were selected as the elements for the Al-Ti-containing HEAs by elemental screening using their binary phase diagrams. AlCrMoTi and AlCrMoTiV HEAs are confirmed as solid solutions with minor ordered B2 phases and have superb specific hardness when compared to that of commercial alloys. The present work demonstrates the desirable possibility for substitution of traditional materials that are applied at intermediate temperature to Al-Ti-containing lightweight HEAs.

Keywords: high-entropy alloys; alloys design; lightweight alloys

1. Introduction

Recently, high-entropy alloys (HEAs) have attracted considerable attention because of their extraordinary properties [1–10], and numerous HEAs have been reported with various compositions [1–10]. Three major HEAs are CoCrFeMnNi alloy which has potential applications in cryogenic environments [5,6], refractory VNbMoTaW alloy for high-temperature structural applications [7,8], and AlCoCrFeNi alloy which maintains high strength up to intermediate temperatures [9,10]. AlCoCrFeNi HEA is relatively lightweight and has excellent specific strength around intermediate temperatures. It is a possible alternative for Ti alloys or wrought superalloys such as Inconel 718 [9,10]. Additional weight reduction can improve the competitiveness of the lightweight HEAs; therefore, we attempted to develop new HEAs that contain lightweight elements, namely Al and Ti.

Most HEAs are developed by a trial-and-error approach based on the effects such as mixing enthalpy and valence electron concentration (VEC) [11,12]. This conventional method is not effective when developing HEAs with a new combination of elements among numerous possibilities. The CALPHAD (Computer Coupling of Phase Diagrams and Thermochemistry) approach may be the best way for designing new HEAs because of its capability in predicting the phase stability [13–16]. However, the application of CALPHAD to the design of new HEAs is challenging because of the lack of a reliable thermodynamic database to cover the entire composition range [13–15]. F. Zhang et al. [17] reported a new approach to design new multi-component FCC HEAs by binary phase diagrams. The FCC single phase formation in the CoCrFeMnNi HEA was predicted using this approach [17]. This method is effective at finding "matching elements" that form a single solid solution and is suitable for designing novel HEAs.

In this work, we sought to design lightweight HEAs, which contain lightweight elements, namely Al and Ti, by using binary phase diagrams. Seven HEAs containing Al and Ti were designed and their mechanical properties were compared with those of commercial alloys.

2. Materials and Methods

The selection of candidate elements was conducted based on their binary phase diagrams. The details regarding the design of the Al-Ti-containing lightweight HEAs are described in Section 3.1.

The aforementioned Al-Ti-containing lightweight HEAs were fabricated by vacuum plasma arc melting (PAM) with high-purity elements. The HEA button ingots were re-melted 4 to 5 times in a melting furnace for the homogenization. The HEA plates were fabricated by vacuum induction melting using a graphite mold. The microstructures of the HEAs were analyzed by optical microscopy (OM), FE-SEM (model: SU-6600, HITACHI), and transmission electron microscopy (TEM, model: Tecnai F20, FEI). The TEM samples were prepared using focused ion beam (FIB). The crystal structure of the material was examined by X-ray diffraction (XRD) measurements on the as-casted material using a MXP21VAHF diffractometer with a CuKα radiation source (model: D/Max-2500VL/PC, RIGAKU). The Vickers hardness tests were carried out using a conventional indenter with a load of 2.94 N for 15 s. A minimum 10 tests were carried out on specimen.

3. Results and Discussion

3.1. Selection of Candidate Elements Based on Their Binary Phase Diagrams

To design lightweight HEAs, we selected Al and Ti as the basic elements owing to their low density. The key idea of designing new HEAs using their binary phase diagrams is finding elements that mix. Therefore, the solubility of Al and Ti should be investigated first. Figure 1 shows the Al-Ti phase diagram [18]. It is very complex, and numerous intermediate ordered phases exist. However, β-Ti (BCC phase) and α-Ti (HCP phase) appear in the Ti-rich region as a solid solution. These solid solution regions suggest the possibility of the formation of single solid solution. Therefore, the lightweight elements, Al and Ti, could be the basic elements for lightweight HEAs.

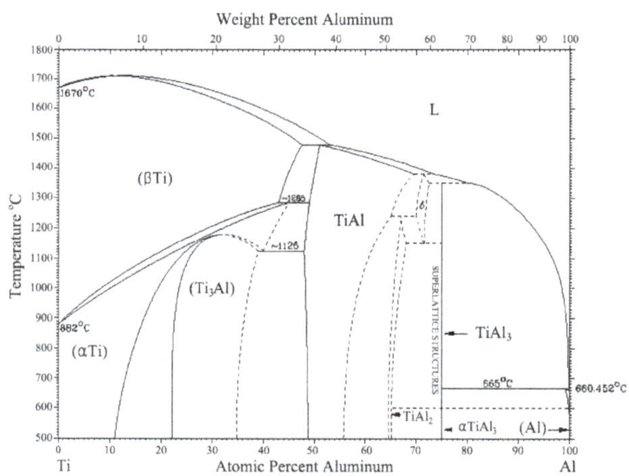

Figure 1. Al-Ti phase diagram [18]. It has solid solution region within a certain temperature and composition range that suggest the possibility of the formation of single solid solution.

The design of the Al-Ti-containing HEAs consisted of three steps. In the first step, the candidate elements were selected from various Ti-X binary phase diagrams wherein "X" represents elements that form homogeneous solid solution with Ti within a certain temperature range [18]. In the second step, a second series of candidate elements were selected from various Y-Al binary phase diagrams, wherein "Y" represents elements that show adequate solubility in Al within a certain composition and temperature range. The final step was the selection of the final candidate elements from the X-Y binary phase diagrams.

From the Ti-X binary phase diagrams, we found 8 candidate elements, Cr, Hf, Mo, Nb, Ta, V, W, and Zr, in which each had a solid solution region in a certain range [18]. Figure 2 shows the Ti-X (X = Cr, Hf, Mo, Nb, Ta, V, W, or Zr) binary phase diagrams. All elements had a large area of solid solution within a certain temperature range, and all the homogeneous solid solution phases had BCC crystalline structure, indicated in green (Figure 2).

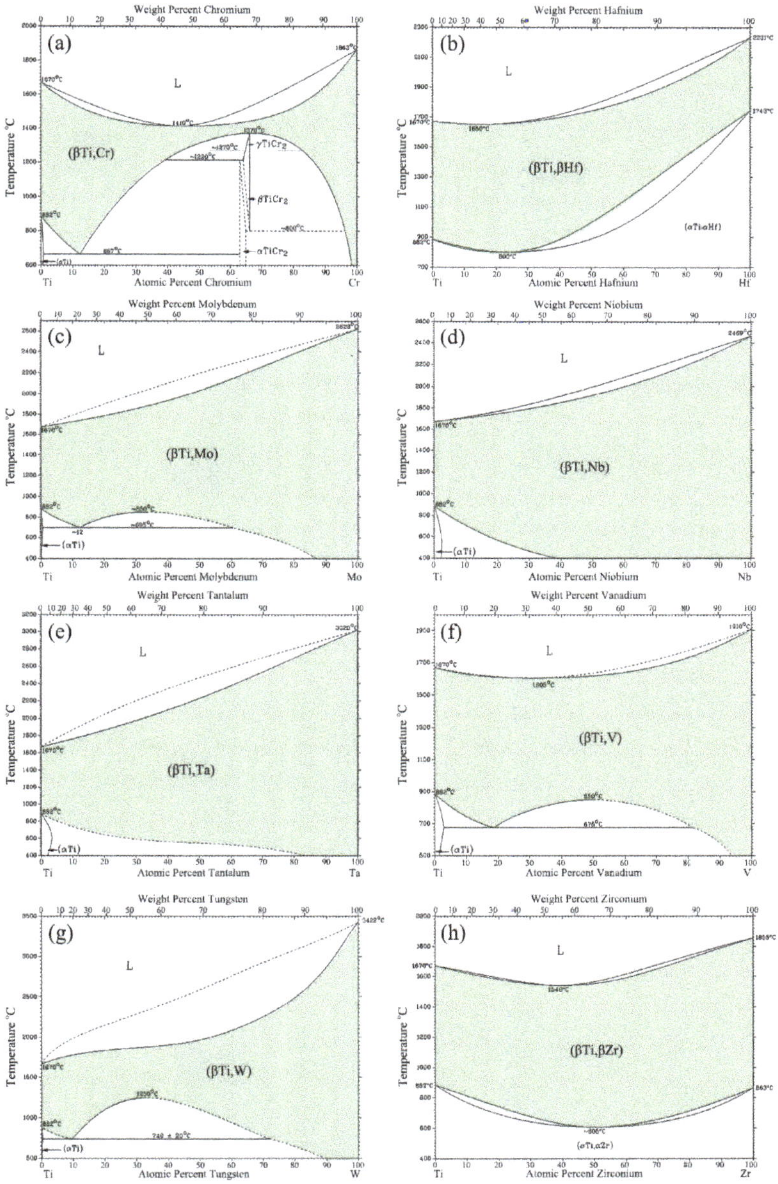

Figure 2. Ti-X phase diagrams where "X" is (**a**) Cr; (**b**) Hf; (**c**) Mo; (**d**) Nb; (**e**) Ta; (**f**) V; (**g**) W; and (**h**) Zr [18].

The next step consisted of finding a second series of candidate elements. Their binary phase diagrams are shown in the supplementary data (Figure S1). 6 candidate elements, Cr, Hf, Mo, Nb, V, and Zr, were chosen from the Y-Al binary phase diagrams. These elements are not form solid solution in all composition ranges, although the solid solutions contain Y's own crystalline structure (BCC) at a certain temperature and composition range. Among the 6 candidates, Nb and Zr were removed because they form solid solution in very restricted range.

From the first and second steps, Cr, Hf, Mo, and V were selected. To select the final candidate elements, the binary phase diagrams between these elements were investigated. Hf-Cr, Hf-Mo, and Hf-V showed complex phase diagrams, which could have possibly formed some intermediate ordered phases (supplementary data, Figure S2). Therefore, Cr, Mo, and V were selected as the elements for the Al-Ti-containing HEAs.

3.2. Microstructure of the Al-Ti-Containing HEAs

Ternary, quaternary, and quinary HEAs were designed by adding the selected elements to Al-Ti. Seven HEAs, numbered #1 to #7, were fabricated and their compositions are detailed in Table 1. The XRD profiles of the Al-Ti-containing HEAs are shown in Figure 3. The AlCrMoTi (#4) and AlCrMoTiV (#7) had a single BCC structure, and ordered BCC peaks appeared for the other HEAs.

Table 1. Valence electron concentration (VEC), atomic size difference (δ), and enthalpy of mixing (ΔH_{mix}) of designed HEAs.

	Alloys	VEC	δ (%)	ΔH_{mix} (kJ/mol)
#1	AlCrTi	4.33	5.49	−20.96
#2	AlMoTi	4.33	1.77	−17.44
#3	AlTiV	4.00	3.42	−21.33
#4	AlCrMoTi	4.75	4.76	−14.00
#5	AlCrTiV	4.50	5.03	−16.75
#6	AlMoTiV	4.50	3.02	−14.25
#7	AlCrMoTiV	4.80	4.51	−12.16

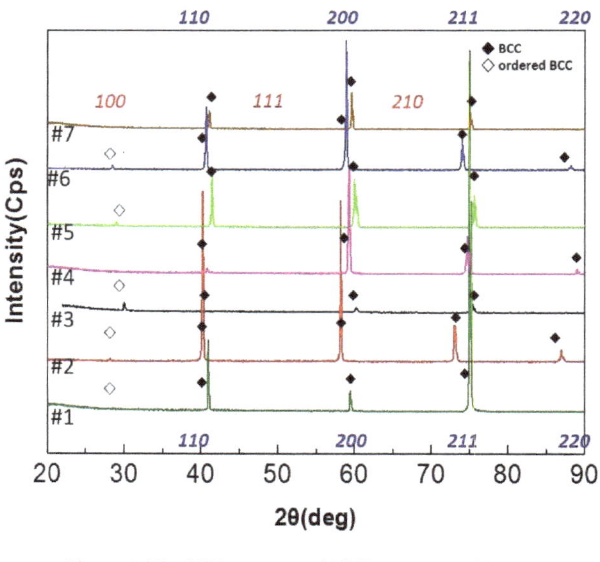

Figure 3. The XRD patterns of Al-Ti-containing HEAs.

The microstructures of AlCrMoTi (#4) and AlCrMoTiV (#7) which had a single BCC structure, are shown in Figure 4a,b. Both showed a dendritic morphology, and the dendritic growth was suppressed by the addition of V. The TEM analysis was performed for AlCrMoTiV (#7), and the results are shown in Figure 4c,d. Figure 4c is bright field image of AlCrMoTiV (#7) and Figure 4d is the diffraction pattern taken from $[001]_{B2/BCC}$ zone axis of Figure 4c. Although the XRD results demonstrate that it is a single BCC structure, diffraction pattern reveals (100) superlattice reflections marked with red-dotted circle that indicate an ordered B2 phase. Nano-scale B2 phases were observed; however, it was not detected by the XRD owing to their small fraction and size. Because AlCrMoTiV (#7) contained minor ordered precipitates, it could be defined as an ordered solid solution [19].

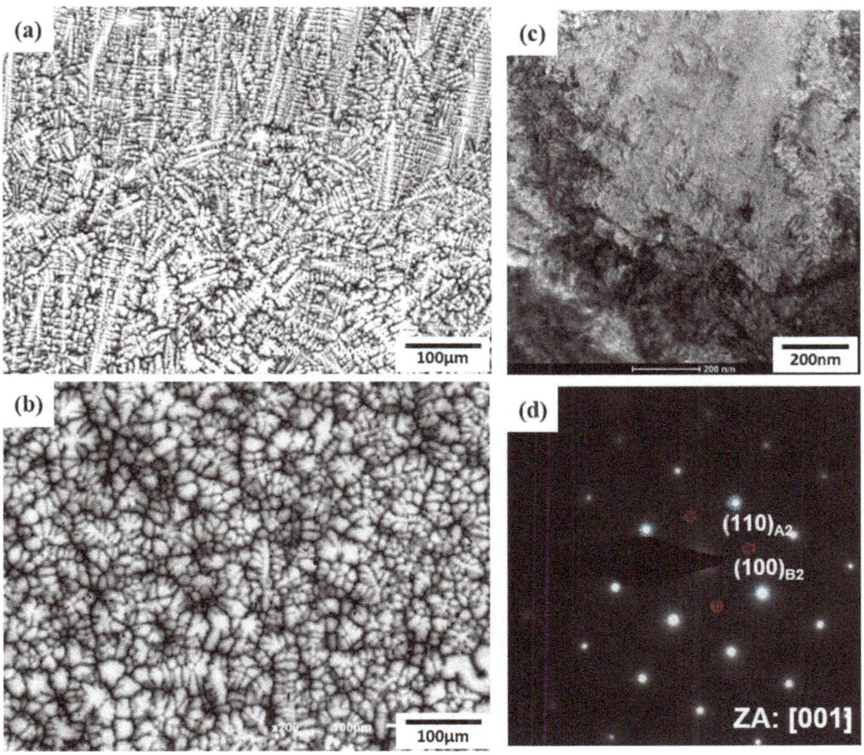

Figure 4. Microstructures of (**a**) AlCrMoTi and (**b**) AlCrMoTiV. Both have dendritic morphology. (**c**) bright field image and (**d**) diffraction pattern of AlCrMoTiV.

One of the physical properties of HEAs is the high-entropy effect [20]. The solid solution is stabilized because of the high configurational entropy [20,21], and this effect increases with an increased number of elements [20]. This trend can be observed in our work. As the number of elements increased from ternary to quinary, the phases progressively became simpler. The high-entropy effect is dominant at high temperatures according to $G = H - TS$, where G is Gibbs free energy, H is enthalpy, T is temperature, and S is entropy [20,21]. Thus, this HEA can exist as a solid solution at elevated temperatures, even though a small fraction of the B2 phase existed in an as-cast state at room temperature.

3.3. Application of Empirical Parameters

The empirical parameters for the HEAs originated from the classic Hume-Rothery rules [22,23]. Guo et al. [24] reported that $-22 \leq \Delta H_{mix} \leq 7$ kJ/mol, and $\delta \leq 8.5\%$ are required for the sole simple phases (i.e., FCC, BCC, and their mixtures, including both ordered/disordered cases). Zhang et al. [19] proposed $-15 \leq \Delta H_{mix} \leq 5$ kJ/mol and $\delta \leq 6.5\%$, whereas Yang et al. [12] proposed $\delta \leq 6.6\%$. In the case of VEC, 8 < VEC and VEC < 6.87 were suggested for the single FCC and BCC structures, respectively [20]. These parameters were statistically determined; therefore, there are differences and exceptions depending on the work [12,19,22,25].

The empirical parameters, valence electron concentration (VEC), atomic size difference (δ), and enthalpy of mixing (ΔH_{mix}) of the designed HEAs were calculated and are shown in Table 1 and Figure 5 [22]. The VEC and δ values satisfied the existing criteria. However, the ΔH_{mix} values were positioned relatively below the alloys which form solid solutions (Figure 5b). All the HEAs satisfy Guo's criterion although only AlCrMoTi (#4), AlMoTiV (#6), and AlCrMoTiV (#7) satisfy Zhang's criterion. The microstructural analysis demonstrates that AlCrMoTi (#4) and AlCrMoTiV (#7) are ordered solid solutions; the empirical parameters of HEAs in present work agree with previous research and support the idea that using the binary phase diagrams can be a solution to screen the proper candidate elements for the design of novel HEAs.

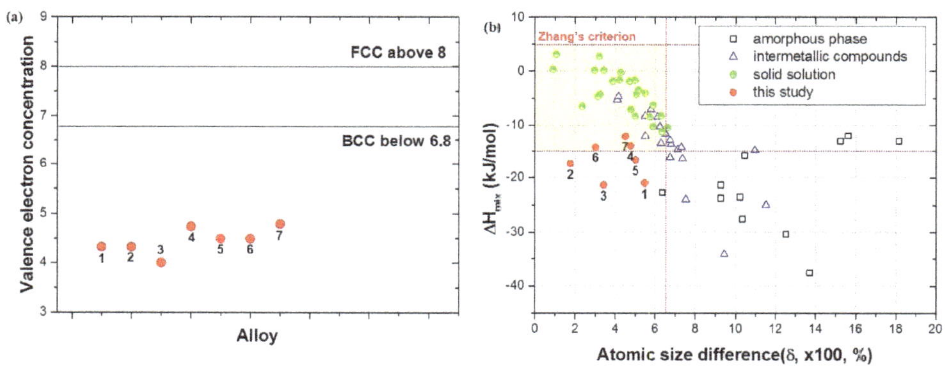

Figure 5. (**a**) VEC and (**b**) $\delta - \Delta H_{mix}$ plot of Al-Ti-containing HEAs. Yellow colored regions highlight Zhang's criterion [19] for single phase to form.

3.4. Specific Hardness of the Al-Ti-Containing HEAs

It has been reported that HEAs have a severe lattice distortion owing to the atomic size differences of elements, which induces solid solution strengthening [25]. Figure 6 shows the relationship between the hardness and δ. The hardness varied across the δ values. Moreover, Figure 6b clearly shows the hardness variation with the addition of Cr, Mo, and V. The hardness increased remarkably with the addition of Cr. The atomic radius of Cr is 1.25 Å which is quite different than the others [24]; this causes further severe lattice distortion.

The hardness of the AlCrMoTi (#4) and AlCrMoTiV (#7) solid solutions were compared with that of other HEAs and commercial alloys and is summarized in Table 2 [26–31]. The hardness of the AlCrMoTi (#4) and AlCrMoTiV (#7) were 606 ± 11 and 556 ± 25 HV, respectively. These values are 17% and 7% higher than that of the AlCoCrFeNi HEA, which is representative of HEAs for intermediate temperature applications [9]. The theoretical densities are also lower than that of the AlCoCrFeNi HEA, so the specific hardnesses were 30% and 19% higher than that of the AlCoCrFeNi HEA. Furthermore, the specific hardness of the HEAs in the present work are considerably higher than that of competitive conventional alloy systems. The AlCrMoTi (#4) showed a 29% improved specific hardness when compared with that of Ti-6Al-4V alloy, which is an alloy that is commonly used for intermediate temperature applications. It is expected that a great improvement in the hardness can accomplish the weight lightening through the gage reduction of machineries despite their higher density when compared to that of the Ti-6Al-4V alloy. Further research on the fine-tuning of the elemental composition for lighter and cheaper alloys as well as a detailed microstructural and high-temperature property analysis could open a new path toward lightweight HEAs for intermediate temperature applications.

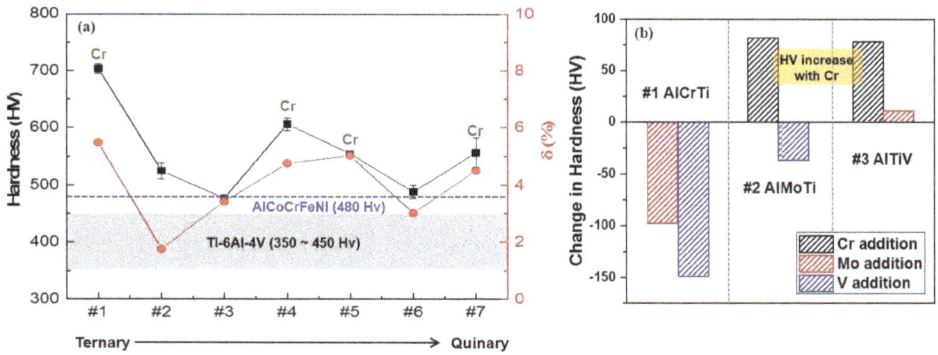

Figure 6. (**a**) Relationship between the hardness and atomic size difference; (**b**) hardness variation with the Cr, Mo, and V addition to ternary HEAs. The hardness increases remarkably with the Cr addition.

Table 2. Comparison of specific hardness of HEAs with other alloys.

Alloys	Hardness (HV)	Theoretical Density (g/cm^3)	Specific Hardness (HV/g/cm^3)
AlCrMoTi (#4)	606	6.01	100.83
AlCrMoTiV (#7)	556	6.00	92.67
AlCoCrFeNi [26]	520	6.70	77.61
AlCoCrCuFeNi [27]	400	7.02	56.98
$Al_{0.5}$CoCrCuFeNiV [28]	640	7.27	88.03
Ti-6Al-4V [29]	346	4.43	78.10
Ti-6242 [30]	339	4.54	74.67
Inconel 718 [31]	355	8.18	43.40

4. Conclusions

In summary, HEAs which contained Al and Ti were designed based on their binary phase diagrams. This approach is powerful for screening candidate elements for novel HEAs. The candidate elements that were selected formed a solid solution within a certain temperature and composition range. The high-entropy effect is enhanced with an increased number of elements; therefore, the AlCrMoTi and AlCrMoTiV HEAs are verified to be solid solutions with a minor ordered B2 phase. These HEAs have a superb specific hardness when compared to that of Ti-6Al-4V and Inconel 718 alloys and show promise as future substitutes for Ti alloys for intermediate temperature structural applications. Furthermore, fine-tuning of the elemental composition of HEAs can lead to the development of novel light-weight HEAs.

Supplementary Materials: The following are available online at http://www.mdpi.com/1099-4300/20/5/355/s1, Figure S1: Y-Al phase diagrams where "Y" is (a) Cr, (b) Hf, (c) Mo, (d) Nb, (e) V, and (f) Zr [18]. Figure S2: (a) Hf-Cr, (b) Cr-Mo, (c) Cr-V, (d) Hf-Mo, (e) Hf-V, and (f) Mo-V binary phase diagrams [18]. Cr, Mo, and V were selected which have simple phase diagrams.

Author Contributions: M.K., K.R.L., J.W.W. and Y.S.N. conceived and designed the experiments; M.K. and K.R.L. performed the experiments and analyzed the data; K.S.L. and Y.S.N. contributed reagents/materials/analysis tools; M.K. wrote the paper.

Acknowledgments: This study was supported financially by Fundamental Research Program (PNK5610) of the Korean Institute of Materials Science (KIMS), and by the Future Material Discovery Project of the National Research Foundation of Korea (NRF) funded by the Ministry of Science, ICT and Future Planning (MSIP) of Korea (NRF-2016M3D1A1023534).

Conflicts of Interest: The authors declare no conflict of interest.

References

1. Cantor, B.; Chang, I.T.H.; Knight, P.; Vincent, A.J.B. Microstructural development in equiatomic multicomponent alloys. *Mater. Sci. Eng. A* **2004**, *375–377*, 213–218. [CrossRef]
2. Manzoni, A.M.; Singh, S.; Daoud, H.M.; Popp, R.; Völkl, R.; Glatzel, U.; Wanderka, N. On the path to optimizing the Al-Co-Cr-Cu-Fe-Ni-Ti high entropy alloy family for high temperature applications. *Entropy* **2016**, *18*, 104. [CrossRef]
3. He, J.Y.; Wang, H.; Huang, H.L.; Xu, X.D.; Chen, M.W.; Wu, Y.; Liu, X.J.; Nieh, T.G.; An, K.; Lu, Z.P. A precipitation-hardened high-entropy alloy with outstanding tensile properties. *Acta Mater.* **2016**, *102*, 187–196. [CrossRef]
4. Chen, Y.; Li, Y.; Cheng, X.; Wu, C.; Cheng, B.; Xu, Z. The Microstructure and Mechanical Properties of Refractory High-Entropy Alloys with High Plasticity. *Materials* **2018**, *11*, 208. [CrossRef] [PubMed]
5. Gali, A.; George, E.P. Tensile properties of high- and medium-entropy alloys. *Intermetallics* **2013**, *39*, 74–78. [CrossRef]
6. Gludovatz, B.; Hohenwarter, A.; Catoor, D.; Chang, E.H.; George, E.P.; Ritchie, R.O. A fracture-resistant high-entropy alloy for cryogenic applications. *Science* **2014**, *345*, 1153–1158. [CrossRef] [PubMed]
7. Han, Z.D.; Chen, N.; Zhao, S.F.; Fan, L.W.; Yang, G.N.; Shao, Y.; Yao, K.F. Effect of Ti additions on mechanical properties of NbMoTaW and VNbMoTaW refractory high entropy alloys. *Intermetallics* **2017**, *84*, 153–157. [CrossRef]
8. Senkov, O.N.; Wilks, G.B.; Scott, J.M.; Miracle, D.B. Mechanical properties of $Nb_{25}Mo_{25}Ta_{25}W_{25}$ and $V_{20}Nb_{20}Mo_{20}Ta_{20}W_{20}$ refractory high entropy alloys *Intermetallics* **2011**, *19*, 698–706. [CrossRef]
9. Lim, K.R.; Lee, K.S.; Lee, J.S.; Kim, J.Y.; Chang, H.J.; Na, Y.S. Dual-phase high-entropy alloys for high-temperature structural applications. *J. Alloys Compd.* **2017**, *728*, 1235–1238. [CrossRef]
10. Lu, Y.; Dong, Y.; Guo, S.; Jiang, L.; Kang, H.; Wang, T.; Wen, B.; Wang, Z.; Jie, J.; Cao, Z.; et al. A promising new class of high-temperature alloys: Eutectic high-entropy alloys. *Sci. Rep.* **2014**, *4*, 6200. [CrossRef] [PubMed]
11. Jin, X.; Zhou, Y.; Zhang, L.; Du, X.; Li, B. A new pseudo binary strategy to design eutectic high entropy alloys using mixing enthalpy and valence electron concentration. *Mater. Des.* **2018**, *143*, 49–55. [CrossRef]
12. Yang, X.; Zhang, Y. Prediction of high-entropy stabilized solid-solution in multi-component alloys. *Mater. Chem. Phys.* **2012**, *132*, 233–238. [CrossRef]
13. Choi, W.M.; Jung, S.; Jo, Y.H.; Lee, S.; Lee, B.J. Design of new face-centered cubic high entropy alloys by thermodynamic calculation. *Met. Mater. Int.* **2017**, *23*, 839–847. [CrossRef]
14. Zhang, C.; Zhang, F.; Chen, S.; Cao, W. Computational thermodynamics aided high-entropy alloy design. *JOM* **2012**, *64*, 839–845. [CrossRef]
15. Gao, M.C.; Alman, D.E. Searching for Next Single-Phase High-Entropy Alloy Compositions. *Entropy* **2013**, *15*, 4504–4519. [CrossRef]
16. Ma, D.; Yao, M.; Pradeep, K.G.; Tasan, C.C.; Springer, H.; Raabe, D. Phase stability of non-equiatomic CoCrFeMnNi high entropy alloys. *Acta Mater.* **2015**, *98*, 288–296. [CrossRef]
17. Zhang, F.; Zhang, C.; Chen, S.L.; Zhu, J.; Cao, W.S.; Kattner, U.R. An understanding of high entropy alloys from phase diagram calculations. *Calphad* **2014**, *45*, 1–10. [CrossRef]
18. ASM Handbook. *Alloy Phase Diagrams*; ASM International: Novelty, OH, USA, 1992.
19. Zhang, Y.; Zhou, Y.J.; Lin, J.P.; Chen, G.L.; Liaw, P.K. Solid-Solution Phase Formation Rules for Multi-component Alloys. *Adv. Eng. Mater.* **2008**, *10*, 534–538. [CrossRef]
20. Tsai, M.H.; Yeh, J.W. High-Entropy Alloys: A Critical Review. *Mater. Res. Lett.* **2014**, *2*, 107–123. [CrossRef]
21. Yeh, J.W. Alloy design strategies and future trends in high-entropy alloys. *JOM* **2013**, *65*, 1759–1771. [CrossRef]
22. Guo, S.; Ng, C.; Lu, J.; Liu, C.T. Effect of valence electron concentration on stability of FCC or bcc phase in high entropy alloys. *J. Appl. Phys.* **2011**, *109*, 103505. [CrossRef]

23. Singh, P.; Sharma, A.; Smimov, A.V.; Diallo, M.S.; Ray, P.K.; Balasubramanian, G.; Johnson, D.D. Design of high-strength refractory complex solid-solution alloys. *npj Comput. Mater.* **2018**, *4*, 16. [CrossRef]
24. Guo, S.; Liu, C.T. Phase stability in high entropy alloys: Formation of solid-solution phase or amorphous phase. *Prog. Nat. Sci.* **2011**, *21*, 433–446. [CrossRef]
25. Yeh, J.W.; Chang, S.Y.; Hong, Y.D.; Chen, S.K.; Lin, S.J. Anomalous decrease in X-ray diffraction intensities of Cu-Ni-Al-Co-Cr-Fe-Si alloy systems with multi-principal elements. *Mater. Chem. Phys.* **2007**, *103*, 41–46. [CrossRef]
26. Ma, S.G.; Zhang, Y. Effect of Nb addition on the microstructure and properties of AlCoCrFeNi high-entropy alloy. *Mater. Sci. Eng. A* **2012**, *532*, 480–486. [CrossRef]
27. Tong, C.J.; Chen, M.R.; Chen, S.K.; Yeh, J.W.; Shun, T.T.; Lin, S.J.; Chang, S.Y. Mechanical Performance of the Al_xCoCrCuFeNi High-Entropy Alloy System with Multiprincipal Elements. *Metall. Mater. Trans. A* **2005**, *36*, 1263–1271. [CrossRef]
28. Chen, M.R.; Lin, S.J.; Yeh, J.W.; Chen, S.K.; Huang, Y.S.; Chuang, M.H. Effect of Vanadium Addition on the Microstructure, Hardness, and Wear Resistance of $Al_{0.5}$CoCrCuFeNi High-Entropy Alloy. *Metall. Mater. Trans. A* **2006**, *37*, 1363–1369. [CrossRef]
29. Niinomi, M. Mechanical properties of biomedical titanium alloys. *Mater. Sci. Eng.* **1998**, *243*, 231–236. [CrossRef]
30. Shen, W.; Soboyejo, A.B.O.; Soboyejo, W.O. Microstructural Effects on Fatigue and Dwell-Fatigue Crack Growth in α/β Ti-6Al-2Sn-4Zr-2Mo-0.1Si. *Metall. Mater. Trans. A* **2004**, *35*, 163–187. [CrossRef]
31. Li, H.Z.; Zeng, H.; Chen, X.Q. An experimental study of tool wear and cutting force variation in the end milling of Inconel 718 with coated carbide inserts. *J. Mater. Process. Technol.* **2006**, *180*, 296–304. [CrossRef]

entropy

MDPI

Article

High Strength and Deformation Mechanisms of $Al_{0.3}CoCrFeNi$ High-Entropy Alloy Thin Films Fabricated by Magnetron Sputtering

Wei-Bing Liao [1], Hongti Zhang [2,3], Zhi-Yuan Liu [4], Pei-Feng Li [5], Jian-Jun Huang [1], Chun-Yan Yu [1,*] and Yang Lu [2,6]

1 College of Physics and Optoelectronic Engineering, Shenzhen University, Shenzhen 518060, China; liaowb@szu.edu.cn (W.-B.L.); huangjj@szu.edu.cn (J.-J.H.)
2 Department of Mechanical Engineering, City University of Hong Kong, Hong Kong, China; zhanght3@shanghaitech.edu.cn (H.Z.); yanglu@cityu.edu.hk (Y.L.)
3 School of Physical Science and Technology, Shanghai Tech University, Shanghai 201210, China
4 College of Mechatronics and Control Engineering, Shenzhen University, Shenzhen 518060, China; zyliu@szu.edu.cn
5 College of Materials Science and Engineering, Shenzhen University, Shenzhen 518060, China; peifengli@szu.edu.cn
6 Nano-Manufacturing Laboratory (NML), City University of Hong Kong Shenzhen Research Institute, Shenzhen 518057, China
* Correspondence: yuchunyan@szu.edu.cn

Received: 23 January 2019; Accepted: 2 February 2019; Published: 4 February 2019

Abstract: Recently, high-entropy alloy thin films (HEATFs) with nanocrystalline structures and high hardness were developed by magnetron sputtering technique and have exciting potential to make small structure devices and precision instruments with sizes ranging from nanometers to micrometers. However, the strength and deformation mechanisms are still unclear. In this work, nanocrystalline $Al_{0.3}CoCrFeNi$ HEATFs with a thickness of ~4 μm were prepared. The microstructures of the thin films were comprehensively characterized, and the mechanical properties were systematically studied. It was found that the thin film was smooth, with a roughness of less than 5 nm. The chemical composition of the high entropy alloy thin film was homogeneous with a main single face-centered cubic (FCC) structure. Furthermore, it was observed that the hardness and the yield strength of the high-entropy alloy thin film was about three times that of the bulk samples, and the plastic deformation was inhomogeneous. Our results could provide an in-depth understanding of the mechanics and deformation mechanism for future design of nanocrystalline HEATFs with desired properties.

Keywords: high-entropy alloys; thin films; hardness; deformation behaviors; nanocrystalline

1. Introduction

It is well known that among all the alloy composition design systems, high-entropy alloys (HEAs) are a brand-new concept based on novel multi-component system composition designs. They contain at least four or five principal metal components and simply form a single face-centered cubic (FCC), body-centered cubic (BCC) or hexagonal close-packed (HCP) phase [1–6]. This novel concept is an important breakthrough of the past 25 years [7,8], as it is completely different from the traditional alloy design concepts in which one major component was selected, and other minor components were added to improve their related physical and chemical performances. It is worth mentioning that HEAs not only have simple phase structures, but also possess many excellent mechanical and physical properties, such as high tensile strength [9–11], good ductility at ambient and cryogenic temperatures [12,13], superior wear and fatigue resistance [14], and strong radiation tolerance [15,16]. These unique features qualify HEAs as potential engineering materials to meet the demanding requirements for complex and harsh environment applications, particularly in the turbine, aerospace, and nuclear industries [17–21]. However, the chemical composition of HEAs contains multiple elements which would naturally raise the cost for industrial application, and limit HEAs extensive development. As a consequence, to reduce the cost for future industrial applications and take full advantage of the above excellent comprehensive properties, HEA thin films (HEATFs) can be efficiently prepared and simultaneously coated on the surface of industrial components, especially for those complex geometry components. In these cases, the HEATFs will play an important role. The initial research of HEATFs is associated with the recent rapid development of HEAs and the high throughput preparation idea [22–26]. As the geometric size and microstructures of the thin films are different from the three-dimensional bulk samples, their performances under loading and service conditions could be completely different [27,28]. So far, HEATFs were verified to have remarkable effects on the hardness [24]. A series work on HEATFs was done not only on the high throughput fabrication but also on the mechanical properties, including the hardness and corrosion properties [29–36]. The previous work has greatly promoted the industrial application of HEATFs. Unfortunately, the related deformation behaviors have not been clearly revealed until now. To facilitate the use of HEATFs and provide a continuous coating technique, the deformation behaviors and reliability of HEATFs merit further investigation. Therefore, in this study we prepared the $Al_{0.3}CoCrFeNi$ HEATFs with a main simple FCC structure by magnetron sputtering, and fabricated nano-scaled pillars on the surface of the thin film by focus ion beams (FIBs), then utilized in situ scanning electron microscopy (SEM) compression to study the deformation behaviors of the HEATFs.

2. Materials and Methods

The target with a composition of $Al_{0.3}CoCrFeNi$ was prepared by metallurgy with high-purity (>99.99%) raw metal materials of aluminum, cobalt, chromium, iron, and nickel. The size of the target is $\phi 76.2 \times 3.175$ mm. The $Al_{0.3}CoCrFeNi$ HEATFs were deposited on silicon wafer substrates by magnetron sputtering. Before putting the target in the vacuum chamber, it was cleaned by argon ion bombardment for about 2 min to remove the oxide or contaminants on the surface. To ensure a uniform deposition a rotation speed of the silicon wafer substrate was set at 2 rpm. The surface roughness of the as-deposited HEATFs was determined by white light interferometry (WLI) using Wyko NT9300 Surface Profiler (Veeco Instruments, Plainview, NY, USA), while the surface morphology and detail nanostructures were characterized by scanning electron microscopy (SEM) and atomic force microscopy (AFM) (Bruker Dimension Icon™, Billerica, MA, USA) with ScanAsyst (Bruker Dimension Icon™, Billerica, MA, USA) at room temperature. To investigate the phase structure of the as-deposited HEATFs, high-energy synchrotron radiation X-ray in transmission mode at 11-ID-C of Advanced Photon Source (APS) was used. The X-ray beam wavelength was 0.117418 Å. The detail microstructures of the as-deposited HEATFs were observed by high-resolution transmission electron microscopy (HRTEM) using a JEOL JEM-2100F instrument (JEOL, Akishima, Tokyo, Japan) operated at 200 kV. The chemical composition was analyzed by the energy dispersive X-ray spectrometer (EDS) equipped in the transmission electron microscopy (TEM). Nanoindentation experiments were performed using a Hysitron TI750 nanoindenter (Hysitron, Inc., Minneapolis, MN, USA) with a Berkovich tip. To avoid any potential effects of the substrate on the experiment, the indentation depth was kept to be less than 10% of the whole thickness of the HEATFs. Micropillars were fabricated out of the $Al_{0.3}CoCrFeNi$ HEAHFs by using a FEI Scios focused ion beam (FIB) (USA) (Thermo Scientific™, Hillsboro, OR, USA) at 30 kV/10pA as the final etching condition. The height of the nanopillars was kept to be less than the thickness of the $Al_{0.3}CoCrFeNi$ HEATFs. The in situ SEM compression tests were conducted at room temperature using a PI 85 PicoIndenter (Hysitron Inc.) with a flat punch diamond tip inside a FEI Quanta 450 FEG (USA) (Thermo Scientific™, Hillsboro, OR, USA), under displacement-control mode and at a strain rate of around 5×10^{-3} s^{-1}. Raw load-displacement data were used to calculate the engineering stress and strain.

3. Results and Discussion

Figure 1a,b show the two-dimensional (2D) and the three-dimensional (3D) surface roughness and profiles of the as-deposited $Al_{0.3}$CoCrFeNi HEATFs, respectively, and Figure 1c,d show the X-profile and Y-profile of the corresponding positions selected on the HEATFs as marked in Figure 1a. It can be clearly seen that there are fine undulating nanostructures on the surface of the $Al_{0.3}$CoCrFeNi HEATFs prepared by the magnetron sputtering deposition technique; however, the entire surface is very flat and smooth, with a roughness R_a of less than 3.5 nm.

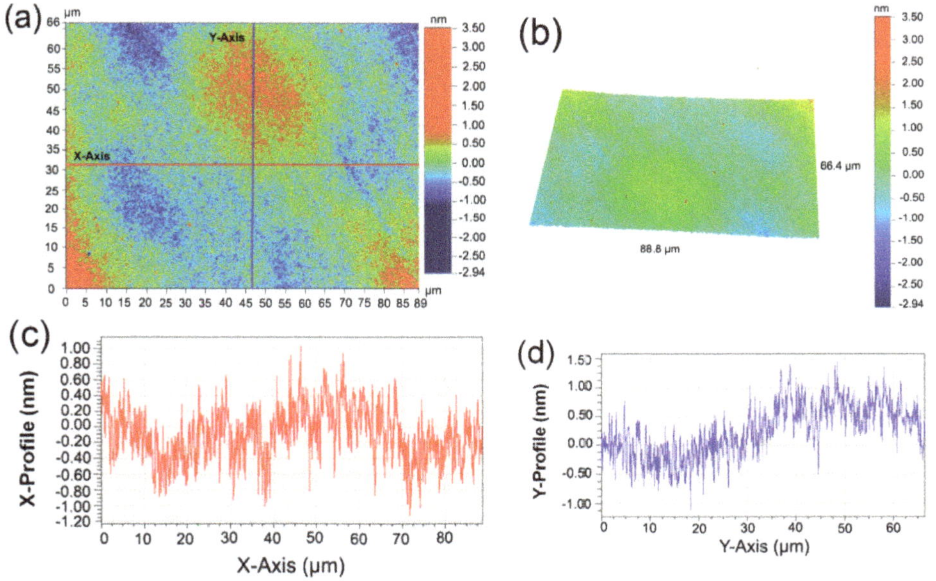

Figure 1. Surface profiles of the $Al_{0.3}$CoCrFeNi high-entropy alloy thin films (HEATFs) characterized by white light interferometry (WLI) technique. (**a**) 2D surface profiles, (**b**) 3D surface profiles, (**c**,**d**): the profiles of the *x*-axis and *y*-axis as marked in (a) respectively.

The SEM surface morphology and the specific fine nanostructures of the as-deposited $Al_{0.3}$CoCrFeNi HEATFs with a magnification of 50,000 times are shown in Figure 2a. It demonstrates that these fine nanostructures are well-knit and compact. The thickness of the HEATFs is about 4 μm, as shown in Figure 2b. To characterize the feature of the HEATFs in more detail, AFM scanning experiments were further conducted. Figure 2c,d show the 2D and 3D AFM images of the surface feature of the $Al_{0.3}$CoCrFeNi HEATFs. Uniform nanostructures are clearly observed, and the heights of these undulating nanostructures were less than 5 nm, which is well consistent with that typically observed by the surface profile. All these experimental data verified that there were a lot of fine nanostructures on the surface of the $Al_{0.3}$CoCrFeNi HEATFs, and the surface was very smooth as a whole, with a roughness R_a of less than 5 nm.

Figure 3a shows the TEM images and the corresponding EDS analysis of the $Al_{0.3}$CoCrFeNi HEATFs. It can be seen that there are a lot of nanocrystalline structures in the HEATFs, with a grain size order of ~10 nm. The elemental distribution of the as-deposited HEATFs is homogenous. The bright and dark places shown in the TEM-EDS images are ascribed to the uneven sample thickness.

Quantitative analysis by EDS confirms that the chemical composition is nearly the same as the composition of the sputtering target, as shown in Table 1.

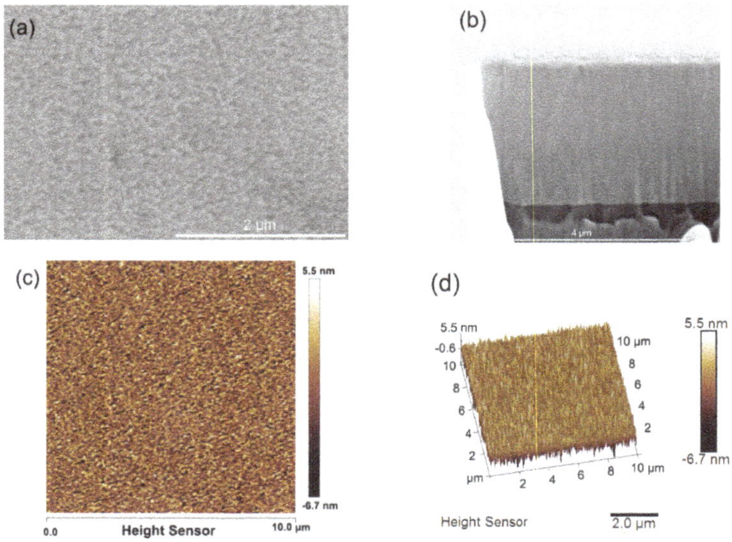

Figure 2. Surface morphologies and microstructures of the $Al_{0.3}CoCrFeNi$ HEATFs characterized in detail by scanning electron microscopy (SEM) and atomic force microscopy (AFM). (**a**) The SEM image of the surface morphologies, (**b**) the cross-section the HEATFs deposited on the silicon substrate, (**c**) the AFM image of the surface structure, and (**d**) 3D surface structures of the HEATFs.

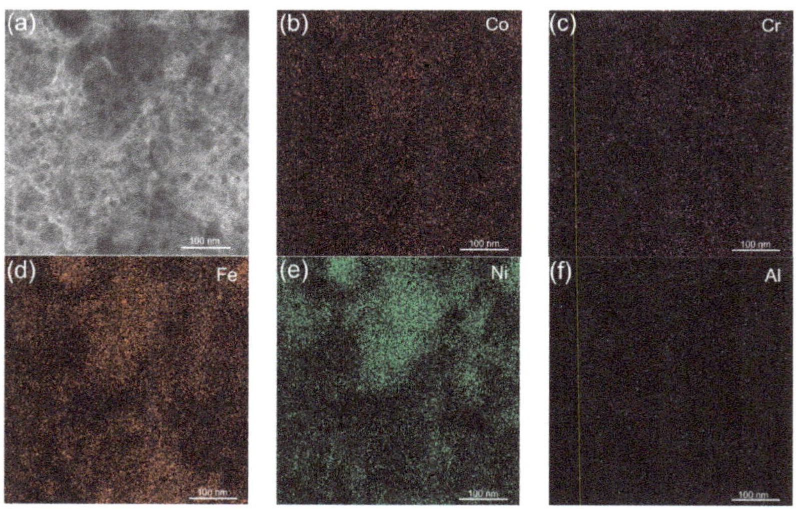

Figure 3. Element distribution mapping for the $Al_{0.3}CoCrFeNi$ HEATFs by TEM-EDS. The top TEM image shows the region analyzed. (**a**) Explanations for subfigure a; (**b**) explanations for subfigure b; (**c**) explanations for subfigure c; (**d**) explanations for subfigure d; (**e**) explanations for subfigure e; (**f**) explanations for subfigure f.

Table 1. The chemical composition of the as-deposited $Al_{0.3}CoCrFeNi$ HEATFs compared with that of the sputtering target.

Elements (at. %)	Al	Co	Cr	Fe	Ni
Nominal target	6.977	23.256	23.256	23.256	23.256
As-deposited HEATFs	6.452	23.436	24.602	23.623	21.887

To obtain the phase structural information of the $Al_{0.3}CoCrFeNi$ HEATFs, high-energy synchrotron radiation X-ray studies were undertaken. Figure 4a shows the X-ray line profiles of the HEATFs. The (111), (200), (220), and (311) phase peaks were observed and identified to be a typical FCC crystalline structure, whilst a small peak appeared before the (111) peak, which means that a minor ordered BCC NiAl type phase structure was in the HEATFs [37]. The corresponding diffraction patterns are exhibited in Figure 4b. The weak continuous rings certify that there are tiny polycrystalline structures in the HEATFs. Interestingly, the diffraction rings of the HEATFs were discrete, with obvious intensity differences, indicating that there were strong textures in the HEATFs. This event could be ascribed to the preferred growth of the thin film induced by the silicon substrate. It should be noted that a rigorous diffraction-intensity-distribution calculation of the solid solution phases responsible for certain orientations in the HEATFs is worthy of a focused topic. However, it is beyond the scope of this work. The synchrotron X-ray experimental results provide cogent evidence that the magnetron sputtering technique is an effect way to prepare the HEATFs with a simple phase structure. Moreover, it could also lead to a wide research range of HEAs, studying the corresponding properties from meso- to nanometer regimes.

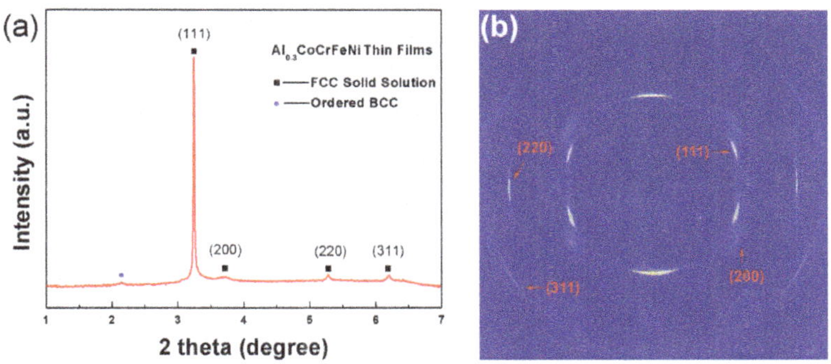

Figure 4. Phase structures of the HEATFs by high-energy synchrotron radiation X-ray. (**a**) synchrotron X-ray line profiles and (**b**) typical diffraction pattern.

Figure 5a shows the nanoindentation properties of the as-deposited HEATFs. Since a series of 4 $\times$ 4 matrix array indentation points were tested in sequence, the average values of elastic modulus and hardness were accurately calculated and were identified to be about 186.01 GPa and 11.09 GPa, respectively. It should be noted that the hardness of the HEATFs is about three times higher than that of the as-cast bulk $Al_{0.3}CoCrFeNi$ HEA sample, but the elastic modulus is nearly the same [38,39]. This enhanced hardness can be ascribed to the nanocrystalline strengthening mechanism which induced the hardening by a large number of grain boundaries observed in Figure 3a. Figure 5b is the typical nanoindentation load-depth curve of the HEATFs. It can be seen that as the loading force increases, the depth of the indenter pressed into the film gradually increases. After unloading, an irreversible depth was retained, indicating a plastic deformation has occurred on the surface of the HEATFs. Figure 5c exhibits the SEM image of the impression mark. The indentation profiles are self-similar. A remarkable pile-up (marked with red arrows) around the indentation can be clearly observed, suggesting that high localized plastic deformation occurred during nanoindentation.

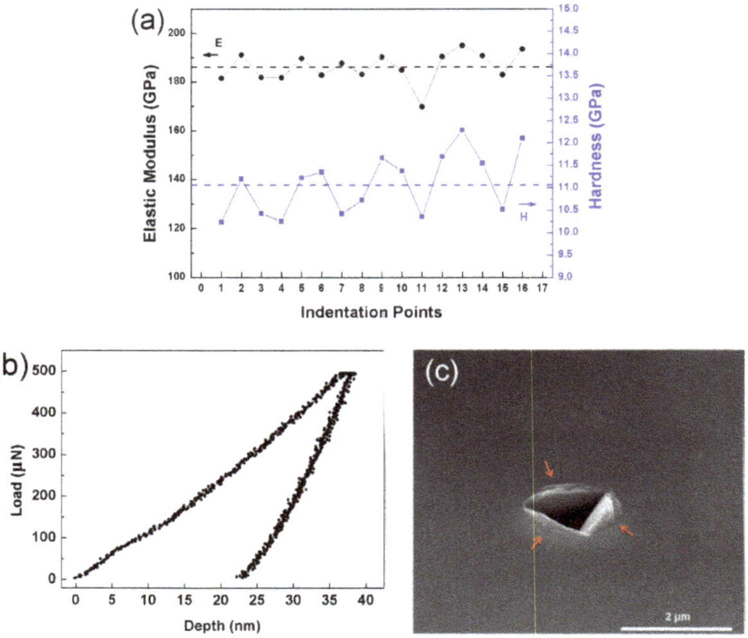

Figure 5. Nanoindentation properties of the as-deposited $Al_{0.3}CoCrFeNi$ HEATFs. (**a**) Elastic modulus and hardness of the HEATFs at different indentation points, (**b**) typical nano-indentation load–depth curve, and (**c**) typical SEM image of the impression mark.

To further characterize the mechanical properties of the $Al_{0.3}CoCrFeNi$ HEATFs, a nanopillar sample with a diameter of 738 nm was fabricated from the HEATFs, and in situ SEM compression tests were conducted on the nanopillar sample, as shown in Figure 6. The entire compression deformation process of the nanopillar can be divided into the following stages. Initially, the nanopillar was deformed elastically, and no significant trace appeared on the surface of the nanopillar, as shown in Figure 6a. Secondly, with the increase of the compressive stress, a large localized metal flow and plastic deformation occurred at the top part of the nanopillar, as marked with a red arrow in Figure 6b. It indicates that the deformation of the $Al_{0.3}CoCrFeNi$ HEA nanopillar was inhomogeneous. After that, it can be observed that a slip was generated at the top part of the nanopillar, which was marked with a red arrow in Figure 6c. The occurrence of the slip is not only related to the plastic deformation, but also has an impact on the work hardening and the serration behavior of the HEAs [40]. The corresponding compression engineering stress–strain curve of the $Al_{0.3}CoCrFeNi$ HEA nanopillar is shown in Figure 6d. The yield strength of the nanopillar is about 1024 MPa, which is also about three times that of the bulk sample [38]. It is consistent with the above experimental results obtained by nanoindentation, as the strength is directly proportion with the hardness. After yielding the nanopillar exhibits work-hardening up to an ultrahigh strength. The compressive strength and corresponding strain were ~2075 MPa and ~11.39%, respectively. Following this, softening dominates until final fracture at a strain of ~12.14%. In general, the compression results further confirmed that the yield strength of the $Al_{0.3}CoCrFeNi$ HEATFs is about three times that of the bulk samples, and the plastic deformation is inhomogeneous.

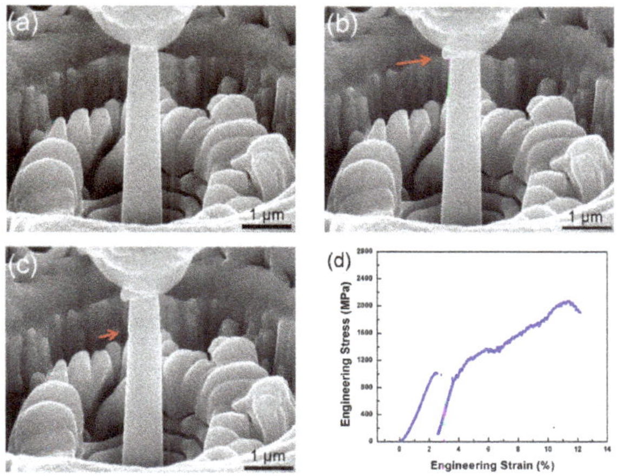

Figure 6. Compression properties for the $Al_{0.3}CoCrFeNi$ pillar with a diameter of 738 nm prepared from the HEATFs. (**a**) Elastic deformation stage, (**b**) localized plastic deformation occurred at the top of the pillar, (**c**) a slip generated at the top part of the pillar, (**d**) a typical compression engineering stress–strain curve of the pillar.

4. Conclusions

In conclusion, the $Al_{0.3}CoCrFeNi$ HEATFs prepared by the magnetron sputtering technique were smooth, with a surface roughness R_a of less than ~5 nm. The chemical composition was homogeneous and amounts of nanocrystallines with a main single FCC phase structure formed in the HEATFs. The hardness of the HEATFs was ~11.09 GPa, and the yield strength of the nanopillar prepared from the HEATFs was ~1024 MPa. Both the hardness and the yield strength were about three times that of the bulk samples. Simultaneously, it was found that the plastic deformation of the HEATFs was inhomogeneous and localized. The present study could provide useful insights in the design and application of HEATFs for functional micro- and nano-devices.

Author Contributions: Formal analysis, W.-B.L. and Z.-Y.L.; Funding acquisition, W.-B.L., C.-Y.Y. and Y.L.; Investigation, W.-B.L., H.Z., P.-F.L., J.-J.H. and C.-Y.Y.; Methodology, W.-B.L. and Y.L.; Supervision, Y.L.; Writing—Original Draft, W.-B.L.; Writing—Review & Editing, W.-B.L.

Funding: This work was supported by the National Natural Science Foundation of China (Grant No. 51801128, 51701125), the Youth Innovation Talent Project of Guangdong Province (Grant No. 2017KQNCX175), the Natural Science Foundation of Shenzhen University (Grant Nos. 2017069, 201553, 827-000180), the Teaching Reform Research Project of Shenzhen University (Grant No. JG2018090), and the Shenzhen Basic Research Project JCYJ20170302142339007. Y. Lu gratefully thanks the funding support from Shenzhen Science and Technology Innovation Committee under the grant JCYJ20170413141157573.

Conflicts of Interest: The authors declare no conflict of interest.

References

1. Yeh, J.W.; Chen, S.K.; Lin, S.J.; Gan, J.Y.; Chin, T.S.; Shun, T.T.; Tsau, C.H.; Chang, S.Y. Nanostructured high-entropy alloys with multiple principal elements: Novel alloy design concepts and outcomes. *Adv. Eng. Mater.* **2004**, *6*, 299–303. [CrossRef]
2. Zhang, Y.; Zuo, T.T.; Tang, Z.; Gao, M.C.; Dahmen, K.A.; Liaw, P.K.; Lu, Z.P. Microstructures and properties of high-entropy alloys. *Prog. Mater. Sci.* **2014**, *61*, 1–93. [CrossRef]
3. Yang, X.; Zhang, Y. Prediction of high-entropy stabilized solid-solution in multi-component alloys. *Mater. Chem. Phys.* **2012**, *132*, 233–238. [CrossRef]
4. Takeuchi, A.; Amiya, K.; Wada, T.; Yubuta, K.; Zhang, W. High-entropy alloys with a hexagonal close-packed structure designed by equi-atomic alloy strategy and binary phase diagrams. *JOM* **2014**, *66*, 1984–1992. [CrossRef]

5. Guo, S.; Ng, C.; Lu, J.; Liu, C.T. Effect of valence electron concentration on stability of fcc or bcc phase in high entropy alloys. *J. Appl. Phys.* **2011**, *109*, 103505. [CrossRef]
6. Ye, Y.F.; Wang, Q.; Lu, J.; Liu, C.T.; Yang, Y. High-entropy alloy: Challenges and prospects. *Mater. Today* **2016**, *19*, 349–362. [CrossRef]
7. Greer, A.L. Confusion by design. *Nature* **1993**, *366*, 303–304. [CrossRef]
8. Cantor, B.; Chang, I.T.H.; Knight, P.; Vincent, A.J.B. Microstructural development in equiatomic multicomponent alloys. *Mater. Sci. Eng. A* **2004**, *375–377*, 213–218. [CrossRef]
9. Lei, Z.; Liu, X.; Wu, Y.; Wang, H.; Jiang, S.; Wang, S.; Hui, X.; Wu, Y.; Gault, B.; Kontis, P.; et al. Enhanced strength and ductility in a high-entropy alloy via ordered oxygen complexes. *Nature* **2018**, *563*, 546–550. [CrossRef]
10. Li, Z.; Pradeep, K.G.; Deng, Y.; Raabe, D.; Tasan, C.C. Metastable high-entropy dual-phase alloys overcome the strength-ductility trade-off. *Nature* **2016**, *534*, 227–230. [CrossRef]
11. He, J.Y.; Wang, H.; Huang, H.L.; Xu, X.D.; Chen, M.W.; Wu, Y.; Liu, X.J.; Nieh, T.G.; An, K.; Lu, Z.P. A precipitation-hardened high-entropy alloy with outstanding tensile properties. *Acta Mater.* **2016**, *102*, 187–196. [CrossRef]
12. Gludovatz, B.; Hohenwarter, A.; Catoor, D.; Chang, E.H.; George, E.P.; Ritchie, R.O. A fracture-resistant high-entropy alloy for cryogenic applications. *Science* **2014**, *345*, 1153–1158. [CrossRef]
13. Liu, J.; Guo, X.; Lin, Q.; He, Z.; An, X.; Li, L.; Liaw, P.K.; Liao, X.; Yu, L.; Lin, J.; et al. Excellent ductility and serration feature of metastable CoCrFeNi high-entropy alloy at extremely low temperatures. *Sci. China Mater.* **2018**. [CrossRef]
14. Chen, P.; Lee, C.; Wang, S.-Y.; Seifi, M.; Lewandowski, J.J.; Dahmen, K.A.; Jia, H.; Xie, X.; Chen, B.; Yeh, J.-W.; et al. Fatigue behavior of high-entropy alloys: A review. *Sci. China Technol. Sci.* **2018**, *61*, 168–171. [CrossRef]
15. Lu, C.; Niu, L.; Chen, N.; Jin, K.; Yang, T.; Xiu, P.; Zhang, Y.; Gao, F.; Bei, H.; Shi, S.; et al. Enhancing radiation tolerance by controlling defect mobility and migration pathways in multicomponent single-phase alloys. *Nat. Commun.* **2016**, *7*, 13564. [CrossRef]
16. Zhang, W.; Liaw, P.K.; Zhang, Y. A novel low-activation $VCrFeTa_xW_x$ (x = 0.1, 0.2, 0.3, 0.4, and 1) high-entropy alloys with excellent heat-softening resistance. *Entropy* **2018**, *20*, 951. [CrossRef]
17. Senkov, O.N.; Wilks, G.B.; Miracle, D.B.; Chuang, C.P.; Liaw, P.K. Refractory high-entropy alloys. *Intermetallics* **2010**, *18*, 1758–1765. [CrossRef]
18. Xia, S.Q.; Yang, X.; Yang, T.F.; Liu, S.; Zhang, Y. Irradiation resistance in $Al_xCoCrFeNi$ high entropy alloys. *JOM* **2015**, *67*, 2340–2344. [CrossRef]
19. Lu, Y.P.; Gao, X.Z.; Jiang, L.; Chen, Z.N.; Wang, T.M.; Jie, J.C.; Kang, H.J.; Zhang, Y.B.; Guo, S.; Ruan, H.H.; et al. Directly cast bulk eutectic and near-eutectic high entropy alloys with balanced strength and ductility in a wide temperature range. *Acta Mater.* **2017**, *124*, 143–150. [CrossRef]
20. Lu, Z.P.; Wang, H.; Chen, M.W.; Baker, I.; Yeh, J.W.; Liu, C.T.; Nieh, T.G. An assessment on the future development of high-entropy alloys: Summary from a recent workshop. *Intermetallics* **2015**, *66*, 67–76. [CrossRef]
21. Luo, H.; Lu, W.; Fang, X.; Ponge, D.; Li, Z.; Raabe, D. Beating hydrogen with its own weapon: Nano-twin gradients enhance embrittlement resistance of a high-entropy alloy. *Mater. Today* **2018**, *21*, 1003–1009. [CrossRef]
22. Feng, X.; Tang, G.; Gu, L.; Ma, X.; Sun, M.; Wang, L. Preparation and characterization of TaNbTiW multi-element alloy films. *Appl. Surf. Sci.* **2012**, *261*, 447–453. [CrossRef]
23. Braeckman, B.R.; Depla, D. Structure formation and properties of sputter deposited $Nb_xCoCrCuFeNi$ high entropy alloy thin films. *J. Alloys Compd.* **2015**, *646*, 810–815. [CrossRef]
24. Liao, W.; Lan, S.; Gao, L.; Zhang, H.; Xu, S.; Song, J.; Wang, X.; Lu, Y. Nanocrystalline high-entropy alloy (CoCrFeNiAl 0.3) thin-film coating by magnetron sputtering. *Thin Solid Films* **2017**, *638*, 383–388. [CrossRef]
25. Marshal, A.; Pradeep, K.G.; Music, D.; Zaefferer, S.; De, P.S.; Schneider, J.M. Combinatorial synthesis of high entropy alloys: Introduction of a novel, single phase, body-centered-cubic FeMnCoCrAl solid solution. *J. Alloys Compd.* **2017**, *691*, 683–689. [CrossRef]
26. Zhang, H.T.; Siu, K.W.; Liao, W.B.; Wang, Q.; Yang, Y.; Lu, Y. In situ mechanical characterization of CoCrCuFeNi high-entropy alloy micro/nano-pillars for their size-dependent mechanical behavior. *Mater. Res. Express* **2016**, *3*, 094002. [CrossRef]

27. Zou, Y.; Maiti, S.; Steurer, W.; Spolenak, R. Size-dependent plasticity in an $Nb_{25}Mo_{25}Ta_{25}W_{25}$ refractory high-entropy alloy. *Acta Mater.* **2014**, *65*, 85–97. [CrossRef]
28. Zou, Y.; Jeffrey, M.W.; Ma, H.; Okle, P.; Spolenak, R. Nanocrystalline high-entropy alloys: A new paradigm in high-temperature strength and stability. *Nano Lett.* **2017**, *17*, 1569–1574. [CrossRef]
29. Gao, L.; Song, J.; Jiao, Z.; Liao, W.; Luan, J.; Surjadi, J.U.; Li, J.; Zhang, H.; Sun, D.; Liu, C.T. High-entropy alloy (HEA)-coated nanolattice structures and their mechanical properties. *Adv. Eng. Mater.* **2017**, *20*, 1700625. [CrossRef]
30. Sheng, W.; Yang, X.; Wang, C.; Zhang, Y. Nano-crystallization of high-entropy amorphous $NbTiAlSiW_xN_y$ films prepared by magnetron sputtering. *Entropy* **2016**, *18*, 226. [CrossRef]
31. Zhang, Y.; Yan, X.-H.; Liao, W.-B.; Zhao, K. Effects of nitrogen content on the structure and mechanical properties of $(Al_{0.5}CrFeNiTi_{0.25})N_x$ high-entropy films by reactive sputtering. *Entropy* **2018**, *20*, 624. [CrossRef]
32. Gao, L.; Liao, W.; Zhang, H.; Surjadi, J.U.; Sun, D.; Lu, Y. Microstructure, mechanical and corrosion behaviors of $CoCrFeNiAl_{0.3}$ high entropy alloy (HEA) films. *Coatings* **2017**, *7*, 156. [CrossRef]
33. Chen, T.K.; Shun, T.T.; Yeh, J.W.; Wong, M.S. Nanostructured nitride films of multi-element high-entropy alloysby reactive DC sputtering. *Surf. Coat. Technol.* **2004**, *188–189*, 193–200. [CrossRef]
34. Braeckman, B.R.; Boydens, F.; Hidalgo, H.; Dutheil, P.; Jullien, M.; Thomann, A.L.; Depla, D. High entropy alloy thin films deposited by magnetron sputtering of powder targets. *Thin Solid Films* **2015**, *580*, 71–76. [CrossRef]
35. Feng, X.B.; Zhang, J.Y.; Wang, Y.Q.; Hou, Z.Q.; Wu, K.; Liu, G.; Sun, J. Size effects on the mechanical properties of nanocrystalline NbMoTaW refractory high entropy alloy thin films. *Int. J. Plasticity* **2017**, *95*, 264–277. [CrossRef]
36. Feng, X.; Fu, W.; Zhang, J.; Zhao, J.; Li, J.; Wu, K.; Liu, G.; Sun, J. Effects of nanotwins on the mechanical properties of AlxCoCrFeNi high entropy alloy thin films. *Scripta Mater.* **2017**, *139*, 71–76. [CrossRef]
37. Li, D.Y.; Li, C.X.; Feng, T.; Zhang, Y.D.; Sha, G.; Lewandowski, J.J.; Liaw, P.K.; Zhang, Y. High-entropy $Al_{0.3}CoCrFeNi$ alloy fibers with high tensile strength and ductility at ambient and cryogenic temperatures. *Acta Mater.* **2017**, *123*, 285–294. [CrossRef]
38. Ma, S.G.; Zhang, S.F.; Qiao, J.W.; Wang, Z.H.; Gao, M.C.; Jiao, Z.M.; Yang, H.J.; Zhang, Y. Superior high tensile elongation of a single-crystal $CoCrFeNiAl_{0.3}$ high-entropy alloy by Bridgman solidification. *Intermetallics* **2014**, *54*, 104–109. [CrossRef]
39. Jiao, Z.M.; Ma, S.G.; Yuan, G.Z.; Wang, Z.H.; Yang, H.J.; Qiao, J.W. Plastic deformation of $Al_{0.3}CoCrFeNi$ and AlCoCrFeNi high-entropy alloys under nanoindentation. *J. Mater. Eng. Perform.* **2015**, *24*, 3077–3083. [CrossRef]
40. Zhang, Y.; Liu, J.P.; Chen, S.Y.; Xie, X.; Liaw, P.K.; Dahmen, K.A.; Qiao, J.W.; Wang, Y.L. Serration and noise behaviors in materials. *Prog. Mater. Sci.* **2017**, *90*, 358–460. [CrossRef]

entropy

MDPI

Article

Effects of Nitrogen Content on the Structure and Mechanical Properties of $(Al_{0.5}CrFeNiTi_{0.25})N_x$ High-Entropy Films by Reactive Sputtering

Yong Zhang [1,*], Xue-Hui Yan [1], Wei-Bing Liao [2] and Kun Zhao [3]

1. State Key Laboratory for Advanced Metals and Materials, University of Science and Technology Beijing, Beijing 100083, China; yanxh_ustb@163.com
2. College of Physics and Energy, Shenzhen University, Shenzhen 518060, China; liaowb@szu.edu.cn
3. Department of Physics, School of Science, Hebei University of Science and Technology, Shijiazhuang 050000, China; zhk3415@163.com

* Correspondence: drzhangy@ustb.edu.cn; Tel.: +86-10-6233-3073

Received: 9 July 2018; Accepted: 19 August 2018; Published: 21 August 2018

Abstract: In this study, $(Al_{0.5}CrFeNiTi_{0.25})N_x$ high-entropy films are prepared by a reactive direct current (DC) magnetron sputtering at different N_2 flow rates on silicon wafers. It is found that the structure of $(Al_{0.5}CrFeNiTi_{0.25})N_x$ high-entropy films is amorphous, with $x = 0$. It transforms from amorphous to a face-centered-cubic (FCC) structure with the increase of nitrogen content, while the bulk $Al_{0.5}CrFeNiTi_{0.25}$ counterpart prepared by casting features a body-centered-cubic (BCC) phase structure. The phase formation can be explained by the atomic size difference (δ). Lacking nitrogen, δ is approximately 6.4% for the five metal elements, which is relatively large and might form a BCC or ordered-BCC structure, while the metallic elements in this alloy system all have a trend to form nitrides like TiN, CrN, AlN, and FeN. Therefore, nitride components are becoming very similar in size and structure and solve each other easily, thus, an FCC (Al-Cr-Fe-Ni-Ti)N solid solution forms. The calculated value of δ is approximately 23% for this multicomponent nitride solid solution. The $(Al_{0.5}CrFeNiTi_{0.25})N_x$ films achieve a pronounced hardness and a Young's modulus of 21.45 GPa and 253.8 GPa, respectively, which is obviously much higher than that of the as-cast $Al_{0.5}CrFeNiTi_{0.25}$ bulk alloys.

Keywords: high-entropy films; phase structures; hardness; solid-solution; interstitial phase

1. Introduction

High-entropy films (HEFs) are a brand-new type of alloy film, which has been developed recently based on the design concept of high-entropy alloys (HEAs) [1]. HEFs can be defined as multiple-component films with high-entropy mixing. Generally, the HEAs are composed of multi-principal-elements (at least five elements with five at a % $\leq$ each element content $\leq$ 35 at %) [2,3]. HEAs feature higher mixing entropy than traditional alloys and tend to form disorder face-centered cubic (FCC) and/or body-centered cubic (BCC) phase structures rather than ordered intermetallic compounds [4–6]. Due to severe lattice distortion and solid solution strengthening, attributable to the multi-components, the HEAs show many excellent mechanical properties, such as high-strength, good ductility, and high-wear and corrosion resistances [1]. Based on a similar scientific concept, HEFs have been designed and investigated gradually and show a great potential for application in the coating industry [7–12]. To date, many excellent properties of HEFs have been discovered and studied, such as high-wear resistance [13,14], high-corrosion resistance [15–17], diffusion-barriers effects [18–20], solar-thermal-conversion effects [10,21], plastic-deformation characteristics [22,23], thermal stabilities [7,24,25], and soft magnetic properties [26]. However, there are few theories that can explain the mechanism of the phase formation of the HEFs.

The phase formation and mechanical properties of $(Al_{0.5}CrFeNiTi_{0.25})N_x$ high-entropy thin films with different N_2 flow rates are studied in this paper. The phase structures and mechanical properties of $Al_xCrFeNiTi_{0.25}$ (x: molar ratio, x = 0, 0.25, 0.5, 0.75, and 1.0) bulks have previously been studied systematically [27]. Herein, the optimal composition of $Al_xCrFeNiTi_{0.25}$ (x = 0.5) alloy is selected as a magnetron-sputtering target to explore the phase formation mechanism and mechanical properties of high-entropy thin films. Significantly, the phase structures of high-entropy thin films transform from amorphous to FCC with the increase of nitrogen content, while the bulk $Al_{0.5}CrFeNiTi_{0.25}$ alloy attains BCC phase structure. Phase formation mechanism is explored from both theoretical calculations and experiments. Concerning high entropy thin films, the ability for solid solution structure formation is first discussed from the atom radius difference (δ). This study can help to provide new insights into understanding the phase formation mechanism of multi-component alloy thin film solids with small atoms such as nitrogen.

2. Materials and Methods

The $(Al_{0.5}CrFeNiTi_{0.25})N_x$ films were deposited on p-type Si (100) wafers by a direct current (DC) magnetron sputtering using non-equal atomic ratio $Al_{0.5}CrFeNiTi_{0.25}$ targets of Φ 60 mm in diameter and 2 mm in thickness. The alloy target was prepared by arc-melting and the smelting was repeated at least five times to ensure uniform mixing of components. Prior to deposition, the Si substrates were cleaned sequentially and rinsed by acetone, alcohol, and distilled water in an ultrasonic bath. Pre-sputtering was an effective way to remove oxide or contaminants on the surface of the target. When the base pressure held at 2.0×10^{-4} Pa, high purity argon was injected into the vacuum chamber and the target was cleaned by argon ion bombardment for 15 min at a power of 100 W. The deposition of the $(Al_{0.5}CrFeNiTi_{0.25})N_x$ films were carried out in an Ar + N_2 mixed atmosphere under a DC power of 100 W with a working pressure of 0.5 Pa. The schematic diagram of reactive sputtering is shown in Figure 1. The metal atoms escaped the surface of the target due to the bombardment of high energy particles. Due to the nitrogen atmosphere, different metal atoms reacted with N-atoms and were deposited on the substrates in the end. During the deposition, the total flow rate of Ar + N_2 was maintained at 30 standard cubic centimeters per minute (sccm) and the ratio was $N_2/(Ar + N_2)$ and R_{N2} was controlled at 0, 10%, 20%, 30%, 40%, and 50%, respectively. The work distance between the substrate and the target was 60 mm and the deposition time was maintained at 60 min. No external heating or bias was used on the substrate during the deposition process.

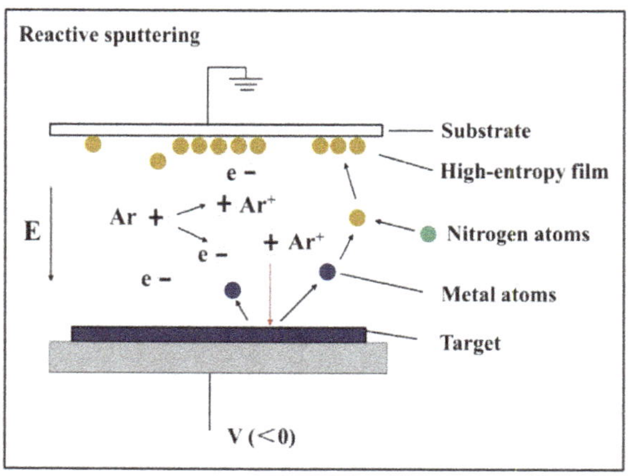

Figure 1. The schematic diagram of reactive sputtering.

The crystal structures of $(Al_{0.5}CrFeNiTi_{0.25})N_x$ films were analyzed by a glancing-incidence (1°) X-ray diffractometer (XRD, BRUKERD8 Discover, Bruker, Karlsruhe, Germany) using the Cu K_α radiation at a scanning rate of 4°/min. The scanning step was 0.02° with a scanning range of 20°–90°. The morphology studies and thickness measurements were carried out using field-emission scanning electron microscopy (SEM, Auriga Field Emission Scanning Electron Microscope, Carl Zeiss, Jena, Germany) equipped with an Energy Dispersive X-ray Spectrometer (EDX) operated at 10 kV. The surface roughness of the coatings were measured by an atomic force microscope (AFM, Veeco DI-3100, Bruker, Beijing, China). The hardness and modulus of the as-deposited films were tested at five points for each sample with a nano-indenter using a Berkovich triangular pyramid indenter. The distance between each indentation was 50 μm. The Poisson's ratio and Elastic modulus of the indenter tip were 0.07 and 1.141×10^6 MPa, respectively. Micrographs of indentations were tested by a laser scanning confocal microscope (LSCM, OLS-4100, Olympus, Tokyo, Japan).

3. Results and Discussion

3.1. Phase Structures

3.1.1. Phase Formation Mechanism Analysis

The XRD patterns of $(Al_{0.5}CrFeNiTi_{0.25})N_x$ films deposited at different R_{N2} are evaluated in Figure 2. An amorphous structure was observed with x = 0. As the N_2 flow rate increased, the phase structures of $(Al_{0.5}CrFeNiTi_{0.25})N_x$ films showed a tendency to crystallize. Using XRD analysis, thin films displayed a simple FCC structure and the grains were nanoscale at about 20 nm. The bulk $Al_{0.5}CrFeNiTi_{0.25}$ alloys counterpart featured a BCC structure [27]. Due to the high-cooling rate, films were far from the equilibrium that could be achieved by bulk alloys, which led to the difference in structure even if they were the same system. Many similar results have been reported for other system film alloys [16,28–31]. While the metallic elements in this alloy system, which all had trends to form nitrides like TiN, CrN, AlN, and FeN, had results that showed only a simple FCC structure in an XRD pattern, rather than a various nitrides phase. It can be inferred that the solid solution occurred between nitrides in high-entropy thin films. Mutual dissolution between carbides and nitrides has been reported in many studies. Figure 3 shows the UC-$ZrC_{0.81}$ pseudo-binary system that showed a miscibility relationship and the substitutional solid solution structure was formed with the carbides as solid solution units [32]. The nitrides also obtained similar results, such as Ti-Al-N, Cr-Al-N, Ti-Cr-N and Ti-Zr-N, Ti-Al-Si-N nitride films [33–37].

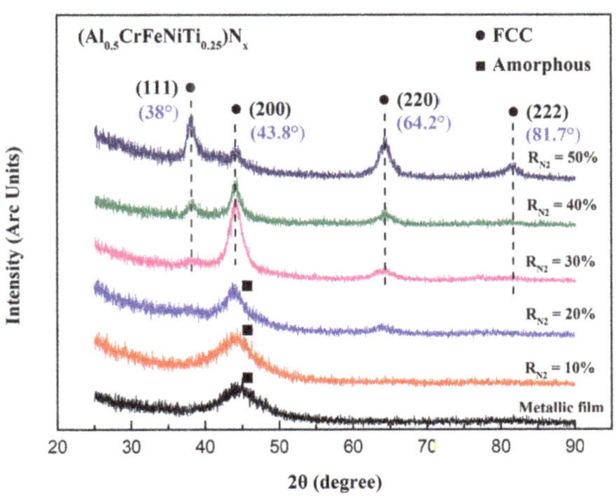

Figure 2. X-ray patterns of the $(Al_{0.5}CrFeNiTi_{0.25})N_x$ films deposited at different R_{N2}.

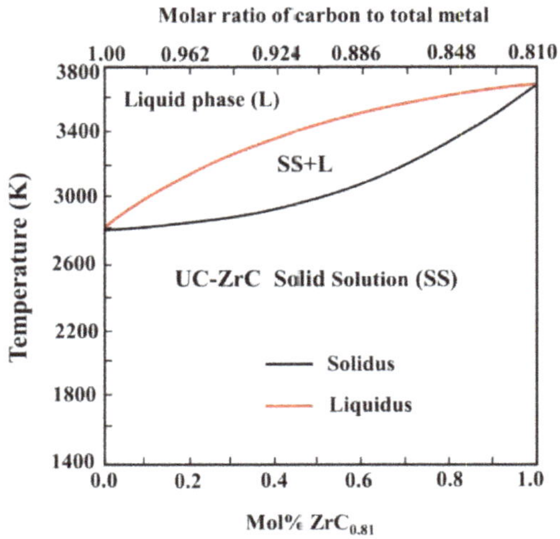

Figure 3. The UC-$ZrC_{0.81}$ pseudo-binary phase diagram. Reproduced with the permission from Reference [34].

A previous study showed that FCC phases were found to have lower δ, whereas BCC phases showed higher δ [38]. Regarding $(Al_{0.5}CrFeNiTi_{0.25})N_x$ with x = 0, δ was approximately 6.4%, which was relatively large, and might form a BCC or ordered BCC structure. Metallic atoms solved each other and formed a BCC solid solution structure (Figure 4a). Accompanying the increase in nitrogen, nitrides formed in $(Al_{0.5}CrFeNiTi_{0.25})N_x$ and were similar in size and structure. The δ between nitrides was relatively small, thus an FCC structure might have formed. When R_{N2} was equal to 10% and 20%, a low concentration of nitrogen existed in the lattice. However, small amounts of nitrogen cannot facilitate lattice reconstruction. Thus, it still maintained an amorphous structure, however, with a higher nitrogen content, the formation of nitrides was promoted that efficiently improved the order of the lattice. The nitride components were becoming more similar in size and structure and solved each other easily, thus, an FCC (Al-Cr-Fe-Ni-Ti)N_x solid solution formed, which is shown in Figure 4b,c.

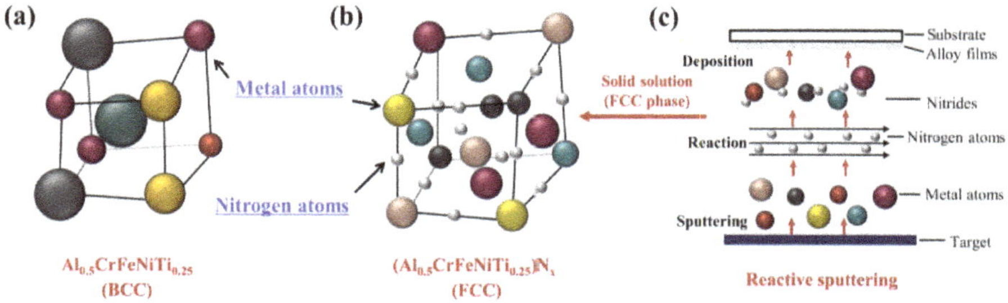

Figure 4. Schematic of lattice structure. (**a**) $Al_{0.5}CrFeNiTi_{0.25}$ bulk; (**b**) $(Al_{0.5}CrFeNiTi_{0.25})N_x$ films; and (**c**) schematic diagram of high-entropy thin film formation.

Figure 5 displays the distribution of component elements of $(Al_{0.5}CrFeNiTi_{0.25})N_x$ thin films deposited at R_{N2} = 50% through EDX, including surface, and line and point scanning. As presented in each element map in Figure 5a, the distribution of component elements was uniform, which indicates a segregate-free characteristic. The results of line and point scanning also reflected the same element distribution characteristics. The internal picture in Figure 5b is the composite surface-scanning of this region and segregate-free characteristics can be observed. Viewing the comparative analysis of the energy spectrum of Point A and Point B in Figure 5c, the distribution of elements was stable and had a slight fluctuation. Moreover, the lattice constant of the FCC solid solution structure was calculated by the Prague formula. The FCC solid solution structure with a lattice constant about 4.093 Å was completely different with nitrides such as TiN, CrN, and so on, which also confirmed the occurrence of a solid solution.

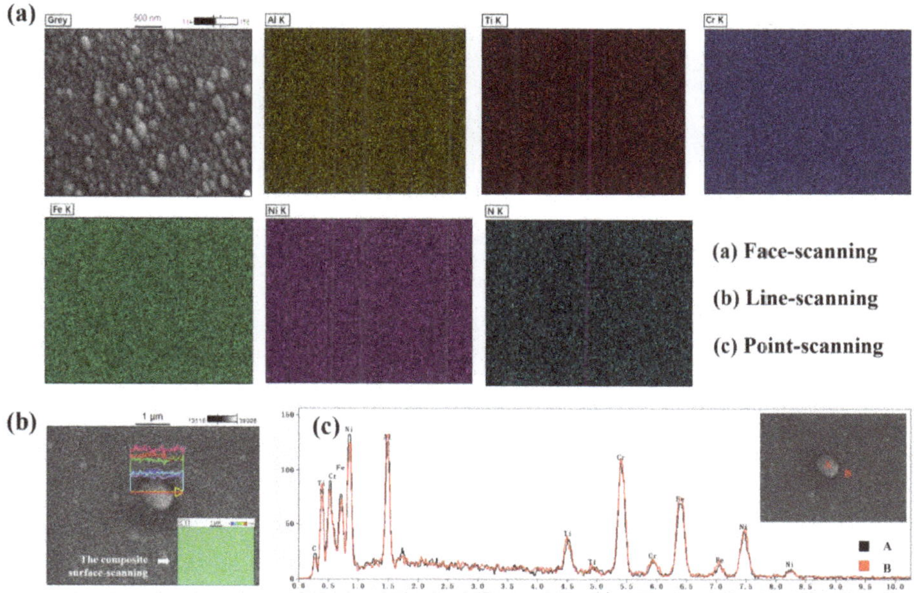

Figure 5. Energy dispersive X-ray spectroscopy of $(Al_{0.5}CrFeNiTi_{0.25})N_x$ high-entropy thin films deposited at R_{N2} = 50%: (**a**) face-scanning; (**b**) line-scanning; the insert picture is the composite surface scanning of this region; and (**c**) point-scanning.

3.1.2. Solid Solution Formation Ability

Regarding high-entropy alloys, the solid-solution formation ability was estimated by the value of Ω and δ [39], which were defined as follows:

$$\Omega = \frac{T_m \Delta S_{mix}}{|\Delta H_{mix}|}$$

$$\delta = \sqrt{\sum_{i=1}^{n} c_i (1 - r_i/\overline{r})^2}$$

where T_m was the melting temperature of the n-elements alloy, ΔS_{mix} was the mixing-entropy of an n-element system, ΔH_{mix} was the mixing-enthalpy of an n-element system, c_i was the atomic percentage of the component, r_i was the atomic radius, and $\overline{r} = \sum_{i=1}^{n} c_i r_i$ was the average atomic radius. Reviewing the previous study, it was concluded that the region where solid-solution structures were formed was in the range of 1.1 to 229.8 for Ω and 0.8% to 6.6% for δ. The radius of component elements in $(Al_{0.5}CrFeNiTi_{0.25})N_x$ are shown in Table 1, and the specific Ω and δ values of $(Al_{0.5}CrFeNiTi_{0.25})N_x$ at different N_2 flow rates are shown in Table 2.

Concerning $(Al\text{-}Cr\text{-}Fe\text{-}Ni\text{-}Ti)N_x$, the δ was described as a function of nitrogen content, as shown in Figure 6a. The analytical curve was calculated based on the atomic ratio of the target, which was the theoretical value. While different deposition yields of component elements can cause fluctuations in the atomic ratio, the experimental values were well matched with theoretical values, as shown by the star points in Figure 6a. To clearly distinguish different phase regions of film alloys in this study, the nitrogen content was also described as a function of δ in Figure 6b. Regarding $Al_{0.5}CrFeNiTi_{0.25}$, δ was about 6.4% located in the SSS zone in Figure 6b, and the BCC solid solution phase structure was stable. When increasing the nitrogen content, the FCC structure formed in the $(Al_{0.5}CrFeNiTi_{0.25})N_x$ system. The calculated value of δ was about 23% for that FCC structure. When the nitrogen content was low, the amorphous structure was stable and located in the middle region.

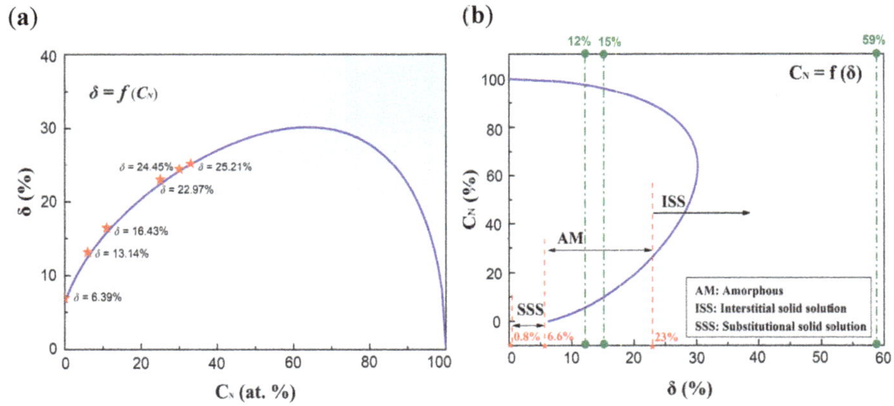

Figure 6. (**a**) Analytical curve of δ as a function of nitrogen content in $(Al_{0.5}CrFeNiTi_{0.25})N_x$ and (**b**) analytical curve of nitrogen content as a function of δ in $(Al_{0.5}CrFeNiTi_{0.25})N_x$ films (12% according to Inoue Principle, 15% and 59% according to the Hume–Rothery Rule).

Table 1. The radius of component elelments in $(Al_{0.5}CrFeNiTi_{0.25})N_x$ films.

Element	N	Al	Ti	Cr	Fe	Ni
Radius (nm)	0.071	0.1434	0.1445	0.1249	0.1241	0.1246

Table 2. The nitrogen content, Ω value, δ value, and the roughness of films at different R_{N2}.

R_{N2} (%)	0	10	20	30	40	50
Nitrogen Content (at %)	0	5.94	10.97	25.10	29.55	32.21
Ω	1.60	1.31	0.44	0.23	0.20	0.19
δ	6.39%	13.14%	16.43%	22.97%	24.45%	25.21%
R_a (nm)	1.375	1.574	3.780	4.542	4.490	5.574

3.2. Deposition Rates

Figure 7 presents the deposition rate of $(Al_{0.5}CrFeNiTi_{0.25})N_x$ as a function of R_{N2} and the exact value of thickness is shown in the internal table. Following the addition of nitrogen, the deposition rate of films was significantly reduced by about 30%. During the deposition, the total flow rate of Ar + N_2 was maintained at 30 sccm. Thus, the density of argon decreased gradually with the increase of N_2 flow rate and resulted in a lower efficiency of argon-ion bombardment of the target. Under higher R_{N2}, the deposition rate continued to slow and maintain a lower sputtering yield. Added to the effect caused by low argon gas density, the ceramicization of the target surface was also an important factor. This was mainly due to the N-containing layer that formed on the target surface, which reduced the conductivity of the target and resulted in a low sputtering yield. This phenomena was "target poisoning" [40,41].

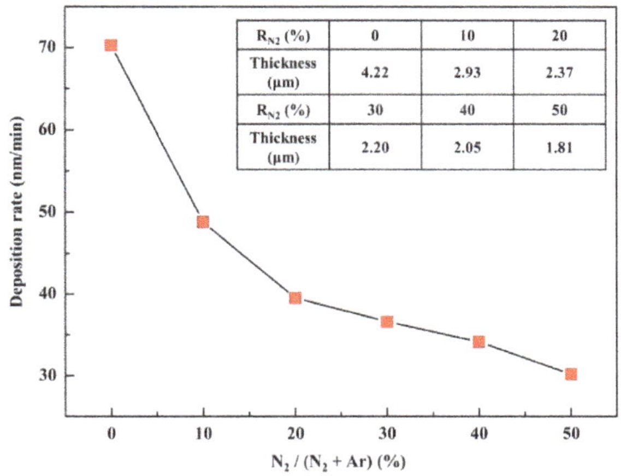

Figure 7. Deposition rates of $(Al_{0.5}CrFeNiTi_{0.25})N_x$ films at different R_{N2}.

3.3. Surface Morphologies

The AFM images and SEM micrographs of the $(Al_{0.5}CrFeNiTi_{0.25})N_x$ films at different R_{N2} are shown in Figure 9. Initially, a very dense and smooth surface with low surface roughness was obtained. Looking at SEM micrographs, it was observed that a small amount of nanostructure precipitation occurred as the R_{N2} increased to 10%. When the N_2 flow rate increased, the number and particle size of the nanostructures were increased significantly. Viewing the AFM images, many needle nanostructures were also observed in the high-entropy thin films. Additionally, the roughness of films was measured by AFM. The value of R_a gradually became larger and the exact value of surface roughness at different R_{N2} is shown in Table 2. The heights of the needle structure in all the HEFs were less than 60 nm, which was consistent with the results since the as-deposited HEFs were smooth and homogeneous, as shown in SEM micrographs.

The surface nanostructures displayed by AFM images were different from each other and the size of the needle structures became larger with the increase of R_{N2}. These nano-scaled needle structures were related to the structure zone models, which had a great impact on the microstructural evolution of thin film. Initially, as R_{N2} was low, the growth of the HEFs effect by nitrogen ions was limited. The nucleation barrier was generally expected to be small, which contributed to the formation of randomly oriented islands for films deposited on the silicon substrates without substrate heating. Accompanying the increase of R_{N2}, a strong driving force was obtained, which facilitated the surface atom diffusion and grain boundary motion. To achieve a stable state, the overall surface and interface energy were minimized through forming the new islands. Islands with lower energy consumed the others. Thus, the roughness of the HEFs increased gradually with the increase of R_{N2}.

3.4. Mechanical Properties

Figure 8a displays the hardness and Young's modulus of $(Al_{0.5}CrFeNiTi_{0.25})N_x$ films as a function of the N_2 flow rate. Through the increase of the nitrogen content, the hardness and Young's modulus of the films showed a significant upward trend. The high-entropy thin films showed the highest hardness and Young's modulus at 21.78 GPa and 253.8 GPa, respectively. Compared with the bulk $Al_{0.5}CrFeNiTi_{0.25}$ alloys (the value of hardness was about 6 GPa), the hardness of the $(Al_{0.5}CrFeNiTi_{0.25})N_x$ HEFs was improved significantly. Considering the load-depth curves of Figure 8b, the manifest and precise change of the hardness and Young's modulus of the HEFs was obtained. When the probe depressed into the same depth, a larger force was required with the increase of the R_{N2}. This was attributed to the fact that solid solution significantly increased the hardness. It is well known that solid solution strengthening increases the yield strength of the material by increasing the stress τ to move dislocations [42]:

$$\Delta\tau = Gb\varepsilon^{3/2}\sqrt{c}$$

where c was the concentration of the solute atoms, G was the shear modulus, b was the magnitude of the Burger's vector, and ε is the lattice strain due to the solute. While increasing the N_2 flow rates, the parameters c and ε were increased correspondingly, which resulted in a higher τ value. Thus, the effect of solid-solution strengthening was enhanced gradually with the increase of R_{N2}. When the R_{N2} = 40% and 50%, the nitrogen content in films increased slowly. The difference in nitrogen content was only 2.66%, as shown in Table 2. The mechanical properties of the HEFs became stable.

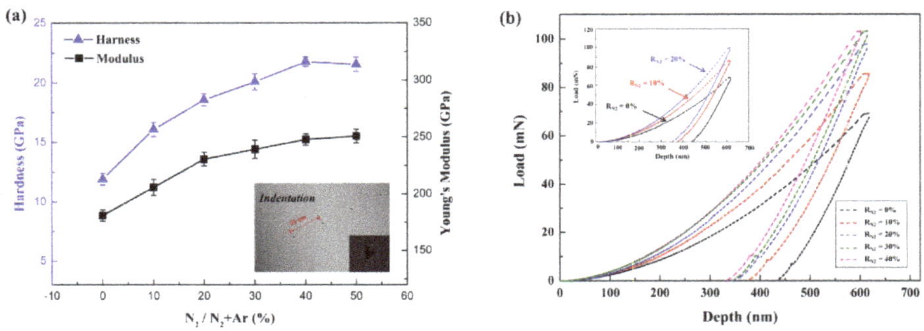

Figure 8. Mechanical properties of $(Al_{0.5}CrFeNiTi_{0.25})N_x$ films deposited at different R_{N2}. (**a**) Hardness and modulus and (**b**) load-depth curve.

4. Discussion

It is well known that the solid solutions can be divided into three types: Substitutional solid solutions; interstitial solid solutions; and vacancy solid solutions. Hume–Rothery rules are a set of basic rules that describe the formation of substitutional and interstitial solid solutions. Regarding a substitutional solid solution, the atomic radius of the solute and solvent atoms must differ by no more than 15%, while solute atoms should have a radius no larger than 59% of the radius of the solvent atoms for an interstitial solid solution [43]. However, the non-metallic elements, such as H, B, C, and N, having a very small atomic radius can form compounds with metal elements. When the ratio of a non-metal atom radius to a metal atom radius (R_x/R_M) is less than 0.59, the compounds have a simple structure such as M_4X, M_2X, MX, MX_2, which is called an interstitial phase. When $R_x/R_M > 0.59$, the compounds have a complex structure such as Fe_3C, thus, there is a interstitial compound [44]. Regarding $(Al_{0.5}CrFeNiTi_{0.25})N_x$, the metallic elements all satisfy the formation rule for an interstitial phase and different nitrides with simple structures can be formed. The metallic elements will tend to be ionic and the ionic radius might well reflect the actual particle. Since most of the elements in this alloy system are transitional metals, their ionic radii are very close to each other. Nitride components are becoming more similar in size and structure and solve each other easily. Thus, the FCC structured phase is formed in $(Al_{0.5}CrFeNiTi_{0.25})N_x$.

The correlation between the composition, processing, microstructures, and the properties of high-entropy thin films can be summarized systematically and is shown in Figure 10. First, the films are prepared by a magnetron sputtering technique and different N_2 flow rates are selected to adjust the composition as well as to change the microstructures. This finds that the nitrogen content in the thin films increases gradually with the increase of the nitrogen content and the structure of films transform from amorphous to FCC. Second, it is further found that the mechanical properties (hardness/Young's modulus) of the high-entropy crystalline films are much better than the high-entropy amorphous films. All told, there is an increasing trend of hardness with higher nitrogen for the high-entropy thin films. Third, due to the difference in the preparation process, the phase structure and the properties of the films are different from the bulk samples, although they are in the same composition system. This study can help enrich the cognition between composition processing, microstructures, and the properties for high-entropy materials, especially for high-entropy thin films.

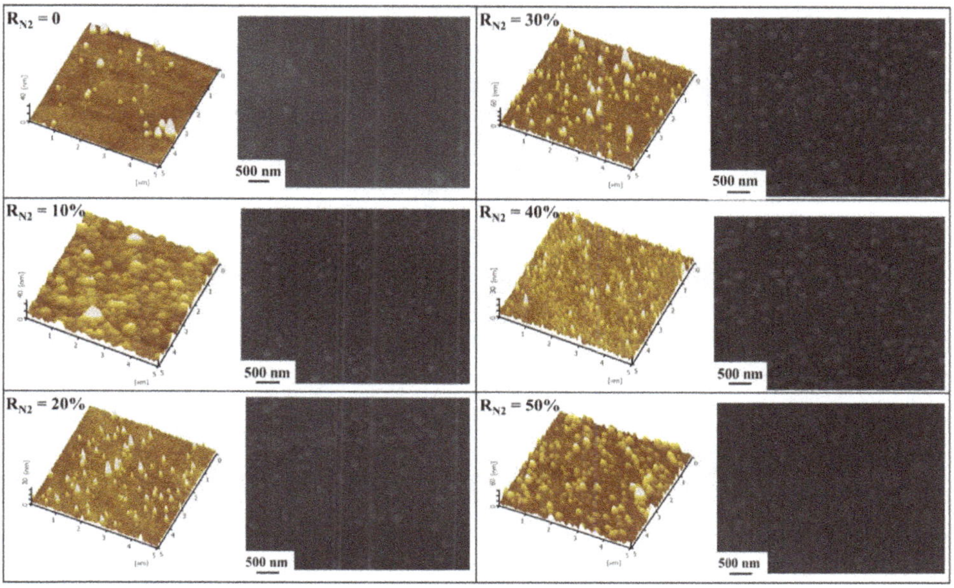

Figure 9. AFM images and SEM micrographs of $(Al_{0.5}CrFeNiTi_{0.25})N_x$ films at different R_{N2}.

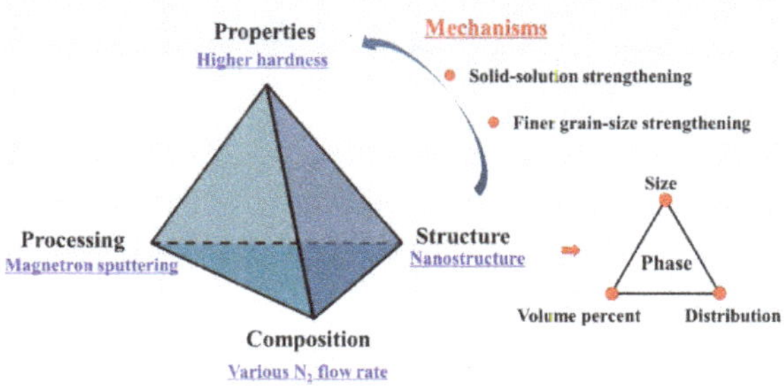

Figure 10. Relationship between composition, processing, properties, and the structures of high-entropy films.

5. Conclusions

The $(Al_{0.5}CrFeNiTi_{0.25})N_x$ high-entropy thin films were deposited on silicon wafers through magnetron sputtering without substrate bias and heating. It was found that the phase structures of the HEFs transformed from an amorphous to an FCC structure with the increase in nitrogen content—the formation of a nitride solid solution was the main reason. Using the XRD analysis, the grain size was nanoscale at about 20 nm. The formation ability of the solid solution phase for the Al-Cr-Fe-Ni-Ti-N system film alloys was discussed regarding δ. When the value of δ was higher than 23%, the solid solution structures were stable in $(Al_{0.5}CrFeNiTi_{0.25})N_x$ thin films. The $(Al_{0.5}CrFeNiTi_{0.25})N_x$ HEFs deposited at R_{N2} = 40% and 50% yielded a maximum hardness and modulus of 21.78 GPa and 253.8 GPa, respectively, which is much higher than the as-cast $Al_{0.5}CrFeNiTi_{0.25}$ bulk alloys. The enhancement in hardness was mainly attributed to solid-solution strengthening and the lattice distortion. Additionally, smaller grain-size was a beneficial factor for increasing hardness.

Author Contributions: X.-H.Y. finished the preparation of high-entropy thin films. X.-H.Y. and W.-B.L. together carried out analyzing and writing. K.Z. provided equipment. Y.Z. offered the theoretical guidance.

Funding: Y. Z. would like to thank the financial supports from the National Science Foundation of China (NSFC, Granted Nos. 51471025 and 51671020). W.-B.L. gratefully acknowledges the financial support from the Natural Science Foundation of Shenzhen University (Grant No. 2017069) and the Shenzhen Basic Research Project JCYJ20170302142339007, and the Youth Innovation Talent Project of Guangdong Province (Grant No. 2017KQNCX175), and the National Natural Science Foundation of China (Grant No. 51801128).

Conflicts of Interest: The authors declare no conflict of interest.

References

1. Zhang, Y.; Zuo, T.T.; Tang, Z.; Gao, M.C.; Dahmen, K.A.; Liaw, P.K.; Lu, Z.P. Microstructures and properties of high-entropy alloys. *Prog. Mater. Sci.* **2014**, *61*, 1–93. [CrossRef]
2. Yeh, J.W.; Chen, S.K.; Lin, S.J.; Gan, J.Y.; Chin, T.S.; Shun, T.T.; Tsau, C.H.; Chang, S.Y. Nanostructured High-Entropy Alloys with Multiple Principal Elements: Novel Alloy Design Concepts and Outcomes. *Adv. Eng. Mater.* **2004**, *6*, 299–303. [CrossRef]
3. Zhang, W.; Liaw, P.K.; Zhang, Y. Science and technology in high-entropy alloys. *Sci. China Mater.* **2018**, *61*, 2–22. [CrossRef]
4. Ma, D.; Grabowski, B.; Körmann, F.; Neugebauer, J.; Raabe, D. Ab initio thermodynamics of the CoCrFeMnNi high entropy alloy: Importance of entropy contributions beyond the configurational one. *Acta Mater.* **2015**, *100*, 90–97. [CrossRef]
5. Wu, Y.D.; Cai, Y.H.; Chen, X.H.; Wang, T.; Si, J.J.; Wang, L.; Wang, Y.D.; Hui, X.D. Phase composition and solid solution strengthening effect in TiZrNbMoV high-entropy alloys. *Mater. Des.* **2015**, *83*, 651–660. [CrossRef]
6. Zhang, Y.; Zhou, Y.J.; Lin, J.P.; Chen, G.L.; Liaw, P.K. Solid-Solution Phase Formation Rules for Multi-component Alloys. *Adv. Eng. Mater.* **2010**, *10*, 534–538. [CrossRef]
7. Hsieh, M.H.; Tsai, M.H.; Shen, W.J.; Yeh, J.W. Structure and properties of two Al–Cr–Nb–Si–Ti high-entropy nitride coatings. *Surf. Coat. Technol.* **2013**, *221*, 118–123. [CrossRef]
8. Huang, P.K.; Yeh, J.W. Effects of substrate temperature and post-annealing on microstructure and properties of (AlCrNbSiTiV)N coatings. *Thin Solid Films* **2009**, *518*, 180–184. [CrossRef]
9. Liu, L.; Zhu, J.B.; Hou, C.; Li, J.C.; Jiang, Q. Dense and smooth amorphous films of multicomponent FeCoNiCuVZrAl high-entropy alloy deposited by direct current magnetron sputtering. *Mater. Des.* **2013**, *46*, 675–679. [CrossRef]
10. Sheng, W.; Yang, X.; Wang, C.; Zhang, Y. Nano-Crystallization of High-Entropy Amorphous NbTiAlSiWxNy Films Prepared by Magnetron Sputtering. *Entropy* **2016**, *18*, 226. [CrossRef]
11. Gao, L.; Song, J.; Jiao, Z.; Liao, W.; Luan, J.; Surjadi, J.U.; Li, J.; Zhang, H.; Sun, D.; Liu, C.T. High-Entropy Alloy (HEA)-Coated Nanolattice Structures and Their Mechanical Properties. *Adv. Eng. Mater.* **2017**, *20*. [CrossRef]
12. Liao, W.; Lan, S.; Gao, L.; Zhang, H.; Xu, S.; Song, J.; Wang, X.; Lu, Y. Nanocrystalline high-entropy alloy (CoCrFeNiAl 0.3) thin-film coating by magnetron sputtering. *Thin Solid Films* **2017**, *638*, 383–388. [CrossRef]
13. Lin, D.; Zhang, N.; He, B.; Zhang, G.; Zhang, Y.; Li, D. Tribological properties of FeCoCrNiAlB x high-entropy alloys coating prepared by laser cladding. *J. Iron Steel Res. Int.* **2017**, *24*, 184–189. [CrossRef]
14. Pogrebnjak, A.; Yakushchenko, I.; Bagdasaryan, A.; Bondar, O.; Krause-Rehberg, R.; Abadias, G.; Chartier, P.; Oyoshi, K.; Takeda, Y.; Beresnev, V. Microstructure, physical and chemical properties of nanostructured (Ti–Hf–Zr–V–Nb) N coatings under different deposition conditions. *Mater. Chem. Phys.* **2014**, *147*, 1079–1091. [CrossRef]
15. Ye, Q.; Feng, K.; Li, Z.; Lu, F.; Li, R.; Huang, J.; Wu, Y. Microstructure and corrosion properties of CrMnFeCoNi high entropy alloy coating. *Appl. Surf. Sci.* **2016**, *396*, 1420–1426. [CrossRef]

16. Hsueh, H.T.; Shen, W.J.; Tsai, M.H.; Yeh, J.W. Effect of nitrogen content and substrate bias on mechanical and corrosion properties of high-entropy films $(AlCrSiTiZr)_{100-x}N_x$. *Surf. Coat. Technol.* **2012**, *206*, 4106–4112. [CrossRef]
17. Shon, Y.; Joshi, S.S.; Katakam, S.; Rajamure, R.S.; Dahotre, N.B. Laser additive synthesis of high entropy alloy coating on aluminum: Corrosion behavior. *Mater. Lett.* **2015**, *142*, 122–125. [CrossRef]
18. Tsai, M.H.; Yeh, J.W.; Gan, J.Y. Diffusion barrier properties of AlMoNbSiTaTiVZr high-entropy alloy layer between copper and silicon. *Thin Solid Films* **2008**, *516*, 5527–5530. [CrossRef]
19. Chang, S.Y.; Li, C.E.; Chiang, S.C.; Huang, Y.C. 4-nm thick multilayer structure of multi-component $(AlCrRuTaTiZr)N_x$ as robust diffusion barrier for Cu interconnects. *J. Alloys Compd.* **2012**, *515*, 4–7. [CrossRef]
20. Chang, S.Y.; Chen, D.S. 10-nm-thick quinary (AlCrTaTiZr)N film as effective diffusion barrier for Cu interconnects at 900 °C. *Appl. Phys. Lett.* **2009**, *94*, 222. [CrossRef]
21. Sheng, W.J.; Yang, X.; Zhu, J.; Wang, C.; Zhang, Y. Amorphous phase stability of $NbTiAlSiN_X$ high-entropy films. *Rare Metals* **2017**, *5*, 1–8. [CrossRef]
22. Feng, X.; Fu, W.; Zhang, J.; Zhao, J.; Li, J.; Wu, K.; Liu, G.; Sun, J. Effects of nanotwins on the mechanical properties of Al_x CoCrFeNi high entropy alloy thin films. *Scripta Mater.* **2017**, *139*, 71–76. [CrossRef]
23. Feng, X.B.; Zhang, J.Y.; Wang, Y.Q.; Hou, Z.Q.; Wu, K.; Liu, G.; Sun, J. Size effects on the mechanical properties of nanocrystalline NbMoTaW refractory high entropy alloy thin films. *Int. J. Plasticity* **2017**, *95*, 264–277. [CrossRef]
24. Cheng, C.Y.; Yeh, J.W. High thermal stability of the amorphous structure of $Ge_xNbTaTiZr$ (x = 0.5, 1) high-entropy alloys. *Mater. Lett.* **2016**, *181*, 223–226. [CrossRef]
25. Feng, X.; Tang, G.; Gu, L.; Ma, X.; Sun, M.; Wang, L. Preparation and characterization of TaNbTiW multi-element alloy films. *Appl. Surf. Sci.* **2012**, *261*, 447–453. [CrossRef]
26. Lin, P.-C.; Cheng, C.-Y.; Yeh, J.-W.; Chin, T.-S. Soft Magnetic Properties of High-Entropy Fe-Co-Ni-Cr-Al-Si Thin Films. *Entropy* **2016**, *18*, 308. [CrossRef]
27. Liu, S.; Gao, M.C.; Liaw, P.K.; Zhang, Y. Microstructures and mechanical properties of Al_x CrFeNiTi 0.25 alloys. *J. Alloys Compd.* **2015**, *619*, 610–615. [CrossRef]
28. Liang, S.C.; Tsai, D.C.; Chang, Z.C.; Sung, H.S.; Lin, Y.C.; Yeh, Y.J.; Deng, M.J.; Shieu, F.S. Structural and mechanical properties of multi-element (TiVCrZrHf)N coatings by reactive magnetron sputtering. *Appl. Surf. Sci.* **2011**, *258*, 399–403. [CrossRef]
29. Lai, C.H.; Lin, S.J.; Yeh, J.W.; Chang, S.Y. Preparation and characterization of AlCrTaTiZr multi-element nitride coatings. *Surf. Coat. Technol.* **2006**, *201*, 3275–3280. [CrossRef]
30. Lin, C.H.; Duh, J.G.; Yeh, J.W. Multi-component nitride coatings derived from Ti–Al–Cr–Si–V target in RF magnetron sputter. *Surf. Coat. Technol.* **2007**, *201*, 6304–6308. [CrossRef]
31. Tsai, D.C.; Huang, Y.L.; Lin, S.R.; Liang, S.C.; Shieu, F.S. Effect of nitrogen flow ratios on the structure and mechanical properties of (TiVCrZrY)N coatings prepared by reactive magnetron sputtering. *Appl. Surf. Sci.* **2010**, *257*, 1361–1367. [CrossRef]
32. Zhang, Y.; Lin, J.P.; Chen, G.L.; Yang, J.X.; Ren, Y.G.; Zhang, S.Z. Uranium carbide powder metallurgy. *Powder Met. Technol.* **1995**, *4*, 303–306. [CrossRef]
33. Zhou, M.; Makino, Y.; Nose, M.; Nogi, K. Phase transition and properties of Ti–Al–N thin films prepared by rf-plasma assisted magnetron sputtering. *Thin Solid Films* **1999**, *339*, 203–208. [CrossRef]
34. Kimura, A.; Kawate, M.; Hasegawa, H.; Suzuki, T. Anisotropic lattice expansion and shrinkage of hexagonal TiAlN and CrAlN films. *Surf. Coat. Technol.* **2003**, *169*, 367–370. [CrossRef]
35. Hörling, A.; Sjölén, J.; Willmann, H.; Larsson, T.; Odén, M.; Hultman, L. Thermal stability, microstructure and mechanical properties of $Ti_{1-x}Zr_xN$ thin films. *Thin Solid Films.* **2008**, *516*, 6421–6431. [CrossRef]
36. Vetter, J.; Scholl, H.; Knotek, O. (TiCr) N coatings deposited by cathodic vacuum arc evaporation. *Surf. Coat. Technol.* **1995**, *74*, 286–291. [CrossRef]
37. Carvalho, S.; Rebouta, L.; Ribeiro, E.; Vaz, F.; Denannot, M.; Pacaud, J.; Rivière, J.; Paumier, F.; Gaboriaud, R.; Alves, E. Microstructure of (Ti, Si, Al) N nanocomposite coatings. *Surf. Coat. Technol.* **2004**, *177*, 369–375. [CrossRef]
38. Zhang, Y.; Lu, Z.P.; Ma, S.G.; Liaw, P.K.; Tang, Z.; Gao, M.C. Guidelines in predicting phase formation of high-entropy alloys. *Mrs Commun.* **2014**, *4*, 57–62. [CrossRef]
39. Yang, X.; Zhang, Y. Prediction of high-entropy stabilized solid-solution in multi-component alloys. *Mater. Chem. Phys.* **2012**, *132*, 233–238. [CrossRef]

40. Lu, C.Y.; Diyatmika, W.; Lou, B.S.; Lu, Y.C.; Duh, J.G.; Lee, J.W. Influences of target poisoning on the mechanical properties of TiCrBN thin films grown by a superimposed high power impulse and medium-frequency magnetron sputtering. *Surf. Coct. Technol.* **2017**, *332*, 86–95. [CrossRef]
41. Arif, M.; Eisenmenger-Sittner, C. In situ assessment of target poisoning evolution in magnetron sputtering. *Surf. Coat. Technol.* **2017**, *324*, 345–352. [CrossRef]
42. Pelleg, J. *Mechanical Properties of Materials*; Springer: New York, NY, USA, 2008; pp. 236–239. ISBN 978-94-007-4341-0.SG.
43. Mizutani, U. The Hume-Rothery rules for structurally complex alloy phases. In *Surface Properties and Engineering of Complex Intermetallics*; World Scientific: Singapore, 2010; pp. 323–399. [CrossRef]
44. Barrett, C.S. *Structure of Metals*; McGraw-Hill Book Company, Inc.: New York, NY, USA, 1943.

entropy

MDPI

Article

Effects of Silicon Content on the Microstructures and Mechanical Properties of (AlCrTiZrV)-Si_x-N High-Entropy Alloy Films

Jingrui Niu [1], Wei Li [1,*], Ping Liu [1], Ke Zhang [1], Fengcang Ma [1], Xiaohong Chen [1], Rui Feng [2] and Peter K. Liaw [2]

[1] School of Materials Science and Engineering, University of Shanghai for Science and Technology, Shanghai 200093, China; njrniujingrui@163.com (J.N.); liuping@usst.edu.cn (P.L.); zhangke@usst.edu.cn (K.Z.); mafengcang@163.com (F.M.); cxh992@163.com (X.C.)

[2] Department of Materials Science and Engineering, The University of Tennessee, Knoxville, TN 37996, USA; fengruisjtu@gmail.com (R.F.); pliaw@utk.edu (P.K.L.)

* Correspondence: liwei176@usst.edu.cn; Tel.: +86-21-552-716-82

Received: 4 December 2018; Accepted: 11 January 2019; Published: 16 January 2019

Abstract: A series of (AlCrTiZrV)-Si_x-N films with different silicon contents were deposited on monocrystalline silicon substrates by direct-current (DC) magnetron sputtering. The films were characterized by the X-ray diffractometry (XRD), scanning electron microscopy (SEM), high-resolution transmission electron microscopy (HRTEM), and nano-indentation techniques. The effects of the silicon content on the microstructures and mechanical properties of the films were investigated. The experimental results show that the (AlCrTiZrV)N films grow in columnar grains and present a (200) preferential growth orientation. The addition of the silicon element leads to the disappearance of the (200) peak, and the grain refinement of the (AlCrTiZrV)-Si_x-N films. Meanwhile, the reticular amorphous phase is formed, thus developing the nanocomposite structure with the nanocrystalline structures encapsulated by the amorphous phase. With the increase of the silicon content, the mechanical properties first increase and then decrease. The maximal hardness and modulus of the film reach 34.3 GPa and 301.5 GPa, respectively, with the silicon content (x) of 8% (volume percent). The strengthening effect of the (AlCrTiZrV)-Si_x-N film can be mainly attributed to the formation of the nanocomposite structure.

Keywords: (AlCrTiZrV)-Si_x-N films; microstructure; mechanical property; nanocomposite structure

---

1. Introduction

Hard coating tools are the fastest developing new techniques and have become the symbol of modern cutting tools. Among them, the most widely-used coatings are binary and ternary systems of traditional metal nitrides or oxides coatings, such as titanium nitrides (TiN), chromium nitrides (CrN), etc. [1]. With the development of the coating science and technology, nanocomposite films with better mechanical properties have become a research hotspot due to their higher mechanical properties, thermal stability, and oxidation resistance. In 1992, Li et al. [2] first prepared TiSiN nanocomposite films with a hardness of over 60 GPa using the physical vapor deposition (PVD), and suggested that the films consisted of TiN and Si_3N_4 phases, in which TiN was crystalline, while Si_3N_4 was amorphous. In 2000, Veprek et al. [3] prepared the TiSiN nanocomposite films with the superhigh hardness of 80–105 GPa, which aroused great interest among researchers all over the world. Since then, researchers have done a great amount of research on the preparation of nanocomposite films and their hardening mechanism, and have proposed the widely-accepted model of nanocrystals encapsulated by the amorphous interfacial phase, which is named as the *nc*-TiN/*a*-Si_3N_4 model (*nc* refers to nanocrystals, and *a* refers to amorphous) [4].

In 2004, Yeh et al. innovatively proposed the concept of high-entropy alloys (HEAs), and Cantor established the equiatomic multicomponent alloys [5,6]. HEAs are a high-entropy solid-solution phase alloy formed by five or more principal elements, each of which has an atomic concentration between 5–35 at.% (atomic percent). The HEAs are not inclined to form intermetallic compounds due to their high-entropy effect, which tend to stabilize the simple body-centered-cubic (BCC), face-centered-cubic (FCC), or hexagonal-close-packed (HCP) solid solution [7–12]. A large number of subsequent studies have shown that HEAs have many superior properties than conventional alloys, such as the high strength, great hardness, strong wear resistance, good fatigue resistance, high oxidation and corrosion resistance [13–28]. With the development of new HEAs, studies of HEA nitride films have also attracted the great interest of many researchers. Feng et al. [29] deposited the (ZrTaNbTiW)N films on the substrate of the Ti_6Al_4V alloy by multi-target magnetron sputtering to investigate the composition, structure, and mechanical properties of the film. They found that the (ZrTaNbTiW)N film showed the BCC and FCC structures, and its mechanical properties were significantly greater than those of ZrTaNbTiW films. Cheng et al. [30] deposited multi-component (AlCrTaTiZr)-Si_x-N films on the monocrystalline Si substrate by reactive radio-frequency (RF) magnetron sputtering and probed the influence of the silicon content on the structures, morphologies, and mechanical properties of the films. The results showed that the incorporation of silicon significantly increased the oxidation resistance of AlCrTaTiZr nitride films, but lowered their hardnesses. Tsai et al. [12] deposited the multi-component (AlCrMoTaTi)N films with different silicon contents by reactive RF magnetron sputtering and studied the effects of silicon contents on the nitride films. Their results showed that the incorporation of silicon led to the lattice distortion of the films, improved the mechanical properties of the films, but reduced the electrical properties of the films. Moreover, Tsai et al. [31] deposited the (AlCrMoTaTi)-Si_x-N films on the Si substrate by the reactive RF magnetron sputtering to systematically study the effects of the silicon content on the oxidation of the films. It was found that the oxidation resistance of the films improved with the increase of the silicon content, which could be attributed to the existence of Al and Si in the films.

Based on the current research situation, it can be seen that there is no clear conclusion about the effects of the Si element on the microstructures and mechanical properties of HEA nitride films. Therefore, this study combines the concepts of the "nanocomposite film" and "HEA nitride films" to prepare (AlCrTiZrV)-Si_x-N films with different silicon contents by the reactive direct-current (DC) magnetron sputtering. The effects of silicon contents on the microstructures and mechanical properties of the films were systematically studied with the expectation of providing the experimental and theoretical basis for the application of the nanocomposite HEA nitride film in the industrial field.

2. Experimental

2.1. Film Preparation

The (AlCrTiZrV)-Si_x-N films with different silicon contents were deposited on monocrystalline Si (100) wafers with a size of 35 mm × 25 mm × 0.5 mm using a JPG–450 multi-target magnetron sputtering system. The wafers were ultrasonically cleaned with the acetone and absolute ethanol for 15 min, and then dried into the sputtering chamber. The target used for sputtering is a self-made compound target with the structure shown in Figure 1. The process of preparing the compound target can be summarized as follows. Both the Si target and the equal-moles AlCrTiZrV target, with the diameter of 75 mm and the purity of 99.99% (volume percent), were cut down 5 slices of 25 equal fan segments by a low speed wire electrical discharge machine (EDM). Then, keeping the 80% (volume percent) of the AlCrTiZrV target unchanged, these fan segments were assembled into the composite target with different silicon contents by changing the number of the Si sector. For example, the composite target shown in Figure 1 has only one Si sector, therefore the Si content (x) of the composite target is 4% (volume percent). According to this method, composite targets with the Si contents (x) of 4%, 8%, 12%, and 16% can be assembled for sputtering.

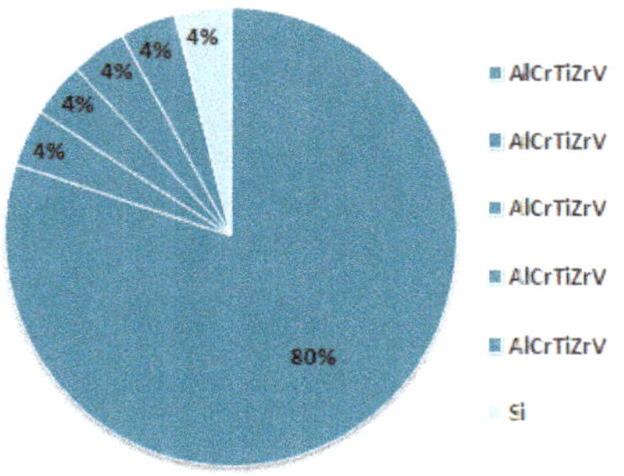

Figure 1. Schematic illustration of the $AlCrTiZrVSi_x$ composite target.

The $AlCrTiZrVSi_x$ compound target was controlled by a DC power supply. The base pressure was pumped down to 3.0×10^{-4} Pa before the deposition. During the experiment, the Ar and N_2 flow rates were both maintained at the 10 standard cubic centimeter per minute (sccm). The working pressure was adjusted to 0.7 Pa, and the sputtering power was 180 W. To improve the homogeneity of films, the substrate was rotated at a speed of 10 r/min. The deposition time was 1.5 h, and finally the (AlCrTiZrV)-Si_x-N films with a thickness of about 1.8 μm were obtained.

2.2. Film Characterization and Measurement

A series of characterization and testing of the deposited (AlCrTiZrV)-Si_x-N films were performed. The structural and phase analyses were conducted on the D8 Advance X-ray diffractometer (XRD, Bruker, Germany) using the CuK_α radiation (λ = 0.15406 nm) with a measurement range of 25° to 90° and the Jade software, respectively. The microstructures of the films were observed by the Quanta FEG450 field emission environmental electron microscope (SEM, FEI, USA) and Tecnai $G^2$20 high-resolution field-emission transmission electron microscope (HRTEM, FEI, USA). The NANO Instrument, a G200 nano indenter (Agilent, USA) with the Berkovich indenter was used to study the mechanical properties. The loading and unloading curves were obtained by accurately recording the change of the loading depth with the load. Then the hardness and elastic modulus were calculated by the Oliver–Pharr model [32]. During the measurement, the loading depth was set to 100 nm, less than 1/10 of the thickness of the films, so as to eliminate the effect of the substrate on the measurements. Each hardness or elastic modulus value was an average of at least 16 measurements.

3. Results

The XRD patterns of the (AlCrTiZrV)-Si_x-N films with different silicon contents are presented in Figure 2. It can be seen that the film without the Si element shows a simple FCC structure with a diffraction peak near $2\theta = 41°$, corresponding to the reflection of the (200) crystal plane, suggesting that an FCC-structured solid-solution phase is formed with good crystallinity, and no other complicated intermetallic compound is developed in the films. The reason for this phenomenon is that, due to the high-entropy effect, the HEA tends to form the stable solid-solution phase rather than intermetallic compounds. In bulk alloys of similar compositions to AlCrTiZrV, several researchers have reported the formation of an order B2 structure [33,34], while in this investigation, the crystal phase of the film without Si is a disordered FCC structure. This phenomenon may contribute to the reaction of the HEA with N_2. The films form the HEA nitride rather than the simple HEA, which leads to the change of the structure of the crystal phase. Combined with the phase analysis of the Jade software, it can be determined that the solid-solution phase is composed of five binary nitrides from Al, Cr, Ti, Zr, and V. This result has been reported in previous works and proven to be thermodynamically stable due to the high mixing entropy effect [35,36].

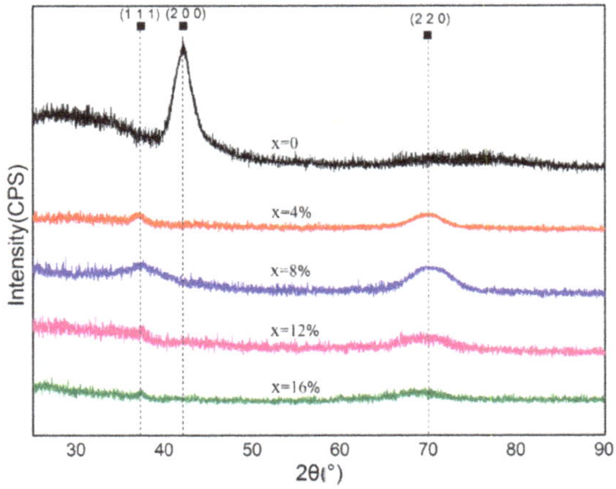

Figure 2. X-ray diffraction (XRD) patterns of (AlCrTiZrV)-Si_x-N films with different silicon contents.

With the incorporation of Si, no obvious diffraction peak can be observed at 2θ = 41°, indicating that the (200) diffraction peaks of films disappear with the incorporation of Si. From the XRD patterns of (AlCrTiZrV)-Si_x-N (x = 4%, 8%, 12%, and 16%) film, it can be seen that the diffraction patterns have two diffraction peaks with low intensities at about 2θ = 37° and 2θ = 70°, corresponding to fcc (111) and (200) peaks, respectively, suggesting that the film crystallinity decreases with the incorporation of the Si element, relative to the (AlCrTiZrV)N film, and the microstructure may be nanocrystalline or amorphous.

Figure 3a shows that the film without the Si element grows in a columnar crystal with the growth direction marked by the arrow. Figure 3b suggests that the (AlCrTiZrV)-$Si_{0.08}$-N film exhibits the amorphous morphology with the growth direction indicated by the arrow. In addition, with the incorporation of Si, the diffraction peaks of the films move to high angles, as presented in Figure 2, suggesting that the lattice parameters of the nitride phase decrease, which may be attributed to the fact that the doped Si atoms can be dissolved into the (AlCrTiZrV)N lattice and occupy some lattice positions. As the radius of the Si atom is smaller than that of other metal elements, the lattice parameters of the nitride phase decrease, which has been confirmed in other studies of ternary Si-contained nitride films [37,38]. Besides, it is clear to see the change of the lattice parameters of the nitride phase according to the values given by Table 1.

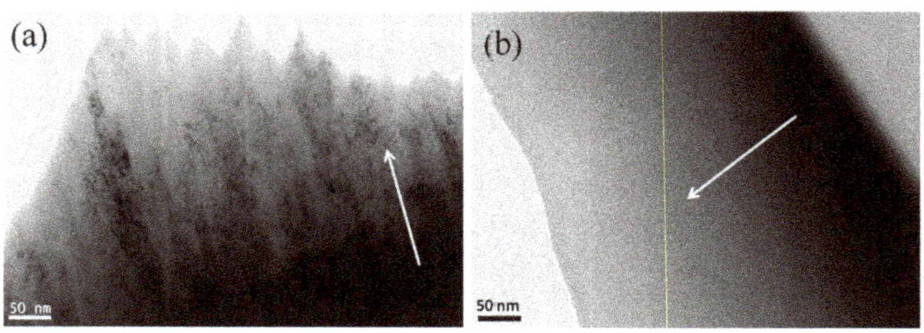

Figure 3. Low-magnification cross-sectional transmission electron microscope (TEM) images of (AlCrTiZrV)-Si_x-N films: (**a**) (AlCrTiZrV)N; (**b**) (AlCrTiZrV)-$Si_{0.08}$-N.

Table 1. The values of the lattice parameters of the (AlCrTiZrV)-Si_x-N films. (d_1 are the calculated values according to the XRD; d_2 are the measurements according to the high-resolution transmission electron microscope (HRTEM)).

(AlCrTiZrV)-Si_x-N	d_1(nm)	d_2(nm)
(AlCrTiZrV)N	0.2018	0.1913
(AlCrTiZrV)-$Si_{0.04}$-N	0.1631	-
(AlCrTiZrV)-$Si_{0.08}$-N	0.1629	0.1652
(AlCrTiZrV)-$Si_{0.12}$-N	0.1627	-
(AlCrTiZrV)-$Si_{0.16}$-N	0.1624	-

The cross-sectional SEM images of the (AlCrTiZrV)-Si_x-N films with different silicon contents are presented in Figure 4. It can be seen that the nitride films are well bonded to the substrates, and there is no obvious micro-gap between them. Moreover, the nitride films are grown uniformly with thickness of about 1.8 μm, and the surface and internal quality of the nitride films are good. All the films have the small surface roughness and a very dense and smooth cross-sectional structure without the visible grain feature. Table 2 shows the calculated grain-sizes values of the (AlCrTiZrV)-Si_x-N films with different silicon contents according to the Scherrer's equation. This trend indicates that the grain sizes of the films reach the nanometer level.

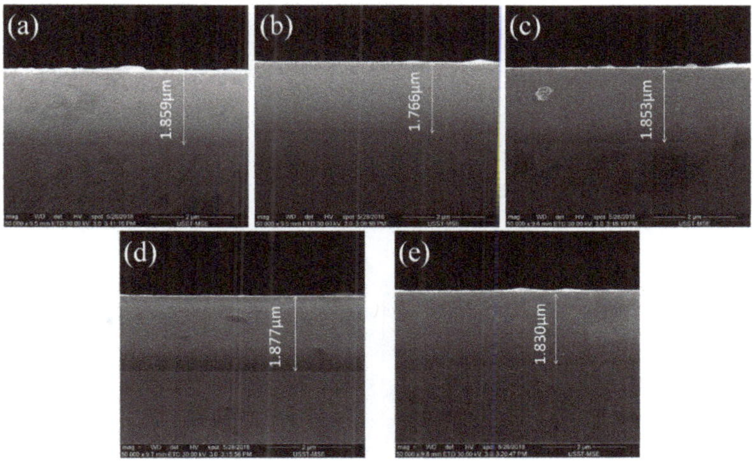

Figure 4. Cross-sectional SEM images of the (AlCrTiZrV)-Si_x-N films. (**a**) (AlCrTiZrV)N; (**b**) (AlCrTiZrV)-$Si_{0.04}$-N; (**c**) (AlCrTiZrV)-$Si_{0.08}$-N; (**d**) (AlCrTiZrV)-$Si_{0.12}$-N; (**e**) (AlCrTiZrV)-$Si_{0.16}$-N.

Table 2. The calculated grain-sizes values of the (AlCrTiZrV)-Si_x-N films.

(AlCrTiZrV)-Si_x-N	D(200)	D(220)
(AlCrTiZrV)N	22.18 nm	-
(AlCrTiZrV)-$Si_{0.04}$-N	-	5.32 nm
(AlCrTiZrV)-$Si_{0.08}$-N	-	4.36 nm
(AlCrTiZrV)-$Si_{0.12}$-N	-	3.20 nm
(AlCrTiZrV)-$Si_{0.16}$-N	-	2.60 nm

The cross-sectional HRTEM images of the (AlCrTiZrV)N and (AlCrTiZrV)-$Si_{0.08}$-N films are presented in Figure 5. Figure 5a,c is the microstructures of the (AlCrTiZrV)N film, in which A, B, and C represent different crystal grains. Figure 5b,d shows the images of the (AlCrTiZrV)-$Si_{0.08}$-N film, where A, B, C, D, E, F, and G denote the nanocrystals inside the films. Figure 5e,f presents the selected-area electron diffraction (SAED) patterns of two films. A comparison of Figure 5a,b indicates that the grains of the (AlCrTiZrV)N film (the areas marked by A, B, and C) are comparatively coarse, reaching tens or hundreds of nanometers, while the grains of the (AlCrTiZrV)-$Si_{0.08}$-N film (inside the yellow dotted line) are much smaller, approximately 2–5 nm, suggesting that the incorporation of the Si element can effectively refine the grain of the film to the nanometer level, which also confirms the previous XRD result in Figure 2. According to Figure 5c,d, it can be seen that there is no other phase existing between the grains of the (AlCrTiZrV)N film (on both sides of the yellow dotted line) from the high-magnification HRTEM, and the grains' contact directly with each other. While the grains of the (AlCrTiZrV)-$Si_{0.08}$-N film do not contact directly, and there are some reticular amorphous interfacial phases between two grains, that is, between two yellow closed dotted lines. It is speculated that the interfacial phase is the amorphous phase and corresponds to the amorphous projection of the XRD pattern. For the SAED patterns images, the FCC structure with the (200) preferred orientation of the (AlCrTiZrV)N film could be seen in Figure 5e, while the SAED patterns of the (AlCrTiZrV)-$Si_{0.08}$-N film in Figure 5f exhibit the nanocrystalline and amorphous features, which are consistent with the XRD analysis of Figure 2.

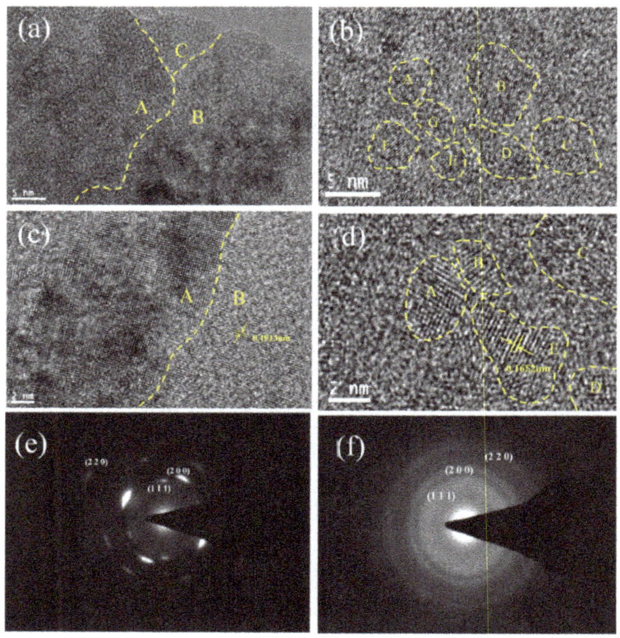

Figure 5. Cross-sectional HRTEM images and selected-area electron diffraction (SAED) patterns of the (**a**,**c**,**e**) (AlCrTiZrV)N and (**b**,**d**,**f**) (AlCrTiZrV)-$Si_{0.08}$-N films: (**a**,**b**) low-magnification HRTEM images; (**c**,**d**) high-magnification HRTEM images; (**e**,**f**) SAED patterns.

The effects of Si contents on the mechanical properties of the (AlCrTiZrV)-Si_x-N films are presented in Figure 6. The hardness and elastic modulus of the (AlCrTiZrV)N film are 30.1 GPa and 274.0 GPa, respectively. With the incorporation of Si, the hardness and elastic modulus of the films both first increase and then decrease. When the Si content (x) is 8%, the hardness and elastic modulus of the film reach the maximum, which are 34.3 GPa and 301.5 GPa, respectively. As the Si content (x) further increases to 12% or 16%, the hardness and elastic modulus of the films decrease and are lower than those of the (AlCrTiZrV)N film. It can be seen that the incorporation of Si has a significant effect on the mechanical properties of the films, which can be attributed to the microstructure change resulting from the incorporation of the Si element.

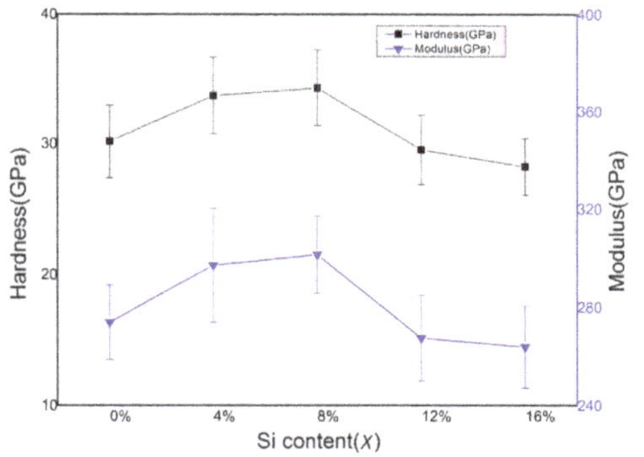

Figure 6. Effect of silicon content on mechanical properties of (AlCrTiZrV)-Si_x-N films.

4. Discussion

4.1. The Formation of the Nanocomposite Structures

The nanocomposite film consisting of two insoluble phases presents a three-dimensional network structure film in which the interfacial phase encapsulates the matrix phase. The difficulty for the formation of the nanocomposite structure within the HEA nitride films lies in the remarkable mixing entropy effect, which enhances the mutual solubility of constituent elements. As a result, the Si atoms may be inclined to be incorporated in the crystalline lattice of the HEA nitride, rather than be segregated as the SiN_x phase, which can serve as the interfacial phase of the nanocomposite structure.

For example, Tsai et al. [30] investigated the effects of the silicon content (0–7.51 at.%) on the (AlCrMoTaTi)N coatings by the reactive-magnetron sputtering, and no formation of the nanocomposite structure had been reported. In particular, the formation of the SiN_x phase was not observed even after the Si content was increased to 7.51 at.%. They attributed the extended solubility of Si to the high entropy effect. Lin et al. [39] reported the formation of the nanocomposite structure in the (AlCrTaTiZr)SiN coatings produced by the reactive RF-magnetron sputtering, and found that the nanocrystalline phase is elongated, rather than equiaxed, which is a typical feature of the nanocomposite films. Moreover, the lattice can be observed within the denoted "amorphous regions", suggesting the probable crystallized feature in these regions. More importantly, the composition distribution had not been provided in these investigations, making it difficult to verify the phase segregation within the coatings.

Compared with the above results, in this investigation, through the high-magnification HRTEM image shown in Figure 5d, the equiaxed regions (the areas marked by A, B, C, D, and E) exhibit an ordered lattice structure, which is the HEA nitride phase, while the areas between two yellow closed dotted lines present an amorphous structure, which is the interfacial phase. Meanwhile, no lattice structure is observed in the amorphous region, indicating that the nanocomposite structure forms within the film resulting from the phase separation between the HEA nitride matrix and SiN_x interfacial phase. Moreover, according to the studies of Prochazka [40], the a-Si_3N_4 phase has the limited solubility in FCC nitrides, and the thermodynamically-driven phase separation will occur during the deposition, which may be the possible reason why the nanocomposite structures could form in the (AlCrTiZrV)-Si_x-N films. Figure 7 is the schematic diagram of nanocomposite structures of the (AlCrTiZrV)-Si_x-N films, in which the (AlCrTiZrV)N matrix phase is a nano-equiaxed crystal, and the SiN_x interfacial phase is a network amorphous phase.

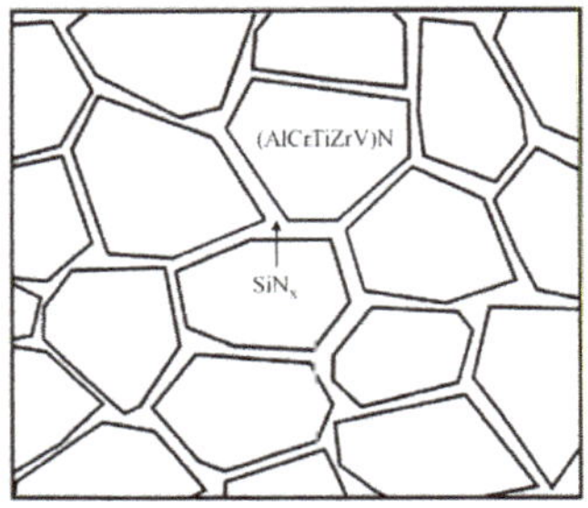

Figure 7. Schematic diagram of the nanocomposite structure of the (AlCrTiZrV)-Si_x-N films.

4.2. The Strengthening Mechanism of the (AlCrTiZrV)-Si_x-N Films

The mechanical properties of the films are improved with the proper addition of the Si element, which can be attributed to the microstructure change due to the Si incorporation. The reasons for improving their mechanical properties can be specifically discussed as follows.

According to the theory of solid-solution hardening [41], the incorporation of the Si element can improve the mechanical properties by causing the lattice distortion of the nitride films. To some extent, Table 1 shows that the incorporation of the Si element causes the lattice distortion of the nitride films. However, compared with the nitride film without Si, the mechanical properties of the nitride films with the high Si content are lower, indicating that the solid-solution strengthening is not the major factor in changing the mechanical properties of the nitride films.

From the classical Hall–Petch relationship [42]:

$$H = H_0 + k_{HP}d^{-1/2}, \quad (1)$$

where H_0 is the intrinsic hardness of the (AlCrTiZrV)N, d is the average grain size, and k_{HP} is the Hall-Petch coefficient of (AlCrTiZrV)N. The reduction of the grain size could lead to the strengthening effect of the HEA-nitride film. According to the HRTEM analysis, it can be seen that with the incorporation of the Si element, the grain sizes of the nitride films significantly decreases to the nanometer scale. Besides, Table 2 shows that the grain sizes of the (AlCrTiZrV)-Si_x-N films reduce with the incorporation of the Si element. Therefore, the effect of the fine-grained strengthening improves the mechanical properties of the nitride films to some extent. However, from the previous analysis, the mechanical properties of the nitride films with the low Si content increase, while the properties with the high Si content further decrease, that is, when the grain sizes reach a minimum, the mechanical properties of the nitride films are worst, suggesting that the fine-grained strengthening mechanism is not the main factor in strengthening the film.

From the HRTEM images and relevant analysis, the nanocomposite structure forms within the HEA nitride films. Combining the strengthening mechanism of the nanocomposite film (*nc*-TiN/*a*-Si_3N_4 model) [4,43,44], with the incorporation of Si, there are many equiaxed crystals with the nanometer size formed in the films, leading to the fact that the generation or multiplication of dislocations cannot happen within the films. At the same time, the thickness of the amorphous interfacial phases formed between the nanocrystals is small, and the crack is difficult to expand in the interfacial phases. Hence, the nitride films present a superhardness effect, and the mechanical properties are improved. Moreover, the quantity of the interfacial phases increases with the increase of the Si content, which makes the grains more refined (as shown in Table 2) and, therefore, effectively strengthens the film. Patscheider et al. [44] pointed out that the hardness of nanocomposite films peaks at the common minimum of the grain size of the crystalline phase and the grain separation by the amorphous phase. Namely, two conditions have to be fulfilled to achieve the maximum hardness for the nanocomposite film: the nanocrystals are approximately 5 nm in size and the mean distance has to be very small, about a few nanometers. Therefore, as shown in Table 2 and Figure 5d, when the average grain size of the crystalline phase is 4.36 nm, and the thickness of the amorphous interfacial phase is approximately 0.5–2 nm, the mechanical properties of the nitride films are the best.

However, if the interfacial phase thickness is too large, the hardness is mainly governed by the properties of the amorphous interfacial phase [44]. In this investigation, when the Si content is 8%, the thickness of the amorphous interfacial phase in the film could reach 2 nm. Because of the phase separation driven by thermodynamics, the (AlCrTiZrV)N matrix phase and the SiN_x interfacial phase are incompatible with each other. The Si only exists in the amorphous region. Thus, when the Si content is above 8%, the thickness of the amorphous interfacial phase will be 3–4 nm or even thicker. While at this time, the grain size of the crystalline phase is only 2–3 nm or even smaller, the hardness of the nitride film is mainly governed by the properties of the amorphous interfacial phase. This trend means that the hardness of the film approaches that of SiN_x. The SiN_x phase is softer than the (AlCrTiZrV)N phase; meanwhile, the crack can expand within the amorphous interfacial phase due to the thickening of the interface, thus leading to the decrease of the mechanical properties. Hence, the mechanical properties of the films with the high Si content are lower than those of the film without Si. Therefore, it is believed that the nanocomposite structure is the main reason for the improvement of the mechanical properties of the HEA nitride films.

Due to the limitation of characterization techniques, however, it is difficult to accurately predict the strengthening effect from the nanocomposite structure in this investigation. As a result, more research should be carried out on the phase-segregation behavior of the Si-containing HEA nitride films to help investigate the strengthening behavior of nanocomposite structures by the advanced techniques, such as atom probe tomography. The relevant modeling and simulations are also needed.

5. Conclusions

(i) The (AlCrTiZrV)N film is a solid-solution phase with the FCC structure and has a (200) preferential orientation. With the incorporation of the Si element, the (200) diffraction peak of (AlCrTiZrV)-Si_x-N films disappeared.

(ii) The (AlCrTiZrV)N film grows with a columnar crystal-growth mode. After the Si element is incorporated, the (AlCrTiZrV)-Si_x-N films exhibit an amorphous fracture morphology. Moreover, the incorporation of the Si element effectively refines the grains of the nitride films, forming nanocrystals with a size of about 2–5 nm, and a large number of amorphous interfacial phases form simultaneously.

(iii) The nanocomposite structure is formed within the (AlCrTiZrV)-Si_x-N films with the incorporation of the Si element, and the mechanical properties are improved. When the Si content (x) is 8%, the hardness and elastic modulus of the (AlCrTiZrV)-Si_x-N films reach 34.3 GPa and 301.5 GPa, respectively. However, the further increase of the Si content will result in the deterioration of the mechanical properties of the (AlCrTiZrV)-Si_x-N films.

Author Contributions: Conceptualization, J.N. and W.L.; Data curation, J.N.; Formal analysis, F.M. and X.C.; Funding acquisition, W.L.; Investigation, J.N.; Methodology, W.L.; Project administration, W.L.; Resources, W.L.; Supervision, W.L.; Validation, P.L. and K.Z.; Visualization, J.N.; Writing – original draft, J.N. and W.L.; Writing—review and editing, R.F. and P.K.L.

Funding: This research was funded by the National Nature Science Foundation of China (NSFC), with the grant number of 51471110. RF and PKL would like to acknowledge the Department of Energy (DOE), Office of Fossil Energy, National Energy Technology Laboratory (DE-FE-0011194) with the program manager, J. Mullen. P.K.L. thanks the support from the National Science Foundation (DMR-1611180 and 1809640) with the program directors, G. Shiflet and D. Farkas.

Conflicts of Interest: The authors declare no conflict of interest.

References

1. Mayrhofer, P.H.; Tischler, G.; Mitterer, C. Microstructure and mechanical/thermal properties of Cr-N coatings deposited by reactive unbalanced magnetron sputtering. *Surf. Coat. Technol.* **2001**, *142*, 78–84. [CrossRef]

2. Li, S.; Shi, Y.; Peng, H. Ti-Si-N films prepared by plasma-enhanced chemical vapor deposition. *Plasma Chem. Plasma Process* **1992**, *12*, 287–297.
3. Veprek, S.; Veprek-Heijman, M.J.G. Industrial applications of superhard nanocomposite coatings. *Surf. Coat. Technol.* **2008**, *202*, 5063–5073. [CrossRef]
4. Veprek, S.; Niederhofer, A.; Moto, K.; Bolom, T.; Mannling, H.-D.; Nesladek, P.; Dollinger, G.; Bergmaier, A. Composition, nanostructure and origin of the ultrahardness in nc-TiN/a-Si_3N_4/a- and nc-$TiSi_2$ nanocomposites with H_v = 80 to ≥ 105 GPa. *Surf. Coat. Technol.* **2000**, *133*, 152–159. [CrossRef]
5. Yeh, J.W.; Chen, S.K.; Lin, S.J.; Gan, J.Y.; Chin, T.S.; Shun, T.T.; Tsau, C.H.; Chang, S.Y. Nanostructured high-entropy alloys with multiple principal elements: Novel alloy design concepts and outcomes. *Adv. Eng. Mater.* **2004**, *6*, 299–303. [CrossRef]
6. Cantor, B.; Chang, I.T.H.; Knight, P.; Vincent, A.J.B. Microstructural development in equiatomic multicomponent alloys. *Mater. Sci. Eng. A* **2004**, *375*, 213–218. [CrossRef]
7. Zhang, Y.; Zuo, T.T.; Tang, Z.; Gao, M.C.; Dahmen, K.A.; Liaw, P.K.; Lu, Z.P. Microstructures and properties of high-entropy alloys. *Prog. Mater. Sci.* **2014**, *61*, 1–93. [CrossRef]
8. Li, W.; Liu, P.; Liaw, P.K. Microstructures and properties of high-entropy alloy films and coatings: A review. *Mater. Res. Lett.* **2018**, *6*, 199–229. [CrossRef]
9. Youssef, K.M.; Zaddach, A.J.; Niu, C.N.; Irving, D.L.; Koch, C.C. A novel low-density, high-hardness, high-entroy alloy with close-packed single-phase nanocrystalline structures. *Mater. Res. Lett.* **2014**, *3*, 95–99. [CrossRef]
10. Gao, M.C.; Yeh, J.W.; Liaw, P.K.; Zhang, Y. *High-Entropy Alloys*, 1st ed.; Springer: Cham, Switzerland, 2016; pp. 35–37.
11. Zhao, Y.J.; Qiao, J.W.; Ma, S.G.; Gao, M.C.; Yang, H.J.; Chen, M.W.; Zhang, Y. A hexagonal close-packed high-entropy alloy: The effect of entropy. *Mater. Des.* **2016**, *96*, 10–15. [CrossRef]
12. Tsai, D.C.; Chang, Z.C.; Kuo, B.H.; Chang, S.Y.; Shieu, F.S. Effects of silicon content on the structure and properties of (AlCrMoTaTi)N coatings by reactive magnetron sputtering. *J. Alloys Compd.* **2014**, *616*, 646–651. [CrossRef]
13. Varalakshmi, S.; Kamaraj, M.; Murty, B.S. Processing and properties of nanocrystalline CuNiCoZnAlTi high entropy alloys by mechanical alloying. *Mater. Sci. Eng. A* **2010**, *527*, 1027–1030. [CrossRef]
14. Shun, T.T.; Hung, C.H.; Lee, C.F. The effects of secondary elemental Mo or Ti addition in Al0.3CoCrFeNi high-entropy alloy on age hardening at 700 °C. *J. Alloys Compd.* **2010**, *495*, 55–58. [CrossRef]
15. Hsu, C.Y.; Juan, C.C.; Wang, W.R.; Sheu, T.S.; Yeh, J.W.; Chen, S.K. On the superior hot hardness and softening resistance of $AlCoCr_xFeMo_{0.5}Ni$ high-entropy alloys. *Mater. Sci. Eng. A* **2011**, *528*, 3581–3588. [CrossRef]
16. Lin, C.M.; Tsai, H.L. Equilibrium phase of high-entropy $FeCoNiCrCu_{0.5}$ alloy at elevated temperature. *J. Alloys Compd.* **2010**, *489*, 30–35. [CrossRef]
17. Tsai, C.W.; Tsai, M.H.; Yeh, J.W.; Yang, C.C. Effect of temperature on mechanical properties of $Al_{0.5}CoCrCuFeNi$ wrought alloy. *J. Alloys Compd.* **2010**, *490*, 160–165. [CrossRef]
18. Hemphill, M.A.; Yuan, T.; Wang, G.Y.; Yeh, J.W.; Tsai, C.W.; Chuang, A.; Liaw, P.K. Fatigue behavior of $Al_{0.5}CoCrCuFeNi$ high entropy alloys. *Acta Mater.* **2012**, *60*, 5723–5734. [CrossRef]
19. Tang, Z.; Yuan, T.; Tsai, C.W.; Yeh, J.W.; Lundin, C.D.; Liaw, P.K. Fatigue behavior of $Al_{0.5}CoCrCuFeNi$ high entropy alloys. *Acta Mater.* **2015**, *99*, 247–258. [CrossRef]
20. Seifi, M.; Li, D.; Yong, Z.; Liaw, P.K.; Lewandowski, J.J. Fracture toughness and fatigue crack growth behavior of as-cast high-entropy alloys. *JOM* **2015**, *67*, 2288–2295. [CrossRef]
21. Thurston, K.V.S.; Gludovatz, B.; Hohenwarter, A.; Laplanche, G.; George, E.P.; Ritchie, R.O. Effect of temperature on the fatigue-crack growth behavior of the high-entropy alloy CrMnFeCoNi. *Intermetallics* **2017**, *88*, 65–72. [CrossRef]
22. Chen, P.Y.; Lee, C.; Wang, S.Y.; Seifi, M.; Lewandowski, J.J.; Dahmen, K.A.; Jia, H.L.; Xie, X.; Chen, B.L.; Yeh, J.W.; et al. Fatigue behavior of the high-entropy alloys: A review. *Sci. China Technol. Sci.* **2018**, *61*, 168–178. [CrossRef]
23. Shi, Y.; Collins, L.; Balke, N.; Liaw, P.K.; Yang, B. In-situ electrochemical-AFM study of localized corrosion of $Al_xCoCrFeNi$ high-entropy alloys in chloride solution. *Appl. Surf. Sci.* **2018**, *439*, 533–544. [CrossRef]
24. Shi, Y.; Collins, L.; Feng, R.; Zhang, C.; Balke, N.; Liaw, P.K.; Yang, B. Homogenization of $Al_xCoCrFeNi$ high-entropy alloys with improved corrosion resistance. *Corros. Sci.* **2018**, *133*, 120–131. [CrossRef]

25. Shi, Y.; Yang, B.; Xie, X.; Brechtl, J.; Dahmen, K.A.; Liaw, P.K. Corrosion of $Al_xCoCrFeNi$ high-entropy alloys: Al-content and potential scan-rate dependentdepedent pitting behavior. *Corros. Sci.* **2017**, *119*, 33–45. [CrossRef]
26. Shi, Y.; Yang, B.; Liaw, P.K. Corrosion-resistant high-entropy alloys: A review. *Metals* **2017**, *7*, 43. [CrossRef]
27. Chuang, M.H.; Tsai, M.H.; Wang, W.R.; Lin, S.J.; Yeh, J.W. Microstructure and wear behavior of $Al_xCo_{1.5}CrFeNi_{1.5}Ti_y$ high-entropy alloys. *Acta Mater.* **2011**, *59*, 6308–6317. [CrossRef]
28. Chen, M.C.; Shi, X.H.; Yang, H.; Liaw, P.K.; Gao, M.C.; Hawk, J.A.; Qiao, J. Wear behavior of $Al_{0.6}CoCrFeNi$ high-entropy alloys: Effect of environments. *J. Mater. Res.* **2018**, *33*, 3310–3320. [CrossRef]
29. Feng, X.; Tang, G.; Ma, X.; Sun, M.; Wang, L. Characteristics of multi-element (ZrTaNbTiW)N films prepared by magnetron sputtering and plasma based ion implantation. *Nucl. Instrum. Methods Phys. Res. Sect. B* **2013**, *301*, 29–35. [CrossRef]
30. Cheng, K.H.; Tsai, C.W.; Lin, S.J.; Yeh, J.W. Effects of silicon content on the structure and mechanical properties of (AlCrTaTiZr)-Si_x-N coatings by reactive RF magnetron sputtering. *J. Phys. D Appl. Phys.* **2011**, *44*, 205405. [CrossRef]
31. Tsai, D.C.; Deng, M.J.; Chang, Z.C.; Kuo, B.H.; Chen, E.C.; Chang, S.Y.; Shieu, F.S. Oxidation resistance and characterization of (AlCrMoTaTi)-Si_x-N coating deposited via magnetron sputtering. *J. Alloys Compd.* **2015**, *647*, 179–188. [CrossRef]
32. Oliver, W.C.; Pharr, G.M. An improved technique for determining hardness and elastic modulus using load and displacement sensing indentation experiments. *J. Mater. Res.* **1992**, *7*, 1564–1583. [CrossRef]
33. Qiu, Y.; Hu, Y.J.; Taylor, A.; Styles, M.J.; Marceau, R.K.W.; Ceguerra, A.V.; Gibson, M.A.; Liu, Z.K.; Fraser, H.L.; Birbilis, N. A lightweight single-phase AlTiVCr compositionally complex alloy. *Acta Mater.* **2017**, *123*, 115–124. [CrossRef]
34. Yurchenko, N.Y.; Stepanov, N.D.; Zherebtsov, S.V.; Tikhonovsky, M.A.; Salishchev, G.A. Structure and mechanical properties of B2 ordered refractory $AlNbTiVZr_x$, (x = 0–1.5) high-entropy alloys. *Mater. Sci. Eng. A* **2017**, *704*, 82–90. [CrossRef]
35. Huang, P.K.; Yeh, J.W. Effects of nitrogen content on structure and mechanical properties of multi-element (AlCrNbSiTiV)N coating. *Surf. Coat. Technol.* **2009**, *203*, 1891–1896. [CrossRef]
36. Huang, P.K.; Yeh, J.W. Effects of substrate temperature and post-annealing on microstructure and properties of (AlCrNbSiTiV)N coatings. *Thin Solid Films* **2009**, *518*, 180–184. [CrossRef]
37. Kim, K.H.; Choi, S.R.; Yoon, S.Y. Superhard Ti-Si-N coatings by a hybrid system ofsystemof arc ion plating and sputtering techniques. *Surf. Coat. Technol.* **2002**, *161*, 243–248. [CrossRef]
38. Pilloud, D.; Pierson, J.F.; Marques, A.P.; Cavaleiro, A. Structural changes in Zr-Si-N films vs. their silicon content. *Surf. Coat. Technol.* **2004**, *180*, 352–356. [CrossRef]
39. Lin, S.Y.; Chang, S.Y.; Chang, C.J.; Huang, Y.C. Nanomechanical properties and deformation behaviors of multi-component $(AlCrTaTiZr)N_xSi_y$ high-entropy coatings. *Entropy* **2014**, *16*, 405–417. [CrossRef]
40. Prochazka, J.; Karvánkova, P.; Veprek-Heijman, M.G.J.; Veprek, S. Conditions required for achieving superhardness of $\geq$ 45 GPa in nc-TiN/a-Si_3N_4 nanocomposites. *Mater. Sci. Eng. A* **2004**, *384*, 102–116. [CrossRef]
41. He, J.L.; Chen, C.K.; Hon, M.H. Micro structure and properties of TiSiN films prepared by plasma-enhanced chemical vapor deposition. *Mater. Chem. Phys.* **1996**, *44*, 9–16. [CrossRef]
42. Liu, W.H.; Wu, Y.; He, J.Y.; Nieh, T.G.; Lu, Z.P. Grain growth and the Hall-Petch relationship in a high-entropy FeCrNiCoMn alloy. *Scr. Mater.* **2013**, *68*, 526–529. [CrossRef]
43. Veprek, S.; Zhang, R.F.; Veprek-Heijman, M.G.J.; Sheng, S.H.; Argon, A.S. Superhard nanocomposites: Origin of hardness enhanc ement, properties and applications. *Surf. Coat. Technol.* **2010**, *2044*, 1899–1906.
44. Patscheider, J.; Zehnder, T.; Diserens, M. Structure-performance relations in nanocomposite coatings. *Surf. Coat. Technol.* **2001**, *146*, 201–208. [CrossRef]

entropy

MDPI

Article

Wear and Corrosion Resistance of $Al_{0.5}$CoCrCuFeNi High-Entropy Alloy Coating Deposited on AZ91D Magnesium Alloy by Laser Cladding

Kaijin Huang [1,2,3,4,*], Lin Chen [2], Xin Lin [3], Haisong Huang [4], Shihao Tang [4] and Feilong Du [4]

1 State Key Laboratory of Materials Processing and Die & Mould Technology, Huazhong University of Science and Technology, Wuhan 430074, China
2 Material Corrosion and Protection Key Laboratory of Sichuan Province, Sichuan University of Science & Engineering, Zigong 643000, China; chenlin3615@163.com
3 State Key Laboratory of Solidification Processing, Northwestern Polytechnic University, Xi'an 710072, China; xlin@nwpu.edu.cn
4 Key Laboratory of Advanced Manufacturing Technology, Ministry of Education, Guizhou University, Guiyang 550025, China; hshuang@gzu.edu.cn (H.H.); shtang@gzu.edu.cn (S.T.); fldu@gzu.edu.cn (F.D.)
* Correspondence: huangkaijin@hust.edu.cn; Tel.: +86-136-0715-0408

Received: 15 November 2018; Accepted: 28 November 2018; Published: 30 November 2018

Abstract: In order to improve the wear and corrosion resistance of an AZ91D magnesium alloy substrate, an $Al_{0.5}$CoCrCuFeNi high-entropy alloy coating was successfully prepared on an AZ91D magnesium alloy surface by laser cladding using mixed elemental powders. Optical microscopy (OM), scanning electron microscopy (SEM), and X-ray diffraction were used to characterize the microstructure of the coating. The wear resistance and corrosion resistance of the coating were evaluated by dry sliding wear and potentiodynamic polarization curve test methods, respectively. The results show that the coating was composed of a simple FCC solid solution phase with a microhardness about 3.7 times higher than that of the AZ91D matrix and even higher than that of the same high-entropy alloy prepared by an arc melting method. The coating had better wear resistance than the AZ91D matrix, and the wear rate was about 2.5 times lower than that of the AZ91D matrix. Moreover, the main wear mechanisms of the coating and the AZ91D matrix were different. The former was abrasive wear and the latter was adhesive wear. The corrosion resistance of the coating was also better than that of the AZ91D matrix because the corrosion potential of the former was more positive and the corrosion current was smaller.

Keywords: laser cladding; high-entropy alloy coating; AZ91D magnesium alloy; wear; corrosion

1. Introduction

High-entropy alloys are a new kind of alloy with excellent properties such as good wear resistance, excellent corrosion resistance, excellent oxidation resistance, low electrical conductivity, low thermal conductivity, and low coefficient of thermal expansion; they were invented by Yeh in 1995 [1–7]. They are composed of five or more elements in the same or an approximately equal molar ratio and have simple BCC and/or FCC solid solution phases. Their microstructure and properties are different from those of traditional alloys such as Fe- and Ni-based alloys or intermetallic compounds such as Ti–Al, Ni–Al, and Fe–Al compounds; this is because of the high entropy effect, severe lattice distortion effect, sluggish diffusion effect, and cocktail effect of the former [5,7].

So far, many methods such as arc melting [8], Tungsten Inert Gas Arc Welding (TIG) [9], Gas Tungsten Arc Welding (GTAW) [10], mechanical alloying [11], DC sputtering [12], thermal spay technology [13], and laser cladding [14–32] have been adopted to prepare different high-entropy alloys or coatings. Among them, the laser cladding method has attracted special attention because of its advantages of rapid heating, rapid cooling, compact coating, and low dilution rate, etc. [19]. At present, the preparation of high-entropy alloy coatings by laser cladding mainly focuses on the microstructure and properties of different matrix materials such as Ti-6Al-4V [14,15], carbon steel [16–21], stainless steel [22–24], tool steel [25], die steel [26], aluminum [27], magnesium [28–30], and magnesium alloys [31,32], and different high-entropy coatings [14–32]. It should be noted that the study of the preparation of high-entropy alloy coatings on magnesium and magnesium alloys by laser cladding was mainly carried out by a few researchers: Yue [28,29,31], Huang [32], and Meng [30]. The possible reason for this is that good-quality laser-clad coatings on magnesium and its alloys are very difficult to obtain because of its low melting point (922K) and low boiling point (1363K).

It is well known that magnesium and its alloys have been widely used in many fields such as the automotive, communication, and aerospace industries, but their poor wear resistance and corrosion resistance hinder their application in those situations where these properties are required. To solve this problem, a method called laser cladding has been developed to improve the wear and corrosion resistance of magnesium and its alloys, and numerous literature reports have confirmed this result [33,34]. Based on the excellent wear resistance and corrosion resistance of high-entropy alloys [1–7] and on the basis of an author's previous research [32], we decided to continue the preparation of new wear-resistant and corrosion-resistant high-entropy alloy coatings by laser cladding on an AZ91D magnesium alloy substrate, which is widely used. To our best knowledge, no studies have reported an $Al_{0.5}CoCrCuFeNi$ high-entropy alloy coating fabricated by laser cladding on an AZ91D magnesium alloy substrate. The reason for studying the $Al_{0.5}CoCrCuFeNi$ alloy is that it has a simple FCC solid solution phase and exhibits excellent wear resistance due to its large work-hardening capacity [35].

In this paper, mixed powders of Cu, Ni, Al, Co, Cr, and Fe were used to fabricate an $Al_{0.5}CoCrCuFeNi$ high-entropy alloy coating on AZ91D magnesium alloys by laser cladding. The microstructure, wear behavior, and corrosion behavior of the coating are shown in detail.

2. Experimental

Laser cladding of mixed powders of Cu, Ni, Al, Co, Cr, and Fe was undertaken on an AZ91D magnesium alloy (Guangdong Dongguan Jubao Magnesium Alloy Material Co., Ltd, Dongguan, China). The sample size used in the laser cladding was 100 mm $\times$ 50 mm $\times$ 10 mm. The powders (Hebei Xingtai Nangong Zhongzhou Alloy Material Co., Ltd, Xingtai, China) had particle sizes of 48–75 μm and purity of 99 wt %. The different element powders were mixed by ball milling (DECO-PBM-V-0.4L, Hunan Changsha Deco Instrument Equipment Co., Ltd, Changsha, China) according to the nominal composition of the $Al_{0.5}CoCrCuFeNi$ high-entropy alloy. The mixed powder from the ball milling was preset on the surface of the AZ91D magnesium alloy using 4 vol % PVA solution. The thickness of the preset layer was about 0.6 mm. The laser cladding experiment was completed under the protection of 99.999% high-purity argon by using a TR050 type CO_2 high-power laser. The optimized laser cladding parameters were as follows: laser power was 3000 W, laser scanning speed was 10 mm/s, laser spot diameter was 4 mm, and laser spot overlap rate was 25%.

The phase structure and microstructure of the coating after laser cladding were identified and observed using an X-ray diffractometer (X' Pert PRO), an optical microscope (Axiovert 200MAT), and a scanning microscope (Quanta 400) with an energy spectrum, respectively. The microhardness of the coating cross section was measured using a microhardness tester (MICROMET 3). The loading force was 100 grams and the loading time was 15 s.

The wear resistance of the coating was evaluated by the dry sliding wear method, in which the size of the sample used for testing was 10 mm $\times$ 10 mm $\times$ 10 mm, and the friction pair used for matching was bearing steel (AISI52100, $HV_{0.1}700$). Figure 1 shows a schematic illustration of the block-on-ring sliding wear tester. The parameters for the dry sliding wear were as follows: the applied load was 98 N, the dry sliding speed was 0.4187 m/s, the dry sliding wear time was 75 mins, and the dry sliding wear distance was 1884 m. The AZ91D magnesium alloy was selected as the experimental material for the dry sliding wear comparison. For each experimental datapoint, the average value of three experimental results was taken as the final data. Wear weight loss was measured using an electronic analytical balance (Bartorius BS110) with an accuracy of 0.1 mg.

The corrosion resistance of the coating was evaluated by testing the corrosion potential polarization curve of a 3.5 wt % sodium chloride solution. A saturated calomel electrode (SCE) was used as the reference electrode and platinum was used as the counter electrode. The test was performed from −2.5 V to 1.5 V with a scanning speed of 1 mV/s.

The surface morphologies of the worn and corroded specimens were observed using a scanning microscope (Quanta 400) with an energy spectrum.

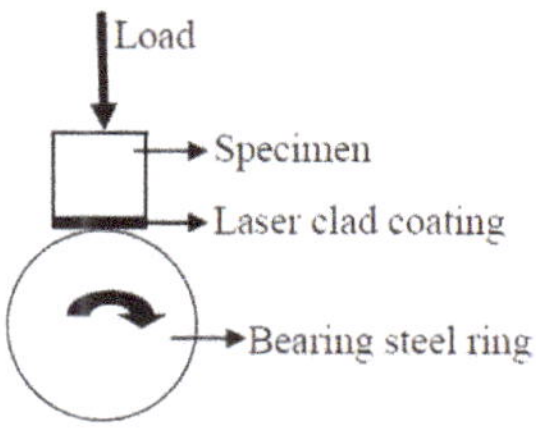

Figure 1. Schematic illustration of the block-on-ring sliding wear tester.

3. Results and Discussion

3.1. Microstructure and Microhardness

Figure 2a shows an optical image of the cross section of the coating/AZ91D substrate sample after laser cladding. It can be seen from Figure 2a that there were basically no defects such as microcracks or pores in the coating. The interface between coating and substrate indicates that they were well combined. In addition, small pieces of AZ91D magnesium alloy substrate were also found in the coating (Figure 2a). This is believed to be due to some partially melted Mg being detached from the substrate and becoming trapped within the rapidly solidifying coating. Figure 2b shows the scanning electron microscopy morphology of position A in Figure 2a. It can be seen from Figure 2b that the coating has typical dendrite microstructure characteristics, and the chemical compositions of the dendrite (DR) and inter-dendrite (ID) are given in Table 1. As shown in Table 1, copper segregates significantly in the inter-dendrite region. In other words, local segregation of copper occurred in the $Al_{0.5}CoCrCuFeNi$ high-entropy alloy coating prepared by laser cladding. These results are similar to the microstructure characteristics of $Al_{0.5}CoCrCuFeNi$ high-entropy alloys prepared by arc melting in the literature [35–38]. The difference is that the dendrite microstructure of the laser cladding coating is fine due to the rapid heating, melting, and rapid solidification of the preset layer during laser cladding.

Table 1. EDS results of the laser-clad high-entropy alloy coating (atom %).

Elements	Al	Co	Cr	Cu	Fe	Ni	Results
DR	9.9	18.7	20.1	12.1	22.0	17.2	poor in Cu element
ID	12.7	5.7	3.5	60.7	3.1	14.3	rich in Cu element
Raw powders	9.0	18.2	18.2	18.2	18.2	18.2	

The copper segregation in the $Al_{0.5}CoCrCuFeNi$ high-entropy alloy can be explained by the mixing enthalpies [39] of different atomic pairs in Table 2. As shown in Table 2, all of the mixing enthalpies of copper with iron, chromium, cobalt, and nickel are positive. This means that copper has a low affinity for these atoms and is easily repelled by them. In the process of laser cladding, different elements in the $Al_{0.5}CoCrCuFeNi$ high-entropy alloy were mixed evenly due to the convection and agitation of the laser cladding pool. However, during the cooling and solidification stage, copper was excluded from the dendrites due to the low affinity of copper atoms with iron, chromium, cobalt, and nickel; thus, copper segregation occurred at the inter-dendrites.

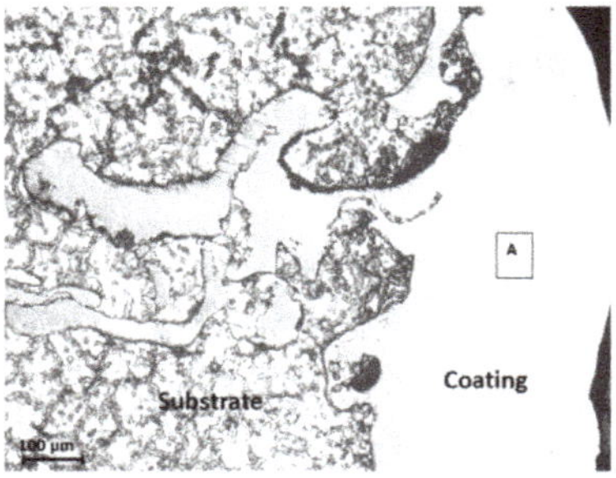

(a) OM image

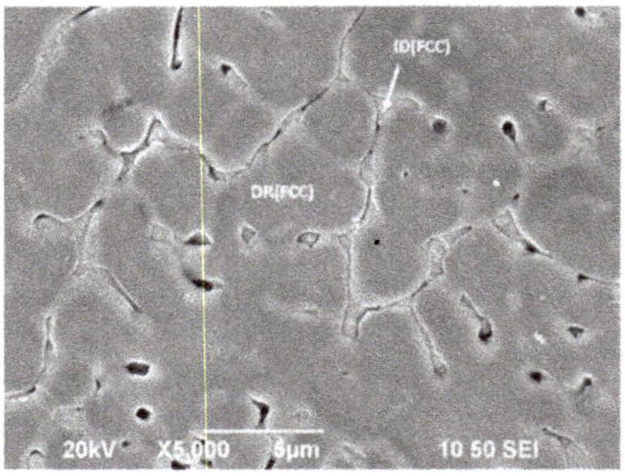

(b) SEM image in position A

Figure 2. Morphology of a laser-clad $Al_{0.5}$CoCrCuFeNi high-entropy alloy coating: (**a**) optical microscope (OM) image; (**b**) SEM image in position A.

Table 2. Values of ΔH_{ij}^{mix} (kJ/mol) for atomic pairs of elements [39].

Mixing Enthalpy	Al	Co	Cr	Cu	Fe	Ni
Al	-	−19	−10	−1	−11	−22
Co		-	−4	6	−1	0
Cr			-	12	−1	−7
Cu				-	13	4
Fe					-	−2
Ni						-

Figure 3 presents the XRD patterns of the $Al_{0.5}$CoCrCuFeNi high-entropy alloy coating prepared by laser cladding. The analytical result confirmed that the phase was a simple FCC solid solution in the laser-clad coating. This result proved that the $Al_{0.5}$CoCrCuFeNi high-entropy alloy coating was successfully fabricated by laser cladding on the AZ91D magnesium alloy based on the mixed powders of Cu, Ni, Al, Co, Cr, and Fe of $Al_{0.5}$CoCrCuFeNi and the optimized laser cladding parameters. This result is consistent with the results reported in the literature [35–38]. Thus, for the laser-clad $Al_{0.5}$CoCrCuFeNi high-entropy alloy coating, both the dendrites and Cu-rich inter-dendrites were of one simple FCC phase (Figure 1b). The formation of a simple solid solution phase rather than intermetallic compounds is mainly attributed to the significant lowering of free energy by the high enthalpy of mixing [40].

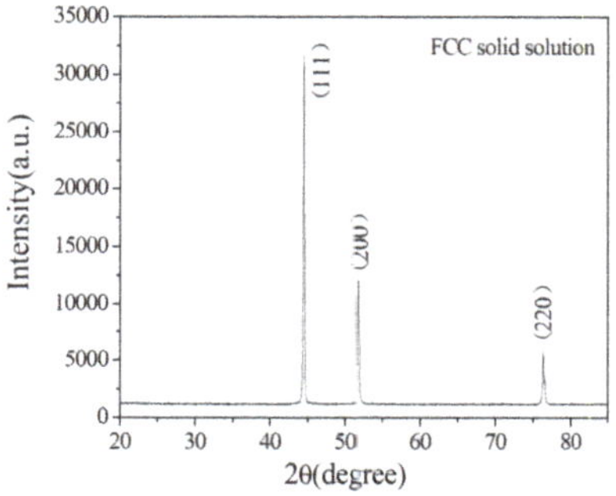

Figure 3. XRD patterns of a laser-clad $Al_{0.5}$CoCrCuFeNi high-entropy alloy coating.

Figure 4 shows the microhardness of the laser-clad $Al_{0.5}CoCrCuFeNi$ high-entropy alloy coating. As can be seen from Figure 4, the microhardness of the coating was $HV_{0.1}365$—about 3.7 times of the AZ91D matrix ($HV_{0.1}98$) and higher than that of the $Al_{0.5}CoCrCuFeNi$ high-entropy alloy (HV_5233) prepared by arc melting [35–37]. The high microhardness of the laser-clad coating was attributed to the solid solution strengthening of different alloy elements and fine grain strengthening of the fine dendritic structure during the cooling process. Obviously, the high microhardness of the coating is more beneficial to increasing the wear resistance of the coating.

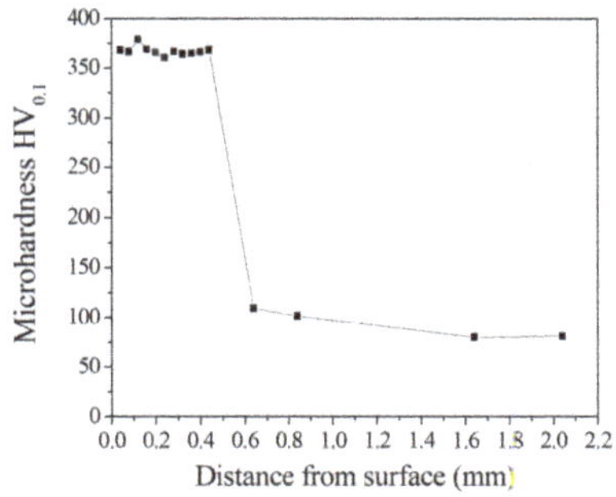

Figure 4. Microhardness of a laser-clad $Al_{0.5}CoCrCuFeNi$ high-entropy alloy coating.

3.2. Wear Properties

Figure 5 shows the dry sliding wear weight loss of different samples. According to Figure 5, the wear resistance of the coating was better than that of the AZ91D matrix, and the wear weight loss was about 2.5 times smaller than that of the AZ91D matrix. This is in accordance with their corresponding microhardness values (Figure 4). In other words, the hardness is higher and the wearability is better. Figure 6 shows the surface morphologies of different samples after dry sliding wear. In Figure 6, both of them present obvious groove characteristics. However, the wear mechanisms of the two samples were also significantly different. The coating showed obvious abrasive wear characteristics, and the AZ91D matrix showed obvious adhesive wear characteristics. Some elongated pockmarks and microcracks (Figure 6a) can be found on the worn surface of the AZ91D matrix. Table 3 shows the EDS results of different positions of different worn samples. It can be seen from Table 3 that AZ91D matrix position 1 and coating position 1 are each matrix materials, although each has a small amount of oxidation due to friction heating. Both AZ91D matrix position 2 and coating position 2 are iron oxide particles, but there are a lot of fine particles on the worn surface of the coating, while the worn surface of the AZ91D matrix is much cleaner.

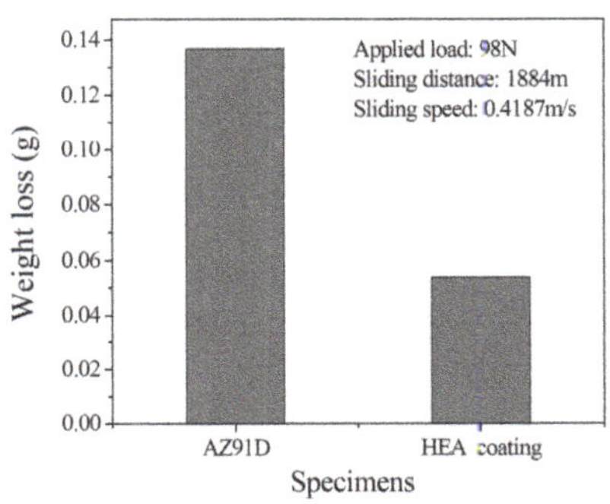

Figure 5. Weight loss of different specimens.

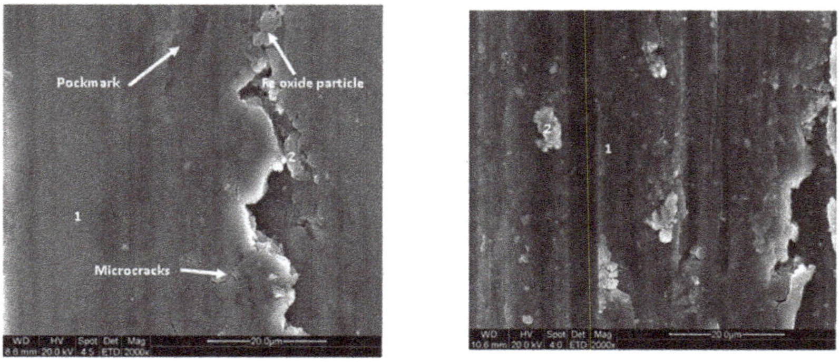

(**a**) AZ91D matrix specimen (**b**) Laser-clad specimen

Figure 6. Worn surfaces of (**a**) an AZ91D matrix specimen and (**b**) a laser-clad specimen.

Table 3. EDS results of worn surfaces of the AZ91D matrix and the laser-clad specimen (wt %).

Samples	Position	C	O	Mg	Al	Zn	Fe	Ni	Cu	Cr	Co	Mn
AZ91D	1	2.2	6.5	83.8	6.8	0.7	-	-	-	-	-	-
AZ91D	2	1.8	44.5	1.3	1.3	0.3	50.8	-	-	-	-	-
Coating	1	1.1	8.9	-	4.7	-	18.7	19.9	12.4	17.3	17.0	-
Coating	2	1.2	33.7	-	3.4	-	34.8	5.3	7.2	7.1	7.3	-
AZ91D	Raw	-	-	90.2	8.9	0.6	-	-	-	-	-	0.3

The causes of the above phenomena should be related to the hardness of the respective matrix materials. Figure 4 shows that the microhardness of the coating was about $HV_{0.1}365$, while the microhardness of the AZ91D matrix was only $HV_{0.1}98$. In addition, the coating is composed of $Al_{0.5}CoCrCuFeNi$ high-entropy alloy, and $Al_{0.5}CoCrCuFeNi$ high-entropy alloy has superior work-hardening characteristics [35], but the AZ91D matrix does not have the same excellent work-hardening capability. As a result, the microhardness of the coating continued to increase when dry sliding wear was applied to the bearing steel friction pair and may even approach or exceed the hardness of the bearing steel friction pair; it would then shear off the steel ring material, and iron wear debris would be smeared over the coating surface. With the extension of the dry sliding wear time, the iron wear debris would be oxidized due to heat generated by friction. Therefore, iron oxide abrasive particles will be used as an abrasive cutting the $Al_{0.5}CoCrCuFeNi$ high-entropy alloy coating. In contrast, the AZ91D matrix did not have this process and only the soft AZ91D matrix was transferred to the hard bearing steel surface, so its wear rate (30.38 μg/s) was much higher than that (11.93 μg/s) of the coating.

3.3. Corrosion Properties

Figure 7 and Table 4 show the potentiodynamic polarization curves and corrosion properties, respectively, of different samples in 3.5 wt % sodium chloride solution. According to the corrosion electrochemical behavior of the two samples and the lack of a passivating region in Figure 7, both the AZ91D matrix and the coating are active dissolved materials. Therefore, using the principle of "the smaller the corrosion current and the higher the corrosion potential, the better the corrosion resistance of the material" to evaluate the corrosion resistance of the active dissolved material, it can be seen that the corrosion resistance of the $Al_{0.5}CoCrCuFeNi$ high-entropy alloy coating prepared by laser cladding was better than that of the AZ91D matrix. This is reflected by the fact that the former had a significantly higher corrosion potential (E_{corr} = −0.998 V) than did the latter (E_{corr} = −1.46 V), and the former also exhibited a lower corrosion current (i_{corr} = 1.60 × 10^{-4} A/cm^2) than did the latter (i_{corr} = 6.20 × 10^{-4} A/cm^2).

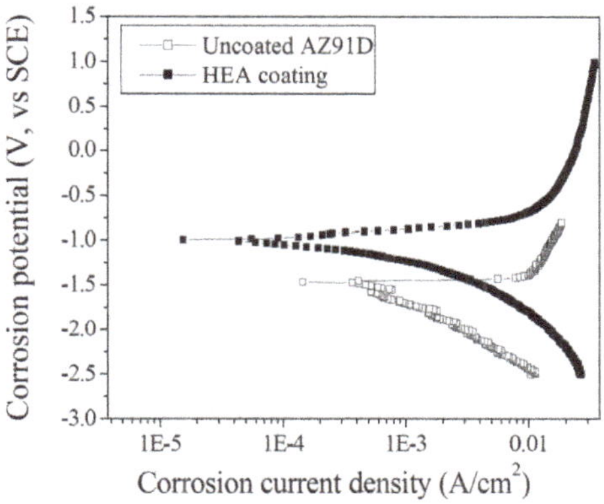

Figure 7. Potentiodynamic polarization curves of different specimens in 3.5 wt % sodium chloride solution.

Table 4. Corrosion parameters of different samples in 3.5 wt % sodium chloride solution.

Specimen	E_{corr} (V)	I_{corr} (A/cm^2)
AZ91D	−1.46	6.20×10^{-4}
Laser-clad coating	−0.998	1.60×10^{-4}

Figure 8 shows the corroded surfaces of two samples. It can be seen from Figure 8a that the magnesium oxide film generated on the AZ91D matrix cannot provide effective protection due to its loose and porous properties and the extremely low equilibrium potential (−2.37 V) of magnesium in the AZ91D matrix (Table 3). The following reactions will occur [41]:

$$Mg(s) + H_2O \rightarrow Mg(OH)_2(s) + H_2(g),$$

$$Mg(OH)_2(s) + Cl^- \rightarrow MgCl_2 + OH^-,$$

$$Mg(s) + Cl^- (aq) \rightarrow MgCl_2.$$

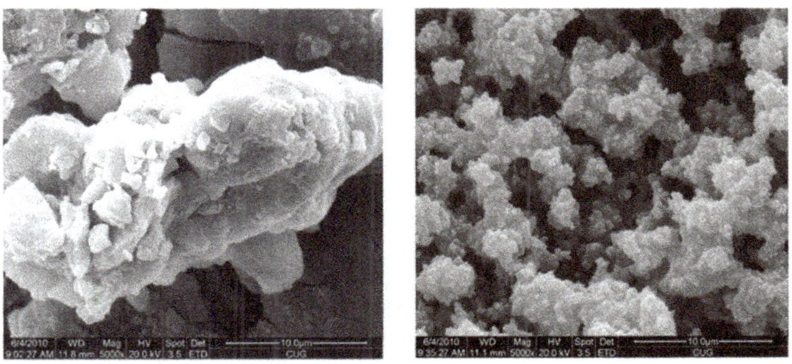

(**a**) AZ91D matrix specimen (**b**) Laser-clad specimen

Figure 8. Corroded surfaces of (**a**) an AZ91D matrix specimen and (**b**) a laser-clad specimen.

Therefore, the corrosion of the AZ91D matrix was severe (Table 4 and Figure 8a).

On the other hand, the corrosion attack on the $Al_{0.5}$CoCrCuFeNi high-entropy alloy coating specimen was much less acute. Figure 8b reveals that the corroded surface has a grainy appearance, indicating that corrosion attacks may mainly occur on dendrite boundaries. It is considered that the presence of Cr, Al, and Ni in the $Al_{0.5}$CoCrCuFeNi high-entropy alloy coating prepared by laser cladding could form dense chromium oxides, aluminum oxides, and nickel oxides at the surface, and this could be an important factor that contributes to the relatively high corrosion resistance of the $Al_{0.5}$CoCrCuFeNi high-entropy alloy coating specimen.

It should be pointed out that the composition of the high-entropy $Al_{0.5}$CoCrCuFeNi selected in this paper is not superior to that of other high-entropy compositions (for example, the one studied in Reference [32]). In this paper, only the high-entropy $Al_{0.5}$CoCrCuFeNi composition is chosen because of its good corrosion resistance.

As for how to choose the composition of the coating to obtain high corrosion resistance, we think that the composition of the high-entropy alloy should be based mainly on whether the same high-entropy alloy prepared by the arc melting method has excellent corrosion performance. If so, it can be selected as a candidate material for laser cladding to prepare a high-entropy alloy coating; otherwise, it should not be selected.

4. Conclusions

(1). An $Al_{0.5}$CoCrCuFeNi high-entropy alloy coating was successfully fabricated on an AZ91D matrix by laser cladding using mixed powders of Cu, Ni, Al, Co, Cr, and Fe. The $Al_{0.5}$CoCrCuFeNi high-entropy alloy coating consisted of fine dendrites with a simple FCC phase.

(2). The dry sliding wear resistance of the $Al_{0.5}$CoCrCuFeNi high-entropy alloy coating prepared by laser cladding was better than that of the AZ91D matrix, and the wear mechanisms of the two materials were different.

(3). The $Al_{0.5}$CoCrCuFeNi high-entropy alloy coating prepared by laser cladding had better corrosion resistance than did the AZ91D matrix in 3.5 wt % sodium chloride solution.

Author Contributions: Investigation—K.H., L.C., H.H., S.T., F.D.; writing—original draft preparation, L.C.; writing—review and editing, K.H.

Funding: This research received no external funding.

Acknowledgments: This work was supported by the Opening Project (2018CL18) of Material Corrosion and Protection Key Laboratory of Sichuan Province, Sichuan University of Science & Engineering, the Open Research Fund Program (XDKFJJ [2016]06) of Key Laboratory of Advanced Manufacturing Technology, Ministry of Education, Guizhou University, and the Fund (SKLSP201733) of the State Key Laboratory of Solidification Processing in NWPU. The authors are also grateful to the Analytical and Testing Center of Huazhong University of Science and Technology.

Conflicts of Interest: The authors declare no conflict of interest.

References

1. Zhang, W.R.; Liaw, P.K.; Zhang, Y. Science and technology in high-entropy alloys. *Sci. China-Mater.* **2018**, *61*, 2–22. [CrossRef]
2. Liu, W.; Liu, P.; Liaw, P.K. Microstructure and properties of high-entropy alloy films and coatings: A review. *Mater. Res. Lett.* **2018**, *6*, 199–229. [CrossRef]
3. Macdonald, B.E.; Fu, Z.; Zheng, B.; Chen, W.; Lin, Y.; Chen, F.; Zhang, L.; Ivanisenko, J.; Zhou, Y.; Hahn, H. Recent progress in high entropy alloy research. *JOM.* **2017**, *69*, 2024–2031. [CrossRef]
4. Shi, Y.Z.; Yang, B.; Liaw, P.K. Corrosion-resistant high-entropy alloys: A review. *Metals.* **2017**, 7. [CrossRef]
5. Zhang, Y.; Zuo, T.T.; Tang, Z.; Gao, M.C.; Dahmen, K.A.; Liaw, P.K.; Lu, Z.P. Microstructure and properties of high-entropy alloys. *Prog. Mater. Sci.* **2014**, *61*, 1–93. [CrossRef]
6. Tsai, M.H.; Yeh, J.W. High-entropy alloys: A critical review. *Mater. Res. Lett.* **2014**, *2*, 107–123. [CrossRef]
7. Yeh, J.W. Recent progress in high-entropy alloys. *Ann. Chim. Sci. Mat.* **2006**, *31*, 633–648. [CrossRef]

8. Chen, M.R.; Lin, S.J.; Yeh, J.W.; Chen, S.K.; Huang, Y.S.; Chuang, M.H. Effect of vanadium addition on the microstructure, hardness, and wear resistance of $Al_{0.5}CoCrCuFeNi$ high-entropy alloy. *Metall. Mater. Trans. A* **2006**, *37*, 1363–1369. [CrossRef]
9. Huo, W.Y.; Shi, H.F.; Ren, X.; Zhang, J.Y. Microstructure and wear behavior of CoCrFeMnNbNi high entropy alloy coating by TIG cladding. *Adv. Mater. Sci. Eng.* **2015**, *2015*. [CrossRef]
10. Abed, H.; Ghaini, F.M.; Shahverdi, H.R. Characterization of Fe49Cr18Mo7B16C4Nb6 high entropy hardfacing layers produced by gas tungsten arc welding (GTAW) process. *Surf. Coat. Technol.* **2018**, *352*, 360–369. [CrossRef]
11. Varalakshmi, S.; Kamaraj, M.; Murty, B.S. Synthesis and characterization of nanocrystalline AlFeTiCrZnCu high entropy solid solution by mechanical alloy. *J. Alloys Compd.* **2008**, *460*, 253–257. [CrossRef]
12. Chen, T.K.; Wong, M.S.; Shun, T.T.; Yeh, J.W. Nanostructured nitride films of multi-element high entropy alloys by reactive DC sputtering. *Surf. Coat. Technol.* **2005**, *200*, 1361–1365. [CrossRef]
13. Huang, P.K.; Yeh, J.W.; Shun, T.T.; Chen, S.K. Multi-principle-element alloys with improved oxidation and wear resistance for thermal spray coating. *Adv. Eng. Mater.* **2004**, *6*, 74–78. [CrossRef]
14. Huang, C.; Zhang, Y.Z.; Shen, J.Y.; Vilar, R. Thermal stability and oxidation resistance of laser clad TiVCrAlSi high entropy alloy coatings on Ti-6A1-4V alloy. *Surf. Coat. Technol.* **2011**, *206*, 1389–1395. [CrossRef]
15. Huang, C.; Zhang, Y.Z.; Vilar, R.; Shen, J.Y. Dry sliding wear behavior of laser clad TiVCrAlSi high entropy alloy coatings on Ti-6Al-4V substrate. *Mater. Des.* **2012**, *41*, 338–343. [CrossRef]
16. Qiu, X.W. Microstructure, hardness and corrosion resistance of $Al_2CoCrCuFeNiTi_x$ high entropy alloy coatings prepared by rapid solidification. *J. Alloys Compd.* **2018**, *735*, 359–364. [CrossRef]
17. Zhao, J.H.; Ma, A.B.; Ji, X.L.; Jiang, J.H.; Bao, Y.Y. Slurry erosion behavior of $Al_xCoCrFeNiTi_{0.5}$ high-entropy alloy coatings fabricated by laser cladding. *Metals.* **2018**, *8*. [CrossRef]
18. Guo, Y.X.; Liu, Q.B.; Zhou, F. Microstructure and properties of $Fe_5Cr_5SiTiCoNbMoW$ coating by laser cladding. *Surf. Eng.* **2018**, *34*, 283–288. [CrossRef]
19. Qiu, X.W.; Zhang, Y.P.; He, L.; Liu, C.G. Microstructure and corrosion resistance of AlCrFeCuCo high entropy alloy. *J. Alloys Compd.* **2013**, *549*, 195–199. [CrossRef]
20. Ye, X.Y.; Ma, M.X.; Cao, Y.X.L.; Liu, W.J.; Ye, X.H.; Gu, Y. The property research on high-entropy alloy $Al_xFeCoNiCuCr$ coating by laser cladding. *Phys. Procedia.* **2011**, *12*, 303–312. [CrossRef]
21. Zhang, H.; Pan, Y.; He, Y.Z.; Jiao, H.S. Microstructure and properties of 6FeNiCoSiCrAlTi high-entropy alloy coating prepared by laser cladding. *Appl. Surf. Sci.* **2011**, *257*, 2259–2263. [CrossRef]
22. Guo, Y.X.; Shang, X.J.; Liu, Q.B. Microstructure and properties of in-situ TiN reinforced laser cladding $CoCr_2FeNiTi_x$ high entropy alloy composite coatings. *Surf. Coat. Technol.* **2018**, *344*, 353–358. [CrossRef]
23. Jiang, L.; Wu, W.; Cao, Z.Q.; Deng, D.W.; Li, T.J. Microstructure evolution and wear behavior of the laser cladded CoFeNi2V0.5Nb0.75 and CoFeNi2V0.5Nb high entropy alloy coatings. *J. Therm. Spray Technol.* **2016**, *25*, 806–814. [CrossRef]
24. Wu, W.; Jiang, L.; Jiang, H.; Pan, X.M.; Cao, Z.Q.; Deng, D.W.; Wang, T.M. Phase evolution and properties of Al2CrFeNiMox high entropy alloys coatings by laser cladding. *J. Therm. Spray Technol.* **2015**, *24*, 1333–1340. [CrossRef]
25. Shang, X.J.; Liu, Q.B.; Guo, Y.X.; Xu, P.; Zhou, F. Laser cladding of a $Fe_5Cr_5SiTiCoNbMoW$ high entropy alloy coating onto a tool steel substrate. *Laser. Eng.* **2018**, *41*, 305–316.
26. Wang, W.M.; Zhang, B.H.; Liu, B.H.; Liu, C.S.; Cui, J.Z. Laser cladding of high entropy alloy on H13 steel. *Rare Metals.* **2014**, *33*, 727–730.
27. Shi, Y.; Ni, C.; Liu, J.; Huang, G.Z. Microstructure and properties of laser clad high entropy alloy coating on aluminum. *Mater. Sci. Technol.* **2018**, *34*, 1239–1245. [CrossRef]
28. Yue, T.M.; Xie, H.; Lin, X.; Yang, H.O.; Meng, G.H. Microstructure of laser re-melted AlCoCrCuFeNi high entropy alloy coatings produced by plasma spraying. *Entropy* **2013**, *15*, 2833–2845. [CrossRef]
29. Yue, T.M.; Xie, H.; Lin, X.; Yang, H.O.; Meng, G.H. Solidification behavior in laser cladding of AlCoCrCuFeNi high entropy on magnesium substrates. *J. Alloys Compd.* **2014**, *587*, 588–593. [CrossRef]
30. Meng, G.H.; Liu, X.T.; Qiu, J.; Wang, B.; Zhao, K. Growth morphology of $CuMg_2$ in laser forming of AlCoCrCuFeNi coating on Mg. *Adv. Eng. Res.* **2017**, *146*, 162–165.
31. Yue, T.M.; Zhang, H. Laser cladding of $FeCoNiCrAlCu_xSi_{0.5}$ high entropy alloys on AZ31 Mg alloy substrates. *Mater. Res. Innovations.* **2014**, *18*, 624–628. [CrossRef]

32. Huang, K.J.; Lin, X.; Wang, Y.Y.; Xie, C.S.; Yue, T.M. Microstructure and corrosion resistance of $Cu_{0.9}NiAlCoCrFe$ high entropy alloy coating on AZ91D magnesium alloy by laser cladding. *Mater. Res. Innovations.* **2014**, *18*, 1008–1011. [CrossRef]
33. Liu, J.L.; Yu, H.J.; Chen, C.Z.; Weng, F.; Dai, J.J. Research and development status of laser cladding on magnesium alloys: A review. *Opt. Laser. Eng.* **2017**, *93*, 195–210. [CrossRef]
34. Yue, T.M.; Su, Y.P.; Yang, H.O. Laser cladding of $Zr_{65}Al_{7.5}Ni_{10}Cu_{17.5}$ amorphous alloy on magnesium. *Mater. Lett.* **2007**, *61*, 209–212. [CrossRef]
35. Chen, M.R.; Lin, S.J.; Yeh, J.W.; Chen, S.K.; Huang, Y.S.; Chuang, M.H. Effect of vanadium addition on the microstructure, hardness, and wear resistance of $Al_{0.5}CoCrCuFeNi$ high-entropy alloy. *Metall. Mater. Trans. A* **2006**, *37*, 1363–1369. [CrossRef]
36. Wu, J.M.; Lin, S.J.; Yeh, J.W.; Chen, S.K.; Huang, Y.S.; Chen, H.C. Adhesive wear behavior of $Al_xCoCrCuFeNi$ high-entropy alloys as a function of aluminum content. *Wear* **2006**, *261*, 513–519. [CrossRef]
37. Chen, M.R.; Lin, S.J.; Yeh, J.W.; Chen, S.K.; Huang, Y.S.; Tu, C.P. Microstructure and properties of $Al_{0.5}CoCrCuFeNiTi_x$ (x = 0–2.0) high-entropy alloys. *Mater. Trans.* **2006**, *47*, 1395–1401. [CrossRef]
38. Tong, C.J.; Chen, Y.L.; Chen, S.K.; Yeh, J.W.; Shun, T.T.; Tsau, C.H.; Lin, S.J.; Chang, S.Y. Microstructure characterization of $Al_xCoCrCuFeNi$ high-entropy alloy system with multiprincipal elements. *Metall. Mater. Trans. A* **2005**, *36*, 881–893. [CrossRef]
39. Takeuchi, A.; Inoue, A. Classification of bulk metallic glasses by atomic size differences, heat of mixing and period of constituent elements and its application to characterization of the main alloying element. *Mater. Trans.* **2005**, *46*, 2817–2829. [CrossRef]
40. Yeh, J.W.; Chen, S.K.; Lin, S.J.; Gan, J.Y.; Chin, T.S.; Shun, T.T.; Tsau, C.H.; Chang, S.Y. Nanostructured high-entropy alloys with multiple principal elements: Novel alloy design concepts and outcomes. *Adv. Eng. Mater.* **2004**, *6*, 299–303. [CrossRef]
41. Staiger, M.P.; Pietak, A.M.; Huadmai, J.; Dias, G. Magnesium and its alloys as orthopedic biomaterials: A review. *Biomaterials.* **2006**, *27*, 1728–1734. [CrossRef] [PubMed]

entropy

MDPI

Article

Fabrication of AlCoCrFeNi High-Entropy Alloy Coating on an AISI 304 Substrate via a $CoFe_2Ni$ Intermediate Layer

Wenyuan Cui [1,*], Sreekar Karnati [1], Xinchang Zhang [1], Elizabeth Burns [2] and Frank Liou [1]

1 Department of Mechanical and Aerospace Engineering, Missouri University of Science and Technology, Rolla, MO 65409, USA; skw92@mst.edu (S.K.); xz25c@mst.edu (X.Z.); liou@mst.edu (F.L.)

2 Department of Metallurgical Engineering, Missouri University of Science and Technology, Rolla, MO 65409, USA; eabpd5@mst.edu

* Correspondence: wcz68@mst.edu; Tel.: +1-573-341-4603

Received: 30 October 2018; Accepted: 19 December 2018; Published: 20 December 2018

Abstract: Through laser metal deposition, attempts were made to coat AlCoCrFeNi, a high-entropy alloy (HEA), on an AISI 304 stainless steel substrate to integrate their properties. However, the direct coating of the AlCoCrFeNi HEA on the AISI 304 substrate was found to be unviable due to cracks at the interface between these two materials. The difference in compositional change was suspected to be the source of the cracks. Therefore, a new transition route was performed by coating an intermediate layer of $CoFe_2Ni$ on the AISI 304 substrate. Investigations into the microstructure, phase composition, elemental composition and Vickers hardness were carried out in this study. Consistent metallurgical bonding was observed along both of the interfaces. It was found that the AlCoCrFeNi alloy solidified into a dendritic microstructure. The X-ray diffraction pattern revealed a transition of the crystal structure of the AISI 304 substrate to the AlCoCrFeNi HEA. An intermediate step in hardness was observed between the AISI 304 substrate and the AlCoCrFeNi HEA. The AlCoCrFeNi alloy fabricated was found to have an average hardness of 418 HV, while the $CoFe_2Ni$ intermediate layer had an average hardness of 275 HV.

Keywords: high-entropy alloy; laser metal deposition; elemental powder; graded material

1. Introduction

As a novel metallic alloy system, high-entropy alloys (HEAs) have received considerable attention in the past decade. The name HEA indicates that the mixing of the principal elements in the alloy leads to a substantial change in entropy. This change in entropy promotes the formation of a simple solid solution instead of complex compounds. One of the extensively studied HEAs is equiatomic AlCoCrFeNi, which shows high hardness, good wear behavior but low tensile ductility [1–7]. As-cast AlCoCrFeNi alloy showed a tensile elongation of 1.0%, while post-heat treatment, the elongation was increased to 11.7% [4]. Wang et al. studied the compressive properties of AlCrFeCoNi HEA prepared by vacuum arc melting. They found that this alloy showed large strain hardening and compressive strength up to 2004 MPa with a 32.7% compressive plasticity [6]. Munitz et al. reported the impact of heat treatment of AlCoCrFeNi HEA, in which the BCC (Body-centered cubic) matrix transformation occurred between 650 and 975 °C. This transformation led to a substantial increase in microhardness [5]. Further modification of this alloy system through the addition of titanium, leading to $AlCoCrFeNiTi_x$ (x = molar ratios), was found to be promising for wear protection [1]. Further, AlCoCrFeNi HEA solidified with dendritic and interdendritic microstructures due to elemental segregation. Dendritic segregation regions were found to be Al- and Ni-rich, while interdendritic areas were Fe- and Cr-rich, and the distribution of Co was uniform. Body-centered cubic (BCC) Fe and Cr precipitates, and B2 (ordered BCC) Al- and Ni-rich matrices were observed in previous studies [2,4,6,8,9]. Most of these studies are based on material fabricated through processes such as casting and arc melting. Unlike these early studies, laser metal deposition (LMD) was implemented in this study.

LMD is capable of fabricating freeform three-dimensional metallic components [10–12] and has been used to fabricate several HEAs [12–16]. Chen et al. fabricated $Al_xCoFeNiCu_{1-x}$ (x = 0.25, 0.5 and 0.75 atom %, respectively) HEAs using elemental powders on the AISI 304 substrate. They reported an increase in hardness with an increase in aluminum content [16]. He et al. used laser cladding to produce $FeCoCrNiAlTi_x$ (x = 0, 0.25, 0.5, 0.75 and 1 atom %, respectively) coating on Q253 steel through the use of elemental powders. Addition of titanium was observed to improve the hardness and wear resistance of the HEA [15]. Similarly, FeCoCrAlCu HEA coating by laser cladding demonstrated good wear resistance under a dry sliding condition [17].

In this paper, the feasibility of coating an AlCoCrFeNi HEA on an AISI 304 stainless steel substrate was investigated. Sole LMD fabrication of AlCoCrFeNi HEA components is very costly due to the need for high-purity (i.e., 99.9%) raw powders of elements such as Co, Cr and Ni. AISI 304 stainless steel, on the other hand, is a low-cost structural material. However, AISI 304 is a soft material with low wear resistance. It is widely used in industrial facilities, transportation equipment and architectural applications. Therefore, by coating AlCoCrFeNi HEA on AISI 304, it can enhance the hardness of AISI 304 structures. This combination of materials could facilitate fabrication of components for applications that require both hardness and wear resistance.

However, direct coating of AlCoCrFeNi HEA on AISI 304 is difficult due to the change in chemistry, thermal expansion and residual stress of the dissimilar materials. For example, the measured coefficient of thermal expansion (CTE, 10^{-6}/K) for AlCoCrFeNi HEA was 9.03 (293–303 K), 12.47 (368–378 K) and 13.54 (423–773 K) [18]. However, the CTE values of AISI 304 were 14.7 (293 K), 16.3 (400 K), 19.5 (700 K) and 20.2 (800 K) [19]. Harihar et al. observed crack formation at the bottom of an AlCoCrFeNi deposit when deposited on an AISI 304 substrate. Due to the brittleness of the deposited material, the deposit broke off from the AISI 304 substrate easily [12]. An extensive network of cracks occurred when a TiVCrAlSi HEA was cladded on a Ti-6Al-4V substrate. This was attributed to the difference between the thermal expansion coefficients and residual stresses associated with the high cooling rate in laser cladding [20].

Therefore, to facilitate the dissimilar material bond, an intermediate layer was necessary and could accommodate the residual stresses and variation in chemistry change [10,21,22]. Intermediate layers of Fe/Cr/V were used between AISI 316 stainless steeland Ti-6Al-4V to facilitate a similar material bond [10]. Currently, there are few studies available identifying the viable intermediate layer between AlCoCrFeNi HEA and AISI 304. In this study, an attempt was made to coat the equiatomic AlCoCrFeNi HEA on the AISI 304 substrate using LMD. The objective was to obtain a strong bond between the two materials. We first demonstrated the issues with direct-coating the HEA onto the substrate. Then we proposed a candidate intermediate material and proved its viability.

2. Materials and Methods

Elemental powders of gas-atomized aluminum (Al), chromium (Cr), cobalt (Co), nickel (Ni) and iron (Fe) from Atlantic Equipment Engineers Inc. were used as precursor materials. These powders, weighed in required ratios, were mixed using a Turbula mixer (Glen Mills Inc., Clifton, NJ, USA) for 1 h to obtain homogeneous blends. Commercially procured AISI 304 bar stock (dimensions: 2.75 inch × 2 inch × 0.25 inch) was used as the substrate material for the deposition. The particle size distribution of the elemental powders stated by the producer is as tabulated in Table 1. Elemental analysis of the elemental powders is listed in Table 2. Elemental compositions (atom %) of the as-blended $CoFe_2Ni$ intermediate layer and AlCoCrFeNi alloy are given in Table 3.

Table 1. Particle size distribution of the precursor elemental powders.

Materials	US Standard Mesh
Al	−100
Co	−100/+325
Cr	−100
Fe	−100
Ni	−100/+325

Table 2. Elemental analysis (atom %) of elemental powders as provided by the manufacturer.

Materials	Al	Cr	Si	Fe	C	Ni	Co	S	Ca
Al	0.88	-	0.07	0.05	-	-	-	-	-
Co	-	-	-	0.002	-	<0.001	~0.99	-	0.001
Cr	-	0.89	-	0.02	0.09	-	-	-	-
Fe	-	-	-	0.99	0.01	-	-	-	-
Ni	-	-	-	0.01	0.05	0.92	-	0.02	-

Table 3. Nominal compositions (atom %) of $CoFe_2Ni$ and AlCoCrFeNi alloy powder blends.

Alloy	Al	Co	Cr	Fe	Ni
$CoFe_2Ni$	0	25	0	50	25
AlCoCrFeNi	20	20	20	20	20

The laser deposition process was performed in an LMD system whose schematic representation is as seen in Figure 1a. The heat source was a 1 kW continuous-wave YAG fiber laser (IPG Photonics, Oxford, MA, USA) with a 2 mm beam diameter. The powders were fed using a vibration X2 powder feed system procured from Powder Motion Labs. The powder was introduced into the melt pool through an alumina tube. A computer numerical control (CNC) table was used to facilitate the movement during the deposition. Argon gas was used to ensure an inert atmosphere and act as a carrier gas to deliver the powder mixture to the melt pool.

In the current setup, the 2 mm spot size is insufficient to attain a large capture efficiency of the powder. This is due to the scatter of the powder flow out of the powder feed tube. This scatter was suspected to vary with individual precursor powder. Therefore, in order to obtain as-deposited compositions that are close to as-blended compositions, the capture efficiency during the deposition process needed to be increased. A trochoidal toolpath (shown in Figure 1b) was designed to create a large enough melt pool to improve capture efficiency during deposition. This toolpath was inspired by "weave"-style toolpaths that are commonly used in welding.

The AISI 304 substrates were cleaned with acetone to remove the impurities such as dirt and oil from the surface. A preheating scan was conducted by running the laser across the substrate surface. To ensure a successful start, the power of the initial five layers of the deposition was carried out at 750 W and 8.5% (3.36 g/min) powder feed rate. The remainder of the deposit was run at a power level of 550 W and 8.5% (3.36 g/min) powder feed rate. The thickness of each layer is 1 mm.

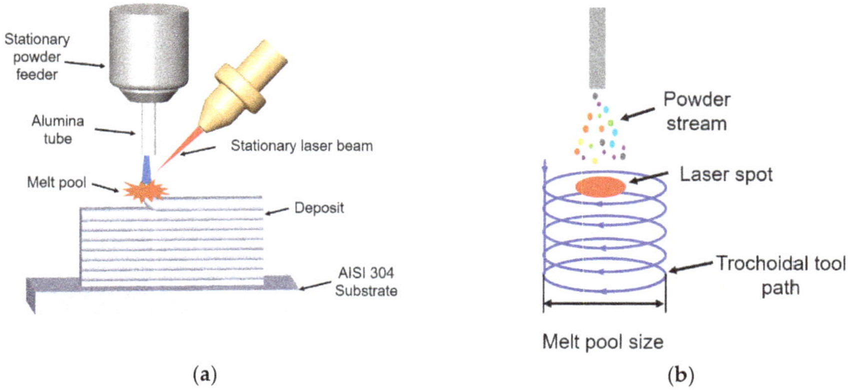

Figure 1. Schematic of the experimental setup, (**a**) laser metal deposition (LMD) system and (**b**) the trochoidal tool path.

After laser deposition, vertical transverse sections of the specimens were cut using a wire electric discharge machine (Hansvedt Industries Inc., Rantoul, IL, USA) and mounted in Bakelite for polishing and etching. The metallographic specimens were first ground using 240, 400, 600 and 800 grit silicon carbide papers and then polished using 15 μm, 9 μm and 3 μm diamond suspensions. The final step of polishing involved 0.05 μm colloidal silica suspension. To reveal the microstructure, the electrolytic etching was carried out in the nitric acid solution (70 mL nitric acid, 30 mL distilled water) at 5 V for 5 seconds. Scanning electron microscopy (SEM), energy dispersive X-ray spectroscopy (EDS) and electron backscatter diffraction (EBSD) were performed on Helios Nanolab 600 SEM (Thermo Fisher Scientific, Waltham, MA, USA). The SEM image was acquired by an Everhart-Thornley detector. The EDS element was analyzed by the factory standardizations provided in the Aztec software. The EBSD step size was selected to be 2.5 μm. EBSD data acquisition and analysis were conducted using Aztec and Channel 5 software, respectively. Grain size was measured by the line intercept method, and the misorientation angle was 10°. Optical microscopy images were collected using a Hirox optical microscope. X-ray diffraction patterns were collected using Philips X'pert MRD using Cu anode. The Vickers hardness was measured using a Struers Duramin hardness tester (Struers Inc., Cleveland, OH, USA) at a 9.8 N load and a 10 s load duration. The reported hardness results were the average of three indentations.

3. Results and Discussions

3.1. Direct Coating of AlCoCrFeNi HEA on AISI 304 Substrate

The direct LMD of the AlCoCrFeNi HEA on the AISI 304 substrate will be discussed first. Figure 2a shows a portion of the vertical transverse section of the HEA deposit near the AISI 304 substrate. An area close to the crack zone, as marked in the dashed-line box, is shown in Figure 2b with high magnification. A network of cracks, mostly transverse and horizontal in orientation, were found to be prevalent. Cracking occurred at the bottom of this HEA deposit. This could be attributed to the mismatch between the thermal expansion coefficients. The CTE of this HEA was reported to be 9.03 (10^{-6}/K, 293–303 K) while the value of AISI 304 was 14.7 (10^{-6}/K, 293 K) [18,19].

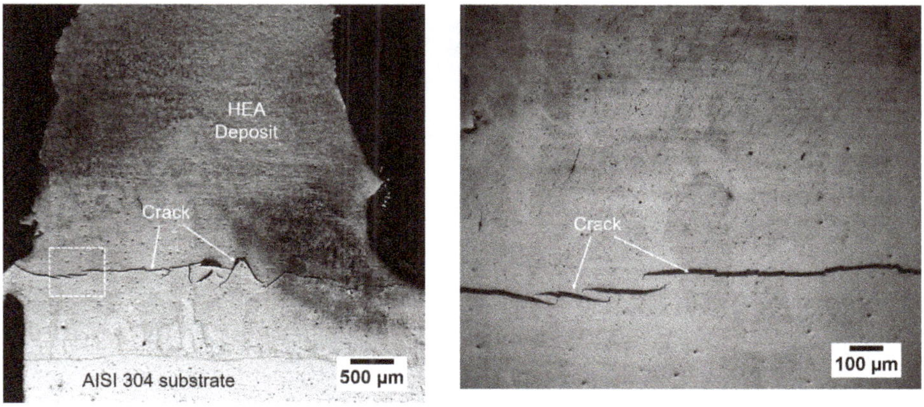

(**a**) (**b**)

Figure 2. (**a**) Optical microscopy image of the vertical transverse cross-section of direct AlCoCrFeNi HEA coating on AISI 304 substrate, (**b**) a high-magnification view of the dashed–line-boxed area in (**a**).

The elemental composition distribution along the interface between the HEA deposit and the AISI 304 substrate is shown in Figure 3. At the bottom of the melted metal, the composition mixing was significant during the laser deposition process (see Figure 3). The bottom of the deposit had high susceptibility of cracking in the transverse cross-section, as seen in Figure 2.

The variation in Vickers hardness across the HEA–AISI 304 direct coating is presented in Figure 4. The average Vickers hardness of the HEA deposits was 412 HV, while that of the substrate was 161 HV. Since the coefficients of thermal expansion are mismatched between HEA and the substrate, residual stresses were developed during the laser deposition process. The AISI 304 substrate had a high elongation rate from 28% to 50% in the temperature range of 300–500 °C [23]. However, the tensile elongations of the AlCoCrFeNi HEA were 1% (as-cast condition) and 11.7% (after heat treatment) [4]. A difference in ductility exists between the substrate and the HEA. Having an intermediate material to bridge these differences was deemed necessary.

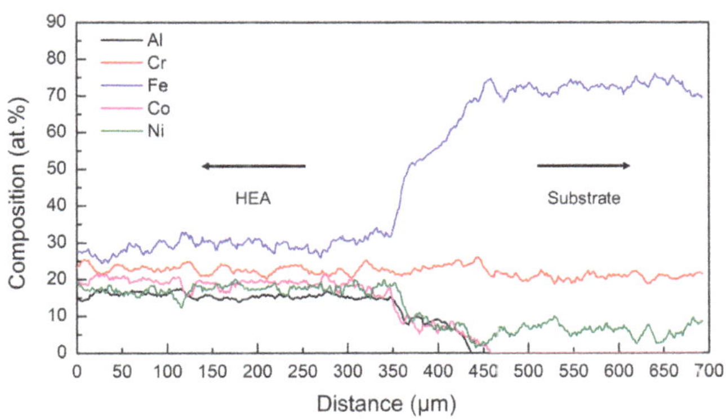

Figure 3. Elemental composition distribution along the interface between the AISI 304 substrate and the HEA deposit.

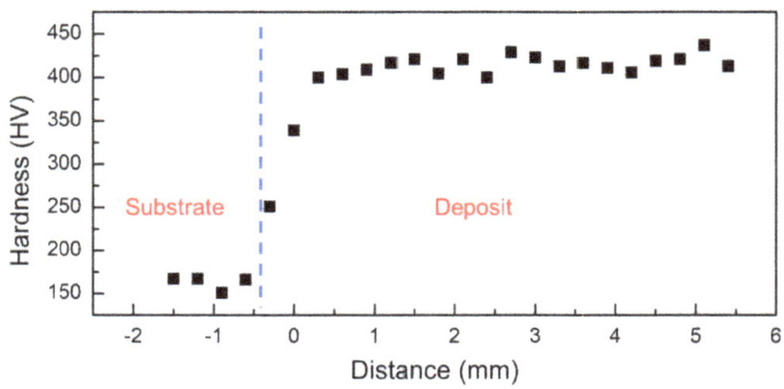

Figure 4. Vickers hardness profile of the direct coating of the AlCoCrFeNi alloy on AISI 304.

3.2. *A New Transition Route*

A blend of Fe, Co and Ni powders was selected as the candidate intermediate material. Since they are among the constituents of the AlCoCrFeNi HEA, no special procurement was needed. A Fe–Co–Ni ternary phase diagram at 1073 K compiled from experimental data is shown in Figure 5 [24]. Fe, Ni and Co have excellent mutual solubility, and no brittle intermetallic phases are expected. From the phase diagram, an atomic composition ratio of Fe, Ni and Co of 50%, 25% and 25%, respectively, was chosen. The selected ratio is expected to bridge the material composition gap between the AlCoCrFeNi HEA and AISI 304. This new transition route, AISI 304 substrate → $CoFe_2Ni$ intermediate layer → AlCoCrFeNi HEA, was then carried out and characterized.

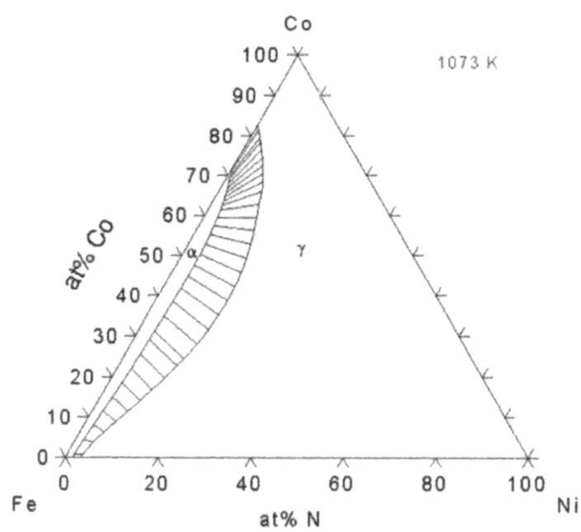

Figure 5. Ternary alloy phase diagram of Fe–Co–Ni at 1073 K [24].

3.3. *AlCoCrFeNi HEA–AISI 304 with an Intermdeiate Layer*

3.3.1. Microstructure

The $CoFe_2Ni$ intermediate layer was coated on the AISI 304 substrate using premixed elemental powder. Then, the AlCoCrFeNi HEA was coated on the intermediate layer by LMD. The intermediate layer composition was theorized to avoid the formation of intermetallic compounds and bridge the large gap in strength differences. Figure 6a,b shows the optical images of etched surfaces of transverse sections of these deposits. Unlike the HEA–AISI 304 direct coating, no apparent cracks were observed, which indicated an improvement in bonding. However, issues of microporosity persisted. A dendrite microstructure was observed along the interface between the intermediate layer and the HEA.

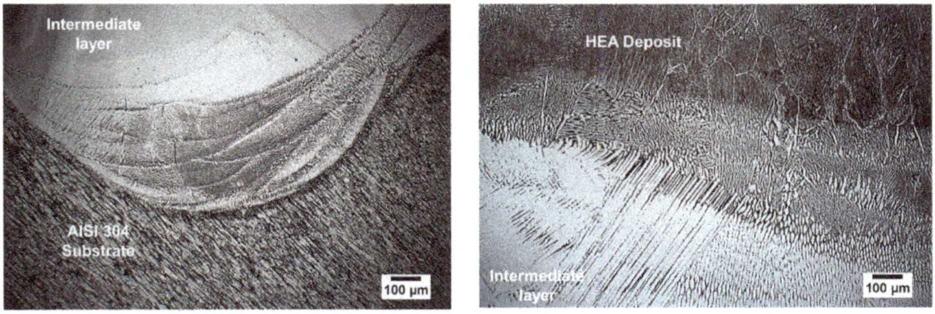

(a) (b)

Figure 6. The optical microstructure of (**a**) the $CoFe_2Ni$ intermediate layer and the AISI 304 substrate and (**b**) the AlCoCrFeNi alloy deposit and the $CoFe_2Ni$ intermediate layer.

A high-magnification secondary electron image of the AlCoCrFeNi HEA deposit is shown in Figure 7, where a two-phase dendritic microstructure was observed. The area fraction of the dendritic microstructure was ~52%, while the interdendritic area fraction was ~48%. The interdendritic region is named A, and the dendritic region is named B. The mean elemental compositions of A and B (average from three arbitrary points) were analyzed by EDS, and the results are listed in Table 4. It is shown that the atomic percentages of Al and Ni were ~29% in A and ~41% in B. The percentages of Fe and Cr were ~54 atom % in A and 43 atom % in B. These results indicate that Fe and Cr were rich in A, while Al and Ni were rich in B. The composition of Co did not show evident differences between A and B. The mixing enthalpies between Fe–Cr, Fe–Ni, Fe–Co, Fe–Al, Cr–Ni, Cr–Co, Cr–Al, Ni–Co, Ni–Al and Co–Al were −1, −2, −1, −11, −7, −4, −10, 0, −22 and −19 kJ/mol, respectively [6,25]. The mixing enthalpy of Al and Ni was higher than other pairs, which indicated that Al and Ni tended to form atomic pairs and segregate. Similar results have been reported for the AlCoCrFeNi HEA, with this microstructure being attributed to the spinodal decomposition [2,4–6,9].

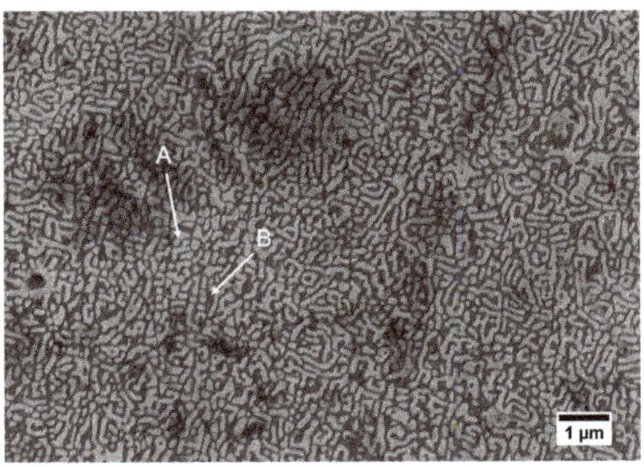

Figure 7. Secondary electron image of the AlCoCrFeNi HEA microstructure at a magnification of 10000.

Table 4. Elemental compositions analyzed by energy dispersive X-ray spectroscopy (EDS)of the AlCoCrFeNi HEA shown in Figure 7.

Elements (atom %)	Al	Co	Cr	Fe	Ni
A	16.2	16.8	23.4	30.2	13.4
B	23.5	15.7	19.4	24.2	17.2

XRD was used to identify the crystal structures of the intermediate layers and the HEA. A transition of the crystal structure was observed from the AISI 304 substrate to the AlCoCrFeNi alloy. The XRD patterns of the AISI 304 substrate, the $CoFe_2Ni$ intermediate layer and the AlCoCrFeNi alloy are shown in Figure 8. The present phases and the corresponding crystallographic information are summarized in Table 5. The peak patterns of FCC were observed in the $CoFe_2Ni$ intermediate layer, while BCC peak patterns were detected in the AlCoCrFeNi alloy. Löbel et al. found BCC and B2 (ordered BCC) phases in $AlCoCrFeNiTi_x$ ($x = 0$) when fabricated via arc melting [1]. A similar result was reported by Shiratori et al., when casting was employed to produce an AlCoCrFeNi HEA [26]. Due to the same basic lattice structure and lattice parameters, the B2 ordered structure is very hard to detect from XRD, as the peak patterns of B2 and BCC are the same [2,9]. However, the evidence of the existence of the B2 phase was found from the EDS analysis above. Previously, an AlCoCrFeNi HEA was reported to also contain the FCC crystal structure with preheating or post-heat treatment [5,13,26]. The FCC structure was not found in this work, which could be because the high cooling rate during LMD inhibited the formation of the FCC crystal structure [5,13,26].

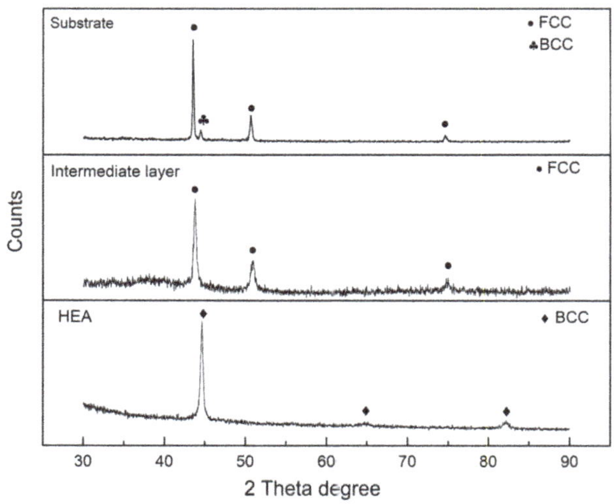

Figure 8. XRD pattern of the AISI 304 substrate, the $CoFe_2Ni$ intermediate layer and the AlCoCrFeNi HEA.

Table 5. Summary of phases detected by XRD analysis for AISI 304, $CoFe_2Ni$ and the AlCoCrFeNi HEA.

Alloy	Lattice	Structure	Space Group	Lattice Parameter (Å)
AISI 304	FCC	Cu	Fm-3m (225)	3.5911
	BCC	Fe	Im-3m (229)	2.87
$CoFe_2Ni$	FCC	Cu	Fm-3m (225)	3.5911
AlCoCrFeNi HEA	BCC	W	Im-3m (229)	2.876

The evolution in chemistry from the intermediate layer to the substrate was characterized by an EDS line scan first. The quantitative results are shown in Figure 9a. The EDS measured results of the AISI 304 substrate (Cr: ~18–19 atom %, Fe: ~70–72 atom %, Ni: ~9–10 atom % in Figure 9a) did not vary from the nominal AISI 304 elemental compositions. Mn (~1–2 atom %) was detected in the AISI 304 substrate by EDS but is not shown in Figure 9. The percentages of Co (~17–22 atom %) and Ni (~21–23 atom %) reduced, while the Fe (~54–56 atom %) content increased from the intermediate layer to the AISI 304 substrate. A small amount of Cr (~3–5 atom %) was present in the intermediate layer, because the substrate was mixed with the intermediate layer. The composition distribution from the HEA to the intermediate layer is shown in Figure 9b. The constituents of the AlCoCrFeNi HEA were detected by EDS (Al: ~16–17 atom %, Co: 19–20 atom %, Cr: ~17 atom %, Fe: ~25 atom %, Ni: ~20–21 atom %). The difference between the as-blended (20 atom %) and as-deposited aluminum (~16–17 atom %) percentages is suspected to be a consequence of inconsistency in capture efficiencies of the constituent powders, and evaporation due to differences in melting point. Al and Cr were present in the intermediate layer as seen in Figure 9b, and their total content was ~4–5 atom %.

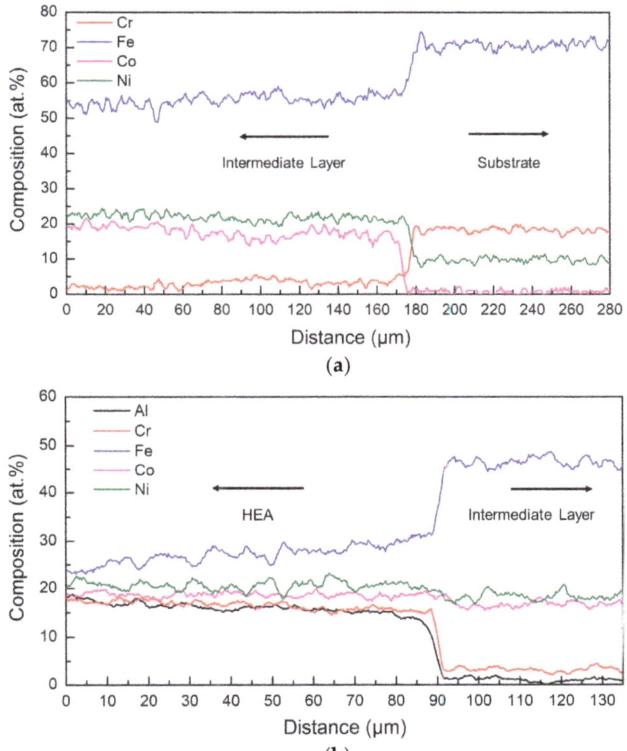

Figure 9. Elemental composition distribution along the boundary, (**a**) $CoFe_2Ni$ intermediate layer and AISI 304 substrate and (**b**) AlCoCrFeNi HEA and $CoFe_2Ni$ intermediate layer.

3.3.2. EBSD

Figure 10a shows the inverse pole figure (IPF) map obtained from the bottom of the HEA section of the specimen. The measured area was approximately 3.4 mm × 1.2 mm of the cross-section parallel to the build direction (BD), which spanned from the left to the right of the specimen. The difference in color indicates the different crystallographic orientations. From Figure 10a, the overall constitution can be classified into two zones—the edge zone (1 and 3) and the middle zone (2). In areas 1 and 3, the grains were observed to be elongated along the build direction (see 1 and 3 in Figure 10a). The distributions of the intercept lengths (using 100 horizontal lines) in different areas are depicted Figure 10b. The median linear intercept for areas 1 and 3 was 72.5 μm, while it was 127.5 μm for area 2. From the linear intercept distribution of area 2, 25% of the intercept values were greater than 300 μm, whereas only 14% of the intercept values were above 300 μm for areas 1 and 3. This grain morphology is likely to be a consequence of deposition toolpath and variation in cooling rate at edges and in the middle [27,28]. Figure 10c,d show the {100}, {110} and {111} pole figures of different areas, which give the distribution of the pole density along the build direction. The pole figure of the areas 1 and 3 (Figure 10c) suggests that the orientations of the grains were close to the <100> direction. However, the grains were random in orientation and did not appear with obvious texture in area 2 (Figure 10d). Further study is necessary to investigate the impact of this toolpath on the grain morphology.

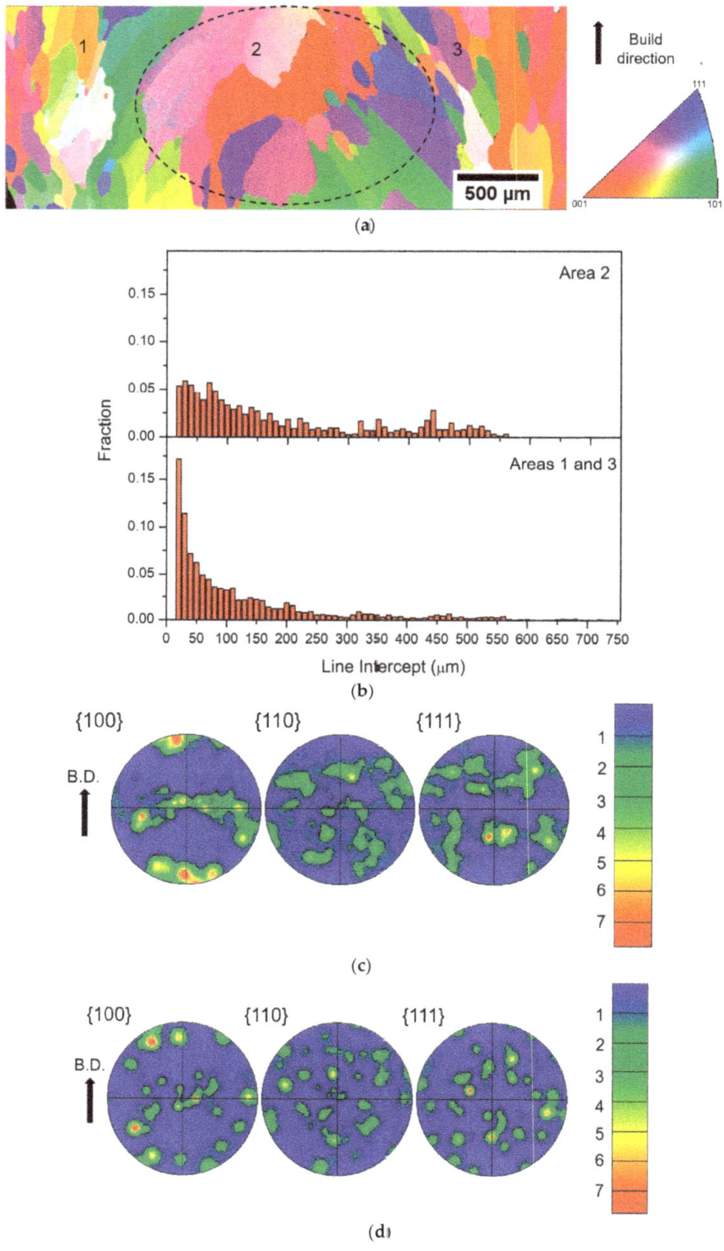

Figure 10. (**a**) Inverse pole figure (IPF)IPF map of the bottom of the HEA section in the specimen; the measured region was approximately 3.4 mm × 1.2 mm, from the left to the right side in the cross-section parallel to the build direction (BD); (**b**) distribution of the intercept length of grains with the bin size of 10 μm; (**c**) pole figure of areas 1 and 3; and (**d**) pole figure of area 2 in (**a**).

3.3.3. Vickers Hardness Analysis

Figure 11 gives the Vickers hardness distribution of the AlCoCrFeNi HEA deposited on the AISI 304 substrate with the $CoFe_2Ni$ intermediate layer. The Vickers hardness of the $CoFe_2Ni$ intermediate layer was around the 275 HV, which could be attributed to the solid solution strengthening. Table 6 lists the Vickers hardnesses of the AlCoCrFeNi HEA, annealed AISI 304, aged Inconel 625, and annealed duplex steel SAF 2205 [29–31]. The average Vickers hardness of the HEA deposit was in the range of 418 HV, because of the second-phase strengthening [4].

According to the XRD results, the AISI 304 substrate and the $CoFe_2Ni$ intermediate layer had an FCC structure, while the AlCoCrFeNi HEA had a BCC structure. The transition from FCC to BCC structure is also expected to enhance the hardness. The high hardness is expected to correlate with good performance in strength and wear resistance [1,16].

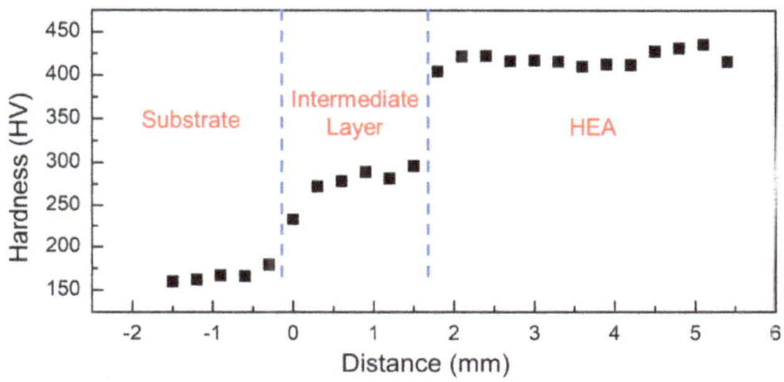

Figure 11. Vickers hardness profile of the AISI 304 substrate—AlCoCrFeNi HEA with the $CoFe_2Ni$ intermediate layer.

Table 6. Vickers hardness of various alloys.

Alloy	Hardness (HV)	Reference
AlCoCrFeNi HEA	418	This work
AISI 304 annealed	188	[29]
Inconel 625 aged	225	[30]
Duplex steel SAF 2205 annealed	290	[31]

4. Conclusions

An AlCoCrFeNi HEA was coated on an AISI 304 substrate by laser metal deposition (LMD) technology. The coating on the substrate without and with the intermediate layer was characterized and discussed. The main conclusions are as follows:

- Cracking was found to be prominent when the AlCoCrFeNi HEA was directly coated on the AISI 304 substrate due to the compositional change between HEA and the substrate.
- Using an intermediate layer of $CoFe_2Ni$ improved the bond. The incorporation of the intermediate layer successfully eliminated crack formation in the deposit.
- XRD patterns revealed a transition of crystal structure from FCC in the AISI 304 substrate to BCC in the AlCoCrFeNi alloy. The evidence of a B2 phase in the AlCoCrFeNi HEA was also found in the EDS analysis results.
- The AlCoCrFeNi alloy fabricated by LMD was found to have an average hardness of 418 HV, while the $CoFe_2Ni$ intermediate layer had an average hardness of 275 HV.

Author Contributions: For this research article, W.C. designed and performed the experiments, analyzed data and wrote the manuscript; S.K. assisted in EDS analysis and manuscript review; X.Z. assisted in SEM analysis; E.B. contributed in EBSD analysis, and F.L. guided the research project.

Acknowledgments: The authors gratefully acknowledge the financial support from NSF (National Science Foundation) grants CMMI-1547042 and CMMI-1625736. The support from the Intelligent Systems Center (ISC) and Materials Research Center (MRC) for the help in sample preparation and materials characterization is also appreciated.

Conflicts of Interest: The authors declare no conflict of interest.

References

1. Löbel, M.; Lindner, T.; Mehner, T.; Lampke, T.; Löbel, M.; Lindner, T.; Mehner, T.; Lampke, T. Influence of Titanium on Microstructure, Phase Formation and Wear Behaviour of AlCoCrFeNiTix High-Entropy Alloy. *Entropy* **2018**, *20*, 505. [CrossRef]
2. Kunce, I.; Polanski, M.; Karczewski, K.; Plocinski, T.; Kurzydlowski, K.J. Microstructural characterisation of high-entropy alloy AlCoCrFeNi fabricated by laser engineered net shaping. *J. Alloys Compd.* **2015**, *648*, 751–758. [CrossRef]
3. Fujieda, T.; Shiratori, H.; Kuwabara, K.; Kato, T.; Yamanaka, K.; Koizumi, Y.; Chiba, A. First demonstration of promising selective electron beam melting method for utilizing high-entropy alloys as engineering materials. *Mater. Lett.* **2015**, *159*, 12–15. [CrossRef]
4. Tang, Z.; Senkov, O.N.; Parish, C.M.; Zhang, C.; Zhang, F.; Santodonato, L.J.; Wang, G.; Zhao, G.; Yang, F.; Liaw, P.K. Tensile ductility of an AlCoCrFeNi multi-phase high-entropy alloy through hot isostatic pressing (HIP) and homogenization. *Mater. Sci. Eng. A* **2015**, *647*, 229–240. [CrossRef]
5. Munitz, A.; Salhov, S.; Hayun, S.; Frage, N. Heat treatment impacts the micro-structure and mechanical properties of AlCoCrFeNi high entropy alloy. *J. Alloys Compd.* **2016**, *683*, 221–230. [CrossRef]
6. Wang, Y.P.; Li, B.S.; Ren, M.X.; Yang, C.; Fu, H.Z. Microstructure and compressive properties of AlCrFeCoNi high entropy alloy. *Mater. Sci. Eng. A* **2008**, *491*, 154–158. [CrossRef]
7. Zhang, Y.; Zuo, T.T.; Tang, Z.; Gao, M.C.; Dahmen, K.A.; Liaw, P.K.; Lu, Z.P. Microstructures and properties of high-entropy alloys. *Prog. Mater. Sci.* **2014**, *61*, 1–93. [CrossRef]
8. Wang, W.-R.; Wang, W.-L.; Wang, S.-C.; Tsai, Y.-C., Lai, C.-H.; Yeh, J.-W. Effects of Al addition on the microstructure and mechanical property of AlxCoCrFeNi high-entropy alloys. *Intermetallics* **2012**, *26*, 44–51. [CrossRef]
9. Manzoni, A.; Daoud, H.; Völkl, R.; Glatzel, U.; Wanderka, N. Phase separation in equiatomic AlCoCrFeNi high-entropy alloy. *Ultramicroscopy* **2013**, *132*, 212–215. [CrossRef]
10. Li, W.; Yan, L.; Karnati, S.; Liou, F.; Newkirk, J.; Taminger, K.M.B.; Seufzer, W.J. Ti-Fe intermetallics analysis and control in joining titanium alloy and stainless steel by Laser Metal Deposition. *J. Mater. Process. Technol.* **2017**, *242*, 39–48. [CrossRef]
11. Syed, W.U.H.; Pinkerton, A.J.; Liu, Z.; Li, L. Coincident wire and powder deposition by laser to form compositionally graded material. *Surf. Coatings Technol.* **2007**, *201*, 7083–7091. [CrossRef]
12. Sistla, H.R.; Newkirk, J.W.; Frank Liou, F. Effect of Al/Ni ratio, heat treatment on phase transformations and microstructure of Al_xFeCoCrNi2−x (x = 0.3, 1) high entropy alloys. *Mater. Des.* **2015**, *81*, 113–121. [CrossRef]
13. Wang, R.; Zhang, K.; Davies, C.; Wu, X. Evolution of microstructure, mechanical and corrosion properties of AlCoCrFeNi high-entropy alloy prepared by direct laser fabrication. *J. Alloys Compd.* **2017**, *694*, 971–981. [CrossRef]
14. Zhang, H.; Pan, Y.; He, Y.-Z. Synthesis and characterization of FeCoNiCrCu high-entropy alloy coating by laser cladding. *Mater. Des.* **2011**, *32*, 1910–1915. [CrossRef]
15. He, B.; Zhang, N.; Lin, D.; Zhang, Y.; Dong, F.; Li, D.; He, B.; Zhang, N.; Lin, D.; Zhang, Y.; et al. The Phase Evolution and Property of FeCoCrNiAlTix High-Entropy Alloying Coatings on Q253 via Laser Cladding. *Coatings* **2017**, *7*, 157. [CrossRef]
16. Chen, X.; Yan, L.; Karnati, S.; Zhang, Y.; Liou, F.; Chen, X.; Yan, L.; Karnati, S.; Zhang, Y.; Liou, F. Fabrication and Characterization of Al_xCoFeNiCu1−x High Entropy Alloys by Laser Metal Deposition. *Coatings* **2017**, *7*, 47. [CrossRef]
17. Zhang, S.; Wu, C.L.; Yi, J.Z.; Zhang, C.H. Synthesis and characterization of FeCoCrAlCu high-entropy alloy coating by laser surface alloying. *Surf. Coat. Technol.* **2015**, *262*, 64–69. [CrossRef]
18. Chou, H.-P.; Chang, Y.-S.; Chen, S.-K.; Yeh, J.-W. Microstructure, thermophysical and electrical properties in AlxCoCrFeNi ($0 \leq x \leq 2$) high-entropy alloys. *Mater. Sci. Eng. B* **2009**, *163*, 184–189. [CrossRef]

19. Shiue, R.K.; Chang, C.T.; Young, M.C.; Tsay, L.W. The effect of residual thermal stresses on the fatigue crack growth of laser-surface-annealed AISI 304 stainless steel: Part I: Computer simulation. *Mater. Sci. Eng. A* **2004**, *364*, 101–108. [CrossRef]
20. Huang, C.; Zhang, Y.; Vilar, R.; Shen, J. Dry sliding wear behavior of laser clad TiVCrAlSi high entropy alloy coatings on Ti–6Al–4V substrate. *Mater. Des.* **2012**, *41*, 338–343. [CrossRef]
21. Li, Y.; Shen, Y.; Hung, C.-H.; Leu, M.C.; Tsai, H.-L. Additive manufacturing of Zr-based metallic glass structures on 304 stainless steel substrates via V/Ti/Zr intermediate layers. *Mater. Sci. Eng. A* **2018**, *729*, 185–195. [CrossRef]
22. Li, W.; Martin, A.J.; Kroehler, B.; Henderson, A.; Huang, T.; Watts, J.; Hilmas, G.E.; Leu, M.C. Fabricating Functionally Graded Materials by Ceramic On-Demand Extrusion with Dynamic Mixing. In Proceedings of the 29th Annual International Solid Freeform Fabrication Symposium, Austin, TX, USA, 13–15 August 2018; pp. 1087–1099.
23. Desu, R.K.; Nitin Krishnamurthy, H.; Balu, A.; Gupta, A.K.; Singh, S.K. Mechanical properties of Austenitic Stainless Steel 304L and 316L at elevated temperatures. *J. Mater. Res. Technol.* **2016**, *5*, 13–20. [CrossRef]
24. Guillermet, A.F. Assessing the thermodynamics of the Fe CO Ni system using a calphad predictive technique. *Calphad* **1989**, *13*, 1–22. [CrossRef]
25. Tang, Z.; Gao, M.C.; Diao, H.; Yang, T.; Liu, J.; Zuo, T.; Zhang, Y.; Lu, Z.; Cheng, Y.; Zhang, Y.; et al. Aluminum Alloying Effects on Lattice Types, Microstructures, and Mechanical Behavior of High-Entropy Alloys Systems. *JOM* **2013**, *65*, 1848–1858. [CrossRef]
26. Shiratori, H.; Fujieda, T.; Yamanaka, K.; Koizumi, Y.; Kuwabara, K.; Kato, T.; Chiba, A. Relationship between the microstructure and mechanical properties of an equiatomic AlCoCrFeNi high-entropy alloy fabricated by selective electron beam melting. *Mater. Sci. Eng. A* **2016**, *656*, 39–46. [CrossRef]
27. Reichardt, A.; Dillon, R.P.; Borgonia, J.P.; Shapiro, A.A.; McEnerney, B.W.; Momose, T.; Hosemann, P. Development and characterization of Ti-6Al-4V to 304L stainless steel gradient components fabricated with laser deposition additive manufacturing. *Mater. Des.* **2016**, *104*, 404–413. [CrossRef]
28. Zhang, X.; Cui, W.; Li, W.; Liou, F. Effects of tool path in remanufacturing cylindrical components by laser metal deposition. *Int. J. Adv. Manuf. Technol.* **2018**, 1–11. [CrossRef]
29. Milad, M.; Zreiba, N.; Elhalouani, F.; Baradai, C. The effect of cold work on structure and properties of AISI 304 stainless steel. *J. Mater. Process. Technol.* **2008**, *203*, 80–85. [CrossRef]
30. Shaikh, M.A.; Ahmad, M.; Shoaib, K.A.; Akhter, J.I.; Iqbal, M. Precipitation hardening in Inconel * 625. *Mater. Sci. Technol.* **2000**, *16*, 129–132. [CrossRef]
31. Calliari, I.; Zanesco, M.; Ramous, E. Influence of isothermal aging on secondary phases precipitation and toughness of a duplex stainless steel SAF 2205. *J. Mater. Sci.* **2006**, *41*, 7643–7649. [CrossRef]

entropy

MDPI

Article

Processing and Characterization of Refractory Quaternary and Quinary High-Entropy Carbide Composite

Hanzhu Zhang and Farid Akhtar *

Division of Materials Science, Luleå University of Technology, 971 87 Luleå, Sweden; hanzhu.zhang@ltu.se
* Correspondence: farid.akhtar@ltu.se

Received: 10 April 2019; Accepted: 28 April 2019; Published: 6 May 2019

Abstract: Quaternary high-entropy ceramic (HEC) composite was synthesized from HfC, Mo_2C, TaC, and TiC in pulsed current processing. A high-entropy solid solution that contained all principal elements along with a minor amount of a Ta-rich phase was observed in the microstructure. The high entropy phase and Ta-rich phase displayed a face-centered cubic (FCC) crystal structure with similar lattice parameters, suggesting that TaC acted as a solvent carbide during phase evolution. The addition of B_4C to the quaternary carbide system induced the formation of two high-entropy solid solutions with different elemental compositions. With the increase in the number of principal elements, on the addition of B_4C, the crystal structure of the HEC phase transformed from FCC to a hexagonal structure. The study on the effect of starting particle sizes on the phase composition and properties of the HEC composites showed that reducing the size of solute carbide components HfC, Mo_2C, and TiC could effectively promote the interdiffusion process, resulting in a higher fraction of a hexagonal structured HEC phase in the material. On the other hand, tuning the particle size of solvent carbide, TaC, showed a negligible effect on the composition of the final product. However, reducing the TaC size from −325 mesh down to <1 μm resulted in an improvement of the nanohardness of the HEC composite from 21 GPa to 23 GPa. These findings suggested the possibility of forming a high-entropy ceramic phase despite the vast difference in the precursor crystal structures, provided a clearer understanding of the phase transformation process which could be applied for the designing of HEC materials.

Keywords: high-entropy ceramic; solid-state diffusion; microstructure; phase evolution; hardness

1. Introduction

High-entropy ceramics (HECs), as a new class of ceramic materials, are developed from the concept of high-entropy alloys (HEAs). In HEAs, multiple principal elemental metals are incorporated to form a single phase alloy or multiphase composites [1,2]. With a similar design concept, HECs consist of multiple principal ceramic compounds such as metallic oxides, nitrides or carbides. In recent work, it has been found that the high entropy effect applies in the HECs system consisting of multiple components [3,4]. Solid solution phases that lack long-range order can be formed attributed to the minimization of Gibbs energy. Christina M. Rost et al. [5] fabricated an entropy stabilized oxide that showed a single-phase face-centered cubic (FCC) structure. Single phase solid solutions have been reported in systems like high entropy borides [6] and high entropy carbides [7,8]. Compared with conventional ceramic materials, single phase HECs show superior mechanical properties contributed by the strain strengthening effect from lattice distortion. For example, the high-entropy carbide fabricated by E. Castle et al. [7] exhibited a single phase FCC structure and enhanced hardness (36.1 GPa) compared to all the component carbides.

Due to the high melting point of ceramic materials, the solid-state pulsed current processing (PCP) is the economically preferable processing route as it offers relatively low sintering temperatures and short sintering times [9]. PCP utilizes pulsed current, directional pressure, and high vacuum during sintering and offers the advantage of providing homogenously densified materials [10,11]. Due to the strong covalent bonding and complex crystal structures of refractory ceramics, the synthesis of high entropy ceramics has been performed with ceramic precursors containing one nonmetal element (for example carbide, nitride or boride) and a similar crystal structure [6,7,12,13] using PCP. The selection approach of the ceramic precursors minimizes the geometrical difference among the starting ceramic precursor materials and achieves a single-phase high-entropy phase during PCP [14].

Our previous work on the high-entropy ceramic composite shows the possibility of forming a single-phase hexagonal structure from carbide precursors with a vast difference in the crystal structures (FCC, hexagonal and rhombohedral) and two nonmetal atoms (carbon and boron), attributing to the independent diffusion of metal and nonmetal atoms during the PCP processing [14]. However, the phase formation rules in a high-entropy ceramic system are still elusive and how B_4C influences the diffusion process remains unclear. In order to develop a better understanding of the formation mechanism of the high-entropy ceramic phase, a systematic study was established and conducted in this work. Quaternary HEC composite was processed from four refractory carbides, HfC, Mo_2C, TaC, and TaC. The phase evolution on B_4C addition was studied by investigating the microstructure and phase composition of the five-component carbides. Moreover, due to the importance of starting particle size in solid-state diffusion [15], HEC composites were fabricated from precursors with different particle sizes. The effects of tuning the particles size of the solvent and the solute carbides on the microstructure, phase composition, and mechanical properties are discussed.

2. Materials and Methods

HfC (<1.25 μm, Sigma Aldrich, Darmstadt, Germany) and (−325 mesh, American Elements, Los Angeles, LA, USA), Mo_2C (−325 mesh, Alfa Aesar, Haverhill, MA, USA) and (2.6 μm, Nanografi, Ankara, Çankaya), TaC (−325 mesh, Alfa Aesar, Haverhill, MA, USA) and (1 μm, US Research Nanomaterials, Houston, TA, USA), TiC (5 μm, H.C. Starck, Munich, Germany) and (2 μm, Alfa Aesar, Haverhill, MA, USA) and B_4C (1–7 μm, Alfa Aesar, Haverhill, MA, USA) were utilized to synthesize high-entropy ceramic composites. Starting materials with different particle sizes were mixed following the designed recipe, with a molar ratio of 2:1:2:2:2. The powder mixture was homogenized in a ball milling machine for two hours, using 4 mm stainless steel balls as a milling media and a powder to ball mass ratio of 1:5. The homogenized powder mixture was filled into a graphite die with a diameter of 10 mm and prepressed before sintering. The sintering was conducted in SPS-530ET (Dr. Sinter Spark Plasma Sintering System, Fuji electronic industrial Co., Ltd., Tsurugashima, Japan) in a glovebox. The samples were heated to 1800 °C with a heating rate of 100 °C/min and then held at 1800 °C for 5 min under vacuum. A uniaxial pressure of 60 MPa was applied during the process.

Before the microstructure characterization and phase identification, the samples were cold mounted in epoxy and polished following standard metallurgical sample preparation procedures. The microstructure was observed using a Scanning Electron Microscope JSM-IT300 (JEOL, Tokyo, Japan) operating at an acceleration voltage of 15 kV. The elemental composition was analyzed using energy dispersive spectrometer (EDS) mounted on JSM-IT300 that was calibrated with Cobalt. X-ray diffraction (XRD) was conducted using an X-ray diffractometer (Empyrean, PANalytical, Malvern, UK) with Cu-Kα radiation (wavelength 0.154 nm). The scanning was performed from 5 to 120° (2 Theta), with a step size of 0.02°. The XRD data was investigated using the software PANalytical X'Pert HighscorePlus with the PDF-4 database. The nanohardness of sintered HEC composites was determined using an MTS NanoIndenter XP with a Berkovich diamond indenter. The measurements were performed with a maximum load of 100 mN at ambient temperature in air. The force-displacement curves of the indenter were recorded. Due to different phase distribution and grain sizes in the microstructure of the PCP samples, the nanohardness values for sample 4-HEC and 5-HEC were obtained by performing at least

5 measurements at specific regions based on the optical microscope mounted on MTS NanoIndenter XP, while the hardness values for sample HEC(+) and HEC(fine) were obtained by performing matrixes indentations and calculating the average value from indentations that were not located in porosity.

3. Results and Discussions

3.1. *Effect of B_4C Addition into (HfMoTaTi)C*

The microstructure of the quaternary HEC composite (4-HEC) sintered from HfC, Mo_2C, TaC, and TiC shows two distinct phases, as shown in Figure 1. The bright phase with a size of 1–5 μm is dispersed uniformly in the dark matrix phase in Figure 1b. Based on the energy-dispersive X-ray spectroscopy (EDS) mapping analysis in Figure 1g, the bright phase is rich in Ta whilst the dark phase contains all constitutional elements Hf, Mo, Ta, Ti and C. Because of the sensitivity of backscattered electron detector (BED) to the atomic number, phases with different densities appear with different contrast in the BED microstructure. The constituent carbides show a vast difference in the densities, from 14.62 g/cm^3 for TaC, 12.2 g/cm^3 for HfC, 9.18 g/cm^3 for Mo_2C to 4.93 g/cm^3 for TiC, therefore the consistent contrast of the dark region in the Figure 1b implies that it represents the quaternary ceramic phase containing all constitutional elements Hf, Mo, Ta, Ti, and C. The average atomic ratio of each phase obtained by performing EDS point analysis on several point locations are listed in Table 1. As EDS lacks the accuracy of quantitative analysis of light elements [16], the atomic ratio of the four metals is normalized to Ta. The results show that the bright phase contains Ta and C as the dominating elements and a minor amount of Mo and trace amount of Hf and Ti, while the dark phase contains all four metal elements with a relatively lower amount of Ta. According to the X-ray diffraction (XRD) data in Figure 1a, the PCP 4-HEC composite consists of two face-centered cubic (FCC) crystal structures with similar lattice parameters (a_1 = 0.4429 nm, a_2 = 0.4399 nm), as marked in the inset in Figure 1a. The BED microstructure, EDS analysis and XRD data, Figure 1 and Table 1, suggest the formation of high-entropy FCC solid solution containing all constitutional elements. In the previous work on high-entropy ceramic $B_4(HfMo_2TaTi)C$ [14], it has been reported that TaC has the lowest metal vacancy formation energy among the precursor carbides, thus it acts as the solvent FCC lattice during the formation of the high-entropy phase [7], i.e., constituent atoms except Ta diffuse into the vacancies in the TaC lattice to form the multicomponent solid solution. Therefore, the Ta-rich phase and the high-entropy phase can be regarded as intermediates in the phase transformation from constitutional carbides to the hexagonal high-entropy ceramic phase. The multicomponent interdiffusion induces TaC lattice distortion, which in this case results in the reduction in the initial lattice parameter of TaC, 0.4460 nm (ICDD reference pattern of TaC: No. 03-065-0282). Based on the quantitative results shown in Table 1, the high-entropy phase with higher content of Hf, Mo, and Ti metal atoms experience intense atomic diffusion compared to the Ta-rich phase. Hence, in the quaternary high-entropy phase, the atomic position exchange between Ta and other metal atoms with similar or smaller atomic radii results in a smaller lattice parameter (0.4399 nm) than the TaC-rich phase (0.4429 nm). Furthermore, the high-entropy phase shows a hardness of 28.4 GPa, which is 23.5% higher than that of the Ta-rich phase (23 GPa) as shown in Figure 2, due to the lattice distortion induced strain strengthening effect. This result is in line with the previous reports on high-entropy materials [6,7].

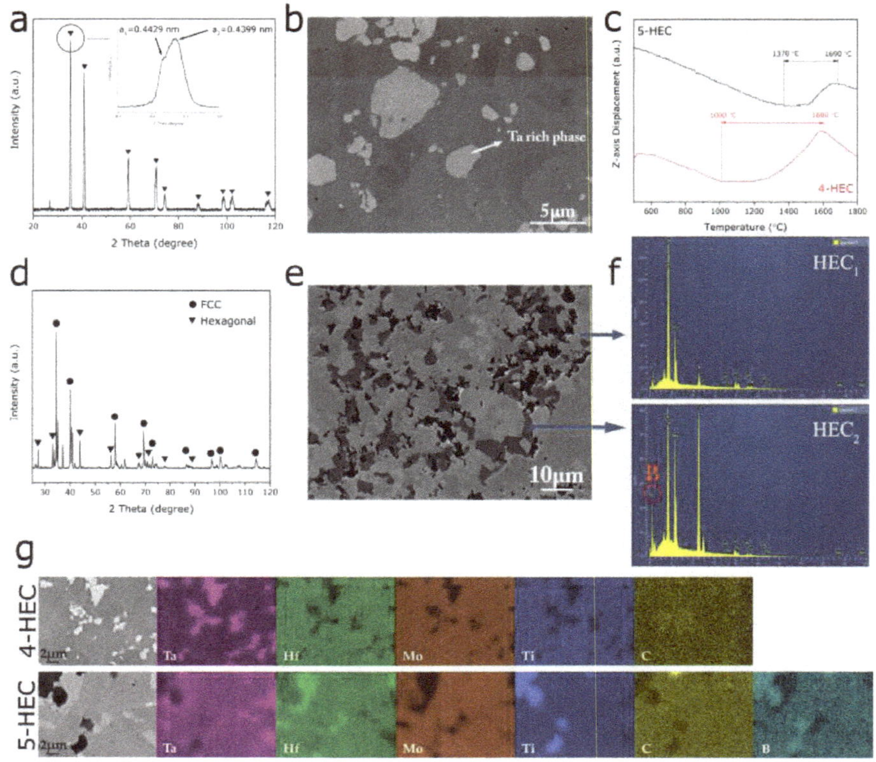

Figure 1. XRD patterns and backscattered electron microstructure of PCP 4-HEC (**a**,**b**) and 5-HEC composite (**d**,**e**), and the EDS mapping analysis (**g**); (**c**) shows the volume change of the material during sintering; (**f**) is the EDS qualitative spectra on different phases of 5-HEC.

Table 1. The quantitative analysis of the metal atom contents, the crystal structure identification of different phases in 4-HEC and 5-HEC.

Compositional Elements	**4-HEC**		**5-HEC/HEC(++)**		
	Ta-Rich Phase (Bright)	**High-Entropy Phase (Dark)**	**Ta-Rich Phase (Bright)**	**HEC_1 (Gray)**	**HEC_2 (Dark)**
Hf	-	2.5	-	0.8	1.7
Mo	Minor	2.4	Minor	1.7	3.3
Ti	-	1.8	-	0.6	7.7
Ta	Major	1	Major	1	1
Crystal structure	FCC	FCC	FCC	FCC	Hexagonal

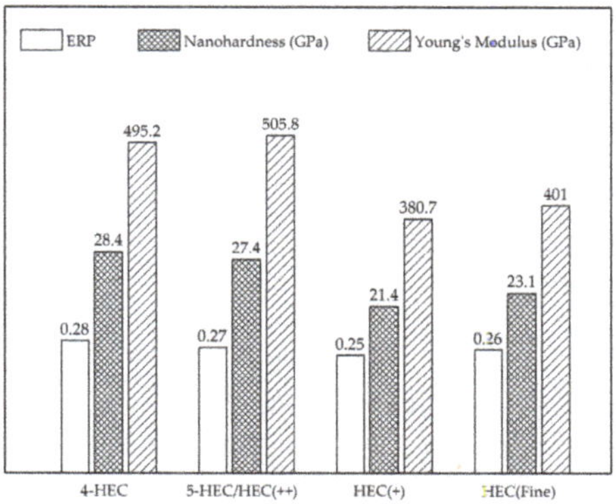

Figure 2. Elastic recovery parameter (ERP), nanohardness and the Young's modulus of the PCP HECs. The nanohardness values refer to the FCC solid solution in 4-HEC, 5-HEC, and average hardness properties for HEC(+) and HEC(Fine).

The curve of Z-axis displacement as a function of temperature was recorded during the PCP. As shown in Figure 1c, the decline of Z-axis refers to the thermal expansion of the material, while the up-climbing region corresponds to the shrinkage of the bulk volume. The reduction of the volume typically refers to the occurrence of sintering phenomenon where the powder material becomes compacted and forms a densified solid mass. Since the sintering temperature of powder material is normally 2/3–3/4 of the melting point [17], the theoretical sintering temperature for current quaternary refractory carbide mixture should be above 1800 °C. Figure 1c shows that the shrinkage of the four-component carbide system 4-HEC takes place from 1000 °C to 1600 °C, while the same phenomenon for the 5-HEC composite was postponed to a higher temperature range 1370 °C–1690 °C. This indicates that the addition of B_4C to the carbide system hinders the solid-state atomic diffusion required for sintering in the multicomponent carbides, leading to a delay of the formation of the high-entropy ceramic phase. A detailed investigation on the sintering behavior of ceramic precursors to form high-entropy ceramic composites will be reported later, elsewhere.

The addition of B_4C to the precursor carbides resulted in the formation of multiple phases during PCP. According to the backscattered electron microstructure in Figure 1e, the PCP 5-HEC composite exhibits three distinct phases. Similar to the 4-HEC, the brightest phase in 5-HEC is rich in Ta, which is coordinated with the fact that Ta has the highest atomic number among the constituent elements, therefore, the Ta-rich phase appears as the brightest phase in the BED microstructure. Two high-entropy solid solutions with different elemental compositions were formed in the five-component system, 5-HEC. The atomic ratio of metal atoms Hf, Mo, Ta and Ti in the gray (HEC_1) and dark phase (HEC_2) in Figure 1e are shown in Table 1, with HEC_2 showing higher content of the solute metal atoms (Hf, Mo, and Ti) in the structure. The higher content of Ti in HEC_2 agrees more with the darker contrast of HEC_2 than HEC_1 in the microstructure, as Ti has the smallest atomic number among the constitutional metal elements. According to the bond dislocation enthalpy (BDE) of transition metal carbides at 298 K [18], Ti-C has the lowest BDE of 423 ± 30 KJ/mol among the solute carbides, while Mo-C_2 and Hf-C have a BDE of 500 and 540 ± 25 KJ/mol, respectively, and the covalent atomic radii vary as Ti < Mo < Hf. These factors might contribute to preferable diffusion of Ti over Mo and Hf during the formation of the multicomponent solid solution, resulting in the metal content ratio in HEC_2 phase as Ti > Mo > Hf in Table 1. A pronounced boron diffraction peak in the EDS pattern was revealed at the HEC_2 phase (Figure 1f), suggesting that HEC_2 experienced a more intensive diffusion of B atoms than HEC_1. Additionally, the diffractions peaks of B_4C were not detected in the XRD diffractogram of the 5-HEC composite in Figure 1d, suggesting the participation of B_4C in the formation of high-entropy solid solutions. However, the diffusion priority of Ti, Mo and Hf was not observed in the high-entropy phase in the 4-HEC composite and HEC_1 phase in 5-HEC, suggesting that the incorporation of B_4C to the transition metal carbides might have promoted the diffusion process of the metal atoms towards a more energetically favorable state.

The XRD data in Figure 1d shows that the 5-HEC composite contains both FCC and hexagonal structured phases. Similar with the 4-HEC, the FCC pattern is generated from diffraction of two FCC crystal structures with similar lattice parameters, including the presence of an FCC Ta-rich phase. Based on the aforementioned discussions, the HEC_1 phase that contains a lower content of the solute atoms should have a closely-matched crystal structure and lattice constant with the solvent TaC lattice due to the reduced atomic position change in the host lattice corresponding to an FCC structure, while the HEC_2 phase corresponds to the hexagonal structure. The HEC_1 phase shows a lattice parameter of 0.4499 nm, which is slightly greater than that of TaC (0.4460 nm). Assuming that the HEC_1 is formed from only transition metal carbides, the lattice parameter should decline as observed in the FCC high-entropy solid solutions in 4-HEC composite (Figure 1a). It is known that the metal-boron bond length is longer than metal-carbon bond length, for example, Ta-C = 2.22 Å [19] and Ta-B = 2.41 Å [20], the expansion of the lattice can be contributed by the addition of B atoms in the formation of an HEC_2 solid solution. The formation of high-entropy solid solutions HEC_1 and HEC_2 that contain all constitutional elements confirms the possibility of processing high-entropy ceramics from a precursor system containing more than one nonmetal atoms and with different crystal structures. The formation of hexagonal structure is likely to be attributed to more B atoms diffusing in the FCC lattice, which induces more severe lattice distortion and consequently leads to the crystal structure change from FCC to a hexagonal structure. In the 5-HEC composite, the FCC structured HEC_1 solid solution shows a nanohardness of 27.4 GPa and Young's modulus of 505.8 GPa, which is close to the FCC solid solution phase in 4-HEC (28.4 GPa and 495.2 GPa for nanohardness and Young's modulus, respectively).

3.2. Effect of Different Starting Particle Sizes

It is well known that the particle size of the precursors is an essential parameter influencing the atomic diffusion and phase evolution during solid-state sintering in PCP [15]. Fine particles promote the solid-state atomic diffusion by reducing the diffusion distance and promote the kinetics of phase transformation [21,22]. To investigate the effect of particle size on the formation of high-entropy ceramics, the same carbide systems with different particle sizes are sintered in PCP (as listed in Table 2). 5-HEC that utilized relative larger particle sizes of solute metal carbides (HfC, Mo_2C, and TiC) is discussed in the previous section and is denoted as HEC(++) in the following discussion. HEC(fine) and HEC(+) contains precursor carbides with the finest particle sizes and a larger particle size of the solvent carbide (TaC), respectively.

Table 2. Precursors with different particle sizes are utilized to study the effect on the phase evolution.

Component	HEC(++)	HEC(+)	HEC(Fine)
B_4C	1–7 μm	1–7 μm	1–7 μm
HfC	−325 mesh	<1.25 μm	<1.25 μm
Mo_2C	−325 mesh	2.6 μm	2.6 μm
TiC	5 μm	2 μm	2 μm
TaC	<1 μm	−325 mesh	<1 μm

The microstructure of the PCP HEC(+) and HEC(fine) show the presence of two phases in Figure 3. According to the compositional mapping analysis in Figure 3d, all four metal elements are distributed in bright and dark phases. The spot-shaped mapping for C is attributed to porosity in the samples, which possibly caused the diamond polishing agents being introduced during the sample preparation procedure. For both PCP HEC(+) and HEC(fine) sample, the bright phase is rich in Ta and Hf while the dark phase contains a greater amount of Mo and Ti. The quantitative analysis results of the selected areas in Table 3 show that these two phases are high-entropy solid solutions with different elemental compositions. The average atomic ratios are normalized to Ta. The bright phase is rich in Ta, while the dark phase has a higher content of other metals (Ti > Mo > Hf ≈ Ta). Based on the discussion about BED microstructure of different transition metal carbides, the dark region with a higher solute metal content refers to more equilibrium composition and a higher extent of phase transformation towards the high-entropy solid solution.

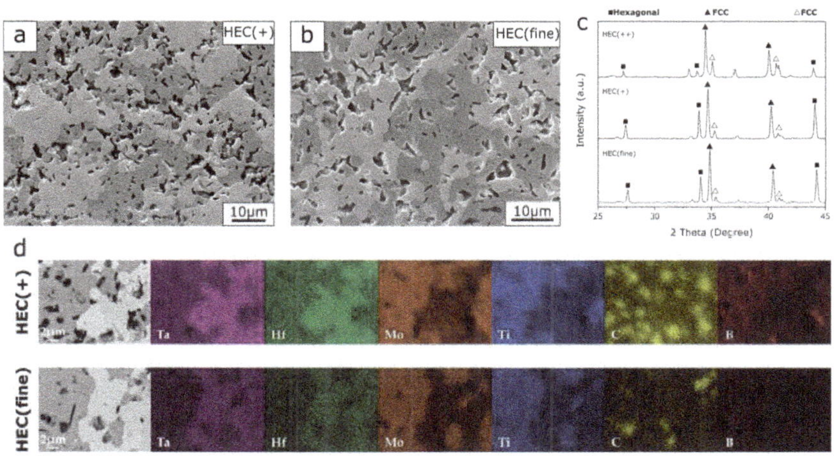

Figure 3. Microstructure (**a**,**b**) and X-ray diffraction phase identification (**c**) and EDS mapping analysis (**d**) of PCP HEC(+) and HEC(fine).

Table 3. The quantitative analysis of the metal atom contents of different phases in HEC(+) and HEC(fine).

Compositional Elements	HEC(+)		HEC(fine)	
	Bright Phase	Dark Phase	Bright Phase	Dark Phase
Hf	0.4	1.2	1	1.1
Mo	0.8	1.9	0.7	2.0
Ti	0.2	2.3	0.8	2.8
Ta	1	1	1	1
Crystal structure	FCC	Hexagonal	FCC	Hexagonal

The XRD patterns of HEC(+) and HEC(fine) show high similarity (Figure 3c). Similar to the diffractogram of HEC(++), the PCP HEC(+) and HEC(fine) composite reveal diffraction patterns from two FCC and one hexagonal structure. Since the elemental mapping from each region with a constant contrast shows a homogenous distribution of constituent elements, a possible reason why the third phase is not distinguishable in the microstructure is that the two FCC phases have indistinguishable contrast in BED images, which suggests that these phases have a similar atomic composition. As the bright phase with less content of foreign atoms (except Ta), it refers to the component with less lattice distortion, therefore it is extrapolated to retain a FCC crystal structure, whilst the dark phase exhibits a hexagonal structure.

For the solid-state phase transformations, the starting particle size has been reported to have a strong influence on the reaction kinetics by tuning the contact area between the solid particles [23–25]. Therefore, it was expected that PCP HEC(fine) with finest starting particle size should have a more promoted phase transformation than HEC(+) and HEC(++) which were fabricated with the same sintering route and sintering conditions. Comparing HEC(fine) with HEC(+), the results show high similarity in the microstructures and phase composition (Figure 3). Both composites consist of a

FCC and a hexagonal crystal structure, with high-entropy solid solutions. With HfC having the lowest formation enthalpy (−1.826 eV) [14] and highest BDE (9.18 g/cm^3) among the precursors [18], the promotion of the solid-state diffusion can be observed by the increased Hf content in the FCC phase in HEC(fine) than HEC(+) (Hf/Ta = 1 and 0.4, respectively) according to the EDS quantitative results in Table 3. On the other hand, HEC(++) has a complicated phase composition compared to HEC(fine). The Ta-rich phase, representing the least diffusion, was observed in HEC(++) (Figure 1e). A lower fraction of the hexagonal structured high-entropy solid solution in HEC(++) composite can be concluded from the microstructure and lower intensity of X-ray diffraction peaks (for example at $2\theta = 44.2°$) in Figure 3c. Therefore, the degree of diffusion is assumed to be in the order of HEC(++) < HEC(+) ≈ HEC(fine). The results suggest that the particle size of solvent carbide TaC is not as essential as the solute carbides in terms of tailoring the phase composition in the multicomponent carbide system. Due to the porosity and small grain size, the nanoindentation testing of HEC(+) and HEC(fine) composites were obtained from 30 indentations, therefore the data represents the overall mechanical properties of the bulk materials instead of each individual high-entropy phase. Although the PCP HEC(fine) and HEC(+) show the same phase composition and similar microstructure, the HEC(fine) composite sintered from precursors with finest grain size shows improved nanoindentation hardness of 23.1 GPa compared to the PCP HEC(+) composite (21.4 GPa). The experimental hardness value is close to the theoretical hardness calculated from the rule of mixture (23.2 GPa). The enhancement of the hardness caused by utilizing small starting particle size during sintering has been reported before [26]. Moreover, Young's modulus of HEC(fine) improves from 380.7 GPa for HEC(+) to 401 GPa, suggesting stronger atomic bonding in the HEC(fine) composite.

4. Conclusions

A high-entropy ceramic (HEC) composite was synthesized from HfC, Mo_2C, TaC, and TiC by pulsed current processing (PCP). The PCP 4-HEC composite contained a high-entropy phase and a Ta-rich phase. Both constituents showed a face-centered cubic (FCC) structure and corresponded to the different extent of phase transformation towards a high-entropy phase. By introducing B_4C in the HEC composite, high-entropy solid solutions that contained all principal elements (including B) were formed. The high-entropy phase with less intense interdiffusion remained FCC structure as the solvent carbide, TaC, while the one with more complete phase transformation experienced crystal structure change from FCC to a hexagonal structure. The results showed the feasibility of synthesizing HEC materials from the multi-principal ceramic system with different crystal structures.

In order to investigate the effect of different particle sizes of solvent and solute components on the phase composition and properties, HEC composites fabricated from precursors with different size of solvent (TaC) and solute carbides (with face-centered cubic (FCC) structure) were PCP consolidated and characterized. The particle size of the solvent carbides was found to be more essential for the interdiffusion process and final phase compositions than that of the solute carbide. HEC composites processed from the fine particle size of solvent carbides HfC, Mo_2C, and TiC showed a higher content of the hexagonal structured HEC phase. On the other hand, reducing the TaC particle size to nanoscale showed negligible influence on the phase composition, but resulted in enhancement of the microhardness from 21.4 GPa to 23.1 GPa.

Author Contributions: Conceptualization, F.A.; methodology, F.A. and H.Z.; formal analysis, H.Z.; investigation, H.Z.; writing—original draft preparation, H.Z.; writing—review and editing, F.A.; supervision, F.A.; project administration, F.A.

Funding: This work was supported by the Swedish Foundation for Strategic Research (SSF) for Infrastructure Fellowship, grant number RIF14-0083.

Conflicts of Interest: The authors declare no conflict of interest.

References

1. Manzoni, A.M.; Glatzel, U. New multiphase compositionally complex alloys driven by the high entropy alloy approach. *Mater. Charact.* **2019**, *147*, 512–532. [CrossRef]
2. Miracle, D.B.; Senkov, O.N. A critical review of high entropy alloys and related concepts. *Acta Mater.* **2017**, *122*, 448–511. [CrossRef]
3. Tsai, M.H.; Yeh, J.W. High-entropy alloys: A critical review. *Mater. Res. Lett.* **2014**, *2*, 107–123. [CrossRef]
4. Gao, M.C.; Yeh, J.-W.; Liaw, P.K.; Zhang, Y. *High-Entropy Alloys*; Springer: Cham, Switzerland, 2016.
5. Rost, C.M.; Sachet, E.; Borman, T.; Moballegh, A.; Dickey, E.C.; Hou, D.; Jones, J.L.; Curtarolo, S.; Maria, J.P. Entropy-stabilized oxides. *Nat. Commun.* **2015**, *6*, 8485. [CrossRef] [PubMed]
6. Gild, J.; Zhang, Y.; Harrington, T.; Jiang, S.; Hu, T.; Quinn, M.C.; Mellor, W.M.; Zhou, N.; Vecchio, K.; Luo, J. High-Entropy Metal Diborides: A New Class of High-Entropy Materials and a New Type of Ultrahigh Temperature Ceramics. *Sci. Rep.* **2016**, *6*, 37946. [CrossRef] [PubMed]
7. Castle, E.; Csanádi, T.; Grasso, S.; Dusza, J.; Reece, M. Processing and Properties of High-Entropy Ultra-High Temperature Carbides. *Sci. Rep.* **2018**, *8*, 8609. [CrossRef] [PubMed]
8. Dusza, J.; Peter, Š.; Girman, V.; Sedlák, R.; Castle, E.G.; Csanádi, T.; Koval, A.; Reece, M.J. Microstructure of (Hf-Ta-Zr-Nb)C high-entropy carbide at micro and nano/atomic level. *J. Eur. Ceram. Soc.* **2018**, *38*, 4303–4307. [CrossRef]
9. Munir, Z.A.; Anselmi-Tamburini, U.; Ohyanagi, M. The effect of electric field and pressure on the synthesis and consolidation of materials: A review of the spark plasma sintering method. *J. Mater. Sci.* **2006**, *41*, 763–777. [CrossRef]
10. Guillon, O.; Gonzalez-Julian, J.; Dargatz, B.; Kessel, T.; Schierning, G.; Räthel, J.; Herrmann, M. Field-assisted sintering technology/spark plasma sintering: Mechanisms, materials, and technology developments. *Adv. Eng. Mater.* **2014**, *16*, 830–849. [CrossRef]
11. Bellosi, A.; Monteverde, F.; Sciti, D. Fast densification of ultra-high-temperature ceramics by spark plasma sintering. *Int. J. Appl. Ceram. Technol.* **2006**, *3*, 32–40. [CrossRef]
12. Zhou, J.; Zhang, J.; Zhang, F.; Niu, B.; Lei, L.; Wang, W. High-entropy carbide: A novel class of multicomponent ceramics. *Ceram. Int.* **2018**, *44*, 22014–22018. [CrossRef]
13. Demirskyi, D.; Borodianska, H.; Suzuki, T.S.; Sakka, Y.; Yoshimi, K.; Vasylkiv, O. Scripta Materialia High-temperature fl exural strength performance of ternary high-entropy carbide consolidated via spark plasma sintering of TaC, ZrC and NbC. *Scr. Mater.* **2019**, *164*, 12–16. [CrossRef]
14. Zhang, H.; Hedman, D.; Feng, P.; Han, G.; Akhtar, F. A high-entropy $B_4(HfMo_2TaTi)C$ and SiC ceramic composite. *Dalt. Trans.* **2019**, *48*, 5161–5167. [CrossRef]
15. Zhang, H.; Feng, P.; Akhtar, F. Aluminium matrix tungsten aluminide and tungsten reinforced composites by solid-state diffusion mechanism. *Sci. Rep.* **2017**, *7*, 12391. [CrossRef]
16. Agarwal, B.K. *X-ray Spectroscopy: An Introduction*; Springer: Berlin/Heidelberg, Germany, 2013; Volume 15, ISBN 3540386688.
17. CDowson, G. *Powder Metallurgy: The Process and Its Products*; Adam Hilger: Bristol, UK, 1990; ISBN 0852740069.
18. Simões, J.A.M.; Beauchamp, J.L. Transition Metal–Hydrogen and Metal–Carbon Bond Strengths: The Keys to Catalysis. *Chem. Rev.* **1990**, *90*, 629–688. [CrossRef]
19. TaC Crystal Structure: Datasheet from "PAULING FILE Multinaries Edition - 2012" in Springer Materials. Available online: https://materials.springer.com/isp/crystallographic/docs/sd_0261184 (accessed on 6 May 2019).
20. TaB2 Crystal Structure: Datasheet from "PAULING FILE Multinaries Edition - 2012" in Springer Materials. Available online: https://materials.springer.com/isp/crystallographic/docs/sd_0260764 (accessed on 6 May 2019).
21. Fick, A. Ueber Diffusion. *Ann. Phys.* **1855**, *170*, 59–86. [CrossRef]
22. Malik, R.; Burch, D.; Bazant, M.; Ceder, G. Particle size dependence of the ionic diffusivity. *Nano Lett.* **2010**, *10*, 4123–4127. [CrossRef]
23. Ferret, R.; Lo, R.; Gutie, M.A. Influence of particle size distribution of precursor oxides on the synthesis of cordierite by solid-state reaction. *Powder Technol.* **2005**, *153*, 34–42.
24. Kim, H.; Kim, J.; Jung, W.; Yoon, D. Effect of starting materials on the properties of solid-state reacted barium titanate powder. *J. Ceram. Process. Res.* **2009**, *10*, 753–757.

25. Buscaglia, M.T.; Bassoli, M.; Buscaglia, V.; Vormberg, R. Solid-State Synthesis of Nanocrystalline $BaTiO_3$: Reaction Kinetics and Powder Properties. *J. Am. Ceram. Soc.* **2008**, *91*, 2862–2869. [CrossRef]
26. Srivatsan, T.S.; Woods, R.; Petraroli, M.; Sudarshan, T.S. An investigation of the influence of powder particle size on microstructure and hardness of bulk samples of tungsten carbide. *Powder Technol.* **2002**, *122*, 54–60. [CrossRef]

entropy

MDPI

Article

Microstructure and Mechanical Properties of Particulate Reinforced NbMoCrTiAl High Entropy Based Composite

Tianchen Li [1], Bin Liu [1,*], Yong Liu [1], Wenmin Guo [2,*], Ao Fu [1], Liangsheng Li [1], Nie Yan [3] and Qihong Fang [4]

1 State Key Laboratory for Powder Metallurgy, Central South University, Changsha 410083, China; litianchencsu@163.com (T.L.); yonliu@csu.edu.cn (Y.L.); iceiceice_isu@hotmail.com (A.F.); lilightsheng@csu.edu.cn (L.L.)
2 College of Mechanical and Energy Engineering, Shaoyang University, Shaoyang 422000, China
3 YuanMeng Precision Technology (Shenzhen) Institute, Shenzhen 518000, China; nieyan725@163.com
4 College of Mechanical and Vehicle Engineering, Hunan University, Changsha 410082, China; fangqh1327@hnu.edu.cn
* Correspondence: binliu@csu.edu.cn (B.L.); wenminguo@hotmail.com (W.G.); Tel.: +86-731-8887-7669 (B.L.); +86-739-530-5016 (W.G.)

Received: 13 June 2018; Accepted: 6 July 2018; Published: 10 July 2018

Abstract: A novel metal matrix composite based on the NbMoCrTiAl high entropy alloy (HEA) was designed by the in-situ formation method. The microstructure, phase evolution, and compression mechanical properties at room temperature of the composite are investigated in detail. The results confirmed that the composite was primarily composed of body-centered cubic solid solution with a small amount of titanium carbides and alumina. With the presence of approximately 7.0 vol. % Al_2O_3 and 32.2 vol. % TiC reinforced particles, the compressive fracture strength of the composite (1542 MPa) was increased by approximately 50% compared with that of the as-cast NbMoCrTiAl HEA. In consideration of the superior oxidation resistance, the P/M NbMoCrTiAl high entropy alloy composite could be considered as a promising high temperature structural material.

Keywords: high entropy alloys; metal matrix composites; mechanical properties

1. Introduction

Recently the concept of high-entropy alloys (HEAs), consisting of four or more principle metallic elements in equiatomic or near-equiatomic ratios, has attracted considerable interest owing to its tendency to form solid solution with outstanding mechanical and functional properties [1–6]. By introducing refractory elements including group IV (Ti, Zr, and Hf), V (V, Nb, and Ta), and VI (Cr, Mo, and W) in HEAs, the developed refractory HEAs (RHEAs) possess high melting temperature, outstanding strength and hardness, high thermal stability, and softening resistance at elevated temperatures, which opens up the possibilities of the alloy developments satisfying structural demands at high temperature.

Since the first WMoTaNb RHEA was reported in 2010, many literatures have been focused on the preparation methods and mechanical properties of RHEAs [7,8]. For example, the BCC WMoTaNb and WMoTaNbV HEAs maintained yield strength of 405 and 477 MPa at 1600 °C respectively, which were much higher than that of Inconel 718 and Haynes 230 [9]. However, most of these RHEAs are generally characterized with high density and poor high temperature oxidation resistance. Such drawbacks are the current bottleneck for utilizing RHEAs as structural materials. In order to improve the oxidation resistance properties of the RHEAs, it was reported that the Ti, Al, Cr, and Si elements are essential for the formation of protective oxide scales. Compared with the traditional alloys, it was widely reported that the RHEAs possessed low diffusivity as well as a wider range of composition design [10]. In addition, it was also reported that the density of RHEAs could be significantly reduced by the additions of light elements, such as Ti [11,12]. Therefore, on the basis of design principles above, a new RHEA with equiatomic composition of NbMoCrTiAl was designed by B. Gorr et al. [13,14]. The as-cast alloy with low density (7.58 g/cm^3) exhibits excellent oxidation resistance. The declining oxidation rate at 1100 °C was mainly due to the formation of Al, Cr-rich oxide layers. However, the compressive strength of the as-cast NbMoCrTiAl alloy was reported to be as low as 1010 MPa and its high temperature strength is insufficient.

As reported in the literature, most of the RHEAs were prepared by the arc melting process. These as-cast alloys generally exhibited a dendritic structure with severe composition segregation [11–13]. Powder metallurgy (P/M), which is considered a high efficiency technique, is widely used to synthesize non-equilibrium materials. The materials characterized with fine grains are without chemical segregation and evaporation of alloying elements caused by arc-melting [15,16]. As reported, P/M WMoTaNbV shows fine grains of micrometer-scale and ultra-high compressive yield strength of 2612 MPa [17]. Therefore, it is an effective way to prepare high performance RHEAs.

Another prospective candidate for the high strength structural materials was the metal matrix composite (MMCs). The high performance was mainly attributed to the interaction between the particles and dislocations, called dispersion strengthening effect and second-phase strengthening effect. As the NbMoCrTiAl matrix is selected, the crucial challenge is to select the suitable strengthening phase and numbers of papers can be used as references [18–21]. The reinforcement particles include carbides (TiC, SiC), oxides (Al_2O_3, Y_2O_3), nitrides, (Si_3N_4, BN), and borides (TiB_2, LaB_6). There are mainly two ways to prepare carbides and oxides strengthened composites. One simple measure is to add them directly, another way is to form in-situ carbides or oxides by introducing C and O elements, which can effectively reduce material defects like pores and cracks. It was reported that the TiC and Al_2O_3 are also commonly used as strengthening phases because of their intrinsic high strength characteristics [22,23].

In this work, in order to further increase the strength of NbMoCrTiAl RHEA, the P/M composite with dispersed carbides and oxides was developed through ball milling and spark plasma sintering (SPS) method. The phase constitutions, microstructures, and mechanical properties of the NbMoCrTiAl HEA composite were investigated in detail.

2. Materials and Methods

Powders of Nb, Mo, Cr, Ti, and Al with purity higher than 99.5 wt. % and particle size of $\leq$45 μm were used as starting materials. The elemental powder was mixed in equiatomic composition and then processed by high energy planetary ball milling for 39 h at 300 rpm in a pure argon atmosphere (YXQM-2L, MITR, Changsha, China). Tungsten carbide vials and balls were utilized as the milling media with a ball-to-powder ratio of 10:1. The empirical addition amount of stearic acid was 3.5 wt. %. The stearic acid was introduced as process control agent (PCA) to prevent cold welding as well as particle agglomeration, and also acted as additive for generating particulate reinforced HEA composite. Subsequently, the as-milled powders were consolidated by SPS (D25/3, FCT, Munich, Germany) under 30 MPa axial pressure at 1700 °C with the heating rate of 100 K/min. During the sintering process, the sample was held at the sintering temperature for 30 min. Finally the sample cooled to room temperature in furnace.

The samples after sintering were analyzed by X-ray diffraction (XRD, D/MAX-2250, Rigaku, Tokyo, Japan) with a Cu Kα radiation at 40 kV and 200 mA. The microstructure of the composite was characterized by field emission scanning electron microscope (FESEM, Quanta FEG250, FEI, Hillsboro, OR, USA). The approximate volume fraction and average grain size of different phases was evaluated by at least ten images using Image Pro software. The phase composition and element distribution were investigated with electron probe microanalysis (EPMA, EPMA-1600, JEOL, Tokyo, Japan). Microhardness of the sample was measured by a Vickers microhardness tester (MicroMet-5104,

Buehler, Lake Bluff, IL, USA) with an indenter load of 100 g and the holding time was 15 s. Cylindrical specimens of compression tests (4 mm in diameter and 6 mm in height) were cut and machined from the sintered bulk. Room temperature compression tests were carried out on an Instron-3369 (INSTRON, Norwood, MA, USA) universal testing machines at strain rate of 10^{-3} s^{-1}.

3. Results

3.1. Microstructure and Phase Evolution after SPS

The general microstructure of the HEA composite is shown in Figure 1. The specimens are highly densified by the reaction sintering process without visible pores. The composite presented three obviously identifiable contrasts, namely light, grey, and black regions (marked as A, B, and C). The volume fraction of the light phase, grey phase, and black phase are 60.8%, 32.2%, and 7.0%, respectively. Since the content of carbon and oxygen cannot be measured accurately by EDS, EPMA experiment was employed to analyze the chemical composition in these areas (Table 1). The results indicate that the region A consists of only 10.66 at. % Ti and 16.79 at. % Al, while Ti and Al atoms are enriched in the region B and C, respectively. Noticeably, the region C in Figure 1 was enriched in Al and O, indicating the formation of Al_2O_3.

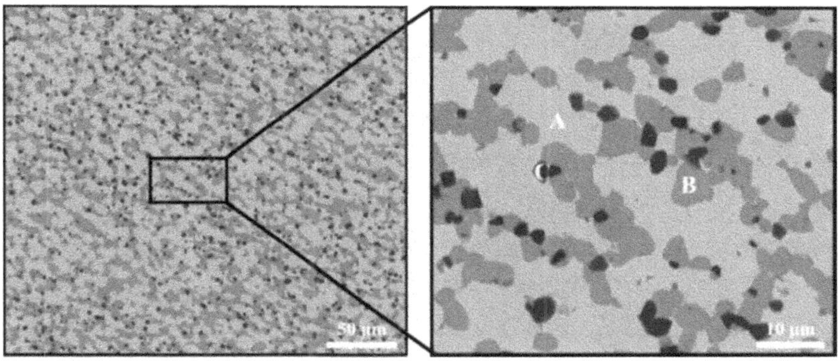

Figure 1. Microstructure of the NbMoCrTiAl composite after spark plasma sintering at 1700 °C for 30 min under a pressure of 30 MPa.

Table 1. Chemical composition of high entropy alloy (HEA) composite determined by electron probe microanalysis (EPMA) analysis.

Region	Nb	Mo	Cr	Ti	Al	O	C
A	20.29	30.95	19.34	10.66	16.79	0.69	1.28
B	15.40	-	-	43.13	-	12.11	29.36
C	-	-	-	-	38.13	61.87	-

The XRD patterns of the RHEA composite were illustrated in Figure 2. Two types of characteristic diffraction peaks could be identified, indicating the present material is primarily composed of BCC crystalline structure (a = 0.315 nm) with diffraction peaks at 2θ = 43.66°, 50.82° 74.67° accompanied with a certain amount of TiC phase (a = 0.435 nm) as the reinforcing phase. Al_2O_3 cannot be detected by X-ray analysis, which was probably ascribed to the limit detection of 5%. In consideration of the EDS analysis results, it could be concluded that the light, grey, and black phase were identified as BCC HEA phase, TiC phase, and Al_2O_3 phase, respectively. The average particle size of TiC and Al_2O_3 were measured to be approximately 4.60 μm and 1.57 μm, respectively.

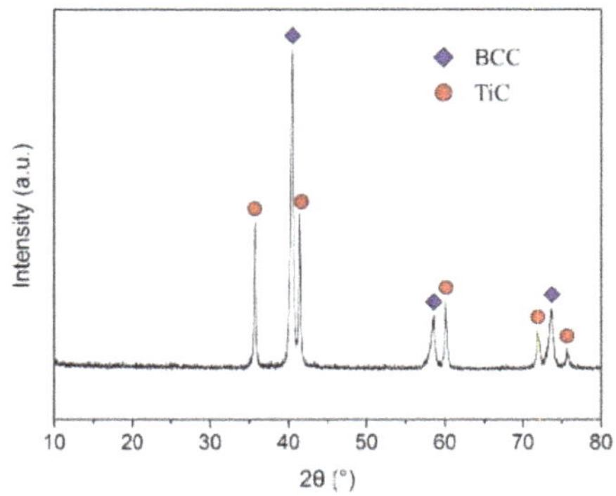

Figure 2. XRD pattern of the NbMoCrTiAl composite.

In order to further investigate the element distribution, the element mapping of the RHEA composite was shown in Figure 3. It was clear that the Mo and Cr elements are enriched almost entirely in the light region (BCC phase). On the contrary, the Ti and C elements are enriched almost entirely in the grey region (TiC phases) with certain amount of Nb, while O and Al are enriched in the black region (Al_2O_3 phase). No apparent debonding between dispersed particles and the HEA matrix could be found, which suggested that the secondary phase grains and the HEA grains possessed good bonding.

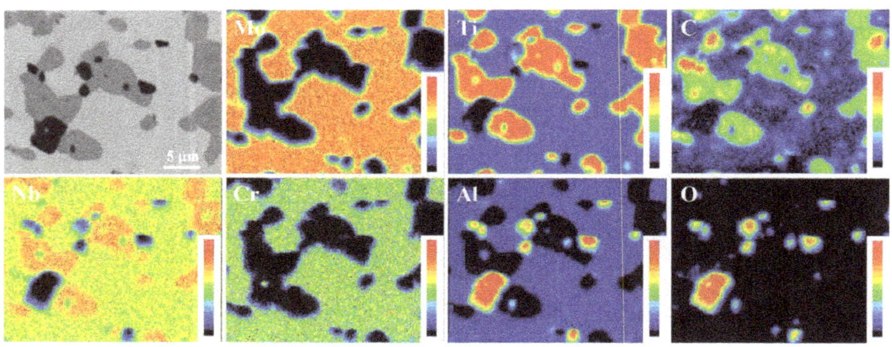

Figure 3. The elemental mapping of the composite obtained by electron probe microanalysis (EPMA).

3.2. Mechanical Properties

The room temperature engineering stress-strain curves of the $Nb_{20}Mo_{20}Cr_{20}Ti_{20}Al_{20}$ composite was given in Figure 4. Despite no ductility, the fracture strength of the composite was 1542 MPa, increasing by more than a half than that of the as-cast alloy (1010 MPa) [13]. The fracture strength was also much higher than that of the traditional WMoTaNb and WMoTaNbV RHEAs [24], which have yield strengths of 1246 MPa and 1058 MPa, respectively, as shown in Figure 4. As a result of the plasticity and oxidation resistance of the NbMoCrTiAl matrix at high temperature, it was concluded that the P/M NbMoCrTiAl HEA composite had relatively superior mechanical properties and could be considered as a potential high temperature structural material [25].

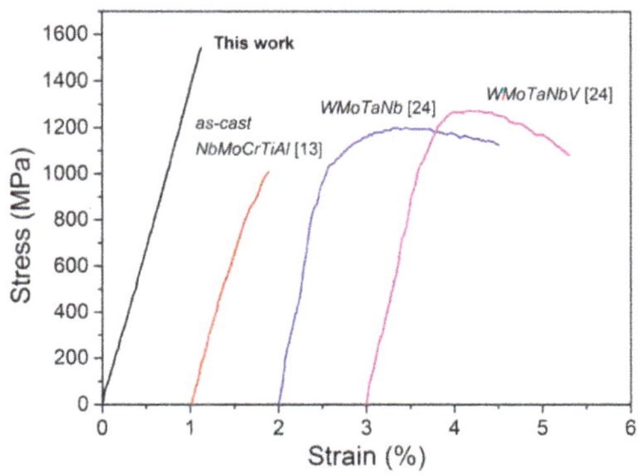

Figure 4. The room temperature engineering stress-strain curves of the composite and typical RHEAs reported in the literatures [13,24].

Microstructure features of the fracture surface of the composite deformed at room temperature are shown in Figure 5. Cleavage fracture is featured by river pattern, which could be clearly seen in Figure 5. This means that the NbMoCrTiAl HEA composite shows typical cleavage fracture mode without ductility. The analysis result is in accordance with the deformation curves above.

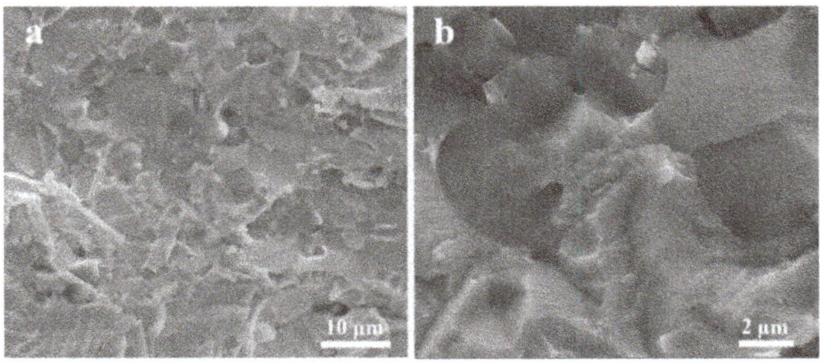

Figure 5. Microstructure features of the fracture surface of the composite deformed at room temperature (**a**) at low magnifications and (**b**) high magnifications.

4. Discussion

4.1. Phase Formation

Although a lack of HEA phase diagrams limit the availability of thermodynamic data, some thermodynamic properties, such as atomic size, melting point, mixing entropy, mixing enthalpy, and valence electron concentration, can be used as criteria to predict the phase formation in the HEA system. According to the empirical rules, BCC solid solutions were generally formed in HEA systems under the following certain conditions. (1) The electron concentration (VEC) was lower than 6.87 [26]; (2) a new parameter combining effects of entropy and enthalpy (Ω) was proposed to be more than 1 and atomic size differences (δ) was less than 6.6% [27]. δ, Ω, and VEC are calculated to be 5.4, 2.92 and 4.80 respectively, so BCC solid solution could be anticipated, which tallied well with the research results produced by H. Chen et al. [13]. Guo et al. [28] further proposed that all ductile alloys have a VEC $\leq$ 4.4, while all brittle alloys have a VEC $\geq$ 4.6. Therefore, the NbMoCrTiAl RHEA should be without ductility and it is indeed a type of brittle RHEA.

As mentioned above, BCC phase, TiC phase, and Al_2O_3 phase can be observed in Figure 1. The actual composition of BCC phase is $Nb_{20.3}Mo_{31.0}Cr_{19.3}Ti_{10.7}Al_{16.7}$, ignoring the contributions made by interstitial elements of carbon and oxygen. According to the actual composition of the BCC high entropy phase, all the parameters calculated for the composition presented in this paper are summarized (δ = 5.6, Ω = 3.46, VEC = 4.96). According to the data, the HEA phase should be of BCC structure, and Figure 2 verifies accuracy of the prediction. As a result, we could confirm that such non-equiatomic phase composition is able to enhance the solid solution especially at high temperature [29].

Combining high energy ball-milling of powder precursors with the following SPS compaction, the HEA composite was successfully synthesized in this work. The addition of stearic acid can be used as an effective way for the preparation of composite materials. During the process of high energy ball-milling, PCA is used to not only prevent cold-welding, but also act as the source of the elements carbon and oxygen. Table 2 shows the values of H_{mix} (kJ/mol) calculated by Miedema's model for atomic pairs between elements with Nb, Mo, Cr, Ti, Al, and C. As shown in Figure 2, this certain concentration of Nb incorporated during grain growth agrees with the large solubility of solute Nb-carbide in solvent TiC mainly because of their largest negative value of Hmix and identical crystal structure. While Mo, Cr, and Al could not dissolve in the TiC due to their relatively weak cohesive force [30].

Table 2. Mixing enthalpies (kJ/mol) of unlike atomic pair [30].

	Nb	Mo	Cr	Ti	Al	C
Nb	-					
Mo	−6	-				
Cr	−7	0	-			
Ti	2	−4	−7	-		
Al	−18	−5	−10	−30	-	
C	−102	−67	−61	−109	−36	-

4.2. Strengthening Mechanism

The composite created a kind of material with highly improved compressive strength, increasing from 1010 MPa to 1542 MPa. This remarkable increase in mechanical properties can be attributed to the following reasons. The most important one is the effect of second phase strengthening. It can be found that introduction of brittle ceramic reinforcements would significantly increase mechanical properties of the material. As listed in Table 3, the yield strength of $(FeCrNiCo)Al_{0.7}Cu_{0.5}$ base HEA reached 630 MPa accompanied by a high strain of 42.7%. With the addition of 10 vol. % TiC, the yield strength increased to 1290 MPa, while plastic strain dropped to 29.2% [31]. FeCoCrNiMn high entropy alloy matrix nanocomposite with addition of SiC was prepared by hot isostatic pressing, which result in the value of yield strength increased to 1600 MPa [18]. In addition, the interphase boundary between the in-situ reinforced particles and the solid solution matrix is free of cracks and pores, resulting in the more possibilities for the design of advanced high strength structural components.

Table 3. Room temperature mechanical properties of HEAs and their composites.

Sample	σ_y (MPa)	σ_{max} (MPa)	ε_p (%)	Hardness (HV)
$(FeCrNiCo)Al_{0.7}Cu_{0.5}$ [31]	630	2270	42.7	313
$(FeCrNiCo)Al_{0.7}Cu_{0.5}$ + 10 vol. % TiC [31]	1290	2444	29.2	580
FeCoCrNiMn [18]	1180	2660	34.5	418
FeCoCrNiMn + 5 wt. % SiC [18]	1600	1930	6.8	545

Another main factor is grain refinement strengthening mechanism. The grain size of the as-homogenized NbMoCrTiAl phase is about 100 μm, while that of the composite is much lower. This is because the PM process is an effective approach to refine grains compared to arc-melting [16]. In addition, the hard phases evenly distributed at the grain boundary, acting as the barrier of grain growth. Praveen et al. [32] reported that the grain growth of the FeCoCrNi composite could be obviously suppressed by two-phase mixture (FCC-HEA and carbide) microstructure.

The third reason is the influence of solution strengthening. It is true that the solution element could have remarkable strengthening effect. In this study, the BCC phase contains 0.69 at. % O and 1.28 at. % C, which may cause significant increase in strength because these two interstitial atoms have a substantially smaller size than any of the metal atoms in the HEA. These interstitial atoms will produce a substantial strain field, with which gliding dislocations will interact. There have been studies on the influence of interstitial elements, such as element C, O, and N. Wang et al. [33,34] reported that the addition of 1.1 at. % carbon to a novel single-phase FCC $Fe_{40.4}Ni_{11.3}Mn_{34.8}Al_{7.5}Cr_6$ HEA not only markedly increased the yield strength from 159 MPa to 355 MPa, but also led to a 25 % increase in the elongation to fracture. An increase in yield strength was also observed in the C-dissolved FeCoCrNi fabricated by selective laser melting [35]. With regard to the carbide, the lattice constant of TiC (0.435 nm) is slightly larger than pure TiC (0.433 nm). The different crystal parameter may result from the solution of atoms with large radius, such as element Nb (r = 0.148 nm) in this composite. Therefore, we may reasonably conclude that solution-strengthening effect resulted in the improved mechanical properties.

5. Conclusions

1. A novel metal matrix composite based on the NbMoCrTiAl high entropy alloy (HEA) was designed by the in-situ formation method. Strengthening phases include TiC particles and Al_2O_3 particles. The volume fraction of the BCC phase, TiC phase, and Al_2O_3 phase were 60.8%, 32.2%, and 7.0%, and the average particle size of TiC and Al_2O_3 were measured to be approximately 4.60 μm and 1.57 μm, respectively.
2. The Mo and Cr elements are enriched almost entirely in the BCC phase. By contrast, the Ti and C elements are enriched almost entirely in the TiC phases with certain amount of Nb, while O and Al are enriched in the Al_2O_3 phase. No apparent debonding between dispersed particles and the HEA matrix could be found, which suggested that the secondary phase grains and the HEA grains possessed good bonding.
3. The composite created a kind of material with highly improved compressive strength (1542 MPa) compared with that of the as-cast NbMoCrTiAl alloy (1010 MPa). On the basis of microstructure features of the fracture surface, typical cleavage fracture mode was found in the composite. The remarkable increase in mechanical properties can be attributed to second phase strengthening effect, grain refinement and solid solution effect.

Author Contributions: Y.L., B.L. and W.G. conceived and designed the experiments T.L. and A.F. prepared the NbMoCrTiAl HEA composite and performed the microstructural characterization of the composite under the supervision of Y.L. and B.L.; the mechanical testing under the supervision of B.L. and W.G. All authors discussed the results and approved the final manuscript.

Funding: This research was funded by the National Natural Science Foundation of China (51771232), the National Key Research and Development Plan of China (2016YFB0700302), the Science and Technology Planning Project of Hunan Province of China (2015SK1002-1), Natural Science Foundation of Hunan Province (2018JJ3477), the State Key Lab of Powder Metallurgy and Project of Innovation, and Entrepreneur Team Introduced by Guangdong Province, China (201301G0105337290).

Conflicts of Interest: The authors declare no conflicts of interest.

References

1. Liu, Y.; Wang, J.; Fang, Q.; Liu, B.; Wu, Y.; Chen, S. Preparation of superfine-grained high entropy alloy by spark plasma sintering gas atomized powder. *Intermetallics* **2016**, *68*, 16–22. [CrossRef]
2. Liu, B.; Wang, J.; Liu, Y.; Fang, Q.; Wu, Y.; Chen, S.; Liu, C.T. Microstructure and mechanical properties of equimolar FeCoCrNi high entropy alloy prepared via powder extrusion. *Intermetallics* **2016**, *75*, 25–30. [CrossRef]

3. Zhang, W.; Liaw, P.K.; Zhang, Y. Science and technology in high-entropy alloys. *Sci. China Mater.* **2018**, *61*, 2–22. [CrossRef]
4. Zhang, Z.; Axinte, E.; Ge, W.; Shang, C.; Wang, Y. Microstructure, mechanical properties and corrosion resistance of CuZrY/Al, Ti, Hf series high-entropy alloys. *Mater. Des.* **2016**, *108*, 106–113. [CrossRef]
5. Xu, J.; Axinte, E.; Zhao, Z.; Wang, Y. Effect of C and Ce addition on the microstructure and magnetic property of the mechanically alloyed fesibalni high entropy alloys. *J. Magn. Magn. Mater.* **2016**, *414*, 59–68. [CrossRef]
6. Shang, C.; Axinte, E.; Sun, J.; Li, X.; Li, P.; Du, J.; Qiao, P.; Wang, Y. CoCrFeNi($W_{1-x}Mo_x$) high-entropy alloy coatings with excellent mechanical properties and corrosion resistance prepared by mechanical alloying and hot pressing sintering. *Mater. Des.* **2017**, *117*, 193–202. [CrossRef]
7. Chen, R.; Qin, G.; Zheng, H.; Wang, L.; Su, Y.; Chiu, Y.; Ding, H.; Guo, J.; Fu, H. Composition design of high entropy alloys using the valence electron concentration to balance strength and ductility. *Acta Mater.* **2018**, *144*, 129–137. [CrossRef]
8. Senkov, O.; Isheim, D.; Seidman, D.; Pilchak, A. Development of a refractory high entropy superalloy. *Entropy* **2016**, *18*, 102. [CrossRef]
9. Senkov, O.N.; Wilks, G.B.; Miracle, D.B.; Chuang, C.P.; Liaw, P.K. Refractory high-entropy alloys. *Intermetallics* **2010**, *18*, 1758–1765. [CrossRef]
10. Miracle, D.B.; Senkov, O.N. A critical review of high entropy alloys and related concepts. *Acta Mater.* **2017**, *122*, 448–511. [CrossRef]
11. Liu, C.M.; Wang, H.M.; Zhang, S.Q.; Tang, H.B.; Zhang, A.L. Microstructure and oxidation behavior of new refractory high entropy alloys. *J. Alloys Compd.* **2014**, *583*, 162–169. [CrossRef]
12. Gorr, B.; Azim, M.; Christ, H.J.; Mueller, T.; Schliephake, D.; Heilmaier, M. Phase equilibria, microstructure, and high temperature oxidation resistance of novel refractory high-entropy alloys. *J. Alloys Compd.* **2015**, *624*, 270–278. [CrossRef]
13. Chen, H.; Kauffmann, A.; Gorr, B.; Schliephake, D.; Seemüller, C.; Wagner, J.N.; Christ, H.J.; Heilmaier, M. Microstructure and mechanical properties at elevated temperatures of a new al-containing refractory high-entropy alloy Nb-Mo-Cr-Ti-Al. *J. Alloys Compd.* **2016**, *661*, 206–215. [CrossRef]
14. Gorr, B.; Mueller, F.; Christ, H.J.; Mueller, T.; Chen, H.; Kauffmann, A.; Heilmaier, M. High temperature oxidation behavior of an equimolar refractory metal-based alloy 20Nb20Mo20Cr20Ti20Al with and without Si addition. *J. Alloys Compd.* **2016**, *688*, 468–477. [CrossRef]
15. Senkov, O.N.; Senkova, S.V.; Woodward, C. Effect of aluminum on the microstructure and properties of two refractory high-entropy alloys. *Acta Mater.* **2014**, *68*, 214–228. [CrossRef]
16. Omori, M. Sintering, consolidation, reaction and crystal growth by the spark plasma system (SPS). *Mater. Sci. Eng. A* **2000**, *287*, 183–188. [CrossRef]
17. Kang, B.; Lee, J.; Ryu, H.J.; Hong, S.H. Ultra-high strength WNbMoTaV high-entropy alloys with fine grain structure fabricated by powder metallurgical process. *Mater. Sci. Eng. A* **2017**, *712*, 616–624. [CrossRef]
18. Rogal, Ł.; Kalita, D.; Tarasek, A.; Bobrowski, P.; Czerwinski, F. Effect of SiC nano-particles on microstructure and mechanical properties of the CoCrFeMnNi high entropy alloy. *J. Alloys Compd.* **2017**, *708*, 344–352. [CrossRef]
19. Guo, N.N.; Wang, L.; Luo, L.S.; Li, X.Z.; Chen, R.R.; Su, Y.Q.; Guo, J.J.; Fu, H.Z. Microstructure and mechanical properties of in-situ MC-carbide particulates-reinforced refractory high-entropy $Mo_{0.5}NbHf_{0.5}ZrTi$ matrix alloy composite. *Intermetallics* **2016**, *69*, 74–77. [CrossRef]
20. Fu, Z.; Koc, R. Ultrafine TiB_2-TiNiFeCrCoAl high-entropy alloy composite with enhanced mechanical properties. *Mater. Sci. Eng. A* **2017**, *702*, 184–188. [CrossRef]
21. Riva, S.; Tudball, A.; Mehraban, S.; Lavery, N.P.; Brown, S.G.R.; Yusenko, K.V. A novel high-entropy alloy-based composite material. *J. Alloys Compd.* **2018**, *730*, 544–551. [CrossRef]
22. Rogal, Ł.; Kalita, D.; Litynska-Dobrzynska, L. CoCrFeMnNi high entropy alloy matrix nanocomposite with addition of Al_2O_3. *Intermetallics* **2017**, *86*, 104–109. [CrossRef]
23. Teber, A.; Schoenstein, F.; Têtard, F.; Abdellaoui, M.; Jouini, N. Effect of SPS process sintering on the microstructure and mechanical properties of nanocrystalline TiC for tools application. *Int. J. Refract. Met. Hard Mater.* **2012**, *30*, 64–70. [CrossRef]
24. Senkov, O.N.; Wilks, G.B.; Scott, J.M.; Miracle, D.B. Mechanical properties of $Nb_{25}Mo_{25}Ta_{25}W_{25}$ and $V_{20}Nb_{20}Mo_{20}Ta_{20}W_{20}$ refractory high entropy alloys. *Intermetallics* **2011**, *19*, 698–706. [CrossRef]

25. Waseem, O.A.; Lee, J.; Lee, H.M.; Ryu, H.J. The effect of ti on the sintering and mechanical properties of refractory high-entropy alloy TixWTaVCr fabricated via spark plasma sintering for fusion plasma-facing materials. *Mater. Chem. Phys.* **2017**, *210*, 87–94. [CrossRef]
26. Guo, S.; Ng, C.; Lu, J.; Liu, C.T. Effect of valence electron concentration on stability of fcc or bcc phase in high entropy alloys. *J. Appl. Phys.* **2011**, *109*, 103505. [CrossRef]
27. Zhang, Y.; Zhou, Y.J.; Lin, J.P.; Chen, G.L.; Liaw, P.K. Solid-solution phase formation rules for multi-component alloys. *Adv. Eng. Mater.* **2008**, *10*, 534–538. [CrossRef]
28. Sheikh, S.; Shafeie, S.; Hu, Q.; Ahlström, J.; Persson, C.; Veselý, J.; Zýka, J.; Klement, U.; Guo, S. Alloy design for intrinsically ductile refractory high-entropy alloys. *J. Appl. Phys.* **2016**, *120*, 164902. [CrossRef]
29. Yeh, J.W. Alloy design strategies and future trends in high-entropy alloys. *JOM* **2013**, *65*, 1759–1771. [CrossRef]
30. Takeuchi, A.; Inoue, A. Classification of bulk metallic glasses by atomic size difference, heat of mixing and period of constituent elements and its application to characterization of the main alloying element. *Mater. Trans.* **2005**, *46*, 2817–2829. [CrossRef]
31. Fan, Q.C.; Li, B.S.; Zhang, Y. The microstructure and properties of (FeCrNiCo)AlxCuy high-entropy alloys and their TiC-reinforced composites. *Mater. Sci. Eng. A* **2014**, *598*, 244–250. [CrossRef]
32. Sathiyamoorthi, P.; Basu, J.; Kashyap, S.; Pradeep, K.G.; Kottada, R.S. Thermal stability and grain boundary strengthening in ultrafine-grained CoCrFeNi high entropy alloy composite. *Mater. Des.* **2017**, *134*, 426–433. [CrossRef]
33. Wang, Z.; Baker, I.; Cai, Z.; Chen, S.; Poplawsky, J.D.; Guo, W. The effect of interstitial carbon on the mechanical properties and dislocation substructure evolution in $Fe_{40.4}Ni_{11.3}Mn_{34.8}Al_{7.5}Cr_6$ high entropy alloys. *Acta Mater.* **2016**, *120*, 228–239. [CrossRef]
34. Wang, Z.; Baker, I.; Guo, W.; Poplawsky, J.D. The effect of carbon on the microstructures, mechanical properties, and deformation mechanisms of thermo-mechanically treated $Fe_{40.4}Ni_{11.3}Mn_{34.8}Al_{7.5}Cr_6$ high entropy alloys. *Acta Mater.* **2017**, *126*, 346–360. [CrossRef]
35. Zhou, R.; Liu, Y.; Zhou, C.; Li, S.; Wu, W.; Song, M.; Liu, B.; Liang, X.; Liaw, P.K. Microstructures and mechanical properties of C-containing FeCoCrNi high-entropy alloy fabricated by selective laser melting. *Intermetallics* **2018**, *94*, 165–171. [CrossRef]

entropy

MDPI

Article

Effect of Binding and Dispersion Behavior of High-Entropy Alloy (HEA) Powders on the Microstructure and Mechanical Properties in a Novel HEA/Diamond Composite

Mingyang Zhang [1], Wei Zhang [1,*], Fangzhou Liu [1], Yingbo Peng [2], Songhao Hu [3] and Yong Liu [1]

1 Powder Metallurgy Research Institute, Central South University, Changsha 410083, China; hugezmy123@gmail.com (M.Z.); lyuhoushyuu@gmail.com (F.L.); yonliu@csu.edu.cn (Y.L.)

2 College of Engineering, Nanjing Agricultural University, Nanjing 210031, China; ybpengnj@njau.edu.cn

3 Henan Huanghe Whirlwind Co., Ltd., Xuchang 461500, China; husonghao2008@outlook.com

* Correspondence: waycsu@csu.edu.cn; Tel.: +86-731-8887-7669

Received: 29 October 2018; Accepted: 30 November 2018; Published: 4 December 2018

Abstract: This study reports the results of the addition of diamonds in the sintering process of a FCC-structured CoCrFeNiMo high-entropy alloy. The effect of raw powder states such as elemental mixed (EM) powder, gas atomization (GA) powder and mechanical alloying (MA) powder on the uniformity of constituent phase was also investigated. Examination of microstructure and evaluation of mechanical properties of the composites depending on the mixing processes were performed. As a result, GA+MA powder composite showed the highest mechanical properties. The experimental results indicated that the powder manufacturing method was an essential parameter to determine the quality of HEA/diamond composites such as the uniformity of phase and binding behavior.

Keywords: high-entropy alloy; diamond; composite; powder metallurgy; spark plasma sintering

1. Introduction

Nowadays, with the rapid development of advanced manufacturing techniques, especially in aerospace, photovoltaics, electronics and communications, there is growing demand for diamond tools [1,2]. Traditionally, diamond tool materials were prepared either by cold-pressing or pressing-and-sintering of a combination of single metal and pre-alloyed powders together with a matrix material, considering the poor thermal stability of diamond and limitations of molding processes, and it is difficult to guarantee the stability and consistency of performance of the matrix materials [3,4]. A new class of materials known as high-entropy alloys (HEAs) have attracted considerable interest by virtue of their outstanding mechanical properties such as high strength, wear and corrosion resistance [5–9]. Unlike a traditional multi-component matrix material, a HEA has a simple crystal structure without any complex phases. HEAs are expected to be adopted increasingly in diamond tools, owing to their impressive features with regard to stability of composition, microstructure, and mechanical properties. FeCoCrNi is considered as one of the most stable HEA alloys [10]. It can be a good matrix material to facilitate the study of the microstructure interface between the HEA matrix and diamond and the mechanical properties of the resulting HEA/diamond composites [11].

At present, the methods of preparing high-entropy alloy powders are mainly gas atomization (GA) and mechanical alloying (MA), and there are also some elemental mixed (EM) methods [12–14]. Different preparation methods have different effects on the shape, particle size and composition uniformity of the HEA powders. The purpose of this study is to investigate the effect of powder state (GA, GA + MA and EM + MA) on the microstructure and mechanical properties of HEA/diamond composites, so as to improve the strength, hardness and wear resistance of the composites and enhance the service life of diamond tool materials.

2. Materials and Methods

The investigated HEA matrix of FeCoCrNiMo with a nominal composition ($Fe_{24.1}Co_{24.1}Cr_{24.1}Ni_{24.1}Mo_{3.6}$ in wt.% was prepared using powder metallurgy. FeCoCrNiMo alloy powders were prepared by gas atomization, and were labeled as GA powders. The GA powders were mechanically milled using conventional planetary ball-milling equipment (the ratio of ball/powders was 10:1, 300 r/min, 20 h), the powders obtained by gas atomization and ball milling were labeled as GA + MA powders. Ball milling powders with five elemental powder mixtures were prepared under the same ball milling conditions, which were labeled as EM+MA powders, as shown in Figure 1. Then these were mixed with 4% (wt.%) synthetic diamond particles (140 mesh) by using a cylinder mixer for 6–8 h. The mixed HEA/diamond powders were then placed in a graphite die of 40-mm diameter and consolidated by using an HPD 25/3 SPS equipment under vacuum (10^{-3} Pa). The SPS process lasted for 20 min, of which the heating time was 10 min, the holding time was 8 min, and the cooling time was 2 min. The sintering temperatures was 950 °C, and the specific pressure was 35 MPa which was selected as SPS parameters for FeCoCrNiMo HEA/diamond composite sintering temperature. Under these conditions, the density, microstructure and properties of the HEA/diamond composite achieved the optimal state (unpublished results). A scanning electron microscope (SEM, FEI, Quanta 250 FEG, Vlastimila Pecha, Czech Republic) equipped with an energy dispersive X-ray (EDX) analyzer was used to investigate the microstructure and chemical compositions of the sintered samples. The phase constitution of the samples was characterized using X-ray diffractometer (XRD, PIGAKV, Rigaku D/MAX-2550 VB+18 Kw, Tokyo, Japan) utilizing Cu Kα radiation. Differential scanning calorimetry (DSC, Netzsh 449C, Selb, Germany) measurements were performed on sintered samples placed in an Al_2O_3 crucible and heated. The bending strength of the samples (size: $12 \times 2 \times 30$ mm^3) was determined by Instron 3369 mechanical testing facility (Instron, Norwood, MA, USA) using the three-point method. The hardness of the alloy was determined using a Buehler 5104 hardness tester (Buehler, Lake Bluff, USA) under a 200-g load for 15 s and was averaged from three measurements. The wearing behavior was measured by HRS-2M high-speed reciprocating line friction test equipment (Lanzhou Zhongke Kaihua Technology Development Co., Ltd., Lanzhou, China). The test parameters were a test time of 15 min, 50N loading, 15 Hz frequency (900 times/min), and 5 mm stroke.

Figure 1. FeCoCrNiMo HEA powders (**a**) GA; (**b**) GA + MA; (**c**) EM + MA.

3. Results and Discussion

3.1. Characterization of the Powders

The three different powders were characterized by SEM and XRD. It can be seen in Figure 1a, the GA powders were nearly spherical and had a large particle size distribution. The average particle diameter is about 45–50 μm. After 20 h of ball milling, the particle size of GA + MA powders and EA + MA powders were smaller than that before ball milling as shown in Figure 1b,c. Both of them became flake or strip shaped after ball milling, the particle size distribution of the powders decreased, and the previously larger spherical particles of GA powders were crushed to form fine particles. The powders with different particle size distribution will form different pore size after sintering [15], thus affecting its sintering densification. The particle size of powders has a great influence on the microstructure and mechanical properties. Ball milling is often used to reduce the particle size in order to speed up the diffusion reaction rate and obtain uniform and fine grains.

From the XRD results of Figure 2a, it can be seen that there was no alloying of GA + MA powders in the milling process and a face-centered-cubic structure (FCC) was maintained. However, the diffraction peaks of the powders were widened and the intensity was obviously reduced after ball milling, indicating that the grain size of the powders was reduced compared with GA powders. Because EM + MA powders were made of element powder by ball milling, mechanical alloying of different metal elements will occur under the action of higher energy in ball milling, resulting in the Cr-rich and Mo-rich area segregation. Moreover, as can be seen from Table 1, EA + MA powders exhibited a great deal of compositional inhomogeneity, especially of the element Cr, which would give rise to large amount of impurity phase after sintering. As shown in Figure 2b, for GA powders, the phase composition after SPS with and without ball milling was basically the same, mainly composed of FCC phase and diamond phase, indicating that there is no impurity formation in the ball milling process, corresponding to Figure 2a. However, for EM + MA powders, due to the inhomogeneity of the powder composition, metastable phases appeared after SPS, like a Cr-rich phase and a Mo-rich phase which were hard and brittle [16].

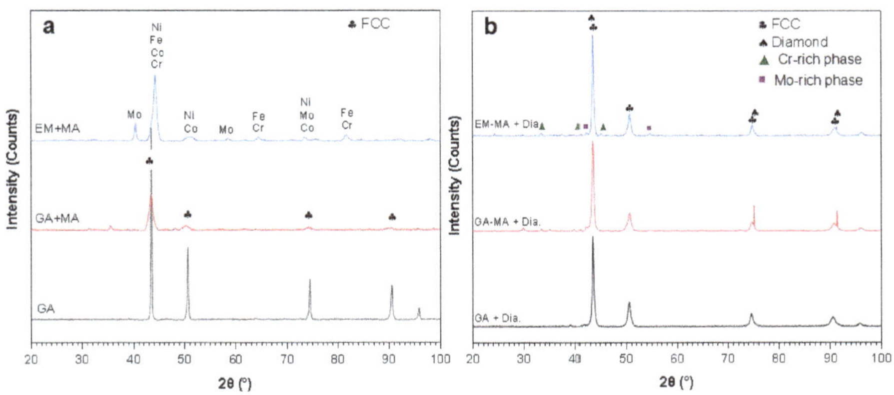

Figure 2. XRD pattern of (**a**) Three kind of FeCoCrNiMo HEA powders; (**b**) HEA/diamond composites after SPS by different powders.

Table 1. Composition of GA, GA + MA and EM + MA powders.

Elements	Fe	Co	Cr	Ni	Mo
Nominal composition (at. %)	24.10	24.10	24.10	24.10	3.6
GA powders	23.83 ± 0.15	22.89 ± 0.23	30.31 ± 0.34	18.53 ± 0.45	4.44 ± 0.23
GA-MA powders	26.70 ± 0.75	22.41 ± 0.50	29.07 ± 0.32	18.13 ± 0.34	3.68 ± 0.67
EM-MA powders	27.17 ± 14.13	15.41 ± 11.10	40.33 ± 29.17	14.07 ± 10.42	3.03 ± 2.00

3.2. Microstructure of HEA/Diamond Composites

Figure 3(a1–c1) show the microstructure of HEA matrix SPSed by GA, GA + MA, and EM + MA powders, respectively. It can be seen that the GA powder SPS sample exhibited a large number of holes left after sintering, and the densification was poor. The density of GA and EM powders increased significantly after ball milling.

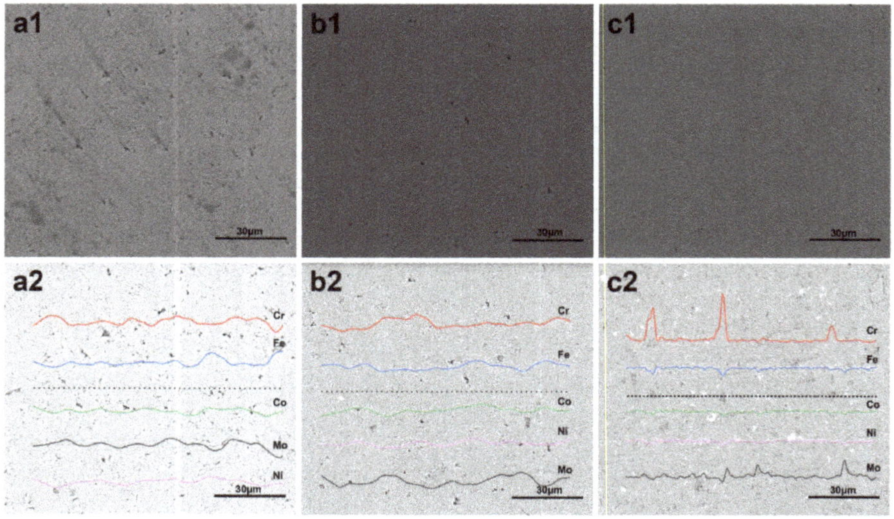

Figure 3. Microstructure (**a1**–**c1**) and element distribution (**a2**–**c2**) of HEA matrix SPSed by (**a**) GA powders, (**b**) GA + MA powders, and (**c**) EM + MA powders.

The average particle size of the powders after ball milling (GA + MA and EM + MA) was smaller than that of the GA powders, the voids between the particles are smaller, and the voids are easy to close in the densification stage during sintering. The larger the specific surface area of the particles are, the greater the sintering driving force is, so the GA + MA and EM + MA powders are easier to achieve sintering densification than the GA powders.

On the other hand, compactness is lower when the powder has a spherical shape than the irregular one after milling. The densification degree also accounts for the green density, prior to hot consolidation stage. The densification parameter is defined as the equation below [17]:

$$\text{Densification parameter} = (\text{Green density} - \text{Apparent density})/(\text{Theoretical density} - \text{Apparent density})$$

The theoretic density is 8.9 g/cm^3 for the HEA/diamond composite. The Green density is 7.32, 7.48 and 7.54 g/cm^3 for GA, GA + MA and EM + MA samples respectively. The apparent density is 4.12, 1.04 and 1.42 g/cm^3 for GA, GA + MA and EM + MA samples respectively. By calculation, the densification parameters are 0.67, 0.81 and 0.82 for the GA, GA + MA and EM + MA samples respectively. It shows that the densification degree of the samples with ball milling is greater than that without ball milling.

The EBSD images in Figure 3(a2–c2) show the element distribution of HEA matrix SPSed by the three kinds of powder. After GA powders and GA + MA powders were sintered, the matrix composition of the HEA/diamond composites was more evenly distributed. However, for EM + MA powders, Cr-rich and Mo-rich phases were bound to occur in the SPS process, according to the XRD pattern in Figure 2. Because elemental segregations of Cr and Mo were generated in the milling process, the components of Mo and Cr elements fluctuated greatly as seen in Figure 3(c2).The ball milling process fines the powder and introduces lattice distortion into the powder particles. The particle size decreases and the surface energy increases, and the repeated deformation of the powders during ball milling leads to a large number of lattice distortions in the particles after ball milling. On one hand, the decrease of particle size increases the surface energy, on the other hand, it shortens the diffusion path, which leads to the rapid completion of the reaction. Moreover, lattice distortion promotes the diffusion process [18], so that the sintering reaction can be completed at a lower temperature.

Figure 4a–c show the microstructures of HEA/diamond composite SPSed by GA, GA + MA and EM + MA powders. After the three kind of powders were sintered, the distribution of diamond particles in the composites was relatively uniform, but in the single diamond particle enlargement images shown in Figure 4d, the diamond grains maintain a well-defined crystalline shape and a complete crystal plane. The HEA matrix SPSed by GA powders were found to be closely packed with diamond. For the GA + MA sample, shown in Figure 4e, the diamond particles became partly spherical as the regular diamond particles were ablated into smaller spheres by the molten HEA matrix. For EM + MA sample, obvious spheroidization occurred in the diamond grains, that was attributed to the graphitization transformation of diamond particles. The sintering pressure, temperature and time of the three kind of powders are the same, but the diamond particles in the composites are graphitized. This is because, by means of mechanical ball milling, the distortion energy is introduced while the powder particles are refined, the reaction temperature is reduced and the reaction time is shortened.

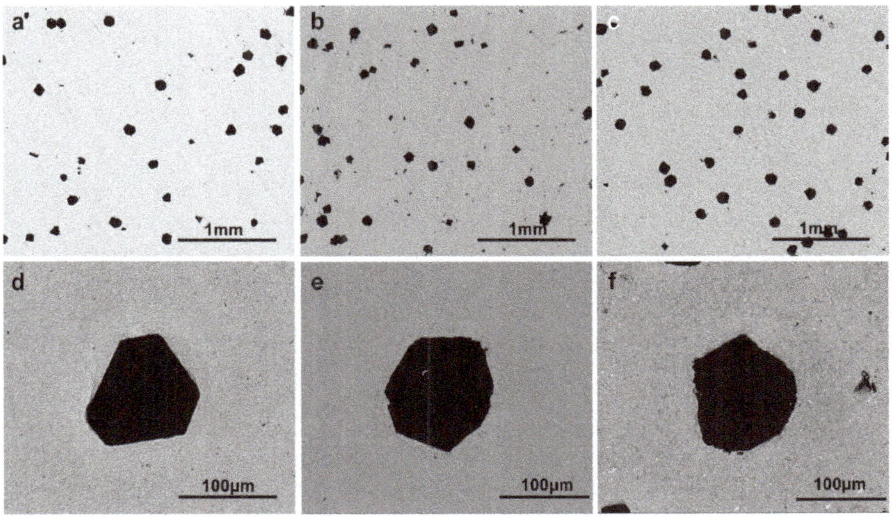

Figure 4. Microstructure of HEA/diamond composite SPSed by (**a**) and (**d**) GA powders; (**b**) and (**e**) GA + MA powders; and (**c**) and (**f**) EM + MA powders.

3.3. Mechanical Properties of HEA/Diamond Composites

3.3.1. Hardness and Bending Strength

Generally, the higher the density of the composites is, the higher the hardness is. The voids between particles of the MA powders are smaller, the voids are easy to be eliminated in the densification stage, and the sintering densification is easier to be realized. As shown in Figure 5, due to the smallest particle size after ball milling, the density of the composite prepared by EM + MA powders is the highest, and the sample prepared by GA powders is the lowest. However, the hardness of EM + MA sample decreases because of the inhomogeneous composition distribution. Considering hardness and density, the GA + MA sample is the best.

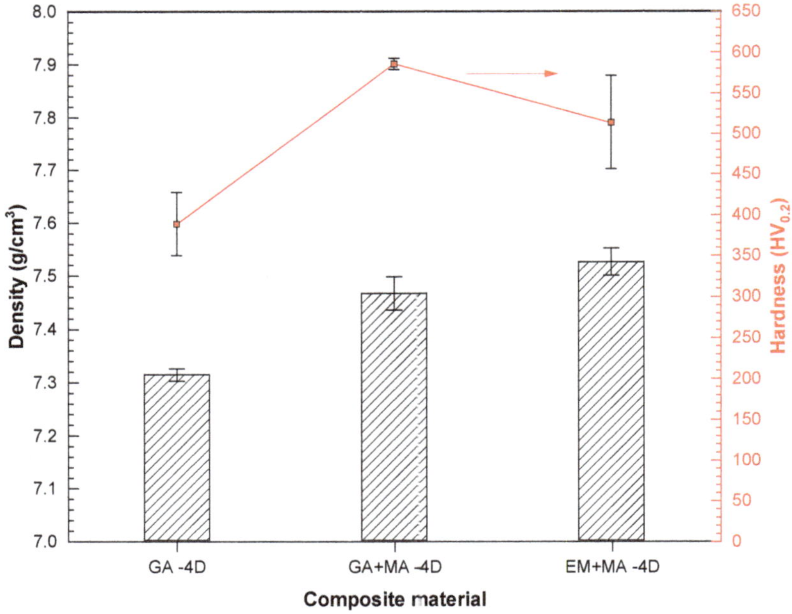

Figure 5. Hardness and density of HEA/diamond composites SPSed by three kinds of powders.

According to the bending stress-strain curves in Figure 6, the bending strength of GA + MA sample is 988 MPa, which is much higher than that of GA and EM + MA samples. The bending strength of GA sample is low is due to the microvoids. The bonding between diamond and matrix is poor, and there is no good metallurgical bonding. There is only a metallurgical bonding and mechanical bonding interface. The bending strength of EM + MA sample is very poor, because of diamond as a hard phase had undergone graphitization and the generation of the segregations like Cr-rich phase and Mo-rich phase. In addition, the elastic slope is different in the curves of the SPSed composites. This phenomenon results from the different density and segregation of samples SPS by different powders. For GA powder, the density is relatively low, microporosity and in HEA matrix results in lower elastic modulus. For EA + MA powders, despite the highest density, a large number of segregation occurred during SPS results in a decline of elastic modulus. GA + MA specimen has relatively higher density and no obvious segregation, so the elastic modulus is high and shows the optimal mechanical properties.

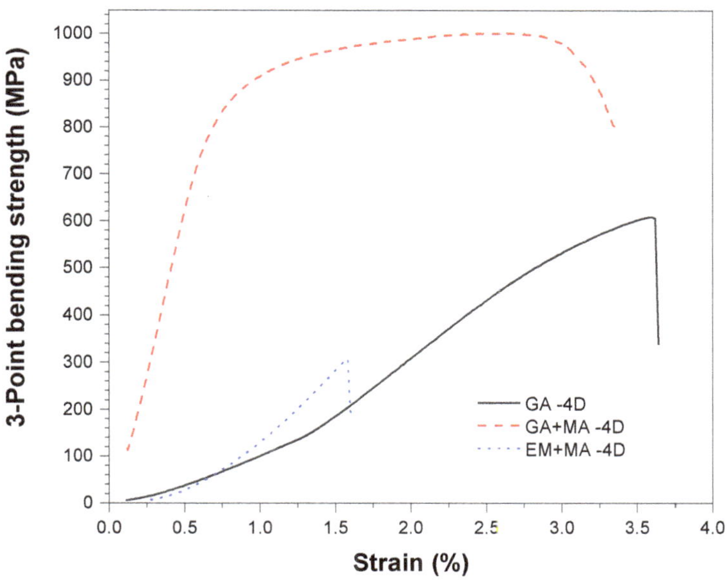

Figure 6. Bending stress-strain curves of HEA/diamond composites SPSed by three kinds of powders.

3.3.2. Wear Properties

The basic type of wear is the combination of abrasive wear and adhesive wear [19]. Generally speaking, the friction failure of diamond tools is diamond shedding or matrix wear. According to the wear mechanism, it can be inferred from Figure 7a,c that the abrasive wear was mainly attributed to abrasive wear and was accompanied by a part of adhesive wear because of the grooves and a small amount of abrasive debris on the surface. However, for GA samples, as shown in Figure 7a, there was an obvious interface between diamond particles and HEA matrix, and signs of breaking off were exhibited. This is because of the metallurgical bonding between diamond and HEA matrix is insufficient, and the loose structure and insufficient bonding will cause the diamond particles to break off from the HEA matrix and fail if friction continues. For the EM + MA sample, as shown in Figure 7c, the surfaces of diamond particles were markedly scratched and deformed after friction, which indicates that the diamond particles are graphitized and softened during sintering, thus losing their wear resistance as abrasive particles. For EM + MA sample, as shown in Figure 7b, although the diamond particles were worn, it had no sign of breaking off and softening. This illustrates that the HEA matrix has enhanced the grinding force to diamond particles, and improved the bonding strength of the interface, thus increasing the service life of the HEA/diamond composite. It is also observed from the wear scratch contour map shown in Figure 8 that the composite sample SPSed by GA + MA powders exhibited the minimum depth of wear scratches.

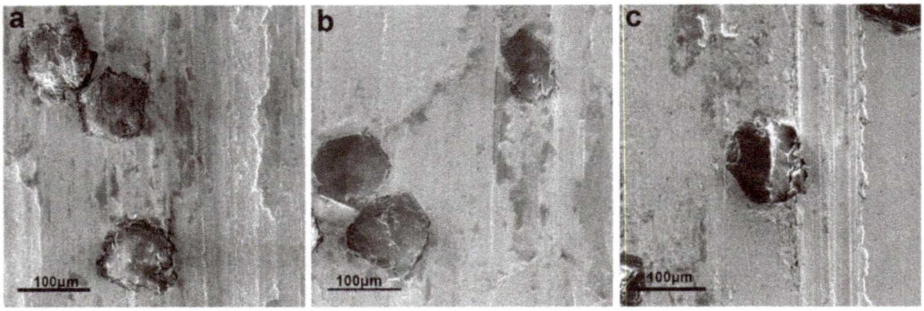

Figure 7. The wearing surface of HEA/diamond composite SPSed by (**a**) GA powders; (**b**) GA + MA powders; and (**c**) EM + MA powders.

Table 2 shows the wear rate of the HEA/diamond composites SPSed by different powders. The holding force of HEA matrix on diamond particles directly determines the performance of the composite. The interfacial bonding strength between diamond and HEA matrix is the key factor affecting the holding force.

Table 2. Wear rate of the HEA/diamond composites SPSed by different powders.

Composites	Wear Rate (mg/min)
GA-dia	0.27
(GA + MA)-dia	0.04
(EM + MA)-dia	0.06

Ball milling can reduce the sintering reaction temperature, improve the alloying degree and densification of the composites, and thus greatly improve the matrix holding force of diamond particles. It can be seen from Table 2 that the wear rate of GA sample is 0.27 mg/min, almost seven times higher than that of GA + MA sample which exhibits minimum worn quality loss. The wear rate of EM + MA sample is higher than GA + MA sample, because graphitization reduces the properties of diamond particles to a certain extent.

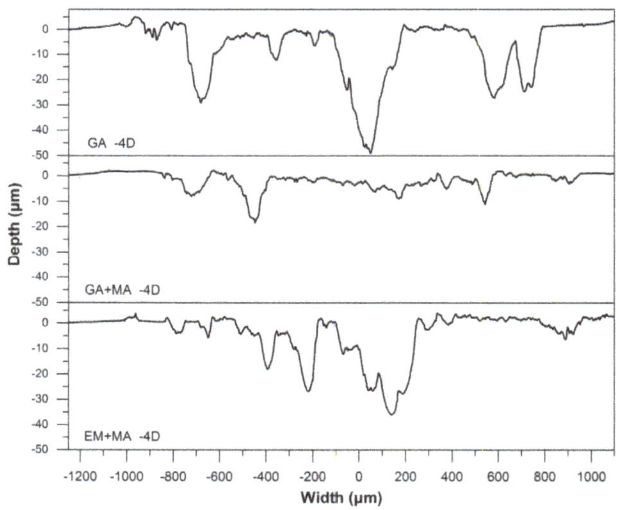

Figure 8. Wear scratch contour map of HEA/diamond composites SPSed by three kinds of powders.

As for the friction coefficient, as illustrated in Figure 9, GA + MA and EM + MA samples have similar frictional behavior characteristics: in the first four seconds, the friction coefficient shows a decline (from about 0.4 to about 0.12), and maintains a stable state (between 0.11 and 0.12) afterward. The friction behavior of GA sample is completely different from that of the other two samples. It enters a stable state at the beginning and the friction coefficient is about 0.05. This is mainly due to the integrity of diamond particles in GA samples, and the friction force mainly comes from the diamond particles, so the friction coefficient is always stable. GA + MA and EM + MA samples were ball milled to reduce the reaction temperature and time, which resulted in an interfacial reaction between diamond and HEA matrix. The HEA matrix is involved in friction. Therefore, the samples of GA + MA and EM + MA undergo a relatively long unstable period in the initial stage of friction. At the same time, the holding force of diamond matrix is improved because of the interfacial reaction, and the friction coefficient increases.

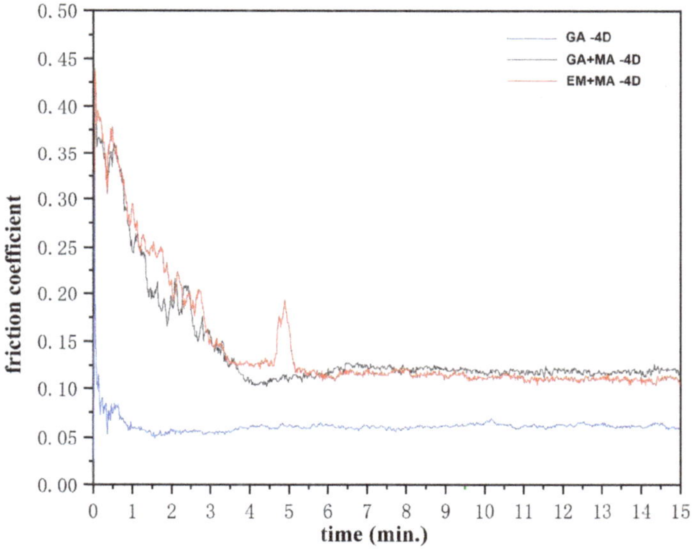

Figure 9. Friction coefficient-time curves of HEA/diamond composites SPSed by three kinds of powders.

4. Conclusions

The phases of GA powders before and after ball milling are basically the same, but the grain size of GA + MA powders decreases significantly. Due to the inhomogeneous distribution of elementary powders, impurity phases occur in EM + MA powders after ball milling. Through the ball milling process, the powder particles are refined and the distortion energy is introduced. The voids between the particles are smaller and easy to close in the densification stage during sintering. The reaction temperature is reduced and the sample is easily densified. The HEA matrix SPSed by GA and GA + MA powders were found to be closely packed with diamond. For the EM + MA samples, graphitization transformation of diamond particles occurred due to the reduction of the reaction temperature. The HEA/diamond composite SPS by GA + MA powders has the optimal hardness, bending strength and wear resistance. Although the EM + MA sample is more dense, the mechanical properties decrease due to the generation of segregations like Cr-rich phase and Mo-rich phase.

Author Contributions: Conceptualization, Y.L. and W.Z.; methodology, W.Z.; validation, F.L., W.Z. and Y.P.; formal analysis, W.Z.; investigation, M.Z. and F.L.; resources, S.H.; data curation, F.L.; writing—original draft preparation, M.Z.; writing—review and editing, W.Z. and Y.P.; project administration, Y.L.; funding acquisition, S.H.

Acknowledgments: The authors wish to acknowledge the financial support of State Key Laboratory of Powder Metallurgy, Central South University, Changsha, China (621011808) and XUCHANG Fellowship Program (XW2017-40).

Conflicts of Interest: The authors declare no conflict of interest.

References

1. Kobaru, Y.; Kondo, E.; Iwamoto, R. Ultra-Precision cutting of single crystal silicon using diamond tool with large top corner radius. *Key Eng. Mater.* **2012**, *523*, 81–86. [CrossRef]
2. Chon, K.S.; Takahashi, H.; Namba, Y. Wear inspection of a single-crystal diamond tool used in electroless nickel turning. *Opt. Eng.* **2014**, *53*, 1709–1717. [CrossRef]
3. Webb, S.W. Diamond retention in sintered cobalt bonds for stone cutting and drilling. *Diam. Relat. Mater.* **1999**, *8*, 2043–2052. [CrossRef]
4. Xu, H.J.; Fu, Y.C.; Xiao, B.; Xu, J.H. Fabrication of monolayer brazed diamond tools with optimum grain distribution. *Key Eng. Mater.* **2004**, *259*, 6–9. [CrossRef]
5. Chuang, M.H.; Tsai, M.H.; Wang, W.R.; Lin, S.J.; Yeh, J.W. Microstructure and wear behavior of AlxCo1.5CrFeNi1.5Tiy high-entropy alloys. *Acta Mater.* **2011**, *59*, 6308–6317. [CrossRef]
6. Senkov, O.N.; Wilks, G.B.; Scott, J.M.; Miracle, D.B. Mechanical properties of Nb25Mo25Ta25W25 and V20Nb20Mo20Ta20W20 refractory high entropy alloys. *Intermetallics* **2011**, *19*, 698–706. [CrossRef]
7. Zhang, Y.; Zuo, T.T.; Tang, Z.; Gao, M.C.; Dahmen, K.A.; Liaw, P.K.; Lu, Z.P. Microstructures and properties of high-entropy alloys. *Prog. Mater. Sci.* **2014**, *61*, 1–93. [CrossRef]
8. Hemphill, M.A.; Yuan, T.; Wang, G.Y.; Yeh, J.W.; Tsai, C.W.; Chuang, A.; Liaw, P.K. Fatigue behavior of Al0.5CoCrCuFeNi high entropy alloys. *Acta Mater.* **2012**, *60*, 5723–5734. [CrossRef]
9. Stepanov, N.D.; Yurchenko, N.Y.; Sokolovsky, V.S.; Tikhonovsky, M.A.; Salishchev, G.A. An AlNbTiVZr0.5 high-entropy alloy combining high specific strength and good ductility. *Mater. Lett.* **2015**, *161*, 136–139. [CrossRef]
10. Gludovatz, B.; Hohenwarter, A.; Catoor, D.; Chang, E.H.; George, E.P.; Ritchie, R.O. A fracture-resistant high-entropy alloy for cryogenic applications. *Science* **2014**, *345*, 1153–1158. [CrossRef] [PubMed]
11. Riva, S.; Tudball, A.; Mehraban, S.; Lavery, N.P.; Brown, S.G.R.; Yusenko, K.V. A novel High-Entropy Alloy-based composite material. *J. Alloy. Comp.* **2018**, *730*, 544–551. [CrossRef]
12. Budin, S.; Almanar, I.P.; Kamaruddin, S.; Maideen, N.C.; Zulkifli, A.H. Modeling of vial and ball motions for an effective mechanical milling process. *J. Mater. Process. Tech.* **2009**, *209*, 4312–4319. [CrossRef]
13. Zhou, P.; Liu, Y.; Yu, Y.; Xiao, D. Phase evolution and mechanical properties of AlCoCrFeNi high entropy alloys by spark plasma sintering. *Mater. Rev.* **2016**, *30*, 95–103.
14. Moravcik, I.; Cizek, J.; Gavendova, P.; Sheikh, S.; Guo, S.; Dlouhy, I. Effect of heat treatment on microstructure and mechanical properties of spark plasma sintered AlCoCrFeNiTi0.5 high entropy alloy. *Mater. Lett.* **2016**, *174*, 53–56. [CrossRef]
15. Bernard, F.; Le Gallet, S.; Spinassou, N.; Paris, S.; Gaffetm, E.; Woolman, J.N.; Munir, Z.A. Dense nanostructured materials obtained by spark plasma sintering and field activated pressure assisted synthesis starting from mechanically activated powder mixtures. *Sci. Sinter.* **2004**, *36*, 155–164. [CrossRef]
16. Tsai, M.H.; Chang, K.C.; Li, J.H.; Tsai, R.C.; Cheng, A.H. A second criterion for sigma phase formation in high-entropy alloys. *Mater. Res. Lett.* **2016**, *4*, 1–6. [CrossRef]
17. Upadhyaya, G.S. *Powder Metallurgy Technology*; Cambridge International Science Publishing: Cambridge, UK, 2002; p. 26.
18. Suryanarayana, C. Mechanical alloying and milling. *Prog. Mater. Sci.* **2001**, *46*, 1–184. [CrossRef]
19. Wu, P.; Du, H.M.; Chen, X.L.; Li, Z.Q.; Bai, H.L.; Jiang, E.Y. Influence of WC particle behavior on the wear resistance properties of Ni-WC composite coatings. *Wear* **2004**, *257*, 142–147. [CrossRef]

entropy

MDPI

Article

Effect of Ti/Ni Coating of Diamond Particles on Microstructure and Properties of High-Entropy Alloy/Diamond Composites

Wei Zhang [1,*], Mingyang Zhang [1], Yingbo Peng [2], Fangzhou Liu [1], Yong Liu [1], Songhao Hu [3] and Yang Hu [4]

1 Powder Metallurgy Research Institute, Central South University, Changsha 410083, China; hugezmy123@gmail.com (M.Z.); lyuhoushyuu@gmail.com (F.L.); yonliu@csu.edu.cn (Y.L.)
2 College of Engineering, Nanjing Agricultural University, Nanjing 210031, China; ybpengnj@njau.edu.cn
3 Henan Huanghe Whirlwind Co., Ltd., Xuchang 461500, China; husonghao2008@outlook.com
4 Yuanmeng Precision Technology (Shenzhen) Institute, Shenzhen 518055, China; yanghu_hust@126.com
* Correspondence: waycsu@csu.edu.cn; Tel.: +86-731-8887-7669

Received: 31 December 2018; Accepted: 5 February 2019; Published: 10 February 2019

Abstract: In this study, an effective way of applying Ti/Ni deposited coating to the surface of diamond single crystal particles by magnetron sputtering was proposed and novel high-entropy alloy (HEA)/diamond composites were prepared by spark plasma sintering (SPS). The results show that the interfacial bonding state of the coated diamond composite is obviously better than that of the uncoated diamond composite. Corresponding mechanical properties such as hardness, density, transverse fracture strength and friction properties of the coated diamond composite were also found to be better than those of the uncoated diamond composite. The effects of interface structure and defects on the mechanical properties of HEA/diamond composites were investigated. The research directions for further improving the structure and properties of high-entropy alloy/diamond composites were proposed.

Keywords: high-entropy alloy; diamond; coating; interface; mechanical properties

1. Introduction

Multi-principal high-entropy alloys break through the traditional alloy design mode based on one kind of alloy element. By optimizing the composition design, excellent performance combinations such as high strength, high hardness, high temperature creep resistance, high temperature oxidation resistance and corrosion resistance can be obtained [1–9]. Based on the excellent properties of high-entropy alloys (HEAs) and diamond, it is of great scientific value and application significance to design a novel high-entropy metal matrix binder for diamond tools and to develop the corresponding theory and technology of heterogeneous multi-phase interface control [10–12].

Under the application conditions, improving the interface state between matrix materials and diamond particles is the key problem to be urgently solved in the research field of diamond tool materials. For diamond tools with an HEA matrix, improving the wettability between the HEA matrix and diamond particles and effectively controlling the reaction products of the HEA/diamond interface are important for improving the interface bonding and overall application performance of diamond tools. Ti, Cr, Mo, V and other strong carbide-forming elements were always selected as coating materials [13–15], which can form a carbide layer on the surface of diamond particles to realize metallurgical bonding between diamond and the matrix. Meanwhile, the direct contact between Fe, Co, Ni and diamond particles can be avoided to prevent the formation of hard and brittle carbides at the interface [16,17]. Therefore, the ideal structure of metallurgical bonding between diamond particles and a metal matrix involves avoiding the formation of carbides.

In this paper, a novel HEA/diamond composite was studied by magnetron sputtering Ti/Ni coating and the spark plasma sintering (SPS) method. A Ti/Ni coating (inner Ti and outer Ni) of diamond particles was realized by magnetron sputtering. An HEA/coated diamond interface was obtained, composed of the solid solution formed by diffusion between the HEA matrix and Ni element in the outer layer of diamond particles. The carbide formed by the reaction of Ti element with diamond particles. On the one hand, this interface can effectively reduce the interface energy between diamond and the HEA matrix and improve the bonding strength. On the other hand, the cracks and micro-holes on the interface after SPS can be filled by generated carbides to improve the density of the composites and improve the strength, toughness and other mechanical properties of the composites.

2. Experimental

Standard MBD4-type synthetic diamond single crystals of 140/170 mesh were boiled and rinsed in HNO_3 and NaOH solutions for surface purification. Diamond was deposited in the magnetron sputtering coating equipment and high purity argon was introduced. The metals Ti and Ni with purity higher than 99.99% were used as targets. The diamond surface was coated by vacuum magnetron sputtering and maintained at room temperature. The vacuum of the reaction chamber was 10^{-4} Pa, the partial pressure of argon was 10^{-1} Pa, the ion deposition rate was 20 nm/min and the diamond was rolled by ultrasonic vibration in the diamond tray to ensure the uniformity of the coating. The thickness was controlled and was about 15 μm as shown in Figure 1. FeCoCrNiMo HEA powders prepared by gas atomization were applied as matrix materials as shown in Figure 1c, in which the average particle size was about 50 μm. Coated and uncoated diamond particles with a mass ratio of 4% were mixed with HEA powder and put into the mixer for 5 hours (mixer speed: 60 r/min). HEA/diamond composites were prepared by SPS (SPS parameters: 950 °C/35 MPa, holding time: 480 s).

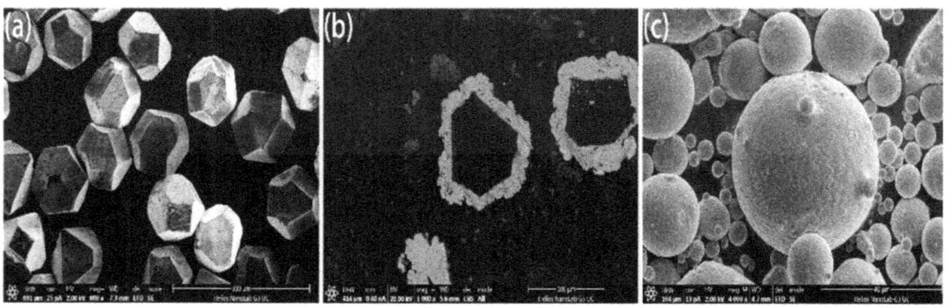

Figure 1. Microstructure of (**a**) uncoated and (**b**) coated diamond particles and (**c**) high-entropy alloy (HEA) powders.

The density of samples was measured by the Archimedes drainage method. The transverse fracture strength of the samples (size: 12 × 2 × 30 mm^3, span: 25 mm, loading rate: 2 mm/min) was determined by the Instron 3369 mechanical testing facility (Instron, Norwood, MA, USA) using the three-point method. The hardness of the alloy was determined using a Vickers hardness tester (200HVS-5, HuaYin, Zhengzhou) under a 200 g load for 15 s and was averaged from five measurements. The wearing behavior was measured by HRS-2M high-speed reciprocating line friction test equipment (Lanzhou Zhongke Kaihua Technology Development Co., Ltd., Lanzhou, China). The test parameters were a test time of 15 min, 50 N loading, 15 Hz frequency (900 times/min) and a 5-mm stroke. A scanning electron microscope (SEM, FEI, Quanta 250 FEG, Vlastimila Pecha, Czech Republic) equipped with an energy dispersive X-ray (EDX) analyzer was used to investigate the microstructure and chemical compositions of the sintered samples. Confocal Raman microscopy was performed on the interface of diamonds/HEA using an inVia Reflex by Renishaw with an Ar laser, using the green line (532 nm, 7.6 mW) with a resolution of 0.5 × 2 μm.

3. Results and Discussion

3.1. Interface of HEA/Diamond Composite

In the SEM analysis of the coated diamond (Figure 2 and Table 1), spot 1 at the edge of the diamond shows all C atoms. Spot 2 is in the transition zone between the diamond and the coated layer. C, Ti and Ni elements can be found in the area near the diamond, of which the content of Ni is relatively small. Spot 3 is in the coated layer, whereas that of Ni element is much more in the outer layer. This is because Ti is a strong carbide-forming element with strong chemical activity and diffusion ability. During the SPS process, Ti reacts with diamond to form a stable, chemically bonded TiC layer. At the same time, because the matrix is a five-element HEA of FeCoCrNiMo, the higher amount of Ni element in the outer layer can form a solid solution interface with the HEA matrix.

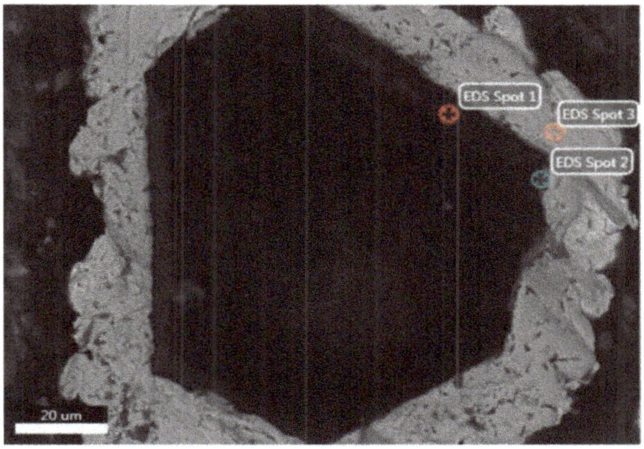

Figure 2. Microstructure of Ti/Ni-coated diamond particles.

Table 1. Chemical composition of different spots in Figure 2.

Composition (wt.%)	C	Ti	Ni	P	O
Spot 1	100	/	/	/	/
Spot 2	59.29	30.64	6.57	3.08	0.42
Spot 3	3.96	0.76	91.09	3.72	0.47

The density of coated diamond increases by about 1/3 due to its multi-layer. This is closer to the density of a high-entropy alloy matrix, which makes the mixture more uniform, reduces the number of voids and indirectly improves the density after sintering. The HEA matrix was found to form a metallurgical bonding with diamond and shows a thin layer with a width of approximately 3 μm at the interface between the diamond and the matrix. This indicates that the metallurgical bonding between the coated diamond particles and the matrix is relatively strong, as shown in Figure 3c,d, while the uncoated diamond resists mechanical bonding to a certain extent, as shown in Figure 3a,b.

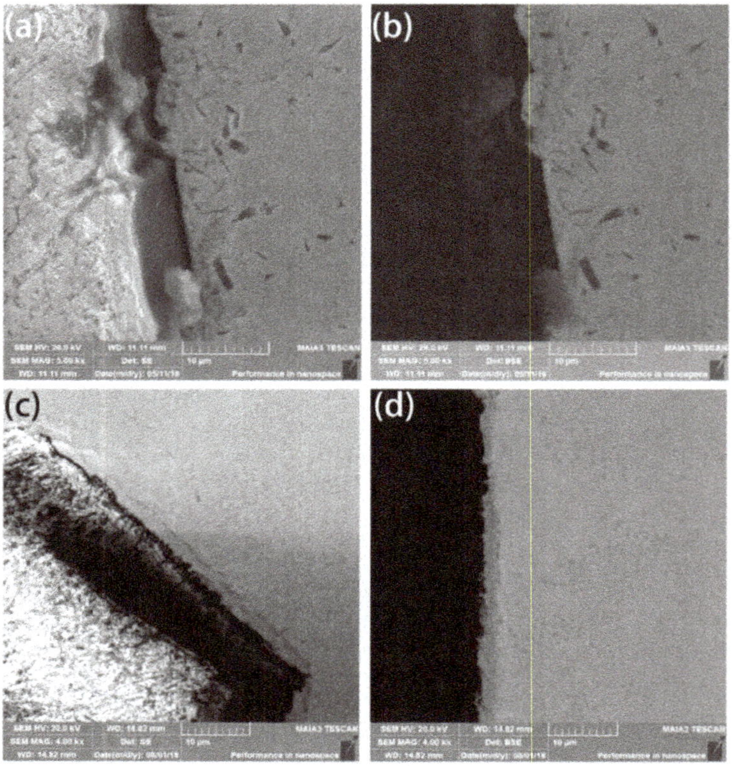

Figure 3. Scanning Electron Microscope (SEM) and Back Scattered Electron (BSE) images of (**a**,**b**) uncoated diamond particles and (**c**,**d**) coated diamond particles.

It was found from the Energy Dispersive Spectrometer (EDS) analysis of the interface of coated diamond/HEA composites that, as shown in Figure 4, there is a high content of Cr and Fe elements between the diamond and the high-entropy alloy matrix, that is, the segregation of Cr, Ti, Ni and Mo elements occurs and the reaction with diamond produces a solid solution layer with a higher bonding strength than that achieved via pure mechanical bonding. It can also be seen from the image that although C atoms aggregate at the interface, they do not extend to the matrix. Although the functional metallic layer has a certain corrosion effect on diamond, it does not destroy the crystalline form of diamond to a large extent. On the contrary, the transition layer restricts the diffusion of carbon atoms to the matrix. It can be seen from Figure 3c,d that the transition layer is different from the non-functional diamond. The smooth bonding surface of diamond composites and the functional diamond composites show a fold structure. This structure increases the bonding area between diamond and the matrix in the transition layer, so that it can be better bonded with the matrix. Moreover, the EDS surface analysis of the interface between coated diamond and the matrix was carried out. As shown in Figure 4b, in accordance with the results of liner scanning, there is obvious segregation at the interface and obvious precipitation of Cr and Mo. This indicates that the interface layer may be a solid solution composed of Cr, Mo, Ti and Ni, indicating that the functionalized layer can form a metallurgical bond with the HEA matrix that will increase the interfacial bonding strength.

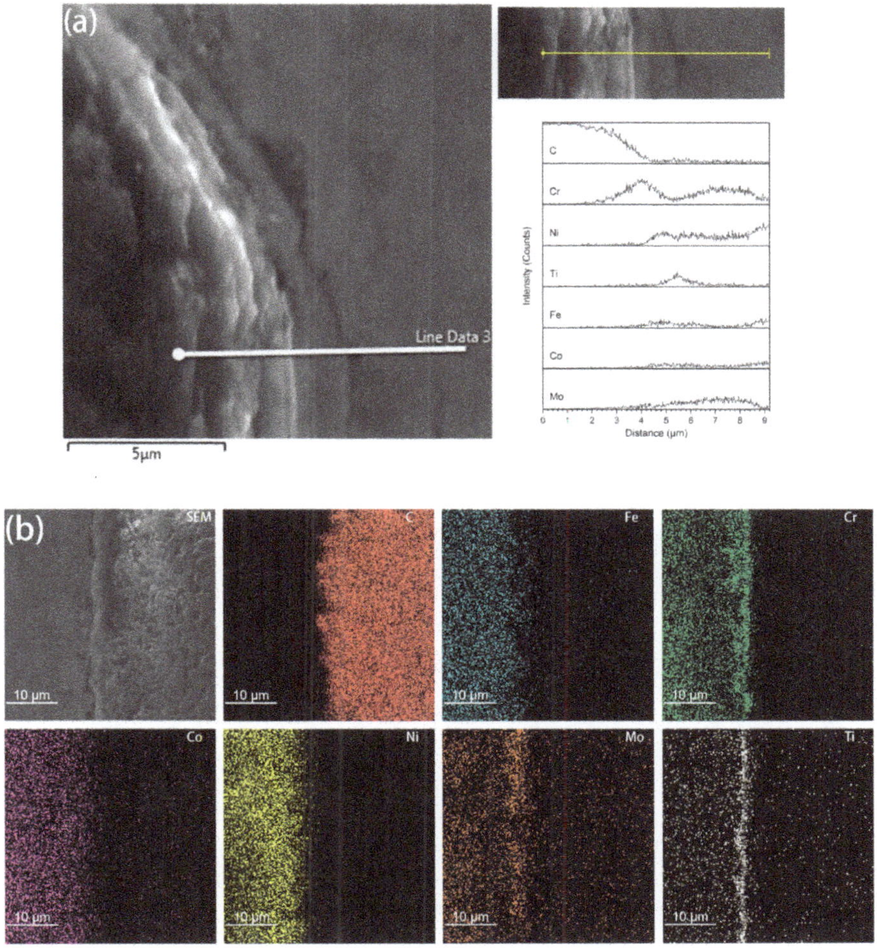

Figure 4. EDS liner scan (**a**) and element mapping (**b**) analysis on the interface of coated diamond/HEA composite.

The interface status of HEA matrix/diamond were also characterized by Raman spectroscopy, as shown in Figure 5. As shown in Figure 5a, there are diamond peaks at 1332 cm^{-1}, graphite peaks at 1350 cm^{-1} (D-band), 1580 cm^{-1} (G-band) and 2700 cm^{-1} (2D-band) and other bonds at the interface of the uncoated diamond specimens. The peak shape is sharp, the G-band strength is more than two times that of D-b (Figure 5b) and the area ratio of ID/IG (Intensity ratio of peak D to peak G) is about 0.52. The interface of the coated diamond specimens only exhibits a diamond peak at 1332 cm^{-1}. For the peaks at 1350 cm^{-1} (D-band) and 1580 cm^{-1} (G-band) (Figure 5c), the peak shape is wide, the G-band strength and D-band strength are close (Figure 5d) and the area ratio of ID/IG is about 2.35, which is obviously higher than that of the uncoated diamond samples. The results show that the graphitization degree of diamond particles in uncoated diamond samples is higher [18] and the other bonds indicate that there are carbides or oxides at the interface [19]. In conclusion, coated diamond can effectively inhibit the graphitization structural transformation of diamond particles in the SPS process and improve the material properties.

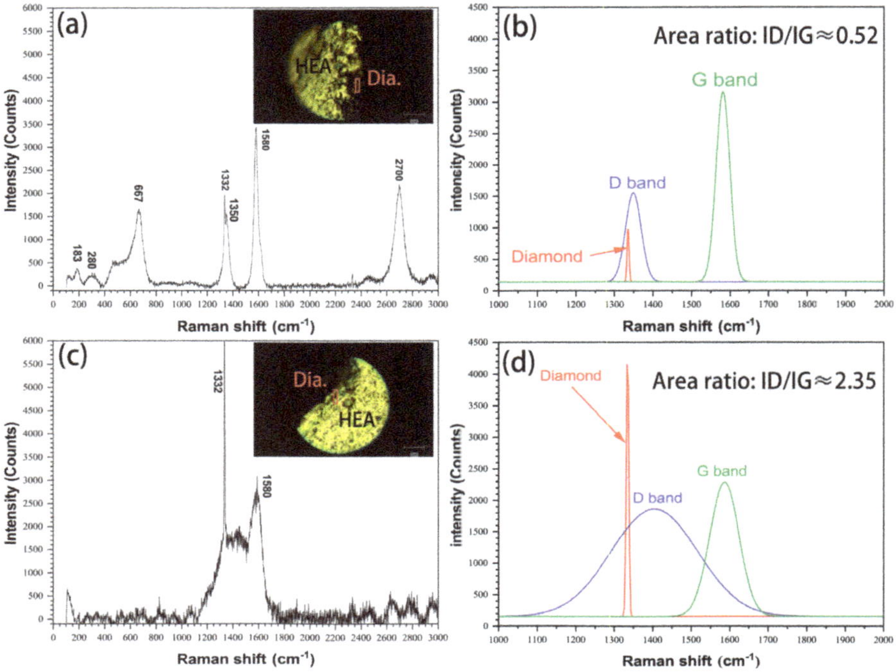

Figure 5. Raman spectra of HEA matrix/diamond interface in (**a**) uncoated diamond and (**c**) coated diamond, as well as the corresponding fitting curves (1000 cm^{-1} ~ 2000 cm^{-1}) (**b**,**d**).

3.2. Microstructure of the HEA/Diamond Composite

The microstructure of the composites fabricated by SPS with coated and uncoated diamond is shown in Figure 6. Uncoated diamond particles are severely ablated. After sintering, the original diamond with a regular shape displays jagged ablation on its surface, which cannot maintain the original excellent crystal form of the diamond as shown in Figure 6a,c. The coated diamond also has a certain degree of ablation, as shown in Figure 6b,c. However, due to the passivation of the coated metal layer, which prevents the corrosion of the matrix to diamond and reduces the diffusion of C element to the matrix, a good crystal form of diamond is maintained. Meanwhile, the coating contains Ti element, while Ti element and C element form carbides and the second-phase carbides disperse in the matrix, thus improving the hardness of the FeCoCrNiMo high-entropy alloy system [20]. Therefore, the friction and wear properties of materials are also affected.

According to the X-ray diffraction (XRD) analysis of the HEA/diamond composite, as shown in Figure 7, only two strong peaks of diamond were detected, because of the low content of diamond and the close distance between the Face-Centred Cubic (FCC)peak and the diamond peak of HEA. However, no carbide and C element were detected. This indicates that there was no large amount of ablation or graphitization of the diamond in coated diamond samples. Moreover, the HEA matrix had an FCC structure and there was no complex multiphase structure and no intermetallic compound found in coated samples in Figure 7. Therefore, the excellent properties of HEA matrix could be maintained. Yet, in uncoated diamond samples, the addition of diamond particles was found to form any complex carbide phase and there existed carbides of Cr according to the powder diffraction files (PDF), such as Cr23C6 (PDF No. 04-004-1672) and Cr7C3 (PDF No. 04-005-9649). These carbides were hard and brittle, which could deteriorate the mechanical properties of the composite.

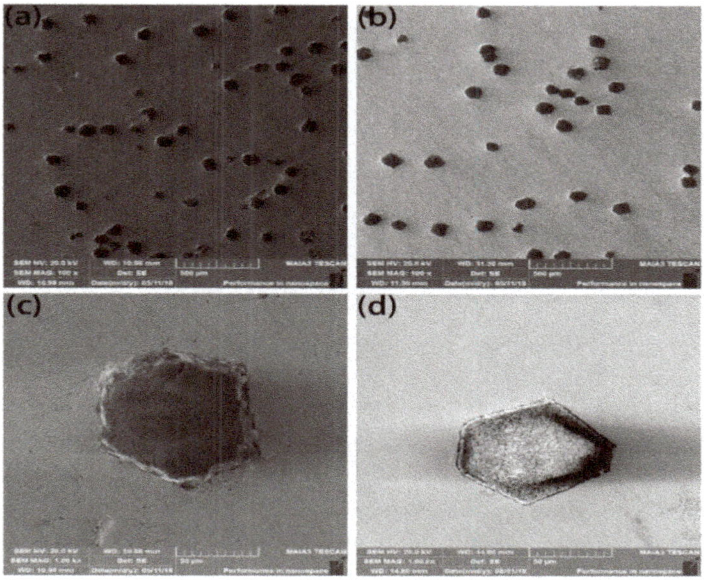

Figure 6. Microstructure of the composites fabricated by spark plasma sintering (SPS) with uncoated (**a**,**c**) and coated diamond (**b**,**d**).

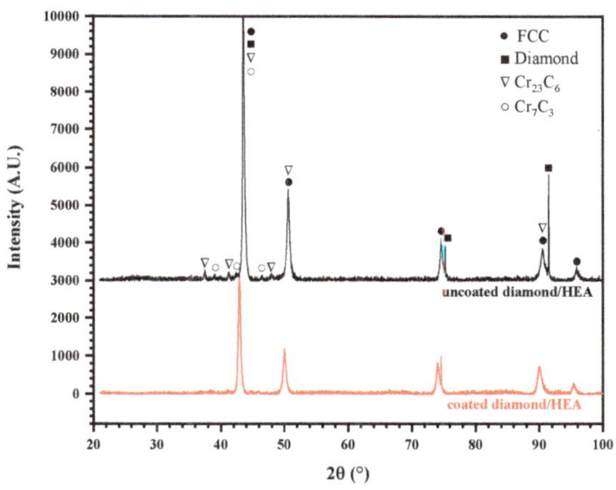

Figure 7. XRD pattern of the composites fabricated by SPS with coated and uncoated diamond.

3.3. Mechanical Properties

From the density of uncoated diamond/HEA composites in Table 2, the density difference between uncoated diamond and HEA powder was found to be large, which not only causes inhomogeneity in the mixing process, but also causes a low density of green compacts. Under such circumstances, it is impossible to eliminate the inhomogeneous distribution of pressure in the sintered block caused by pores, which ultimately results in the insufficient density of composites. The surface coating of diamond can improve the density of diamond particles, make the mixture more uniform, reduce the number of pores and increase the sintering density. At the same time, the activation of diamond can promote the diffusion reaction between diamond and the matrix and produce good metallurgical bonding. According to the results of hardness tests (Table 2), it was found that the hardness of the samples of coated diamond composites is about 30 HV higher than the that of uncoated samples, because the graphitization of diamond particles (Figure 4c) in the uncoated diamond composites reduces the hardness of the matrix.

The fracture strength of coated diamond composites is 745 MPa, which is far greater than that of uncoated diamond samples, as shown in Table 2. The reason for the low fracture strength of uncoated diamond composites is the low density caused by the micro-voids in the materials, the graphitization of diamond as a hard phase and the poor bonding between diamond and the HEA matrix. It does not produce good metallurgical bonding, but is only a metallurgical bonding and mechanical bonding interface.

Table 2. Density, hardness and transverse fracture strength of the composites fabricated by SPS with coated and uncoated diamond.

Composite Samples	Uncoated Diamond	Coated Diamond
Density(g/cm^3)	7.072 ± 0.053	7.240 ± 0.046
$HV_{0.2}$	342.2 ± 38.5	370.4 ± 34.8
Transverse fracture strength (MPa)	~550	~750

Friction and wear performance is an important property of diamond tools. Comparing the SEM images after friction from the wear mechanism, as shown in Figure 8, there were ravines and trace debris on the surface of the both coated and uncoated diamond composites. It can be inferred that the wear mechanism is abrasive wear with partially adhesive wear. The scratches on the surface of uncoated diamond composites were obvious and diamond shedding occurred. Through the analysis of the exfoliated parts in Figure 8a,b, it was found that although the diamond and the matrix have a certain degree of metallurgical bonding, due to the loose structure and inadequate bonding, the diamond will fall off from the matrix and fail. Compared with the coated diamond composite, the coated diamond can be worn without showing signs of falling off (Figure 8c,d). This shows that the Ti/Ni coating can enhance the control ability of the HEA matrix over diamond particles, improve the bonding strength and thus enhance the service life of diamond composites.

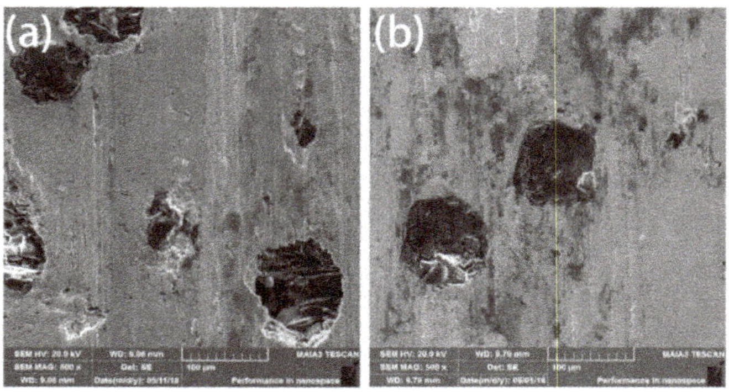

Figure 8. The wearing surface of the composites fabricated by SPS with uncoated (**a**) and coated (**b**) diamond.

The friction coefficients of the two kinds of samples fabricated by SPS ranged from 0.04 to 0.06 and maintained a very stable trend, as shown in Figure 9. After 15 min of friction, the wear of the uncoated diamond samples was 0.0144 g, while the wear of the coated diamond samples was only 0.0127 g. This shows that the friction and wear properties of the two samples are both excellent, which also confirms that the hardness of the two samples is not very different. The reason for the smaller friction coefficient is that the diamond in the sample is relatively complete and bare. The agate ball only rubs on the diamond surface, while the strong control of the matrix on diamond keeps the friction coefficient in a stable linear trend. That is, in the actual friction process, diamond plays a major role in friction, while the HEA matrix holds on the diamond. The basic type of wear is the combination of abrasive wear and adhesive wear. In general, the friction failure of diamond tools is the shedding of diamond or the wear of matrix; however, the surface coating of diamond can effectively avoid the shedding of diamond particles.

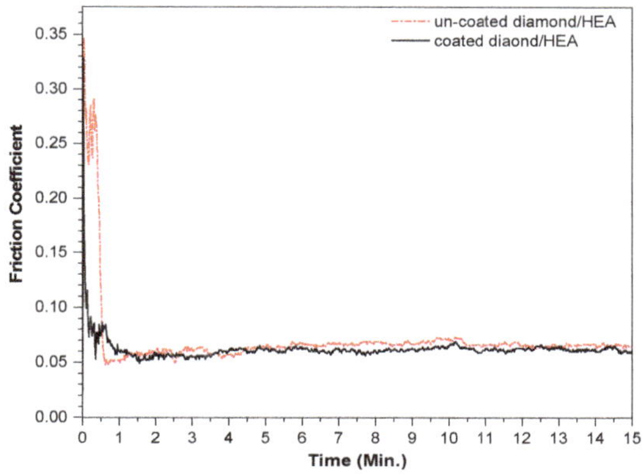

Figure 9. The friction coefficient curve of the composites fabricated by SPS with uncoated and coated diamond.

4. Conclusions

Diamond single crystal particles with Ti/Ni coatings were obtained by magnetron sputtering. The diamond gained more than 30% weight and the coating thickness was 10–20 μm.

The interfacial diffusion and solid solution formation in coated diamond composites led to the increase of interfacial bonding strength. The interface bonding strength and mechanical properties of the composites without diamond coating decreased, due to the occurrence of graphitization at the interface and the formation of carbides.

Compared with the composite with uncoated diamond, the density, hardness, transverse fracture strength and interfacial bonding properties of the composite with coated diamond were significantly improved.

The wear resistance of the coated composites is obviously higher than that of the uncoated composites, because the coating on the diamond particles maintains a relatively complete crystal form in the SPS process and has a higher bonding strength with the HEA matrix. This improves the holding force of the matrix and gives full play to the excellent properties of the HEA.

Author Contributions: Conceptualization, W.Z. and Y.L.; Methodology, W.Z.; Validation, Y.P.; Formal Analysis, W.Z. and Y.P.; Investigation, M.Z. and F.L.; Resources, Y.L. and S.H.; Data Curation, M.Z.; Writing—Original Draft Preparation, M.Z. and F.L.; Writing—Review & Editing, W.Z. and Y.P.; Project Administration, W.Z. and Y.L.; Funding Acquisition, Y.L., S.H. and Y.H.

Funding: The authors wish to acknowledge the financial support of State Key Laboratory of Powder Metallurgy (CSU 621011808), the XUCHANG Fellowship Program (grant number: XW2017-40), the project of Innovation and Entrepreneur Team Introduced by Guangdong Province (201301G0105337290) and the Special Funds for Future Industrial Development of Shenzhen (No. HKHTZD20140702020004).

Conflicts of Interest: The authors declare no conflict of interest.

References

1. Yeh, J.W.; Chen, S.K.; Lin, S.J.; Gan, J.Y.; Chin, T.S.; Shun, T.T.; Tsau, C.H.; Chang, S.Y. Nanostructured high-entropy alloys with multiple principal elements: Novel alloy design concepts and outcomes. *Adv. Eng. Mater.* **2004**, *6*, 299–303. [CrossRef]

2. Guo, N.N.; Wang, L.; Luo, L.S.; Li, X.Z.; Chen, R.R.; Su, Y.Q.; Guo, J.J.; Fu, H.Z. Hot deformation characteristics and dynamic recrystallization of the MoNbHfZrTi refractory high-entropy alloy. *Mater. Sci. Eng. A* **2016**, *651*, 698–707. [CrossRef]
3. Chou, Y.L.; Wang, Y.C.; Yeh, J.W.; Shih, H.C. Pitting corrosion of the high-entropy alloy Co1. 5CrFeNi1. 5Ti0. 5Mo0. 1 in chloride-containing sulphate solutions. *Corr. Sci.* **2010**, *52*, 3481–3491. [CrossRef]
4. Moravcik, I.; Gouvea, L.; Hornik, V.; Kovacova, Z.; Kitzmantel, M.; Neubauer, E.; Dlouhy, I. Synergic strengthening by oxide and coherent precipitate dispersions in high-entropy alloy prepared by powder metallurgy. *Scr. Mater.* **2018**, *157*, 24–29. [CrossRef]
5. Chuang, M.H.; Tsai, M.H.; Wang, W.R.; Lin, S.J.; Yeh, J.W. Microstructure and wear behavior of AlxCo1. 5CrFeNi1. 5Tiy high-entropy alloys. *Acta Mater.* **2011**, *59*, 6308–6317. [CrossRef]
6. Nene, S.S.; Sinha, S.; Frank, M.; Liu, K.; Mishra, R.S.; McWilliams, B.A.; Cho, K.C. Unexpected strength–ductility response in an annealed, metastable, high-entropy alloy. *Appl. Mater. Today* **2018**, *13*, 198–206. [CrossRef]
7. Menou, E.; Tancret, F.; Toda-Caraballo, I.; Ramstein, G.; Castany, P.; Bertrand, E.; Gautier, N.; Pedro Eduardo Jose Rivera Díaz-Del-Castillo. Computational design of light and strong high entropy alloys (HEA): Obtainment of an extremely high specific solid solution hardening. *Scr. Mater.* **2018**, *156*, 120–123. [CrossRef]
8. Tian, Y.Z.; Sun, S.J.; Lin, H.R.; Zhang, Z.F. Fatigue behavior of CoCrFeMnNi high-entropy alloy under fully reversed cyclic deformation. *J. Mater. Sci. Technol.* **2019**, *35*, 334–340. [CrossRef]
9. Long, Y.; Liang, X.; Su, K.; Peng, H.; Li, X. A fine-grained NbMoTaWVCr refractory high-entropy alloy with ultra-high strength: Microstructural evolution and mechanical properties. *J. Alloy. Comp.* **2019**, *780*, 607–617. [CrossRef]
10. Riva, S.; Tudball, A.; Mehraban, S.; Lavery, N.P.; Brown, S.G.R.; Yusenko, K.V. A novel High-Entropy Alloy-based composite material. *J. Alloys. Comp.* **2018**, *730*, 544–551. [CrossRef]
11. Zhang, M.; Zhang, W.; Liu, Y.; Liu, B.; Wang, J. Fecocrnimo high-entropy alloys prepared by powder metallurgy processing for diamond tool applications. *Powder Metal.* **2018**, *61*, 123–130. [CrossRef]
12. Loginov, P.A.; Sidorenko, D.A.; Shvyndina, N.V.; Sviridova, T.A.; Churyumov, A.Y.; Levashov, E.A. Effect of Ti and TiH2 doping on mechanical and adhesive properties of Fe-Co-Ni binder to diamond in cutting tools. *Int. J. Refract. Met. Hard Mater.* **2019**, *79*, 69–78. [CrossRef]
13. Liu, X.; Lu, P.; Wang, H.; Ren, Y.; Tan, X.; Sun, S.; Jia, H. Morphology and structure of Ti-doped diamond films prepared by microwave plasma chemical vapor deposition. *Appl. Surf. Sci.* **2018**, *442*, 529–536. [CrossRef]
14. Wang, L.; Li, J.; Che, Z.; Wang, X.; Zhang, H.; Wang, J.; Kim, M.J. Combining Cr pre-coating and Cr alloying to improve the thermal conductivity of diamond particles reinforced Cu matrix composites. *J. Alloy. Comp.* **2018**, *749*, 1098–1105. [CrossRef]
15. Chang, R.; Zang, J.; Wang, Y.; Yu, Y.; Lu, J.; Xu, X. Preparation of the gradient Mo layers on diamond grits by spark plasma sintering and their effect on Fe-based matrix diamond composites. *J. Alloy. Comp.* **2017**, *695*, 70–75. [CrossRef]
16. Rommel, D.; Scherm, F.; Kuttner, C.; Glatzel, U. Laser cladding of diamond tools: Interfacial reactions of diamond and molten metal. *Surf. Coat. Technol.* **2016**, *291*, 62–69. [CrossRef]
17. Lu, J.; Cao, Z.; Qi, F.; Qian, F.; Zhang, W. Evolution of interface carbide diamond brazed with filler alloy containing Cr. *Diam. Relat. Mater.* **2018**, *90*, 116–125. [CrossRef]
18. Inzoli, F.; Dellasega, D.; Russo, V.; Ghezzi, F.; Passoni, M. Early stages of diamond growth on substrates with different carbon diffusivity. *Diam. Relat. Mater.* **2017**, *80*, 69–75. [CrossRef]
19. Sidorenko, D.A.; Zaitsev, A.A.; Kirichenko, A.N.; Levashov, E.A.; Kurbatkina, V.V.; Loginov, P.A.; Rupasov, S.I.; Andreev, V.A. Interaction of diamond grains with nanosized alloying agents in metal–matrix composites as studied by Raman spectroscopy. *Diam. Relat. Mater.* **2013**, *38*, 59–62. [CrossRef]
20. Cheng, H.; Chen, W.; Liu, X.; Tang, Q.; Xie, Y.; Dai, P. Effect of Ti and C additions on the microstructure and mechanical properties of the FeCoCrNiMn high-entropy alloy. *Mater. Sci. Eng. A* **2018**, *719*, 192–198. [CrossRef]

entropy

MDPI

Review

Coherent Precipitation and Strengthening in Compositionally Complex Alloys: A Review

Qing Wang [1], Zhen Li [2], Shujie Pang [3], Xiaona Li [1,*], Chuang Dong [1] and Peter K. Liaw [4,*]

1. Key Laboratory of Materials Modification by Laser, Ion and Electron Beams (Ministry of Education), School of Materials Science and Engineering, Dalian University of Technology, Dalian 116024, China; wangq@dlut.edu.cn (Q.W.); dong@dlut.edu.cn (C.D.)
2. School of Mechanical Engineering, Dalian University of Technology, Dalian 116024, China; lizhen@dlut.edu.cn
3. Key Laboratory of Aerospace Materials and Performance (Ministry of Education), School of Materials Science and Engineering, Beihang University, Beijing 100191, China; pangshujie@buaa.edu.cn
4. Department of Materials Science and Engineering, The University of Tennessee, Knoxville, TN 37996, USA

* Correspondence: lixiaona@dlut.edu.cn (X.L.); pliaw@utk.edu (P.K.L.); Tel.: +86-411-84708389 (X.L.)

Received: 29 October 2018; Accepted: 14 November 2018; Published: 15 November 2018

Abstract: High-performance conventional engineering materials (including Al alloys, Mg alloys, Cu alloys, stainless steels, Ni superalloys, etc.) and newly-developed high entropy alloys are all compositionally-complex alloys (CCAs). In these CCA systems, the second-phase particles are generally precipitated in their solid-solution matrix, in which the precipitates are diverse and can result in different strengthening effects. The present work aims at generalizing the precipitation behavior and precipitation strengthening in CCAs comprehensively. First of all, the morphology evolution of second-phase particles and precipitation strengthening mechanisms are introduced. Then, the precipitation behaviors in diverse CCA systems are illustrated, especially the coherent precipitation. The relationship between the particle morphology and strengthening effectiveness is discussed. It is addressed that the challenge in the future is to design the stable coherent microstructure in different solid-solution matrices, which will be the most effective approach for the enhancement of alloy strength.

Keywords: precipitation; strengthening; coherent microstructure; conventional alloys; high entropy alloys

1. Introduction

Precipitation strengthening with intermetallic compounds is the most effective approach for the enhancement of alloy strength in engineering structural materials, compared with solid-solution strengthening, grain-boundary strengthening, and work hardening [1–3]. Especially at high temperatures (HTs), the precipitation strengthening is indispensable due to the prominent long-time microstructural stabilities caused by second-phase precipitates in the solid-solution matrix [4–8]. Among them, the coherent ordered phases, such as $L1_2$-Ni_3Al (*cP*4-Cu_3Au) of the face-centered-cubic (FCC) solid solution [4,5], and B2-NiAl (*cP*2-ClCs) of the body-centered-cubic (BCC) solid solution [9–13], are crucial for the HT creep-resistant properties of alloys due to the perfect coherency between the ordered phase and the solid-solution matrix. It should be pointed out that the precipitation strengthening is related not only to the macroscopic properties of precipitates, but also to their microstructural morphologies. For instance, the prominent creep-resistant property of Ni-based superalloys at up to 85% of the insipient melting temperature (as high as 1100 °C) is primarily attributed to the special microstructure of spherical or cuboidal $L1_2$-Ni_3Al nanoprecipitates coherently-precipitated into the FCC matrix [5].

In order to meet the service-performance requirements, including mechanical strength, corrosion- and oxidation-resistant properties, etc., several solute elements are generally added to alloy or minor-alloy the solvent matrix constituted of one or two primary elements in conventional engineering structural materials [14]. From the viewpoint of element species, most of high-performance metallic materials, including Al alloys, Mg alloys, Cu alloys, stainless steels, and Ni superalloys, are all compositionally-complex alloys (CCAs), resulting in a uniform microstructure of diverse second-phase particles distributed in their solid-solution matrix. Recently, another kind of newly-developed CCAs are not based on one or two solvent elements, but based on the equimolar or near-equimolar mixing of multi-principal elements, which are also named high-entropy alloys (HEAs) [15–20]. HEAs have attracted more attention due to their unique properties resulted from simple crystalline structures, such as FCC, BCC, close-packed hexagonal (HCP), and their derivatives ($L1_2$, B2, etc.) [19–22]. Thus, they can also be regarded as a special kind of solid-solution alloys, similar to conventional engineering alloys.

Therefore, the present work will comprehensively generalize the precipitation behavior and precipitation strengthening in CCAs, including conventional engineering alloys and high-entropy alloys, where the coherent precipitation will be specially emphasized. The morphology evolution of second-phase particles and precipitation strengthening mechanisms will be illustrated firstly. Then, the relationship between the particle morphology (shape and particle size) and strengthening effectiveness in diverse CCAs will be discussed, respectively. Finally, several thoughts on the coherent precipitation to design and develop high-performance CCAs in the future will be suggested.

2. Equilibrium Morphology of a Misfitting Particle

In the absence of elastic stress, the equilibrium morphology of a second-phase particle embedded into a matrix is established solely by the particle-matrix interfacial energy and its dependence on crystallographic orientation [23]. However, experimentally, the particle morphology, which arises during a diffusional phase transition in many alloys, was often not the shape that minimizes the total interfacial energy. In most cases, the presence of the lattice misfit between the second phase and the matrix can induce an elastic stress field, which in turn, affects the particle morphology [24]. For instance, in Ni-based superalloys with a microstructure of ordered $L1_2$-Ni_3Al particles coherently-embedded into the FCC solid-solution matrix, the particles were observed to undergo changes in shape from spheres to cuboids, and then to plates with increasing particle size [25,26], or even fission into smaller particles once they reach a critical size [27–29]. Actually, the particle equilibrium shape is determined by minimizing the total energy E_t, the sum of interfacial energy, E_i, and elastic energy, E_e, at a constant particle volume, in which the E_i and E_e scale with the surface area and the volume of the particle, respectively [30,31]. So, the equilibrium shape of a misfitting particle is dependent on the particle size. That is to say, the particle shape should tend towards the shape that minimizes the E_i at a smaller particle size, and towards the shape that minimizes the E_e at a larger size. Furthermore, with increasing particle size, the elastic energy plays an increasingly important role in setting the shape since it is the driving force of the particle growth and coarsening.

The relative importance of the elastic energy and the interfacial energy can be evaluated through the characteristic parameter L [32,33], i.e., $L = \varepsilon_2 C_{44} r/s$, where ε is the lattice misfit between the particle and matrix phases, C_{44} is the elastic constant of the matrix, r is the average particle size, and s is the average specific interfacial energy. Figure 1 gives the morphology evolution of the particle with the parameter L [33], from which it is found that a small L usually corresponds to a spherical particle, which can transform to ellipsoidal or cuboidal shape when L increases. At a much larger L, the fourfold symmetry of cuboidal particles will be broken and some low symmetric shapes, like plates or needles, will begin to appear due to the elastic anisotropy. In addition, when the particle sizes are comparable, the lattice misfit ε will play the key role in determining the particle shape, since the parameter L is proportional to both the lattice misfit ε and the particle size r. Apparently, the particle morphology has a profound effect on the mechanical properties of alloys, which will be discussed in the following diverse alloy systems, respectively.

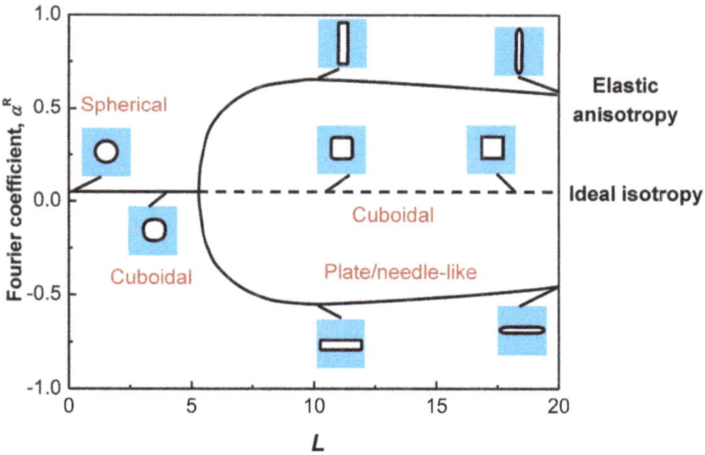

Figure 1. Particle morphology evolution with the characteristic parameter L, in which the vertical axis is the Fourier coefficient a^R to represent the energy of different particle shapes. It shows that the bifurcation from the four-fold symmetric cuboid to the two-fold symmetric shapes (plate or needle) occurs at a critical value (L = 5.6) [33].

3. Precipitation Strengthening Mechanisms

The precipitation strengthening mechanisms can be divided into two categories [1–3], the dislocation shearing mechanism and the Orowan dislocation bypassing mechanism, depending on the interaction between moving dislocations and precipitates. The one leading to a smaller strength increment is the operative mechanism. The dislocation-shearing mechanism is generally active when the precipitates are coherent with the matrix, and the particle size is small, while the Orowan bypassing mechanism dominates when the coherent particle size exceeds a critical value or when the particles are incoherent with the matrix. For the shearing mechanism, three factors contribute to the increase in yield strength, coherency strengthening ($\Delta\sigma_{CS}$), modulus mismatch strengthening ($\Delta\sigma_{MS}$), and order strengthening ($\Delta\sigma_{OS}$). The former two ($\Delta\sigma_{CS}$ and $\Delta\sigma_{MS}$) occur before the dislocation shears the particle and the latter ($\Delta\sigma_{OS}$) during shearing. Thereof, the larger value of ($\Delta\sigma_{CS}$ + $\Delta\sigma_{MS}$) or $\Delta\sigma_{OS}$ is expected to be the total strength increment from the shearing mechanism. The equations available to calculate these strength increments caused by both dislocation shearing and bypassing are as follows [34–39]:

$$\Delta\sigma_{CS} = M \times \alpha_\varepsilon \times (G\varepsilon_c)^{\frac{3}{2}} \times (\frac{rf}{0.5Gb})^{\frac{1}{2}} \tag{1}$$

$$\Delta\sigma_{MS} = M \times 0.0055(\Delta G)^{\frac{3}{2}} \times (\frac{2f}{G})^{\frac{1}{2}} \times (\frac{r}{b})^{\frac{3m}{2}-1} \tag{2}$$

$$\Delta\sigma_{OS} = M \times 0.81 \times \frac{\gamma_{apb}}{2b} \times (\frac{3\pi f}{8})^{\frac{1}{2}} \tag{3}$$

$$\Delta\sigma_{orowan} = M \times \frac{0.4Gb}{\pi\sqrt{1-v}} \times \frac{\ln(2\sqrt{\frac{2}{3}}r/b)}{\lambda_p},\ \lambda_p = 2\sqrt{\frac{2}{3}}r(\sqrt{\frac{\pi}{4f}}-1) \tag{4}$$

where M = 2.73 for BCC structure and M = 3.06 for FCC structure (Taylor Factor) [1], α_ε = 2.6 (a constant) [35,36], m = 0.85 (a constant) [37,38], $\varepsilon_c = 2\varepsilon/3$ [2,35,36], the constrained lattice misfit. G and ΔG are the shear-modulus of the matrix and the shear modulus mismatch between precipitates and matrix, respectively; b is the Burgers vector; r and f are the average size and the volume fraction of precipitates, respectively; γ_{apb} is the anti-phase boundary energy of precipitates; v is the Poisson ratio; and λ_p is the inter-precipitate spacing.

Since the shearing and bypassing mechanisms occur concurrently and are independent to each other, the strengthening is determined by the smaller of $\Delta\sigma_{shearing}$ or $\Delta\sigma_{orowan}$. In other words, the softer mechanism initiates the plastic deformation. Ideally, the largest yield strength increment could be reached when $\Delta\sigma_{\text{shearing}} = \Delta\sigma_{orowan}$ at a critical particle size r_0 with a fixed f [1]. Figure 2 shows the variation tendency of the yield strength increment with the particle size by competing the dislocation shearing and bypassing mechanisms, in which the maximum strength increment reaches at the critical r_0 when the volume fraction f is fixed.

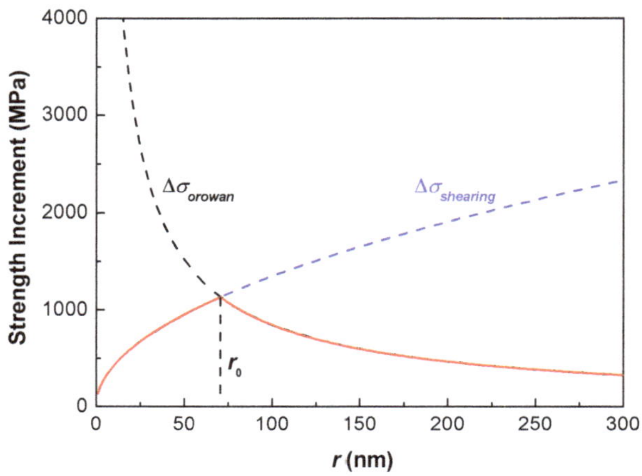

Figure 2. The variation tendency of yield strength increment with the particle size [1–3], in which the strength increments caused by dislocation shearing mechanism ($\Delta\sigma_{shearing}$) and bypassing mechanism ($\Delta\sigma_{orowan}$) are shown, and the maximum increment reaches at a critical particle size r_0.

4. Precipitate Morphology and Precipitation Strengthening in CCAs

In this section, the precipitation behavior and precipitation strengthening effects in each CCA system are generalized in details. Typical alloy systems with precipitation strengthening include Ni-based superalloys, Al alloys, Mg alloys, Cu alloys, stainless steels, and high-entropy alloys. The overviews are elaborated as follows.

4.1. Ni-Based Superalloys

Ni-based superalloys exhibit the most outstanding mechanical properties (especially the creep-resistance), corrosion- and oxidation-resistant properties at elevated temperatures among all the conventional structural materials. Their excellent properties are benefited from their specially coherent microstructures of spherical or cuboidal $L1_2$-γ' nanoprecipitates into FCC-γ solid solution [4–6]. Especially the coherent precipitation of cuboidal $L1_2$-γ' particles in single-crystal superalloys is responsible for the necessary strength at much higher temperatures near to the melting point [40,41]. However, the single-crystal superalloys with similar compositions often possess different creep-resistant properties, even containing cuboidal γ' precipitates with a comparable particle size.

Figure 3 exhibits the creep curves at 1100 °C/137 MPa of TMS-138 (Ni-6Co-3Cr-3Mo-6W-6Al-6Ta-0.1Hf-5Re-2Ru, wt.%) and TMS-75(+Ru) (Ni-12Co-3Cr-2Mo-6W-6Al-6Ta-0.1Hf-5Re-1.5Ru, wt.%) alloys, in which the microstructural evolutions during the creep process are also shown [40]. Both superalloys have similar compositions with a minor difference in the amounts of Co, Mo, and Ru. The particle sizes of cuboidal γ' nanoprecipitates in these two alloys are comparable, being r = 230 ± 30 nm (TMS-138) and r = 245 ± 25 nm (TMS-75(+Ru)), respectively, with a volume fraction of about f = 65% in experiments. But the lattice misfit ε between γ and γ' phases are different in both alloys, being $\varepsilon = -0.33\%$ in TMS-138 and $\varepsilon = -0.16\%$ in TMS-75(+Ru) at 1100 °C, respectively, in which the lattice misfit is calculated with the equation of $\varepsilon = 2(a_{\gamma'} - a_{\gamma})/(a_{\gamma'} + a_{\gamma})$ ($a_{\gamma'}$ and a_{γ}: the lattice constants of γ' and γ phases, respectively). Remarkably, TMS-138 possesses a longer creep life and a lower minimum creep rate, which is attributed to its larger γ/γ' lattice misfit ε. Specifically, in the primary creep stage, such as the time t = 2 h, the larger misfit stress caused by the larger ε in TMS-138 drives the loops of matrix dislocations to move by cross-slip through the matrix channels, while in TMS-75(+Ru), the dislocations move by climbing around the γ' cuboids due to the insufficient driving force by the relatively-smaller ε (as seen in the microstructures in Figure 3). With increasing creep time to t = 60 h, there are many superdislocations in γ' cuboids in TMS-75(+Ru), while many perfect dislocation networks on the γ/γ' interface are formed in TMS-138, which can effectively prevent the gliding dislocations in the γ channels from cutting the rafted γ/γ' structure. More importantly, the larger lattice misfit can result in the denser γ/γ' interfacial dislocation networks. Clearly, the dislocation networks are much denser in TMS-138 after rupture, which is the key for the small minimum creep rate.

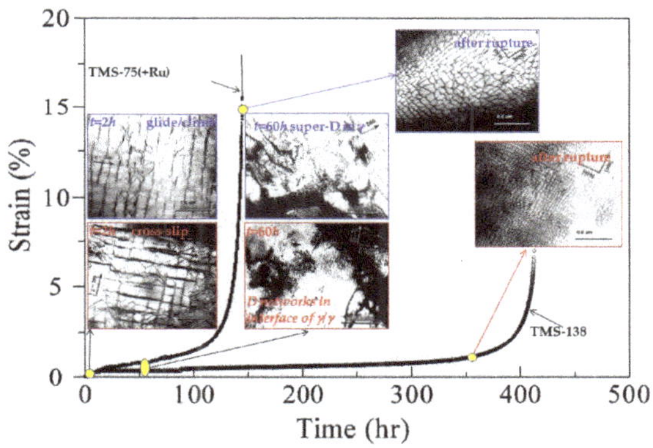

Figure 3. Creep curves at 1100 °C/ 137 MPa of TMS-75(+Ru) and TMS-138 superalloys, in which the microstructures at different creep stages (primary stage, steady state and after rupture) are also presented [40].

Not only the precipitate morphology, but also the strengthening effect of coherent precipitation are closely related to the lattice misfit between the ordered phase and solid-solution phase. We calculated the yield strength increments given by shearing and bypassing mechanisms according to the Equations (1)–(4) with the particle size r, in which the dominant ($\Delta\sigma_{CS} + \Delta\sigma_{MS}$) presents the strength increment caused by the shearing mechanism. The used parameters for strength increment calculations are M = 3.06, α_{ε} = 2.6, G = 81 GPa, ΔG = 4 GPa, m = 0.85, b = 0.254 nm, γ_{apb} = 0.12 J/m^2, and v = 0.35, respectively. Figure 4 gives the variation tendencies of ($\Delta\sigma_{CS} + \Delta\sigma_{MS}$) and $\Delta\sigma_{orowan}$ as a function of the particle size r for TMS-138 and TMS-75(+Ru) superalloys. It was found that the optimal particle size r_0 corresponding to the maximum strength increment given by the theoretical calculation in TMS138 is r_0 = 193 nm, which is consistent with the experimental size of $r = 230 \pm 30$ nm. While the experimental particle size ($r = 245 \pm 25$ nm) is far away from its optimal r_0 = 406 nm in TMS-75(+Ru), indicating that the strengthening effect does not reach the maximum. Hence, the TMS-138 superalloy exhibits a much higher strength and a better creep resistance due to a larger lattice misfit. Therefore, in the case of coherent precipitation, the control of the lattice misfit between the ordered phase and its parent solid solution is significant to develop high-performance CCAs.

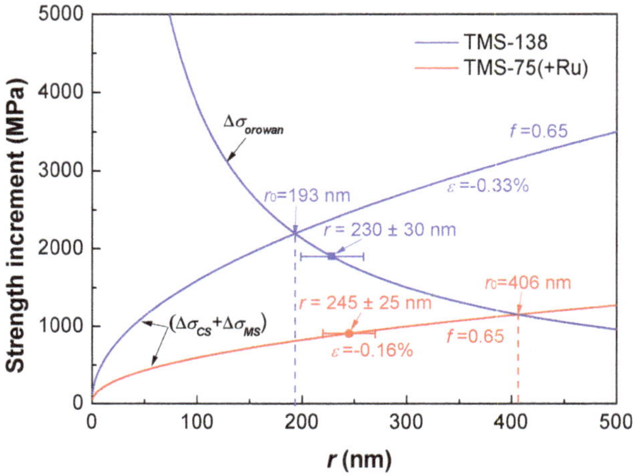

Figure 4. The variation tendency of ($\Delta\sigma_{CS}$ + $\Delta\sigma_{MS}$) and $\Delta\sigma_{orowan}$ with the particle size r of TMS-138 and TMS-75(+Ru) superalloys, in which the optimal particle size r_0 from the calculation and the experimentally-measured r are also marked for each alloy.

4.2. Al-Based Alloys

Al alloys have been used widely as engineering structural materials due to their high specific strength, among which the high-strength Al-Zn-Mg-Cu series of alloys (7000 series) are extensively applied into aeronautical fields [42–46]. The Al-Cu binary system is a well-studied precipitation-strengthening system, since it forms the basis for many types of age-hardening alloys with technological importance [47]. The precipitation sequence during the aging process, Al SS → G.P. zone → θ″-Al_3Cu → θ′-Al_2Cu → θ-Al_2Cu, was often taken as a model for describing the fundamentals of precipitation strengthening. The coherent Guinier-Preston (G.P.) zone consisting of a single layer of pure Cu atoms was firstly precipitated from the FCC-Al solid solution (SS) matrix. Then, Al-Cu clusters with a stoichiometrical Al_3Cu (θ″) were formed, which is also coherent with the FCC matrix. It could transform into the metastable θ′-Al_2Cu phase with a body-centered-tetragonal structure, which is the main strengthening phase, but semi-coherent with the matrix. Finally, the metastable θ′-Al_2Cu would transform into the equilibrium tetragonal θ′-Al_2Cu phase, which is incoherent with the matrix [48]. In this case, the coherent relationship between the precipitated phase and the matrix could be destroyed with prolonging the aging time, which eventually leads to the formation of coarse precipitates, as a final result of weakening the strengthening effect, compared with the coherent fine precipitates. Figure 5 shows the effect of the particle morphology on the hardness variation with the aging time of Al-0.8Mg-0.79Si (wt.%, 6061) alloy [49]. The peak hardness reaches at 175 °C aging for 4 h, corresponding to the semi-coherent precipitation of the needle β″ (a monoclinic structure with a stoichiometrically MgSi) nanoparticles with a size of about 10~15 nm. Once the needle β″ nanoparticles are transformed to relatively-coarse rod β′ particles (a hexagonal structure with a stoichiometrically $Mg_{1.7}Si$) at 200 °C aging for 20 h, the hardness will decrease due to the incoherency of β′ and the matrix [49].

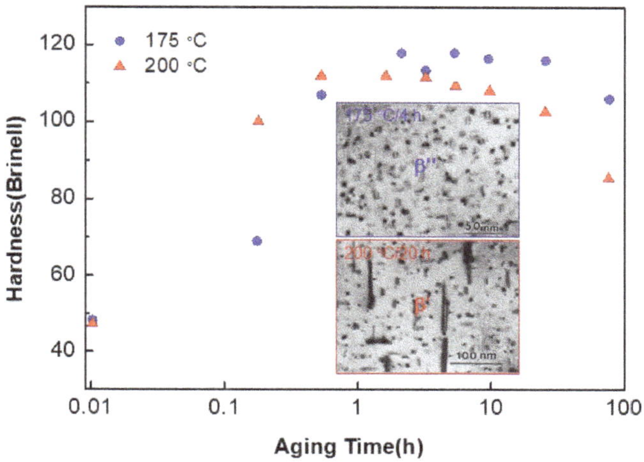

Figure 5. Variation of hardness of the 6061 Al alloy with the aging time at both 175 °C and 200 °C, in which the microstructures at peak aging and over aging are also shown [49].

Recently, the coherent precipitation of ordered $L1_2$-Ni_3M (M = Sc, Er, Zr, etc.) in the disordered FCC matrix of Al-Zr-Sc-Er alloy systems has attracted more attention since it can provide significant strengthening to a temperature of about 300 °C. Such Al alloys are excellent candidates for some high-temperature automotive and aerospace applications [50–52]. Supersaturated Al-Sc binary alloys generally possess high strength due to the coherent precipitation of $L1_2$-Al_3Sc nanoparticles [53,54]. Based on it, the addition of Zr can form coarsening-resistant $L1_2$-Al_3(Sc,Zr) lobed-cuboids consisting of a Sc-enriched core surrounded by a Zr-enriched shell in Al-0.06Sc-0.06Zr (wt.%) alloy [45,55] (as seen in Figure 6). More interestingly, the Er further substitution for Zr can form spheroidal Al_3(Sc,Zr,Er) nanoprecipitates with a core/double-shell structure consisting of an Er-enriched core surrounded by a Sc-enriched inner shell and a Zr-enriched outer shell in the Al-0.06Sc-0.04Zr-0.02Er (wt.%) alloy (Figure 6), which are stable and difficult to be coarsened even at a higher temperature of 400 °C for 64 days [42]. Resultantly, it is due to the particle size that renders the two alloys with a remarkable difference in microhardness, as shown in Figure 6. The particle size of lobed-cuboidal Al_3(Sc,Zr) precipitates is about 25 nm in the former alloy, while the spheroidal Al_3(Sc,Zr,Er) nanoprecipitates with a particle size of 3~8 nm result in a drastic improvement of the alloy strength, from a peak harness of 243 MPa in the former alloy to 451 MPa in the later one at 400 °C aging. Therefore, more and more interests have been focused on the coherent precipitation in Al alloys to develop new light-weight materials that can be applied in high-temperature environments (>300 °C).

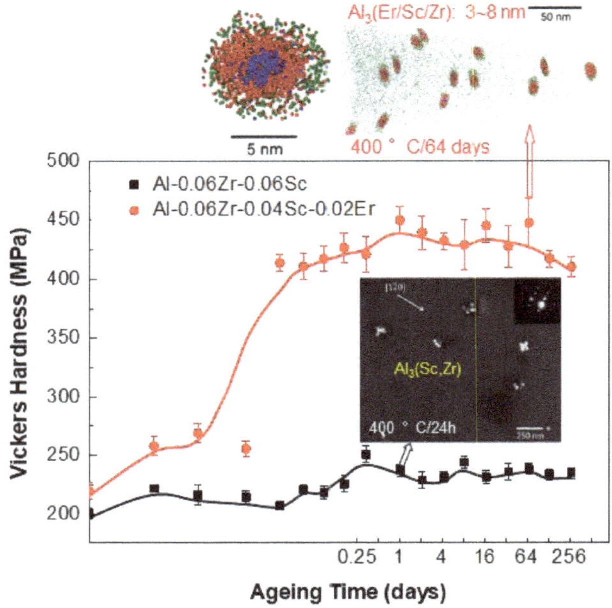

Figure 6. Variation of Vickers hardness of Al-Zr-Sc-Er alloys with the aging time at 400 °C, in which the precipitate morphologies of coherent $Al_3(Sc,Zr)$ after aging 24 h and $Al_3(Sc,Zr,Er)$ after aging 64 days are also presented [50].

4.3. Mg-Based Alloys

Mg alloys with a close-packed-hexagonal (HCP) matrix are the lightest among all the commonly-used structural materials and have great potentials for application in the automotive, aircraft, aerospace, and electronic industries [56]. Their useful mechanical properties were generally achieved via age-hardening process to form high-strength precipitates, which is similar to the precipitation in Al alloys. Actually, it is more difficult to keep the coherency between the precipitates and the matrix due to the HCP structure of the matrix. The structure, morphology, and orientation of precipitates, precipitation sequence, and hardening response in various Mg alloy systems have been generalized, in which the effects of precipitate shapes on strengthening and the rational design of microstructures for higher strengths were also emphasized [57,58]. For the most widely-used Mg-Al-based alloys, such as AZ91 (Mg-8.7Al-0.7Zn-0.1Mn, wt.%), the final stable incoherent β-$Mg_{17}Al_{12}$ phase with a BCC structure precipitates directly from the supersaturated HCP-Mg solid solution without any coherent G.P. zones, which could not result in an appreciable strengthening response due to the existence of relatively-coarse plate/lath-like β particles [59–61]. For most of the high-strength Mg alloy series, the precipitation usually follows the sequence of Mg SS $\rightarrow$ G.P. zone $\rightarrow$ coherent metastable phase $\rightarrow$ incoherent stable phase [62–71]. A typical Mg-Gd(-Y)-based Mg-15Gd-0.5Zr (wt.%) alloy was taken for an instance [72]. Firstly, the coherent G.P. zone precipitated from the HCP-Mg solid solution at the initial stage of aging at 250 °C. Then the metastable β''-Mg_3Gd phase with an ordered DO_{19} structure of the HCP solid solution appeared after aging for 0.5 h, which keeps the perfect coherent orientation with the matrix and exhibits a hexagonal prism morphology. Another metastable orthorhombic β'-Mg_3Gd phase with a lenticular particle morphology would substitute for the β'' when the aging time increased to 8 h, resulting in a peak strengthening. Figure 7 shows the microhardness variation with the aging time, from which an obvious improvement of microhardness is attributed to the precipitation of metastable β'' and β' phases. Further prolonging the aging time, β' will transform into a FCC β_1-Mg_3Gd with a DO_3-$L2_1$ structure, and then to the final stable β-Mg_5Gd with a FCC structure, which can weaken the strengthening response due to the plate-like morphology of β_1 and β particles [72,73].

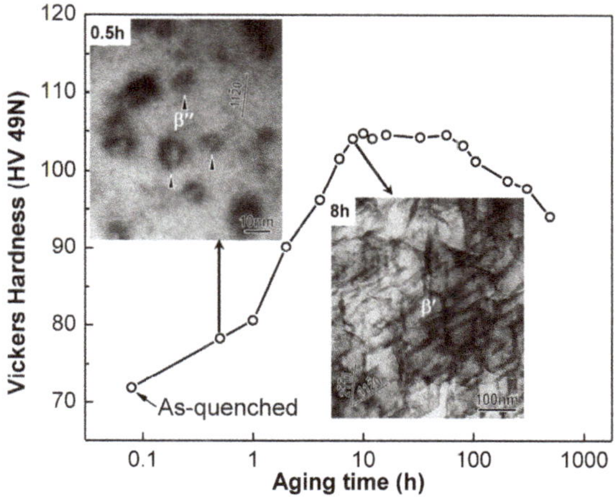

Figure 7. Variation of Vickers hardness of the Mg-15Gd-0.5Zr alloy with the aging time at 250 °C, in which the morphologies of coherent β″ and β′ precipitates are also shown [72,73].

It is emphasized that the DO_3-$L2_1$ β_1 phase often exists in many high-strength Mg alloys, which is a highly-ordered superstructure of the BCC solid solution, consisting of eight BCC unit cells. If the HCP-Mg matrix is changed to a BCC structure, the coherency will be achieved between the $L2_1$ phase and BCC-Mg solid solution. It is fascinating that the ordered $L2_1$-Li_2MgAl phase was coherently-precipitated into the BCC-Mg matrix in the recently-reported Mg-11Li-3Al (wt.%) alloy, which renders the alloy with high strength, good ductility, and excellent corrosion resistance [74]. Figure 8 compared the room-temperature mechanical tensile properties (yield strength and elongation to fracture) of Mg-11Li-3Al alloy with several traditional Mg alloys. It is found that this BCC-based alloy exhibits not only a higher strength, but also a much better ductility with an elongation to fracture of about 27%, which is attributed to the coherent precipitation of spherical $L2_1$ nanoparticles with a particle size of 2~20 nm (as shown in Figure 8).

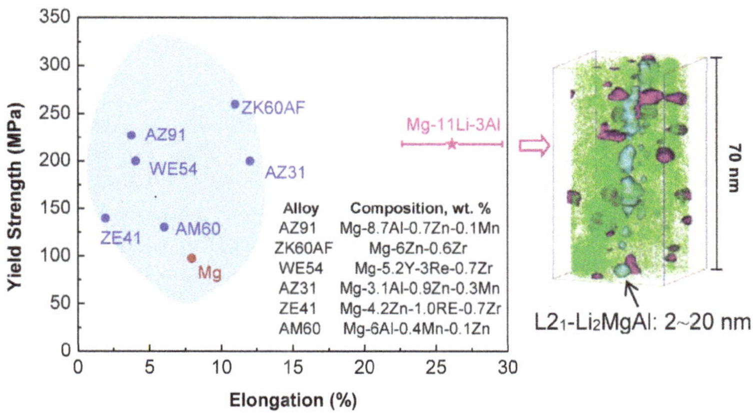

Figure 8. Comparison of the mechanical properties (yield strength and elongation) of the BCC-based Mg-11Li-3Al alloy in traditional HCP Mg alloys, in which the spherical $L2_1$ nanoparticles of the former BCC Mg alloy are also presented [74].

4.4. Cu-Based Alloys

The precipitates in various Cu alloys generally show different phase structures, in which the coherent precipitation of metastable ordered $L1_2$ in the FCC-Cu solid solution matrix can appear in the commonly-used Cu-Ni-Sn alloy system [75–77]. The model alloy is the Cu-15Ni-8Sn (wt.%, C72900), in which the $(Cu,Ni)_3Sn$ precipitates have four crystal structures, FCC-DO_3, tetragonal DO_{22}, ordered $L1_2$, and orthorhombic δ. The phase evolution sequence from high to low temperatures is similar to that in aged Al alloys for different times. When aging at above 550 °C, discontinuous and intragranular DO_3-γ precipitates (plate-like shape) appeared in the FCC matrix, as seen in Figure 9a. Spinodal decomposition often occurred during the early stage of the decomposition below ~500 °C, followed by the DO_{22} ordering (as seen in Figure 9b) and then $L1_2$ ordering (the inset of Figure 9c) [77]. With decreasing aging temperature, spherical $L1_2$ nanoparticles could be coherently-precipitated into the FCC matrix, compared with the rod-like DO_{22} precipitates. The particle morphology can affect the mechanical property of this alloy, as seen in Figure 9c, being the variation tendency of the tensile yield strength with the aging time [77–79]. It was found that the coherent precipitation of spherical $L1_2$ nanoprecipitates corresponds to the highest strength.

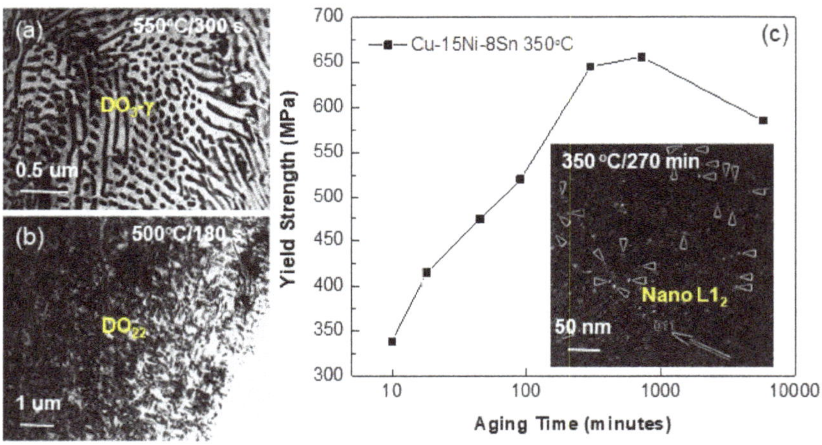

Figure 9. Morphologies of ordered DO_3-(**a**), DO_{22}-(**b**), and $L1_2$-$(Cu,Ni)_3Sn$ precipitates (**c**), as well as the variation tendency of the tensile yield strength of Cu-15Ni-8Sn alloy with the aging time at 350 °C (**c**) [77,78].

4.5. Fe-Based Stainless Steels

It is well known that the most common strengthening precipitates are carbides, being $Cr_{23}C_6$ and MC (M = Nb, Ti, V, etc.) [7,8]. For some special stainless steels (SSs), there also exist other kinds of precipitates, such as Laves phases (Fe_2M), Ni_3M, B2-NiAl, σ-FeCr, and Z-CrNbN, to strengthen the FCC or BCC matrix [7]. For instance, austenitic SSs for the use in high-temperature (600~800 °C) and oxidation environment are generally strengthened by MC, $Cr_{23}Cr_6$, Z, Fe_2M, or B2-NiAl [80,81]. However, it is noted that these phases are not coherent with the FCC austenite matrix, which can lead to the coarsening of second phase precipitates, as a final result of softeness or embrittlement (the latter mainly caused by the σ phase). Very interestingly, another kind of high-temperature ferritic SSs with a coherent microstructure of ordered B2 phase precipitation into BCC matrix have been developed [12,13,82–88]. As shown in Figure 10a, the Fe-6.5Al-10Ni-10Cr-3.4Mo-0.25Zr-0.005B (wt.%, FBB8) alloy exhibits a prominent creep resistance at 700 °C, better than the conventional P92, P122, T122, and 12CR steels [12,84]. It is attributed to the coherent precipitation of spherical B2 nanoparticles into BCC ferritic matrix (Figure 10b), similar to that in FCC Ni-based superalloys [5]. The further addition of 2 wt.% Ti into FBB8 induces another ordered $L2_1$-Ni_2AlTi phase of the BCC solid solution, forming a coherent microstructure with cuboidal B2/L2$_1$ hierarchical precipitates (Figure 10c) [89,90]. We calculated the lattice misfit between BCC and B2/L2$_1$ phases with the formulas of $\varepsilon = 2(a_{B2} - a_{BCC})/(a_{B2} + a_{BCC})$ and $\varepsilon = 2(a_{L21} - 2a_{BCC})/(a_{L21} + 2a_{BCC})$, in which a_{B2}, a_{L21}, and a_{BCC} are the lattice constants of B2, L2$_1$, and BCC solid solution phases, respectively. It was found that the cuboidal precipitation of B2/L2$_1$ phases mainly resulted from the larger lattice misfit of $\varepsilon = 0.7\%$ between the BCC and L2$_1$ phases in the Ti-modified FBB8, while the smaller value of $\varepsilon = 0.06\%$ promotes the formation of spherical B2 precipitates in FBB8. More importantly, it is due to the cuboidal precipitation that further improves the creep-resistant property of the Ti-modified FBB8 alloy at 700 °C (Figure 10a), possessing a better creep life than FBB8.

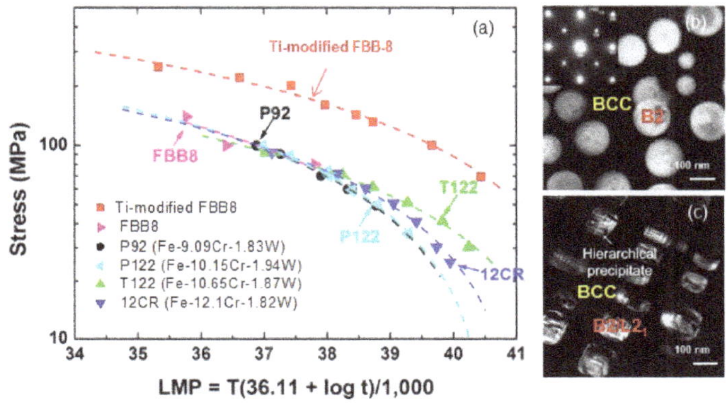

Figure 10. (**a**) Larson-Miller parameters (LMP) for FBB8, Ti-modified FBB8, and several conventional steels (P92, P122, T122, and 12CR) [89]; (**b**,**c**) morphologies of spherical B2 nanoprecipitates in FBB8 [12] and cuboidal L21 nanoprecipitates in Ti-modified FBB8 [89,90], respectively.

Moreover, the precipitation of intermetallic compound Ni_3M in the BCC martensite matrix can make maraging stainless steels with a much higher strength (tensile yield strength of σ_y = 1.2~1.5 GPa) [91–94]. Very recently, the coherent precipitation of spherical B2 nanoparticles with a particle size of 3~5 nm, rather than the Ni_3M, in the BCC martensite rendered a Fe-17Ni-6.2Al-2.3Mo-0.48Nb-0.37C-0.05B (wt.%) steel with a superhigh strength (σ_y > 1.9 GPa) [9], as shown in Figure 11. It is attributed to the smaller lattice misfit ($\varepsilon = 0.17\%$) that permits the dislocations cutting through the B2 nanoparticles, in which the high anti-phase boundary (APB) energy can promote the ordering strengthening by increasing the dislocation shear resistance in the particles.

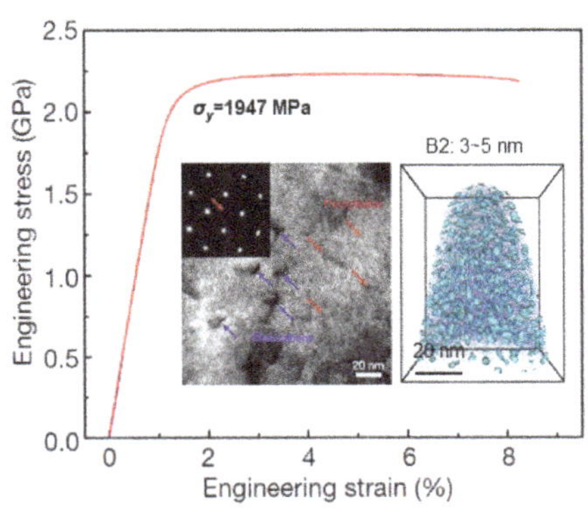

Figure 11. Tensile engineering stress-strain curve of the aged Fe-17Ni-6.2Al-2.3Mo-0.48Nb- 0.37C-0.05B steel, in which the microstructures of B2 nanoparticles with a size about 3~5 nm and dislocations are also shown [9].

4.6. High-Entropy Alloys

High-entropy alloys (HEAs) with equimolar or near-equimolar mixing of multiple elements have been found in diverse alloy systems and have attracted more attention due to their interesting properties and associated scientific understandings. Especially, CoCrFeNi-based HEAs [95–104], composed of late transition metals (LTMs), were widely investigated due to their exceptional mechanical properties and potential industrial applications. For instance, the single FCC CoCrFeNiMn HEA displays an excellent damage tolerance with higher tensile strength and remarkable fracture toughness than traditional engineering stainless steels at cryogenic temperatures down to −196 °C, which resulted from the microstructural diversity caused by the twinning-induced plasticity (TWIP) effect [98].

A further addition of Al or Ti into these LTMs-based HEAs can produce phase transformations [105–131], and then lead to microstructural diversities due to the strong interactions between Al/Ti and LTMs, resulting in an enhancement of strength. It is primarily attributed to the coherent precipitation of intermetallic phases, such as $L1_2$-Ni_3Al from the FCC matrix [125,126], and B2-NiAl [110,118–124] and $L2_1$-Ni_2AlTi [107–109] from the BCC matrix. As example, the $(NiCoFeCr)_{94}Ti_2Al_4$ (at.%) HEA possesses a special coherent microstructure with fine spherical $L1_2$-(Ni_3(Al,Ti)) nanoprecipitates in the FCC matrix, resulting in a significant strength improvement with a yield strength over 1 GPa [125]. A newly-developed kind of high entropy Ni-based alloys, such as $Ni_{48.6}Al_{10.3}Co_{17}Cr_{7.5}Fe_{9.0}Ti_{5.8}Ta_{0.6}Mo_{0.8}W_{0.4}$ (at.%), exhibit a higher high-temperature hardness resulted from the γ'-$L1_2$ precipitation-strengthening of γ-FCC matrix, where the coherent γ/γ' microstructure can be thermodynamically stable after aging from 700 °C to 1100 °C for at least 500 h [126]. It is noted that a small amount of Al addition in HEAs (e.g., $Al_{0.3}$FeCoNiCr [115–117] and $Al_8Co_{17}Cr_{17}Cu_8Fe_{17}Ni_{33}$ [127]) generally renders spherical $L1_2$-Ni_3Al particles precipitated in the FCC matrix, resulting in high strength and good ductility, similar to that in Ni-based superalloys [5,132]. With increasing Al content, not only the FCC matrix of Al_xFeCoNiCr series of HEAs transforms into the BCC phase, but also the precipitates change from $L1_2$-Ni_3Al to B2-NiAl, as a result of an enhancement of strength drastically [114–118]. In addition, fixing Al content, the mutation of transition metals can also change the phase transition from FCC to BCC, as evidenced by the two HEAs of the FCC-based $Fe_{36}Co_{21}Cr_{18}Ni_{15}Al_{10}$ (at.%) and the BCC-based $Fe_{36}Mn_{21}Cr_{18}Ni_{15}Al_{10}$ (at.%) HEA with cuboidal B2 nanoprecipitates [109,123]. Furthermore, a minor addition of Ti (4 at.%) into these two HEAs leads to the formation of $L2_1$-Ni_2AlTi phase, rather than the B2, but the $L2_1$ precipitates exhibit different morphologies, being plate-like shape in the former alloy and cuboidal shape in the latter, respectively [109].

Experimentally, it is difficult to obtain cuboidal or spherical morphology of coherent B2 or $L2_1$ precipitates in BCC-based HEAs, which is primarily attributed to a large lattice misfit between BCC and B2 phases caused by the large composition difference. Thus, a weave-like microstructure of BCC and B2/$L2_1$ always occurred in these BCC-based HEAs, since it is sensitive to the Al/Ti content, such as the AlFeCoNiCr HEA, leading to a serious brittleness [114]. Massive efforts have been done to research for the cuboidal or spherical B2/$L2_1$ precipitation in various systems through adjusting both Al and transition metals [107–109,118–124,128,129]. It is fascinating that spherical or cuboidal B2/$L2_1$ nanoprecipitates are coherently-existed not only in Al/Ti-LTM HEAs (e.g., $Fe_{34}Cr_{34}Ni_{14}Al_{14}Co_4$ at.% [121]), but also in refractory HEAs consisted of Al and early transition metals [111,128,129]. For instance, the coherent precipitation of spherical B2 nanoparticles in BCC matrix improves the room- temperature compressive ductility on a large extent of refractory $Al_{0.5}NbTa_{0.8}Ti_{1.5}V_{0.2}$Zr HEA, while maintains high yield strength at both room and elevated temperatures [129].

Table 1 lists the mechanical properties (yield strength σ_y and ductility δ) at room and elevated temperatures of some precipitation-strengthened HEAs, including $L1_2$-strengthed FCC-based HEAs and B2/$L2_1$-strenthened BCC-based HEAs. The mechanical properties of commercial boiler steel HR3C ($Fe_{54.73}Cr_{24.01}Ni_{20.6}C_{0.05}Nb_{0.37}N_{0.24}$, wt.%) [133,134], newly-developed Ti-modified FBB8 ferritic stainless steel [89,90], and commercial Ni-based polycrystalline superalloy Inconel 718

($Ni_{53}Fe_{18.5}Cr_{19}Nb_{5.1}Mo_{3.0}Ti_{0.9}Al_{0.5}$ wt.%) [135,136] were also listed in Table 1 for reference. Compared with the HR3C austenitic stainless steel strengthened by MC-type carbides and Z-NbCrN nanoparticles, the FCC-based HEAs containing coherent spherical $L1_2$ nanoprecipitates generally exhibit higher yield strengths at both room and elevated temperatures, in which the particle size must exceed a certain value to ensure high strength according to the strengthening mechanism. In fact, the high-temperature strength has been improved in the newly-developed Ti-modified FBB8 ferritic stainless steel due to the coherent precipitation of cuboidal $L2_1$ nanoparticles [89,90]. The higher room-temperature strengths of BCC-based HEAs with cuboidal $B2/L2_1$ precipitation are comparable to that of Ni-based Inconel 718 superalloy. Especially for the Al-contained refractory HEAs [128,129], the yield strengths at both room and elevated temperatures (up to 1200 °C) are all much higher than that of Inconel 718, which will be softened at the temperature above 900 °C.

Table 1. Data summary for some typical precipitation-strengthened HEAs, including alloy composition, matrix phase, particle size and morphology of precipitated phase at room temperature (RT), and mechanical properties (yield strength σ_y and ductility δ) at both room and elevated temperatures. The mechanical properties of commercial HR3C steel and Inconel 718 superalloy, as well as Ti-modified FBB8, are also listed for reference.

Alloys	Matrix Phase	Particle Size (nm) and Morphology of Precipitated Phase at RT	Mechanical Properties		
			Temp. (°C)	σ_y (MPa)	δ (%)
$(NiCoFeCr)_{94}Ti_2Al_4$ (at.%) [125] (tensile test)	FCC	25 Spherical $L1_2$	RT	1005	17
$Ni_{48.6}Al_{10.3}Co_{17}Cr_{7.5}Fe_{9.0}Ti_{5.8}Ta_{0.6}Mo_{0.8}W_{0.4}$ (at.%) [126] (microhardness test)	FCC	287 Spherical $L1_2$	RT 500 700 900	1230 1128 1056 918	- - - -
$Fe_{36}Co_{21}Cr_{18}Ni_{15}Al_{10}$ (at.%) [123] (compressive test)	FCC	90 plate-like B2/BCC	RT 400 600	250 155 150	>50 >50 >50
$Al_{0.3}FeCoNiCr$ [117] (tensile test)	FCC	35 Spherical $L1_2$	RT	160	60.8
$Al_{0.5}CoCuCrFeNi$ [113,131] (tensile test)	FCC	200~400 Spherical $L1_2$	RT 600 700	1290 425 186	5.7 2.8 5.4
$Al_8Co_{17}Cr_{17}Cu_8Fe_{17}Ni_{33}$ (at.%) [127] (tensile test)	FCC	<20 Spherical $L1_2$	RT 500	357 315	9 0.7
$Fe_{36}Mn_{21}Cr_{18}Ni_{15}Al_{10}$ (at.%) [123] (tensile test)	BCC	160 Cuboidal B2	RT 400 500 600	750 640 515 310	2.5 20 42 55
$Fe_{35}Mn_{20}Cr_{17}Ni_{12}Al_{12}Ti_4$ (at.%) [109] (compressive test)	BCC	190 Cuboidal $L2_1$	RT 400 600	1280 1100 355	31 >50 >50
$Fe_{35}Co_{20}Cr_{17}Ni_{12}Al_{12}Ti_4$ (at.%) [109] (compressive test)	BCC	65 Plate-like $L2_1$	RT 400 600 800	1420 1285 795 285	18 24 >50 >50
$Al_{0.7}NiCoFe_2Cr$ [119] (tensile test)	BCC	50~90 Cuboidal B2	RT 650 700	1085 454 108	8.2 6.4 5.5
$Al_{0.7}NiCoFeCr_2$ [120] (compressive test)	BCC	80~120 Cuboidal B2	RT	1718	26.5
$(Al_2Ti)_{0.7}NiCoFeCr_2$ [107] (compressive test)	BCC	50~60 Cuboidal $L2_1$	RT	1808	35
$AlMo_{0.5}NbTa_{0.5}TiZr$ [128] (compressive test)	B2	10~55 Cuboidal & plate-like BCC	RT 600 800 1000 1200	2000 1870 1597 745 250	10 10 11 >50 >50
$Al_{0.5}NbTa_{0.8}Ti_{1.5}V_{0.2}Zr$ [129] (compressive test)	BCC	50 Spherical B2	RT 600	1345 1423	38 16.2

Table 1. *Cont.*

Alloys	Matrix Phase	Particle Size (nm) and Morphology of Precipitated Phase at RT	Mechanical Properties		
			Temp. (°C)	σ_y (MPa)	δ (%)
HR3C $Fe_{54.73}Cr_{24.01}Ni_{20.6}C_{0.05}Nb_{0.37}N_{0.24}$ (wt.%) [132,133] (tensile test)	FCC	25~30 Spherical Z-NbCrN phase	RT	368	48
			650	180	46
Ti-modified FBB8 $Fe_{67.85}Al_{6.5}Ni_{10}Cr_{10}Mo_{3.4}Ti_2Zr_{0.25}B_{0.005}$ (wt.%) [89,90] (In-situ tensile test)	BCC	84~138 Cuboidal $L2_1$	RT	1138 MPa, converted from *HV* value	
			700	230	7.5
Inonel 718 $Ni_{53}Fe_{18.5}Cr_{19}Nb_{5.1}Mo_{3.0}Ti_{0.9}Al_{0.5}$ (wt.%) [134,135] (tensile test)	FCC	15~25 Spherical γ'-$L1_2$ and γ''-Ni_3Nb	RT 600 700 800 900	1206 1048 979 629 262	17.8 17.1 20.5 28.0 83.0

In our recent work, we obtained spherical or cuboidal B2 nanoprecipitates in the BCC matrix of Al-TM HEAs with the composition formula of Al_2M_{14} (=$Al_{0.7}M_5$ in molar fraction), in which Al is fixed and M represents different mutations of Ni, Co, Fe, and Cr [119,120]. The Al_2M_{14} was designed with the guide of a cluster formula approach through mutating the combinations of TMs, rather than the Al. It is noted that the addition of a much more amount of BCC stabilizers (Fe and Cr) can favor to the formation of the BCC/B2 structures without any FCC phase in alloys. More significantly, the cuboidal B2 nanoprecipitation in alloys with M = $NiCoFe_2Cr$ and M = $NiCoFeCr_2$ is strongly attributed to a moderate lattice misfit (ε ~0.4%) between BCC and B2 phases, as seen in Figure 12a [120]. It is due to the cuboidal B2 nanoparticles in the BCC matrix that produces a prominent mechanical property with higher strength (σ_y = 1.1~1.7 GPa) and good ductility. Besides, these cuboidal B2 nanoprecipitates in these two alloys are very stable and could not be coarsened even after a long-time aging at 773 K for 1080 h [137]. In addition, the Ti substitution for Al in Al_2M_{14} HEAs can change the phase structures of ordered precipitates, from B2-NiAl to $L2_1$-Ni_2AlTi. Furthermore, a minor amount of Ti substitution with a ratio of Al/Ti $\geq$ 2/1 can still keep the cuboidal morphology of $L2_1$ nanoprecipitates for the achievement of a higher strength (σ_y = 1.8 GPa) due to a moderate lattice misfit (Figure 12b) [107,108]. Figure 12c shows the variation tendencies of ($\Delta\sigma_{CS}$ + $\Delta\sigma_{MS}$) and $\Delta\sigma_{orowan}$ with the particle size r in typical $(Al,Ti)_2M_{14}$ alloys at a fixed f. The used parameters for yield strength increment calculations are M = 2.73, α_ε = 2.6, G_{BCC} = 83 GPa, G_{B2} = 80 GPa, G_{L21} = 73.6 GPa, m = 0.85, b = 0.254 nm, $(\gamma_{apb})_{B2}$ = 0.25 J/m^2, $(\gamma_{apb})_{L21}$ = 0.04 J/m^2, and v =0.3, respectively. The optimal particle size r_0, corresponding to the largest strength increment, is calculated by the equation of ($\Delta\sigma_{CS}$ + $\Delta\sigma_{MS}$) = $\Delta\sigma_{orowan}$ in light of the precipitation strengthening mechanism. It is noted that when the particle size of B2 or $L2_1$ nanoprecipitates exceeds a certain value (about 40 nm), the ($\Delta\sigma_{CS}$ + $\Delta\sigma_{MS}$), rather than the $\Delta\sigma_{OR}$, will dominates the dislocation-shearing mechanism, compared with the phenomenon in the above-mentioned superhigh strength Fe-17Ni-6.2Al-2.3Mo-0.48Nb-0.37C-0.05B (wt.%) alloy [9]. More importantly, it can be demonstrated that the higher strength of $(Al,Ti)_2M_{14}$ HEAs is primarily attributed to the fact that the experimental particle size (50~120 nm) of B2 or $L2_1$ cuboids is close to the optimal size r_0 for the maximum strength increment from theoretical calculations.

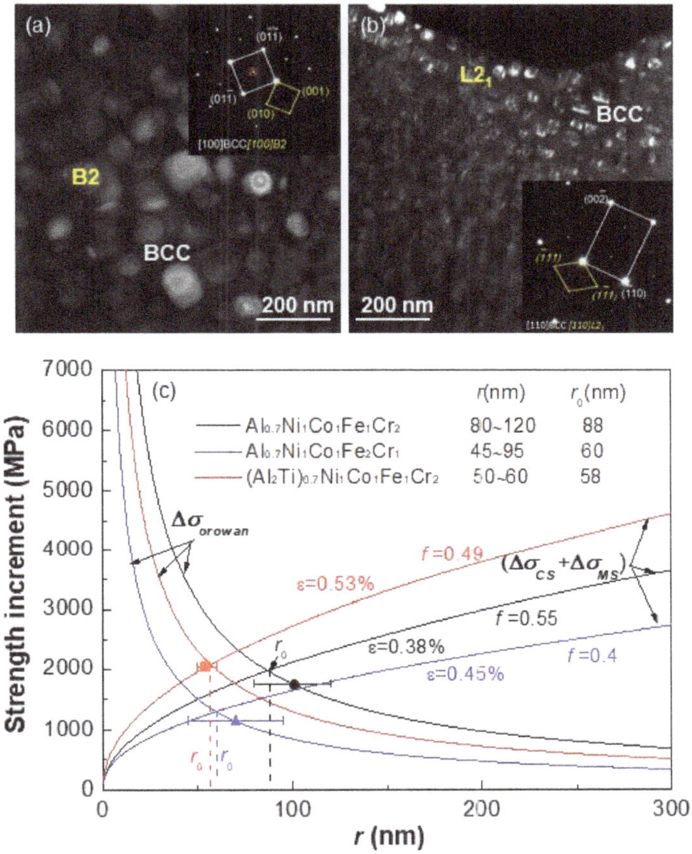

Figure 12. Computations of ($\Delta\sigma_{CS}$ + $\Delta\sigma_{MS}$) and $\Delta\sigma_{orowan}$ as a function of particle size r for the present $Al_{0.7}Ni_1Co_1Fe_1Cr_2$, $Al_{0.7}Ni_1Co_1Fe_2Cr_1$, and $(Al_2Ti)_{0.7}Ni_1Co_1Fe_1Cr_2$ HEAs (**c**), in which the morphologies of the coherent cuboidal B2 (**a**) and L2$_1$ (**b**) precipitates are also presented [107,108,120]. The optimal particle size r_0 from the calculation and the experimentally-measured r are also marked for each alloy.

5. Thoughts on the Coherent Precipitation Strengthening

The coherent precipitation of ordered L1$_2$-γ' nanoprecipitates into the FCC-γ matrix renders the Ni-based superalloys with prominent mechanical properties at elevated temperatures close to melting points, which is hard to realize in other conventional alloy systems. The main reason is that in most conventional alloy systems, such as Al alloys, Mg alloys, and Cu alloys, the finally-stable precipitated phases are not the ordered superstructures of their parent solid solutions. In fact, the precipitation sequence of G.P. zones $\rightarrow$ metastable coherent ordered phases $\rightarrow$ stable non-coherent intermetallic phases often appears during the aging process, in which only the coherent precipitation corresponds to the peak strength. Therefore, many researchers have been exploring how to maintain the long-term stability of coherent precipitates through adjusting the amount of alloying elements or changing element species. Take the Mg alloys for instance. Based on the Mg-Al binary alloys without any ordered phases, the addition of a superlarge-size element Ca can form an ordered G.P. zone on the basal plane of the HCP matrix, i.e., clusters induced by Ca atoms, which can improve the creep resistance, but cannot enhance the strength [65]. So, in order to further improve the alloy strength, Mg-Gd-, Mg-Y-, and Mg-Nd-based alloys have been developed, in which several typical alloys, such as

Mg-18.2Gd-1.9Ag-0.3Zr, Mg-6Y-4.9Zn, and Mg-10Gd-5.7Y-1.6Zn-0.7Zr (wt.%), possess a much higher ultimate tensile strength of above 400 MPa [57]. An obvious feature in these alloy systems is that the interplanar distance of G.P. zones (d = 0.37 nm) is getting closer to the lattice constant (a~0.32 nm) of the basal plane of HCP-Mg matrix, compared with that (d = 0.556 nm) in Mg-Al-Ca alloys. Hence, the much better coherency between the G.P. zones and the Mg matrix promotes the formation of ordered superstructure (DO_{19}-Mg_3(Gd,Nd)) of the HCP solid solution, resulting in a higher strength. From the viewpoint of alloying element species, it can be found that the key factor for the better coherency is that the super large-size elements (Gd, Y, etc.) and small-size elements (Zn, Ag, etc.) must be added simultaneously to balance the interplanar distance of G.P. zones close to the lattice constant of Mg matrix. Till recently, the HCP-Mg matrix was changed into the BCC-Mg matrix in Mg-Li-Al system, in which the coherent precipitation of ordered superstructure of $L2_1$-Li_2MgAl improves the tensile ductility, besides the higher strength of alloys [74]. It demonstrates sufficiently the important role of coherent precipitation in the development of high-strength structural materials.

In addition, in the case of coherent precipitation, the morphology of the coherent precipitates is also important to the mechanical properties of alloys, which has been identified by the presence of cuboidal $L1_2$-γ' nanoprecipitates in Ni-based single-crystal superalloys [5]. It is known that the precipitate morphology (shape and size) is primarily controlled by the lattice misfit between the ordered phase and its parent solid solution. A moderate lattice misfit for cuboidal nanoprecipitates should be achieved by mutating the lattice constants of these two phases simultaneously. For the γ/γ' coherent microstructure, it is relatively easy to adjust the lattice misfit because there exists a relatively-small composition difference between γ and γ' phases. In contrast, it is difficult to adjust the lattice misfit between the B2 or $L2_1$ phase and the BCC matrix rationally in conventional alloy systems due to the relatively-larger composition difference between these two phases. In most cases, there often exhibits a weave-like microstructure induced by spinodal decomposition. The recent development of high-entropy alloys shows a bright insight to achieve the ideal (cuboidal) coherent microstructure since the phase compositions can be adjusted within a wide range in multi-principal alloy systems. Actually, all of the Ni-based single-crystal superalloys are concentratedly-complex alloys, generally containing more than ten elements, which favors the mutation of lattice misfit. So, some new BCC-based superalloys with superhigh strength and/or better creep resistance have been developed recently, which is derived from the coherent precipitation of spherical/cuboidal B2 or $L2_1$ nanoprecipitates in the BCC matrix [9,89,90,108,120].

It is emphasized that whether the formation of coherent phases or the morphology of coherent precipitates is considered, both are closely related to the chemical composition. However, the rational matching of solute elements in the ordered phase and the solid solution was seldom considered when designing alloy compositions, since the solute distribution in the parent solid solution has not been clear until now. Actually, the local chemical structural units (chemical short range orders, CSROs) induced by solute elements are crucial to the stability of parent solid solution at high temperatures [138,139], which decides the precipitation of ordered phase during the aging or as-cast state. The recently-proposed cluster-plus-glue-atom model has defined such local chemical structure units in solid solutions [120,140–142], in which the cluster is the nearest-neighbor polyhedron centered by a solute atom having the strong interaction with the base solvent atoms to represent the strongest CSRO, and some other solute atoms (i.e., glue atoms) with weak interactions are certainly required to fill the space between the clusters to balance the atomic-packing density. Thus, a composition formula of [cluster] (glue atom) x (x being the glue-atom number) can be obtained from the cluster model. So, the phase compositions of ordered phase and their parent solid solution could be considered in light of the chemical structural units, respectively. Combined with the volume fraction of coherent precipitates, it would provide a new approach of composition design to develop high-performance compositionally-complex alloys with coherent precipitation.

6. Conclusions

In the present work, the precipitation behavior and precipitation strengthening in compositionally-complex alloys were generalized comprehensively, including high-performance conventional engineering materials (Ni superalloys, Al alloys, Mg alloys, Cu alloys, and stainless steels), and newly-developed high entropy alloys. The morphology evolution of second-phase particles and precipitation strengthening mechanism were introduced firstly. Then, the precipitation behaviors in diverse compositionally-complex alloy systems are illustrated, respectively. After discussing the relationship between the particle morphology and strengthening effectiveness, alloys with the coherent microstructure of the ordered phase precipitated in the disordered solid solution matrix were specially emphasized, since they exhibit prominent mechanical properties (superhigh strength/toughness and excellent high-temperature creep resistance). The universal feature existed in all compositionally-complex alloys is the coherent precipitation, which will be the most effective approach for the enhancement of alloy strength.

Author Contributions: Conceptualization, P.K.L. and Q.W.; methodology, Q.W., Z.L. and C.D.; validation, X.L. and P.K.L.; formal analysis, Q.W. and S.P.; investigation, X.L.; resources, C.D.; writing—original draft preparation, Q.W.; writing—review and editing, C.D. and P.K.L.

Funding: This research was funded by the National Key Research and Development Plan, grant number 2017YFB0702400; the Science Challenge Project, grant number TZ2016004; the National Magnetic Confinement Fusion Energy Research Project, grant number 2015GB121004.

Acknowledgments: P.K.L. would like to acknowledge the Department of Energy (DOE), Office of Fossil Energy, National Energy Technology Laboratory, grant number DE-FE-0011194; the U.S. Army Research Office project, grant number W911NF-13-1-0438; the National Science Foundation, grant number DMR-1611180 and 1809640.

Conflicts of Interest: The authors declare no conflict of interest.

References

1. Hosford, W.F. *Mechanical Behavior of Materials*, 1st ed.; Cambridge University Press: New York, NY, USA, 2005.
2. Argon, A. *Strengthening Mechanisms in Crystal Plasticity*; Oxford University Press: Oxford, UK, 2007.
3. Brown, L.M.; Ham, R.K. Dislocation-particle Interactions. In *Strengthening Methods in Crystals*; Kelly, A., Nicholson, R.B., Eds.; Elsevier: Amsterdam, The Netherlands, 1971.
4. Giamei, A.F.; Anton, D.L. Rhenium additions to a Ni-base superalloy: Effects on microstructure. *Metall. Trans. A* **1985**, *16*, 1997–2005. [CrossRef]
5. Reed, R.C. *The Superalloys: Fundamentals and Applications*; Cambridge University Press: New York, NY, USA, 2006.
6. Zhou, H.; Okada, I.; Ro, Y.; Koizumi, Y.; Kobayashi, K. Thermomechanical fatigue behavior of the third-generation, single-crystal superalloy TMS-75: Deformation structure. *Metall. Mater. Trans. A* **2004**, *35*, 1779–1787. [CrossRef]
7. Lo, K.H.; Shek, C.H.; Lai, J.K.L. Recent developments in stainless steels. *Mater. Sci. Eng. R* **2009**, *65*, 39–104. [CrossRef]
8. Zhou, Y.H.; Liu, Y.C.; Zhou, X.S.; Liu, C.X.; Yu, J.X.; Huang, Y.; Li, H.J.; Li, W.Y. Precipitation and hot deformation behavior of austenitic hear-resistant steels: A review. *J. Mater. Sci. Technol.* **2017**, *33*, 1448–1456. [CrossRef]
9. Jiang, S.H.; Wang, H.; Wu, Y.; Liu, X.J.; Chen, H.H.; Yao, M.J.; Gault, B.; Ponge, D.; Raabe, D.; Hirata, A.; et al. Ultrastrong steel via minimal lattice misfit and high-density nanoprecipitation. *Nature* **2017**, *544*, 460–464. [CrossRef] [PubMed]
10. Jiao, Z.B.; Luan, J.H.; Zhang, Z.W.; Liu, C.T. High-strength steels hardened mainly by nanoscale NiAl precipitates. *Scr. Mater.* **2014**, *87*, 45–48. [CrossRef]
11. Jiao, Z.B.; Luan, J.H.; Miller, M.K.; Yu, C.Y.; Liu, C.T. Effects of Mn partitioning on nanoscale precipitation and mechanical properties of ferritic steels strengthened by NiAl nanoparticles. *Acta Mater.* **2015**, *84*, 283–291. [CrossRef]

12. Teng, Z.K.; Miller, M.K.; Ghosh, G.; Liu, C.T.; Huang, S.; Russell, K.F.; Fine, M.E.; Liaw, P.K. Characterization of nanoscale NiAl-type precipitates in a ferritic steel by electron microscopy and atom probe tomography. *Scr. Mater.* **2010**, *63*, 61–64. [CrossRef]
13. Vo, N.Q.; Liebscher, C.H.; Rawling, J.S.; Asta, M.; Dunand, D.C. Creep properties and microstructure of a precipitation-strengthened ferritic Fe-Al-Ni-Cr alloy. *Acta Mater.* **2014**, *71*, 89–99. [CrossRef]
14. Aggen, G.; Allen, M. *ASM Handbook Volume I, Properties and Selection: Irons, Steels, and High-Performance Alloys*; The Materials Information Company: New York, NY, USA, 2018.
15. Senkov, O.N.; Miller, J.D.; Miracle, D.B.; Woodward, C. Accelerated exploration of multi-principal element alloys with solid solution phases. *Nat. Commun.* **2015**, *6*, 6529–6535. [CrossRef] [PubMed]
16. Ye, Y.F.; Wang, Q.; Lu, J.; Liu, C.T.; Yang, Y. High-entropy alloy: challenges and prospects. *Mater. Today* **2016**, *19*, 349–362. [CrossRef]
17. Miracle, D.B.; Senkov, O.N. A critical review of high entropy alloys and related concepts. *Acta Mater.* **2016**, *122*, 488–511. [CrossRef]
18. Yeh, J.W.; Chen, S.K.; Lin, S.J.; Gan, J.Y.; Chin, T.S.; Shun, T.T.; Tsau, C.H.; Chang, S.Y. Nanostructured high-entropy alloys with multiple principal elements: Novel alloy design concepts and outcomes. *Adv. Eng. Mater.* **2004**, *6*, 299–303. [CrossRef]
19. Diao, H.; Xie, X.; Sun, F.; Dahmen, K.A.; Liaw, P.K. Mechanical Properties of High-Entropy Alloys. In *High-Entropy Alloys: Fundamentals and Applications*; Gao, M.C., Yeh, J.W., Liaw, P.K., Zhang, Y., Eds.; Springer International Publishing: Gewerbestrasse, Switzerland, 2016.
20. Zhang, Y.; Zuo, T.T.; Tang, Z.; Gao, M.C.; Dahmen, K.A.; Liaw, P.K.; Lu, Z.P. Microstructures and properties of high-entropy alloys. *Prog. Mater. Sci.* **2014**, *61*, 1–93. [CrossRef]
21. Zhang, Y.; Zhou, Y.J.; Lin, J.P.; Chen, G.; Liaw, P.K. Solid-solution phase formation rules for multi-component alloys. *Adv. Eng. Mater.* **2008**, *10*, 534–538. [CrossRef]
22. Youssef, K.M.; Zaddach, A.J.; Niu, C.; Irving, D.L.; Koch, C.C. A novel low-density, high-hardness, high-entropy alloy with close-packed single-phase nanocrystalline structures. *Mater. Res. Lett.* **2014**, *2*, 95–99. [CrossRef]
23. Cahn, J.W.; Kalonji, G. Symmetry in Solid State Transformation Morphologies. In Proceedings of the International Conference on Solid-solid Phase Transformations, Whistler, BC, Canada, 28 June–3 July 1982.
24. Khachaturyan, A.G. *Theory of Structural Transformations in Solids*; Wiley: New York, NY, USA, 1983.
25. Ardell, A.J.; Nicholson, R.B. On the modulated structure of aged Ni-Al alloys: with an Appendix On the elastic interaction between inclusions by J. D. Eshelby. *Acta Metall.* **1966**, *14*, 1295–1309. [CrossRef]
26. Doi, M.; Miyazaki, T.; Wakatsuki, T. The effects of elastic interaction energy on the γ' precipitate morphology of continuously cooled nickel-base alloys. *Mater. Sci. Eng.* **1985**, *74*, 139–145. [CrossRef]
27. Miyazaki, T.; Imamura, H.; Kozakai, T. The formation of "γ' precipitate doublets" in Ni_3Al alloys and their energetic stability. *Mater. Sci. Eng.* **1982**, *54*, 9–15. [CrossRef]
28. Kaufman, M.J.; Voorhees, P.W.; Johnson, W.C.; Biancaniello, F.S. An elastically induced morphological instability of a misfitting precipitate. *Metall. Trans. A* **1989**, *20*, 2171–2175. [CrossRef]
29. Glatzel, U.; Feller-Kniepmeier, M. Calculations of internal stresses in the γ/γ' microstructure of a nickel-base superalloy with high volume fraction of γ'-phase. *Scripta Metall.* **1989**, *23*, 1839–1844. [CrossRef]
30. Johnson, W.C.; Coorhees, P.W. Elastically-induced precipitate shape transitions in coherent solids. *Solid State Phenom .* **1992**, *23*, 87–103. [CrossRef]
31. Khachaturyan, A.G.; Semenovskaya, S.V.; Morris, J.W., Jr. Theoretical analysis of strain-induced shape changes in cubic precipitates during coarsening. *Acta Metall.* **1988**, *36*, 1563–1572. [CrossRef]
32. Voorhees, P.W.; Mcfadden, G.B.; Johnson, W.C. On the morphological development of second-phase particles in elastically-stressed solids. *Acta Metall. Mater.* **1992**, *40*, 2979–2992. [CrossRef]
33. Thompson, M.E.; Su, C.S.; Voorhees, P.W. The equilibrium shape of a misfitting precipitate. *Acta Metall. Mater.* **1994**, *42*, 2107–2122. [CrossRef]
34. Ardell, A.J. Precipitation hardening. *Metall. Trans. A* **1985**, *16*, 1985–2135. [CrossRef]
35. Gerold, V.; Haberkorn, H. On the Critical Resolved Shear Stress of Solid Solutions Containing Coherent Precipitates. *Phys. Stat. Sol.* **1966**, *16*, 675–684. [CrossRef]
36. Jansson, B.; Melander, A. On the critical resolved shear stress from misfitting particles. *Scr. Metall.* **1978**, *12*, 497–498. [CrossRef]

37. Melander, A.; Persson, P.A. The strength of a precipitation hardened AlZnMg alloy. *Acta Metall.* **1978**, *26*, 267–278. [CrossRef]
38. Nembach, E. Precipitation hardening caused by a difference in shear modulus between particle and matrix. *Phys. Stat. Sol.* **1983**, *78*, 571–581. [CrossRef]
39. Gleiter, H.; Hornbogen, E. Theorie der wechselwirkung von versetzungen mit kohärenten geordneten zonen. *Phys. Stat. Sol.* **1965**, *12*, 235–250. [CrossRef]
40. Zhang, J.X.; Wang, J.C.; Harada, H.; Koizumi, Y. The effect of lattice misfit on the dislocation motion in superalloys during high-temperature low-stress creep. *Acta Mater.* **2005**, *53*, 4623–4633. [CrossRef]
41. Ding, Q.Q.; Li, S.X.; Chen, L.Q.; Han, X.D.; Zhang, Z.; Yu, Q.; Li, J.X. Re segregation at interfacial dislocation network in a nickel-based superalloy. *Acta Mater.* **2018**, *154*, 137–146. [CrossRef]
42. Andersen, S.J.; Zandbergen, H.W.; Jansen, J.; Traeholt, C.; Tundal, U.; Reiso, O. The crystal structure of the β'' phase in Al-Mg-Si alloys. *Acta Mater.* **1998**, *46*, 3283–3298. [CrossRef]
43. Murayama, M.; Hono, K. Pre-precipitate clusters and precipitation processes in Al-Mg-Si alloys. *Acta Mater.* **1999**, *47*, 1537–1548. [CrossRef]
44. Fuller, C.B.; Murray, J.L.; Seidman, D.N. Temporal evolution of the nanostructure of Al(Sc,Zr) alloys: Part I-Chemical compositions of Al(ScxZrx) precipitates. *Acta Mater.* **2005**, *53*, 5401–5413. [CrossRef]
45. Sha, G.; Cerezo, A. Early-stage precipitation in Al-Zn-Mg-Cu alloy (7050). *Acta Mater.* **2004**, *52*, 4503–4516. [CrossRef]
46. Marlaud, T.; Deschamps, A.; Bley, F.; Lefebvre, W.; Baroux, B. Influence of alloy composition and heat treatment on precipitate composition in Al-Zn-Mg-Cu alloys. *Acta Mater.* **2010**, *58*, 248–260. [CrossRef]
47. Gayle, F.W.; Goodway, M. Precipitation hardening in the first aerospace aluminum alloy: the wright flyer crankcase. *Science* **1994**, *266*, 1015–1507. [CrossRef] [PubMed]
48. Biswas, A.; Siegel, D.J.; Wolverton, C.; Seidman, D.N. Precipitates in Al-Cu alloys revisited: Atom-probe tomographic experiments and first-principles calculations of compositional evolution and interfacial segregation. *Acta Mater.* **2011**, *59*, 6187–6204. [CrossRef]
49. Edwards, G.A.; Stiller, K.; Dunlop, G.L.; Couper, M.J. The precipitation sequence in Al-Mg-Si alloys. *Acta Mater.* **1998**, *46*, 3893–3904. [CrossRef]
50. Booth-Morrison, C.; Dunand, D.C.; Seidman, D.N. Coarsening resistance at 400 °C of precipitation-strengthened Al-Zr-Sc-Er alloys. *Acta Mater.* **2011**, *59*, 7029–7042. [CrossRef]
51. Wen, S.P.; Gao, K.Y.; Huang, H.; Nie, W.Z.R. Precipitation evolution in Al-Er-Zr alloys during aging at elevated temperature. *J. Alloy. Compd.* **2013**, *574*, 92–97. [CrossRef]
52. Li, H.Y.; Bin, J.; Liu, J.J.; Gao, Z.H.; Lu, X.C. Precipitation evolution and coarsening resistance at 400 °C of Al microalloyed with Zr and Er. *Scr. Mater.* **2012**, *67*, 73–76. [CrossRef]
53. Marquis, E.A.; Seidman, D.N. Nanoscale structural evolution of Al_3Sc precipitates in Al(Sc) alloys. *Acta Mater.* **2001**, *49*, 1909–1919. [CrossRef]
54. Seidman, D.N.; Marquis, E.A.; Dunand, D.C. Precipitation strengthening at ambient and elevated temperatures of heat-treatable Al(Sc) alloys. *Acta Mater.* **2002**, *50*, 4021–4035. [CrossRef]
55. Clouet, E.; Nastar, M.; Barbu, A.; Sigli, C.; Martin, G. An Atomic and Mesoscopic Study of Precipitation Kinetics in Al-Zr-Sc Alloys. *Adv. Eng. Mater.* **2006**, *8*, 1228–1231. [CrossRef]
56. Avedesian, M.M.; Baker, H. *ASM Specialty Handbook: Magnesium and Magnesium Alloys*; ASM International: Headquarters, OH, USA, 1999.
57. Nie, J.F. Precipitation and hardening in magnesium alloys. *Metall. Mater. Trans. A* **2012**, *43*, 3891–3939. [CrossRef]
58. Wang, D.S.; Amsler, M.; Hegde, V.I.; Saal, J.E.; Issa, A.; Zhou, B.C.; Zeng, X.Q.; Wolverton, C. Crystal structure, energetics, and phase stability of strengthening precipitates in Mg alloys: A first-principles study. *Acta Mater.* **2018**, *158*, 65–78. [CrossRef]
59. Duly, D.; Zhang, W.Z.; Audier, M. High-resolution electron microscopy observations of the interface structure of continuous precipitates in a Mg-Al alloy and interpretation with the O-lattice theory. *Philos. Mag. A* **1995**, *71*, 187–204. [CrossRef]
60. Nie, J.F.; Xiao, X.L.; Luo, C.P.; Muddle, B.C. Characterisation of precipitate phases in magnesium alloys using electron microdiffraction. *Micron* **2001**, *32*, 857–863. [CrossRef]
61. Zhang, M.; Zhang, W.Z.; Ye, F. Interpretation of precipitation crystallography of $Mg_{17}Al_{12}$ in a Mg-Al alloy in terms of singular interfacial structure. *Metall. Mater. Trans. A* **2005**, *36*, 1681–1688. [CrossRef]

62. Geng, J.; Gao, X.; Fang, X.Y.; Nie, J.F. Enhanced age-hardening response of Mg-Zn alloys via Co additions. *Scr. Mater.* **2011**, *64*, 506–509. [CrossRef]
63. Hono, K.; Mendis, C.L.; Sasaki, T.T.; Oh-Ishi, K. Towards the development of heat-treatable high-strength wrought Mg alloys. *Scr. Mater.* **2010**, *63*, 710–715. [CrossRef]
64. Bettles, C.J.; Gibson, M.A.; Venkatesan, K. Enhanced age-hardening behaviour in Mg-4 wt.% Zn micro-alloyed with Ca. *Scr. Mater.* **2004**, *51*, 193–197. [CrossRef]
65. Homma, T.; Nakawaki, S.; Oh-ishi, K.; Hono, K.; Kamado, S. Unexpected influence of Mn addition on the creep properties of a cast Mg-2Al-2Ca (mass%) alloy. *Acta Mater.* **2011**, *59*, 7662–7672. [CrossRef]
66. Wilson, R.; Bettles, C.J.; Muddle, B.C.; Nie, J.F. Precipitation hardening in Mg-3 wt% Nd (-Zn) casting alloys. *Mater. Sci. Forum* **2003**, *419*, 267–272. [CrossRef]
67. Nishijima, M.; Hiraga, K.; Yamasaki, M.; Kawamura, Y. Characterization of β' Phase Precipitates in an Mg-5 at% Gd Alloy Aged in a Peak Hardness Condition, Studied by High-Angle Annular Detector Dark-Field Scanning Transmission Electron Microscopy. *Mater. Trans.* **2006**, *47*, 2109–2122. [CrossRef]
68. Nie, J.F.; Muddle, B.C. Characterisation of strengthening precipitate phases in a M-Y-Nd alloy. *Acta Mater.* **2000**, *48*, 1691–1703. [CrossRef]
69. Zhang, M.X.; Kelly, P.M. Morphology and crystallography of $Mg_{24}Y_5$ precipitate in Mg-Y alloy. *Scr. Mater.* **2003**, *48*, 379–384. [CrossRef]
70. Hilditch, T.; Nie, J.F.; Muddle, B.C. The Effect of Cold Work on Precipitation in Alloy WE54. In *Magnesium Alloys and Their Applications*; Mordike, B.L., Kainer, K.U., Eds.; Werkstoff-Informat: Frankfurt, Germany, 1998; pp. 339–344.
71. Nie, J.F.; Oh-ishi, K.; Gao, X.; Hono, K. Solute segregation and precipitation in a creep-resistant Mg-Gd-Zn alloy. *Acta Mater.* **2008**, *56*, 6061–6076. [CrossRef]
72. Gao, X.; He, S.M.; Zeng, X.Q.; Peng, L.M.; Ding, W.J.; Nie, J.F. Microstructure evolution in a Mg-15Gd-0.5Zr (wt.%) alloy during isothermal aging at 250 °C. *Mater. Sci. Eng. A* **2006**, *431*, 322–327. [CrossRef]
73. Nishijima, M.; Hiraga, K. Structural changes of precipitates in an Mg-5at%Gd alloy studied by transmission electron microscopy. *Mater. Trans.* **2007**, *48*, 10–15. [CrossRef]
74. Xu, W.Q.; Birbilis, N.; Sha, G.; Wang, Y.; Daniels, J.E.; Xiao, Y.; Ferry, M. A high-specific-strength and corrosion-resistant magnesium alloy. *Nature* **2015**, *14*. 1229–1236. [CrossRef] [PubMed]
75. Carisa, J.; Li, D.Q.; Stephens, J.J., Jr.; Lewandowskia, J.J. Microstructural effects on tension behavior of Cu-15Ni-8Sn sheet. *Mater. Sci. Eng. A* **2010**, *527*, 769–781. [CrossRef]
76. Spooner, S.; Lefevre, B.G. The effect of prior deformation on spinodal age hardening in Cu-15Ni-8Sn alloy. *Metall. Trans. A* **1980**, *11*, 1085–1093. [CrossRef]
77. Zhao, J.C.; Notis, M.R. Spinodal decomposition, ordering transformation, and discontinuous precipitation in a Cu-15Ni-8Sn alloy. *Acta Mater.* **1998**, *46*, 4203–4218. [CrossRef]
78. Lefevre, B.G.; D'annessa, A.T.; Kalish, D. Age hardening in Cu-15Ni-8Sn alloy. *Metall. Trans. A* **1978**, *9*, 577–586. [CrossRef]
79. Hermann, P.; Morris, D.G. Relationship between microstructure and mechanical properties of a spinodally decomposing Cu-15Ni-8Sn alloy prepared by spray deposition. *Metall. Mater. Trans. A* **1994**, *25*, 1403–1412. [CrossRef]
80. Sun, Z.Q.; Edmondson, P.D.; Yamamoto, Y. Effects of Laves phase particles on recovery and recrystallization behaviors of Nb-containing FeCrAl alloys. *Acta Mater.* **2018**, *144*, 716–727. [CrossRef]
81. Yamamoto, Y.; Brady, M.P.; Lu, Z.P.; Maziasz, P.J.; Liu, C.T.; Pint, B.A.; More, K.L.; Meyer, H.M.; Payzant, E.A. Creep-resistant, Al_2O_3-forming austenitic stainless steels. *Science* **2007**, *316*, 433–436. [CrossRef] [PubMed]
82. Zhu, S.M.; Tjong, S.C.; Lai, K.L. Creep behavior of a β'(NiAl) precipitation strengthened ferritic Fe-Cr-Ni-Al alloy. *Acta Mater.* **1998**, *46*, 2969–2976. [CrossRef]
83. Stallybrass, C.; Schneider, A.; Sauthoff, G. The strengthening effect of (Ni, Fe)Al precipitates on the mechanical properties at high temperatures of ferritic Fe-Al-Ni-Cr alloys. *Intermetallics* **2005**, *13*, 1263–1271. [CrossRef]
84. Teng, Z.K.; Liu, C.T.; Ghosh, G.; Liaw, P.K.; Fine, M.E. Effects of Al on the microstructure and ductility of NiAl-strengthened ferritic steels at room temperature *Intermetallics* **2010**, *18*, 1437–1443. [CrossRef]
85. Teng, Z.K.; Ghosh, G.; Miller, M.K.; Huang, S.; Clausen, B.; Brown, D.W.; Liaw, P.K. Neutron-diffraction study and modeling of the lattice parameters of a NiAl-precipitate-strengthened Fe-based alloy. *Acta Mater.* **2012**, *60*, 5362–5369. [CrossRef]

86. Teng, Z.K.; Zhang, F.; Miller, M.K.; Liu, C.T.; Huang, S.; Chou, Y.T.; Tien, R.H.; Chang, Y.A.; Liaw, P.K. New NiAl-strengthened ferritic steels with balanced creep resistance and ductility designed by coupling thermodynamic calculations with focused experiments. *Intermetallics* **2012**, *29*, 110–115. [CrossRef]
87. Huang, S.; Gao, Y.; An, K.; Zheng, L.; Wu, W.; Teng, Z.; Liaw, P.K. Deformation mechanisms in a precipitation-strengthened ferritic superalloy revealed by in situ neutron diffraction studies at elevated temperatures. *Acta Mater.* **2015**, *83*, 137–148. [CrossRef]
88. Sun, Z.; Song, G.; Ilavsky, J.; Liaw, P.K. Nano-sized precipitate stability and its controlling factors in a NiAl-strengthened ferritic alloy. *Sci. Rep.* **2015**, *5*, 16081. [CrossRef] [PubMed]
89. Song, G.; Sun, Z.Q.; Poplawsky, J.D.; Xu, X.D.; Chen, M.W.; Liaw, P.K. Primary and secondary precipitates in a hierarchical-precipitate-strengthened ferritic alloy. *J. Alloy. Compd.* **2017**, *706*, 584. [CrossRef]
90. Song, G.; Sun, Z.Q.; Poplawsky, J.D.; Gao, Y.; Liaw, P.K. Microstructural evolution of single Ni_2TiAl or hierarchical NiAl/Ni_2TiAl precipitates in Fe-Ni-Al-Cr-Ti ferritic alloys during thermal treatment for elevated-temperature applications. *Acta Mater.* **2017**, *127*, 1–16. [CrossRef]
91. Hattestrand, M.; Nilsson, J.O.; Stiller, K.; Liu, P.; Andersson, M. Precipitation hardening in a 12%Cr-9%Ni-4%Mo-2% Cu stainless steel. *Acta Mater.* **2004**, *52*, 1023–1037. [CrossRef]
92. Schober, M.; Schnitzer, R.; Leitner, H. Precipitation evolution in a Ti-free and Ti-containing stainless maraging steel. *Ultramicroscopy* **2009**, *109*, 553–562. [CrossRef] [PubMed]
93. Xu, W.; Rivera-Diaz-del-Castillo, P.E.J.; Yan, W.; Yang, K.; Martin, D.S.; Kestens, L.A.I.; Zwaag, S. A new ultrahigh-strength stainless steel strengthened by various coexisting nanoprecipitates. *Acta Mater.* **2010**, *58*, 4067–4075. [CrossRef]
94. Ifergane, S.; Sabatani, E.; Carmeli, B.; Barkay, Z.; Ezersky, V.; Beeri, O.; Eliaz, N. Hydrogen diffusivity measurement and microstructural characterization of Custom 465 stainless steel. *Electrochim. Acta* **2015**, *178*, 494–503. [CrossRef]
95. Liu, W.H.; Wu, Y.; He, J.Y.; Nieh, T.G.; Lu, Z.P. Grain growth and the Hall-Petch relationship in a high entropy alloy FeCrNiCoMn. *Scr. Mater.* **2013**, *68*, 526–529. [CrossRef]
96. Li, Z.; Pradeep, K.G.; Deng, Y.; Raabe, D.; Tasan, C.C. Metastable high-entropy dual-phase alloys overcome the strength-ductility trade-off. *Nature* **2016**, *534*, 227–230. [CrossRef] [PubMed]
97. Raabe, D.; Tasan, C.C.; Springer, H.; Bausch, M. From High-entropy alloys to high-entropy steels. *Steel Res. Int.* **2015**, *86*, 1127–1138. [CrossRef]
98. Gludovatz, B.; Hohenwarter, A.; Catoor, D.; Chang, E.H.; George, E.P.; Ritchie, R.O. A fracture-resistant high-entropy alloy for cryogenic applications. *Science* **2014**, *345*, 1153–1158. [CrossRef] [PubMed]
99. Schuh, B.; Mendez-Martin, F.; Volker, B.; George, E.P.; Clemens, H.; Pippan, R.; Hohenwarter, A. Mechanical properties, microstructure and thermal stability of a nanocrystalline CoCrFeMnNi high-entropy alloy after severe plastic deformation. *Acta Mater.* **2015**, *96*, 258–268. [CrossRef]
100. Tsai, K.Y.; Tsai, M.H.; Yeh, J.W. Sluggish diffusion in Co-Cr-Fe-Mn-Ni high-entropy alloys. *Acta Mater.* **2013**, *61*, 4887–4897. [CrossRef]
101. Owen, L.R.; Pickering, E.J.; Playford, H.Y.; Stone, H.J.; Yucker, M.G.; Jones, N.G. An assessment of the lattice strain in the CrMnFeCoNi high-entropy alloy. *Acta Mater.* **2017**, *122*, 11–18. [CrossRef]
102. Otto, F.; Yang, Y.; Bei, H.; George, E.P. Relative effects of enthalpy and entropy on the phase stability of equiatomic high-entropy alloys. *Acta Mater.* **2013**, *61*, 2628–2638. [CrossRef]
103. Cantor, B.; Chang, I.T.H.; Knight, P.; Vincent, A.J.B. Microstructural development in equiatomic multicomponent alloys. *Mater. Sci. Eng. A* **2004**, *375*, 213–218. [CrossRef]
104. Zhang, Z.J.; Mao, M.M.; Wang, J.; Gludovatz, B.; Zhang, Z.; Mao, S.X.; George, E.P.; Yu, Q.; Ritchie, R.O. Nanoscale origins of the damage tolerance of the high-entropy alloy CrMnFeCoNi. *Nat. Commun.* **2015**, *6*, 10143. [CrossRef] [PubMed]
105. Na, Y.S.; Lim, K.R.; Chang, H.J.; Kim, J. Effect of trace additions of Ti on the microstructure of AlCoCrFeNi-based high entropy alloy. *Sci. Adv. Mater.* **2016**, *8*, 1984–1988. [CrossRef]
106. Li, D.Y.; Li, C.X.; Feng, T.; Zhang, Y.D.; Shang, G.; Lewandowski, J.J.; Liaw, P.K.; Zhang, Y. High-entropy $Al_{0.3}$CoCrFeNi alloy fibers with high tensile strength and ductility at ambient and cryogenic temperatures. *Acta Mater.* **2017**, *123*, 285–294. [CrossRef]
107. Li, C.L.; Ma, Y.; Hao, J.M.; Wang, Q.; Pang, S.J.; Dong, C.; Liaw, P.K. Effect of Ti substitution for Al on the cuboidal nanoprecipitates in $Al_{0.7}NiCoFeCr_2$ high-entropy alloys. *J. Mater. Res.* **2018**, *33*, 3266–3275. [CrossRef]

108. Li, C.L.; Ma, Y.; Hao, J.M.; Yan, Y.; Wang, Q.; Dong, C.; Liaw, P.K. Microstructures and mechanical properties of body-centered-cubic $(Al,Ti)_{0.7}(Ni,Co,Fe,Cr)_5$ high entropy alloys with coherent $B2/L2_1$ nanoprecipitation. *Mater. Sci. Eng. A* **2018**, *737*, 286–296. [CrossRef]
109. Stepanov, N.D.; Shaysultanov, D.G.; Tikhonovsky, M.A.; Zherebtsov, S.V. Structure and high temperature mechanical properties of novel nonequiatomic Fe-(Co,Mn)-Cr-Ni-Al-(Ti) high entropy alloys. *Intermetallics* **2018**, *102*, 140–151. [CrossRef]
110. Niu, S.Z.; Kou, H.C.; Guo, T.; Zhang, Y.; Wang, J.; Li, J.S. Strengthening of nanoprecipitations in an annealed $Al_{0.5}CoCrFeNi$ high entropy alloy. *Mater. Sci. Eng. A* **2016**, *671*, 82–86. [CrossRef]
111. Senkov, O.N.; Senkova, S.V.; Woodward, C. Effect of aluminum on the microstructure and properties of two refractory high-entropy alloys. *Acta Mater.* **2014**, *68*, 214–228. [CrossRef]
112. Singh, S.; Wanderka, N.; Murty, B.S.; Glatzel, U.; Banhart, J. Decomposition in multi-component AlCoCrCuFeNi high-entropy alloy. *Acta Mater.* **2011**, *59*, 182–190. [CrossRef]
113. Hemphill, M.A.; Yuan, T.; Wang, G.Y.; Yeh, J.W.; Tsai, C.W.; Chuang, A.; Liaw, P.K. Fatigue behavior of high entropy alloys. *Acta Mater.* **2012**, *60*, 5723–5734. [CrossRef]
114. Wang, Y.P.; Li, B.S.; Ren, M.X.; Yang, C.; Fu, H.Z. Microstructure and compressive properties of AlCrFeCoNi high entropy alloy. *Mater. Sci. Eng. A* **2008**, *491*, 154–158. [CrossRef]
115. Kao, Y.F.; Chen, T.J.; Chen, S.K.; Yeh, J.W. Microstructure and mechanical property of as-cast, -homogenized, and -deformed $Al_xCoCrFeNi$ ($0 \leq x \leq 2$) high-entropy alloys. *J. Alloy. Compd.* **2009**, *488*, 57–64. [CrossRef]
116. Wang, W.R.; Wang, W.L.; Yeh, J.W. Phases, microstructure and mechanical properties of $Al_xCoCrFeNi$ high-entropy alloys at elevated temperatures. *J. Alloy Compd.* **2014**, *589*, 143–152. [CrossRef]
117. Shun, T.T.; Du, Y.C. Microstructure and tensile behaviors of FCC $Al_{0.3}CoCrFeNi$ high entropy alloy. *J. Alloy. Compd.* **2009**, *479*, 157–160. [CrossRef]
118. Ma, Y.; Jiang, B.B.; Li, C.L.; Wang, Q.; Dong, C.; Liaw, P.K.; Xu, F.; Sun, L. The BCC/B2 morphologies in $Al_xNiCoFeCr$ high-entropy alloys. *Metals* **2017**, *7*, 57. [CrossRef]
119. Wang, Q.; Ma, Y.; Jiang, B.B.; Li, X.N.; Shi, Y.; Dong, C.; Liaw, P.K. A cuboidal B2 nanoprecipitation-enhanced body-centered-cubic alloy $Al_{0.7}CoCrFe_2Ni$ with prominent tensile properties. *Scr. Mater.* **2016**, *120*, 85–89. [CrossRef]
120. Ma, Y.; Wang, Q.; Jiang, B.B.; Li, C.L.; Hao, J.M.; Li, X.N.; Dong, C.; Nieh, T.G. Controlled formation of coherent cuboidal nanoprecipitates in body-centered cubic high-entropy alloys based on $Al_2(Ni,Co,Fe,Cr)_{14}$ compositions. *Acta Mater.* **2018**, *147*, 213–225. [CrossRef]
121. Zhou, Y.; Jin, X.; Zhang, L.; Du, X.Y.; Li, B.S. A hierarchical nanostructured $Fe_{34}Cr_{34}Ni_{14}Al_{14}Co_4$ high-entropy alloy with good compressive mechanical properties. *Mater. Sci. Eng. A* **2018**, *716*, 235–239. [CrossRef]
122. Zhang, L.; Zhou, D.; Li, B.S. Anomalous microstructure and excellent mechanical properties of $Ni_{35}Al_{21.67}Cr_{21.67}Fe_{21.67}$ high-entropy alloy with BCC and B2 structure. *Mater. Lett.* **2018**, *216*, 252–255. [CrossRef]
123. Shaysultanov, D.G.; Salishchev, G.A.; Ivanisenko, Y.V.; Zherebtsov, S.V.; Tikhonovsky, M.A.; Stepanov, N.D. Novel $Fe_{36}Mn_{21}Cr_{18}Ni_{15}Al_{10}$ high entropy alloy with bcc/B2 dual-phase structure. *J. Alloy. Compd.* **2017**, *705*, 756–763. [CrossRef]
124. Stepanov, N.D.; Shaysultanov, D.G.; Chernichenko, R.S.; Tikhonovsky, M.A.; Zherebtsov, S.V. Effect of Al on structure and mechanical properties of Fe-Mn-Cr-Ni-Al non-equiatomic high entropy alloys with high Fe content. *J. Alloy. Compd.* **2019**, *770*, 194–203. [CrossRef]
125. He, J.Y.; Wang, H.; Huang, H.L.; Xu, X.D.; Chen, M.W.; Wu, Y.; Liu, X.J.; Nieh, T.G.; An, K.; Lu, Z.P. A precipitation-hardened high-entropy alloy with outstanding tensile properties. *Acta Mater.* **2016**, *102*, 187–196. [CrossRef]
126. Tsao, T.K.; Yeh, A.C.; Kuo, C.M.; Murakami, H. On the superior high temperature hardness of precipitation strengthened high entropy Ni-based alloys. *Adv. Eng. Mater.* **2017**, *19*, 1600475. [CrossRef]
127. Daoud, H.M.; Manzoni, A.; Volkl, R.; Wanderka, N.; Glatzel, U. Microstructure and tensile behavior of $Al_8Co_{17}Cr_{17}Cu_8Fe_{17}Ni_{33}$ (at. %) high-entropy alloy. *JCM* **2013**, *65*, 1805–1814. [CrossRef]
128. Senkov, O.N.; Isheim, D.; Seidman, D.N.; Pilchak, A.L. Development of a refractory high entropy superalloy. *Entropy* **2016**, *18*, 102. [CrossRef]
129. Soni, V.; Senkov, O.N.; Gwalani, B.; Miracle, D.B.; Banerjee, R. Microstructural design for improving ductility of an initially brittle refractory high entropy alloy. *Sci. Rep.* **2018**, *8*, 8816. [CrossRef] [PubMed]

130. Tsai, C.W.; Tsai, M.H.; Yeh, J.W.; Yang, C.C. Effect of temperature on mechanical properties of $Al_{0.5}CoCrCuFeNi$ wrought alloy. *J. Alloy. Compd.* **2010**, *490*, 160–165. [CrossRef]
131. Zhang, W.R.; Liaw, P.K.; Zhang, Y. Science and technology in high-entropy alloys. *Sci. China Mater.* **2018**, *61*, 2–22. [CrossRef]
132. Wang, X.G.; Liu, J.L.; Jin, T.; Sun, X.F. The effects of ruthenium additions on tensile deformation mechanisms of single crystal superalloys at different temperatures. *Mater. Des.* **2014**, *63*, 286–293. [CrossRef]
133. Wang, B.; Liu, Z.D.; Cheng, S.C. Microstructure Evolution and Mechanical Porperties of HR3C Steel during Long term Aging at High Temperature. *J. Iron Steel Res. Int.* **2014**, *21*, 765–773. [CrossRef]
134. Zhang, Z.; Hu, Z.F.; Tu, H.Y. Microstructure evolution in HR3C austenitic steel during long-term creep at 650 °C. *Mater. Sci. Eng. A* **2017**, *681*, 74–84. [CrossRef]
135. Physical Constants and Thermal Properties. Available online: http://www.specialmetals.com/assets/smc/documents/alloys/inconel/inconel-alloy-718.pdf (accessed on 14 November 2018).
136. Chaturvedi, M.C.; Han, Y. Effect of particle size on the creep rate of superalloy Inconel 718. *Mater. Sci. Eng.* **1987**, *89*, L7–L10. [CrossRef]
137. Ma, Y.; Hao, J.M.; Wang, Q.; Zhang, C.; Li, C.L.; Dong, C. Microstructural stability of coherent cuboidal B2 particles in precipitation-strengthened body-centered-cubic $Al_{0.7}NiCoFeCr_2$ high-entropy alloy. *J. Mater. Sci* **2018**. submitted.
138. Cowley, J.M. Short- and long-range order parameters in disordered solid solutions. *Phys. Rev.* **1960**, *120*, 1648–1657. [CrossRef]
139. Reinhard, L.; Schönfeld, B.; Kostorz, G.; Bührer, W. Short-range order in α-brass. *Phys. Rev. B* **1990**, *41*, 1727–1734. [CrossRef]
140. Ma, Y.; Wang, Q.; Li, C.L.; Santodonato, L.J.; Feygenson, M.; Dong, C.; Liaw, P.K. Chemical short-range orders and the induced structural transition in high-entropy alloys. *Scr. Mater.* **2018**, *144*, 64–68. [CrossRef]
141. Hong, H.L.; Wang, Q.; Dong, C.; Liaw, P.K. Understanding the Cu-Zn brass alloys using a short-range-order cluster model: significance of specific compositions of industrial alloys. *Sci. Rep.* **2014**, *4*, 7065. [CrossRef] [PubMed]
142. Ma, Y.P.; Dong, D.D.; Dong, C.; Luo, L.J.; Wang, Q.; Qiang, J.B.; Wang, Y.M. Composition formulas of binary eutectics. *Sci. Rep.* **2015**, *5*, 17880. [CrossRef] [PubMed]

entropy

MDPI

Article

Microstructure and Room Temperature Mechanical Properties of Different 3 and 4 Element Medium Entropy Alloys from HfNbTaTiZr System

Jiří Zýka [1,*], Jaroslav Málek [1], Jaroslav Veselý [1], František Lukáč [2,3], Jakub Čížek [3], Jan Kuriplach [3] and Oksana Melikhova [3]

1 UJP PRAHA a.s., Nad Kamínkou 1345, 156 10 Praha 5—Zbraslav, Czech Republic; malek@ujp.cz (J.M.); vesely@ujp.cz (J.V.)

2 Department of Materials Engineering, Institute of Plasma Physics CAS, Za Slovankou 1782, 1820 Praha 8, Czech Republic; lukac@ipp.cas.cz

3 Faculty of Mathematics and Physics, Charles University, V Holešovičkách 2, 18000 Praha 8, Czech Republic; jakub.cizek@mff.cuni.cz (J.C.); jan.kuriplach@mff.cuni.cz (J.K.); oksana.melikhova@mff.cuni.cz (O.M.)

* Correspondence: zyka@ujp.cz; Tel.: +420-227-180-364

Received: 12 December 2018; Accepted: 15 January 2019; Published: 26 January 2019

Abstract: Refractory high entropy alloys (HEA) are promising materials for high temperature applications. This work presents investigations of the room temperature tensile mechanical properties of selected 3 and 4 elements medium entropy alloys (MEA) derived from the HfNbTaTiZr system. Tensile testing was combined with fractographic and microstructure analysis, using scanning electron microscope (SEM), wavelength dispersive spectroscope (WDS) and X-Ray powder diffraction (XRD). The 5 element HEA alloy HfNbTaTiZr exhibits the best combination of strength and elongation while 4 and 3 element MEAs have lower strength. Some of them are ductile, some of them brittle, depending on microstructure. Simultaneous presence of Ta and Zr in the alloy resulted in a significant reduction of ductility caused by reduction of the BCC phase content. Precipitation of Ta rich particles on grain boundaries reduces further the maximum elongation to failure down to zero values.

Keywords: refractory high entropy alloys; medium entropy alloys, mechanical properties; microstructure

1. Introduction

High entropy alloys (HEAs) attract attention of a growing number of scientists and researchers. A concept of mixing of 5 or more elements in equimolar or near equimolar concentrations is used in order to explore central regions of multicomponent alloy phase space [1]. This approach is driven by possibility to get stable solid solution microstructure of the alloy with favourable mechanical properties (high strength and good ductility) as well as other physical properties, for example, good corrosion resistance [2]. Microstructure stability shall be provided by high configurational entropy of the system, supposing random arrangement of alloying elements [3,4]. Although increased configurational entropy may in principle stabilize solid solution its stabilizing effect is usually found insufficient to counteract driving forces for formation of intermetallic phases [5,6].

Since the pioneering work of Yeh [4], original concept has been evolved and widened. The term HEA is connected mainly with approach utilizing high configurational entropy to get single phase solid solution of multiple elements. Multiple principal element alloys (MPEA) represent a broader approach when the main motivation is exploring of a vast composition space of multi-principal element alloys without primary concern about the magnitude of the configurational entropy [1]. MPEAs with multi-phase microstructures are denoted complex concentrated alloys (CCA) [1].

Different elements are used to fabricate HEAs. Although the number of possible 4 or 5 elements combinations is enormous five basic metallic HEA groups can be distinguished: FCC HEAs based on the Cantor alloy, BCC refractory metal HEAs, light element HEAs, HCP HEAs and precious functional HEAs [7]. Ceramic HEAs have been established as well [8,9].

Our attention was attracted by the refractory HEA group, which is composed of elements from IV, V and VI groups of the periodic table of elements [10]. These elements are characterized by prevailing good mutual miscibility and high melting points. The melting point of titanium (1668 °C) is the lowest among them. Therefore HEAs composed of these elements are intended for high temperature applications. Moreover some of these elements are biocompatible [11]. Therefore some of these alloys can be attractive for bioimplant related materials research. Note that β-Ti alloys used in bioimplant research are usually composed of this group of elements [12]. Several biocompatible HEAs have been already studied [13–16]. Also, hard ceramic composites were produced from the metals from IV, V and VI groups of the periodic table [17].

Original definition of HEAs with single phase solid solution microstructure and good room temperature mechanical properties, especially ductility, is beneficial also for use in human medicine. Inspiration can be found also in other materials used for production of bioimplants which are used or have been used in other parts of material research, for example, TiAl6V4 in aerospace, CoCrMo in aero engines.

Our first attempt in research of HEAs was therefore related to HfNbTaTiZr refractory metal alloy [18]. Our research was inspired by Senkov [19] and confirmed our expectations. Ingots produced by vacuum arc melting possessed single phase solid solution microstructure, with a high room temperature tensile strength and ductility.

Some of the elements used in the HfNbTaTiZr alloy are very similar to each other with very similar chemical behaviour; Nb is chemically similar to Ta and Zr to Hf. Therefore a question has emerged whether it is necessary to use all of these elements in order to get random solid solution and good mechanical properties (high strength and sufficient ductility).

The design of the present experiment was as follows. First, we reduced the original HfNbTaTiZr alloy system by removing Hf, which is almost identical to Zr from the chemical point of view, then we distracted Ta, which is very similar to Nb; thus getting NbTiZr alloy with equiatomic concentrations. Then we produced also other 3 element combinations containing Ti, namely TaTiZr and NbTaTi. Based on the observed results; 4 element alloys containing Ti and Ta with different Nb to Zr ratio were produced as well to explore the influence of chemical composition on elongation.

NbTaTiZr [14] and NbTiZr [20,21] studies have already been reported. No reports about TaTiZr and NbTaTi were found in literature.

One can notice that set of the produced element combinations is not complete. The motivation for element selection was as follows. Titanium was used in all combination because it is element with the lowest density. Hafnium was not used since it is very reactive in ambient atmosphere and it is hard to get Hf free of impurities. Both hafnium and tantalum are high density and high price element; thus less suitable for practical use.

2. Materials and Methods

Experimental alloys were prepared by vacuum arc melting in water cooled copper crucible. Chemical purity of inserted elements was 99.9%. Final cast ingot has approximately 100 mm in length, 30 mm in width, 10 mm in height and 400 g in weight. Casting was performed 8x times and flipped for each melt to mix the elements thoroughly and suppress chemical heterogeneity. All investigated alloys were prepared at the same equipment under the same conditions. Nor HIP, nor other heat treatment was applied to the cast material.

Tensile test bodies 5 mm in diameter and 25 mm of measured length were strained with the strain rate of 2×10^{-4} s^{-1} using Instron 1185 testing machine equipped with video extensometer.

Metallographic sections perpendicular to the length of the casting were prepared. Vickers hardness HV30 was measured using Zwick ZHU 250 Topline universal testing machine.

Fractography analysis of broken specimen surfaces was carried out on a scanning electron microscope (JEOL JSM 7600F).

Microstructure was examined by scanning electron microscope (FEI Quanta 200F) equipped with wavelength dispersive X-ray spectroscope (WDS) for local chemical composition analysis.

In order to determine phase composition and structures of phases in the alloys, powder X-ray diffraction (PXRD) analysis was performed. A Bruker D8 Discover diffractometer, CuKα radiation and 1D detector were used. Lattice parameters and phase composition were determined by Rietveld analysis of PXRD diffraction patterns using TOPAS V5 code [22].

Ab-initio quantum Monte Carlo (MC) simulation was performed to obtain atomic configuration of the 5 element HfNbTaTiZr alloy. Hf, Nb, Ta, Ti, Zr ions were distributed randomly in equimolar concentration into a 250 atom BCC supercell. This initial state was relaxed with respect to ion positions until minimum of the total free energy was reached. The equilibrium configuration corresponding to minimum free energy was obtained using a Metropolis MC algorithm at 300 K (room temperature). The details of the simulation and more complete results will be published elsewhere.

3. Results

3.1. Chemical Composition

Chemical compositions of investigated alloys in atomic percentages are given in Table 1. Important HEA related parameters are listed in Table 2. Valence electron concentration—VEC, mixing enthalpy$-\Delta H_{mix}$, difference in atomic radii$-\delta$, Ω parameter, mixing entropy$-S_{mix}$ are calculated in accordance with literature [1], where T_m is a calculated average melting point. According to entropy-based definition HEAs are characterized by the configurational entropy higher than 1.61 *R* [1]. Hence, according to this definition, only HfNbTaTiZr alloy is high entropy alloy. The other alloys studied in the present work are medium entropy alloys.

Table 1. Chemical composition of investigated alloys in at.%, including important properties of individual elements.

	Ti	Ta	Nb	Zr	Hf
molar mass	47.867	180.94	92.9	91.22	91.224
density (g cm^{-3})	4.51	7.14	7.8	6.51	6.5
melting temp. (K)	1941	3269	2741	2125	2495
atomic radius (Å)	1.47	1.43	1.43	1.62	1.6
VEC	4	5	5	4	4
HfNbTaTiZr	0.2	0.2	0.2	0.2	0.2
NbTaTiZr	0.25	0.25	0.25	0.25	0
NbTaTi	0.333	0.333	0.333	0	0
TaTiZr	0.333	0.333	0	0.333	0
NbTiZr	0.333	0	0.333	0.333	0
$Nb_{1.5}TaTiZr_{0.5}$	0.25	0.25	0.375	0.125	0
$Nb_{0.5}TaTiZr_{1.5}$	0.25	0.25	0.125	0.375	0

Table 2. Important high entropy alloy (HEA) related parameters if investigated alloys calculated using data from Table 1.

	VEC	ΔH_{mix} [kJ/mol]	δ [%]	T_m [K]	Ω	S_{mix} [J/(mol K)]	S_{mix} [R*J/(mol K)]	
HfNbTaTiZr	4.4	2.72	5.51	2514.20	12.37	13.38	1.61	HEA
NbTaTiZr	4.5	2.50	5.26	2519.00	11.61	11.53	1.39	MEA
NbTaTi	4.662	1.33	1.31	2647.68	18.17	9.13	1.10	MEA
TaTiZr	4.329	1.77	5.43	2442.56	12.57	9.13	1.10	MEA
NbTiZr	4.329	2.66	5.43	2266.73	7.78	9.13	1.10	MEA
$Nb_{1.5}TaTiZr_{0.5}$	4.625	2.13	4.20	2596.00	13.42	10.98	1.32	MEA
$Nb_{0.5}TaTiZr_{1.5}$	4.375	2.38	5.67	2442.00	11.29	10.98	1.32	MEA

3.2. Mechanical Properties

Tensile curves of investigated alloys performed at room temperature are shown in Figure 1. Results of tensile tests performed and hardness measurements performed at room temperature are given in Table 3.

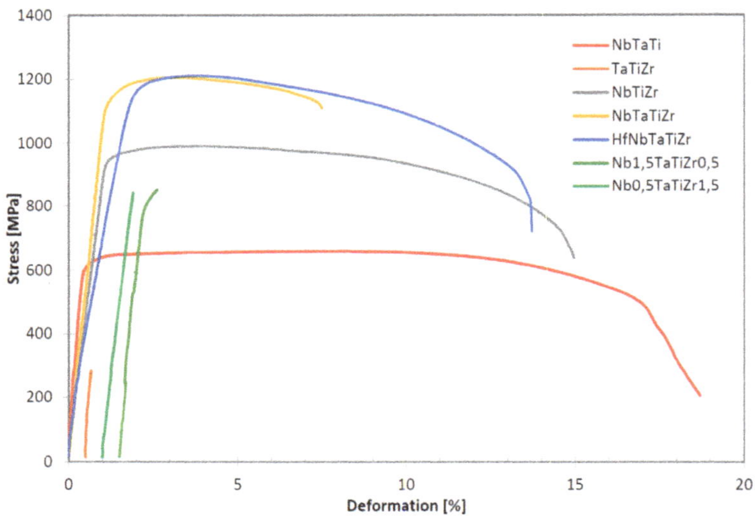

Figure 1. Tensile curves of investigated alloys at room temperature. Curves for brittle alloys TaTiZr, $Nb_{1.5}TaTiZr_{0.5}$ and $Nb_{0.5}TaTiZr_{1.5}$ were shifted horizontally to make them visible.

Table 3. Room temperature mechanical properties of the alloys studied.

Alloy	$Rp_{0.2}$ [MPa]	R_m [MPa]	A [%]	E [GPa]	HV30
HfNbTaTiZr	1155	1212	12.3	59	359
NbTaTiZr	1144	1205	6.4	98	358
NbTaTi	620	683	18.5	143	246
TaTiZr	-	284	0	157	485
NbTiZr	956	991	14.2	88	295
$Nb_{1.5}TaTiZr_{0.5}$	822	852	0.33	127	294
$Nb_{0.5}TaTiZr_{1.5}$	-	843	0	93	489

The 5 element HEA HfNbTaTiZr alloy has the best tensile properties among the alloys studied, both regarding the strength and elongation. The 4 element NbTaTiZr alloy has similar strength and hardness as HfNbTaTiZr but elongation is reduced to 6.4%. The NbTaTi has low strength, half of that for HfNbTaTiZr but the highest elongation. The NbTiZr alloy has strength values below 1000 MPa but elongation is relatively high (14.2%). The TaTiZr alloy exhibited brittle behaviour, breaking before reaching the yield point. The $Nb_{0.5}TaTiZr_{1.5}$ alloy performed very similarly, breaking also before reaching yield point but a little bit later than TaTiZr alloy. The $Nb_{1.5}TaTiZr_{0.5}$ alloy reached the yield point (822 MPa) but the elongation was 0.33% only.

Highest hardness (HV30 = 485) values were measured in the most brittle alloys TaTiZr and $Nb_{0.5}TaTiZr_{1.5}$. Hardness of the 4 and 5 element equimolar alloy was very similar to each other, lower by $\approx$130 HV compared to the brittle alloys. Hardness of the NbTiZr and $Nb_{1.5}TaTiZr_{0.5}$ are slightly lower than 300 HV. The most ductile alloy NbTaTi exhibited the lowest hardness of 246 HV.

3.3. Fractographic Analysis

Fracture surfaces of specimen broken during tensile test were subjected to fractographic analysis, see Figure 2. Both transgranular and intergranular ductile fracture can be found on fracture surfaces of the NbTiTaZr alloy, see Figure 2a–c. Elongated particles were observed on the intergranular ductile fracture surface. Ductile dimples of different diameters were found on the transgranular fracture surface of NbTaTi alloy, see Figure 2d. Transgranular ductile fracture was found also in the case of NbTiZr alloy (Figure 2f). On the other hand, fracture surface of TaTiZr alloy is dominantly flat and brittle and containing number of needle like particles, see Figure 2e. River-like pattern is present as well (not shown). $Nb_{1.5}TaTiZr_{0.5}$ specimen fracture surfaces are composed of transgranular ductile fracture and intergranular fracture, see Figure 2g. A lot of particles are present on the intergranular part of the fracture surface. $Nb_{0.5}TaTiZr_{1.5}$ fracture surfaces (Figure 2h–i) are similar to those of $Nb_{1.5}TaTiZr_{0.5}$ and NbTaTiZr alloy but without presence of particles on the intergranular part of the fracture surface.

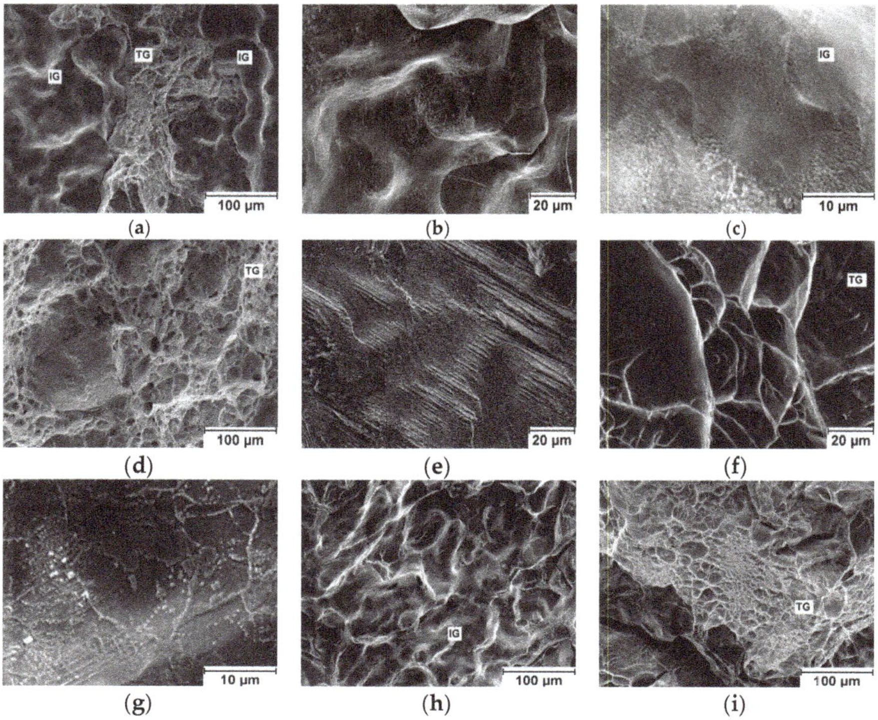

Figure 2. Scanning electron microscopy (SEM) image of investigated fracture surfaces: (**a**), (**b**), (**c**) NbTaTiZr alloy; (**d**) NbTaTi alloy (**e**) TaTiZr alloy; (**f**) NbTiZr alloy; (**g**) $Nb_{1.5}TaTiZr_{0.5}$ alloy; (**h**), (**i**) $Nb_{0.5}TaTiZr_{1.5}$ alloy; where TG denotes transgranular ductile fracture and IG intergranular ductile fracture.

3.4. Microstructure

Microstructure of the investigated alloys is shown in Figure 3. Grains size was estimated by light microscopy on metallographic specimens with mirror-like polished and slightly etched surface using the linear intercept procedure [23]. It revealed grain size around 0.5 mm which is similar for all alloys. All alloys, except of NbTiZr alloy, exhibited dendritic segregation.

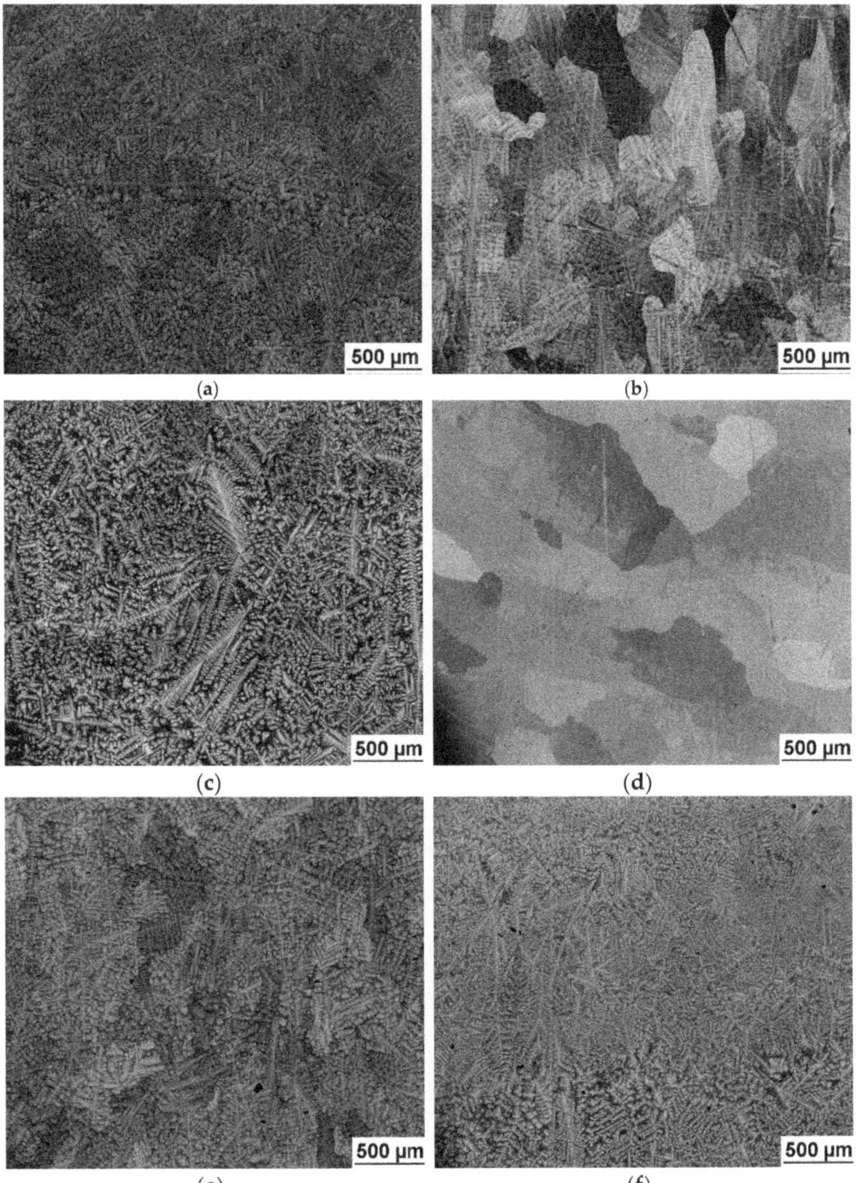

Figure 3. SEM image of investigated alloys microstructure: (**a**) NbTaTiZr alloy; (**b**) NbTaTi alloy; (**c**) TaTiZr alloy; (**d**) NbTiZr alloy; (**e**) $Nb_{1.5}TaTiZr_{0.5}$ alloy; (**f**) $Nb_{0.5}TaTiZr_{1.5}$ alloy.

Small submicron precipitates were found on grain boundaries of TaTiZr, $Nb_{1.5}TaTiZr_{0.5}$ and $Nb_{0.5}TaTiZr_{1.5}$ alloys, see Figure 4. WDS analysis of these precipitates revealed that they are rich in Ta. For example the average Ta concentration in the precipitates in TaTiZr alloy is (52 ± 1) at.% while the Ta content in the matrix of this alloy was found to be (38 ± 1) at.%.

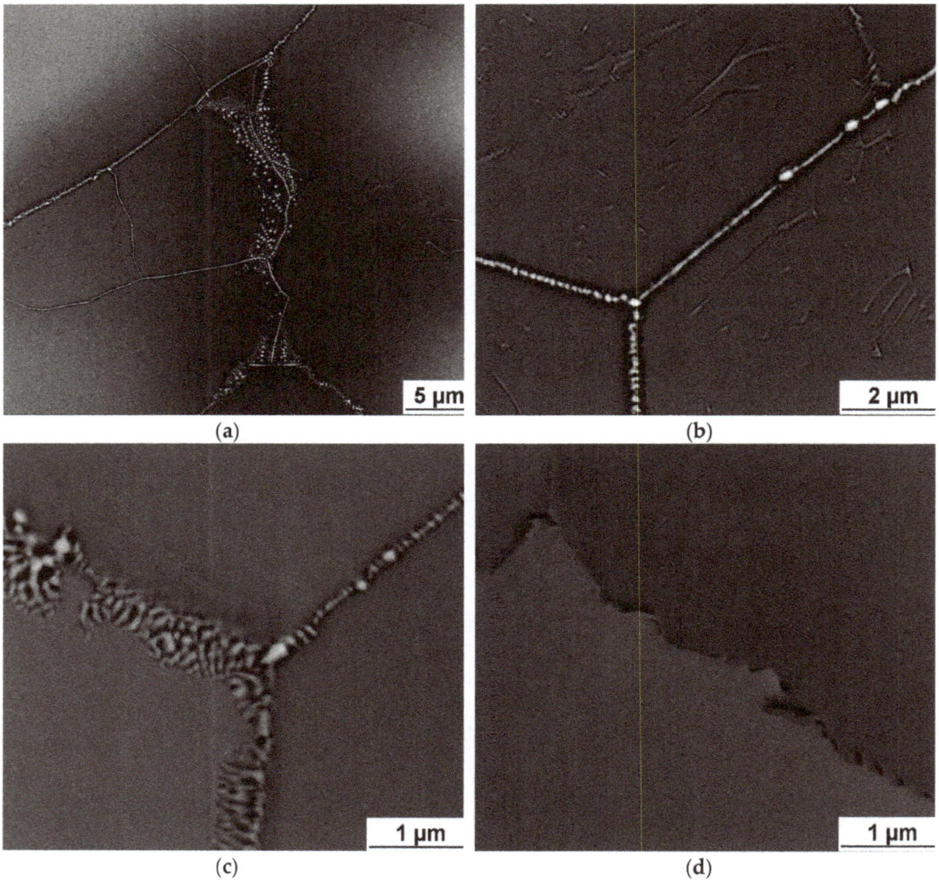

Figure 4. SEM image of investigated alloys microstructure: (**a**), (**b**) TaTiZr alloy; (**c**) $Nb_{1.5}TaTiZr_{0.5}$ alloy; (**d**) $Nb_{0.5}TaTiZr_{1.5}$ alloy.

Figure 5 illustrates existence of two phases. One of them appears brighter because of higher average Z number while the second one appears darker due to lower average Z number. Phase separation into these two phases occurs on the length scale of ~20 µm. The only exception is NbTiZr alloy, which shows no phase segregation, see Figure 5d. WDS line analysis showed enhanced concentration of Zr and Ti in the dark phase and Ta and Nb in the bright phase, see Table 4.

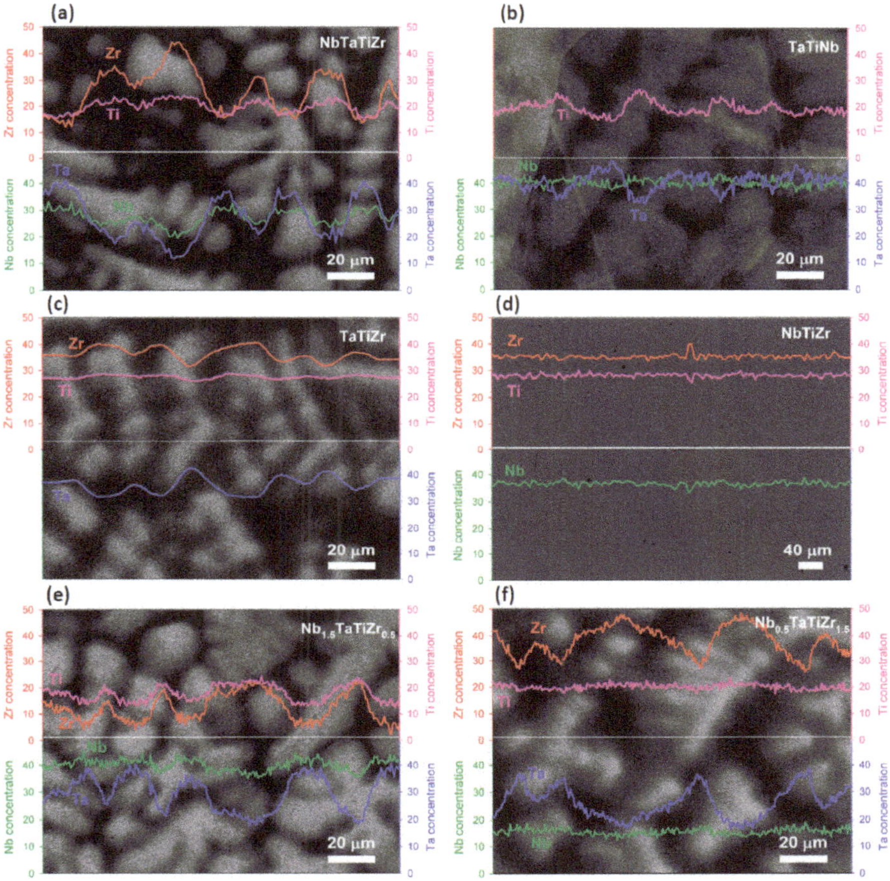

Figure 5. WDS line analysis of: (**a**) NbTaTiZr alloy; (**b**) TaTiNb alloy; (**c**) TaTiZr alloy; (**d**) NbTiZr alloy; (**e**) $Nb_{1.5}TaTiZr_{0.5}$ alloy; (**f**) $Nb_{0.5}TaTiZr_{1.5}$ alloy.

Table 4. Chemical composition in atomic % of bright and dark phase in investigated alloys. Uncertainties of concentrations (one standard deviation) are given in parenthesis.

Alloy	Phase	Ti	Zr	Nb	Ta
$Nb_{0.5}Ta\ TiZr_{1.5}$	Bright	0.19(1)	0.28(1)	0.16(1)	0.37(2)
	Dark	0.22(2)	0.42(3)	0.15(2)	0.21(2)
	nominal	0.25	0.375	0.125	0.25
$Nb_{1.5}Ta\ TiZr_{0.5}$	Bright	0.13(1)	0.07(1)	0.41(3)	0.39(3)
	Dark	0.22(2)	0.22(2)	0.36(2)	0.20(2)
	nominal	0.25	0.125	0.375	0.25
NbTaTiZr	Bright	0.17(1)	0.16(1)	0.32(2)	0.35(22)
	Dark	0.25(2)	0.35(2)	0.23(2)	0.17(1)
	nominal	0.25	0.25	0.25	0.25
NbTaTi	Bright	0.23(2)		0.28(1)	0.49(2)
	Dark	0.42(2)		0.28(1)	0.30(1)
	nominal	0.333		0.333	0.333
TaTiZr	Bright	0.32(2)	0.16(2)		0.52(2)
	Dark	0.40(3)	0.35(3)		0.25(1)
	nominal	0.333	0.333		0.333

3.4.1. XRD Analysis

XRD patterns in Figure 6 exhibit peaks in positions corresponding to reflection of a BCC phase but with different extent of broadening. HfNbTaTiZr [18], NbTaTi and NbTiZr alloys consist of single BCC phase. Rietveld refinement of XRD patterns suggested existence of two BCC phases with slightly different lattice parameters in case of NbTaTiZr and TaTiZr alloys (Table 5). This corresponds to the previously observed presence of the dendritic microstructure in the feedstock powder particles [24]. The regions that solidify earlier were enriched in Nb, Ta, thereby triggering a measurable change in the respective lattice parameters of the BCC regions as compared to the interdendritic regions with increased Zr, Ti content. Rietveld refinement suggested even three BCC phases in case of $Nb_{1.5}TaTiZr_{0.5}$, $Nb_{0.5}TaTiZr_{1.5}$ with similar lattice parameters (Table 5). Grain size cannot be calculated from XRD data since the broadening of XRD reflections caused by finite grain size was found to be negligible. It means the average grain size of the alloys studied is higher than 100 nm which is consistent with metallographic and SEM observations. Dominant source of peak broadening is chemical heterogeneity, that is, local changes of lattice parameter due to spatial variations of chemical composition.

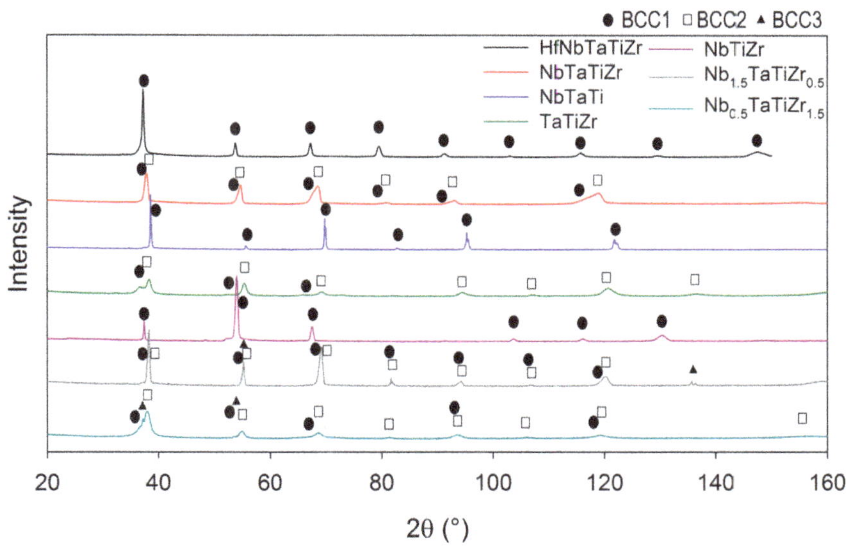

Figure 6. X-ray diffraction (XRD) patterns of investigated alloys. Reflections of the BCC phases are marked by labels.

Table 5. Phase composition of investigated alloys. The lattice parameter (a) and phase content were obtained using Rietveld refinement fitting.

Alloy	BCC1		BCC 2		BCC3	
	a [Å]	[%]	a [Å]	[%]	a [Å]	[%]
HfNbTaTiZr	3.4089(1)	100				
NbTaTiZr	3.3509(8)	60.15	3.380(2)	39.85		
NbTaTi	3.29685(7)	100				
TaTiZr	3.446(1)	29.08	3.3184(2)	70.92		
NbTiZr	3.3969(1)	100				
$Nb_{1.5}TaTiZr_{0.5}$	3.3220(5)	28.25	3.334(2)	71.22	3.3273(2)	0.53
$Nb_{0.5}TaTiZr_{1.5}$	3.451(5)	19.64	3.3395(3)	77.97	3.4121(4)	2.39

Note: Error of the last digit is shown in the parentheses.

3.4.2. Monte Carlo Simulation of Microstructure

Ab-initio MC simulation was performed to evaluate the microstructure stability of investigated alloys. Simulation was performed for the 5 element HfNbTaTiZr alloy, because it contains all elements considered in investigated alloys.

Hf, Nb, Ta, Ti, Zr ions were distributed randomly in equimolar concentration into a 250 atom BCC supercell. This initial state was relaxed with respect to ion positions until minimum of the total free energy was reached. The equilibrium configuration corresponding to minimum free energy was obtained using a Metropolis MC algorithm at 300 K (room temperature). The details of the simulation and more complete results will be published elsewhere. Figure 7 shows the equilibrium atomic configuration corresponding to the minimum of total energy. The most apparent effect is a rather one-dimensional Ta object ('wire') along the <100> direction. The Ta wire is surrounded predominantly by Nb ions. The rest of the simulation box is filled up by the mixture of Ti, Zr and Hf. The latter two elements appear to be well separated from the Ta and Nb region. Hence, this preliminary result indicates inhomogeneity of the HfNbTaTiZr alloy. Such inhomogeneities–though they need to be yet verified experimentally–could affect physical properties of the alloy.

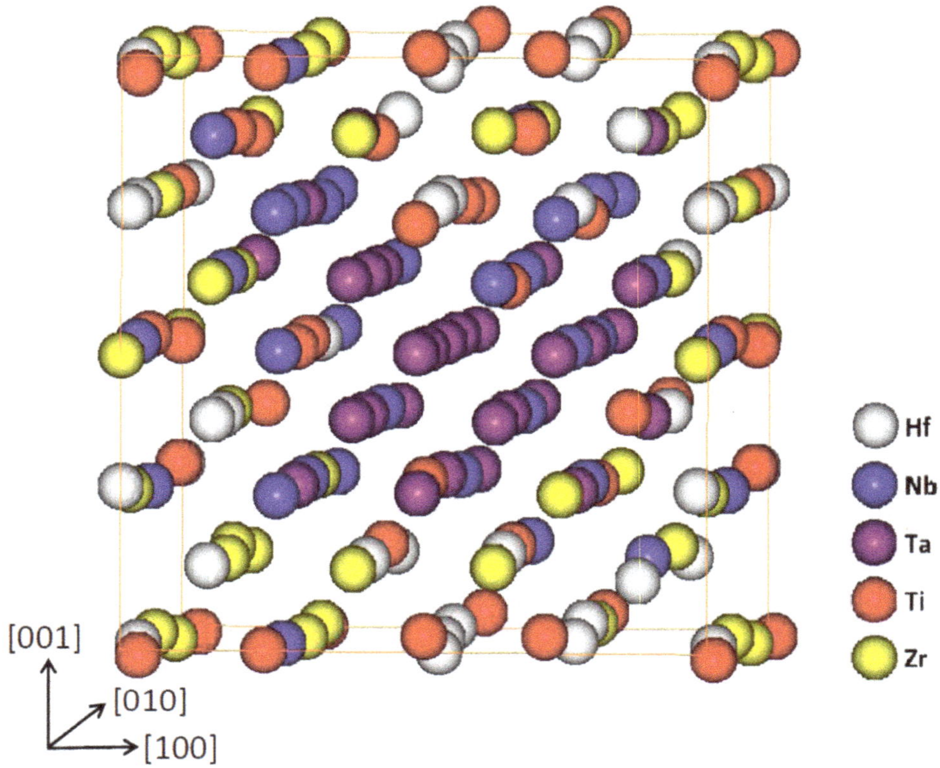

Figure 7. Equilibrium atomic configuration of simulated HfNbTaTiZr alloy.

4. Discussion

Mechanical properties tests have revealed that combination of Ta and Zr reduces elongation. Microstructure analysis revealed existence of dark and bright phases because of Ta-Zr segregation. This kind of dendritic segregation was reported elsewhere [25–27]. It was shown [25] that Ta with Nb segregates during solidification to the solid and Zr with Ti to the liquid.

Figure 8a shows relation between the elongation to failure, A and the total atomic fraction of Ta and Zr. All alloys studied exhibit linear relationship of A on the Ta + Zr concentration. The only exception is $Nb_{1.5}TaTiZr_{0.5}$ alloy not following the linear relationship because, it is brittle despite of relatively low Ta + Zr content. In the latter alloy Nb probably plays similar role as tantalum. Figure 8b shows the content of the BCC1 phase as a function of the total concentration of is Ta and Zr. Obviously it obeys similar linear relationship with the net concentration of Ta and Zr as the elongation.

Intergranular nanosized precipitates were found on grain boundaries of brittle alloys with zero or almost zero elongation, namely TaTiZr, $Nb_{1.5}TaTiZr_{0.5}$ and $Nb_{0.5}TaTiZr_{1.5}$. XRD analysis revealed 2 BCC phases (BCC1, BCC2) which are probably caused by the microsegregation [24]. In two cases small amount of third BCC phase (BCC3) was detected, it can be connected with the intergranular precipitation. However, it was not investigated in detail in the present study.

Usually relations between the misfit parameter δ [1] and strength or hardness are reported, since rising δ shall indicate higher solid solution strengthening. Since in the present case we have single phase solid solution only in 3 alloys out of 7 investigated, the relation between hardness and δ is more complicated than simple linear dependence. Indeed if we exclude two most brittle alloys, indicated by red symbols in Figure 9a, the dependence of hardness and δ becomes rather close to the linear relationship. In elongation and δ relationship NbTaTi alloy, the most ductile one, destroys possible correlation as well, Figure 9b.

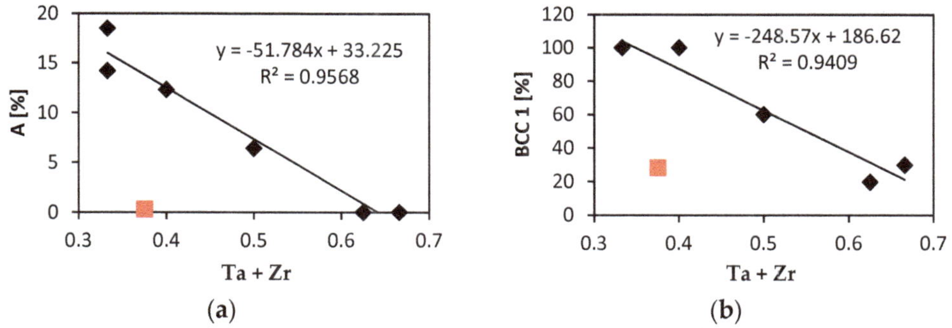

Figure 8. (**a**) Relation of the total elongation to failure A on the sum of Ta and Zr atomic concentration, (**b**) the concentration of the BCC1 phase plotted as a function of the sum of Ta and Zr content. Data for $Nb_{1.5}TaTiZr_{0.5}$ alloy are indicated by red symbols and were excluded from linear regression.

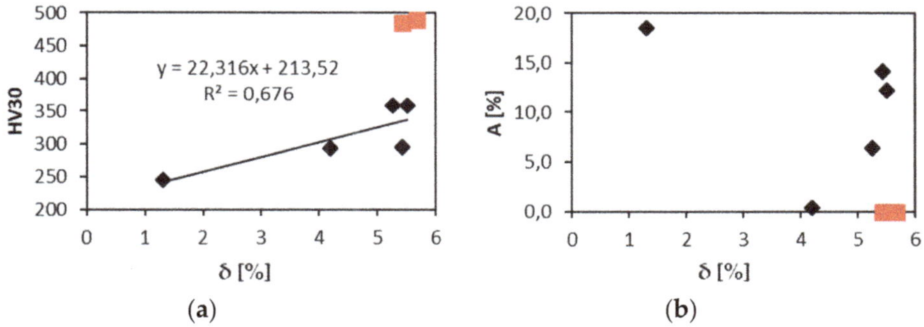

Figure 9. Relation of δ and: (**a**) HV30; (**b**) elongation.

A relation between the elongation and the VEC parameter was reported in case of refractory HEAs [28]. Alloys with VEC lower than 4.4 shall be ductile, alloys with VEC higher than 4.6 shall be brittle. Figure 10a shows relation between elongation of investigated alloys and their VEC parameters. VEC values are close to the boundary value of 4.5. Brittle alloys TaTiZr (VEC = 4.329), $Nb_{0.5}TaTiZr_{1.5}$ (VEC = 4.375) and ductile alloy NbTaTi (VEC = 4.663) do not follow the reported rule. However, these alloys are MEAs and not HEAs. Brittle alloys with VEC lower than 4.5 are not single phase systems but contain 2 or 3 BCC phases and intergranular precipitates. NbTaTi is ductile because of absence of Ta-Zr combination in the alloy. The NbTiZr alloy exhibits similar behaviour. The HfNbTaTiZr alloy is also ductile, despite the fact that it contains Ta-Zr combination. Although MC simulation showed segregation of Ta, any mark of such segregation, was detected in experiment. When plotting the total elongation A as a function of the BCC1 phase content one can observe a clear linear relationship, see Figure 10b. Hence, there is a positive correlation between the total elongation and the BCC1 phase content. This is not surprising since both the total elongation and the BCC1 phase content decrease with Ta + Zr concentration, see Figure 8. The only exception is $Nb_{1.5}TaTiZr_{0.5}$ alloy which does not follow this trend.

The presence of other BCC phases influences also fracture mechanisms and fracture surfaces of tensile specimens.

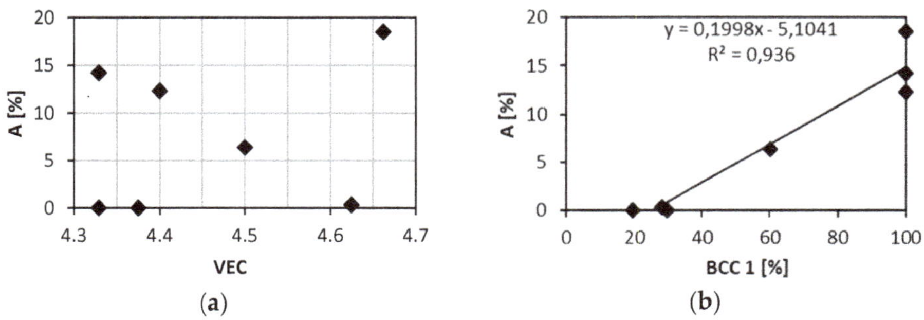

Figure 10. Relation of the elongation to failure and: (**a**) the VEC parameter; (**b**) the content of BCC 1 phase.

It is important to investigate the question of existence of 2 or more BCC phases in our alloys. In HEA related research a lot of work has been done to establish connection between chemical composition or related parameters and microstructure, especially existence of single phase or intermetallics or amorphous microstructure. Satisfying $\delta < 0.066$ and $\Delta H_{mix} > -11.6$ kJ/mol, however, is necessary but not sufficient conditions to form solid solutions in HEAs. Checking the binary phase diagrams among constituent elements can give some further guidance in designing solid solutions forming HEAs [28]. All our alloys satisfy these two conditions but some of them are not single phase. Thus binary diagram analysis is needed.

Mixing enthalpies of element pairs relevant to the investigated alloys is shown in Table 6. Some mixing enthalpies are zero (Ti-Zr, Ti-Hf, Zr-Hf and Ta-Nb – pairs of elements from the same group) or almost zero (Ti-Ta, Ti-Nb), which is ideal for solid solution forming. Other mixing enthalpies are little bit higher (Ta-Zr, Ta-Hf, Nb-Zr, Nb-Hf – pairs from different but neighbouring groups). Above zero mixing enthalpies can cause existence of miscibility gaps in respective binary diagrams. Binary equilibrium phase diagrams of pairs with higher mixing enthalpies are shown in Figure 11. No intermetallic phases are present but large miscibility gaps can be found. Although in case of Ta-Zr diagram miscibility gap is in the range between 800 °C and 1780 °C, see Figure 10a, we found two BCC phases at room temperature.

Table 6. Mixing enthalpies of element pairs relevant to the investigated alloys [29].

Ti-Ta	Ti-Nb	Ti-Zr	Ti-Hf
1	2	0	0
-	Ta-Nb	Ta-Zr	Ta-Hf
-	0	3	3
-	-	Nb-Zr	Nb-Hf
-	-	4	4
-	-	-	Zr-Hf
-	-	-	0

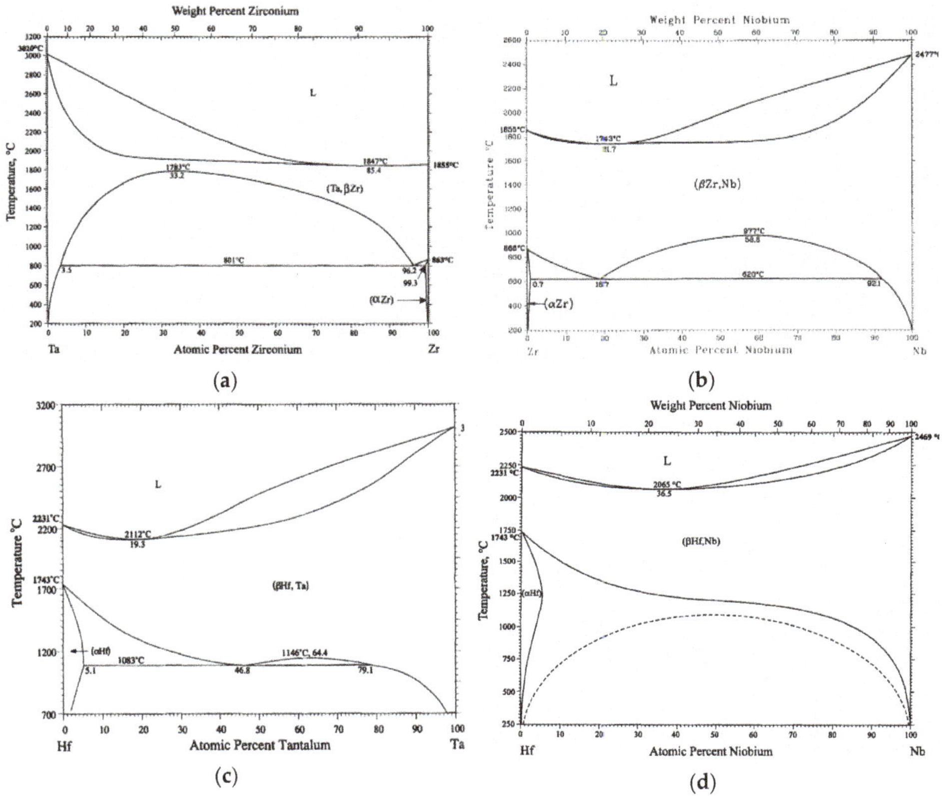

Figure 11. Binary phase diagrams: (**a**) Ta-Zr. Reprinted from [30] with permission of Springer Nature; (**b**) Nb-Zr. Reprinted from [31] with permission of Springer Nature; (**c**) Ta-Hf. Reprinted from [32] with permission of Springer Nature; (**d**) Nb-Hf. Reprinted from [33] with permission of Springer Nature.

It is not clear whether presence of BCC phases in our less ductile alloys is due to stabilizing by rapid cooling during vacuum arc melting or due to differences between binary alloy and 3, 4 or 5 element alloys.

Existence of two BCC phases was reported in similar RHEA system with Mo and V instead of Ta and Nb [34]. CALPHAD simulation of the HfNbTaTiZr was performed, too [35]. A mixture of BCC and HCP phases with Ta Zr segregation was predicted. Experimental observation was performed on cold deformed and heat treated alloy in combination with high pressure torsion (HPT) and isothermal annealing in the range 300 °C–1100 °C and BCC + HCP phases were identified. Combination of two BCC phases and HCP was detected in a HfNbTaTiZr specimen annealed at 500 °C for 100 h. On contrary we observed only a single BCC phase in as-cast state.

System HfNbTaTi, similar to our NbTaTiZr alloy was calculated by CAPLHAD method [36]. Also combination of BCC + HCP phases was found to be stable at room temperature the BCC2 phase appeared between 750 °C and 1000 °C.

In Ref. [25] equimolar NbTaTiZr was modified by adding Ti and removing Ta and results similar to the present work were observed. The elongation to failure increased with rising titanium content. Dendritic segregation came from solidifying, segregation of Ta and Zr and intergranular precipitates were formed after 1200 °C/8 h annealing. But only single BCC phase was found in the as-cast state. Stabilizing of BCC phase by reduction of Ta content and rising Ti content was proposed [25]. A clear correlation between strength or elongation and the misfit parameter δ were established. It is probably due to single phase microstructure.

Table 7 shows comparison of tensile properties of the alloys studied in the present work with MEAs reported in literature. There is a good agreement in ultimate tensile strength of NbTaTiZr alloy but referred rupture strength is higher and elongation much lower than in case alloy studied in this work.

Elongation, A calc., based on tantalum and zirconium combined content using equation from Figure 8a was calculated. There is a good correlation in higher Ti contents in case of alloys referred. But there is important difference in lower Ti content, although single phase solid solution for all compositions was reported [25].

Table 7. Tensile properties of present and reported medium entropy alloys (MEAs) [25].

Alloy	Ta + Zr	HV30	$Rp_{0.2}$ [MPa]	R_m [MPa]	A [%]	A calc. [%]	Ref.
HfNbTaTiZr	0.4	359	1155	1212	12.3	12.51	[This work]
NbTaTiZr	0.5	358	1144	1205	6.4	7.33	[This work]
NbTaTi	0.333	246	620	683	18.5	15.98	[This work]
TaTiZr	0.666	485	-	284	0	−1.26	[This work]
NbTiZr	0.333	295	956	991	14.2	15.98	[This work]
$Nb_{1.5}TaTiZr_{0.5}$	0.375	294	822	852	0.33	13.81	[This work]
$Nb_{0.5}TaTiZr_{1.5}$	0.625	489	-	843	0	0.86	[This work]
NbTaTiZr	0.5	-	* 1190	* 1190	* 0	7.33	[25]
$NbTa_{0.8}Ti_{1.2}Zr$	0.45	-	* 1100	* 1110	* 1	9.92	[25]
$NbTa_{0.6}Ti_{1.4}Zr$	0.4	-	* 1030	* 1070	* 2.5	12.51	[25]
$NbTa_{0.4}Ti_{1.6}Zr$	0.35	-	910	1040	18	15.10	[25]
$NbTa_{0.2}Ti_{1.8}Zr$	0.3	-	790	920	22	17.69	[25]

* Values estimated from the figure [25]

This difference corresponds with important change in elongation with Ti fraction. No clear explanation of this phenomenon was given [25]. Comparing experimental details of the present study with Ref. [25] one can recognize that dimensions of ingots were different. Ingots of alloys studied in the present work were larger and thereby solidifies and cools slower than ingots in Ref. [25]. Therefore NbTaTiZr alloy containing 100% of the BCC1 phase reported in Ref. [25] was likely due to higher cooling rate; while in the present work the same alloy exhibits a mixture of BCC1 and BCC2 phases, see Table 4. On the contrary in Ref. [25] the NbTaTiZr is referred as brittle, we measured 6.4% elongation, see Table 3, however, fracture mechanism has changed from transgranular with ductile dimples to intergranular, see Figure 2. Different interstitials element content may cause the difference in deformation of nominally same alloy.

In this study we use the same furnace and the same size of ingots in case of all investigated alloys. Therefore it is solely the effect of Ta + Zr content what affects the stability of the BCC1 phase. The cooling rate was high enough to produce 100% BCC1 microstructure in case of 5 element HfNbTaTiZr alloy but not in the 4 element alloy NbTaTiZr.

Intensive precipitation at grain boundaries was observed in the brittle TaTiZr alloy, see Figure 4. This alloy contains highest combined content of Ta + Zr and a low amount of the BCC1 phase.

Similar microstructures were found in Refs. [37,38]. These works investigated the influence of middle temperature annealing on the 5 element HEA HfNbTaTiZr in deformed and homogenisation annealed state. Combination of BCC and HCP phases were detected, at specific conditions (longer annealing temperatures and longer times) Ta rich BCC and Zr rich HCP phase precipitates were formed. Similarity of the microstructures and phases suggests that these effects have similar origin, namely Ta+Zr content, partial decomposition of the BCC1 phase to the BCC2 because of relatively high Ta-Zr mixing enthalpy and resulting miscibility gap.

The 5 element HfNbTaTiZr alloy was studied in Refs. [35,36]. It has lower Ta + Zr content than NbTaTiZr, $Nb_{0.5}TaTiZr_{1.5}$ and TaTiZr studied in the present work. Ingots used in Refs. [35,36] were smaller compared to that prepared in the present work, thus gaining higher cooling rate. The ingots were also annealed for homogenization and water quenched, thus metastable BCC1 solid solution was obtained. However, subsequent mid temperatures annealing caused decomposition of the BCC1 solid solution.

In our study, alloys are destabilized by higher Ta + Zr content and the cooling rate was slower than in other studies, therefore the BCC1 phase decomposition took place already during cooling. The MC simulation revealed segregation of Ta in the equilibrium state of the HfNbTaTiZr alloy. This supports the idea of Ta + Zr destabilizing HEAs also during ageing annealing. Thus single phase BCC1 structure is metastable in the 5 element HfNbTaTiZr HEA alloy at room temperature and can be preserved after high temperature annealing by rapid cooling. The role of rapid cooling has been discussed in the review [1].

The alloys containing the BCC2 phase are free of Hf. An HCP phase was predicted and confirmed experimentally in Hf containing alloys [35–38]. It means that Hf acts as an element stabilizing HCP phase.

Further research shall be focused on detail investigations of the BCC2 and BCC3 phases and intergranular precipitates detected in some alloys. Destabilising effect of Ta + Zr on the BCC phase stability during manufacturing ingots and ageing annealing shall be verified and utilised in future research. Stabilising effect of Hf on the HCP phase formation during mid temperature ageing shall be verified as well. Influence of the cooling rate on phase composition of HEAs shall be investigated, too.

Further investigation of the NbTiZr MEA alloy with promising results shall be performed.

Obtained results can be used in evolution of other variants of the investigated system HfNbTaTiZr, for example, for reduction of high temperature oxidation [39] or in research of ageing behaviour of similar RHEAs, especially containing both Ta and Zr.

5. Conclusions

Investigations of mechanical properties and microstructure of different MEAs derived from the HfNbTaTiZr alloy system has been performed. It has been shown that ductility is significantly reduced by simultaneous presence of Ta and Zr in the alloy, which leads to reduction of the BCC1 phase content. These elements segregate during solidifying. Because of the miscibility gap in the Ta-Zr equilibrium phase diagram two BCC phases are found after cooling to room temperature. Clear correlation between the BCC1 phase content and the total elongation to failure of the alloy was found. Precipitation of Ta rich precipitates on grain boundaries reduces the elongation to almost zero values. No correlation was found between the tensile mechanical properties and microstructure related parameters in investigated set of alloys. The 5 element HEA alloy exhibits the best combination of strength and elongation. 4 and 3 element MEAs have lower strength to various extent. Some of them are ductile (NbTaTiZr, NbTaTi, NbTiZr) but some of them are brittle (TaTiZr, $Nb_{1.5}TaTiZr_{0.5}$, $Nb_{0.5}TaTiZr_{1.5}$) depending on microstructure.

Obtained results on microstructure stability and related mechanical properties can be useful also for long term aging annealing of HEA alloys. The Monte Carlo simulation performed pointed out, that single solid solution of 5 element HfNbTaTiZr alloy is a metastable state at room temperature. Stabilizing effect of Hf on the HCP phase formation during mid temperature aging was proposed.

Author Contributions: Conceptualisation, J.Z.; Resources, J.V. and J.M.; Investigation, J.Z., F.L., O.M. and J.K.; Formal analysis, J.Z.; Writing—original draft, J.Z., J.C. and J.M.; Writing—review and editing, J.Z. and J.C.; supervision, J.C.

Acknowledgments: This research was funded by the Czech Science Foundation, project 17-17016S.

Conflicts of Interest: The authors declare no conflict of interest.

References

1. Miracle, D.B.; Senkov, O.N. A critical review of high entropy alloys and related concepts. *Acta Mater.* **2016**, *122*, 448–511. [CrossRef]
2. Zhang, W.; Liaw, P.K.; Zhang, Y. Science and technology in high-entropy alloys. *Science China Materials, January* **2018**, *61*, 2–22. [CrossRef]
3. Cantor, B.; Chang IT, H.; Knight, P.; Vincent AJ, B. Microstructural development in equiatomic multicomponent alloys. *Mater. Sci. Eng. A* **2004**, *377*, 213–218. [CrossRef]
4. Yeh, B.J.; Chen, S.; Lin, S.; Gan, J.; Chin, T.; Shun, T.; Tsau, C. Nanostructured High-Entropy Alloys with Multiple Principal Elements: Novel Alloy Design Concepts and Outcomes. *Adv. Eng. Mater.* **2004**, *6*, 299–303. [CrossRef]
5. Schön, C.G.; Duong, T.; Wang, Y.; Arróyave, R. Probing the entropy hypothesis in highly concentrated alloys. *Acta Mater.* **2018**, *148*, 263–279. [CrossRef]
6. Otto, F.; Yang, Y.; Bei, H.; George, E.P. Relative effects of enthalpy and entropy on the phase stability of equiatomic high-entropy alloys. *Acta Mater.* **2018**, *61*, 2628–2638. [CrossRef]
7. Li, Z.; Ludwig, A.; Savan, A.; Springer, H.; Raabe, D. Combinatorial metallurgical synthesis and processing of high-entropy alloys. *J. Mater. Res.* **2018**, *33*, 3156–3169. [CrossRef]
8. Tsai, M.; Yeh, J.; Tsai, M.; Yeh, J. High-Entropy Alloys: A Critical Review High-Entropy Alloys: A Critical Review. *Mater. Res. Lett.* **2014**, *2*, 107–123. [CrossRef]
9. Rost, C.M.; Sachet, E.; Borman, T.; Moballegh, A.; Dickey, E.C.; Hou, D.; Jones, J.L.; Curtarolo, S.; Maria, J.P. Entropy-stabilized oxides. *Nat. Commun.* **2015**, *6*, 1–8. [CrossRef]
10. Senkov, O.N.; Miracle, D.B.; Chaput, K.J. Development and exploration of refractory high entropy alloys—A review. *J. Mater. Res.* **2018**, *33*, 3092–3128. [CrossRef]
11. Niinomi, M.; Nakai, M.; Hieda, J. Development of new metallic alloys for biomedical applications. *Acta Biomater.* **2012**, *8*, 3888–3903. [CrossRef] [PubMed]
12. Niinomi, M. Recent Research and Development in Titanium Alloys for Biomedical Applications and Healthcare Goods Recent research and development in titanium alloys for biomedical applications and healthcare goods. *Sci. Technol. Adv. Mater.* **2003**, *4*, 445–454. [CrossRef]
13. Wang, S.; Xu, J. TiZrNbTaMo high-entropy alloy designed for orthopedic implants: As-cast microstructure and mechanical properties. *Mater. Sci. Eng. C* **2017**, *73*, 80–89. [CrossRef] [PubMed]
14. Todai, M.; Nagase, T.; Hori, T.; Matsugaki, A.; Sekita, A.; Nakano, T. Scripta Materialia Novel TiNbTaZrMo high-entropy alloys for metallic biomaterials. *SMM* **2017**, *129*, 65–68. [CrossRef]
15. Li, H.F.; Xie, X.H.; Zhao, K.; Wang, Y.B.; Zheng, Y.F.; Wang, W.H.; Qin, L. In vitro and in vivo studies on biodegradable CaMgZnSrYb high-entropy bulk metallic glass. *Acta Biomater.* **2013**, *9*, 8561–8573. [CrossRef] [PubMed]
16. Nnamchi, P.S. First principles studies on structural, elastic and electronic properties of new Ti-Mo-Nb-Zr alloys for biomedical applications. *JMADE* **2016**, *108*, 60–67. [CrossRef]
17. Pogrebnjak, A.D.; Beresnev, V.M.; Jurga, S. Structural and mechanical characterization of (TiZrNbHfTa) N/WN multilayered nitride coatings. *Mater. Lett.* **2018**, *229*, 364–367. [CrossRef]
18. Zyka, J.; Malek, J.; Pala, Z.; Andrsova, I.; Vesely, J. Structure And Mechanical Properties Of TaNbHfZrTi High Entropy Alloy. In Proceedings of the METAL 2015—24rd International Conference on Metallurgy and Materials, Brno, Czech Republic, 3–5 June 2015; pp. 3–8.
19. Senkov, O.N.; Scott, J.M.; Senkova, S.V.; Miracle, D.B.; Woodward, C.F. Microstructure and room temperature properties of a high-entropy TaNbHfZrTi alloy. *J. Alloys Compd.* **2011**, *509*, 6043–6048. [CrossRef]
20. Senkov, O.N.; Zhang, C.; Pilchak, A.L.; Payton, E.J.; Woodward, C.; Zhang, F. CALPHAD-aided development of quaternary multi-principal element refractory alloys based on NbTiZr. *J. Alloys Compd.* **2019**, *783*, 729–742. [CrossRef]
21. Senkov, O.N.; Rao, S.; Chaput, K.J.; Woodward, C. Acta Materialia Compositional effect on microstructure and properties of NbTiZr-based complex concentrated alloys. *Acta Mater.* **2018**, *151*, 201–215. [CrossRef]
22. Coelho, A.A. *TOPAS*; Version 5; Computer Software; Coelho Software: Brisbane, Australia, 2016.
23. *Standard Test Methods for Determining Average Grain Size*; ASTM E112-13; ASTM: West Conshohocken, PA, USA, 2013.

24. Lukac, F.; Dudr, M.; Musalek, R.; Klecka, J.; Cinert, J.; Cizek, J.; Chraska, T.; Cizek, J.; Melikhova, O.; Kuriplach, J.; et al. Introduction, Spark plasma sintering of gas atomized high-entropy alloy HfNbTaTiZr. *J. Mater. Res.* **2018**, *33*, 3247–3257. [CrossRef]
25. Nguyen, V.T.; Shi, Z.; Song, T. Compositional design of strong and ductile (tensile) Ti-Zr-Nb-Ta medium entropy alloys (MEAs) using the atomic mismatch approach. *Mater. Sci. Eng. A* **2019**, *742*, 762–772. [CrossRef]
26. Maiti, S.; Steurer, W. Structural-disorder and its effect on mechanical properties in single- phase TaNbHfZr high-entropy alloy. *Acta Mater.* **2016**, *106*, 87–97. [CrossRef]
27. Couzinié, J.P.; Dirras, G.; Perrière, L.; Chauveau, T.; Leroy, E.; Champion, Y.; Guillot, I. Microstructure of a near-equimolar refractory high-entropy alloy. *Mater. Lett.* **2014**, *126*, 285–287. [CrossRef]
28. Sheikh, S.; Shafeie, S.; Hu, Q.; Ahlström, J.; Persson, C.; Veselý, J.; Zýka, J. Alloy design for intrinsically ductile refractory high-entropy alloys Alloy design for intrinsically ductile refractory high-entropy alloys. *J. Appl. Phys.* **2016**, *120*. [CrossRef]
29. Takeuchi, A.; Inoue, A. Classification of Bulk Metallic Glasses by Atomic Size Difference, Heat of Mixing and Period of Constituent Elements and Its Application to Characterization of the Main Alloying Element. *Mater. Trans.* **2005**, *46*, 2817–2829. [CrossRef]
30. Okamoto, H. Ta-Zr (Tantalum-Zirconium). *J. Phase Equilib.* **1996**, *17*, 555. [CrossRef]
31. Okamoto, H. Nb-Zr (Niobium-Zirconium). *J. Phase Equilib.* **1992**, *13*, 577. [CrossRef]
32. Okamoto, H. Hf-Ta (Hafnium-Tantalum). *J. Phase Equilib.* **1996**, *17*, 382. [CrossRef]
33. Okamoto, H. Hf-Nb (Hafnium-Niobium). *J. Phase Equilib.* **1998**, *19*, 288. [CrossRef]
34. Wu, Y.D.; Cai, Y.H.; Chen, X.H.; Wang, T.; Si, J.J.; Wang, L.; Wang, Y.D.; Hui, X.D. Phase composition and solid solution strengthening effect in TiZrNbMoV high-entropy alloys. *Mater. Des.* **2015**, *83*, 651–660. [CrossRef]
35. Schuh, B.; Volker, B.; Todt, J.; Schell, N.; Perriere, L.; Li, J.; Couzinie, J.P.; Hohenwarter, A. Thermodynamic instability of a nanocrystalline, single-phase TiZrNbHfTa alloy and its impact on the mechanical properties. *Acta Mater.* **2018**, *142*, 201–212. [CrossRef]
36. Coury, F.G.; Butler, T.; Chaput, K.; Saville, A.; Copley, J.; Foltz, J.; Mason, P.; Clarke, K.; Kaufman, M.; Clarke, A. Phase equilibria, mechanical properties and design of quaternary refractory high entropy alloys. *Mater. Des.* **2018**, *155*, 244–256. [CrossRef]
37. Stepanov, N.D.; Yurchenko, N.Y.; Zherebtsov, S.V.; Tikhonovsky, M.A.; Salishchev, G.A. Aging behavior of the HfNbTaTiZr high entropy alloy. *Mater. Lett.* **2018**, *211*, 87–90. [CrossRef]
38. Chen, S.Y.; Tong, Y.; Tseng, K.; Yeh, J.; Poplawsky, J.D.; Wen, J.G.; Gao, M.C.; Kim, G.; Chen, W.; Ren, Y.; et al. Phase transformations of HfNbTaTiZr high-entropy alloy at intermediate temperatures. *Scr. Mater.* **2019**, *158*, 50–56. [CrossRef]
39. Sheikh, S.; Bijaksana, M.K.; Motallebzadeh, A.; Shafeie, S.; Lozinko, A.; Gan, L.; Tsao, T.; Klement, U.; Canadinc, D.; Murakami, H.; et al. Accelerated oxidation in ductile refractory high-entropy alloys. *Intermetallics* **2018**, *97*, 58–66. [CrossRef]

entropy

MDPI

Article

Effects of Mo, Nb, Ta, Ti, and Zr on Mechanical Properties of Equiatomic Hf-Mo-Nb-Ta-Ti-Zr Alloys

Ko-Kai Tseng [1], Chien-Chang Juan [1], Shuen Tso [2], Hsuan-Chu Chen [2], Che-Wei Tsai [1,2] and Jien-Wei Yeh [1,2,*]

1 Department of Materials Science and Engineering, National Tsing Hua University, Hsinchu 30013, Taiwan; zouts2@gmail.com (K.-K.T.); d943554@oz.nthu.edu.tw (C.-C.J.); chewei@mx.nthu.edu.tw (C.-W.T.)

2 High Entropy Materials Center, National Tsing Hua University, Hsinchu 30013, Taiwan; a8701787017@gmail.com (S.T.); xiphofly@gmail.com (H.-C.C.)

* Correspondence: jwyeh@mx.nthu.edu.tw; Tel.: +886-3-5719558

Received: 30 November 2018; Accepted: 21 December 2018; Published: 25 December 2018

Abstract: Nowadays refractory high-entropy alloys (RHEAs) are regarded as great candidates for the replacement of superalloys at high temperature. To design a RHEA, one must understand the pros and cons of every refractory element. However, the elemental effect on mechanical properties remains unclear. In this study, the subtraction method was applied on equiatomic HfMoNbTaTiZr alloys to discover the role of each element, and, thus, HfMoNbTaTiZr, HfNbTaTiZr, HfMoTaTiZr, HfMoNbTiZr, HfMoNbTaZr, and HfMoNbTaTi were fabricated and analyzed. The microstructure and mechanical properties of each alloy at the as-cast state were examined. The solid solution phase formation rule and the solution strengthening effect are also discussed. Finally, the mechanism of how Mo, Nb, Ta, Ti, and Zr affect the HfMoNbTaTiZr alloys was established after comparing the properties of these alloys.

Keywords: high-entropy alloys; refractory high-entropy alloys; alloys design; elevated-temperature yield strength; solid solution strengthening effect

---

1. Introduction

Refractory elements, including Rhenium (Re), Molybdenum (Mo), Niobium (Nb), Tantalum (Ta), and Tungsten (W) [1] are very important for improving mechanical properties in advanced alloys such as Titanium alloys and Nickel-base superalloys. Generally, elements with a melting point higher than Titanium (Ti), such as Chromium (Cr), Hafnium (Hf), Osmium (Os), Ruthenium (Ru), Vanadium (V), and Zirconium (Zr) are also classified as refractory elements. In Titanium alloys, refractory elements, especially Mo and W, are all beta stabilizers and possess a strong solid solution strengthening characteristic [2]. Nb forms Ni_3Nb γ phase in Nickel-base superalloys such as Inconel 718 [3]. The improved creep resistance of the sixth generation superalloy TMS-238 mainly results from Re and Ru additions [4]. Refractory alloys also play significant roles in industry. In the second generation nuclear power plants, the most-used materials for fuel cladding are Zircoloy 2 and Zircoloy 4. The Niobium alloy C103 is used for the nozzle extension of satellites [5]. Refractory alloys are also known as having the great potential for elevated-temperature applications because of their high strength at elevated temperature. According to thermodynamics, thermal efficiency of a turbine engine could be enhanced by increasing the turbine inlet temperature. However, the melting point of Nickel-base superalloy limits the application itself above 1200 °C. Therefore, it is necessary to develop new refractory alloys, especially for applications at temperatures higher than 1200 °C.

In 2004, Professor Yeh and his group published the concept of high-entropy alloys (HEAs) [6]. He defined high-entropy alloys (HEAs) as having five or more major elements beneath 5–35 at.%, and minor elements below five at.%. In 2006, he also established four core effects of HEAs [7]:

High-entropy, severe-lattices-distortion, sluggish-diffusion, and the cocktail effect. With the new concepts, scientists are able to develop and expand alloys without restrictions [8–12]. Some HEAs have been found to have attractive properties on diffusion [13], oxidation [14], corrosion [15], fatigue [16], creep [17], fracture toughness [18], and elevated-temperature strength [19]. High-entropy superalloys (HESAs) [20], eutectic HEAs (EHEAs) [21], light-weight HEAs (LWHEAs) [22], refractory high-entropy alloys (RHEAs) [23], etc. have been proposed and attract increased attention. In 2010, Senkov and Miracle first published RHEAs, MoNbTaW and MoNbTaVW [23,24]. These two alloys possess body-centered cubic (BCC) structure and have excellent elevated-temperature yield strength which is around 400 MPa at 1600 °C. But their density is much higher, and the room temperature compressive ductility is very low. In 2012, they published the equiatomic composition of HfNbTaTiZr, which possesses excellent compressive ductility up to 50%, lower density but poor elevated-temperature yield strength [25,26]. From then on, there were over 150 papers published concerning RHEAs [19].

There are some researches working on the addition of Al [27,28], Mo [29], Ti [30], V [31], or Zr [32], but few of them focus on the elevated-temperature mechanical performance and the overall effect of the constituent elements. To understand the elemental effect on mechanical properties, equiatomic HfMoNbTaTiZr alloy is firstly designed as a base alloy by adding the high modulus refractory element Mo to HfNbTaTiZr. Then the subtraction method is used to analyze each elemental effect in the equiatomic HfMoNbTaTiZr. Thus, in this study, six alloys HfMoNbTaTiZr, HfNbTaTiZr, HfMoTaTiZr, HfMoNbTiZr, HfMoNbTaZr, and HfMoNbTaTi are investigated and compared on their microstructure. Further, the compressive properties at room temperature and at elevated temperature are investigated. By comparing HfMoNbTaTiZr and HfNbTaTiZr, the influence of the addition of Mo can be understood. Likewise, the influence of the addition of Nb, Ta, Ti, or Zr can be articulated. In addition, promising compositions were found and further improved design is suggested. The solid solution phase formation rule and the solid solution strengthening effect will be discussed.

2. Materials and Methods

The experimental Hf-Mo-Nb-Ta-Ti-Zr alloy series was prepared by vacuum-arc melting. The purity of raw elements including Hf, Mo, Nb, Ta, and Zr was 99.9 wt.%, and that of Ti was 99.99 wt.%. The melting points of each constituent element used in alloys are shown in Table 1. These pure metals were stacked together in the sequence of low melting point to high melting point from bottom to top. The stacked metals were melted together in a water-cool copper mold and solidified therein. The ingot of each alloy was flipped and re-melted, at least, four times to improve the chemical homogeneity. The crystal structure of the alloy samples taken from the portion near the copper mold was examined with the Shimadzu XRD-6000 X-ray diffractometer (SHIMADZU CORPORATION, Kyoto, Japan), operated at 30 kV and 20 mA with a scanning rate of 4°/min from 20° to 100°. JEOL JSM-5410 (JEOL Ltd., Tokyo, Japan) scanning electron microscope (SEM) and JXA-8500F FE-EPMA (JEOL Ltd., Tokyo, Japan) was used to analyze the samples in backscattering electron (BSE) mode. Energy dispersive spectrometry (EDS) was also used to confirm the chemical compositions. The cylindrical samples for the compression test were 3.6 mm in diameter and 6 mm in height. The room temperature compression tests were conducted with Instron 4468 (INSTRON, Norwood, MA, USA) universal testing machine, and the high temperature compression tests were performed on Gleeble-3500 (DYNAMIC SYSTEMS INC, Poestenkill, NY, USA) thermal–mechanical simulator. All tests were examined at the crosshead speed of 0.36 mm/min, which imposed the strain rate of 10^{-3} s^{-1} on the samples.

Table 1. Results of scanning electron microscope-energy dispersive spectrometry (SEM-EDS) analysis (at. %). Nominal composition means the designed composition. DR means the dendritic region. ID means the interdendritic region.

	Element	Hf	Mo	Nb	Ta	Ti	Zr
HfMoNbTaTiZr	Nominal	16.7	16.7	16.7	16.7	16.7	16.7
	DR	14.3	18.4	19.5	24.4	12.4	10.8
	ID	21.1	13.6	12.3	9.9	18.3	24.7
HfNbTaTiZr	Nominal	20.0	-	20.0	20.0	20.0	20.0
	DR	18.5	-	22.4	27.4	18.2	13.5
	ID	22.6	-	17.5	12.8	20.2	26.9
HfMoTaTiZr	Nominal	20.0	20.0	-	20.0	20.0	20.0
	DR	20.6	21.4	-	23.9	18.3	15.7
	ID	24.4	16.2	-	11.0	21.2	27.1
HfMoNbTiZr	Nominal	20.0	20.0	20.0	-	20.0	20.0
	Overall	20.8	20.6	19.7	-	19.2	19.7
HfMoNbTaZr	Nominal	20.0	20.0	20.0	20.0	-	20.0
	DR	18.5	20.8	21.7	24.5	-	14.5
	ID	27.0	15.6	13.5	9.9	-	34.0
HfMoNbTaTi	Nominal	20.0	20.0	20.0	20.0	20.0	-
	DR	15.5	22.7	19.5	25.7	16.6	-
	ID	30.7	16.9	17.6	10.7	24.1	-

3. Results

Figure 1a–f are BSE images of experimental Hf-Mo-Nb-Ta-Ti-Zr alloys. A typical dendritic structure is observed. Their compositions as obtained from SEM-EDS are shown in Table 1. It is noted that the dendritic area is rich in Ta and Mo which have the highest two melting points. This is expected since high melting point elements tend to crystalize first during solidification. By contrast of dendritic structure is poor in HfMoNbTiZr alloy as observed in the BSE image. This means the partition between dendrite and interdendrite is small and the coring phenomenon was less obvious. This is due to the subtraction of the highest melting point element Ta which would solidify first with a Ta-rich solid solution.

Figure 2 shows the X-Ray diffraction patterns. The main phase of the Hf-Mo-Nb-Ta-Ti-Zr alloy series is a BCC disordered solid solution. The asymmetry of (200) and (211) peaks are shown in the diffraction pattern results from the cored dendritic structure. The composition variation of dendritic and interdendritic areas causes a little difference in the lattice constant. The lattice constants of the Hf-Mo-Nb-Ta-Ti-Zr alloy series listed in Table 2 are in the range of 3.305 to 3.400 Å calculated by the Nelson–Riley extrapolation function. Referring to the phase diagrams of each binary alloy between Hf, Mo, Nb, Ta, Ti, and Zr, Hf-Mo, and Mo-Zr binary alloys form Mo_2Hf and Mo_2Zr, respectively, in a certain range of composition even at high temperature. However, these two intermetallic compounds or others do not show up in Hf-Mo-Nb-Ta-Ti-Zr alloys, which means the high entropy effect has a significant benefit in forming simple BCC solid solution in this alloy system especially at high temperature.

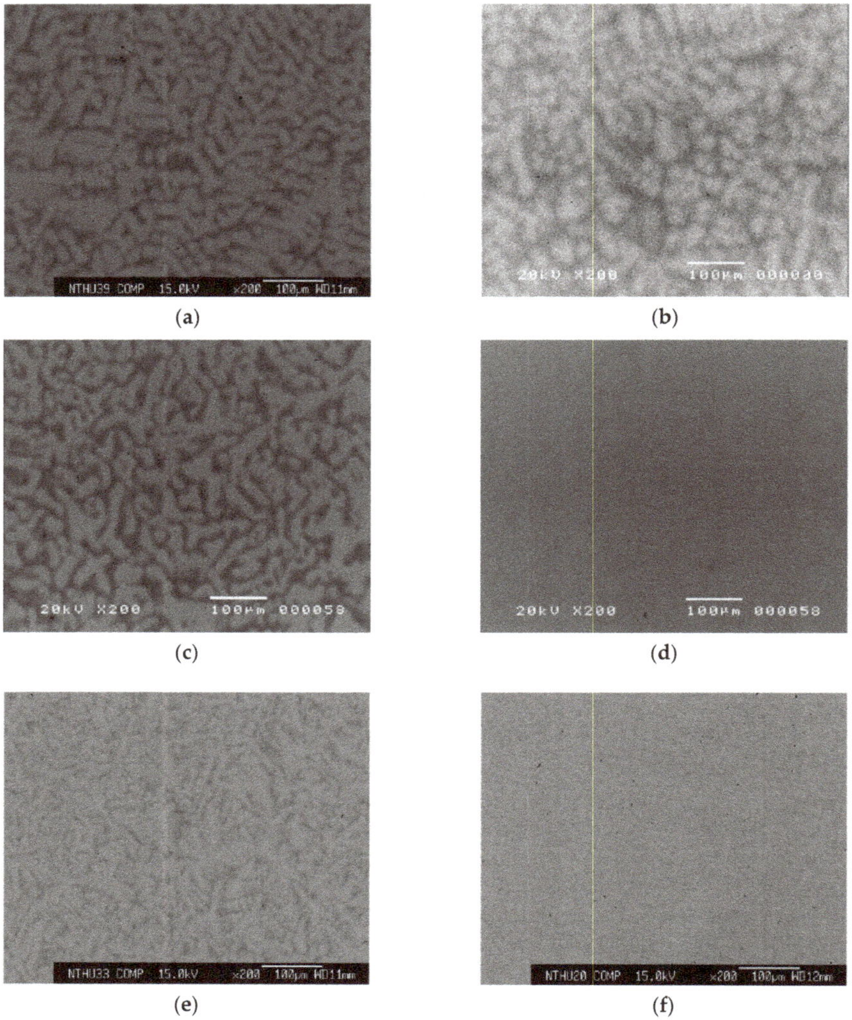

Figure 1. Backscattering electron (BSE) images of (**a**) HfMoNbTaTiZr, (**b**) HfNbTaTiZr, (**c**) HfMoTaTiZr, (**d**) HfMoNbTiZr, (**e**) HfMoNbTaZr, and (**f**) HfMoNbTaTi. All the alloys show the dendritic structure except HfMoNbTiZr.

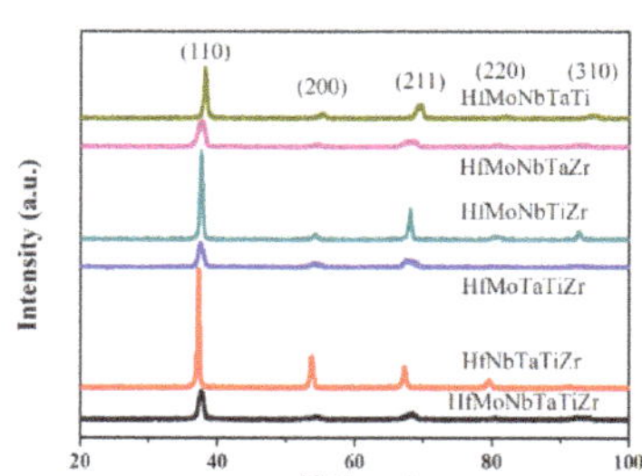

Figure 2. X-ray diffraction patterns of Hf-Mo-Nb-Ta-Ti-Zr alloy series.

Table 2. The lattice constants (Å) of the Hf-Mo-Nb-Ta-Ti-Zr alloy series. Cal. means the value calculated from Vegard's Law. Exp. means the value calculated by Nelson–Riley extrapolation function based on X-ray diffraction pattern.

	HfMoNb TaTiZr	HfNbTa TiZr	HfMoTa TiZr	HfMoNb TiZr	HfMoNb TaZr	HfMoNb TaTi
Cal.	3.361	3.404	3.373	3.373	3.378	3.317
Exp.	3.345	3.400	3.364	3.369	3.347	3.305

Figure 3 shows the compression test results of Hf-Mo-Nb-Ta-Ti-Zr alloy series. At room temperature, the yield strength of HfMoNbTaTiZr alloy was 1512 MPa, and ultimate strength was 1828 MPa when the strain was 11%. The compression tests for HfMoNbTaTiZr alloy were also conducted at 800 °C, 1000 °C, and 1200 °C, respectively. At 800 °C, the yield strength of HfMoNbTaTiZr alloy was 1007 MPa and ultimate strength was 1489 MPa when the strain was 19%, which shows obvious work hardening. At 1000 °C and 1200 °C, the results of yield strength were 814 MPa and 556 MPa, respectively, but the strength kept decreasing from the yield point to the end of the test, showing the work softening behavior. No crack was observed at 1000 °C and 1200 °C.

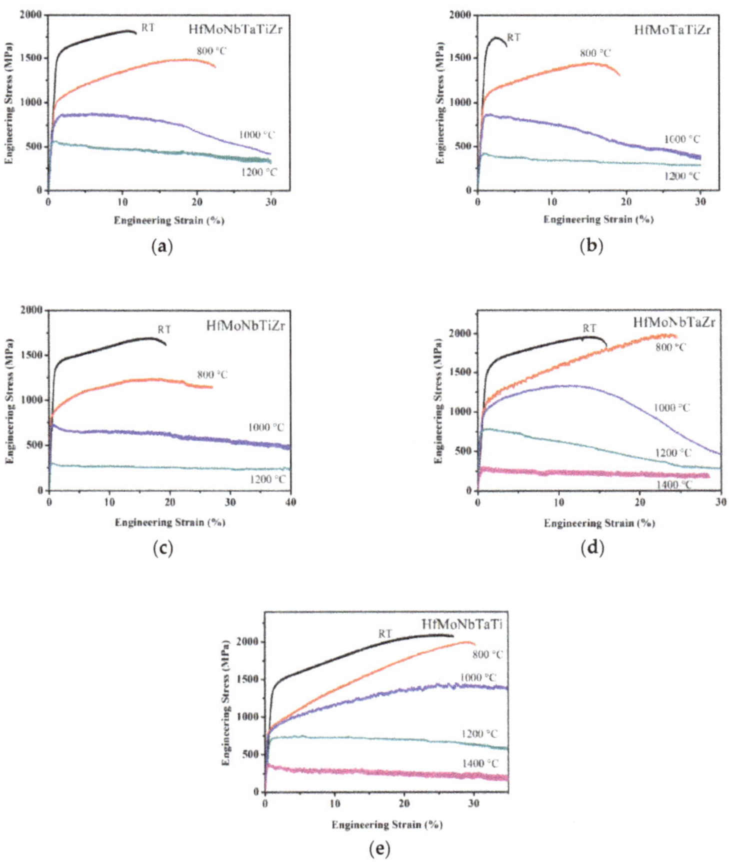

Figure 3. Engineer compressive stress–strain curve of (**a**) HfMoNbTaTiZr, (**b**) HfMoTaTiZr, (**c**) HfMoNbTiZr, (**d**) HfMoNbTaZr, and (**e**) HfMoNbTaTi.

When an element is removed from HfMoNbTaTiZr, the behavior is changed. The subtraction of Nb gives HfMoTaTiZr alloy. At room temperature, the yield strength of HfMoTaTiZr alloy was 1600 MPa, and the ultimate strength was 1743 MPa when the strain was 3%. At 800 °C, the yield strength was 1045 MPa, and the ultimate strength was 1446 MPa when the strain was 23%. As for the HfMoNbTaTiZr alloy, the stress–strain curve of HfMoTaTiZr alloy shows obvious work hardening effect. The results of yield strength were 855 MPa and 404 MPa at 1000 °C and 1200 °C, respectively. The strength kept decreasing from the yield point to the end of the test. No crack was observed at 1000 °C and 1200 °C.

The subtraction of Ta from HfMoNbTaTiZr gives HfMoNbTiZr. At room temperature, the yield strength of alloy was 1351 MPa, and the ultimate strength was 1698 MPa when the strain was 17%. This alloy performs with better toughness than HfMoNbTaTiZr and HfMoTaTiZr does at room temperature. At 800 °C, the yield strength was 829 MPa and the ultimate strength was 1244 MPa when the strain was 18%. At 1000 °C, yield strength was 721 MPa, and the strength kept decreasing when the strain increased. At 1200 °C, the yield strength was 301 MPa. The strength of the alloy was almost constant after the yield point to the end of the test. The strain softening effect was balanced by the strengthening effect.

Ti was removed in sequence. At room temperature, the yield strength of HfMoNbTaZr alloy was 1524 MPa, and the ultimate compress strength was 1963 MPa when the strain was 13.5%. At 800 °C, the compressive yield strength was 1005 MPa, and the ultimate strength was 1991 MPa when the strain was 24%. As with HfMoNbTaTiZr, there is an obvious work hardening effect shown at 800 °C in the stress–strain diagram. The yield strength was 927 MPa at 1000 °C. There was still a work hardening effect at the beginning of the test. The ultimate strength was 1336 MPa when the strain was 11%, but the strength decreased drastically to 464 MPa at the end of the test. At 1200 °C, the yield strength was 694 MPa, but the strength decreased to 289 MPa when the test stopped at 30% strain. At 1400 °C, the yield strength was 278 MPa, and the strength barely decreased during the test. Except for 800 °C, no fracture was observed during the test at elevated temperatures.

Finally, Zr was removed. At room temperature, the yield strength of HfMoNbTaTi alloy was 1369 MPa, and the ultimate compress strength was 2094 MPa when the strain was 25%. The yield strength at 800 °C was 822 MPa, and the ultimate strength was 1998 MPa when the strain was 29%. An obvious work hardening effect can be observed in the stress–strain curve. At 1000 °C, the yield strength was 778 MPa, and there was still work hardening effect at the beginning of the test until the ultimate strength 1454 MPa, at 27.5% strain. The results of yield strength were 699 MPa and 367 MPa at 1200 °C and 1400 °C, respectively. Both stress–strain diagrams show a steady decrease in strength after yield points. No fracture was observed at 1000 °C, 1200 °C, and 1400 °C.

The Tables 3 and 4 summarize the results of compressive tests. Comparing the performance at room temperature, HfMoTaTiZr alloy has the best yield strength 1600 MPa; HfMoNbTaTi has the highest 27% fracture strain; comprehensively, HfMoNbTaTi has the best mechanical properties (yield strength 1369 MPa and 27% fracture strain). The fracture strain increases from 4% for the HfMoTaTiZr alloy to 12% for HfMoNbTaTiZr. The presence of Ta can increase the yield strength but decrease the toughness. Ti seems to have no significant influence on strength, but it has a negative effect on toughness. Zr increases the strength; however, it strongly deteriorates the toughness at room temperature, because the fracture strain decreases from 27% for the HfMoNbTaTi alloy to 12% for HfMoNbTaTiZr. The presence of Mo, which has the highest shear modulus, increases the yield strength at room temperature from 929 MPa to 1512 MPa significantly. Nevertheless, the toughness decreases tremendously, fracture strain declines from > 50% to 12%. However, the presence of Nb decreases the yield strength only slightly from room temperature to 1000 °C and increases the yield strength by 38% at 1200 °C, but largely improves the room temperature fracture strain. This indicates that more Nb could be added for higher ductility.

Table 3. The room temperature compressive yield strength and fracture strain of the Hf-Mo-Nb-Ta-Ti-Zr alloy series.

	HfMoNb TaTiZr	HfNbTa TiZr [25]	HfMoTa TiZr	HfMoNb TiZr	HfMoNb TaZr	HfMoNb TaTi
Yield strength (MPa)	1512	929	1600	1351	1524	1369
Fracture strain (%)	12	> 50	4	20	16	27

Table 4. The elevated temperature compressive yield strength (MPa) of the Hf-Mo-Nb-Ta-Ti-Zr alloy series.

Temperature (°C)	HfMoNb TaTiZr	HfNbTa TiZr [25]	HfMoTa TiZr	HfMoNb TiZr	HfMoNb TaZr	HfMoNb TaTi
800	1007	535	1045	829	1005	822
1000	814	295	855	721	927	778
1200	556	92	404	301	694	699
1400	N. A.	N. A.	N. A.	N. A.	278	367

The melting point of the elements in the alloy affects the strength performance at elevated temperature. For instance, at 1200 °C, HfMoTaTiZr alloy and HfMoNbTiZr alloy which have lower melting-point elements (Ta, Mo, Nb) have less strength; HfMoNbTaZr alloy and HfMoNbTaTi alloy which have higher melting-point elements (Ti, Zr) have better strength. Moreover, all the alloys with Mo present have much better strength than HfNbTaTiZr does. Therefore, Mo makes a significant contribution to strength at elevated temperature.

The elevated temperature yield strength versus temperature of Hf-Mo-Nb-Ta-Ti-Zr alloys is shown in Figure 4a. Except at 800 °C, the strength of the above mentioned Hf-Mo-Nb-Ta-Ti-Zr alloys is better than the commercial nickel base superalloys, CMSX-4 and Inconel 718. Additionally, Hf-Mo-Nb-Ta-Ti-Zr alloys also have better resistance to softening at elevated temperature. Figure 4b is the specific strength of Hf-Mo-Nb-Ta-Ti-Zr alloys, HfNbTaTiZr alloy, CMSX-4, and Inconel 718 at different temperatures. Below 900 °C, CMSX-4 and Inconel 718 perform better; at 1000 °C, Hf-Mo-Nb-Ta-Ti-Zr alloys, except for HfNbTaTiZr, are better than CMSX-4 and Inconel 718. At temperatures above 1200 °C, HfMoNbTaTi alloy has the highest specific strength.

Comprehensively speaking, Hf-Mo-Nb-Ta-Ti-Zr alloys have a potential application at elevated temperature.

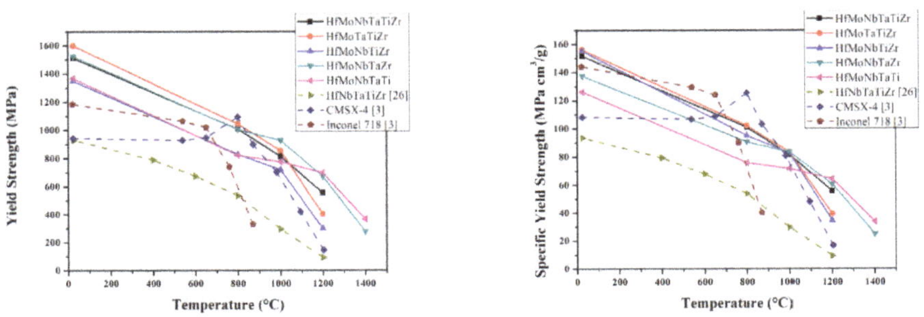

(a) (b)

Figure 4. (**a**) Elevated temperature yield strength and (**b**) elevated temperature specific yield strength versus temperature between Hf-Mo-Nb-Ta-Ti-Zr alloy series, CMSC-4, and Inconel 718 [3]. The elevated temperature yield strength of HfNbTaTiZr is from Reference [26].

4. Discussion

4.1. Phase Formation Rule

The solid solution phase formation rules were checked with the microstructure and crystal structure of the alloys in this study. First are the criterion based on thermodynamic parameters and atomic size parameter [33,34]. The thermodynamic parameters are mixing entropy ΔS_{mix}, mixing enthalpy ΔH_{mix}, and Ω, respectively:

$$\Delta S_{mix} = -\sum c_i \ln c_i \tag{1}$$

$$\Delta H_{mix} = \sum 4\Delta H_{ij} c_i c_j,\ i \neq j \tag{2}$$

$$\Omega = \frac{T_m \Delta S_{mix}}{\Delta H_{mix}} \tag{3}$$

where R is the gas constant, c_i is the atomic percentage of the element i, c_j is the atomic percentage of the element j, ΔH_{ij} is the enthalpy of the binary liquid state of elements i and j at an equiatomic composition from the Miedema's model [35,36], and T_m is the melting point of the alloy defined by rule of mixing:

$$T_m = \sum c_i T_{m,i} \tag{4}$$

where $T_{m,i}$ is the melting point of the element i. The atomic size parameter is atomic size difference δ:

$$\delta = \sqrt{\sum c_i (1 - \frac{r_i}{\bar{r}})^2} \tag{5}$$

where r_i is the atomic radius of element i. $\bar{r}$ is the average radius of the alloy defined by rule of mixing.

$$\bar{r} = \sum c_i r_i \tag{6}$$

The second criterion determining crystal type is related to electronic parameters, valence electron concentration VEC [37], and the third criterion determining Laves phase is related to Allen electronegativity difference $\Delta\chi_{Allen}$ [38]:

$$VEC = \sum c_i VEC_i \tag{7}$$

$$\Delta\chi_{Allen} = \sqrt{c_i (1 - \frac{\chi_i^{Allen}}{\bar{\chi}})^2} \tag{8}$$

where VEC_i is the valence electron concentration of the element i [39], χ_i^{Allen} is the electronegativity of the element i from Allen et al. [40], and $\bar{\chi}$ is the average electronegativity of the alloy defined by rule of mixing:

$$\bar{\chi} = \sum c_i \chi_i^{Allen} \tag{9}$$

The ranges in the three criterions for predicting the phases and crystal structure of HEAs are (1) Disorder solid solution phase forms when Ω >1.1 and δ < 6.6% [34]; (2) face-centered cubic (FCC) is stable when VEC > 8, and BCC is stable when VEC < 6.87 [37]; and (3) Laves phase forms when $\Delta\chi_{Allen}$ > 7% and δ > 5% in HEAs [41]. All the criterions mentioned above are established through statistical approach, so there is still some error especially in the boundary condition. The properties of pure elements Hf, Mo, Nb, Ta, Ti, and Zr are listed in Table 5. The BCC atomic radii of Hf, Ti, and Zr are Goldschmidt radii since all the alloys are a BCC structure. All the parameters of six alloys are calculated and listed in Table 6. From Table 5, one can observe Hf and Zr possess the largest atomic radius and smallest electronegativity, and Mo possesses the smallest atomic radius and biggest electronegativity. Furthermore, from the experiment results, all the alloys at the as-cast state form a single BCC disorder solid solution phase and no Laves phase is observed. This indicates that the formation of a single solid solution phase is consistent with criteria (1) and (2) but not consistent with criterion (3). Base on criterion (3), Laves phase might form in all the alloys except HfNbTaTiZr which is at the margin. It is necessary to check the criterion for Laves phase formation because Mo_2Hf or Mo_2Zr might form according to the Hf-Mo or Mo-Zr binary phase diagram. The result shows that criterion 3 is not fulfilled in the present alloy series. The minimum value 7% seems to be lower. One can observe that the Ω parameter values of these alloys are much higher than 1.1 and the VEC values are significantly lower than 6.87. This demonstrates that the high entropy effect is significant in enhancing the formation of a solid solution when mixing enthalpy and strain energy is small.

Table 5. Various data of the properties of Hf, Mo, Nb, Ta, Ti, and Zr. HCP means hexagonal close-packing.

ΔH_{ij} (kJ/mol)	Hf	Mo	Nb	Ta	Ti	Zr
Hf	-	−4	4	3	0	0
Mo	−4	-	−6	−5	−4	−6
Nb	4	−6	-	0	2	4
Ta	3	−5	0	-	1	3
Ti	0	−4	2	1	-	0
Zr	0	−6	4	3	0	-
r_i (nm)	0.159 (HCP) 0.155 (BCC)	0.136	0.143	0.143	0.147 (HCP) 0.142 (BCC)	0.162 (HCP) 0.157 (BCC)
$T_{m,i}$ (K)	2506	2896	2750	3290	1941	2128
χ_i^{Allen}	1.16	1.47	1.41	1.34	1.38	1.32
VEC_i	4	6	5	5	4	4
G (GPa)	30	120	38	69	44	33

Table 6. The values of thermodynamics, atomic size, and electronic parameters of the Hf-Mo-Nb-Ta-Ti-Zr alloy serious.

	HfMoNb TaTiZr	HfNbTa TiZr	HfMoTa TiZr	HfMoNb TiZr	HfMoNb TaZr	HfMoNb TaTi
ΔH_{mix} (kJ)	−0.9	2.7	−1.9	−1.6	−1.1	−1.4
ΔS_{mix} (J)	14.9	13.4	13.4	13.4	13.4	13.4
T_m (K)	2585.2	2523.0	2552.2	2444.2	2714.0	2676.6
Ω	43.3	12.4	17.8	20.4	32.4	24.9
δ	6.3%	5.5%	6.7%	6.7%	6.9%	5.4%
VEC	4.7	4.4	4.6	4.6	4.8	4.8
$\Delta\chi_{Allen}$	7.2%	6.6%	7.6%	7.8%	7.8%	7.8%

4.2. Solution Hardening Mechanism

The solid solution strengthening effect is calculated to examine the yield strength of the alloys in this study. From the experiment results, all the alloys possess a single phase of BCC disorder solid solution. It is valuable to use the yield strength of the alloys to check the solution strengthening mechanism. The solution strengthening mechanism of HEAs was proposed by Senkov et al. and then modified by Yao et al. [25,42]. The solution strengthening value $\Delta\sigma_i$ contributed by element *i* is:

$$\Delta\sigma_i = AGf_i^{4/3}c_i^{2/3} \quad (10)$$

where A is a material-dependent dimensionless constant of the order of 0.04, G is the shear modulus of the alloy, and f_i is the mismatch parameter of element *i* related to shear modulus and atomic size:

$$f_i = \sqrt{\delta_{G,i}{}^2 + \alpha^2\delta_{r,i}{}^2} \quad (11)$$

where $\delta_{G,i}$ and $\delta_{r,i}$ are the modulus mismatch parameter and atomic radius mismatch parameter, respectively as Equations (12) and (13). The value of α depends on the type of dislocation. For mixed dislocation, the value is designated to be nine.

$$\delta_{G,i} = \frac{9}{8}\sum c_j \delta_{G,ij} \tag{12}$$

$$\delta_{r,i} = \frac{9}{8}\sum c_j \delta_{r,ij} \tag{13}$$

where $\delta_{G,ij}$ and $\delta_{a,ij}$ are the differences between elements i and j in shear modulus and atomic radius, respectively as Equations (14) and (15). Nine is the number of atoms in the i-centered cluster in the BCC lattice, eight is the number of atoms neighboring with the center atom i.

$$\delta_{G,ij} = \frac{2(G_i - G_j)}{(G_i + G_j)} \tag{14}$$

$$\delta_{r,ij} = \frac{2(r_i - r_j)}{(r_i + r_j)} \tag{15}$$

where G_i and G_j are the shear modulus of element i and j, respectively, and r_j is the atomic radius of element j. Eventually, the solution strengthening $\Delta\sigma$ contributed by all the alloying elements is obtained by summation of $\Delta\sigma_i$. The calculated yield stress σ_c is the summation of the yield stress, σ_m, by rule of mixing and $\Delta\sigma$.

$$\Delta\sigma = \left(\sum (\Delta\sigma_i)^{3/2}\right)^{2/3} \tag{16}$$

$$\sigma_c = \sigma_m + \Delta\sigma \tag{17}$$

As the shear modulus of the HfMoNbTaTiZr alloy system is still lacking, we reasonably use the rule of mixing to calculate it since the modulus relates to the interatomic potential energy well:

$$G_m = \sum c_i G_i \tag{18}$$

The calculated results are listed in Table 7 and compared in Figure 5. One can observe that the σ_m of all the alloys is small and almost the same. In addition, the trends of $\Delta\sigma$ and σ_c are consistent with $\sigma_{0.2}$. This means that high yield strength of this alloy series all comes from solution strengthening effect despite there being some deviation, about 30%, between calculated values and experimental values. It is interesting to note that Mo, with the smallest atomic radius and the largest shear modulus, interacts frequently with other elements, thus, the HfNbTaTiZr alloy possesses the smallest yield stress without the addition of Mo. Ti, having the average atomic radius and the average shear modulus, interacts slightly with other elements, and, thus, HfMoNbTaZr alloy possess the largest $\Delta\sigma$ without the addition of Ti. As for the deviation between σ_c and $\sigma_{0.2}$, it might be due to the overestimated shear modulus. Young's modulus of HfNbTaTiZr is 81 GPa reported by Juan et al. [43], and, thus, the shear modulus can be calculated to be 31 GPa. This value is obviously smaller than the average shear modulus 43 GPa of HfNbTaTiZr. This implies that all the average shear moduli might be overestimated. If we calculated the shear modulus from the experimental $\sigma_{0.2}$, by assumption that σ_c equals to $\sigma_{0.2}$, the result is shown in the G_{cal} column of Table 7. From the figure, the trend of calculated shear modulus G_{cal} is consistent with the trend of average shear modulus although significantly smaller than G_m by ~30%. One can observe that the calculated shear modulus of HfNbTaTiZr 32 GPa is in a good agreement with the literature [43]. This indicates that severe lattice distortion in HEAs has a strong solution hardening effect, but Young's modulus was effectively lower. This is reasonable and could be related to its effect on lattice constant [11]. In the alloy series, NiCo, NiCoFe, NiCoFeCr, and NiCoFeCrMn, the lattice constant of real crystal structure has more deviation from that predicted by Vegard's law. The increased lattice constants as compared with ideal average lattice constant indicates that lattice distortion has the effect to expand the lattice. Thus, the interatomic bonding strength is effectively lower and the shear modulus is simultaneously lower. However, in the present alloy series, the calculated lattice constants based on Vegard's law are larger than the experimental lattice constants measured from X-ray diffraction patterns as listed in Table 2, especially with the addition of Mo. This is because Mo has a strong interaction with other elements to reduce the bond length according to ΔH_{ij} in Table 5. Al has the same effect and is reported in Reference [27]. However, there needs to be more research in the future to confirm the reasons for the reduced shear modulus.

Table 7. Comparisons of G_m, $\Delta\sigma$, σ_m, σ_c, $\sigma_{0.2}$, and G_{cal} of the present alloy series.

	G_m (GPa)	$\Delta\sigma$ (MPa)	σ_m (MPa)	σ_c (MPa)	$\sigma_{0.2}$ (MPa)	G_{cal} (GPa)
HfMoNbTaTiZr	55	1669	260	1929	1512	41
HfNbTaTiZr	43	938	225	1163	929 [26]	32
HfMoTaTiZr	60	1918	264	2182	1600	41
HfMoNbTiZr	53	1683	278	1961	1351	33
HfMoNbTaZr	58	1948	273	2221	1524	37
HfMoNbTaTi	60	1610	256	1866	1369	41

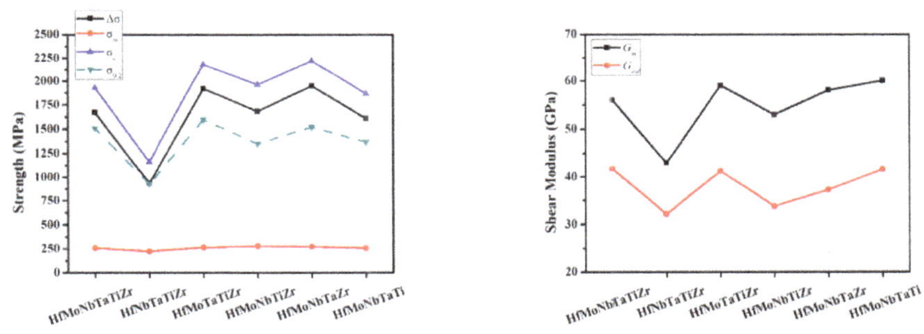

(a) (b)

Figure 5. The trend of Hf-Mo-Nb-Ta-Ti-Zr alloy series: (**a**) $\Delta\sigma$, σ_m, σ_c, and $\sigma_{0.2}$, and (**b**) G_m and G_{cal}.

5. Conclusions

The equiatomic HfMoNbTaTiZr alloy was chosen to analyze the effect of each constituent elemental by the subtraction method and, thus, HfMoNbTaTiZr, HfNbTaTiZr, HfMoTaTiZr, HfMoNbTiZr, HfMoNbTaZr, and HfMoNbTaTi were studied. Among these alloys, HfMoNbTaTi has the best mechanical performance, that is, 27% compressive strain at room temperature and yield strength 367 MPa at 1400 °C. HfMoNbTaTi has great potential for elevated-temperature applications. As the alloy system does not contain very expensive elements such as Re and Ru, it is cost competitive for high-temperature applications like Nb-Hf-Ti alloys in space vehicles. Further modification of composition and/or anti-oxidation coatings are still required for high-temperature applications in the air.

According to the experiment results, the effects of Mo, Nb, Ta, Ti, and Zr on mechanical properties of equiatomic Hf-Mo-Nb-Ta-Ti-Zr alloys were described. For higher room-temperature strength, one should add an element which interacts frequently with the alloy, such as Mo. For higher elevated-temperature strength, one should add the elements which possess high melting points, such as Mo, Nb, or Ta. One should add more Nb for higher ductility. With Ti or Zr addition, the elevated-temperature strength and the density decreases. All these elemental effects could also be applied to all other RHEAs systems, but more research is required to confirm this premise.

The solid solution phase formation rule and the solid solution strengthening effect of RHEAs have been discussed. The high entropy effect of the present alloys is significant in enhancing the formation of a solid solution. The shear modulus of RHEAs is smaller than that predicted from mixture rule by about 30%. This reduction is attributable to severe lattice distortion.

Author Contributions: K.-K.T. did the writing-original draft preparation, writing-review & editing, formal analysis, and the investigation. C.-C.J. did the methodology and the data curation. S.T. and H.-C.C. did the writing-original draft preparation. C.-W.T. and J.-W.Y. did the writing-review & editing, supervision, project administration, and funding acquisition.

Funding: This work was financially supported by the "High Entropy Materials Center" from the Featured Areas Research Center Program within the framework of the Higher Education Sprout Project by the Ministry of Education (MOE) and from the Project MOST 107-3017-F-007-003 by Ministry of Science and Technology (MOST) in Taiwan.

Acknowledgments: The authors acknowledged Woei-Ren Wang (Industrial Technology Research Institute, Tainan, Taiwan), who provided all the high temperature compressive test with Gleeble 3500.

Conflicts of Interest: The authors declare no conflict of interest.

References

1. International, A.S.M.; Handbook, C. *Properties and Selection: Nonferrous Alloys and Special-Purpose Materials*, ASM Handbook, 10th ed.; ASM International: Materials Park, OH, USA, 1990; Volume 2.
2. Smith, W.F. *Structure and Properties of Engineering Alloys*, 2nd ed.; McGraw-Hill: New York, NY, USA, 1993; 630p.
3. Reed, R.C. *The Superalloys: Fundamentals and Applications*; Cambridge University Press: Cambridge, UK, 2006. [CrossRef]
4. Kawagishi, K.; Yeh, A.C.; Yokokawa, T.; Kobayashi, T.; Koizumi, Y.; Harada, H. Development of an Oxidation-Resistant High-Strength Sixth-Generation Single-Crystal Superalloy TMS-238. *Superalloys* **2012**. [CrossRef]
5. Satya Prasad, V.V.; Baligidad, R.G.; Gokhale, A.A. Niobium and Other High Temperature Refractory Metals for Aerospace Applications. In *Aerospace Materials and Material Technologies: Volume 1: Aerospace Materials*; Prasad, N.E., Wanhill, R.J.H., Eds.; Springer: Singapore, 2017; pp. 267–288.
6. Yeh, J.W.; Chen, S.K.; Lin, S.J.; Gan, J.Y.; Chin, T.S.; Shun, T.T.; Tsau, C.H.; Chang, S.Y. Nanostructured high-entropy alloys with multiple principal elements: Novel alloy design concepts and outcomes. *Adv. Eng. Mater.* **2004**, *6*, 299–303. [CrossRef]
7. Yeh, J.W. Recent progress in high-entropy alloys. *Annales de Chimie Science des Matériaux* **2006**, *31*, 633–648. [CrossRef]
8. Miracle, D.B.; Senkov, O.N. A critical review of high entropy alloys and related concepts. *Acta Mater.* **2017**, *122*, 448–511. [CrossRef]
9. Cheng, C.Y.; Yang, Y.C.; Zhong, Y.Z.; Chen, Y.Y.; Hsu, T.; Yeh, J.W. Physical metallurgy of concentrated solid solutions from low-entropy to high-entropy alloys. *Curr. Opin. Solid State Mater.* **2017**, *21*, 299–311. [CrossRef]
10. Yeh, J.W. Alloy Design Strategies and Future Trends in High-Entropy Alloys. *JOM* **2013**, *65*, 1759–1771. [CrossRef]
11. Gao, M.C.; Yeh, J.W.; Liaw, P.K.; Zhang, Y. *High-Entropy Alloys: Fundamentals and Applications*, 1st ed.; Springer International Publishing: Cham, Switzerland, 2016.
12. Murty, B.S.; Yeh, J.W.; Ranganathan, S. *High Entropy Alloys*; Butterworth-Heinemann: Boston, MA, USA, 2014.
13. Tsai, K.Y.; Tsai, M.H.; Yeh, J.W. Sluggish diffusion in Co–Cr–Fe–Mn–Ni high-entropy alloys. *Acta Mater.* **2013**, *61*, 4887–4897. [CrossRef]
14. Shen, W.J.; Tsai, M.H.; Tsai, K.Y.; Juan, C.C.; Tsai, C.W.; Yeh, J.W.; Chang, Y.S. Superior Oxidation Resistance of $(Al_{0.34}Cr_{0.22}Nb_{0.11}Si_{0.11}Ti_{0.22})_{50}N_{50}$ High-Entropy Nitride. *J. Electrochem. Soc.* **2013**, *160*, C531–C535. [CrossRef]
15. Shi, Y.; Yang, B.; Liaw, P. Corrosion-Resistant High-Entropy Alloys: A Review. *Metals* **2017**, *7*, 43. [CrossRef]
16. Chen, P.; Lee, C.; Wang, S.Y.; Seifi, M.; Lewandowski, J.J.; Dahmen, K.A.; Jia, H.; Xie, X.; Chen, B.; Yeh, J.W.; et al. Fatigue behavior of high-entropy alloys: A review. *Sci. China Technol. Sci.* **2017**, *61*, 168–178. [CrossRef]
17. Tsao, T.K.; Yeh, A.C.; Kuo, C.M.; Kakehi, K.; Murakami, H.; Yeh, J.W.; Jian, S.R. The High Temperature Tensile and Creep Behaviors of High Entropy Superalloy. *Sci. Rep.* **2017**, *7*, 12658. [CrossRef] [PubMed]

18. Gludovatz, B.; Hohenwarter, A.; Catoor, D.; Chang, E.H.; George, E.P.; Ritchie, R.O. A fracture-resistant high-entropy alloy for cryogenic applications. *Science* **2014**, *345*, 1153–1158. [CrossRef] [PubMed]
19. Senkov, O.N.; Miracle, D.B.; Chaput, K.J.; Couzinie, J.-P. Development and exploration of refractory high entropy alloys—A review. *J. Mater. Res.* **2018**, 1–37. [CrossRef]
20. Yeh, A.C.; Tsao, T.K.; Chang, Y.J.; Chang, K.C.; Yeh, J.W.; Chiou, M.S.; Jian, S.R.; Kuo, C.M.; Wang, W.R.; Murakami, H. Developing New Type of High Temperature Alloys–High Entropy Superalloys. *Int. J.Metall. Mater. Eng.* **2015**, *1*. [CrossRef]
21. Lu, Y.; Dong, Y.; Guo, S.; Jiang, L.; Kang, H.; Wang, T.; Wen, B.; Wang, Z.; Jie, J.; Cao, Z.; et al. A promising new class of high-temperature alloys: Eutectic high-entropy alloys. *Sci. Rep.* **2014**, *4*, 6200. [CrossRef] [PubMed]
22. Kumar, A.; Gupta, M. An Insight into Evolution of Light Weight High Entropy Alloys: A Review. *Metals* **2016**, *6*, 199. [CrossRef]
23. Senkov, O.N.; Wilks, G.B.; Miracle, D.B.; Chuang, C.P.; Liaw, P.K. Refractory high-entropy alloys. *Intermetallics* **2010**, *18*, 1758–1765. [CrossRef]
24. Senkov, O.N.; Wilks, G.B.; Scott, J.M.; Miracle, D.B. Mechanical properties of $Nb_{25}Mo_{25}$Ta25W_{25} and $V_{20}Nb_{20}Mo_{20}Ta_{20}W_{20}$ refractory high entropy alloys. *Intermetallics* **2011**, *19*, 698–706. [CrossRef]
25. Senkov, O.N.; Scott, J.M.; Senkova, S.V.; Miracle, D.B.; Woodward, C.F. Microstructure and room temperature properties of a high-entropy TaNbHfZrTi alloy. *J. Alloys Compd.* **2011**, *509*, 6043–6048. [CrossRef]
26. Senkov, O.N.; Scott, J.M.; Senkova, S.V.; Meisenkothen, F.; Miracle, D.B.; Woodward, C.F. Microstructure and elevated temperature properties of a refractory TaNbHfZrTi alloy. *J. Mater. Sci.* **2012**, *47*, 4062–4074. [CrossRef]
27. Yurchenko, N.; Stepanov, N.; Tikhonovsky, M.; Salishchev, G. Phase Evolution of the Al*x*NbTiVZr (x = 0; 0.5; 1; 1.5) High Entropy Alloys. *Metals* **2016**, *6*, 298. [CrossRef]
28. Lin, C.M.; Juan, C.C.; Chang, C.H.; Tsai, C.W.; Yeh, J.W. Effect of Al addition on mechanical properties and microstructure of refractory Al_xHfNbTaTiZr alloys. *J. Alloys Compd.* **2015**, *624*, 100–107. [CrossRef]
29. Juan, C.C.; Tseng, K.K.; Hsu, W.L.; Tsai, M.H.; Tsai, C.W.; Lin, C.M.; Chen, S.K.; Lin, S.J.; Yeh, J.W. Solution strengthening of ductile refractory $HfMo_x$NbTaTiZr high-entropy alloys. *Mater. Lett.* **2016**, *175*, 284–287. [CrossRef]
30. Waseem, O.A.; Lee, J.; Lee, H.M.; Ryu, H.J. The effect of Ti on the sintering and mechanical properties of refractory high-entropy alloy Ti_xWTaVCr fabricated via spark plasma sintering for fusion plasma-facing materials. *Mater. Chem. Phys.* **2017**. [CrossRef]
31. Wu, Y.D.; Cai, Y.H.; Chen, X.H.; Wang, T.; Si, J.J.; Wang, L.; Wang, Y.D.; Hui, X.D. Phase composition and solid solution strengthening effect in TiZrNbMoV high-entropy alloys. *Mater. Des.* **2015**, *83*, 651–660. [CrossRef]
32. Yurchenko, N.Y.; Stepanov, N.D.; Zherebtsov, S.V.; Tikhonovsky, M.A.; Salishchev, G.A. Structure and mechanical properties of B2 ordered refractory AlNbTiVZr*x* (x = 0–1.5) high-entropy alloys. *Mater. Sci. Eng. A* **2017**, *704*, 82–90. [CrossRef]
33. Zhang, Y.; Zhou, Y.J.; Lin, J.P.; Chen, G.L.; Liaw, P.K. Solid-Solution Phase Formation Rules for Multi-Component Alloys. *Adv. Eng. Mater.* **2008**, *10*, 534–538. [CrossRef]
34. Yang, X.; Zhang, Y. Prediction of high-entropy stabilized solid-solution in multi-component alloys. *Mater. Chem. Phys.* **2012**, *132*, 233–238. [CrossRef]
35. Boer, F.R.d. *Cohesion in Metals: Transition Metal Alloys*; North-Holland: Amsterdam, The Netherlands, 1988; 758p.
36. Takeuchi, A.; Inoue, A. Classification of bulk metallic glasses by atomic size difference, heat of mixing and period of constituent elements and its application to characterization of the main alloying element. *Mater. Trans.* **2005**, *46*, 2817–2829. [CrossRef]
37. Guo, S.; Ng, C.; Lu, J.; Liu, C.T. Effect of valence electron concentration on stability of fcc or bcc phase in high entropy alloys. *J. Appl. Phys.* **2011**, *109*, 103505. [CrossRef]
38. Poletti, M.G.; Battezzati, L. Electronic and thermodynamic criteria for the occurrence of high entropy alloys in metallic systems. *Acta Mater.* **2014**, *75*, 297–306. [CrossRef]
39. Guo, S.; Liu, C.T. Phase stability in high entropy alloys: Formation of solid-solution phase or amorphous phase. *Prog. Nat. Sci: Mater. Int.* **2011**, *21*, 433–446. [CrossRef]
40. Mann, J.B.; Meek, T.L.; Knight, E.T.; Capitani, J.F.; Allen, L.C. Configuration Energies of the d-Block Elements. *J. Am. Chem. Soc.* **2000**, *122*, 5132–5137. [CrossRef]

41. Yurchenko, N.; Stepanov, N.; Salishchev, G. Laves-phase formation criterion for high-entropy alloys. *Mater. Sci. Technol.* **2016**, *33*, 17–22. [CrossRef]
42. Yao, H.W.; Qiao, J.W.; Hawk, J.A.; Zhou, H.F.; Chen, M.W.; Gao, M.C. Mechanical properties of refractory high-entropy alloys: Experiments and modeling. *J. Alloy. Compd.* **2017**, *696*, 1139–1150. [CrossRef]
43. Juan, C.C.; Tsai, M.H.; Tsai, C.W.; Hsu, W.L.; Lin, C.M.; Chen, S.K.; Lin, S.J.; Yeh, J.W. Simultaneously increasing the strength and ductility of a refractory high-entropy alloy via grain refining. *Mater. Lett.* **2016**, *184*, 200–203. [CrossRef]

entropy

MDPI

Article

A Novel Low-Activation $VCrFeTa_xW_x$ ($x = 0.1, 0.2, 0.3, 0.4$, and 1) High-Entropy Alloys with Excellent Heat-Softening Resistance

Weiran Zhang [1], Peter K. Liaw [2] and Yong Zhang [1,3,*]

1 State Key Laboratory for Advanced Metals and Materials, University of Science and Technology Beijing, Beijing 100083, China; zhangweiranruth@163.com
2 Department of Materials Science and Engineering, The University of Tennessee, Knoxville, TN 37996, USA; pliaw@utk.edu
3 Beijing Advanced Innovation Center for Materials Genome Engineering, University of Science and Technology Beijing, Beijing 100083, China
* Correspondence: drzhangy@ustb.edu.cn

Received: 15 November 2018; Accepted: 6 December 2018; Published: 11 December 2018

Abstract: The microstructure, Vickers hardness, and compressive properties of novel low-activation $VCrFeTa_xW_x$ (x = 0.1, 0.2, 0.3, 0.4, and 1) high-entropy alloys (HEAs) were studied. The alloys were fabricated by vacuum-arc melting and the characteristics of these alloys were explored. The microstructures of all the alloys exhibited a typical morphology of dendritic and eutectic structures. The $VCrFeTa_{0.1}W_{0.1}$ and $VCrFeTa_{0.2}W_{0.2}$ alloys are essentially single phase, consisting of a disordered body-centered-cubic (BCC) phase, whereas the $VCrFeTa_{0.2}W_{0.2}$ alloy contains fine, nanoscale precipitates distributed in the BCC matrix. The lattice parameters and compositions of the identified phases were investigated. The alloys have Vickers hardness values ranging from 546 $HV_{0.2}$ to 1135 $HV_{0.2}$ with the x ranging from 0.1 to 1, respectively. The $VCrFeTa_{0.1}W_{0.1}$ and $VCrFeTa_{0.2}W_{0.2}$ alloys exhibit compressive yield strengths of 1341 MPa and 1742 MPa, with compressive plastic strains of 42.2% and 35.7%, respectively. $VCrFeTa_{0.1}W_{0.1}$ and $VCrFeTa_{0.2}W_{0.2}$ alloys have excellent hardness after annealing for 25 h at 600–1000 °C, and presented compressive yield strength exceeding 1000 MPa with excellent heat-softening resistance at 600–800 °C. By applying the HEA criteria, Ta and W additions into the VCrFeTaW are proposed as a family of candidate materials for fusion reactors and high-temperature structural applications.

Keywords: low-activation high-entropy alloys (HEAs); high-temperature structural alloys; microstructures; compressive properties; heat-softening resistance

1. Introduction

With the rapid development of human civilization, the demand for energy is increasing and the fossil fuel sources are running out. As nuclear energy produces more energy with less pollution, in the long run, nuclear energy will be the next major energy source after fossil fuels, such as coal and oil, to meets human needs [1–3]. However, with fast-growing nuclear power technology, people have higher requirements for the reliability and safety of the nuclear-power [4,5]. The structural materials for new commercial fusion nuclear reactors operate in a harsh environment that is high temperature and chemically reactive, and experiences time-varying stress and intense neutron radiation [4,6], while required to be environmentally-friendly (reduced activation properties). This has motivated worldwide research and development (R&D) on advanced nuclear power systems. Therefore, the exploitation of novel and advanced materials that meet the requirements of these severe conditions will be a key issue for the development of new commercial reactors in the future [4].

Reduced activation or low-activation materials means that the main source of radioactivity after neutron irradiation is short- or medium-lived radioactive elements [7]. The challenge is managing the radioactive waste after shutting down the reactor, and fusion will lose its advantage of being a cleaner energy. Hence, it is important to choose specific materials for the reliable operation of these reactors [8]. The reduced activation ferritic/martensitic (RAFM) steels [9–11], such as Eurofer 97, China low activation martensitic (CLAM), and F82H, are considered to be the original candidate blanket structural materials and/or first wall for future fusion-power devices due to their excellent thermophysical properties, high thermomechanical capabilities, low-activation property, and resistance to neutron irradiation. RAFM steels have been developed using modified (8–12)CrMoVNb type ferritic martensitic steels by replacing Nb, Mo, and Ni with W, Mn, and Ta to achieve the low activation properties [1,7]. However, the operating temperature limit of RAFM steels is currently about 550 °C, which limits the overall thermodynamic efficiency of the power plant [12]. In order to widen the operating temperature window for fusion reactors, several alternative advanced materials options are being pursued. These alternatives include oxide-dispersion-strengthened (ODS) ferritic steels [5], vanadium alloys, and silicon carbide fiber-reinforced silicon carbide matrix composites [13,14]. Existing materials struggle to meet the requirements of fusion reactors operating in extreme environments, such as higher temperatures and stronger neutron irradiation [13,15–17]. Therefore, the first task for the development of fusion energy is to develop high-performance materials. According to the requirements of fusion reactors for short- or medium-lived radioactive materials and existing alloys, such as RAFM steels, ODS steels, and vanadium alloys, we summarize the low activation elements and high activation elements in Table 1.

Table 1. Low activation elements and high activation elements.

Low activation element	Fe	Ti	Cr	V	Ta	Zr	W	Mn	Si	Al	B	C	N	O
High activation element	Nb	Ni	Co	Cu	Mo									

High-entropy alloys (HEAs) are new materials developed in the field of metals in the past decade [18–20]. The term HEAs signifies unconventional alloy systems composed of at least four principal elements, and the atomic percent of each composed element is between 5 at. % (atomic percent) and 35 at. %, which benefits the formation of single-phase solid-solution on the simple underlying face-centered-cubic (FCC), body-centered-cubic (BCC), and hexagonal-close-packing (HCP) structures compared with intermetallics [19,21,22]. HEAs are strongly contrasted with conventional alloys, which are usually based on one or two major elements, and the addition of trace amounts of alloying elements mostly leads to the formation of new phases [4,23]. HEAs can have high hardness [24–27], great creep resistance [28,29], good irradiation resistance [4,6,16,30], good structural stability [28,31–34], and excellent high-temperature strength [28,31,35]. These advantages make HEAs specifically suitable for high-temperature [13] and irradiation applications [14,30]. Zhang et al. [15] reported that $Al_xCoCrFeNi$ (x = 0.1, 0.75, and 1.5) shows great phase stability and swelling resistance under heavy ion irradiation at room temperature to high displacement per atom (dpa). They thought that this trend is due to the severe lattice distortion and sluggish diffusion, which are unique to HEAs. MoNbHfZrTi [36] shows a high compressive yield strength of 1719 MPa and 1575 MPa in as-cast and as-homogenized states at room temperature, respectively. This alloy has high compressive yield strength at elevated temperatures (825 MPa at 800 °C and 187 MPa at 1200 °C) and shows a drop in flow stress after yielding.

The study of HEAs has expanded from the central region to the surroundings of the phase diagram, which means that the research is changing from examining equiatomic single-phase solid-solution alloys to non-equimolar multi-phase solid-solution alloys [9,22,37,38]. According to the concept of HEAs, the reduced activation elements of Fe, Cr, V, W, and Ta were chosen to form several non-equimolar and an equiatomic TaWFeCrV HEAs. However, W is a candidate element for plasma-oriented components in commercial energy-fusion reactors, as it can increase strength and reduce the brittle transition temperature [3,39–41]. Cr reinforces the corrosion resistance of the alloys [5,42,43]. V and Ta can improve creep properties, reduce grain size, and enhance the toughness and strength of the alloys [44,45]. Fe has excellent ductility [46] and is inexpensive. In this study, the V–Cr–Fe–Ta–W alloys were prepared by vacuum-arc melting. Their alloying behaviors, microstructures, and mechanical properties were investigated in detail.

2. Experimental Procedures

Alloy ingots with the nominal composition of $VCrFeTa_xW_x$ (x: molar ratio; x = 0.1, 0.2, 0.3, 0.4, and 1 denoted by T0.1, T0.2, T0.3, T0.4, and T1, respectively) were prepared by vacuum arc melting with high purity elements (the purity of each elements was better than 99.9 wt. %) in a Ti-gettered high-purity argon atmosphere. The nominal chemical compositions of the obtained five alloys are listed in Table 2. These ingots were re-melted at least 5 times in order to achieve compositional homogeneity, and each sample weight was around 50 g. The produced alloys were annealed at 800 °C for 25 h in a high-purity argon furnace and cooled down in water.

Table 2. Nominal compositions of experimental alloys (Nominal compositions, at. %).

Alloy	Identification	Fe	Cr	V	Ta	W
$VCrFeTa_{0.1}W_{0.1}$	T0.1	31.2	31.2	31.2	3.2	3.2
$VCrFeTa_{0.2}W_{0.2}$	T0.2	29.4	29.3	29.3	6	6
$VCrFeTa_{0.3}W_{0.3}$	T0.3	27.8	27.7	27.7	8.4	8.4
$VCrFeTa_{0.4}W_{0.4}$	T0.4	26	26	26	11	11
VCrFeTaW	T1	20	20	20	20	20

The analyzed samples were cut from the middle part of the as-cast alloys to attain flat surfaces for the microstructure study. The exposed surfaces were subsequently ground and polished with the standard polishing process. The crystal structures of as-cast and annealed samples were identified by X-ray diffraction (XRD) (Bruker D8, Karlsruhe, Germany.) with Cu Kα radiation generated at 40 kV and 40 mA and with the scanning angles (2θ) ranging from 30° to 100° at a step of 0.0102° and 10 s dwell time per step. A scanning electron microscope (SEM) (ZEISS SUPRA 55, Jena, Germany.) was used for microstructural analyses, and experimental compositions were analyzed by the energy-dispersive spectroscopy (EDS) (ZEISS SUPRA 55, Jena, Germany.).

Mechanical properties were studied in terms of compression and micro-hardness in air at room temperature. The cylindrical test specimens with a diameter of 3 mm and height of 6 mm were cut from the middle of the ingots for compressive tests, which were performed on a computer-controlled electronic universal testing machine. The high-temperature compressive performances were studied on a Gleeble machine. The heating rate was set to 20 °C/s, and the holding time was five minutes, then air cooled. The initial strain rate for all compressive tests was 10^{-3} s^{-1}. The Vickers micro-hardness tests were carried out with a load of 200 g and a 15 s dwelling time with at least 12 tested points for each test specimen. SEM was used to observe the fracture surfaces of the samples after compressive tests.

3. Results

3.1. Structural Characterization

The XRD patterns of as-cast alloys are shown in Figure 1. Only one BCC1 crystal structure (the Fe-, Cr-, and V-rich phase) had lattice parameters of 0.2935 nm and 0.2937 nm, which were identified in T0.1 and T0.2 alloys in the as-cast state according to Bragg's law, respectively. With the increase in Ta and W content, the structures became rather complex. Both BCC1 and BCC2 (W-rich phase) solid-solution structures, together with Laves (Fe_2Ta-type) phases, were found in these alloys. When the molar ratio of Ta and W increased to 0.3, BCC2 appeared with a lattice parameter of 0.3174 nm. The lattice parameter of the BCC1 phase was 0.2947 nm with a peak width much wider than those of the T0.1 and T0.2 alloys. However, there was a weaker peak of the Laves phase around 47°. The phase composition of T0.4 is similar to T0.3, with the only difference in the strength. The peak (110) for BCC1 is broader than that of the T0.3, and the peak intensity is weaker. The peak of BCC2 phase with the lattice parameter of 0.3178 nm is stronger, both in the peak width and intensity compared with T0.3. Despite this trend, T0.4 is mainly composed of BCC1 with a lattice parameter of 0.2963 nm. When the alloy reaches an equimolar ratio, VCrFeTaW, which is denoted T1, is composed of two BCC phases with lattice parameters of 0.2962 nm for BCC1 and 0 3166 nm for BCC2.

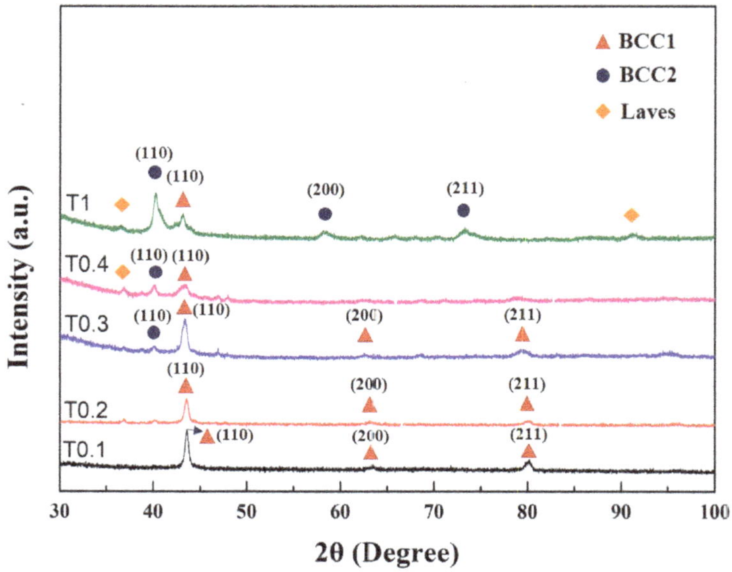

Figure 1. X-ray diffraction (XRD) patterns of the as-cast $VCrFeTa_xW_x$ (x = 0.1, 0.2, 0.3, 0.4, and 1) alloys.

Comparing the X-ray diffraction parameters of each sample, the lattice parameter of BCC1 first rises and then falls, as shown in Figure 1. When the contents of Ta and W reached 0.4 molar, the lattice constants of BCC1 and BCC2 were the largest. The lattice-constant changing trend of BCC2 is similar to that of BCC1. The difference is that the decreasing degree of BCC2 is greater than BCC1. It can be seen from the XRD pattern that the Laves phase has obvious diffraction peaks when x reaches 0.4. The intensity of the diffraction peak of BCC1 decreased significantly but still dominated. The phase composition and lattice constant of each alloy are listed in Table 3. We explain this phenomenon in two ways. Considering the atomic radius, the atomic radii of Ta and W are 1.48 Å and 1.41 Å, respectively, which are larger than the atomic radii of other elements (Fe: 1.24 Å, Cr: 1.25 Å, and V: 1.32 Å). When the contents of Ta and W were less than 0.4, they could be dissolved in the solid solution of BCC1 to a certain degree. Since the atomic radii of Ta and W are larger than those of other elements, the lattice distortion of the solid-solution matrix, BCC1, and the lattice constant of BCC1 increased, which inevitably appeared. Thereby, the solid-solution lattice-strain energy increased. When the content of Ta exceeds 0.4, the solid solubility of the Ta element in the solid-solution matrix of BCC1 has reached saturation. The addition of excess Ta induces the precipitation of the Laves phase. The precipitate of the second phase mitigates the lattice distortion of the solid-solution matrix to a certain extent. Hence, the lattice strain energy of the solid solution is released. An interesting phenomenon can be seen from the XRD pattern: when x is 1, the peak of the Laves phase is not enhanced by the previous analysis, but is slightly weakened. This feature may be related to the phase of BCC2, which is rich in W.

Table 3. Phase compositions of the VCrFeTaxWx alloy system and lattice constants of the solid-solution phase.

Alloy	Phase Composition	Lattice Constant (nm)
$VCrFeTa_{0.1}W_{0.1}$	BCC1	0.2935
$VCrFeTa_{0.2}W_{0.2}$	BCC1	0.2937
$VCrFeTa_{0.3}W_{0.3}$	BCC1	0.2947
	BCC2	0.3174
$VCrFeTa_{0.4}W_{0.4}$	BCC1	0.2963
	BCC2	0.3178
	Laves	-
VCrFeTaW	BCC1	0.2962
	BCC2	0.3166
	Laves	-

When x increased from 0.3 to 0.4, due to the precipitated Laves phase, the lattice constant of BCC2 significantly increased. When x increased to 1, the lattice constant of W drastically reduced, which may be related to the fact that W is a dominant phase, and elements, such as Fe, Cr, and V, having a small atomic radius, are dissolved in W.

3.2. Microstructures and Chemical Compositions

The microstructures of the as-cast and as-polished $VCrFeTa_xW_x$ samples captured using SEM exhibited multiple phases. Figure 2 presents the microstructures of the samples. All the alloys exhibited a typical cast dendrite (DR) structure. The EDS component analyses of the alloys are provided in Table 4. W and V were mainly present in the DR regions, Fe and Ta were concentrated in the inter-dendritic (IR) regions, whereas Cr was more evenly distributed. The distribution of elements could mainly be explained by the mixing enthalpy between elements and the difference in melting points. During the solidification of the alloys, W and V, with higher melting points, first solidified to form DR. The mixing enthalpy (Table 5) between Ta and Fe, −15 KJ/mol, which is the minimum for the selected alloy system, indicates that Ta and Fe combined together more easily than other elemental pairs in the process of solidification due to their great intermetallic compounding ability. Due to the low melting point of Fe, Fe and Ta formed in IR after the formation of DR. Ta and Fe formed a Laves phase in the solid-solution matrix composed of W and V. The microstructures of the alloy system will be described in detail later.

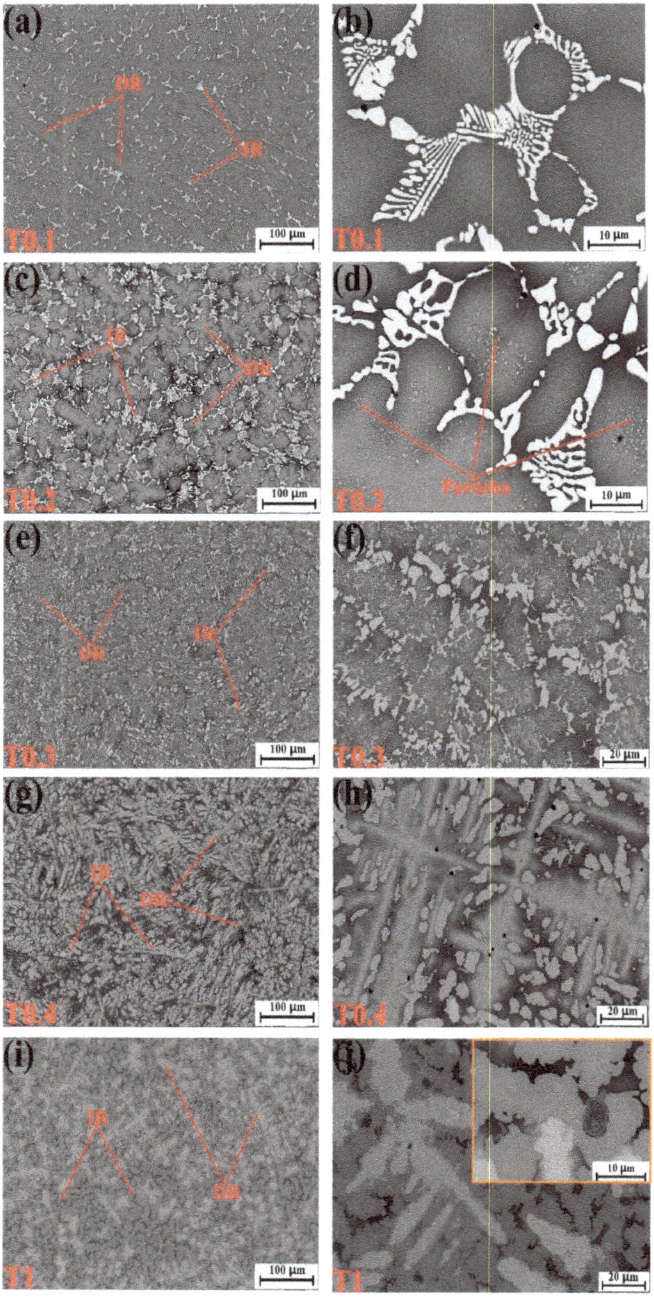

Figure 2. Scanning electron microscope (SEM) backscatter electron images of the as-cast alloys. (**a**,**b**) $VCrFeTa_{0.1}W_{0.1}$; (**c**,**d**) $VCrFeTa_{0.2}W_{0.2}$; (**e**,**f**) $VCrFeTa_{0.3}W_{0.3}$; (**g**,**h**) $VCrFeTa_{0.4}W_{0.4}$; and (**i**,**j**) VCrFeTaW.

Table 4. Chemical compositions in different regions of various alloys by energy-dispersive spectroscopy (EDS) (at. %).

Alloy	Region		V	Cr	Fe	Ta	W
$VCrFeTa_{0.1}W_{0.1}$		Overall	33.29 ± 0.29	30.94 ± 0.17	30.75 ± 0.22	2.54 ± 0.27	2.47 ± 0.02
	IR	White	19.66 ± 0.30	22.26 ± 0.35	37.57 ± 0.52	20.51 ± 0.79	/
	DR	Gray	35.98 ± 0.30	33.32 ± 0.35	24.80 ± 0.52	2.20 ± 0.24	3.70 ± 0.40
$VCrFeTa_{0.2}W_{0.2}$		Overall	31.15 ± 0.21	29.97 ± 0.31	28.58 ± 0.40	5.01 ± 0.59	5.29 ± 0.08
	IR	White	15.98 ± 0.18	23.30 ± 0.21	38.75 ± 0.32	21.96 ± 0.49	/
	DR	Gray	35.57 ± 0.21	33.72 ± 0.19	21.95 ± 0.21	2.76 ± 0.31	6 ± 0.07
	DR	Black	32.41 ± 0.18	26.72 ± 0.16	39.49 ± 0.24	1.39 ± 0.11	/
$VCrFeTa_{0.3}W_{0.3}$		Overall	30.27 ± 0.20	28.54 ± 0.19	28.52 ± 0.52	6.21 ± 0.37	6.36 ± 0.04
	IR	White	10.33 ± 0.25	22.54 ± 0.62	42.89 ± 0.61	24.24 ± 0.14	/
	DR	Gray	24.75 ± 0.28	31.51 ± 0.59	31.36 ± 0.60	/	12.38 ± 0.10
	DR	Black	44.96 ± 0.36	23.82 ± 0.54	25.74 ± 0.65	5.48 ± 0.07	/
$VCrFeTa_{0.4}W_{0.4}$		Overall	27.54 ± 0.20	27.57 ± 0.21	27.19 ± 0.26	9.09 ± 0.43	8.62 ± 0.07
	IR	White	15.19 ± 0.28	23.59 ± 0.66	36.01 ± 0.63	25.20 ± 0.15	/
	DR	Gray	28.49 ± 0.54	26.94 ± 0.66	25.51 ± 0.62	/	19.06 ± 0.13
	DR	Black	28.88 ± 0.28	28.43 ± 0.52	36.17 ± 0.60	/	6.52 ± 0.07
VCrFeTaW		Overall	22.12 ± 0.34	18.99 ± 0.71	20.99 ± 0.65	19.00 ± 0.82	18.90 ± 0.76
	DR	White	13.08 ± 0.50	7.76 ± 0.59	/	/	79.16 ± 0.39
	IR	Gray	14.64 ± 0.30	22.23 ± 0.41	30.96 ± 0.63	32.22 ± 0.18	/
	IR	Black	39.21 ± 0.60	25.32 ± 0.69	35.48 ± 0.76	/	/

Table 5. The formation enthalpies between elements (KJ/mol).

	Fe	Cr	V	Ta	W
Fe	-	−1	−7	−15	0
Cr	-	-	−2	−7	1
V	-	-	-	−1	−1
Ta	-	-	-	-	−7
W	-	-	-	-	-

Figure 2a–d correspond to the microstructures of the T0.1 and T0.2 alloys. From these figures, we found that there is a small amount of the second phase in the white regions in the figures, which is the Laves phase according to the EDS results. The volume fraction of the Laves phase is lower than the XRD detection limit, as no obvious Laves phase peak was observed in the XRD pattern, as presented in Figure 1. It can be seen from the figures that they exhibit a similar DR microstructure. The IR structure of the Laves phase combined with the primary BCC1 phase in the DR region formed eutectic structures and is composed of a network-like appearance. We found that the DR size of T0.2 is less than T0.1, indicating that the addition of Ta and W can effectively reduce the size of the DR crystal grains, which is consistent with the width of the diffraction peak of T0.2, which is wider than T0.1 in the XRD pattern. The eutectic structure regions of the T0.2 alloy are significantly larger than those of the T0.1 alloy. The plate-like Laves phase thickens with increasing Ta content, meaning that the volume fraction of the Laves phase is increasing with the Ta content. This trend is consistent with the appearance of a less pronounced Laves phase around 36° on the XRD curve of T0.2 (Figure 1). From the magnified view of T0.2 in Figure 2b, there is a black area in the middle of the gray and white regions. Based on the EDS results, we found that the area is enriched by V and Fe, which was the BCC1 solid solution. On the gray DR, there are many fine precipitate particles, which is consistent with the increase in the BCC1 lattice constant.

When the content of Ta and W reached 0.3, the microstructure of the alloy still exhibited a typical DR structure (Figure 2e,f). Combined with the EDS results, the DR structure is formed by gray regions of solid-solution phases and precipitated particles of the Laves phase. The Laves phase in the IR region, with the BCC1 primary phase of the black regions in the V-rich area, formed a eutectic structure.

The Laves-phase volume fraction increased significantly, which is consistent with the XRD results (Figure 1).

For the microstructure of the T0.4 alloy, shown in Figure 2g,h, the Laves phase volume fraction increased relative to T0.1, T0.2, and T0.3 alloys shown in Figure 2a–c, respectively. The Laves phases grew into a bar-like shape distributed in the matrix (Figure 2h). Combined with XRD and EDS results, the DR regions consist of a gray W-rich area of BCC2 and a black Fe–Cr–V-rich area of BCC1. The eutectic structure formed by the precipitated Laves peak of the IR structure and the primary BCC1. The distortion of the matrix was released, and the lattice constant of BCC1 reached the maximum owing to the complete precipitation of the Laves phase, which was also observed for BCC2.

The T1 alloy in Figure 2i,j had a similar appearance to T0.4, and the microstructure was a typical DR crystal structure. The white region of DR is distinctly fishbone-like shaped and is enriched in W, whereas a portion of V is a solid dissolved in the W matrix. Due to the smaller atomic radius of the V solid solution in the W matrix, the lattice constant of W decreased, which is consistent with the XRD pattern in Figure 1. The IR structure presents a hypereutectic structure, and the primary phase is not the BCC1 in the black regions, but the Laves phase in the gray regions, whereas, in the highlighted region, Laves and BCC1 phases are present in eutectic structure. These trends indicate that the additions of Ta and W not only influence the phase composition of the alloy system, but also change the microstructure of the alloy system. The addition of Ta promoted regular changes in the microstructures of the BCC1 alloy system, and the transition from the hypoeutectic structure ($x = 0.4$) to hypereutectic structures ($x = 1$) occurred.

3.3. Mechanical Properties

3.3.1. Mechanical Properties at Room Temperature

Figure 3a shows a histogram of the microhardness of the alloys at room temperature and the average microhardness values are listed in Table 6. It can be seen from Figure 3a that the microhardness of the alloy system increases linearly with the increase in the Ta and W contents, from 564 $HV_{0.2}$ of T0.1 to 1135 $HV_{0.2}$ for the equimolar alloy, T1. The hardness enhanced with increasing x due to: (1) the Laves phase belongs to the intermetallic compound, and the hardness is high and (2) the addition of Ta and W caused the lattice distortion of the alloy matrix to increase continuously, and the hardness of the alloy also improved. The relationship between Ta and W content, x, and microhardness (y_{HV}) is fitted in Figure 3b, which can be expressed as $y_{HV} = 553.1 + 609.2x$, where y is the microhardness of the alloy. The linear correlation coefficient, R, is 0.942, meaning that relationship is linear.

Table 6. Mechanical properties of the as-cast VCrFeTaxWx (x = 0.1, 0.2, 0.3, 0.4, and 1) alloys.

Alloy	Vickers Hardness ($HV_{0.2}$)	$\sigma_{0.2}$ (MPa)	σ_{bc} (MPa)	ε_p (%)
$VCrFeTa_{0.1}W_{0.1}$	564	1341	2917	42.2 (Not broken)
$VCrFeTa_{0.2}W_{0.2}$	673	1742	3265	35.7 (Not broken)
$VCrFeTa_{0.3}W_{0.3}$	726	/	701	/
$VCrFeTa_{0.4}W_{0.4}$	886	1580	1767	/
VCrFeTaW	1135	/	1501	/

$\sigma_{0.2}$: yield strength; σ_{bc}: ultimate compressive strength; and ε_p: plastic-strain limit.

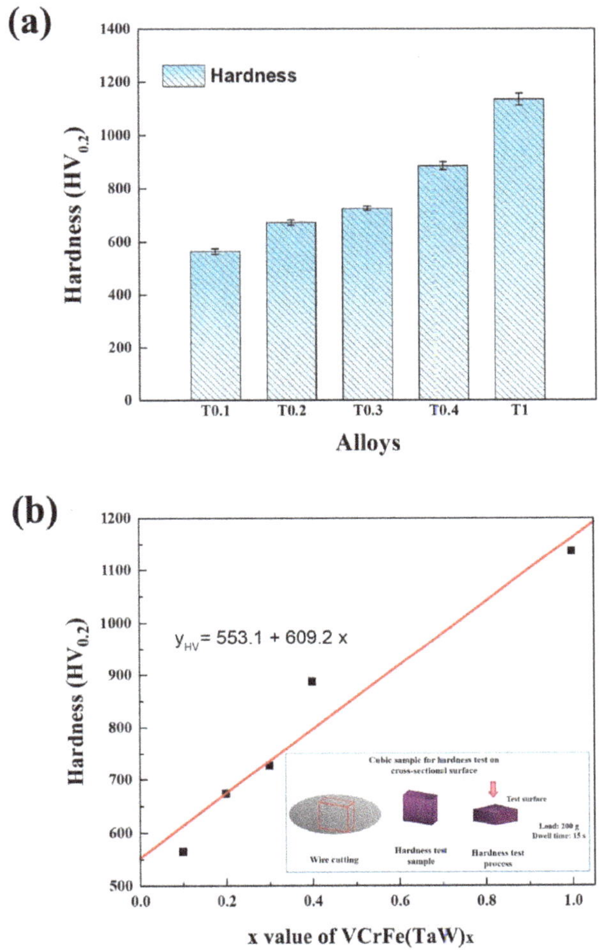

Figure 3. (**a**) Hardness histogram and standard deviation of the as-cast $VCrFeTa_xW_x$ (x = 0.1, 0.2, 0.3, 0.4, and 1) alloys at room temperature; (**b**) The hardness curve of the as-cast $VCrFeTa_xW_x$ (x = 0.1, 0.2, 0.3, 0.4, and 1) alloys as a function of the Ta and W content, and hardness test process.

The compressive engineering stress-strain curves are shown in Figure 4, and the compressive properties, such as yield strength $\sigma_{0.2}$, fracture strength σ_{bc}, and plastic strain limit ε_p, are summarized in Table 6. Evidentially, the Ta and W content had a very pronounced effect on the compressive behavior of the alloys. It can be seen that the T0.1 and T0.2 alloys showed excellent compressive properties, with yield strength, fracture strength, and plastic strain values in T0.1 and T0.2 of 1341 MPa, 2917 MPa, and 42.2%; and 1742 MPa, 3265 MPa, and 35.7%, respectively. Compared with the compressive properties of T0.1 and T0.2, both the yield strength and fracture strength increased, which is a trade-off with plastic strain. This trend is mainly due to the presence of a large amount of dispersed fine precipitates in the T0.2 alloy and an increase in the lattice distortion, resulting in an increase in the strength of the alloy. The plastic strain decreased due to the increase in the volume fraction of the second phase. It can be seen that the compressive properties substantially decrease with the Ta and W content changing from 0.3 to 1 mainly because of the increase in the volume fraction and size of the Laves phase as the Ta content increases. The compressive performance of T0.3 is worse than T0.4 and T1, probably due to the precipitated particles growing in the matrix.

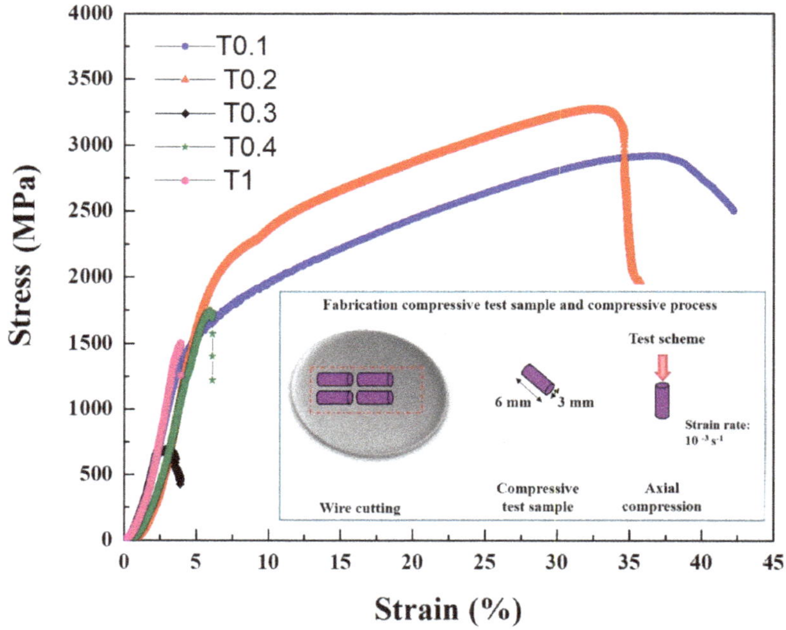

Figure 4. The compressive stress-strain curves of the as-cast VCrFeTa_xW_x (x = 0.1, 0.2, 0.3, 0.4, and 1) alloys at room temperature with a diameter of 3 mm.

The fracture-surface morphologies of the as-cast alloys after compressive tests at room temperature are depicted in Figure 5. Figure 5a,b show the side sections of T0.1 and T0.2 alloys, respectively. It can be seen from the figures that the alloys are not crushed, and the macroscopic appearance are waist-drum shaped. The picture of the T0.1 alloy in Figure 5a shows that the fracture surface presents severe plastic deformation, indicating its superior room-temperature plasticity and exhibiting a certain plasticity fracture. In Figure 5b, the T0.2 alloy exhibits a fracture angle of nearly 45°, and approximately parallel sliding steps of unevenness can be observed in the fracture profile. The fracture surface of the T0.3 alloy is shown in Figure 5c, which has a distinct cleavage step due to the cleavage fracture. T0.4 and T1 alloys have similar fracture patterns, and the fracture surfaces are relatively flat, demonstrating obvious tear and river patterns, which are typical quasi-cleavage fractures, in Figure 5d,e, respectively. This trend is consistent with the compressive strength of T0.4 and T1 being better than T0.3.

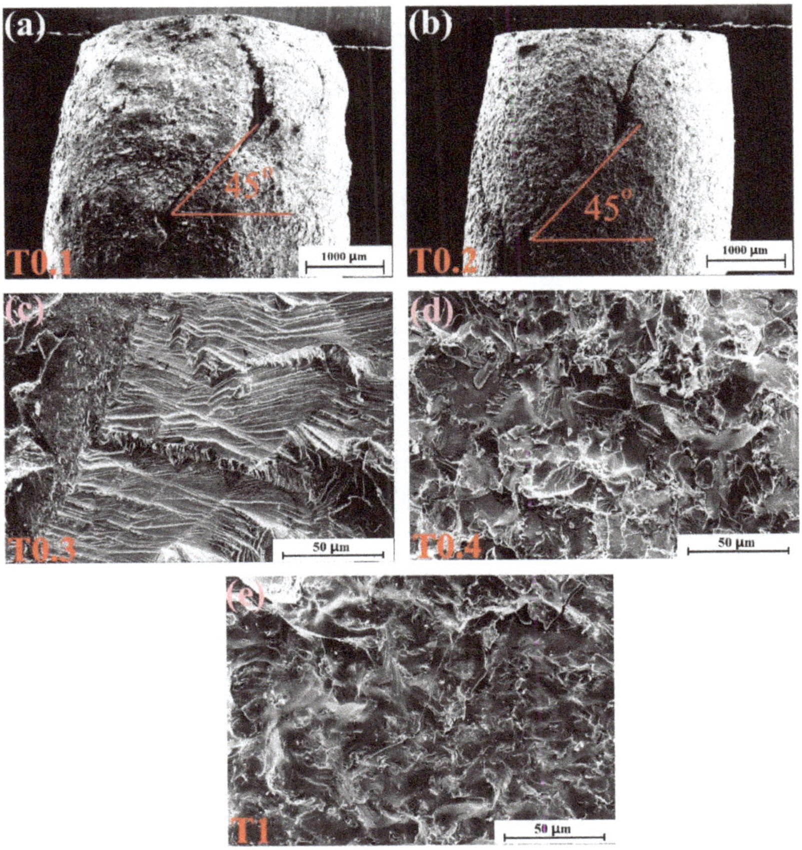

Figure 5. SEM micrographs of the fracture surfaces of VCrFeTa$_x$W$_x$ (x = 0.1, 0.2, 0.3, 0.4, and 1) alloys at room temperature: (**a**) VCrFeTa$_{0.1}$W$_{0.1}$; (**b**) VCrFeTa$_{0.2}$W$_{0.2}$; (**c**) VCrFeTa$_{0.3}$W$_{0.3}$; (**d**) VCrFeTa$_{0.4}$W$_{0.4}$; and (**e**) VCrFeTaW.

3.3.2. Mechanical Properties at High-Temperature

Since T0.1 and T0.2 alloys exhibit excellent room-temperature strength, we also examined their high-temperature properties. Annealing at 600 °C, 800 °C, and 1000 °C for 25 h did not cause alloy softening. The hardness of the annealed T0.1 alloys are 605 HV$_{0.2}$/ 600 °C and 621 HV$_{0.2}$/800 °C, and T0.2 alloys are 721 HV$_{0.2}$/600 °C and 762 HV$_{0.2}$/800 °C, respectively, obviously higher than those of as-cast alloys presented in Table 7, indicating that the T0.1 and T0.2 alloys have the great softening resistance. Table 7 and Figure 6 show the compressive engineering stress-strain curves of T0.1 and T0.2 alloys from room temperature to 1000 °C. The strength increased and the ductility decreased with the yield strength at 600 °C. The strength decreased and the ductility increased at 800 °C. At 1000 °C, the strength of T0.1 and T0.2 alloys dropped significantly, indicating a certain degree of softening, which is also reflected in the hardness at this temperature listed in Table 7. Surprisingly, the yield strength, fracture strength, and plastic strain at 800 °C for T0.1 and T0.2 are 1019 MPa, 1289 MPa, and 50% and 1033 MPa, 1260 MPa, and 40.6%, respectively.

Table 7. Mechanical properties of T0.1 and T0.2 alloys at 600–1000 °C.

Alloy	Temperature (°C)	$\sigma_{0.2}$(MPa)	σ_{bc} MPa	ε_p (%)	Vickers Hardness ($HV_{0.2}$)
$VCrFeTa_{0.1}W_{0.1}$	600	1234	2158	28.1	605
	800	1019	1289	> 50	621
	1000	371	421	> 50	538
$VCrFeTa_{0.2}W_{0.2}$	600	1657	2316	22.6	721
	800	1033	1260	40.6	762
	1000	182	253	> 50	665

Note: $\sigma_{0.2}$: yield strength, σ_{bc}: ultimate compressive strength, and ε_p: plastic-strain limit.

(a)

(b)

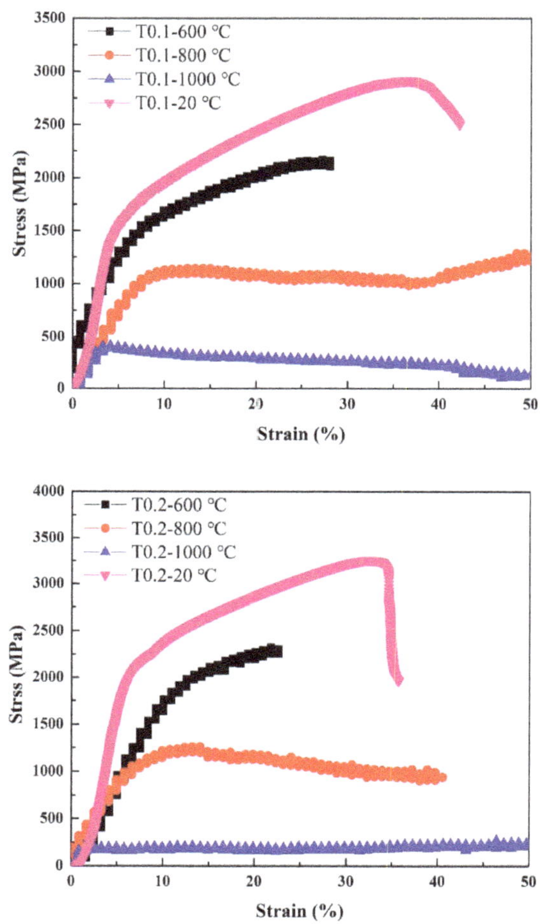

Figure 6. The compressive stress-strain curves of (**a**) $VCrFeTa_{0.1}W_{0.1}$ and (**b**) $VCrFeTa_{0.2}W_{0.2}$ alloys at different temperatures with a diameter of 3 mm.

4. Discussion

4.1. Phase Selection

From the results, the phase structures of the $VCrFeTa_xW_x$ HEAs are not as simple as presented in the XRD patterns in Figure 1. The various criteria were analyzed in order to further understand the phase formation of the alloy in this system.

Studies have shown that HEAs are prone to form solid solutions with an FCC structure or a BCC [19,21,47]. Hume-Rothery rules [21,48,49] govern the criteria for the formation of solid solutions in the binary alloy system, which include the crystal structure factor, atomic-size factor, valence-electron-concentration factor, and chemical electronegativity factor. As a special kind of the solid-solution alloy, HEAs have many components, complicating distinguishing solutes or solvents. Hence, it is difficult to study them using traditional methods [22,50]. Recent investigations extended the Hume-Rothery rules for explaining the criteria for the formation of the solid-solution structure in the HEA area with the aid of empirical relationships.

Zhang et al. [48,51] proposed three parameters affecting the formation of the HEA solid-solution phase: atomic-size difference (delta, δ), mixing enthalpy (ΔH_{mix}), and mixing entropy (ΔS_{mix}), to predict the phase formation in HEAs, amorphous metallic glasses, and intermetallic compounds. These calculation methods are detailed as follows:

$$\Delta H_{mix} = \sum_{i=1,i\neq j}^{n} \Omega_{ij} c_i c_j$$

$$\Omega_{ij} = 4\Delta H_{AB}^{mix}$$

$$\Delta S_{mix} = k\ln w = -R\sum_{i=1}^{n}(c_i \ln c_j)$$

$$\delta = \sqrt{\sum_{i=1}^{n} c_i (r_i - \bar{r})^2}$$

$$\bar{r} = \sum_{i=1}^{n} c_i r_i$$

where n is the number of the involved elements in an alloy, ΔH_{AB}^{mix} is the mixing of the enthalpy of binary equiatomic AB alloys, Ω_{ij} is the regular melt-inter-action parameter between the i-th and j-th elements, R is the gas constant, c_i and r_i are the atomic percentage and atomic radius of the i-th element, respectively, and $\bar{r}$ is the average atomic radius. They concluded that solid solutions tend to form in the region delineated by $\delta > 6.6\%$, -15 KJ/mol $\leq \Delta H_{mix} \leq 5$ KJ/mol, and 11 J/(K·mol) $\leq \Delta S_{mix}$ $\leq$16.5 J/(K·mol).

To be better understand the criteria of HEAs, a new parameter Ω [21,51] was proposed to correlate the relative contribution of the change in ΔH_{mix} and ΔS_{mix}, expressed as:

$$\Omega = \frac{T_m \Delta S_{mix}}{|\Delta H_{mix}|}$$

$$T_m = \sum_{i=1}^{n} c_i (T_m)_i$$

where $(T_m)_i$ is the melting temperature of the i-th component. By analyzing the phase formation using the parameters, Ω and δ, of various reported multicomponent alloys, new parameters for forming solid-solution phases in HEAs were suggested [48]: $\Omega \geq 1.1$ and $\delta \leq 6.6\%$.

HEAs often form solid solutions of an FCC or BCC phase, and the above criteria can effectively predict whether the alloy can form a solid-solution structure, but it is impossible to predict whether the solid-solution structure of the alloy is an FCC or BCC phase. Guo et al. [52] proposed the relationship between the valence-electron concentration (VEC) and solid-solution stability. The VEC in a multi-component system is:

$$\mathrm{VEC} = \sum_{i=1}^{n} c_i(\mathrm{VEC})_i$$

where $(\mathrm{VEC})_i$ is the VEC of the *i*-th element. From published experimental results [37,52], the limitation that they suggested were: FCC phases occur at VEC $\geq$ 8.0, BCC phases at VEC < 6.87, and a mixture of FCC and BCC phases at 6.87 $\leq$ VEC < 8.

For the $VCrFeTa_xW_x$ alloys in the present work, we studied the phase-formation rules according to the above parameters. The specific results are shown in Table 8. It can be seen from the table that as the contents of Ta and W increase, the δ of the alloy system grows from 3.59% to 5.41%, as plotted in Figure 7a. This indicates that the degree of the lattice distortion caused by the atomic arrangement is increasing, which is consistent with the lattice constants of the two phases when x is between 0.1 and 0.4 (the lattice constant decreases in the case of the equimolar HEA, which is related to the Laves phase being the primary phase, and V is distributed into a solid-solution with W). Despite the change in δ and ΔH_{mix} with the increase in the Ta and W content, their values meet the requirement for forming solid solutions, as plotted in Figure 7a,b, indicating that the mixing entropy effect of the alloy is stronger than that of the mixing enthalpy. The tendency of the alloy to form solid-solutions is improved, indicating that the effect of the mixing of entropy on the solid solution formation strengthens. The VEC in the studied alloys is shown in Figure 7c. With increasing Ta and W addition, the value of VEC decreased from 6.28 to 6, which meets the BCC-forming requirement (6.87 $\leq$ VEC < 8), demonstrating that the BCC phase is stable in the $VCrFeTa_xW_x$ alloy system.

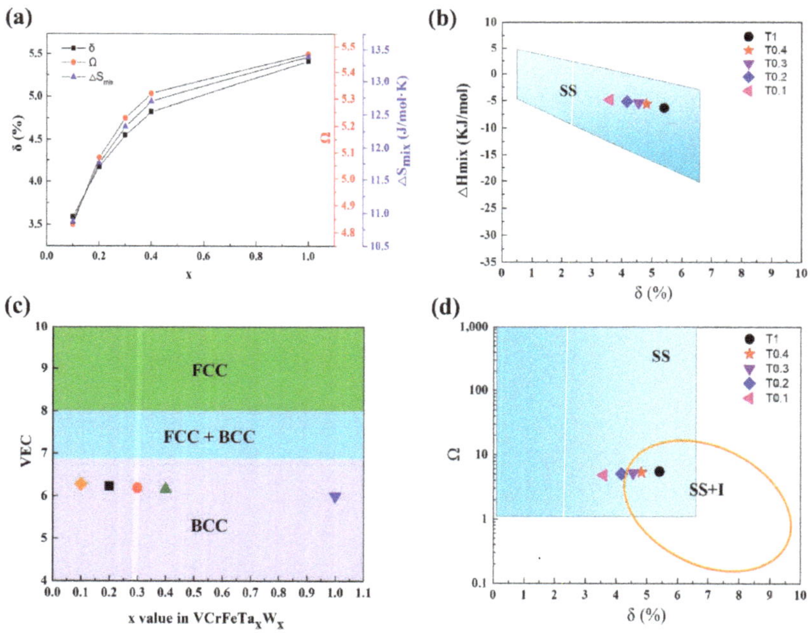

Figure 7. (**a**) The curves of δ, Ω, and ΔS_{mix} as a function of the Ta and W content for $VCrFeTa_xW_x$ (x = 0.1, 0.2, 0.3, 0.4, and 1) alloys; (**b**) The relationship between parameters δ and ΔH_{mix} for the as-cast $VCrFeTa_xW_x$; (**c**) The relationship between the valence-electron concentration (VEC), and Ta and W content of $VCrFeTa_xW_x$ alloys; (**d**) The relationship between the parameters δ and Ω, for the as-cast $VCrFeTa_xW_x$. (SS: Solid-Solutions; I: Intermetallics compound; SS + I: Solid-Solutions + Intermetallics compound).

Table 8. ΔH_{mix}, ΔS_{mix}, Ω, δ, valence-electron concentration (VEC), the theoretical density, and the melting points for $VCrFeTa_xW_x$ (x = 0.1, 0.2, 0.3, 0.4, and 1) alloys.

Alloy	δ	ΔH_{mix} (KJ/mol)	ΔS_{mix} (J/mol·K)	Ω	VEC	ρ_{theor} (g/cm^3)	T_m (K)
$VCrFeTa_{0.1}W_{0.1}$	3.59	−4.83	10.87	4.83	6.28	7.85	2147.7
$VCrFeTa_{0.2}W_{0.2}$	4.17	−5.15	11.75	5.08	6.23	8.58	2227.3
$VCrFeTa_{0.3}W_{0.3}$	4.55	−5.41	12.32	5.23	6.19	9.19	2297.0
$VCrFeTa_{0.4}W_{0.4}$	4.82	−5.63	12.70	5.32	6.16	9.72	2360.1
VCrFeTaW	5.41	−6.4	13.38	5.47	6	11.81	2631.8

With the increase in the Ta and W content, although both ΔS_{mix} and Ω ($\geq$ 1.1) increased (Figure 7a), from Figure 7d, T0.3, T0.4, and T1 HEAs are in the SS + I (SS: Solid-solution, I: Intermetallics compound) region. The mixing of the enthalpy promotes the formation of intermetallic compounds, whereas the Ta–Fe binary system has the most negative mixing of enthalpy (−15 kJ/mol) listed in Table 5, indicating that the bonding force between these two elements is the strongest. The Fe_2Ta phase formation occurs, which is consistent with the XRD patterns in Figure 1.

4.2. Ta and W Effects at Room Temperature

The microstructure analysis of the $VCrFeTa_xW_x$ alloys with different Ta and W contents, performed in the current research, revealed several features. First, an increase in the Ta and W content substantially decreased the BCC1 matrix phase of the alloys (Figure 1). Conversely, the volume fraction of BCC2 and Laves phases increase. Second, an addition of Ta and W resulted in the formation of intermetallic phases, namely the Laves phase of Fe_2Ta-type. The Laves phase can be associated with a highly negative-enthalpy of the intermetallic-phases formation [19,37] (Table 6). Third, with increasing Ta and W contents, the change of precipitation was the most obvious and most significant phenomenon. When x = 0.1, the Laves precipitation phase has a lamellate shape and forms an eutectic structure with the matrix. When x = 0.2, the Laves-precipitation phase is thicker, and still forms an eutectic structure with the matrix, and fine particles, most of which are the Laves phase precipitated on the matrix. When x = 0.3–1, the precipitation phase gradually aggregates and grows. These results indicate that the addition of Ta and W not only changes the phase composition of the alloy system, but also varies the microstructure of the alloy system. However, the addition of Ta promoted the regular Laves phase microstructure change of the BCC1 alloy system, and a transition from the hypoeutectic structure ($x \leq 0.4$) to hypereutectic structure (x = 1) occurred. This indicates that Ta can promote the eutectic transformation of the Fe–Cr–V alloy system.

The Ta and W contents have a prominent effect on the mechanical properties of the $VCrFeTa_xW_x$ alloys, as expected given the pronounced changes in the microstructure. With the addition of Ta and W, the hardness of the alloy increases linearly, while the plasticity and strength of the studied alloys exhibited a complex dependence. In the HEA solid-solution phase, various atoms randomly occupy the lattice position of the crystal. Each atom is surrounded by other kinds of atoms, and all atoms can be regarded as solute atoms or solvent atoms [37,53]. In addition, the types of atoms vary in size, causing severe lattice distortion in the solid solution, which in turn leads to high solid-solution strengthening [46,54,55], thereby increasing the strength and hardness of the alloy, especially the HEAs of the BCC structure [24,56–58]. Conversely, W obviously promotes the formation of BCC2, and Ta prefers forming Laves phases. The mixing enthalpy between Ta and Fe is −15 kJ/mol (Table 5), which is the lowest for the $VCrFeTa_xW_x$ system, meaning that Ta and Fe combine together quite easily compared with the other element pairs during solidification owing to their great compatibility. The hardness values of the $VCrFeTa_xW_x$ alloys increase with increasing x. As the content of Ta and W increased, the yield strength increased from T0.1 to T0.2. However, in the $VCrFeTa_xW_x$ (x = 0.3, 0.4, and 1) alloys, the yield strength deteriorated quickly with an increase in x (Figure 4). This phenomenon can be associated with a large volume fraction of Laves phases, and the size of Laves particles increases with the change in x. Notably, the strengthening contribution of the precipitation particles to the BCC phase alloys was prominent, which has been previously studied [59]. In the present study, the fine second phase particles of the Laves phase even had a positive effect on the strength, as the T0.2 alloy has an excellent yield strength and plasticity.

As engineering materials, the yield strength of HEAs is an important parameter for the design of a component. The dependence of the room-temperature compressive yield strength of the reported HEAs [13,24,29,56–72] and the studied alloys in the present work are plotted in Figure 8. The T0.1 and T0.2 HEAs possess relatively-high yield strength, whereas they all exhibit excellent compressive strains. T0.1 and T0.2 have a yield strength of 1341 MPa and 1742 MPa with plastic strains of 42.2% and 35.7%, respectively. They exhibit excellent plasticity, as they could not be broken in the compressive tests. The refractory HEAs reported in papers exhibit high yield strength, while their plastic strains are lower than 15%; or the reported refractory HEAs show remarkable plastic strains more than 50%, while their yield strengths are not remarkable. If we set the coordinates of the alloys in Figure 8 at (10%, 1600 MPa), only the T0.2 alloy is in the first quadrant. Therefore, we concluded that the BCC HEAs developed so far still have insufficient compressive strain at room temperature, or exhibit low strength. It is noteworthy that T0.2 studied in the current work not only possesses a compressive strain up to 35.7%, but also has a yield strength as high as 1742 MPa. In other words, T0.2 may have excellent ductility, which makes this kind of HEA better than other HEAs for engineering applications. More significantly, the T0.2 alloy, as a low-activation HEA, has obvious performance advantages over the previously-reported low-activation alloys, indicating that it has potential as a candidate material for fusion reactors.

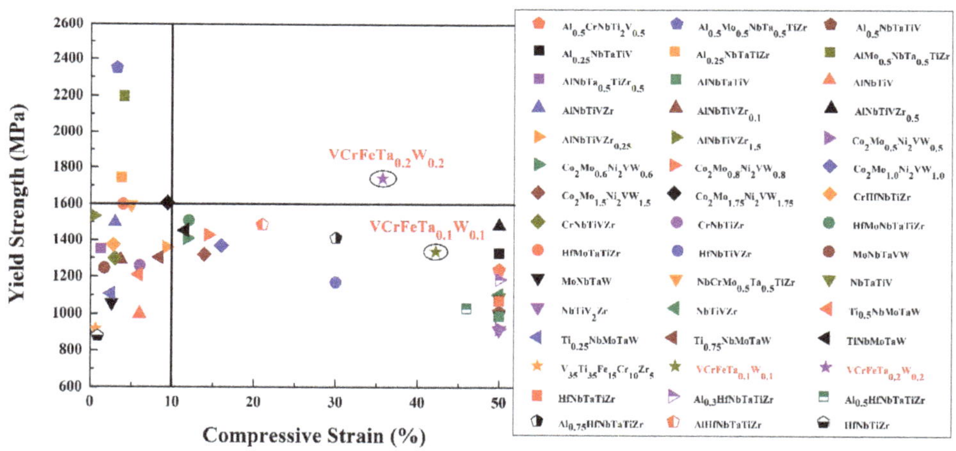

Figure 8. The map of compressive yield strength and ductility combinations of various refractory high-entropy alloys [13,24,29,56–72] at room temperature. Initial strain rates range from 10^{-4} to 10^{-3} s^{-1}.

4.3. Heat-Softening Resistance

Typically, annealing leads to alloy softening due to the rapid diffusion of atoms, resulting in internal stress relaxation caused by alloy defects at high temperatures [28,29]. However, due to the high mixing entropy and the obvious sluggish diffusion effect, HEAs have good high-temperature stability and great resistance to high temperature softening [28,35,37]. According to the Gibbs free energy formula, with the increase in temperature, the effect of entropy is more obvious than that of enthalpy [21]. The high mixing-entropy increases the solid solubility of the alloy, which is conducive to the formation of the solid-solution structure. Because of the sluggish diffusion effect, the alloy easily forms a supersaturated solid solution and a fine precipitated phase. Annealing at elevated temperatures can partially release the internal stress and dissolve atoms. For the studied alloys in the present work, the increase in hardness after annealing from 600 °C to 800 °C is due to solid-solution strengthening. Although the microhardness of the T0.1 and T0.2 alloys decreased at 1000 °C, the reduction was not large compared to the hardness values at room temperature. For traditional alloys, tempering after quenching tends to result in significant softening. However, HEAs have significant advantages in this regard. HEAs, such as NbMoTaW [56], NbMoTaWV [56], and AlNbTiV [68], with a typical BCC structure have been reported to show that, although the alloys have poor compressive strains at room temperature, the plasticity of the alloys increases as the temperature rises. As shown in Figure 9, from room temperature to 800 °C, the yield strength of the T0.1 and T0.2 alloys are significantly higher than those of superalloys, such as Inconel 718 and Haynes 230. Although the strength of MoNbHfZrTi is higher than those of the T0.1 and T0.2 alloys at room temperature, its compression plasticity is only 10.1%, far lower than the T0.1 and T0.2 alloys. However, the strengths of T0.1 and T0.2 alloys exceed the reported high-entropy alloys include MoNbHfZrTi at 600–800 °C [36,68]. Thereby, the HEAs exhibit excellent resistance to high-temperature softening. As an alternative alloy for future fusion reactors with the low activation, the target alloys in the present work have certain advantages at high temperatures compared to CLAM, ODS, and V–4Cr–4Ti, as presented in Figure 9.

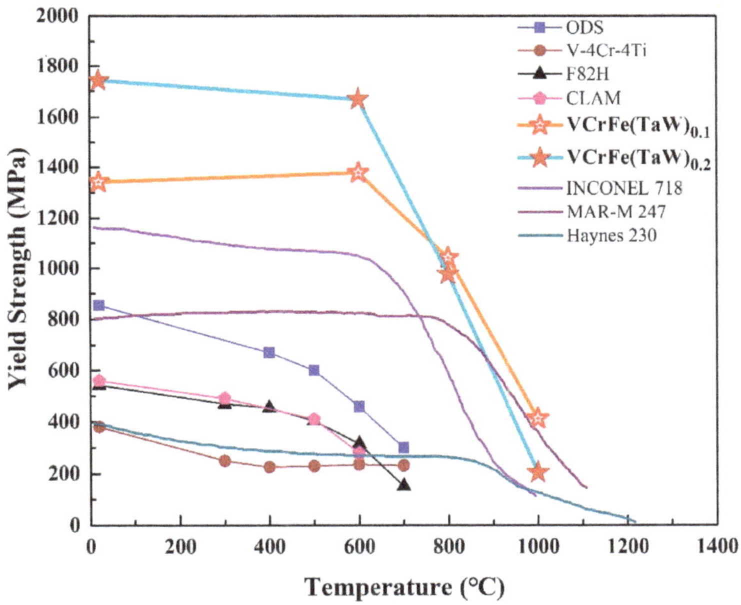

Figure 9. Temperature dependence of the compressive yield strength of superalloys [37], reported HEAs [36,68], and low activation alloys [13].

5. Summary

Novel low-activation HEAs $VCrFeTa_xW_x$ ($x = 0.1, 0.2, 0.3, 0.4$, and 1) were fabricated by vacuum arc melting, and their microstructure and mechanical properties were studied in the present work. The investigated alloys exhibited a relatively simple microstructure and promising properties. Based on the obtained results and discussions, our conclusions are as follows:

(1) The microstructures of all investigated alloys exhibited a typical dendritic and eutectic structure, with $VCrFeTa_{0.1}W_{0.1}$ and $VCrFeTa_{0.2}W_{0.2}$ presenting a mainly BCC1 (a VCrFe-rich region) solid solution and Laves phases (an Fe_2Ta-type). $VCrFeTa_{0.3}W_{0.3}$, $VCrFeTa_{0.4}W_{0.4}$, and VCrFeTaW contain BCC1 and BCC2 (W-rich region) solid solutions and Laves phases.

(2) The Vickers hardness of the alloys increased with increasing Ta and W contents. The hardness values of $VCrFeTa_{0.1}W_{0.1}$, $VCrFeTa_{0.2}W_{0.2}$, $VCrFeTa_{0.3}W_{0.3}$, $VCrFeTa_{0.4}W_{0.4}$, and VCrFeTaW are 546 $HV_{0.2}$, 673 $HV_{0.2}$, 726 $HV_{0.2}$, 886 $HV_{0.2}$, and 1135 $HV_{0.2}$, respectively. This feature is attributed to the solid-solution strengthening and the increased amount of Laves and BCC2 phases.

(3) The augmented Ta and W contents increased the compressive strength but decreased the plastic strain of T0.1 and T0.2 alloys. The T0.1 and T0.2 alloys exhibited compressive yield strengths of 1341 MPa and 1742 MPa, with plastic strains of 42.2% and 35.7%, respectively. The solid-solution strengthening of the BCC matrix and the formation of hard Laves phases precipitated in particles are two main factors contributing to alloy strengthening. With the addition of Ta and W, the compressive performance deteriorated sharply due to the increase in the volume fraction and the growth of the Laves phase.

(4) By applying the atomic ratios strategy and the criteria for disordered solid solutions, the alloy system of $VCrFeTa_{0.1}W_{0.1}$, $VCrFeTa_{0.2}W_{0.2}$, $VCrFeTa_{0.3}W_{0.3}$, $VCrFeTa_{0.4}W_{0.4}$, and VCrFeTaW fully met the VEC, δ-Ω, and δ-$\triangle H_{mix}$ criteria. Studies have shown that HEAs maintain stable phase structures and properties after high-temperature annealing. This means that the design of high-temperature structural materials using the HEA concept could promote the development of these alloys for use in extreme high-temperature extreme such as blades, engines, aerospace, fusion reactors and other applications.

(5) The high-temperature mechanical properties of T0.1 and T0.2 alloys were examined. After annealing 25 h at 600–1000 °C, the T0.1 and T0.2 alloys maintained high hardness. The compressive yield strengths of the T0.1 and T0.2 alloys are promising with heat-softening resistance at 600–800 °C. The yield strengths of T0.1 and T0.2 alloys were much higher than 1000 MPa at 800 °C. This study provides guidance for further development of the current concepts to produce refractory high-temperature structural materials, which are candidate alloys for fusion reactors.

Author Contributions: W.Z. prepared the high-entropy alloys and finished the manuscript. Y.Z. and P.K.L. offered the theoretical guidance. All authors contributed to the general discussion.

Funding: Y.Z. would like to thank the financial supports from the National Science Foundation of China (NSFC, Granted Nos. 51471025 and 51671020). P.K.L. would like to acknowledge the National Science Foundation (DMR-1611180 and 1809640), and Department of Energy (DOE) Office of Fossil Energy, National Energy Technology Laboratory (NETL) (DE-FE-0011194), and the U.S. Army Office Project (W911NF-13-1-0438).

Conflicts of Interest: The authors declare no conflict of interest.

References

1. Knaster, J.; Moeslang, A.; Muroga, T. Materials research for fusion. *Nat. Phys.* **2016**, *12*, 424. [CrossRef]
2. Zinkle, S.J.; Was, G. Materials challenges in nuclear energy. *Acta Mater.* **2013**, *61*, 735–758. [CrossRef]
3. Kanpara, S.; Khirwadkar, S.; Belsare, S.; Bhope, K.; Swamy, R.; Patil, Y.; Mokariya, P.; Patel. N.; Patel, T.; Galodiya, K. Fabrication of tungsten & tungsten alloy and its high heat load testing for fusion applications. *Mater. Today Proc.* **2016**, *3*, 3055–3063.
4. Yan, Z.; Liu, S.; Xia, S.; Zhang, Y.; Wang, Y.; Yang, T. He behavior in Ni and Ni-based equiatomic solid solution alloy. *J. Nuclear Mater.* **2018**, *505*, 200–206. [CrossRef]
5. Zhang, G.; Zhou, Z.; Mo, K.; Wang, P.; Miao, Y.; Li, S.; Wang, M.; Liu, X.; Gong, M.; Almer, J. The microstructure and mechanical properties of Al-containing 9Cr ODS ferritic alloy. *J. Alloys Compd.* **2015**, *648*, 223–228. [CrossRef]
6. Jin, K.; Lu, C.; Wang, L.M.; Qu, J.; Weber, W.J.; Zhang, Y.; Bei, H. Effects of compositional complexity on the ion-irradiation induced swelling and hardening in Ni-containing equiatomic alloys. *Scr. Mater.* **2016**, *119*, 65–70. [CrossRef]
7. Tan, L.; Katoh, Y.; Tavassoli, A.A.F.; Henry, J.; Rieth, M.; Sakasegawa, H.; Tanigawa, H.; Huang, Q. Recent status and improvement of reduced-activation ferritic-martensitic steels for high-temperature service. *J. Nuclear Mater.* **2016**, *479*, 515–523. [CrossRef]
8. Tan, L.; Snead, L.L.; Katoh, Y. Development of new generation reduced activation ferritic-martensitic steels for advanced fusion reactors. *J. Nuclear Mater.* **2016**, *478*, 42–49. [CrossRef]

9. Huang, Q.; Team, F. Development status of clam steel for fusion application. *J. Nuclear Mater.* **2014**, *455*, 649–654. [CrossRef]
10. Abdou, M.; Morley, N.B.; Smolentsev, S.; Ying, A.; Malang, S.; Rowcliffe, A.; Ulrickson, M. Blanket/first wall challenges and required R&D on the pathway to DEMO. *Fus. Eng. Des.* **2015**, *100*, 2–43.
11. Alimov, V.K.; Hatano, Y.; Yoshida, N.; Watanabe, H.; Oyaidzu, M.; Tokitani, M.; Hayashi, T. Surface modification and sputtering erosion of reduced activation ferritic martensitic steel F82H exposed to low-energy, high flux deuterium plasma. *Nuclear Mater. Energy* **2016**, *7*, 25–32. [CrossRef]
12. Emmerich, T.; Qu, D.; Vaßen, R.; Aktaa, J. Development of W-coating with functionally graded W/EUROFER-layers for protection of First-Wall materials. *Fus. Eng. Des.* **2018**, *128*, 58–67. [CrossRef]
13. Xian, X.; Zhong, Z.; Zhang, B.; Song, K.; Chen, C.; Wang, S.; Cheng, J.; Wu, Y. A high-entropy V35TI35FE15Cr10Zr5 alloy with excellent high-temperature strength. *Mater. Des.* **2017**, *121*, 229–236. [CrossRef]
14. Xia, S.Q.; Yang, X.; Yang, T.F.; Liu, S.; Zhang, Y. Irradiation resistance in AlxCoCrFeNi high entropy alloys. *JOM* **2015**, *67*, 2340–2344. [CrossRef]
15. Xia, S.Q.; Gao, M.C.; Yang, T.F.; Liaw, P.K.; Zhang, Y. Phase stability and microstructures of high entropy alloys ion irradiated to high doses. *J. Nuclear Mater.* **2016**, *480*, 100–108. [CrossRef]
16. Zhao, Y.; Lee, D.-H.; Seok, M.-Y.; Lee, J.-A.; Phaniraj, M.; Suh, J.-Y.; Ha, H.-Y.; Kim, J.-Y.; Ramamurty, U.; Jang, J.-I. Resistance of cocrfemnni high-entropy alloy to gaseous hydrogen embrittlement. *Scr. Mater.* **2017**, *135*, 54–58. [CrossRef]
17. Hosemann, P.; Frazer, D.; Fratoni, M.; Bolind, A.; Ashby, M. Materials selection for nuclear applications: Challenges and opportunities. *Scr. Mater.* **2018**, *143*, 181–187. [CrossRef]
18. Yeh, J.W.; Chen, S.K.; Lin, S.J.; Gan, J.Y.; Chin, T.S.; Shun, T.T.; Tsau, C.H.; Chang, S.Y. Nanostructured high-entropy alloys with multiple principal elements: Novel alloy design concepts and outcomes. *Adv. Eng. Mater.* **2004**, *6*, 299–303. [CrossRef]
19. Zhang, Y.; Zuo, T.T.; Tang, Z.; Gao, M.C.; Dahmen, K.A.; Liaw, P.K.; Lu, Z.P. Microstructures and properties of high-entropy alloys. *Progr. Mater. Sci.* **2014**, *61*, 1–93. [CrossRef]
20. Cantor, B.; Chang, I.; Knight, P.; Vincent, A. Microstructural development in equiatomic multicomponent alloys. *Mater. Sci. Eng. A* **2004**, *375*, 213–218. [CrossRef]
21. Gao, M.C.; Yeh, J.W.; Liaw, P.K.; Zhang, Y. *High-Entropy Alloys: Fundamentals and Applications*; Springer: Berlin, Germany, 2016.
22. Zhang, W.; Liaw, P.K.; Zhang, Y. Science and technology in high-entropy alloys. *Sci. China Mater.* **2018**, *61*, 2–22. [CrossRef]
23. MacDonald, B.; Fu, Z.; Zheng, B.; Chen, W.; Lin, Y.; Chen, F.; Zhang, L.; Ivanisenko, J.; Zhou, Y.; Hahn, H. Recent progress in high entropy alloy research. *JOM* **2017**, *69*, 2024–2031. [CrossRef]
24. Senkov, O.N.; Miracle, D.B.; Chaput, K.J.; Couzinie, J.P. Development and exploration of refractory high entropy alloys—A review. *J. Mater. Res.* **2018**, *33*, 1–37. [CrossRef]
25. Han, Z.D.; Luan, H.W.; Liu, X.; Chen, N.; Li, X.Y.; Shao, Y.; Yao, K.F. Microstructures and mechanical properties of tixnbmotaw refractory high-entropy alloys. *Mater. Sci. Eng. A* **2018**, *712*, 380–385. [CrossRef]
26. Sarker, P.; Harrington, T.; Toher, C.; Oses, C.; Samiee, M.; Maria, J.-P.; Brenner, D.W.; Vecchio, K.S.; Curtarolo, S. High-entropy high-hardness metal carbides discovered by entropy descriptors. *Nat. Commun.* **2018**, *9*, 4980. [CrossRef]
27. Bagdasaryan, A.A.; Pshyk, A.V.; Coy, L.E.; Konarski, P.; Misnik, M.; Ivashchenko, V.I.; Kempiński, M.; Mediukh, N.R.; Pogrebnjak, A.D.; Beresnev, V.M.; et al. A new type of (TiZrNbTaHf)N/MoN nanocomposite coating: Microstructure and properties depending on energy of incident ions. *Compos. B Eng.* **2018**, *146*, 132–144. [CrossRef]
28. Zou, Y.; Ma, H.; Spolenak, R. Ultrastrong ductile and stable high-entropy alloys at small scales. *Nat. Commun.* **2015**, *6*, 7748. [CrossRef]
29. Schuh, B.; Völker, B.; Maier-Kiener, V.; Todt, J.; Li, J.; Hohenwarter, A. Phase decomposition of a single-phase AlTiVNb high-entropy alloy after severe plastic deformation and annealing. *Adv. Eng. Mater.* **2017**, *19*, 1600674. [CrossRef]
30. Kumar, N.A.P.K.; Li, C.; Leonard, K.J.; Bei, H.; Zinkle, S.J. Microstructural stability and mechanical behavior of fenimncr high entropy alloy under ion irradiation. *Acta Mater.* **2016**, *113*, 230–244. [CrossRef]

31. Zou, Y.; Wheeler, J.M.; Ma, H.; Okle, P.; Spolenak, R. Nanocrystalline high-entropy alloys: A new paradigm in high-temperature strength and stability. *Nano Lett.* **2017**, *17*, 1569–1574. [CrossRef]
32. Feng, X.; Zhang, J.; Xia, Z.; Fu, W.; Wu, K.; Liu, G.; Sun, J. Stable nanocrystalline nbmotaw high entropy alloy thin films with excellent mechanical and electrical properties. *Mater. Lett.* **2018**, *210*, 84–87. [CrossRef]
33. Sathiyamoorthi, P.; Basu, J.; Kashyap, S.; Pradeep, K.; Kottada, R.S. Thermal stability and grain boundary strengthening in ultrafine-grained cocrfeni high entropy alloy composite. *Mater. Des.* **2017**, *134*, 426–433. [CrossRef]
34. Zhang, C.; Zhang, F.; Diao, H.Y.; Gao, M.C.; Tang, Z.; Poplawsky, J.D.; Liaw, P.K. Understanding phase stability of Al-Co-Cr-Fe-Ni high entropy alloys. *Mater. Des.* **2016**, *109*, 425–433. [CrossRef]
35. Rao, J.C.; Diao, H.Y.; Ocelík, V.; Vainchtein, D.; Zhang, C.; Kuo, C.; Tang, Z.; Guo, W.; Poplawsky, J.D.; Zhou, Y. Secondary phases in AlxCoCrFeNi high-entropy alloys: An in-situ tem heating study and thermodynamic appraisal. *Acta Mater.* **2017**, *131*, 206–220. [CrossRef]
36. Guo, N.N.; Wang, L.; Luo, L.S.; Li, X.Z.; Su, Y.Q.; Guo, J.J.; Fu, H.Z. Microstructure and mechanical properties of refractory MoNbHfZrTi high-entropy alloy. *Mater. Des.* **2015**, *81*, 87–94. [CrossRef]
37. Miracle, D.B.; Senkov, O.N. A critical review of high entropy alloys and related concepts. *Acta Mater.* **2017**, *122*, 448–511. [CrossRef]
38. Xing, Q.; Ma, J.; Wang, C.; Zhang, Y. High-throughput screening solar-thermal conversion films in a pseudobinary (Cr, Fe, V)–(Ta, W) system. *ACS Comb. Sci.* **2018**, *20*, 602–610. [CrossRef]
39. Hasegawa, A.; Fukuda, M.; Yabuuchi, K.; Nogami, S. Neutron irradiation effects on the microstructural development of tungsten and tungsten alloys. *J. Nuclear Mater.* **2016**, *471*, 175–183. [CrossRef]
40. Waseem, O.A.; Lee, J.; Lee, H.M.; Ryu, H.J. The effect of ti on the sintering and mechanical properties of refractory high-entropy alloy TixWTaVCr fabricated via spark plasma sintering for fusion plasma-facing materials. *Mater. Chem. Phys.* **2018**, *210*, 87–94. [CrossRef]
41. Rieth, M.; Dudarev, S.; De Vicente, S.G.; Aktaa, J.; Ahlgren, T.; Antusch, S.; Armstrong, D.; Balden, M.; Baluc, N.; Barthe, M.-F. Recent progress in research on tungsten materials for nuclear fusion applications in europe. *J. Nuclear Mater.* **2013**, *432*, 482–500. [CrossRef]
42. Xiao, D.H.; Zhou, P.F.; Wu, W.Q.; Diao, H.Y.; Gao, M.C.; Song, M.; Liaw, P.K. Microstructure, mechanical and corrosion behaviors of AlCoCuFeNi-(Cr,Ti) high entropy alloys. *Mater. Des.* **2017**, *116*, 438–447. [CrossRef]
43. Shon, Y.; Joshi, S.S.; Katakam, S.; Shanker Rajamure, R.; Dahotre, N.B. Laser additive synthesis of high entropy alloy coating on aluminum: Corrosion behavior. *Mater. Lett.* **2015**, *142*, 122–125. [CrossRef]
44. Dong, Y.; Zhou, K.Y.; Lu, Y.P.; Gao, X.X.; Wang, T.M.; Li, T.J. Effect of vanadium addition on the microstructure and properties of AlCoCrFeNi high entropy alloy. *Mater. Des.* **2014**, *57*, 67–72. [CrossRef]
45. Salishchev, G.A.; Tikhonovsky, M.A.; Shaysultanov, D.G.; Stepanov, N.D.; Kuznetsov, A.V.; Kolodiy, I.V.; Tortika, A.S.; Senkov, O.N. Effect of Mn and V on structure and mechanical properties of high-entropy alloys based on CoCrFeNi system. *J. Alloys Compd.* **2014**, *591*, 11–21. [CrossRef]
46. Li, Z.M.; Pradeep, K.G.; Deng, Y.; Raabe, D.; Tasan, C.C. Metastable high-entropy dual-phase alloys overcome the strength–ductility trade-off. *Nature* **2016**, *534*, 227–230. [CrossRef]
47. Diao, H.Y.; Feng, R.; Dahmen, K.A.; Liaw, P.K. Fundamental deformation behavior in high-entropy alloys: An overview. *Curr. Opin. Solid State Mater. Sci.* **2017**, *21*, 252–266. [CrossRef]
48. Zhang, Y.; Zhou, Y.J.; Lin, J.P.; Chen, G.L.; Liaw, P.K. Solid-solution phase formation rules for multi-component alloys. *Adv. Eng. Mater.* **2008**, *10*, 534–538. [CrossRef]
49. Pickering, E.J.; Jones, N.G. High-entropy alloys: A critical assessment of their founding principles and future prospects. *Int. Mater. Rev.* **2016**, *61*, 183–202. [CrossRef]
50. Lu, Z.P.; Wang, H.; Chen, M.W.; Baker, I.; Yeh, J.W.; Liu, C.T.; Nieh, T.G. An assessment on the future development of high-entropy alloys: Summary from a recent workshop. *Intermetallics* **2015**, *66*, 67–76. [CrossRef]
51. Yang, X.; Zhang, Y. Prediction of high-entropy stabilized solid-solution in multi-component alloys. *Mater. Chem. Phys.* **2012**, *132*, 233–238. [CrossRef]
52. Guo, S.; Ng, C.; Lu, J.; Liu, C. Effect of valence electron concentration on stability of fcc or bcc phase in high entropy alloys. *J. Appl. Phys.* **2011**, *109*, 103505. [CrossRef]
53. Zhang, F.; Wu, Y.; Lou, H.B.; Zeng, Z.D.; Prakapenka, V.B.; Greenberg, E.; Ren, Y.; Yan, J.Y.; Okasinski, J.S.; Liu, X.J. Polymorphism in a high-entropy alloy. *Nat. Commun.* **2017**, *8*, 15687. [CrossRef]

54. Maiti, S.; Steurer, W. Structural-disorder and its effect on mechanical properties in single-phase TaNbHfZr high-entropy alloy. *Acta Mater.* **2016**, *106*, 87–97. [CrossRef]
55. Huang, H.; Wu, Y.; He, J.; Wang, H.; Liu, X.; An, K.; Wu, W.; Lu, Z. Phase-transformation ductilization of brittle high-entropy alloys via metastability engineering. *Adv. Mater.* **2017**, *29*, 1701678. [CrossRef]
56. Senkov, O.N.; Wilks, G.B.; Scott, J.M.; Miracle, D.B. Mechanical properties of $Nb_{25}Mo_{25}Ta_{25}W_{25}$ and $V_{20}Nb_{20}Mo_{20}Ta_{20}W_{20}$ refractory high entropy alloys. *Intermetallics* **2011**, *19*, 698–706. [CrossRef]
57. Senkov, O.N.; Woodward, C.F. Microstructure and properties of a refractory NbCrMo0.5Ta0.5TiZr alloy. *Mater. Sci. Eng. A* **2011**, *529*, 311–320. [CrossRef]
58. Juan, C.C.; Tsai, M.H.; Tsai, C.W.; Lin, C.M.; Wang, W.R.; Yang, C.C.; Chen, S.K.; Lin, S.J.; Yeh, J.W. Enhanced mechanical properties of HfMoTaTiZr and HfMoNbTaTiZr refractory high-entropy alloys. *Intermetallics* **2015**, *62*, 76–83. [CrossRef]
59. Senkov, O.; Jensen, J.; Pilchak, A.; Miracle, D.; Fraser, H.J. Compositional variation effects on the microstructure and properties of a refractory high-entropy superalloy $AlMo_{0.5}NbTa_{0.5}TiZr$. *Mater. Des.* **2018**, *139*, 498–511. [CrossRef]
60. Senkov, O.N.; Wilks, G.B.; Miracle, D.B.; Chuang, C.P.; Liaw, P.K. Refractory high-entropy alloys. *Intermetallics* **2010**, *18*, 1758–1765. [CrossRef]
61. Senkov, O.N.; Senkova, S.V.; Miracle, D.B.; Woodward, C. Mechanical properties of low-density, refractory multi-principal element alloys of the Cr–Nb–Ti–V–Zr system. *Mater. Sci. Eng. A* **2013**, *565*, 51–62. [CrossRef]
62. Senkov, O.N.; Senkova, S.V.; Woodward, C.; Miracle, D.B. Low-density, refractory multi-principal element alloys of the Cr–Nb–Ti–V–Zr system: Microstructure and phase analysis. *Acta Mater.* **2013**, *61*, 1545–1557. [CrossRef]
63. Senkov, O.N.; Woodward, C.; Miracle, D.B. Microstructure and properties of aluminum-containing refractory high-entropy alloys. *JOM* **2014**, *66*, 2030–2042. [CrossRef]
64. Lin, C.M.; Juan, C.C.; Chang, C.H.; Tsai, C.W.; Yeh, J.W. Effect of Al addition on mechanical properties and microstructure of refractory Al*x*HfNbTaTIZr alloys. *J. Alloys Compd.* **2015**, *624*, 100–107. [CrossRef]
65. Wu, Y.D.; Cai, Y.H.; Wang, T.; Si, J.J.; Zhu, J.; Wang, Y.D.; Hui, X.D. A refractory $Hf_{25}Nb_{25}Ti_{25}Zr_{25}$ high-entropy alloy with excellent structural stability and tensile properties. *Mater. Lett.* **2014**, *130*, 277–280. [CrossRef]
66. Senkov, O.N.; Jensen, J.K.; Pilchak, A.L.; Miracle, D.B.; Fraser, H.L. Compositional variation effects on the microstructure and properties of a refractory high-entropy superalloy AlMo0.5NbTa0.5TiZr. *Mater. Des.* **2017**, *139*, 498–511. [CrossRef]
67. Stepanov, N.D.; Yurchenko, N.Y.; Panina, E.S.; Tikhonovsky, M.A.; Zherebtsov, S.V. Precipitation-strengthened refractory Al0.5CrNbTi2V0.5 high entropy alloy. *Mater. Lett.* **2017**, *188*, 162–164. [CrossRef]
68. Yurchenko, N.Y.; Stepanov, N.D.; Zherebtsov, S.V.; Tikhonovsky, M.A.; Salishchev, G.A. Structure and mechanical properties of B2 ordered refractory AlNbTiVZrx (x = 0–1.5) high-entropy alloys. *Mater. Sci. Eng. A* **2017**, *704*, 82–90. [CrossRef]
69. Sheikh, S.; Shafeie, S.; Hu, Q.; Ahlström, J.; Persson, C.; Veselý, J.; Zýka, J.; Klement, U.; Guo, S. Alloy design for intrinsically ductile refractory high-entropy alloys. *J. Appl. Phys.* **2016**, *120*, 3445. [CrossRef]
70. Yang, X.; Zhang, Y.; Liaw, P.K. Microstructure and compressive properties of NbTiVTaAlx high entropy alloys. *Procedia Eng.* **2012**, *36*, 292–298. [CrossRef]
71. Senkov, O.N.; Zhang, F.; Miller, J.D. Phase composition of a CrMo0.5NbTa0.5TiZr high entropy alloy: Comparison of experimental and simulated data. *Entropy* **2013**, *15*, 3796–3809. [CrossRef]
72. Jiang, H.; Zhang, H.; Huang, T.; Lu, Y.; Wang, T.; Li, T. Microstructures and mechanical properties of Co2MoxNi2VWx eutectic high entropy alloys. *Mater. Des.* **2016**, *109*, 539–546. [CrossRef]

entropy

Article

Magnetic Properties and Microstructure of $FeCoNi(CuAl)_{0.8}Sn_x$ ($0 \le x \le 0.10$) High-Entropy Alloys

Zhong Li [1], Chenxu Wang [1], Linye Yu [1,2], Yong Gu [1,3], Minxiang Pan [1], Xiaohua Tan [1,*] and Hui Xu [1,*]

1 Institute of Materials Science, School of Materials Science and Engineering, Shanghai University, Shanghai 200072, China; hanying880205@163.com (Z.L.); wcx123581@163.com (C.W.); 17721232010@163.com (L.Y.); gy2018@hznu.edu.cn (Y.G.); panminxiang@cjlu.edu.cn (M.P.)

2 Shanghai Marine Diesel Engine Reserch Institute, Shanghai 201108, China

3 Qianjiang College, Hangzhou Normal University, Hangzhou 310036, China

* Correspondence: tanxiaohua123@shu.edu.cn (X.T.); huixu8888@shu.edu.cn (H.X.); Tel.: +86-21-5633-7032 (X.T.); +86-21-5633-7887 (H.X.)

Received: 9 October 2018; Accepted: 8 November 2018; Published: 13 November 2018

Abstract: The present work exhibits the effects of Sn addition on the magnetic properties and microstructure of $FeCoNi(CuAl)_{0.8}Sn_x$ ($0 \le x \le 0.10$) high-entropy alloys (HEAs). The results show all the samples consist of a mixed structure of face-centered-cubic (FCC) phase and body-centered-cubic (BCC) phase. The addition of Sn promotes the formation of BCC phase, and it also affects the shape of Cu-rich nano-precipitates in BCC matrix. It also shows that the Curie temperatures (T_c) of the FCC phase and the saturation magnetization (M_s) of the $FeCoNi(CuAl)_{0.8}Sn_x$ ($0 \le x \le 0.10$) HEAs increase greatly while the remanence (B_r) decreases after the addition of Sn into $FeCoNi(CuAl)_{0.8}$ HEA. The thermomagnetic curves indicate that the phases of the $FeCoNi(CuAl)_{0.8}Sn_x$ ($0 \le x \le 0.10$) HEAs will transform from FCC with low T_c to BCC phase with high T_c at temperature of 600–700 K. This work provides a new idea for $FeCoNi(CuAl)_{0.8}Sn_x$ ($0 \le x \le 0.10$) HEAs for their potential application as soft magnets to be used at high temperatures.

Keywords: high-entropy alloys (HEAs); phase constitution; magnetic properties; Curie temperature; phase transition

1. Introduction

Since the first report of high-entropy alloys (HEAs) in 2004 [1], researchers have shown an increased interest in the study of HEAs. HEAs are the definition of alloys that are typically composed of more than 5 principal elements, which have broken the traditional alloy design concept based on 1 or 2 principal elements [2]. In contrast with the conventional alloys, HEAs predominantly trend to form an amorphous structure [3,4] or a simple solid solution with body-centered-cubic (BCC) phase [5,6], face-centered-cubic (FCC) phase [7–9] or a mixture of them [10,11], which is attributed to the high mixing entropy of HEAs [12,13]. The unique design concept and the significant mixing entropy effect of HEAs give them potential application in many high-entropy structural and functional materials. For example, HEAs have huge potential for use in jet-engine turbines, thin-film resistors, heat- or wear-resistant parts, functional coatings, and electronic products [13,14]. Recently, the good magnetic properties of HEAs capture the increasing interest in this new field of materials [15,16]. It is worth noting that many HEAs [17–20] consist of several ferromagnetic elements, such as Fe, Co, and Ni, and they also have a good comprehensive mechanical properties, which make them have great application potential in soft magnetic materials. HEAs with good magnetic and mechanical properties are expected to be used in electric motors, electromagnets, and magnetic recording. Liu et al. [21,22] reported the $FeCoNi_{0.25}Al_{0.25}$ HEA exhibits a high saturation magnetization (M_s = 101.0 emu/g) and a low coercivity (H_c = 268 A/m). Zuo et al. [23] found CoNiMnGa HEA shows a low saturation magnetostriction coefficient and a high Curie temperature (T_c). In our previous work [24], the $FeCoNi(CuAl)_{0.8}$ HEA consisting of BCC and FCC phases shows good magnetic and mechanical properties, and it was found that BCC phases show a higher M_s than that of FCC phases for the $FeCoNi(CuAl)_{0.8}$ HEA. The other work of our group [25] found a minor amount of Ga addition into $FeCoNi(CuAl)_{0.8}$ HEA can promote the formation of BCC phase and improve the M_s of the alloy, and the value of remanence (B_r) and coercivity (H_c) also increases. It was reported that the addition of Sn can hinder the formation of FCC phase [26] and promote the formation of BCC phase [27]. Therefore, in this work, a minor amount of Sn was added into the $FeCoNi(CuAl)_{0.8}$ HEA hoping to get a higher volume fraction of BCC phase.

In this work, the $FeCoNi(CuAl)_{0.8}Sn_x$ ($0 \leq x \leq 0.10$) HEAs were studied from phase constitutions to microstructure and magnetic properties. It was found that these HEAs show high M_s, low B_r, and high T_c, which indicate their potential application as soft magnetic materials. This paper offers a good method for designing future high-performance soft magnetic materials.

2. Materials and Methods

The $FeCoNi(CuAl)_{0.8}Sn_x$ ($0 \leq x \leq 0.10$) HEAs were prepared via arc-melting the constituent elements of 99.99% purity using a water-cooled Cu crucible. The alloys were sucked into a 100 × 10× 2 mm water-cooled Cu mold after remelting four times. X-ray diffraction (XRD, D/max-2500 V, Rigaky Corporation, Tokyo, Japan) was used to characterized the crystal structures of the $FeCoNi(CuAl)_{0.8}Sn_x$ ($0 \leq x \leq 0.10$) HEAs at a scan speed of 1°/min. Scanning electronic microscopy (SEM, Hitachis-3400N) was used to observe the morphology of the samples. Transmission electron microscope (TEM, JEM-2100F, JEOL, Ltd., Tokyo, Japan) was employed for the microstructure of HEAs. Dual-beam focused ion beam (FIB, FEI Helios 600i, Hillsboro, OR, USA) was used to prepare the TEM samples. The high angle annular dark field (HAADF) images were performed by a scanning transmission electron microscope with energy dispersive spectrometer (STEM/EDS, JEM-2100F, JEOL, Ltd., Tokyo, Japan). The M_s and thermomagnetic curves was obtained from vibrating sample magnetometer (VSM, Lakeshore 7407, Westerville, OH, USA). The coercivity (H_c), hysteresis losses (P_u), remanence (B_r), initial permeability (μ_i), and maximum permeability (μ_{max}) were obtained from hysteresis curves (DC) test system (HCTS, FE-2100SD, Forever elegance, Hunan, China) using a 27 × 9 × 1.7 mm rectangular sample at a magnetic field of 25 kA/m. The thermal stability was analyzed by differential scanning calorimeter (DSC, DIAMOND) at a heating rate of 10 K/min.

3. Results and Discussion

3.1. X-ray Diffraction

Figure 1 shows the XRD patterns of the $FeCoNi(CuAl)_{0.8}Sn_x$ ($0 \leq x \leq 0.10$) HEAs. It is found that all of these HEAs consist of a mixed structure of FCC phase and BCC phase. A few diffraction peaks of unknown phases appear in the XRD patterns for $x \geq 0.02$ and it is especially obvious for $x \geq 0.06$. Here, $I_{(111)FCC}$ and $I_{(110)BCC}$ were used to denote the diffraction intensity of the strongest peak of (111) for FCC phase and (110) for BCC phase, respectively. Therefore, the relative content of FCC and BCC phases can be expressed as $I_{(110)BCC}/I_{(111)FCC}$. Table 1 shows the ratio of $I_{(110)BCC}/I_{(111)FCC}$. For x = 0, the ratio of $I_{(110)BCC}/I_{(111)FCC}$ is 0.38. This means the FCC phase in the $FeCoNi(CuAl)_{0.8}$ HEA is a dominant phase. When the content of Sn increases from 0.02 to 0.10, the ratio of $I_{(110)BCC}/I_{(111)FCC}$ increases from 0.69 to 10.53, suggesting that the content of BCC phase increases rapidly. Based on XRD patterns, the lattice parameters of FCC and BCC phases can be calculated and shown in Table 1. It is seen that the lattice parameters of FCC and BCC phases all increase as x increases from 0 to 0.04, then they decrease and remain almost stable as x further increases. The decrease of lattice parameters of FCC and BCC phases may be due to the precipitation of these unknown phases.

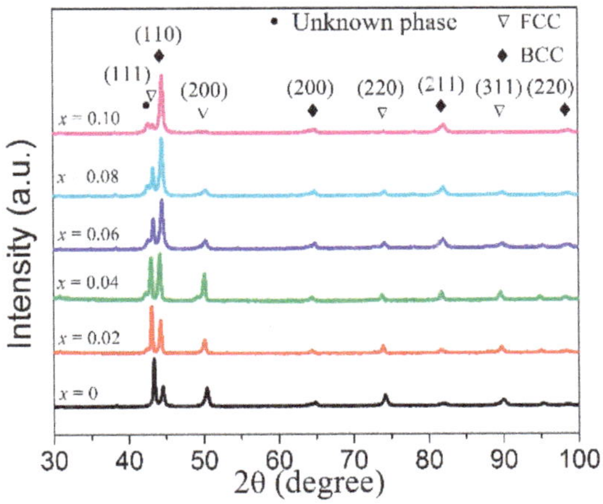

Figure 1. XRD patterns of $FeCoNi(CuAl)_{0.8}Sn_x$ ($0 \leq x \leq 0.10$) high-entropy alloys (HEAs).

Table 1. The ratio of $I_{(110)BCC}/I_{(111)FCC}$ and lattice parameters of $FeCoNi(CuAl)_{0.8}Sn_x$ ($0 \leq x \leq 0.10$) HEAs.

x	$I_{(110)BCC}/I_{(111)FCC}$	a_{FCC} (nm)	a_{BCC} (nm)
0	0.38	0.3588	0.2856
0.02	0.69	0.3634	0.2891
0.04	1.11	0.3642	0.2894
0.06	1.81	0.3614	0.2876
0.08	2.40	0.3612	0.2879
0.10	10.53	0.3616	0.2876

3.2. *Magnetic Properties*

Figure 2 shows hysteresis loops of $FeCoNi(CuAl)_{0.8}Sn_x$ ($0 \leq x \leq 0.10$) HEAs measured by VSM. Figure 3 shows the magnetization curves and hysteresis loops of $FeCoNi(CuAl)_{0.8}Sn_x$ ($0 \leq x \leq 0.10$) HEAs measured by HCTS. The corresponding magnetic parameters obtained from Figures 2 and 3 are shown in Table 2. As can be seen from Table 2, the value of M_s increases from 78.6 Am^2/kg to 88.8 Am^2/kg as x increases from 0 to 0.10, which increases almost 13 percent. To make it easier to compare, the corresponding magnetic parameters as well as the ratio of $I_{(110)BCC}/I_{(111)FCC}$ as a function of x are shown in Figure 4. It is obvious that the M_s, coercivity (H_c), and hysteresis losses (P_u) increase while the initial permeability (μ_i) and maximum permeability (μ_{max}) have a generally decreasing trend with increasing x. These results may be due to the increase of the volume fraction of BCC phase and the decrease of the volume fraction of FCC phase, which is in agreement with that of reported $FeCoNi(CuAl)_{0.8}Ga_x$ ($0 \leq x \leq 0.08$) HEAs [25]. However, the value of remanence (B_r) decreases after the addition of Sn into $FeCoNi(CuAl)_{0.8}$ HEA, and this is completely different from the effect of Ga, which contributes to the increase of B_r [25]. A careful analysis of these magnetic parameters shows the decrease of B_r may be due to the rapid decrease of permeability.

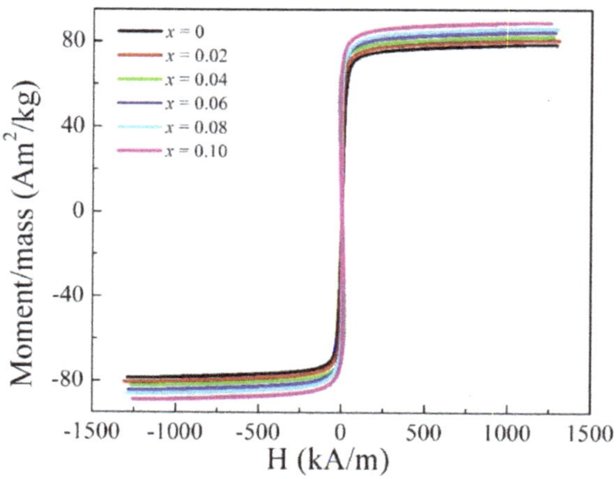

Figure 2. Hysteresis loops of FeCoNi(CuAl)$_{0.8}$Sn$_x$ ($0 \leq x \leq 0.10$) HEAs measured by VSM.

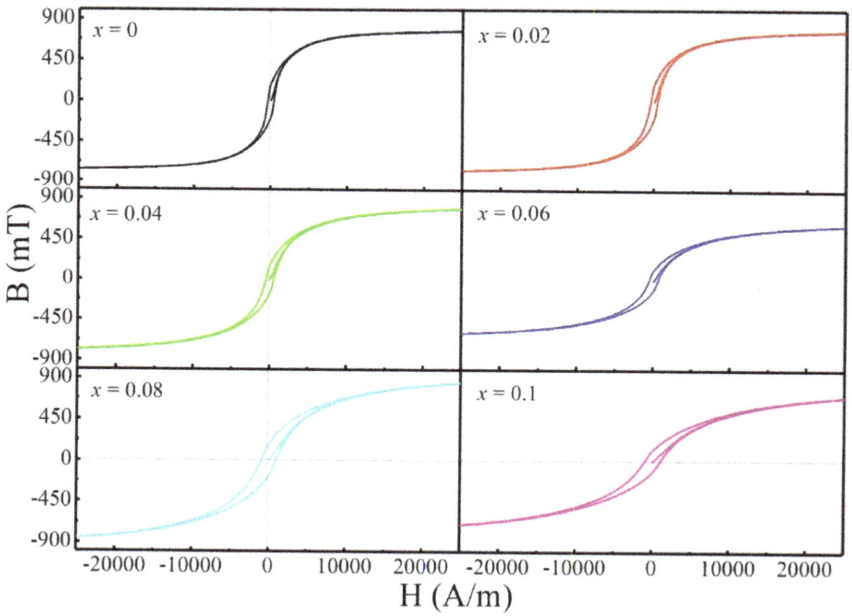

Figure 3. Magnetization curves and hysteresis loops of FeCoNi(CuAl)$_{0.8}$Sn$_x$ ($0 \leq x \leq 0.10$) HEAs measured by HCTS.

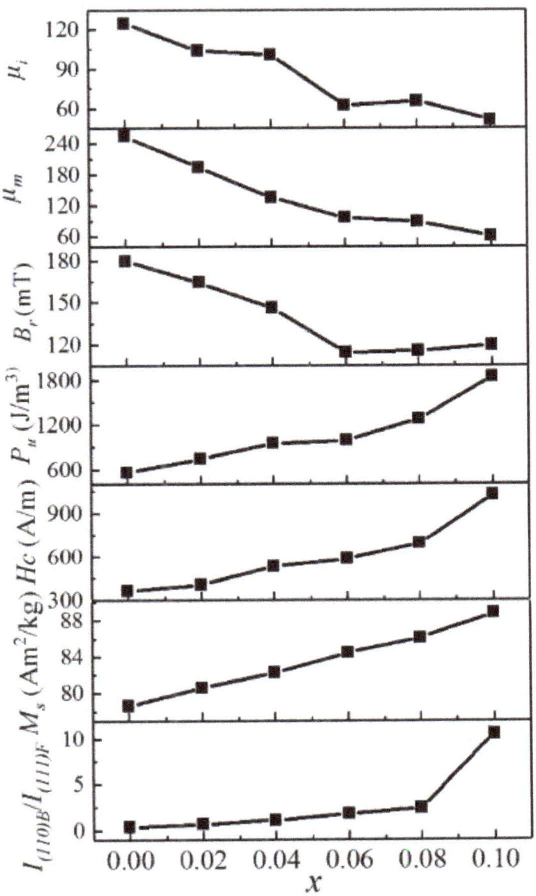

Figure 4. The ratio of $I_{(110)BCC}/I_{(111)FCC}$ and magnetic properties as a function of x for $FeCoNi(CuAl)_{0.8}Sn_x$ ($0 \leq x \leq 0.10$) HEAs.

Table 2. Magnetic parameters of $FeCoNi(CuAl)_{0.8}Sn_x$ ($0 \leq x \leq 0.10$) HEAs.

x	M_s (Am^2/kg)	B_r (mT)	H_c (A/m)	P_u (J/m^3)	μ_m	μ_i
0	78.6	179.5	362.0	558.6	254.3	124.8
0.02	80.6	164.5	404.5	738.2	193.6	103.9
0.04	82.2	145.9	529.8	947.4	135.4	100.9
0.06	84.4	114.2	580.7	988.5	96.15	62.44
0.08	86.1	115.4	685.0	1275	87.69	65.55
0.10	88.8	119.5	1020	1848	60.86	50.94

The temperature dependence of magnetization for $FeCoNi(CuAl)_{0.8}Sn_x$ ($0 \leq x \leq 0.10$) HEAs measured at an applied magnetic field of 1 T is shown in Figure 5. It can be seen that the magnetization of the $FeCoNi(CuAl)_{0.8}$ HEA decreases first with increasing temperature, and it has hit bottom of 28.9 Am^2/kg when the temperature is 630.4 K, then it increases quickly and reaches a constant value of about 52 Am^2/kg at 706 K. It is worth noting that the magnetization does not reduce to 0 at its lowest point. That means a magnetic phase with higher Curie temperature exists in the alloy. As can be seen from Figure S1, there is one phase transformation peak [25] for $FeCoNi(CuAl)_{0.8}$ HEA, at which the phase transforms from FCC to BCC phase. Therefore, it can be concluded that the Curie temperatures of BCC phase are obvious higher than that of FCC for FeCoNi(CuAl)$_{0.8}$ HEA. That means the FCC phase exhibits paramagnetic behavior and a disordered magnetic structure while the BCC phase still shows ferromagnetic behavior with the increase of temperature. It also agrees well with our previous study [25]. Therefore, the magnetization of the alloy will increase when the temperature is above 630.4 K and below 706 K. The Curie temperature (T_c), corresponding to the ferromagnetic to paramagnetic state transition of FCC phase for FeCoNi(CuAl)$_{0.8}$ HEA, is indicated by arrow in Figure 5. For ease of comparison, the Curie temperatures of FCC phase for other FeCoNi(CuAl)$_{0.8}$Sn$_x$ ($0.02 \leq x \leq 0.10$) HEAs are also shown in Figure 5. For $x = 0.04$, the outline of the curve is similar to that of the Sn-free alloy, but the Curie temperature of FCC phase increases significantly and reaches 634.9 K. For $x \geq 0.08$, the Curie temperature of FCC phase continues to increase. However, there is only little change for the value of magnetization as the temperature continues to increase. This is because only a small number of FCC phase transform to BCC phase according to the results of XRD in Figure 1 and DSC curves in Figure S1. Therefore, it can be concluded that the addition of a minor amount of Sn can obviously increase the Curie temperature of FCC phase. The phases of the FeCoNi(CuAl)$_{0.8}$Sn$_x$ ($0 \leq x \leq 0.10$) HEAs will transform from FCC with low T_c to BCC phase with high T_c at temperature of 600–700 K, which leads to the increase of magnetization. This provides a new idea for FeCoNi(CuAl)$_{0.8}$Sn$_x$ ($0 \leq x \leq 0.10$) HEAs for their potential application as soft magnets to be used at high temperature.

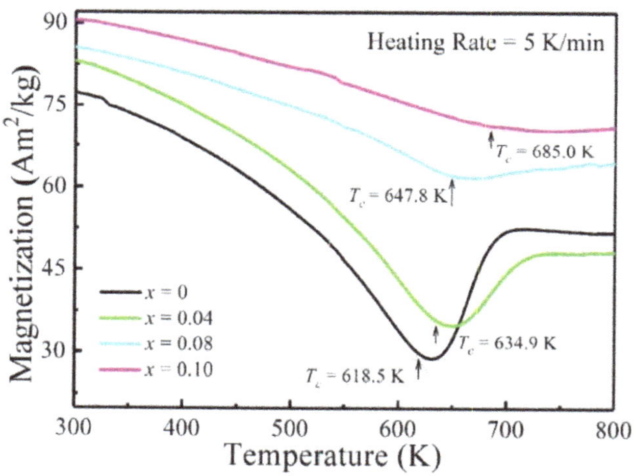

Figure 5. The thermomagnetic curves of FeCoNi(CuAl)$_{0.8}$Sn$_x$ ($0 \leq x \leq 0.10$) HEAs.

3.3. Microstructure

The SEM backscattered-electron (SEM-BSE) microstructures of the FeCoNi(CuAl)$_{0.8}$Sn$_x$ ($0 \leq x \leq 0.10$) HEAs are displayed in Figure 6. In Figure 6a, two obviously identifiable contrasts are found in the Sn-free alloy, which can be identified as dendritic regions and interdendritic regions (marked as DR and IR, respectively). According to our previous studies [24,25], the DR and IR region can be confirmed to be FCC and BCC phase, respectively. For $x \geq 0.02$, regions with strong contrast appear between the DR and IR regions, namely, the phase boundary regions, which can be marked as PB region. Moreover, the volume fraction of DR gradually decreases while the volume fraction of IR and PB increase with the increase of x. In addition, it is worth noting that the DR phases are almost invisible in the FeCoNi(CuAl)$_{0.8}$Sn$_{0.10}$ HEA (Figure 6f). The above results are in good agreement with the XRD results. Therefore, we can conclude that the addition of Sn in FeCoNi(CuAl)$_{0.8}$Sn$_x$ ($0 \leq x \leq 0.10$) HEAs can promote the formation of BCC phases and PB phases.

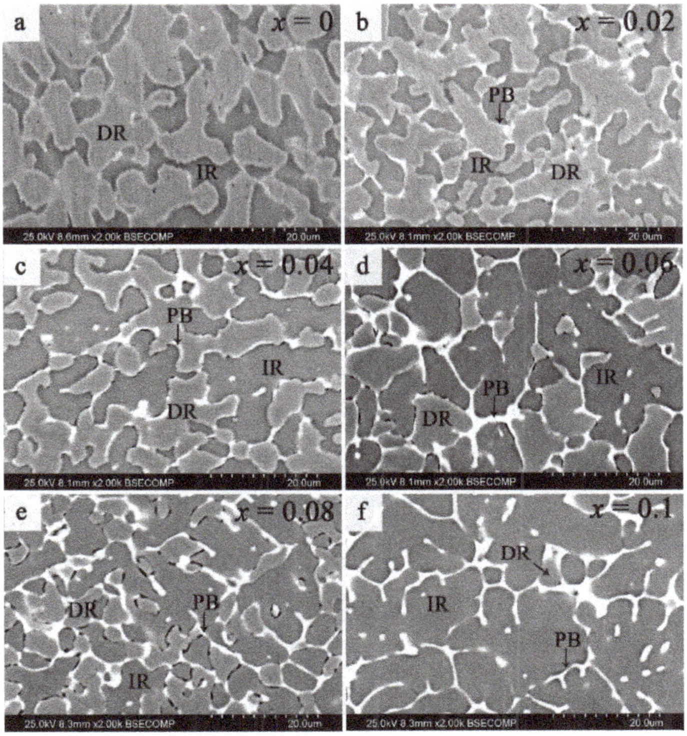

Figure 6. Typical SEM-BSE images of FeCoNi(CuAl)$_{0.8}$Sn$_x$ ($0 \leq x \leq 0.10$) HEAs.

In order to get more details of the FeCoNi(CuAl)$_{0.8}$Sn$_x$ ($0 \leq x \leq 0.10$) HEAs, the structures of the FeCoNi(CuAl)$_{0.8}$ and FeCoNi(CuAl)$_{0.8}$Sn$_{0.10}$ HEAs were further analyzed by TEM and the results are shown in Figure 7. Figure 7a1 confirms that FeCoNi(CuAl)$_{0.8}$ HEA consists of two kinds of phases which are named as DR (dendritic region) and IR (interdendritic region) according to the results revealed by the typical SEM-BSM images (Figure 6). The selected-area-electron-diffraction (SAED) patterns of Figure 7a2,a3 suggest the FCC crystal structure of DR and the BCC crystal structure of IR in FeCoNi(CuAl)$_{0.8}$ HEA. Meanwhile, it can be seen from Figure 7a1 that the surface of FCC phase is very smooth while the surface of BCC phase is much harsh. The high-magnification bright-field image of BCC phase for FeCoNi(CuAl)$_{0.8}$ HEA is shown in Figure 7a4 and displays that the BCC phase contains a large number of nanoscale precipitates which distribute homogeneously in the BCC matrix. Moreover, the average size of nanoscale precipitates is 20 ± 5 nm. After the addition of Sn into the FeCoNi(CuAl)$_{0.8}$ HEA, DR and IR regions can also be found in the FeCoNi(CuAl)$_{0.8}$Sn$_{0.10}$ HEA in Figure 7b1. Similarly, the former one can be indexed as FCC phase while the latter one is BCC phase according to the diffraction calibration in Figures 7 and 7. In addition, it sees that two new regions (marked as A and B, respectively) are observed in Figure 7b1. We can infer that they are the source of the unknown phase peak in the XRD pattern. As shown in Figure 7b4, the high-magnification bright-field image of BCC phase for FeCoNi(CuAl)$_{0.8}$Sn$_{0.10}$ HEA is apparently different from that of the Sn-free alloy. The shape of the nano-precipitates is rod-like, and their density is lower than that in FeCoNi(CuAl)$_{0.8}$ HEA. At the same time, the nanoprecipitates with an average length cf about 100 nm and a width of about 30 nm are larger than that in FeCoNi(CuAl)$_{0.8}$ HEA.

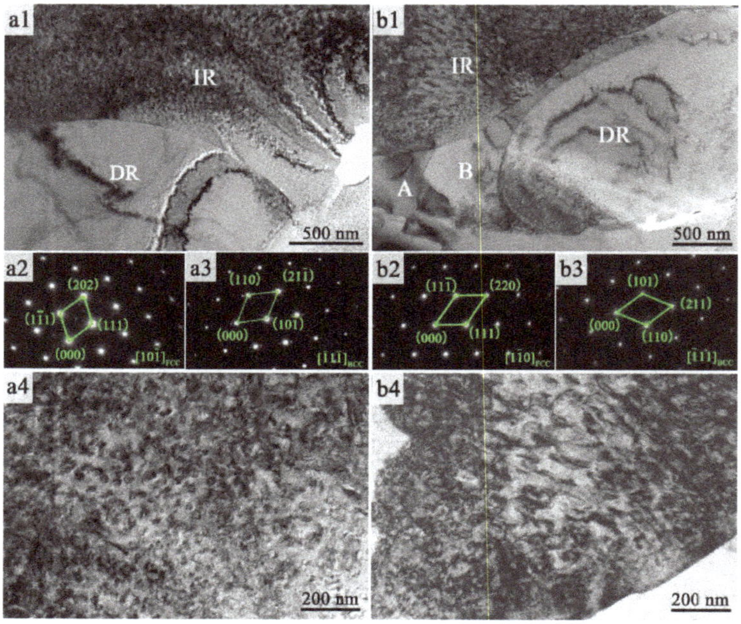

Figure 7. TEM images of $FeCoNi(CuAl)_{0.8}$ HEA: (**a1**) bright-field image; (**a2**) SAED pattern of DR region; (**a3**) SAED pattern of IR region; (**a4**) high-magnification bright-field image of IR region. TEM images of $FeCoNi(CuAl)_{0.8}Sn_{0.10}$ HEA: (**b1**) bright-field image; (**b2**) SAED pattern of DR region; (**b3**) SAED pattern of IR region; (**b4**) high-magnification bright-field image of IR region.

The high angle annular dark field (HAADF) image and element mappings of Fe, Co, Ni, Cu, and Al for $FeNiCo(CuAl)_{0.8}$ HEA measured by STEM-EDS technique are displayed in Figure 8. As can be seen from Figure 8a, the microstructure of the $FeNiCo(CuAl)_{0.8}$ HEA is similar to that shown in Figure 7a1. It is worth noting that a phase boundary region with a width of about 13 nm can also be found in Figure 8a. We can confirm that these nanoprecipitates in the BCC region and the phase boundary region are rich in Cu (Figure 8e). Figure 8b–f suggests that the distribution of Fe and Co is very uniform in the alloy, while Ni and Al are enriched in the BCC region. The formation of the Cu-rich phase boundary regions in the as-cast $FeNiCo(CuAl)_{0.8}$ HEA may be caused by the following three reasons. First, the melting point of Cu is lower than that of Fe, Co, and Ni [28], thus it may solidify after Fe, Co, and Ni when the temperature decreases. Second, the mixing enthalpies [28] (Table S1) between Cu and Fe, Co, and Ni are 13, 6, and 4 kJ/mol, respectively, meaning Cu is more likely to be repelled by other elements to form the Cu-rich phase boundary region. Finally, and most importantly, Cu and other elements are completely soluble at high temperature, but a large amount of Cu will precipitate out due to the rapid decrease of Cu solubility during casting. As for the formation of Cu-rich nano-precipitate in the BCC phase, it is due to the great difference in crystal structure between copper and the BCC phase matrix as well as the decrease of Cu solubility in BCC phase and the positive mixing enthalpies of Cu with Fe, Co, Ni, and Sn.

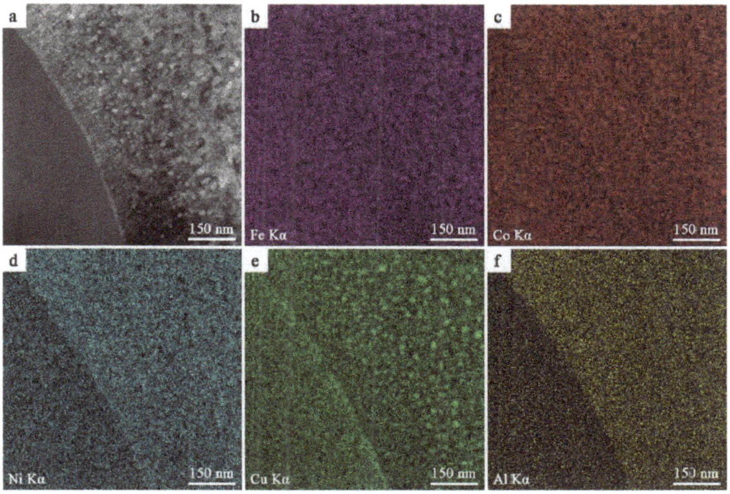

Figure 8. High angle annular dark field (HAADF) image (**a**) and elemental mapping images of $FeCoNi(CuAl)_{0.8}$ HEA for Fe-Kα (**b**), Co-Kα (**c**), Ni-Kα (**d**) Cu-Kα (**e**) and Al-Kα (**f**).

Figure 9 shows the HAADF image and elemental mapping images of $FeCoNi(CuAl)_{0.8}Sn_{0.10}$ HEA. Compared with that of $FeCoNi(CuAl)_{0.8}$ HEA, the microstructure shown in Figure 9a is more complex. From Figure 9a, a large number of Cu-rich nanoprecipitates with rod-like shape can be found in BCC phase. The shape of the precipitates changes to spherical as they approach the phase boundary region, and their size also gradually decreases. Figure 9b–g suggests that the distribution of Fe and Co is uniform in FCC and BCC phase regions, however, little Fe and Co can be found in the region between FCC and BCC phase. Moreover, the distribution of Ni and Al is enriched in BCC regions. One of the interesting things is that Cu and Sn segregate in the region between FCC and BCC phase. Besides, in the region where Sn is enriched, the content of Ni is also very high. Therefore, we can deduce that the unknown phase shown in the XRD patterns is composed of two phases, and one of it is rich in Cu, the other one is rich in Ni and Sn.

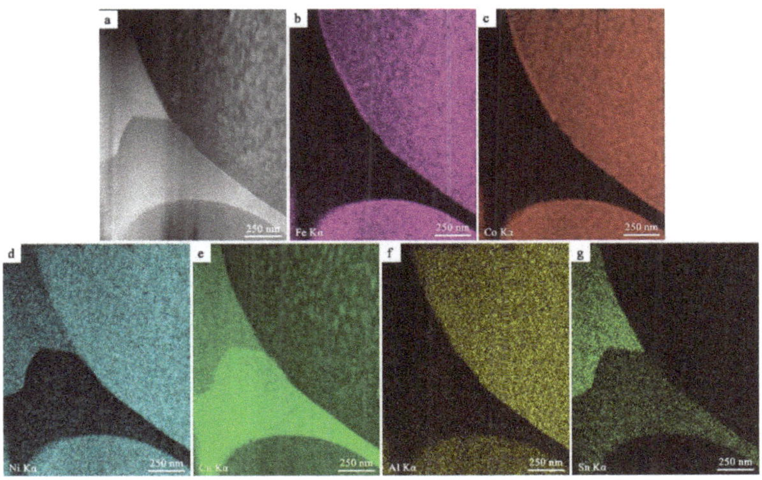

Figure 9. HAADF image (**a**) and elemental mapping images of $FeCoNi(CuAl)_{0.8}Sn_{0.10}$ HEA for Fe-Kα (**b**), Co-Kα (**c**), Ni-Kα (**d**) Cu-Kα (**e**), Al-Kα (**f**) and Sn-Kα (**g**).

4. Conclusions

In this work, the FeCoNi(CuAl)$_{0.8}$Sn$_x$ ($0 \leq x \leq 0.10$) HEAs were prepared by vacuum arc-melt casting. Effects of Sn content on the phase constitution and magnetic properties were studied. All the samples are composed of FCC and BCC phases, whereas some unknown phases appear with the addition of Sn. The addition of Sn promotes the formation of BCC phase, and it also affects the shape of Cu-rich nanoprecipitates in the BCC matrix. Moreover, the M_s increases greatly while the remanence (B_r) decreases with the increasing of x for FeCoNi(CuAl)$_{0.8}$Sn$_x$ ($0 \leq x \leq 0.10$) HEAs. The addition of Sn can obviously increase the Curie temperature of the FCC phase. The phase of alloys with a mixture of FCC and BCC will transform from FCC to BCC phase at high temperature, leading to an increase of magnetization. They can be used as new soft magnetic materials at high temperatures.

Supplementary Materials: The following are available online at http://www.mdpi.com/1099-4300/20/11/872/s1, Figure S1: DSC curves of FeCoNi(CuAl)$_{0.8}$Sn$_x$ ($0 \leq x \leq 0.10$) HEAs. Table S1: The melting points (K) [28] of different elements and the mixing enthalpies (kJ/mol) [29] between two elements.

Author Contributions: H.X. and X.T. conceived and designed the experiments; L.Y. and Z.L. prepared the FeCoNi(CuAl)$_{0.8}$Sn$_x$ ($0 \leq x \leq 0.10$) HEAs; L.Y. and Z.L. performed the microstructural characterization of the samples; Z.L. and C.W. prepared the original draft; Z.L., X.T. and H.X. revised the manuscript; All authors discussed, analyzed the results and approved the final manuscript.

Funding: This research was funded by the National Natural Science Foundation of China (grant number 51471101 and U1531120).

Acknowledgments: The authors thank Jianchao Peng, Pengfei Hu, Na Min and Xue Liang of the Instrumental Analysis & Research Center, Shanghai University, China for their assistance in TEM samples preparation and measurements.

Conflicts of Interest: The authors declare no conflict of interest.

References

1. Yeh, J.W.; Chen, S.K.; Lin, S.J.; Gan, J.Y.; Chin, T.S.; Shun, T.T.; Tsau, C.H.; Chang, S.Y. Nanostructured high-entropy alloys with multiple principal elements novel alloy design concepts and outcomes. *Adv. Eng. Mater.* **2004**, *6*, 299–303. [CrossRef]
2. Murty, B.S.; Yeh, J.W.; Ranganathan, S. *High-Entropy Alloys*; Elsevier: Amsterdam, The Netherlands, 2014. [CrossRef]
3. Wei, R.; Sun, H.; Chen, C.; Tao, J.; Li, F. Formation of soft magnetic high entropy amorphous alloys composites containing in situ solid solution phase. *J. Magn. Magn. Mater.* **2018**, *449*, 63–67. [CrossRef]
4. Xu, J.; Axinte, E.; Zhao, Z.F.; Wang, Y. Effect of C and Ce addition on the microstructure and magnetic property of the mechanically alloyed FeSiBAlNi high entropy alloys. *J. Magn. Magn. Mater.* **2016**, *414*, 59–68. [CrossRef]
5. Lin, C.M.; Juan, C.C.; Chang, C.H.; Tsai, C.W.; Yeh, J.W. Effect of Al addition on mechanical properties and microstructure of refractory Al$_x$HfNbTaTiZr alloys. *J. Alloys Compd.* **2015**, *624*, 100–107. [CrossRef]
6. Borkar, T.; Chaudhary, V.; Gwalani, B.; Choudhuri, D.; Mikler, C.V.; Soni, V.; Alam, T.V.; Ramanujan, R.; Banerjee, R. A combinatorial approach for assessing the magnetic properties of high entropy alloys: Role of Cr in AlCo$_x$Cr$_{1-x}$FeNi. *Adv. Eng. Mater.* **2017**, *19*, 1700048. [CrossRef]
7. Xu, X.D.; Liu, P.; Guo, S.; Hirata, A.; Fujita, T.; Nieh, T.G.; Liu, C.T.; Chen, M.W. Nanoscale phase separation in a fcc-based CoCrCuFeNiAl$_{0.5}$ high-entropy alloy. *Acta Mater.* **2015**, *84*, 145–152. [CrossRef]
8. Tsai, M.H.; Yuan, H.; Cheng, G.; Xu, W.; Tsai, K.Y.; Tsai, C.W.; Jian, W.W.; Juan, C.C.; Shen, W.J.; Chuang, M.H.; et al. Morphology, structure and composition of precipitates in Al$_{0.3}$CoCrCu$_{0.5}$FeNi high-entropy alloy. *Intermetallics* **2013**, *32*, 329–336. [CrossRef]
9. Li, R.; Niu, P.; Yuan, T.; Cao, P.; Chen, C.; Zhou, K. Selective laser melting of an equiatomic CoCrFeMnNi high-entropy alloy: Processability, non-equilibrium microstructure and mechanical property. *J. Alloys Compd.* **2018**, *746*, 125–134. [CrossRef]
10. Wang, W.R.; Wang, W.L.; Wang, S.C.; Tsai, Y.C.; Lai, C.H.; Yeh, J.W. Effects of Al addition on the microstructure and mechanical property of Al$_x$CoCrFeNi high-entropy alloys. *Intermetallics* **2012**, *26*, 44–51. [CrossRef]

11. Chou, H.P.; Chang, Y.S.; Chen, S.K.; Yeh, J.W. Microstructure, thermophysical and electrical properties in $Al_xCoCrFeNi$ ($0 \leq x \leq 2$) high-entropy alloys. *Mater. Sci. Eng. B* **2009**, *163*, 184–189. [CrossRef]
12. Zhang, Y.; Zuo, T.T.; Tang, Z.; Gao, M.C.; Dahmen, K.A.; Liaw, P.K.; Lu, Z.P. Microstructures and properties of high-entropy alloys. *Prog. Mater. Sci.* **2014**, *61*, 1–93. [CrossRef]
13. Gao, M.M.C.; Yeh, J.W.; Liaw, P.K.; Zhang, Y. *High-Entropy Alloys: Fundamentals and Applications*, 1st ed.; Springer International Publishing: Cham, Switzerland, 2016. [CrossRef]
14. Lim, X.Z. Metal mixology stronger, tougher, stretchier: With a simple new recipe, metallurgists are creating a generation of alloys with remarkable properties. *Nature* **2016**, *533*, 306–307. [CrossRef] [PubMed]
15. Zhang, Y.; Zuo, T.T.; Cheng, Y.Q.; Liaw, P.K. High-entropy alloys with high saturation magnetization, electrical resistivity, and malleability. *Sci. Rep.* **2013**, *3*, 1455. [CrossRef] [PubMed]
16. Zaddach, A.J.; Niu, C.; Oni, A.A.; Fan, M.; LeBeau, J.M.; Irving, D.L.; Koch, C.C. Structure and magnetic properties of a multi-principal element Ni–Fe–Cr–Co–Zn–Mn alloy. *Intermetallics* **2016**, *68*, 107–112. [CrossRef]
17. Cheng, H.; Liu, X.; Tang, Q.; Wang, W.; Yan, X.; Dai, P. Microstructure and mechanical properties of $FeCoCrNiMnAl_x$ high-entropy alloys prepared by mechanical alloying and hot-pressed sintering. *J. Alloys Compd.* **2019**, *775*, 742–751. [CrossRef]
18. Alijani, F.; Reihanian, M.; Gheisari, K. Study on phase formation in magnetic FeCoNiMnV high entropy alloy produced by mechanical alloying. *J. Alloys Compd.* **2019**, *773*, 623–630. [CrossRef]
19. Kilmametov, A.; Kulagin, R.; Mazilkin, A.; Seils, S.; Boll, T.; Heilmaier, M.; Hahn, H. High-pressure torsion driven mechanical alloying of CoCrFeMnNi high entropy alloy. *Scripta Mater.* **2019**, *158*, 29–33. [CrossRef]
20. Gao, M.C.; Miracle, D.B.; Maurice, D.; Yan, X.; Zhang, Y.; Hawk, J.A. High-entropy functional materials. *J. Mater. Res.* **2018**, *33*, 3138–3155. [CrossRef]
21. Li, P.P.; Wang, A.D.; Liu, C.T. Composition dependence of structure, physical and mechanical properties of $FeCoNi(MnAl)_x$ high entropy alloys. *Intermetallics* **2017**, *87*, 21–26. [CrossRef]
22. Li, P.P.; Wang, A.D.; Liu, C.T. A ductile high entropy alloy with attractive magnetic properties. *J. Alloys Compd.* **2017**, *694*, 55–60. [CrossRef]
23. Shang, C.; Axinte, E.; Ge, W.; Zhang, Z.; Wang, Y. High-entropy alloy coatings with excellent mechanical, corrosion resistance and magnetic properties prepared by mechanical alloying and hot pressing sintering. *Surf. Interfaces* **2017**, *9*, 36–43. [CrossRef]
24. Zhang, Q.; Xu, H.; Tan, X.H.; Hou, X.L.; Wu, S.W.; Tan, G.S.; Yu, L.Y. The effects of phase constitution on magnetic and mechanical properties of $FeCoNi(CuAl)_x$ ($x = 0-1.2$) high-entropy alloys. *J. Alloys Compd.* **2017**, *693*, 1061–1067. [CrossRef]
25. Li, Z.; Xu, H.; Gu, Y.; Pan, M.X.; Yu, L.Y.; Tan, X.H.; Hou, X.L. Correlation between the magnetic properties and phase constitution of $FeCoNi(CuAl)_{0.8}Ga_x(0 \leq x \leq 0.08)$ high-entropy alloys. *J. Alloys Compd.* **2018**, *746*, 285–291. [CrossRef]
26. Liu, L.; Zhu, J.B.; Zhang, C.; Li, J.C.; Jiang, Q. Microstructure and the properties of FeCoCuNiSnx high entropy alloys. *Mater. Sci. Eng. A* **2012**, *548*, 64–68. [CrossRef]
27. Zuo, T.; Gao, M.C.; Ouyang, L.; Yang, X.; Cheng, Y.; Feng, R.; Chen, S.; Liaw, P.K.; Hawk, J.A.; Zhang, Y. Tailoring magnetic behavior of CoFeMnNiX (X = Al, Cr, Ga, and Sn) high entropy alloys by metal doping. *Acta Mater.* **2017**, *130*, 10–18. [CrossRef]
28. Takeuchi, A.; Inoue, A. Classification of bulk metallic glasses by atomic size difference, heat of mixing and period of constituent elements and its application to characterization of the main alloying element. *Mater. Trans.* **2005**, *46*, 2817–2829. [CrossRef]
29. Dean, J.A. *Lange's Handbook of Chemistry*; McGraw-Hill Inc.: New York, NY, USA, 1999.

entropy

MDPI

Article

Effect of Co and Gd Additions on Microstructures and Properties of FeSiBAlNi High Entropy Alloys

Sicheng Zhai, Wen Wang, Juan Xu, Shuai Xu, Zitang Zhang and Yan Wang *

School of Materials Science and Engineering, University of Jinan, Jinan 250022, China; zscshr123321@163.com (S.Z.); ww_nwc@163.com (W.W.); xujuanMISS@163.com (J.X.); 17862915273@163.com (S.X.); 13796829730@163.com (Z.Z.)

* Correspondence: mse_wangy@ujn.edu.cn; Tel: +86-531-8276-5473; Fax: +86-531-8797-4453

Received: 10 April 2018; Accepted: 19 June 2018; Published: 22 June 2018

Abstract: FeSiBAlNi (W5), FeSiBAlNiCo (W6-Co), and FeSiBAlNiGd (W6-Gd) high entropy alloys (HEAs) were prepared using a copper-mold casting method. Effects of Co and Gd additions combined with subsequent annealing on microstructures and magnetism were investigated. The as-cast W5 consists of BCC solid solution and FeSi-rich phase. The Gd addition induces the formation of body-centered cubic (BCC) and face-centered cubic (FCC) solid solutions for W6-Gd HEAs. Whereas, the as-cast W6-Co is composed of the FeSi-rich phase. During annealing, no new phases arise in the W6-Co HEA, indicating a good phase stability. The as-cast W5 has the highest hardness (1210 HV), which is mainly attributed to the strengthening effect of FeSi-rich phase evenly distributed in the solid solution matrix. The tested FeSiBAlNi-based HEAs possess soft magnetism. The saturated magnetization and remanence ratio of W6-Gd are distinctly enhanced from 10.93 emu/g to 62.78 emu/g and from 1.44% to 15.50% after the annealing treatment, respectively. The good magnetism of the as-annealed W6-Gd can be ascribed to the formation of Gd-oxides.

Keywords: high entropy alloys; elemental addition; annealing treatment; magnetic property; microhardness

1. Introduction

Recently, a new concept was proposed for high entropy alloys (HEAs), which has aroused wide attention and interest [1–3]. Generally, HEAs with equiatomic or near-equiatomic alloying elements mainly consist of face-centered cubic (FCC), body-centered cubic (BCC), or hexagonal closed-packed (HCP) solid solutions, and some intermetallic or amorphous phases. Owing to the special phase structure, HEAs usually possess excellent mechanical properties [4,5] and corrosion resistance [6], especially magnetic properties [7–9]. Several studies have reported that additions of certain elements into HEAs could induce the transformation of crystalline structures and further affect the related properties of HEAs [9–11]. The addition of Al, Ga, and Sn to the CoFeMnNi HEA induced the phase transition from FCC to ordered BCC phases, and further led to the significant improvement of the saturation magnetization (M_s) [9]. The microstructural evolution of $(FeCoNiCrMn)_{100-x}Al_x$ HEA system transformed from the initial single FCC structure to final single BCC structure as Al concentration increased from 0 to 20 at. % [10]. Both the tensile fracture and yield strength were enhanced with increasing Al concentration. The HEAs phases are metastable in thermodynamics, therefore, would transform to the stable microstructure after subsequent annealing, which obviously affects the properties of HEAs to some degree [3,8,12–15]. Annealing under a given condition can lead to the occurrence of phase transition from a FCC to BCC phase for $FeCoNi(CuAl)_{0.8}$ HEAs, resulting in a substantial increase in the M_s from 78.9 Am^2/kg to 93.1 Am^2/kg [15].

Recently, our group has prepared a series of as-milled FeSiBAlNi-based HEA powders using a mechanical alloying (MA) process [11,14,16]; these displayed an interesting microstructural evolution and magnetic properties. In the present study, the equiatomic FeSiBAlNiM (M = Co, Gd) HEAs were fabricated by a copper-mold spray casting technique. The effects of Co and Gd additions and subsequent annealing treatment on the microstructures, microhardness, and magnetism of the FeSiBAlNi HEAs were systematically investigated.

2. Experimental

The ingots of FeSiBAlNi, FeSiBAlNiCo, and FeSiBAlNiGd HEAs (denoted as W5, W6-Co, and W6-Gd, respectively) were prepared by an arc melting technique. The melting of these ingots was repeated at least five times to ensure the composition homogeneity in a Ti-gettered high-purity argon atmosphere. Then the ingots were remelted and made into 8 mm diameter rods by copper mold spray casting in an argon atmosphere. Then, the rods were annealed at given temperatures for two hours and cooled inside the furnace in the argon atmosphere. The annealing temperatures were set as two segments denoted as T_I and T_{II} in a low and high temperature region, respectively. There are 600 and 1000 °C for W5 HEA, 600 and 1000 °C for W6-Co HEA, and 650 and 1050 °C for W6-Gd HEA. The relatively high annealing temperatures selected for W6-Gd are attributed to the higher melting point (T_m) than that of the other two samples, as shown in the differential scanning calorimetry (DSC) curves.

Microstructural characterization of the as-cast and as-annealed HEAs were conducted by X-ray diffraction (XRD, Rigaku D8 Advance, Bruker, Germany) using Cu Kα radiation, field emission scanning electron microscopy (FESEM, QUANTA FEG 250 operated at 15 kV, Japan) coupled with energy dispersive spectrometry (EDS). The working distance used in SEM measurements was less than 10 mm. The thermal properties were analyzed by differential scanning calorimetry (DSC, TGA/DSC1, Mettler-Toledo, Greifensee, Switzerland) used under a continuous flow (30 mL/min) high-purity argon atmosphere at a heating rate of 10 K/min scanned from room temperature to 1400 °C. Microhardness of the tested HEAs was determined by a Vickers hardness tester (HV-10B), with a load of 200 g and a duration time of 15 s. The HV measurement for every tested sample was repeated ten times in order to obtain the average values. The coercive force (H_c), M_s, and remanence ratio (M_r/M_s, M_r: remanence) were determined by an alternating gradient magnetometer (AGM) at room temperature with a maximum applied field of 14000 Oe.

3. Results and Discussion

The XRD patterns of the as-cast W5, W6-Co, and W6-Gd HEAs are shown in Figure 1a. The as-cast W5 HEA consists of BCC1 (a = 4.475 Å) solid solution and FeSi-rich phase. The XRD pattern of the as-cast W6-Co HEA mainly displays the FeSi-rich phase of solution of other principal elements. In addition, other phase peaks may overlap with FeSi-rich phase peaks. Compared with the W5 and W6-Co HEAs, the effect of Gd addition on phase composition presents an obvious difference. The as-cast products of W6-Gd HEA exhibit the formation of new BCC2 (a = 4.484 Å) and FCC solid solutions. However, the FeSi-rich phase doesn't appear. The phase products of the as-cast W5, W6-Co, and W6-Gd HEAs are presented in Table 1.

Figure 1b shows the DSC curves of the as-cast HEAs. The T_m value of W6-Co HEA is 1129 °C, which is lower than that of W5 (1152 °C). However, the Gd addition increases the value of T_m, which reaches 1185 °C.

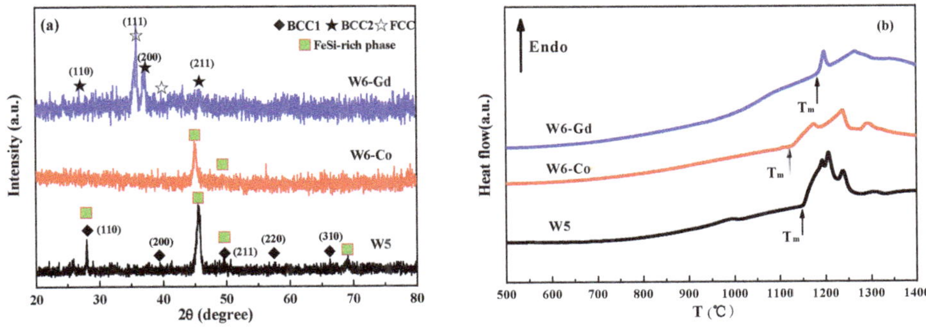

Figure 1. XRD patterns (**a**) and differential scanning calorimetry (DSC) curves (**b**) of the as-cast W5, W6-Co, and W6-Gd HEAs.

Table 1. Phase products of as-cast and as-annealed W5, W6-Co, and W6-Gd high entropy alloys (HEAs) at T_I and T_{II}, identified distinctly from XRD patterns.

HEAs	As-Cast	As-Annealed	
		T_I	T_{II}
W5	BCC1+FeSi-rich	BCC3+FeSi-rich	BCC3+FeSi-rich
W6-Co	FeSi-rich	FeSi-rich	FeSi-rich
W6-Gd	BCC2+FCC	BCC2 + FCC + AlNi + AlGd + Gd-oxide	BCC2 + FCC + AlNi + AlGd + Gd-oxide

To further investigate the difference in morphologies and compositions caused by the Co and Gd additions, FESEM coupled with EDS analysis was carried out and is presented in Figure 2 and Table 2. As shown in Figure 2a, a larger number of polygonous light-grey phases distribute dispersedly in the matrix, as well as the irregular black phases with small size. The inset of Figure 2a-1 reveals one rhombic grain with edge sizes less than 8 μm. It needs to be noted that metalloid B as a light element can't be accurately measured. Moreover, the B content of some samples is very small in most regions, therefore, they are omitted in the present study. According to the EDS results, matrix (A) contains more Al and Ni elements, and a certain amount of Fe and Si elements. Furthermore, there is a partial FeSi-rich phase in region (A) and the black region (B) is enriched with Fe and Si elements. This indicates that the FeSi-rich phase mainly exists in region (B), presenting an evenly distribution in the matrix. The rhombic grain (C) is mainly composed of Fe and B elements (instead, region (C) contains more B element above 10 at. %.). The Co addition induces the refinement of the precipitated grains (Figure 2b). The inset in Figure 2b-1 shows that the larger number of dark regions (D) with the smaller size exhibit a uniform distribution state and enrich Fe and Si elements. The gray region (E) is rich in the Ni element, and the bright-grey region (F) is poor in the Al element. Moreover, the component ratio of Fe and Si in regions (E) and (F) is close to 1:1. Although several phases appear in the SEM images, no peaks other than the FeSi-rich phase can be seen in the XRD results of as-cast W6-Co HEA (Figure 1). It suggests that the precipitates probably have a similar lattice constant and crystal structure concerning the matrix [17]. Moreover, it could be a complex compositional fluctuation in the as-cast W6-Co HEA [18]. Figure 2c and inset (Figure 2c-1) present that the as-cast W6-Gd HEA consists of coarse rod-like dendrites as the FCC-matrix phase. They are rich in each principal element with near-equiatomic ratio except for Al element (region (G)). However, the Al element segregates in the interdendritic grains ((H): dark- and deep-grey regions) corresponding to the BCC2 phase with less contents. Usually the precipitation pathways in HEAs can be very complex, and it is a particularly challenging topic, which remains to be studied [19].

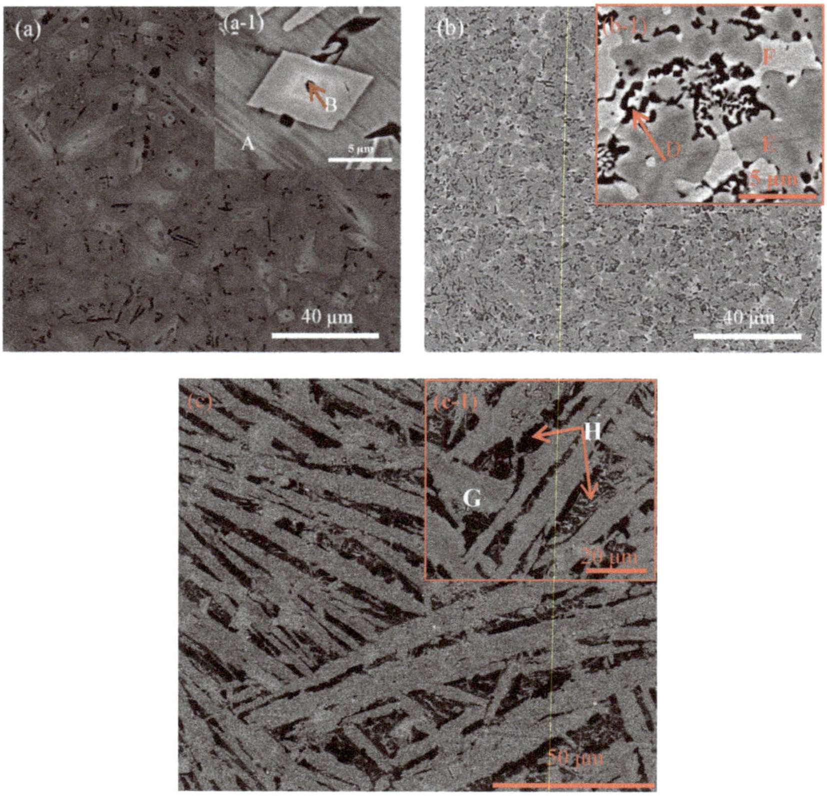

Figure 2. Field emission scanning electron microscopy (FESEM) micrographs of the as-cast high entropy alloys (HEAs): (**a**) W5; (**b**) W6-Co; and (**c**) W6-Gd. The insets of (a-1), (b-1), and (c-1) are the partial enlargements corresponding to (**a**), (**b**), and (**c**), respectively.

Table 2. Chemical compositions of representative regions for W5, W6-Co, and W6-Gd HEAs obtained by energy dispersive spectrometry (EDS).

Regions	Elements (at. %)						
	Fe	Si	B	Al	Ni	Co	Gd
A	11.36	10.01	-	36.29	42.34	-	-
B	39.6	33.8	-	12.95	13.65	-	-
C	80.9	1.65	10.97	2.6	3.87	-	-
D	41.15	27.85	-	6.38	10.49	14.13	-
E	11.42	12.51	-	17.59	34.78	23.7	
F	24.72	22.07	-	2.1	27.66	23.45	
G	24.5	27.9	-	0.48	17.41	-	29.71
H	14.57	2.55	-	51.32	11.06	-	20.5

Figure 3 shows the XRD patterns of the as-annealed W5, W6-Co, and W6-Gd HEAs at different temperatures; their annealing products are also listed in Table 1. After annealing at T_I, the annealed products of W5 HEAs consist of a new BCC3 (a = 4.033 Å) solid solution with a FeSi-rich phase. However, the contents of the FeSi-rich phase obviously decrease compared to that of the as-cast state. Moreover, the BCC1 solid solution disappears (Figure 3a). Via annealing at T_{II} the two obtained phases still exist, but the diffraction peak intensity becomes strong, indicating the further growth and coarsening of the grains. Being distinct from the W5 HEA, no new phase transformation occurs in the W6-Co HEA after annealing at T_I and T_{II}, indicating that the W6-Co HEA possesses a good thermal stability (Figure 3b). It suggests that the Co addition leads to the transformation from the metastable characteristic of W5 HEA to a more stable state in thermodynamics. However, the inset presents that the main diffraction peak of the W6-Co HEA shifts to the lower angle with the increased annealing temperature, suggesting a serious lattice distortion caused by the expansion of the lattice. Figure 3c reveals the formation of new phases of AlNi, AlGd, Gd-oxides, besides the primary BCC2 and FCC solid solutions for the as-annealed W6-Gd HEA at T_I. The as-annealed products are unchanged at T_{II}, except for the increased phase amounts of Gd-oxides. Moreover, compared with the W5 and W6-Co HEA, the highest T_m value of the W6-Gd HEA may be attributed to the high T_m values of precipitated intermetallic compounds of AlNi (1638 °C) and AlGd (1200 °C).

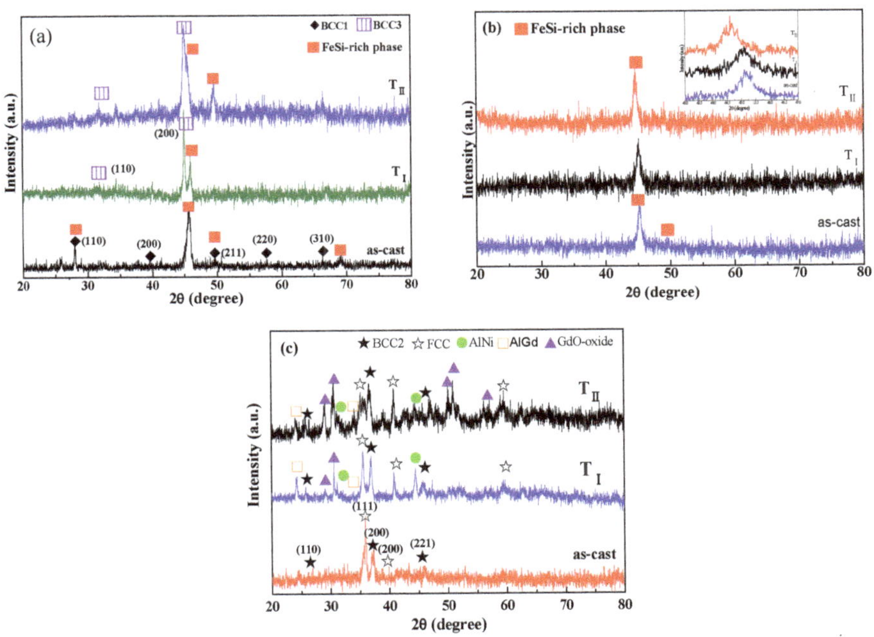

Figure 3. XRD patterns of the as-annealed HEAs at different temperatures (T_I and T_{II}): (**a**) W5, (**b**) W6-Co, and (**c**) W6-Gd.

Figure 4 shows the Vickers hardness (HV) of the as-cast and as-annealed HEAs. The W5 HEA displays the highest HV among the tested HEAs, and the as-cast W5 HEA possesses the highest hardness of 1210 HV. The additions of Co and Gd cause the decline of HV values, and the as-annealed W6-Gd HEA (T_{II}) displays the maximal decline of HV (738). It suggest that the annealing treatment plays a negative effect on the HV of the as-cast samples, which is in agreement with Salishchec's results [20]. With the increased annealing temperature, the internal stress of the as-cast HEAs gradually decreases as well as the microstructural coarsening. The effect of solid solution strengthening became the smaller, and strain softening was revealed in HEAs [21]. Compared with W6-Co and W6-Gd HEAs, the FeSi-rich phase, as the second strengthening phase in the W5 HEAs (especially the as-cast one) evenly distributes in the BCC solid solution matrix, which can contribute to the high HV values.

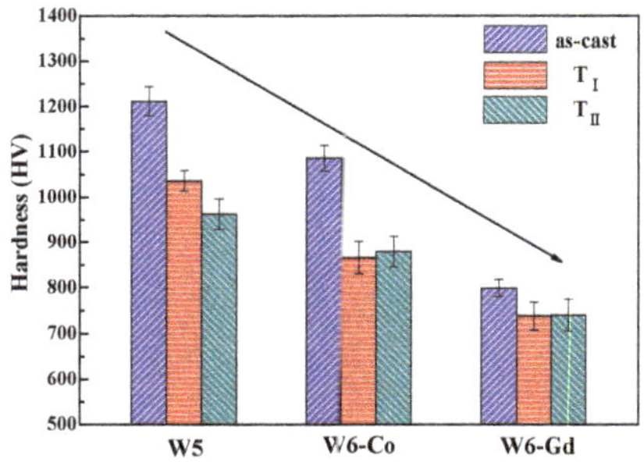

Figure 4. Vickers hardness of the as-cast and as-annealed W5, W6-Co, and W6-Gd HEAs.

The mass magnetization (M) as a function of the magnetic field intensity (H) for the as-cast and as-annealed samples was tested. The H_c, M_s, and M_r/M_s of these HEAs are shown in Figure 5. All H_c values of the tested HEAs are in the range from 10 to 180 Oe (Figure 5a), indicating the soft magnetism nature of these HEAs. It suggests that the annealing treatment induces a weak decrease of H_c values for the W5 and W6-Co HEAs, but the H_c of W6-Gd HEA becomes large after annealing. The H_c is mainly affected by impurity, deformation, crystallite size, and stress, and the subsequent heat-treatment process [22]. Therefore, the H_c values of as-annealed W5 and W6-Co HEAs are slightly lower than those of the as-cast samples, suggesting that the former possess a little larger average crystallite size according to the well-known coercivity-crystal size relationship [8]. Moreover, the origin of the lower H_c can be attributed to the low number density of domain-wall pinning sites [23]. The as-annealed products of W6-Gd HEA contain complex phase compositions, and display the inhomogeneous characteristics, which obviously work towards the pinning effect of domain wall movement. Therefore the H_c values can be enhanced after the annealing treatment.

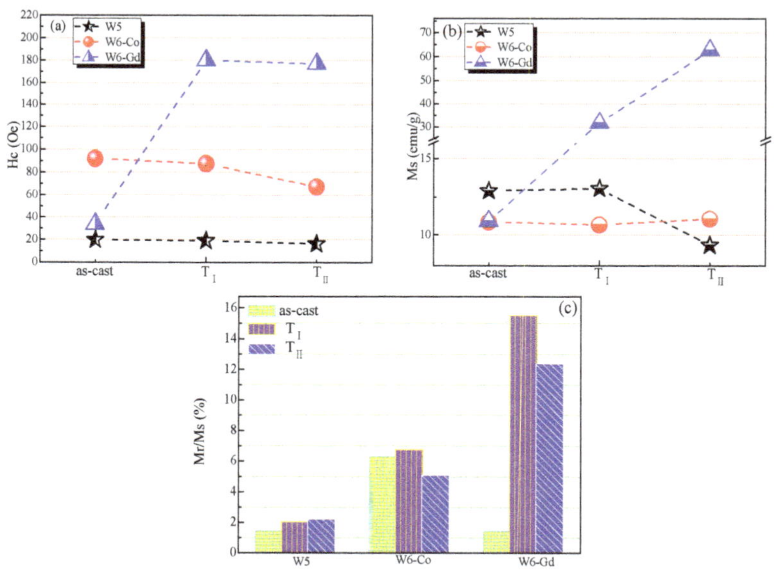

Figure 5. Magnetic properties of the as-cast and as-annealed W5, W6-Co, and W6-Gd HEAs: H_c (**a**), M_s (**b**) and M_r/M_s (**c**).

The variations of M_s are exhibited in Figure 5b. In the as-cast state, there is no distinct difference in M_s for all the tested samples, and W5 HEA emerges with a slightly higher M_s of 12.91 emu/g. After annealing at T_I, the M_s of W5 HEA remains nearly unchanged, whereas declines with a reduction of 27.7% at T_{II}. There is no obvious magnetism changes revealed for the as-cast and as-annealed W6-Co HEAs, and the M_s values become stabilized at about 11 emu/g. This stability of M_s is resulted from the stable phase characteristic of W6-Co HEA in the annealing stage. Unlike W5 and W6-Co HEAs, the magnetism of W6-Gd HEA is enhanced during annealing treatment. The M_s value of the as-annealed W6-Gd HEA increased from 10.93 emu/g to 31.91 emu/g at T_I, and further up to 62.78 emu/g at T_{II}, suggesting increased soft magnetic properties.

From Figure 5c, the M_r/M_s values of as-cast W5 and W6-Co HEAs are similar to their as-annealed states, which depend on their similar phase compositions. The as-annealed products of W6-Gd HEAs are significantly different from the as-cast one, and the M_r/M_s values are enhanced from 1.44% (as-cast) to 15.5% (at T_I). Moreover, the as-annealed W6-Gd HEAs reveal the highest M_r/M_s values among the tested samples, indicating a better soft magnetism.

Residual stress exists in the as-cast HEAs which can deteriorate the soft magnetic properties. Appropriate heat treatment can induce stress relief, which is beneficial to improve the soft magnetic properties [24]. Therefore, except for the as-annealed W5 HEA at T_{II}, the soft magnetic properties of the tested HEAs are properly improved by the structural relaxation through stress-relief annealing [25]. Notably, it suggests that M_s of the as-annealed W6-Gd HEA at T_{II} is about five times higher than that obtained for the as-cast one. Moreover, the magnetic properties are strongly dependent on the microstructure of the materials. The microstructure contribution to magnetism arises from morphology: properties such as magnetic anisotropy, magnetostriction, coercivity, and volume fraction of the precipitates. The decrease in M_s for as-annealed W5 at T_{II} can be related to the enhanced density of grain boundaries and the increase of volumetric fraction of BCC3 solid solutions around FeSi-rich phases, which reduce the magnetic moment. According to the effect of phase compositions on magnetic properties, the increase in M_s for the as-annealed W6-Gd HEA can be ascribed to the formation of Gd-oxides. Moreover, M_s is enhanced by increasing the contents of Gd-oxides after elevating the annealing temperature.

4. Conclusions

The phase composition, microstructures, microhardness, and magnetic properties of as-cast and as-annealed W5, W6-Co, and W6-Gd HEAs have been investigated. The as-cast and as-annealed W6-Co HEAs maintain the same phase compositions, and are composed of single FeSi-rich phases, indicating the stable phase characteristic. The addition of Gd obviously enhances T_m (1185 °C) compared with W5, owing to the exhibition of AlNi and AlGd with high melting points. As-cast W5 possesses the highest hardness of 1210 HV, which is attributed to the uniform distribution of the FeSi-rich phase in the matrix. All the tested HEAs display soft magnetic properties. Moreover, the M_s and M_r/M_s values of W6-Gd were enhanced from 10.93 emu/g to 62.78 emu/g and from 1.44% to 15.50% via the annealing process, respectively. It suggests that Gd-oxides are beneficial to the enhancement of magnetic properties in W6-Gd.

Author Contributions: Conceived and designed the experiments, S.Z. and S.X.; Reviewed relevant studies and literature, W.W.; Performed the experiments, J.X. and Z.Z.; Analyzed the data, S.X., W.W. and Y.W.; Wrote the paper with the help of the rest of the authors, S.Z. and Z.Z.; Provided feasible advices and critical revision of the manuscript, Y.W.; All authors have read and approved the final manuscript.

Funding: This research was funded by the National Natural Science Foundation of China (No. 51671095).

Conflicts of Interest: The authors declare no conflict of interest.

References

1. Yeh, J.W.; Chen, S.K.; Lin, S.J.; Gan, J.Y.; Chin, T.S.; Shun, T.T.; Tsau, C.H.; Chang, S.Y. Nanostructured High-Entropy Alloys with Multiple Principal Elements: Novel Alloy Design Concepts and Outcomes. *Adv. Eng. Mater.* **2004**, *6*, 299–303. [CrossRef]
2. Cantor, B.; Chang, I.T.H.; Knight, P.; Vincent, A.J.B. Microstructure development in equiatomic multicomponent alloys. *Mater. Sci. Eng. A* **2004**, *375–377*, 213–218. [CrossRef]
3. Zhang, Y.; Zuo, T.T.; Tang, Z.; Gao, M.C.; Dahmen, K.A.; Liaw, P.K.; Lu, Z.P. Microstructures and properties of high-entropy alloys. *Prog. Mater. Sci.* **2014**, *61*, 1–93. [CrossRef]
4. Dong, Y.; Zhou, K.Y.; Lu, Y.P.; Gao, X.X.; Wang, T.M.; Li, T.J. Effect of Vanadium Addition on the Microstructure and Properties of AlCoCrFeNi High Entropy Alloy. *Mater. Des.* **2014**, *57*, 67–72. [CrossRef]
5. Wu, Z.; Bei, H.; Pharr, G.M.; George, E.P. Temperature dependence of the mechanical properties of equiatomic solid solution alloys with face-centered cubic crystal structures. *Acta Mater.* **2014**, *81*, 428–441. [CrossRef]
6. Chen, Y.Y.; Duval, T.; Hung, U.D.; Yeh, J.W.; Shih, H.C. Microstructure and electrochemical properties of high entropy alloys-a comparison with type-304 stainless steel. *Corros. Sci.* **2005**, *47*, 2257–2279. [CrossRef]
7. Ji, W.; Wang, W.M.; Wang, H.; Zhang, J.Y.; Wang, Y.C.; Zhang, F.; Fu, Z.Y. Alloying behavior and novel properties of CoCrFeNiMn high-entropy alloy fabricated by mechanical alloying and spark plasma sintering. *Intermetallics* **2015**, *56*, 24–27. [CrossRef]
8. Li, P.P.; Wang, A.D.; Liu, C.T. A ductile high entropy alloy with attractive magnetic properties. *J. Alloys Compd.* **2017**, *694*, 55–60. [CrossRef]
9. Zuo, T.T.; Gao, M.C.; Ouyang, L.Z.; Yang, X.; Cheng, Y.Q.; Feng, R.; Chen, S.Y.; Liaw, P.K.; Hawk, J.A.; Zhang, Y. Tailoring magnetic behavior of CoFeMnNiX (X = Al, Cr, Ga, and Sn) high entropy alloys by metal doping. *Acta Mater.* **2017**, *130*, 10–18. [CrossRef]
10. He, J.Y.; Liu, W.H.; Wang, H.; Wu, Y.; Liu, X.J.; Nieh, T.G.; Lu, Z.P. Effects of Al addition on structural evolution and tensile properties of the FeCoNiCrMn high-entropy alloy system. *Acta Mater.* **2014**, *62*, 105–113. [CrossRef]
11. Xu, J.; Axinte, E.; Zhao, Z.; Wang, Y. Effect of C and Ce addition on the microstructure and magnetic property of the mechanically alloyed FeSiBAlNi high entropy alloys. *J. Magn. Magn. Mater.* **2016**, *414*, 59–68. [CrossRef]
12. Zhuang, Y.X.; Xue, H.D.; Chen, Z.Y.; Hu, Z.Y.; He, J.C. Effect of annealing treatment on microstructures and mechanical properties of FeCoNiCuAl high entropy alloys. *Mater. Sci. Eng. A* **2013**, *572*, 30–35. [CrossRef]
13. Zhu, X.; Zhou, X.; Yu, S.; Wei, C.; Xu, J.; Wang, Y. Effects of annealing on the microstructure and magnetic property of the mechanically alloyed FeSiBAlNiM (M = Co, Cu, Ag) amorphous high entropy alloys. *J. Magn. Magn. Mater.* **2017**, *430*, 59–64. [CrossRef]
14. Wang, J.; Zheng, Z.; Xu, J.; Wang, Y. Microstructure and magnetic properties of mechanically alloyed FeSiBAlNi (Nb) high entropy alloys. *J. Magn. Magn. Mater.* **2014**, *355*, 58–64. [CrossRef]
15. Zhang, Q.; Xu, H.; Tan, X.H.; Hou, X.L.; Wu, S.W.; Tan, G.S.; Yu, L.Y. The effects of phase constitution on magnetic and mechanical properties of $FeCoNi(CuAl)_x$ (x = 0–1.2) high-entropy alloys. *J. Alloys Compd.* **2017**, *693*, 1061–1067. [CrossRef]
16. Xu, J.; Shang, C.; Ge, W.; Jia, H.; Liaw, P.K.; Wang, Y. Effects of elemental addition on the microstructure, thermal stability, and magnetic properties of the mechanically alloyed FeSiBAlNi high entropy alloys. *Adv. Powder Technol.* **2016**, *27*, 1418–1426. [CrossRef]
17. Tsai, M.H.; Yuan, H.; Cheng, G.; Xu, W.; Tsai, K.Y.; Tsai, C.W.; Jian, W.W.; Juan, C.C.; Shen, W.J.; Chuang, M.H.; et al. Morphology, structure and composition of precipitates in $Al_{0.3}CoCrCu_{0.5}FeNi$ high-entropy alloy. *Intermetallics* **2013**, *32*, 329–336. [CrossRef]
18. Chen, C.; Pang, S.; Cheng, Y.; Zhang, T. Microstructure and mechanical properties of $Al_{20-x}Cr_{20+0.5x}Fe_{20}Co_{20}Ni_{20+0.5x}$ high entropy alloys. *J. Alloys Compd.* **2016**, *659*, 279–287. [CrossRef]
19. Zhang, Y.; Zuo, T.; Cheng, Y.; Liaw, P.K. High-entropy alloys with high saturation magnetization, electrical resistivity, and malleability. *Sci. Rep.* **2013**, *3*, 1455. [CrossRef]
20. Salishchev, G.A.; Tikhonovsky, M.A.; Shaysultanov, D.G.; Stepanov, N.D.; Kuznetsov, A.V.; Kolodiy, I.V.; Tortika, A.S.; Senkov, O.N. Effect of Mn and V on structure and mechanical properties of high-entropy alloys based on CoCrFeNi system. *J. Alloys Compd.* **2014**, *591*, 11–21. [CrossRef]

21. Zhu, Z.G.; Ma, K.H.; Yang, X.; Shek, C.H. Annealing effect on the phase stability and mechanical properties of $(FeNiCrMn)_{(100-x)}Co_x$ high entropy alloys. *J. Alloys Compd.* **2017**, *695*, 2945–2950. [CrossRef]
22. Sun, G.F.; Qiang, W.J. *Magnetic Material*; Chemical Industry Press: Beijing, China, 2007.
23. Bitoh, T.; Makino, A.; Inoue, A. Origin of low coercivity of $(Fe_{0.75}B_{0.15}Si_{0.10})_{100-x}Nb_x$(x = 1–4) glassy alloys. *J. Appl. Phys.* **2006**, *99*, 08F102. [CrossRef]
24. Wei, R.; Tao, J.; Sun, H.; Chen, C.; Sun, G.W.; Li, F.S. Soft magnetic $Fe_{26.7}Co_{26.7}Ni_{26.6}Si_9B_{11}$ high entropy metallic glass with good bending ductility. *Mater. Lett.* **2017**, *197*, 87–89. [CrossRef]
25. Kong, F.L.; Chang, C.T.; Inoue, A.; Shalaan, E.; Al-Marzouki, F. Fe-based amorphous soft magnetic alloys with high saturation magnetization and good bending ductility. *J. Alloys Compd.* **2014**, *615*, 163–166. [CrossRef]

entropy

MDPI

Article

Evaluation of Radiation Response in CoCrFeCuNi High-Entropy Alloys

Yang Wang [1,2], Kun Zhang [1,2,*], Yihui Feng [2,3], Yansen Li [1,2], Weiqi Tang [1,2] and Bingchen Wei [1,2,*]

1. Key Laboratory of Microgravity (National Microgravity Laboratory), Institute of Mechanics, Chinese Academy of Sciences, Beijing 100190, China; wangyang2@imech.ac.cn (Y.W.); liyansen@imech.ac.cn (Y.L.); tangweiqi16@mails.ucas.ac.cn (W.T.)
2. School of Engineering Science, University of Chinese Academy of Sciences, Beijing 100049, China
3. State Key Laboratory of Nonlinear Mechanics, Institute of Mechanics, Chinese Academy of Sciences, Beijing 100190, China; fengyh@lnm.imech.ac.cn

* Correspondence: zhangkun@imech.ac.cn (K.Z.); weibc@imech.ac.cn (B.W.); Tel.: +86-108-254-4108 (K.Z.); +86-108-254-4118 (B.W.)

Received: 10 October 2018; Accepted: 26 October 2018; Published: 31 October 2018

Abstract: CoCrFeCuNi high-entropy alloys (HEAs) prepared by arc melting were irradiated with a 100 keV He^+ ion beam. Volume swelling and hardening induced by irradiation were evaluated. When the dose reached 5.0×10^{17} ions/cm^2, the Cu-rich phases exhibited more severe volume swelling compared with the matrix phases. This result indicated that the Cu-rich phases were favorable sites for the nucleation and gathering of He bubbles. X-ray diffraction indicated that all diffraction peak intensities decreased regularly. This reduction suggested loosening of the irradiated layer, thereby reducing crystallinity, under He^+ ion irradiation. The Nix-Gao model was used to fit the measured hardness in order to obtain a hardness value H_0 that excludes the indentation size effect. At ion doses of 2.5×10^{17} ions/cm^2 and 5.0×10^{17} ions/cm^2, the HEAs showed obvious hardening, which could be attributed to the formation of large amounts of irradiation defects. At the ion dose of 1.0×10^{18} ions/cm^2, hardening was reduced, owing to the exfoliation of the original irradiation layer, combined with recovery induced by long-term thermal spike. This study is important to explore the potential uses of HEAs under extreme irradiation conditions.

Keywords: high-entropy alloy; ion irradiation; hardening behavior; volume swelling

1. Introduction

High-entropy alloys (HEAs) have recently drawn increased interest because of their distinct compositions, microstructures, and flexible properties. In contrast to conventional alloys, HEAs are composed of more than five principal elements at equal or nearly equal atomic percentages (at.%). HEAs exerts four primary effects: (1) high-entropy effect; (2) sluggish diffusion effect; (3) severe lattice distortion effect; and (4) cocktail effect [1–4]. These effects render HEAs more likely to form a simple solid-solution structure rather than an intermetallic compound, which confers distinct properties on HEAs, including high fatigue strength [5], high hardness [6], good abrasion resistance and corrosion resistance, high breaking strength at low temperatures [7–10], and good softening resistance at elevated temperatures [11]. The use of HEAs as structural materials under extreme environments has been proposed owing to their desirable mechanical properties and thermodynamic stability.

Previous studies on the irradiation effects of HEAs have mostly focused on irradiation resistance and phase stability. Zhang et al. [12] evaluated the effects of irradiation on $Al_xCoCrFeNi$ (x = 0.1, 0.75, and 1.5) HEAs under 3 MeV Au^+ ion irradiation at room temperature. Results indicated that compared with conventional nuclear materials, single-phase HEAs based on the face-centered cubic

(FCC) structure in the $Al_xCoCrFeNi$ system showed improved radiation resistance; in addition, volume swelling in the $Al_xCoCrFeNi$ alloys in the ascending order was FCC < FCC + BCC < BCC (body-centered cubic). Yang et al. [13] recently reported that with temperature increased from 523 K to 923 K, the irradiation-induced defect density of $Al_{0.1}CoCrFeNi$ HEAs decreased, whereas the size of the defect increased. Meanwhile, irradiation led to Ni and Co enrichment as well as Fe, Cr, and Al depletion in dislocation loops and dislocation regions. T. Nagase. et al. [14] showed that as-sputtered CoCrFeMnNi HEAs maintained good irradiation resistance within a wide temperature range from 298 to 773 K without grain coarsening under fast electron irradiation. Jin et al. studied a Ni-based multicomponent alloy under 3 MeV Ni ion irradiation at 773 K [15,16] and found that irradiation-induced volume swelling decreased with an increase in the number of elements in the disordered solid solution under identical irradiation conditions. This finding suggested that irradiation-induced volume swelling was also strongly affected by compositional complexity.

In the last decade, FeCoNi-based HEAs, as one of the successful HEAs, have attracted more and more attention, especially because of its mechanical properties and microstructure evolution [17]. M. Klimova et al. [18] studied the microstructure and mechanical properties evolution of the Al-, C-containing CoCrFeNiMn-type high-entropy alloy during cold rolling. They reported that rolling resulted in an increase in strength and a decrease in ductility of the Al-, C-containing CoCrFeNiMn-type alloy. Feng et al. [19] found that the short range order is positive between Al-Al, Al-Si, Si-Si pairs and negative between Ni-Al, Co-Si, Fe-Co, Ni-Si, and Fe-Si pairs, which leads to an increase in the elastic modulus by sacrificing ductility and isotropy. In addition, the appropriate doping of Y_2O_3 as a reinforcement phase in CoCrFeMnNi HEAs could increase both the room temperature tensile strength and the wear-resistance [20].

With regard to the irradiation response of HEAs, there are not many researches on the evolution of mechanical properties of FeCoNi-based HEAs after irradiation. Meanwhile, most studies focused on the effect of composition on irradiation-induced swelling and vertical inhomogeneity of radiation damage, and the difference in irradiation response between the phase and phase boundary has rarely been reported. Whether lateral inhomogeneity of irradiation response exists in the HEAs, especially in single-phase FCC structure-based HEAs, has yet to be determined. In the present study, a typical single-phase FCC structure-based CoCrFeCuNi HEA was selected, and the key objective is to determine the irradiation response exerted by He^+ ion irradiation. The present study provides an initial examination of the fundamental irradiation behavior of an HEA material, thus offering an insight into the potential of this family of materials for application under extreme environments.

2. Materials and Methods

CoCrFeCuNi HEAs with a diameter of 5 mm and a length of 100 mm were prepared by arc melting a mixture of pure metal (purity > 99.9% wt %) in a Ti-gettered high-purity argon atmosphere. These ingots were remelted at least four times to prevent chemical heterogeneity and were eventually drop-cast into a copper mold. The cast bar was cut into thin pieces, each with a thickness of 2 mm, and then mechanically polished to a mirror finish. Irradiation experiments were conducted in a BNU-400 kV electrostatic accelerator in a test chamber, applying pressure near or below 10^{-5} torr. The polished samples were irradiated with a 100 keV He^+ ion beam at fluences of 2.5×10^{17} ions/cm^2, 5.0×10^{17} ions/cm^2, and 1.0×10^{18} ions/cm^2 at a normal angle at room temperature.

Figure 1 shows the results for ion distribution, irradiation damage (displacement per atom or dpa), and concentration of He atoms with the increases in depth calculated using the Ion Distribution and Quick Calculation of Damage mode in the SRIM (Stopping and Range of Ions in Matter) 2008 code [21]. As shown in Figure 1a, when a large number of incident ions enter the HEAs, they form a spatial influence area rather than a specific path owing to collision cascade. The total number of ions is 99,999, and the effective depth of ion influence is ~560 nm. The dpa is determined by the sum of the predicted Co, Cr, Cu, Fe, and Ni vacancy concentrations and recoil events within 0 nm to ~560 nm from the surface. According to Figure 1b,c, the three curves in each figure linearly change with increased ion doses, implying that the dpa and concentration of He atoms are positively correlated with the total doses of incident ions. Equations (1) and (2) present the method used to extract the irradiation damage (total vacancies produced/atom = dpa) and He atom concentration (in at.%) from SRIM output files [21].

$$\overbrace{\left(\frac{vacancies}{ions \times \text{Å}}\right)}^{\text{vacancy.txt}} \times \left(\frac{10^8\left(\frac{\text{Å}}{cm}\right) \times \text{Fluence}\left(\frac{ions}{cm^2}\right)}{8.705 \times 10^{22}\left(\frac{atoms}{cm^3}\right)}\right) = \text{dpa}, \tag{1}$$

$$\overbrace{\left(\frac{atoms/cm^3}{atoms/cm^2}\right)}^{\text{range.txt}} \times \left(\frac{\text{Fluence}\left(\frac{ions}{cm^2}\right)}{8.705 \times 10^{22}\left(\frac{atoms}{cm^3}\right)}\right) \times 100 = \left(\frac{ions}{atom}\right) \times 100 = \text{at\% ion}, \tag{2}$$

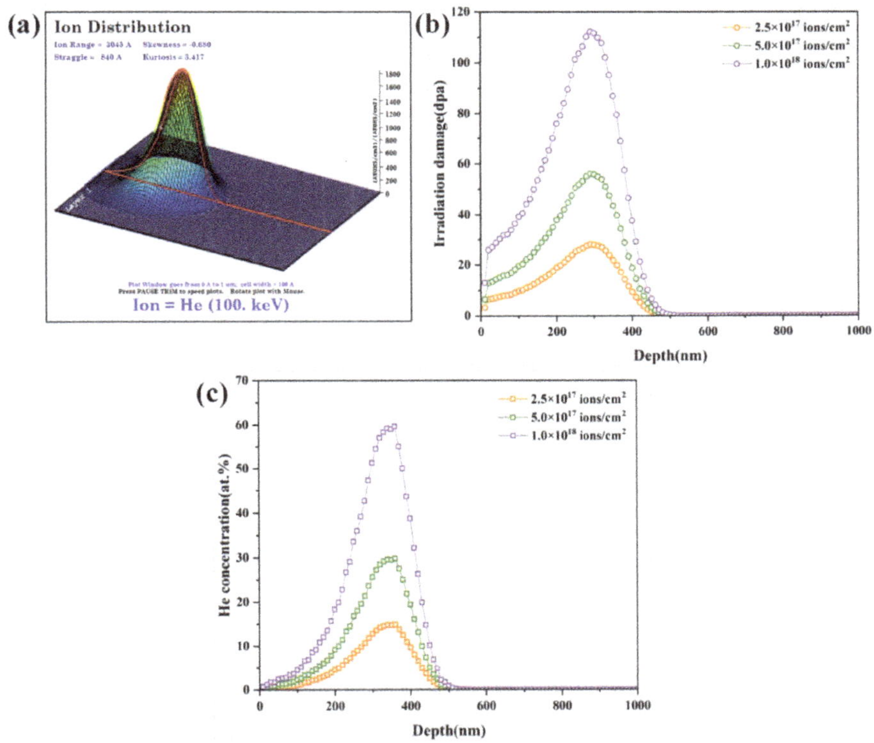

Figure 1. SRIM simulations of CoCrFeCuNi HEAs irradiated with 100 keV He^+ ions. (**a**) Ion distribution; (**b**) irradiation damage (dpa); and (**c**) concentration of He atoms with depth.

In addition, the number of atomic displacements reached the peak at a depth of approximately 290 nm, which corresponded to the projected range of the He^+ ions in CoCrFeCuNi HEAs. The peak concentrations of the He atoms were approximately 14.9%, 29.8%, and 59.6% under different doses.

The phase structures of pristine and irradiated samples were analyzed by slow-scan X-Ray diffraction (XRD, Philips PW 1050 diffractometer with CuKα radiation, Amsterdam, Netherlands) at 45 kV voltage, 0.01 step size, and 200 mA current for phase analysis with improved accuracy. The microstructure and compositions of pristine and irradiated samples were analyzed by scanning electron microscopy (SEM, HITACHI S-4800, Tokyo, Japan) taken at a working voltage of 20.0 kV and a working distance of 14.3 and 15.0 mm, with energy-dispersive X-ray spectroscopy (EDX, HORIBA X-Max, Kyoto, Japan).

Nanoindentation experiments on pristine and irradiated samples were conducted using the Agilent Technologies Nano Indenter G200 with a stand Berkovich diamond indenter at room temperature (29.0 °C). The hardness measurement parameter included a strain rate of ~0.05 s^{-1}, frequency of ~45 Hz, and harmonic displacement of ~2.0 nm. Each nanoindentation test ran to a maximum of 1000 nm into the indented surface, and all hardness measurement positions were in the matrix phases. Specifically, the sample irradiated at fluences of up to 1.0×10^{18} ions/cm^2, the hardness measurement positions were located apart from the exfoliation of the damaged areas, that is, on the remaining part underneath. The geometry of the tip was standardized from the indentation on fused silica with the same indentation depth, and each sample was subjected to 5 indents with a space wider than 50 μm.

3. Results

3.1. Microstructural Characterization

The XRD patterns of the pristine and irradiated CoCrFeCuNi HEA samples at different doses are presented in Figure 2. The figure reveals that the samples remained fully crystalline, and the diffraction peaks of CoCrFeCuNi HEAs in both pristine and irradiated samples at 43°, 50°, and 74° are from the typical FCC phases corresponding to (111), (200), and (220) lattice planes, respectively. These findings are similar to those in previous studies [22,23]. Figure 2 shows that no obvious phase decomposition occurs under any irradiation condition, which is consistent with high configurational entropy playing a significant role in single-phase stability [24]. A closer examination reveals that each major peak contains two slightly separated peaks. The left peak of the major split peaks is attributed to the Cu-rich phases, whereas the right peak is attributed to the matrix phases. These findings are consistent with the following EDX results. The separated major peaks are hardly differentiated at low angles because no significant difference in average atomic radius between the Cu-rich phases and matrix phases in CoCrFeCuNi HEA is indicated. Moreover, the diffraction peaks of the irradiated HEAs shift to the left, meaning a slight lattice expansion happens. Meanwhile, the intensity of the diffraction peaks decreases with the increasing of irradiation dose, owing to the increased concentration of defects in the near surface of HEAs.

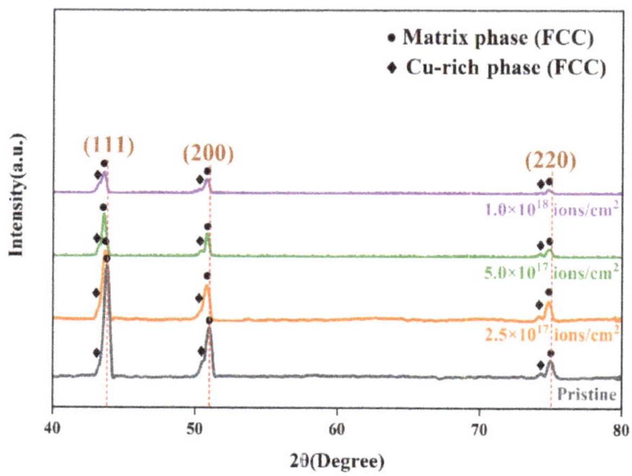

Figure 2. XRD patterns of pristine and He^+ ion-irradiated CoCrFeCuNi HEAs at different doses.

Irradiation has been known to enhance diffusion in solids by creating defects, such as vacancies and interstitials. These defects fill the irradiated layer with voids, eventually leading to a decrease in crystallinity. Irradiation-induced defects generally lead to structural damage in the HEAs, such as cluster of small defects and dislocation loops [25]. According to Makinson et al. [26], the peak diffraction intensity of X-rays depends on the concentration of defects in a material. Meanwhile, peak intensity decreases with an increase in defect concentration, and vice versa. In the present study, the defect concentration induced by He^+ ion irradiation increases with an increase in irradiation dose of up to 1.0×10^{18} ions/cm^2, leading to a decrease in peak intensity. Moreover, the low solubility of helium in CoCrFeCuNi HEAs leads to its precipitation into bubbles and voids [27], further decreasing the peak intensities.

3.2. Surface Topography

The SEM images reveal the surface morphology of pristine and irradiated CoCrFeCuNi HEAs at different irradiation doses, as shown in Figure 3a–d. Table 1 shows the components of different regions in the microstructures of pristine and irradiated CoCrFeCuNi alloys estimated by EDX spectroscopy. The analysis was conducted on five pristine and five irradiated regions randomly chosen from the samples. The EDX maps showing the elemental distribution of Cr, Co, Fe, Cu, and Ni on the HEAs surface at different doses, are shown in Figure 4a–d, respectively. The EDX maps show matrix phases enriched with Cr, Fe, Co, and Ni, as well as Cu-rich phases at the phase boundaries with fair amounts of Ni, and no obvious diffusion occurs in the HEA components after irradiation, implying that the composition of HEA is stable after irradiation to some extent. Nevertheless, the distribution of the elements is very homogeneous in both the matrix phases and Cu-rich phases. Moreover, both the matrix phases and Cu-rich phases consist of only one simple phase, which is consistent with previous studies [28]. According to Figure 3b–d, when the irradiation dose increases to 5.0×10^{17} ions/cm^2, the volume swelling of the Cu-rich phases markedly increases in severity. Meanwhile, no significant change is observed in the matrix phases, suggesting the inhomogeneity of radiation resistance on the surface perpendicular to the ion incident direction. The evolution of volume swelling after irradiation is discussed in the subsequent section.

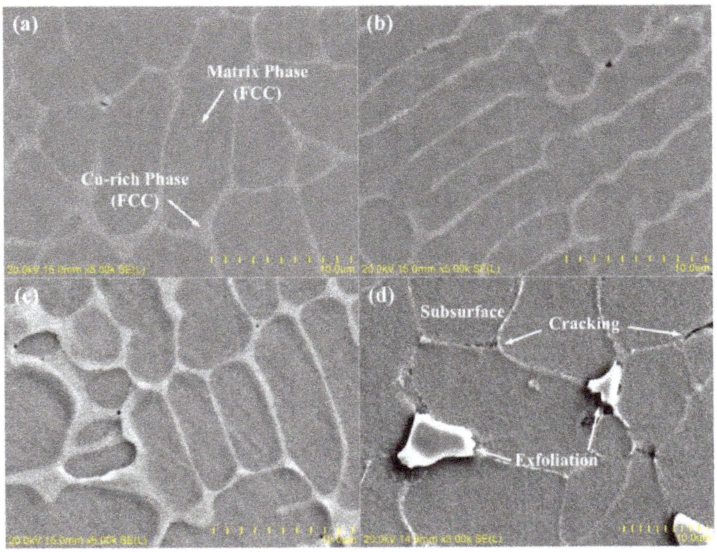

Figure 3. SEM images of CoCrFeCuNi HEAs. (**a**) Pristine and (**b–d**) He^+ ions irradiated at fluences of 2.5×10^{17} ions/cm^2, 5.0×10^{17} ions/cm^2, and 1.0×10^{18} ions/cm^2.

Table 1. Components of different regions in the microstructure of CoCrFeCuNi HEAs (at. %).

Condition	Region	Co	Cr	Fe	Cu	Ni
Nominal	—	20	20	20	20	20
Pristine	Matrix Phase	22.87 ± 0.52	23.08 ± 0.24	22.67 ± 0.46	10.75 ± 0.62	20.46 ± 0.50
	Cu-rich Phase	5.60 ± 0.57	6.26 ± 0.85	6.53 ± 0.62	68.86 ± 2.15	12.76 ± 0.48
2.5×10^{17}	Matrix Phase	22.83 ± 0.47	23.41 ± 0.19	22.59 ± 0.32	10.64 ± 0.11	20.38 ± 0.31
ions/cm^2	Cu-rich Phase	6.29 ± 0.71	7.11 ± 0.49	6.92 ± 0.91	69.39 ± 3.18	10.29 ± 0.62
5.0×10^{17}	Matrix Phase	22.66 ± 0.65	23.23 ± 0.38	23.06 ± 0.43	10.61 ± 0.36	20.43 ± 0.59
ions/cm^2	Cu-rich Phase	4.69 ± 0.32	5.19 ± 1.24	5.19 ± 0.92	74.09 ± 3.06	10.85 ± 0.72
1.0×10^{18}	Matrix phase	23.55 ± 0.12	23.30 ± 0.18	22.85 ± 0.75	10.68 ± 0.29	19.62 ± 0.61
ions/cm^2	Cu-rich phase	9.84 ± 0.99	11.33 ± 1.50	11.00 ± 0.92	51.66 ± 3.93	16.19 ± 0.77

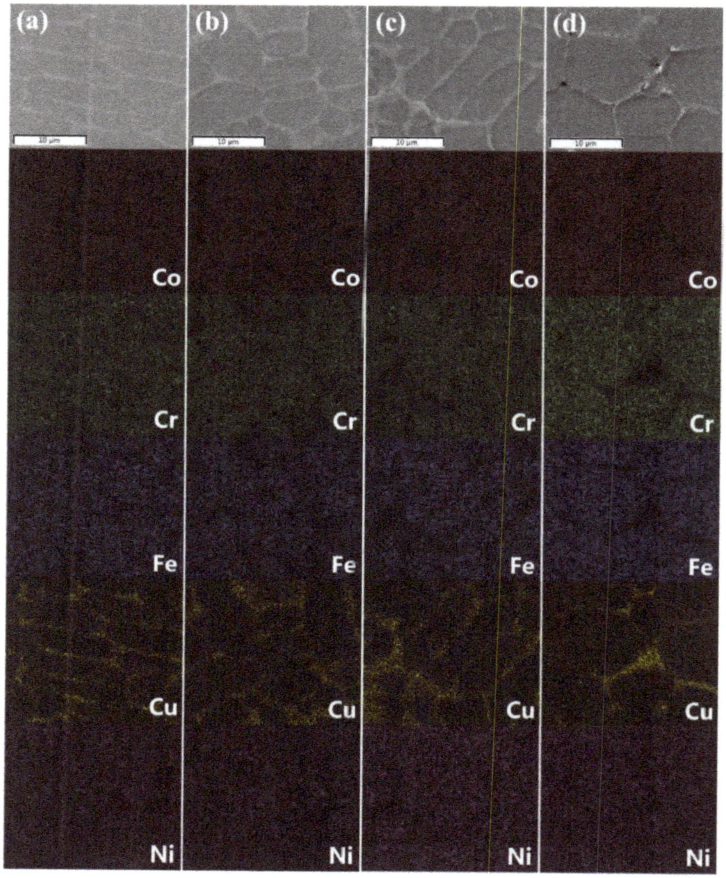

Figure 4. Energy-dispersive X-ray spectroscopy (EDX) maps showing elemental distribution of CoCrFeCuNi HEAs. (**a**) Pristine and (**b–d**) He$^+$ ions irradiated at fluences of 2.5×10^{17} ions/cm^2, 5.0×10^{17} ions/cm^2, and 1.0×10^{18} ions/cm^2.

However, at a dose of 1.0×10^{18} ions/cm^2, cracking and exfoliation occur, and the subsurface becomes visible after the surface flakes off. Wei et al. [29] reported a similar occurrence on $Cu_{48}Zr_{48}Al_4$ bulk metallic glass composites induced by He$^+$ ion irradiation at the same dose. When high-energy He$^+$ ions are bombarded at a target material, most ions take electrons from the matrix to form helium atoms, which can cause distortion, elastic rebound, and stress in the lattice [30]. With increasing irradiation fluence, irradiation-induced helium gas pressure develops inside the matrix. At a fluence of 1.0 × 10^{18} ions/cm^2, excessive pressure of the helium gas was developed inside the irradiated area, which eventually led to cracking and exfoliation [31,32], as shown in Figure 3d.

3.3. Mechanical Behavior

A nanoindentation test was conducted to reveal the average microhardness of the HEA surface and thus investigate the irradiation-induced hardening behavior. Figure 5 shows typical (a) depth profiles of nanoindentation hardness and (b) dependence of the H_{irr}/H_{unirr} ratio on the indentation depth of pristine and irradiated CoCrFeCuNi HEAs at different irradiation doses. The irradiated region and substrate region are consistent with the dpa profiles (see Figure 1b). Figure 5a shows that the pristine samples decrease in hardness with increasing indent depth at the indentation depth of h > 50 nm. Such depth-dependent hardness behavior is regarded as the indentation size effect (ISE) [33]. By contrast, hardness increases with increasing indent depth at the depth of $h < 50$ nm is considered as reverse ISE, which is usually caused by uncertainty factors from the surface and the geometrical shape of the Berkovich indenter. Therefore, to exclude uncertainty factors from the surface, we ignored the data in regions shallower than 100 nm in this study. Compared with pristine samples, all irradiated samples exhibited different degrees of hardening at $h > 100$ nm. To more clearly characterize the degree of increment, the dependence of the H_{irr}/H_{unirr} ratio on the indentation depth of all samples is exhibited in Figure 5b. Notably, the normalized nanoindentation hardness reaches a peak with an increase in indentation depth. The same peak position on all curves is located at a depth of ~146 nm (the dotted transition line). For the shallower region before the transition line, H_{irr}/H_{unirr} increases with an increase in indentation depth, which is similar to the results for other metals [25] irradiated with similar ions. However, for the deeper region after the transition line, H_{irr}/H_{unirr} decreases with an increase in indentation depth.

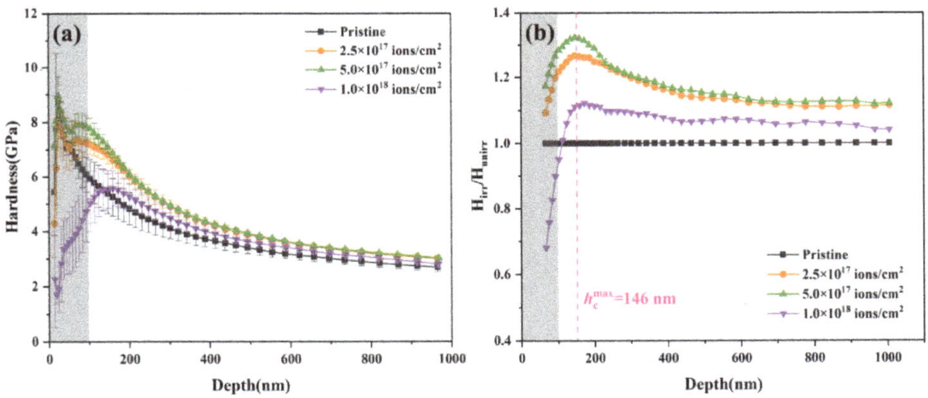

Figure 5. (**a**) Typical depth profiles of nanoindentation hardness and (**b**) dependence of the H_{irr}/H_{unirr} ratio on the indentation depth of pristine and He+ ion-irradiated CoCrFeCuNi HEAs at different irradiation doses.

An irradiation-induced hardened layer forms in the HEAs after He^+ ion implantation. When indented into the sample surface, the hemispheric influence zone beneath the indenter can reach 4 to 10 times the indentation depth. As the indenter presses deeper, if the radius of the influence zone is less than the thickness of the hardened layer, the damage grade effect is exerted, and the hardening ratio increases; otherwise the softer substrate effect is exerted, and the hardening ratio decreases [34].

To explain the ISE, Nix and Gao [35] developed a model based on the concept of geometrically necessary dislocations. This model predicts the hardness-depth profile by using the following equation.

$$H = H_0 \left(1 + \frac{h^*}{h}\right)^{1/2}, \tag{3}$$

where H is the hardness, h is the indentation depth, H_0 is the hardness at an infinite depth (i.e., macroscopic hardness), and h^* is the length that characterizes the depth dependence of hardness for a given material and indenter geometry, which depends on the material and shape of the indenter tip (i.e., statistically stored dislocation density) [36]. To aid the discussion, the hardness data are plotted as H^2 versus $1/h$ in Figure 6. H_0 is the square root of the intercept for the linear fitting of the hardness data in the near-surface region. The value of H_0 excludes the size effect and can be used as a parameter to characterize the hardening effect in HEAs under irradiation.

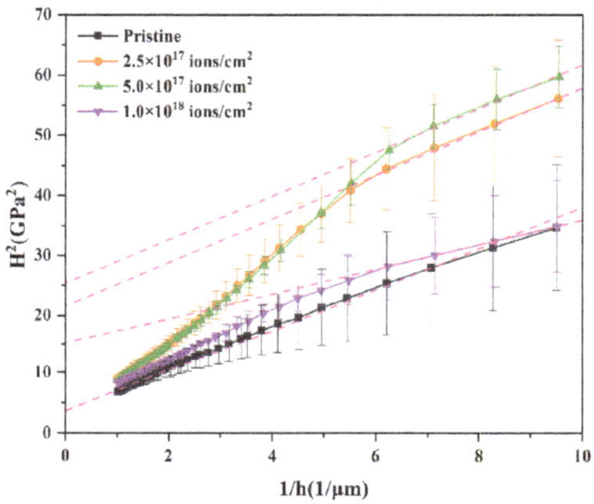

Figure 6. Plot of H^2 versus $1/h$ for the pristine and irradiated HEAs at different doses.

In Figure 6, the pristine sample shows good linearity in the range of 100 nm < h < 1000 nm. Regardless, the irradiated samples exhibit bilinearity with a shoulder at the depth of about 150 nm consisting of the depth of the transition line, (see Figure 6). Table 2 lists H_0 and h^* for the pristine samples and irradiated samples calculated by least squares fitting of hardness data in the 100 nm < h < 1000 nm range according to Equation (3) [37].

Table 2. Calculated H_0 and h^* based on the Nix–Gao model.

Materials	Irradiation Doses (ions/cm^2)	H_0 (GPa)	h^* (nm)
CoCrCuFeNi HEAs	0	1.948	902
	2.5×10^{17}	4.779	152
	5.0×10^{17}	5.063	137
	1.0×10^{18}	3.934	130

The data in Table 2 shows that the calculated H_0 first increases and then decreases with an increase in irradiation dose, which is consistent with the hardness results. The calculated h^* decreases systematically, implying a smaller size effect during irradiation, which is consistent with previous studies [37].

4. Discussion

The schematic in Figure 7 illustrates the swelling effect, as well as the hardening behavior, of Cu-rich phases during He$^+$ irradiation. With regard to the formation of He bubbles, nuclear loss-induced collisional damage, such as dislocations, voids, and interfaces, act as sinks for point defects. Dislocations can partly absorb more interstitials than vacancies. A large number of vacancies are consequently generated in the HEAs, and vacancy clusters grow and trap He to form He bubbles. For CoCrFeCuNi HEA, the Cu-rich phases appeared as defect-rich phase boundary regions in the matrix phases, similar to grain boundary, which are also favorable nucleation sites of He bubbles [38]. With the ion dose increasing to 5.0 × 10^{17} ions/cm^2, the He concentration increasingly rises and reaches near-saturation [39] in the matrix phases. Thus, He bubbles tend to migrate and gather toward the Cu-rich phases, resulting in the increased severity of volume swelling, as shown in Figure 7. We also reported that the irradiation response is inhomogeneous along the ion incident direction because of a perpendicular local shear stress [40]. Cu-rich phases may act as local low-stress regions because of the more loosely packed He atoms. He bubbles may thus preferentially migrate toward the Cu-rich phases to unload the stress, resulting in the increased severity of volume swelling [41,42]. This factor can be potentially influence the volume swelling of Cu-rich phases.

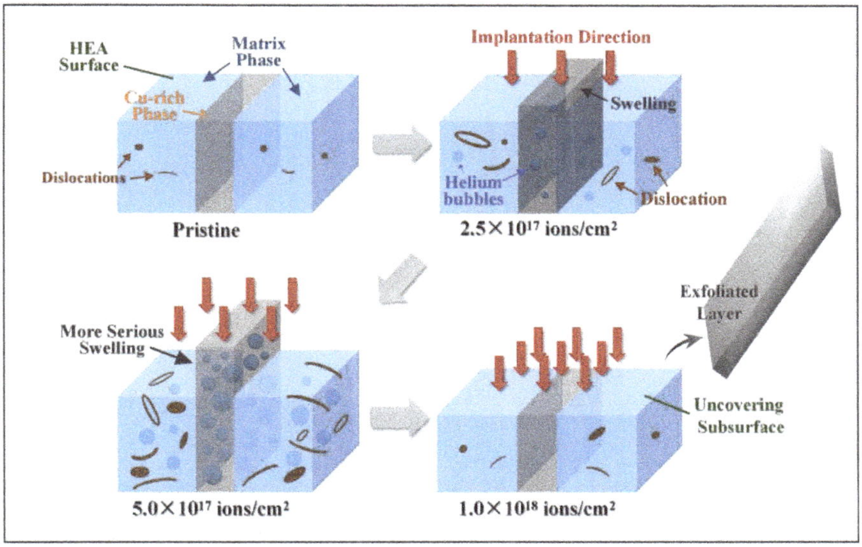

Figure 7. Schematic of radiation response in CoCrFeCuNi HEAs.

At ion doses ranging from 2.5 × 10^{17} ions/cm^2 to 5.0 × 10^{17} ions/cm^2, HEAs exhibit marked hardening behavior, which can be attributed to obstacles to dislocation glide, proposed by Orowan and Seeger [43,44]. In general, irradiation-induced defect clusters and/or dislocation loops, as well as He bubbles, pin dislocation lines and impede dislocation glide, causing HEAs to harden [45]. However, as previously mentioned, as the irradiation dose increases, numerous He bubbles migrate and gather toward the Cu-rich phases to form cavities, which may induce swelling and softening of the Cu-rich phases. While the irradiation-induced dislocations continue to increase in the matrix phases, the overall result for the HEA ultimately reflects the hardening effect.

When the ion dose increases to 1.0 × 10^{18} ions/cm^2, the hardness of the HEA slightly decreases. To avoid surface irregular interference, we selected the flatter subsurface for hardness testing. Consistent with the SEM micrograph in Figure 3d, at the highest dose, the uncovering of the subsurface after exfoliation corresponds to a new irradiation plane, except that it is only affected by diminished irradiation effects, and therefore the hardness decreases. In addition, recovery [46–48] might occur during irradiation. The thermal-spike effect during long-term irradiation can lead to annealing of defects and annihilation of helium bubbles, thus reducing the hardening effect. We hypothesize that the weakened irradiation effect on the uncovering subsurface, combined with recovery, causes softening of the HEA.

5. Conclusions

CoCrFeCuNi HEAs prepared by arc melting were irradiated with a 100 keV He^+ ion beam at fluences of 2.5×10^{17} ions/cm^2, 5.0×10^{17} ions/cm^2, and 1.0×10^{18} ions/cm^2 at room temperature. XRD results proved that all diffraction peak intensities along the (111), (200), and (220) lattice planes decreased regularly, indicating that the irradiated layer was filled with voids, leading to a decrease in crystallinity. When the irradiation dose increased to 5.0×10^{17} ions/cm^2, the irradiation-induced swelling became increasingly severe in the Cu-rich phases, whereas no significant change occurred in the matrix phases. This difference suggested that the degree of swelling varied between the two phases under the same irradiation condition. This finding contributed to the migration and gathering of the He bubble-induced noncompact structure and the lateral inhomogeneity of radiation damage. At a higher dose, 1.0×10^{18} ions/cm^2, cracking and exfoliation occurred revealing the subsurface after the surface flaked off mainly from the excessive pressure of the He gas. Nanoindentaion was performed to indicate that hardening induced by He^+ ion irradiation occurred in both irradiated samples. The depth-dependent hardness behavior was explained by the Nix-Gao model. At ion doses of 2.5×10^{17} ions/cm^2 and 5.0×10^{17} ions/cm^2, the HEAs exhibited an obvious hardening behavior, which can be explained by dislocation-dominated hardening effect. When the ion dose reached 1.0×10^{18} ions/cm^2, the hardening effect decreased probably because of the weakened irradiation effect on the uncovered subsurface, combined with the long-term thermal-spike effect-induced recovery.

Author Contributions: Conceptualization, K.Z. and Y.W.; software, W.T.; validation, B.W., K.Z. and Y.W.; formal Analysis, Y.W.; investigation, Y.W. and Y.L.; resources, B.W.; data curation, Y.W. and Y.F.; writing-original draft preparation, Y.W.; writing-review & editing, Y.W. and K.Z.; visualization, Y.W.; supervision, B.W. and K.Z.; project administration, B.W.

Funding: This research was funded by the National Natural Science Foundation of China (Grant No. 51401028, No. 51271193, No. 11402277, No. 11790292) and the Strategic Priority Research Program of the Chinese Academy of Sciences (Grant No. XDB22040303).

Conflicts of Interest: The authors declare no conflict of interest.

References

1. Zhang, Y.; Zuo, T.T.; Tang, Z.; Gao, M.C.; Dahmen, K.A.; Liaw, P.K.; Lu, Z.P. Microstructures and properties of high-entropy alloys. *Prog. Mater. Sci.* **2014**, *61*, 1–93. [CrossRef]
2. Yeh, J.-W.; Chen, S.-K.; Lin, S.-J.; Gan, J.-Y.; Chin, T.-S.; Shun, T.-T.; Tsau, C.-H.; Chang, S.-Y. Nanostructured high-entropy alloys with multiple principal elements: Novel alloy design concepts and outcomes. *Adv. Eng. Mater.* **2004**, *6*, 299–303. [CrossRef]
3. Zhang, Y.; Zhou, Y.J.; Lin, J.P.; Chen, G.L.; Liaw, P.K. Solid-solution phase formation rules for multi-component Alloys. *Adv. Eng. Mater.* **2008**, *10*, 534–538. [CrossRef]
4. Yang, X.; Zhang, Y. Prediction of high-entropy stabilized solid-solution in multi-component alloys. *Mater. Chem. Phys.* **2012**, *132*, 233–238. [CrossRef]
5. Tang, Z.; Yuan, T.; Tsai, C.-W.; Yeh, J.-W.; Lundin, C.D.; Liaw, P.K. Fatigue behavior of a wrought $Al_{0.5}$CoCrCuFeNi two-phase high-entropy alloy. *Acta. Mater.* **2015**, *99*, 247–258. [CrossRef]
6. Senkov, O.N.; Senkova, S.V.; Woodward, C.; Miracle, D.B. Low-density, refractory multi-principal element alloys of the Cr–Nb–Ti–V–Zr system: Microstructure and phase analysis. *Acta. Mater.* **2013**, *61*, 1545–1557. [CrossRef]
7. Huang, P.-K.; Yeh, J.-W.; Shun, T.-T.; Chen, S.-K. Multi-principal-element alloys with improved oxidation and wear resistance for thermal spray coating. *Adv. Eng. Mater.* **2004**, *6*, 74–78. [CrossRef]
8. Hemphill, M.A.; Yuan, T.; Wang, G.Y.; Yeh, J.W.; Tsai, C.W.; Chuang, A.; Liaw, P.K. Fatigue behavior of $Al_{0.5}$CoCrCuFeNi high entropy alloys. *Acta. Mater.* **2012**, *60*, 5723–5734. [CrossRef]

9. Chuang, M.-H.; Tsai, M.-H.; Wang, W.-R.; Lin, S.-J.; Yeh, J.-W. Microstructure and wear behavior of $Al_xCo_{1.5}CrFeNi_{1.5}Ti_y$ high-entropy alloys. *Acta. Mater.* **2011**, *59*, 6308–6317. [CrossRef]
10. Gludovatz, B.; Hohenwarter, A.; Catoor, D.; Chang, E.H.; George, E.P.; Ritchie, R.O. A fracture-resistant high-entropy alloy for cryogenic applications. *Science.* **2014**, *345*, 1153. [CrossRef] [PubMed]
11. Lu, Y.; Dong, Y.; Guo, S.; Jiang, L.; Kang, H.; Wang, T.; Wen, B.; Wang, Z.; Jie, J.; Cao, Z.; Ruan, H.; Li, T. A promising new class of high-temperature alloys: Eutectic high-entropy alloys. *Sci. Rep.-UK* **2014**, *4*, 6200. [CrossRef] [PubMed]
12. Xia, S.Q.; Yang, X.; Yang, T.F.; Liu, S.; Zhang, Y. Irradiation resistance in $Al_xCoCrFeNi$ high entropy alloys. *JOM* **2015**, *67*, 2340–2344. [CrossRef]
13. Yang, T.; Xia, S.; Guo, W.; Hu, R.; Poplawsky, J.D.; Sha, G.; Fang, Y.; Yan, Z.; Wang, C.; Li, C.; et al. Effects of temperature on the irradiation responses of $Al_{0.1}CoCrFeNi$ high entropy alloy. *Scripta Mater.* **2018**, *144*, 31–35. [CrossRef]
14. Nagase, T.; Rack, P.D.; Noh, J.H.; Egami, T. In-situ TEM observation of structural changes in nano-crystalline CoCrCuFeNi multicomponent high-entropy alloy (HEA) under fast electron irradiation by high voltage electron microscopy (HVEM). *Intermetallics* **2015**, *59*, 32–42. [CrossRef]
15. Jin, K.; Lu, C.; Wang, L.M.; Qu, J.; Weber, W.J.; Zhang, Y.; Bei, H. Effects of compositional complexity on the ion-irradiation induced swelling and hardening in Ni-containing equiatomic alloys. *Scripta Mater.* **2016**, *119*, 65–70. [CrossRef]
16. Jin, K.; Sales, B.C.; Stocks, G.M.; Samolyuk, G.D.; Daene, M.; Weber, W.J.; Zhang, Y.; Bei, H. Tailoring the physical properties of Ni-based single-phase equiatomic alloys by modifying the chemical complexity. *Sci. Rep.* **2016**, *6*, 20159. [CrossRef] [PubMed]
17. Tsai, M.-H.; Yeh, J.-W. High-Entropy Alloys: A Critical Review. *Mater. Res. Lett.* **2014**, *2*, 107–123. [CrossRef]
18. Klimova, M.; Stepanov, N.; Shaysultanov, D.; Chernichenko, R.; Yurchenko, N.; Sanin, V.; Zherebtsov, S. Microstructure and mechanical properties evolution of the Al, C-containing CoCrFeNiMn-type high-entropy alloy during cold rolling. *Materials* **2017**, *11*, 53. [CrossRef] [PubMed]
19. Feng, W.; Qi, Y.; Wang, S. Effects of short-range order on the magnetic and mechanical properties of FeCoNi $(AlSi)_x$ high entropy alloys. *Metals* **2017**, *7*, 482. [CrossRef]
20. Liu, X.; Yin, H.; Xu, Y. Microstructure, mechanical and tribological properties of oxide dispersion strengthened high- entropy alloys. *Materials* **2017**, *10*, 1312. [CrossRef] [PubMed]
21. Egeland, G.W.; Valdez, J.A.; Maloy, S.A.; McClellan, K.J.; Sickafus, K.E.; Bond, G.M. Heavy-ion irradiation defect accumulation in ZrN characterized by TEM, GIXRD, nanoindentation, and helium desorption. *J. Nucl. Mater.* **2013**, *435*, 77–87. [CrossRef]
22. Zhu, Z.G.; Ma, K.H.; Wang, Q.; Shek, C.H. Compositional dependence of phase formation and mechanical properties in three CoCrFeNi-(Mn/Al/Cu) high entropy alloys. *Intermetallics* **2016**, *79*, 1–11. [CrossRef]
23. Ma, Y.; Peng, G.J.; Wen, D.H.; Zhang, T.H. Nanoindentation creep behavior in a CoCrFeCuNi high-entropy alloy film with two different structure states. *Mat. Sci. Eng. A Struct.* **2015**, *621*, 111–117. [CrossRef]
24. Kumar, N.A.P.K.; Li, C.; Leonard, K.J.; Bei, H.; Zinkle, S.J. Microstructural stability and mechanical behavior of FeNiMnCr high entropy alloy under ion irradiation. *Acta Mater.* **2016**, *113*, 230–244. [CrossRef]
25. Odette, G.R.; Alinger, M.J.; Wirth, B.D. Recent Developments in Irradiation-Resistant Steels. *Annu. Rev. Mater. Res.* **2008**, *38*, 471–503. [CrossRef]
26. Makinson, J.D.; Lee, J.S.; Magner, S.H.; Angelis, R.J.D.; Weins, W.N.; Hieronymus, A.S. X-ray Diffraction Signatures of Defects in Nanocrystalline Materials. *Adv. X-Ray Anal.* **2000**, *42*, 407–411.
27. Zhang, X.; Mei, X.; Zhang, Q.; Li, X.; Wang, Y.; Wang, Y. Study of irradiation damage induced by He^{2+} ion irradiation in $Ni_{62}Ta_{38}$ metallic glass and W metal. *Nucl. Instrum. Methods Phys. Res. Sect. B* **2017**, *406*, 548–554. [CrossRef]
28. Oh, S.M.; Hong, S.I. Microstructure and mechanical properties of equitomic CoCrFeCuNi high entropy alloy. *Key Eng. Mater.* **2018**, *765*, 149–154. [CrossRef]
29. Wei, Y.; Zhang, K.; Wei, B.; Zhao, Z.; Yuan, J. Microstructural evolution and mechanical properties in $Cu_{48}Zr_{48}Al_4$ bulk metallic glass composites induced by He^+ ion irradiation. *Nucl. Instrum. Methods Phys. Res. Sect. B* **2018**, *428*, 17–23. [CrossRef]
30. Ghauri, I.M.; Afzal, N. Effects of neutron irradiation on the stress relaxation rate in Al–Cu–Mg alloy. *J. Phys. D Appl. Phys.* **2007**, *40*, 6044. [CrossRef]

31. Rafique, M.; Afzal, N.; Ahmad, R. Impact of 18 MeV He+ ions on the morphological and structural properties of pure Fe. *Mater. Res. Express* **2017**, *4*, 096504. [CrossRef]
32. Zhang, X.; Mei, X.; Zhang, Q.; Li, X.; Qiang, J.; Wang, Y. Damage induced by helium ion irradiation in Fe-based metallic glass. *J. Nucl. Mater.* **2017**, *490*, 216–225. [CrossRef]
33. Pharr, G.M.; Herbert, E.G.; Gao, Y. The indentation size effect: A critical examination of experimental observations and mechanistic interpretations. *Annu. Rev. Mater. Res.* **2010**, *40*, 271–292. [CrossRef]
34. Liu, X.; Wang, R.; Ren, A.; Jiang, J.; Xu, C.; Huang, P.; Qian, W.; Wu, Y.; Zhang, C. Evaluation of radiation hardening in ion-irradiated Fe based alloys by nanoindentation. *J. Nucl. Mater.* **2014**, *444*, 1–6. [CrossRef]
35. Nix, W.D.; Gao, H. Indentation size effects in crystalline materials: A law for strain gradient plasticity. *J. Mech. Phys. Solids* **1998**, *46*, 411–425. [CrossRef]
36. Was, G.S. Fundamentals of radiation materials science: Metals and alloys. *Mater. Today* **2007**, *10*, 52.
37. Kasada, R.; Takayama, Y.; Yabuuchi, K.; Kimura, A. A new approach to evaluate irradiation hardening of ion-irradiated ferritic alloys by nano-indentation techniques. *Fusion Eng. Des.* **2011**, *86*, 2658–2661. [CrossRef]
38. Yu, K.Y.; Liu, Y.; Sun, C.; Wang, H.; Shao, L.; Fu, E.G.; Zhang, X. Radiation damage in helium ion irradiated nanocrystalline Fe. *J. Nucl. Mater.* **2012**, *425*, 140–146. [CrossRef]
39. Nagata, S.; Tsuchiya, B.; Sugawara, T.; Ohtsu, N.; Shikama, T. Helium and hydrogen trapping in W and Mo single-crystals irradiated by He ions. *J. Nucl. Mater.* **2002**, *307–311*, 1513–1516. [CrossRef]
40. Zhang, K.; Hu, Z.; Zhao, Z.; Wei, B.; Li, Y.; Wei, Y. Whiskers growth and self-healing in Ti-based metallic glasses during ion irradiation. *Appl. Surf. Sci.* **2018**, *437*, 176–180. [CrossRef]
41. Van Dillen, T.; Polman, A.; Onck, P.R.; van der Giessen, E. Anisotropic plastic deformation by viscous flow in ion tracks. *Phys. Rev. B* **2005**, *71*, 024103. [CrossRef]
42. Volkert, C.A. Density changes and viscous flow during structural relaxation of amorphous silicon. *J. Appl. Phys.* **1993**, *74*, 7107–7113. [CrossRef]
43. Orowan, E. *Symposium on Internal Stresses in Metals and Alloys*; Institute of Metals: London, UK, 1948.
44. Seeger, A.K. On the theory of radiation damage and radiation hardening. In Proceedings of the Second United Nations International Conference on the Peaceful Uses of Atomic Energy, Geneva, Switzerland, 1–13 September 1958; pp. 250–273.
45. Ou, X.; Anwand, W.; Kögler, R.; Zhou, H.-B.; Richter, A. The role of helium implantation induced vacancy defect on hardening of tungsten. *J. Appl. Phys.* **2014**, *115*, 123521. [CrossRef]
46. Gao, F.; Bacon, D.J.; Howe, L.M.; So, C.B. Temperature-dependence of defect creation and clustering by displacement cascades in α-zirconium. *J. Nucl. Mater.* **2001**, *294*, 288–298. [CrossRef]
47. Fabritsiev, S.A.; Pokrovsky, A.S. Effect of irradiation temperature on microstructure, radiation hardening and embrittlement of pure copper and copper-based alloy. *J. Nucl. Mater.* **2007**, *367–370*, 977–983. [CrossRef]
48. Miyazawa, T.; Nagasaka, T.; Kasada, R.; Hishinuma, Y.; Muroga, T.; Watanabe, H.; Yamamoto, T.; Nogami, S.; Hatakeyama, M. Evaluation of irradiation hardening of ion-irradiated V–4Cr–4Ti and V–4Cr–4Ti–0.15Y alloys by nanoindentation techniques. *J. Nucl. Mater.* **2014**, *455*, 440–444. [CrossRef]

MDPI
St. Alban-Anlage 66
4052 Basel
Switzerland
Tel. +41 61 683 77 34
Fax +41 61 302 89 18
www.mdpi.com

Entropy Editorial Office
E-mail: entropy@mdpi.com
www.mdpi.com/journal/entropy

www.ingramcontent.com/pod-product-compliance
Lightning Source LLC
Chambersburg PA
CBHW041931260926
51850CB00018B/48
* 9 7 8 3 0 3 9 4 3 6 1 9 4 *

www.ingramcontent.com/pod-product-compliance
Lightning Source LLC
Chambersburg PA
CBHW041936260926
51850CB00018B/210
9783725812851